AF214173

Advances in Energy Storage

Advances in Energy Storage

Latest Developments from R&D to the Market

Edited by Andreas Hauer

ZAE-Bayern
Garching, Germany

WILEY

This edition first published 2022
© 2022 John Wiley & Sons Ltd

All rights reserved. No part of this publication may be reproduced, stored in a retrieval system, or transmitted, in any form or by any means, electronic, mechanical, photocopying, recording or otherwise, except as permitted by law. Advice on how to obtain permission to reuse material from this title is available at http://www.wiley.com/go/permissions.

The right of Andreas Hauer to be identified as the author of the editorial material in this work has been asserted in accordance with law.

Registered Offices
John Wiley & Sons, Inc., 111 River Street, Hoboken, NJ 07030, USA
John Wiley & Sons Ltd, The Atrium, Southern Gate, Chichester, West Sussex, PO19 8SQ, UK

Editorial Office
The Atrium, Southern Gate, Chichester, West Sussex, PO19 8SQ, UK

For details of our global editorial offices, customer services, and more information about Wiley products visit us at www.wiley.com.

Wiley also publishes its books in a variety of electronic formats and by print-on-demand. Some content that appears in standard print versions of this book may not be available in other formats.

Limit of Liability/Disclaimer of Warranty
While the publisher and authors have used their best efforts in preparing this work, they make no representations or warranties with respect to the accuracy or completeness of the contents of this work and specifically disclaim all warranties, including without limitation any implied warranties of merchantability or fitness for a particular purpose. No warranty may be created or extended by sales representatives, written sales materials or promotional statements for this work. The fact that an organization, website, or product is referred to in this work as a citation and/or potential source of further information does not mean that the publisher and authors endorse the information or services the organization, website, or product may provide or recommendations it may make. This work is sold with the understanding that the publisher is not engaged in rendering professional services. The advice and strategies contained herein may not be suitable for your situation. You should consult with a specialist where appropriate. Further, readers should be aware that websites listed in this work may have changed or disappeared between when this work was written and when it is read. Neither the publisher nor authors shall be liable for any loss of profit or any other commercial damages, including but not limited to special, incidental, consequential, or other damages.

Library of Congress Cataloging-in-Publication Data
Names: Hauer, Andreas, 1962- editor.
Title: Advances in energy storage : latest developments from R&D to the market / edited by Andreas Hauer.
Description: Hoboken, NJ : John Wiley & Sons, 2022. | Includes bibliographical references and index.
Identifiers: LCCN 2021020792 (print) | LCCN 2021020793 (ebook) | ISBN 9781119239352 (hardback) | ISBN 9781119760146 (pdf) | ISBN 9781119760108 (epub) | ISBN 9781119239390 (ebook)
Subjects: LCSH: Energy storage.
Classification: LCC TJ165 .A39 2022 (print) | LCC TJ165 (ebook) | DDC 621.31/26--dc23
LC record available at https://lccn.loc.gov/2021020792
LC ebook record available at https://lccn.loc.gov/2021020793

Cover image: © geniusksy/Shutterstock; © onurdongel/Getty Images; © zhongguo/Getty Images
Cover design by Wiley

Set in 9.5/12.5pt STIXTwoText by Integra Software Services Pvt. Ltd, Pondicherry, India

Printed and bound by CPI Group (UK) Ltd, Croydon CR0 4YY

C9781119239352_140322

Contents

List of Contributors

Fouzia Achchaq
Institut de Mécanique et d'Ingénierie (UMR CNRS 5295), Université de Bordeaux, CNRS, Talence, FR

Thomas Beikircher
Bavarian Center for Applied Energy Research (ZAE Bayern), Divison Energy Storage, Garching, Germany

Andreas Belderbos
Department of Mechanical Engineering, University of Leuven (KU Leuven), Heverlee, Belgium
EnergyVille Research Institute, Genk, Belgium

Ronnie Belmans
Department of Electrical Engineering, University of Leuven (KU Leuven), Heverlee, Belgium
EnergyVille Research Institute, Genk, Belgium

Pierre Bérest
LMS, Ecole Polytechnique, Palaiseau, France

Léonard E.A. Berlouis
Department of Pure and Applied Chemistry, University of Strathclyde, Glasgow, UK

Tom Brijs
Department of Electrical Engineering, University of Leuven (KU Leuven), Heverlee, Belgium
EnergyVille Research Institute, Genk, Belgium

Benoît Brouard
Brouard Consulting SAS, Paris, France

Declan Bryans
Department of Pure and Applied Chemistry, University of Strathclyde, Glasgow, UK

Luisa F. Cabeza
GREiA Research Group, Universitat de Lleida, Lleida, Spain

Jianyun Cao
School of Materials, Faculty of Science and Engineering, University of Manchester, Manchester, UK

Shijie Cheng
State Key Laboratory of Advanced Electromagnetic Engineering and Technology, Huazhong University of Science and Technology, Wuhan, China
Department of Mechanical Engineering, Tsinghua University, Beijing, China

Ningwei Justin Chiu
KTH-Royal Institute of Technology, Dept. of Energy Technology, Stockholm, Sweden

Jong-Hoo Choi
Universität zu Köln, Department für Chemie, Köln, Germany

Andrew Cruden
Energy Technology Group, Engineering Sciences, Southampton University, UK

Alvaro de Gracia
GREiA Research Group, Universitat de Lleida, Lleida, Spain

Elena Palomo del Barrio
CIC energiGUNE, Álava, Spain

Christian Doetsch
Fraunhofer Institute for Environmental Safety and Energy Technology UMSICHT, Oberhausen, Germany

David J. Evans
Retired – formerly British Geological Survey, Keyworth, Nottingham, UK

Jiakun Fang
State Key Laboratory of Advanced Electromagnetic Engineering and Technology, Huazhong University of Science and Technology, Wuhan, China

Thomas Gaal
Axpo Solutions AG, Baden, Switzerland

Jaume Gasia
GREiA Research Group, Universitat de Lleida, Lleida, Spain

Frederik Geth
Department of Electrical Engineering, University of Leuven (KU Leuven), Heverlee, Belgium
EnergyVille Research Institute, Genk, Belgium

Andrew J. Goodwin
Chief Engineer, KTSI (UK) Design Centre Limited, UK

Diego Gordon
University of Florida, Department of Mechanical and Aerospace Engineering Gainesville, USA

Anna Grevé
Fraunhofer Institute for Environmental, Safety, and Energy Technology UMSICHT, Oberhausen, Germany

Ulf Groos
Fraunhofer Institute for Solar Energy Systems, ISE, Freiburg, Germany

Stefan Gschwander
Fraunhofer ISE, Freiburg, Switzerland

Peter J. Hall
Electrical Engineering, Sheffield University, UK

Andreas Hauer
ZAE Bayern, Bavarian Center for Applied Energy Research, Garching, Germany

Thomas Haussmann
Fraunhofer ISE, Freiburg, Switzerland

Thomas Hildinger
e-netz ag Switzerland

Martin R, Jiminez
Department of Pure and Applied Chemistry, University of Strathclyde, Glasgow, UK

Stefanie Kemmer
e-netz ag Switzerland

David Kerr
Department of Pure and Applied Chemistry, University of Strathclyde, Glasgow, UK

Kris Kessels
VITO, Mol, Belgium
EnergyVille Research Institute, Genk, Belgium

Michael J. King
The Hydrodynamics Group, LLC, Edmonds, Washington, USA

Lingbin Kong
State Key Laboratory of Advanced Processing and Recycling of Non-Ferrous Metals, School of Materials Science and Engineering, Lanzhou University of Technology, China

Eberhard Lävemann
Bavarian Center for Applied Energy Research (ZAE Bayern), Garching, Germany

Robert Langer
Philipps-Universität Marburg, Marburg, Germany
Martin-Luther-Universität Halle, Halle, Germany

Perry Y. Li
Center for Compact and Fluid Power (CCEFP), University of Minnesota, Minnesota, USA

Jingdong Li
State Key Laboratory of Advanced Electromagnetic Engineering and Technology, School of Electrical and Electronic Engineering, Huazhong University of Science and Technology, Wuhan, China

Xiaohong Li
Renewable Energy Group, College of Engineering, Mathematics and Physical Sciences, University of Exeter, UK

Yuxiang Liao
State Key Laboratory of Advanced Electromagnetic Engineering and Technology, School of Electrical and Electronic Engineering, Huazhong University of Science and Technology, Wuhan, China

Maocheng Liu
State Key Laboratory of Advanced Processing and Recycling of Non-Ferrous Metals, School of Materials Science and Engineering, Lanzhou University of Technology, Lanzhou, China

Yang Liu
State Key Laboratory of Advanced Electromagnetic Engineering and Technology, Huazhong University of Science and Technology, Wuhan, China

Louis Londe
Geostock, France

Eric Loth
Center for Compact and Fluid Power (CCEFP), University of Minnesota, USA

Viktoria Martin
KTH-Royal Institute of Technology, Dept. of Energy Technology, Stockholm, Sweden

Jennifer M. Maxwell
Department of Pure and Applied Chemistry, University of Strathclyde,
Glasgow, UK

Dylan McCord
University of Florida Department of Mechanical and Aerospace Engineering
Gainesville, USA

Laia Miró
GREiA Research Group, Universitat de Lleida, Lleida, Spain

Jon M. Mitxelena
Department of Pure and Applied Chemistry, University of Strathclyde,
Glasgow, UK

George Moridis
The Hydrodynamics Group, LLC, Edmonds, Washington, USA

Richard W. Newark
CEO, Kinetic Traction Systems Incorporated (KTSI)

Tabea Obergfell
Fraunhofer ISE, Freiburg, Switzerland

Masaya Okumaya
Nagoya University, Nagoya, Japan

Halime Ö. Paksoy
Department of Chemistry, Cukurova University, Adana, Turkey

Claudia Pollak-Reibenwein
Lead Engineer/Department Penstock & Gates, ANDRITZ HYDRO GmbH, Linz,
Austria

Martin H.G. Prechtl
Universität zu Köln, Department für Chemie, Köln, Germany
Instituto Superior Técnico, University of Lisbon, Lisboa, Portugal

Chao (Chris) Qin
Center for Compact and Fluid Power (CCEFP), University of Minnesotaa, USA

Timing Qu
Department of Mechanical Engineering, Tsinghua University, Beijing, China

Christoph Rathgeber
Bavarian Center for Applied Energy Research (ZAE Bayern), Garching, Germany

Li Ren
State Key Laboratory of Advanced Electromagnetic Engineering and Technology, School of Electrical and Electronic Engineering, Huazhong University of Science and Technology, Wuhan, China

Arnaud Réveillère
Geostock, France

Alenka Ristić
Department of Inorganic Chemistry and Technology, National Institute of Chemistry, Ljubljana, Slovenia

Matthias Rottmann
Bavarian Center for Applied Energy Research (ZAE Bayern), Divison Energy Storage, Garching, Germany

Manfred Sallaberger
ANDRITZ HYDRO AG, Switzerland

Jonathan Scheffe
University of Florida Department of Mechanical and Aerospace Engineering Gainesville, USA

Jing Shi
State Key Laboratory of Advanced Electromagnetic Engineering and Technology, Huazhong University of Science and Technology, Wuhan, China

Terrence W. Simon
Center for Compact and Fluid Power (CCEFP), University of Minnesota, USA

Daan Six
VITO, Mol, Belgium
EnergyVille Research Institute, Genk, Belgium

Per Alex Soerensen
PlanEnergi, Skoerping, Denmark

Pål-Tore Storli
The Waterpower Laboratory, Dept. of Energy and Process Engineering, Norwegian University of Science and Technology, Trondheim, Norway

Yuejin Tang
State Key Laboratory of Advanced Electromagnetic Engineering and Technology, Huazhong University of Science and Technology, Wuhan, China

Helena Teschner
Politics and Markets, BVES – German Energy Storage Association, Berlin, Germany

James D. Van de Ven
Engineering, Center for Compact and Fluid Power (CCEFP), University of Minnesota, USA

Dominic van der Waals
Universität zu Köln, Department für Chemie, Köln, Germany

Wim Van Helden
AEE INTEC, Gleisdorf, Austria

Frank C. Walsh
Electrochemical Engineering Laboratory, Energy Technology Research Group, Faculty of Engineering and the Environment, University of Southampton, UK

Rutao Wang
State Key Laboratory of Advanced Processing and Recycling of Non-Ferrous Metals, School of Materials Science and Engineering, Lanzhou University of Technology, Lanzhou, China

Jinyu Wen
State Key Laboratory of Advanced Electromagnetic Engineering and Technology, Huazhong University of Science and Technology, Wuhan, China

Robin J. White
Sustainable Catalytic Materials Group, Division Hydrogen Technologies, Fraunhofer Institute for Solar Energy Systems ISE, Freiburg, Germany

Ying Xu
State Key Laboratory of Advanced Electromagnetic Engineering and Technology, School of Electrical and Electronic Engineering, Huazhong University of Science and Technology, Wuhan, China

Guo Yan
Northwest Institute for Nonferrous Metal Research, Xi'an, China

Kun Yan
State Key Laboratory of Advanced Electromagnetic Engineering and Technology, Huazhong University of Science and Technology, Wuhan, China

Sinian Yan
State Key Laboratory of Advanced Electromagnetic Engineering and Technology, Huazhong University of Science and Technology, Wuhan, China

Wei Yao
State Key Laboratory of Advanced Electromagnetic Engineering and Technology, School of Electrical and Electronic Engineering, Huazhong University of Science and Technology, Wuhan, China

Nada Zamel
Fraunhofer Institute for Solar Energy Systems, ISE Freiburg, Germany

Dr. Thomas Zell
Adama Makhteshim Ltd., Beer Sheva, Israel

Lihui Zhang
State Key Laboratory of Advanced Electromagnetic Engineering and Technology, School of Electrical and Electronic Engineering, Huazhong University of Science and Technology, Wuhan, China

Weibin Zhang
State Key Laboratory of Advanced Processing and Recycling of Non-Ferrous Metals, School of Materials Science and Engineering, Lanzhou University of Technology, Lanzhou, China

Guangjin Zhao
Laboratory for Grid Waste Treatment and Resource Recycle Technology, State Grid Corporation of China, Zhengzhou, China

Xiao Zhou
State Key Laboratory of Advanced Electromagnetic Engineering and Technology, Huazhong University of Science and Technology, Wuhan, China

1

Energy Storage Solutions for Future Energy Systems

Andreas Hauer

ZAE Bayern, Germany

1.1 The Role of Energy Storage

Wherever energy is available but not immediately used, energy storage can be utilized. Energy storage technologies help to absorb energy and release it at a later time (or in a different place) when it is needed. Hence, energy storage makes surplus energy usable, and is, therefore, equivalent to energy sources like fossil fuels and their market competition. Conventional energy resources – crude oil, natural gas, and coal, for example – share many energy storage qualities while not being renewable. They allow for a wide variety of storage methods at high energy densities, for example 40 GJ/m^3 for crude oil and coal. Considering these facts, energy storage technologies need to have either similar technical characteristics or different advantages of an economic or ecologic nature.

Electric mobility – may it be cars, public transportation or any other kind of vehicle – is a prime example of direct competition between a storage system and fossil fuels. The storage, i.e. the battery, is supposed to take in energy from, for ecological reasons, renewable sources and deliver it whenever the consumer deems fit. At the same time, the storage needs to make the consumer physically independent of these energy sources. Besides a purely economic comparison of vehicles fitted with combustion and electric engines, environmental issues like air quality might cause a crucial bias toward energy storage.

The cost of storage usually adds to the total energy cost. The cost for stored energy must not be significantly higher than the cost of energy supplied directly to the consumer. Prices, however, fluctuate with actual demand. Storage is economically most attractive when energy can be obtained at low cost and provided at a higher price during a peak in demand.

With an increasing share of renewable energy, future energy systems will also have an increased need for balancing supply and demand. Fluctuating renewable energy sources combined with energy storage systems are able to provide demand adapted energy. This application will become the most relevant.

1.2 The Definition of Energy Storage

In many cases it is not necessary to define the topic of a paper, since it is quite obvious to everyone what is actually discussed. In the case of energy storage this is not valid. Energy storage compared to other components within an energy system, whether it be on the production or the consumer side, is rather complex. It is connected to an energy

Advances in Energy Storage: Latest Developments from R&D to the Market, First Edition.
Edited by Andreas Hauer.
© 2022 John Wiley & Sons, Ltd. Published 2022 by John Wiley & Sons, Ltd.

source and to an energy consumer at the same time. The condition for charging might differ from the conditions for discharging. Even the energy form – electrical, mechanical, chemical, or thermal – might change within the storage process.

Therefore three definitions are given here: One of the energy storage process itself, one of the energy form stored, and one concerning the relation between the energy storage system and the application [1].

1.2.1 What is an Energy Storage?

An energy storage system can absorb energy and deliver it at a later point in time. The storage process itself consists of three stages: Charging storage, and discharging, as shown in Figure 1.1. After discharging, storage can be recharged.

This definition of energy storage excludes fossil fuel storage. For fossil fuel storage only the storage and the discharging step is taken into account. The charging, which takes millions of years is neglected! And it puts a focus on the possibilities of energy storage systems that can be charged again and again.

Also of importance is the distinction between the three stages since each of them is afflicted with losses and can be optimized independently. Minimizing these losses leads to high overall storage efficiency.

1.2.2 What is Actually Stored?

The form of energy (electrical, thermal, mechanical, chemical) that is taken up by an energy storage system is usually the same as the one that is delivered.

However, in many cases the charged type of energy has to be transformed for storage (e.g. pumped hydrostorage or batteries). It is re-transformed for the discharging. In some energy storage systems the transformed energy type is delivered (e.g. Power-to-Gas or Power-to-Heat).

This working definition explains how energy storage components store a certain form of energy. In many cases the storage process consists of several energy transformations. A pumped hydroplant for example converts electricity into potential mechanical energy by pumping water to a higher elevation. The energy will be converted into kinetic energy of flowing water and finally back into electricity by a turbine. Looking at this storage as a black box, electrical energy is charged to the system and later electrical energy is discharged from the system (Figure 1.2).

Finally the definition includes storage technologies like power-to-gas or power-to-heat, which deliver a different energy form to the one used in charging. This really expands the definition of energy storage. Electricity can be stored and synthetic gas or heat/cold can be discharged. Through this, energy storage systems are able to link different energy sectors like electricity and mobility or heat/cold, which opens very promising fields of application.

Figure 1.1 The three steps of a storage process – charging, storage, and discharging.

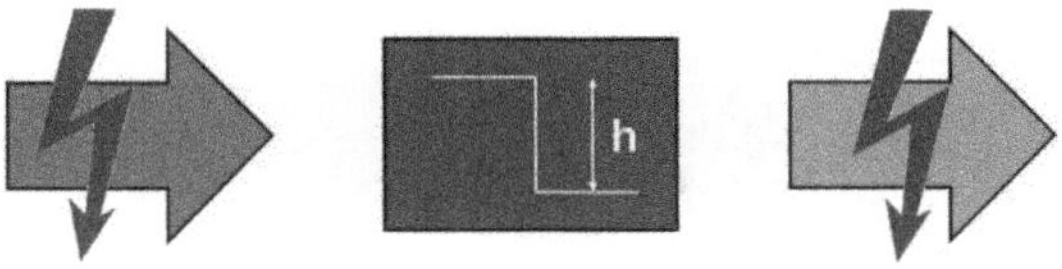

Figure 1.2 Electricity input, conversion to potential mechanical energy and final electricity output at a pumped hydroplant.

1.2.3 Energy Storage System and Its Application

The relationship between energy storage systems and their application is of the highest importance:

> The technical and economical requirements for an energy storage system are determined by its actual application within the energy system. Therefore any evaluation and comparison of energy storage technologies is only possible with respect to this application.

The application determines the technical requirements (e.g. type of energy, storage capacity, charging/discharging power) as well as the economic environment (e.g. expected pay-back time, price for delivered energy).

As an example Figure 1.3 shows the different application possibilities of one storage technology (power-to-gas). Synthetic hydrogen can be provided to the mobility sector either as fuel or can be converted to electricity before delivery. Both discharging possibilities offer different technical and economic benefits.

Unlike any other component in our energy system, energy storage has to satisfy the requirements of the system on two sides: The energy source and the consumer. This high degree of connectivity and complexity makes it impossible to discuss the benefits of an energy storage operation without referring to an actual use case. Comparison of different storage technologies is not possible without comparison of the operating context.

1.2.4 Energy and Power Storage

Looking at actual applications for energy storage systems, there are two categories requiring definition: The storage of energy and the storage of power. Power is energy per time and describes how fast a certain amount of energy can be charged or discharged. Many applications require a high power input or output, e.g. fast power reserve for balancing the electrical grid. These storage units are described by their maximum power. Also their economic properties, like investment cost, are related to

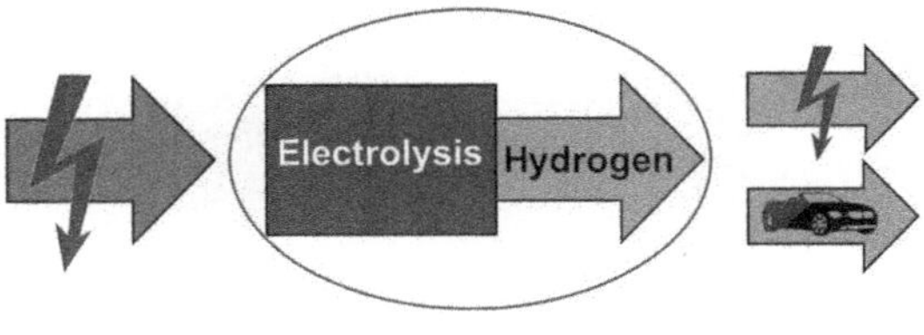

Figure 1.3 Power-to-gas storage with different outputs of hydrogen to the electricity and mobility sectors.

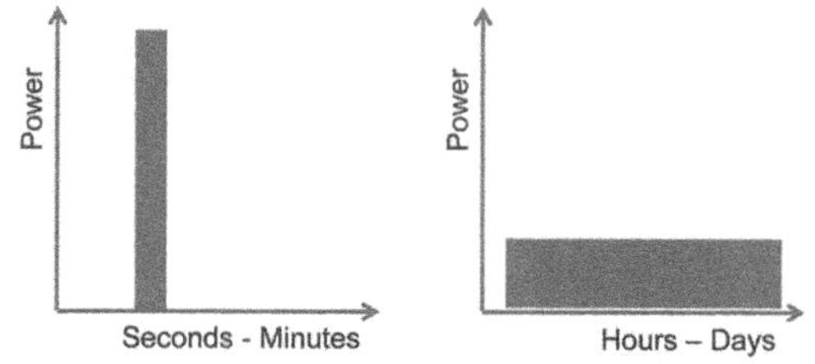

Figure 1.4 Dynamic behavior of power (left) and energy (right) storage systems [2].
Source: Hauer, A.,"Energy Storage – Definitions, Properties, and Economics", Latin America Public-Private Partnerships Workshop on Energy Storage for Sustain-able Development (UN), Rio de Janeiro, Brazil,16, 2015.

their power. Such systems are schematically shown on the left-hand side of Figure 1.4. Here high power is delivered for a short time.

Energy storage systems with a characteristic like the diagram on the right-hand side of Figure 1.4 – a long charging and discharging process at lower power – focus on energy input and output. Their main parameter is the amount of energy stored and delivered, not so much how fast the energy can be provided. Typical applications are long-term storage, e.g. seasonal storage of heat, where the charging process lasts over the summer in order to provide heating throughout the winter.

In order to evaluate energy storage systems correctly it has to be clearly identified which category the actual system belongs to.

1.2.5 Temporal Mismatch between Supply and Demand

Storage can be used wherever there is a quantitative mismatch between the supply and demand of energy in need of compensation. Without storage the consumer would have to accept all the energy produced or the supplier would have to deliver only the amount needed, neither scenario is desirable. Figure 1.5 shows two cases in which either the supply or demand of energy is stable while the other value fluctuates over time.

In Figure 1.5 (a) energy is supplied by a conventional base load plant while the fluctuations in demand are compensated by energy storage so the plant does not need to react to them and can at all times be operated within its optimal range. Since, in this mode of operation, the storage covers peaks in demand, the highest possible prices can

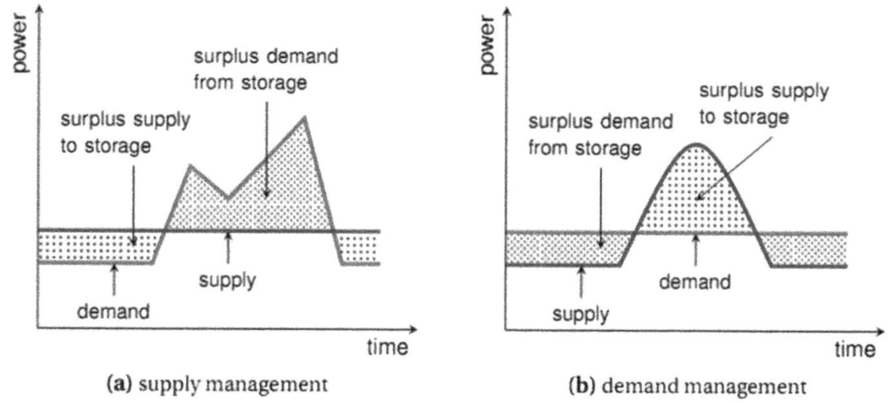

Figure 1.5 Supply and demand management [3].
Source: Hauer, A., Hiebler, S., Reuß, M.: Wärmespeicher, BINE Fachbuch, 5. Vollständig überarbeitete Auflage.

be obtained for the supplied energy. Another example (Figure 1.5 (b)) would be a cogeneration unit, fuelled by biogas, storing heat while maintaining a constant power feed rate. The thermal energy from the storage is then made available as needed.

In example (b) the energy supply fluctuates as in, for example, renewables like photovoltaics (PV), wind, or solar thermal energy. In this case, the energy storage supplies enough energy to cover the base load (constant demand). Energy storage, in such a scenario, serves as a means to integrate renewables into the local energy system.

The future reality of the energy sector will include temporal variation of energy supply due to a growing contribution of renewables, as well as of energy consumption and, therefore, an ever growing necessity for the application of efficient energy storage.

The function of an energy storage can also be defined by positioning it on the supply or demand side of the market. On the supply side, storage is designed to take in energy, which is a typical application in renewable energies or the utilization of industrial waste heat. The main goal is the storage of energy, e.g. electricity from PV or wind, which would otherwise be lost. Storage on the consumer side is designed to compensate for fluctuations in demand to level out supply to the consumer. In this case, the demand profile is uncoupled from supply.

Storage on the demand side is usually decentralized and found on site with the consumer. These are often relatively small units since they are adapted to the consumer's needs. Storage on the supply side, on the other hand, is normally large, where the centralized storage systems are directly linked to production units.

1.3 Technologies for Energy Storage

1.3.1 How Can Energy be Stored?

While physics differentiate between different forms of energy, energy technology differentiates between different forms of delivery to the consumer: electrical current, mechanical (kinetic) energy, chemical energy carriers, or thermal energy. These forms differ in their percentage of actual usable energy. While electric current can usually be put to use at 100%, the efficiency of the conversion process needs to be taken into account when using kinetic energy (e.g. in a flywheel generator) or chemical energy (e.g. hydrogen). For thermal energy, as long as it is not used directly as heat or cold, the laws of thermodynamics apply for any further conversion. The measure for practical usability of theoretical working capacity in this context is exergy [4].

1.3.1.1 Electrical/Electrochemical Energy Storages

The most commonly known way of storing electricity is a battery. It is an example of electrochemical energy storage, since the electricity is stored in a reversible chemical reaction. But there are also ways to store electricity directly: either in super-capacitors by charge separation on extremely thin metal capacitor plates – so called super-caps – or in super-conducting magnetic energy storages – so called SMES – within the magnetic field created by direct current in a super-conducting coil at very low temperatures.

The future integration of renewable electricity from solar and wind power plants can partly be accomplished through electrochemical storage systems. The demand for electrochemical storages (accumulators, redox flow batteries) and hybrid systems consisting of batteries and super-capacitors with high power density and long service life will constantly rise over the coming years together with the share of energy from decentral, fluctuating sources, pushing the stationary application of these technologies. Simultaneously, the development of high-performance batteries for mobility applications will gain importance. Future challenges will be a user-friendly cost structure for electrochemical storages and the development of user-oriented system solutions. This is especially important for stationary, electrochemical energy storage that is fed fluctuatingly by photovoltaic and wind plants. The evident development here is the rerouting of surplus wind energy into large battery systems and integration of these into a service energy system designed to efficiently stabilize the voltage and frequency of the distribution grid.

1.3.1.2 Mechanical Energy Storage

Mechanical energy storage systems convert electrical energy into mechanical potential energy. During discharge the mechanical energy is converted back into electrical energy.

Mechanical energy storage systems are characterized by their large storage capacities and electrical power and they are currently the most economical solution for storing large amounts of electrical energy.

Pumped storage hydropower stations (PSH) store electrical energy by converting it into potential energy of the height of a water reservoir; water is pumped from a lower level to an upper level storage reservoir. In order to discharge the storage, water flows in the opposite direction through water turbines, which drive electricity generators.

Compressed air energy storages (CAES) convert electrical energy into potential energy of compressed air having been compressed and stored in underground caverns or pressure vessels. In discharge, the compressed air is expanded through gas turbines. The storage efficiency of CAES depends on the efficiencies of the compression and expansion. The air entering the gas turbines during discharge has to be heated. In "adiabatic CAES" the heat generated during the compression process when charging the CAES can be stored in a separate heat storage and later on be used to heat up the expanding air.

Flywheel energy storage systems convert electrical energy into rotational energy of a flywheel. During charging, the flywheel is accelerated by an electrical motor thus the rotational speed increases. During discharge the flywheel drives the electrical motor used as a generator, which leads to a decreasing rotational speed of the flywheel. Latest generation flywheels are hence equipped with magnetic bearings and operated in vacuum housings. Storage capacities and electrical power depend on the moment of inertia of the flywheel and its rotational speed and therefore on the strength of the rotating materials and the bearing.

1.3.1.3 Chemical Energy Storage

Energy can also be stored by chemical conversion. For this purpose energy is used to produce chemical energy carriers with high energy density and nearly infinite storage periods. Most investigated energy carriers for conversion of electricity (or heat) are hydrogen and methane.

One way to generate hydrogen is by using excess electrical energy from renewables for electrolysis of water. In the process of electrolysis water is split up into oxygen and hydrogen, which can then be stored. Hydrogen can also be generated by thermally splitting water but high temperatures are required for this process. During surplus energy periods the generation of hydrogen is therefore a possible solution for the storage of energy.

The reconversion of electricity can follow in fuel cells or in gas turbines. Due to a relatively low energy density with respect to the volume, hydrogen has to be compressed to reach the energy density levels of other fuels. After compression, hydrogen can also be used in transport. Hydrogen can be stored in caverns, or engineered tanks. It can also be fed to the natural gas network but only to a certain amount since other natural gas applications tolerate only small amounts of hydrogen.

Hydrogen can also be used to synthesize methane. In the process of methanization, CO_2 and H_2 are combined to form methane (CH_4) and water. An advantage is the versatility of methane. Fed into the existing natural gas infrastructure (with no limits) it can be used in all fields of the energy sector. The reconversion of electricity via gas turbines or fuel cells is a potential option to bridge longer periods of insufficient renewable power supply [5]. When compressed or liquefied, synthetic methane (from renewable electricity by electrolysis) can reach energy densities that make it suitable for mobility applications. Synthetic methane, like conventional natural gas, can also be used for heating purposes.

Important in this context is the fact that the hydrogen and methane are produced from renewable electricity and through the processes of electrolysis and methanization the electricity is converted into a fuel in which the energy is "stored." After the charging and storage steps, discharging can happen as electricity or gas for mobility or heating purposes. These synthetic gases have to be clearly distinguished from fossil fuels.

1.3.1.4 Thermal Energy Storage

Another important instrument for the efficient use of energy is the storage of thermal energy. Possible fields of application for thermal energy storage are buffer storage systems for power adjustment in energy supply, energy storage for different temporal ranges from short-term to seasonal, and storage in thermally differing environments ranging from low temperature storage for water heating to high temperature storage in solar thermal energy generation (concentrated solar power). Thermal storage allows for stationary and mobile use. A significant positive influence on energy efficiency is to be expected particularly in waste heat recovery. In principle, thermal energy can be stored as sensible or latent heat as well as in thermochemical processes.

Thermal energy storage systems may store "cold" and "hot." These terms differentiate in the question whether the temperature of the thermal energy supplied is greater than (= hot) or less than (= cold) ambient temperature. A cold storage differs from a hot storage in that it releases heat instead of taking it in during its "storage" process.

The temperature difference of the hot (or cold) delivered to the ambient environment describes the energy content of the storage. But this difference tends to decrease over time due to thermal losses. Only thermochemical processes can overcome this limitation.

1.3.1.5 Properties of Energy Storage Systems

In order to describe the different energy storage technologies some technical properties need to be defined:

- *Capacity*: The storage capacity gives the possible amount of energy stored within and finally delivered from the storage unit. The capacity is linked to the energy density of the storage. The energy density is a function of the capacity: the stored energy related to the volume or the mass of the storage.
- *Power*: The power – energy per time – defines how fast the energy storage can be charged and discharged. Charging and discharging power can differ in most storage processes. The power density gives the power (charging or discharging) of a certain unit related to its volume or mass.
- *Efficiency*: The efficiency of the energy storage is the ratio between the energy provided to the consumer and the energy charged to the storage. Usually the efficiency is below 100%, due to losses over the storage period or inefficiencies of the charging and discharging processes or within any conversion process performed by the storage system.
- *Storage period*: The storage period defines the length of time that the charged energy is stored until discharge. This time can go from seconds (or less!) to months. The storage period is given by the application and limited by the losses of the system.
- *Cost*: The cost of an energy storage system can refer to either the capacity (money per energy, e.g. US $/kWh^{-1}) or the power (money per power, e.g. US $/ kW^{-1}) installed. The cost of the energy (or power) delivered from the storage system is strongly dependent on the number of cycles per time for an actual application

The properties listed here are interdependent. An energy storage system intended for high power output is designed in a different way compared to a system optimized for high storage capacity. Capacity and power are in this case complementary. Another example would be an efficient long-term thermal energy storage with high cost for the insulation of the system. Here efficiency, storage period, and cost are related.

So far the properties were discussed as properties of storage technology. In Section 3.2 it will become clear that the actual application of any storage system influences very strongly all these properties. In general, energy storage systems are linked to their application much more intensely than any other components within the energy system! They are not only delivering a certain energy form or consuming it. They have to do both and match supply and demand requirements.

Table 1.1 shows a list of storage technologies and their properties. However, such a list is not suitable for comparing technologies when they are used for different applications.

The price per energy discharged or finally delivered to the user in US $/kWh can only be calculated for an actual application, because parameters like the number of storage cycles per time or price of replaced energy in the application influence the price substantially. In order to quantify typical prices for delivered energy from storage systems, relevant reference systems would be required for each application. Without operating conditions given by such reference systems a comparison of storage technologies is not possible.

1.3.2 Structure of Energy Storage Technologies

The different ways to store energy can be described and structured by the physical (or chemical) storage effect and the energy form that is actually stored. Basically three

Table 1.1 Energy storage technologies and their properties [6–15].

Storage Technology	Storage Mechanism	Power	Capacity	Storage Period	Density		Efficiency	Lifetime		Cost	
		MW	MWh	Time	$kWht^{-1}$	$kWhm^{-3}$	%	#cycles	a	$\$kW^{-1}$	$\$kWh^{-1}$
Lithium ion (Li Ion)	Electrochemical	0.005–100	0.004–130	min-weeks	30–350	150–600	85–97	200– 25 000	5–25	900–400	240–3800
Sodium sulfur (NAS) battery	Electrochemical	0.05–50	0.4–300	h-weeks	100–150	80–300	75–90	2000–5000	10–20	350–400	100–500
Lead-acid battery	Electrochemical	0.001–50	0.001–40	Min-days	10–50	40–100	70–92	50–4500	5–17	300–600	130–600
Redox/Flow battery	Electrochemical	0.005–200	0.2–800	h-month	10–85	15–60	65–85	2000–13 000	5–20	430–2500	150–2800
Compressed air energy storage (CAES)	Mechanical	0.003–1000	0.002–3000	h-month	30–60	0.5–6	40–60 (–90with heat stroage)	8000–30 000	20–100	400–2700	2–250
Pumped hydroenergy storage (PHES)	Mechanical	10–500	500–8000	h-days	0.5–1.5	0.5–15	70–85	10 000–50 000	40–100	600–4300	5–100
Hydrogen	Chemical	0.3–60	n.a.	day-months	100–10 000 (only gas: 33 000)	3 (only gas@1bar)	20–66	20 000	5–20	500–5500	2–15
Methane	Chemical	n.a.	n.a.	day-months	1389.9 (gas only)	9.96 (gas only)	30–38	n.a	n.a.	n.a.	n.a.

Table 1.1 (Continued)

Sensible storage-Water	Thermal	<10	<100	hour-year	18537	<60	50–90	~5000	–	0.1–13
Phase change materials (PCM)	Thermal	<10	<10	hour-week	50–150	<120	75–90	~5000	–	13–65
Thermochemical storage (TCS)	Thermal	<1	<10	hour-week	120–250	120–250	80–100	~3500	–	10–130

Note: While most of the technical properties seem to be valid for longer times, the cost values are quite volatile. Here the development is difficult to predict and can be rapid. There are two values given: Cost per storage capacity in US $/kWh and per discharging power in US $/kW.

Sources: Based on Quinnel, J.; Hauer, A.; Laevemann, E.: Energy Storage Technologies - Caharac-teristics, Comparison and Synergies. In: Stolten, D.; Scherer, V. (Editors): Tran-sition to Renewable Energy Systems. Weinheim: Wiley-VCH, 2013. P. 800; Stenzel, Peter, et al. "Database development and evaluation for techno-economic assessments of electrochemical energy storage systems." Energy Conference (ENERGYCON), 2014 IEEE International. IEEE, 2014; Ferreira, Helder Lopes, et al. "Characterisation of electrical energy storage technologies." Energy 53 (2013): 288-298; Kempener, Ruud, and Eric Borden. "Battery storage for renewables: Market status and technology outlook." International Renewable Energy Agency, Abu Dhabi (2015): 32; Moseley, Patrick T., and Jürgen Garche, eds. Electrochemical energy storage for renewable sources and grid balancing. Newnes, 2014; [11] Luo, Xing, et al. "Overview of current development in electrical energy storage technologies and the application potential in power system operation." Applied energy 137 (2015): 511-536; Zakeri, Behnam, and Sanna Syri. "Electrical energy storage systems: A com-parative life cycle cost analysis." Renewable and Sustainable Energy Reviews 42 (2015): 569-596; Jeffrey W. Braithwaite, William L. Auxer: Handbook of Batteries. 2. Auflage. McGraw-Hill, 2002, ISBN 0-07-135978-8, Kapitel 40: Sodium-Beta Batteries, S. 40.1–40.31; US department of Energy Global Energy Storage Database and Sterner M. and Stadler I., Energiespeicher: Bedarf, Technologien, Integration. Springer Vieweg, 2014

forms of energy – electrical, chemical, and thermal – can be stored. Energy storage technologies for this purpose can be subdivided into numerous technical solutions. Figure 1.6 shows a structure of storage technologies.

Furthermore, there are energy storage technologies located between the three main technologies – storage of electrical, chemical, and thermal energy. Power-to-gas-to-power and power-to-heat-to-power use chemical and thermal storage technologies in order to provide storage of electrical energy. Technologies like power-to-heat or power-to-gas (not explicitly shown in Figure 1.6) can be defined as thermal or chemical storage following the physical storage effect they are based upon. In these cases only the energy source electricity is converted into heat/cold or hydrogen for storage.

1.4 Applications for Energy Storage

1.4.1 List of Applications

Due to the vast diversity of energy storage applications, it is not possible to compile a complete list of possible applications. Table 1.2 shows a number of relevant storage applications. The applications are structured following the two pillars of the German "Energie wende" – the integration of renewable energies and the increase of energy efficiency [1]. This list is also incomplete and gives only a first hint of the possibilities of energy storage in any future energy system. The fields of application mentioned can be served by different storage technologies. The list gives an impression that *all* different technologies will play their role in the future. There will not be a one-technology-fits-all solution!

1.4.2 Energy Storage Configurations and New Fields of Application

Energy storage can only be evaluated within an actual application. Therefore it is important to discuss energy storage configurations – a configuration consisting of an energy source, energy storage, and energy user – rather than isolated storage technologies.

This is obvious for energy storage applications like pumped hydroplants within the electricity grid or battery storages for optimizing the self-consumption of private PV installations – applications well known to the public. In these cases the technical and economic environment of the energy storage, the energy storage configuration, is understood to be part of the discussion.

There are countless, maybe yet unknown, possibilities for energy storage configurations. By way of example are two examples of energy storage applications that are now a commercial product and which were literally unknown a few years ago:

- A thermochemical heat storage realized by a zeolite-water reaction has been integrated in a dishwasher. It is charged while the water is heated at the beginning of the washing process and delivers heat for the drying at the end of the process. The thermal energy storage reduces the electricity consumption by about 25% compared to the most efficient dishwasher without such a storage system [16].
- The so called "PowerBooster" is basically a battery, which can be charged at low power and discharged at much higher power. The battery is applied for enhancing the charging infrastructure of electric vehicles. The PowerBooster provides high capacity for the charging process while energy is recharged at the available network

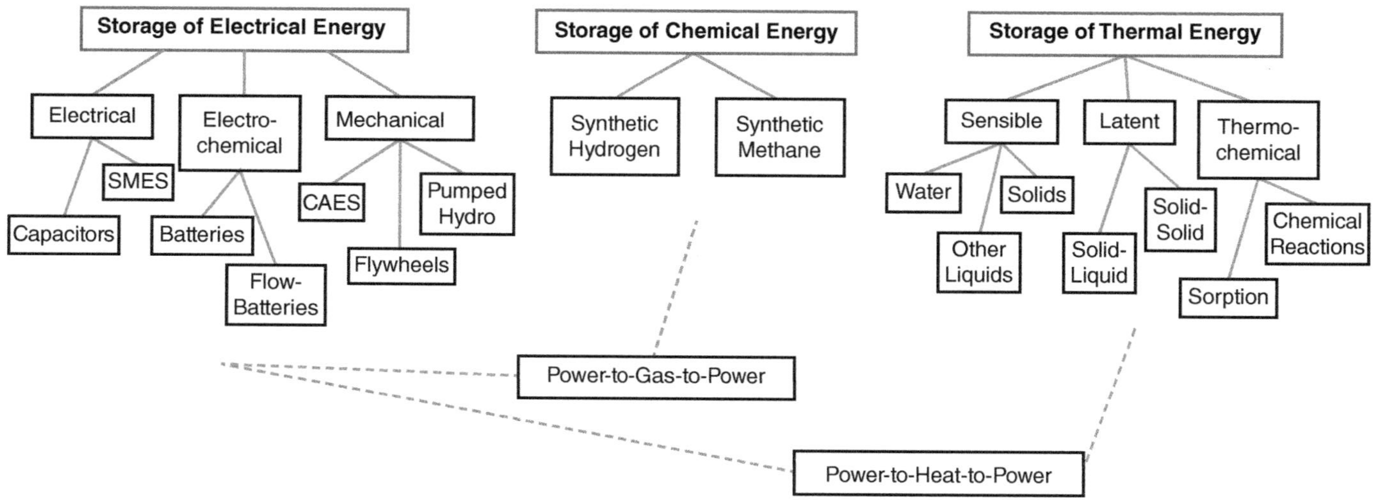

Figure 1.6 Structure of energy storage technologies following the physical storage effect.

Table 1.2 List of relevant energy storage applications following the structure of the German "Energiewende."

Renewable Energies	Energy Efficiency
Integration of Renewable Electricity	**Industrial Processes**
• Grid Stability	• Waste Heat Utlization
– Frequency regulation	• Recuperation of Mechanical Energy
– Voltage support	
– Transmission and distribution congestion relief	
– Black start	
• Grid balancing	**Buildings**
– Fast power reserve	• Heating and Cooling
– Peak shaving	– Day/night-balancing
– Self-consumption, off-grid	– Summer/winter-balancing
• Demand side integration	**Electricity production**
– Dispatchable load	• Fossil Thermal Power Plants
– Power-to-gas	• Heat Utilization of CHP
– Power-to-heat	
Integration of Renewable Thermal Energy	**Mobility**
• Concentrated Solar Power	• Propulsion
• Solar-thermal Process Heat	• Heating/Air Conditioning
• Solar-thermal Heating and Cooling	

connection point at low power. This not only saves time, but also eliminates the need for elaborate medium-voltage systems, building cost subsidies, or expensive network expansion [17].

Within an activity of the energy storage program of the International Energy Agency (IEA), the impact of distributed energy storage solutions for future energy systems was investigated. Distributed energy storage is located on the consumer side. For example in an electricity grid they would be placed on the distribution grid level. Over 90 different distributed energy storage (DES) configurations have been identified. Most of them are commercially available (TRL $\geq$ 9). About 40% of the DES configurations are realized by electrical energy storage, 40% by thermal energy storage, and 20% by power-to-gas installations. In general all types of energy storage contribute to DES solutions. Almost all DES configurations operated today are optimized for their local application. Grid (electrical and thermal) services provided by DES configurations are expected to present additional benefits and will play an increasingly important role in the future [18].

The concept of sector coupling offers also new opportunities for the implementation of energy storage. Sector coupling basically describes the distribution of renewable electricity (from wind and PV) from the electricity sector to the heat/cold and mobility sectors. However the mismatch in supply from renewable sources and demand is not solved by this concept. Only the implementation of sufficient storage capacity can solve this problem. Energy storage in this context faces new challenges and opportunities. For thermal energy storage in the heating and cooling sector electricity with 100% exergy content now offers the possibility of high temperature storage. For the mobility sector, power-to-gas or power-to-fuel solutions offer the possibility for renewables to enter, economically, much more valuable markets.

All these new developments influence the research and development of energy storage technologies, which have to adapt to the ever changing requirements and possibilities. Therefore the field of energy storage will certainly remain innovative and challenging in the future.

References

[1] Bundesverband Energiespeicher BVES, https://www.bves.de

[2] Hauer, A. (2015). Energy storage – definitions, properties, and economics. In: *Latin America Public – Private Partnerships Workshop on Energy Storage for Sustainable Development (UN)*, 16–17 April, Rio de Janeiro, Brazil.

[3] Hauer, A., Hiebler, S., and Reuß, M. (2013). *Wärmespeicher*, BINE Fachbuch, 5. Vollständig überarbeitete Auflage.

[4] Baehr, H.D. and Kabelac, S. (2006). *Thermodynamik – Grundlagen Und Technische Anwendungen*. Berlin: Springer.

[5] Hauer, A., Specht, M., and Sterner, M. (2010). Energiespeicher – steigerung der energieeffizienz und integration erneuerbarer energien. In: *Forschungsverbund Erneuerbare Energien (FVEE), Berlin (Hrsg.): Forschung für das Zeitalter Erneuerbarer Energien. Jubiläumstagung des FVEE 11–12 October*, p. 148, Proceedings.

[6] Quinnel, J., Hauer, A., and Laevemann, E. (2013). Energy storage technologies - Characteristics, comparison and synergies. In: *Transition to Renewable Energy Systems* (eds. D. Stolten and V. Scherer), 800. Weinheim: Wiley-VCH.

[7] Stenzel, P., Baumann, M.J., Fleer, J., et al. (2014). Database development and evaluation for techno-economic assessments of electrochemical energy storage systems. In: *Energy Conference (ENERGYCON), 2014 IEEE International*. IEEE.

[8] Ferreira, H.L., Garde, R., Fulli, G., et al. (2013). Characterisation of electrical energy storage technologies. *Energy* 53: 288–298.

[9] Kempener, R. and Borden, E. (2015). *Battery storage for renewables: Market status and technology outlook*, p. 32. Abu Dhabi: International Renewable Energy Agency.

[10] Moseley, P.T. and Garche, J. (eds). (2014). *Electrochemical Energy Storage for Renewable Sources and Grid Balancing*. Oxford: Newnes.

[11] Luo, X., Wang, J., Dooner, M., et al. (2015). Overview of current development in electrical energy storage technologies and the application potential in power system operation. *Applied Energy* 137: 511–536.

[12] Zakeri, B. and Syri, S. (2015). Electrical energy storage systems: A comparative life cycle cost analysis. *Renewable and Sustainable Energy Reviews* 42: 569–596.

[13] Braithwaite, J.W. and Auxer, W.L. (2002). *Handbook of Batteries*. 2. Auflage. McGraw-Hill, ISBN 0-07-135978-8, Chapter 40: Sodium-Beta Batteries, S. 40.1–40.31

[14] US department of Energy Global Energy Storage Database, https://www.energystorageexchange.org

[15] Sterner, M. and Stadler, I. (2014). *Energiespeicher: Bedarf, Technologien, Integration*. Berlin: Springer Vieweg

[16] https://wiki.bsh-group.com/en/wiki/Zeolite_drying

[17] https://www.ads-tec.de/en/energy-storage/industrial-infrastructure/powerbooster/description.html

[18] http://www.eces-desire.org

Part I

Electrochemical, Electrical, and Super Magnetic Energy Storages

2

An Introduction to Electrochemistry in Modern Power Sources

Frank C. Walsh,[1,] Andrew Cruden,[1] and Peter J. Hall[2]*

[1] *Energy Technology Group, Engineering Sciences, Southampton University*
[2] *Electrical Engineering, Sheffield University*
* *Author for correspondence*

2.1 Introduction

The history of electrochemical devices for energy conversion goes back to the earliest battery, a pile of closely separated silver and zinc discs, due to Volta (1799) and Faraday's laws of electrolysis relating electrical energy to the change in active electrochemical materials (1833/34). The principle of a hydrogen-oxygen fuel cell can be independently attributed to Schönbein (1838) and Bacon (1839). The concept of a redox flow battery can be traced back to Kangro (1949) while super-capacitors were developed by, for example GEC in the early 1950s. The benefits of electrochemical devices include ease of measurement (current flow being monitored as the rate of electron movement in an external circuit), control, and the ability to scale, while drawbacks include materials degradation and unwanted reactions. It is important to consider the following aspects of electrochemical energy conversion:

a) The nature of electrode reactions, including their reactants, products, and phase changes.
b) Types of cell and charge carriers within them.
c) The balanced benefits and drawbacks of electrochemical energy storage.
d) The thermodynamic (theoretical energy change) and kinetic (practical) characteristics of cells.
e) Fundamental expressions describing cell performance.

2.2 Electrode Reactions

An extremely diverse range of chemistry is possible, including solid, liquid, and gaseous states and their interconversion. Common examples in electrochemical galvanic cells include:

a) Surface phase transformation, e.g. oxidation of nickel oxide to nickel oxyhydroxide at the anode of a Ni-Cd battery:

$$\mathrm{Ni(OH)_2 + OH^- - e^- = NiO(OH) + H_2O} \qquad E^o = 0.49\,\mathrm{V}\ vs.\ \mathrm{SHE} \qquad (2.1)$$

Advances in Energy Storage: Latest Developments from R&D to the Market, First Edition.
Edited by Andreas Hauer.
© 2022 John Wiley & Sons, Ltd. Published 2022 by John Wiley & Sons, Ltd.

b) Fuel oxidation, e.g. hydrogen oxidation at the anode of a hydrogen-oxygen fuel cell:

$$H_2 - 2e^- = 2H^+ \qquad E^o = 0V\ vs.\ SHE \tag{2.2}$$

c) Oxygen reduction, e.g. in a metal-air battery under neutral/alkaline conditions:

$$O_2 + 2H_2O + 4e^- = 4OH^- \qquad E^o = 0.40V\ vs.\ SHE \tag{2.3}$$

or at the cathode of a hydrogen-air proton exchange membrane fuel cell, under acidic conditions:

$$O_2 + 4H^+ + 4e^- = H_2O \qquad E^o = 1.23V\ vs.\ SHE \tag{2.4}$$

d) Metal dissolution, e.g. zinc dissolution at the anode of a Zn-air battery:

$$Zn + 4OH^- - 2e^- = Zn(OH)_4^{2-} \qquad E^o_{cell} = -1.25V\ vs.\ SHE \tag{2.5}$$

e) Extraction of an intercalated Li^+ ion, from the positive electrode of a lithium ion battery:

$$Li - e^- = Li^+ \qquad E^o_{cell} = -3.04V\ vs.\ SHE \tag{2.6}$$

f) Hydrogen adsorption into a solid catalyst such as finely divided palladium in electrochemical hydrogen storage:

$$H_{2(g)} = 2H_{ads} \tag{2.7}$$

g) A change in the oxidation state of a metal in a soluble ion, as in the all vanadium redox flow battery at the positive electrode:

$$VO_2^+ + e^- = VO^{2+} \qquad E^o = 1.00V\ vs.\ SHE \tag{2.8}$$

or, at the negative electrode:

$$V^{2+} - e^- = V^{3+} \qquad E^o = -0.255V\ vs.\ SHE \tag{2.9}$$

2.3 Electrochemical Cells

A definition sketch of an electrochemical cell, consisting of an anode and a cathode in ionic and electronic contact, is shown in Figure 2.1.

Commonly, the electronically conducting electrodes are metallic or carbon although ceramic polymeric and composite electrodes are increasingly important. The electrochemical cell is comprised of an anode and a cathode in contact with an electrolyte. The electrolyte is an ionic conductor and may be (i) an aqueous solution or gel, (ii) a non-aqueous (ionic liquid) solution, (iii) a molten salt, or (iv) a solid state (polymeric or ceramic) ionic conductor. The cell may be undivided, with one electrolyte common to both electrodes, or divided by an ionically conductive material, typically a microporous polymer or ceramic or else an ion-exchange polymer.

The anode is the electrode at which oxidation of a reactant occurs at the electrode/electrolyte interface via electron loss; at the cathode, electron gain at the electrode/electrolyte interface leads to reduction. The net cell reaction is a corresponding redox process. The desired (or primary) electrode reactions often have competition from secondary ones, resulting in a real charge capability below the ideal.

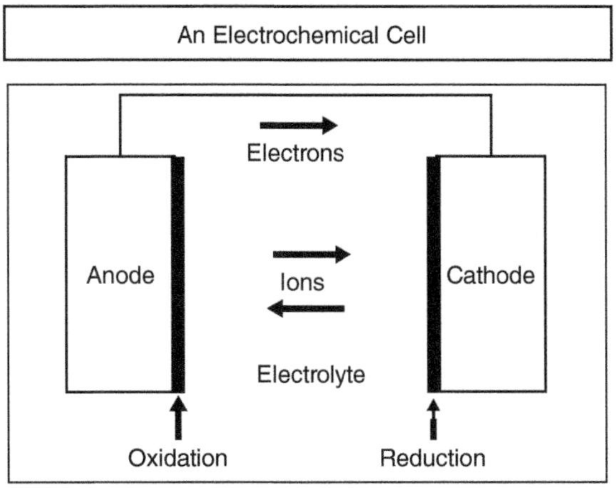

Figure 2.1 A fundamental electrochemical cell, showing unidirectional electronic current flowing in the external circuit and bidirectional ionic current flow in the electrolyte.

As a whole, the cell has several notable features:

1. It is electrically neutral.
2. Electrical current flows bidirectionally, as cations and anions, in the electrolyte.
3. Electrical current flows unidirectionally, as electrons (e^-) in the external circuit.
4. The reactions take place locally at the electrode/electrolyte interfaces, resulting in a potential difference, E_{cell}, in the outer circuit between the electrodes.

Examples of cell reactions include:

a) In a proton exchange membrane, PEM hydrogen-oxygen fuel cell, reactions (2.2) and (2.4) can be coupled to give the combustion of hydrogen by oxygen to water:

$$H_2 + 0.5O_2 = H_2O \quad E^o_{cell} = 1.23V \tag{2.10}$$

b) In a discharging zinc-air cell, electrode reactions (5) and (3) may be coupled to give:

$$Zn + O_2 = Zn\left(OH\right)_4^{2-} + 2OH^- + 2H_2O \quad E^o_{cell} = 1.65V \tag{2.11}$$

c) In a discharging all vanadium redox flow cell, reactions (2.8) and (2.9) may be coupled to show that vanadium ions having oxidation states of V and II undergo reaction to vanadium ions in the IV and II oxidation state:

$$VO_2^+ + V^{2+} = VO^{2+} + V^{3+} \quad E^o_{cell} = 1.255V \tag{2.12}$$

In practice, it is common to electrically connect a number of cells in series to realize a battery having a sufficiently high potential difference.

2.4 The Case for Electrochemical Power Sources

Electrochemical power conversion has the benefits of:

a) Direct electrical supply from controlled chemical change.
b) A controllable electrical output that is readily turned off by open circuiting the discharging cell.
c) A predictable open-circuit cell potential difference and reasonable charge capacity, together with

d) the capability of clean recharging in the case of secondary cells (i.e. those where the electroactive materials can be reformed by external supply of direct current, at least at the source of the electrical supply).

A more critical appraisal might include several limitations of cells, including:

a) Availability of active materials and electrolytes.
b) Active materials can have very different densities and current efficiencies, leading to varying (specific or volumetric) energy capability and charge capacity.
c) Limited materials stability, e.g. due to corrosion, passivation, or phase transformation.
d) Overcharging leading to overheating or gassing.
e) Restricted recycling and environmental remediation.
f) Irreversibility.

2.5 The Thermodynamics of Electrochemical Cells

The ideal electrochemical cell supports a redox cell reaction comprised of a reduction at the cathode and an oxidation at the anode. The thermodynamically available energy of an electrochemical cell at equilibrium (open-circuit or zero current) is related to the Gibbs free energy change for the cell reaction, ΔG_{cell} that is directly related to the equilibrium cell potential, $E_{e,cell}$

$$\Delta G_{cell} = -zFE_{e,cell} \tag{2.13}$$

where F is the Faraday constant and z is the number of electrons interchanged in each electrode reaction.

In galvanic cells, the electrochemical cell reaction has a negative ΔG_{cell} and electrical energy is generated spontaneously, as in batteries and fuel cells (Figure 2.2). In contrast, electrolytic cells, having a positive ΔG_{cell}, require an input of electrical energy to drive the electrochemical reaction, as in secondary batteries during charging.

By convention, the cell potential (cell voltage) is expressed as the difference between the potentials of the right-and left-hand side electrodes, at open-circuit under equilibrium conditions:

$$E_{e,cell} = E_{e,R} - E_{e,L} \tag{2.14}$$

If the cathode is placed on the right-hand side,

$$E_{e,cell} = E_{e,C} - E_{eA} \tag{2.15}$$

2.6 The Actual Cell Voltage: Thermodynamic, Electrode Kinetic, and Ohmic Losses

Away from equilibrium, when current flows, the cell potential difference can be expressed as:

$$E_{cell} = E_{e,cell} - \sum |\eta| - \sum |IR| \tag{2.16}$$

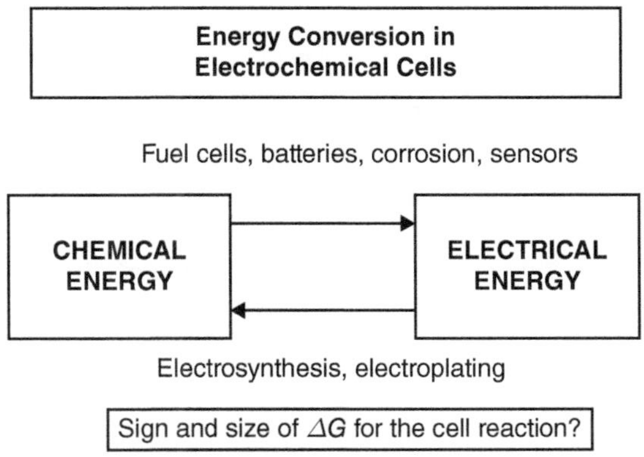

Figure 2.2 Energy conversion in electrochemical cells, showing the different direction of chemical to electrical energy conversion in galvanic and electrolytic cells.

The first term on the right-hand side is set by choice of electrode reactions and thermodynamics, the second one by electrode kinetics and mass transport to and from the electrodes, and the third term by ohmic losses. Expanding the expression gives:

$$E_{cell} = E_{e,cell} - |\eta_{c,act}| - |\eta_{a,act}| - |\eta_{c,conc}| - |\eta_{a,conc}| - \sum |IR| \qquad (2.17)$$

(The modulus of values is taken to allow for negative cathode currents values, which indicate a reduction process). where $\eta_{c,act}$ and $\eta_{a,act}$ are the activation (charge transfer) polarization at the cathode and anode, respectively, representing kinetic charge transfer limitations. These terms dominate at low current densities and can be minimized by selecting appropriate catalysts for the desired reaction and by operating at a higher temperature. The concentration polarizations (which occur under convective diffusion) terms $\eta_{c,conc}$ and $\eta_{a,conc}$ are important at high current densities, since they cause a potential drop due to mass transport limitations of electroactive species reaching, or leaving, the electrode surface and can be minimized with high surface area electrodes and high mass transport flow regimes or turbulence promoters. The *IR* term is the sum of the electronic and ionic resistances across the cell and comprises all ohmic drops in the system, such as those in the electrolyte(s), membrane, electrodes, current collectors, and external electrical contacts. It is important to minimize energy losses in cells by reducing: (i) overpotentials at the electrodes by suitable choice of catalytic electrode materials and (ii) the resistances within electrodes, across the electrolyte(s) and through the membrane. For an electrolytic cell, such as a secondary cell on charge (an electrolytic cell), minimizing potential losses in the cell minimizes the cell potential required; for a galvanic cell, such as a fuel cell or flow battery on discharge, minimizing potential losses ensures that the maximum cell potential is available on discharge. Equation (2.14) is illustrated graphically in Figure 2.3 by considering the components of cell potential in the case of a galvanic cell.

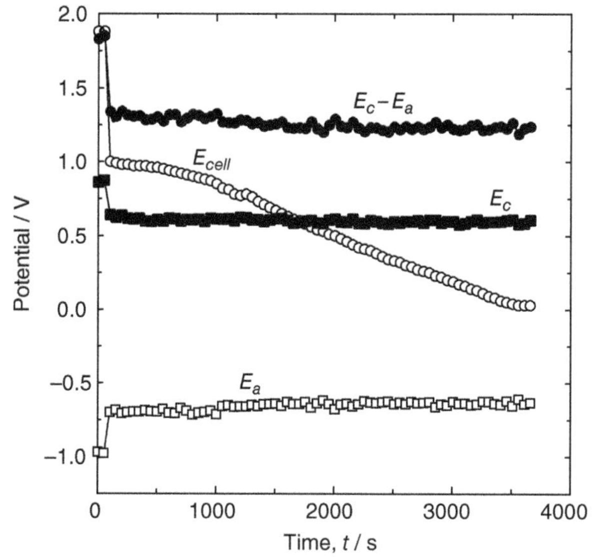

Figure 2.3 Components of cell potential with distance between the electrodes, showing the cell potential, E_{cell} to consist of components due to electrode polarization, η at the electrode/electrolyte interfaces and ohmic drop, *IR* in the electrolyte. The cell components in Equation 2.14 are illustrated graphically for a borohydride-peroxide fuel cell at 20°C discharging at a constant current of 2 A (corresponding to a current density of 31.2 mA cm^{-2}). The cathode potential, E_c, anode potential, E_a, their difference, (E_c-E_a) and the cell potential, E_{cell} are shown as a function of time. After [1].
Source: C. Ponce-de-León, F.C. Walsh, A. Rose, J.B. Lakeman, D.J. Browning, R.W. Reeve, A direct borohydride - acid peroxide fuel cell, 64 (2007) 441.

2.7 Faraday's Laws and Charge Capacity

The electrical charge, q flowing through a cell due to the movement of z electrons is quantitatively related to the amount of chemical change, Δn by Faraday's laws of electrolysis:

$$q = \Delta n.z.F \tag{2.18}$$

where the electrical charge, q in the time interval $t_2 - t_1$, is:

$$q = \int_{t1}^{t2} I.dt \tag{2.19}$$

The Faraday constant, F is the charge associated with a mole of electrons. If $t_2 = 0$, the overall charge is involved; otherwise, the charge in the time interval (t_2-t_1) is involved.

In the case of constant current, I flowing for a time t, Equation (2.6) simplifies to:

$$q = \Delta n.z.F\,/\,t \tag{2.20}$$

The current efficiency, ϕ is ideally unity but is typically < 1 due to the occurrence of secondary reactions at the electrode surface. In practice, the rate of charge of a cell will be limited by the rate of one of the electrode reactions.

2.8 The Performance of Cells: Charge Capacity and Specific Energy Capability

The charge capacity of a battery depends on the electrical charge that can be achieved from a known mass of active material, *w*. From Equation (2.6)

$$q = \Delta n.z.F \,/\, w \tag{2.21}$$

The specific energy capability, E_s, i.e. the electrical energy available from a known mass of active material developing an electrical power *P*, is

$$E_s = \frac{1}{w} \cdot \int_{t_1}^{t_2} P.\, dt \tag{2.22}$$

It is essential to declare whether the mass is that of the active material alone or if it includes the electrolyte, internal electrode connectors, and container.

In a classical redox flow battery, energy is stored as soluble redox species and the volumetric energy change, E_V is:

$$E_V = \frac{1}{V} \cdot \int_{t_1}^{t_2} P.dt \tag{2.23}$$

where *V* is the constant electrolyte volume (cm^3) and *P* is the power (W).

2.9 Types of Electrochemical Device for Energy Conversion

A *battery*, (e.g. the lead-acid car battery) is self-contained; the electrolyte is always inside the cell and power is stored inside the electrode structures. Unlike primary batteries, secondary batteries can be recharged by the input of an external direct current power supply. In a *fuel cell*, (e.g. the hydrogen-oxygen fuel cell, the fuel, and oxidant (often gases) are stored externally to the cell and electrical energy is generated at the electrode/electrolyte interface. In a *redox flow cell* (e.g. the all vanadium cell), liquid electrolytes are stored externally to the cell and electrical energy is stored as soluble species in the electrolytes. Electrochemical double layer capacitors or *super-capacitors* are considered have a high power density, short charge and discharge time and long lifespan. Despite their recent development, such qualities make super-capacitors complementary to fuel cells and rechargeable batteries, despite their energy density being much lower than rechargeable batteries.

Table 2.1 attempts a general comparison of four types of electrochemical power source, a static battery, a redox flow battery, a fuel cell, and a super-capacitor. The charging and discharging of batteries is relatively slow; degradation of the composition or structure of active materials can occur over time. Batteries have a moderate power density and lose their ability to retain energy with time. In super-capacitors, energy is stored electrostatically on the surface of the active electrode materials and does not involve electrochemical reactions. Super-capacitors can be charged quickly, leading to a very high power density and do not easily lose their storage capabilities over time.

Table 2.1 A general comparison of four types of electrochemical power source.

Property	Static battery	Redox flow battery	Fuel cell	Super-capacitor
Reversibility	Primary cells discharge only; secondary cells charge–discharge cycling.	Charge–discharge cycling.	Usually designed for discharge only; regenerable fuel cells can perform limited charge–discharge cycling.	Usually designed for limited charge–discharge cycling.
Nature of charge storage	As activity of Faradaic materials on electrodes.	As activity of Faradaic materials flowing to electrodes.	As activity of Faradaic materials introduced to electrodes.	As double layer capacitance at electrode/electrolyte interface.
Reactant location and nature	Internal, as active material on electrode surfaces.	External as soluble species in the electrolyte tank-pump-cell flow loop.	External as fuel (anodic reactant) or oxidant, commonly oxygen in air (cathodic reactant).	Internal, as active material on high surface area electrode surfaces.
Charging time	Typically 10–100 hours.	Typically 10–100 hours.	Not applicable.	Typically 1–100 seconds.
Energy storage capability	40–125 $W\,h\,kg^{-1}$ (based on active materials).	10–50 $W\,h\,L^{-1}$ (based on stored electrolyte).	250–1500 $W\,h\,kg^{-1}$ (based on active materials).	3–850 $W\,h\,kg^{-1}$ (based on active materials).

They can last for millions of charge/discharge cycles. The main shortcoming of super-capacitors is their low energy density, particularly compared to batteries. The cost of super-capacitor materials often exceeds that of battery materials due to the increased difficulty in creating high-performing super-capacitor materials, such as graphene and precious metal oxides. Recent advances in creating new super-capacitor materials and improving material production methods may soon bridge the energy density gap for some applications although super-capacitors tend to be appreciably larger or heavier than batteries for a given charge capacity.

Energy devices can be divided into those involving power quality and energy management. The former refers to charge–discharge cycles on the short timescale (seconds to minutes) and includes sag compensation, power smoothing, grid stabilization, and frequency regulation. The latter concerns charge-discharge cycles on a longer timescale (minutes to hours), including load levelling, load following, power balancing, peak shaving, and time shifting, plus improving the grid load factor. A UPS (uninterruptable power supply) is an energy storage service with intermediate features. Depending on the type of device, operating times range from fractions of a second, with response times in the order of milliseconds, to several hours. Power output ranges from a few kW to many MW as shown in Figure 2.4(a).

Figure 2.4(b) shows a general (Ragone) plot of energy density vs power density for various batteries and super-capacitors, showing the characteristic domains of devices.

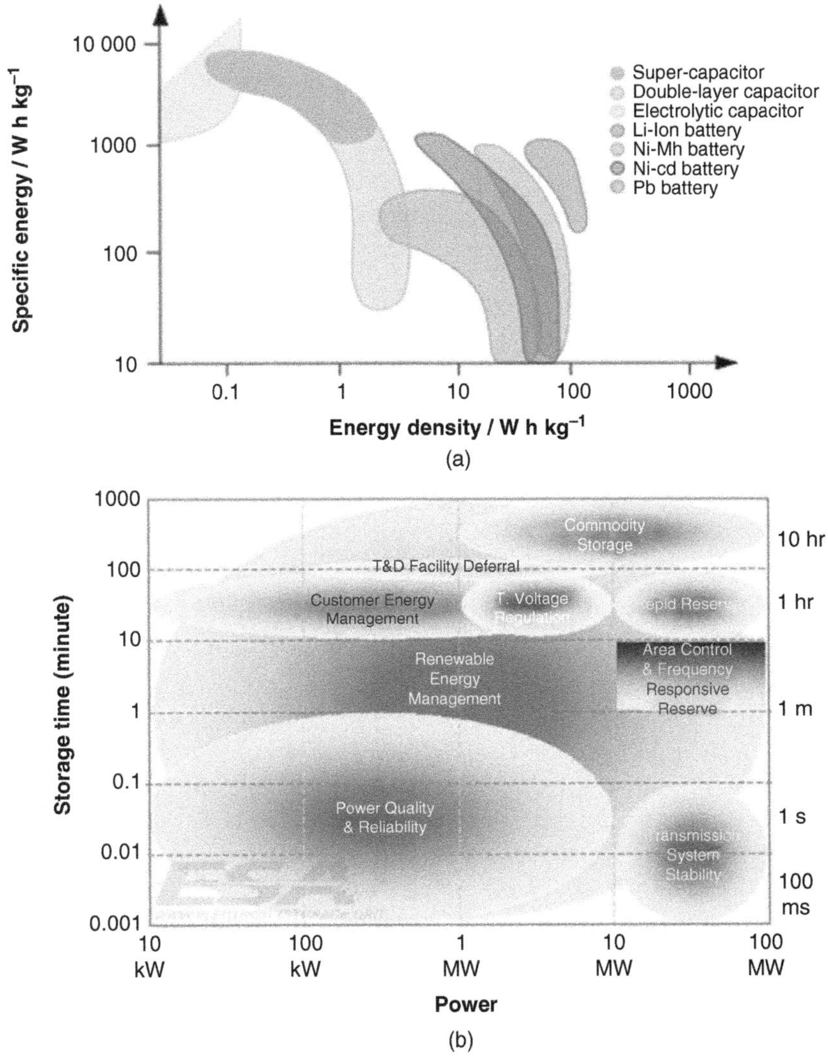

Figure 2.4 Specific energy density vs specific power density (Ragone) plot, showing the domain of various batteries and super-capacitors as electrochemical energy devices. a) After [2]. b) Specific energy density vs power output of various devices. After [3]

2.4a: Source: Elcap/Wikimedia. https://commons.wikimedia.org/wiki/File:Power_vs_energy_density.png. Public Domain

2.4b: Source: P. Alotto, M. Guanieri, F. Moro, Redox flow batteries for the storage of renewable energy: A review, Renewable and Sustainable Energy Reviews, 29 (2014) 325–335.

This type of chart continues to evolve due to improvements in materials and construction together with the advent of hybrid devices.

It is important to rationalize performance and facilitate comparisons by presenting normalized figures of merit. In the case of batteries, the specific charge capacity is often used:

$$q_s = q / w \tag{2.24}$$

and it is important to state if the mass, w, refers to one electrode, both electrodes, or the total mass of the battery, including electrolyte and case.

In the case of redox flow batteries, the charge capacity most usefully refers to the total electrolyte volume in the reservoirs, V_T, i.e. the sum of electrolyte volumes in the positive and negative electrolyte holding tanks; the volumetric charge capacity is:

$$q_v = q / V_T \qquad (2.25)$$

For all batteries, it is essential to refer performance to a known state of charge, a state of charge (SOC) of 100% corresponding to fully charged and 0% indicating a completely discharged battery.

For super-capacitors, it is common to consider the capacitance per unit mass of active material, w, i.e. the specific capacitance:

$$C_s = C / w \qquad (2.26)$$

For practical devices, it is common to use units as follows:

a) Electrical charge, A h
b) Volumetric charge capacity, A h m^{-3}
c) Energy density, W h m^{-2}

where 1 A h = 3600 A s – 3600 C.

Figure 2.5 shows another Ragone plot, highlighting the typical discharge time characteristics of batteries and super-capacitors. Super-capacitors and batteries are both used in some energy storage applications. Traditional capacitors store energy through electrostatic charging at their electrode – electrolyte interfaces under an applied potential. This process allows capacitors to deliver energy very rapidly, but only in small amounts. In contrast, batteries store energy through electrochemical reactions that typically occur throughout the entire bulk of active materials within electrodes, allowing comparatively large amounts of energy to be stored, albeit with a slower delivery than from capacitors. This trade-off between power and energy is illustrated in the Ragone plot in Figure 2.6.

Hybrid devices aim to provide a combination of the characteristics of different power devices, albeit with limitations due to the compromises. For example, a combination super-capacitor and battery can achieve rapid charging then store energy as active electrode material, allowing Faradaic discharge, as in a battery, albeit at lower rates and over a smaller number of cycles than a super-capacitor.

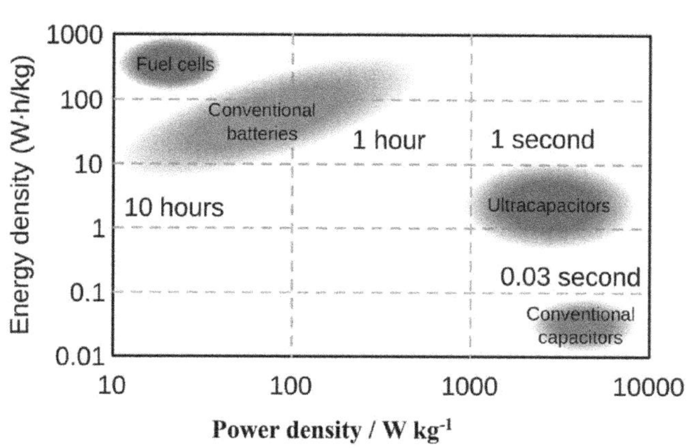

Figure 2.5 A Ragone plot showing the typical discharge time characteristics of batteries and super-capacitors as electrochemical energy devices. After [4].

Source: Stan Zurek, Supercapacitors Chart, Wikimedia Commons, 2017. Available at https://commons.wikimedia.org/wiki/File:Supercapacitors_chart.svg. Licensed under CC BY-SA 3.0.

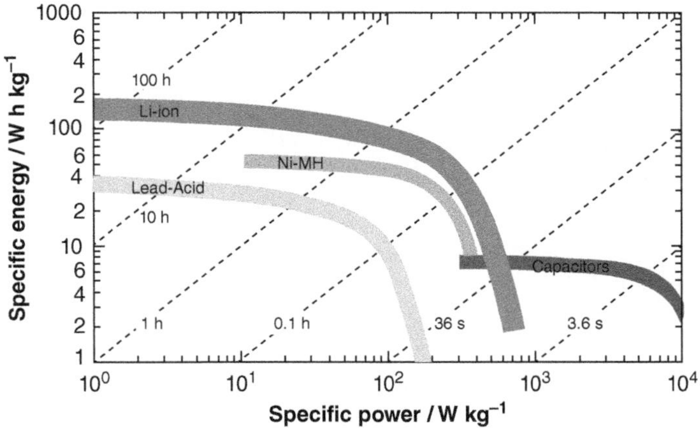

Figure 2.6 A Ragone plot indicating the energy–power trade-off for batteries and super-capacitors. For electric vehicles, the vertical axis is linked to acceleration while the horizontal axis determines range. After [5] Adapted with permission from B.D. McCloskey, Expanding the Ragone plot: pushing the limits of energy storage, *J. Phys. Chem. Lett.* 6 (2015) 3592–3593. © (2015) American Chemical Society and [6].

Acknowledgments

The authors gratefully acknowledge UK EPSRC funding for a Centre for Doctoral Training in Energy Storage and its Applications (EP/L016818/1).

Nomenclature

Symbol	Meaning	Units
A	electrode area	cm^2
C	capacitance	F
Cs	specific capacitance	$F\,g^{-1}$
E^o	standard electrode potential	V
E_{cell}	cell voltage (cell potential)	V
E	energy	J
F	Faraday constant	$C\,mol^{-1}$
I	electrical current	A
j	current density	$A\,cm^{-2}$
n	amount of material undergoing electrochemical change	mol
P	electrical power	W
q	electrical charge	C
q_V	volumetric electrical charge	$C\,m^{-3}$
t	time	s
w	mass of active material	g
z	number of electrons (stoichiometric number)	dimensionless
Greek		
ϕ	current efficiency (current yield)	dimensionless

Subscripts

A	anode
C	cathode
cell	cell
e	equilibrium

Superscripts

o	standard

Abbreviations

e^-	the electron
RFB	redox flow battery
SHE	standard hydrogen electrode
SOC	state of charge

References

[1] Ponce-de-león, C., Walsh, F.C., Rose, A., et al. (2007). A direct borohydride – acid peroxide fuel cell. *Journal of Power Sources* 164: 441.

[2] Alotto, P., Guanieri, M., and Moro, F. (2014). Redox flow batteries for the storage of renewable energy: A review. *Renewable and Sustainable Energy Reviews* 29: 325–335.

[3] Electricity Storage Association http://www.electricitystorage.org/technology/tech_archiv,e/technologycomparisons (accessed 20 April 2017).

[4] Sandia. https://commons.wikimedia.org/wiki/File:Super-capacitors_chart.svg#filelinks Stan Zurek (raster), CC BY-SA 3.0 http://creativecommons.org/licenses/by-sa/3.0/, via Wikimedia Commons (accessed 20 April 2017).

[5] McCloskey, B.D. (2015). Expanding the Ragone plot: Pushing the limits of energy storage. *Journal of Physical Chemistry Letters* 6: 3592–3593.

[6] Srinivasan, V. (2011). The three laws of batteries (and a bonus zeroth law). https://gigaom.com/2011/03/18/thethreelawsofbatteriesandabonuszerothlaw (accessed 20 April 2017).

Further Reading

Arenas-Martínez, L.F., Ponce De León, C., and Walsh, F.C. (2017). Engineering aspects of the design, construction and performance of modular redox flow batteries for energy storage. *Journal of Energy Storage* 11: 119–153.

Bagotsky, V.S., Skundin, A.M., and Volfkovich, Y.M. (2015). *Electrochemical Power Sources: Batteries, Fuel Cells, and Super capacitors*. Hoboken, NJ: Wiley.

Bard, A.J. and Faulkner, L. (2001). *Electrochemical Methods Fundamentals and Applications*, 2nd ed. New York: Wiley.

Conway, B.E. (1999). *Electrochemical Super-capacitors. Scientific Fundamentals and Technological Applications*. Boston, MA: Springer.

Larminie, J. and Dicks, A. (2003). *Fuel Cell Systems Explained*, 2nd ed. Chichester: Wiley.

Leung, P.K., Li, X., Ponce De Léon, C., et al. (2012). Progress in flow batteries, remaining challenges and their applications in energy conversion and storage. *RSC Advances* 2: 10125–10156.

Menictas, C., Skyllas-Kazacos, M., and Lim, T.M. (eds.). (2015). *Advances in Batteries for Large- and Medium-scale Energy Storage: Applications in Power Systems and Electric Vehicles*. Oxford: Woodhead.

Pletcher, D. (2009). *A First Course in Electrode Processes*, 2nd ed. Cambridge: Royal Society of Chemistry.

Pletcher, D. and Walsh, F.C. (1990). *Industrial Electrochemistry*, 2nd ed. London: Chapman and Hall.

Srinivasan, S. (2006). *Fuel Cells: From Fundamentals to Applications*. Boston, MA: Springer.

Vincent, C.A. and Scrosati, B. (1997). *Modern Batteries 2nd Edition: An Introduction to Electrochemical Power Sources*. Oxford: Butterworth-Heinemann.

Winter, M. and Brodd, R.J. (2004). What are batteries, fuel cells, and super-capacitors? *Chemical Reviews* 104: 4245–4269.

3

Standalone Batteries for Power Backup and Energy Storage

*Declan Bryans, Martin R Jiminez, Jennifer M Maxwell, Jon M Mitxelena, David Kerr, and Léonard E A Berlouis**

Department of Pure and Applied Chemistry, University of Strathclyde, Glasgow G1 1XL, UK

3.1 Introduction

Due to the increase in the world population and widespread industrial development, the global energy demand and so energy generation has increased significantly over recent decades. Nearly 70% of this energy is currently being generated through the burning of fossil fuels. To reduce CO_2 emissions and so global warming, much research has been directed toward developing new energy technologies with a small or zero carbon footprint. Energy generation from renewable sources has been perceived as a means of achieving this goal. However, the intermittency of these sources, especially when connected directly to the electricity grid, can be problematic for ensuring the reliability of the supply, to improve power quality, frequency, and voltage. An energy storage system (ESS) is an elegant approach to overcome the intermittency issue as well as play a role in controlling electricity costs. Battery energy storage systems (BESS) represent one option for complementing renewable energy sources by managing the intermittency in their energy production. Just as for other ESSs, this technology can be used for grid stability as well as energy management. However, the modular nature of BESS means that it can be designed to meet the specific requirements of the given application with no restrictions for installation sites, unlike pumped hydro ES or compressed air energy storage (ES). This allows BESS to support remote communities with poor grid connectivity by providing better energy security and better use of their local renewable sources.

3.2 Standalone Battery Technologies

Batteries have multiple common applications ranging from portable devices (e.g. mobile phones) to medical devices (e.g. pacemakers). However, this section focuses on some common examples of battery systems that are utilized as large-scale, standalone systems used in energy management and grid stabilization.

Advances in Energy Storage: Latest Developments from R&D to the Market, First Edition.
Edited by Andreas Hauer.
© 2022 John Wiley & Sons, Ltd. Published 2022 by John Wiley & Sons, Ltd.

3.2.1 Lead-acid Battery

3.2.1.1 Battery Fundamentals

The lead-acid battery is the oldest example of rechargeable battery having been developed by Gaston Planté in 1859. This battery type has a very large power density with a low construction cost making them attractive to the automotive industry. The battery has been under constant development in order to improve safety controls and the low energy density of this system. The half-cell reactions that occur during the discharge cycle are:

$$Pb + SO_4^{2-} \rightarrow PbSO_4 + 2e^- \qquad E^0 = -0.36\ V \tag{3.1}$$

$$PbO_2 + 4H^+ + SO_4^{2-} + 2e^- \rightarrow PbSO_4 + 2H_2O \qquad E^0 = +1.69\ V \tag{3.2}$$

at the anode and cathode, respectively [1]. Thus, during the discharge process, the anodic oxidation causes the sulfate ions, from the sulfuric acid, to react with the generated Pb^{2+} ions, forming a $PbSO_4$ precipitate on the negative electrode (the anode). The cathodic reduction at the positive electrode (cathode) causes the lead dioxide (Pb(IV)) again to form $PbSO_4$ (Pb(II)). This results in both electrodes becoming sulfated, which when combined with the change in acidity from the formation of water and consumption of sulfate ions, causes the voltage to change during the discharge process. Theoretically, this cell reaction can produce 167 W h kg^{-1} [2].

The reverse of these processes occurs during charging of the battery. From the consumption and release of sulfhate anions and protons during charge/discharge, the sulfuric acid concentration within this electrolyte can fluctuate between 0.5 M and 1.0 M. Hydrogen evolution and oxygen evolution can also occur at the electrodes during charge (secondary reactions). Figure 3.1 shows the reactions occurring within this battery.

From all commercially available batteries that use an aqueous solution as an electrolyte, the lead-acid has the highest overall cell voltage of 2.05 V. Monitoring the state of charge (SoC) of this battery system is possible due to the fluctuations in the sulfuric acid concentration through measuring the specific gravity of the electrolyte.

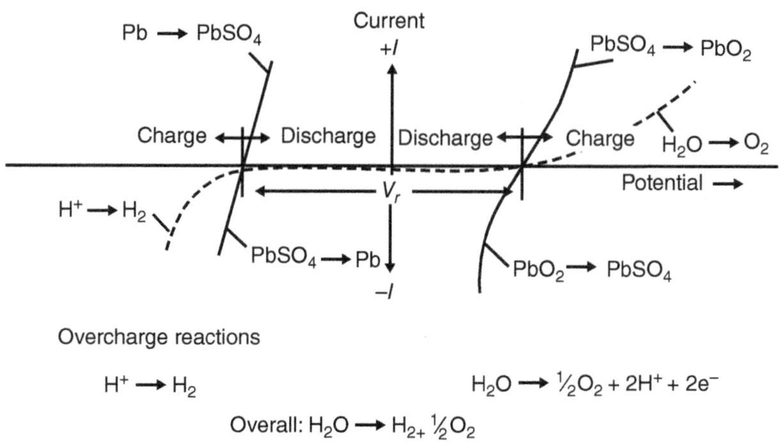

Figure 3.1 I-V characteristics of the reactions in the lead-acid battery [1].

The current collectors for lead-based batteries are typically produced by lightweight materials to maximize the system's energy density. Due to the hazards associated with the possible evolution of hydrogen and oxygen gas in the cell, the flooded lead-acid battery has been replaced with a sealed lead-acid battery. This system is known as the valve regulated lead-acid (VRLA) battery, which recombines the evolved gases into water, so negating the risk of explosion and prolonging the life of this system. The choice of material components in the battery is dependent on the application of this system, e.g. uninterruptible power supply (UPS) uses thicker electrodes to achieve deeper discharges [2].

However, several issues still persist with this technology. The corrosion of the lead alloy occurs at the positive plate during charge and this gets progressively worse over time. This phenomenon occurs due to the high potential at both electrodes that are exposed to the sulfuric acid, and the corrosion rate increases as electrolyte concentration decreases. A porous coating is formed on the grid and this cannot be reversed, resulting in permanent failure. The incorporation of some additives such as lead-calcium–antimony alloys as solid electrode materials in VRLA batteries inhibits this corrosion and prevents the creation of a passivation barrier between the electrode and the electrolyte.

The loss of electrolyte causes the system to dehydrate and results in loss of contact with the active material. The loss of this electrolyte originates mainly from gas evolution or electrolyte evaporation. The VRLA is designed to counteract the gas evolution. However, during discharge, where $PbSO_4$ is formed, the resulting non-conductive material contributes to the increase in the cell resistance, causing higher temperatures and leading to electrolyte evaporation. This necessitated maintenance procedures to be put in place to rehydrate the battery at set time intervals in order to regain some of the lost energy capacity. $PbSO_4$ can also accumulate on the positive electrode after charging instead of being converted to PbO_2. Thus, the positive electrode compresses and does not regain its initial morphology because of its softness. This causes the active material/current collector interactions to decrease resulting in a loss in voltage during discharge and in system efficiency. This is avoided by using lead-calcium alloys and the addition of tin to the support grids to reinforce the PbO_2.

One of the most common causes of lead-acid battery failure derives from sulfation. During high rates of discharge and prolonged storage at low SoC, $PbSO_4$ crystals slowly accumulate and grow as a stable coating on the negative electrode surface, leading to an efficiency loss as the lead cannot be properly reformed from these large crystals. The operation of the battery at low temperatures also influences this sulfation process. The $PbSO_4$ crystals impede sulfate diffusion from the electrolyte, which inhibits the cycling efficiency and so leads to a loss in capacity. The initial morphology of the electrodes is key to mitigating this effect, e.g. the presence of macropores in the deposits increase the surface area, ion movement, and the conductivity of the active materials. Carbon can also be added to impede this sulfation effect at the negative electrode. During discharge, the presence of the carbon increases conductivity, distributing the current more uniformly along the plate and delaying crystals accumulation. If sulfation could be suppressed it would result in a significant increase to cycle life and efficiency.

Two VRLA types exist, classified by their mode of operation on how to manage the gas space, *viz.*, gel and absorptive glass-mat (AGM). The AGM models have the electrolyte totally absorbed within the separators located at the positive electrode, which are made of glass mat fibers. Thus, electrolyte leakage is avoided. The gel electrolyte is a static layer, with some channels where gas can pass. Here, deeper discharge endurances and longer cycle-life are achieved, resulting in gel-type VRLA being implemented as the preferred lead-acid technology for large-scale BESS. Many of the typical battery failures are avoided due to the compacted design of the gel-type VRLA, i.e. there is no need for any electrolyte preservation (e.g. free of periodic water addition to the electrolyte for rehydration), acid stratification, and electrolyte leakage. There are no serious safety issues, and homogeneity is achieved within the battery giving a more uniform reaction throughout. Lead-calcium-antimony and lead-calcium alloys are used as positive and negative plate grids respectively to prevent failures and increase the system efficiency. The structure of the gel-type VRLA can be seen in Figure 3.2.

Another variation is the lead-acid "Ultrabattery™," which utilizes an ultracapacitor along with the lead-sponge plate of the battery [1]. This type of Ultrabattery™ model could be defined as a hybrid electric asymmetric ultracapacitor, as a parallel-connected ultracapacitor exists within it, which can store 100 times more energy than conventional capacitors and provides instant power. The current (Faradaic and capacitive) is shared between the negative electrodes and so is protected from high-rates of discharge. As a result, the efficiency, lifetime, and high-charge acceptance under a partial SoC are improved compared to conventional lead-acid batteries. The Ultrabattery™ allows for improved cycle-life and endurance for high depths of discharge (DoD) with both cycle-life and capacity improvement. This battery type is unique in the market with its powerful discharge density rate output, having an immediate response. With the addition of selected carbon material to the negative electrode forming alloys, the battery acquires a longer cycle-life. These two surface layers, where the reactions occur rapidly and in a reversible way, can achieve 100% DoD over 1000 cycles. The thickness of these carbon additive electrodes also enhances the energy capacity of the battery.

For these batteries, with high recyclability, the capital cost is the lowest of all batteries. Lead-acid batteries can perform under high temperature conditions and their maturity has helped to improve and demonstrate their reliability and safety in ESS

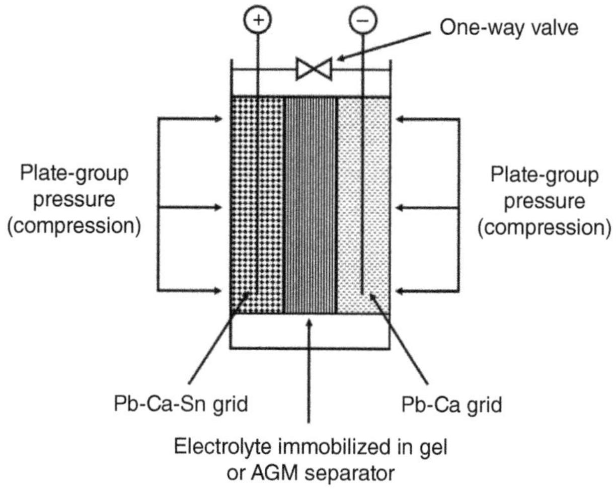

Figure 3.2 Main components of the VRLA [1].

installations. According to the International Renewable Energy Agency (IRENA), current lead-acid installations costs are below US $200 per kW h [3]. These characteristics have allowed the technology to continue to be a popular choice in the automotive industry as well as being the battery of choice for a series of large-scale standalone installations. This is particularly so in the US where a series of projects are currently underway.

3.2.1.2 Large-scale Projects

The Notrees Wind Farm, in Texas, has the biggest lead-acid BESS implemented in the world [4] (Figure 3.3). Energy is obtained from a 153 MW wind farm. The project is led by the largest electricity power company in the US, Duke Energy, along with the support from the Electric Power Research Institute (EPRI) and the Xtreme Company. The battery delivers a ca. 36 MW output.

A key characteristic of the facility is its fast response when it is required, thus optimizing dispatch, grid reliability, and managing energy supply/demand. It provides an excellent model for other companies due to the reliability and efficiency of the facility. In a few milliseconds, the BESS can translate from a standby status to its maximum output. The BESS is hosted in a substation of 6000 square meters, next to the wind station, with both systems connected to the grid. The battery technology used here is the Advanced-VRLA/Ultrabattery™ type, which has a high performance and cycle-life with a successful grid stability and fast response of energy storage when demand is high; suitable for grid management. The BESS has a capacity of 24 MW h. Data from 2015 indicates that the project's net revenue has increased by 35%, yielding US $2.8M gross profit and it is expected to pay back Duke Energy's investment within nine years. The entire system consists of 1200 individual Powercell packs, each providing 1 MW h energy storage and a bidirectional inverter/charger PCS. The batteries are collocated in two parallel rows, each providing 500 kW h of energy storage. Both racks are connected to the PCS; 24 parallel packs in total. By 2015, the plant had stored 340 033 MW h of energy with an average efficiency of 60%. In December 2017 however, Younicos [5] completed a replacement programme of the installed lead-acid batteries at this site with advanced Samsung SDI lithium-ion batteries [6].

The BESS installation at Metlakatla, Alaska, involving GNB Industrial Power and Exide Technologies has substituted fossil-fuel-derived energy and typical hydro units from the sea, decreasing energy costs from US $5M to $200k in 11 years [7]. A 2800 m^2 building hosts the entire BESS. The Exide company employed 378 GNB Absolyte™ VRLA batteries of 1.4 MW h capacity in this project. The BESS collects energy from the excess of hydroelectric generators, operating continuously and is maintained under partial state of charge. It had an eight-year lifespan and 1200 cycles of total discharge capacity, rated at 1 MW. Satisfactory results were obtained after the first eight years of the cell, as it was maintained in good condition with no sulfation observed and it remained relatively free of corrosion.

The Kaheawa and Kawailoa wind farm is one of the largest renewable energy projects existing on Maui, Hawaii and it has been operating since 2006. This 30 MW facility provides energy to 11 000 homes annually and consists of 34 wind turbines in total, each at 1.5 MW. Thanks to the 10 MW BESS used, 126 000 tons of CO_2 annually are avoided. The turbines are connected to the electrical substation by underground

Figure 3.3 Lead-acid installation at the Notrees Windfarm 6000 m^2 substation, West Texas [4].

electrical power lines. This substation absorbs excess wind energy. The installation is supported by Younicos, which implemented a lead-acid battery of high capability [8]. The battery was capable of increasing or decreasing output when required. The entire EES consist of a PCS integrated with lead-acid batteries. The facility though has recently been upgraded, with the lead-acid batteries replaced by advanced Samsung SDI lithium-ion batteries [6].

The Pennsylvania–New Jersey–Maryland (PJM) Interconnection Project regulates the services to the grid and the frequency on peak demand, matching supply of electricity to demand in the PJM Interconnection from solar PV system and wind farms. The BESS here consists of the Deka Ultrabattery™ technology, providing continuous frequency regulation at 3 MW power to the grid. At the same time, the BESS is employed for peak demand management [9].

3.2.2 Lithium-ion Battery

3.2.2.1 Battery Concepts

Lithium batteries provide high energy and power densities, due to the low density (0.534 g/cm^3 at 20ºC) and large electrochemical potential (E^0 = –3.045 V) of lithium. This element reacts violently with water, so lithium metal is not used in modern lithium-based batteries. Instead, Li-ion batteries are used with intercalated graphite for the negative electrode and different lithiated transition metal oxides, such as $LiCoO_2$ (LCO), for the positive electrode. The active materials used in both the positive and the negative electrodes have different Gibb's Free energy for Li^+ intercalation and so, depending on the nature of the cathode material, the cell voltage varies. A lithium salt is used as the electrolyte, dissolved in an organic solvent (Figure 3.4). The reaction between the charged anode material and organic solvents forms a protective layer (solid electrolyte interphase [SEI]), which allows lithium-ion transport and protects the electrolyte from further reactions [10].

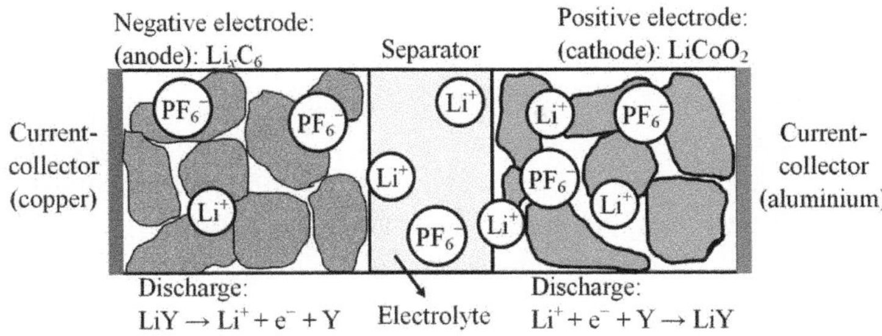

Figure 3.4 Basic design of a LCO Li-ion cell with $LiPF_6$ as the lithium salt [10].

During discharge, the intercalated lithium ions in the carbon travel through the electrolyte and intercalate into the positive electrode, as represented by:

$$xLiC_6 \rightarrow xLi^+ + xe^- + xC_6 \tag{3.3}$$

$$Li_{1-x}CoO_2 + xLi^+ + xe^- \rightarrow LiCoO_2 \tag{3.4}$$

During the first charge, the lithiated positive electrode provides the necessary Li^+ ions for intercalation into the negative electrode. The performance of the positive electrode depends mostly on the composition and crystalline structure of the electroactive material. As noted above, transition metal oxides are commonly used as active materials, and many of them form one of these three crystalline structures: olivine lattice, layered structure, or spinel lattice. Commercially, lithium cobalt oxide (layered) and lithium manganese oxide (spinel) are the most used active-materials.

3.2.2.2 Battery Components

Introduced in 1991, by the Japanese Sony Company, the lithium cobalt oxide ($LiCoO_2$, also known as LCO) system was the first mass produced component for the positive electrode for the Li-ion battery. While charging, up to half of the lithium is extracted from the positive electrode and correspondingly, half of the Co^{3+} is oxidized to Co^{4+}, in order to balance charges. It provides a potential between 3.5 V and 4.3 V vs Li/Li^+, a long cycle-life, and is safe enough for commercial use. Despite cobalt being environmentally hazardous, this layered compound is extensively used in small commercial Li-ion batteries for portable applications (e.g. mobile phones).

The $LiMn_2O_4$ (LMO) has a potential of 4.0 V *vs* Li/Li^+. Despite the energy density being lower than for LCO, the 3D spinel structure allows a faster insertion of Li^+ ions. Due to steric reasons solvent molecules cannot intercalate in the positive electrode, and this leads to a lower electrode decomposition comparing to LCO. The high stability and performance, safety, low toxicity, and non-polluting character of LMO have made this system a really good alternative to LCO, especially for large-scale implementation. However, poor cycling and storage stability are observed at high temperatures, and the dissolution of Mn^{2+} in the electrolyte can result in the poisoning of the negative electrode.

Carbon-based materials are mostly used in commercial cells for the negative electrode, as they provide high stability and more negative redox potentials. Among the carbon-based materials, graphite (natural or artificial) and amorphous carbon are most often employed. Graphite consists of carbon atoms that are arranged in hexagonal layers between which the Li^+ ions intercalate. However, because of its structure, some organic solvents tend to co-intercalate with lithium ions, reducing the maximum charge capacity. During charge (intercalation), the distance between adjacent layers increases by 10%. Amorphous carbon is known for its high level of structural disorder. It provides good electrode cycling stability, power, and better safety than graphite, although with a lower capacity and higher cost.

The electrolyte allows ionic conductivity between electrodes and it must remain stable, non-poisonous, and non-flammable over a large temperature range. Most commercial cell electrolytes are liquids with solvated Li^+ ions or polymers that allow good ion mobility. Liquid electrolytes typically consist of a lithium salt dissolved in an organic solvent. During the first charge, the organic solvent is decomposed around the negative electrode and the SEI is formed. This interphase, which does allow Li^+ ion transport, prevents further electrolyte decomposition. The most common organic solvents are propylene carbonate (PC), ethylene carbonate (EC), and gamma-Butyrolactone. Additives are usually used to improve the efficiency of the electrolyte such as SEI forming improvers, positive electrode protecting agents, and safety protecting agents.

Although Li-ion batteries are generally safe, combustion may occur due to leakage of the liquid electrolyte and/or short-circuit of the electrodes. Solid electrolytes avoid this safety issue by providing a safe ion conduction mechanism, where no liquid substance is involved [11]. These electrolytes can be inorganic compounds or solid polymer electrolytes (SPE). SPEs are mostly based on polyethylene oxide (PEO) – one of the few polymers capable of dissolving lithium salts, particularly, the highly conducting lithium bis(trifluoromethanesulfonyl)imide (LiTFSI) electrolyte. Inorganic solid electrolytes can be either crystalline or amorphous, where Li^+ ions move through empty spaces in the structure. These electrolytes tend to have a lower conductivity than traditional liquid electrolytes, yet they are not as flammable as the liquid ones. Lithium lanthanum zirconate ($Li_7La_3Zr_2O_{12}$) and lithium phosphorus oxynitride ($Li_{2.9}PO_{3.3}N_{0.46}$) are good examples of inorganic solid electrolytes.

The purpose of the separator is to prevent electronic contact between the positive and the negative electrode, i.e. to block the direct flow of electrons between them and permit only ionic conduction. A Li-ion cell separator must be permeable (porous) and compatible with the electrolyte. It should also be as thin as possible to minimize Ohmic losses, and its pores should be well distributed in order to get optimum ion transport. Thanks to the chemical design and structure of the separator, the cell automatically shuts down, due to the separator interrupting ion transport through the closing of pores, when the cell temperature inside reaches a critical value. Small commercial batteries use a single-layered polyethylene separator, whereas bigger industrial batteries usually use triple-layered separators, such as polypropylene/polyethylene/polypropylene.

3.2.2.3 Safety and Costs

Safety is a primary concern for a Li-ion battery. Prevention and protection systems can be found in nearly every Li-ion battery. These include electronic protections, and mechanical and electric design to try to prevent any possible hazard, while other mitigation systems try to reduce the consequences of any accidents or failures that may occur [12].

Despite its initial high cost, Li-ion is now also emerging as an alternative for large-scale applications. Building such batteries involves taking extra safety measures. Fire and explosions can be extremely hazardous on this scale, whereas environmental issues must also be considered, i.e. in case of leakage, as although some chemicals may not be very harmful to humans, they can cause major damage to nearby fauna and flora. The risks of fires, explosions, and gas venting are very high with these high energy density systems and so, due care must be taken in the design and engineering of these cells.

Thermal runaways generally occur when an exothermic reaction goes out of control. This arises from a series of events such as (i) the breakdown of the SEI due to the high temperature with the negative electrode reacting with the electrolyte in a rapid and uncontrolled manner, (ii) heat from SEI decomposition favoring the reaction between intercalated lithium and the organic solvent, releasing highly flammable hydrocarbon gases, (iii) at around 130°C the polymer separator decomposes, provoking a short-circuit between positive and negative electrodes, (iv) heat from the electrolyte decomposition causes the breakdown of the metal oxide electrode (positive electrode), releasing oxygen, which now enables the burning of the hydrocarbon gases leading to total breakdown of the cell by fire and/or explosion. The thermal runaway risk is measured by a "Reactivity level," a scale in which values from 0 to 7 are given depending on the potential risk of the cell. Here, the overall chemistry and cell conditions, including environmental conditions, are considered [12]. At very high reactivity levels (5–7), cells can heat fast enough to catch fire or explode. Avoiding thermal runaway is a complex problem but suppressing or reducing the activity of active materials is a common way to improve the thermal runaway tolerance of the cell.

Availability of materials, processing cost, battery assembly, and future ancillary costs are key factors in large-scale applications, yet policy and possible environmental impacts must also be considered [13]. Battery storage capacity is directly related to the amount of active material used. Both raw material costs as well as production costs must be considered, the latter including manufacturing high purity electrodes and electrolytes, battery assembly, current collectors, and safety features. Lithium carbonate costs less than $5/kg, and other materials such as cobalt and manganese oxide are also relatively inexpensive [10]. It is the ancillary materials, current collectors, electrodes, and extremely thin separators that make these batteries so expensive. It is estimated that only 40% of the price of a conventional Li-ion battery is determined by the materials, whereas ancillary costs represent 60% of the total cost. Manufacturing highly pure organic solvents is also expensive, and the electrolyte conductivity on dissolving the lithium salts is relatively low compared to aqueous electrolytes. Lower conductivity means that the separators must be thinner in order to achieve a good performance. Lastly, cost can also be increased due to geopolitical reasons since importation costs can vary considerably depending on the relationship between different governments. Nevertheless, the Li-ion battery manufacturing cost has decreased considerably over the past decade, and an even larger decline (up to 70 %) is expected within the next 15 years.

3.2.2.4 Large-scale Projects

The Tehachapi Wind Energy Storage Project [14], built in 2014, is located at Southern California Edison's Monolith Substation in Tehachapi, California (Figure 3.5).

Figure 3.5 Tehachapi Wind Energy Storage Plant, Southern California [14]. Batteries located in 6300 square foot building with PCS in two 40-foot containers.

This project includes an 8 MW/32 MW h BESS. This project has been used as a demonstration project of a Li-ion BESS connected directly to the grid and in the integration of intermittent resources, in this case, wind. The battery storage devices and technology were developed and manufactured by LG Chem [15]. The battery is divided in to four sections, each section containing 151 battery racks connected in parallel. Each rack contains 18 battery modules connected in series and a battery management system (BMS). Each section contributes 2 MW of power. The total cost was of the project US $49.5 M. However, apart from storing wind energy, this plant also fulfils various other functions such as grid stabilization, decreasing transmission losses, and diminishing congestion.

The Feldheim Regional Regulating Power Station located in Brandenburg, Germany functions as a regulating power station at a regional level [16]. It was opened by German wind turbine manufacturer Enercon and wind developer Energiequelle and commissioned in September 2015. With a total power output of 10 MW and 10.8 MW h of capacity, it is the largest of its type in Europe. Inside the plant there are 3360 Li-ion storage modules, designed and manufactured by LG Chem [17]. This plant provides load-balancing power to the local grid, relieving it in case of oversupply and supplying power when required, thanks to the energy previously obtained from nearby wind turbines. Its total cost was US $14.3M.

An energy storage system was installed in 2013 and commissioned in 2015 at Nishi-Sendai Transformer Substation in Sendai City, Japan by the Tohoku-Electric Power Company [18]. It comprises of a Li-ion battery system, with a capacity of 20 MW h and a nominal output of 20 MW, although a peak output power of 40 MW can be achieved for a short period of time. It has stabilized the grid frequency fluctuation caused by the introduction of new solar power facilities in the area. The system has been designed to verify the frequency control logics and analyze the improvements achieved in combining the BESS and thermal storage with renewable energy. The performance of the battery itself, in terms of charge/discharge levels and efficiency losses has been successfully assessed. Indeed, in February 2016, Toshiba announced that their second Li-ion battery system, rated 40 MW/40 MW h, supplied to the Tohoku Electric Power

Company had started operation at the Minami-Soma Substation Project (in the Fukushima prefecture) to verify the improvement of supply–demand balance with large-capacity power storage systems [19].

EDF Renewables, the subsidiary of the giant French energy utility company Électricité de France (EDF) have developed and built a large Li-ion battery storage plant in McHenry, Illinois [20]. The project started operation in December 2015 added 20 MW/8.5 MW h of energy storage to the PJM Interconnection regional transmission organization and is intended to serve in both the regulation and capacity markets. The battery and power electronics were provided by BYD America and consisted of 11 containerized units.

One of the world's largest Li-ion battery installations has been carried out by the San Diego Gas & Electric in partnership with AES Energy storage at Escondido in California. This was deployed in February 2017 and consists of a 30 MW/120 MW h storage system, to help SDG&E achieve its clean energy and low carbon emission goals in grid stabilization and providing green energy to its customers [21].

3.2.3 Redox Flow Batteries

3.2.3.1 Battery Concepts

Redox flow batteries (RFBs), of particular interest in this study, can provide a solution to large-scale energy storage and a more efficient link between energy production and energy demand [22–23]. This type of battery system has the advantage of lower cost, rapid response, low level of self-discharge, and is considered to have a much safer operation compared to other battery systems such as the sodium sulfur and lithium ion batteries [24–25]. Additionally, as with all battery systems, it has the advantage of being more flexible and mobile in relation to non-electrochemical technologies, such as pumped hydro and compressed air storage. The latter large-scale energy management technologies are restrained by the suitability of the terrain whereas batteries, such as RFBs, can be readily installed anywhere [26].

The RFB typically consists of two external reservoirs that store the electrolyte containing the electroactive species, as shown in Figure 3.6. By controlling the total volume and concentration of the electroactive species the energy storage density of this storage system is controlled [27]. The stacks, which consist of two electrodes separated by a separator or membrane with the electrolyte flowing either side, control the power density of the battery. As with all energy storage applications, high power and energy capacities are often required to meet electrical demand. RFBs achieve this through stacking unit cells together in an electrical series to increase the voltage and then these stacks can be electrically connected in parallel to obtain high power levels. Figure 3.7 shows the typical set up of "bipolar" electrodes in order to improve the system's feasibility by reducing the weight and volume of the system while maximizing the voltage and current output [22].

The vast range of RFBs available can be categorized in three types: aqueous, non-aqueous, and hybrid. However, only aqueous flow batteries have been developed for large-scale application and the focus of this work will be on two of these systems. Note that aqueous RFBs are constrained to the limits where water electrolysis occurs with hydrogen evolution occurring below −0.44 V *vs* SHE

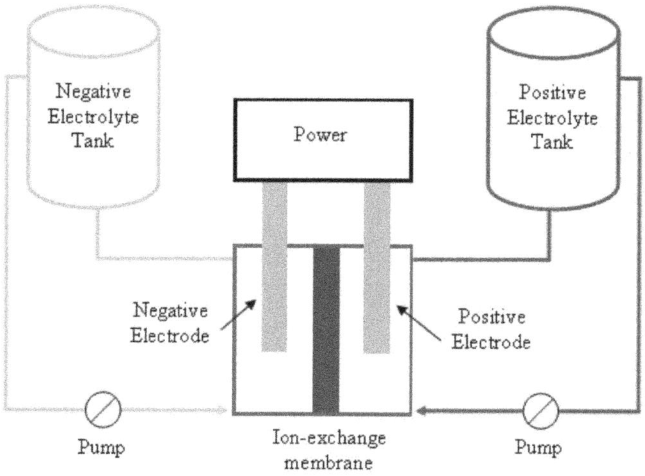

Figure 3.6 Typical structure of a redox flow battery.

(on carbon electrodes) and oxygen evolution above 1.23 V *vs* SHE. The engineering aspects of RFBs are covered in [28].

3.2.3.2 Vanadium Redox Flow Battery (VRFB)

As noted above, aqueous RFBs are among the most developed, with numerous flow battery systems already being demonstrated [29–30]. As the name suggests, these are systems with a water-based electrolyte. Although they have a high power density, these batteries have low energy densities and high material costs. Several developments, such as that of the mixed acid electrolyte [31] in the vanadium redox flow battery have yielded higher energy densities, but they are still struggling to compete with other battery technologies. The all-vanadium RFB is the most iconic and commercially available of all the RFBs. Developed by M. Skyllas-Kazacos et al. in 1988, the vanadium RFB has seen considerable development both at the fundamental and industrial level [26, 32–33]. One of its advantages is its resilience to electroactive species cross-membrane contamination. Since the same element is used on both sides of the cell, should the electroactive species cross over, the electrolytes can simply be regenerated through remixing and controlled electrolysis without destruction of any of the materials or the requirement for the system to undergo complicated separation treatment. The Generation I – VRFB [34] employed concentrated sulfuric acid as the supporting electrolyte. This VRFB gives the following reactions during discharge [35]:

Anode:

$$V^{2+} \rightarrow V^{3+} + e^{-} \quad E^0 = -0.26 \text{ V } vs. \text{ SHE} \tag{3.5}$$

Cathode:

$$VO_2^{+} + 2H^{+} + e^{-} \rightarrow VO^{2+} + H_2O \quad E^0 = 1.00 \text{ V } vs. \text{ SHE} \tag{3.6}$$

Overall:

$$V^{2+} + VO_2^{+} + 2H^{+} \rightarrow V^{3+} + VO^{2+} + H_2O \quad E^0 = 1.26 \text{ V} \tag{3.7}$$

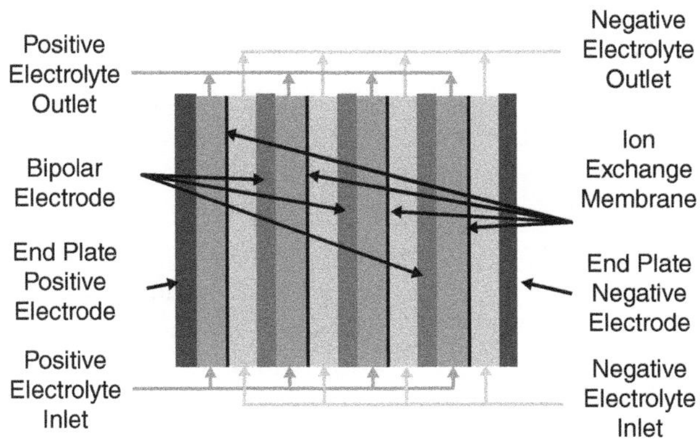

Figure 3.7 Bipolar plates of stacked redox cells [20].

As indicated above, this gives an overall open circuit voltage of 1.26 V under standard conditions. The energy density is limited by the concentration at which the different vanadium ions can stay within solution. Currently, this is 2 M $VOSO_4$ in 2.0 M sulfuric acid. Above this concentration the VO_2^+ (V(V)) ions can precipitate out as V_2O_5, especially when temperatures are above 40°C, Conversely, the V^{2+}/V^{3+} exists as solid vanadium oxide at temperatures below 10°C [36]. This limits most practical examples of these batteries to operate at a temperature range between 10°C and 40°C and a concentration of less than 2 M. Such concentrations give an open circuit of 1.6 V when fully charged. The VRFB has the ability to undergo numerous charge–discharge cycles leading to the system having a better level led out cost of electricity (a measure of economic value over the potential lifetime of the technology), despite the vanadium having a high cost. On top of this, the system also has a 70–90% energy efficiency due to fast kinetics and can be over charged or undergo deep discharge with no lasting damage to the system. Nevertheless, it is worth noting that when the cell is overcharged, side reactions leading to hydrogen evolution can occur at the cathode:

$$2H^+ + 2e^- \rightarrow H_2 \tag{3.8}$$

Gas evolution can affect the flow of the electrolyte, create imbalance in the electrolyte, increase the cell resistance, and alter the pH of the solution, which can affect the performance of the proton-exchange membrane. This gas evolution is kept to a minimum by increasing the overpotential of this reaction through the use of carbon electrodes.

The main drawbacks of the Generation I model (*viz.* the high cost of vanadium, low power density, and temperature limits) was tackled in the Generation II model. This system replaced the VO^{2+}/VO_2^+ of the half-cell with the polyhalide couple $Br^-/ClBr_2^-$ [37]:

$$V^{2+} + 0.5\ Br_2 \rightarrow V^{3+} + Br^- \quad E^0 = 1.35\ V \tag{3.9}$$

Using a mixed acid electrolyte of HBr and HCl to ensure the availability of bromide ions on the positive side, the battery's energy density almost doubled to ca. 50 W h kg^{-1} while achieving an energy efficiency of 80% [34]. The concentration of vanadium could be increased to 3 M, with the precipitation of V^{4+}/V^{5+} no longer a factor in this system, so increasing the energy density. Furthermore, the total volume required on the positive half-cell could be halved to match that of the negative half-cell, reducing the total volume of electrolyte in the system by a quarter while still maintaining an increased energy density. However, work was still required in the need of more stable electrode materials, cheaper ion-exchange membranes, and solving the toxic vapour issue from the electrogenerated bromine. The latter was tackled by employing two common complexing agents, *N*-methyl-*N*-ethylpyrrolidinium bromide and *N*-methyl-*N*-ethylmorpholinium bromide, which could complex with the generated bromine and form an immiscible phase in the aqueous electrolyte, so reducing the vapour pressure of the bromine. However, the additional cost of the complexing agents, alongside the existing cost of the membrane material and the vanadium electrolyte, led it to be considered inferior overall to the advantages of the all vanadium RFB.

This led to the Generation III VRFB breakthrough, which used the same reactions as the Generation I model. By employing a mixed sulfuric and hydrochloric acid electrolyte, the energy capacity increased by more than 70% over the original Generation I model and even surpassed the Generation II model. The increased energy capacity stems from the acidic mixture being able to stabilize the electrolyte. Studies have indicated that the VO^{2+} species concentration is determined by the solubility of $VOSO_4$ and the V^{3+} by the solubility of both $V_2(SO_4)_3$ and VOCl. The V^{2+} is stable in both the mixed acid electrolyte and sulfuric acid electrolyte. In addition, the mixed electrolyte also improves the effective temperature range in which the battery can operate, from −5°C to 50°C. This comes from the formation of $VO_2Cl(H_2O)_2$ at elevated temperatures, which maintains its solubility as opposed to forming the insoluble V_2O_5. This removes the need for expensive temperature controls and cooling systems. This version gives an energy efficiency of 87%. These have been implemented into the market by various providers such as RedT Energy (UK), Prudent Energy [38], UniEnergy Technologies (US), and Sumitomo (Japan) for large-scale implementations.

3.2.3.2.1 Large-scale Projects: Sumitomo Electric Industries Ltd.

Sumitomo Electric Industries Ltd. (SEI) have been involved with VRFBs since 1985 in partnership with Kansai Electric Power Company in Japan [39]. They deployed their first large-scale system in 2001, a 3 MW unit that could function as a UPS, delivering 3 MW for 1.5 s at a liquid crystal display manufacturing plant or for peak shaving, 1.5 MW for 1 h. There have been many other VRFB installations by SEI over the past two decades or so and one of their latest has been the 60 MW h (15 MW for 4 h) system at the Minami-Hayakita substation of Hokkaido Electric Power Company in 2015 [40]. This demonstration project verified the VRFB's performance in all the functional demands placed on it but it was clear that cost had to be driven down in order to make the system more appealing for widespread use. Their latest development is with San Diego Gas and Electric Utility with a 2 MW/8 MW h installation (Figure 3.8) to assess the effectiveness of the integration of the battery with a renewable energy facility to improve flexibility and grid stability [41].

The outcome of this four-year project, commissioned in 2017, will monitor the voltage frequency, power outage support, peak shifting on providing power equivalent to 1000 homes in the area.

Figure 3.8 Sumitomo Electric Industries VRFB Installation in San Diego. [41] VRFB units in shipping containers.

Dalian, a city in China, experiences adverse weather conditions that result in frequent power outages. UniEnergy Technologies (UET) and Rongke Power are partnering in the manufacturing of the world's largest battery storage system, which will consist of 10 20 MW/80 MW h VRFBs [42]. The battery will be there to provide electrical energy and enhance grid stabilization for this region in China. In addition, this large-scale battery will form an additional load centre and provide black-start capabilities in the event of an emergency. The battery will be installed in two phases. The first phase is set to be completed at the end of 2017 and the second phase will be completed at the end of 2018. Once fully commissioned by year 2020, it will be able to peak shave 8% off the expected load.

Another company producing VRFBs is called redT Energy Storage [43]. redT is a company who have been researching and developing energy storage technology for the past 15 years or so. Their VRFBs have a lifetime of up to 25 years with 100% depth of discharge and a long storage duration. They produce different sizes of the VRFB with many different power ratings and capacities, ranging from 5 kW/20 kW h to 60 kW/300 kW h modules.

The CellCube [44] was created by the company "Gildemeister Energy Solutions." Research first started in this area in 1999 but the first field trials did not happen until 2004. After the first flow system was introduced in 2008, Gildemeister started to participate in the production and manufacturing of the CellCube system in 2010. With low maintenance, intelligent sensors, and control functions it is clear that the CellCube is one of the most promising VRFBs on the market today. The battery has a capability of 100% discharge depth and an energy efficiency of about 80%. There are three different storage sizes that can be purchased and these are 400 kW h, 800 kW h, and 1,600 kW h systems. All these batteries have a normal discharge power output of 200 kW. The battery can be used for grid support, backup, wind and solar parks, and re-powering applications. The CellCube has a projected lifetime of 20 years and is currently used all over the world at more than 8000 installations.

3.2.3.3 Zinc Bromine Redox Flow Battery (ZBFB)

The zinc bromine hybrid RFBs have been shown to yield high energy and power densities with lower costs but lose the ability to scale up power and energy independent of each other as when charged, zinc deposits on the electrode as opposed to remaining in the electrolyte [45]. Zinc is an attractive material for use in the flow battery due to its natural abundance and low cost [46]. Additionally, zinc also has a large negative potential that allows for the system to have a high power density when coupled with another redox couple for the positive electrode side. During the charge–discharge cycle of the redox flow battery the zinc reaction is [22]:

$$Zn^{2+} + 2e^- \rightarrow Zn \qquad E^0 = -0.76\ \text{V}\ \textit{vs.}\ \text{SHE} \tag{3.10}$$

The Br_2/Br^- redox couple is used as the positive electrode reaction in the zinc–bromide, polysulfide bromide, and vanadium–bromide batteries as it offers a large positive potential resulting in a high cell voltage for the battery system. At the positive electrode, the reaction is:

$$Br_2 + 2e^- \rightarrow 2Br^- \qquad E^0 = +1.09\ \text{V}\ \textit{vs.}\ \text{SHE} \tag{3.11}$$

Here the bromide converts into bromine liquid. However, bromine liquid is still very volatile (vapour pressure of 202 mmHg at 298 K) and will result in some Br_2 vapour being formed, meaning that some of the overall battery efficiency will be lost due to this gas escaping into any available head space in the reservoirs. Additionally, bromine is extremely toxic by inhalation (LC50 750 ppm 1 h (mouse)) [47]. This system also has other issues, such as the growth of dendritic zinc and cross over of bromine. Despite these issues, the zinc bromine RFB is the second most commercially developed flow battery after the VRFB.

To counteract the bromine volatility issue, as with the Generation II VRFB, a series of additives are employed to complex with the electrogenerated bromine and hold it within the positive electrolyte. These additives are typically quaternary ammonium compounds that capture the Br_2 and complex this to higher polybromide forms (e.g. Br_3^-, Br_5, $^-$or Br_7^-) [48]. Normally, the $Q^+\ Br_x^-$ yields an immiscible liquid phase, which requires these battery systems to have additional pumping procedures for the positive electrolyte during discharge. There has been a great deal of interest in synthesis compounds that would reduce the self-discharge of the zinc bromine battery. One of the first papers to relate it to batteries was from F. Rallo et al. [49], where this method was proposed as a way to reduce the solubility of the bromine in the aqueous phase by using low molecular weight tetraalkylammonium halides and perchlorates to form the barely soluble polyhalides [49]. The formation of these liquid-phase polybromides was investigated by D.J. Eustace, where unsymmetrical substituted cyclic quaternary ammonium bromides (QBr) were used [50]. Three were specifically investigated, *viz.*, *N*-ethyl-*N*-methylmorpholinium bromide, *N*-chloromethyl-*N*-methylpyrrolidinium bromide, and *N*-methoxymethyl-*N*-methylpiperidinium bromide.

All these compounds were found to create an immiscible bromine-containing oily phase, which was modelled as a micelle-like process, where the dominant phase separation process is controlled by the polyhalide ion activity. From the complexing agents used *N*-methoxymethyl-*N*-methylpiperidinium bromide was found to be unstable whereas *N*-ethyl-*N*-methylmorpholinium bromide has been successful in several zinc–bromine commercial systems. However, these systems require additional pumping and valve controls to ensure the full dispersion of the immiscible phase, as shown in Figure 3.9.

There exists, nevertheless, a considerable interest in developing novel additive compounds that could lead to improvements in the cycling efficiencies, kinetics, cost of materials, or physical nature of complexed polybromide phase achieved [51–52]. Very few papers focus on the immiscible phase itself other than giving an analysis of its impact on the electrokinetics [53]. However, the mass transport issues created by having the secondary phase have been recognised by Yang et al. [54]. In that study, surfactants were employed to break up the immiscible phase to improve its dispersion within the aqueous electrolyte. Using a small quantity of a polysorbate (polysorbate 20), they managed to increase the dispersion of the polybromide complex phase within the aqueous phase, leading to an increase in the coulombic efficiency.

The second major issue, which affects all zinc systems, is the formation of dendrites. These dendrites are finger like projections that protrude from the negative electrode surface when the battery is being charged and zinc is being deposited. These dendrites can potentially reduce the coulombic efficiency (if they break off) or drastically reduce the RFBs lifespan (should they pierce the membrane and cause a short-circuit). A high rate of zinc electrodeposition and dissolution is usually desirable for cost and performance related parameters, such as for high power and keeping pumping costs low. Therefore, the morphology of such depositions are key to improving the current zinc systems and make them more efficient and cost effective [55].

Additives, such as polyethylene glycol (PEG, MW = 200) [55], are commonly used to inhibit the growth of these dendrites. PEG has a good stability and is resistant to chemical degradation during the charge/discharge cycles of the zinc-bromine battery. Ethanol has also been employed in a zinc iodine RFB to successfully inhibit the growth of these dendrites [56]. This led to a finer grain of deposits being produced and this

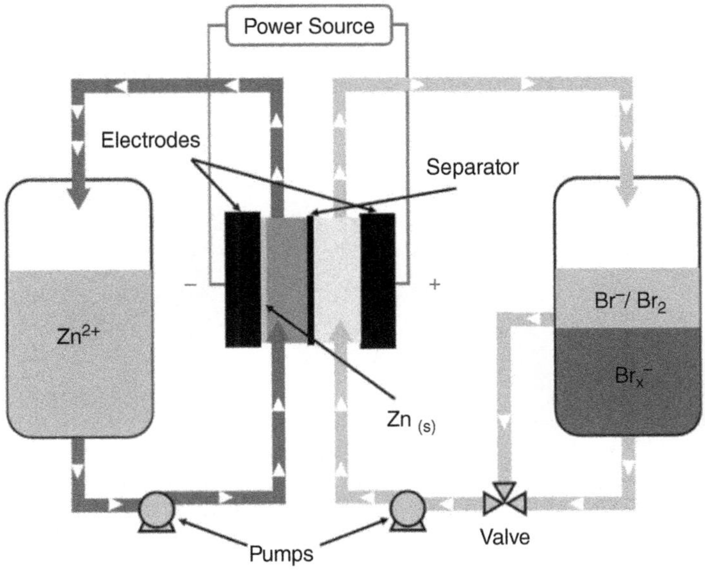

Figure 3.9 Schematic for the ZBFB, showing 2-phase electrolyte on the positive side.

effect is believed to be the outcome of the ethanol coordination with the zinc ions allowing for the plating overpotential to be increased, due to a lower plating exchange current density. Although this work focused primarily on the zinc iodine RFB, the outcome is of relevance to the other aqueous zinc-based hybrid RFBs.

3.2.3.3.1 Large-scale Projects

The company Primus Power [57] has designed a low-cost zinc/bromine flow battery called the EnergyPod-2 (Figure 3.10). The EnergyPod-2 has a lifetime of 20 years, and can supply power for up to five hours (a discharge capacity of 125 kW h at a rated power of 25 kW). The efficiency of the battery is 70% and the depth of discharge can be at 100%. The EnergyPod-2 can operate at temperatures between −5°C and 40°C. However, an extended hot-and-cold-weather package can be purchased and that battery can operate in temperatures between −30°C and 50°C.

A major undertaking by Primus Power has been in the Republic of Kazakhstan. Kazakhstan desires to generate up to 30% of its total energy from renewable sources by the year 2030. Currently 70% of the total energy generation is from coal alone. Kazakhstan produced 20 GW of power in the year 2015 and this figure is expected to rise to 28 GW by the year 2030. The country will also have installed 4.6 GW of wind energy and 0.5 GW of solar. A major reason for working with Primus Power is that with their technology, the membranes in the batteries do not need to be replaced thus reducing the overall cost. Primus Power raised US $25M in funding to bring the 25 MW/100 MW h zinc bromine flow battery system into the Republic of Kazakhstan in order to achieve their goal of 50% renewable energy usage by 2050 [58].

Australian energy storage company Redflow Limited manufacture the ZCELL and the ZBM2 systems [59]. The ZCELL is mainly used for small-scale application such as the home or office. However, the ZBM2 is used for industrial and grid-scale energy storage. The scalable 10 kW h battery unit can deliver 100% discharge without

Figure 3.10 Primus Power EnergyPod-2 zinc bromine battery [57]. Pod dimensions are 1.8 m × 2.1 m × 2.2 m.

causing any damage to the battery at 3 kW (5 kW peak). It operates at 48 V and has a warranted stack lifetime of 10 years or 36.5 MW h discharge. The Redflow system was involved in the Townsville Queensland Solar City transformation project between 2006 and 2013. In this study, sustainable models for electricity supply were assessed through the use of distributed solar technologies, energy storage, energy efficiency, load management, smart meters, and cost-reflective pricing combined in large-scale grid connected urban sites [60]. In 2015 it was announced that Ergon Energy [61], which provides electricity to over 733 000 customers in the state of Queensland, would be using the RedFlow ZBM2 flow battery as a large-scale storage system. The BESS would be linked with community-scale PV solar panels. The BESS consisted of 48 ZBM2 units all contained in a shipping container. The battery produces 100 kW of power and has a storage capacity of 480 kW h. In October 2017, it was announced that the Queensland Government of Australia were to invest $40M in schemes to improve access to both solar energy and energy storage systems [62], starting from March 2018. The Government's aim was then to generate 50% of its electricity from renewable sources by 2030. Redflow batteries were also to be deployed at five specific sites in South Africa from October 2018 to provide standby power to remote mobile phone towers in the event of a power outage [63]. The key factor here was that the Redflow batteries undergo no self-discharge once charged and, when energy is required, the battery can come to full operational status in the same timeframe as a standby generator.

3.2.4 Sodium-Sulfur Battery

3.2.4.1 Battery Fundamentals

The sodium-sulfur (NAS) battery is a high temperature rechargeable secondary battery that operates between 300°C and 350°C. The ceramic β-alumina, Al_2O_3 has a dual purpose in this battery, as it acts as both electrolyte and separator between the two half-cells. These ceramic oxides are known for their structure of closely packed conduction slabs and loosely packed layers, which is ideal for use in batteries as it allows optimum conduction between the interface of slabs and layers [64]. The downside to β-alumina is the need for high temperatures ≥ 300 °C, to maintain its function as an electrolyte and so maintain the conductivity for sodium ions. As this reaction is carried at such high temperatures both reactants, sulfur and sodium, are molten [65].

The electrochemical reaction during discharge occurs at the negative electrode where the molten sodium is the active material.

$$Na_{(s)} \rightarrow Na^+_{(aq)} + e^- \tag{3.12}$$

The oxidized sodium ion is then transported through the solid β-alumina electrolyte membrane to the cathode. The cathode is slightly more complicated, as there is a sequence of products produced based on the level of discharge of the battery. The first stage of battery discharge happens at 350°C when the sulfur content of the cathode is between 78% and 100%. This sulfur-rich phase contains molten sodium sulfide and forms two immiscible layers, due to the difference in the densities of these species (Figure 3.11). As both reactants are in liquid (molten) phase, conductivity remains optimal. The sulfide composition of this phase mainly consists of disodium pentasulfide (Na_2S_5). Therefore, the cathodic reaction at the positive electrode can be represented as

$$xS_{(l)} + 2e^- \rightarrow S^{2-}_{x(l)} \tag{3.13}$$

where x refers to the stoichiometry that is related to the DoD. When x is large, the DoD of the battery is low (or high state of charge) and as x decreases then the DoD for the battery increases towards 100%. The overall electrochemical reaction can be represented as:

$$2Na_{(l)} + xS_{(l)} \rightarrow Na_2S_{(l)} \tag{3.14}$$

To understand the workings of the NAS battery, it is useful to consider the Gibbs phase rule. This states that "for a heterogeneous system in equilibrium at a definite temperature and pressure, the number of degrees of freedom, F, is equal to the difference in the number of components, C, and the number of phases, P. The equilibrium is not influenced by external factors [66]." This can be expressed by the following equation:

$$F = C - P + 2 \tag{3.15}$$

The phase rule can be applied to the positive electrode of this battery as there are two components and two phases. This can be substituted into Equation 3.15 shown above. This gives the fully charged battery two degrees of freedom, *viz.* temperature and pressure. With no remaining degree of freedom, the cell potential remains constant at 2.075 V. As the battery discharges, the polysulfide chain falls from S_5^{2-} to eventually S_2^{2-}. Initially though, the composition changes from Na_2S_5 to Na_2S_4 and subsequently, the stoichiometry falls to Na_2S_3. At this point the composition is only one phase, liquid, as can be seen from the phase diagram of sodium polysulfide, reproduced in Figure 3.11 [67]. Thus, the degree of freedom increases to three and the cell potential is now no longer constant. Thus, between the composition of Na_2S_5 to Na_2S_3, the cell potential falls from 2.075 V to 1.74 V.

Inside the battery, the pressure can be assumed constant. In this scenario, the equation can be altered accordingly to take this into consideration:

$$F = C - P + 1 \tag{3.16}$$

This decreases the initial discharge degrees of freedom to one, where the temperature is now the only external factor that can be changed without resulting in a change in the number of phases formed. Equation 3.16 can also be applied to the latter stages of discharge. For optimum efficiency, the battery must be operated at a temperature of 350°C, where the resistivity (ρ), of β-alumina is 5–6 Ω cm which is close to that of aqueous electrolytes. From the phase diagram in Figure 3.11, it can be seen that the fully discharged battery leads to the formation of solid Na_2S_2. This is detrimental to the longevity of the battery as it then cannot be recharged. This is due to increased resistance at the positive electrode [68]. The battery system has now been engineered to limit the open circuit potential to 1.78 V to minimize the risk of solid disodium disulfide forming.

The design of the NAS battery is different to that other batteries on the market as it has a tubular structure. Figure 3.12 shows molten sulfur in the outer ring connected to the positive terminal. The magnification of the interface shows how sodium ions are

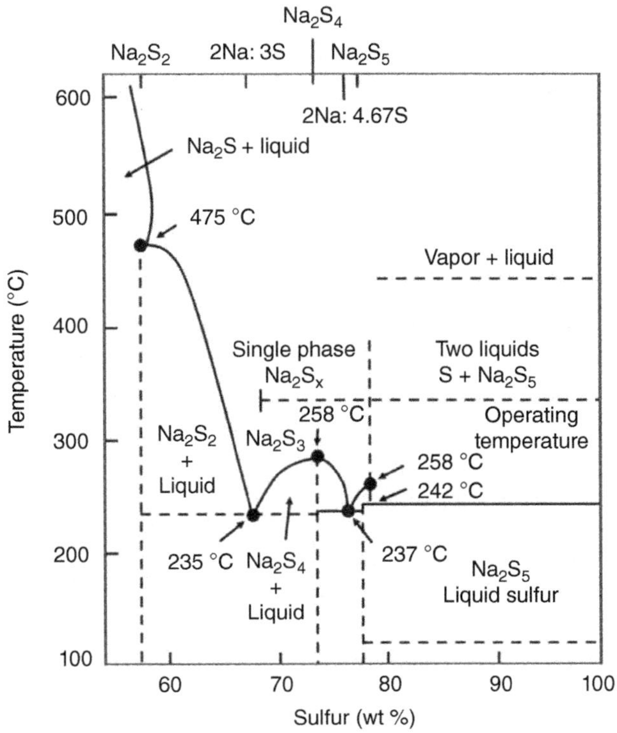

Figure 3.11 Phase diagram for sodium polysulfide [67].

transported through the β-alumina electrolyte into the molten sulfur during discharge. The positive electrode contains carbon, which enhances the electrical conductivity and can carry out the electrochemical reactions effectively [69]. During charging the electrochemical reactions are reversed, sodium ions are reduced at the negative electrode whereas the polysulfide ions are oxidized.

This design has been deemed the most favorable as it allows a higher volumetric energy density, which is beneficial as the battery can possess a high capacity and have long operating times. This design also increases the longevity of the battery as sulfur is

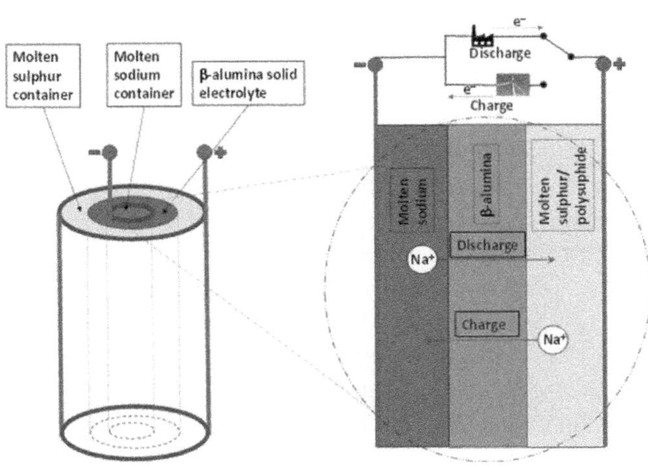

Figure 3.12 Schematic design of NAS battery [69].

a better thermal insulator than sodium and so, less heat is lost to the outside environment. This reverse configuration design is also beneficial as sodium is encased in the centre with the ceramic–alumina surrounding it. This is particularly useful as the sodium is highly volatile and reactive and difficult to store but the β-alumina ensures that it is not exposed to any sources of air and/or water. Limiting the DoD to 90% also prohibits the production of corrosive Na_2S_3. The outer casing then only needs protection from corrosion and the ability to withstand high temperature without rupturing. An inexpensive material such as aluminium or steel can be used on the outside of the battery. As most metals can withstand high temperatures it does not affect the choice of metal or alloy and it is mainly down to weight, cost, and maintenance. The NAS battery has an expected lifespan of between 20 and 25 years if regularly maintained and run at optimal conditions. If the battery is run to full DoD, the battery is capable of running for 2500 cycles whereas the number of cycles almost doubles to 4500 if the depth of discharge is limited to 90% [65].

The main concern for this battery is thermal management that, if left unmonitored with a fault in the module, could cause a fire. This is due to the high operating temperature and exothermic reaction if the reactive chemicals were to come in contact within the cell. This hazard cannot be completely eradicated but safeguards can be put in place to ensure the battery is operated with minimal fire risk as far as is reasonably practicable. The battery should have cooling systems in place to cool down the battery should overheating occur.

The cost to install and run this ESS varies with the amount of capacity required. As an example, using a capacity for the bulk energy storage on the megawatt scale, a 30 MW h with a power of 50 MW has a total cost of between US $3100 and US $3300 per kW and a cost of US $520 to US $550 per kW h [65]. On the other hand, utility T&D (transmission and distribution) grid support with a much lower capacity of 7.2 MW h and power of 1 MW has a wider range of costs, between US $3200 and US $4000 per kW and US $445 to US $555 per kW h. The utilization of the NAS battery over other competing batteries is down to its high-energy density and high energy efficiency of ~ 70%–85%. To supply a large amount of energy, many cells must be connected and packed in a "vacuum thermal enclosure," which is known as a battery module. The cells are packed in a "series-parallel" configuration within the thermal enclosure. This gives the desired power and voltage of the battery unit. The capacity of the battery enclosure and module can be changed to meet energy demand.

3.2.4.2 Large-scale Projects

During the early 1980s Tokyo Electric Power Company (TEPCO) chose NGK's NAS® technology for implementation as it had the potential to shift the dependence on pumped hydro energy storage to dispersed utility energy storage [70]. Indeed, since 1984, NGK and TEPCO have cooperatively piloted the development of the NAS battery in Japan and by 2004, the company had over 100 projects implemented with a total output and capacity of approximately 80 MW and 625 MW h, respectively [70]. To date, NGK has installed NAS batteries in over 200 locations worldwide, with a total output of over 600 MW and a total storage capacity of over 4.2 GW h [71]. The NGK NAS four-container series-connected unit is shown in Figure 3.13, with a total 5.8 MW h capacity and maximum output of 1.0 MW.

One of the largest NAS installations in operation is the 50 MW/300 MW h system at the Buzen storage battery substation, Fukuoka, Japan, balancing supply from solar

Figure 3.13 The NGK NAS four-container unit (each container has 1.45 MW h capacity with maximum 250 kW output). Dimensions 6.1 m × 5.6 m × 5.2 m, weight 84 tonnes [72].

and non-solar generated power and demand by the Kyushu Electric Power Company [73]. It was commissioned in March 2016 to stabilize the extra solar generated power. At Rokkasho Village, Aomori, Japan, the 51 MW wind farm by Futamata Wind Development Company employed NKG NAS batteries to store and supply the generated electricity to the national grid. The 34 MW/245 MW h NAS battery system was commissioned in 2008. At the Oki Islands, Shimane, Japan, the energy storage system was commissioned in September 2015 and operates with a rated output of 4.2 MW and a capacity of 25.2 MW h. This plant is interesting as a Li-ion battery is also used alongside the NAS battery. Thus, although the particular focus on this project is to utilize the NAS battery for both large and small power fluctuations, the system is also able to accommodate rapid response due to its integral operation with Li-ion battery technology.

The first NAS systems in North America was the 2 MW/14 MW h battery in Bluffton, Ohio, installed and commissioned in 2009 by the S&E Electric Company for the American Electric Power Company [74]. This was installed on a 48 km-long 34.5 kV feeder and the 'islanded' system could serve customers downstream if power to the feeder was interrupted. Since then, numerous other stations have been commissioned and are operating, ranging from a 1 MW/6 MW h in British Columbia, Canada [75] to a 2 MW/14 MW h at Vaca Dixon, California, and a 4 MW/ 28 MW h at Yerba Buena, San Jose, California [76] with a total combined capacity of ~125 MW h of energy. The batteries' usage in these stations can vary from improving the reliability of power in remote locations in Texas to the 4 MW/24 MW h ESS, substituting in and reducing the part load emissions from diesel generators on Catalina Island, Southern California.

The largest NAS station in Europe is situated in the Campania Region of Italy. This is a 34.8 MW/250 MW h system commissioned in 2015 to allow excess wind storage in the southern part of Italy to be stored rather than forcing the curtailment of the turbines [77]. One of NGK's notable installation can be found in the Middle East, in Abu Dhabi, UAE. The NGK ESS comprises of 15 systems in total across 10 sites, totalling 108 MW/648 MW h, with each system capable of storing electrical energy for 6 h. A smaller demonstration project has also been installed for the Dubai Electricity and Water Authority, with a 1.2 MW/7.2 MW h NAS battery storage linked to its large desert solar park to stabilize solar fluctuations, time shifting, and frequency control [78]. It has also been recently announced that NGK won the order from the Ministry of Energy in Mongolia to install its NAS batteries (600 kW/3.6 MW h) in Mongolia alongside solar power generation facilities. This will improve the country's energy security as well as forming part of its Upscaling Renewable Energy Project, reducing its CO_2 emissions [79].

3.3 Comparisons

The four BESSs discussed represent the best candidates for large-scale implementations to support renewable energy sources, energy security in remote communities, energy management, or grid stabilization. Each of the technologies offers unique benefits and drawbacks, depending on the operating conditions required. Table 3.1 compares some key performance indicators for the battery systems examined.

3.4 Conclusions

The Li-ion battery is the most well-known of all the systems due to its application in many domestic items. Additionally, it is also the most common battery system found in electric vehicles due to its high power and energy densities coupled with relatively low production costs. However, this system presents some significant risks, requiring stringent control parameters to negate these. In addition, as the number of cycles increases, the morphology of the lithiated electrodes changes causing this system to have one of the lowest service lives compared to the other batteries. The lead-acid is the most mature technology of those compared giving it a sense of good reliability in terms of what it can deliver. Additionally, coupled with the abundance of the raw materials, this allows this system to have the lowest installation cost of all the other systems. However, that may balance itself out as a result of the low service life and number of cycles it is able to achieve. This system though is > 95% recyclable. It is to be noted though that at least two large-scale storage projects involving the lead-acid

Table 3.1 Key performance indicators for the five battery systems discussed.

Parameters	Pb-Acid	Li-ion	VRFB	ZBFB	NAS
Cycle life (no. of cycles)	~ 1250	> 10 000	> 10 000	>10 000	~ 3000
Energy density ($W h L^{-1}$)	100	200–500	15–30	40	150–300
Energy efficiency (%)	70	80–90	80	80	70–85
Service life (years)	6–15	15–20	20–25	20–25	15–25
Discharge time (hours)	~ 9	~ 4	~ 15	~ 10	~ 6

battery have been upgraded in recent years to Li-ion technology as the large-scale operations and safety of the latter systems become better understood.

Redox flow batteries offer a long service life and high energy efficiency. However, they are also the most expensive. The VRFB uses expensive materials for the electrolyte and membrane separator. The ZBFB raises the generally lower energy density typical of flow battery systems and lowers the cost due to the more abundant electrolyte material and cheaper separator options. However, the advantage of decoupling the energy capacity and power output are lost due to its hybrid nature, with zinc plating onto the electrode surface during charge. These systems do offer a high number of cycles and a very long service life that, when coupled with the longer discharge times, makes them well suited for energy arbitrage applications. The NAS battery also offers a long service life with a high energy density. With good energy efficiency over an appreciable discharge time, this allows the battery to maintain its market competitiveness. However, with the battery requiring high operating temperatures to ensure the salts and electrodes remain in the molten state, this increases the operating costs and also displays a lower number of cycles with greater risks.

References

[1] Sauer, D.U. (2009). Secondary batteries – lead–acid systems I Lifetime determining processes. In: J. Garche (ed.), *Encyclopedia of Electrochemical Power Sources*, 805–815. Amsterdam: Elsevier. doi: 10.1016/B978-044452745-5.00137-4.

[2] Takehara, Z. (2006). On the reaction in the lead-acid battery (special review article by the 2005' Gaston Planté Medal recipient). *Journal of Power Sources* 158 (2): 825–830. doi: 10.1016/j.jpowsour.2005.11.014.

[3] IRENA. (2017). *Electricty Storage and Renewables: Costs and Markets to 2030.* Available at: https://www.irena.org/publications/2017/Oct/Electricity-storage-and-renewables-costs-and-markets (accessed 9 November 2018).

[4] Duke Energy. (2013, November). *Technology Performance Report: Duke Energy Notrees Wind Storage Demonstration Project*, 1–67.

[5] Younicos. (2017). *Notrees Windfarm Battery Upgrade.* Available at: https://www.windpowerengineering.com/business-news-projects/younicos-upgrades-battery-storage-duke-energys-notrees-wind-farm-texas (accessed 9 November 2018).

[6] Samsung SDI. (2018). *ESS Batteries by Samsung SDI Top Safety & Reliability Solutions.* Available at: http://www.samsungsdi.com/upload/ess_brochure/201803_SamsungSDIESS_EN.pdf.

[7] May, G.J., Davidson, A., and Monahov, B. (2018). Lead batteries for utility energy storage: A review. *Journal of Energy Storage* 15 (3): 145–157. doi: 10.1016/j.est.2017.11.008.

[8] Younicos. (2018). *New Batteries for Hawaiian Wind Farm.* Available at: https://www.aggreko.com/en-gb/news/2018/global-news/january/new-batteries-for-hawaiian-wind-farm (accessed 9 November 2018).

[9] Ecoult. (2012). *PJM Regulation Services Project.* Available at: https://www.ecoult.com/media1/videos/15-news/68-east-penn-and-ecoult-launch-pjm-regulation-services-project-with-doe-support-to-demonstrate-the-advantages-of-deka-Ultrabattery--technology (accessed 9 November 2018).

[10] Kurzweil, P. (2015). Lithium battery energy storage: State of the art including lithium–air and lithium–sulfur systems. In: *Electrochemical Energy Storage for Renewable Sources and Grid Balancing* (ed. P.T. Moseley and J. Garche), 269–307. Amsterdam: Elsevier. doi: 10.1016/B978-0-444-62616-5.00016-4.

[11] Maranas, J.K. (2012). Solid polymer electrolytes. *Dynamics of Soft Power*: 123–143. doi: 10.1007/978-1-4614-0727-0_5.

[12] Rosewater, D. and Williams, A. (2015). Analyzing system safety in lithium-ion grid energy storage. *Journal of Power Sources* 300: 460–471. doi: 10.1016/j.jpowsour.2015.09.068.

[13] Chemical Sciences Roundtable. (2012). Critical materials in large-scale battery applications. In: (ed. D. Friedman, T. Masciangioli, and S. Olson), *The Role of the Chemical Sciences in Finding Alternatives to Critical Resources*. National Academies Press, Washington, DC, pp. 37–44. https://doi.org/10.17226/13366

[14] SouthernCaliforniaEdison. (2015). *Tehachapi Wind Energy Storage Project*, 714–895. Available at: https://www.osti.gov/servlets/purl/1349233 (accessed 16 November 2021).

[15] LG. (2018). *LG Chem*. Available at: http://www.lgchem.com/global/main (accessed 9 November 2018).

[16] Feldheim. (2015). *Feldheim Regional Power Station*. Available at: http://www.energystorageexchange.org/projects/1446 (accessed 9 November 2018).

[17] LGChem. (2018). *LG Chem ESS Battery*. Available at: http://www.lgchem.com/global/ess/ess/product-detail-PDEC0001 (accessed 9 November 2018).

[18] Kenji Kaneko. (2015). *Tohoku-Electric Power Begins Operation of 20,000kW Storage Battery System*. Available at: https://tech.nikkeibp.co.jp/dm/english/NEWS_EN/20150223/405564/?ST=msbe (accessed 9 November 2018).

[19] Toshiba. (2016). *BESS for Tohoku Electric Power Company*. Available at: https://www.toshiba.co.jp/sis/en/topics/2016/20160226.htm (accessed 9 November 2018).

[20] EDF. (2015). *McHenry Storage Project*. Available at: https://www.edf-re.com/project/mchenry-storage-project (accessed 9 November 2018).

[21] Goldman, S. (2017). *AES Energy Storage*. Available at: http://blog.fluenceenergy.com/sdge-and-aes-energy-storage-unveil-worlds-largest-battery-storage-installation (accessed 22 November 2018).

[22] Leung, P., Li, X., León, C.P. de, et al. (2012). Progress in redox flow batteries, remaining challenges and their applications in energy storage. *RSC Advances* 2 (27): 10125–10156. doi: 10.1039/C2RA21342G.

[23] Amstutz, V., Toghill, K.E., Powlesland, F., et al. (2014). Renewable hydrogen generation from a dual-circuit redox flow battery. *Energy & Environmental Science* 7 (7): 2350–2358. doi: 10.1039/C4EE00098F.

[24] Sezer, H., Aygun, M., Mason, J.H., et al. (2015). A computational model for sodium sulfur battery analysis H. Sezer. *ECS Transactions* 69 (1): 91–100.

[25] Peng, P. and Jiang, F. (2016). Thermal safety of lithium-ion batteries with various cathode materials: A numerical study. *International Journal of Heat and Mass Transfer* 103: 1008–1016. doi: 10.1016/j.ijheatmasstransfer.2016.07.088.

[26] Skyllas-Kazacos, M., Chakrabarti, M.H., Hajimolana, S.A., et al. (2011). Progress in flow battery research and development. *Journal of the Electrochemical Society* 158 (8): R55. doi: 10.1149/1.3599565.

[27] Zhao, H., Wu, Q., Hu, S., et al. (2014). Review of energy storage system for wind power integration support. *Applied Energy* 137: 545–553. doi: 10.1016/j.apenergy.2014.04.103.

[28] Arenas, L.F., Ponce de León, C., and Walsh, F.C. (2017). Engineering aspects of the design, construction and performance of modular redox flow batteries for energy storage. *Journal of Energy Storage* 11: 119–153. doi: 10.1016/j.est.2017.02.007.

[29] Rahman, F. and Skyllas-Kazacos, M. (2009). Vanadium redox battery: Positive half-cell electrolyte studies. *Journal of Power Sources* 189 (2): 1212–1219. doi: 10.1016/j.jpowsour.2008.12.113.

[30] Kear, G., Shah, A.A., and Walsh, F.C. (2012). Development of the all-vanadium redox flow battery for energy storage: A review of technological, financial and policy aspects. *International Journal of Energy Research* 36 (11): 1105–1120. doi: 10.1002/er.1863.

[31] Li, L., Kim, S., Wang, W., et al. (2011). A stable vanadium redox-flow battery with high energy density for large-scale energy storage. *Advanced Energy Materials* 1 (3): 394–400. doi: 10.1002/aenm.201100008.

[32] Rychcik, M. and Skyllas-Kazacos, M. (1988). Characteristics of a new all-vanadium redox flow battery. *Journal of Power Sources* 22 (1): 59–67. doi: 10.1016/0378-7753(88)80005-3.

[33] Bryans, D., Berlouis, L., Spicer, M., et al. (2018). Characterisation of a 200 kW/400 kWh vanadium redox flow battery. *Batteries* 4 (4): 54. doi: 10.3390/batteries4040054.

[34] Skyllas Kazacos, M. and Kazacos, G. (2010). Recent advances with UNSW vanadium-based redox flow batteries. *International Journal of Energy* Research 34 (2): 182–189. doi: 10.1002/er.

[35] Ding, C., Zhang, H., Li, X., et al. (2013). Vanadium flow battery for energy storage: prospects and challenges. *Journal of Physical Chemistry Letters* 4 (3): 1281–1294.

[36] Yang, Z., Zhang, J., Kintner-Meyer, M.C.W., et al. (2011). Electrochemical energy storage for green grid. *Chemical Reviews* 111 (5): 3577–3613. doi: 10.1021/cr100290v.

[37] Skyllas-Kazacos, M. (2003). Novel vanadium chloride/polyhalide redox flow battery. *Journal of Power Sources* 124 (1): 299–302. doi: 10.1016/S0378-7753(03)00621-9.

[38] Prudent Energy. (2012) *Prudent Energy VRB Energy Storage Solutions*. Available at: http://www.pdenergy.com/pdfs/GillsOnions-PrudentEnergy-WhitePaper-July2012.pdf (Accessed: 16 April 2021).

[39] Sumitomo Electric. (2018). *Sumitomo Electric Redox Flow Battery*. Available at: https://global-sei.com/products/redox (accessed 9 November 2018).

[40] Monai, T., Suzaki, H., and Yano, K. (2017). Large-scale power conditioning system for grid battery storage system with redox flow battery having world's highest capacity class of 60 MW h. In *Energy Creation and Social Infrastructure Solutions Contributing to Creation of Sustainable Societies* (Ed. Eguchi, N.) *Fuji Electric Review* 63 (1): 41–47. ISSN 0429–8284.

[41] Sumitomo Electric. (2017). *Sumitomo Electric Industries California*. Available at: https://global-sei.com/company/press/2017/03/prs025.html (accessed 22 November 2018).

[42] redT. (2018). *redT Energy Storage Technology*. Available at: https://redtenergy.com (accessed 9 November 2018).

[42] UET. (2016). *200 MW/800 MWh VRFB with Rongke Power*. Available at: http://www.uetechnologies.com/news/71-200mw-800mwh-energy-storage-station-to-be-built-with-rongke-power-s-vanadium-flow-battery (accessed 9 November 2018).

[43] NGK. (2018). *NGK Sodium-Sulphur Battery*. Available at: https://www.ngk.co.jp/nas/case_studies (accessed 9 November 2018).

[44] CellCube. (2018). *CellCube Advanced Energy Storage Solutions*. Available at: https://www.cellcubeenergystorage.com (accessed 9 November 2018).

[45] Zhang, L., Zhang, H., Lai, Q., et al. (2013). Development of carbon coated membrane for zinc/bromine flow battery with high power density. *Journal of Power Sources* 227: 41–47. doi: 10.1016/j.jpowsour.2012.11.033.

[46] Lim, H.S., Lackner, A.M., and Knechtli, R.C. (1977). Zinc-bromine secondary battery. *Journal of the Electrochemical Society* 124 (8): 1154. doi: 10.1149/1.2133517.

[47] Withers, R.M.J. and Lees, F.P. (1986). The assessment of major hazards: The lethal toxicity of bromine. *Journal of Hazardous Materials*. 13, 279–299. https://doi.org/10.1016/0304-3894(86)85002-6

[48] Noack, J., Roznyatovskaya, N., Herr, T., et al. (2015). The chemistry of redox-flow batteries. *Angewandte Chemie – International Edition* 54 (34): 9776–9809. doi: 10.1002/anie.201410823.

[49] Rallo, F. and Silvestroni, P. (1972). Insoluble polybromides of quaternary ammonium salts and their phase-transitions in the presence of aqueous bromine-bromide solution. *Journal of the Electrochemical Society* 119 (11): 1471. doi: 10.1149/1.2404025.

[50] Eustace, D.J. (1980). Bromine complexation in zinc-bromine circulating batteries. *Journal of the Electrochemical Society* 127 (3): 528–532. doi: 10.1149/1.2129706.

[51] Rajarathnam, G.P., Easton, M., Schneide, M., et al. (2016). The influence of ionic liquid additives on zinc half-cell electrochemical performance in zinc/bromine flow batteries. *RSC Advances* 6 (33): 27788–27797. doi: 10.1039/C6RA03566C.

[52] Bryans, D., Amstutz, V., Girault, H., et al. (2017). Synthesis and characterisation of novel additives for use in the hybrid ZnBr 2 flow battery D. Bryans. *ECS Transactions* 77 (1): 33–36.

[53] Kautek, W., Conradi, A., Fabjan, C., et al. (2001). In situ FTIR spectroscopy of the Zn-Br battery bromine storage complex at glassy carbon electrodes. *Electrochimica Acta* 47 (5): 815–823. doi: 10.1016/S0013-4686(01)00762-9.

[54] Yang, J.H., Yang, H.S., Ra, H.W., et al. (2015). Effect of a surface active agent on performance of zinc/bromine redox flow batteries: Improvement in current efficiency and system stability. *Journal of Power Sources* 275: 294–297. doi: 10.1016/j.jpowsour.2014.10.208.

[55] Banik, S.J. and Akolkar, R. (2013). Suppressing dendrite growth during zinc electrodeposition by PEG-200 additive. *Journal of the Electrochemical Society* 160 (11): D519–D523. doi: 10.1149/2.040311jes.

[56] Li, B., Nie, Z., Vijayakumar, M., et al. (2015). Ambipolar zinc-polyiodide electrolyte for a high-energy density aqueous redox flow battery. *Nature Communications* 6: 1–8. doi: 10.1038/ncomms7303.

[57] PrimusPower. (2017, August). How advances in long-duration, low-cost energy storage are making possible the creation of self-sufficient, high-resilience micro-grids, 1–6. Available at: https://primuspower.com/assets/pdf/Self-Sufficient-High-Resilience-Micro-Grids.pdf (Accessed 16 November 2021).

[58] Kazakhstan. (2015). *Flow Batteries to Kazakhstan*. Available at: https://www.kazakhembus.com/content/primus-power-raises-25m-bring-flow-batteries-kazakhstan (accessed 9 November 2018).

[59] RedFlow Ltd. (2013). RedFlow limited – Energy storage solutions. Available at: http://www.redflow.com.

[60] TownsvilleSolarCity. (2013). *Townsville Solar City Final Report* 2006–2013. Available at: https://www.ergon.com.au/__data/assets/pdf_file/0010/310996/Townsville-Solar-City-Final-Report_2006_2013.pdf (accessed 16 November 2021).

[61] ErgonEnergy. (2015). *Redflow Queensland*. Available at: https://www.energymatters.com.au/renewable-news/redflow-battery-demonstration-em5240 (accessed 9 November 2018).

[62] Hutchins, M. (2017). *Queensland to Offer No-interest Loans for Solar Systems*. Available at: https://www.pv-magazine.com/2017/10/25/queensland-to-offer-no-interest-loans-for-solar-systems (accessed 9 November 2018).

[63] Redflow. (2018). *Redflow South Africa*. Available at: https://redflow.com/tag/south-africa (accessed 9 November 2018).

[64] Lu, X., Xia, G., Lemmon, J.P., et al. (2010). Advanced materials for sodium-beta alumina batteries: Status, challenges and perspectives. *Journal of Power Sources* 195 (9): 2431–2442. doi: 10.1016/j.jpowsour.2009.11.120.

[65] Moseley, P.T. and Rand, D.A.J. (2014). *High-Temperature Sodium Batteries for Energy Storage, Electrochemical Energy Storage for Renewable Sources and Grid Balancing*. Amsterdam: Elsevier B.V. doi: 10.1016/B978-0-444-62616-5.00015-2.

[66] Singh, D., Deshwal, B., and Vats, S.K. (2010). *Comprehensive Engineering Chemistry*. Delhi, India: I.K International Pvt Ltd.

[67] Sangster, J. and Pelton, A. (1997). The Na-S (Sodium-Sulfur) system. *Journal of Phase Equilibria* 18: 89. doi: https://doi.org/10.1007/BF02646762.

[68] Hueso, K.B., Armand, M., and Rojo, T. (2013). High temperature sodium batteries: Status, challenges and future trends. *Energy and Environmental Science* 6 (3): 734–749. doi: 10.1039/c3ee24086j.

[69] Dunn, B., Kamath, H., and Tarascon, J.-M. (2011). Electrical energy storage for the grid: A battery of choices. *Science* 334 (6058): 928–935. doi: 10.1126/science.1212741.

[70] Bito, A. (2005). Overview of the sodium-sulfur battery for the IEEE stationary battery committee. *IEEE Power Engineering Society General Meeting, 2005*, 2346–2349. doi: 10.1109/PES.2005.1489556.

[71] NGK Insulators. (2020). NAS worldwide battery installations. Available at: https://www.ngk-insulators.com/en/product/nas-solutions.html (accessed 15 April 2021).

[72] NGK Insulators. (2021). NGK NAS Container type battery system. Available at: https://www.ngk-insulators.com/en/product/nas-configurations.html (accessed 28 April 2021).

[73] NGK Insulator. (2016). *Buzen City, Fukuoka, Japan*. Available at: https://www.ngk.co.jp/nas/case_studies/buzen Buzen substation, Fukuoka, Japan. Available at: https://www.ngk-insulators.com/en/product/nas-solutions02.html (accessed 28 April 2021).

[74] S&C Electric Company. (2009). *Energy Storage for Microgrids*. Available at: https://www.sandc.com/en/solutions/energy-storage-for-microgrids/ (Accessed 16 May 2021).

[75] British Columbia Hydro. (2015). *Project Outreach Report: Energy Storage and Demand Response for Improved Reliability in an Outage-prone Community*. Available at: https://www.bchydro.com/content/dam/BCHydro/customer-portal/documents/projects/energy-in-bc/bch-energy-storage-project-outreach-report.pdf (accessed 28 April 2021).

[76] Thalman, J.E. (2015). *PG & E Battery Energy Storage Pilot Projects: Safety Plans and Lessons Learned*. Available at: http://www.cpuc.ca.gov/WorkArea/DownloadAsset.aspx?id=3072.

[77] IRENA. (2019). *Utility-scale Batteries: Innovation Landscape Brief, International Renewable Energy Agency, Abu Dhabi*. Available at: https://www.irena.org/-/media/Files/IRENA/Agency/Publication/2019/Sep/IRENA_Utility-scale-batteries_2019.pdf. (Accessed 16 November 2021).

[78] MiddleEastUtilities. (2018). *NGK Dubai Battery Storage*. Available at: https://www.utilities-me.com/power/11500-dubai-to-trial-large-scale-battery-storage (accessed 9 November 2018).

[79] NGK Insulators. (2021). *NGK Wins Order for Renewable Energy Project in Mongoli*. Available at: https://www.advancedbatteriesresearch.com/articles/23403/ngk-wins-order-for-renewable-energy-project-in-mongolia#:~:text=NGK%20Insulators%20Ltd%20based%20in,Ministry%20of%20Energy%20of%20Mongolia (accessed 16 November 2021).

[80] RedFlow Ltd. (2013). RedFlow Limited – Energy storage solutions. Available at: http://www.redflow.com.

4

Environmental Aspects and Recycling of Battery Materials

Guangjin Zhao

Laboratory for Grid Waste Treatment and Resource Recycle Technology
Zhengzhou, China

4.1 Introduction

At the beginning of the 1990s lithium-ion batteries (LIBs) were regarded as the most promising energy or power sources for mobile phones, laptops, and other portable devices [1–2], which has been proven true today. LIBs have developed rapidly and different types of chemistry for battery components, especially for cathode materials, have successfully been introduced. For instance, LIBs have totally replaced NiCd batteries as well as NiMH batteries in mobile phones as they offer a lot of advantages compared to other rechargeable battery systems, such as high operating voltage, high specific energy, high power densities, and a long life-time [1].

Furthermore, for LIBs, development is currently going on in larger applications such as energy storage, partly or fully powered electric vehicles (EVs), industrial vehicles, lifts, cranes, harbor machines, mining vehicles, boats, and submarines [2]. These have recently received much attention due to rapidly growing high energy demands [3–4]. Among those applications listed, power LIBs in particular are attracting much more attention, since they can enable all types of EVs to maintain traffic safety and longer driving distances.

The wide use of EVs will produce a sustained huge demand on power batteries [5]. For instance, in 2017, the application amount of LIB market for new energy vehicles in China exceeded 44 GWh. In 2018, the production of LIBs used in electric vehicles was 65 GWh. And the shipment quantity of power LIBs reached 71 GWh and 80 GWh in 2019 and 2020, respectively, as shown in Figure 4.1.

The widespread and constantly increasing use of LIBs will lead to an increasing number and weight of spent LIBs [6], which have to be recycled with regard to environmental and economical sustainability [1].

For example, it is expected that the accumulative amount of spent power LIBs in China will reach 350 000 and 450 000 tons in 2025 and 2035, respectively, which will exert enormous pressure on the environment. Furthermore, according to the statistics of the State Grid Corporation of China (SGCC), by the end of 2015, over 100 MWh spent energy storage LIBs had been produced, including 88.44 MWh phosphoric-acid-iron ($LiFePO_4$) batteries, 9.74 MWh lithium-manganese-oxide ($LiMn_2O_4$) batteries, and 2.77 MWh ternary-ternary composite material ($Li(Ni_xCo_yMn_Z)O_2$) batteries. Besides, the energy storage LIBs in service are more than 1000 MWh. Therefore, in the

Advances in Energy Storage: Latest Developments from R&D to the Market, First Edition.
Edited by Andreas Hauer.
© 2022 John Wiley & Sons, Ltd. Published 2022 by John Wiley & Sons, Ltd.

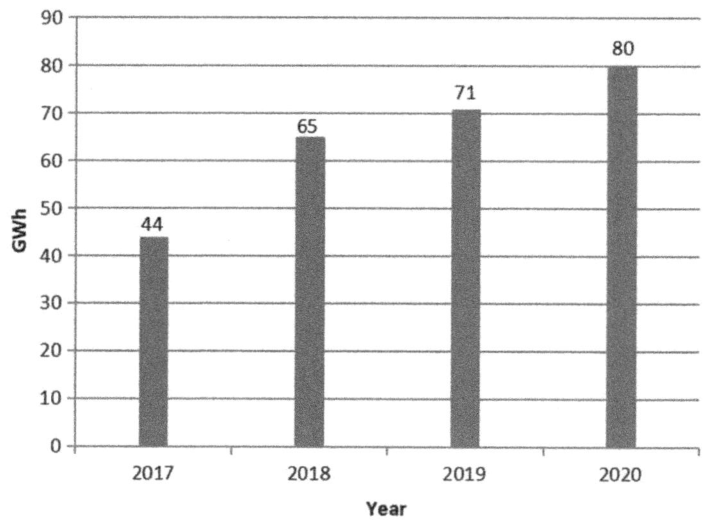

Figure 4.1 LIB market for new energy vehicles in China.

coming years, there will be hundreds of megawatt-hours of retired power and energy storage LIBs per year.

With the retirement of a large number of LIBs from energy storage systems, EVs, and other application scenarios, their cumulative effects on health, the environment, and safety are multiplied, which should not be ignored [7]. On the one hand, recycling of LIBs has become increasingly important because their safe disposal may become a serious problem due to the presence of flammable and toxic elements or compounds, although spent LIBs are not generally classified as dangerous waste [8–9]. Unlike other batteries, LIBs may explode during the recycling process due to radical oxidation of the lithium metal produced from battery metals. So there are two problems to be solved: disposal of harmful waste and prevention of explosion during recycling of LIB waste [10].

On the other hand, some economic benefits could be achieved in the recovery of major components from LIBs [11–12]. Spent LIBs also represent a valuable waste material in the recovery of the metals present (Co, Li, Mn and Ni) or their compounds [13], which can be re-used to make batteries, but belong to non-renewable resources, so recycling of spent batteries can result in excellent economic benefits [11–12].

Therefore, if the spent batteries are not recycled and disposed effectively, it will not only do great harm to the environment but will also cause a waste of resources. In a word, recycling waste LIBs to achieve resource recovery and reuse of metal materials in them will reduce the total amount development and use of raw materials, thereby reducing the damage to the ecological environment and improving the utilization rate of resources [14].

We can conclude that recycling of waste LIBs can meet important practical needs, evidently reducing battery costs and promoting the application of EVs [15].

Nowadays, LIBs have been widely used in many scenarios, it seems that this technology has reached a high level of maturity and confidence that enables it to be pushed toward more demanding applications [16]. However, current LIBs fall short of the desired specifications for EVs and the storage of electrical energy generated by wind and solar power [5]. To meet these increasing demands, intensified research is required to develop novel technologies and materials for next-generation LIBs with dramatically improved performances, including improved specific energy and volumetric energy density, cyclability, charging rate, stability, and safety [17].

In the past two decades, the field of LIBs has been very interesting, important, and attractive for large scientific and technology communities interested in materials science, electrochemistry, chemical engineering, computational chemistry, solid-state chemistry, physics, etc., because these battery systems are complicated and their development requires multi-disciplinary skills and efforts [16]. It is much more important to investigate and develop each single component for LIBs: new electrode materials [18], solution species, new separators, and even cases require rigorous studies of the correlation among composition, morphology structure [19], surface chemistry, intrinsic electrochemical behaviour, and thermal stability.

Of three of the key components (cathode, anode, and electrolyte) of LIBs, cathode material is usually the most expensive with the highest weight in the battery, and from the viewpoint of current LIB technology, the cell voltage and capacities are mainly determined by the cathode material, which is also the limiting factor for the Li transportation rate. The developments of cathode materials therefore become extremely crucial and have receive much attention in recent decades [18].

Battery safety and sustainable batteries should also receive their deserved emphasis and attention in the future. In fact, safety considerations are not less important than performance because high energy density means high safety risks [16].

Finally, new battery concepts have to be further developed to go beyond LIBs [20].

4.2 Classical Batteries

The device that converts chemical energy into electrical energy is called an electrochemical power source, usually known as electrochemical power source or electrochemical battery. In 1800, Volta placed the zinc and silver plates close together, added the woolen cloth and leather absorbed with salty water between them and thus made the first real electrochemical power source in the world [21].

In 1859, 59 years later, the lead acid battery was invented by Plante. As yet, it is still widely used in all aspects of life. Later, in 1868, Leclanch invented the zinc manganese dioxide dry battery, namely Zn-MnO_2 dry battery. Jungner and Edison invented the alkaline secondary batteries Cd-Ni and Fe-Ni respectively in 1899 and 1901. These inventions had far-reaching significance for the development of electrochemical power sources. Over the next few decades, the Zn-Ag battery, the Cd-Ni battery, the Ni-MH battery, and LIBs appeared in succession and have gradually been industrialized.

Modern LIBs generally use transition metal oxides with lithium in their lattice structure (e.g. lithium cobalt oxide ($LiCoO_2$) and $LiMn_2O_4$) as the active cathode materials, lithium salts dissolved in an organic solvent or gel (e.g. $LiPF_6$ in ethylene carbonate) as the electrolyte, and, typically, graphite as the active anode material [22]. Between the cathode and anode electrodes are separator membranes with uniform pores for preventing electronic contact between those two electrodes. During discharge, lithium atoms ionize and deintercalate from the anode, diffuse through the electrolyte, and intercalate in the lattice structure of the transition metal oxides in the cathode; electrons flow through an external circuit from anode to cathode. During charging, the reverse process occurs.

There are various ways of classification based on different LIB types. As divided by shapes, there are: cuboid and cylindrical; as divided by cathode materials: $LiCoO_2$, $LiFePO_4$, $LiMn_2O_4$, lithium nickel oxide ($LiNiO_2$), and Li ($Ni_xCo_yMn_Z$)O_2; as divided by applications, there are communication applications, energy storage applications, power applications, for example.

The cathode material is the most critical raw material in an LIB. In the current market, cathode materials used are mainly $LiCoO_2$, $LiFePO_4$, $LiMn_2O_4$, $LiNiO_2$, and $Li(Ni_xCo_yMn_Z)O_2$. And, the commercialized power battery mainly uses three categories of materials: $LiFePO_4$, $LiMn_2O_4$, and Li ($Ni_xCo_yMn_Z$) O.

It appears that graphite will remain the most important anode material for LIBs. $Li_4Ti_5O_{12}$ is also an important anode material for special applications such as low-temperature performance and very prolonged cycling when the energy density is less important. Finally, Li-Si has the potential to become an important practical anode material [16].

4.3 Summary

Firstly, the rising and international scope and importance of LIBs was reviewed in our work. Then, based on our research results, we investigated the environmental hazards posed by incremental spent LIBs in detail. And finally, the important economical aspect of resourceful recycling of spent LIBs wa also introduced.

For the purpose of a greater understanding of the history of electrochemical batteries, we give a development milestone for various electrochemical batteries, including zinc-air, lead-acid, nickel-cadmium, and metal hydride.

In particular, modern developments of various lithium cells, as widely used in a range of fields, were introduced in detail. The challenges including diverse chemistry, materials, complex and surface structures for LIBs were also reviewed.

Additionally, based on our experimental data, environmentally friendly recycling technology routes for spent LIBs, for example, the "Three Wastes" treatment and prevention of secondary pollution during the recycling process, was be reviewed. A novel automated battery recycling method and its related equipment will be described in the section "Perspectives."

4.4 Future Perspectives

As we known, LIB recycling technology mainly experiences three development stages: first, recycling technology for the cathode materials of small-sized $LiCoO_2$ batteries (mainly used for mobile phone, laptop, digital devices, etc.); second, recycling technology for the cathode materials of lithium-ion power batteries; third, comprehensive resource utilization technology for lithium-ion power batteries.

(1) Recycling technology for the cathode materials of small-sized $LiCoO_2$ batteries

The process of recycling of small-sized $LiCoO_2$ batteries is roughly divided into three steps: Preprocessing of battery, separation of active substances, and current collector, recycling, and reusing of cobalt [23]. The study mainly focuses on the technology required for the recycling of cathode materials, which can be divided into four categories: Physical method, chemical method, biological method, and direct synthesis method [24–26].

(2) Recycling technology for the cathode materials of lithium-ion power batteries

This technology mainly relates to the recycling method for cathode materials of small-sized $LiCoO_2$ batteries. First, preprocessing the spent batteries, including disassembling the batteries, the classification of electrodes, separation of active substances, and then, recycling and reuse of the active substances, including acid solution and the selective precipitation to manganese (nickel) metal and lithium [6].

As a $LiFePO_4$ (LFP) battery does not include cobalt and other precious metals the recycling of certain elements in those LFP batteries has a lower economic efficiency. Therefore, currently consideration for the disposal of spent LFP batteries uses the Solid State Method for its regeneration.

(3) Comprehensive resource utilization technology for lithium-ion power batteries

Recently, we started a research program supported by SGCC for investigating the recycling technology and processes for lithium-ion power batteries. Our study mainly focus on the recycling of lithium-ion power batteries according to the 3R rules: Recycle, reuse, reduce.

Currently, the reuse of lithium-ion power batteries involves using them in energy storage systems [5, 15].

In the recycling process, for spent LFP batteries that have a soft pack, we explore the pretreatment technique and recycling technology for the constituent materials of LFP batteries considering the characteristics of a simple process, a low cost, and the rate of energy consumption, as described below:

(1) The battery is cut and crushed using the die cutting method. (2) Following the magnetism of the LFP, the positive current collectors are separated from the negative current collectors and separator membranes using magnetic separation equipment. (3) The separator membranes and negative current collectors are separated by rinsing with water. (4) The positive and negative current collectors are then put into separating devices, respectively, for the purpose of separating the cathode and anode materials from current collectors utilizing water, and using appropriate external forces such as wipe-and-stir as assistance at the same time. It is possible to recycle, directly, the current collector, separator membranes, anode materials, and cathode materials, using physical and chemical mixing methods, to obtain the ferric and lithium compounds for reusing.

Environmentally friendly technology is required in the whole recycling process for spent LIBs, and are based on our experimental data, as are "Three Wastes" treatment and prevention for secondary pollution during the recycling battery process (see Figure 4.2).

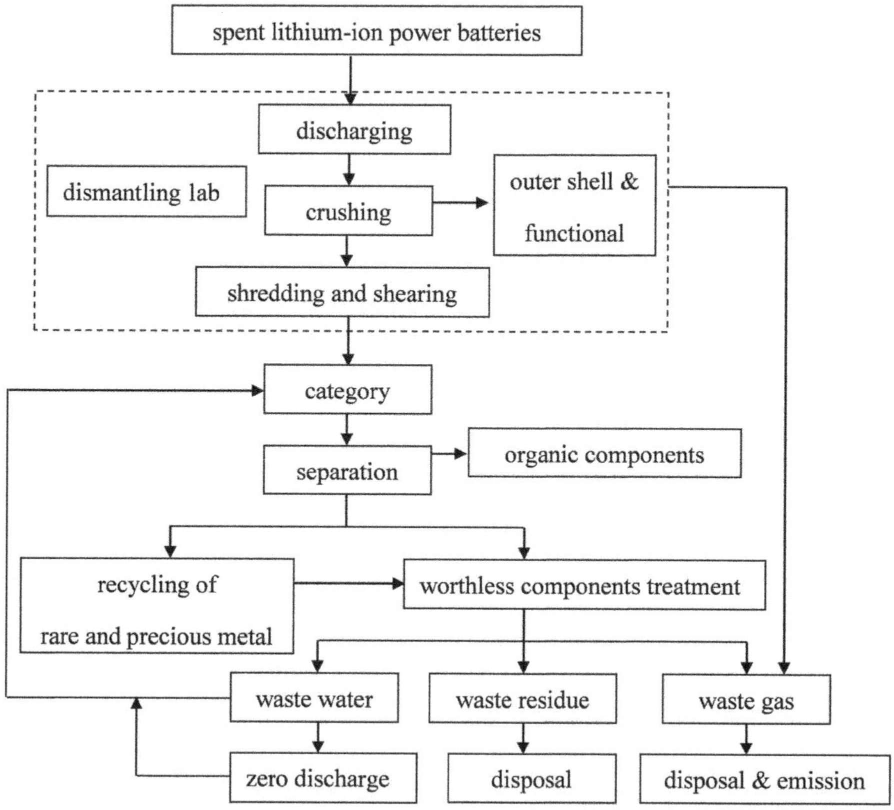

Figure 4.2 The technical route for recycling lithium-ion power batteries based on our research results.

(1) "Three Wastes" Treatment in Recycling of LFP Power Batteries

Some gases are produced during the recycling process of LFP power batteries, and some solid and liquid wastes are also produced. In order to avoid secondary pollution, "Three Wastes" monitoring methods and processing systems must be set up and strictly adhered to in order to realize the recycling of spent LIBs environmentally friendly.

Waste gases treatment: It is known from the recycling steps of LFP power batteries that the electrolytic solution will volatilize during cutting, breaking up, and solution immersing into positive or negative electrodes. So the cutting, breaking-up process chamber (to avoid the potential risks of breakage, and to make it convenient for collecting the exhaust gas produced), and the airtight transmitting procedure should be set up separately. The exhaust system should be installed at the top of each process to collect and recycle the volatilized electrolytic solution (alkaline solution absorption treatment or organic solvent absorption and purification). After the cathode materials are peeled off, the acid solution will absorb the HF hydrolyzed by $LiPF_6$ through heating and soaking. In addition, as the Li-ion solution is alkaline the anode is processed by water solution, so it can also absorb the electrolytic solution $LiPF_6$.

Waste water "zero discharge": During the purification process, the positive electrode produces an alkaline aqueous solution, which can be used for dissolving and absorbing the electrolyte in a waste gas treatment, and can also be used for soaking and washing the anode. The emphasis is that the alkaline solution does not affect the copper foil current collector, and the lithium contained in the anode also shows alkaline when it dissolves in the solution. When the Li-ion content concentration in the cleaning solution at the anode reaches a certain level, the cleaning solution can mix with the solution of the anode, the Li_2CO_3 precipitation reaction starts. As a result the "zero discharge" of waste water can be realized through the whole recycling process.

Waste solid residue "zero discharge": During the recycling of the LFP battery, the shell, aluminum foil current collector, cathode materials, and copper foil are all recycled for reuse. And all of the separator membranes, anode materials, and conductive agents will be treated harmlessly, which confirms the waste solids "zero discharge."

(2) Prevention for secondary pollution during recycling battery process

Before recycling, some waste LIBs may have been damaged or decayed due to external reasons. Such pollutants as waste gas, waste liquid, and waste residues will be produced during the entire process of spent battery recycling. And if mishandled, these pollutants will cause serious environment pollution. So, during the recycling process, it is highly necessary to have secondary pollution prevention measures and to have controlling equipment in the appropriate place, and then formulate a complete set of secondary pollution prevention and control measures.

Wet-processing metallurgy produces much less waste gas and residues than that of pyrometallurgy, and the components contained in them are simpler to some extent. Considering the possible secondary pollution during the recycling process, such as power over discharge, fire, and even explosion caused by electrolyte leakage and short-circuit. It is essential to keep batteries dry and to avoid high ambient temperatures during storage. Storage racks and equipment must be made of acid and alkali resistant and anticorrosive materials. Floors and walls must resist corrosion, fire, and explosion while the collection, storage, and production sites must have appropriate of fire-fighting facilities. Toxic gases must be collected, absorbed, and disposed together using a negative pressure exhaust system. And it is important to monitor the exhaust gases in real time and discharge to ensure relevant standards are met.

For the purpose of preventing secondary pollution, during the battery recycling process, strictly controlling the acid used in production, developing appropriate standards for use, and paying attention to the storage area ventilation, and so on, should be strictly regulated. The acid used must be stored separately from waste battery materials. The storage area must have access to water and fire-fighting equipment and should be installed with safety shower and eye-wash equipment. Working staff should wear protection suits, gas masks, and protective goggles, etc. Also, concentrated acid use precautions, acid spill emergency treatment, and poisoning first-aid measures, etc. should be formulated.

During the wet recycling of cathode and anode materials, acid and alkali corrosion resistant equipment must be used, appropriate leakage-prevention emergency measures must be made, and the field operators must wear appropriate protective clothing and equipment. Formulate waste water recycling regulations, monitor waste water ion concentration, and purify waste water with a too high ion concentration by the hyperfiltration method for future use, so as to realize zero discharge of waste water. Carry out centralized storage of solid products by category, prohibit random stacking, timely pack and ship the products, leaving no waste behind.

4.5 Future Developments

For further optimizing the recycling technology, we reported a novel automated battery recycling method [27], which mainly uses a physical method for the separation and collection of components of LIBs including electrode materials, separator membranes, current collector metals, plastic diaphragms, battery shells and so on. Figure 4.3 shows a schematic diagram of the separation and collection method for the full range of component materials of spent LIBs. We can see from it that the method is characterized by mechanical crushing, vacuum separation, vibration screening, specific gravity separation, flow separation, and other simple dry processes and means. Based on the separation and collection methods for the full range of component materials of spent lithium-ion battery, the system requires a variety of mechanical equipment.

Using this method no chemical reagent is introduced during the separation and collection, and the whole process is fully automated. Throughout the process, electrolyte and dust are collected and disposed of.

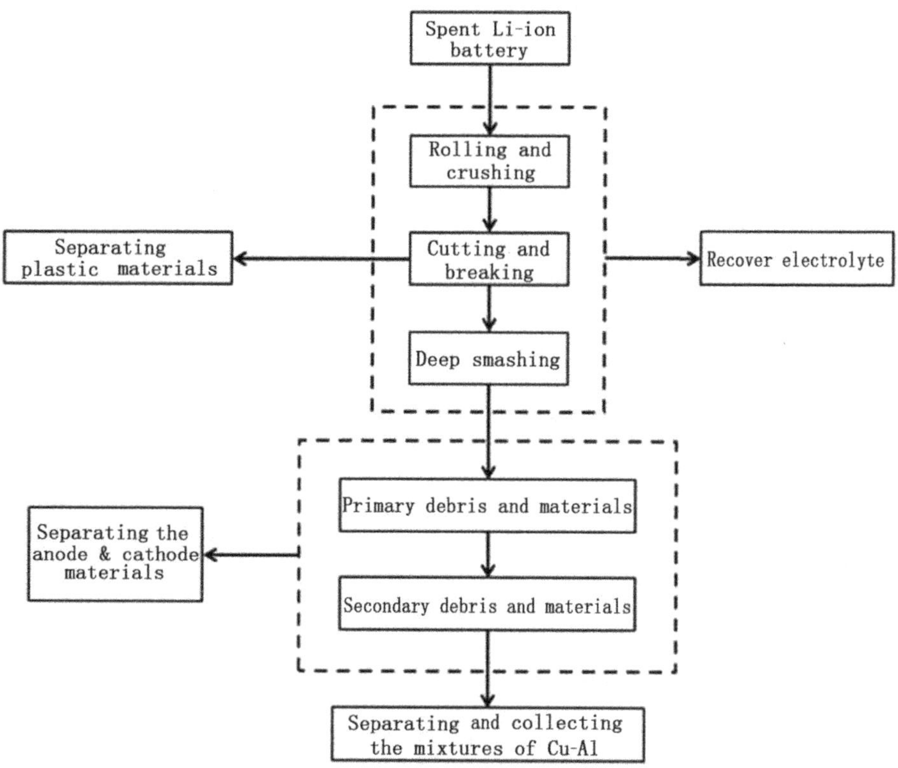

Figure 4.3 Schematic diagram of the separation and collection method for the full range of component materials of spent LIB.

Following the diagram the process is divided into the following steps: (i) Rolling and crushing of spent lithium-ion batteries is to obtain the primary debris and materials; (ii) Cutting and breaking the primary debris and materials into the secondary debris and materials; (iii) Putting the secondary debris and materials through the plastic materials separating and collecting device to separate plastic materials from electrode materials and current collectors and other materials; (iv) Deeply smashing the secondary debris materials; (v) Using the electrolyte absorption and collection device to remove the electrolyte during the battery rolling and crushing, cutting and breaking, and deep smashing; (vi) Initially separating and collecting the electrode materials; (vii) Secondary separation and collection of the electrode materials; (viii) Separation and collection of the mixtures of Cu-Al particles of the current collector; (ix) Separation and collection of the anode and cathode materials.

References

[1] Georgi-Maschlera, T., Friedricha, B., Weyheb, R., et al. (2012). Development of a recycling process for Li-ion batteries. *Journal of Power Sources* 207: 173–182.

[2] Väyrynen, A. and Salminen, J. (2012). Lithium ion battery production. *The Journal of Chemical Thermodynamics* 46: 80–85.

[3] Guo, Y.G., Hu, J.S., and Wan, L.J. (2008). Nanostructured materials for electrochemical energy conversion and storage devices. *Advanced Materials* 20: 2878–2287.

[4] Manthiram, A., Vadivel Murugan, A., Sarkar, A., and Muraliganth, T. (2008). Nanostructured electrode materials for electrochemical energy storage and conversion. *Energy & Environmental Science* 1: 621–638.

[5] Zhao, G.J., Wu, W.L., Qiu, W.B., et al. (2012). Secondary use of PHEV and EV lithium-ion batteries in stationary applications as energy storage system. *Advanced Materials Research* 528: 202–205.

[6] Zhao, G.J., Wu, W.L., Qiu, W.B., et al. (2012). Recycling opportunities for lithium-ion power batteries. *Advanced Materials Research* 518–523: 3441–3443.

[7] Zhao, G.J., Wu, W.L., and Guo, Y. (2013). The possibility of using oxide cathode materials of spent lithium-ion power batteries for carbon dioxide capture from fossil fuel plant. *Advanced Materials Research* 779–780: 52–55.

[8] Castillo, S., Ansart, F., Laberty-Robert, C., et al. (2002). Advances in the recovering of spent lithium battery compounds. *Journal of Power Sources* 112: 247–254.

[9] Bankole, O., Gong, C., and Lei, L. (2013). Battery recycling technologies: Recycling waste lithium ion batteries with the impact on the environment in-view. *Journal of Environment & Ecology* 4 (1): 14–28.

[10] Lee, C.K. and Rhee, K.I. (2002). Preparation of $LiCO_2$ from spent lithium ion batteries. *Journal of Power Sources* 109: 17–21.

[11] Conard, B.R. (1992). The role of hydrometallurgy in achieving sustainable development. *Hydrometallurgy* 30: 1–28.

[12] Zhong, H., Li, J., and Chai, L. (2001). Study on technology for recycling cobalt from Li-ion secondary battery anode scrap of Al-Co film. *Rare Metals and Cemented Carbides* 1 (144): 1–3.

[13] Kondás, J., Jandováa, J., and Nemeckova, M. (2006). Processing of spent Li/MnO_2 batteries to obtain Li_2CO_3. *Hydrometallurgy* 84: 247–249.

[14] Xua, J.Q., Thomasb, H.R., Francis, R.W., et al. (2008). A review of processes and technologies for the recycling of lithium-ion secondary batteries. *Journal of Power Sources* 177: 512–527.

[15] Zhao, G.J., (2017). *Reuse and Recycling Of Lithium-Ion Power Batteries*. Singapore: Wiley.

[17] Deng, D. (2015). Li-ion batteries: Basics, progress, and challenges. *Energy Science and Engineering* 3 (5): 385–418.

[16] Etacheri, V., Marom, R., Elazari, R., et al. (2011). Challenges in the development of advanced Li-ion batteries: A review. *Energy & Environmental Science* 4: 3243–3262.

[19] Verma, P., Mairel, P., and Novák, P. (2010). A review of the features and analyses of the solid electrolyte interphase in Li-ion batteries. *Electrochimica Acta* 55: 6332–6341.

[20] Goodenough, J.B. (2013). Evolution of strategies for modern rechargeable batteries. *Accounts of Chemical Research* 46 (5): 1053–1061.

[21] Shi, P.F. (2006). *Chemical Power Technology*. Harbin: Harbin Institute of Technology Press.

[22] Sonoca, A. and Jeswieta, J. (2014). A review of lithium supply and demand and a preliminary investigation of a room temperature method to recycle lithium ion batteries to recover lithium and other materials. *Procedia CIRP* 15: 289–293.

[23] Li, L., Dunn, J.B., Zhang, X.X., et al. (2013). Recovery of metals from spent lithium-ion batteries with organic acids as leaching reagents and environmental assessment. *Journal of Power Sources* 233: 180–189.

[25] Hu, X.F. (2009). Reclamation of waste lead-acid battery. *Nonferrous Metals Engineering & Research*. 30 (6): 30–32 (Chinese).

[26] Yang, J.K., Zhu, X.F., and Liu, W.C. (2009). Review of recovery technology for lead paste from spent lead acid battery. *Modern Chemical Industry* 29 (3): 32–37 (Chinese).

[28] Xu, B., Qian, D.N., Wang, Z.Y., et al. (2012). Recent progress in cathode materials research for advanced lithium ion batteries. *Materials Science and Engineering R Reports – Lausanne* 73 (5–6): 51–65.

[24] Zhang, Z.J., Qi, G.S., Li, D.H., et al. (2004). Study on cleaner production process of reclaim waste storage cell containing lead and acid. *Environmental Protection Science* 30 (1): 27–29 (Chinese).

[27] Zhao, G.J., Tang, G.P., and Wu, W.L. (2018). An equipment for separation and collection of full components for spent lithium-ion battery. 2016.05, ZL 2016 1 0314509.2.

5

Supercapacitors for Short-term, High Power Energy Storage

Lingbin Kong,[a] Maocheng Liu,[a] Jianyun Cao,[b] Rutao Wang,[a] Weibin Zhang,[a] Kun Yan,[a] Xiaohong Li,[c] and Frank C. Walsh[d]*

[a] *State Key Laboratory of Advanced Processing and Recycling of Non-Ferrous Metals, School of Materials Science and Engineering, Lanzhou University of Technology, Lanzhou 730050, People's Republic of China*
[b] *Department of Materials, School of Natural Sciences, Faculty of Science and Engineering, University of Manchester, Manchester, M139PL, United Kingdom*
[c] *Renewable Energy Group, College of Engineering, Mathematics and Physical Sciences, University of Exeter, Penryn Campus, Cornwall TR10 9FE, United Kingdom*
[d] *Electrochemical Engineering Laboratory, Energy Technology Research Group, Faculty of Engineering and the Environment, University of Southampton, Highfield, Southampton, SO17 1BJ, United Kingdom*

5.1 Introduction

With rapid growth in global population and economy, energy conversion has become one of the great challenges of the twenty-first century. This is due not only to depletion of traditional energy resources (e.g. fossil fuels) but also to associated environmental damage. We have to face the major challenges of an energy crisis and environmental pollution [1–2]. Renewable and clean alternative energy sources (wind, solar, hydro, etc.) are crucial to enable a sustainable development of our economy [3]. To make effective and efficient use of these energy resources, advanced, low-cost, and environmentally friendly energy storage systems must be developed.

Among various energy storage and conversion systems, including solar cells, fuel cells, lithium-ion batteries (LIBs) and electrochemical capacitors (ECs), ECs are becoming a more important energy storage technology that has attracted much attention in the past three decades. Figure 5.1(a) compares the Ragone plots (specific power versus specific energy) of LIBs, conventional capacitors, and ECs [4]. It is well known that LIBs were first introduced in 1990 by Sony then numerous efforts have been devoted to improving their performance. In spite of the high energy density of LIBs (180 W h kg^{-1}), they usually suffer from a sluggish power delivery or uptake. As a result, LIBs cannot meet the performance demands when fast and short-term energy storage and delivery are required [1]. In this case, ECs having high specific power would be an ideal solution.

ECs, which are also known as supercapacitors or ultracapacitors, have attracted considerable research interests in supplementing, or even replacing LIBs for energy storage, mainly due to their high specific power (10 kW kg^{-1}), long cycle-ife (>100 000 cycles), rapid dynamics of charge propagation, superior reversibility, and low maintenance costs [2, 3, 5]. Based on the charge storage mechanism, ECs can be generally classified into two categories: electrochemical double-layer capacitors (EDLCs) and "pseudo-capacitors" [6–8].

Advances in Energy Storage: Latest Developments from R&D to the Market, First Edition.
Edited by Andreas Hauer.
© 2022 John Wiley & Sons, Ltd. Published 2022 by John Wiley & Sons, Ltd.

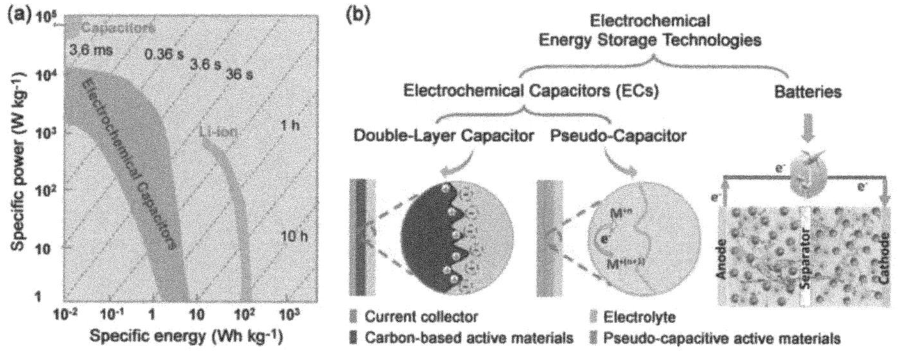

Figure 5.1 a) Comparison of Ragone plots (specific power versus specific energy) and charge–discharge times for lithium-ion battery, conventional capacitors, and electrochemical capacitors. b) Schematic comparison of the fundamental mechanisms of electrochemical energy storage in double-layer capacitors, pseudo-capacitors, and batteries. Reproduced with permission from Springer Nature Customer Service Centre GmbH. *Graphene-Based Composites for Electrochemical Energy Storage* by Liu, J. Copyright (2017) Springer Nature.

Becker, [9] first proposed electricity storage and delivery via the electrochemical double-layer capacitance. The main energy storage mechanism in EDLCs is through reversible charge accumulation in the electrical double layer at the interface between electrode and electrolyte (Figure 5.1b) [4]. An EDLC cell often consists of two activated carbon electrodes and a separator, immersed in the electrolyte solution. During the charging process, the negative electrode attracts cations from the electrolyte, while anions are accumulated on the surface of the positive electrode. The electrode/electrolyte interface can be treated as a conventional capacitor; herein, the capacitance is dependent on the specific surface area (SSA) of the electrode, the type of electrolyte, and the effective thickness of the double layer (the Debye length), which can be calculated from:

$$C = \varepsilon_r \varepsilon_0 A_s / d \tag{5.1}$$

where A_s ($m^2\ g^{-1}$) is the SSA of the electrode, which for an EDLC is the active surface area of activated carbon; ε_0 is the dielectric constant of vacuum ($F\ m^{-1}$); ε_r (a dimensionless constant) is the relative dielectric constant of the electrolyte solvent; and d (m) is the effective thickness of the electrical double layer [4, 10]. Currently, electrode materials are mainly made from porous carbon materials with high SSA and well-controlled pore structures, such as carbon nanotubes (CNTs), activated carbon (AC), mesoporous carbon (MC), templated carbon, carbon xerogels, carbon onions, graphene, carbide-derived carbon (CDC), and porous carbon microspheres [1]. Research has shown that a number of key factors, such as SSA, pore diameters, pore volume, electrical conductivity, particle size, hetero-atoms, ash content, and surface wettability, have a great effect on the electrochemical performance of the porous carbon materials [2, 6]. Generally, carbon-based materials can provide high power density and excellent cycling stability. However, the relatively low specific capacitance ($< 200\ F\ g^{-1}$) of carbon materials restricts the further development of high-energy density EDLCs.

"Pseudo-capacitors" chemically store electric charge via fast and reversible faradaic redox (oxidation/reduction) reactions at the surface/subsurface of the electrode as shown in Figure 5.1b [4]. The Faradaic charge transfer process leads to a potential-dependent charge accumulation (or release), which means that the derivative dq/dV

can be equivalent to a capacitance. Over the past 30 years, many pseudo-capacitive materials have been systematically studied, including metal oxides (e.g. RuO_2, MnO_2, MoS_2, Fe_3O_4) and conducting polymers (e.g. polypyrrole and polyaniline) [1, 11]. As a typical example, RuO_2 has been widely studied due to its high electrical conductivity and high specific capacitance. The charge storage mechanism is based on electro-adsorption of protons on the surface of RuO_2 in acidic solution, and leads to a capacitive behavior with ion adsorption following a Frumkin-type isotherm [11]. Compared with the double-layer capacitance of carbon materials, metal oxides and conducting polymers generally possess higher specific capacitance due to the pseudo-capacitive redox reaction. However, their practical application is restricted by limited surface area, high toxicity, low conductivity, high cost, inferior specific power, and poor cycling stability.

As a rising and competitive energy storage system, ECs possess outstanding high power delivery and uptake ability, with excellent cycling capability. The main market targeted by ECs manufacturers in the short term is likely to be in the transportation arena, including electric vehicles, aircrafts, and metro trains. In the near future, ECs have a great potential to become one of the energy storage systems for applications such as portable electronics and transportation.

5.2 Electrode Materials

Electrode materials are key component that determinesthe overall performance of supercapacitors [12]. They can be classified into carbons, conductive polymers, metal oxides, and other metal compounds.

5.2.1 Carbons

The energy storage mechanism of carbon materials is mainly based on the EDLC. The specific capacitance (C) of carbons is proportional to the surface area (A) of the carbons and inversely proportional to the thickness (d) of the electrical double layer according to Equation (5.1). Since the thickness of the electrical double layer is usually a constant and determined by the type and size of electrolyte ions or solvated ions, increasing the SSA by tailoring the nanoporous structure is the main way to improve the specific capacitance of the carbon materials. In addition, the wettability of the electrolyte solution with the surface of porous carbon should be considered since only the surface infiltrated by the electrolyte solution contributes to the formation of the electrical double layer. The surface area that is accessible to the electrolyte ions is defined as the effective specific surface area (ESSA) [13–14]. This means that the specific capacitance of the carbons should be proportional to ESSA.

Due to the high SSA and relatively low cost, activated carbon powder and activated carbon fiber cloth have been used as active materials in the commercialized EDLCs. These carbons could be derived from different carbon-enriched organic precursors by carbonization in an inert atmosphere. The directly carbonized materials have a low SSA and less porosity, which result in poor specific capacitance. Thus, a subsequent physical or chemical activation process is necessary to increase the SSA and pore volume [15]. The activated carbons could reach a high specific capacitance of 250–350 F g^{-1} and excellent cycling stability up to 100 000 cycles [1].

Nano-casting is another important approach to fabricating mesoporous carbon. This synthetic procedure involves: (i) impregnation of organic carbon precursors (sucrose, glucose, furfuryl alcohol, pitch, or thermosetting resin) into the porous structure of templates, (ii) carbonization of precursors at a high temperature in an inert atmosphere, and (iii) removal of the templates. After the removal of templates, the porous structure of the templates is perfectly copied to the templated carbons. The commonly used templates can be divided into hard templates (mesoporous silica, anodic aluminum oxide, and other mesoporous inorganic materials) and soft templates (e.g. various tri-block copolymers and surfactants). This nano-casting method could precisely control the porous structure at the nanometer scale by selection of suitable templates [16]. It is effective in fabrication of well-designed and precisely ordered mesoporous carbon materials, such as CMK-3 [17] and CMK-8 [18]. Generally, ordered mesoporous carbon has a large surface area, narrow pore size distribution, high electron conductivity, and fast ionic transportation, and is a competitive electrode material for EDLCs.

To simplify the complicated fabrication process of nano-casting, a facile method has been proposed recently that uses direct carbonization of porous polymers or block copolymers. As an example, nanoporous nitrogen-enriched carbon was prepared through a simple carbonization of well-defined poly-(n-butyl acrylate) (PBA)-b-poly-acrylonitrile (PAN) block copolymer precursors. The PAN acts as the precursor of nitrogen-enriched carbon while the PBA is the sacrificial block (source of mesoporosity) [19] (Figure 5.2). In this process, the pore size could be precisely controlled by varying the lengths of PBA blocks and the porosity could be adjusted through varying the PAN/PBA ratios. This nitrogen-enriched hierarchically porous carbon exhibited an unusually high capacitance per unit surface area (>30 μF cm^{-2}), which may be due to the pseudo-capacitance of nitrogen atoms within the materials. A similar method was used to fabricate highly structure-controllable mesoporous carbons (HSCMCs) by carbonizing PAN-poly-methylmethacrylate (PMMA) diblock copolymers [20–21]. By precisely adjusting and controlling the molecular weight or volume fraction of the

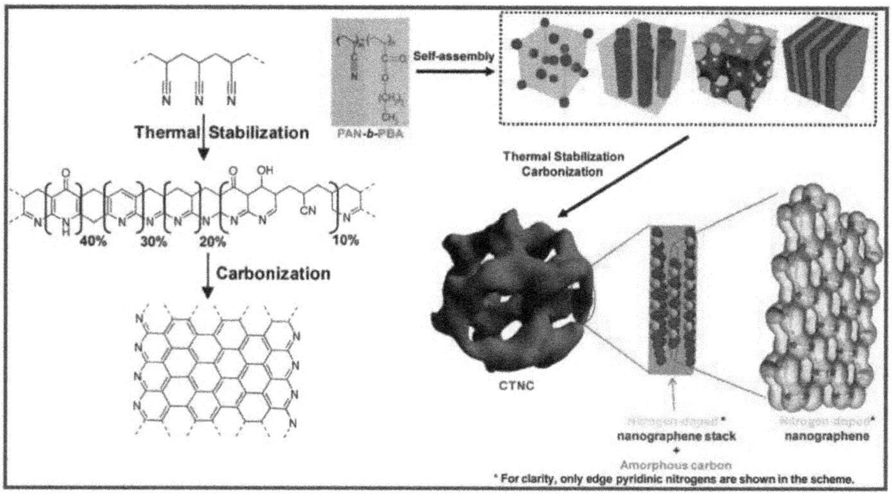

Figure 5.2 Fabrication scheme for the nanoporous nitrogen-enriched carbon by carbonization of PBA-b-PAN block copolymers. Reproduced with permission from [19]. Copyright American Chemical Society.

sacrificial block, the pore size of HSCMCs could be designed precisely on a molecular scale. The HSCMCs exhibited well-controlled mesopore sizes in the range 5.96–17.42 nm and high specific surface areas of 428–213 $m^2\ g^{-1}$. The HSCMC is an ideal material to study the relationship between porous structure properties (SSA, pore size distribution, and pore volume) and electrical double-layer capacitance (specific capacitance and rate capability). The results indicate that an HSCMC with an optimal mesoporous size of 13.68 nm could achieve the highest specific capacitance of 254 $F\ g^{-1}$ at a current density of 0.5 $A\ g^{-1}$ and an excellent rate capability (78% capacitance retention as the current density increased from 0.5 to 5 $A\ g^{-1}$), as well as superior cycle stability (96% of the capacitance was retained after 10 000 cycles). The above studies point to new ways to fabricate structure-controllable mesoporous carbons and provide a platform for the study of the effects of porous structure on the electrical double-layer capacitive performance.

Since the first report in 2004, graphene has been considered as a promising material for supercapacitors. Graphene has an ultrahigh theoretical SSA (2630 $m^2\ g^{-1}$), extraordinary electronic, mechanical, thermal, and optical properties [22]. However, two obstacles inhibit its commercial application in supercapacitors, namely, the serious aggregation and restacking of graphene sheets and challenges associated with the large-scale production of high-quality graphene at an acceptable cost [1, 23], sometimes including expensive starting materials.

A well-defined porous structure is the key factor for achieving a high electrical double-layer capacitive performance. As described above, the specific capacitance of carbons is proportional to the ESSA of the carbons. The ESSA of porous carbons is mainly determined by the total SSA, pore size distribution, the type and size of ion in the electrolyte, and the wettability of electrolyte on the surface of the carbon materials [24]. Generally, the size of electrolyte ions (< 1 nm) is smaller than the pore size of the mesopores (>2 nm). The solvation of ions should also be considered as ions are generally solvated in electrolyte and the solvated ions have a larger size than the bare ions. So, the ESSA of mesoporous carbons is also determined by the wettability of electrolyte on the surface of carbons. It is also important to appreciate that solvated ions can be relatively large and may experience problems in penetrating into small pores in the electrode. There are two ways to improve the wettability of carbon with the electrolyte solution. The first method is chemical modification and activation by acids (e.g. H_2SO_4 and HNO_3) [25–27], alkalis (e.g. KOH) [28], or gases (e.g. CO_2, O_2, and water vapor) [29]. After chemical modification, the SSA, pore size distribution, and pore volume of the carbons will be modified due to chemical corrosion. Additionally, oxygen and nitrogen-containing functional groups (e.g. -OH, -COOH, -CO, and $-NO_2$) can also be formed on the carbon surface [30]. These hydrophilic surface groups can greatly enhance the ESSA of carbons by improving surface wettability. The second method is the grafting of hydrophilic polymers onto the surface of carbons. The hydrophilic molecular chains could enhance the surface wettability leading to improved ESSA. The ESSA is mainly determined by the porous structure and surface wettability of the carbon materials, the appropriate porous structure and high surface wettability will introduce a larger ESSA and higher electrical double layer capacitive performances.

Another effective way to improve the specific capacitance of carbon is to introduce pseudo-capacitive functionalities. For example, chemical modification of carbon with HNO_3 and KOH is usually employed to introduce oxygen and nitrogen functional groups thus extra pseudo-capacitance. These oxygen and nitrogen-containing groups (e.g. -OH, -COOH, -CO, and -NO_2) store electric charge according to the following redox reactions [31]:

$$>C{-}OH \longleftrightarrow >C{=}O + H^+ + e^- \tag{5.2}$$

$$-COOH \longleftrightarrow -COO + H^+ + e^- \tag{5.3}$$

$$>C{=}O + e^- \longleftrightarrow >C{-}O^- \tag{5.4}$$

$$>C{-}NO_2 + e^- \longleftrightarrow >C{-}N^-O_2 \tag{5.5}$$

The pseudo-capacitance from oxygen and nitrogen-containing surface groups is possibly proportional to their atomic contents in the carbon materials. Theoretically, increasing the contents of the surface groups could lead to high specific capacitance. However, high contents of the surface groups will decrease the electrical conductivity of the carbons, which results in poor rate capability. Therefore, it is important to adjust the type and concentration of surface groups to obtain a high specific capacitance while maintaining excellent rate capability.

In summary, electrical double-layer capacitance of carbon materials could be improved by the suitable design of porous structures, improvement in surface wettability, and introduction of pseudo-capacitance.

5.2.2 Conducting Polymers

While electrical double-layer capacitors deliver high power density and long cycle-life, their specific capacitance and energy density are still much lower than those of batteries. To overcome these limitations, pseudo-capacitive materials with high specific capacitance have been proposed as electrode materials. Conducting polymers undergo fast and reversible redox reaction during charge–discharge process. They have been considered as promising materials for pseudo-capacitors due to their high electric conductivity, low cost, and environmental friendliness [32–34]. The charge storage mechanism of the conducting polymers is mainly based on the reversible electrochemical doping/de-doping, which corresponds to the charge–discharge process [35].

In the past two decades, conducting polymers including polyaniline (PANI) [36], polypyrrol (PPY) [37], polythiophene (PTH) [38], and their derivatives have been investigated thoroughly as materials for pseudo-capacitors. They exhibit high specific capacitance in both aqueous and organic electrolytes. The specific capacitances of conducting polymers in aqueous electrolytes are generally higher than those in organic electrolytes. However, the organic electrolytes provide a larger potential window than aqueous electrolytes due to their higher decomposition voltage, this is important for the conducting polymers to provide a high energy density. The major challenge for conducting polymers is the poor cycling stability resulted from the swelling and shrinking of polymer chains during ion insertion/de-insertion process [39]. The poor cycling stability inhibits the commercial application of conducting polymers. Nevertheless, due to their elastic nature, conducting polymers are promising in the design and fabrication of a flexible supercapacitor that can be widely used in the future for advanced flexible electronic devices [3].

5.2.3 Metal Oxides/Hydroxides

Metal oxides/hydroxides, which mainly store energy by fast Faradaic processes, are promising pseudo-capacitive materials for developing high power and high energy supercapacitors. Since the Faradaic reactions (electrosorption and redox processes) take place at the surface and subsurface of the electrode materials, metal oxides/ hydroxides show much higher rate capability and thus higher power density than battery materials, in which the redox reactions occur in the bulk of the material. In addition, the Faradaic reaction of metal oxides/hydroxides leads to higher specific capacitance than that of carbon materials. The pseudo-capacitance of metal oxides/ hydroxides is strongly dependent on their crystallinity, microstructure, morphology, electronic conductivity, and hydrated water content, etc. [40].

RuO_2, first reported as a pseudo-capacitive material in 1971, shows ideal pseudo-capacitive behavior in aqueous H_2SO_4 electrolytes. This is because the high electrical conductivity of $RuO_2{\cdot}xH_2O$ allows the rapid electron transport while the presence of structural water enables rapid proton transportation. The rapid electron and proton transportation in $RuO_2{\cdot}xH_2O$ lead to fast Faradaic reactions at high rates. Considering the high theoretical specific capacitance (< 1715 F g^{-1}), wide potential window, long cycle-life, and excellent chemical stability, RuO_2 is an ideal material for pseudo-capacitors [41]. The energy storage mechanism of RuO_2 is based on the following reaction [42–43]:

$$RuO_x(OH)_y + \delta H^+ + \delta e^- \leftrightarrow RuO_{x-\delta}(OH)_{y+\delta} \tag{5.6}$$

The high cost of RuO_2 limits its practical application. Many other transition metal oxides/hydroxides with low cost have been investigated as alternative pseudo-capacitive materials, including NiO [44–45], CoO_x [46–47], MnO_x [48–50], V_2O_5 [51], and $NiCo_2O_4$ [52–53].

MnO_x appears to be a competitive material to replace RuO_2 due to its low cost and high capacitance. It shows promising pseudo-capacitive properties in neutral alkali metal salt electrolytes. The main energy storage mechanism for MnO_x is based on reversible redox reactions involving the exchange of protons and/or cations with the electrolyte and the transitions between different oxidation states. The reaction mechanism can be expressed as [54–55]:

$$MnO_\alpha(OC)_\beta + \delta C^+ + \delta e^- \leftrightarrow MnO_{\alpha-\delta}(OC)_{\beta+\delta} \tag{5.7}$$

where C represents protons or alkali metal cations. MnO_x also exhibits pseudo-capacitive properties in alkaline electrolyte, showing higher specific capacitance than that in neutral/alkali metal-salts electrolyte. However, MnO_x is unstable in alkaline electrolyte with a poor cycling stability.

NiO and Co_3O_4 can also provide a high performance as pseudo-capacitive materials in alkaline electrolytes [50, 56]. The reported specific capacitance of NiO reached 1776 F g^{-1}, which is higher than that of RuO_2 [57]. However, the rate capability and cycle stability of NiO must be further improved befor possible commercial application. Co_3O_4 has better rate capability and cycle stability than NiO, with relative smaller specific capacitance (746 F g^{-1}) [56].

Other metal oxides, including MoO_3 [58] and V_2O_5 [51], have also been investigated as pseudo-capacitive materials. However, their poor specific capacitances prevent them being competitive in practical applications. To combine the advantages of each type of metal oxides, binary metal oxides ($A_xB_yO_z$, where A and B represent the metal elements having low and high oxidation state, respectively) were employed as pseudo-capacitive materials, including MCo_2O_4, $MMoO_4$, MWO_4, and $M_3V_2O_8$ (M = Ni or Co) [59–60]. These binary metal oxides possess multiple oxidation states that enable multiple redox reactions, leading to higher electrochemical performance than those of single component metal oxides. It is reported that element A plays a significant role in the generation of pseudo-capacitance, while B has a minor influence on the capacitances of the binary metal oxides. Herein, elements with high pseudo-capacitance can be employed as element A [62–63]. For example, $M_3V_2O_8$ (M = Ni or Co) shows the highest specific capacitance among previous reported binary metal oxides [60].

Design of nanocomposites based on metal oxides is another approach to improve the pseudo-capacitive properties of the metal oxides. For example, Co_3O_4 nanowire-MnO_2 ultrathin nanosheet core/shell arrays [61] exhibited better pseudo-capacitive properties than the individual components, where the excellent electrochemical performance can be ascribed to the synergetic contribution from the porous Co_3O_4 nanowire core and the ultrathin MnO_2 shell, as well as the ordered array geometry. Hierarchical $MnMoO_4/CoMoO_4$ nanowires [62] and $CoMoO_4/NiMoO_4$ $x H_2O$ bundles [59] also exhibited higher specific capacitance and better rate capability than either component. $Ni_3V_2O_8/Co_3V_2O_8$ nanocomposites [60] inherited the structural characteristics and combined the pseudo-capacitive benefits of both $Ni_3V_2O_8$ and $Co_3V_2O_8$, showed higher specific capacitance than $Co_3V_2O_8$ and superior rate capability as well as better cycling stability than $Ni_3V_2O_8$. Other composites such as ZnO/MoO_3 core/shell nanocables [63], vertically aligned V_2O_5-TiO_2 nanotube arrays [64] and $Ni_2P/Co_3V_2O_8$ nanocomposite [65] exhibited better pseudo-capacitance than the individual metal compound. The main reason for the enhanced performance is that the composites combine the advantages of the both metal oxides and also the synergistic effect between different components.

While some metal oxides and their composites show high specific capacitance, they suffer from poor electrical conductivity. Poor electrical conductivity is unfavorable for the materials to provide excellent rate capability. The low electrical conductivity not only introduces large IR potential drop when the electrode is charged/discharged at high current density, but also prevents the rapid migration of electrons at high charge–discharge rate, therefore introducing electrochemical polarization and reducing the utilization efficiency of active materials. A common strategy for improving electrical conductivity is to add mesoporous carbon or graphene to improve electrical conductivity. Examples include MnO_2/C [66], reduced graphene oxide/carbon nanotubes/NiO composite [67], hierarchical carbon nanotubes and graphene nanosheets supported MnO_2 nanoneedles [68]. Such carbon enhanced composites showed a better rate capability than metal oxides due to improved electrical conductivity.

5.2.4 Other Metal Compounds

Metal oxides suffer from low rate capability, which seriously restricts their practical applications. This mainly originates from their poor electrical conductivity which confines fast electron transport during the charge–discharge process. Although the design and fabrication of metal oxide composites by adding mesoporous carbon or graphene could improve rate capability, these carbon materials do not significantly contribute to electrochemical capacitance and compromise the high specific capacitance of metal oxides. Therefore, the improvement of electrical conductivity in the composites is limited because the content of the carbons must be kept at a low value. A more effective way to improve the performance of pseudo-capacitive materials is to improve the electrical conductivity of the metal oxides/compounds involved.

Currently, identification and synthesis of other metal compounds with high electrical conductivity as pseudo-capacitive materials has attracted certain research attentions. Compounds including metal sulfides [20, 69], metal selenides [34, 70] and metal phosphides [71–73] have been reported. Metal sulfides offer redox reactions similar to metal oxides, but have a much lower optical band gap in energy and higher conductivity, and have thus become a new type of highly electrical conductive and pseudo-capacitive material. They not only inherit the high specific capacitance of metal oxides but also can provide a higher rate capability than metal oxides due to better electrical conductivity. For instance, copper sulfides (CuS, Cu_7S_4 and Cu_9S_5) [80] show excellent high rate performance, good stability, and high specific capacitance. Uniform NiS_2 hollow nanoprisms [74] and Ni_3S_2 nanoparticles [75] exhibit excellent pseudo-capacitive performance. Highly conductive $NiCo_2S_4$ urchin-like nanostructures [76], $NiCo_2S_4$ nanosheet arrays with self-decorated nanoneedles [77], and $CoNi_2S_4$/graphene nanocomposite [70] were reported as high performance pseudo-capacitive materials.

In the periodic table of elements, selenium is located in the same family as oxygen and sulfur, therefore metal selenides can also serve as pseudo-capacitive materials. For example, three-dimensional hierarchical $GeSe_2$ [34] nanostructures were synthesized and achieved a specific capacitance of 300 F g^{-1} at a current density of 1 A g^{-1}. SnSe was reported to have a specific capacity of 228 F g^{-1} at 0.5 A g^{-1} [79]. However, these single-phase metal selenides show a relatively low specific capacitance compared with oxides and sulfides. Bimetallic Ni-Co selenides [76] with different Ni/Co ratios have been synthesized as materials for high power energy storage, showing a high specific capacitance (535 F g^{-1} at a current density of 1 A g^{-1}) and improved rate capability (the capacitance retention is 63% after a 50-fold increase in current density), combined with excellent cycling stability (the capacity retention is 98% after 2000 cycles). These studies indicate that metal selenides are potential materials for high performance pseudo-capacitors, but further investigations are needed in this research area.

Transition metal phosphides (MP_X, where M = Fe, Co, Ni, etc.) are binary compounds consisting of metal and phosphorus. They have a high electrical conductivity close to that of metals (shown in Figure 5.3), thus they could be served as a new type of high electrical conductive pseudo-capacitive material [33, 80]. For instance, Ni_2P nanoparticles [81] on reduced graphene oxide exhibited excellent specific capacitance (2266 F g^{-1})

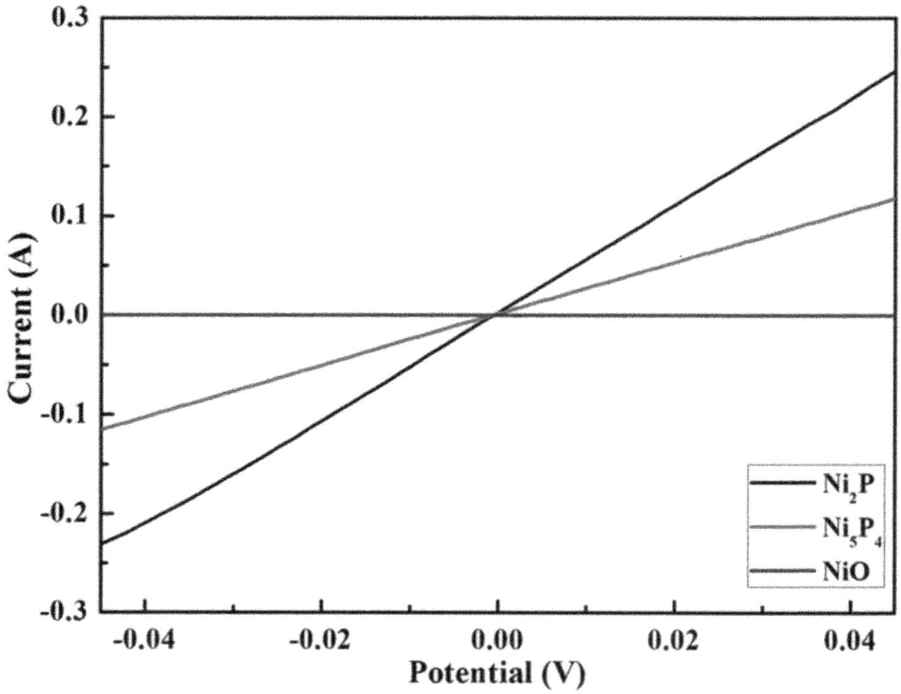

Figure 5.3 Current-potential curves of Ni_2P, Ni_5P_4, and NiO. Reproduced with permission from [71]. Copyright John Wiley and Sons.

and cycling stability. Au/$Ni_{12}P_5$ core/shell nanoparticles [82] showed good specific capacitance (806.1 F g^{-1}) and superior cycling stability (a retention of 91.1% after 500 charge-discharge cycles). Ni_2P and Ni_5P_4 [72] were also investigated as electrochemical supercapacitor materials. Ni_2P exhibits a high specific capacitance of 843.25 F g^{-1} while the value for Ni_5P_4 is 801.5 F g^{-1}. The Ni_2P exhibited excellent cycle stability (with the specific capacitance retention rate of 96% after 1000 cycles). This study suggests that the electrical conductivity of Ni_2P and Ni_5P_4 are much higher than that of NiO, which is very important for materials to obtain excellent rate capability.

5.3 Supercapacitor Devices

A supercapacitor consists of two electrodes separated by an ion-permeable separator with an electrolyte ionically connecting the electrodes [13]. The initial type of supercapacitor is of symmetric design in which both positive and negative electrodes are the same material and approximately the same mass [42, 83]. Aqueous electrolyte was first used in a supercapacitor and later replaced by an organic electrolyte to increase the cell voltage. Later, asymmetric designs were introduced by using different materials as the positive and negative electrodes, [42, 84–85]. The evolution of different types of supercapacitor leads to their increasing popularity for electronics, transportation, power grid, and military uses [42, 13, 85–86].

5.3.1 Symmetric Supercapacitors

A symmetric supercapacitor consists of two electrodes that are usually equal in capacitance and use the same mechanisms for energy storage (shown in Figure 5.4). The energy storage mechanisms are usually related to EDLC [87–88]. The aqueous electrolyte first used in symmetric supercapacitors was replaced later by non-aqueous electrolytes (organic and ionic electrolytes) for high cell voltage [89]. The most commonly used electrode materials for symmetrical supercapacitors are carbon materials of different types, including activated carbon, graphene, CNTs, carbon aerogel, and carbide-derived carbon [90–94].

(1) Aqueous symmetric supercapacitors

The prototype supercapacitors developed by SOHIO, in the US in the 1960s used active carbon for both positive and negative electrodes, each with approximately the same mass to keep the same capacitance values [95]. The primary supercapacitors working in aqueous electrolytes were intended to replace aluminum electrolytic capacitors. Later, aqueous symmetric supercapacitors based on the pseudo-capacitive materials were developed, such as RuO_2, IrO_2, and conducting polymers [13, 42, 83]. However, most of these aqueous symmetric supercapacitors are not fully commercialized due to the high cost of RuO_2 and IrO_2, as well as the limited cycle-life of conductive polymers. Aqueous solutions of acids, bases, and salts can be used as electrolytes for symmetric supercapacitors. Sulfuric acid (3.7 mol dm^{-3}) and potassium hydroxide (27.5% wt) are usually the best choices of electrolytes due to their excellent conductivity [42, 86, 89]. Furthermore, the aqueous symmetric device is one of the most widely used cell

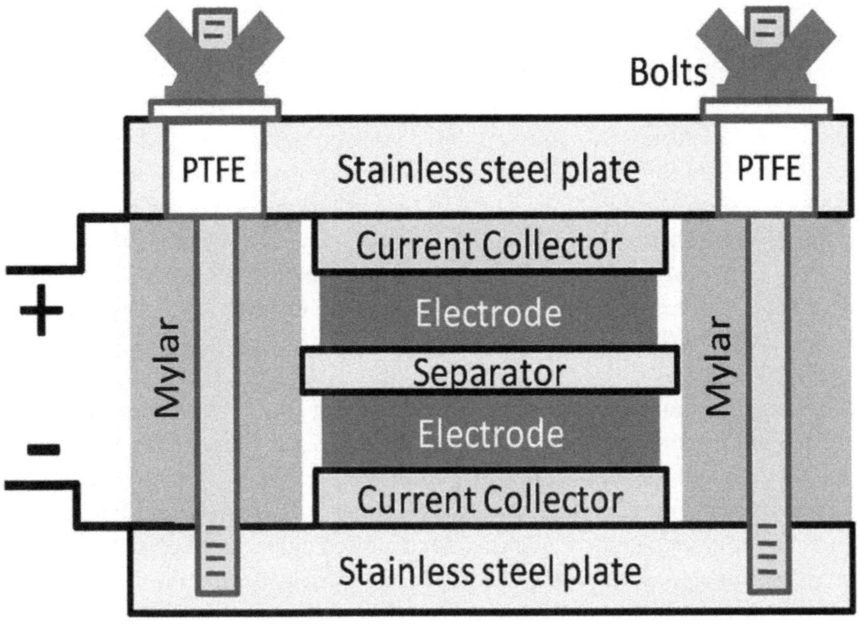

Figure 5.4 Schematic of two-electrode configuration of an EDLC unit cell. Reproduced with permission from [88]. Copyright American Chemical Society.

configurations for carbon materials because assembly of these devices is simple and convenient without the need for any special protective atmosphere [89].

It should be noted that the aqueous electrolyte has a limited operating cell voltage (up to ca. 1.2 V) leading to a relatively low specific energy according to:

$$E = 0.5CV^2 \tag{5.8}$$

where E is the specific energy, C is the specific capacitance, V is the cell voltage. Nevertheless, researchers from both academic and industrial areas have devoted great efforts to improve the energy/power performance of aqueous symmetric supercapacitor for the purpose of commercial application. Frackowiak and Beguin [90] proposed that the maximum cell voltage is essentially limited by an irreversible anodic oxidation at the positive electrode. By performing a controlled chemical oxidation of the carbon surface using hydrogen peroxide, the irreversible anodic oxidation can be minimized, leading to a maximum operating cell voltage of 1.9 V with an excellent stability during 10,000 charge–discharge cycles [96]. Recently, [77] prepared a nitrogen-doped mesoporous carbon with robust redox reactions at nitrogen-associated functionalities, leading to a capacitance of 855 F g^{-1} in aqueous electrolytes. The as-assembled aqueous symmetric supercapacitors with working cell voltage of 1.6 V can offer a high specific energy of 41 W h kg^{-1} (19.5 W h L^{-1}) in 2 M Li_2SO_4 (pH = 1.8) electrolyte.

(2) Non-aqueous symmetric supercapacitors

Non-aqueous symmetric supercapacitors are similar to aqueous symmetric ones but use an organic electrolyte solution or ionic liquid rather than an aqueous electrolyte [97–98]. The organic electrolyte is typically an ammonium salt (e.g. tetraethylmethyl ammonium tetrafluoroborate ($TEMABF_4$)) dissolved in an organic solvent such as propylene carbonate (PC) or acetonitrile (AN), which allows operation at higher unit cell voltages (2.3-2.7 V) [42, 83, 86, 97–98].

However, supercapacitors based on organic electrolytes have several disadvantages. Firstly, the dielectric constant of the organic solvent is lower than that of aqueous electrolytes. Acetonitrile, the commonly used organic solvent, is marked as highly flammable and harmful by inhalation in the European Union. Secondly, the larger size of organic ions cannot enter all the pores in the carbon materials, leading to partial utilization of electrode material [91, 98]. As a result, there is a significant decrease in the capacitance per mass of electrode material when compared to aqueous electrolyte [97]. Thirdly, the power performance decreases due to the lower ionic conductivity of the organic electrolyte compared to that of aqueous electrolytes, especially at low temperatures [42]. Indeed, the freezing point of organic electrolytes should always be considered. The device must be carefully and tightly packaged in a protective atmosphere since trace quantities of water in the electrolyte can result in a short cycle-life and reduced working cell voltage. In practice, exceeding the working cell voltage limit (2.7 V) will trigger undesired side reactions (such as gas evolution and passive film formation on the electrode surface) and cause serious damage to the cell [83]. Despite all these disadvantages, substantially increased energy density remains a great advantage for organic electrolytes based symmetrical supercapacitors [83, 97]. The energy density of this type of supercapacitor can reach up to 8 W h kg^{-1}, which is much higher than that of aqueous symmetric supercapacitors (Table 5.1). Organic electrolyte based

Table 5.1 Summary of parameters of electrolytic capacitors, electrochemical capacitors, and the Li-ion battery.

Cell type	Electrolyte	Cell voltage (V)	Specific energy density (W h kg^{-1})	Temperature range (°C)
Aluminum electrolytic capacitors	Polymer or metal oxide	4–550	0.01–0.3	–40 to 125
AC//AC SCs	H_2SO_4	0–1.2	0.5–1	–20 to 60
AC//AC SCs	$TEMABF_4$/PC	0–2.7	3–6	–40 to 65
AC//AC SCs	Ionic Liquids	0–4	5–12	–40 to 200
$Ni(OH)_2$//AC SCs	KOH	1.6	8–12	–20 to 60
PbO_2//AC SCs	H_2SO_4	2.2	25	–20 to 60
LTO//AC SCs	Li salt in organic solution	1.4–2.8	6–2	–40 to 60
Graphite//AC SCs	Li salt in organic solution	2–4	8–20	–10 to 60
Li-ion battery	Li salt in organic solution	2.5–4.2	100–250	–20 to 60

symmetric supercapacitors are amongst the most popular commercial devices today and have been widely used in rail transportation, smart power grid, portable electronics, and the military field [97].

Recently, ionic liquids (ILs), which are generally composed of a bulky, asymmetric organic cation and weakly coordinating inorganic/organic anion, have attracted significant attention as electrolytes for supercapacitors [2, 98–111]. ILs are solvent-free room-temperature liquid electrolytes, generally non-flammable, and have high chemical and thermal stabilities [101, 104, 108]. A careful selection of both the cation and anion can realize a supercapacitor with a working cell voltage up to 4 V [90, 103]. Aprotic ILs, which are based on pyrrolidinium and imidazolium cations, have been recognized as potential electrolytes for supercapacitors [103–108]. Generally, pyrrolidinium-based ionic liquids have a larger working potential window and higher stability against oxidation than that of imidazolium cation-based ILs. On the contrary, the imidazolium-based ionic liquids usually show lower viscosities and higher conductivities, meaning faster charging – discharging and lower internal resistance properties than that of pyrrolidinium-based ILs [102, 108]. Herein, supercapacitors made of pyrrolidinium-based ionic liquid display higher energy but lower power densities than that of those imidazolium based ILs supercapacitors [2, 108].

In 2010, Liu et al. [107] reported that graphene-based supercapacitors using 1-ethyl-3-methylimidazolium tetrafluoroborate ($EMIMBF_4$) as an electrolyte can operate with a cell voltage of 4 V and exhibit a high energy density of 85.6 W h kg^{-1} (based on the active materials). Whilst ILs are attractive for the realization of supercapacitors with high cell voltages, for commercial application, they face great challenges due to their high resistance and poor rate capability caused by the high viscosity and low ionic conductivity of ILs. The power density of IL-based supercapacitors is often limited, especially at low temperatures [100, 102, 104, 108].

To further increase the ionic conductivity and reduce viscosity, ILs dissolved in organic solvents such as PC, AN, or g-butyrolactone have been investigated [100, 103, 110–111]. This strategy makes it possible to achieve ion concentrations in solvents higher than those with conventional solid ammonium salts. A supercapacitor featuring BF_4- based ILs in AN displays a wider working potential than conventional organic electrolytes [103]. However, this approach loses the main advantage offered by ILs. The safety and working temperature regimes of these electrolytes is inferior to that of solvent-free systems, which are not feasible with commercial systems [102]. Clearly, some drawbacks such as limited power densities, high inner resistance, undesired side reactions and high cost, need to be solved before the practical use of ILs in commercial cells [2, 102, 106].

5.3.2 Asymmetric (Hybrid) Supercapacitors

In asymmetric/hybrid systems, two distinct electrodes that differ considerably in specific capacitance and/or charge storage mechanisms are paired and separated by a membrane separator [42, 89, 83]. The asymmetric supercapacitors combine the characteristics of EDLCs and Faradaic devices (pseudo-capacitor), have increased the capacity due to the Faradaic redox reactions, while maintain the fast charge–discharge ability of the EDLCs [97]. The use of a Faradaic material with a fixed thermodynamic potential can minimize the self-discharge of asymmetric supercapacitors. In addition, Faradaic materials with a large capacity lead to the full utilization of the double-layer capacitance in the carbon electrode.

Moving from the symmetric to asymmetric supercapacitor design provides a choice for increasing cell voltage and correspondingly energy density via full utilizing the different working potential windows of positive and negative electrode materials. Full use of the potential window can be achieved by proper cell design, including mass balance, potential accommodation, and charge balance [83, 89]. The two electrode materials of an asymmetric supercapacitor should be mass balanced to achieve an optimal potential distribution in the ultimate asymmetrical device, thereby ensuring that the individual electrodes remain in their respective potential windows for a stable operation [89]. Such advantages ensure that asymmetric supercapacitors have the potential to bridge the energy/power density gap between conventional symmetrical supercapacitor and batteries.

(1) Aqueous asymmetric supercapacitors

Most aqueous asymmetric supercapacitors are made of metal oxides and carbon electrodes. Metal oxides, which store energy by pseudocapacitive (e.g. MnO_2 and RuO_2) or battery-like mechanisms (e.g., $Ni(OH)_2$, PbO_2, $LiMn_2O_4$), are usually used as the positive electrode material; carbon materials, which storage energy based on electric double-layer capacitance (e.g. activated carbon, graphene, CNTs, and aerogel carbon), are usually used as the negative electrode [13, 42, 89, 112–113]. Aqueous solutions of acids, salts, and alkaline are often used as the electrolytes [42, 83, 89]. The fabrication and packaging of hybrid cells are simple as a controlled atmosphere is not required (no need for a glove box or rigorous dry environment). In the aqueous asymmetric configuration, H_2 and O_2 evolution occurring at the carbon-based negative electrode and battery-like or pseudo-capacitive positive electrode, respectively, can be suppressed by the high overpotentials, extending the effective working voltage up to 2 V [89, 114–115]. Hence, the hybrid cell can achieve a significantly higher specific energy than that of symmetric supercapacitors using aqueous electrolytes. Furthermore, the use of aqueous electrolytes facilitates the high power performance due to the high ionic conductivity. Additionally, the low cost and high safety of a water-based electrolyte leads to greater advantages over energy-storage technologies based on non-aqueous electrolytes (e.g. non-aqueous supercapacitor and Li-ion batteries) [83, 89].

The early commercial design of aqueous asymmetric supercapacitors, produced by ESMA, Troitsk (Russia), were based on a PbO_2 or a $Ni(OH)_2$ battery-like positive electrode in an aqueous acidic or alkaline electrolyte, respectively, paired with an AC negative electrode [112–113]. The $Ni(OH)_2$/AC asymmetric supercapacitor can work in a cell voltage range of 0.8–1.6 V, but should not be charged/discharged at voltages below 0.9 V due to the occurrence of unwanted corrosion process [116]. The PbO_2/AC asymmetric supercapacitor can be charged and discharged in a cell voltage window ranging from 0.8 to 2.2 V. Higher operating voltages (>2.2 V) lead to water decomposition, while a lower operating voltage (<0.8 V) can cause irreversible electrochemical reactions [89, 113, 116–117]. These commercial asymmetric supercapacitors have typical specific energies ranging from ca. 12 W h kg^{-1} for $Ni(OH)_2$/AC and ca. 25 W h kg^{-1} for PbO_2/AC (Table 5.1). These specific energy values are higher than those of commercial symmetric supercapacitor (3–8 W h kg^{-1}).

However, some drawbacks exist in these commercial aqueous asymmetric supercapacitors, e.g. the limited ability to withstand a short-circuit or a reversed polarity for $Ni(OH)_2$/AC, the toxicity of lead, and the use of the strong acidic or alkali electrolytes [89]. Recently, MnO_2 based asymmetric supercapacitors have been recognized as an alternative due to their low cost, high theoretical specific capacity (1350 F g^{-1}), and the use of a mild, neutral electrolyte [89]. Importantly, MnO_2 is a typical pseudocapacitive material, indicating that the power density is potentially higher than those of battery-like materials (e.g. $Ni(OH)_2$ and PbO_2).The specific energy of MnO_2 based devices can reach up to ca. 28 W h kg^{-1}. However, MnO_2-based aqueous asymmetric supercapacitors remain at an early stage of development. Further improvement in cycling stability could potentially lead to the practical application and commercialization of MnO_2-based asymmetric system.

(2) Non-aqueous asymmetric supercapacitors

The configuration of non-aqueous asymmetric supercapacitor is similar to that of the aqueous asymmetric supercapacitor. Generally, the non-aqueous asymmetric cells are made of two distinct electrodes (Figure 5.5): (1) a battery-type (e.g. $Li_4Ti_5O_{12}$ (lithium-titanate-oxide (LTO)), graphite, or Li_2MnO_4) or a pseudo-capacitive (e.g. metal oxides, metal nitrides, and conductive polymer) electrode that relies on a Faradaic mechanism for charge storage; (2) a carbon electrode with high surface area, where charge is stored primarily as double-layer capacitance [69, 83, 118–119]. The major difference between a non-aqueous and an aqueous asymmetric supercapacitor is the electrolyte. A non-aqueous asymmetric supercapacitor usually uses a Li or Na salt dissolved in organic solvents as electrolyte solutions, which allow operation at high unit cell voltages of 2.7–4.5 V.

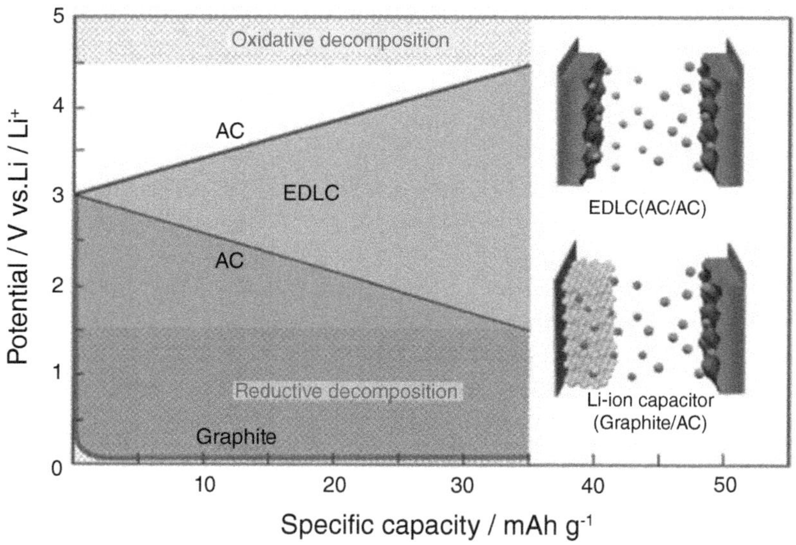

Figure 5.5 Typical voltage profiles for an EDLC cell and an HLIC cell. An EDLC cell has a symmetric configuration and utilizes AC for both positive and negative electrodes. The HLIC cell is featured by a Li-ion battery type electrode coupled with a EDLC electrode. Compared to EDLCs, HLIC can achieve higher voltages over 4.0 V thus the energy density can be enhanced more than three-fold. Reproduced with permission from [83]. Copyright RSC Publishing.

In these non-aqueous asymmetric systems, the hybrid Li-ion capacitors (HLICs) have attracted widespread attention [69, 83, 119–120]. HLICs were first proposed in the early 1980s by Kanebo Co., Japan (Japan patent No. 56-92626). The common HLICs use a hybrid electrochemical energy storage device that combines a lithium battery-like anode and a double-layer capacitive cathode [83].

LTO is a good option for the negative electrode due to the advantages of minimum volume change during the lithium intercalation process, high columbic efficiency, a high theoretical capacity of 175 mA h g^{-1}, and good safety without solid electrolyte interface (SEI) formation and gas evolution [83]. In 2001 Amatucci et al. [121] first introduced the LTO as negative electrodes in HLICs and a 500 F prototype HLICs deice with an energy density of 11 W h kg^{-1} was reported [122–125]. However, the conventional LTO suffers from poor power capability due to its inherent low Li^+ diffusion coefficient ($<10^{-6}$ cm^2 s^{-1}) and poor electronic conductivity ($<10^{-13}$ Ω^{-1} cm^{-1}), thus greatly limiting the rate performance of HLICs. Hence, reducing the particle size to nanoscale was recognized as the most useful method to ensure a high utilization of LTO at high power density. This is because a much smaller particle size can greatly reduce the diffusion path of lithium ion and facilitate ionic transport. Furthermore, post-treatments (e.g. carbon coating) or the introduction of conductive scaffolds, such as carbon nanofiber (CNF), CNTs, graphene, are effective in increasing the conductivity and stability of LTO. Two research group [124, 126] reported a HLIC system, in which the anode was featured by LTO nanoparticles that were homogeneously located on CNF, by coupling with an AC cathode, showed a high energy density of 13 W h L^{-1} and a power density of 4.7 kW L^{-1}.

The anode for HLICs was originally made of LTO, but this has now been replaced by graphitic carbon [125, 127–129]. This is due to that the voltage plateau of LTO-based materials is ca. 1.5 V *vs* Li/Li^+, thus the as-assembled hybrid supercapacitor has a limited working voltage lower than 3.2 V, resulting in a restricted energy density. In contrast, the graphite negative electrode with a theoretical capacity of 372 mA h g^{-1} undergoes the Li-ion insertion at a low potential (close to 0 V *vs.* Li/Li^+). As a result, the graphite based HLICs can be operated at a high working voltage up to 4.5 V (Figure 5.5). Such a high working voltage enables the HLICs to achieve a high energy density of more than 30 W h kg^{-1}.

Graphite-based HLICs were mainly developed in early 2000 by Japanese companies, including JM Energy Corp., Fuji Heavy Industries, Asahi Kasei FDK Corp., Advanced Capacitor Technologies (ACT), and Taiyo Yunden Co., Ltd. FDK announced its EneCapTen lithium-ion capacitor in 2007 [83, 124]. The EneCapTen performed well at high temperatures (up to 80°C), offered a maximum cell voltage of 4.0 V, a maximum capacitance of 2,000 F, and an energy density of up to 25 W h L^{-1} (14 W h kg^{-1}). These values are 3–4 times higher than those of FDK's elder EDLCs [130]. JM Energy Corp. started the mass production of lithium-ion capacitors in 2008 with a monthly output capacity of 25 000 units. Two series of products with capacitance of 1100 F and 2200 F, respectively, were initially produced, which can offer a volumetric energy density of 21–25 W h L^{-1}, and gravimetric density of 12–14 W h kg^{-1} [131].

Several drawbacks restrict the further application of graphite HLICs devices. Graphite HLICs devices exhibit good performance at a high temperature of 60 °C but show poor low-temperature performance, possibly due to metal deposition on the graphite [83]. Lithium pre-doping of the negative electrode to form a solid electrolyte interphase layer is crucial to successful fabrication of device [83, 125, 128–129]. This process is expensive and is inefficient. Metallic lithium is currently used as the lithium source and must be handled with caution to minimize risks due to the violent reaction of lithium metal with water from moisture, which could cause an explosion and fire [132]. Finally, an improved energy density can be achieved by enlarging the working voltage but the poor power capability of graphite materials would also greatly limit the power density of HLICs [83, 122, 126]. HLICs based on other types of battery materials, such as Si and Fe_3O_4, exhibit high working voltages and high energy density, but still suffer from low rate capability, short cycle-life, and a large volume expansion of the anode [133–134].

The use of the battery-like materials causes a limited power density of HLICs. The imbalance between the kinetics of the cathode and the anode is a challenge that needs to be solved for high-performance HLICs. This is because anode kinetics using the Faradaic redox reaction are far more sluggish than that of a cathode using electric-double-layer capacitance. Such challenges could be solved by the development of high-power anode materials to narrow the kinetic gap between two electrodes. Pseudocapacitive materials, including Nb_2O_5 [135], MoS_2 [136], MoO_3 [137], $H_2Ti_6O_{13}$ [138], $TiNb_2O_7$ [118], vanadium nitride [33], and MXenes [139] store energy commonly associated with surface and near surface reversible redox reactions and/or insertion/de-insertion reactions with little structural change. These pseudocapacitive materials exhibit high power capability, which shows the potential to bridge the kinetic gap between anode and cathode. Although most of the HLICs based on pseudo-capacitive materials exhibit the high energy density and enhanced power density, they are still limited by low working cell voltages (<3.5 V) or short cycle-life.

In view of the low cost and abundant sodium resource, asymmetric supercapacitors based on sodium salts dissolved in organic solvents, also called hybrid Na-ion capacitors (HNICs), have attracted wide research attention [140]. The radius of the Na ion is larger than that of Li, indicating slow kinetics for Na ion diffusion in bulk and low columbic efficiency. Thus, energy and power densities of HNICs are lower than that of HLICs. Further advances in the electrode materials having a suitable Na ion transport pathway for fast Na intercalation, should continue to drive the development and commercialization of Na - ion based asymmetric supercapacitors.

5.4 Conclusions

5.4.1 Materials

(1) Electrodes

Carbon is one of the most studied materials for supercapacitor electrodes, which is suitable for all kinds of electrolytes and cell configurations (symmetric or asymmetric). Carbon materials may be found in various forms with reasonably adaptable morphology and diverse structures. The high specific capacitance obtained from various carbons opens potential applications of supercapacitors in power back-up systems, electric vehicles, portable electronic devices, etc. In the future, more attention should be focused on low-cost carbons that can be obtained from simpler carbonization processes without any activation.

Metal oxides can possess larger specific capacitance than carbons. They store electric charges by Faradaic reactions within appropriate potential windows. To maximize their specific capacitance and enable fast kinetics during the charge–discharge process, metal oxides need to have a large specific surface area, high electrical conductivity, preferable microstructure, and morphology. In addition, a good cycling stability is crucial for the practical application of metal oxides. For future applications, metal oxides will possibly replace or complement carbons in many cases to maximize the energy and power performance of supercapacitors.

Conducting polymers that have high electrical conductivity at doped state, large specific capacitance, and adjustable redox activity through chemical modification, show great potential as electrode materials for supercapacitors. However, the swelling and shrinking of conducting polymers during the electrochemical doping/de-doping process leads to a poor cycling stability. Therefore, more efforts are needed to improve the electrochemical stability of conducting polymers, such as composites with carbon materials.

(2) Electrolytes

So far, the aqueous electrolytes (acidic, alkali, and neutral) are the most-used electrolytes for supercapacitors. It is well known that aqueous acidic and alkali electrolytes have higher electrical conductivities than organic electrolytes, leading to better electrochemical performance at higher power densities. However, the maximum operating voltage is limited to 2 V due to the confined electrochemical stability of water. Furthermore, the aqueous acidic solution is aggressive to electrode materials and conductive current collectors, causing further challenges for the aqueous electrolytes.

Organic electrolytes have been adopted for manufacturing of commercial supercapacitors, mainly due to their wider operating voltage than those of aqueous electrolytes.

However, the specific capacitances of electrode materials measured in organic electrolytes are generally smaller than that of aqueous electrolyte. In addition, organic electrolytes show relatively lower ionic conductivity, and higher costs than aqueous ones. Since most of the organic solvent is flammable (e.g. AN), the safety issue is another major challenge for organic electrolytes.

Recently, room-temperature ionic liquids have been considered as a promising class of electrolytes due to their non-volatile, poory combustiblity, and heat-resistant properties.

Moreover, room-temperature ionic liquids remain stable at increased temperatures and provide a wider working cell voltage than that of aqueous and organic electrolytes. However, the ionic conductivity of ionic liquids is generally lower than that of organic electrolytes, especially at low temperatures, greatly reducing the power performance of supercapacitors.

5.4.2 Devices

The emergence of supercapacitors could potentially solve the dilemma between power density and energy density in energy systems. For the further development of supercapacitors research attention should focus on the following aspects:

- Development of electrode materials with good performance and low cost to reduce the equivalent internal resistance thus improves the energy density and power density of supercapacitors.
- Better combinations of electrode materials and electrolytes to increase the operating cell voltage of supercapacitors.
- Reasonable evaluation and prediction of the performance and life expectancy of supercapacitors under actual working conditions.
- Broadenig the operating temperature range of supercapacitors, enabling their ability to work stably under extreme environments.

In the past four decades, great progress has been achieved in the development of supercapacitors. However, the performance and price performance ratio of supercapacitors have to be further improved to compete with other energy storage systems. Fortunately, with the development of solar and wind energy, supercapacitors with the ability of fast charge–discharge and long cycle-life, show great opportunity to act as temporary electric energy storage system to smooth out the inconsistencies of solar and wind energy. By understanding the fundamental science behind supercapacitors, their performance will be continuously improved, and the applications will be further expanded, leading to bright market prospects.

5.5 Outlook

5.5.1 The Importance of Materials

(1) Flexibility

With the development of wearable and portable electronic products, including flexible display, electronic paper, flexible batteries, etc., flexible and portable energy storage devices will have a large market demand in the near future. As one of the potential flexible energy storage devices, supercapacitors will have a great opportunity. In addition, combining 3D macrostructure with secondary 3D nanostructure could be another research direction to improve the performance of the flexible supercapacitors. Furthermore, constructing high performance composite electrodes based on flexible 3D macrostructure will be a good option to achieve a flexible supercapacitor with a high energy density and power density.

(2) Stretchability

Stretchable electronics is an emerging field. In stretchable electronics, circuits can be built or embedded in stretchable substrates, which have the capability to adapt large levels of strain without significant degradation in their electronic performance. It has received continuous attention for its potential in biomedical devices, wearable electronics, electronic paper displays, sensor embedded artificial skins, power source devices, etc. As an essential component of stretchable electronic products, energy storage and conversion units should be capable of accommodating a large strain while retaining the performance to power these highly stretchable devices. Therefore, great efforts should be made to develop stretchable energy storage and conversion devices, including highly stretchable supercapacitors.

(3) Miniaturization

In recent years, electronic devices are constantly moving towards miniaturization, which requires smaller energy storage components. On-chip micro-supercapacitors made of interdigitated electrodes are a promising option for powering up these miniature electronic devices. Potential applications of micro-supercapacitors will be, for example, micro-sensors and radio frequency identification tags.

(4) Operating conditions and environment

Increases in temperature will initiate unwanted chemical/electrochemical reactions and accelerate the degradation of supercapacitors. Supercapacitors working at a high overvoltage can experience severe problems, including a rapid loss of capacitance and an increase in electrical resistance. It is imperative to develop new types of supercapacitors that can work under extreme conditions, such as high and low temperatures, high charging voltage, and high intensity of magnetic field.

(5) Catastrophic damage

Catastrophic damage can significantly compromise the performance of a supercapacitor. If the supercapacitor experiences an accident and natural disaster then an entire energy storage system can be interrupted. It is important that supercapacitor development includes rigorous testing of devices under such adverse conditions.

References

[1] Yan, J., Wang, Q., Wei, T., et al. (2014). Recent advances in design and fabrication of electrochemical supercapacitors with high energy densities. *Advanced Functional Materials* 4 (4): 1300816.

[2] Zhong, C., Deng, Y., Hu, W., et al. (2015). A review of electrolyte materials and compositions for electrochemical supercapacitors. *Chemical Society Reviews* 44 (21): 7484–7539.

[3] Xie, K. and Wei, B. (2014). Materials and structures for stretchable energy storage and conversion devices. *Advanced Materials* 26 (22): 3592–3617.

[4] Liu, J. (2017). *Graphene-Based Composites for Electrochemical Energy Storage.* Singapore: Springer.

[5] Tarascon, J.M. and Armand, M. (2001). Issues and challenges facing rechargeable lithium batteries. *Nature* 414: 359–367.

[6] Dai, L., Chang, D., Baek, J.B., et al. (2012). Carbon nanomaterials for advanced energy conversion and storage. *Small* 8 (8): 1130–1166.

[7] Ellis, B.L., Knauth, P., and Djenizian, T. (2014). Three-dimensional self-supported metal oxides for advanced energy storage. *Advanced Materials* 26 (21): 3368–3397.

[8] Lin, T., Chen, I., Liu, F., et al. (2015). Nitrogen-doped mesoporous carbon of extraordinary capacitance for electrochemical energy storage. *Science* 350 (6267): 1508–1513.

[9] Becker et al. (1957). Firstly proposed electricity storage and delivery via the electrochemical double-layer. U.S. Patent 2800616.

[10] Deng, Y., Xie, Y., Zou, K., et al. (2016). Review on recent advances in nitrogen-doped carbons: Preparations and applications in supercapacitors. *Journal of Materials Chemistry A* 4 (4): 1144–1173.

[11] Zhi, M., Xiang, C., Li, J., et al. (2013). Nanostructured carbon–metal oxide composite electrodes for supercapacitors: A review. *Nanoscale* 5: 72–88.

[12] Xia, X., Zhang, Y., Chao, D., et al. (2015). Tubular TiC fibre nanostructures as supercapacitor electrode materials with stable cycling life and wide-temperature performance. *Energy & Environmental Science* 8: 1559–1568.

[13] Simon, P. and Gogotsi, Y. (2008). Materials for electrochemical capacitors. *Nature Materials* 7: 845–854.

[14] Zhang, L., Yang, X., Zhang, F., et al. (2013). Controlling the effective surface area and pore size distribution of sp2 carbon materials and their impact on the capacitance performance of these materials. *Journal of the American Chemical Society* 135 (15): 5921–5929.

[15] Wood, K.N., O'Hayre, R., and Pylypenko, S. (2014). Recent progress on nitrogen/carbon structures designed for use in energy and sustainability applications. *Energy & Environmental Science* 7 (4): 1212–1249.

[16] Ania, C.O., Khomenko, V., Raymundo-Piñero, E., et al. (2007). The large electrochemical capacitance of microporous doped carbon obtained by using a zeolite template. *Advanced Functional Materials* 17 (11): 1828–1836.

[17] Jun, S., Joo, S.H., Ryoo, R., et al. (2000). Synthesis of new, nanoporous carbon with hexagonally ordered mesostructure. *Journal of the American Chemical Society* 122 (43): 10712–10713.

[18] Gierszal, K.P., Jaroniec, M., Kim, T.W., et al. (2008). High temperature treatment of ordered mesoporous carbons prepared by using various carbon precursors and ordered mesoporous silica templates. *New Journal of Chemistry* 32 (6): 981–993.

[19] Zhong, M., Kim, E.K., McGann, J.P., et al. (2012). Electrochemically active nitrogen-enriched nanocarbons with well-defined morphology synthesized by pyrolysis of self-assembled block copolymer. *Journal of the American Chemical Society* 134 (36): 14846–14857.

[20] Yang, Z.H., Zhu, X., Wang, K., et al. (2015). Preparation of NiCo2S4 flaky arrays on Ni foam as binder-free supercapacitor electrode. *Applied Surface Science* 347: 690–695.

[21] Yan, K., Kong, L.B., Shen, K.W., et al. (2016). Facile preparation of nitrogen-doped hierarchical porous carbon with high performance in supercapacitors. *Applied Surface Science* 364: 850–861.

[22] Novoselov, K.S., Geim, A.K., Morozov, S.V., et al. (2004). Electric field effect in atomically thin carbon films. *Science* 306 (5696): 666–669.

[23] Li, D., Muller, M.B., Gilje, S., et al. (2008). Processable aqueous dispersions of graphene nanosheets. *Nature Nanotechnology* 3 (2): 101–105.

[24] Dutta, S., Bhaumik, A., and Wu, K.C.W. (2014). Hierarchically porous carbon derived from polymers and biomass: Effect of interconnected pores on energy applications. *Energy & Environmental Science* 7 (11): 3574–3592.

[25] Zhao, L., Fan, L.Z., Zhou, M.Q., et al. (2010). Nitrogen-containing hydrothermal carbons with superior performance in supercapacitors. *Advanced Materials* 22 (45): 5202–5206.

[26] Liu, M.C., Kong, L.B., Kang, L., et al. (2014). Synthesis and characterization of M3V2O8 (M = Ni or Co) based nanostructures: A new family of high performance pseudocapacitive materials. *Journal of Materials Chemistry A* 2 (14): 4919–4926.

[27] Li, M.L., Sun, G.Y., Yin, P.P., et al. (2013). Controlling the formation of rodlike V2O5 nanocrystals on reduced graphene oxide for high-performance supercapacitors. *ACS Applied Materials & Interfaces* 5 (21): 11462–11470.

[28] He, X., Li, R., Qiu, J., et al. (2012). Synthesis of mesoporous carbons for supercapacitors from coal tar pitch by coupling microwave-assisted KOH activation with a MgO template. *Carbon* 50 (13): 4911–4921.

[29] Li, J.J., Hu, Y.X., Liu, M.C., et al. (2016). Mechanical alloying synthesis of Ni3S2 nanoparticles as electrode material for pseudocapacitor with excellent performances. *Journal of Alloys and Compounds* 656: 138–145.

[30] Liu, M.C., Kong, L.B., Lu, C., et al. (2012). Waste paper based activated carbon monolith as electrode materials for high performance electric double-layer capacitors. *RSC Advances* 2 (5): 1890–1896.

[31] Li, H., Xi, H.A., Zhu, S., et al. (2006). Preparation, structural characterization, and electrochemical properties of chemically modified mesoporous carbon. *Microporous and Mesoporous Materials* 96 (1–3): 357–362.

[32] Zhang, Y., Feng, H., Wu, X., et al. (2009). Progress of electrochemical capacitor electrode materials: A review. *International Journal of Hydrogen Energy* 34 (11): 4889–4899.

[33] Wang, R.T., Lang, J.W., Zhang, P., et al. (2015). Fast and large lithium storage in 3D porous VN nanowires-graphene composite as a superior anode toward high-performance hybrid supercapacitors. *Advanced Functional Materials* 25 (15): 2270–2278.

[34] Wang, X.F., Liu, B., Wang, Q.F., et al. (2013). 3D nanostructures: Three-dimensional hierarchical GeSe2 nanostructures for high performance flexible all-solid-state supercapacitors. *Advanced Materials* 25 (10): 1369.

[35] Ryu, K.S., Wu, X., Lee, Y.G., et al. (2003). Electrochemical capacitor composed of doped polyaniline and polymer electrolyte membrane. *Journal of Applied Polymer Science* 89 (5): 1300–1304.

[36] Ryu, K.S., Kim, K.M., Park, N.G., et al. (2002). Symmetric redox supercapacitor with conducting polyaniline electrodes. *Journal of Power Sources* 103 (2): 305–309.

[37] Zhang, J., Kong, L.B., Wang, B., et al. (2009). In-situ electrochemical polymerization of multi-walled carbon nanotube/polyaniline composite films for electrochemical supercapacitors. *Synthetic Metals* 159 (3–4): 260–266.

[38] Di Fabio, A., Giorgi, A., Mastragostino, M., and Soavi, F. (2001). Carbon-poly (3-methylthiophene) hybrid supercapacitors. *Journal of the Electrochemical Society* 148 (8): A845–A850.

[39] González, A., Goikolea, E., Barrena, J.A., et al. (2016). Review on supercapacitors: Technologies and materials. *Renewable and Sustainable Energy Reviews* 58: 1189–1206.

[40] Wang, G., Zhang, L., and Zhang, J. (2012). A review of electrode materials for electrochemical supercapacitors. *Chemical Society Reviews* 41 (2): 797–828.

[41] Das, R.N., Liu, B., Reynolds, J.R., et al. (2009). Engineered macroporosity in single-wall carbon nanotube films. *Nano Letters* 9 (2): 677–683.

[42] Conway, B.E. (1999). *Electrochemical Supercapacitors: Scientific Fundamentals and Technological Applications*. New York: Kluwer Academic.

[43] Ramani, M., Haran, B.S., White, R.E., et al. (2001). Synthesis and characterization of hydrous ruthenium oxide-carbon supercapacitors. *Journal of the Electrochemical Society* 148 (4): A374–A380.

[44] Liang, K., Tang, X.Z., and Hu, W.C. (2012). High-performance three-dimensional nanoporous NiO film as a supercapacitor electrode. *Journal of Materials Chemistry* 22 (22): 11062–11067.

[45] Kim, S.I., Lee, J.S., Ahn, H.J., et al. (2013). Facile route to an efficient NiO supercapacitor with a three-dimensional nanonetwork morphology. *ACS Applied Materials & Interfaces* 5 (5): 1596–1603.

[46] Yuan, C.Z., Yang, L., Hou, L.R., et al. (2012). Growth of ultrathin mesoporous Co3O4 nanosheet arrays on Ni foam for high-performance electrochemical capacitors. *Energy & Environmental Science* 5 (7): 7883–7887.

[47] Zhang, F., Yuan, C.Z., Lu, X.J., et al. (2012). Facile growth of mesoporous Co3O4 nanowire arrays on Ni foam for high performance electrochemical capacitors. *Journal of Power Sources* 203: 250–256.

[48] Gao, H.C., Xiao, F., Ching, C.B., et al. (2012). High-performance asymmetric supercapacitor based on graphene hydrogel and nanostructured MnO2. *ACS Applied Materials & Interfaces* 4 (5): 2801–2810.

[49] Lu, X.H., Zhai, T., Zhang, X.H., et al. (2012). WO3-x@Au@MnO2 core-shell nanowires on carbon fabric for high-performance flexible supercapacitors. *Advanced Materials* 24 (7): 938.

[50] Rakhi, R.B., Chen, W., Cha, D., et al. (2012). Substrate dependent self-organization of mesoporous cobalt oxide nanowires with remarkable pseudocapacitance. *Nano Letters* 12 (5): 2559–2567.

[51] Li, M., Liu, C., Cao, H., et al. (2014). KOH self-templating synthesis of three-dimensional hierarchical porous carbon materials for high performance supercapacitors. *Journal of Materials Chemistry A* 2 (36): 14844–14851.

[52] Wei, T.Y., Chen, C.H., Chien, H.C., et al. (2010). A cost-effective supercapacitor material of ultrahigh specific capacitances: Spinel nickel cobaltite aerogels from an epoxide-driven sol-gel process. *Advanced Materials* 22 (3): 347.

[53] Huang, L., Chen, D.C., Ding, Y., et al. (2013). Nickel-cobalt hydroxide nanosheets coated on NiCo2O4 nanowires grown on carbon fiber paper for high-performance pseudocapacitors. *Nano Letters* 13 (7): 3135–3139.

[54] Pang, S.C., Anderson, M., and Chapman, T.W. (2000). Novel electrode materials for thin-film ultracapacitors: Comparison of electrochemical properties of sol-gel-derived and electrodeposited manganese dioxide. *Journal of the Electrochemical Society* 147 (2): 444–450.

[55] Hu, C.C. and Tsou, T.W. (2002). Ideal capacitive behavior of hydrous manganese oxide prepared by anodic deposition. *Electrochemistry Communications* 4 (2): 105–109.

[56] Kong, D.S., Wang, J.M., Shao, H.B., et al. (2011). Electrochemical fabrication of a porous nanostructured nickel hydroxide film electrode with superior pseudocapacitive performance. *Journal of Alloys and Compounds* 509 (18): 5611–5616.

[57] Gao, Y.Y., Chen, S.L., Cao, D.X., et al. (2010). Electrochemical capacitance of Co3O4 nanowire arrays supported on nickel foam. *Journal of Power Sources* 195 (6): 1757–1760.

[58] Jiang, J.B., Liu, J.L., Peng, S.J., et al. (2013). Facile synthesis of alpha-MoO3 nanobelts and their pseudocapacitive behavior in an aqueous Li2SO4 solution. *Journal of Materials Chemistry A* 1 (7): 2588–2594.

[59] Liu, M.C., Kong, L.B., Lu, C., et al. (2013). Design and synthesis of CoMoO4-NiMoO4·xH2O bundles with improved electrochemical properties for supercapacitors. *Journal of Materials Chemistry A* 1 (4): 1380–1387.

[60] Liu, M.C., Kong, L.B., Zhang, P., et al. (2012). Porous wood carbon monolith for high-performance supercapacitors. *Electrochimica Acta* 60: 443–448.

[61] Liu, J.P., Jiang, J., Cheng, C.W., et al. (2011). Co3O4 nanowire@MnO2 ultrathin nanosheet core/shell arrays: A new class of high-performance pseudocapacitive materials. *Advanced Materials* 23 (18): 2076.

[62] Mai, L.Q., Yang, F., Zhao, Y.L., et al. (2011). Hierarchical MnMoO4/CoMoO4 heterostructured nanowires with enhanced supercapacitor performance. *Nature Communications* 2 (1) 381.

[63] Li, G.R., Wang, Z.L., Zheng, F.L., et al. (2011). ZnO@MoO3 core/shell nanocables: Facile electrochemical synthesis and enhanced supercapacitor performances. *Journal of Materials Chemistry* 21 (12): 4217–4221.

[64] Yang, Y., Kim, D., Yang, M., et al. (2011). Vertically aligned mixed V2O5-TiO2 nanotube arrays for supercapacitor applications. *Chemical Communications* 47 (27): 7746–7748.

[65] Hu, Y.M., Liu, M.C., Hu, Y.X., et al. (2016). One-pot hydrothermal synthesis of porous nickel cobalt phosphides with high conductivity for advanced energy conversion and storage. *Electrochimica Acta* 215: 114–125.

[66] Kim, I.T., Kouda, N., Yoshimoto, N., et al. (2015). Preparation and electrochemical analysis of electrodeposited MnO2/C composite for advanced capacitor electrode. *Journal of Power Sources* 298: 123–129.

[67] Bai, Y., Du, M., Chang, J., et al. (2014). Supercapacitors with high capacitance based on reduced graphene oxide/carbon nanotubes/NiO composite electrodes. *Journal of Materials Chemistry A* 2 (11): 3834–3840.

[68] Hsieh, C.T., Tzou, D.Y., Lee, W.Y., et al. (2016). Deposition of MnO2 nanoneedles on carbon nanotubes and graphene nanosheets as electrode materials for electrochemical capacitors. *Journal of Alloys and Compounds* 660: 99–107.

[69] Nguyen, V.H., Lamiel, C., and Shim, J.J. (2015). Hierarchical mesoporous graphene@Ni-Co-S arrays on nickel foam for high-performance supercapacitors. *Electrochimica Acta* 161: 351–357.

[70] Chen, H.C., Jiang, J.J., Zhang, L.,et al. (2013). Highly conductive NiCo2S4 urchin-like nanostructures for high-rate pseudocapacitors. *Nanoscale* 5 (19): 8879–8883.

[71] Wang, D., Kong, L.B., Liu, M.C., et al. (2015). An approach to preparing Ni–P with different phases for use as supercapacitor electrode materials. *Chemistry – A European Journal* 21 (49): 17897–17903.

[72] Wang, D., Kong, L.B., Liu, M.C., et al. (2015). Amorphous Ni–P materials for high performance pseudocapacitors. *Journal of Power Sources* 274: 1107–1113.

[73] Chen, X.J., Cheng, M., Chen, D., et al. (2016). Shape-controlled synthesis of Co2P nanostructures and their application in supercapacitors. *ACS Applied Materials & Interfaces* 8 (6): 3892–3900.

[74] Dai, Z.Y., Zang, X.X., Yang, J., et al. (2015). Template synthesis of shape-tailorable NiS2 hollow prisms as high-performance supercapacitor materials. *ACS Applied Materials & Interfaces* 7 (45): 25396–25401.

[75] Li, J., Liu, K., Gao, X., et al. (2015). Oxygen-and nitrogen-enriched 3D porous carbon for supercapacitors of high volumetric capacity. *ACS Applied Materials & Interfaces* 7 (44): 24622–24628.

[76] Chen, H.C., Chen, S., Fan, M.Q., et al. (2015). Bimetallic nickel cobalt selenides: A new kind of electroactive material for high-power energy storage. *Journal of Materials Chemistry A* 3 (47): 23653–23659.

[77] Lin, L.Y., Liu, J.L., Liu, T.M., et al. (2015). Growth-controlled NiCo2S4 nanosheet arrays with self-decorated nanoneedles for high-performance pseudocapacitors. *Journal of Materials Chemistry A* 3 (34): 17652–17658.

[78] Du, W.M., Wang, Z.Y., Zhu, Z.Q., et al. (2014). Facile synthesis and superior electrochemical performances of CoNi2S4/graphene nanocomposite suitable for supercapacitor electrodes. *Journal of Materials Chemistry A* 2 (25): 9613–9619.

[79] Zhang, C., Yin, H., Han, M., et al. (2014). Two-dimensional tin selenide nanostructures for flexible all-solid-state supercapacitors. *ACS Nano* 8 (4): 3761–3770.

[80] Wang, Y., Liu, F.Y., Ji, Y., et al. (2015). Controllable synthesis of various kinds of copper sulfides (CuS, Cu7S4, Cu9S5) for high-performance supercapacitors. *Dalton Transactions* 44 (22): 10431–10437.

[81] An, C.H., Wang, Y.J., Wang, Y.P., et al. (2013). Facile synthesis and superior supercapacitor performances of Ni2P/rGO nanoparticles. *RSC Advances* 3 (14): 4628–4633.

[82] Duan, S.B. and Wang, R.M. (2014). Au/Ni12P5 core/shell nanocrystals from bimetallic heterostructures: In situ synthesis, evolution and supercapacitor properties. *NPG Asia Materials* 6: e122.

[83] Naoi, K., Ishimoto, S., Miyamoto, J.I., et al. (2012). Second generation 'nanohybrid supercapacitor': Evolution of capacitive energy storage devices. *Energy & Environmental Science* 5 (11): 9363–9373.

[84] Conway, B.E. (1991). Transition from "supercapacitor" to "battery" behavior in electrochemical energy storage. *Journal of the Electrochemical Society* 138 (6): 1539–1548.

[85] Lu, M., Béguin, F., and Frąckowiak, E. (2013). *Supercapacitors: Materials, Systems, and Applications Materials Science*. Weinheim: Wiley-VCH. DOI:10.1002/9783527646661

[86] Burke, A. (2000). Ultracapacitors: Why, how, and where is the technology? *Journal of Power Sources* 91: 37–50.

[87] Pandolfo, A.G. and Hollenkamp, A.F. (2006). Carbon properties and their role in supercapacitors. *Journal of Power Sources* 157: 11–27.

[88] Stoller, M.D., Park, S., Zhu, Y., et al. (2008). Graphene-based ultracapacitors. *Nano Letters* 8 (10): 3498–3502.

[89] Long, J.W., Bélanger, D., Brousse, T., et al. (2011). Asymmetric electrochemical capacitors—Stretching the limits of aqueous electrolytes. *MRS. Bulletin* 36 (7): 513–522.

[90] Frackowiak, E. and Béguin, F. (2001). Carbon materials for the electrochemical storage of energy in capacitors. *Carbon* 39 (6): 937–950.

[91] Futaba, D.N., Hata, K., Yamada, T., et al. (2006). Shape-engineerable and highly densely packed single-walled carbon nanotubes and their application as super-capacitor electrodes. *Nature Materials* 5 (12): 987–994.

[92] Chmiola, J., Yushin, G., Gogotsi, Y., et al. (2006). Anomalous increase in carbon capacitance at pore sizes less than 1 nanometer. *Science* 313 (5794): 1760–1763.

[93] Miller, J.R., Outlaw, R.A., and Holloway, B.C. (2010). Graphene double-layer capacitor with AC line-filtering performance. *Science* 329 (5999): 1637–1639.

[94] Bonaccorso, F., Colombo, L., Yu, G., et al. (2015). Graphene, related two-dimensional crystals, and hybrid systems for energy conversion and storage. *Science* 347 (6217): 1246501.

[95] Miller, J. (2007). A brief history of supercapacitors. Battery & Energy Storage Technology. Available at: https://www.bestmag.co.uk/content/brief-history-supercapacitors (accessed 17 November 2021).

[96] Gao, Q., Demarconnay, L., Raymundopinero, E., et al. (2012). Exploring the large voltage range of carbon/carbon supercapacitors in aqueous lithium sulfate electrolyte. *Energy & Environmental Science* 5 (11): 9611–9617.

[97] Gogotsi, Y. and Simon, P. (2011). True performance metrics in electrochemical energy storage. *Science* 334 (6058): 917–918.

[98] Deschamps, M., Gilbert, E., Azais, P., et al. (2013). Exploring electrolyte organization in supercapacitor electrodes with solid-state NMR. *Nature Materials* 12 (4): 351–358.

[99] Chmiola, J., Largeot, C., Taberna, P., et al. (2010). Monolithic carbide-derived carbon films for micro-supercapacitors. *Science* 328 (5977): 480–483.

[100] Balducci, A., Dugas, R., Taberna, P.L., et al. (2007). High temperature carbon–carbon supercapacitor using ionic liquid as electrolyte. *Journal of Power Sources* 165 (2): 922–927.

[101] Lazzari, M., Mastragostino, M., and Soavi, F. (2007). Capacitance response of carbons in solvent-free ionic liquid electrolytes. *Electrochemistry Communications* 9 (7): 1567–1572.

[102] Zhu, Q., Song, Y., Zhu, X., et al. (2007). Ionic liquid-based electrolytes for capacitor applications. *Journal of Electroanalytical Chemistry* 601 (1–2): 229–236.

[103] Zhu, Y., Murali, S., Stoller, M.D., et al. (2011). Carbon-based supercapacitors produced by activation of graphene. *Science* 332 (6037): 1537–1541.

[104] Arbizzani, C., Biso, M., Cericola, D., et al. (2008). Safe high-energy supercapacitors based on solvent-free ionic liquid electrolytes. *Journal of Power Sources* 185 (2): 1575–1579.

[105] Huang, J., Sumpter, B.G., and Meunier, V. (2008). A universal model for nanoporous carbon supercapacitors applicable to diverse pore regimes, carbon materials, and electrolytes. *Chemistry – A European Journal* 14 (22): 6614–6626.

[106] Armand, M., Endres, F., MacFarlane, D.R., et al. (2009). Ionic-liquid materials for the electrochemical challenges of the future. *Nature Materials* 8 (8): 621–629.

[107] Liu, C., Yu, Z., Neff, D., et al. (2010). Graphene-based supercapacitor with an ultrahigh energy density. *Nano Letters* 10 (12): 4863–4868.

[108] Brandt, A., Pohlmann, S., Varzi, A., et al. (2013). Ionic liquids in supercapacitors. *MRS. Bulletin* 38 (07): 554–559.

[109] Feng, G., Li, S., Presser, V., et al. (2013). Molecular insights into carbon supercapacitors based on room-temperature ionic liquids. *Journal of Physical Chemistry Letters* 4 (19): 3367–3376.

[110] Zhang, L., Zhang, F., Yang, X., et al. (2013). Porous 3D graphene-based bulk materials with exceptional high surface area and excellent conductivity for supercapacitors. *Scientific Reports* 3: 1408.

[111] Xu, Y., Lin, Z., Zhong, X., et al. (2014). Holey graphene frameworks for highly efficient capacitive energy storage. *Nature Communications* 5: 4554.

[112] Brikez, A.M. (1999). Method of manufacturing a battery electrode. US Patent 5,972,056. Applicant: ESMA & ELIT, Russia.

[113] Razoumov, S., Litvinenko, S., and Beliakov, A. (2001) Asymmetric electrochemical capacitor and method of making. U.S. Patent 6,222,723.

[114] Wang, Y.G. and Xia, Y.Y. (2006). Hybrid aqueous energy storge cells using activated carbon and lithium-intercalated compounds. *Journal of the Electrochemical Society* 153 (2): A450–A454.

[115] Zhang, J.T., Jiang, J.W., Li, H.L., et al. (2011). A high-performance asymmetric supercapacitor fabricated with graphene-based electrodes. *Energy & Environmental Science* 4 (10): 4009–4015.

[116] Belyakov, A.I., Dashko, O.G., Kazarov, V.A. et al. (2001). Capacitor with dual electric layer. US Patent 6,195,252.

[117] Beliakov, A.I. (1993). Russian supercapacitors to start engines. Batteries International.

[118] Aravindan, V., Sundaramurthy, J., Jain, A., et al. (2014). Unveiling TiNb2O7 as an insertion anode for lithium ion capacitors with high energy and power density. *ChemSusChem* 7 (7): 1858–1863.

[119] Ma, Y.F., Chang, H.C., Zhang, M., et al. (2015). Graphene-based materials for lithium-ion hybrid supercapacitors. *Advanced Materials* 27 (36): 5296–5308.

[120] Amatucci, G.G., Badway, F., Pasquier, A.D., et al. (2001). An asymmetric hybrid nonaqueous energy storage cell. *Journal of the Electrochemical Society* 148 (8): A930–A939.

[121] Pasquier, A.D., Plitz, I., Gural, J., et al. (2003). Characteristics and performance of 500F asymmetric hybrid advanced supercapacitor prototypes. *Journal of Power Sources* 113 (1): 62–71.

[122] Pasquier, A.D., Plitz, I., Menocal, S., et al. (2003). A comparative study of Li-ion battery, supercapacitor and nonaqueous asymmetric hybrid devices for automotive applications. *Journal of Power Sources* 115 (1): 171–178.

[123] Naoi, K. (2010). "Nanohybrid capacitor": The next generation electrochemical capacitors. *Fuel Cells* 10 (5): 825–833.

[124] Naoi, K., Naoi, W., Aoyagi, S., et al. (2013). New generation "nanohybrid supercapacitor." *Accounts of Chemical Research* 46 (5): 1075–1083.

[125] Sivakkumar, S.R. and Pandolfo, A.G. (2012). Evaluation of lithium-ion capacitors assembled with pre-lithiated graphite anode and activated carbon cathode. *Electrochimica Acta* 65: 280–287.

[126] Khomenko, V., Raymundo-Pinero, E., and Beguin, F. (2008). High-energy density graphite/AC capacitor in organic electrolyte. *Journal of Power Sources* 177 (2): 643–651.

[127] Cao, W.J. and Zheng, J.P. (2012). Li-ion capacitors with carbon cathode and hard carbon/stabilized lithium metal powder anode electrodes. *Journal of Power Sources* 213: 180–185.

[128] Aravindan, V., Gnanaraj, J., Lee, Y.S., et al. (2014). Insertion-type electrodes for nonaqueous Li-ion capacitors. *Chemical Reviews* 114 (23): 11619–11635.

[129] Cao, W.J. and Zheng, J.P. (2013). The effect of cathode and anode potentials on the cycling performance of Li-ion capacitors. *Journal of the Electrochemical Society* 160 (9): A1572–A1576.

[130] Egashira, M. and Ueda, K. (2008). Lithium-ion capacitor using Lithium Triazolate as electrolyte salt. http://www.greencarcongress.com/2009/01/fdk-to-begin-ma.html (accessed 17 November 2021).

[131] Hideki, S., Yukinori, H., Osamu, H., et al. (2006). Lithium-ion capacitor predoping and high performance. Subaru technical review. http://www.greencarcongress.com/2008/07/jm-energy-intro.html.

[132] Park, M.S., Lim, Y.G., Kim, J.H., et al. (2011). A novel lithium-doping approach for an advanced lithium ion capacitor. *Advanced Energy Materials* 1 (6): 1002–1006.

[133] Zhang, F., Zhang, T.F., Yang, X., et al. (2013). A high-performance supercapacitor-battery hybrid energy storage device based on graphene-enhanced electrode materials with ultrahigh energy density. *Energy & Environmental Science* 6 (5): 1623–1632.

[134] Yi, R., Chen, S.R., Song, J.X., et al. (2014). High-performance hybrid supercapacitor enabled by a high-rate Si-based anode. *Advanced Functional Materials* 24 (47): 7433–7439.

[135] Augustyn, V., Come, J., Lowe, M.A., et al. (2013). High-rate electrochemical energy storage through Li+ intercalation pseudocapacitance. *Nature Materials* 12 (6): 518–522.

[136] Cook, J.B., Kim, H.S., Yan, Y., et al. (2016). Mesoporous MoS2 as a transition metal dichalcogenide exhibiting pseudocapacitive Li and Na-ion charge storage. *Advanced Energy Materials* 6 (9): 1501937.

[137] Brezesinski, T., Wang, J., Tolbert, S.H., et al. (2010). Ordered mesoporous alpha-MoO3 with iso-oriented nanocrystalline walls for thin-film pseudocapacitors. *Nature Materials* 9 (2): 146–151.

[138] Wang, Y.G., Hong, Z.S., Wei, M.D., et al. (2012). Layered H2Ti6O13-nanowires: A new promising pseudocapacitive material in non-aqueous electrolyte. *Advanced Functional Materials* 22 (24): 5185–5193.

[139] Come, J., Naguib, M., Rozier, P., et al. (2012). A non-aqueous asymmetric cell with a Ti2C-based two-dimensional negative electrode. *Journal of the Electrochemical Society* 159 (8): A1368–A1373.

[140] Ding, J., Wang, H.L., Li, Z., et al. (2015). Peanut shell hybrid sodium ion capacitor with extreme energy-power rivals lithium ion capacitors. *Energy & Environmental Science* 8 (3): 941–955.

6

Overview of Superconducting Magnetic Energy Storage Technology

Jing Shi, Xiao Zhou, Yang Liu, Li Ren, Yuejin Tang, and Shijie Chen

State Key Laboratory of Advanced Electromagnetic Engineering and Technology, Huazhong University of Science and Technology, Wuhan, China

6.1 Introduction

The demand for energy storage technologies is increasing rapidly in a modern power grid as the developments in electric power and the progress in renewable energy power generations. Superconducting magnetic energy storage (SMES) is unique among the technologies proposed for energy storage for electric utilities in that there is no conversion of the electric energy, which is stored directly as a circulating current in a large superconducting magnet. The research about SMES for the power grid began in the 1970s. The original concept of SMES is to level out the daily and weekly cyclic power variations of a power grid. A research verified that the rapid response capability is available for stabilizing system oscillations, in which the interaction between a superconducting magnet and a power grid through a multiphase Graetz bridge was studied [1–2].

Although it seems infeasible technically and economically to manufacture a large superconducting magnet for levelling out the daily load variations of a modern power grid, it has been confirmed that SMES has unique advantages in cycle-life and response speed, and can compensate active and reactive power independently [3–4]. Because of these features, SMES is a promising technology for enhancing the stability of the power grid, upgrading the quality of power supply, and levelling out the random output fluctuation of wind or solar power generations. This chapter illustrates the principle and structure of SMES, summarizes its development status, and discusses the developing trends as well as identifying research topics for developing SMES from the point of view of manufacturing and operating SMES.

6.2 The Principle of SMES

6.2.1 The Configuration of SMES

A general configuration of SMES used in the power grid is shown in Figure 6.1. SMES is composed of a superconducting magnet, a cryogenic system, protection system, and power conditioning system(PCS) that include converter, controller, and monitoring system. The transformer in Figure 6.1 is not a component of SMES, it is only used to match the voltage between SMES and the power grid.

Advances in Energy Storage: Latest Developments from R&D to the Market, First Edition.
Edited by Andreas Hauer.
© 2022 John Wiley & Sons, Ltd. Published 2022 by John Wiley & Sons, Ltd.

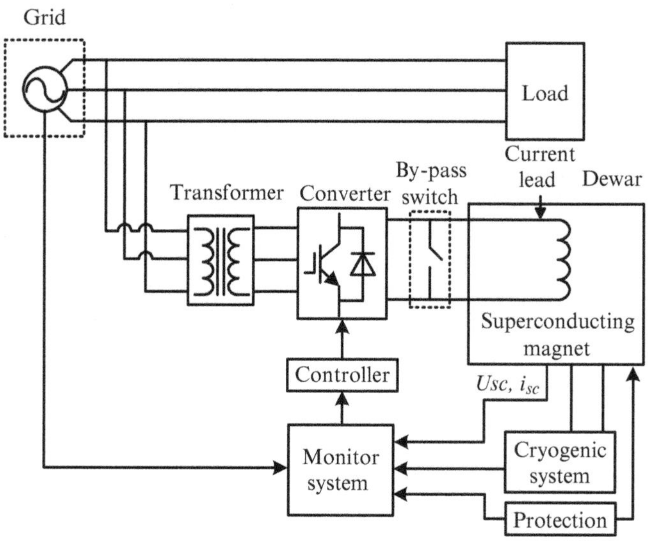

Figure 6.1 The general structure of SMES.

SMES stores electromagnetic energy in the superconducting magnet, which is connected to the power grid through a bidirectional converter. Charging the magnet means absorbing energy from the grid, and discharging the magnet equals the release of energy to the grid. Monitoring systems collect information from the grid and SMES, and give out control orders to the converter and protection system. The converter together with the controller and monitor is also known a power conditioning system. The cryogenic system is used to keep the magnet in a superconducting state. Protection measures are usually used to guard SMES, especially the magnet.

6.2.2 The Components of SMES

(1) Superconducting magnet

A superconducting magnet(coils) is the core component of SMES. The energy stored in SMES can be calculated from Equation (6.1).

$$W_{SMES} = \frac{1}{2} L_{SMES} I_{SMES}^2 \tag{6.1}$$

Here, L_{SMES} is the inductance of the magnet and I_{SMES} is the current in the coils. The value of L_{SMES} affects not only the output voltage and power but also the response speed of the SMES.

Superconducting magnets for SMES can be manufactured as solenoid coils or toroidal arranged coils. The solenoid type has a simple structure, but larger stray magnetic field, while the toroidal type has a smaller stray magnetic field, but more complicated structure. Because the current carrying capacity of the superconducting wire is sensitive to the magnetic field the primary issue of designing a superconducting magnet is whether the magnetic field can fulfil the requirements of the superconducting wires. In addition, while designing a superconducting magnet, many factors should be considered, such as electric parameters matching the demands of the power grid, the mechanical strength of the superconducting wires, operational reliability, magnet protection, and cooling methods.

Although the superconducting magnet has no joule losses when operated under a stable DC current, there are losses while the SMES is compensating power to the grid because a changing current or magnetic field induces AC loss in superconducting wires and eddy loss in the support parts made of conductive materials.

(2) Cryogenic system

The cryogenic system is utilized to maintain a low temperature environment for keeping magnets in a superconducting state. This is also called the cooling system. The cryogenic system directly impacts the performance of SMES, such as the thermal stability of the superconducting magnet, and the cost and reliability of the SMES. Generally, immersion cooling and conducting cooling can be adopted to cool the superconducting magnet [3, 5–6]. Conducting cooling does not require a cryogenic medium, but its thermal transfer speed is lower than that of immersion cooling.

For a low temperature superconducting (LTC) magnet, the cryogenic medium is usually liquid helium (4.2 K). If a high temperature superconductor (HTc) is used, a cooling system of 20–30 K is needed. Due to higher operation temperature, which means lower costs and a higher cooling efficiency, HTc SMES is more promising for practical applications.

(3) Power conditioning systems

PCS are composed of a converter, a controller, and a monitoring system. The magnet can realize active and reactive power compensation in four quadrants quickly and independently through the PCS.

The converter can be divided into two basic typologies: current source converter (CSC), and voltage source converter (VSC), as shown in Figure 6.2 [7–8]. Since the superconducting magnet has inherent current source characteristics, it can be directly connected to the DC side of CSC, and then the magnet needs to be connected to the DC side of VSC through a DC-DC chopper. Nowdays, the converter is usually controlled by Pulse-width Modulation (PWM), which can obtain a lower harmonic level and more sensitive results than a conventional trigger phase control.

The controller is composed of an outer loop and an inner loop control. The outer loop controller provides operation orders of active and reactive power to the inner

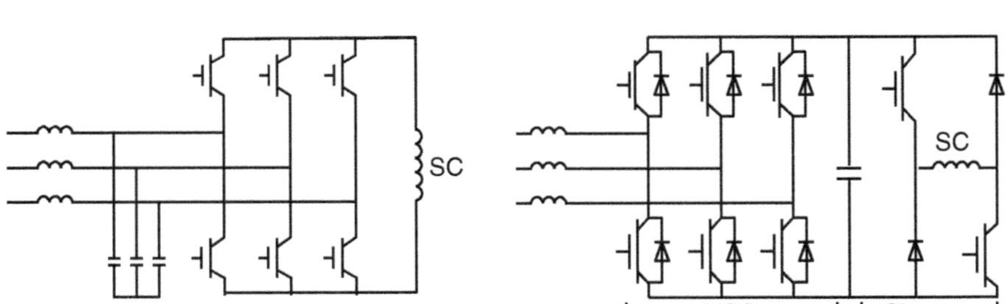

Figure 6.2 Circuit topology of power conditioning systems.

loop. Then, the inner loop controller generates switch trigger signals to the converter, thus realizing power compensation to the grid.

The monitoring system supervises the status of SMES and gives out control orders to the converter and protection system based on the information collected from the grid and SMES. Control strategies are already set according to the operation objective of SMES.

(4) Protection system

In order to improve the security and reliability of SMES, protection system are necessary for the magnet, the converter, and the cooling system. Usually, the cooling system has faults occurring more readily than other component because it is related to cryogenic and vacuum technologies. But the most important and difficult protection is that of the magnet. The magnet can be damaged by quenching if the energy stored in it is not transferred out or cut down on time. Unfortunately, it is difficult to detect quench quickly.

6.3 Development Status of SMES

6.3.1 SMES with LTc Superconductor

The concept of storing energy with superconducting magnet was proposed in 1968. The first SMES was built in the early 1970s. It consisted of a superconducting magnet and a three-phase AC/DC Graetz-bridge. It marked the beginning of applying SMES in power grid [1, 3, 9]. As the power electronic technology developed, PWM controlled converter were used in SMES [10–11].

Before a HTC superconducting conductor was available for practical application, research on SMES progressed with low temperature condition (LTC) superconducting wires of NbTi, mainly in the US and Japan. A summary of LTC SMES projects is shown in Table 6.1. LTC SMES has been verified available for enhancing transient stability, damping low-frequency oscillation, up-grading power quality, and to be used as UPS.

The largest energy storage capacity of LTC SMES has reached 100 MJ [12], and commercial products of LTC SMES with a capacity of megajoules (MJ) class was declared available in the early 1990s. However, there is no evidence that LTC SMES has been used widely in power grids. It is noteworthy that the reliability of the cooling system with liquid helium did not give users sufficient confidence, accompanied the expensive cost of SMES.

Table 6.1 LTc-SMES projects.

Institute	Conductor	Parameters and main objects	References
LANL (US)	NbTi	30 MJ/10 MW; For power oscillation (0.35 Hz).	Field operation. [3]
Kyushu Electric Power (Japan)	NbTi	3.6 MJ/1 MW; For stabilization of load fluctuation.	Field experiment. [13]

Table 6.1 (Continued)

AMSC (US)		2.4 MJ/1.4 MVA; For power quality.	Operated for 3 years. [14]
KERI (Korea)	NbTi	1 MJ/300 kVA; For UPS.	System experiment. [15]
ISTEC (Japan)	NbTi	500 kWh/100 MW; For load compensation.	CICC conductor. [16]
Florida State University (US)	NbTi	100 MJ/5 MW; For load fluctuation compensation.	Field operation. [12]
Chubu Electric Power (Japan)	NbTi	5 MJ/5 MVA; For power quality.	Field experiment. [17]
THU (China)	NbTi	0.3 MJ; voltage compensation;	System experiment. [18]
KERI (Korea)	NbTi	3 MJ/750 kVA; For power quality.	Tested as UPS. [19]
Ansaldo Ricerche Spa (Italy)	NbTi	2.6 MJ/1 MW; For sensitive load.	Magnet test. [20]
Chubu Electric Power (Japan)	NbTi	20 MJ/10 MVA; For load fluctuation.	Field experiment. [21]

6.3.2 SMES with HTc Superconductor

HTC has made rapid progress during the past 20 years particularly on superconducting tapes and HTC SMES has also made great progress in the twenty-first century. Table 6.2 summarizes research projects on HTC SMES, in which the magnets are made of Bi2212, Bi2223, or YBCO tapes. Although all of this research verified that HTC SMES can realize all the same functions as LTC SMES in a power grid the energy storage capacity is not large enough to fulfil the needs of a power grid.

SMES with a HTC superconducting magnet can operate at 20–30 K. Compared with LTC SMES, a higher operating temperature may bring merits over a simpler cooling system and have a cheaper operating cost. HTC SMES is promising for the future, because the development of a HTC superconducting conductor is still making rapid progress.

Another new concept that has been proposed [22], this combines the use of liquid hydrogen (LH_2) as the bulk energy carrier with much faster and efficient SMES than in SMES in which MgB_2 is used.

6.3.3 Simulation Research about the Application of SMES in a Power Grid

Besides the research projects described, there are much research into the effect of applying SMES by simulation analysis, such as using D-SMES to improve the voltage stability of a transmission system [31], applying SMES in an electric vehicle (EV)

Table 6.2 HTc-SMES projects.

Institute	Conductor	Parameters and main objects	References
ACCEL (Germany)	Bi2223	150 kJ; UPS.	Theoretical design. [23]
Chubu Electric Power (Japan)	Bi2212	1 MJ; For instantaneous voltage dips.	Field experiment. [5]
HUST (China)	Bi2223	35 kJ/7.5 kW; For transient stability.	Field experiment. [6]
KERI (Korea)	Bi2223	36.5 kJ; Fundamental research.	Magnet test. [24]
IEECAS (China)	Bi2223	30 kJ; Fundamental research.	Magnet test. [25]
KERI (Korea)	Bi2223	600 kJ; For voltage compensation.	Magnet test. [26]
DGA (France)	Bi2212	800 kJ; Pulsed power generator.	Magnet test. [27]
ISTEC (Japan)	YBCO	2 GJ/100 MVA; For load fluctuation compensation.	System design [28]
IEECAS (China)	Bi2223	1 MJ/0.5 MW; Power quality improvement.	Field experiment. [29]
HUST (China)	YBCO/Bi	150 kJ; For stability of power grid.	Field experiment. [30]

charging station [32], in hybrid energy storage system [33–34], in wind energy generation plant, as well as to evaluate the stability of a power grid [35].

6.4 Development Trend of SMES

6.4.1 Promising Ways to Develop SMES

(1) HTc SMES

As described in Section 6.3, LTC SMES based NbTi is almost a mature superconducting equipment as the 100MJ LTC SMES has been developed and has completed field experiment/demonstration operation. However, LTC SMES has not been widely used until now. It is noted that the cost, security, and reliability are three main factors affect its application. In addition, other energy storage technologies are progressing with high speed, so SMES has strong competitive opponents. Considering the NbTi conductor technology is mature, LTC SMES with NbTi has little space for upgrading its technique and economic performances. Therefore, HTC SMES is the main route for development, including using MgB_2 operated at 20 K [22]. Although HTC conductors are currently more expensive than LTC conductors, there is still development space for HTC conductors; the higher temperature of HTC SMES offers an advantage.

(2) Modular SMES and distributed SMES

Due to the difficulties of developing a large-scale HTC magnet, the concept of a dispersal module type SMES is proposed. It means that a number of small-scale HTC magnets, even multiple SMES are collectively controlled as a larger scale one.

According to the conditions and requirements of the power grid, a number of SMES can be located at different sites and co-controlled: the concept of distributed SMES.

Modular SMES and distributed SMES have features of layout flexibility, high reliability, easy expansion, and so on.

(3) Hybrid energy storage

Because of the urgent requirements for energy storage techniques, the development of different types of battery with high performance is progressing very fast. Batteries can realize large energy storage capacity easier and cheaper than SMES, but its cycle-life and response speed is relatively poor when compared with SMES. A promising application method is to combine SMES with a battery [33–34]. The controller lets the SMES respond to fast and frequent needs, while the battery bears the slow changes and longer response times. Figure 6.3 shows the topology of a power electronic converter for SMES and a battery [34]. Both the SMES and the battery are connected to the DC bus of a grid-side DC/AC converter through their respective bidirectional DC/DC

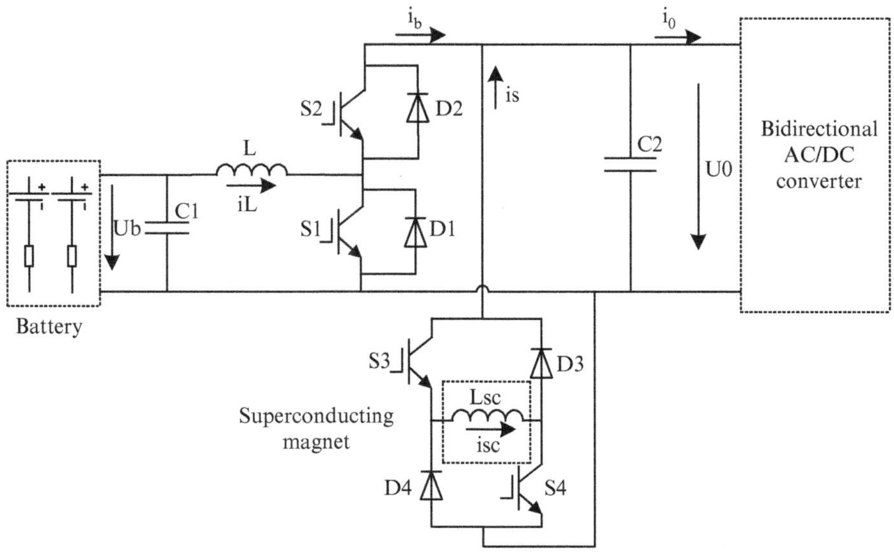

Figure 6.3 Topology of cooperated hybrid energy storage system with SMES and battery.

converter. SMES generally undertakes rapid or short duration transient or dynamic power compensation; the battery with a large capacity is used for long-period power compensation and energy regulation.

Another hybrid energy storage concept has been proposed that combines the use of liquid hydrogen (LH_2) as the bulk energy carrier with much faster and efficient SMES [22]. LH_2 with its high volumetric energy density is a prime candidate for large-scale stationary energy storage. But it is unrealistic to balance load or supply fluctuations due to not only the losses related to the re-conversion into electricity but also to the response times of the flow control. HTC SMES is proposed for this purpose, which could be operated in the LH_2 bath. With this approach the cryogenics-related costs for SMES are widely cut.

6.4.2 Promising Applications of SMES

6.4.2.1 Near-term Applications

At the present stage, due to the high cost and limited current-carrying capability of HTC tapes, the capacity of HTC SMES is not large. So, it is better to focus the application of SMES on the quick response characteristic instead of on energy storage capacity.

(1) Enhancing power grid transient stability

When the power grid is subjected to a large disturbance, SMES acts as a flexible and controllable active and reactive power source to damp the system oscillation, and thus enhance the transient stability of the power grid. The response time of SMES is less than 10 ms, faster than the generator's controller.

(2) Improving power quality

With the development of precision manufacturing, information technology, and digital technology, power quality is receiving increasing attention. Power quality problems include many factors, such as harmonic, three-phase voltage asymmetry, long-time or instantaneous voltage fluctuation, and instantaneous or short-time power supply interruption.

SMES can play an important role in improving power quality systems with its fast response speed as well as the ability to compensate active and reactive power independently. Not only can SMES compensate for the change in voltage, but it also can be used as an uninterrupted power supply.

6.4.2.2 Long-term Applications

If 10 MJ ~ GJ HTc SMES can be realized as the progress of HTC superconducting tapes, SMES can be utilized to damp low-frequency oscillations in the power grid, supply the power grid reserve, and compensate unbalanced power in renewable energy generation.

(1) Damping low-frequency oscillations in power grid

The reason for low-frequency oscillation is the negative damping effect of the power grid. This often occurs at a weak link, on a long-distance, and on a heavy-load transmission line; it is more prone to occur under the condition of a fast and high magnification excitation system. A 30 MJ/10 MW HTc SMES has been installed in a 500 kV transmission line near the western coast of the US [3], which successfully damped the emerging 0.35 Hz low-frequency oscillation that occurred and improved the stability of the transmission line. So, the HTC SMES with large capacity can also be utilized to damp low-frequency oscillations of the power grid.

(2) Compensating peak load

Some electric loads are intermittent or pulsed, such as trains and a lifter/crane. If the capacity of the load is larger enough, turning it on or off of the device may affect the performance of the connected power grid. SMES can be used to compensate power to relax the impact of peak load. Also, SMES is expected to level out the daily load curve of the power grid. However, to achieve that a very large capacity is needed. Therefore, it is a good candidate for compensating peak loads, which can cope with pumped hydro power generation.

(3) Improving the stability of renewable energy generation

The most efficient way to utilize renewable energy is by generating electric power and connecting it to the power grid. However, some renewable energy sources, such as solar energy, wind energy, and ocean energy are influenced by natural climate with intermittent, randomness, and non-smoothness, which will aggravate the power fluctuations in a power grid, and even affect the stable operation of the grid. SMES can smooth the output power of the renewable energy power generation with its fast power regulation ability and efficient energy storage ability. Thus SMES can effectively suppress problems such as voltage fluctuation and flicker caused by such power sources, and improve controllability and stability. For this application, SMES is also available now for use in collaboration with battery banks [34].

6.5 Research Topics for Developing SMES

6.5.1 Key Problems Concerned with SMES Components

(1) Superconducting wires

The experiences of the past 40 years indicate that it is difficult to apply widely SMES cooled with liquid helium in a power grid, so it is necessary to develop SMES with a higher operation temperature. For high temperature superconductors, BSCCO and YBCO are successfully manufactured to be usable conductors, but with tape form. This gives some hopes to HTC SMES. However, the characteristics of HTC superconducting tapes are still not good enough for large-scale SMES. Their anisotropic electromagnetic characteristics make the critical current in a HTC magnet non-uniform. Their fragile and weak mechanical properties make it difficult to get twisted large current conductors and manufacture large-scale magnets because they cannot bear a strong electromagnetic force. An HTC conductor with enhanced mechanical properties, lower cost, and better performance is a key topics in developing HTC SMES. There are many researchers trying to make twisted HTC large current conductors with isotropic performance.

Other new superconducting wires such as Nb_3Al, MgB_2 are also under development. They can be used in SMES with some special application methods.

(2) Insulation of the superconducting magnet

The PCS interfaces the magnet and power grid, which controls energy transfer of SMES. The output voltage of PCS with PWM converter is pulsed square waves with different pulse widths. Over-voltage and non-uniform distribution may be induced on the magnet when the pulsed voltage at a steep rising edge is applied, because the magnet has an LC distributed equivalent circuit [35–36]. On the other hand, a dynamic or transient current in the magnet induces overvoltage too. Dynamical current appears while the SMES is compensating power to the power grid, quench occurs in the magnet, or some other fault situations. It is necessary to develop cryogenic temperature insulation materials and design optimal insulation structures for the magnet.

(3) Optimal design method for a superconducting magnet

In order to manufacture SMES with high technique performances and workable economic cost (including operation cost), the design of the magnet should be optimized under multiple operation conditions. It is necessary to consider the dynamic characteristics of SMES while designing the magnet for SMES [37–38]. Designing a superconducting magnet for SMES is a complex issue, which covers many disciplines, such as superconductivity, electric field, magnetic field, structural mechanics, heat transfer and cryogenics, power grid, electric circuit theory, and low temperature insulation. In addition, optimal design relates to computer programming, optimization algorithm, and optimization software development, etc.

(4) Cooling system

The cooling system has a serious effect on the performance of an SMES, such as (operating cost), security, reliability as well as its expected length of life. Although developing cooling systems mainly depends on the progress of cryogenic and vacuum techniques, some work can also be done on designing, manufacturing, and operating SMES. For example, the cooling ability must be optimized together with the magnet considering the heat produced by dynamic current when the SMES operates in a power compensating state [38–39].

(5) Power condition system

One of the limits for developing larger scale SMES comes from the PCS, more specifically, from the converter. In order to develop SMES with a large capacity, high power density, fast response characteristics, the PCS needs to address the following aspects: high voltage and high current power electronic devices; serial-parallel technology of power electronic devices and modular converter; and the control strategy of the PCS.

6.5.2 Key Problems Concerned with SMES Operation

(1) Selecting optimal site to locate the SMES

SMES can be applied in the power grid to solve problems such as transient stability, voltage stability, and power quality. For different problems, the parameters and control strategies of the SMES are different. For other the sensitive points for different problems are diverse according to the configuration of the power grid. Setting the SMES at the most sensitive node can get the best response and highest operation efficiency. The site selection of SMES is a comprehensive topic, it involves power grid planning, dynamic performances of power grid and SMES, and technical and economic performances of SMES.

(2) Status evaluation of SMES

To ensure the security and reliability of the SMES, it is important to measure all of the key technical parameters related to the operation of SMES and evaluate the status based on these parameters. In addition, as the output capability of SMES has a close relationship to the status of the superconducting magnet, such as the energy stored in the magnet (can be expressed by the current) and the temperature of the magnet. If the maximum output can be controlled based on the online status evaluation of SMES, the security of the SMES may be improved greatly [40]. To achieve this goal, it is necessary to study the interaction between the dynamic characteristics of superconducting magnets and the power grid.

(3) Coordinated operation with existing control equipment

In order to perform better both technically and economically, it is essential that the SMES coordinates with other existing control equipment. For example, if an SMES is used to increase the transient stability of the power grid, the SMES can be coordinately controlled with the excitation controller of the generator. SMES and the excitation controller share their responsibilities according to the their characteristics. Another coordination operation issue is that SMES needs to cooperate with other energy storage devices such as the battery bank, which is the way of hybrid energy storage method. In a hybrid energy storage system, SMES responds the orders of rapid and high-frequency power changes, while the battery bank serves the power fluctuations with a low speed but a long time cycle. With such a control strategy, the SMES does not need large energy storage capacity, and the battery bank can improve its cycle-life.

6.6 Conclusions

SMES has progressed to be distinctive energy storage equipment, which can compensate active and reactive power independently with a fast response speed, and has no limit on charge and discharge cycle-life. SMES promises that it can cooperate well with the competitive opponents in electric energy storage in the future power grid.

Considering the experiences of LTC SMES, the developing focus should be put on HTC SMES. However, expensive cost and unconfirmed reliability have restricted the application of SMES in the power grid widely. At present, there are still many technical difficulties in developing large-scale HTC SMES. Currently available HTC superconducting tapes are difficult to produce to support SMES with an energy storage capacity of 10 MJ ~ 1 G, which is expected in the modern power grid. Also, the reliability and cycle-life of the cooling system do not provide users with enough confidence to use SMES.

In order to develop SMES, three application models are proposed. The first one uses a modular superconducting magnet or modular SMES. The second one is distributed SMES, where several SMESs are installed at different nodes but work together. The third is to enable SMES to cooperate with other competitive energy storage devices.

Some important problems to be researched in the development of SMES are identified and include essential ones such as developing a superconducting conductor with a higher performance, and application problems such as developing effective online methods to evaluate the operation status of SMES.

References

[1] Peterson, H.A., Mohan, N., and Boom, R.W. (1975). Superconductive energy storage inductor-convertor units for power systems. *IEEE Transactions on Power Apply System* 94 (4): 1337–1348.

[2] Hassenzahl, W.V. (1983). Superconducting magnetic energy storage. *Proceeding of the IEEE* 71 (9): 1089–1098.

[3] Rogers, J.D., Schermer, R.I., Miller, B.L. et al. (1983). 30-MJ superconducting magnetic energy storage system for electric utility transmission stabilization. *Proceedings of the IEEE* 71 (9): 1099-1107.

[4] Ali, M.H., Wu, B., and Dougal, R.A. (2010). An overview of SMES applications in power and energy systems. *IEEE Transactions on Sustainable Energy* 1 (1): 38–47.

[5] Shikimachi, K., Moriguchi, H., Hirano, N., et al. (2005). Development of MVA class HTS SMES system for bridging instantaneous voltage dips. *IEEE Transaction on Applied Superconductivity* 15 (2): 1931–1934.

[6] Shi, J., Tang, Y., Zhou, Y., et al. (2007). Development of a conduction-cooled HTC SMES. *IEEE Transaction on Applied Superconductivity* 17 (3): 3846–3852.

[7] Zhang, C.Z. and Ooi, B. (1993). Multimodular current-source SPWM converters for a superconducting magnetic energy storage system. *IEEE Transaction on Power Electronics* 8 (3): 250–256.

[8] Iglesias, I.J., Acero, J., and Bautista, A. (1995). Comparative study and simulation of optimal converter topologies for SMES system. *IEEE Transaction on Applied Superconductivity* 5 (2): 254–257.

[9] Luongo, C.A. (1996). Superconducting storage systems: An overview. *IEEE Transactions on Magnetics* 32 (4): 2214–2223.

[10] Murakami, Y., Ise, T., and Fujita, Y. (1988). Controlled characteristics of a superconducting AC coil by pulse width modulated current. *IEEE Transactions on Magnetics* 24 (2): 903–906.

[11] Hassenzahl, W.V. (2000). Superconducting magnetic energy storage. *Power Engineering Review IEEE* 17 (11): 607–612.

[12] Luongo, C.A., Baldwin, T., Ribeiro, P., et al. (2003). A 100 MJ SMES demonstration at FSU-CAPS. *IEEE Transactions on Applied Superconductivity* 13 (2): 1800–1805.

[13] Hayashi, H., Honda, K., Kanetaka, H., et al. (1999). Results of tests on components and the system of lkWh/lMW module-type SMES. *IEEE Transaction on Applied Superconductivity* 9 (2): 313–317.

[14] Cerulli, J. (1999). Operational experience with a superconducting magnetic energy storage device at Brookhaven National Laboratory, New York. *IEEE Power Engineering Society* 99: 1247–1252.

[15] Seong, K.C., Kim, H.J., Kim, S.W., et al. (2002). Design and test of a 1-MJ SMES system. *IEEE Transaction on Applied Superconductivity* 12 (1): 391–394.

[16] Hanai, S., Shimada, M., Tsuchihashi, T., et al. (2003). Design and test results of CIC conductor for a cost reduced 100 MW/500 kWh SMES. *IEEE Transaction on Applied Superconductivity* 13 (2): 1810–1813.

[17] Nagaya, S., Hirano, N., Kondo, M., et al. (2004). Development and performance results of 5 MVA SMES for bridging instantaneous voltage dips. *IEEE Transaction on Applied Superconductivity* 14 (2): 699–704.

[18] Jiang, X., Zhu, X., Cheng, Z., et al. (2005). A 150 kVA/0.3 MJ SMES voltage sag compensator for distribution system. *IEEE Transactions on Applied Superconductivity* 15 (2): 1903–1906.

[19] Kim, H.J., Seong, K.C., Cho, J.W., et al. (2006). 3 MJ/750kVA SMES system for improving power quality. *IEEE Transaction on Applied Superconductivity* 16 (2): 574–577.

[20] Ottonello, L., Canepa, G., Albertelli, P., et al. (2006). The largest Italian SMES. *IEEE Transaction on Applied Superconductivity* 16 (2): 602–607.

[21] Katagiri, T. (2009). Field test result of 10 MVA/20 MJ SMES for load fluctuation compensation. *IEEE Transaction on Applied Superconductivity* 19 (3): 1993–1998.

[22] Sander, M., Gehring, R., and Neumann, H. (2013). LIQHYSMES – A 48 GJ toroidal MgB2-SMES for buffering minute and second fluctuations. *IEEE Transactions on Applied Superconductivity* 23: 5700505.

[23] Kreutz, R., Salbert, H., Krischel, D., et al. (2003). Design of a 150 kJ high-Tc SMES (HSMES) for a 20 kVA uninterruptible power supply system. *IEEE Transaction on Applied Superconductivity* 13 (2): 1860–1862.

[24] Kim, J.H., Kim, W.S., Hahn, S.Y., et al. (2005). Characteristic test of HTS pancake coil modules for small-sized SMES. *IEEE Transaction on Applied Superconductivity* 15 (2): 1919–1922.

[25] Wang, Q., Dai, Y., Song, S., et al. (2008). A 30 kJ Bi2223 high temperature superconducting magnet for SMES with solid-nitrogen protection. *IEEE Transaction on Applied Superconductivity* 18 (2): 754–757.

[26] Kim, W.S., Kwak, S.Y., Lee, J.K., et al. (2006). Design of HTS magnets for a 600 kJ SMES. *IEEE Transaction on Applied Superconductivity* 16 (2): 620–623.

[27] Tixador, P., Bellin, B., Deleglise, M., et al. (2007). Design and first tests of a 800 kJ HTS SMES. *IEEE Transaction on Applied Superconductivity* 17 (2): 1967–1972.

[28] Shikimachi, K., Hirano, N., Nagaya, S., et al. (2009). System coordination of 2 GJ class YBCO SMES for power system control. *IEEE Transaction on Applied Superconductivity* 19 (3): 2012–2018.

[29] Guo, W., Xiao, L., Dai, S., et al. (2011). Control and test of a 0.5 MVA/1 MJ SMES. *IEEE Transaction on Applied Superconductivity* 22 (3): 5700804.

[30] Ren, L., Xu, Y., Zuo, W., et al. (2015). Development of a movable HTS SMES system. *IEEE Transaction on Applied Superconductivity* 25 (4): 5701109.

[31] Ross, M., Borodulin, M., and Kazachkov, Y. (2001). Using D-SMES devices to improve the voltage stability of a transmission system. *IEEE Proceedings: Transmission and Distribution Conference and Exposition* 2: 1144–1148.

[32] Liu, Y., Tang, Y., Shi, J., et al. (2015). Application of small-sized SMES in an EV charging station with DC bus and PV system. *IEEE Transaction on Applied Superconductivity* 25 (3): 5700406.

[33] Ise, T., Kita, M., and Taguchi, A. (2005). A hybrid energy storage with a SMES and secondary battery. *IEEE Transaction on Applied Superconductivity* 15 (2): 1915–1918.

[34] Wang, S., Tang, Y., Shi, J., et al. (2015). Design and advanced control strategies of a hybrid energy storage system for the grid integration of wind power generations. *IET Renewable Power Generation* 9 (2): 89–98.

[35] Arsoy, A.B., Wang, Z., and Liu, Y. (1999). Transient modeling and simulation of a SMES coil and the power electronics interface. *IEEE Transaction on Applied Superconductivity* 9 (4): 4715–4724.

[36] Shirai, Y. and Nitta, T. (2002). On-line evaluation of power system stability by use of SMES. *IEEE Conference on Transmission and Distribution* 3: 1548–1553.

[37] Xu, Y., Tang, Y., Ren, L., et al. (2014). Numerical simulation of coupled heat transfer, stress and electromagnetic properties in a SMES magnet. *IEEE Transaction on Applied Superconductivity* 24 (3): 4901005.

[38] Xu, Y., Tang, Y., Ren, L., et al. (2015). An evaluation method for small-scale conduction cooled SMES cryogenic cooling system based on thermal analysis. *Cryogenics* 71: 30–38.

[39] Xu, Y., Tang, Y., Ren, L., et al. (2016). Numerical simulation and experimental validation of a cooling process in a 150-kJ SMES magnet. *IEEE Transaction on Applied Superconductivity* 26 (6): 5700507.

[40] Liu, Y., Tang, Y., and Xu, Y. (2015). Status evaluation method for SMES used in power grid. *IEEE Transactions on Applied, Superconductivity* 25 (5): 5701310.

7

Key Technologies of Superconducting Magnets for SMES

Ying Xu, Li Ren, Jing Shi, and Yuejin Tang

State Key Laboratory of Advanced Electromagnetic Engineering and Technology, School of Electrical and Electronic Engineering, Huazhong University of Science and Technology, Wuhan, China

7.1 Introduction

A whole SMES mainly consists of a cryogenic system, a vacuum dewar, a superconducting magnet, a current lead, a converter, a data acquisition system, and a quench protection system. The magnet is the core because it plays the role of energy storage. Obviously, the design of the magnet determines the energy density of the SMES, which may be regarded as the level of a SMES to some degree. Besides, the SMES magnet will not be in a superconducting state when its operation conditions, such as current, stress, and temperature and so on, change because of the feature of superconducting material used for manufacturing the magnet. Therefore, ensuring the safety and stability of SMES in operation through the design of the magnet is also a primary point to consider. Succinctly, the SMES magnet decides the level of SMES and whether the SMES can operate normally. As for the other parts of SMES, they must be coordinated with the magnet during the design process. In other words, the magnet is also the basis of an SMES.

7.1.1 Key Parameters of SMES Magnets

The design of the SMES magnet focuses mainly on the following parameters: energy storage W_m, inductance L, magnet voltage V, current I_{op}, and output power P. The relationship between each parameter and magnet operating characteristics is described.

7.1.1.1 Stored Energy W_m and Inductance L

Energy of SMES is stored in the magnetic field. The energy in the magnet is defined as in Equation (7.1):

$$W_m = \frac{1}{2}\int_V \vec{H} \cdot \vec{B} dV \tag{7.1}$$

Where W_m is the stored energy; $\vec{H}$, the magnetic field intensity; $\vec{B}$, the magnetic flux density; and V, the volume in the region of the magnetic field. So the storage energy, W_m, of the SMES magnet is given by Equation (7.1).

Advances in Energy Storage: Latest Developments from R&D to the Market, First Edition. Edited by Andreas Hauer.
© 2022 John Wiley & Sons, Ltd. Published 2022 by John Wiley & Sons, Ltd.

While the energy stored in an SMES magnet (of inductance, L) when current, I_{op}, is passing via it is:

$$W_m = \frac{1}{2} L I_{\text{op}}^2 \tag{7.2}$$

After the storage energy, W_m, is determined, the inductance of the magnet is given by Equation (7.2).

7.1.1.2 Critical Current I_c and Operation Current I_{op}

In the process of designing a SMES magnet, calculating the critical current Ic of the magnet is quite an important step. It is worth mentioning that the critical current of the magnet must be predicted with great accuracy. The critical current of superconducting tapes used for a winding magnet is affected by some factors, such as the magnetic force, the stress, and the strain, which should be taken into account when calculating the critical current of the magnet. Analyzing the influence of the magnetic force and strain on the critical current of the magnet is the main task, in which the former is to determine the maximum current allowed or the magnet and the latter is to examine whether superconducting tapes can meet the demands of mechanical stability under this allowed current.

During the process of calculating the critical current, the first step is acquiring the tape's electromagnetic property described as the J_c-B-θ curve and contrasting it with the magnetic force extracted from the magnet's model at different angles. And then the reference critical current values of the magnet at each angle can be obtained. The minimum of these current values can be regarded as the preliminary critical current of the SMES magnet. The operation current I_{op} of a superconducting magnet must be lower than its critical current. Thus the operation current I_{op} is set to 0.6 to 0.8 times that of the critical current to ensure a magnet's safety in actual application.

7.1.1.3 Voltage U and the Output Power P

SMES exchanges power with the grid through a DC (Direct Current) link voltage. DC link voltage is the voltage U across the magnet. The output power of the SMES is the product of the DC link voltage and the magnet's current. The energy variation and power can be described with Equations (7.3) and (7.4):

$$\Delta E = Pt \tag{7.3}$$

$$P = U I_{\text{op}} \tag{7.4}$$

Where t is the discharge period, U is the voltage of the DC link, and I_{op} is the current of the magnet. ΔE and P are the energy variation and the discharge power, respectively. As shown in Equation (7.4), the output power is determined by U and I.

7.1.2 Structures of SMES Magnets

The configurations of commonly used SMES magnets are single solenoid, multiple solenoid, and toroidal magnet, as shown in Figure 7.1. Each configuration has its own electromagnetic and mechanical characteristics. Single solenoid magnet is the simplest in structure and most widely used magnet. A multiple parallel arrangement of

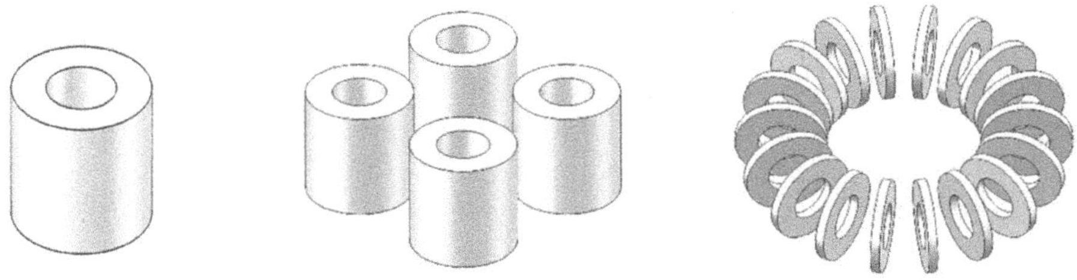
(a) (b) (c)

Figure 7.1 Three types of configurations of SMES magnets: (a) Single solenoid; (b) Multiple solenoid; (c) Toroidal magnet.

single solenoids becomes a multiple solenoid, which can reduce the stray field. A toroidal magnet forms a closed magnetic field, with a small stray magnetic field. Currently, accounting for magnets in use and magnets in production, the single solenoid is the most widely used. Multiple solenoids and toroidal magnets are commonly applied to large-scale SMES devices because of their minimal stray magnetic field.

Usually, a superconducting magnet is made up of superconducting coils, which is no different to the coils of an electrical machine and transformer. It's called a magnet rather than a coil since the only function for superconducting coils, in most applications, is just producing a magnetic field. It functions exactly the same as an electromagnet and a permanent magnet and can obtain an even higher magnetic field intensity. The structures for commonly used SMES magnets are: single solenoid, multi-solenoid, and toroidal magnet. The single solenoid magnet is suitable for small-scale SMES systems, and the multi-solenoid and the toroidal structure magnet are suitable for middle-scale and large-scale SMES systems.

The structure of a single solenoid magnet (Figure 7.1(a)) is the simplest and is the most widely used magnet in the industry. Magnetic fields produced by a single solenoid magnet are approximately parallel to the axis of the magnet. In order to improve the uniformity of magnetic field distribution, the layout of the magnet winding can be designed as high multi-layer windings. Measures like compensation winding can also be used. A single solenoid magnet has the advantage of a simple structure, a high material utilization rate, and high energy storage efficiency. The main drawback is that the external leakage of magnetic field is larger, which will affect the surrounding environment.

A multiple parallel arrangement of the single solenoid becomes a multi-solenoid magnet system, shown as Figure 7.1(b). Due to the electromagnetic coupling relationship between every solenoid magnet, a multi-solenoid magnet is normally treated as one single magnet when considering current flow changes. An equivalent current in the reverse direction is loaded on adjoining groups of magnets to cancel out the reduced the magnetic field leakage, meanwhile, the capacity of the magnet is maintained, but the energy storage efficiency is low. Simulation results show that the magnetic leakage field decreases with the decrease of ratio of height to radius of the magnet, but the critical current is affected simultaneously.

The structure of a toroidal magnet is shown in Figure 7.1(c). This kind of magnet forms a closed magnetic field, with external leakage being very small, which could help to avoid the adverse effects on the surrounding environment and expand the application of SMES. Since it is difficult to make a complete winding of a toroidal magnets, generally multiple solenoids are placed annularly instead. In this arrangement the coil can be set as a round shape, a D shape, or other geometric shape depending on the optimization of electromagnetic characteristics, stress design, easy installation layout, and the convenience for supporting, etc. The D-shape coil has a better bearing capacity than the round shape coil, but it has more complicated structure and higher manufacturing cost.

7.2 The Development of SMES Magnets

As a new type of energy storage technology, SMES has been under study for electric utility energy storage applications since the early 1970s [1]. With the development of cryogenic technology and superconducting conductors, the capacity of LTS (Low Temperature Superconducting) SMES can reach 100 MJ. Although the capacity of HTS (High Temperature Superconducting) SMES is not as large as LTS SMES, HTS tapes are also competitive in SMES applications as it is very attractive to use HTS tapes to build SMES systems to achieve larger energy densities. This opens up more opportunities for the design of HTS energy storage magnets. Some examples of SMES units follow.

7.2.1 LTS SMES

Los Alamos Scientific Laboratory (LASL) developed the concept of installing a small SMES unit for the purpose of providing system damping similar to that now available through modulation of the Pacific HVDC (High-voltage Direct Current) intertie [2–4]. The coil is immersion cooled in liquid helium at 4.5 K and the stored energy, when fully charged, is to be 30 MJ, even though a maximum of only 9.1 MJ is to be exchanged with the AC (Alternative Current) power system on each cycle. The magnet contains 20 NbTi double pancakes with an inductance of 2.4 H, the current drop from 5 kA to 4.17 kA during the magnet discharge.

The Center for Advanced Power System (CAPS), as an outgrowth of the National High Magnetic Field Laboratory (NHMFL), built a new facility in Tallahassee, Florida, to house the new program. An important feature of the facility will be the ability to simulate pulsed power conditions up to the rating of the test bed, the network will include energy storage in the form of a 100 MJ SMES that will serve not only as a pulsed power supply during tests, but also as a field demonstration of SMES technology at a scale not achieved before and meaningful for both military and utility applications [5]. The SMES system can provide 100 MW peak and 50 MW oscillatory power with 100 MJ of stored energy, typical frequencies in which significant amounts of energy are exchanged with the electrical grid are in the range of 0.2 to 3 Hz. The base line for the coil design is a NbTi cable-in-conduit conductor (CICC), with a rated voltage of 24 kV, and operating at a nominal temperature of 4.5 K. The number of double pancakes is 44 and the inductance is 10.8 H. Figure 7.2 shows the 100 MJ solenoid superconducting coil system in the CAPS laboratory.

An SMES with an output 5 MW has been developed, by Chubu Electric Power, to bridge instantaneous voltage dips. The field examination is executed by setting up this system at the new and large liquid crystal factory in Japan. The developed SMES

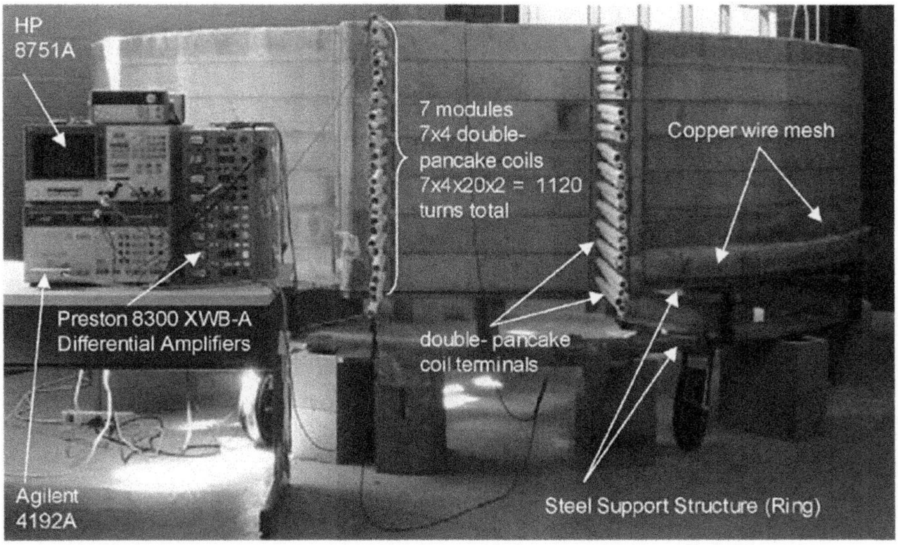

Figure 7.2 100MJ solenoid superconducting coil system in the CAPS laboratory.

Figure 7.3 Four-pole solenoid superconducting coil system.

can store up to 7.34 MJ of magnetic energy in superconducting coils using the NbTi conductor [6]. The operating current, rated voltage, and inductance of the magnet is 2.7 kA, 2.5 kV, and 2.08 H, respectively. The magnet is cooled by a boiling liquid helium pool and works at 4.2 K. The test results confirmed the performance of bridging instantaneous voltage dips by SMES, the long-term drive reliability, and the standby loss characteristic. Figure 7.3 shows Four-pole solenoid superconducting coil system.

In the Chinese Academy of Sciences, a large-scale superconducting magnet with four parallel solenoids was designed, fabricated, and tested for high storage energy density SMES (see Figure 7.3) [7]. The superconducting magnet stores 2 MJ of energy with a current of 490 A and a peak magnetic field of 5.4 T, the inductance of the magnet is 16.644 H. Two GM (Gifford-McMahon) cryo-coolers cool the whole system to realize zero evaporation of liquid helium and keep the normal operation with a bath temperature of 4.2 K. High temperature superconducting current leads of Bi2223 are used and cooled through one GM cryocooler. The height of the superconducting magnet is 0.612 m. The maximum diameter of the magnet is limited to 1.0 m for economic use of and to improve the storage energy density. A special structure is needed to support the attractive forces between the solenoids. A monolithic NbTi/Cu conductor is used to improve the mechanical stability.

7.2.2 HTS SMES

In Japan, a SMES system of MVA class for bridging instantaneous voltage dips has been developed using Bi2212 wire [6, 8–9]. The Bi2212 wire has high-performance conductive characteristics that do not deteriorate at a low temperature in high magnetic fields beyond 10 T. A total of 18 unit coils were stacked to make a coil system (outer diameter: 700 mm; height: 554 mm; inductance: 7.87 H; stored energy: 984 kJ) and installed into an SMES system of 1 MVA for bridging instantaneous voltage dips. An indirect cooling system of recondensing helium by using two compact cryo-coolers (cooling capacity: 4.3 W at 4.3 K) was adopted to cool the coils. Additionally, HTS power leads and an 80K shield were individually cooled by single-stage cryo-coolers.

Chubu Electric Power developed an SMES system for load fluctuation compensation of a system with an output capacity of 100 MVA and a storage capacity of the 2 GJ class, which is expected to meet power system control requirements [10–11]. The SMES system includes a toroidal type YBCO coil consisting of 180 compact, high magnetic field multi-unit coils, a conduction cooling method is used for the superconducting coil by the latest single-stage GM type cooler (4.8 kW @ 20 K). In order to reduce power loss from the cryo-cooler, the high temperature end of the high-temperature superconducting current lead is cooled by the cryo-cooler, and a thermal anchor is installed close to the liquid nitrogen temperature. Figure 7.4 shows an overview of the system.

A 35 kJ/7 kW conduction-cooled HTS SMES has been developed by R&D (Research and Development) Center of Applied Superconductivity in Huazhong University of Science and Technology [12]. The magnet of the SMES is a solenoid-type magnet composed of 32 Bi2223/Ag double pancakes, and cooled to about 20 K by a conduction-cooled method. Up to the present, a conduction-cooled HTS hybrid magnet has been developed for a 600 V/150 kJ/100 kW SMES system used for power conditioning [13–14]. The solenoid-type HTS magnet consists of 17 double-pancake coils. Coils 6 and coils 12–17 are wound with a Type-HT-CA BSCCO tape. Coils 7–11 are wound with an Amperium 4.8 YBCO tape. The magnet is cooled by conduction cooling method.

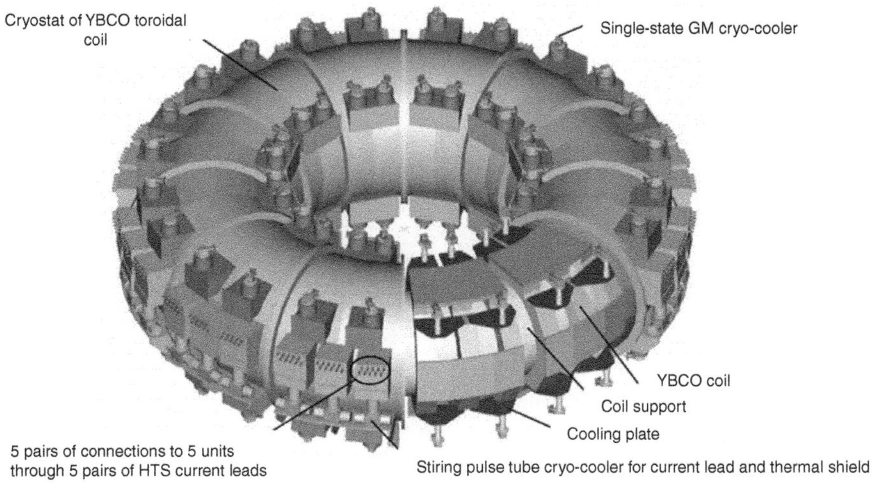

Figure 7.4 Overview of the 100 MVA/2 GJ class YBCO SMES. [9]

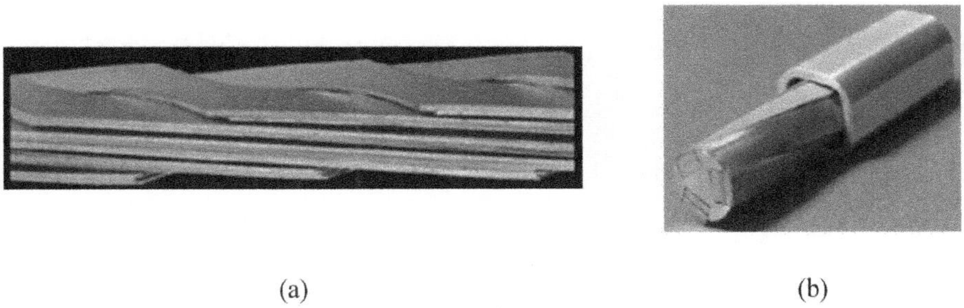

Figure 7.5 11 Two kinds of high current density cables with coated conductor. (a) High IC class CC ROEBEL cable. (b) Three helical channel CICC composed of 20-YBCO-tape cable in each groove in a copper diameter 9.53 mm (Total 60 YBCO tapes).

7.3 Considerations in the Design of SMES Magnets

The design of an SMES magnet involves many disciplines, making it a typical multi-disciplinary integration issue, and it is related to the fields of electric field, magnetic field, structural mechanics, heat, circuit theory, and low temperature insulation. In addition, the optimization design of the superconducting coil also includes computer programming, adaptive optimization algorithm, and development of optimization software. According to the classical electromagnetic formula, considering the required performance parameters, general electromagnetic design of superconducting magnets can be executed. With the progress of computers and the development of the electromagnetic field analysis and optimization design software, the design of superconducting magnet is generally finished by software like ANSYS or COMSOL. Therefore, special considerations related to the design of superconducting magnet electromagnetism follow:

7.3.1 The Current-carrying Capacity

The critical current of superconductor is influenced by the amplitude and direction of the external magnetic field. For the design of a superconducting magnet, the magnetic field distribution on the superconducting tape is essential. Special attention should also be paid to the effect on the critical current of the larger magnetic field direction, which is parallel to the magnetic field of the superconductor's *c* axis in a high temperature superconducting magnet. For Bi-based tape, magnetic field paralleled to the *c* axis is perpendicular to the magnetic tape surface. Due to the high temperature superconductor lattice generated direction tilt or when the consistency is bad, the greatest magnetic field influence on the critical current may deviate from the strip on the surface of the flat vertical direction, such circumstances have been reported in the sample of the second generation of high temperature superconducting tape (Y-coated conductor) experiments. Figure 7.5 shows two kinds of high current density cables with coated conductor.

On the other hand, the current capacity of a single HTS tape is limited, which is unable to meet the requirement of a large current. Large-scale SMES systems requir development of high current density cables capable of carrying high currents. The current capacity of HTS conductors can be increased with a parallel arrangement of the tapes. As the existing HTS tape has a strong anisotropy, cabling methods for HTS tapes have been developed.

7.3.2 Mechanical Properties

The critical current of HTS tape is related to the mechanical stress on the tape. After the mechanical stress reaches a certain limited value, critical current decreases dramatically. When winding coils, the bending radius of the tape also affects the critical current. The mechanical properties of low temperature superconductors is relatively better, with a smaller bending radius. Due to its ceramic properties, mechanical performance of high temperature superconductor is poorer, with a larger bending radius. In particular the lateral bending of HTS tape is limited.

Measures have been taken to optimize mechanical stress on the wires. Figure 7.6 shows the experimental device and schematic illustration of force-balanced coil. A superconducting force-balanced coil (FBC), which is geometrically one-tenth the size of a 100 MJ class SMES coil, has been developed. The FBC is a helically wound hybrid coil of toroidal field coils and a solenoid [15]. This coil configuration can minimize the required mass of the structure for induced electromagnetic forces. The FBC model with an outer diameter of 0.53 m will have 270 kJ magnetic energy at the critical magnetic field of 7.1 T using NbTi strands. The critical coil current and self-inductance are 552 A and 1.8 H, respectively.

The FBC can control the distribution of the mechanical stresses caused by electromagnetic forces and significantly reduce the required mass of the structure for the same magnetic energy compared with the conventional coil configurations. Large-scale SMES coils require special considerations for induced electromagnetic forces to limit allowable tensile stresses, such as bedrock supports, indicating that the FBC will be a feasible options for large-scale SMES.

(a) (b)

Figure 7.6 Force-balanced coil: (a) A small experimental device; (b) Schematic illustration.

7.3.3 AC Loss and the Cooling Design

There is no resistance and no loss under DC current in superconductors, but AC loss exists under alternating current and alternating magnetic field. The design of superconducting magnet should consider the application purpose and current flow. AC loss in a superconducting magnet is related to the magnetic field acting on the superconducting tape. The amplitude, direction, and changing form of a magnetic field will all affect the value of AC loss.

Superconducting tape itself is made from a variety of composite materials, thermal properties are difficult to obtain accurately. In the design process, if we cannot evaluate the heat transfer process of the superconducting magnet, the thermal stability analysis of the SMES magnet will incur a lot of errors. In the process of operating it is possible that SMES will suffer a hit from an uncontrolled overcurrent, a pulse current will generate instantaneous high heat in the magnet. If the heat cannot be taken away quickly, the superconducting magnet may quench, and the magnet may even burn. Therefore, it is important to analyze the heat transfer characteristics and thermal stability of superconducting magnets.

The design of the SMES cooling system is associated with current-carrying capacity, AC loss, stability, and other related technical parameters, and also with operation efficiency and reliability of the SMES system. Due to current and magnetic field variation during operation, the AC losses may cause heat generation in the SMES magnet. Therefore, the power exchange processes should be simulated to make sure that the capacity of the cooling system could meet these requirements (see ESH055). Ensuring that the internal heat is taken away quickly and effectively is a guarantee of the technical performance of magnets. The efficiency and reliability of the cooling system should be also promised.

7.3.4 Insulation Design

During the power exchange process, the voltage across the HTS magnet is produced via switching operations of the PCS semiconductor devices and is characterized by a steep rising edge and high frequency. Through the high frequency transient model, PWM voltage is not distributed evenly between coils. This transient overvoltage can endanger the insulation and even lead to SMES operation failure. To this effect, understanding the electromagnetic transient interaction between the HTS magnet and the PCS is critical in regards to the successful design of SMES magnet insulations.

The lumped parameter network model of the magnet, an example being the transformer transient model, composed of self-inductances, mutual couplings, and series and shunt capacitances is typically used to analyze the voltage distribution on the HTS magnet. In order to reduce the calculation amount, several turns or several disks are typically lumped into a basic unit. Though not fully accurate, the lumped parameter network model can be utilized as a reference.

Experimental study can also contribute to our understanding of the transient behaviors of SMES magnets. Considering the complicated control system, protection system, and cooling system, experiments on a practical VSC SMES are labor intensive and time consuming. A relatively small experiment platform was built to more elegantly and efficiently simulate the working status of a real SMES system.

Experiment and simulation on a five series connection stacked HTS magnet show that the transient overvoltage due to power switches may be twice the value of the steady voltage; the transient overvoltage of the double-pancake increases from the middle towards either end of the magnet; the transient overvoltage decreased with an increase in IGBT action time.

Insulation design should refer to the transient overvoltage rather than steady-state voltage. For the solenoid magnet, insulation should be strengthened at both ends of the coils, which suffer high transient over-voltage. Transient voltage characteristics are not exactly the same when the configuration of a magnet changes. For different magnets and different applications, appropriate modeling analysis should be carried out so as to provide a reliable insulation design.

7.3.5 The Optimization Design and the Field-circuit Coupling Design

Generally, the optimization parameters of a magnet include energy storage, magnet current, length of HTS tapes, magnet volume, and magnetic field leakage. There are many kinds of optimization schemes to meet the single performance standard, but the results may be different from the point of view of considering the dynamic characteristics of SMES. If the magnet needs a more precise optimization scheme, the parameters of the SMES can be optimized for the power exchange between the SMES and the power grid or other devices. The optimization parameters include the AC loss of the magnet and the temperature rise of the magnet. After a lot of optimization, the optimized design scheme is more targeted, and the magnet performance is better.

Generally, there are special requirements on the magnetic field configuration in its application circumstances. So, the primary goal of magnet design optimization is to obtain satisfying configuration, strength, evenness, and gradient of the magnetic field. Next, under a sufficient safety margin, economical efficiency needs to be considered, such as the least number of superconducting tapes and the least magnet volume. Then, some comprehensive factors should be considered, such as the size of the magnetic field leakage and its influence on the surrounding environment.

The current-carrying capacity of superconducting tapes is associated with the intensity of the background magnetic field, the mechanical stress, and the temperature of superconducting tapes. Therefore, the current-carrying capacity, which is the basic electromagnetic characteristic, is a multi-coupling problem, directly related to magnetic field, stress field, and temperature field. At the same time, insulation is an indispensable part in electromagnetic magnets. The design and the manufacture process of the winding must consider the influence of the electric field intensity and distribution. Superconducting properties such as AC loss, thermal stability, quench, and quench propagation are closely related to its efficiency, reliability, and so on. The design of superconducting magnet should also be associated with the operation state of the magnets, considering dynamic electromagnetic parameters like the current rate and the current distribution in the superconductors. Overall, the design and the operation of the whole system will be combined for comprehensive optimization, namely the field-circuit coupling design. Selecting an SMES magnet as an example, the influence factors of magnet design are analyzed, with their relationships shown in Figure 7.7.

After determining the wire type, the environment temperature of the cryogenic system, and the magnetic field distribution, the main problems and influencing factors of the multi-field coupling analysis are:

(1) The shape of the magnet: requirement for magnetic field configuration; magnetic field distribution on the tape and the influence of magnetic field on critical current and AC loss; the influence of the AC loss; stress distribution and its effect on the critical current, etc.
(2) The magnet wire packing ratio (The cross section of average current density): insulation level; the cooling efficiency; wire thermal stability; etc.
(3) Heat loss: low temperature container adiabatic level; current lead heat leakage; AC loss on superconducting wires; etc.

The main problems and influencing factors of the field-circuit coupling analysis are:

(1) Magnet AC loss: dynamic current morphology; intensity and frequency of dynamic power compensation.
(2) The magnet quench protection: amplitude and changing rate of dynamic current; overcurrent expectations; fault in the power control system and power system; etc.

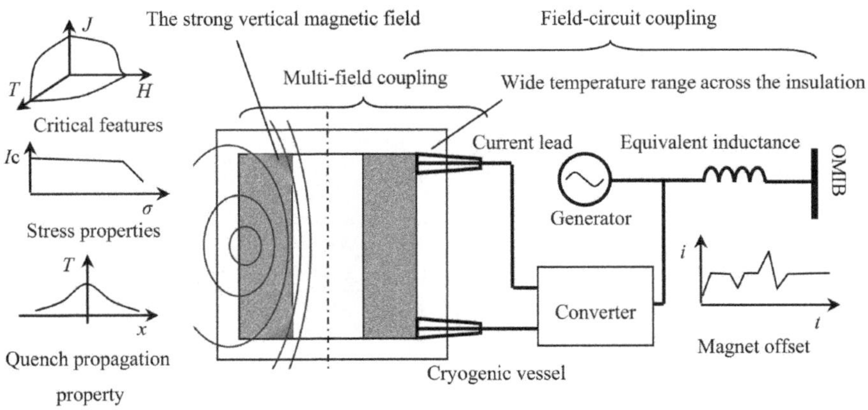

Figure 7.7 Main factors for multi-field coupling and field-circuit coupling design in SMES.

(3) Magnets and current lead insulation level: over voltage waveform, the amplitude of the peak; frequency of occurrence in power system; and so on.
(4) It is obvious that a number of factors considered are relevant in the multi-field coupling design and field-circuit coupling design.

7.4 Current Leads of SMES Magnets

7.4.1 Classification of Current Leads

For superconducting devices, current lead is one of the main heat sources, which is the main factor that causes the low temperature medium boiling or power consumption of refrigerators. For most superconducting devices the cost for maintaining cryogenic circumstances decides the operation cost of the whole system. The main purpose of optimizing current lead is to reduce the heat leakage of superconducting devices to the external environment so as to improve the economic benefits.

Figure 7.8 presents different classifications of current lead. According to the material type, it can be divided into full-metal current lead, high temperature superconductor (HTS) current lead, and peltier current lead. According to the cooling method, it can be divided into conduction-cooled, vapor-cooled, and gas-cooled current lead.

The schematic views of three kinds of current lead are shown in Figure 7.9. The current lead made of regular metal is also called a one-dimensional lead. This type of lead mainly uses copper and its alloys, and the heat leakage is relatively high though it has a simple structure and is easy to fabricate. Binary or mixed current lead, some of whose materials are replaced by high temperature superconducting materials, can eliminate the joule heat when its operating temperature is under the critical temperature of its superconducting state. In addition, the thermal conductivity of a high temperature superconductor is very low (the average thermal conductivity of Bi2223 is a third of stainless steel from 4 K to 77 K), which reduces the conduction of heat flow along the wire from a high to a low temperature area. Therefore, a current lead made of a high temperature superconductor can significantly reduce the heat leakage. At present, the application of binary current leads has become the mainstream in low temperature superconducting magnets. For high temperature superconducting magnets operating in liquid nitrogen temperature, the effect of using a binary current lead is not obvious due to the narrow

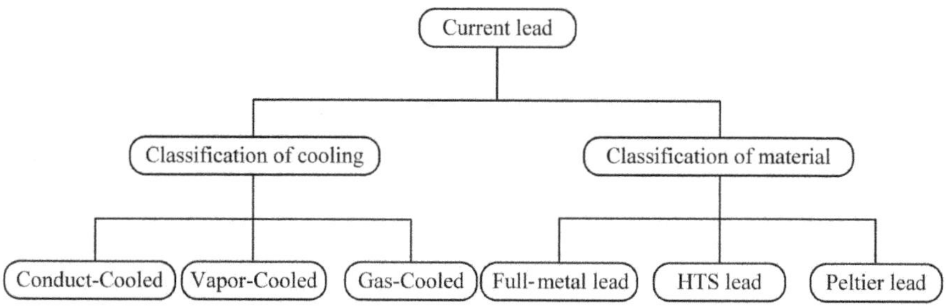

Figure 7.8 Classification of current leads.

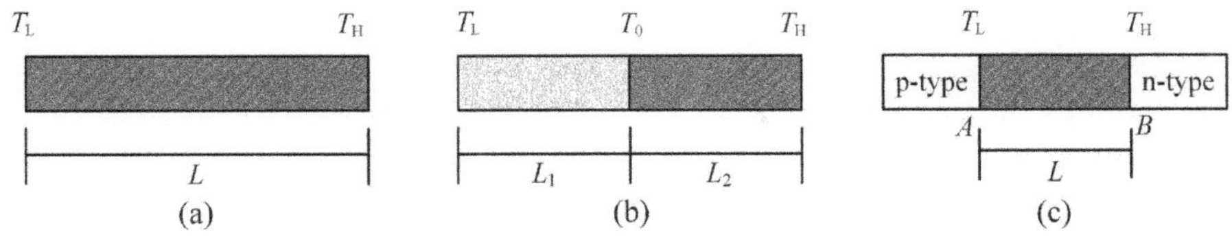

Figure 7.9 Schematic views of three kinds of current leads. (a) Full metal current lead. (b) Binary current lead. (c) Peltier current lead.

temperature zone of a high temperature superconductor over 77 K. HTS materials used for current leads are: large Bi2212 tubing and rod in burning process, sedimentary Bi2223, melt texture Y123, HTS tapes with high thermal conductivity substrate (such as Bi2223 with the substrate of silver-gold, Y123 coating conductor), etc. In 1996, Yamaguchi proposed a new type of current lead composed of thermoelectric materials for a superconducting magnet system, called a peltier lead. It is widely believed that a peltier lead can reduce 16%–26% of heat leakage. The effect of reducing heat leakage of a peltier current lead is pretty good when the current lead is conduction cooled. In 2006, Japanese scholars succeeded in applying a peltier current lead to a 5 T superconducting magnet system, and the experimental results showed that the peltier lead has the potential to significantly reduce the heat leakage of a superconducting magnet system.

7.4.2 The Method of Designing Current Leads

In the superconducting magnet system, the heat leakage of the current lead consists of conduction heat leakage, joule heat, and heat exchange between low temperature gas and current lead, where conduction heat leakage is caused by the temperature difference between the two ends of the current lead and joule heat is produced by the resistance of the lead itself. Generally speaking, the heat exchange between low temperature gas and current lead can be ignored for theoretical calculation and optimization, and only conduction heat leakage and joule heat are taken into consideration. Given a constant current, there is a corresponding minimum heat leakage when the conduction heat leakage equals joule heat in ideal conditions. In a specific superconducting magnet system, it is necessary to consider the actual structure and dimension of the superconducting magnet system when optimizing the material type, length, and cross-section area of the current lead.

The design of the current lead should include thermal design and insulation design. Under the condition of guaranting the insulation level, reducing the heat loss is the main goal of the design.

7.4.2.1 Thermal Design of Current Leads

The upper part of the current lead is in a normal temperature, and the lower part is connected with the lower temperature of the superconducting component, and the heat is transferred from the normal temperature to the low temperature through the lead wire. When the current is passed through the lead wire, the joule heat is generated by the resistance of the lead conductor, and the joule heat is also penetrates through to the low temperature end due to the heat conduction.

Due to the fact that thermal conductivity and resistivity of the conductor as well as the thermal conductivity of insulation materials are functions of temperature, the mechanism of heat transfer and heat transfer coefficient between low temperature mediums and the conductors are not only associated with the temperature of mediums themselves, but also associated with temperature difference between them and conductors. The early lead design needs to simplify a number of minor factors. With the development of computer technology and numerical computing technology, it is now generally using numerical analysis to optimize the heat flux distribution on the current leads. Even so, it is still challenging to consider all relative factors and all non-linear parameters of the temperature characteristics, so it is necessary to make an appropriate simplification.

7.4.2.2 Insulation Design of Current Leads

Current lead connections are either at room temperature or low temperature, the electrical insulation is involved in normal temperature insulation and low temperature insulation. In normal temperature, the insulation design can be referenced to the conventional current lead insulation design method of electrical devices. At the low temperature end we should focus attention to the low temperature properties of the insulating material. We should also be aware of the transition design of the normal temperature insulation and cold insulation between low temperature and room temperature, as well as the performance of the insulating material under temperature change.

7.4.2.3 The Structure Design of Air Cooled Current Leads

In the optimization design of air cooled current leads, it is necessary to carry on the special treatment on the structure. We should reduce the heat loss point of view to design the structure, such as heat can be taken away by the evaporation of low temperature medium gas, temperature shield segment should be set in the middle of the lead.

The research on conventional metal lead has formed a relatively mature theory system in foreign countries. And the calculation method together with structure optimization for conventional metal leads has also been perfected, with small current levels of current lead moving towards commercialization. In a domestic situation, the research is still in the theoretic state of optimization and analysis, as well as at single prototype development.

Theoretically, the minimum heat leakage of a metal current lead is $1.04\ \text{W}\cdot\text{kA}^{-1}$. In the condition of liquid helium cooling with sufficient heat exchange, the optimization heat leakage level of pure copper and brass wire are $1.07\ \text{W}\cdot\text{kA}^{-1}$ and $1.08\ \text{W}\cdot\text{kA}^{-1}$, respectively. Table 7.1 gives the minimum heat leakage and the required electric power of cryo-generator for different leads after optimization.

Table 7.1 Optimization design of pure copper, brass, and stainless steel wire through liquid nitrogen cooling.

Current lead optimization design in 77 K		pure copper	brass	stainless steel
Air-cooled lead	Heat leakage Q_L ($\text{W}\cdot\text{kA}^{-1}$)	23.2	25	25.5
	Electric power P_e ($\text{W}\cdot\text{kA}^{-1}$)	≈350	≈370	≈380

7.4.3 Cases of Current Leads

Table 7.2 is a typical example of superconducting current lead development. In recent years, the binary current lead has become the mainstream. For superconducting device whose cold end temperature is below 30 K, the binary lead can reduce the heat leakage, and the heat penetration can be reduced to 1/5 ~ 1/7 of a conventional metal lead. The following is a brief introduction of three kinds of binary current leads.

7.4.3.1 1 kA Binary Gas-cooled Current Lead [16–17]1 kA Binary Gas-cooled Current Lead [16–17]

The Fuji electric power company in Japan developed a pair of 1 kA gas-cooled binary current leads for a 3.6 MJ/1 MW SMES system, which is shown in Figure 7.10. The low temperature section of the current lead consists of Bi2212 round bulks, with copper used in the high temperature section. The current lead is cooled by 4.5 K helium gas flowing along the whole current lead with a rated flow rate of 0.05 $g\cdot s^{-1}$. The test results indicated that the heat load in a steady state at a rated current of 1 kA was 0.035 W, well below the specified value of 0.1 W.

7.4.3.2 1.1 kA Binary Conduction-cooled Current Lead [18]

Ansaldo-CRIS and CESI in Italy designed and manufactured a pair of 1.1 kA resistive/HTS current leads cooled by a closed-cycle refrigerator for a 1 MJ/1 MW SMES, which is presented in Figure 7.11. The low temperature part of the current lead is made of Bi2223/AgAu, and the high temperature part adopts copper. The whole current lead is directly cooled by a refrigerator, which can provide around 100 W at the temperature of 60 K to keep the upper part of the HTS material at a temperature below 70 K.

Table 7.2 A typical example of superconducting current lead.

institution	parameters	Temperature of low end	status
Eurus Energy America	2 kA, heat leakage < 0.1 $W\cdot kA^{-1}$	liquid-helium cooling, 4.2 K	commercialized
AMSC	10~20 kA, heat leakage < 0.1 $W\cdot kA^{-1}$	liquid-helium cooling, 4.2 K	commercialized
Technical institute of physics in Germany	60 kA, heat leakage < 0.1 $W\cdot kA^{-1}$	liquid-helium cooling, 4.2 K	development stage
Sumitomo (Japan)	10~20 kA, heat leakage < 0.1 $W\cdot kA^{-1}$	liquid-helium cooling, 4.2 K	commercialized
International hot nuclear research center	70 kA, heat leakage < 0.48 $W\cdot kA^{-1}$	gaseous-helium cooling, 80 K	completed in lab
Northwest Institute for Non-ferrous Metal Research	1 kA	liquid-nitrogen cooling, 77 K	commissioning
The Chinese Academy of Sciences (Plasma Physics)	15 kA	liquid-helium cooling, 4.2 K	commissioning

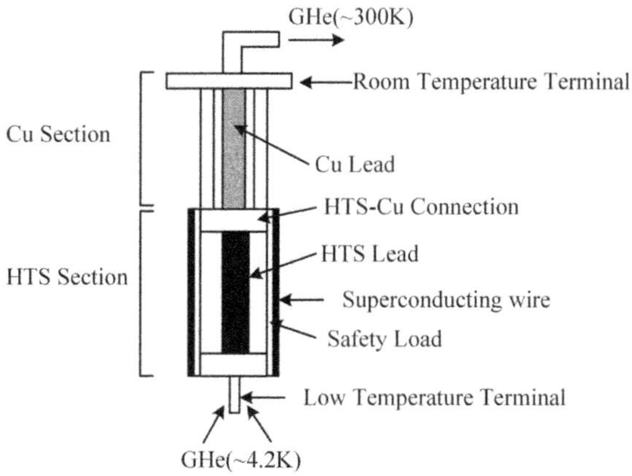

Figure 7.10 Schematic view of 1 kA gas-cooled current lead. [16]

7.4.3.3 52/68 kA HTS Current Lead [19–20]

The Institute of Plasma Physics, Chinese Academy of Science (ASIPP) designed and fabricated a pair of 52/68 kA HTS current leads for the ITER (International Thermonuclear Experimental Reactor) superconducting magnet, shown in Figure 7.12. The current lead consists of an LTS module, an HTS module, and a copper transition module. The HTS module includes 1080 Bi2223/AgAu tapes and cylinder binary shunt. The copper part of the current lead is cooled by 50 K helium gas to keep the HTS module at a temperature below 65 K and reduce heat leakage from 65 K warm end to 5 K cold end.

7.5 Quench Protection for SMES Magnets

Quenching of superconducting magnet is the process of electromagnetic energy being transferred into heat energy. The quench will induce the issues of overheat and overvoltage. Moreover, the temperature and volume of the low temperature medium will increase, which will induce excessive pressure in cryogenic vessels. Therefore, once the quench happens effective protection measures must be taken to meet the following requirements: (1) Reducing the energy released by the winding in the normal area to prevent overheating in superconducting tapes; (2) Reducing the voltage of the winding in the normal area to prevent turn-to-turn insulation breakdown; (3) Reducing the energy released in the low temperature container to prevent evaporation of the low temperature medium.

Quench protection has two different types. The first method is to quickly transfer the energy in the magnet to the external environment and release it through the resistance element, this is called external protection. The second method, which is called inner protection or self-protection, is to artificially propagate the normal area to all magnets instantaneously in order to prevent a high local temperature rise. The former is also called active protection and the latter called passive protection.

The active protection for magnet quench is shown in Figure 7.13. In Figure 7.13(a), R_e is the resistance for energy-releasing, R_q is the resistance of normal state after

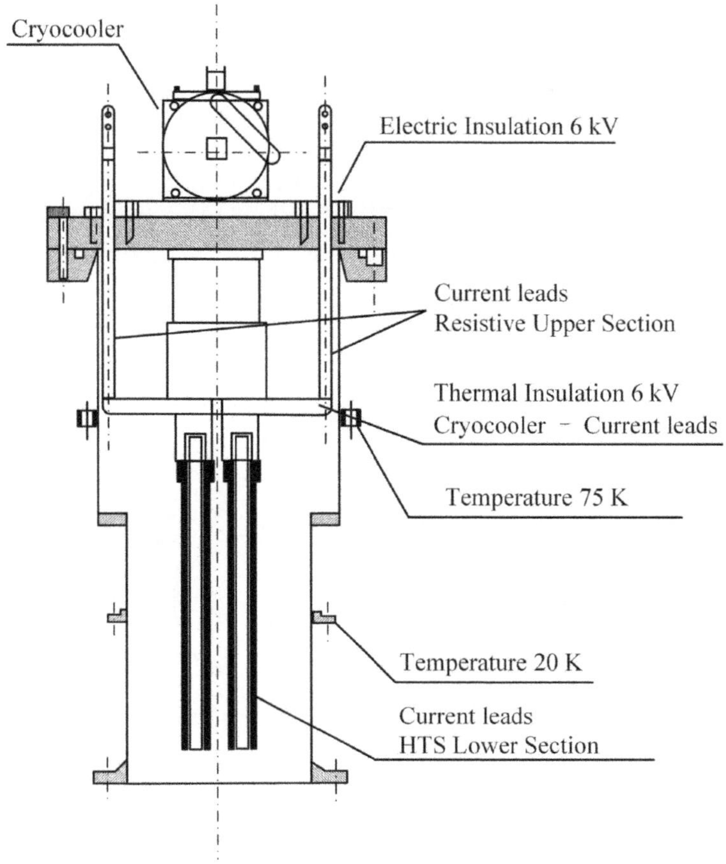

Figure 7.11 Schematic view of 1.1 kA conduction-cooled current lead. [11]

quenching. Once the quench signal has been detected, the switch K_1 is controlled to be closed, and the energy in the magnet can be transfered to the energy-releasing resistance. When cutting off the magnet's power supply, parallel resistance (non-linear or linear resistance) should be set to protect the magnet from overvoltage damage. For power sources, overvoltage protection measures should also be taken by establishing a new current path for power supply, as shown in the figure as power protection resistance R_p.

For a superconducting magnet with a small energy storage, quench protection can adopt simple method, as shown in Figure 7.13(b). The thyristor is connected in parallel to the magnet, without the energy-releasing resistance. Once the quench occurs, the power supply is cut off and the magnet current gradually decays in the short-circuit thyristor loop. At the same time, energy consumes through the thyristor and the resistance R_q.

Passive protections include the segmented resistance method, the inductive coupling method, and the heater method. The key point is to prevent quench point (normal state area) concentrating on one fixed position, which makes the local temperature too high and damages magnets. The segmented resistance protection method is shown in Figure 7.14. When one segment of the coil quenches, the

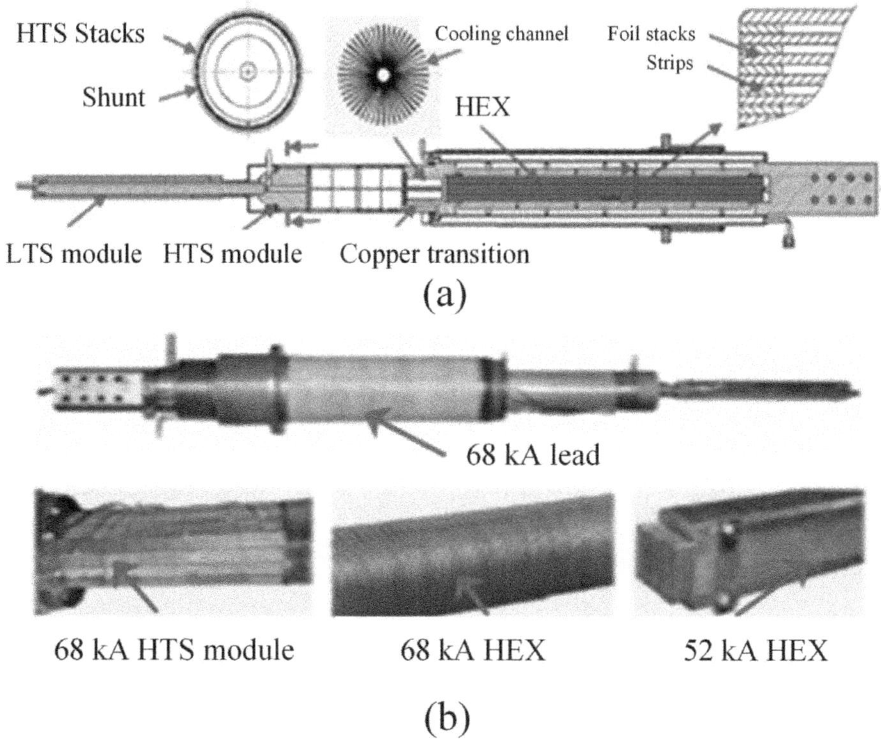

Figure 7.12 52/68 kA HTS current lead. (a) Schematic view of the current lead. (b) Photo of the current lead. [20]

existence of normal state resistance impels the power flow to the shunt resistor, and reduces the burden of coil itself. Due to the magnetic coupling relationship between coil sections, current drop in one coil section may cause other section current increase. This can speed up other coil section quenches. Then quenches spread rapidly through the magnet, which reduces the risk of the magnet being damaged.

For effective quench protection, sensitive quench detection is essential. There are many types of quench detection methods, including voltage detection, ultrasonic

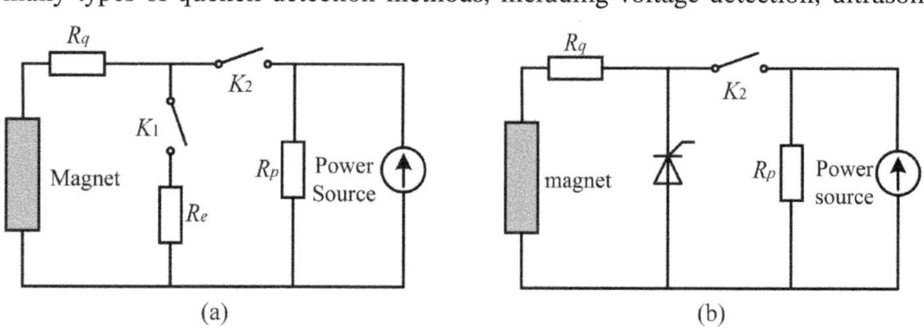

Figure 7.13 The basic form of superconducting magnet quenches, active protection. (a) Discharge through energy-releasing resistance. (b) Short-circuit by thyristor.

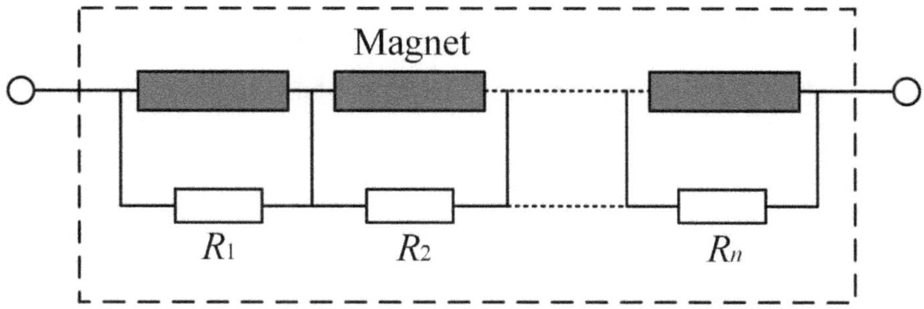

Figure 7.14 Superconducting magnet segmented resistance protection.

detection, and detection of temperature rise at low temperature, pressure and velocity. The physical quantity of temperature, pressure, and velocity is related to the heat transference of the low temperature medium, which has the characteristic of time delay. So, the most direct, rapid, and accurate detection method is still the voltage detection method. Practical application processes should choose the protection scheme and protection device parameters of superconducting magnets according to the highest temperature that the magnets are allowed to reasch. Accurately calculating local temperature rises of a superconducting magnet quench point is difficult, because it is related to current value and the quench resistance, which is associated with many factors such as the wire structure, cooling conditions, and current form. For the voltage detection method, to measure the voltage between turns is difficult due to the installation of voltage sensors at every turn. Furthermore, the issue of electromagnetic noise is hard to solve. Therefore, it is feasible to detect quench by detecting voltage at every section of the whole magnet. Bridge circuits are often used due to its high detection accuracy. As shown in Figure 7.15, the magnet is divided into two sections that form a bridge circuit with external resistors connected in parallel respectively. When the coil is in a superconducting state the galvanometer displays zero in value; when quench occurs, there will be an imbalance in the circuit due to resistance changing inside the superconducting magnet, which causes a current flow through the galvanometer. Quench can be detected by changes in the galvanometer. In addition, the bridge circuit method can also be extended by adding more sections in the magnet.

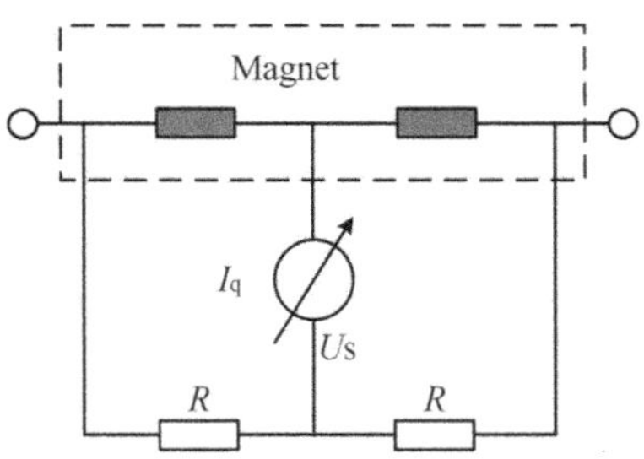

Figure 7.15 Bridge circuit detection method.

7.6 Summary

From the technical level, low temperature SMES is more mature, but the room for improvement of low temperature superconducting wire is limited, the storage energy density is difficult to increase substantially. The operating temperature of LTS wire cannot increase. The cost of the cooling system and the operating cost restrict its commercial development.

The technology of the HTS wire is less mature than that of LTS wire, and the current carrying capacity of a single commercial HTS tape can't meet the demand of large-capacity and high-powered SMES. The price of HTS wire is much higher than the LTS wire. However, there are still a lot of room for improvement on the performance of HTS wire, and there are more choices of cooling methods. With the development of high-temperature superconducting technology, a high-temperature SMES system is more likely to achieve a breakthrough in the future.

In addition to wire performance, price constraints and complex refrigeration system also hinder the SMES application. Magnet protection is a great challenge, and there is not a perfect method to guarantee the safe and stable operation of SMES in complex conditions. These issues require further analysis and resolution.

References

[1] Boom, R.W. and Peterson, H.A. (1972). Superconductive energy storage for power systems. *IEEE Transactions on Magnetics* 8 (3): 701–703.

[2] Rogers, J.D., Schermer, R.I., Miller, B.L., et al. (1983). 30-MJ superconducting magnetic energy storage system for electric utility transmission stabilization. *Proceedings of the IEEE* 71 (9): 1099–1107.

[3] Boenig, H.J. and Hauer, J.F. (1985). Commissioning tests of the Bonneville Power Administration 30 MJ superconducting magnetic energy storage unit. *IEEE Transactions on Power Apparatus and Systems* PAS-104 (2): 302–312.

[4] Rogers, J.D., Boening, H.J., and Hauer, J.F. (1985). Operation of the 30-MJ SMES system in the Bonneville Power Administration electrical grid. *IEEE Transactions on Magnetics* 21 (2): 752–755.

[5] Luongo, C.A., Baldwin, T., Ribeiro, P., et al. (2003). A 100 MJ SMES demonstration at FSU-CAPS. *IEEE Transactions on Applied Superconductivity* 13 (2): 1800–1805.

[6] Nagaya, S., Hirano, N., Kondo, M., et al. (2004). Development and performance results of 5 MVA SMES for bridging instantaneous voltage dips. *IEEE Transactions on Applied Superconductivity* 14 (2): 699–704.

[7] Wang, Q., Dai, Y., Zhao, B., et al. (2010). Development of large-scale superconducting magnet with very small stray magnetic field for 2 MJ SMES. *IEEE Transactions on Applied Superconductivity* 20 (3): 1352–1355.

[8] Nagaya, S., Hirano, N., Shikimachi, K., et al. (2004b). Development of MJ-class HTS SMES for bridging instantaneous voltage dips. *IEEE Transactions on Applied Superconductivity* 14 (2): 770–773.

[9] Shikimachi, K., Moriguchi, H., Hirano, N., et al. (2005). Development of MVA class HTS SMES system for bridging instantaneous voltage dips. *IEEE Transactions on Applied Superconductivity* 15 (2): 1931–1934.

[10] Shikimachi, K., Hirano, N., Nagaya, S., et al. (2009). System coordination of 2 GJ class YBCO SMES for power system control. *IEEE Transactions on Applied Superconductivity* 19 (3): 2012–2018.

[11] Ueda, H., Ishiyama, A., Shikimachi, K., et al. (2010). Stability and protection of coils wound with YBCO bundle conductor. *IEEE Transactions on Applied Superconductivity* 20 (2): 1320–1323.

[12] Shi, J., Tang, Y., Zhou, Y., et al. (2007). Development of a conduction-cooled HTS SMES. *IEEE Transactions on Applied Superconductivity* 17 (3): 3846–3852.

[13] Ren, L., Xu, Y., Liu, H., et al. (2015). The experimental research and analysis of a HTS SMES hybrid magnet. *IEEE Transactions on Applied Superconductivity* 25 (3): 1–5.

[14] Ren, L., Xu, Y., Zuo, W., et al. (2015). Development of a movable HTS SMES system. IEEE T*ransactions on Applied Superconductivity* 25 (4): 1–9.

[15] Nomura, S., Kasuya, K., Tanaka, N., et al. (2008). Experimental results of a 7-T force-balanced helical coil for large-scale SMES. *IEEE Transactions on Applied Superconductivity* 18 (2): 701–704.

[16] Bohno, T., Tomioka, A., Nose, S., et al. (1997). Development of HTS current leads for 1 kWh/1 MW module type SMES system. II. Manufacturing and testing of prototype leads. *IEEE Transactions on Applied Superconductivity* 7 (2): 688–691.

[17] Konno, M., Sakaki, K., Uede, T., et al. (1997). Development of HTS current leads for 1 kWh/1 MW module type SMES system. I. Design study. *IEEE Transactions on Applied Superconductivity* 7 (2): 684–687.

[18] Mariani, M., Ariante, M., Matrone, A., et al. (2002). Cryocooler cooled HTS current leads for a 1 MJ/1 MW-Class SMES system. *IEEE Transactions on Applied Superconductivity* 12 (1): 1293–1296.

[19] Bauer, P., Bi, Y., Cheng, A., et al. (2010). Test results of 52/68 kA trial HTS current leads for ITER. *IEEE Transactions on Applied Superconductivity* 20 (3): 1718–1721.

[20] Zhou, T., Ding, K., Liu, C., et al. (2015). R&D on 52-kA HTS current lead at ASIPP. *IEEE Transactions on Applied Superconductivity* 25 (2): 4802704.

8

Testing Technologies for Developing SMES

Jing Shi, Yuxiang Liao, Lihui Zhang, Ying Xu, Li Ren, Jingdong Li, and Yuejin Tang

State Key Laboratory of Advanced Electromagnetic Engineering and Technology, School of Electrical and Electronic Engineering, Huazhong University of Science and Technology, Wuhan, China

8.1 Introduction

The development of superconducting magnetic energy storage (SMES) systems has made great progress, and several SMES systems have been developed successfully and tested in a laboratory setting or power grid in recent years. Before the SMES systems is put into use in the power grid, a series of strict performance tests are needed to promise the stability, security, and performance of the SMES system. Thus, this chapter mainly discusses the testing method of SMES systems. The testing considers the key components of an SMES system involving superconductivity, the test methods of High Temperature Superconducting (HTS) tape, superconducting coils, and the SMES system are presented respectively.

8.2 HTS Tape Property Test Method

8.2.1 HTS Tapes Critical Current Measurement

The critical current of HTS tapes is related to temperature, the magnitude and direction of the external magnetic field, stress, and the non-uniform degree of the material. So, tests of HTS tapes mainly include an anisotropic test, a tensile properties test, a bending characteristic test, and the critical current uniformity of the test strip.

Anisotropy measurements of HTS tapes generally adopt fourwire-measurement methods, the measuring principle is shown in Figure 8.1. HTS tapes are fixed to a rotatable sample holder, and the extraction voltage signal line is welded on the surface of the tapes. The magnitude of critical current could be calculated by through-flow test according to quench criterion of 1 μV/cm. Maintaining the temperature of the test sample, and varying the amplitude of the external magnetic field and the angle between magnetic field and the tapes surface, the measurement of the anisotropy of superconducting tapes under temperature can be achieved [1].

Advances in Energy Storage: Latest Developments from R&D to the Market, First Edition.
Edited by Andreas Hauer.
© 2022 John Wiley & Sons, Ltd. Published 2022 by John Wiley & Sons, Ltd.

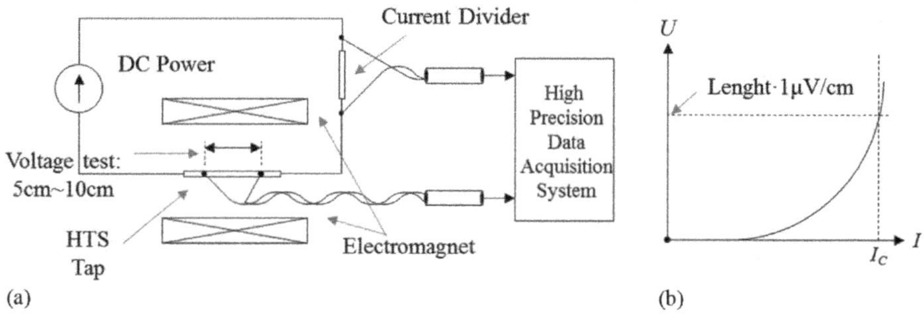

Figure 8.1 HTS tapes critical current measurement schematic: (a) Measurement schematic; (b) Critical current.

For BSCCO-tapes and YBCO-tapes', Its superconducting layer lattice structure will change when they suffer mechanical stress, then the critical current of the tapes will reduce. The stress characteristic test [2] consists of the tensile stress test and the bending stress test. Figure 8.2 shows the tensile properties of superconducting tapes measuring principle and measurement results. Through increasing the tension gradually, the critical current under different fixed tensions can be measured with the four-wire method. Then reducing the tension, the degree of recovery of the critical current can be observed. Repeat the tests until the strip current generates greater attenuation.

The bending stress characteristics test is similar to the tensile stress characteristics test, the difference is that superconducting tapes with different bending radiI are fixed on a special sample holder in the bending stress test, and the superconducting tape is mainly affected by bending stresses. After each bending stress test, we also need to test the recovery performance, that is, measuring the critical current of the superconducting tapes after they have been straightened.

Current distribution measurement of superconducting tapes consists of a non-contact method and a contact method. The contact method requires installation of

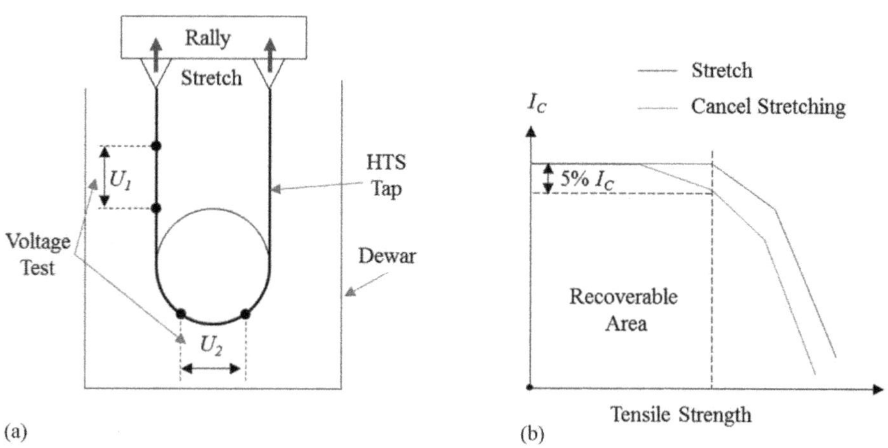

Figure 8.2 Tensile properties test schematic of high-temperature superconducting tapes: (a) Measurement schematic; (b) Rally to restore relations with critical current.

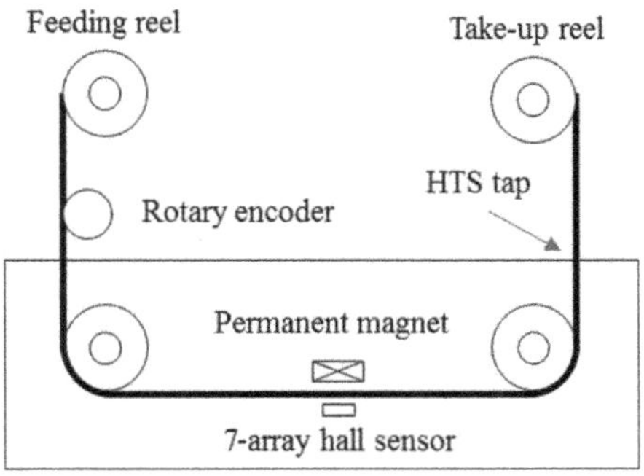

Figure 8.3 HTS tape measurement with current distribution characteristic diagram.

several contact probes on the surface of superconducting tapes, which may slow the measurement speed and damage the superconducting tapes. The non-contact method [3–7] uses the magnetic measurement method which utilizes multi-hall array probes. Scanning the entire superconducting tape through the wheel dial obtains a current distribution from the calculation of superconducting tape, Figure 8.3 shows the measurement schematic.

8.2.2 AC Loss Measurement of High Temperature Superconducting Tapes

According to the measuring principle, AC loss measurement methods are divided into thermal methods, magnetic methods, and electrical measurement methods.

Thermal methods include temperature differential method and calorimetry. The temperature differential method is measure the temperature difference between the upstream and downstream fluid to calculate the superconductor's AC losses by forcing the coolant (liquid nitrogen or liquid helium) flow through the superconductor. Calorimetry calculates superconductor's AC losses by measuring the rate of evaporation of the liquid coolant. Magnetic measurement method calculates the loss of magnetization per unit volume of the sample by measuring the induced voltage around the pickup coil. Electrical measurement method [8] detects the AC loss through a lock-in amplifier. Figure 8.4 shows the schematic diagram of the electrical measurement. The lock-in amplifier measures the terminal voltage of superconducting tapes after locking the current phase. The voltage component of the current direction is the size equivalent to the resistance voltage AC losses generated, multiplying this component and the current gives the AC loss.

All these methods have their advantages and disadvantages. The thermal method has features of low sensitivity, time-consuming, and require complicated equipment, but they are not affected by ambient electromagnetic noise, so the measurement range is broader. The magnetization method has high sensitivity, but it can only measure the static hysteresis loss of unit volume superconductors. The electrical measurement method is sensitive to external disturbances and the experimental environment, but has a higher sensitivity and short measurement time, so it is widely used.

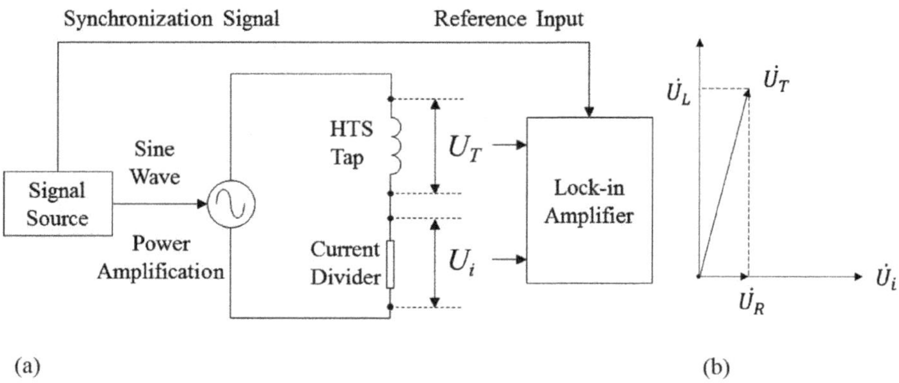

Figure 8.4 AC loss of superconducting tape measure schematic: (a) AC loss measurement circuit diagram; (b) Voltage vector diagram.

8.3 Magnet Coils Experimental Methods

8.3.1 AC Loss Measurements of the Superconducting Coil

AC loss measurement of the high temperature superconducting coil is also divided into magnetic measurement methods, thermal measurement method, and electrical measurement method. It is similar to the principle of AC loss measurement of high temperature superconducting tapes, so the first two methods are not specifically described. Compared with superconducting tape, the inductance value of a superconducting coil is larger. Therefore, measuring programs need to be adjusted when using the electrical measurement method for loss measurement.

The principle of AC loss measurement for a high temperature superconducting coil is shown in Figure 8.5. Because of the larger inductance of the superconducting coil, the terminal voltage is much larger than the voltage measurement range of a lock-in amplifier. At this point, it is necessary to use a resistor divider and a

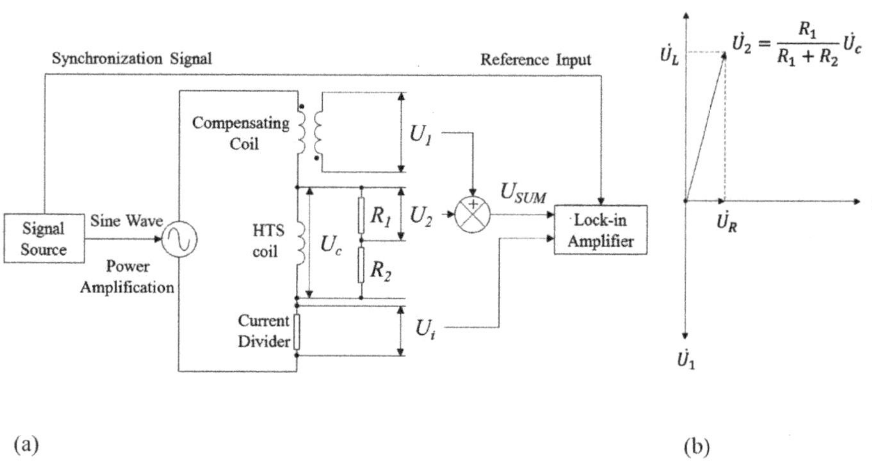

Figure 8.5 AC loss superconducting coil measurement schematic: (a) AC loss measurement circuit diagram (b) Voltage vector diagram.

compensation coil. Dividing resistors shrink the size of the superconducting coil terminal voltage proportionally, and the compensation coil offsets the induced voltage of the superconducting coil. According to the actual measurement object parameters, voltage divider resistors and the compensation coil ratio between primary and secondary side are designed legitimately, thus meeting the measurement accuracy and the lock-in amplifier range.

8.3.2 SMES Superconducting Magnet's Inductance Measurement

Superconducting magnet's inductance is influenced by temperature, current frequency, current size, current distribution, winding capacitance, eddy current loss, and other parameters. The inductance parameter measurement of superconducting magnet are different under different conditions. Generally, inductance measurement can be measured in low frequency conditions. Inductance of superconducting magnet can be measured by an LCR meter, an impedance analyzer, etc. The oscilloscope can also be used, as shown in Figure 8.6. Through a slow increasing current in the superconducting magnet the terminal voltage of the superconducting magnet is measured according to the basic formula $L = U\Delta t/\Delta I$, then the inductance can be calculated.

8.3.3 SMES Superconducting Magnet's Insulation Measurement

The method of measuring the insulation of superconducting magnets consists of the impedance measurement method and the high-voltage insulation measurement method. Impedance measurement methods use an LCR meter to measure the resistance, inductance, resistance to ground, and other parameters. The basic parameters will be quite different from the normal value if under serious insulation condition. The high-voltage insulation measurement method uses the withstanding voltage tester. First, a suitable location is selected for connecting the withstanding voltage tester with the superconducting magnets and the ground, then the withstanding voltage tester is operated to obtain the breakdown voltage. The breakdown voltage should be greater than the design parameters of the superconducting magnet, and a safety margin should be left. If to the contrary, there is a superconducting magnet insulation problem that needs to be detected.

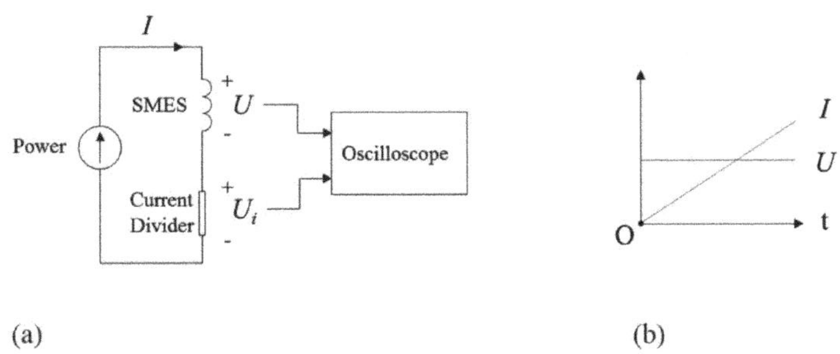

Figure 8.6 Superconducting magnet's inductance measurement method schematic: (a) Energization measurement schematic; (b) Voltage and current waveforms.

8.4 SMES Test

The test of SMES is divided into four parts: (1) superconducting magnets and cryogenic system test, (2) PCS test, (3) the SMES power regulation characteristic test, and (4) the dynamic simulation experiment of power system. In these tests, the SMES is considered as a whole device, and the quality inspection of each part of SMES in the process of the SMES production is not included.

8.4.1 Preparation Work

Several items should be prepared before the SMES test:

(1) Selection, layout, and calibration of temperature probe for the superconducting magnet and cryogenic system

In the process of testing, the temperature characteristics of the superconducting magnet and the cryogenic system are very important to the stability and operating efficiency of the whole device, so it should be monitored all the time. If the type of temperature probe or the location of the arrangement is inappropriate it will have a great impact on the accuracy of the measurement and on the performance of the SMES.

(2) Data acquisition system

In the process of testing, some parameters need to be recorded, such as voltage and current of the power system, and the voltage, current, resistance, temperature, the magnetic field of superconducting magnet, AC/DC side voltage, and the current of the converter. These data, some of which are crucial parameters for SMES control, some are the necessary parameters for SMES state monitoring, are essential for the operation and technical performance analysis of SMES. Different sampling frequencies and recording periods can be set according to the diverse target. Before the experiment, the data should be collected and verified.

(3) Inspection of cryogenic system

Checking the operating condition of the cryogenic system includes vacuum degree, temperature, working condition of the cooling machine, and cooling water. For immersion-cooling SMES, cold media liquid level, pressure, and other parameters should be included.

(4) The confirmation of storage temperature of the magnet

Before the test of the magnet, whether the magnet can work steadily at the set temperature should be confirmed.

8.4.2 Test of the Magnet

The performance of the magnet directly affects the effectiveness of the dynamic compensation of SMES. The test of the magnet is a key part of the SMES test.

(1) Parameter measurement of the cryogenic system

Before the magnet test, the operating loss of the cryogenic system should be measured. The relationship between the temperature and the vacuum degree and the stability margin of the thermal disturbance should be checked.

(2) Basic electromagnetic parameter measurement of magnets

A conventional LRC meter is used to test the inductance and resistance of the magnet. The maximum magnetic field of the magnet is measured by a hall probe, and the basic electromagnetic parameters of the magnet are measured by a tramegger.

(3) Insulation characteristics test of superconducting magnets

The insulation level of the superconducting magnet should be detected to confirm whether the insulation is strong enough under the rated frequency and voltage operation in the power system, which comprises a magnet (dewar, a magnet bracket) insulation, wire-to-wire insulation, insulation between wires, and a cryogenerator. Once the superconducting magnet is made it is difficult to measure the internal turn insulation of the superconducting magnet. Therefore, the insulation characteristics of the magnet should be carried out during the manufacture phase [9–10].

(4) Rated current flow test

This is to test the current flow capability and verify the electromagnetic thermal stability of the magnet at different current levels and change rates. Current, voltage (especially the resistive voltage), the magnet temperature, the maximum magnetic field, and the magnet external dissipation field should be recorded.

A superconducting magnet DC test wiring diagram is shown in Figure 8.7. Parallel resistors R1 and R2 protect the DC power source and the magnet respectively. The resistance value of R1 and its current capacity should be able to meet the requirements of the maximum parameters of the DC current source, while the resistance value and capacity of R2 it should meet the rated current of the magnet. When the DC current changes, the magnet will induce voltage on the magnet. A compensating coil could be used to decrease the influence of induced voltage. When the change rate of the current alters, the quench current of the superconductor will be different. The magnet of the

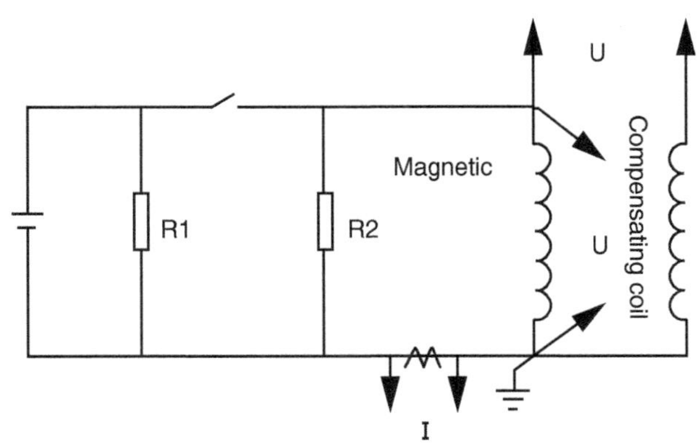

Figure 8.7 Superconducting magnet DC test wiring diagram.

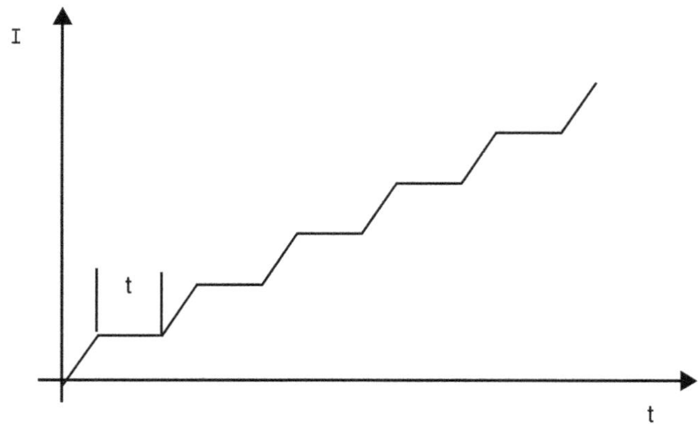

Figure 8.8 Field current rising waveform diagram.

SMES must be able to withstand a certain dynamic current. In the process of the experiment, the test can be carried out many times by using a low-rising rate. On the other hand, it is necessary to get the maximum current rise rate that the magnet can bear in order to meet the needs of dynamic compensation in the power system. The current flow test includes a uniform velocity current rising test and a constant speed current rising test.

Changing the excitation velocity and current stable operation time, the changes of temperature in different current levels held steady for a period of time should be observed. As the example shown in Figure 8.8, current rises at the rate of 0.02 A/s and maintains 10 min for every 20 A. Finally the magnet should be operated for 10 min at the rated current flow. The magnet temperature and the magnetic field of the magnets of each part should be recorded in the process at all levels of stable through-current. Once the unexpected voltage change occurs, the test should end immediately.

(5) Magnet critical current measurement

Increasing the current until quench is detected, the current of the magnet is the critical current and should be recorded. The experiment is to establish the maximum current that the magnets can bear. In order to avoid damaging the magnets, this test should be carried out in well-protected conditons. The rise rate of the current should also be restricted. The high rise rate of current will not only influence the accuracy of the measurement, but also will probably damage the magnets.

(6) The performance of magnets under a different magnetic charge rate test

When the power system needs power compensation, SMES should supply power compensation, and the charge and discharge rate varies with the requirement of the power system. Therefore, the magnet operation characteristics under different magnetic charge rates should be tested [11–12].

This experiment should be completed under DC power source and converter disparately. The test process is shown in Figure 8.9.

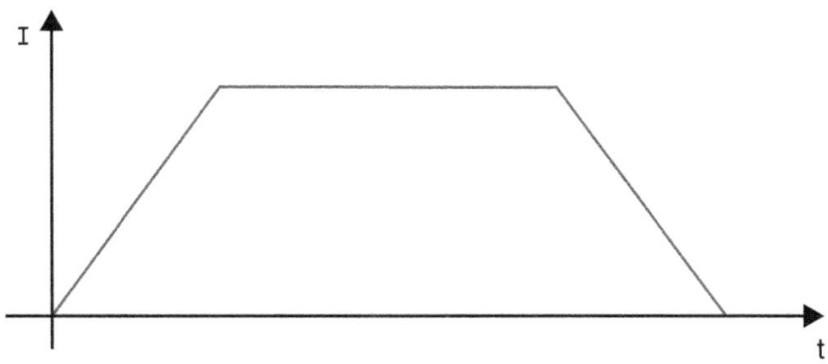

Figure 8.9 Current rate change schematic. Rising current at a certain rate–stable operation–decreasing current at a certain rate.

Several groups of experiments should be completed at different current rise rates, from 0.02 A/s to 1 A/s.

For experimental purposes:

1) Test the influence of different current change rates on the magnet temperature.
2) Inspect the magnet temperature recovery features.
3) Inspect the controllability of the converter in magnetization and demagnetization.

We can calculate the inductance value according to the charge and discharge time constant, and compare it with inductance value measure by LRC meter.

(7) AC loss measurement of the magnet

The operation current in SMES is the DC current. In the process of dynamic power compensation, the current will rise or fall repeatedly during charging and discharging process. The magnet current (duration, amplitude, frequency) varies with the operation of the power system. Through spectrum analysis, high frequency current is detected. The change of the magnet current will produce eddy current loss in the conductive magnet supports. Therefore, the losses of SMES include AC losses in superconducting wires and the eddy current losses in the support component. It is difficult and unnecessary to distinguish between these two losses.

There are two common methods to measure the AC loss of magnet: the calorimetry and electrical method.

For calorimetry method, we calculate the AC loss by measuring the volume of evaporation gas. In order to improve the accuracy it is necessary to eliminate the effects of liquid nitrogen or liquid helium caused by current lead and cryogenic vessel leakage.

The electrical method calculates the power consumed by the coil directly. Figure 8.10 shows the circuit diagram of AC loss measurement. It is measured by using a phase-locked amplifier and Rogowski coil. The output signal of the Rogowski coil is taken as a reference signal (which can be regarded as pure inductance signal). The "standard" signal (pure resistance voltage signal or pure inductance voltage signal) is taken as a reference signal to adjust the measurement signal. Thus, we can separate the resistance voltage component and the inductance voltage component. The device takes full advantage of high measurment accuracy and phase resolution of the phase-locked amplifier, which accurately measures the voltage loss.

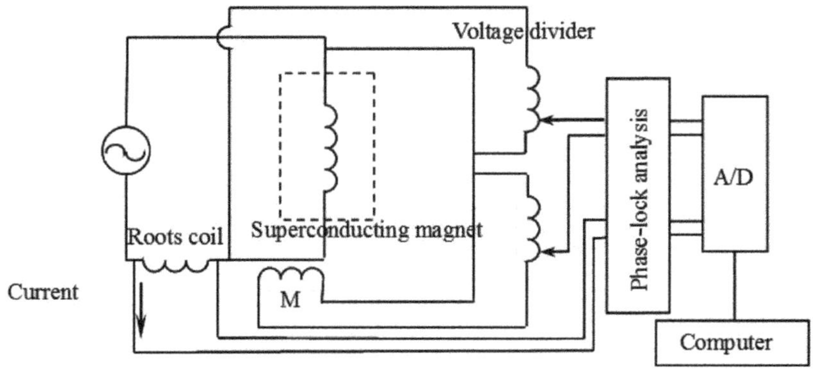

Figure 8.10 Magnet AC loss measuring circuit.

In order to evaluate, roughly, the loss characteristics of superconducting magnets, we can measure the input power at the end of the magnet at different frequencies's of current and find the relationship between the loss and the frequency. It is worth noting that the capacity of the superconducting magnet for a high frequency current may be less than that of low frequency current. Therefore, this measurement method is difficult to take where the magnet is working in a rated magnetic field intensity. In order to actually evaluate the magnet loss under the rated conditions we need to simulate the dynamic power compensation current through the converter control.

(8) Testing the frequency response of magnets

Superconducting magnets exchange power with the power grid through a power conditioning system. In order to achieve the four quadrant power regulation between the SMES and the system, power conditioning systems generally adopt high frequency power switching devices. The metal parts outside the superconducting magnet may resonate with power switching devices in the power conditioning system. Therefore, it is necessary to test the frequency response characteristics [13] of the superconducting magnet. Figure 8.11 shows the schematic diagram of superconducting magnet frequency response measurement.

8.4.3 Test of Power Regulation Characteristic

A power regulation test confirms whether the converter could control the magnet current and also tests that the communication between the converter and the main monitoring system is normal. During the performance test of the power regulation system the temperature of the magnet should be monitored and analyzed at the same time.

(1) Magnetizing and release magnetism test

The magnet has undergone magnetizing–stable operation–releasing magnetism. This process can be carried out throughout the testing process. Under different operating conditions, the change of temperature caused by magnetizing or releasing magnetism should be detected.

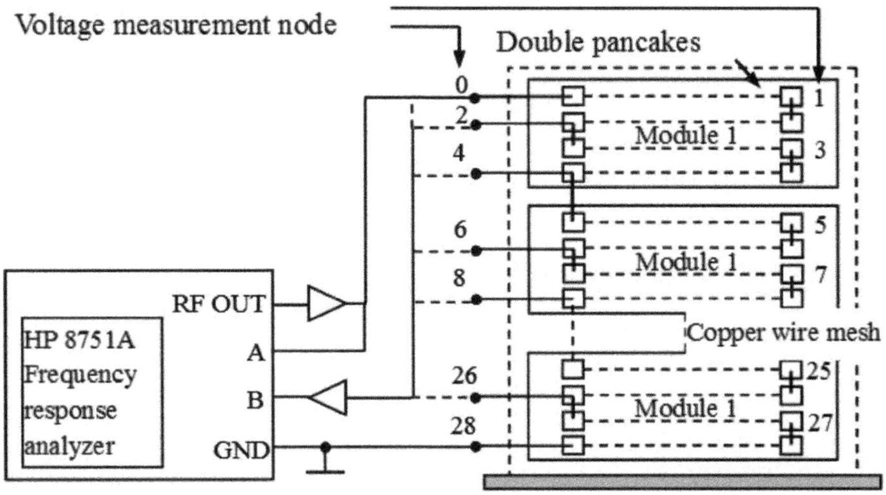

Figure 8.11 Superconducting magnet frequency response measurement schematic diagram [16].

(2) Long operation test under different current levels

Long operation characteristics under different current levels should be recorded, such as: voltage change, current change, magnet change, temperature change.

(3) Active power control test

As mentioned, the power regulation test has two purposes: (i) to verify the power exchange ability between the magnet and the system, and (ii) to check out the stability of the magnet in the process of operation. The power regulation test generally includes:

1) magnetizing—hold-- + P(power switching cycle and duration)—release magnetism
2) magnetizing—hold-- -P(power switching cycle and duration)—release magnetism
3) magnetizing—hold-- + P、 -P(power switching cycle and duration)—release magnetism

The test should be done at different power switching cycle durations, power exchange durations, and reactive powers. The ability of the exchange of active power between the system and the magnets should be verified, and the influence of the power switching cycle and the duration on magnet temperature changes need to be detected at the same time.

(4) Reactive power control test

Similar to the active power control test, the reactive power control test includes:

1) magnetizing—hold-- + Q(power switching cycle and duration)—release magnetism
2) magnetizing—hold-- -Q(power switching cycle and duration)—release magnetism
3) magnetizing—hold-- + Q、 -Q(power switching cycle and duration)—release magnetism

The test should be done at different power switching cycle durations power exchange durations, and active powers. The ability of the exchange of active power between the system and the magnets should be verified, and the influence of the power switching cycle and the duration on magnet temperature changes need to be detected at the same time.

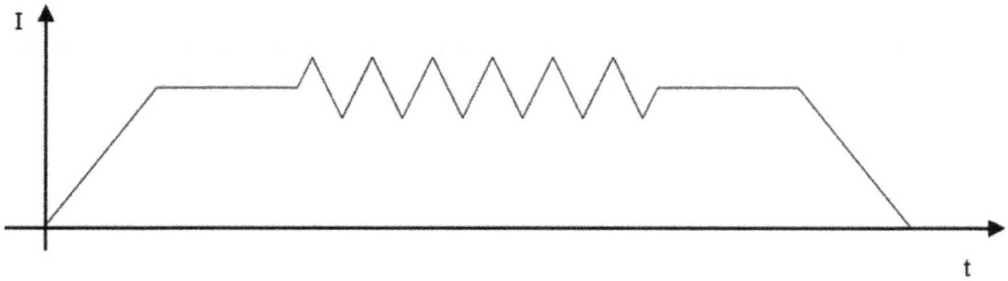

Figure 8.12 Substation circulation flow process diagram.

(5) Four quadrant power regulation test

This test aims to verify the four quadrant power regulation ability of SMES, and to detect temperature changes of the magnet during four quadrant dynamic operation. The process of the test is: magnetizing–hold–four quadrant power regulation–releasing magnetism.

The test should be done at different power switching cycle durations, power exchange durations, and different amplitudes of power regulation. The ability of the exchange of active power between system and magnets should be verified, and the influence of the power switching cycle and the duration on magnet temperature changes need to be detected at the same time.

(6) The losses of power compensation and magnet temperature rising test

Figure 8.12 shows the current change process of equivalent power compensation. Absorbing power and releasing power by the control of the converter, the discharge current can be achieved. The amplitude or frequency of current change could be set randomly with the converter control instruction. This experiment assesses the dynamic

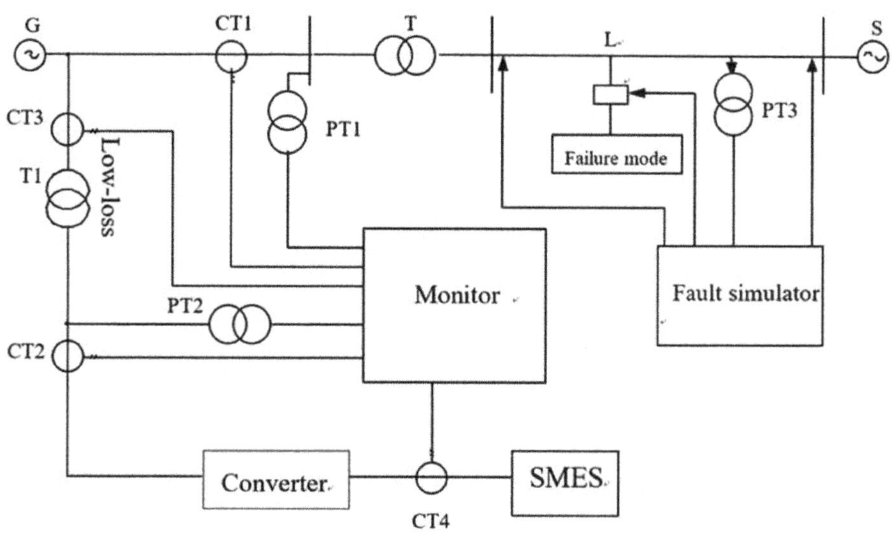

Figure 8.13 Dynamic model test system.

loss performance of the magnet by measuring the input power of the magnet and monitoring any temperature changes of the magnet [14–15].

8.4.4 Response Characteristic of an SMES System Test

The test of an SMES system response characteristic is to check out the effect of the control strategy and verify the response speed and power compensation ability of the SMES. As a practical device it is difficult to simulate accidents or voltage fluctuation situations in a power system frequently, so we can use the special simulation platform. In other words, we can test through a dynamic simulation experiment of power system. The system structure of the SMES dynamic simulation experiment is shown in Figure 8.13. Keeping the export power of the generator constant, the SMES's dynamic response to the system should be recorded in different short-circuit times.

8.5 Conclusions

SMES is a new power device that requires multi-disciplinary cooperation. It involves physics, materials, electrical engineering, cryogenic engineering, automatic control theory, and several other disciplines. From the perspective of electrical engineering, it includes several second-level disciplines such as superconducting magnet technology, power electronics, and power systems' stability and control technology. Thus, experiments involving SMES are complex and extensive.

References

[1] Kosse, A.I., Prokhorov, A.Y., Khokhlov, V.A., et al. (2008). Measurements of the magnetic field and temperature dependences of the critical current in YBCO films and procedures for an appropriate theoretical model selection. *Superconductor Science and Technology* 21 (7): 075015.

[2] Shin, H.S., Kim, K.H., Dizon, J.R.C., et al. (2005). The strain effect on critical current in YBCO coated conductors with different stabilizing layers. *Superconductor Science and Technology* 18 (12): S364.

[3] Bentzon, M.D. and Vase, P. (1999). Critical current measurements on long BSCCO tapes using a contact-free method. *IEEE Transactions on Applied Superconductivity* 9 (2): 1594–1597.

[4] Grimaldi, G., Bauer, M., and Kinder, H. (2001). Continuous reel-to-reel measurement of critical currents of coated conductors. *Applied Physics Letters* 79 (26): 4390–4392.

[5] Furtner, S., Nemetschek, R., Semerad, R., et al. (2004). Reel-to-reel critical current measurement of coated conductors. *Superconductor Science and Technology* 17 (5): S281–S284.

[6] Wang, Y., Lu, Y., Xu, X., et al. (2007). Detecting and describing the inhomogeneity of critical current in practical long HTS tapes using contact-free method. *Cryogenics* 47 (4): 225–231.

[7] Kim, H.S., Oh, S.S., Lee, N.J., et al. (2010). Nondestructive measurement of critical current distribution of Sm-BCO coated conductor using hall probe. *IEEE Transactions on Applied Superconductivity* 20 (3): 1537–1540.

[8] Hughes, T., Yang, Y., Beduza, C., et al. (1994). Measurements of self-field AC losses in PbBi-2223 Ag-sheathed tapes. *Physica C: Superconductivity* 235–240 (235): 3423–3424.

[9] Choi, J.H., Kwag, D. S., Cheon, H.G., et al. (2007). The insulation characteristics of coil-cryocooler for conduction cooled HTS SMES. *IEEE Transactions on Applied Superconductivity* 17 (2): 1517–1520.

[10] Choi, J.H., Kwag, D.S., Cheon, H.G., et al. (2008). The electrical insulation design of 600 kJ conduction cooled HTS SMES. *IEEE Transactions on Applied Superconductivity* 18 (2): 1394–1397.

[11] Kim, J.H., Kim, W.S., Hahn, S.Y., et al. (2005). Characteristic test of HTS pancake coil modules for small-sized SMES. *IEEE Transactions on Applied Superconductivity* 15 (2): 1919–1922.

[12] Tixador, P., Deleglise, M., Badel, A., et al. (2008). First tests of a 800 kJ HTS SMES. *Transactions on Applied Superconductivity* 18 (2): 774–778.

[13] Park, M.J., Kwak, S.Y., Kim, W.S., et al. (2007). AC loss and thermal stability of HTS model coils for a 600 kJ SMES. *IEEE Transactions on Applied Superconductivity* 17 (2): 2418–2421.

[14] Ren, L., Zuo W., Shi, X., et al. (2015). Development of a movable HTS SMES system. *IEEE Transactions on Applied Superconductivity* 25 (4): 1–9.

[15] Ren, L., Xu, Y., Liu, H., et al. (2015). The experimental research and analysis of a HTS SMES hybrid magnet. *IEEE Transactions on Applied Superconductivity* 25 (3): 1–5.

[16] Source: Steurer, M., & Hribernik, W. (2005). *Frequency Response Characteristics of a 100 MJ SMES Coil—Measurements and Model Refinement. IEEE Transactions on Appiled Superconductivity*, 15(2), 1887–1890.

9

Superconducting Wires and Tapes for SMES

Yuejin Tang,[1] Ying Xu,[1] Sinian Yan,[1] Feng Feng,[2] and Guo Yan[3]

[1] State Key Laboratory of Advanced Electromagnetic Engineering and Technology, Huazhong University of Science and Technology, Wuhan, China
[2] Division of Advanced Manufacturing, Tsinghua Shenzhen International Graduate School, Tsinghua University, Shenzhen, China
[3] Northwest Institute for Nonferrous Metal Research, Xi'an, China

9.1 Introduction

In 1911, Kammerlingh Onnes found that the resistance of mercury (Hg) dropped to zero abruptly below 4.15 K. After that, scientists in the world discovered that many materials can transit to zero resistance while cooling them to a certain low temperature. It is considered that those materials transited to a special phase below a critical temperature (Tc), which was called the superconducting state (SC state), and those materials were classed as superconductors. Alongside zero resistance, some special properties of superconductors were discovered that are different to their properties in a normal conducting state (normal state), and the properties identified while in the superconducting state were labeled superconductivity.

Until 1986, the Tc of superconductors was very low, the highest one being 23.2 K of Nb_3Ge. In 1986, Bednorz and Muller found that $La_{1-x}BaYCuO_y$ showed superconductivity at 35 K, and this work opened a new path to find a superconductor from copper oxide compounds (cuprate). In a few years, a lot of cuprate superconductors were found, and achieved the highest critical temperature of 134 K with $HgBa_2Ca_2Cu_3O_9$.

Entering the twenty-first century (2001), a new MgB_2 (magnesium diboride) was found to be a superconductor with a Tc of 40 K, and in 2008 an iron-pnictide of $LaFeAsO_{1-x}F_x$ was found presenting superconductivity at 26 K. Not long after many iron-pnictide superconductors were found, which have a critical temperature of higher than 40 K. Now, more than 1000 materials have been found to be superconductors; some of them are shown in Table 9.1.

Superconductors can be classed as type I and type II according to the electro-magnetic characteristics. Type I superconductors have a very low critical magnetic field, and cannot be applied to practical use in the field of electric power. All superconducting wires or tapes for manufacturing electric equipment are made of type II superconductors. In applied superconductivity, superconductors were classed as Low and High Temperature (LTS and HTS), divided by the temperature 25 K. Although a lot of materials were found to be superconductors, there are only a few types of superconductors that have currently been made into usable conductors, they are marked with * in Table 9.1. Superconductors used in electric equipment have two cross-section shapes,

Advances in Energy Storage: Latest Developments from R&D to the Market, First Edition.
Edited by Andreas Hauer.
© 2022 John Wiley & Sons, Ltd. Published 2022 by John Wiley & Sons, Ltd.

Table 9.1 Some well known superconductors and their critical parameters.

Material		Tc(0) K	Hc(0) T	Material		Tc(0) K	Hc2(0) T
Elemental metal	Nb	9.29	0.1944	Alloy	*NbTi	9.5	11.5 (4.2 K)
	Tc	8.2	0.141	Metal compounds	*Nb_3Sn	18.2	24 (4.2 K)
	Pb	7.2	0.0803		Nb_3Al	18.9	32 (4.2 K)
	Ta	4.48	0.083		Nb_3Ga	20.3	34 (4.2 K)
	Hg-α	4.153	0.0412		V_3Ga	16.8	21 (4.2 K)
	Sn	3.722	0.0309		Nb_3Ge	23.2	37 (4.2 K)
	In	3.4	0.0293		*MgB_2	40.0	–
	Al	1.183	0.0104	Metal oxide compounds	*$Bi_2Sr_2CaCu_2O_8$	85	22//533 (0 K)
	Ga	1.09	0.0059		*$Bi_2Sr_2Ca_2Cu_3O_{10}$	110	
	Mo	0.92	0.0098		*$YBa_2Cu_3O_7$	92	122//674 (0 K)
	Zn	0.85	0.0052		$Tl_2Ba_2Ca_2Cu_3O_9$	110	42//1400 (0 K)
	Zr	0.55	0.0047		$HgBa_2Cu_3O_9$	134	
	Ti	0.420.39	0.0056		$LaFeAsO_{1-x}F_x$	26	
	W	0.016	0.0001		SmOFeAs	55	>50(20 K)

*Materials that have been manufactured to be wire or tape for practical application.

wire and tape, according to the manufacture process. Here we use "superconducting wire" generically for superconducting wire and tape unless it needs to be specified.

In order to understand, design, manufacture, and operate Superconducting Energy Storage Systems (SMES), it is necessary to know some superconductivities and the characteristics of superconducting wires.

9.2 A Brief Explanation of Superconductivity

9.2.1 Zero Resistance and the Messiner Effect

In order to confirm whether a material is a superconductor, two properties have to be observed, namely zero electrical resistance and perfect diamagnetism. Zero resistance is the most promising property to apply to a superconductor in strong current equipments for decreased energy loss. It is verified, through experiments with a superconductor ring, that the resistivity in the superconducting state can be lower than $10^{-24\Omega}$/cm. For engineering applications, if the resistivity is smaller than $10^{-12}\Omega$/cm, the superconductor is considered to be in a superconducting state.

Perfect diamagnetism means that a superconducting material does not permit an externally applied magnetic field to penetrate its interior. This, was discovered by W. Messiner and R. Ochsenfeld in 1933, and is called the Messiner Effect.

However, London Equations illustrate that magnetic flux can enter the superconductor according to Equation (9.1).

$$\nabla^2 H = \lambda_L^{-2} H \tag{9.1}$$

Here, H is the strength of the magnetic field applied on a superconductor, and λ_L is the magnetic penetrate depth.

If a superconductor occupying the region $x>0$, and a magnetic field H_1 lying parallel to the superconductor surface, e.g. H//Z, the resolution of Equation 9.1 can be written as Equations (9.2) and (9.3).

$$H(x) = H_1 e^{-x/\lambda_L} \tag{9.2}$$

$$\lambda_L = \left(\frac{\Lambda}{\mu_0}\right)^{1/2} = \left(\frac{m_s}{\mu_0 n_s e_s^2}\right)^{1/2} \tag{9.3}$$

Here, m_s,n_s, and e_s are the mass, the density, and the electric charge of superconducting electronics in superconductors respectively. Figure 9.1 shows this penetration schematially, where the shaded area is the superconductor.

9.2.2 Critical Parameters of a Superconductor

The superconducting state can only be maintained below a certain temperature, i.e. Tc, and heating a superconductor can destroy its superconductivity. This means superconducting equipment can only operate below its Tc.

The superconducting state can only be kept below a limited magnetic field strength, and this field is defined as critical field H_c, which means that superconductivity can be destroyed by magnetic field. H_c is related to temperature as shown in Equation (9.4).

$$H_c(T) = H_c(0)[1 - (T / T_c)^2 \tag{9.4}$$

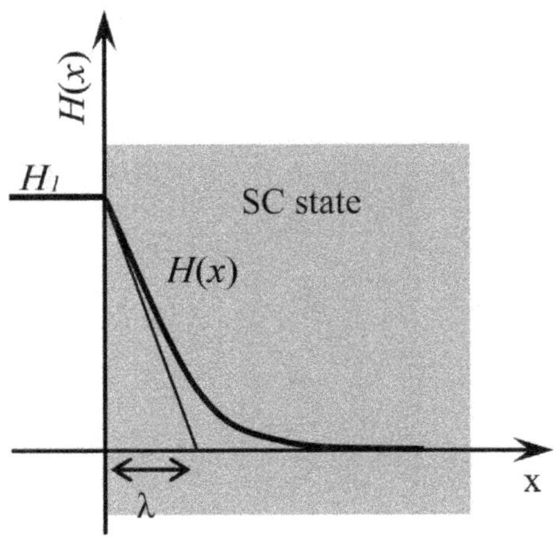

Figure 9.1 The distribution of flux penetrating a superconductor.

with $H_c(0)$ values for elements typically being less than 10^3Oe.

Another critical parameter of a superconductor is current densityJ_c. For an engineering application, a critical current I_c is more convenient than J_c. Considering the relationship between current and magnetic field, it is easy to understand that *Ic* relates with the magnetic field applied to the superconductor, not only the amplitudes, but also to the directions of current and field. I_c is also dependent on temperature as shown in Equation (9.5).

$$I_c(T) = I_c(0)\left\{1-(T/T_{co})^2\right\} \tag{9.5}$$

A superconductor can only maintain a superconducting state below the three critical parameters as shown in Figure 9.2, and any critical parameter is affected by the other two. This makes the design of practical superconducting electric equipment more complex than a normal conductor.

9.2.3 Type I and Type II Superconductors

Superconductors are classed as type I and type II, the differences between the two types mainly appears in the characteristics of superconductors under the stress of a magnetic field. Here we do not illustrate why there are type I and type II, but point out that the difference is observable when an external magnetic field is applied.

While an external magnetic field *He* is applied, the density of magnetic flux *B* in a material is given by Equations (9.6) and (9.7).

$$B = \mu_0(H_e + M) \tag{9.6}$$

$$M = \chi H_e \tag{9.7}$$

where M and χ are the magnetization and magnetic permeability of the material. Figure 9.3 compares the *M–H* curves of type I and type II superconductors. For a type I superconductors, the Meissner Effect shows that $\chi = -1$, while the field is larger than H_c, the superconducting state is destroyed and the superconductor transits to normal state, $M = 0$. For the type II superconductor, while the external magnetic field reaches H_{c1}, a lower value than H_c, fluxes begin penetrating the superconductor and M begins to decrease, but M does not decrease to zero until H_e reaches to H_{c2}, which is larger than H_c. This phenomena shows that type II transits from the superconducting state to normal gradually from H_{c1} to H_{c2}. Between H_{c1} and H_{c2} the superconductor is of mixed state; both normal and superconducting

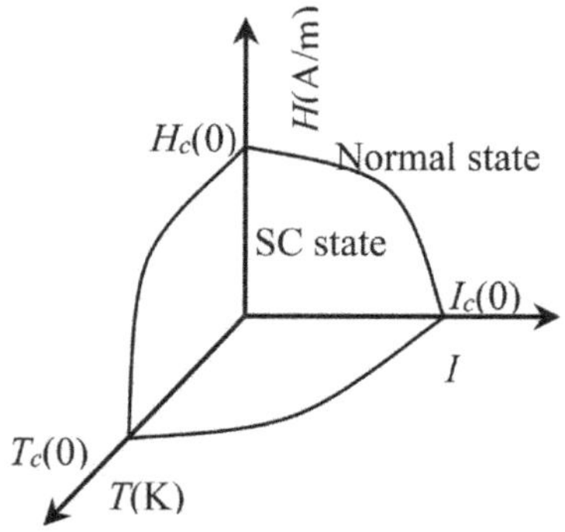

Figure 9.2 Relations between I_c, H_c, and T_c.

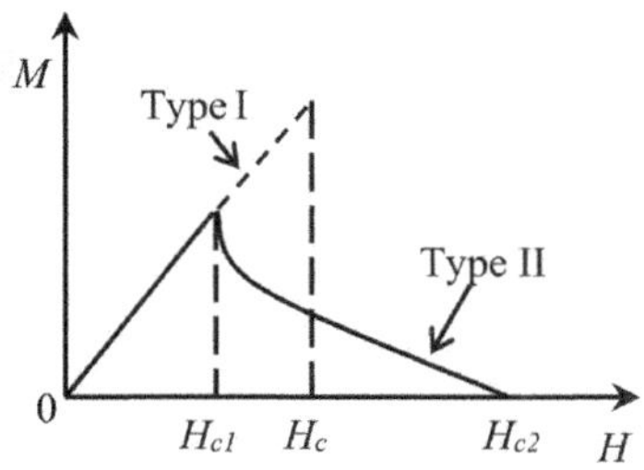

Figure 9.3 Comparison of magnetization between type I and type II superconductor.

state present together. The distribution of fluxes in a mixed state is shown in Figure 9.4, the shaded areas are normal state cores that are widely dispersed, and each flux of φ_0 penetrates through a normal core surrounded by a current like vortex, φ_0 is called flux quantum with a value of 2.07×10^{-15}Wb. As the external field increases, the number of normal cores increase until it fulfils the superconductor at H_{c2}. H_{c1} and H_{c2} are named as the lower and upper critical magnetic fields respectively.

As the H_c of the type I superconductor is very low, it is not suitable to be applied to electric power applications, such as SMES or transformers. Although H_{c1} of the type II superconductor is very low it can maintain a part of the superconducting state until a very high H_{c2} is achieved. All superconducting wires used to manufacture electric power equipment are made of a type II superconductor.

9.2.4 Flux Motion and AC Loss

9.2.4.1 Flux Flow and Flux Pinning

Considering a transport current runs through a type II superconductor in a mixed state as shown in Figure 9.5, the Lorentz force will be produced due to the interaction between current and fluxes. Therefore, fluxes will be moved by the Lorentz force. The magnetic flux motion results in the fluxes cutting off the normal state cores and so losses will be produced in the superconductor.

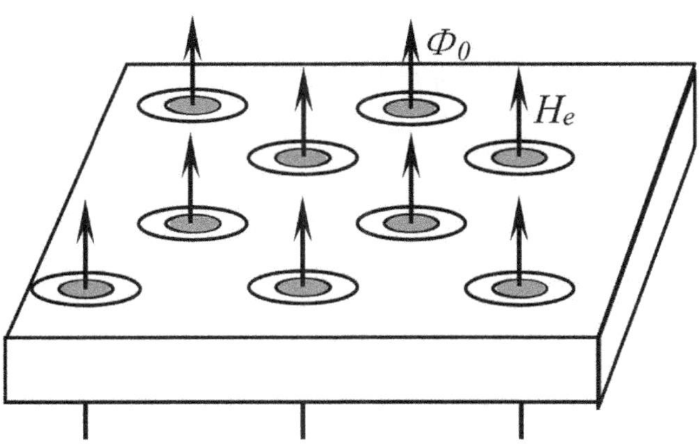

Figure 9.4 Distribution of fluxes in mixed state of type II superconductor under magnetic field.

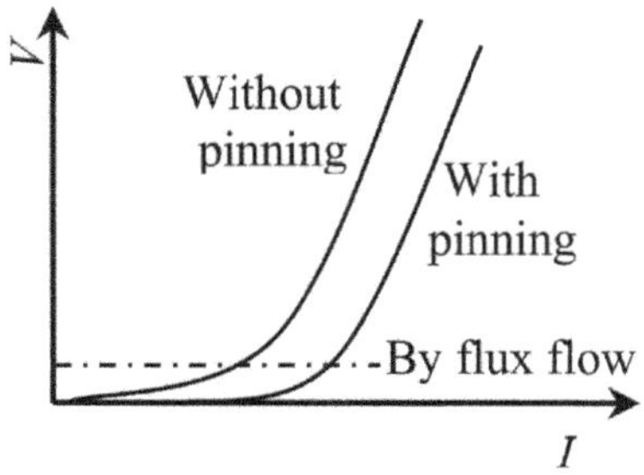

Figure 9.5 Compassion of V-I curve with or without pinning.

Although flux flow loss is very small, a continuous loss makes the superconducting state unstable. In order to apply the superconductor in practical applications, flux motion must be limited. Flux pinning is an effective way for stopping flux flow.

Flux pinning is usually produced by the defects in a practical superconductor, which include various lattice defects, grain boundaries, dislocations, non-superconducting precipitates, etc. Those defects act as pinning centers because they have a lower energy potential for magnetic flux than that of the superconductor state. Fluxes that go out or off those pinning centers have to get energy larger than the potential, and that means that there is pinning force F_P in a pinning center. The pinning force can prevent flux motion while it is larger than the Lorentz force F_L and so allows the superconductor to transport a larger current with zero resistance than that without pinning, as shown in Figure 9.5.

In order to increase the ability for transporting the current of a superconductor, it is necessary to make pinning centers during the manufacture of a practical superconductor wire.

9.2.4.2 Inreversible Magnetization and AC Loss

Because of the existance of the pinning force, fluxes entering or exiting pinning centers consume energy, and that allows the magnetization curve to become irreversible as seen in Figure 9.6.

Pinning force is a function of the magnetic field and temperature, if the magnetic field strength is larger than H_i, the magnetization becomes reversible as shown in Figure 9.6. Also, if the temperature is higher than T_i, the magnetization become reversible too. H_i and T_i are named irreversible magnetic field and irreversible temperature, they are lower than H_c and T_c of the superconductor respectively. Irreversible

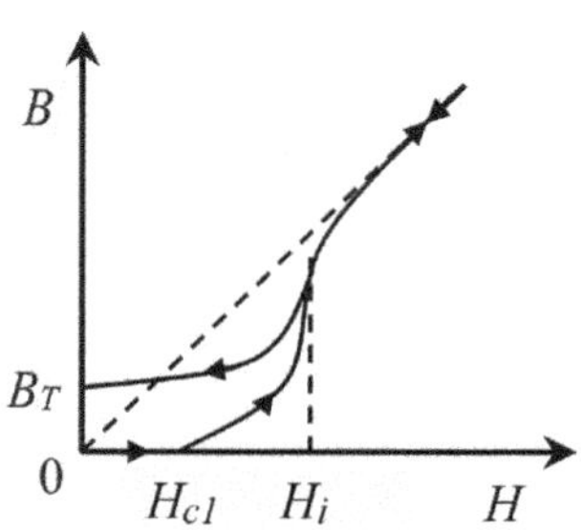

Figure 9.6 Irreversible magnetization curve with pinning force.

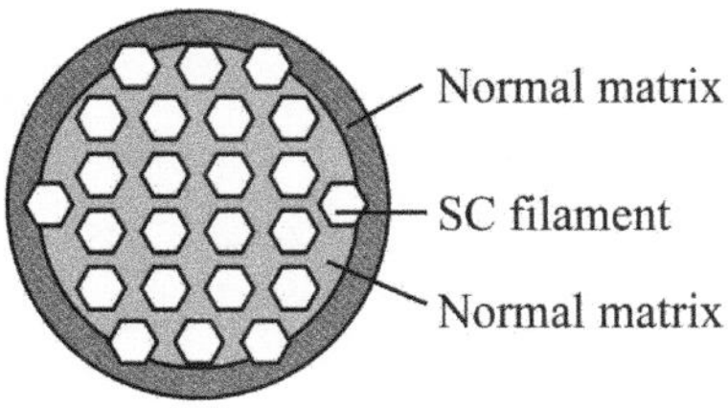

Figure 9.7 General construction of the cross-section of a superconducting wire.

magnetization means there are losses when a superconducting wire operates under a changed magnetic field or current. This kind of loss is named hysteresis loss.

The superconducting wire must work at $H<H_i$ and $T<T_i$ to decrease AC losses in the wire.

Research has verified that to decrease hysteresis loss/per volume, the superconducting filament should be very thin. For practical use, the superconducting wire has to carry a large current, and that allows the superconducting wire to be a composite, that is, a usable superconducting wire is made of a combination of twisted filaments separated with a normal metal matrix, its cross section is shown in Figure 9.7.

In the construction shown in Figure 9.7, if the wire operates under changing fields, then losses are generated in the nomal metal matrix, which is called coupling loss.

AC loss in a superconducting wire includes hysteresis loss and coupling loss.

It is one of the important challenges to technology reduce AC loss in superconducting wire, which will operate under a changing electro magnetic environment (magnetic field or current). For SMES, AC losses are generated while it compensates energy to the grid, while the current in the SC magnet is a dynamical one.

9.2.5 Stability of Superconducting Wires

For a practical application of a superconducting wire, the transition of superconducting state to normal state must be prevented, otherwise the superconducting equipment could be damaged.

In a superconducting wire or equipment these transitions can be initiated by a disturbance that produces heat pulses locally in the superconducting wire resulting in a normal zone seed. Disturbances may come from thermal, mechanical, electrical, and magnetic changes.

While a normal zone occurs in a superconducting wire, heat Q_R is produced by the transport current and the resistance, and transfers as shown in Figure 9.8, a part of the heat (Q_T) is consumed by heating the wire, Q_k is transferred along the wire by thermal conduction, and Q_c is cooled by cryogenic media. If Q_R-Q_T-Q_k-Q_c>0, the normal zone will expand and let the superconducting wire be quenched, otherwise the normal zone shrinks and the wire has thermal stability.

In order to increase thermal stability, it is necessary to add some normal metal as a stabilizer to produce a practical superconducting wire as shown in Figure 9.8. The stabilizer in a superconducting wire can increase the thermal stability by decreasing the temperature rise and shielding some current from the superconductor.

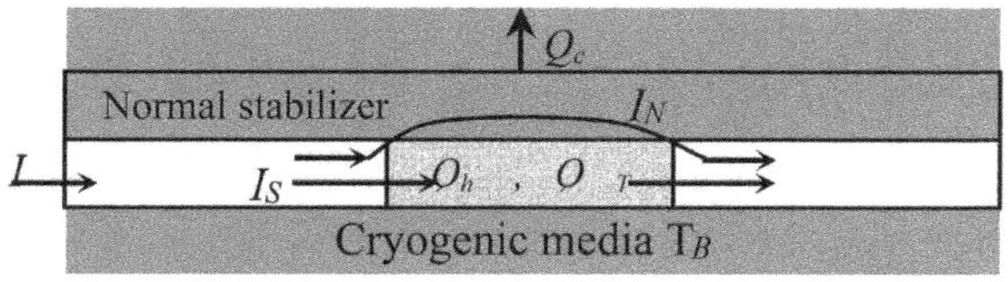

Figure 9.8 Current shared by normal stabilizer.

9.2.6 Key Paramaters for Evaluating a Superconducting Wire

As described, a practical superconducting wire must be made of a type II superconductor, and must address the following technical problems:

- Prevent flux motion by making pinning centres.
- Decrease AC loss by making the superconductor a thin filament.
- Combine a lot of filaments to carry a large current, which are separated with a normal matrix.
- Add a normal metal stabilizer to increase stability.

Incorporating those requirements, a practical superconducting wire is usually of the structure shown in Figure 9.7. However, until now it has been difficult to construct such a structure for some HTS superconductors, which can only be made to be in the form of tapes.

There are some key parameters for evaluating the quality of a superconducting wire.

Critical current. The critical current of a superconducting wire depends on the magnetic field strength (amplitude and angle) acted on and the temperature of the wire. Also the critical current shall degrade under the affect of strains.

n value. During the process of a superconductor transiting from superconducting state to normal state, the voltage–ample characteristics is almost $V \propto I^n$. For different samples of the same wire, n values reflect their uniformities, the larger n has better uniformity. For samples of different wires, the n values reflect their speeds of quench propagation, also their thermal stabilities, a smaller n value equates to a slower quench speed, and better stability.

AC loss. AC loss is an important parameter when the superconductor wires operate under a changing electro magnetic environments. For an application with a higher change rate of current or magnetic field, a thinner superconducting filament is needed. However, HTS wires are more difficult to produce with a thin filament than LTS.

Thermal stability. As described in Section 2.5, there are many factors affecting the stability of superconducting wire or magnets. The **ratio of copper to superconductor** is usually used to reflect the stability of a superconducting wire. The safest condition is incorporating sufficient stabilizer (copper or alumina) in the wire to share current entirely while the superconducting filaments quench perfectly.

Mechanic strength. Superconducting wire usually bears an electromagnetic force while it is operating. The limits of pulling strain and bending strain must be known while using a superconducting wire, which also affects the critical current.

9.3 Wires Made from LTc Superconductors

9.3.1 NbTi

NbTi is an alloy of niobium and titanium that performs as a type II superconductor with T_c of 9.5 K(at 0 T) and H_{c2} of 11.5 T(at 4.2 K). NbTi has good mechanical properties, and can be easily fabricated into wire through the extruding and drawing process. An NbTi strand is usually made of a number of thin NbTi filaments in an aluminium or copper matrix. NbTi wire is a mature commercial product. For different applications, the NbTi wire can be manufactured with different parameters through changing the diameter of the NbTi filament, the twist angle of the filaments, the number of filaments in a strand, and the number of strands in a completed conductor, as well as the ratio of Cu:NbTi. For example, for a NbTi strand with diameter of 0.73 mm, Cu:NbTi of 2.3, the critical current density reaches 339A at 5 T/4.2 K [1]. This NbTi strand is manufactured for ITER poloidal field magnets and a number of strands are twisted together in order to pass a large current.

The shape of NbTi can be different according to different usages, such as the Rutherford cable [2] or cable-in-conduit conductors (CICCs) [3]. The earliest SMES of 30MJ was made of NbTi wire with maximum operating current of 4.9 kA [4].

NbTi wire needs to operate at 4.2 K, usually cooled with liquid helium and the capacity of the carrying current decreases fast as the external magnetic field increases. Although SMES of 30MJ with NbTi wires were demonstrated in 1983 LTS SMES have not been applied in commercial equipment until now. Low T_c and H_c of NbTi can be considered two important reasons.

9.3.2 Nb_3Sn

Nb_3Sn is an intermetallic compound of niobium (Nb) and tin (Sn), used industrially as a type II superconductor with T_c of 18.2 K (at 0 T) and H_{c2} of 22 T (at 4.2 K).

Mechanically, Nb_3Sn is extremely brittle and thus can not be easily drawn into a wire. To overcome this, wire manufacturers typically draw down composite wires containing ductile precursors. The internal-tin process includes separate alloys of Nb, Cu, and Sn-2%Ti. The bronze process contains Nb in a copper–tin bronze matrix. Typically, with both processes the strand is drawn to the final size and coiled into a solenoid or cable before heat treatment. It is only during the heat treatment that the Sn reacts with the Nb to form the brittle, superconducting niobium–tin compound. The powder-in-tube (PIT) process is also used. Some Nb_3Sn wires can be wound after heat treatment [5]. Recently, some research suggest the react-and-wind process in the manufacture of Nb_3Sn magnets is possible by using Cu–Nb reinforced bronze Nb_3Sn wires. Now a record non-copper critical current density of Nb_3Sn superconducting wires are claimed of about 3000 A/mm^2 at 12 T and 4.2 K [6–7].

Because both the manufacturing and heat-treatment process are a little complicated, it is more expensive manufacturing a superconducting magnet with Nb_3Sn superconducting wire than NbTi, but Nb_3Sn can withstand magnetic field intensity values up to 20 Tesla (T), whereas NbTi can withstand only up to roughly 10 T. In order to obtain a magnetic field of higher than 10 T, Nb_3Sn is a potential candidate. For example, Nb_3Sn wires are used in accelerators, nuclear fusion reactors, etc. [5–8], but there is not an example in SMES.

9.4 Wires or Tapes Made from HTc Superconductors

The first copper oxide compound (cuprate) superconductor was created in 1986. After that a series of cuprate superconductors were made, some of which have a critical temperature higher than 77 K. This means that the cuprate superconductor can maintain superconductivity with liquid nitrogen, enable its cooling system to be more economical than at using liquid helium. The concept of the HTS superconductor was identified after cuprate superconductors were discovered. In the following decades, tremendous efforts were made to develop usable HTS wires, and a lot of progress has been achieved on HTS wire industrialization.

Three types of cuprate superconductors have been fabricated into practically usable conductors; they are $Bi_2Sr_2Ca_2Cu_3O_{10}$ (BSCCO-2223), $REBa_2Cu_3O_{7-x}$ (REBCO-123 for short, where RE represents rare earth elements, one well used is $YBa_2Cu_3O_7$ (YBCO-123)), and $Bi_2Sr_2CaCu_2O_8$ (BSCCO-2212). Both BSCCO-2223 and REBCO-123 are in tape form, while BSCCO-2212 can be made as a round wire.

BSCCO-2223 and YBCO-123 have been widely utilized during the recent decades, and both could be superconductive under liquid nitrogen temperature (77 K under standard atmospheric pressure) and self-field. Dependent on the sequence of industrialization, they were labeled first and second generation (1G and 2G) HTS wires, respectively. The 1G wire is made from the $(Bi, Pb)_2Sr_2Ca_2Cu_3O_{10+y}$ (Bi-2223 or BSCCO-2223 with T_c of about 110 K, while 2G wire is mae from the REBCO-123 with T_c of about 92 K.

BSCCO-2212 can be made as a round wire, it presents good performances at 4.2 K, but it has a very low current carrying capacity in a magnetic field at 77 K.

The superconductivity of MgB_2 was discovered in 2001, its critical temperature is 40 K, and upper critical field H_{c2} (parallel to ab planes) is ~ 14.8 T, (perpendicular to ab planes) ~ 3.3 T, in thin films up to 74 T, in fibres up to 55 T. It essentially has applications at temperatures below 25 K. As the critical temperature is higher than 25 K, MgB_2 can be classed as an HTS superconductor. MgB2 wire has been rapidly developing in recent years, and has been used in MRI applications.

For iron-based superconductors, which were discovered in 2008, short wires and tape samples had been made successfully [9].

9.4.1 BSCCO-2223/Ag Tapes

The wires using BSCCO-2223 superconductor are generally manufactured with the "powder-in-tube" (PIT) or "silver sheath" method, which consists of filling the raw material powder into silver tubes, deforming into a wire tapes, and sintering the polycrystalline superconductor [10].

In the PIT process, a mixture of precursor powders with mainly $Bi_2Sr_2Ca_1Cu_2O_{8+x}$, lead rich phases, and alkaline earth cuprates is packed into the silver tubes. The composite tube is drawn to a desirable diameter, typically 1 mm, which is then bundled and packed into another silver alloy tube to form a multifilament structure. After consecutive deformation, including multi-pass drawing and rolling, multi-filament green tapes can be produced, with a typical dimension of 4 mm in width and 0.2–0.3 mm in thickness. This green tape does not contain any BSCCO-2223 phases. Complex heat treatment processes with intermediate deformation are be carried out to transfer the precursor powder in the filaments into almost pure BSCCO-2223 grains. It should be noticed that the BSCCO-2223 tape using Ag-Au sheath has a low heat conduction coefficient at the temperature region around 4 K to 77 K. It is a very promising material for HTS current leads that can sustain a large current and break heat conduction as well.

Plenty of producers appeared during the past two decades, such as AMSC, Sumitomo Electric, INNOST, EHTS, and Trithor. But some of them have quit the market since the 2G wire of YBCO coated conductor entered the market successfully. Sumitomo Electric are still in production and its products, performance currently meet the highest parameters [11].

BSCCO-2223/Ag tapes are available as industrial productions in km lengths. The typical critical current (I_c) can reach to 200 A at 77 K and self-field with a dimension of about 4 mm × 0.02 mm. The critical stress is around 100 MPa to 400 MPa depending on the different reinforced materials. The I_c of BSCCO-2223/Ag tapes at 77 K can be easily suppressed under perpendicular fields, but their I_c at 77 K under parallel fields is comparable to and even higher than that of 2G tapes. Furthermore, I_c of BSCCO-2223/Ag tapes at below 20 K and high fields is also comparable to that of 2G tapes. These properties make the BSCCO-2223/Ag tape a candidate for HTS cables at 77 K, but for SMES, motor, and generator at below 30 K.

9.4.2 REBCO Coated Conductors

The critical current of BSCCO-2223 tape degreases fast while there is an external magnetic field. 2G HTS wire has a higher irreversibility field (H_i), at which J_c vanishes, compared with BSCCO-2223. Thus 2G wire has a better in-field performance, enlarging its application fields. Additionally, there is less noble metal in the architecture of 2G wire, leading to a low cost potential.

However, the fabrication of 2G wire is somewhat difficult due to its complicated architecture, which mainly includes a metallic substrate, several buffer layers, an HTS layer, and a silver stabilizer layer. The 2G wire is fabricated by coating the above layers on to the substrate, hence it is also named as coated conductor. The metallic substrate, which is generally a nickel-based alloy or stainless steel, could enhance the mechanical strength of the 2G wire. The silver stabilizer layer could protect the HTS layer. The buffer layers could prevent element diffusion between metallic substrate and HTS layer, and the topside buffer layer, also known as the cap layer, could serve as the template for the epitaxial growth of the biaxially textured HTS layer. Buffer layers are different not only in the materials but also the number of layers according to different manufactures [12–13]. A copper or stainless steel layer could also be laminated on to the 2G wire to further enance the mechanical strength.

The manufacturing of 2G wires is under development in many countries. Architecture with different material combinations and deposition methods is used by these companies. Most of these manufacturers could provide 2G wires with the pieces in lengths of hundreds meters or over 1 km, and some of them could achieve an annual output of several hundred km. A typical I_c of the wires is over 300 A/cm-width at 77 K, self-field [14].

Since the manufacturing techniques are still under development, the above architecture and data might vary in the future years. To have an up-to-date knowledge of the latest 2G wires please refer to the websites shown in Table 9.2.

Table 9.2 The websites of some companies manufacturing 2G wire.

Compannies	Website
AMSC (US)	http://www.amsc.com/solutions-products/hts_wire.html
Bruker (German)	https://www.bruker.com/products/superconductors-and-metal-composite-materials/superconductors
Fujikura (Japan)	http://www.fujikura.co.jp/eng/products/newbusiness/superconductors
SCSTC (China)	http://www.china-superconductor.com/page/cpxg/index.php
SHSCTEC (China)	http://www.shsctec.com/Pages/SpecialProductList.aspx?ID=61&cl=cn
SuNAM (Korea)	http://www.coatedconductor.com
SuperOx (Russia & Japan)	http://www.superox.ru/en/products/974-vtspprovod
Superpower (US)	http://www.superpower-inc.com/content/2g-hts-wire
THEVA (German)	http://theva.com/superconductors/theva-pro-line

9.4.3 BSCCO-2212

$Bi_2Sr_2CaCu_2O_8$(BSCCO-2212) has a critical temperature of about 85 K. Although it has a very low current carrying capacity in a magnetic field at 77 K, it has a good high field property in the temperature range between 4.2 and 20 K. BSCCO-2212 can be manufactured as a round wire [15], and the critical current shows no anisotropy with respect to an applied field.

Several successful methods of achieving high current density conductors have been demonstrated, including powder-in-tube, dip-coating, tape casting, electrophoretic coating, and spray coating.

Conductors of this material have been made in long lengths of wire with uniform properties without breakage. The versatility of BSCCO-2212 principally provides a potential for wide engineering use where the operating temperature is lower than 20 K.

Oxford Superconductor Technologies (OST) has measured the I_c of an OST round BSCCO-2212 conductor with a diameter of 0.7 mm at different temperatures and differed magnetic fields. The J_c drops dramatically at higher temperatures [16]. OST has produced a BSCCO-2212 round strand with a critical current density higher than $300 A/mm^2$ at 4.2 K and 15 T [17].

The performance of BSCCO-2212 is excellent below 20 K and is promising in high field magnets. Multi-filament tapes of BSCCO-2212 have been made for application in high field magnets [18] as well as in SMES [19]. However, at temperatures higher than 20 K its performance is limited. From this point of view, BSCCO-2212 was eclipsed by BSCCO-2223 wires.

9.4.4 Research on Larger Current HTS Conductors

BSCCO-2212 can be made as a round wire or structured to be composite cables [18, 20], Although BSCCO-2223 or REBCO superconductors are in the form of tapes, the critical current of a single tape is not large enough for manufacturing electric equipments such as SMES, transformers, or generators, on a commercial scale, which usually have a capacity of MW ~ GW and operating current of kA ~ MA. Presently, HTS SMES with a storage capacity of 1 MW is available with BSCCO-2223, but the magnets cooled down to 4.2 K were made of double pancakes wound with two parallel tapes, and the end of the magnet wound with four parallel tapes [21]. An unbalanced current share presents very easily in parallel superconducting branches, while the current is a dynamic one, because the resistance is very low and the current share rate is decided by reactances. In order to develop a large-scale HTS SMES with storage capacity of hundreds MW or GW, it is better that HTS tapes are assembled as cables that are able to transport kA to MA currents. In addition, to minimize losses and homogenize current distribution among tapes, the tapes in the cable have to be transposed.

There are three main difficulties in the production of BSCCO-2223 or YBCO-123 to produce a larger current cable:

- The tape form is difficult to twist and transpose.
- The fragile mechanical property is difficult to twist and transpose.
- The crystal structures of cuprate superconductors have sheet-like two-dimensional CuO_2 square lattices where superconductivity occurs. This particular structure results in an anisotropic property, both on I_c and H_c. The current transport capacity of BSCCO-2223 and YBCO-123 is weaker under a perpendicular magnetic field than a parallel one. If N tapes are assembled to a larger cable with twisting and transposing, the total current capacity should be less than that of a single tape times N.

In recent years, much research has resulted in a larger current cable with transposed REBCO tapes. Typical ways proposed and researched are simple twisted [22], stacked and twisted [23], and the Roebel cable [24].

9.4.5 MgB_2

Due to the low cost of its constituent elements, MgB_2 has more promise for low-cost superconducting wire. MgB_2 wire has been researched by Hyper Tech Research (US), Columbus Superconductors (Italy), Hitachi (Japan), as well as Northwest Institute for Nonferrous Metal Research (China).

MgB_2 wires can be produced through the powder-in-tube ex-situ and in-situ processes. In the in-situ variant, a mixture of boron and magnesium is poured into a metal tube, which is reduced in diameter by conventional wire drawing. The wire is then heated to the reaction temperature to form MgB_2 inside. In the ex-situ variant, the tube is filled with MgB_2 powder, reduced in diameter, and sintered at 800°C to 1000°C.

A cross-section of a typical Hyper Tech wire and its performance in magnetic field have been measured [25]. This wire is manufactured in 1 to 4 km lengths with 7 and 19 filaments. Typical J_c values are 1×10^5 A cm^2 at 5 T, 4.2 K and 2×10^5 A cm^2 at 1 T, 20 K. The I_c at 1 T and 20 K is approximately 200 A.

MgB_2 wire is very versatile and could have many potential applications in devices that can operate at temperatures below 25 K, under a low to medium field. Some of the possible applications include transformers, inductors, reactors, motors, generators, fault current limiters, and magnetic resonance imaging (MRI) systems. In 2010, a

1 MW-5s SMES with MgB_2 was designed [26], but to date there has not been a perfect SMES manufactured using MgB_2 wires.

9.5 Discussion

The advantages of SMES come mainly from two properties of superconductor. The first one is zero resistance, which makes the superconducting magnet able to store energy with very low losses while the magnet storage has a stable current. The second is the high current carrying density of superconducting wires, which make the superconducting magnet have a high energy storage density. However, cooling power is consumed to maintain the magnet in a superconducting state. In addition, the magnet will produce AC loss while the current is changing, such as when an SMES is absorbing or discharging energy, that presents during the process of compensating the power demand of the gird. In order to reduce the operating cost, the superconducting wire in an SMES is required to be able to operate at higher temperatures, bearing higher magnetic fields, and carrying higher current density, with less AC loss. Unfortunately, all superconducting wires are not good enough for commercial-level SMES.

The manufacturing technology of NbTi wire is almost mature, but it has to be operated at 4.2 K and H_c is not so high. Although a capacity of 30MJ SMES with NbTi was tested early in 1983 [4], SMES with NbTi has not been commercially applied until now. It is significant that the main obstacle comes from the low operating temperature and critical magnetic field. Nb_3Sn as well as Nb_3Al [27] have higher critical magnetic fields than NbTi, but they also need to be operated at 4.2 K, and there are no reports about using them in SMES yet.

It is of considerable importance that promising SMES should be made of HTS superconducting wires, which can be operated at a higher temperature, and so with less operating cost.

Both BSCCO-2212 and BSCCO-2223 have been used to manufacture SMES, but they need to be cooled to below 30 K for BSCCO-2223, and 4.2 K for BSCCO-2212.

YBCO-123 is also used in SMES, and it may be that it can operate at a higher temperature than BSCCO-2223. In recent years, the properties of 2G HTS wires have been improved significantly. However, the price becomes a severe problem for massive applications of 2G HTS wires. To lower the cost of wire production, several key problems and the corresponding approaches are under investigation.

MgB_2 and iron-based superconductors [9] are expected to be cheaper superconducting wires compared with BSCCO and YBCO, but a critical temperature lower than 40 K may be a problem for decreasing the operating costs.

Fortunately, developing spaces for HTS superconducting wires exists both in the technology of manufacture wires and the demand from the grid. For REBCO and MgB_2 wires performances are continuously being upgraded. Also, the demand for energy storage with a fast response speed is becoming more urgent in the power grid. Together with the progress on HTS wires, cryogenic techniques, and the techniques for increasing the reliability of SMES, superconducting magnetic energy storage is promising in the future electric power system and other application fields such as pulsed power generation.

References

[1] Muzzi, L., et al. (2011). Test results of a NbTi wire for the ITER poloidal field magnets: A validation of the 2-pinning components model. *IEEE Transactions on Applied Superconductivity* 21 (3): 2132–2137.

[2] Kim, H.J., et al. (2006). 3 MJ/750kVA SMES system for improving power quality. *IEEE Transactions on Applied Superconductivity* 16 (2): 574–577.

[3] Luongo, C.A., et al. (2003). A 100 MJ SMES demonstration at FSU-CAPS. *IEEE Transactions on Applied Superconductivity* 13 (2): 1800–1805.

[4] Rogers, J.D., et al. (1983). 30-MJ superconducting magnet energy storage system for electric utility transmission stabilization. *Proceeding of the IEEE* 71 (9): 1099–1107.

[5] Zhang, P.X., et al. (2010). Fabrication and characterization of internal Sn and bronze-processed Nb_3Sn strands for ITER application. *Superconductor Science and Technology* 23 (2010): 034013.

[6] Parrell, J.A., et al. (2003). High field Nb_3Sn conductor development at Oxford Superconducting Technology. *IEEE Transactions on Applied Superconductivity* 13 (2): 3470–3473.

[7] Freda, R,. et al. (2015). Performance test of superconducting wires subject to heavy deformations. *IEEE Transactions on Applied Superconductivity* 25 (3): 4800904.

[8] Holik, E.F., et al. (2016). Fabrication and analysis of 150-mm-aperture Nb3Sn MQXF coils. *IEEE Transactions on Applied Superconductivity* 26 (4): 4000907.

[9] Ma, Y. (2015). Development of high-performance iron-based superconducting wires and tapes. *Physica C* 516 (2015): 17–26.

[10] Ayai, N., et al. (2006). Achievement of high-temperature superconducting wire with critical current exceeding 200 A. *SEI Technical Review* 63, December 2006: 58–64.

[11] Simitomo Electric. (2016). http://www.sei.co.jp/super/hts.

[12] Rupich, M.W., et al. (2013). Second generation wire development at AMSC. *IEEE Transactions on Applied Superconductivity* 23 (3): 6601205.

[13] Superpower. (2016). http://www.superpower-inc.com/content/2g-hts-wire.

[14] Fleshler, S., et al. (2014). Development status of AMSC Amperium® Wire, 11th European conference on applied superconductivity (EUCAS2013). *Journal of Physics: Conference Series* 507 (2014): 022005.

[15] Miller, A.T., et al. (1991). Micro-filamentary conductor based on Bi(2212). *IEEE Transactions On Magnetics* 27 (2): 909–911.

[16] Barzi, E., et al. (2006). High temperature superconductors for high field superconducting magnets. *Advances in Cryogenic Engineering* 52 (416): 2006.

[17] Huang, Y., et al. (2014). Bi-2212 round wire development for high field applications. *IEEE Transactions on Applied Superconductivity* 24 (3): 6400205.

[18] Byeon, J., et al. (2008). Bi-2212 structured cable for high-field solenoid inserts. *IEEE Transactions on Applied Superconductivity* 18 (2): 513–515.

[19] Shikimachi, K., et al. (2005). Development of MVA class HTS SMES system for bridging instantaneous voltage dips. *IEEE Transactions on Applied Superconductivity* 15 (2): 1931–1934.

[20] Sumiyoshi, F., et al. (2006). Inter-strand coupling losses in Bi-2212 Rutherford cables. *IEEE Transactions on Applied Superconductivity* 16 (2): 1055–1058.

[21] Zhang, J., et al. (2011). The electromagnetic analysis and structural design of a 1 MJ HTS magnet for SMES. *IEEE Transactions on Applied Superconductivity* 21 (3): 1344–1347.

[22] Marzi, G.D., et al. (2015). Characterization of the critical current capabilities of commercial REBCO coated conductors for an HTS cable-in-conduit conductor. *IEEE Transactions on Applied Superconductivity* 25 (3): 6602804.

[23] Celentano, G., et al. (2014). Design of an industrially feasible twisted-stack HTS cable-in-conduit conductor for fusion application. *IEEE Transactions on Applied Superconductivity* 24 (3): 4601805.

[24] Fleiter, J., et al. (2013). Electrical characterization of REBCO roebel cables. *Superconductor Science and Technology* 26 (2013): 065014.

[25] Tomsic, M., et al. (2007). Development of magnesium diboride (MgB_2) wires and magnets using in-situ strand fabrication method. *Physica C* 456 (2007): 203–208.

[26] Morandi, A., et al. (2016). Design and performance of a 1MW-5s high temperature superconductor magnetic energy storage system. *Superconductor Science and Technology* 29 (2016): 015014.

[27] Banno, N., et al. (2011). Fabrication of long-length Nb_3Al wire by the metastable solid-solution-strand restacking method. *IEEE Transactions on Applied Superconductivity* 21 (3): 2517–2520.

Further Reading

Kalsi, S. S. (2011). *Application of High Temperature Superconductors to Electric Power Equipment*. Hoboken, NJ: John Wiley & Sons, Inc.

Poole, C. P., Jr., Farach, H. A., Vreswick, R. J., et al. (2007). *Superconductivity*, 2e. San Diego: Elsevier Ltd.

Seeber B. (1998). *Handbook of Applied Superconductivity*. Bristol and Philadelphia, PA: Institute of Physics Publishing.

10

Cryogenic Technology

Li Ren, Ying Xu, and Yuejin Tang

State Key Laboratory of Advanced Electromagnetic Engineering and Technology, School of Electrical and Electronic Engineering, Huazhong University of Science and Technology, Wuhan, China

10.1 Introduction

The cryogenic system plays an important role in superconducting technology. The progress of cryogenic technology resulted in the discovery of superconductivity. In 1908, Onnes liquefied helium gas to obtain a 4.2 K temperature and then discovered the superconducting phenomenon. To date, the cryogenic temperature is not only an essential condition to maintain the superconducting state of superconductors, but it also determines technical performance, economic performance, security, and the stability of superconducting devices.

10.1.1 Function of Cryogenic for SMES

SMES is mainly composed of five parts: the magnet; data collection and the quench protection subsystem (DCQPS); the cryogenic subsystem (CS); the monitoring and control subsystem (MCS); and the power conditioning subsystem (PCS). Figure 10.1 shows the structure of a SMES.

The cryogenic subsystem keeps the superconducting magnet in a superconducting state and makes it work in a stable way at the operating temperature. The operating temperature is related to the system cost and the cooling efficiency. For low temperature superconducting magnets, liquid helium (4.2 K) is usually used as the cryogenic liquid. For large superconducting magnets, superfluid helium can be used to improve the cooling capacity and efficiency. For high temperature superconducting magnets, other cryogenic liquids or cryogenic gas can be used. However, for a small-scale superconducting magnet, cryo-cooler-conduction cooling is also a practical method.

10.1.2 Cool-down Method of Superconducting Magnets

There are two basic methods of cooling down a superconducting magnet: conduction cooling through a cryo-cooler or immersion cooling through cryogenic liquid or gas. The former uses solid thermal conduction by connecting the magnet directly to the cryo-cooler. The latter uses heat exchange of the magnet into a cryogenic medium through immersion, such as cryogenic liquid or gas. It is more difficult to

Advances in Energy Storage: Latest Developments from R&D to the Market, First Edition.
Edited by Andreas Hauer.
© 2022 John Wiley & Sons, Ltd. Published 2022 by John Wiley & Sons, Ltd.

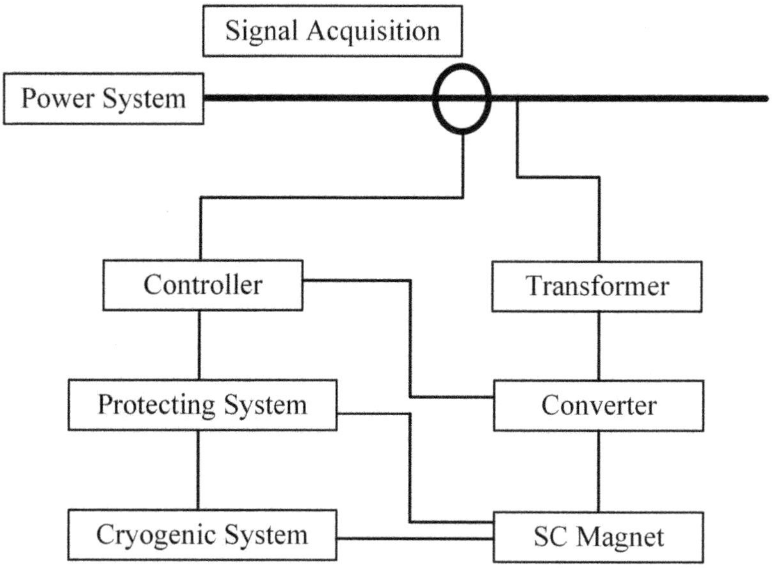

Figure 10.1 The basic structure of an SMES.

eliminating heat at lower temperatures than that at higher temperatures, so reducing the heat source and suppressing the thermal intrusion from the ambient surroundings are key issues to address to improve operating efficiency and reduce the operating cost.

10.2 Cryogens

Generally, gases satisfy the ideal state Equation (10.1):

$$PV = nRT \tag{10.1}$$

Where P is the pressure (Pa); V is the volume (m^3); n is the amount of substance (mol); R is the molar gas constant that equals 8.314 J mol^{-1} K^{-1}; T is the absolute temperature (K).

Lowering the temperature of the gas can turn it into a liquid state, and further cooling into a solid state. Conversely, heating can melt a solid, and then evaporate it. Matter exists in three states: solid, liquid, and gas. A phase diagram shows the relationship between the thermodynamic parameters to express different states of matter. Different thermodynamic parameters, produce different phase diagrams. Owing to more engagement with the physical parameters of pressure P and temperature T in daily life, the P-T phase diagram is most widely used and most intuitive. The P-T phase diagram of gas is shown in Figure 10.2.

In Figure 10.2 SB is the dividing line of liquid and gas, namely the vaporization line. SD is the dividing line between solid and liquid, namely the melting line. SE is the demarctionline of solid sublimation to gas. The intersection point S is a three

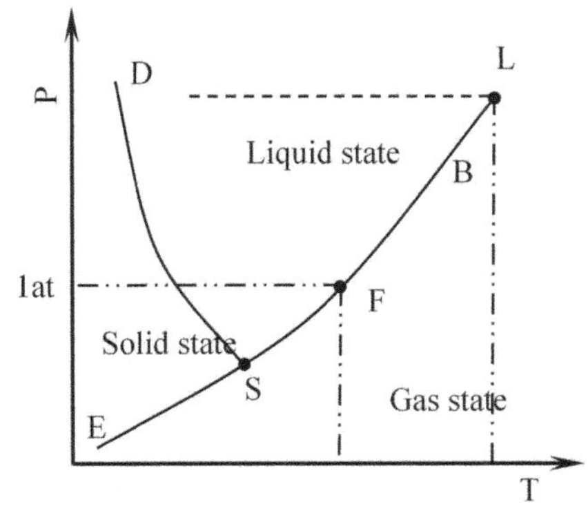

Figure 10.2 P-T phase diagram of gas.

phase point, which corresponds to definite pressure and temperature and is the only state of coexistence of gas, liquid, and solid. Any point on the three lines means a balance of two states, which is called saturation. In Figure 10.2 point L is called the critical point, and the gas cannot be in a liquid or gas state after the pressure reaches a certain value.

The heat that is absorbed or released in the process of a phase transition, with the same temperature, is called latent heat. Latent heat during the different phase change procedures has different names. From solid to liquid, the latent heat is called melting latent heat. When a cryogenic media converts to gas by boiling or evaporation, it will absorb a certain quantity of heat, which is called vaporization heat or latent heat of vaporization. The method of using cryogenic media for cooling an object is to use a liquid's latent heat of vaporization. The more latent heat, the less amount of liquid required to cool the object.

10.2.1 Cryogenic Media

The boiling point of cryogenic media used for cooling superconducting devices must be below the critical temperature of superconducting material (the safety margin also needs to be considered). At 1 atm, there are some gases that have a boiling point below 100 K, such as He, H_2, Ne, N_2, CO, F_2, Ar, O_2. Table 10.1 shows the boiling points and freezing points of these gases. The critical temperatures of NbTi and Nb_3Sn are between 10 K to 20 K, while the critical temperatures of Bi and Y superconductors are between 80 K to 110 K. Therefore, the cryogenic media that can be used in superconducting devices are mainly He, H_2, Ne, N_2. Their basic physical parameters are shown in Table 10.2. The insulation strength is related to the voltage, the electrode, and the gap distance, and it is generally non-linear. So the insulation strength in Table 10.2 is an approximate range.

H_2 is divided to n-H_2 and p-H_2. The spin directions of two atomic nucleuses of n-H_2 are the same, but the spin directions are opposite for p-H_2. At room temperature, 75% of H_2 are n-H_2, and only 25% of H_2 are p-H_2. However, 99.79% of H_2 are p-H_2 for liquid H_2 in 1 atm. Tables 10.1 and 10.2 mainly take account of the data of p-H_2. Considering

Table 10.1 The boiling point and freezing point of different gases (1 atm).

Type	^{3}He	^{4}He	H_2	Ne	N_2	CO	F_2	Ar	O_2
Boiling point (K)	3.19	4.22	20.28	27.09	77.35	81.62	85.24	87.29	90.19
Freezing point (K)	-	-	13.81	24.57	63.15	69.1	55.2	63.85	54.36

Table 10.2 Basic physical parameters of several cryogenic media.

Physical parameters		Unit	^{4}He	H_2	Ne	N_2
Molecular weight			4.003	2.016	20.18	28.01
Gas constant R		J kg^{-1} K^{-1}	2077.2	4124.6	441.95	296.81
Boiling point (1 atm)	Temperature	K	4.224	20.28	27.09	77.35
	Liquid density	kg m^{-3}	124.98	70.792	1204.8	808.61
	Gas density	kg m^{-3}	16.89	13.4	9.584	4.612
	Vaporization heat	kJ kg^{-1}	20.416	443.0	86.98	198.64
	Specific heat capacity of liquid	kJ kg^{-1} K^{-1}	2.37	9.66		2.063
	Thermal conductivity of liquid	mW m^{-1} K^{-1}	186.6	118.5	113.2	148.5
	Dielectric constant		1.047	1.231	1.188	1.431
	Insulation strength	kV cm^{-1}	~300	~700		~800
Critical point	Temperature	K	5.201	32.9	44.40	126.20
	Pressure	MPa	0.227	1.28	2.653	3.40
Three-phase point	Temperature	K	2.177	13.81	24.55	63.15
	Pressure	kPa	5.04	7.04	43.35	12.53
Gas density (273 K, 1 atm)		kg m^{-3}	0.179	0.09		1.25
Sensible heat (boiling point to 300 K)		kJ kg^{-1}	4010	3510	282.4	233.9

safety and economy, the cryogenic media usually used in superconducting devices are mainly liquid He and liquid N_2.

10.2.2 Helium (He)

Helium is an inert gas with two kinds of isotopes. One is ^{3}He and its relative atomic mass is 3. The other one is ^{4}He whose relative atomic mass is 4. The proportion of ^{3}He obtained from separating natural gas is very small, so we generally refer to ^{4}He as He.

We can get solid He only under the high pressure of 25 atm. In Figures 10.2 and 10.3, the point that shows 2.17 K on the saturation line is called the λ point. At a temperature below λ point, He is called He-II, At a temperature beyond the λ point, He is called He-I. These two kinds of He have different characteristics. The specific heat capacity is not continuous near 2.17 K. Based on the λ point, we can get two curves of specific heat capacity. It is why we call 2.17 K the λ point.

One of the properties of He-II is the superfluidity. He-II is not sticky, and it can still flow in capillaries under no pressure difference. He-II can overcome gravity and extend to the top in the form of a film. Thus, it is very difficult to use a vacuum pump to reduce the steam pressure in the container because the He-II film will flow along the vessel vacuum tube to reach the high temperature region and to be evaporated with the increasing steam pressure.

Another characteristic of He-II is its heat transfer characteristics. Thermal conductivity of He-II is very high, higher than that of copper. If there is a heat pulse, heat will spread in the form of a wave in He II, and He II will immediately restore to its original temperature, which is similar to the spread of sound waves and acoustics. So He-II is also known as the second sound wave.

He-II also has the characteristics of superfluidity, a large specific heat capacity, and a strong cooling capacity. Large superconducting magnets often need forced cooling. Hence, superfluid helium is a powerful candidate.

10.2.3 Nitrogen (N_2)

Liquid nitrogen is a transparent, odorless, tasteless liquid. Figure 10.3 shows the P-T phase diagram of nitrogen. Its boiling point is 77 K at 1 atm.

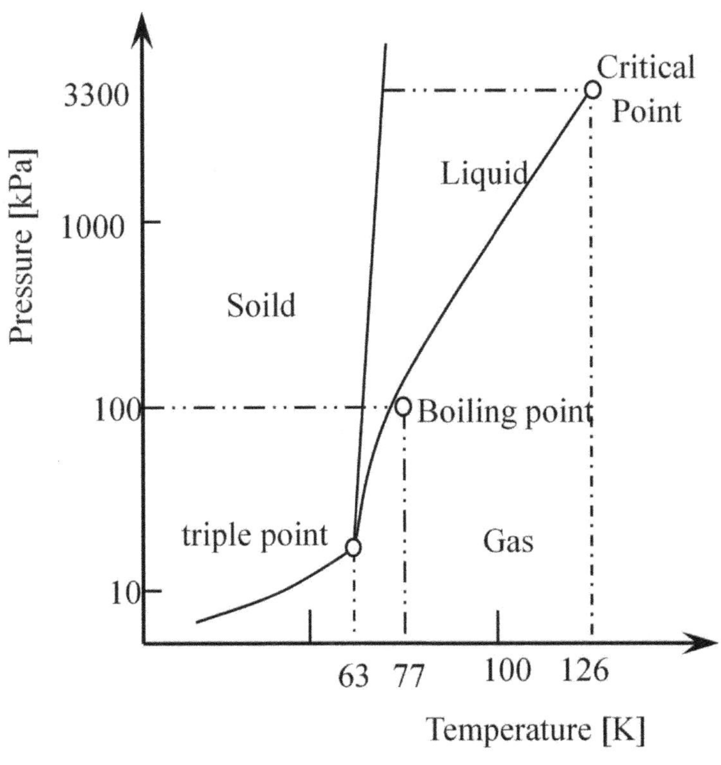

Figure 10.3 P-T phase diagram of nitrogen.

Liquid nitrogen is widely used in superconducting technology, in three main ways:

(1) For high temperature superconductors with critical temperatures above 77 K, liquid nitrogen can be used as a cryogenic media.
(2) Generally, liquid helium is 10 to 20 times the price of liquid nitrogen. Owing to the cheaper price and much larger latent heat than He, liquid nitrogen is usually used to precool the object for saving costs when the object needs to be cooled by liquid helium. Precooling with liquid nitrogen to 77 K can complete more than 90% of the cooling task. In practical applications, precooling by liquid nitrogen can also provide an opportunity to investigate the reliability of the devices. In general, more than 90% of cooling shrinkage of an object is completed above 77 K, so precooling by liquid nitrogen can test the ability of the superconducting devices to withstand cooling shrinkage stress.
(3) The third application is to provide a heat shielding layer for liquid helium, namely forming a shielding layer around 80 K to prevent direct invasion of heat from 300 K to 4.2 K. In order to obtain the same amount of cooling capacity, the cost of energy at liquid helium temperature is much larger than that at liquid nitrogen temperature, so using a shielding layer can greatly improve the energy efficiency of the cryogenic system. At the same time, liquid nitrogen is much cheaper than liquid helium, so using liquid nitrogen to prevent heat from invading the liquid helium region can also reduce the operating costs of the cryogenic system.

10.3 Cryo-cooler

State change of matter under certain conditions can lead to temperature changes. For the cryocooler, the fundamental principle of obtaining a low temperature is how to make use of thermodynamic cycle process that makes working fluids absorb heat constantly through circulation state changes from the object to be cooled. Cooling with gas as the working fluid is achieved by following three processes: (i) compressing the gas and removing heat, (ii) utilizing the gas expansion to absorb heat, and (iii) isolating the compression unit and expansion unit through a heat exchanger or a regenerator and separating heat. It is available to different thermodynamic cycles, such as isentropic expansion and isenthalpic expansion by designing different cycling conditions and processes in the compression and expansion links. Cryocoolers with different principles can be manufactured by using different thermodynamic cycles.

Currently, the cryocoolers that can be used in superconducting power applications are the Stirling refrigerator, the Gifford-McMahon (GM) refrigerator, and the pulse tube refrigerator.

10.3.1 Stirling Refrigerator

The Stirling refrigerator has a compact structure and a large cooling capacity at liquid nitrogen temperature. The Stirling refrigerator is driven by a crank linkage and has the advantages of high efficiency, fast cooling rate, compact structure, and a wide range of cooling temperature, etc. The refrigeration power range can reach the kilowatt range, but the lubricating oil that is required in moving parts would pollute the working fluid

Table 10.3 Hectowatt/kilowatt Stirling refrigerator at the liquid nitrogen temperature.

Product type	Input power (kW)	Weight (kg)	Refrigerating capacity	Relative carnot efficiency (%)
Stirling C&R LPC-01	12	850	500 W @ 65 K	15
Stirling C&R LPC-02	25	1500	1300 W @ 65 K	18
Stirling C&R LPC-04	60	3750	2800 W @ 65 K	17
Stirling C&R LPC-08	135	7500	5800 W @ 65 K	16
Stirling C&R SPC-01	11	600	1050 W @ 80 K	26
Stirling C&R SPC-04	45	1255	1500 W @ 80 K	27
GEA (Germany)	2.6	70	150 W @ 77 K	17
			235 W @ 109 K	16
Aisin Seki Model SC	14	–	1000 W @ 77 K	20
Aisin Seki Model SS	6	–	400 W @ 77 K	20

and affect the cooling performance. A free-piston Stirling refrigerator avoids excessive vibration and working. However, the technology is still immature and the cost is high. Therefore, the practical application is still limited when a large cooling capacity is needed. So far, the crank rod driven Stirling refrigerator has a good development prospect in applications requiring a large cooling capacity because of its outstanding cooling performance and manufacturing cost advantages. Table 10.3 shows the current parameter hectowatt/kilowatt levels of the main products in the Stirling refrigerator at liquid nitrogen temperature.

10.3.2 GM Refrigerator

The characteristic of the GM refrigerator is the separation of the cold head and compressor, the low speed of the cold head inside the moving parts, small vibration, easy operation, long service life, and so on. The working temperature of the GM refrigerator ranges from liquid helium temperature to room temperature. If there is no special requirement for weight and size, the GM refrigerator can be applied widely in space, materials, electronics, communications, and medical application, and so on. It has a unique advantage in the temperature region at 20 K with a large cooling capacity. Now, single-stage, two-stage, and three-stage GM refrigerators have been developed.

According to the operating temperature of the GM refrigerator, it can be divided into three categories:

(1) Liquid helium temperature (@4.2 K): The cooling capacity is 0.5 W and cooling time is generally less than 60 minutes. This kind of GM refrigerator is mainly used

for small superconducting magnets and can also be used for cooling superconducting electronic devices and the recondensation of evaporated liquid helium in large magnet systems. In addition, it can be used in ultrahigh vacuum condensing pumps, small helium making machines, superconducting magnet cooling of Magnetic Resonance Imaging (MRI), and cryogenic sources in laboratories.

(2) Liquid hydrogen temperature (@10–20 K): The cooling capacity is 0.5 W–40 W, and cooling time is generally less than 60 minutes. This kind of GM refrigerator is mainly used as a cold-background cryogenic source of large space environmental simulation equipment in the aerospace sector. It can also be used for, for example, cooling large or medium high temperature superconducting (HTS) magnets and motors; the cryogenic system of high temperature superconducting Maglev trains; a cryogenic source of large and medium-sized pumps; a cooling high energy physics research liquid hydrogen target system, and large high temperature superconducting filter. All these applications require the refrigerator to provide a larger amount of cooling.

(3) Liquid nitrogen temperature (@77 K): The cooling capacity is 5 W–320 W and the cooling time is generally 15–20 minutes. This kind of GM refrigerator is mainly used for precooling of large- and medium-sized cryogenic systems. It can also be used for cooling a superconducting quantum interference device and small liquid - nitrogen-making machine, etc. In the mid to late 1980s, the discovery of magnetic cold accumulating materials of high specific heat in the liquid helium temperature region resulted in the development of the GM refrigerator for the region of temperature for liquid helium. Now, the research of the GM refrigerator for the region of temperature for liquid helium has been divided into two categories: one is to further improve the refrigeration efficiency and capacity, the other is to expand the application fields.

10.3.3 Pulse Tube Refrigerator

The pulse tube refrigerator was first proposed and developed by Gifford and McMahon in 1963. The pulse tube refrigerator has the characteristics of no moving parts in the low temperature region, a simple structure, and reliable operation. But the refrigeration efficiency is not high and it has a small cooling capacity. According to the differences of driven methods of pulse tube refrigerator, it can be divided into four categories: Stirling type, GM type, thermo-acoustic driven type, and thermo–acoustic Stirling heat type. Depending on the different phase modulation methods, it can be divided into several categories.

The Stirling type pulse tube refrigerator is usually referred to as a non-valve pulse tube refrigerator, and it is driven by a non-valve compressor, and the compressor is directly connected to the high-temperature end of the regenerator. Its operating frequency is determined by the compressor, and it is usually between 30 and 120 Hz.

The GM type pulse tube refrigerator uses an electromagnetic valve or rotary valve connecting the refrigerator and compressor; its gas distribution system is similar to that of the GM refrigerator. The GM type pulse tube refrigerator's operating frequency is determined by the switching speed of the rotary valve, which depends on the frequency controller. Generally, its working frequency is 1–2 Hz, so it is also called the low-frequency pulse tube refrigerator. It is easy for the GM type pulse tube refrigerator to obtain a large refrigerating capacity in very low temperatures, and it has important applications in: the semiconductor industry, medical equipment, and scientific research. Currently, the GM type pulse tube refrigerator has achieved commercial applications, but it remains to be improved in terms of refrigeration efficiency and reliability.

The thermo-acoustic driven pulse tube refrigerator is a type of driving pulse tube refrigerator using a thermo-acoustic engine instead of the traditional mechanical compressor. The thermo-acoustic driven refrigerator has the following two prominent advantages: 1) the whole machine has no moving parts, completely eliminating mechanical wear and tear problems, and 2) a thermo-acoustic engine that can be driven by heat energy, such as solar energy and gas, and so on, especially since using low-grade thermal energy can not only improve the thermodynamic efficiency of the whole system, but it also has more practical significance for the occasional lack of electric energy.

10.3.4 Development Trends

According to the current technical level, a Stirling refrigerator is suitable for a large cooling capacity and high temperature area, such as liquid nitrogen. The GM refrigerator is suitable for the temperature range required from liquid helium to liquid nitrogen, and the pulse tube refrigerator can be used in small devices, such as superconducting devices in applications of superconducting electronics. Magnetic annealing cooling, dilution refrigeration obtaining temperature at mK or less, with μW cooling capacity, can be used for superconducting property research, but has not reached the level of application in superconducting power devices. For a large superconducting device, a large cooling capacity is required, and a single refrigerator cannot meet the requirements, and although more than one refrigerator to cool could be used, using a large cryogenic liquefier, helium liquefier, and nitrogen liquefier is more reasonable.

A refrigerator can use a multi-stage refrigeration mode to achieve lower temperatures; for example a single-stage Stirling refrigerator's temperature range is 173 ~ 73 K, two-stage is 12 ~ 15 K, and three-stage is 7 K. Multi-stage refrigeration adopts a different refrigeration principle, a large helium liquefier is generally a multi-level integration of the multiple refrigeration principle.

Refrigerator technology is still rapidly developing. For example, the Q drive company claims that its 2S262 K Stirling refrigerator has reached 1 kW/77 K; GM refrigerator's maximum cooling capacity at 77 K was only tens of watts in 2005, but there has been a refrigerator commodity of 280 W/80 K in 2011; In 2003, the parameter of the pulse tube refrigerator was still mW/77 K, but there has been a 10 W/77 K commercial product (Sumitomo) in 2011.

10.4 Cryogenic System

10.4.1 Cryogenic System of Large-scale Magnet

It is difficult to use the described simple ways to cool large superconducting equipment, or the superconducting devices need long running time and it is also not economical that open-cooling loses a lot of low-temperature liquid. It is necessary to adopt

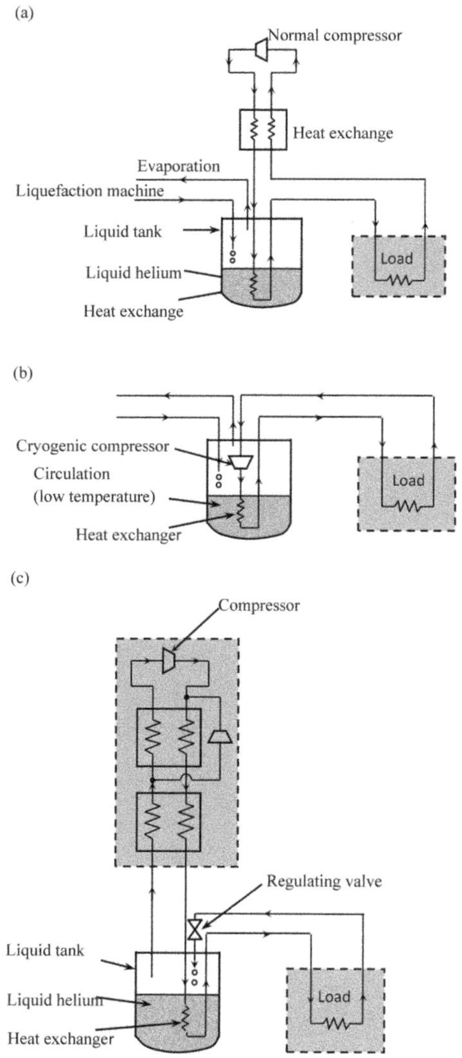

Figure 10.4 Three ways of supercritical forced cooling.
(a) Normal compressor, subject to liquefaction machine. (b) Cryogenic compressor, subject to liquefaction machine. (c) Owned liquefaction system.

a closed-cycle cooling system in large installations. Figure 10.4(a) is a cooling system of a large superconducting device cooled by liquid helium, which consists of a liquefaction machine, liquid dewar, and a boil-off gas recovery system. The cooling process is divided into four steps:

(1) Cool from room temperature to the vicinity of 80 K.The device is slowly cooled to about 80 K, transporting helium into it through heat exchange by a liquid helium compressor without an expander.
(2) Start the expander, and inject liquid helium into the liquid Dewar.
(3) Cool from 80 K to about 10 K. The liquefaction machine used as a refrigerator injects cooling helium into the device and cools the device to about 10 K. In this

case, if the cooling capacity is too drastic, it may cause uneven cooling of the internal device, and produce thermal stress resulting in local damage of the device. Slow cooling homogeneity should be considered.

(4) When the device is cooled to about 10 K, start injecting liquid helium from the upper dewar flask into the device, and the evaporated helium is processed by liquefaction or the gas recovery system. Initially, the liquid helium should be added from the bottom of the superconducting magnet, and it can be added from the top when the magnet's temperature is substantially uniform.

10.4.2 Forced Cooling by Supercritical Helium

In some cases, it is necessary to use forced circulation cooling, such as the common hollow conductor of a large magnet, which requires a cooling medium passing the conductor inside, then, uses pumps and compressors to force the circulation of cooling medium. In the case of helium refrigerant, pressure, and temperature changes may lead to a phase change during the cycle, thereby causing a large pressure loss, affecting the cooling effect. Therefore, helium forced circulation generally uses supercritical helium, and there are usually three ways as shown in Figure 10.4.

In Figure 10.4(a), the cooling medium's temperature increases to room temperature after a through load, then make sit into the heat exchanger by compression to cool again in a room temperature zone. The advantage of this approach is that the compressor works at room temperature, and has easy maintenance and management, but the thermal efficiency is not high. Because the density of room temperature and the cryogenic cooling medium vary widely, it requires a large capacity compressor under normal temperatures to maintain a certain cold flow. Figure 10.4(b) is similar to (a), but we use a cryogenic compressor instead of the cooling medium reverting to normal. The configurations shown in Figures 10.4(a) and 10.4(b) need an additional liquefaction machine that provides supercritical liquid helium. Figure 10.4(c) uses a liquefaction system so that the compressor of the liquefaction machine provides supercritical helium with the pressure of forced circulation, and cryogenic liquid helium produced by the JT expansion makes supercritical helium cool down, which forces circulation to bypass the heat exchanger. In this way, the pressure of the supercritical helium must be under the inverted curve of helium.

In closed-cycle cooling, once the circulation failure has appeared the temperature of the load will increase. Reliability of the pumps and compressors is very important, and they must be equipped with adequate safeguards.

10.4.3 Conduction-cooled Method

With the advance in cryogenic technology, you can use the refrigerator to cool the object directly by conduction cooling down to 77 K, 40 K, 20 K, and 4.2K. However, the cooling capacity of the refrigerator at temperatures below 77 K is not high. The refrigerator itself is relatively expensive, so, the conduction-cooled method is generally used for small superconducting devices. However, this method is beneficial for simplifying the operation and enhancing reliability because there is no need for a cooling liquid. Hence, it is an important direction in the development of superconducting cooling technology.

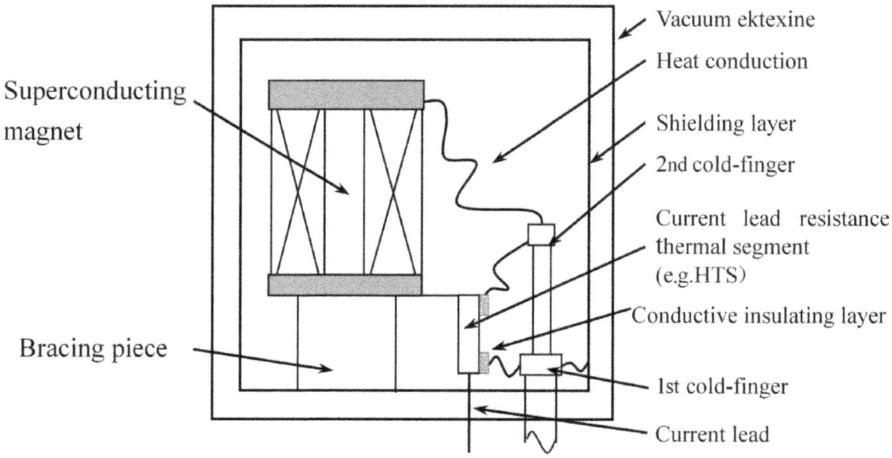

Figure 10.5 Structure principle of indirect cooling magnet cryogenic system.

Figure 10.5 is a schematic diagram of a superconducting magnet conduction-cooled by two cryo-coolers. The magnet is cooled by a single-stage GM cryocooler 1# while the radiating shield and the hybrid current leads are cooled by a two-stage GM cryocooler 2#.

10.5 Vacuum Technology

Vacuum technology is usually used in superconducting and cryogenic devices. Vacuum is measured in units of pressure, and the SI unit of pressure is the Pascal (Pa). Additionally, vacuum is often measured in atm (atmospheric pressure), τ (torr, named after Torricelli), and mmHg. A torr is equal to the displacement of a millimeter of mercury (mmHg) in a manometer with 1 torr equaling 133.3223684 Pa above absolute zero pressure. Hence the relationship between them is: $1\ \tau = 1$ mmHg $\approx 1.33 \times 10^2$ Pa. And 1 atm is equal to 760 mmHg in a manometer. The degree of vacuum is represented in Table 10.4.

10.5.1 Vacuum Pump

We can obtain the vacuum by using a vacuum pump. There are many types of vacuum pump. The mechanical pump and diffusion pump are commonly used. The diffusion

Table 10.4 Pressure ranges of each quality of vacuum in different units.

The levels of vacuum	Vacuum		
Low vacuum	10^5~10^3 Pa	760~10 mmHg	760~10 τ
Middle vacuum	10^3~10^{-1} Pa	10~10^{-3} mmHg	10~10^{-3} τ
High vacuum	10^{-1}~10^{-5} Pa	10^{-3}~10^{-7} mmHg	10^{-3}~10^{-7} τ
Ultra high vacuum	10^{-5}~10^{-9} Pa	10^{-7}~10^{-11} mmHg	10^{-7}~10^{-11} τ
Extreme high vacuum	<10^{-9} Pa	<10^{-11} mmHg	<10–11τ

pump can reach a higher vacuum degree than a mechanical pump, but it cannot be used alone because it needs a pre-stage vacuum upon reaching 1 mmHg. A good pre-vacuum is necessary for a diffusion pump to obtain a high vacuum. There are some other pumps that can achieve a high vacuum, such as a molecular pump, a condensate pump, and an ion pump. The degrees of vacuum that these pumps can reach are shown in Table 10.5.

The vacuum suction rate is tightly related to the form of the exhaust pipe. Supposing Q is gas flow rate, $(P_2 - P_1)$ is the pressure difference of the pipe end, the flow capacity of the pipe which is denoted by c is defined as:

$$C = \frac{Q}{P_2 - P_1} \tag{10.2}$$

The conductance is proportional to the radius of the conduit under high vacuum conditions ($c \propto r$), so coarse vacuum piping should be selected.

Table 10.5 Classification and working ranges of different vacuum pumps.

Types of pump			**Working pressure range(mmHg)**
Mechanical vacuum pump	Oil-sealed vacuum pump	Rotary vane vacuum pump	$760\sim10^{-2}$ (Single-stage), $760\sim10^{-4}$ (Double-stage)
		Piston vacuum pumps	$760\sim10^{-2}$ (Single-stage), $760\sim10^{-4}$ (Double-stage)
		Stator pump	$760\sim10^{-2}$
	Reciprocating pumps		$760\sim3$, $760\sim10^{-2}$ (Double-stage)
	Turbomolecular pump		$10^{-2}\sim10^{-10}$
	Mechanical booster		$30\sim10^{-4}$
	Water ring vacuum pump		$760\sim100$
Vapor stream vacuum pump	Diffusion pump	Oil diffusion pump	$10^{-3}\sim10^{-7}$, $10^{-3}\sim10^{-4}$ (with water-cooled baffle)
		Mercury diffusion pump	$10^{-4}\sim10^{-7}$
	Jet pump	Steam jet pump	$760\sim100$ (Single-stage), $200\sim20$ (Double-stage)
		Oil booster	$1\sim10^{-4}$
	Molecular sieve adsorption pump		$10^{-2}\sim10^{-3}$ (Single-stage), 10^{-4} (Double-stage)
	Sublimation pump		$10^{-3}\sim10^{-11}$
	Getter ion pump	Ballistic titanium pump	$10^{-4}\sim10^{-11}$
		Sputter ion pump	$10^{-2}\sim10^{-12}$
	Condensate pump		$10^{-3}\sim10^{-12}$

10.5.2 Measurement of Vacuum

In superconducting devices, the vacuum uses the basic adiabatic method. The degree of vacuum is associated with the capability of preventing the heat from invading superconducting devices, which will affect the efficiency and reliability of the superconducting device. It is very important to ensure the safety of superconducting devices by monitoring the degree of vacuum accurately.

According to the principle of measuring vacuum, there are two kinds of vacuum gauges. The one is called "absolute vacuum gauge," which means that it can measure the pressure of gas directly, such as the U-shaped tube vacuum gauge, which measures the displacement of liquid level in the U-shaped tube and the elastic deformation vacuum gauge, which measures the elastic deformation of the elastic element under pressure. The other is called "relative vacuum gauge," which means we can use some physical parameters or physical phenomena that are related to the pressure of gas to calculate the pressure of the gas. For example, according to the phenomenon that heat conduction corresponds to stress in low pressure, we can make a resistive thermocouple vacuum gauge and vacuum gauge; according to the phenomenon that low pressure gas heat radiation corresponds to stress in low pressure, we can make a heat radiation gauge; according to the phenomenon that the ionization current that corresponds to the density of particles produced when gas molecules collide with energetic molecules, we can make an ionization gauge, and so on.

In general, the measurement of the vacuum in superconducting applications requires relative vacuum gauges. Usually the relative vacuum gauge consists of a vacuum gauge sensor, and measurement components for controlling and display. A high-vacuum gauge is only applicable in a high-vacuum range, and it will be damaged in a low-vacuum range. Therefore, the measurement of a vacuum generally requires a different vacuum gauge in different vacuum ranges, such as using a pointer vacuum gauge (elastic deformation vacuum gauge) at a low-vacuum range, then when the vacuum range is reaches the high-vacuum range, an ionization vacuum gauge may be needed to begin to measure the degree of vacuum.

There are many types of ionization vacuum gauge, such as the cold-cathode type, hot-cathode magnetron type, and ionization type. The vacuum range for the heat insulation in superconducting devices is generally between 10^{-2} and 10^{-5} Pa. The ionization gauge has a wide measuring range, it is commonly used in superconducting applications.

10.6 An Evaluation Method for Conduction-cooled SMES Cryogenic Cooling Systems

An SMES system consists of a superconducting magnet, a cryogenic cooling system, and a power conditioning system. The cryogenic cooling system is of great importance to the safe operation of SMES. Recent developments and advances in cryogenic technology have made the conduction-cooled method widely applied to HTS SMES cryogenic cooling systems. The conduction-cooled method is an appropriate cooling method for countries and regions where is difficult to obtain helium. The following are the requirements for the SMES cryogenic cooling system: (i) The system can operate safely at the target temperature; (ii) When SMES exchanges power with the power grid, the temperature rise is acceptable and the heat can be effectively taken away; (iii) The cryogenic cooling system is cost-effective. In order to achieve the three requirements simultaneously, the suitability of the capacity of cryogenic cooling system should be estimated. Due to current and magnetic field variation during the operation, the AC losses may cause heat generation in the SMES magnet. Therefore, the power exchange processes should be simulated to make sure that the capacity of the cryogenic cooling system could meet these requirements. For different applications, the requirements of the current rate and the duration of power change can be very different.

10.6.1 Definition of Factor δ

As mentioned in Section 10.1, AC losses including eddy current losses may cause heat generation in the SMES magnet during operation. After obtaining the temperature characteristic of the SMES by thermal analysis, some quantitative parameters should be used to evaluate the capabilities of the cryogenic cooling system. A lot of field tests concerning SMES have been carried out to select quantifiable parameters. Based on the temperature variation when an SMES exchanges power with the grid, two parameters are selected: ΔT_{temp} and Δt. ΔT_{temp} is the temperature rise when the power exchange finishes. Δt is the required recovery time that cooling SMES to the steady state. When SMES is connected to the power grid, the magnet is excited to an operating current, and then the SMES is in standby state. At this moment, according to the orders, the SMES can release its energy to the power grid. The operating current is called a steady-state current and is represented by I_0. We selected three parameters to describe the characteristics of the current. The schematic diagram of the current and the temperature variation in SMES is shown in Figure 10.6. I_0 and I_d are steady-state current and dynamic current, respectively. D is the discharging time.

GM cryo-coolers have been widely used for conduction-cooling systems to cool the magnet, the current leads, and the radiation shield. In order to assess the technological and economic validity of the chosen cryogenic scheme, a new technical and economical factor δ is defined. The following paragraph will detail the factor δ of the cryo-cooler.

When the SMES exchanges power with the grid, the temperature of the magnet is not supposed to rise too much, which means the lower ΔT_{temp} is the better. When the SMES finishes the power exchange, the magnet is expected to be cooled down as quickly as possible. This means the smaller Δt is the better. Therefore, ΔT_{temp} and Δt are used to show the technical performance indicators of the cryogenic cooling system. In addition, the cost of the cryogenic cooling system is regarded as an economical evaluation standard. The cheaper the cryogenic cooling system is, the better. Based on the above consideration, we define the technical and economical factor δ of the cryocooler by:

$$\delta = \frac{1}{\Delta t \cdot \Delta T_{\text{temp}}} \cdot \frac{1}{f(P_{\text{M}})} \tag{10.3}$$

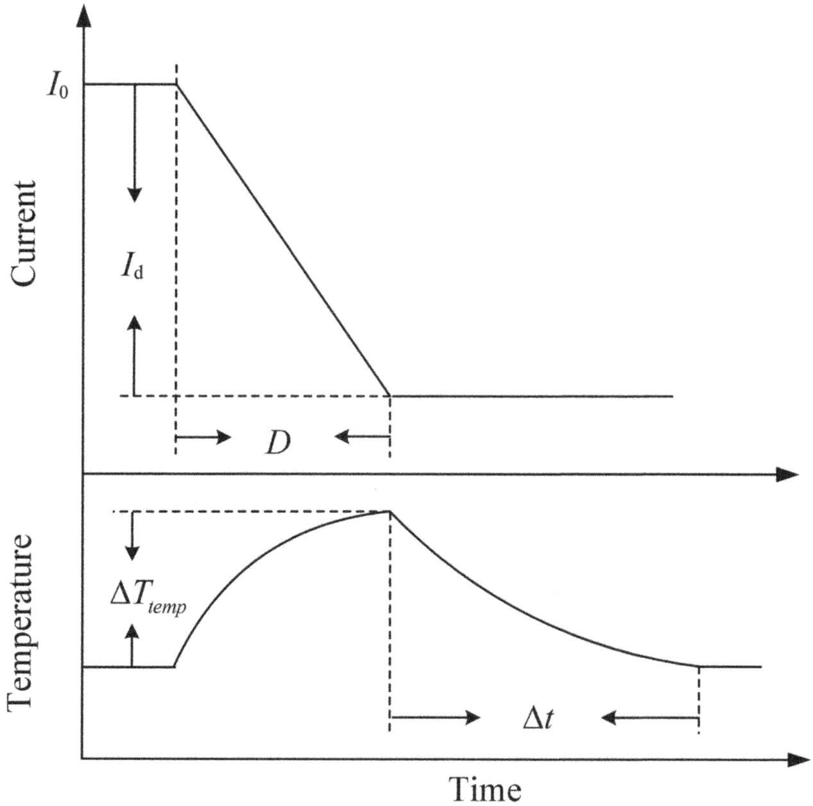

Figure 10.6 Schematic diagram of the current and temperature variation in SMES.

where P_M is the cooling power of the cryocooler at 20 K. $f(PM)$ is the cost of the cryo-cooler. The greater the value of δ is, the more technical and more economical the cryo-cooler. To simplify the correlation between the cost and the cooling power, it is assumed that the cost of a cryo-cooler is of linear correlation with the cooling power. $f(PM)$ can be approximated by $k_1P_M+H\cdot e\cdot k_2P_M+M$, where k_1P_M and $H\cdot e\cdot k_2P_M$ are the cost and the energy charge of cryocoolers, respectively. k_1 is 714 US\$ kW^{-1}. H is the operating life of cryo-coolers, which is supposed to be 15 (years) × 300 (days) × 24 (hours). e is the price of one kilowatt-hour electricity, which is 0.113 US\$ kWh^{-1}. k_2P_M is the input power of the cryo-cooler. It is assumed that the input power is 160 times greater than its cooling power. Since the unit of P_M is watt and the price of electricity is calculated as kilowatt, k_2 is 160 × 10^{-3}. M is the maintenance costs, which is US\$50 000.00 for 15 years.
In most cases, Equation (10.4) could meet the design requirements of the cryogenic cooling system. While in some particular cases, formula (4) is proposed as a supplementary to Equation (10.4).

$$\delta = 0 \text{ (If } \Delta t \text{ doesn't meet operating requirements)} \tag{10.4}$$

Compared with Δt, temperature rise ΔT_{temp} is not as sensitive to the cooling power. A more effective solution to bring down ΔT_{temp} is reducing the eddy current loss and the AC loss.

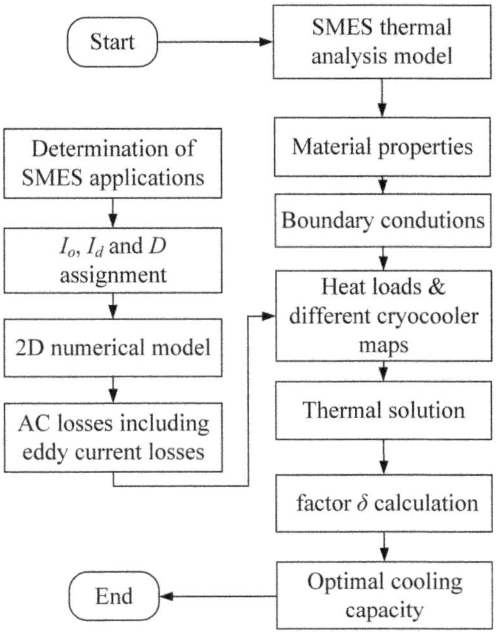

Figure 10.7 Flow chart of an evaluation of the technical and economical characteristics of a cryogenic cooling system.

10.6.2 Evaluation Procedure

The evaluation procedure of the technical and economical aspects of a cryogenic cooling system involves four parts: Firstly, a thermal analysis model is established based on the finite-element method. Secondly, the thermal material properties and boundary conditions are assigned to the SMES model. After determining the SMES applications, parameters I_o, I_d, and D are used to calculate the AC losses including eddy current losses by a 2D numerical model. Thirdly, the losses are transferred to the thermal analysis model as heat sources. Then, different cryocooler load maps are assigned to the model. Finally, we compare the simulation results and determine the most suitable cooling capacity. Figure 10.7 shows the flow chart of an evaluation of the technical and economical characteristics of a cryogenic cooling system.

10.7 Case Study

10.7.1 Circulating Liquid Helium Cooling System

A circulating liquid helium cooling system is shown in Figure 10.8 [1], where the cryogenic system is divided into two loops: primary and secondary loop. In the primary loop the heat load of the magnet is taken by the supercritical helium and is dumped in the LHe bath. The liquid helium from the first heat exchanger (HEX 1) flows through the cold compressor (CC) pump before going to HEX 2 and the flow of liquid helium is controlled by two valves.

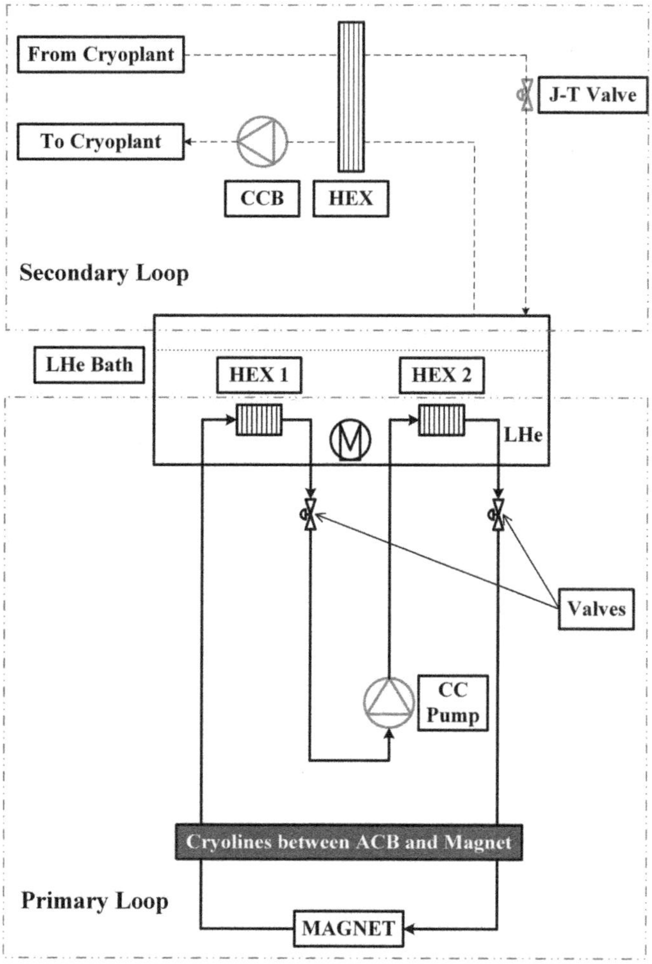

Figure 10.8 A circulating liquid helium cooling system. [1]

In the secondary loop the helium comes from the cryoplant and goes to the heat exchanger for pre-cooling and then through the J-T valve. The liquid helium gets stored in the LHe bath. The vapor generated inside the LHe bath goes to the heat exchanger where it is heated and then through the cold compressor, from where it goes back to the cryoplant. Then a new loop starts.

10.7.2 Cryo-cooler Cooled System

The conduction cooling scheme of the 150-kJ/100-kW SMES developed by the Huazhong University of Science and Technology is shown in Figure 10.9 [2]. It is cooled by conduction cooling using two GM cryocoolers. One is the AL325 (single stage) with a CP1110 compressor from Cryomech Inc., and the other CH-210 (two stages) with an F70 compressor from Sumitomo Electric Ind. Ltd. The AL325 GM cryocooler is designed to cool the magnet. Copper braided wires are used to connect the cold head of the cryo-cooler and the magnet, providing a flexible connection between the cryocooler and the magnet. The CH-210 GM cryocooler adopts the same connection method to cool the current lead and the radiation shield.

The HTS coil bobbins and the conduction cooling plates are made of brass and copper (OFHC), respectively. Two main conduction cooling plates are installed at the top and the bottom of the magnet. The thickness of the top one is 15 mm and the bottom one 10 mm. The top one is connected to the AL325 GM cryocooler by copper braided wires. In addition, there is a conduction cooling plate, whose thickness is 1 mm, between each two adjacent coils. The refrigeration capacity from the AL325 conducts through the main conduction cooling plates and 16 cooling rods made of copper (OFHC) to each conduction cooling plate between any two coils. The 16 cooling rods are installed along the circumference evenly to make a uniform distribution of temperature and improve the efficiency of cooling. Figure 10.10 shows the actual magnet.

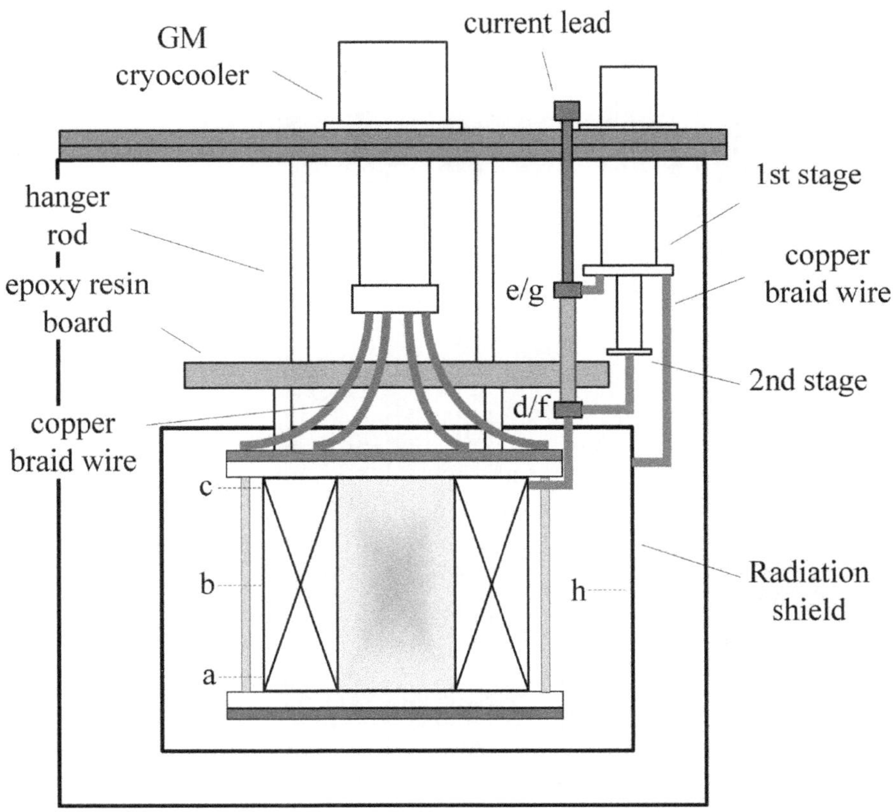

Figure 10.9 The conduction cooling scheme of the 150-kJ/100-kW SMES.[2]

10.7.3 Cryo-cooler and Liquid-nitrogen/Gas-helium Combined Cooling System

The Korean Electric Research Institute (KERI) has developed a cryo-cooler and liquid-nitrogen (LN_2)/gas-helium (GHe) combined cooling system of a 600 kJ HTS SMES [3]. The schematic diagram of the cooling system is shown in Figure 10.11. The system is composed of two GM cryo-coolers, a LN_2/GHe heat exchanger that provides a fast cool down, and 16 copper conduction bars (cooling plates).

The cooler's first stage is cooling the current leads and radiation shield through braided wire and the second stage is cooling the simulation coils through conduction bars. The LN_2/GHe heat exchanger provides a fast cool-down in the initial cooling. LN_2 or GHe flows in through the supply tube and cools the two semicircular copper blocks and flows out. This heat exchanger reduces the second stage heat load of the cryo-cooler in the initial cool-down.

Figure 10.10 The 150-kJ/100-kW SMES magnet.

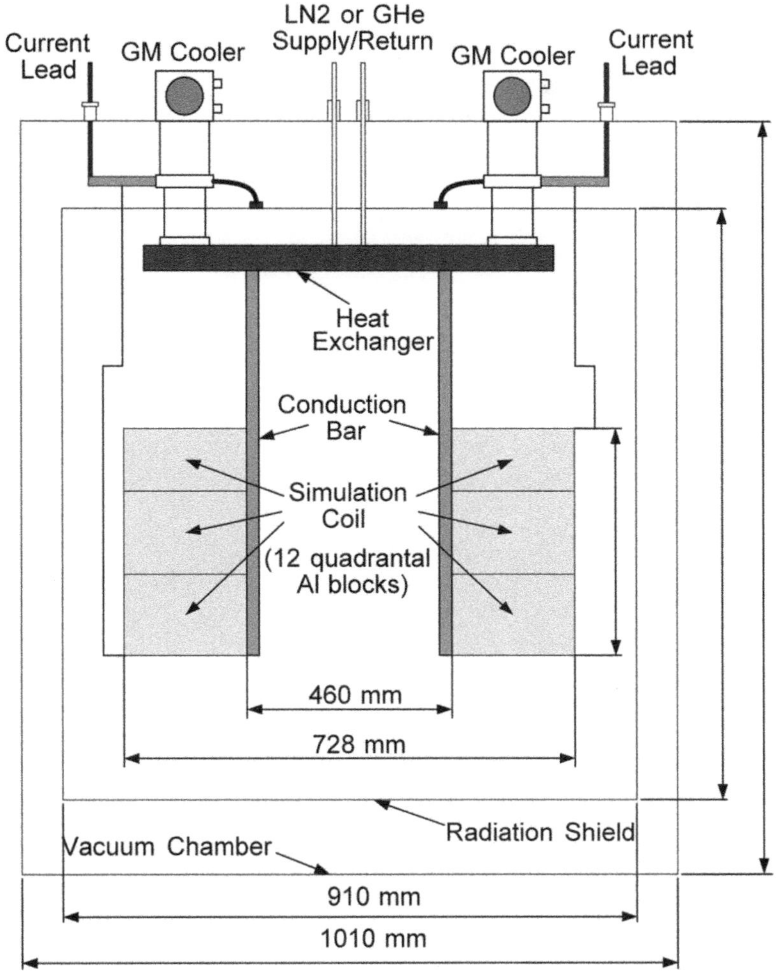

Figure 10.11 The schematic diagram of the KERI cooling system. [3]

References

[1] Vaghela, H., Sarkar, B., Bisht, A., et al. (2015). Thermal performance analysis of cryogenic system for cooling of superconducting magnets at 4K temperature level. *5th Nirma University International Conference on Engineering (NUiCONE).*

[2] Xu, Y., Tang, Y., Ren, L., et al. (2016). Numerical simulation and experimental validation of a cooling process in a 150-kJ SMES magnet. *IEEE Transactions on Applied Superconductivity* 26 (6): 5700507.

[3] Yeom, H.K., Park, S.J., Hong, Y.J., et al. (2008). An experimental study of the conduction cooling system for the 600 kJ HTS SMES. *IEEE Transactions on Applied Superconductivity* 18 (2): 741–744.

Further Reading

Flynn, T. (2005). *Cryogenic Engineering*. Boca Raton, FL: CRC Press.

Jousten, K. (2016). *Handbook of Vacuum Technology*. Weinheim, Germany Wiley-VCH.

Timmerhaus, K.D. and Reed, R.P. (2006). *Cryogenic Engineering: Fifty Years of Progress*. London: Springer.

Yoshimura, N. (2014). *Vacuum Technology: Practice for Scientific Instruments*. London: Academic Press.

11

Control Strategies for Different Application Modes of SMES

Jiakun Fang, Wei Yao, Jinyu Wen, and Shijie Cheng*

State Key Laboratory of Advanced Electromagnetic Engineering and Technology,
Huazhong University of Science and Technology
** Author for correspondence*

11.1 Overview of the Control Strategies for SMES Applications

Energy storage has been receiving increasing attention due to its important role in smart grids and accommodating renewable energies. Different storage technologies have been proposed, including potential in pumped hydro storage, the kinetic energy stored in the flywheel, and chemical energy in ultra-capacitors. Among these storage techniques, superconducting magnetic energy storage (SMES) has featured with a high efficiency, a high power density, and a fast response, and is a potential solution for dealing with power imbalance in a power system, enhancing stability, and improving power quality.

SMES stores electrical energy in a large superconductor with a large inductance and a very small resistance. Typically the energy stored in a superconductor E is associated with its inductance L and the current I. The energy in an SMES can be also measured by the magnetic field (B in Tesla). It is connected to the power system via an AC/DC converter, namely a power conditioning system (PCS). The converter can achieve a bi directional power exchange between the power system and the superconducting coil. Because of the high efficiency (over 95%) and fast response at the milliseconds level, there are a multitude of different applications for SMES. As a result, the control strategy varies with the objectives.

Generally, the control strategies for SMES can be divided into groups according to the application modes. Electromagnetic control can be implemented in the PCS to improve the power quality including supporting the transient voltage dip, reducing the harmonic distortion, and enhancing the reliability of the power supply. Early studies investigated the feasibility of using SMES to improve the power quality of an islanded power system. Further, a kJ-class, 20 kW HTS SMES has been demonstrated to compensate the dynamic power and voltage fluctuation caused by wind power [1]. In [2], the unified power quality conditioner (UPQC) concept is proposed with a superconducting coil embedded. Although the topology of the UPQC is different from conventional SMES with two converters interfacing the DC inductance and AC power grid, the capability of SMES to is still to be demonstrated. In [3], hybrid energy storage with SMES is simulated in the DC grid to compensate the voltage sags. A set-membership affine projection algorithm-based adaptive control strategy for the SMES to smooth the wind farm output and to suppress the harmonics is detailed in [4]. These controllers are embedded in the PCS of the SMES.

Advances in Energy Storage: Latest Developments from R&D to the Market, First Edition.
Edited by Andreas Hauer.
© 2022 John Wiley & Sons, Ltd. Published 2022 by John Wiley & Sons, Ltd.

For a longer time scale, the upper-level controller of the SMES is mainly associated with the electromechanical dynamics of the power system. The use of SMES to damp the low-frequency oscillations in power systems started with the Bonneville Power Administration (BPA) in the late 1980s. The SMES installed at the Tacoma transmission system has been used to suppress the fluctuation of the Pacific AC intertie. The low-frequency oscillation results from the poorly damped oscillation modes among the generators' shafts and/or disturbances such as faults, sudden load changes, and line outages. It is represented as oscillations of the active power, system frequency, bus voltage, etc. within the 0.1 ~ 2.5 Hz range. Being applied to damp the low-frequency oscillations, SMES acts like a flexible AC transmission system (FACTs) device. The major advantage is that SMES can output not only reactive power like a static synchronous compensator (STATCOM) but also the active power, which is proven to be more effective in low-frequency oscillation damping [5]. Traditionally, the proportional–integral (PI) or lead–lag controllers have been employed, which can provide a satisfactory damping performance for a specific operating condition. However, the fixed parameters may degrade performance when the system operating condition changes. Efforts have been made to make the SMES-based damping controller robust to operating conditions caused by load change and/or generation output variation. Later various control theories, such as H_∞ theory [6] and fuzzy logic [7], have been adopted in the design of SMES-based power oscillation damping controller. Based on the physical characteristics of the power system, an energy-function-based design is proposed in [8]. In [9], a model-free design procedure has been proposed, which can obtain an approximation of the power system model through model identification techniques.

From the literature identified, one can see that the design of the controller plays a vital role in the damping performance of the closed-loop system. So in this chapter, we give two examples of this topic, addressing two different control issues in Section, 11.2 and 11.3, respectively. To implement these control strategies in a real-world application, we give the design procedure of the monitoring and control unit in an SMES system in Section 11.4.

11.2 Robust Control for SMES in Coordination with Wind Generators

11.2.1 Problem Formulation: Stability Issues Brought by Renewable Sources

Climate change, fossil resource depletion, policy incentives as well as higher public awareness in term of sustainability have promoted the deployment of renewable generators, typically wind and solar. By the end of 2015, the total capacity of the installed wind turbines had reached 432.4 GW worldwide. In some countries, such as Denmark, wind power may produce significant portion, sometimes even 100% percent, of electricity in a specific time period.

Those large-scale wind farms with capacities of over several hundreds of MW have brought significant influences on the power system stability. The effect of wind power can be categorized into three primary mechanisms: first, synchronous generators are replaced by wind generators with different dynamic behaviors thus system dynamics change; second, the controllers of wind generators interacting with synchronous generators such that damping torques are partly influenced; third, altering the dispatch of conventional generation and the profile of power flow, thereby exerting an influence on the power system operating conditions. For the first case, the wind generators usually have much smaller shafts than coal-fired synchronous generators in central power plants. Hence, when the synchronous generators are gradually replaced by wind generators, system inertia significantly reduces. Additionally, for the second aspect, modern wind turbines are connected to the grid with power electronics interfaces. Not only the dynamic behavior of different types of wind generators differ, but also the control strategies for the converters influence dynamic performance. They can either negatively or positively contribute to the oscillation damping with complementary control strategies for the converters. The third mechanism results from the fact that the power generated from wind farms leads to network flow changes and hence influences the network damping characteristics.

To analyze such effects, research has been initiated by the modeling of wind generators and integration to the power system. Later investigation of practical cases shows that the variable-speed wind generators such as double-fed induction generators (DFIGs) have less, or can contribute a little to the system damping performances. However, there is also claim that the wind farm has a detrimental influence. According to existing literature, it can be seen that the views about the impact of wind generators on power system damping are controversial, mainly because the causes are too complex and there is a lack of a system-level index to quantify its damping performance. Later, probabilistic methods have been used to derive the probabilistic density function (PDF) of the critical eigenvalues from the distribution of the wind power output.

Although the wind generators with a localized controller have marginal effects on the power system dynamic performance, the wind generators, especially those power electronics interfaced ones, have the potential to provide additional damping to the power system. The control strategies have been designed to unlock such potential. With wide-area measurements system (WAMS) and properly designed controllers, wind power plants are capable of damping inter-area oscillations. In coordination with terminal converters of the high-voltage direct current (HVDC), the flexible alternative current transmission system (FACTS), etc., it is conceivable that the wind generators' capability of fast controlling the active and reactive power to improve power system stability could be used. Energy storage is commonly used in corporation with wind farms. Those electromagnetic energy storages with fast responses can not only be used to mitigate the fluctuations of the wind farm output but can also be used to improve system stability. So the coordination of the wind farm and SMES is promising in the enhancement of future power system stability.

In this Section, we first illustrate the above phenomenon via sample cases. A systematic method for analying how DFIG-based wind farms can influence the system small-signal stability is proposed. Then we describe a robust damping controller to coordinate the DFIG and SMES. The proposed control strategy can handle the system uncertainties.

11.2.2 System Modeling and Analysis

The dynamic model of a power system is usually described by a set of differential-algebraic equations. Figure 11.1 provides an overview of the power system model with a wind generator and SMES. The component blocks, including synchronous

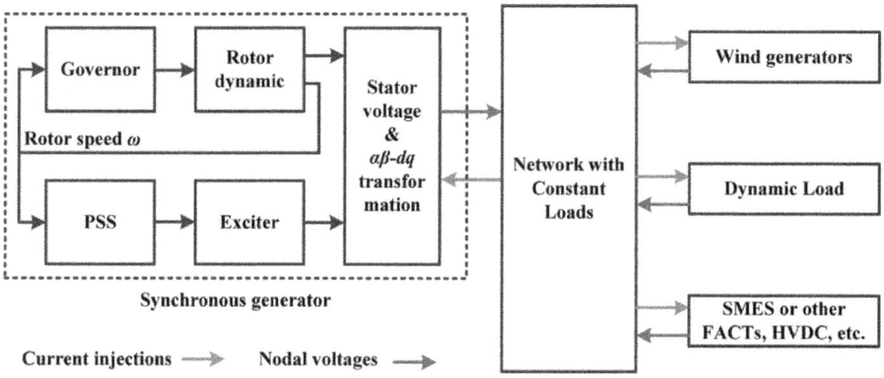

Figure 11.1 Power system model with wind generator and SMES.

generators, wind generators, dynamic load, SMES, FACTs, etc., embeds the differential equations inside. The network with a constant load is represented by a set of algebraic equations. These blocks interact with the current injections, and nodal voltage signals iteratively in every simulation step. Since this chapter mainly focuses on the system-level low-frequency stability, only positive sequence networks are included, and it is assumed that the power system remains in a quasi-steady-state.

In this chapter, the benchmark power system used in the following subsections is studied as an example. A modified 4-machine 11-bus system from [10] is used as the test case. A detailed system structure can be found in [11]. At the terminal of the DFIG, the capacitor banks are installed to provide the reactive power so that the power factor of DFIG i increased to 1.0.

In the studied system, the synchronous generators are modeled according to [10]. And the schematic diagram for the SMES and DFIG model can be found in [11]. A simplified third order SMES model is used here, which can be divided into three parts: the upper-level control, converter dynamic, and the DC current dynamic. The modulation ratio M and firing angle α are modeled as two first order lags. The upper-level controller consists of the damping controller and the voltage restorer. The damping controller collects remote signals from the power system and gives active power reference to the converter while the voltage restorer measures the terminal bus voltage and gives a reactive power reference to the converter.

The DFIG is modeled as a third order asynchronous generator and two-mass rotor, derived from the basic equations of the asynchronous machine, neglecting the electromagnetic behaviors in the stator,

$$\dot{\vec{E}}' = -js\vec{E}' - \frac{R_r}{L_r}\left[\vec{E}' - j(L_s - X')\vec{I}_s - j\frac{L_m}{R_r}\vec{U}_r\right] \tag{11.1}$$

where $\overrightarrow{E'} = e_d + je_q$ is the vector of electrical potential components. The parameters L_s, R_s, L_m, L_r, R_r are the machine stator reactance, stator resistance, mutual reactance, rotor reactance, and rotor resistance, respectively. With stator flux oriented controlled exciting voltage $\overrightarrow{U_r} = u_{dr} + ju_{qr}$, the DFIG's active and reactive power is decoupled.

The drive train is represented by a two-mass model, with a flexible coupling between them, which represents the gearbox,

$$\begin{cases} 2H_M\dot{\omega}_M = T_M - T_G - D_M\omega_M \\ 2H_G\dot{\omega}_r = T_G - T_E - D_G\omega_r \\ \dot{\theta}_s = \omega_0(\omega_M - \omega_r) \end{cases} \tag{11.2}$$

where ω_M and ω_r are the rotor speed of the windmill and the generator, while T_M and T_E are mechanical torque of the turbine and electrical torque of the generator. $T_G = K_S\theta_s - D_s(\omega_r - \omega_M)$ represents the damping characteristic of the two masses. The low voltage power logic (LVPL) is also implemented in the DFIG model.

Summarizing the models of the synchronous generators, SMES and DFIG, the studied system can be built under the arrangement shown in Figure 11.1. Next the small-signal-stability according to the pre-fault system operating conditions is analyzed. The small signal stability analysis is a common way of assessing power system stability subject to small disturbances. Such analysis is usually used to assist in the design of the damping controllers. Preliminaries of the small-signal stability are well elaborated [10].

In the modified power system, since the DFIG is modeled with the two-mass drive train, there are in total five shafts and hence four oscillation modes lying in the paradigm of low-frequency oscillation. Under a typical operating condition, whose power flow is exactly the same with [10], these four modes are analyzed and the results are shown in Table 11.1.

The *participating factor* is the term to measure the correlation between the kth state x_k and the ith eigenvalue λ_i, where the v_{ki} and u_{ki} are the elements from ith left eigenvector v_i and right eigenvector u_i. The larger participating factors indicate that the oscillation mode is most relevant for certain generator(s). From Table 11.1, it can be see that in modes 1, 2, and 3 the DFIG has a much smaller P_{ki} than other conventional generators, while mode 4 dominates. So modes 1 and 3 are the ordinary low-frequency oscillations between the two synchronous generators. In mode 2, the DFIG acts as a constant voltage source, while mode 4 is the low-frequency oscillation between the two-mass of the drive train inside the DFIG.

In the above study, the wind farm is represented by the equivalent of a single wind generator in a low frequency oscillation study. This is a trade-off between simulation accuracy and computational burden. The wind, however, is constantly fluctuating, which brought two major challenges. First, the system operating condition varies with

Table 11.1 Mode analysis of the power system.

Mode no.	Frequency(Hz)	Damping Ratio	Participating Factor $p_{ki} = \frac{u_{ki}v_{ik}}{v_i^T u_i}$			
			G1	G2(WT)	G3	G4
1	0.6889	0.0282	**0.3823**	0.0137	0.0811	0.0610
2	0.4051	0.0981	**0.1443**	0.0032	**0.1985**	**0.1842**
3	1.1000	0.1020	0.0002	0.0000	**0.2434**	**0.3008**
4	0.9089	0.1464	0.0101	**0.4886**	0.0004	0.0002

the output of the wind generator. Second, when the wind turbines are cutting in and out, the parameters of the equivalent generator varies. So in this section, we analysis how these variations can influence small-signal stability. We propose the solution in the following section.

Regarding the variation of the parameters of the equivalent wind generator, we divide its parameters into two categories: the electrical ones and the mechanical ones. The electrical parameters include the stator inductance, rotor inductance, while the mechanical parameters are the inertias, the stiffness of the shaft, etc. We let these parameters vary from 50% to 150% of its original value. Figure 11.2 illustrates the trajectories of the low-frequency modes along with the electrical and mechanical parameter changes, respectively. From Figure 11.2(a), we can see that the changes of the electrical parameter has little impact on the low-frequency modes. Figure 11.2(b) shows that the change of the mechanical parameters has an influence on mode 3 and 4. Note that here mode 4 involves the DFIG only while mode 3 is conventional low-frequency oscillation associated with a synchronous generator 3 and 4. So a conclusion can be drawn that the parameter variation of the equivalent wind generator can influence low-frequency oscillation only when its mechanical parameters change with the operating conditions of the wind farm, i.e. in power system small-signal stability analysis; we pay particular attention to the mechanical parameters while the electrical parameters are, comparatively, not so important.

When the wind fluctuates, the power flow in the network also varies, which leads to different linearization results. In Figure 11.2 (c), the real power output of the DFIG

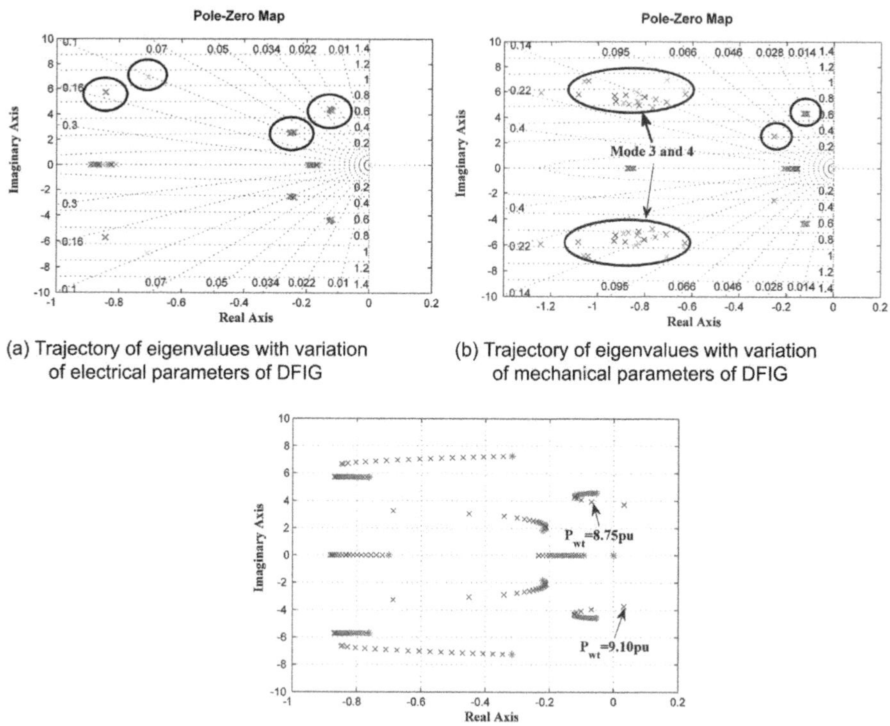

(a) Trajectory of eigenvalues with variation of electrical parameters of DFIG

(b) Trajectory of eigenvalues with variation of mechanical parameters of DFIG

(c) Trajectory of eigenvalues along with penetration level of wind farm

Figure 11.2 Eigenvalues affected by variation of parameters of DFIG and its output.

increases from 2.8 pu to 9.1 pu. The penetration level increases from 30% to 50% in area I. The red star point in the figure is the initial operating condition. In this aspect of power flow, the fluctuating wind power influences all the low-frequency oscillation modes. The eigenvalues of modes 2, 3, and 4 move to the left in the complex plane, while the trajectory of mode 1 moves in the opposite direction. When the real power of the DFIG exceeds 9.0 pu (900 MW in this case), the eigenvalue moves to the right half in the complex plane.

Though the above analysis is carried out on a small test system, the conclusion can be drawn that the wind power fluctuation can not only change the system load flow, but also has a complicated impact on the internal dynamic behavior of the wind farm, or the equivalent wind generator. In control theory, this stochastic nature can be summarized as the uncertainties of the plant, i.e. the dynamic system does not behave as a linearized plant. So in the Section 11.2.3 a robust control strategy coordinating wind generator and SMES is proposed.

11.2.3 Robust Coordinative Control Strategy

In order to achieve a stable closed-loop system, we have to properly design the controller according to the plant so that all the eigenvalues are placed on the left half of the complex plane. State-feedback control can place the eigenvalues precisely on the complex plane, but in power systems, most of the states are hard to measure and collect in the control center. WAMS enables power engineers to access remote signals via the via phasor measurement units (PMUs). In this perspective, the output feedback control is preferred in practice. On the other hand, the previously-mentioned uncertainties require the designed controller to be robust. In this section, the H_2 / H_∞ mixed sensitivity robust control is employed to handle this problem.

The system is augmented as Figure 11.3, penalizing the error signal *e*, control signal *u* and the output signal *y* with weighing functions $W_1(s), W_2(s)$ and $W_3(s)$. Here the open-loop plant *P*(*s*) represents the linearized power system model built in the previous section. And *K*(*s*) represents the controller, to be designed. The control signal *u* includes the active and reactive power references for both DFIG and SMES, while the output *y* is selected remote signals from the WAMS. z_2 and z_∞ are artificial outputs used to measure the H_2 and H_∞ performance, respectively. Usually, W_1 is designed as a high-pass filter to ensure the robustness of the controller *K*(*s*); W_2 is selected as a high-pass filter or a constant so that the control is efficient in power. W_3 is a low-pass filter to mitigate the effects from the disturbance. In this work, the weighing functions are selected as shown in Table 11.2.

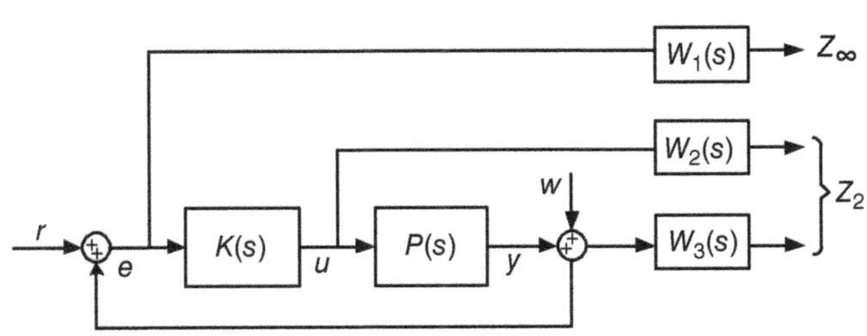

Figure 11.3 Multi-objective damping controller synthesis configuration.

Table 11.2 Weighing functions for mixed sensitivity.

W_1	$W_1 = \frac{0.5s}{s+50}$
W_2	$W_2 = 0.005$
W_3	$W_3 = \frac{0.1s+1}{s+20}$

Then the augmented plant $P(s)$ in Figure 11.3 becomes

$$P(s) = \begin{bmatrix} W_1(s) & -W_1(s)P(s) \\ 0 & W_2(s) \\ 0 & W_3(s)P(s) \\ I & -P(s) \end{bmatrix}$$

with the state space form of

$$P(s) \sim \begin{cases} \dot{x} = Ax + B_1 w + B_2 u \\ z_\infty = C_1 x + D_{11} w + D_{12} u \\ z_2 = C_2 x + D_{21} w + D_{22} u \\ y = C_y x + D_{yw} w + D_u \end{cases} \tag{11.3}$$

The aim is to design an output feedback controller

$$K(s) \sim \begin{cases} \dot{x}_k = A_k x_k + B_k y \\ u = C_k x_k + D_k y \end{cases} \tag{11.4}$$

so that the closed-loop system has its pole on the desired place on the complex plane with reasonable H_2 / H_∞ performance to handle exogenous disturbances. Together with the plant $P(s)$, the closed-loop system can be presented in a compact format:

$$\begin{cases} \dot{x}_c = A_c x_c + B_c w \\ z_\infty = C_{c1} x_c + D_{c1} w \\ z_2 = C_{c2} x_c + D_{c2} w \\ y = C_c x + D_1 w \end{cases} \tag{11.5}$$

where $x = \begin{bmatrix} x \\ x_k \end{bmatrix}$, $A_c = \begin{bmatrix} A + B_2 D_k C & B_2 C_k \\ B_k C_y & A_k \end{bmatrix}$, $B_c = \begin{bmatrix} B_2 + B_2 D_k D_1 \\ B_k D_1 \end{bmatrix}$, $C_{c1} = \begin{bmatrix} C_1 + D_{12} D_k C_y & D_{12} C_k \end{bmatrix}$, $C_{c2} = \begin{bmatrix} C_2 + D_{22} D_k C_y & D_{22} C_k \end{bmatrix}$, $D_{c1} = D_{11} + D_{12} D_k D_1$, $D_{C2} = D_{21} + D_{22} D_k D_1$. With the closed-loop system (11.5), the requirements to the controller $K(s)$ can be transferred to linear matrix inequalities (LMIs) as in the following.

(1) Pole placement. In the complex plane, a region D can be described by the LMI $D = \left\{ s \in C : L + sM + \bar{s}M^T < 0 \right\}$. For a specific system matrix A, if it satisfies that all the eigenvalues of A lies in region D, there exists a positive symmetric matrix X that

$$L \otimes X + M \otimes (AX) + M^T \otimes (AX)^T < 0 \tag{11.6}$$

where $\otimes$ means the *Kronecker* product of two matrices.

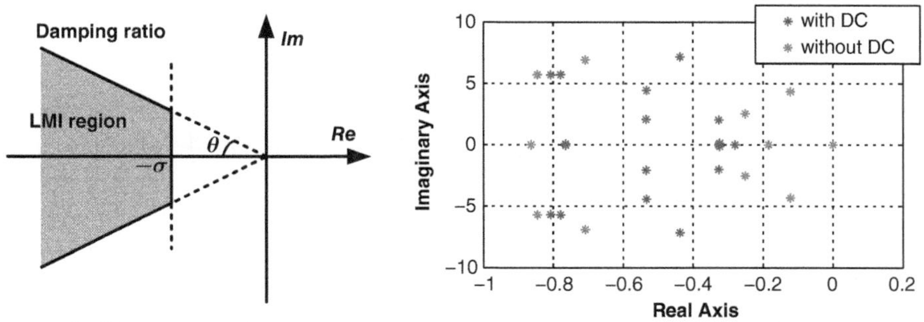

(a) LMI region for pole placement (b) Linearization of the closed-loop power system

Figure 11.4 Desired and resulting pole placement.

(2) H_2 performance. The two-norm of the transfer function from disturbance signal ω to z_2 is less than the given constant η_0.

$$\begin{bmatrix} A_c X + X A_c^T & B_c \\ B_c^T & -I \end{bmatrix} < 0$$

$$\begin{bmatrix} Q & C_{c2} X \\ X C_{c2}^T & X \end{bmatrix} < 0 \tag{11.7}$$

$$\text{Trace}(Q) < \eta_0^2$$

(3) H_∞ performance. The infinit-norm of the transfer function from disturbance signal ω to z_∞ is less than the given constant γ.

$$\begin{bmatrix} A_c X + X A_c^T & B_c & X C_{c1}^T \\ B_c^T & -I & D_{c1}^T \\ C_{c1} X & D_{c1} & -\gamma^2 I \end{bmatrix} < 0 \tag{11.8}$$

By adopting positive matrix X and symmetrical matrix Q, the multi-objective controller satisfying the above mentioned performance requirements can be found by solving the following LMI:

$$\min_{\gamma, X, Q} \{\alpha\gamma + \beta \text{Trace}(Q)\} \tag{11.9}$$

st. (11.6), (11.7), and (11.8)

In this work, the LMI region (or the D region in Section 11.2.2) is designed as a union of a sector area and a half plane area, as shown in Figure 11.4 (a). Placing the eigenvalues in the sector area makes the closed-loop system have a desired damping ratio, while the left half plane guarantees system stability. The shaded part in Figure 11.4 (a) indicating $\sigma = -0.25$ and $\theta = 85°$ is desired.

The coordinated damping controller between the DFIG and the SMES is studied in this section. The inputs of the plant are selected as the reference value of the SMES active power output and the supplementary control signal added to the maximum power point tracking control of the DFIG. By solving the LMI (11.9), the closed-loop performance is shown in Figure 11.4 (b). With the designed damping controller, the real parts of all the eigenvalues are less than -0.25.

11.2.4 Simulation, Observations, and Conclusion

Figure 11.5 shows the system performances under 30% and 50% wind power penetration, respectively. The three-phase fault occurs at time 1.0s. The fault location lies in the middle of the transmission line. 50 ms later after the fault, the line is tripped off with a successful re-closure at the time 2.1s.

With the coordinated damping control of the DFIG and SMES, low-frequency oscillation is dampened effectively. The blue line in these figures illustrates the controller designed without considering robustness, i.e. solving the LMI only (11.6) instead of (11.9). Under this operating condition, the robust damping controller performs a little worse than the conventional one. The robust damping controller sacrificed some of its dynamic performance, under the typical operating condition, for a better robustness. In Figure 11.5 (d), (e), and (f), the penetration level has been increased to 50%, though the conventional controller can still dampen the swing, the robust controller performed better.

To quantitatively assess the system's dynamic performance during the transient process, the integral of the time multiplied by the absolute error (ITAE) is calculated by (11.10)

$$J_{\mathrm{ITAE}} = \int_0^{t_{sim}} \frac{\left(|\delta_1 - \delta_4| + |\delta_2 - \delta_4| + |\delta_3 - \delta_4|\right)}{3} t dt \tag{11.10}$$

A smaller J_{ITAE} indicates less deviation of synchronousness and a shorter time to reach a new steady state. Hence a better performance and therefore the system is more stable. Figure 11.6 shows the performance comparison through the ITAE index defined in (11.10) with different operating conditions: a closed-loop system with properly designed damping controller surely performs better than the open-loop system; but the robust damping controller is much less sensitive to the change in the operating conditions.

11.3 Anti-windup Compensation for SMES-Based Power System Damping Controller

11.3.1 Major Concern on the Capacity of SMES

In recent years, the novel power system stabilizer based on an energy storage system, as the potential solution to mitigate both power generation and consumption fluctuations, which threaten power system stability, draws much attention, especially when taking intermittent renewable energy generation into account. SMES with the features of high efficiency, high power density, and fast response is the potential solution to the power imbalance in the power system to enhance stability and improve power quality. Unfortunately, the cost of an SMES system is much higher than conventional FACTs devices such as STATCOM and increases exponentially with its capacity. However, the investment in SMES is still considerable, though it is promising in power system applications. The methodologies for optimal sizing of SMES are investigated. This work shows a simple phenomenon that the system performance degrades with a decrease of SMES capacity. In Section 11.2, the damping controller for SMES in coordination with

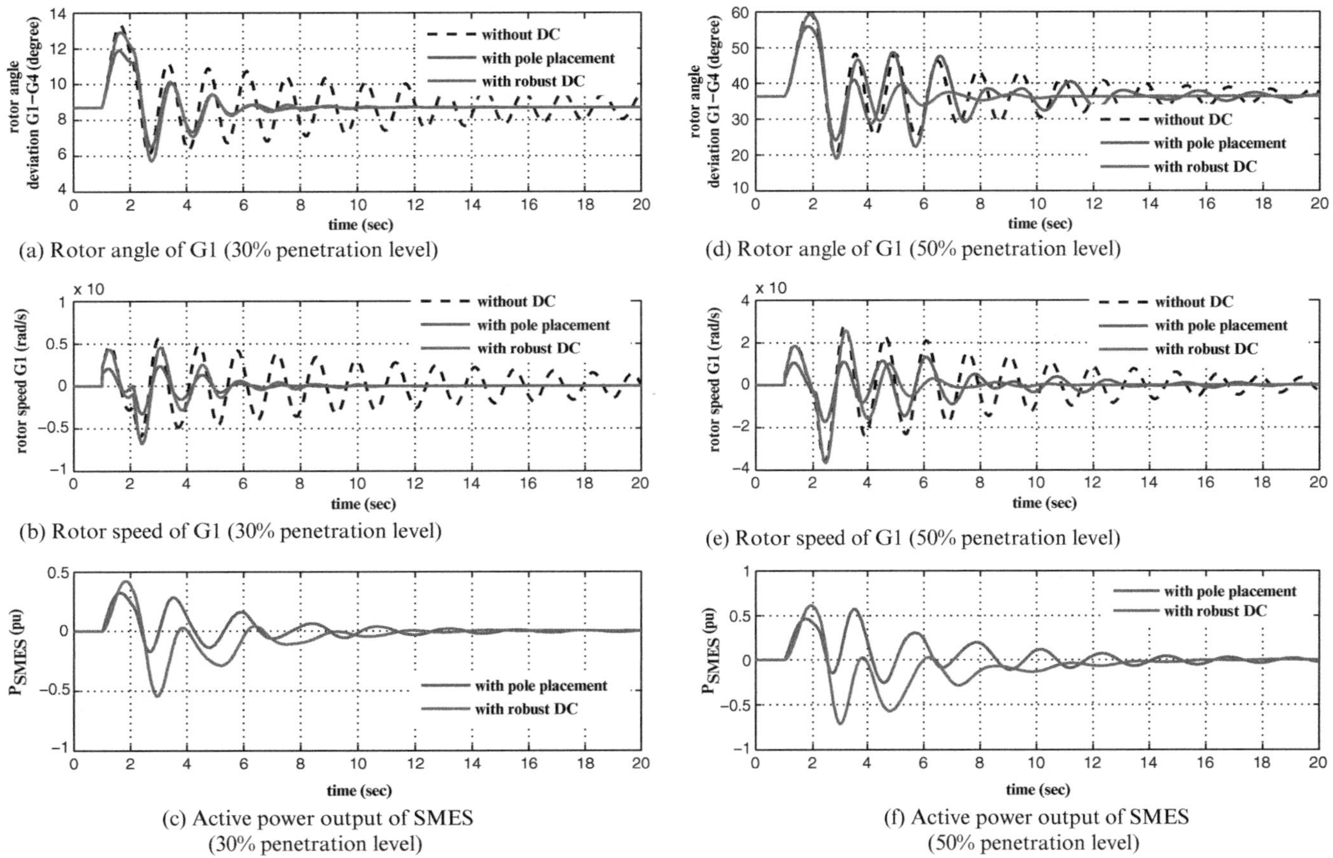

Figure 11.5 Closed-loop system response.

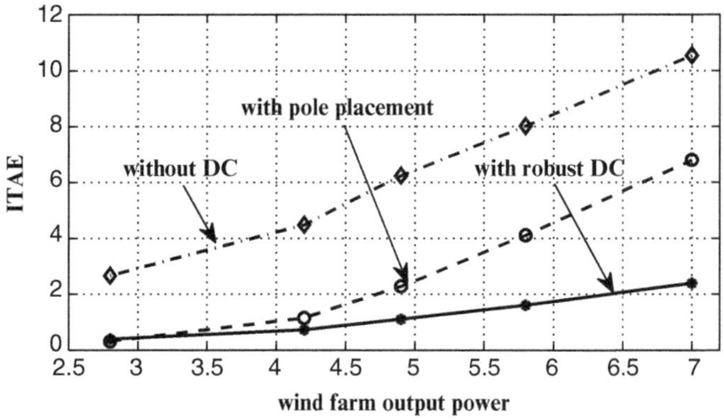

Figure 11.6 Performance under different penetration levels.

DFIG-based wind farm was designed. In this section, we deal with the situation where the capacity of the SMES is less than the reference value produced by the upper-level controller. This leads to saturation non-linearities.

From the control point of view, a linear system with a hyperbolic equilibrium is globally stable. If the magnitude constraints on control input are considered, however, the system may not be globally stable. The saturation problems can be traced back to the 1940s when the control systems consisted of analog controllers. It was believed at that time that it was unreasonable to sacrifice the small signal behavior of the control system to guarantee large signal performance. Later when the control systems became more and more digitalized, researchers began to look into the so-called "windup" problem, namely the performance or even stability loss after the saturation limits were reached by the actuators. Then the "anti-windup" strategies were naturally considered in order to mitigate the saturation-induced malfunctioning of the control systems. Early literature involving anti-windup compensation heralded various work on "intelligent integrators," most of which were tricky fixes proposed by engineers, specific to certain problems. Most of these approaches have a lack of formal stability guarantees.

The alternative solution to saturation is anti-windup compensation, which is a two-step procedure in controller design. First, the unconstrained controller is designed without considering saturation. And then when the actuator reaches its output bounds, feedback is used to adjust the controller's output to meet the system requirements. In this section, the anti-windup compensation scheme is applied to the SMES-based power system damping controller. First, the saturation problem is formulated, and then the design procedure is given. Time-domain simulation demonstrates that the designed compensator is effective if, and only if, the saturation phenomenon happens.

11.3.2 Problem Formulation

The reason for degradation is the output power limitation of SMES, which can be mathematically described as the actuator saturation in a dynamic power system. These saturation non-linearities are ubiquitous since the physical actuator, such as SMES, is subject to saturation at its maximum and minimum limits. For brevity, the SMES model is simplified to a second order model,

$$\begin{cases} \dot{P}_{ES} = -\dfrac{1}{T_c} P_{ES} + \dfrac{1}{T_c} u_p \\ \dot{Q}_{ES} = -\dfrac{1}{T_c} Q_{ES} + \dfrac{1}{T_c} u_q \end{cases} \tag{11.11}$$

And only the active power of SMES is considered as the system input. The rotor angle of number 1 generator is adopted as the system's feedback signal for the damping controller. The residues method is used to design the damping controller in the form of

$$K(s) = K_d \frac{T_w s}{1 + T_w s} \left[\frac{1 + sT_1}{1 + sT_2} \right]^m \tag{11.12}$$

The residue-based method is quite a common way to design a damping controller. There are m lead–lag compensators, a washout, and a steady state gain K_d in the linear controller $K(s)$

$$m = \text{ceil}\left[\frac{\pi - \arg(R_k)}{\pi / 3} \right] \tag{11.13}$$

where R_k is the residues of the open-loop system. The phase α for each lead–lag compensator can be calculated as

$$\alpha = \frac{1 - \sin(\phi / m)}{1 + \sin(\phi / m)}$$

Then, with $m = 2$, $T_1 = \dfrac{1}{\omega_i \sqrt{\alpha}} = 0.5791$, $T_2 = \alpha T_1 = 0.0922$. The trajectory of eigenvalues for the closed-loop system can be obtained, from which the steady state gain $K_d = 0.5$ is selected. When $K_d = 0.5$, the damping ratio of dominant oscillation mode 1 improves from 0.0282 to 0.1276 while mode 2 decreases from 0.096 to 0.078. The other 2 modes remain nearly unchanged.

The dynamic model of a power system excluding the SMES based damping controller can be linearized at an equilibrium point as follows:

$$\Sigma_p \sim \begin{cases} \dot{x}_p = A_p x_p + B_1 w + B_2 u \\ y = C_2 x_p + D_{21} w \\ z = C_1 x_p + D_{11} w + D_{12} u \end{cases} \tag{11.14}$$

where $x_p \in R^{n}{}_p$ is the state variables, $u \in R^{n}{}_u$ the control inputs, $w \in R^{n}{}_w$ the disturbances, $y \in R^{n}{}_y$ the measurement outputs, and $z \in R^{n}{}_z$ the controlled outputs. A is the state matrix, B the input matrix, and C the output matrix. w reflects disturbances to the system.

11.3.3 Anti-windup Compensation Scheme

A vector of saturation functions $\mathbf{sat}(\cdot)$ indicates a series of decentralized actuators' saturations $\left[sat_1(u_1), sat_2(u_2), \ldots, sat_m(u_m)\right]^T$. For each $\text{sat}_i(u_i)$,

$$\text{sat}_i(u_i) = \begin{cases} \bar{u}_{0(i)} & u_i > \bar{u}_{0(i)} \\ u_i & \underline{u}_{0(i)} \le u_i \le \bar{u}_{0(i)} \\ \underline{u}_{0(i)} & u_i < \underline{u}_{0(i)} \end{cases} \tag{11.15}$$

Then, a system with saturation can be expressed as

$$\tilde{\sum}_p \sim \begin{cases} \dot{x}_p = A_p x_p + B_1 w + B_2 \mathbf{sat}(u) \\ y = C_2 x_p + D_{21} w \\ z = C_1 x_p + D_{11} w + D_{12} \mathbf{sat}(u) \end{cases} \tag{11.16}$$

If the actuator saturation exists, the performance of the closed-loop control system designed without considering actuator saturation may seriously deteriorate. Even though replacing the controller may solve this problem, it is preferable to implement simple design methods for the controller, such as lead–lag type and PID controller. On the other hand, linear control theory has been extensively applied to a power system for decades. Hence, a two-step procedure is preferred and studied in this chapter. Firstly the linear controller is designed without taking saturation into consideration, and then the anti-windup scheme is used to reduce the influence of actuator saturation.

The anti-windup compensation is a two-step procedure. Firstly, the state space form of linear controller

$$\Sigma_c \sim \begin{cases} \dot{x}_c = A_c x_c + B_{cy} y + B_{cw} w \\ u = C_c x_c + D_{cy} y + D_{cw} w \end{cases} \tag{11.17}$$

is designed to achieve the closed-loop performance without saturation. In this work, Σ_c is a damping controller for SMES. The closed-loop system's performance is degraded by the saturations according to the capacity of SMES to output active and reactive power. Hence, a correction term proportional to $q = u - \text{sat}(u)$ is added to the controller Σ_c, which is

$$\tilde{\Sigma}_c \sim \begin{cases} \dot{x}_c = A_c x_c + B_{cy} y + B_{cw} w + \eta_1 \\ u = C_c x_c + D_{cy} y + D_{cw} w + \eta_2 \end{cases} \tag{11.18}$$

where $\eta = \left[\eta_1^T \ \eta_2^T\right] = -\Lambda q$ is the additional part for anti-windup compensation. As in the anti-windup scheme shown in Figure 11.7, for $-\Lambda$ being a static gain vector, the anti-windup control's (AWC) effect takes place as soon as the saturation occurs and quits as soon as the saturation disappears. Hence, the AWC has no influence on the original controller designed with linear control algorithm. One can see from Figure 11.7, if no saturation occurs, there is no difference between u and $\text{sat}(u)$, then the output of AWC η

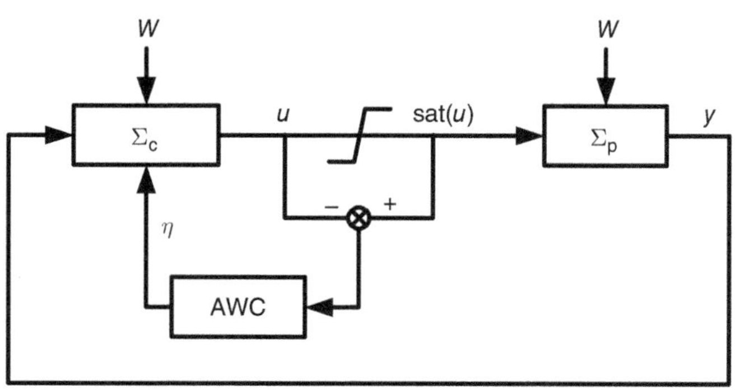

Figure 11.7 Schematic diagram for anti-windup control.

becomes zero immediately, the AWC will have no effect on the linear controller. According to (11.18), the dimension of η and Λ are $n_y + n_u$ and $(n_y + n_u) \times n_u$, respectively.

In industrial applications, the controller and actuator's output are usually monitored. The correction to the controller's states and output with a static gain is easy to implement. Hence, this anti-windup scheme is quite practical only if the AWC is properly designed.

Earlier works on anti-windup design try to minimize the effect of saturation in a direct way by reducing the difference between the input and output of the actuators, which has to modify the controller's output even when there is no saturation happening. Modern anti-windup aims to reduce the effect of saturation indirectly by trying to improve closed-loop stability and performances. In the event of actuator saturation, a signal based on the difference between the linear controller's output and the saturated actuator's output is produced and feedback to the linear controller. When there is no such saturation, the original controller remains unaffected.

The problem of anti-windup design lies in the anti-windup gain matrix Λ to meet various performance indices. In [12], the design objective is to minimize the L_2 gain from the disturbance w to the controlled output z. By selecting $x = \left[x_p^T \; x_c^T\right]^T$, the closed-loop system connects (11.16) and (11.17) in series in the linear fractional transformation format. For a decentralized saturation function, Δ is a diagonal matrix with each element 1(·)-sat(·). A diagonal scaling $n_u \times n_u$ matrix W is introduced to reduce conservatism. After equivalent transformation of the block diagram, the system $\tilde{\Sigma}$ becomes

$$\tilde{\Sigma} \sim \begin{cases} \dot{x} = Ax + \begin{bmatrix} B_w & (B_q - B_\eta \Lambda)W^{-1} \end{bmatrix} \begin{bmatrix} w \\ \tilde{q} \end{bmatrix} \\ \begin{bmatrix} z \\ u \end{bmatrix} = \begin{bmatrix} C_z \\ C_u \end{bmatrix} x + \begin{bmatrix} D_{zw} & (D_{zq} - D_{z\eta}\Lambda)W^{-1} \\ D_{uw} & (D_{uq} - D_{u\eta}\Lambda)W^{-1} \end{bmatrix} \begin{bmatrix} w \\ \tilde{q} \end{bmatrix} \end{cases} \tag{11.19}$$

where $x = \begin{bmatrix} x_p \\ x_c \end{bmatrix}$, $A = \begin{bmatrix} A_p + B_2 D_{cy} C_2 & B_2 C_c \\ B_{cy} C_2 & A_c \end{bmatrix}$,

$$B_w = \begin{bmatrix} B_1 + B_2(D_{cy}D_{21} + D_{cw}) \\ B_{cy}D_{21} + B_{cw} \end{bmatrix}, \; B_q = \begin{bmatrix} -B_2 \\ 0 \end{bmatrix}, \; B_\eta = \begin{bmatrix} 0 & B_2 \\ I & 0 \end{bmatrix},$$

$$B_\eta = \begin{bmatrix} 0 & B_2 \\ I & 0 \end{bmatrix}, \; C_z = \begin{bmatrix} C_1 + D_{12}D_{cy}C_2 & D_{12}C_c \end{bmatrix},$$

$$D_{zw} = D_{11} + D_{12}D_{cy}D_{21} + D_{12}D_{cw}, \; D_{zq} = -D_{12},$$

$$C_u = \begin{bmatrix} D_{cy}C_2 & C_c \end{bmatrix}, \; D_{uw} = D_{cy}D_{21} + D_{cw}, \; D_{uq} = 0, \; D_{u\eta} = \begin{bmatrix} 0 & I \end{bmatrix}.$$

Then (11.19) is a standard linear time invariant system with $[\omega \quad \tilde{q}]^T$ as input and $[z \quad u]^T$ as output. By the virtue of bounded real lemma and Schur's formula for the determinant of a partitioned matrix, H_∞ synthesis is performed to guarantee that the L_2 gain of transfer function from input $[\omega \quad \tilde{q}]^T$ output $[z \quad u]^T$ is less than γ. Then the following LMI is derived. Technically, $M = W^{-2}$ becomes a search variable for a better optimization.

$$\begin{bmatrix} AQ+QA^T & * & * & * & * \\ MB_q^T - X^T B_\eta^T & -M & * & 0 & * \\ C_u Q & D_{uq}M - D_{u\eta}X & -M & * & 0 \\ B_w^T & 0 & D_{uw}^T & -\gamma I & * \\ C_z Q & D_{zq}M - D_{z\eta}X & 0 & D_{zw} & -\gamma I \end{bmatrix} < 0 \tag{11.20}$$

If the LMI (11.20) is feasible, then the AWC can be obtained from $\Lambda = XM^{-1}$. And the L_2 gain is less than γ [13]. By using the Matlab function *MINCX*, γ can be optimal minimized under the constraint of LMI (20).

11.3.4 Simulation Validation

The three-phase fault takes place at the midpoint of the line, lasting 0.12s, and then successful reclosure takes place after 1s. Comparing the solid and dotted lines in Figure 11.8(a), one can see that the effectiveness of the designed damping controller. When the capacity of SMES is set to 2 MW (0.02 pu), the system's performance is degraded. Figure 11.8(b) compares the saturated and unsaturated output of SMES, indicating the serious saturation occurred during SMES control.

Using the method in Section 3.3, the AWC is designed as Λ = [0.0011 0 0.0002 -0.9364]^T. With a properly designed AWC, the degradation of system performance by the actuator saturation can be reduced as shown in Figure 11.9. A better performance has been achieved with the AWC.

Figure 11.9 shows an unstable case to demonstrate the essentiality of the AWC. The three-phase fault clearing time is extended to 0.2s, while the capacity of SMES is still 2 MW. The saturated damping controller cannot stabilize the system without AWC. These results indicate that the supplementary anti-windup compensation has the potential to deal with the saturation and thus, improve the stability of the power system.

The simulation results also indicate that the AWC acts only when saturation occurs. In Figure 11.9(b), one can observe the difference of controller's outputs with and without the correction by the AWC. Figure 11.9(c), however, shows nearly the same output of SMES in the linear zone when no saturation occurred. This means the AWC only takes effect in the non-linear zone of the system.

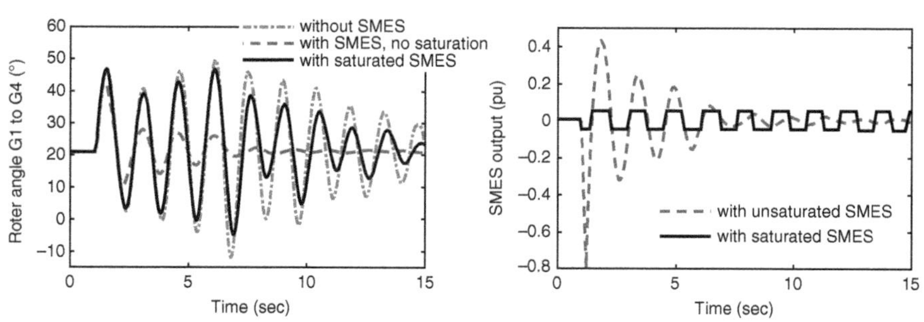

(a) system performance with non-saturated and saturated SMES

(b) SMES output power under saturation

Figure 11.8 System performance degraded by saturation of SMES.

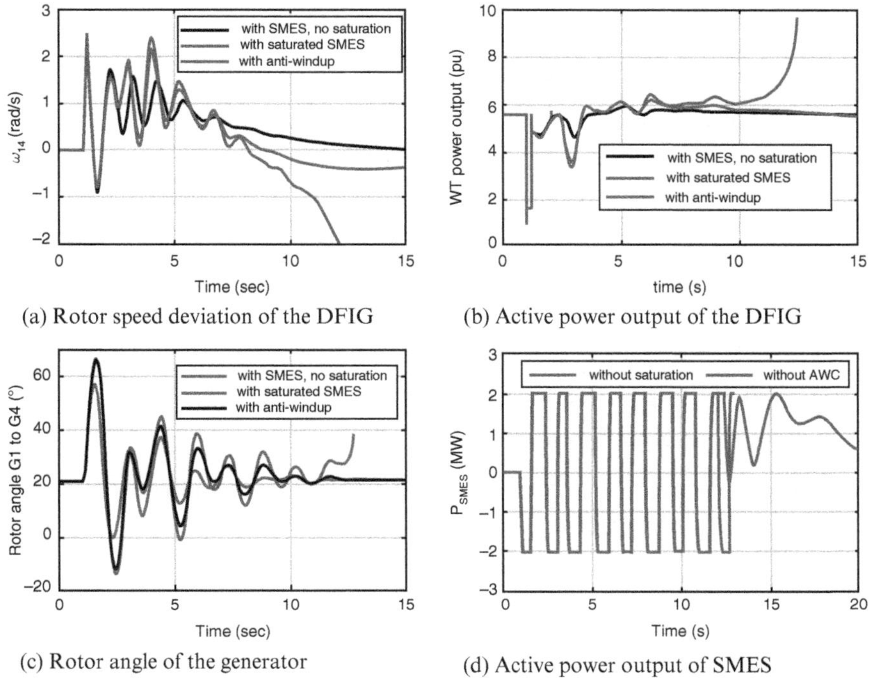

(a) Rotor speed deviation of the DFIG
(b) Active power output of the DFIG
(c) Rotor angle of the generator
(d) Active power output of SMES

Figure 11.9 System performance improved by AWC, unstable cases.

Figure 11.10 compares system's J_{ITAE} under the same fault from the stable case, but different capacities of SMES. The palest bars in the figure indicate that SMES outputs exactly as the controller output without magnitude limited. So when there is no saturation, J_{ITAE} keeps constant with the change of SMES capacity. This figure shows that under the same capacity of SMES, system performance is better with the AWC loop in the control. Although the AWC cannot entirely eliminate the adverse effects of the saturation to the system, it can be an effective compensation to the linear controller. On the other hand, with the increase of SMES capacity, the difference between the controller's and SMES's output is smaller. Hence the AWC has less influence on the dynamic system.

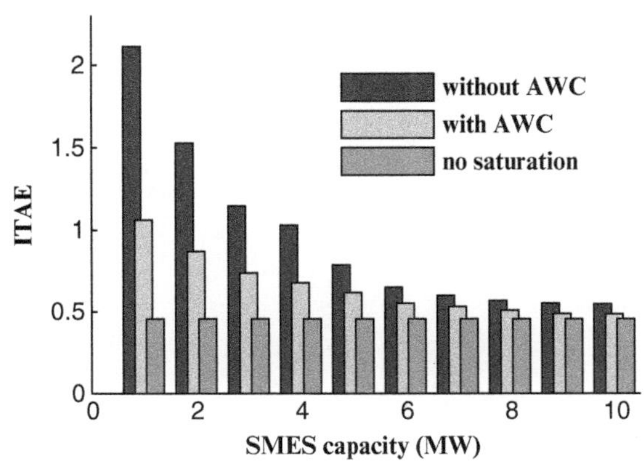

Figure 11.10 Performance comparison under different saturation levels.

11.4 Monitoring and Control Unit of SMES

The SMES system usually consists of following subsystems: (1) the power conditioning subsystem (PCS) interfacing the DC superconductor and AC power grid; (2) the cryogenic refrigeration subsystem to initially cool and maintain the superconductor at its working temperature; (3) the monitoring and control unit (MCU); (4) the superconducting coil and its containing dewar. Among the subsystems in SMES, the MCU plays a vital role in embedding the control algorithms such as those mentioned in the previous Sections 11.2 and 11.3. The MCU also functions to record the operating conditions of the SMES and the experimental results. In this section, we introduce real-world instance of MCU for SMES, including its design, implementation, and testing.

11.4.1 General Functionalities of the MCU for SMES

Generally, the MCU in the SMES system is used to measure the desired signals indicating the operation conditions of both the SMES and its connected power system, and command the PCS to control the power exchange between superconductor and power grid. The MCU consists of four major parts according to the different functions: the human-machine interface (HMI), the data acquisition (DAQ), the upper-level controller (UC), and the waveform recording (WR). The HMI is a personal computer (PC), communicating with the PCS to send out commands from the operator and display information of the SMES. DAQ collects real-time information of the SMES and its connected power system and gives such information to the UC and/or WR. There are mainly two categories of signals: binary signals, and analog signals. The binary signals mainly indicate the switching status, e.g. open/closed of the relays. The analog signals indicate those quantities such as active power, voltage, current. Communication and multi-channel data acquisition are needed. Taking the measurements as input, UC calculates the active and reactive power references based on the embedded upper-level controller and sends them to the PCS. WR simply records some of the signals from the DAQ and saves it for re-use. The schematic diagram of the MCU for SMES can be seen in Figure 11.11.

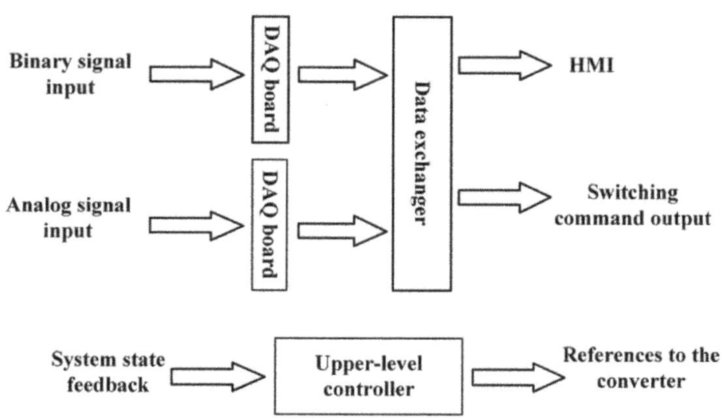

Figure 11.11 Schematic diagram of the MCU for SMES.

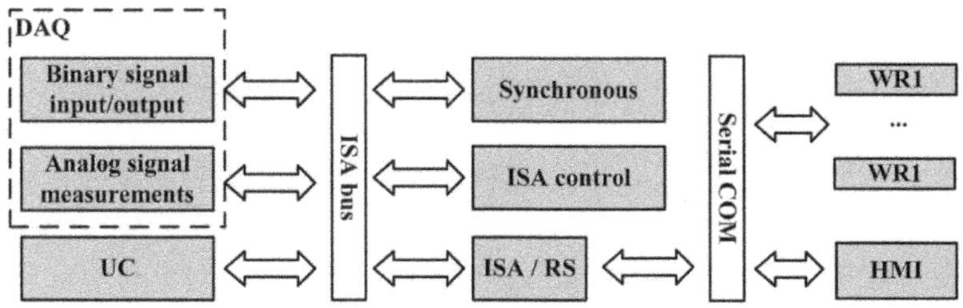

Figure 11.12 ISA and serial communication within the MCU.

11.4.2 Design and Implementation

In this section, we will briefly introduce the schematic design process of the MCU for SMES. The technical content of this section is mainly based on the practical instance of SMES developed by Huazhong University of Science and Technology [11]. For this SMES, a modular-based design procedure has been adopted. As shown in Figure 11.12, the communication for the MCU is based on serial communication and Industry Standard Architecture (ISA). Several ISA interfaced cards are developed according to their different functions. The binary signal input/output, analog signal measurements, and the UC communicate with the ISA bus, while the WR and HMI are connected to the serial communication network. The HMI is also connected to the serial communication network, developed with the Visual C++ programming platform so that it can easily be used on different PCs. Each add-in in the diagram is equipped with a dynamic random access memory (DRAM) for data storage and buffering.

The UC, embedded in the MCU, consists of two control loops for active and reactive power. One adopts the active power deviation, ΔP_e as the input while the other adopts the voltage deviation as the input. Dead bands are set to 0.1 kW and 10 V, respectively.

11.4.3 Laboratory and Field Tests

The bias of active power, ΔP_e instead of the rotor speed derivation, $\Delta\omega$, is adopted as the feedback signal of the upper-level controller. In the field test of the 30 kJ SMES system in the hydro-power plant of Kyushu Electric Power Co., the $\Delta\omega$ control mode [14] is used for system stabilization. The rotor speed derivation, however, is usually hard to measure in power systems. So here, the SMES is connected to the terminal of the generator. So the local signals such as the derivation of electrical power of the generator output are measured and used as the feedback signal for the upper-level controller.

The dynamic experiment of the SMES for improving power system stability is carried out. A single machine infinite bus (SMIB) system is configured. The specifications are listed in [11]. In the first experiment, line 1 is tripped during a steady state. Figure 11.13(a) illustrates the power oscillation with and without the SMES. Comparing the dotted and solid lines, one can see that the power oscillation is dampened and the system stability is improved by the SMES.The response of the SMES to a three-phase short-circuit is studied in the second case.

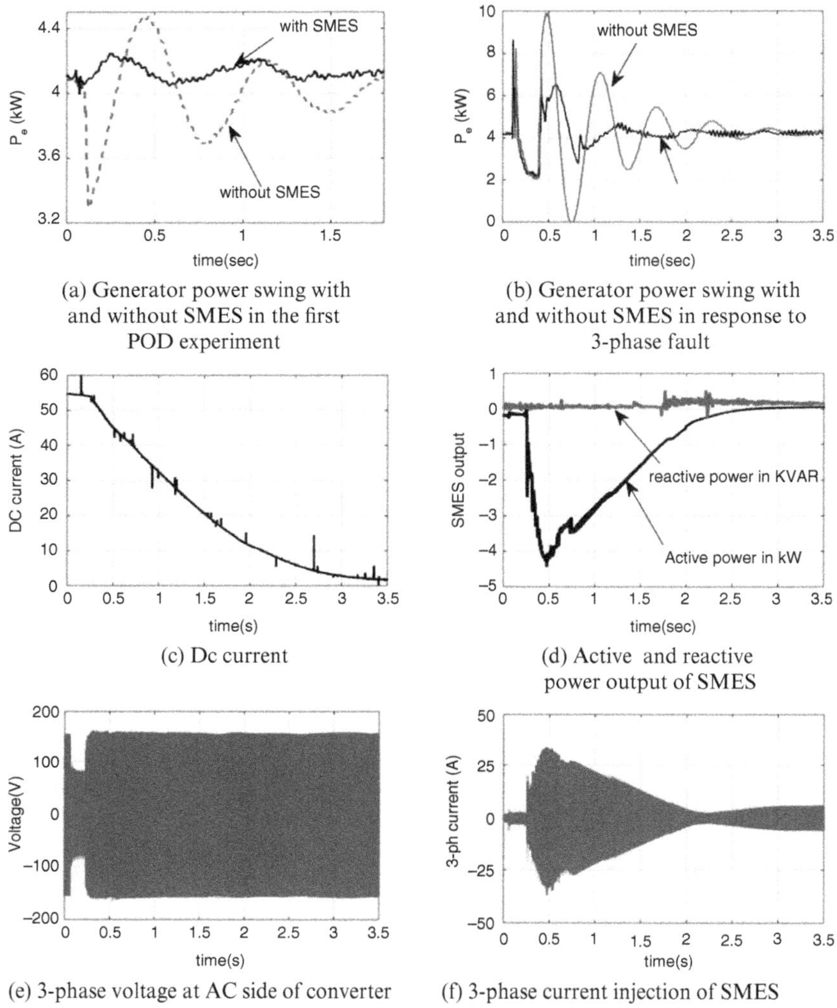

Figure 11.13 Dynamic system response under 3-phase fault.

The fault occurs at the middle of line 2, lasting for 300 ms, emulated by three-phase relay whose reclosure angle can be controlled accurately. Since line 1 had been out of service before the fault, this disturbance causes a much more severe power oscillation. Figure 11.13(b) depicts the process. Due to the very low voltage at the generator terminal where SMES is connected, the power chasing control scheme of the SMES, extracting or absorbing active power during a fault, can easily trigger the over current protection of the CSC. The SMES resumes the system stability control when the generator's terminal voltage recovers to 90% of the nominal value according to MCU's measurement. Therefore, during the period from 0.2 to 0.4 seconds, the dotted and solid lines coincide. After the time of 0.4 seconds, the dynamic stability of the AC power system is enhanced by the SMES as shown in Figures 11.13(c) to 11.13(f).

The field test of the SMES has been carried out in the Laohukou hydropower plant in Hubei province. Figure 11.14 shows the power system for the field experiment; the generator is connected to the nearest substation through a 1.7 km 10 kV transmission

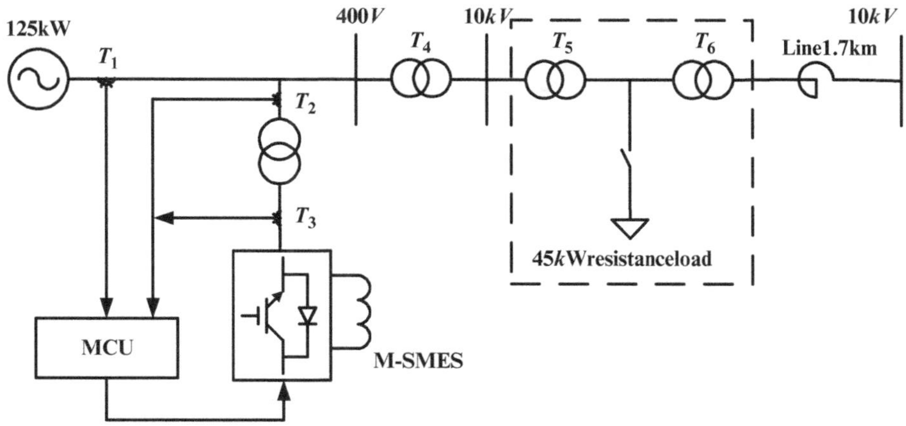

Figure 11.14 The field experiment circuit [11].

line. T_5 and T_6, being the same type of transformers as the step-up transformer T_4, are added to provide a suitable total inductance between the generator and the substation, which makes the system weaker so that the power oscillation can be observed more easily. The disturbance is created by suddenly closing the circuit breaker to connect a 45 kW resistance load between T_5 and T_6 as shown in Figure 11.14.

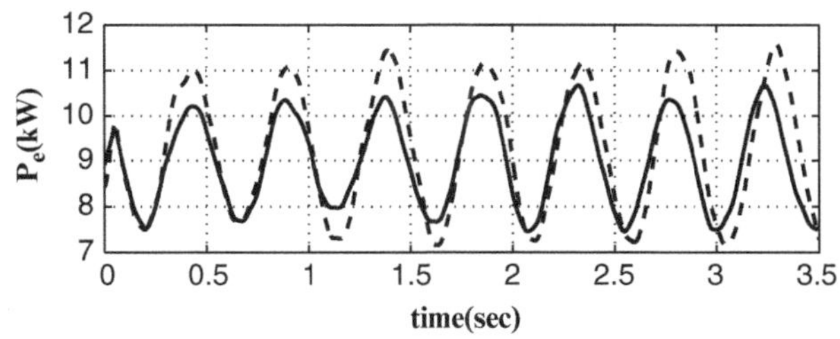

(a) Power oscillation with (solid line) and without (dash line) SMES.

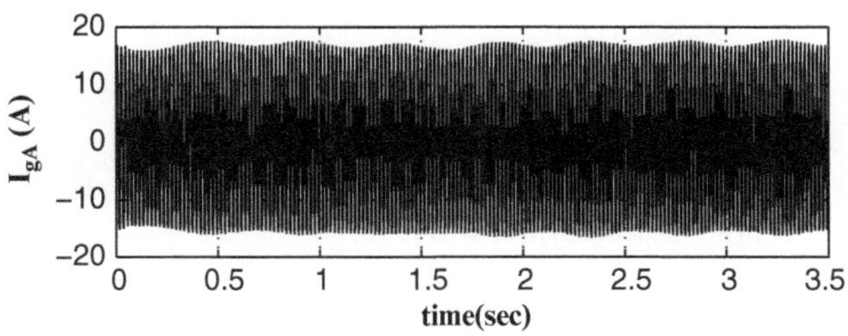

(b) 1 phase generator current with SMES.

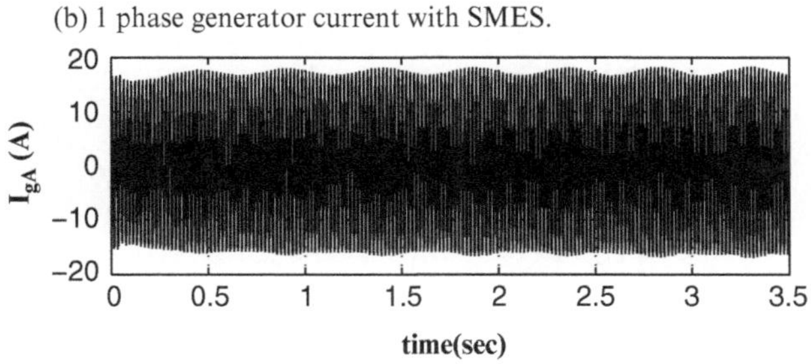

(c) 1 phase generator current without SMES.

Figure 11.15 Field test results, the generator is working at the point of 10 kW.

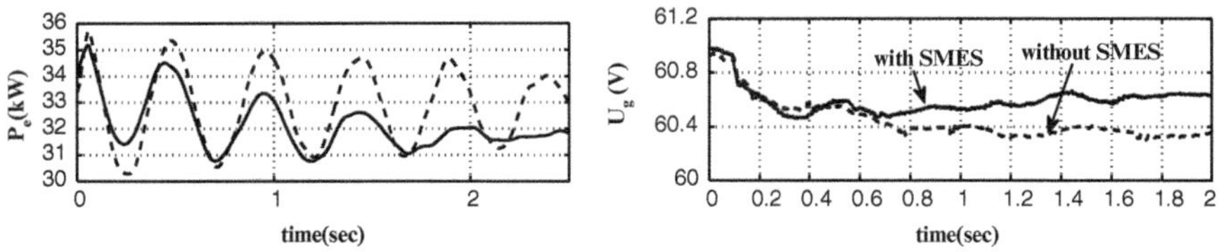

(a) Generator power oscillation with (solid line) and without (dash line) SMES at the working point of 33.5kW.

(b) The RMS voltage at the terminal of the generator

Figure 11.16 Field test results.

The power swing can be caused by the load disturbance when the circuit breaker is closed. Since the SMES has a relatively small capacity, the 125 kW generator is adjusted to operate at a low power production of 9 kW. The SMES is set to output active power only. From Figure 11.15, one can see that the power oscillation caused by 45 kW disturbance is suppressed by the SMES effectively.

Another identical test is conducted, but the generator output is adjusted to 33.5 kW. Figure 11.16 shows the active power fluctuation of the generator. The SMES achieved the best damping effect in this test and stabilized the generator in four swings. The fluctuation continues more than six cycles without SMES.

Another purpose of the field test is to verify that the SMES's ability to improve power quality. Laohukou hydropower plant is located at the end of the power system with severe voltage fluctuations. In this test, the test circuit remains the same as Figure 11.14, but the generator stops running. Only the SMES is connected to the low-side of T_4 directly. When a 45 kW load is suddenly connected, the bus voltage drops, as shown by the dotted lines in Figure 11.16(b). The SMES is activated after 500 ms to boost the bus voltage. Although the SMES has a limited capacity compared to the generator and system, and the bus voltage is difficult to return to the steady-state value, it is apparent that the SMES improved the voltage and restored it to closer to the steady-state value.

11.5 Conclusion

This chapter has focused on the design and implementation of the control strategies for the SMES used in damping power system oscillations. Two practical issues are addressed: first the increasing penetration of the renewable energy results in the variance of operating condition and uncertainties in system parameters; secondly, the capacity limits of the SMES lead to the saturation non-linearities. To handle the fluctuations and uncertainties, the H_2 / H_∞ mixed robust control algorithm is introduced. And the anti-windup compensator is designed to improve system stability under the situation that saturation happens. Finally, the implementation of the MCU for a SMES system was introduced. Laboratory and field test results were given.

References

[1] Zhu, J., Yuan, W., Qiu, M., et al. (2015). Experimental demonstration and application planning of high temperature superconducting energy storage system for renewable power grids. *Applied Energy*, 137: 692–698.

[2] Kinhal, V.G., Agarwal, P., and Gupta, H.O. (2011). Performance investigation of neural-network-based unified power-quality conditioner. *IEEE Transactions on Power Delivery* 26 (1): 431–437.

[3] Chen, X.Y., Jin, J.X., Xin, Y., et al. (2014). Integrated SMES technology for modern power system and future smart grid. *IEEE Transactions on Applied Superconductivity* 5 (24): 3801605.

[4] Hasanien, H.M. (2014). A set-membership affine projection algorithm-based adaptive-controlled SMES units for wind farms output power smoothing. *IEEE Transactions on Sustainable Energy* 5 (4): 1226–1233.

[5] Du, W., Wang, H.F., Cao, J., et al. (2012). Robustness of an energy storage system-based stabiliser to suppress inter-area oscillations in a multi-machine power system. *Generation, Transmission & Distribution, IET* 6 (4): 339–351.

[6] Wan, Y. and Zhao, J. (2013). $H\infty$ control of single-machine infinite bus power systems with superconducting magnetic energy storage based on energy-shaping and backstepping. *IET Control Theory and Applications* 5 (7): 757–764.

[7] Hemeida, A.M. (2010). A fuzzy logic controlled superconducting magnetic energy storage, SMES frequency stabilizer. *Electric Power Systems Research* 80 (6): 651–656.

[8] Shi, J., Tang, Y., Xia, Y., et al. (2012). Energy function based SMES controller for transient stability enhancement. *IEEE Transactions on Applied Superconductivity* 22 (3): 5701304.

[9] Yao, W., Jiang, L., Fang, J., et al. (2016). Adaptive power oscillation damping controller of superconducting magnetic energy storage device for interarea oscillations in power system. *International Journal of Electrical Power and Energy Systems*, 78: 555–562.

[10] Prabha, K. (1994). *Power System Stability and Control*. New York: McGraw-Hill.

[11] Fang, J., Wen, J., Wang, S., et al. (2013). Laboratory and field tests of movable conduction-cooled high-temperature SMES for power system stability enhancement. *IEEE Transactions on Applied Superconductivity* 23 (4): 5701607.

[12] Wu, X. and Lin, Z. (2012). On immediate, delayed and anticipatory activation of anti-windup mechanism: Static anti-windup case. *IEEE Transactions on Automatic Control* 57 (3): 771–777.

[13] Sajjadi Kia, S. and Jabbari, F. (2009). Modified anti-windup compensators for stable plant. *IEEE Transactions on Automatic Control* 54 (8): 1934–1939.

[14] Irie, F., Takeo, M., Sato, S., et al. (1992). A field experiment on power line stabilization by a SMES system. *IEEE Transactions on Magnetics* 28 (1): 426–429.

Part II

Mechanical Energy Storage and Pumped Hydro Energy Storage

12

Overview of Pumped Hydro Resource

Pål-Tore Storli

The Waterpower Laboratory, Dept. of Energy and Process Engineering, Norwegian University of Science and Technology, Alfred Getz' veg 4, 7491 Trondheim

12.1 Pumped Hydro Storage Basic Concepts

Pumped Hydro Storage (PHS, sometimes also called Pumped Storage Power (PSP) or Pumped Hydro Energy Storage (PHES)) is the storage of electrical energy as mechanical energy. The pumping operation of the unit is converting electrical energy into rotating mechanical energy, which is further converted into hydraulic mechanical energy. At the outlet of the pumping unit this energy is consisting of pressure, kinetic, and potential energy, and water flows towards an upper reservoir where the intent is to store the energy. When the water enters this reservoir the remaining kinetic energy is lost when the water settles, so eventually the mechanical energy is stored as pressure and potential energy in the upper reservoir where the water has settled. The energy is regenerated to electricity when the stored water is pulled by gravity through a turbine, similar to how a conventional hydro power plant works.

The basic concept involves one or several pumps (single or multi-stage) and one or several turbines operating between the upper and lower reservoirs. The pumps and turbines can have separate motors and generators (Quaternary set), or they can be mounted on a single long shaft, which is connected to a single motor/generator (Ternary set). Alternatively, the pumps and turbines can be replaced with a Reversible Pump Turbine (RPT), which operates as a pump when rotation is in one direction and as a turbine when rotating in the opposite direction (Binary set). Binary sets are most common because of the reduced installation cost, since one machine is capable of performing the action for both pumping and generating modes. This RPT option is limited to certain height differences between the upper and lower reservoirs, because the highest differences between reservoir levels require Pelton turbines for electrical power generation, and because a Pelton turbine cannot be operated in reverse, to act as a pump [1]. The details in the different types of machines and layouts of the mechanical equipment of a PHS will be the foci of several chapters in Part II: Chapters 13, 14, 15, and 16.

12.1.1 PHS Schematic Drawing

As Figure 12.1 shows, the pumped storage hydro plant is operated by cycling water between two reservoirs. After a pumping cycle, the upper reservoir is filled with water and used for storing potential energy. When conditions are right, the generating cycle

Advances in Energy Storage: Latest Developments from R&D to the Market, First Edition.
Edited by Andreas Hauer.
© 2022 John Wiley & Sons, Ltd. Published 2022 by John Wiley & Sons, Ltd.

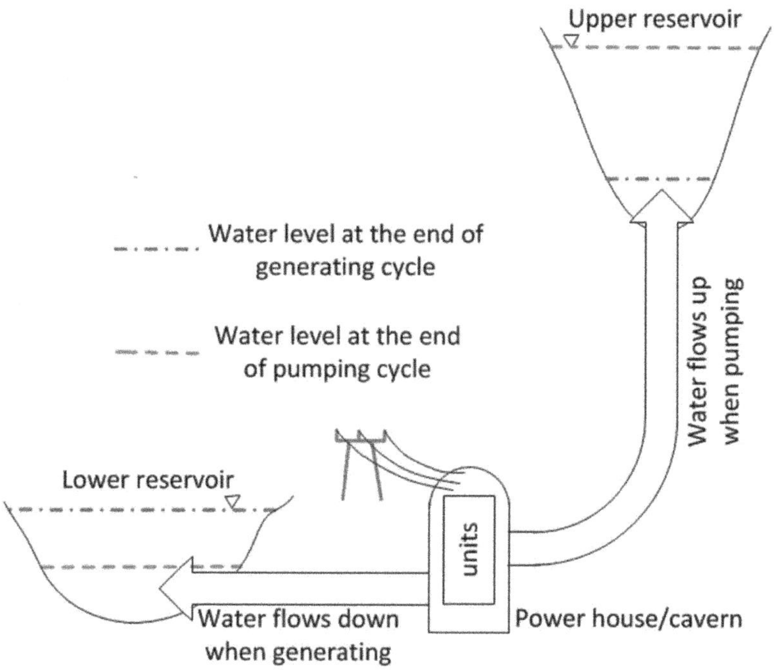

Figure 12.1 Operating principles of PHS (ClimatechWiki.org).

is initiated, and water is drained through an electricity generating unit. The upper reservoir will eventually be emptied, and the pumping cycle can be started again, when conditions are right for pumping. What defines "right" conditions varies for different PHS plants, but usually some consideration regarding electricity prices plays a major role.

A PHS plant is a large installation with big structures distributed over, often, large distances. There are a lot of construction and civil works needed to finalize and ultimately commission a PHS plant. The technical solutions and scales are highly site dependent, and is outside the scope of this article.

12.1.2 Pumping and Generating Cycles

When generating, electric energy is produced and sold, resulting in an income. When pumping, electric energy is consumed and must be bought, clearly leading to a cost. The reason why PHS (and in principle other electric energy storage technologies) have emerged is the fact that at some times electricity is traded at a low price, and sometimes at a high price. The reason for this price difference is that electrical energy is traded in markets, where buyers and sellers finally agree on a price based on balancing expectations for future supply and demand. As in all such markets, the price goes down if supply increase and/or demand goes down, or vice versa. This means that a net income can be generated by buying electrical energy when the price is low, and sold when the price is high. For this to be possible, the energy needs to be stored, and this storage is what energy storage technologies provide. This strategy of operation is called price arbitrage. It uses the difference in prices to generate a net income. Prices are the results of markets finding the balance between prices the producers are willing to sell for, and the prices consumers are willing to pay. This "willingness" differs a lot, both from the producers' point of view and from the consumers' point of view. Seasonal changes in temperatures and availability of power from sources that have seasonal cycles lead to a price that has a seasonal profile. Pumping in the low-price season and using the stored energy for generation in the high-price season is often called energy arbitrage. On a shorter timescale, there are weekly price profiles resulting from the consumption being higher in the weekdays than in the weekends, due to offices and shops being closed in weekends, etc. Some PHS have their operations tuned to match with this profile, and pump during low prices at weekends and generate during the weekdays. Ultimately, we end up at a timescale with unit "hour," because the market with the shortest timescale has this temporal resolution. This market takes care of the differences within the day, typically characterized with a high mid-day peak consumption (thus high price) period in some markets, or with two peaks typically in the morning and in the afternoon in other markets. The market mechanisms resulting in these prices due to the characteristics of production and consumption will lead to an operation that is very good at keeping the system in balance on an hourly timescale, at all times. Traditionally, this price arbitrage operation has been sufficient for generating a net income for the PHS plant operators. However, this is no longer considered to be the case [1]. Fortunately, for the operators, PHS plants are in possession of several other properties that provide the basis for sound and economic operation.

In the electrical system, there must always be a balance between the electricity consumed, produced, and stored. The previous section explained the balance that was found on an hourly basis by assuming (with statistical confidence) a consumption profile, and then balancing this with the right amount of production and storage. The actual consumption of electrical energy will for most of the time deviate somewhat from the assumed consumption. The necessary power needed to provide the instantaneous balance due to these deviations are available from contracts made in the markets that trade the short-term capability to produce power, so-called ancillary services (ancillary services: "those services necessary to support the transmission of electric power from seller to purchaser given the obligations of control areas and transmitting utilities within those control areas to maintain reliable operations of the interconnected transmission system," Wikipedia). The buyers in these markets are the Transmission System Operators (TSOs) that are responsible for maintaining stable operation in the transmission system. PHS have many flexible features that can help provide such ancillary services, such as (adopted from [2]):

- Providing contingency reserve; helping to provide a new balance if production or consumption changes pattern by starting to generate or stopping to pump, respectively.
- Providing load following; PHS is capable of fast ramping rates and can change the power generated quickly. Variable speed units can also provide load following to a certain degree in pumping mode.
- Frequency regulation; can help shift the system frequency by changing production on other units slightly, finding a new equilibrium between overall production and consumption at a new system frequency.

- Spinning reserve; operating at low power, ready for swift power ramp up.
- Synchronous condenser mode; some PHS can be synchronized on the grid with water evacuated from the rotating parts and the unit is operated as a voltage control unit for improving the state of the grid.
- Hydraulic short-circuit; operating the pump and turbine in a closed loop, ready to switch off the unit not needed for imminent operation and swiftly change to either pumping mode or turbine mode.

Providing such services is highly important for the stability of the electrical grid, and can be priced accordingly. Even when paying for the energy needed to pump, the pumping action can be turned into a net income mode of operation. The flexibility of PHS units in both operational modes and timescales are the key to providing such services.

At some physical locations there are interfaces where the PHS plant is connected to the grid, typically where the grid is connected to the PHS plant transformer. On the PHS side of this interface the plant operator has to decide on how to manage the plant and the connection to the grid; with all the possibilities and limitations intrinsic to the specific plant. On the other side of the interface, the grid operator has to decide on similar management problems.

12.1.3 PHS Basic Math. Calculation

The energy storage in a PHS power plant is by potential energy due to an increase in height at which a mass is positioned above a reference level. This is equivalent to the solid body potential energy increasing when mass is lifted against the pull of gravity. The analogy can be seen in Figure 12.2.

Figure 12.2 shows the similarities between solid body potential energy storage and fluid potential energy storage (PHS); namely that it is the increased center of gravity that is representative of the total energy stored in the mass that has been lifted. It also shows to the differences; the shape of the volume of the solid body is fixed whereas the shape

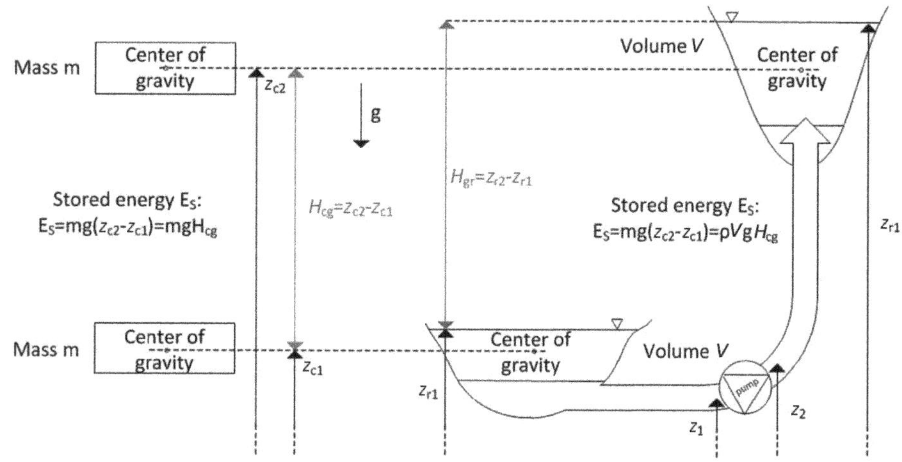

Figure 12.2 Basic storage mechanism for PHS.

of the volume of the water changes. This means that the difference in water levels between the upper and lower reservoir is possibly changing differently to the difference in levels for the center of gravity. This leads to a common misconception saying that the stored energy is represented by the difference in elevation of the water levels when the water volume is stored in the upper reservoir. It is the difference in elevation of the *center of gravity* of the pumped volume that represents the energy storage.

Consider a fluid parcel of fixed volume being lifted at the beginning of a pumping operation; the lower reservoir is filled, and the upper reservoir is empty, and the fluid parcel is lifted the height corresponding to the difference in the reservoir levels and gets an increase in energy directly proportional to this height. But as the pumping continues, fluid parcels must be lifted higher and higher because the lower and upper reservoir levels decrease and increase, respectively. The consecutive parcels need increasingly more and more energy as they are lifted, and if the rate of parcels lifted is constant, the rate of energy needed, *power*, increases. The average energy increase in lifting a number of parcels with a combined volume V is by considering the difference in the position of the center of gravity for the volume V before and after lifting. A pump with variable speed capabilities is able to maintain a constant discharge through it despite changing reservoir levels, corresponding to a constant rate of fluid parcels being lifted as in the fixed-fluid parcel example above. This pump would have to increase the *instantaneous* power that is used for pumping because this is dependent on the difference in water levels, which is increasing. We return to this using mathematical expressions later in this chapter.

A pump interacts with a fluid by reducing the mechanical energy of the fluid at its inlet and increasing the mechanical energy of the fluid at its outlet so that water drains into the pump inlet and flows away from the pump outlet; the fluid is pumped. The flow going into the inlet must be the same as the flow leaving the outlet by applying mass conservation principles. The mechanical energy (in unit meter, m) of a fluid is called "head", often denoted by H. It is expressed as the sum of static pressure and dynamic pressure, both converted to unit meter, and elevation, z, parallel to the gravitational acceleration, over some arbitrary reference [3], the terms on the right-hand side of Equation (12.1), respectively:

$$H = \frac{p}{\rho g} + \frac{c^2}{2g} + z \tag{12.1}$$

Where p is the static pressure (kgm^{-1}s^{-2}), ρ is the density (kgm^{-3}), g is the gravitational constant (ms^{-2}), c is the velocity (ms^{-1}), and z is the elevation (m). If subscript 1 and subscript 2 is used for the inlet and outlet of a pump, respectively, the difference in the head H between outlet and inlet is described by

$$\Delta H = \frac{p_2}{\rho g} + \frac{c_2^2}{2g} + z_2 - \left(\frac{p_1}{\rho g} + \frac{c_1^2}{2g} + z_1\right) = \frac{\Delta p}{\rho g} + \frac{\Delta\left(c^2\right)}{2g} + \Delta z = H_p \tag{12.2}$$

This increase in head is what the pump has transferred to the fluid when going through the pump, often called the "lifting height", denoted by H_p. The hydraulic power P_h (kgm^{2}s^{-3}) put into this flowing fluid is described by Equation (12.3) [3]:

$$P_h = \rho g Q H_p \tag{12.3}$$

Where Q (m^3s^{-1}) is the discharge through the pump. Assuming no losses in flow between the reservoirs and through the pump, the Energy equation (Bernoulli equation including the contribution of the pump) can be used to show that the lifting height of the pump must be equal to the gross head, H_{gr}, which is the instantaneous difference in elevation between the water levels in the reservoirs see (Figure 12.2 [3]). However, for real flows where friction is involved, the pump must also overcome these frictional losses, so the lifting height of the pump is equal to:

$$H_p = H_{gr} + h_l \tag{12.4}$$

Where h_l is the instantaneous total head loss (both from friction and singular losses) in the flow from the lower to the upper reservoirs (m). The hydraulic power added to the flow will therefore be equal to:

$$P_h = \rho g Q\left(H_{gr} + h_l\right) \tag{12.5}$$

Energy is the time integral of power, and the hydraulic energy E_h (kgm^2s^{-2}) put into the flow is expressed by integration of Equation (12.5):

$$E_h = \int P_h dt = \rho g \int Q\left(H_{gr} + h_l\right) dt \tag{12.6}$$

However as explained earlier, the lifting height H_{gr} increases due to the reservoir levels changing, in a highly site-specific manner linked to the surface area of the reservoirs. To complicate matters, the discharge Q through the pump is a function of the lifting height $H_{gr} + h_l$. This is what is commonly called "the pump characteristic"; the relationship between the flow through a pump and the lifting height the pump is able to generate. To complicate matters even further, the head loss h_l is a function of the discharge though the system, which is equal to the discharge Q through the pump. The time and mutual dependency of these variables are needed if a detailed investigation of the energy that is put into the flow by the pump is required. The energy that is stored is however somewhat easier to evaluate, because the losses are, after all, not stored:

$$E_{\text{stored}} = E_h - E_{\text{loss}} = \rho g \int Q\left(H_{gr} + h_l - h_l\right) dt = \rho g \int Q H_{gr} dt \tag{12.7}$$

Still, there are a lot of things that have to be known to evaluate this integral, and the appendix shows one example of this evaluation. However, we still know that the stored energy also can be found by the solid body analogy used at the beginning of this chapter, so:

$$E_{stored} = \rho g \int Q H_{gr} dt = \rho g V H_{cg} \tag{12.8}$$

12.1.3.1 Round-trip Efficiency (RTE)

We have now described the energy conversion chain from mechanical energy put into the flow by the pump to the stored energy in the reservoir. But the energy conversion chain for a PHS plant starts at the grid connection. Due to losses in the pump, we have more mechanical rotation energy, E_m, available than that which is put into the flow. And due to losses in the pump motor/drive, we have to put more electrical energy $E_{e,m}$, into the motor than what ends up as E_m. Due to losses in the transformation of voltage between the grid and the motor/drive of the pump we need to used more energy E_e, from the grid than what end up as $E_{e,m}$.

The details in these energy conversions are outside the scope of this chapter, but the flow of energy has to move through these steps: Starting with electrical energy E_e from the grid; to electrical energy $E_{e,m}$ on the terminals of the motor; to mechanical energy E_m on the rotor; to hydraulic energy E_h transferred by the pump to the flowing fluid; to the energy stored E_{stored}.

The concept of efficiency is handy when describing energy conversion chains. The efficiency, often denoted by η, is defined as the ratio of the energy that goes in, to the energy that comes out of, an energy conversion process.

$$\eta = \frac{E_{out}}{E_{in}} \tag{12.9}$$

We are now able to describe the energy conversion chain from electrical energy taken from the grid to energy stored in the upper reservoir

$$E_{e,fromgrid} = \frac{E_{e,m}}{\eta_{transf.}} = \frac{E_m}{\eta_{transf.}\eta_{motor}} = \frac{E_h}{\eta_{transf.}\eta_{motor}\eta_{pump}} = \frac{E_{stored}}{\eta_{transf.}\eta_{motor}\eta_{pump}\eta_{pipes}} \tag{12.10}$$

The subscript indicates which component or link in the energy conversion chain the efficiency relates to. By using the definition of efficiency, we can describe the combined efficiency of the pumping cycle as

$$\eta_{pumping} = \frac{E_{stored}}{E_{e,\ from\ grid}} = \eta_{transf.}\eta_{motor}\eta_{pump}\eta_{pipes} \tag{12.11}$$

This is however only half of the energy conversion chain of a PHS because the water must complete full cycle in order for power to be produced. This generating mode energy conversion also has losses involved, and can be described in a similar way by using efficiencies:

$$\eta_{generating} = \frac{E_{e,to\ grid}}{E_{stored}} = \underset{\text{turbine mode}}{\eta_{pipes}}\ \underset{\text{turbine mode}}{\eta_{hydraulic\ machine}}\ \underset{\text{turbine mode}}{\eta_{electric\ machine}}\ \underset{\text{turbine mode}}{\eta_{transformers}} \tag{12.12}$$

The overall quality of the conservation of energy for water going full cycle is called Round-trip Efficiency (RTE). It is quite simply the two generation mode efficiencies multiplied with each other:

$$\eta_{RTE} = \frac{E_{e,to\ grid}}{E_{e,from\ grid}} = \frac{E_{stored}}{E_{e,\ from\ grid}}\frac{E_{e,\ to\ grid}}{E_{stored}} = \eta_{pumping}\eta_{generating} \tag{12.13}$$

At some locations the climatic conditions lead to evaporation of the water in the upper reservoir reducing the amount of stored water. In these cases, an efficiency $\eta_{storage}$, can be included to take into account the loss of volume of stored water:

$$\eta_{RTE} = \eta_{pumping}\eta_{storage}\eta_{generating} \tag{12.14}$$

For all steps in the energy conversion chain of a pumped storage power plant the efficiency is high. This is seen in the energy flow diagram in Figure 12.3. Losses depend mainly on component efficiencies but also on friction losses and singular losses (valves, losses due to change of cross section areas or flow direction) in the waterways, and inefficient recovery of kinetic energy into pressure energy at the reservoirs. The high efficiency of turbines, pumps, motors, generators, and electrical equipment leads to

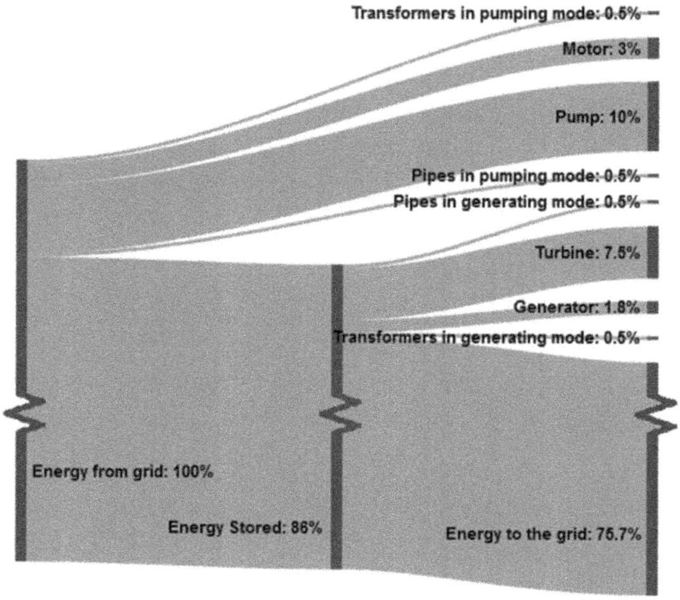

Figure 12.3 Energy flow diagram (Figures from Escher Wyss).

the overall RTE efficiency of existing PHS plants to be in the range of 70 to 80%. Modern PHS plants like Goldisthal in Germany (1060 MW, 2003), have a RTE between 80 and 85%.

When applicable, the reversible pump turbine can reduce investment costs because it is a single machine doing the job of two machines. However, due to the physics involved it is not possible to design a machine that will work in an optimal way as both a pump and in reverse as a turbine. In practice, the design must be made according to the pumping mode. This means that in turbine mode, the unit is operating at head lower than its intended design. The consequence is that the efficiency of the RPT in turbine mode is lower than for a turbine designed to operate between the same two reservoirs. This will reduce RTE, compared to PSHs having pumps and turbines, rather than RPTs. The problem is that in turbine mode the unit is spinning too fast for optimal energy conversion. This problem has been solved by modern RPTs having variable speed capabilities. These units are spinning slower in generating mode, than in pumping mode. The variable speed capabilities are also providing other benefits. In pumping mode, the power used for pumping can be chosen much more at will than for fixed speed RPTs, highly useful for PHS plants that have a large variation in lifting height. Actually, this was the main motivation for using variable speed machines at PHS plants in the first place. The previously mentioned Goldisthal has variable speed capabilities, partly resulting in the high RTE of the plant.

12.1.3.2 Energy Storage Calculation

As a calculation example, we can consider an upper storage volume, which is shaped like an inverted pyramidal frustum with bottom base of 45 000 m^2, a top base of 50 000 m^2, and a height of 10 m. This volume is approximately 47 5000 m^3, and the center of gravity is 4.91 m below the upper surface. When in the lower reservoir, the volume has a different shape; for the sake of this example let us say it is a rectangular body which is higher than the sides of the square base area; 100 meters high and base sides of approximately 69 m. The difference in the center of gravity for the two positions for the volume, H_{cg}, is 400 m, the layout is seen in Figure 12.4.

The stored energy in lifting this volume from the lower to the upper reservoir can be found using Equation (12.8):

$$E_{stored} = \rho g V H_{cg} = 1000 \cdot 9.81 \cdot 475000 \cdot 400 Ws = 517.8 MWh \tag{12.15}$$

If the difference in elevation between the reservoir levels *after* the volume has been pumped to the upper reservoir (erroneously) is used, the result becomes

$$E_{\substack{\text{stored}\\ \text{incorrectly}}} = \rho g V H_{\text{gr,ultimately}} = 1000 \cdot 9.81 \cdot 475000 \cdot (400 + 50 + 4.91) Ws = 588.8 MWh \tag{12.16}$$

The difference in these two figures show the dangers of calculating incorrectly, although this case had rather unrealistic reservoir shapes. However, there can be a substantial difference between the upper and lower reservoir levels, as well as the lower reservoir having a larger surface area than the upper reservoir. If the two reservoirs were to change position, the calculation example could be rather realistic, and the numbers would be the same as used in this example.

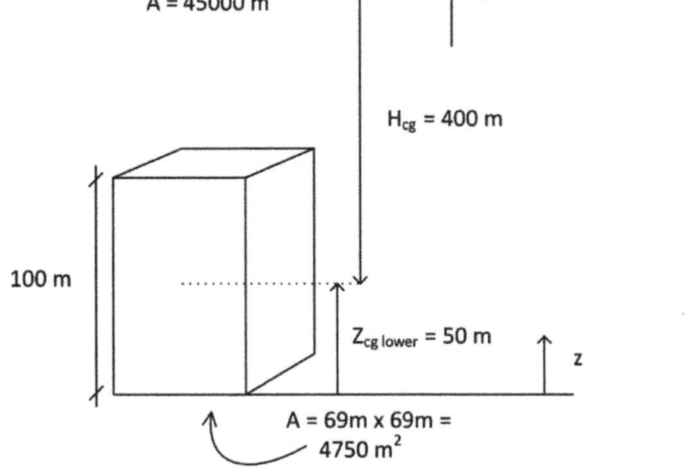

Figure 12.4 Layout of calculation example.

The power that can be extracted from this energy storage for the duration of 10 hours is 51.78 MW, and using typical efficiency figures (from Figure 12.3) a turbine generator designed for approximately 48 MW of rotating mechanical power should be installed. This is found using the turbine mode equivalent of Equation (12.3), including the turbine efficiency:

$$\begin{aligned} P_{m,turbine} &= P_{stored}\eta_{pipes}\eta_{turbine} = \frac{E_{stored}}{duration}\eta_{pipes}\eta_{turbine} \\ &= \frac{517.8MWh}{10h}\cdot 0.995\cdot 0.925 = 47.66MW \end{aligned} \tag{12.17}$$

The pump can be imagined having to replace the stored energy on the remaining 14 hours of the day, which means the power added to the energy storage should be 37 MW. The motor driving the pump must be designed to deliver 41.3 MW of rotating mechanical power to the pump impeller:

$$\begin{aligned} P_{\text{m,pump}} = \frac{P_{\text{h}}}{\eta_{\text{pump}}} = \frac{P_{\text{stored}}}{\eta_{\text{pump}}\eta_{\text{pipes}}} = \frac{E_{\text{stored}}/duration}{\eta_{\text{pump}}\eta_{\text{pipes}}} = \frac{517.8MWh/14h}{0.9\cdot 0.995} \\ = 41.3MW \end{aligned} \tag{12.18}$$

12.1.3.3 Head–Storage Capacity Relationship

The head–storage capacity relationship is relating the head to the storage. As discussed previously, increasing the head means that the water level increases in the upper reservoir, and decreases in the lower reservoir. Since practically all reservoirs have an increasing surface area at higher water levels, and most lower reservoirs have larger surface areas than the upper reservoirs, pumping water will gradually result in a slower change in the reservoir levels since the water is gradually being deposited to a larger surface area. The available storage volume at higher heads is therefore bigger than the available storage volume at lower heads. This feature is important when it comes to planning the operation of the power plant, making sure one doesn't buy or sell less or more energy than that which the PHS plant can cycle at a given time. Quite often this relationship is assumed linear, even if in most cases it is more asymptotic. If detailed relationships are needed, piecewise linear assumption can be applied, but has to be used with caution.

12.1.4 Sub-types of PHS

Several sub-types of PHS exist and are in operation [4]. Within the subtypes there are also a large variety in layouts, technologies used and practical solutions. The main subtypes are commonly called closed loop, semi-open loop, and open loop.

12.1.4.1 Closed Loop

The closed-loop PHS operates with a water volume that is finite because it is separated from natural waterways, as seen in Figure 12.5 and 12.6. In essence, two gigantic pools at different elevations are the upper and lower reservoir. The biggest closed-loop PHS plant in the world is Dinorwig PHS in Wales, UK (1728 MW). This PHS can be started up very quickly, and in 16 seconds it can go from 0 to 1728 MW in generation [5].

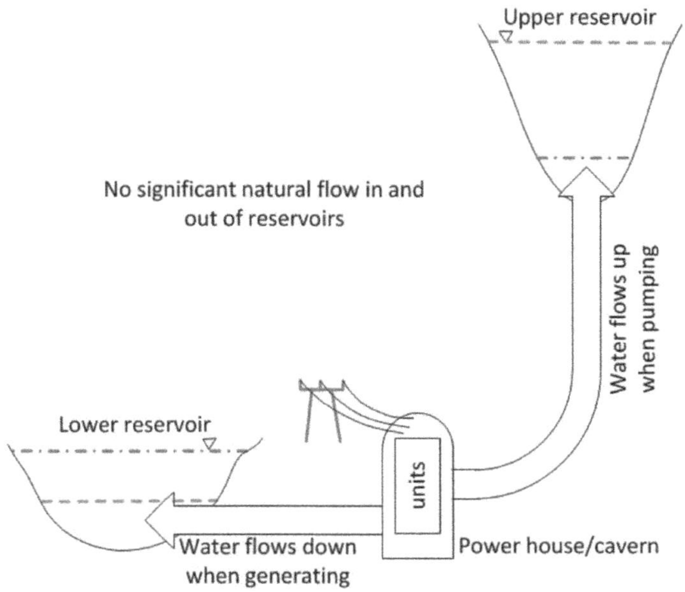

Figure 12.5 Schematics of a closed-loop PHS.

12.1.4.2 Semi-open Loop

A semi-open PHS has a natural and continuous flow through one of the reservoirs, most often the lower one, as in Figure 12.7. Okinawa's Yanbaru seawater PHS seen in Figure 12.8 is a special type of semi-open system that has the ocean as the lower reservoir.

Figure 12.6 Turlough Hill, closed-loop PHS in Ireland (DOE, Sandia National Laboratories, Strategen Consulting). DOE (Department of Energy U.S).

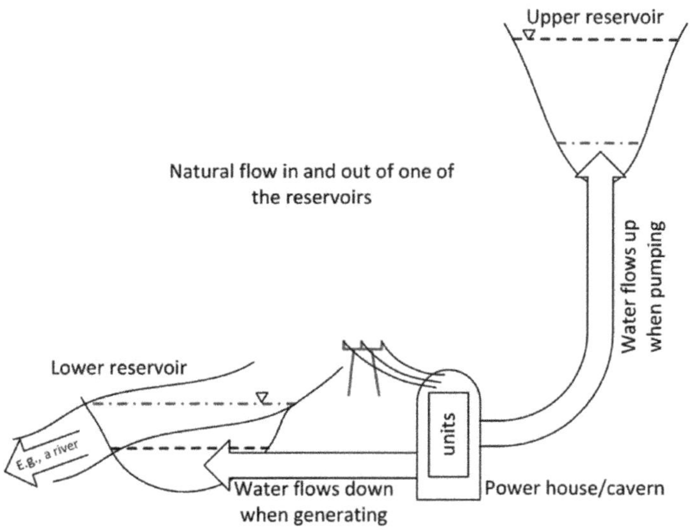

Figure 12.7 Schematics of semi-open PHS.

Another special type of semi-closed PHS is the project Muuga in Estonia, expected to be commissioned in 2028.[1] It uses the sea as the upper reservoir, and old underground caverns originating from granite excavation as the lower reservoir [1]. It is not uncommon to skip this classification and use only closed or open as classification groups for PHS.

12.1.4.3 Open Loop

An open system has a natural and continuous flow through both reservoirs. The layout might be as in Figure 12.9, which depicts a system often referred to as "pump-back" system. In natural waterways, conventional hydro plants might be present as well.

Figure 12.8 Yanbaru seawater PHS, (DOE, Sandia National Laboratories, Strategen Consulting) DOE. (Department of Energy U.S).

1 https://tyndp.entsoe.eu/tyndp2018/projects/storage_projects/1004

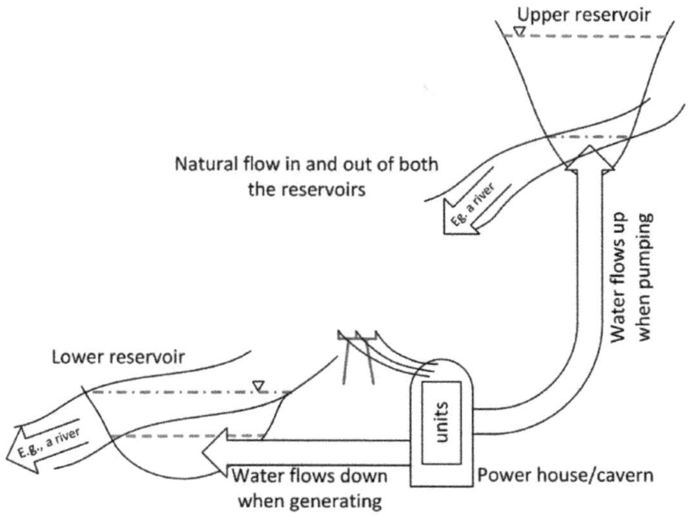

Figure 12.9 Schematics of open loop PHS.

However, the turbines might also be connected to the same conduit system as the pumps, as in the case of Saurdal PHS in Norway in Figure 12.10, its big reservoir can be seen in Figure 12.11.

12.1.5 PHS A Complex and Multidisciplinary Project

The different sub-types of PHS have emerged as a result of local geographical limitations for the establishment of a PHS plant, and the calculation example shows which

Figure 12.10 The machine hall of Saurdal PHS, Norway, the two furthermost units are Reversible Pump Turbines (Statkraft).

Figure 12.11 The upper reservoir of Saurdal PHS, a three-year energy storage reservoir and Norway's biggest (Statkraft).

considerations can be made regarding dimensioning of the units with respect to operational strategies and determining the powers for the design of the hydraulic-mechanic equipment. However, a complete techno-economic analysis will have to be performed to find the optimal layout and operation of the units, and this is a formidable task. The techno-economic analysis must take into account what is actually a possible project seen from the environmental, social, and political aspect. Ultimately, many iterations at all levels (technological, economical, societal, political) have to be performed before arriving at a solution that is ready to take one step closer to the construction phase. If a PHS project is not complex and multidisciplinary enough as it is, the topic of electricity markets emerges as a big uncertainty; the PHS plant is constructed for many decades of operation, but will be at the mercy of the future and rather uncertain electrical energy markets in which it sells its services.

12.2 Historic Perspective

Predicting the future is difficult, but collecting information about things that have happened can be quite difficult as well. Details of the historical development of PHS is surprisingly difficult to find and validate. The reason is that many PHS plants have been expanded, abandoned, replaced, and renamed, and so on. So, lists of PHS's existing and in operation at a given historic year in a given country is very difficult to establish. The following section aims to identify milestones and trends in the development, as well as where the majority of developments have occurred. The majority of the information regarding the year of commissioning and installed power presented for the period from 1950 to today have been found in the DOE Global Energy Storage Database (Department of Energy, US and Sandia Laboratories).

12.2.1 Before and Around 1900

PHS has existed for a long time as the first PHS plants were used in Italy and Switzerland in the 1880s and 1890s. The first recorded such system dates back to 1882 in Zurich, Switzerland, when a reciprocating pump was used to store energy. In 1894, a centrifugal pump of about 50 kW was used in Luino, Italy, to store water at a head of circa 70 meters and this was used to drive a spinning mill. In 1909 the first commercial pumped storage plant was built near Schaffhausen, Switzerland, with a capacity of 1500 kW [6]. It made use of a separate pump and turbine.

12.2.2 From 1920 to 1960

Rudolf-Fettweis-Werk with the plant Schwarzenbachwerk in Germany was the first large-scale pumped storage plant in Europe commissioned in 1926. Two Pelton turbines generating 46 MW and a variable-speed (likely two-speed) pump rated at 20 MW was installed [7, 8]. Pumped storage grew rapidly in the 1930s and 1940s especially in Europe, with a total installed capacity of approximately 630 MW in 1951, with Germany alone accounting for approximately 450 MW. In Europe, Germany and Austria had the most PSP developments, followed by Switzerland.

In Germany the Koepchenwerk/Herdecke plant (132 MW) was completed in January 1930 [9, 10]. Niederwartha PHS near Dresden was constructed making use of two 20 MW pumps and two 22.4 MW turbines [11], and was racing Herdecke for completion, losing by a few months [12]. In March 1930 its last unit was on grid. The commissioning of Waldshut, Reisach, Happurg, Geesthacht, and Tanzmühle in the 1950s (150 MW in 1951, 99 MW in 1955, 160 MW in 1958, 120 MW in 1958, and 35 MW in 1959) completed a period of high activity in Germany.

Austria developed, amongst others, Rodundwerk 1 (150 MW, 1943) and ended the period with Kaprun Oberstufe (112 MW, commissioned 1956) and Lunerseewerk (230 MW, commissioned 1958).

In Switzerland, Etzelwerk (137 MW) was commissioned in 1937, a power plant that generated electricity at a frequency of 16.66 Hz, used to power electrical engines on trains. Gougra PHS was commissioned late in the period (164 MW, 1958).

In France, Lac Noir PHS was commissioned in 1933, but an accident struck on 4 January 1934, when a pipe burst. Several people were killed in this accident (*L'Ouest-Éclair*, 6 January 1934). Two PHS plants were commissioned in France in the 1950s; Alrance (11 MW, 1951) and Argental PHS (48 MW), but the majority of the French PHS was developed later.

Around 1930 the reversible pump turbine made its entry, Pedreira was the first site in the world to have a reversible pump turbine installed, around 1937 [13, 14]. This original unit was replaced in 1953 by a 15 MW unit [15]. It was however theorized that there was a 25 MW limit to how big reversible units could be.

The United States started to develop PHS in this period, with Rocky River (31 MW) being the first in 1930 [16]. From the paper describing the project: "It takes 1.63 times as much energy to raise the water into the reservoir as is obtained from the same water when it flows back to generate electric power. The condition which makes the project economically sound and not a senseless dream is the fact that the stored water is used at seasons of the year and hours of the day when it has a high value, while the energy used for the pumping is obtained at times of the year and in the night hours from water that would otherwise spill over the dams or from steam plants that have to be kept in service anyway. Expressed differently, the pumping is done with low-cost off-peak energy while the energy from the stored water can be released at peak-load period when it has a high value." Almost 90 years later, the identical argument for PSP is being made, even if providing ancillary services are considered necessary for economic sound operation. In the same paper, it is stated that 41 pumped storage plants have been built or are being built in Europe, a number that has been very difficult to verify, pointing to the challenges faced when trying to find historical data.

In 1954, Flatiron Power and Pumping Plant was completed, having two Francis turbines, and a single, two-speed reversible pump turbine. The dam was mainly constructed due to irrigation needs. The pump turbine had a special design that gave no regulation of the turbine power [17]. Two years later, in 1956, the reversible pump-turbine at Hiwassee Dam was installed. The dam built was for flood control and power generation in the 1930s, and unit 2 (59.5 MW) was the first unit in the United States to be used solely for energy storage, and was used for periods of daily system peak loads [15]. The previously theorized maximum value for an RPT of 25 MW was put to the test and not found true at Hiwassee dam, and in 1956 the company Allis Chalmers found it was possible to have a true reversible unit operating on a single shaft. This was the breakthrough of reversible pump turbines [2]. Still today all current US PHS units are RPTs.

12.2.3 From 1960 to 2000

Since the 1960s, world PHS capacity expanded as a way to balance low and high electricity demand in grids with nuclear or coal fired electricity generation facilities that operated at a constant load.

In Europe, France started developing their PHS potential, the biggest plant being the huge Grand Maison (1820 MW commissioned in 1987). Italy started developing, with Chiotas (Entracque) as the biggest (1184 MW, commissioned in 1981), followed in size by Roncovalgrande (commissioned in 1973), Edolo (commissioned in 1985), and Domenico Cimarosa (commissioned in 1991), all at 1000 MW.

In this period Germany still had a high pace with several large units like Markersbach A (1050 MW commissioned in 1979) and Wehr (910 MW commissioned in 1976). Switzerland also had high activity with many units, but no big units. All of the UK PHS plants were built and completed between 6191 and 1982, the last being Dinorwig PHS with a massive 1728 MW capacity. The small country of Luxembourg commissioned Vianden Pumped Storage Power Plant (1096 MW, 1964), and Slovakia commissioned Čierny Váh Pumped Storage Power Plant (735 MW) in 1981. The Czech Republic had a few projects, the biggest being Dlouhé Stráně Pumped Storage Power Plant (650 MW, commissioned in 1996)

North America had a great development in this period, primarily in United States, commissioning 35 plants starting with Lewinston Pump-Generating plant in 1961 (240 MW). The real boom was in the 1970s, with 17 plants commissioned, including two plants with a generating capacity of more than 1500 MW, Raccoon Mountain Pumped Storage Plant (1650 MW, 1978) and Ludington Pumped Storage (1870 MW, 1973). In the 1980s, nine plants were commissioned, but only one with more capacity than 1500 MW; Bath County Pumped Storage Station, originally at 2100 MW (1985), became, at the time, the world's biggest with respect to installed power.

Central and South America has had very little PHS activity. Argentina commissioned two PHS plants, one in 1984, the Los Reyunos Pumped Hydro Storage (224 MW) and one in 1986, the Rio Grande-Cerro Pelado Hydroelectric Complex (750 MW). Brazil's single PHS plant, Pedreira Pumped Hydro Power Station (the world first PHS plant to use an RPT), was increased in size from 15 MW to 20 MW in 1988 [18].

Africa has developed little of its hydro power potential. Still, PHS plants exist. South Africa constructed three plants in this period, Drakensberg (1000 MW, commissioned in 1981), Palmiet Pumped Storage Scheme (400 MW, commissioned in 1988), and the first one commissioned in 1979, Steenbras Dam Pumped Storage Scheme (180 MW)

In Asia and Oceania, Australia developed several PHS plants, five plants were commissioned from 1966 to 1977, three of them in 1977. The biggest being the Tumuth Hydroelectric Power Station 3 (1500 MW, commissioned in 1973), Poatina Power Station (300 MW, commissioned in 1977), and Wivenhoe Power Station (500 MW, commissioned in 1984). South Korea commissioned three PHS plants; Cheongpyeong (400 MW 1979), Samrangjin (600 MW, 1985), and Muju (600 MW, 1995). Taiwan commissioned two large PHS plants; Minghu Pumped Storage (1008 MW, 1985) and Minglan Dam (1600 MW) in 1994. Thailand had some activity, the biggest being Srinagarind Hydroelectric Power Plant (720 MW, commissioned in 1991)

India also commissioned a few PHS plants, one of them a big plant, Srisailam PHS, 1670 MW, commissioned in 1981.

Japan developed most of its current PHS capacity from the early 1960s to 2000, becoming one of the countries with most PHS capabilities installed. Japan started a tremendous construction of PHS from the early 1960s to 2000. More than 30 PHS were constructed, the majority being over 500 MW, with almost 10 being over 1000 MW, amongst these the two giants Okumino (1500 MW, commissioned in 1995) and Okutataragi (1932 MW, 1998). The rather special pumped seawater storage power station at Okinawa, Yanbaru seawater PHS (30 MW) was commissioned in 1999.

China is another Asian PHS flower that was an even later bloomer than Japan, starting to increase its PHS capacity in the 1990s. Some seven PHS plants were constructed with five of them in the 1990s. Shisanling PHS (800 MW, 1997) being the biggest, but the true giants were still to come.

12.2.4 After 2000

Today, in grids with a significant penetration of renewable energy technologies, PHS is increasingly being installed to balance times of low and high electricity supply from variable wind and solar electricity generators and to assist in keeping the grid stable by providing ancillary services. As an indication of the importance of keeping the grid stable, the PHS Kops II in Austria was equipped with the possibility to operate in a so-called "hydraulic short-circuit". This operation uses the pump to power the turbine directly, and the water moves in a closed loop, obviously wasting energy. However, since both pump and turbine are on-line and operating filled with water, changing operational mode to either purely pumping mode or generating mode, is really fast. This allows for fast ramping of power either from the grid, or to the grid, depending on what the need is [19].

Starting with the last country from the previous period, China entered the new millennia with the commissioning of two large PHS plants; Guangzhou PHS (2400 MW, 2000), claiming the, at the time, world record for installed power, and Tianhuanping (1836 MW, 2000). Including these two, China has commissioned, or has under construction, 27 PHS plants with a total capacity of more than 30500 MW of PHS capacity. To put this into perspective, this is close to identical to the installed hydro power capacity of Norway, the sixth largest producer of electricity from hydro power in the world.

In Japan, the development of PHS halted towards the year 2000, with one PHS plant commissioned, Kazunogawa PHS (800 MW) in 2000. However, after the Fukushima Nuclear power plant accident in March 2011, Japan has installed 10 GW of wind power and other renewables. This explains the boost of PHS development in Japan seen after 2011, with at least six new plants at different stage of construction [20]. In 2011, the large Omarugawa PHS was commissioned (1200 MW). Kazunogawa had a pump expansion in 2014 of 400 MW, and Kyogoku had three units at different years of commission, each at 200 MW, in 2014, 2015 and 2020. The Kannagawa Pumped Hydro Plant units 1-6 have been commissioned during 2010's, with the last units commissioned January 1st 2020. All units combined the power is 2840 MW, second only to Bath PSP in the world. These units are pushing the limit for head, and are amongst the first to use splitter blade design in the RPT runner, claiming that turbine mode efficiency have been increased by 1.5% [21].

India has continued to develop PHS, but at a faster pace than before, with more than six PHS plants commissioned in this period, the biggest one being Sardar Sarovar Pumped Storage Power Station, rated at 1450 MW, commissioned in 2006.

In the United States only one project has been commissioned in this period, the small Olivenhain-Hodges (40 MW) in 2012, however between 2004 and 2009 Bath County Pumped Storage Station had its six units upgraded, reclaiming the current world record with the new plant capacity of 3000 MW. The Eagle Mountain project (1300 MW) and Iowa Hill Pumped Storage (1000 MW) are the biggest ones that have been announced. A few smaller projects have been announced as well. The announced Maysville PHS (1000 MW) recently got its permit rejected, due to a lack of project progress.

In Europe, several projects have been commissioned. In Germany, Goldisthal (1060 MW, commissioned in 2004) and Waldeck (140 MW, commissioned in 2010). In Italy, Pont Ventoux-Susa Hydroelectric Plant (158 MW) was commissioned in 2005. Austria commissioned five plants; Kops II (525 MW) in 2009, Koralpe (50 MW), Feldsee (140 MW) and Limberg II (480 MW) in 2011, and Reisseck II (430 MW) in 2014. Poland commissioned Dychów (90 MW) in 2005. In Portugal, two plants were commissioned: Alqueva (520 MW) in 2004 and Frades I (191.6 MW) in 2005. Russia commissioned Zagorsk PSP I (1200 MW) in 2003. Slovenia commissioned Avče (185 MW) in 2010. Spain commissioned the huge La Muela (2000 MW) in 2013. Switzerland commissioned the Linth-Limmern Expansion in 2017 (1000 MW), and plans for Lago Bianco PHS exists (1000 MW). In Portugal, Frades II (778 MW) and Venda Nova III (736 MW) was commissioned in 2016. It seems that most of the planned projects in Europe are expansions of already developed sites.

In Africa, Morocco constructed one PHS, Afourer Pumped Storage Scheme, (465 MW), commissioned in 2004, and have announced another one, Abdelmoumen Pumped Power Station (350 MW). In South Africa, the Ingula Pumped Storage Scheme had all its four units in operation in 2016 (1332 MW).

Figure 12.12 shows the development of the global installed PHS capacity from 1980 to 2012 (IEA). The increase in Asia and Oceania after 1990 is due to the development in China and Japan. "America" could have been substituted with "the United States", since there is very little PHS elsewhere in North America. A stagnation is observed in both Europe and North America, as the best project sites that provide a lot of installed capacity have already been developed.

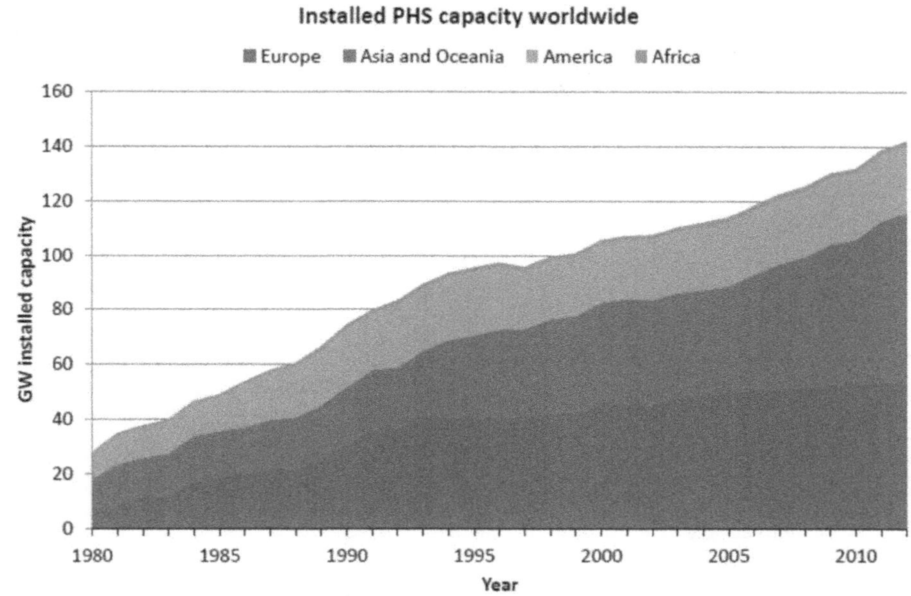

Figure 12.12 Worldwide installed PHS capacity 1980–2012 (Data from [13] and [14]).

12.3 Worldwide Installed Base

As with the historical data, complete and up to date worldwide statistics are difficult to find. Different statistics have to be compiled to form a relatively complete picture. The statistics presented in this section are taken from the DOE Global Energy Storage Database (Department of Energy, US and Sandia Laboratories). This database is not complete, and the author knows of a few missing entries, however, the data is considered to be useful for the purpose of giving an overview of worldwide installations.

With over 145 GW, (in 2015, IHA) the more than 300 PHS power plants represent nearly 99% of worldwide installed electrical storage capacity which is about 3% of global generation capacity. An overview of the number of plants by continent and country can be seen in Table 12.1.

12.4 The Future for PHS

In a world desperately trying to avoid the worst climate changes due to global warming, PHS seems to be a technology equipped with all the needs of a future electrical energy system. The technology is mature, however it is not without interesting and fundamental problems that could lead to technological improvement, if tackled. The technology provides energy storage on many different scales, providing electrical energy storage and generation on many different time-scales needed to have an instantaneous balance between over all production, consumption, and storage in an electrical grid. Furthermore, the technology provides a significant amount of rotating inertia to alternating current grids, a property that is highly important for stabilizing such grids. The draw back is that the technology requires favorable geographical conditions; water and elevation in an area suited for such large-scale construction work. A lot of research is

Table 12.1 Global installed base by continent and country, data aggregated from DOE Global Energy Storage Database.

Continent	Country	No. of plants	Installed PHS Capacity (MW)	Miscellaneous
N. and S. America	Argentina	2	974	43 plants of close to 23,760 MW installed power. Smallest PHS Plants: Flatiron 8.5 MW and O'Neil, 25 MW. Largest PHS Plants: Bath County, 3000 MW and Ludington, 1870 MW. All US units are reversible pump-turbines.
	Brazil	1	20	
	Canada	1	174	
	United States	38	22 560	
Europe	Austria	18	4680	171 plants of close to 61059 MW installed power. The smallest PHS plants are the Danish GO development Pumped Hydro Demonstration Project at only 6 kW, and Lepenica PSP Plant in Croatia, at 1.14 MW. The former is a project using a large bladder with soil on top as the upper reservoir, the soil giving the effect of additional pressure mimicking an elevated reservoir. The biggest plants are Dinorwig (1728 MW) and Grand Maisson (1820 MW), but the two close PHS plants Zagrosk-1 and Zagorsk-2 in Russia has a combined power of 2000 MW. The Linthal 2015 (Linth-Limmern expansion) of 1000 MW is not included in the entries for Switzerland, as is has status "under construction" in the DOE database, however it was inaugurated in September 2016.
	Belgium	2	1307	
	Boznia and Hercegovina	1	420	
	Bulgaria	3	1052	
	Croatia	3	281	
	Czech Republic	3	1145	
	Denmark	1	0.006	
	France	10	5812	
	Germany	28	6528	
	Greece	2	699	
	Ireland	1	292	
	Italy	18	7070	
	Lithuania	1	900	
	Luxembourg	1	1096	
	Norway	4	967	
	Poland	6	1745	
	Portugal	12	3546	
	Romania	1	53	
	Russia	5	2226	
	Serbia	1	614	
	Slovakia	4	1017	
	Slovenia	1	185	
	Spain	22	7990	
	Switzerland	16	5427	
	Ukraine	3	3173	
	United Kingdom	4	2828	

Table 12.1 (Continued)

Continent	Country	No. of plants	Installed PHS Capacity (MW)	Miscellaneous
Asia and Oceania	Australia	6	2541	107 plants with a capacity of more than 80 012 MW. Smallest plant is Okawachi PHP Station at 1.28 MW in Japan. Largest Huizhou PSP Station at 2448 MW in China. Japan has the world largest installed base of Units with Doubly-fed Induction Machines or DFIM adjustable speed machines. Japan numbers include 2 units of total 2,08 GW which has a status "operational" in the DOE database, but a commissioning date around 2020. Israel will commission its first PHS plant Gilboa at 300 MW in 2018.
	China	34	32 000	
	India	10	6772	
	Iran	1	1040	
	Japan	43	28 251	
	Philippines	1	709	
	South Korea	7	4700	
	Taiwan	2	2608	
Africa	Morocco	1	465	5 plants with a capacity of 3377 MW. The smallest being Steenbras Dam Pumped Storage Scheme of 180 MW in South Africa. The largest plant is also in South Africa, Ingula Pumped Storage Scheme at 1332 MW.
	South Africa	4	2912	

being performed on developing technical solutions for PHS that can make other sites possible for use; underground PHS using old mines or other cavities as upper and/or lower reservoirs, as well as investigating seawater options. But even if the technological challenges are faced and overcome, there is still the aspect of getting the PHS investment to be an economic sound action. The lack of sufficient transmission capacity can be a problem for some locations; possible PHS locations might be far away from consumers and generation units, so the energy must be cycled back and forth over long distances. If transmission capacity is insufficient, the construction of new transmission lines adds a great cost to the project. In some electricity markets, there can be a challenge returning the investment since capital costs of PHS investments are high and energy prices are, in general, relatively low. Creating markets that push the energy sector in the desired direction is a continuous challenge.

If finding historical and current data is a challenge, finding information regarding the future development of the construction of PHS plans is even more difficult. There seem to be a lot of projects in some stage of planning, but how close to actual construction they are is uncertain. One uncertain but announced project in Canada, Revelstoke, is planned for 4000 GW of power, claiming the world record by a large margin. This announced new world record indicates that PHS is a technology with a bright future, in the emerging new regime of energy supply from renewable sources. PHS is said to experience a renaissance. Since many of the best sites have already been developed, a lot of activity is seen, and expected to continue, in projects connected to expansion, rehabilitation, upgrading, and uprating of existing PHS plants.

What will the future look like? What disruptive energy technologies will emerge along the way? Some kind of fusion technology [22, 23]? Synthetic photosynthesis [24, 25]? Nobody knows, and this makes the energy industry a tremendously interesting field! Whatever the future looks like, it is likely to need some large-scale storage technology that can take care of discrepancies between energy production and energy consumption on many time scales. PHS is an excellent example of such large-scale energy storage technology.

References

[1] Pérez-Díaz, J.I., Cavazzini, G., Blázques, F., et al. (2014). Technological developments for pumped-hydro energy storage. https://www.eera-set.eu/wp-content/uploads/Technological-Developments-for-Pumped-Hydro-Energy-Storage_EERA-report-2014.pdf (accessed 21 April 2017).

[2] MWH. (2009). Technical analysis of pumped storage and integration with wind power in the pacific northwest. http://www.hydro.org/wp-content/uploads/2011/07/PS-Wind-Integration-Final-Report-without-Exhibits-MWH-3.pdf (accessed 20 April 2017).

[3] Cengel, Y.A. and Cimbala, J.M. (2014). *Fluid Mechanics, Fundamentals and Applications*. New York: McGraw Hill Education.

[4] Wänn, A., Leahy, P., Reidy, M., et al. (2012). Environmental performance of existing energy storage installations. StoRe, Deliverable D **3**. https://www.researchgate.net/profile/Eamon_Mckeogh/publication/267860326_Environmental_Performance_of_Existing_Energy_Storage_Installations/links/55e5a7bd08aede0b57368c19.pdf (accessed 21 April 2017).

[5] Du, P. and Lu, N. (2014). *Energy Storage for Smart Grids: Planning and Operation for Renewable and Variable Energy Resources (VERs)*. New York: Academic Press.

[6] Munoz-Hernandez, G.A. and Jones, D.I. (2012). *Modelling and Controlling Hydropower Plants*. Berlin: Springer Science & Business Media.

[7] EnBW Energie Baden-Württemberg AG. (2009). Wasser ist Energie: Wasserkraft bei der EnBW. https://www.enbw.com/media/konzern/docs/energieerzeugung/broschuere-wasserkraft-bei-der-enbw.pdf (accessed 20 April 2017).

[8] EnBW Energie Baden-Württemberg AG. (2011). Das Rudolf-Fettweis-Werk in Forbach: Ein Pumpspeicherkraftwerk auf dem Weg in die Zukunft. https://www.enbw.com/media/konzern/docs/energieerzeugung/anzeige_infoveranstaltung_psw_forbach.pdf (accessed 20 April 2017).

[9] Hartmann, E., Helmer, F., and Latrille, W. (1987). Replacement of the existing pumped storage plant Koepchenwerk at Herdecke/Germany by a new 150 MW pump-turbine plant. *Water Power and Dam Construction, Conference on Upgrading and Refurbishing Power Plants*. Strasbourg.

[10] RWE Power AG. (2006). Power: Perspectives 2006. http://www.rwe.com/web/cms/mediablob/en/1761208/data/1761186/1/rwe-power-ag/power-perspectives/blob.pdf (accessed 20 April 2017).

[11] VDE Dresden. (2014). 85 Jahre PSW Niederwartha. https://www.vde-dresden.de/resource/blob/666676/13fbdcd237cea921136140cd11128913/85-jahre-psw-niederwartha—download-data.pdf (accessed 20 April 2017).

[12] Weinhold, K. (2014). ENSO-Blog Energiegeschichte: 85 Jahre Pumpspeicherwerk Niederwartha. http://www.enso-blog.de/energiegeschichte-85-jahre-pumpspeicherwerk-niederwartha (accessed 15 February 2017).

[13] DOE "Pedreira PSP", Department of Energy, USA and Sandia National Laboratories. (n.d.). DOE Global Energy Storage Database. http://www.energystorageexchange.org/projects/874 (accessed 20 April 2017) Strategen Consulting LLC.

[14] IEA International Energy Agency. (2014). Energy storage technology roadmap, technological annex. https://www.iea.org/media/freepublications/technologyroadmaps/AnnexA_TechnologyAnnexforweb.pdf (accessed 20 April 2017).

[15] TETSESME. (1981), The east tennessee section of the American Society of Mechanical Engineers, Hiwassee dam unit 2 reversible pump-turbine (1956). https://www.asme.org/getmedia/4a637087-db16-4eb1-8240-e2a1a40e9d8c/67-Hiwassee-Dam.aspx (accessed 21 April 2017).

[16] Heslop, P.L. (1928). A hydro electric plant that pumps its own water supply, the rocky river hydro-electric plant. *44th Annual Meeting, Conneticut Society of Civil Engineers.*

[17] Denton, S. and Britt, H. (1955). Pumped storage and hydro generation at Flatiron power plant. *Transactions of the American Institute of Electrical Engineers. Part III: Power Apparatus and Systems* 74 (3): 614–618.

[18] Schwery, A., Fass, E., Henry, J.M., et al. Evolution of pump storage technology. http://lme.epfl.ch/files/content/sites/lme/files/users/wetter/private/pdf_divers/cigre/posters/EvolutionPump.pdf (accessed 20 April 2017).

[19] Mitteregger, A. and Penninger, G. (2008). Austrian pumped storage power stations supply peak demands. *World Pumps* 2008 (500): 16–21.

[20] Fairley, P. (2015). IEEE spectrum, "a pumped hydro energy-storage renaissance." 2015. http://spectrum.ieee.org/energy/policy/a-pumped-hydro-energystorage-renaissance (accessed 21 April 2017).

[21] Peltier, R. (2006). Kannagawa hydropower plant, Japan. http://www.powermag.com/kannagawa-hydropower-plant-japan (accessed 29 March 2017).

[22] Hornfeld, H. (2016). Strategic opportunities in fusion energy. *Journal of Fusion Energy* 35 (1): 102–106.

[23] Sykes, A., Costley, A., Windsor, C., et al. (2016). Compact fusion energy based on the spherical tokamak. *26th IAEA Fusion Energy Conference- IAEA CN-234.* http://www.firefusionpower.org/IAEA_2016_EXP3-36_Costley.pdf (accessed 21 April 2017).

[24] Aresta, M. and Dibenedetto, A. (2014). Synthetic photosynthesis for the conversion of large volumes of carbon dioxide into energy-rich molecules: Saving fossil fuels by recycling carbon. *Advances in Organometallic Chemistry and Catalysis: The Silver/Gold Jubilee International Conference on Organometallic Chemistry Celebratory Book*, 513–527. Wiley Online Library.

[25] Burki, S.J. (2017). *Technology and Development. Rising Powers and Global Governance: Changes and Challenges for the World's Nations*. New York: Palgrave Macmillan, 235–252.

Appendix

This appendix describes the development of equations for power delivered to the pumped flow, and the energy put into the flow as a result of pumping over some duration. We will assume no energy losses, so friction is neglected in waterways and in the pump, or it is the stored energy we find, as indicated by Equation (12.7). Units and notations are the same in this Appendix as in the main article.

Pumped fluid energy is the time integral of power put into the fluid flow:

$$E_{\mathrm{h}} = \int P_{\mathrm{h}} dt = \int \rho g Q H dt = \rho g \int Q(t) H(t) dt \tag{A1}$$

In which the flow and head are functions of time, respectively; Q = Q(t), H = H(t). Let us omit this notation in the following and use Q and H. First, let us describe the flow as a function of time. It can be expressed as the time integral of dQ/dt from a starting time t_0, up to the time of interest t, adding to the initial flow:

$$Q = \int_{t=0}^{t} \frac{dQ}{dt} dt + Q_0 \tag{A2}$$

dQ/dt can be expanded as:

$$\frac{dQ}{dt} = \frac{\partial Q}{\partial H} \frac{\partial H}{\partial t} \tag{A3}$$

The first partial derivative of Q with respect to H is the instantaneous slope of the pump characteristic;

$$\frac{\partial Q}{\partial H} = \left(\frac{\partial H}{\partial Q} \right)^{-1} = -C^{-1} = -\frac{1}{C} \tag{A4}$$

where C is a positive constant for a stable pump characteristic. Let us for simplicity assume that this is a constant for the entire pumping action (This will imply a time varying power which we will see that we get, but this is the power *put into the flow*, discarding any efficiency of the pump and possibly a constant power input to a pump and pump drive). Piecewise linear integration can be applied if the variation in C is significant over the duration of the pumping action.

The second partial derivative of H with respect to time is dependent on how fast the water levels change when the flow moves water from the lower to the upper reservoir. There is a water-level reduction in the lower reservoir because water is removed, and the time rate of this change is the instantaneous flow divided by the surface area A_{l} of the lower reservoir. But the same water makes the level increase in the upper reservoir, and the time rate of this increase is the instantaneous flow divided by the surface area A_{u} of the upper reservoir. The total time rate of head increase seen by the pump is these two added together:

$$\frac{\partial H}{\partial t} = \frac{Q}{A_{\mathrm{u}}} + \frac{Q}{A_{\mathrm{l}}} = \left(\frac{1}{A_{\mathrm{u}}} + \frac{1}{A_{\mathrm{l}}} \right) Q = \frac{Q}{A_{\mathrm{eq}}} \tag{A5}$$

Where A_{eq} have been introduced as an equivalent reservoir area for the sake of expression simplicity.

Substituting Equation (A3), (A4), and (A5) into Equation (A2) gives:

$$Q = -\frac{1}{C}\int_{t=0}^{t}\frac{Q}{A_{eq}}dt + Q_0 \tag{A6}$$

Differentiating Equation (A6) with respect to time (or directly combining Equation (A5) and (A6) as in Equation (A4)) we get:

$$\frac{dQ}{dt} = -\frac{Q}{A_{eq}C} \tag{A7}$$

Separating variables and integrating to solve for Q gives:

$$[\ln Q]_{Q_0}^{Q} = -\frac{1}{A_{eq}C}(t - t_0) \tag{A8}$$

Which rearranged gives:

$$Q = Q_0 e^{-\frac{(t-t_0)}{A_{eq}C}} \tag{A9}$$

We have now solved for Q, and we need to solve for H, we go back to Equation (A5) and substitute the newly found Equation (A9) for Q:

$$\frac{\partial H}{\partial t} = \frac{Q_0 e^{-\frac{(t-t_0)}{A_{eq}C}}}{A_{eq}} \tag{A10}$$

Integrating with respect to time gives:

$$H = \frac{Q_0}{A_{eq}}\int_{t_0}^{t} e^{-\frac{(t-t_0)}{A_{eq}C}} dt + H_0 =$$

$$\frac{Q_0}{A_{eq}} e^{\frac{t_0}{A_{eq}C}}\int_{t_0}^{t} e^{-\frac{t}{A_{eq}C}} dt + H_0 = -CQ_0 e^{\frac{t_0}{A_{eq}C}}\left(e^{-\frac{t}{A_{eq}C}} - e^{-\frac{t_0}{A_{eq}C}}\right) + H_0$$

$$H = -CQ_0\left(e^{-\frac{(t-t_0)}{A_{eq}C}} - 1\right) + H_0 \tag{A11}$$

Inserting expressions for head and flow into the equation for power to get:

$$P_h = \rho g H Q = \left(-CQ_0\left(e^{-\frac{(t-t_0)}{A_{eq}C}} - 1\right) + H_0\right)\cdot Q_0 e^{-\frac{(t-t_0)}{A_{eq}C}}$$

$$P_h = \left(CQ_0^2 + H_0 Q_0\right)e^{-\frac{(t-t_0)}{A_{eq}C}} - CQ_0^2 e^{-\frac{2(t-t_0)}{A_{eq}C}} \tag{A12}$$

Integrating Equation (A12) to find the expression for energy E_h:

$$E_h = \int_{t_0}^{t} P_h dt = \rho g \int_{t_0}^{t} QH dt = \rho g \int_{t_0}^{t} \left(\left(CQ_0^2 + H_0 Q_0 \right) e^{-\frac{(t-t_0)}{A_{eq}C}} - CQ_0^2 e^{-\frac{2(t-t_0)}{A_{eq}C}} \right) dt =$$

$$\rho g \left(\left(CQ_0^2 + H_0 Q_0 \right) \int_{t_0}^{t} e^{-\frac{(t-t_0)}{A_{eq}C}} dt - CQ_0^2 \int_{t_0}^{t} e^{-\frac{2(t-t_0)}{A_{eq}C}} dt \right) =$$

$$\rho g \left(-\left(CQ_0^2 + H_0 Q_0 \right) A_{eq} C \left(e^{-\frac{(t-t_o)}{A_{eq}C}} - 1 \right) + CQ_0^2 \frac{A_{eq}C}{2} \left(e^{-\frac{2(t-t_o)}{A_{eq}C}} - 1 \right) \right) \tag{A13}$$

13

Pumped Storage Machines – Motor Generators

Stefanie Kemmer and Thomas Hildinger

ewz ag (company), Switzerland

The purpose of hydro power plants – specifically of pumped storage hydro power plants – had changed in former decades from pure energy storages into dynamic grid support.

In general, these plants must satisfy two main aspects:

- fast and flexible drive operation characteristics and
- regulation of power input and output.

To apply these two essential requirements there are different types of electrical machines that can possibly be chosen for hydro power plants – either fixed-speed or variable-speed machines.

Fixed-speed drive systems can be connected directly to the electrical grid since they provide a constant frequency. Usually those drive systems consist of a synchronous machine and are installed in hydro power plants operating at a constant speed level.

While the power input is controlled by adjusting the turbines distributer, there are some challenges concerning the variable power output. Generally, it can be realized either by means of hydraulic or by means of electrical regulation. A variable-speed machine enables variable power input into the electrical grid as well as variable power supply out of the electrical grid. Thus, energy deficits and surpluses are more easily compensated by pumped storage hydro power plants with variable-speed drives.

There are a few power plants with hydraulic operation system, such as Kopswerk II in Montafon, Austria or Wehr in Germany. The machinery train is oriented vertically in Kops and horizontally in Wehr; in both cases it carries a ternary system consisting of a motor generator, a Pelton turbine, and a pump set. The variability of power input enables the so called hydraulic short-circuit. If a power surplus arises, which is not sufficient to drive the pump set, the turbine gets activated as well. Electrical power generation is controlled by adjusting the distributer of the turbine.

Aside from the flexibility of the power input there are relatively high losses caused by the operation of two hydraulic systems, which usually should be prevented at any time.

For the realization of an electrical variable-speed drive system there are two different methods: on the one hand it is possible to install a conventional synchronous machine which is connected to the grid via a voltage source inverter (VSI). On the other hand, the operation of a doubly fed induction machine is sufficient, as already used in wind power plants. The converter is placed between the rotor and the power grid.

Advances in Energy Storage: Latest Developments from R&D to the Market, First Edition.
Edited by Andreas Hauer.
© 2022 John Wiley & Sons, Ltd. Published 2022 by John Wiley & Sons, Ltd.

Whereas the frequency in the power network remains constant, the machinery train consisting of a reversible pump turbine and generator is able to drive variable speeds.

13.1 Synchronous Machine Fixed Speed

13.1.1 Operating Principle and Components

The synchronous machine can be illustrated in a single-phase equivalent circuit diagram as in Figure 13.1.

In generator mode the air gap of the synchronous machine is affected by the stator current. This reaction and the associated voltage drop are summarized in main reactance X_{1h}. The electrical field of the stator coils caused by self-induction is not linked with the rotor field and is expressed with $X_{1\sigma}$. Resistance R_1 represents the occurring current heat losses of the stator coils.

The rotor circuit does not find any consideration in the single-phase equivalent circuit diagram since the synchronous speed of the rotary field and rotor does not induce any voltage.

The following equation can be derived from Figure 13.1:

$$\underline{U}_{\mathrm{p}} = \underline{U}_1 + R_1 \cdot \underline{I}_1 + j\,(X_{1\sigma} + X_{1h}).\,\underline{I}_1 \tag{13.1}$$

To simplify Equation (13.1) we can summarize X_{1h} and $X_{1\sigma}$ as the so-called synchronous reactance X_{d}. Additionally the ohmic part of the stator reactance R_1 is negligible as soon as we are talking about larger machines. Therefore, we can change the stator voltage equation and the corresponding single-phase-equivalent circuit diagram as in Figure 13.2.

$$\begin{aligned} \underline{U}_{\mathrm{p}} &= \underline{U}_1 + R_1 \cdot \underline{I}_1 + j\underbrace{(X_{1\tilde{A}} + X_{1\mathrm{h}})}.\underline{I}_1 \\ \underline{U}_{\mathrm{p}} &= \underline{U} + j\left(X_{\mathrm{d}}\right) \cdot \underline{I}_1 \end{aligned} \tag{13.2}$$

The main feature that distinguishes a synchronous machine from an induction machine is the phase shift operation by which inductive or capacitive reactive power in electrical grids can be compensated. If the machine runs in idle mode and the excitation current gets increased, the synchronous generated voltage increases at the same time. A current is created to balance the voltage difference. This current is 90° ahead of the terminal voltage U. That reflects the behavior of a condenser and the state is called over excitation.

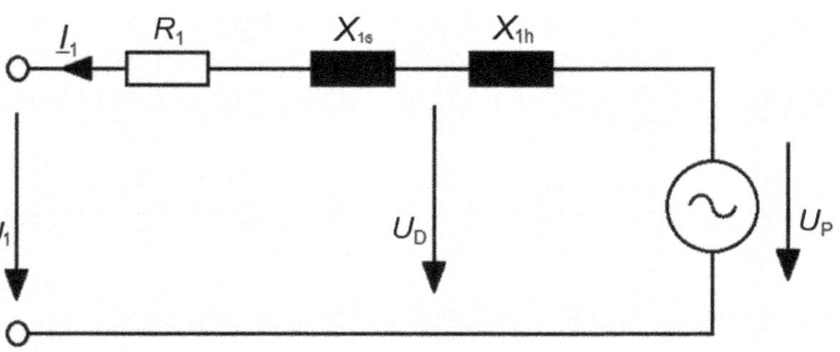

Figure 13.1 Single-phase-equivalent circuit diagram of a synchronous machine.

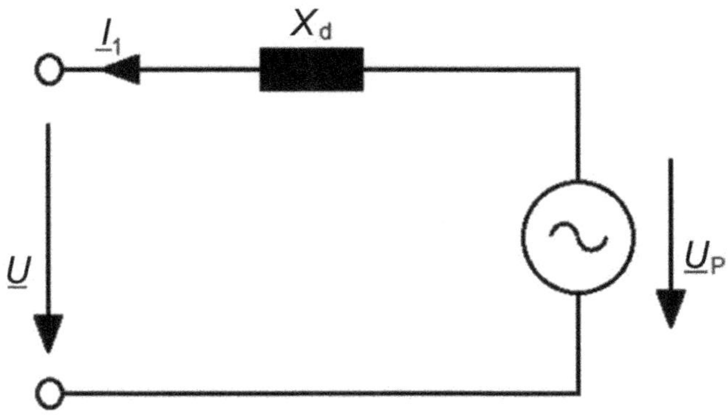

Figure 13.2 Simplified single-phase-equivalent circuit.

In rare instances an inductive reactive power output is necessary, which means a decrease of the excitation current and consequently a decrease of the synchronous generated voltage. Thus, the emerging current for the voltage compensation lags 90° behind the terminal voltage and the machine perform as a coil. This mode is called under excitation.

With the help of these two phase shift modes the cosρ can be enhanced essentially. Both modes are pictured in Figure 13.3.

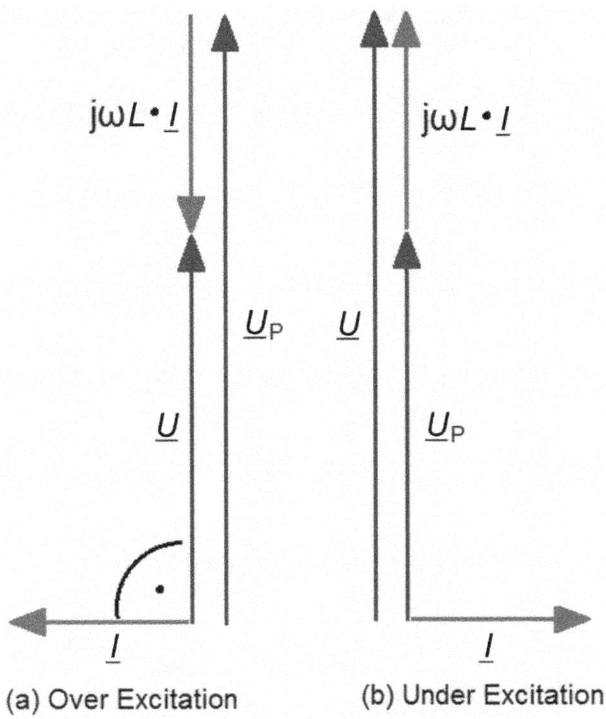

Figure 13.3 Over and under excitation of a synchronous machine.

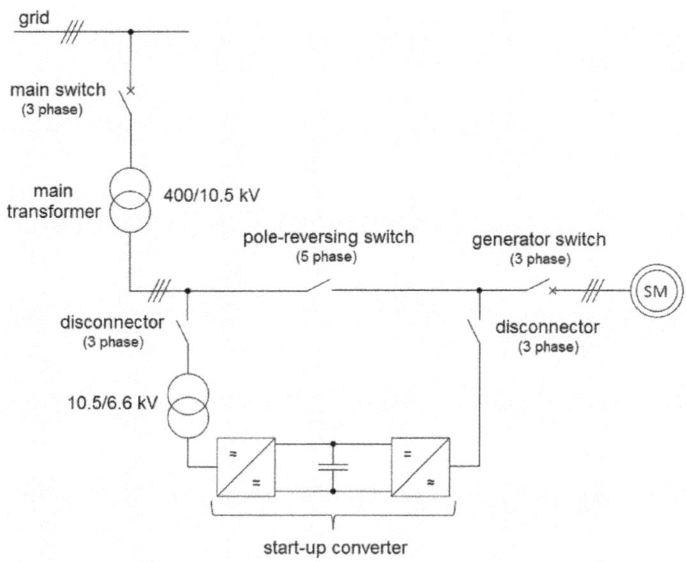

Figure 13.4 Conventional synchronous machine with bypass.

In practice, interconnection of an ordinary synchronous machine to the grid looks as shown in Figure 13.4.

Figure 13.4 shows the connection including a start-up converter. After starting the machine up to a certain speed, the bypass will open. For pump mode the pole-reversing switch will close appropriately. Figure 13.5 illustrates another interconnection possibility without a start-up converter and pole-reversing switch. Instead, a voltage source inverter is installed to enable acceleration of the machine as well as the switch-over into pump mode.

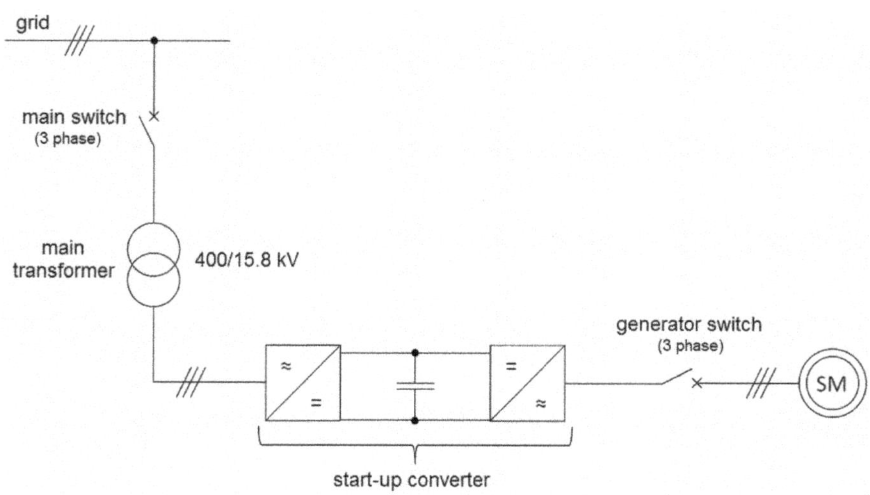

Figure 13.5 Conventional synchronous machine without a bypass.

13.1.2 Excitation System

The functional principle of a synchronous machine operates on the same principle as an induction machine with a rotating field generated in the stator. Therefore, the stators of both machine types are similar. The difference is due to the rotor. It contains a closed magnifying system, whereas the number of poles corresponds with those of the stator field. This magnifying system can be generated in two different ways: electrical or with a permanent magnet. The operating behavior is based on the excitation system.

- Electrical Excitation system

So called salient and non-salient pole machines need an electrical excitation system to feed the rotor coils with a variable DC-current flow. This means two different electrical networks for only one machine – a three-phase power system for the stator and a DC-system for the rotor.

In several cases, a smaller synchronous machine attached to the drive shaft with a rectifier in serial is more common. The fast adaption of the excitation voltage and a relatively high speed are beneficial to the user. Additionally, only one electrical system is necessary for this way of excitation. A drawback, on the other hand, is the use of slip rings, which are needed. It means shorter maintenance intervals and thus higher costs.

For large conventional synchronous machines, the static excitation system is preferred. The main advantages are fast-acting excitation, which means an extremely fast field current adjustment for changing operational conditions and the extended stability limits caused by the steady state stability. By using an internal free control and power electronic system a high control speed can be achieved.

- Permanent Magnet System

Commonly used in smaller machines, permanent magnets are used to excite the rotor system. This drive mode brings some essential benefits:
  - no excitation supply is necessary,
  - the higher density of electromagnetic energy permits smaller constructions, and
  - higher efficiency means lower temperature increase.

As a disadvantage it should be mentioned that the excitation field cannot be varied making the operation in the field weakening range impossible.

13.1.3 Converters for Grid Connection

Generally, converters are utilized to convert an AC-current via a DC-link into an AC-current with different frequency. In the past few years, converter technology has been enhanced significantly.

For each business case there is an adequate technology. Usually, converters that are already available on the market are integrated. Otherwise, a new development is needed, which results in increasing costs.

The dimensions of the converter system, namely the necessary power electronics, depend on the maximum current load – the higher the maximum current load, the larger the converter's dimensions. Understandably, that brings higher construction costs and space requirements with it. The three main layout criteria are as follows:

- high efficiency,
- large power variations in pump mode, and
- requirements concerning the start-up process.

S-called Cycloconverters are frequently used as start-up converters for conventional synchronous machines.

There are step-up and step-down Cycloconverters, depending whether the frequency is to be increased or decreased. Commonly used are step-down converters to reduce the frequency. They consist of a relatively large number of thyristors, in which the triggering of the electronics is very complex.

A disadvantage is the fixed power factor. The constant reactive power is a result of the phase controlled modular. If the reactive power is manipulated by varying the firing angle, the active power is modified adequately. In that way the phase shift between current and voltage remains unchanged. Additionally, high stress peaks can emerge, which affect the live time of the installed electrical motor negatively. Because of the high harmonic content filters are necessary to protect the power grid as well as all other electrical components from improper high stresses.

13.1.4 Power Chart

For every electrical machine there is a power chart that illustrates the safe operating area. In Figure 13.6 you can find an example of a synchronous machine's power chart.

The chart can be explained as follows:

If the machine is operating in its nominal point, which is marked with "nominal current" and the turbine valve is choked while the excitation current is constant,

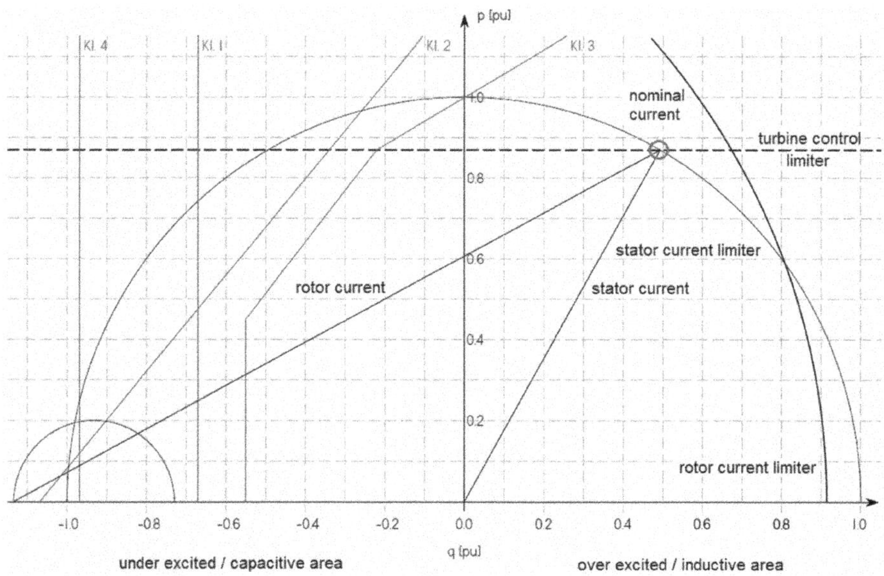

Figure 13.6 Power chart of a synchronous machine.

the operating point is moving semi-circular in the direction of the q-axis. Hence, the reactive power output increases, whereas the active power output decreases progressively. If the turbine valve gets closed entirely, the operating point is located on the q-axis, which means a maximum reactive power output. In this case there is no active power supply to the grid. When decreasing the excitation current with an unchanged turbine valve, originating from the nominal point, the operation point is shifted on the horizontal axis in the left direction. Meanwhile the mechanical power remains constant, which results in a constant active power output, whereas the inductive reactive power is decreased. If the operating point is located on the p-axis, reactive power is neither absorbed nor fed. By a further decrease of the excitation current reactive power becomes negative, which means that the generator absorbs reactive power. At the same time the polar wheel angle increases up to the stability limit of theoretically 90° (for a salient pole machine). A maximum of the capacitive reactive power is reached. By reducing the turbine valve the active power decreases.

To avoid an overload of the rotor and its excitation system there are delimiters installed that limit the operation within the over excitation area. The active power is limited by the maximum drive capacity of the turbine and is commonly between 80 and 90% of the generator's nominal power. In Figure 13.3 this limit is represented by a dashed horizontal line. The under excited area is limited by the machine current or the stability limit of the synchronous machine which is – as already mentioned – at a polar wheel angle of around 90°. If this limit is reached the machine tilts over and stops immediately.

The four lines marked Kl 1 to Kl 4 in Figure 13.6 indicate the parameters of the protection devices. Line K3 illustrates the regulation of the polar wheel angle. If this line is exceeded, an alarm occurs. After reaching line K1 respectively K2 the under excitation protection interferes which causes the shutdown within 30 seconds if the excitation current remains too low. When lines K3, K2, and K1 are exceeded very fast, so that line K4 gets passed as well, the machine stops immediately to avoid the loss of synchronization.

In the opposite direction the excitation current can increase so that the line of the rotor current or stator current limiter gets passed. In this case an alarm occurs primarily. After a certain time the temperature rises inadmissibly so that the thermal element is activated, which leads into a shutdown of the generator.

13.1.5 Load Change (P/M/n – Curve)

In this chapter, we study the dynamic performance of a synchronous machine in turbine mode, operating at a fixed speed. Figure 13.7 shows a load change, which is expressed by the straight black line. That means an increase of active power issued by the synchronous machine. Since the rotational speed has to remain constant to keep up with the grids frequency, the horizontal line shows an approximately horizontal progress.

The unstable line (light gray) – which expresses the actual active power – tries to follow the new set point as fast as possible. Examining the moment directly after the load change there is a short decrease initially before increasing in an unstable manner. That illustrates a pressure drop, which occurs after opening the gate valves before the

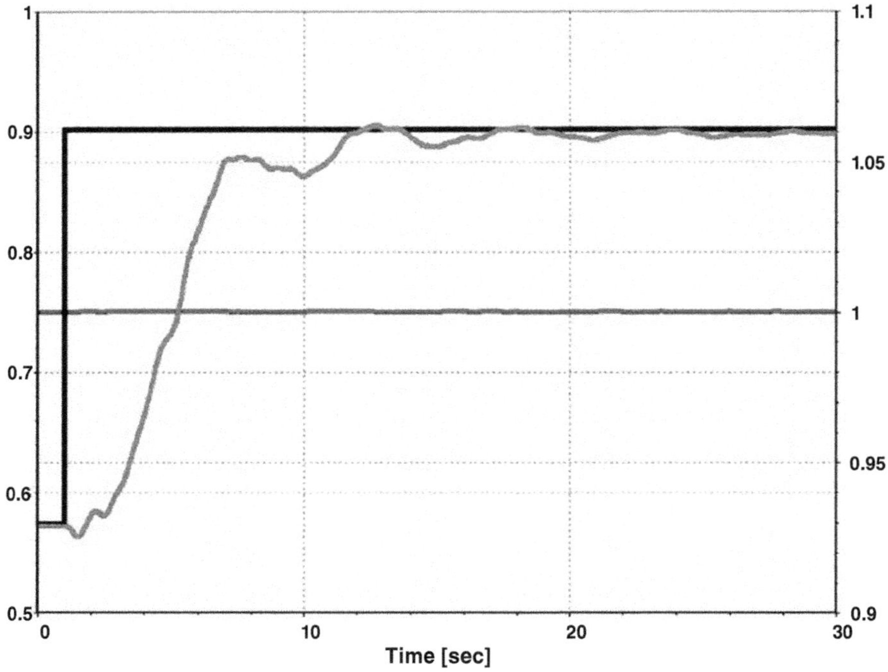

Figure 13.7 Synchronous machine – active power and speed in turbine mode.

water flow accelerates. Since the power depends on the pressure and water flow, there is a short drop of actual power. The transient time is approximately 60 seconds until oscillations are controlled.

If the active power set point gets decreased the process has to be passed through in the opposite direction. The gate valves get closed, which can cause so called water hammer when closing the valves to fast. The resulting shock waves generate high pressure vibrations that cause loud noises and ultimately mechanical damages. This must be prevented by sufficient measures.

In pump mode a controlled variation of active power is impossible since there is no scope of influence regarding the input power.

13.1.6 Advantages/Disadvantages

There are several appreciable advantages concerning the Synchronous machine, operating on fixed speed:

- cost efficient caused by less material input;
- refitting into a variable speed is possible;
- compact construction; and
- proved and highly reliable technology since several decades.

On the other hand one should consider the disadvantages as well:

- only fixed speed in pump mode possible, and
- start-up converter most commonly necessary in motor (pump) operation,

13.2 Doubly fed Induction Machine Adjustable Speed (DFIM)

13.2.1 History

For more than 20 years, scientists in Japan have been researching in the field of variable speed drives. Thus, one can find several electrical machines in hydro power plants with an adjustable speed range in operation. In 1990 the first machine had been launched in Yagisawa with a rated power of 85 MVA, constructed by Toshiba. The system contains a Cycloconverter whereas IEGT (Injection Enhanced Gate Transistor) inverters are preferred nowadays.

The latest variable speed machine in Europe is called Frades II and is installed in Portugal. The constructor Voith provided two pump turbines with a rated power of 380 MW each, two doubly feed induction motor generators with a rated power of 433 MVA each, a frequency converter and the control technology as well as hydraulic steel components; speed range is from 350 rpm to 381 rpm. The system had been set to go on grid in year 2016.

In Switzerland the largest hydro power plant projects are called Nant de Drance with a rated speed of 6 × 157 MW and a speed range of 428.6 rpm + 7 – 14 % and Linthal 2015 with a rated speed of 4 × 255 MW and 500 rpm +/- 6 %. Both are currently under construction and the first pump turbine was connected to the grid in 2020.

Until the present day, statements about the quality of the two variable speed drives in pumped storage hydro power plants are quite different, providing a good reason to analyze both concepts. All regulations must be strictly adhered to, while efficiency, costs, maintenance, and other aspects are observed and evaluated.

This way we want to clarify the operating principles of the fully fed synchronous machine and the doubly fed induction machine. Conclusively, the main criteria for the decision between these two drive methods will be mentioned. In addition, we identify advantages and disadvantages that will help choosing a variable speed system.

13.2.2 Operating Principle and Components

The electrical machine is connected to the grid via the stator, while the rotor is electrically excited via slip rings. Thus, the system can be controlled in both directions at variable speed by varying the rotor voltage. The speed range is relatively narrow with a maximum of 20%, which is caused by the converter that is dimensioned for the slip power only.

Active power can be controlled in turbine and in pump mode, whereas reactive power is controlled via the VSI in the rotor circuit. The dynamic operation can be described as follows: the polar wheel angle and the electrical active power follow the mechanical power with a certain delay, which is controlled via the turbine regulator. The polar wheel angle and the stator voltage are measured during which the control of the other electrical variables occurs. A deviation between mechanical and electrical power generates a change in speed and slip. In practice the electrical power is adjustable very quickly while the turbine regulator alters the speed quite slowly.

The balance of active power of a lossless DFIM can be described as in (13.3).

$$P_\delta = P_{mech} + P_2 = P_{mech} + s \times P_1 \quad \text{where} \tag{13.3}$$

P_1 = active power stator side
P_2 = active power rotor side
P_δ = transferred air gap power
P_{mech} = mechanical power
s = slip

This equation shows that the speed is adjusted by the slip and the rotor frequency.

The converter in the rotor circuit is constructed almost identically to one in a fully fed synchronous machine but operates in a different way. The rotor is fed via the VSI that corresponds with a controlled voltage source. The pulse inverter on the grid side is an active controlled grid converter. Via the converter on the machine side pulse inverter voltage pulses are guided from the intermediate voltage circuit in sinus-PWM-method to the rotor circuit. This causes a sinusoidal current of variable frequency and amplitude. If the speed is over-synchronous the frequency becomes negative, which corresponds with a reversal of the rotating field. Equation (13.4) illustrates the frequency of the rotor current:

$$f_r = P_{mech} - (n_{(r)} \times p)/120 \tag{13.4}$$

P_{mech} = rotor frequency
P_{grid} = grid frequency
n_r = rotor speed
n = number of poles

Thus, the rotor has to carry a three-phase winding system which is fed by a three-phase current of a variable frequency depending on the speed. The limit of the rotor power is given by the maximum thermal load. The converter is constructed depending on the slip power, which must be transferred. A large speed range compared to the synchronous speed requires a large slip power, which increases costs and size of the VSI.

The performance of the machine is determined by the power electronics. Both the rotor side and the grid side of the VSI must be constructed based on IGBTs (insulated-gate bipolar transistor) or MOSFETs (metal-oxide-semiconductor field-effect transistor) to enable a power flow in both directions. If the generator is operating in under excited mode the rotor has to be fed while in over excited mode the power flow occurs in the opposite direction. Eventually a system is able to operate in a so called four quadrants drive mode.

The greatest advantage of the Doubly Fed Induction Machine is the converters compact construction. As already mentioned, it has to provide the slip power for the rotor which is about 20–30 % of the machine rated power. Plant operators achieve more flexibility by operating a DFIM although there are limits concerning the speed range. Figure 13.8 shows the circuit principle.

To provide a safety start-up and change over, there is a integrated switch – the so-called pole-reversing switch. Generally, the plant operator has to decide individually if switches and protection systems are necessary. It mainly depends on the safety requirements.

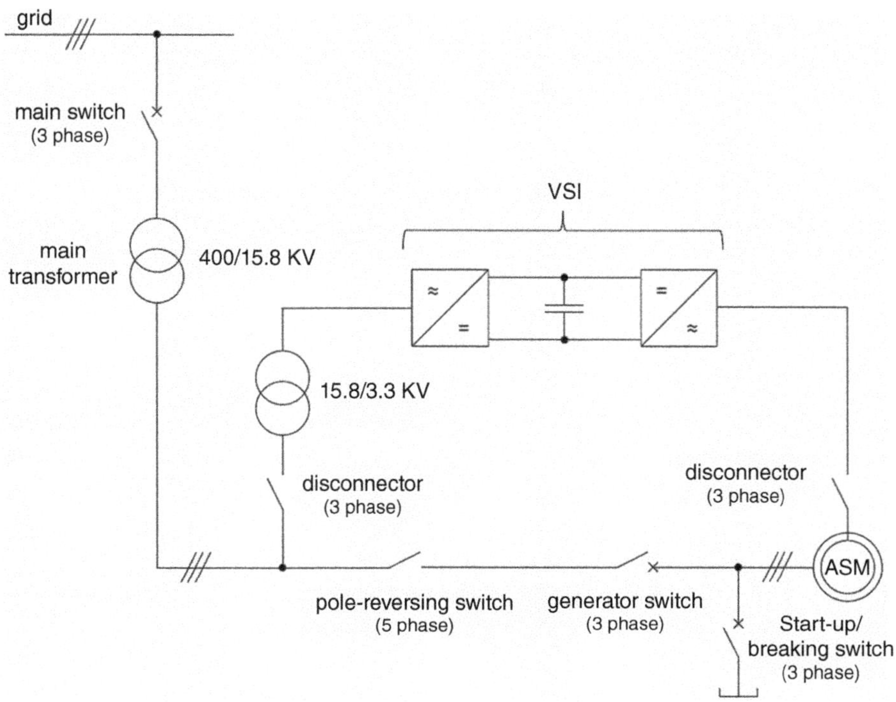

Figure 13.8 DFIM – Circuit principle.

13.2.3 Converters for Grid Connection

By means of a three-level converter the line-to-line voltage can be adjusted in five steps: U_{dc}, $U_{dc}/2$, 0, $-U_{dc}/2$ and $-U_{dc}$ whereas U_{dc} stands for the voltage of the intermediate circuit. The system usually is installed in high voltage equipments.

To reach output voltages of 15 kV to 20 kV converters can theoretically be connected in series. By implementing a three-level converter one can reduce the stress of the machine winding's isolation and extend life expectancy of the electrical machine. This is benefitted by the flat increase and decrease of the converter output voltage and the thus lower voltage peaks.

The converter is separated in two sections: transformer side and machine side, which can be constructed based on IGBTs or MOSFETs as already mentioned. If higher voltage peaks are expected it can also be constructed with IGCTs. The power electronics must have a good electric strength compared to other converter types since the relatively high intermediate voltage is switched by the power electronics parts. Switching losses should be very low since there are numerous switching processes while operating a DFIM. Additional filters help to smooth currents and voltages while space requirements have to be considered as well. Whereas a Cycloconverter requires a transformer with two secondary taps, the three-level Converter needs a transformer with one secondary tap only.

With a multi-layer converter, the advantages of a three-level converter can be extended while the disadvantages are minimized. By cascading several converters harmonics can be reduced which results in saving filters.

Another converter technology is called LCI (Load-Commutated Inverter), this has an intermediate current circuit, consisting of a smoothing choke instead of a capacity. In general, LCI are based on SCRs – a special thyristor type. By igniting via the gate, the thyristor becomes conductive. As the converter types before, the input AC current gets smoothed and on the output side converted into an AC voltage with varying frequency. The SCRs turn-on and turn-off instantly depending on the rotating motor which is why the SCR are called load-commutated. This requires a relatively complex control causing high costs. Thus, the LCI technology is not very efficient and reliable compared to other PWM controls.

13.2.4 Load Chance (P/M/n – Curve)

Additionally, the reaction of a DFIM after a load change is illustrated by a simulation result to that shown in Figure 13.9. The steady black line in Figure 13.9 shows the set point of the active power, which steps from 0.57 pu up to 0.9 pu. One can observe that the speed collapses for a short period of time, which is visualized by the light gray uneven line. This is explained by the increased need of active power that is obtained by the speed. Since the system usually strives to keep the nominal speed, the turbines gates will be opened as fast as possible to speed up the machine to its nominal value. This process is executed after around 25 seconds. If the active power is increased in pump mode one can see curve progressions similar to those in Figure 13.9.

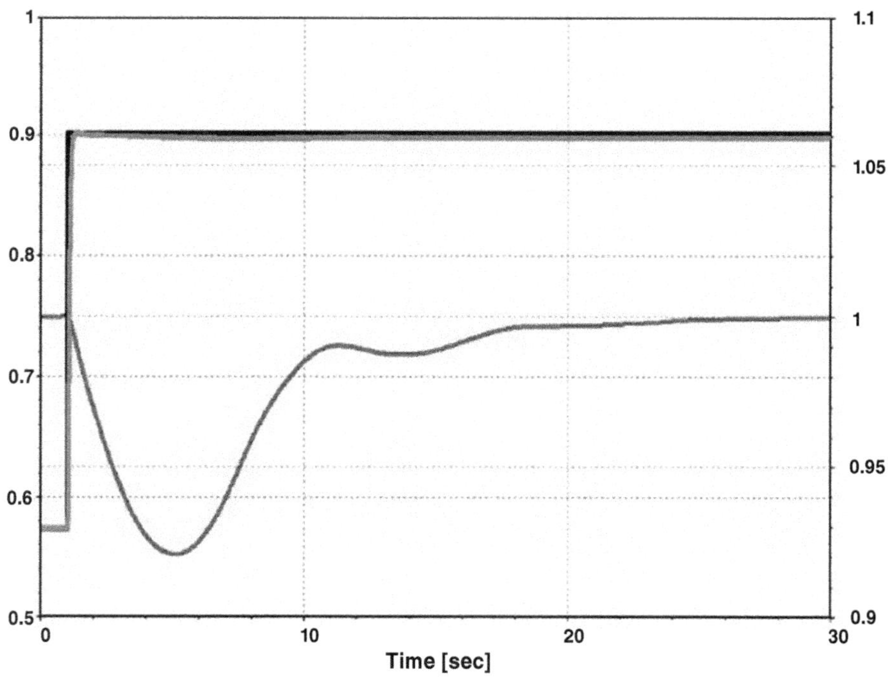

Figure 13.9 DFIM – active power and speed in turbine mode.

13.2.5 Advantages/Disadvantages

One of the main advantages is the cost saving since the VSI of a doubly fed induction machine is very small compared to a FFIM due to the lower rated power needed to feed the rotor. This fact results into lower space requirements.

Understandably, there is no start-up converter necessary but an additional start-up switch for the synchronization process right after speeding up the machine and a breaking switch to stop the machine. To run the DFIM in both directions – both in generator and in motor mode – a pole-reversing switch has to be a part of the entire system. The complex construction of the rotor results in a higher amount of maintenance.

13.2.6 Comparison of Doubly Feed Induction Machine (DFIM) with Fixed Speed Synchronous Machine

As already shown, the load change performance of a variable-speed machine is much better than of a fixed-speed machine.

This means a greater contribution to the grid stability, since variable-speed machine are able to respond much more quickly to load changes in the grid, which happens continuously.

In Figure 13.10 the operation areas of a fixed-and a variable-speed machine are mapped. Concerning the operation area of the variable-speed machine, one has to consider cavitation inside the turbine, the maximum pumping power, the head of water, and instabilities of the entire system. Still there are many operation points on which the variable-speed machine can be run, whereas there is only one point for the operation of a synchronous machine.

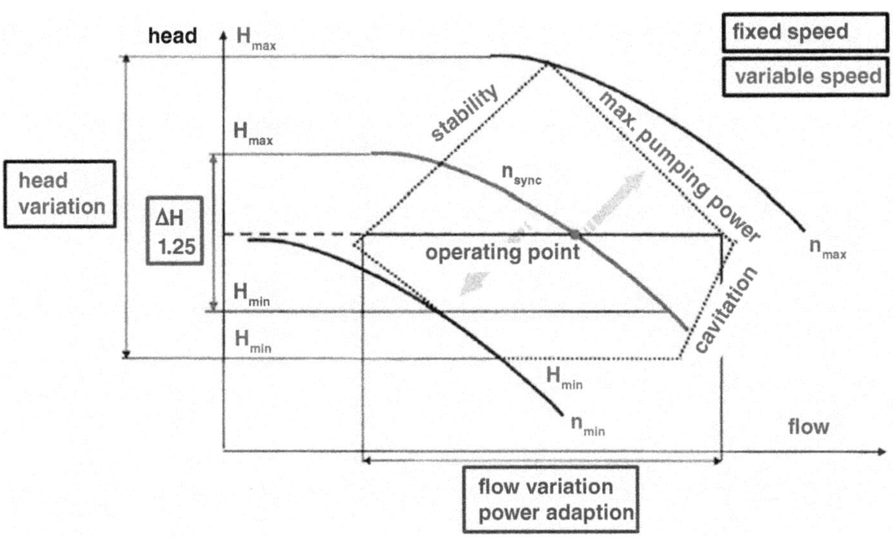

Figure 13.10 Comparison fixed speed *vs* variable speed machine – operating area in pump mode.

13.3 Synchronous Machine Adjustable Speed (FFIM)

13.3.1 Operating Principle and Components

A fully fed synchronous machine is composed of a conventional synchronous machine and a voltage source inverter, whereas the VSI is installed between the stator and the grid. The excitation system of the rotor is fed by a DC current.

There are two different circuitry systems as shown in Figures 13.11 and 13.12. The first one shows the system without any redundancy, which means that the machine can be run via the VSI only. In Figure 13.12 one can see the bypass parallel to the VSI, which results in a redundancy. If the full-size converter breaks down, the machine can continuously be run via the bypass. Although the higher construction components cause higher material costs.

Theoretically, this drive system provides a large speed range from zero up to the rated speed. In practice the speed range is limited by the turbine, which can be driven at very low speeds.

The operating principle of this machine's drive mode can be explained as follows: The DC excited rotor, which carries a one-phase winding system, is powered by the turbine via a mechanical shaft. Thus, a rotating field arises and hence a synchronous generated voltage is induced in the three-phase stator winding. The frequency of the voltage and current depend on the rotor speed and thus the turbine and can be calculated as in Equation (13.5)

$$f = (n \times pp)/60 \tag{13.5}$$

where: f = frequency
n = speed
pp = pole pair

To decouple the frequency of the speed the VSI with a voltage intermediate circuit is connected between the stator and the grid. It converts the machine-side input voltage

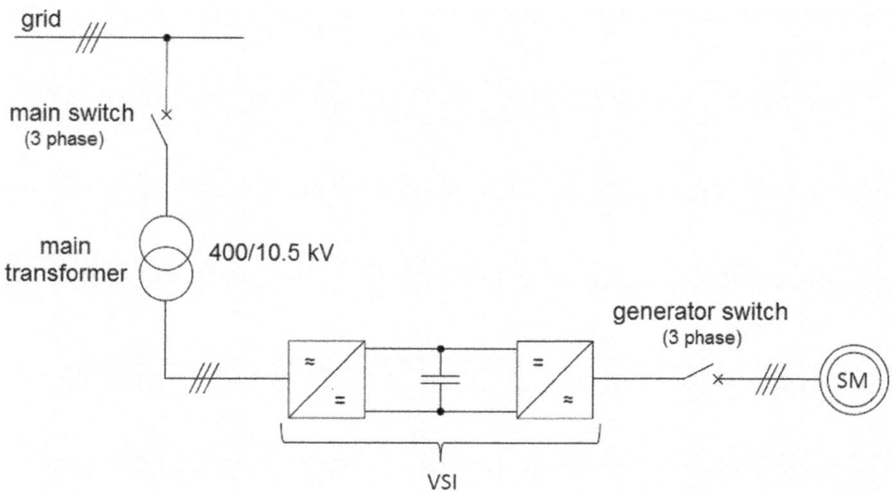

Figure 13.11 FFIM – circuit principle without a bypass.

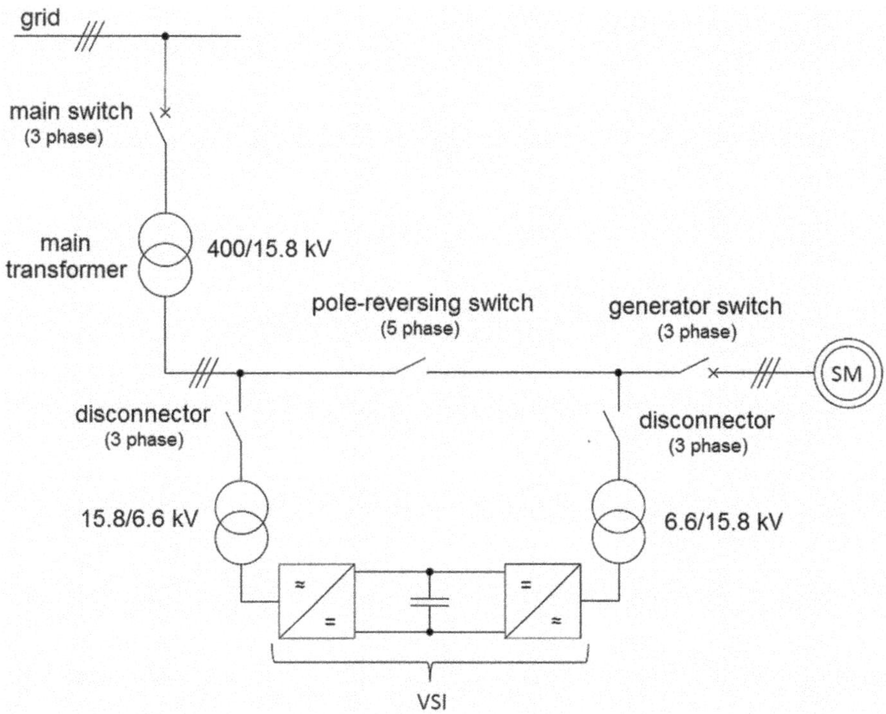

Figure 13.12 FFIM – circuit principle with a bypass.

into a grid-side output voltage with an arbitrary frequency, advisably the grid frequency. This enables the connection to the grid.

The amplitude is directly controlled by the dc excitation or the rotor winding which is coupled with the stator current, illustrated in Equation (13.6).

$$\underline{S} = m \times \underline{U}_s \times \underline{I}_s^* = P + jQ \tag{13.6}$$

$\underline{I}_s^*$ = complex conjugate stator current
m = number of windings
P = active power
Q = reactive power
$\underline{S}$ = complex apparent power
$\underline{U}_s$ = complex conjugate stator voltage (phase-to-ground)

With the VSI generation of both inductive reactive power by under excitation and capacitive reactive power by over excitation is possible, which are freely adjustable on both sides of the converter, while the amplitude depends on the converters dimensions. It basically operates like a both-side electrical source with independent frequency and voltage amplitude. The current can be regulated individually and is only limited by the maximum rated current of the VSI.

Since operation in all four operation quadrants is possible via the converter, there is no need for a pole-reversing switch. It must be integrated in a system with a bypass parallel to the VSI to enable pump and turbine mode without a converter and the converters losses.

13.3.2 Converters for Grid Connection

As already mentioned, the fully fed synchronous machine is connected to the grid via a full-size converter. Usually, a three-level converter topology is favorably installed, since this system smooths the current amplitude. Thus, the THD (Total Harmonic Distortion) and efficiency are enhanced essentially. Moreover, power electronics are stressed less.

13.3.3 Advantages/Disadvantages

A great advantage is the longtime experience of operating synchronous machines, where reliability and efficiency had been proven over decades. The large flexibility concerning speed range also means a huge advantage. Since almost maximum torque is given for the start-up process from standstill up to rated speed, start-up and switching time are minimal.

An existing power plant with a conventional fixed-speed synchronous machine can be converted into a fully fed system, enabling variable speed mode.

The advantage of the variable speed in pump mode is decimated by the higher losses inside the VSI. Compensation can be reached by an excellent efficiency of the hydraulic facility.

The main disadvantage is the relatively high capital charges and efficiency losses caused by the converter. Furthermore, the space requirements of a fully fed synchronous machine are relatively high, since the full size converter itself is considerably large.

There is no pole-reversing switch or separate start-up converter necessary, remember the equivalent circuit without bypass. As already mentioned, the bypass enables the synchronous machine in pump and turbine mode to run without the VSI and the emerging VSI losses. The redundancy structure offers emergency operation and thus higher security standards.

13.3.4 Comparison of DFIM and FFIM

To choose the appropriate drive system in a pumped storage hydro power plant there are different aspects to be considered in each case. Prioritization depends on the customer's requirements. The following criteria are relevant in the decision:

1. **Switching time**

   Basically there is no exact time to switch an electrical machine from pump mode into turbine mode and vice versa, since there are many factors of influence:
   - rated power of the converter (proportional to the machines rated power);
   - inertia of the machine (larger volume respectively larger diameter results in a higher moment of inertia);
   - installation depth: if natural dewatering is not possible, an additional hydraulic system is necessary to steam out the water before the start-up process in motor mode; and

- speed of hydraulic/mechanic, which is given by the material – the faster the water flow, the harder water hammers.

Those criteria are very specific for each single project, which constitutes the individuality. Principally, these aspects result from the customer's needs and the geographic facts. In addition, there are some regulations that have to be met. Thus, the drive system has to be adjusted accordingly. Usually, the fastest control is preferably installed to enable flexibility and fast availability.

2. Dimensions

The stator of a doubly fed induction machine usually has a longer shaft caused by the slip ring collector, but the diameter is a bit smaller. Moreover, the internal electrical field in a DFIM determines larger dimensions and at the same time a heavier weight, which is positive for the flywheel effect. Thus, active power is more readily available. Since the natural flywheel of an induction machine is more than three times higher than of a synchronous machine, the dimensions of a FFIM have to be larger to enable the same centrifugal moment [1].

The constructions of the stators are almost identical, but the rotors look very different. The synchronous machine contains a single-phase winding to be fed by a DC current, whereas the DFIM carries a three-phase winding system to be fed by a three-phase current of variable frequency. There are additional slip rings to transfer the three-phase current [2].

Since the rotor of the DFIM consists of embedded poles, the synchronous machine has a rotor with salient poles. Thus, the DFIM is the heavier one.

Besides the electrical machine there are reasonable dimension differences between the corresponding converters. Caused by the rated power and the speed range, the VSI of a DFIM is quite small since it only has to feed the machines rotor with approximately 30% of the machines rated power as already mentioned. Whereas the VSI in the stator circuit of a variable speed synchronous machine is much larger – in average about three times. It has to carry the entire machine output.

3. Maintenance

Operating characteristics of the electrical and hydraulic machine and auxiliary systems like filters, pumps, relay, etc. are usually evaluated weekly. Critical values like temperature and availability of sensors are measured permanently by the machine controls.

In general, maintenance intervals are determined individually depending of the operation, condition, and availability of the machine. Typically, machines are maintained between every one to three years. There are several exemplary services operations required:

Annual: Visual inspection of turbine, motor-generator, and auxiliary systems
Annual: Analyses of bearing oil
2 years: Maintenance of turbine regulation
3 years: Changing bearing oil and clean oil cooler
3–5 years: Electrical check of high voltage windings
5–10 years: If necessary, replacement of tube and gaskets

Servicing of highly demanded machine parts, auxiliaries and switches is specified by operating cycles. Variable speed machine sets usually have shorter service intervals. For example, the converters have to be maintained every year. Additionally, the carbon brushes of the induction machine have to be replaced frequently. Thus, new and modern plants need an annual service.

Due to the slip rings and the complicated rotor construction, the doubly fed induction machine is more maintenance intensive than a fully fed synchronous machine. Down times are almost identical.

4. Speed range

Ultimately there is a significant difference between both variable speed concepts concerning the speed range.

Experiences say that a DFIM usually is economically dimensioned for speed ranges +/- 10% of the rated speed. If not considering these limits, the machine volume and the necessary converter are getting colossally large and costly. However, a FFIMs speed range is much larger.

References

[1] Schwery, A., Fass, E., Henry, J.-M., et al. (2005). Pump storage power plants – Alstom's long experience and technological innovation. *Hydro 2005*, Villach (Austria).

[2] Henning, H.,Ködding, L., and Hildinger, T. (n.d.) Key aspects to design of variable speed motor-generators – Asynchtonous (DFIM) and synchronous (SMFI) electric machinery options for pumped storage schemes. Paper, Voith.

14

Pumped Storage Machines – Ternary Units

Manfred Sallaberger and Thomas Gaal

14.1 Ternary Units

14.1.1 Introduction

Pumped storage power plants are well-established systems for energy storage. During pump operation energy is stored by pumping up water from a lower reservoir to an upper reservoir. In turbine mode the water passes from the upper reservoir to the lower reservoir through the turbine and energy is generated. In addition to this classical way of energy storage, pumped storage power plants are also able to contribute to the grid and frequency control. In combination with quick transition times from energy supply to energy consumption, pumped storage plants are used for frequency stabilization. Such pumped storage power plants can be realized by separate arrangements of pump together with a motor and turbine together with a generator. In ternary units the generator-motor, the turbine, and the pump are arranged along the same shaft line. Therefore, only one electric machine is acting as a motor and generator. In pumped storage plants with reversible pump-turbines the pump and the turbine function are combined into one hydraulic machine. With this concept one electric machine and one hydraulic machine is needed. The concept of ternary units has its advantages and is widely used especially for high-head pumped storage plants.

The increasing contribution of renewable energy to the electrical grid has given new challenges and opportunities to pumped storage plants. In addition to conventional energy storage further functions can be taken by pumped storage:

- energy storage,
- frequency regulation,
- voltage regulation, and
- reactive power compensation.

The increased use of these services from the grid also leads to special requirements to pumped storage. High flexibility in operation and short transition times for load changes and change of operation mode from pump to turbine and vice versa are required. Further the regulation of power in pump operation is needed. This will lead to new concepts of pumped storage like operation in hydraulic short-circuit. Operating with variable speed in combination with frequency converters will provide this flexibility. The combination of these concepts with ternary units further increases their benefits.

Advances in Energy Storage: Latest Developments from R&D to the Market, First Edition.
Edited by Andreas Hauer.
© 2022 John Wiley & Sons, Ltd. Published 2022 by John Wiley & Sons, Ltd.

14.1.2 System of Pumped Storage Plant with Ternary Units

The pumped storage power plants with ternary units consist of a complex system of reservoirs and tunnels (see Figure 14.1). For storage, the pump of the ternary unit is used to pump up water from the lower to the upper reservoir. For electricity generation water passes through the turbine from the upper to the lower reservoir.

On the headrace side from the intake structure at the upper reservoir water passes the intake gate, entering the pressure tunnel. During the load changes and changing of operation from turbine to pump and vice versa the water flow in the pressure tunnel is decelerated abruptly and has to be accelerated in the opposite direction again. In order to avoid too high a pressure rise in the system a surge tank with throttling device is arranged between pressure tunnel and pressure shaft. The surge tank is acting as a balancing reservoir – a certain amount of water can enter the surge tank and thus the pressure fluctuations in the entire headwater are reduced. The water passes down to the high head manifold via the pressure shaft (penstock) and the main inlet valve to

Figure 14.1 System of pumped storage power plant Kopswerk II (Source: Illwerke).

the turbine and pump in the power house. In many cases this is realized as an underground powerhouse hosting the hydroelectric units and a transformer cavern nearby. The tailrace system with distribution pipes, surge tank, and tailrace tunnel ends in the lower reservoir in the outlet and intake structure.

14.1.3 Arrangement and Machine Concepts of Ternary Units

In ternary units the generator-motor, the turbine, and the pump are arranged along the same shaft line. As turbine and pump are turning in the same sense of rotation no change of direction is needed for change of operation mode from energy generation to energy consumption or vice versa. In vertical units normally the pump is located at the lower end in order to utilize the maximum counter pressure at the pump inlet. The various selections of the machines, arrangements of bearings, and the methods for coupling determine the specific behavior and characteristics of the entire machine set.

14.1.3.1 Turbine and Pump Type

Depending on head and flow in ternary units Francis or Pelton turbines are applied. The Francis turbine offers higher efficiency at peak and in the upper range of power compared to the Pelton. The Pelton turbine has the advantage of better efficiency at part load and enables operation at very low load. Besides hydraulic performance other criteria like maintenance or silt-loaded water may impact the selection of the machines.

For the pump, single- or multi-stage pumps are used depending on the head. In most cases single flow pumps are applied, in cases of high flow rate double flood pumps may also be installed.

14.1.3.2 Shaft Arrangement

For the shaft arrangement horizontal as well as vertical arrangements are used for ternary units. The first units were designed in a horizontal arrangement; for higher power output the vertical arrangement is preferred. The limits for horizontal design have been continuously pushed towards higher power output. For horizontal units Pelton turbines with up to three nozzles are realized, Pelton turbines with higher numbers of nozzle, with up to six are installed vertically.

The pumped storage plant Obervermuntwerk II realized by Illwerke in Austria (www.illwerkevkw.at), with a power of 180 MW, is designed in horizontal arrangement. One of the criteria to apply horizontal design was the size and cost of the cavern. According to Illwerke [1]; www.illwerkevkw.at) in the case of Obervermuntwerk II the costs for the cavern for the horizontal arrangement were considerably lower than for a vertical design.

14.1.3.3 Machine Arrangement

For horizontal ternary units in most cases the generator-motor is in the center with the turbine and pump on each side. Various arrangements of the machines are possible for vertical ternary units. In many cases the generator is on top and then the turbine and pump are arranged. In other arrangements the turbine is on top followed by the generator and the pump at the bottom. In these cases the waterproof sealing between the turbine and generator has to be ensured.

14.1.3.4 Coupling

The machines of the ternary units may either be arranged on the same shaft, fixed coupled, or connected to each other by switchable couplings.

Between generator-motor and turbine normally fixed coupling or a shifting coupling is used. In general, with fixed coupling higher ventilation losses occur. With the shifting clutch of the turbines the turbine runner can be disengaged in pump and phase-shifting mode and thus ventilation losses can be reduced.

The synchronizing self-switching coupling is a mechanical coupling that automatically couples two rotating components as soon as they are at a similar rotating speed. Both the driving part and the driven part have to be brought to similar rotating speed, then a mechanism adjusts the speed of the driven part and coupling. With the integrated gear coupling the torque is transmitted as long as both parts are at a synchronous speed.

For ternary units this device can be used for the coupling of the turbine. While the generator-motor is rotating at a synchronous speed the turbine is started up by water. As soon as both generator-motor and turbine have similar rotational speed the synchronizing coupling adjusts the speed and couples the two machines. After coupling the gear coupling is transmitting the torque during continuous operation.

The connection of the pump to the generator-motor may be applied as fixed coupling or a coupling that is switchable in standstill, like a curved tooth coupling. By means of a hydraulic torque converter the pump can even be coupled to the generator-motor during rotation.

The hydraulic torque converter is a particular piece of hydraulic machinery that is used for synchronizing and coupling of the pump to the generator-motor. It again consists of a pump on the driving shaft and a turbine runner on the powered shaft and is operated with water that is circulating between the pump and turbine runner in a closed loop. During the start up the converter is filled with water and the pump runner drives the water in an internal loop through the turbine runner. Thus, the turbine runner is accelerated. As soon as both driving and driven shaft are close to the same speed a gear coupling is engaged and the torque is transmitted via the gear coupling. The pump and turbine runner are de-watered again and thus ventilation losses only occur by turning in air. The advantages of a hydraulic torque converter are short synchronization time and low ventilation losses. Further the low water consumption is fed from the tail water reservoir thus no energy losses occur.

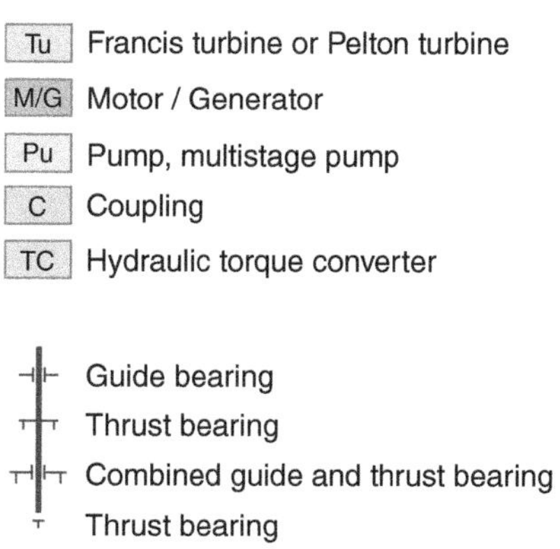

Figure 14.2 Nomenclature.

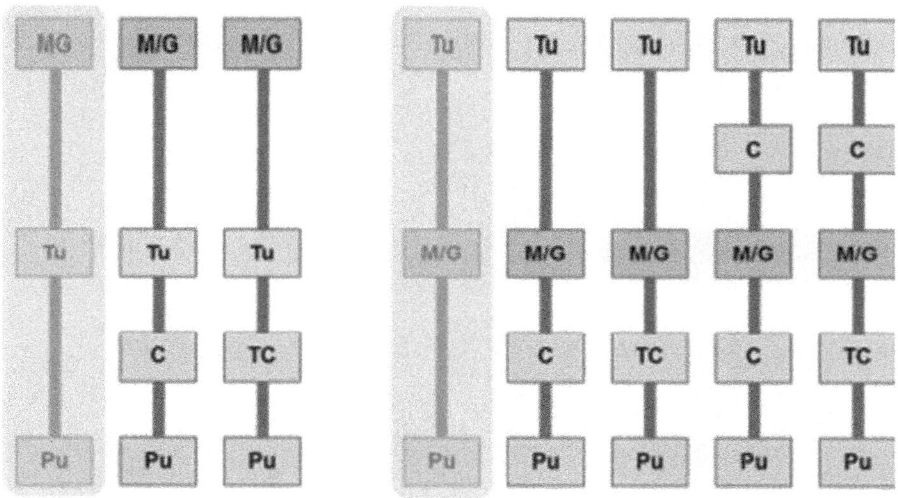

Figure 14.3 Example of vertical machine arrangements of ternary units.

14.1.3.5 Bearing Arrangement

In the arrangement of ternary units the type and selection of the coupling is closely connected to the number and arrangements of the bearings (Figures 14.2 and 14.3). For each detachable part of the shaft line a separate thrust bearing and at least two guide bearings will be necessary. An overview of typical arrangements of vertical ternary units is given in Figure 14.3.

14.1.3.6 Machine Selection for Ternary units

The main parameters for the hydraulic machinery of ternary units are head, discharge, rotational speed, and power as a result of these quantities. Based on these parameters the turbine and pump are selected.

One determining parameter for the selection of the turbine is the gross head of the entire pumped storage plant. For highest heads Pelton turbines will be the more technically feasible and useful application (see Figure 14.4). For medium and high head installations Francis turbines as wells a Pelton turbines will be feasible. Currently, the Francis turbine of the pumped storage plant Haeusling in Austria, with a head of 734 m, is still the one with the highest heads worldwide (see Figure 14.5). As the overlapping head range with both types is possible, further criteria like setting level, rotating speed, or operational behavior may be determining factors.

The Pelton turbine offers operation down to a very low part load at reasonable efficiency. Pelton turbines normally a certain amount of overhang of the turbine to the tail water level is needed, therefore the setting level of the turbine needs to be higher than the tail water level. In exceptional cases Pelton turbines can be arranged to a certain level below the tail water level like in the pumped storage plant Kopswerk II. Then the water level within the Pelton turbine has to be lowered by air pressurizing the entire inner casing of the turbine.

The advantage of the Francis turbine is higher peak efficiency, the efficiency of the Francis turbine at part load normally will be lower than for the Pelton. Currently, there are developments ongoing for Francis turbines to enable operation down to very

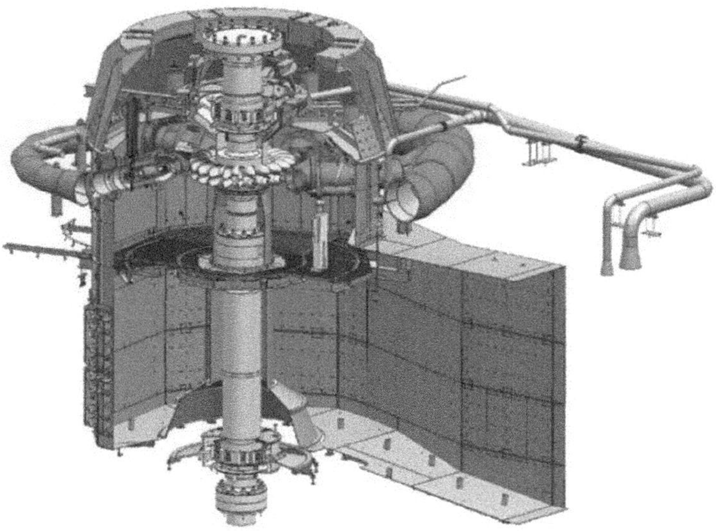

Figure 14.4 3D view of the five jet Pelton turbine at Veytaux II (FMHL+), Alpiq, Switzerland (Source: Andritz Hydro).

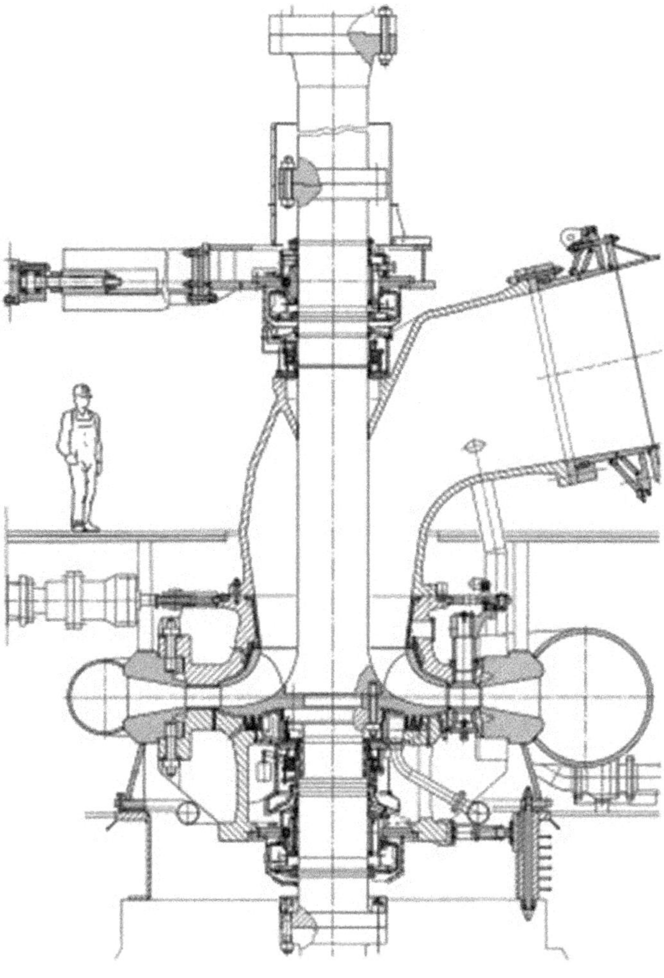

Figure 14.5 Cross section of the Francis turbine at Haeusling, Verbund, Austria (Source: Andritz Hydro).

low part load by means of suitable profile design. On the other hand the Francis turbine needs a certain submergence, the setting level needs to be below the tail water level, which is easily achieved in ternary units, because the setting level for the pump requires even more submergence.

As in ternary units the turbine, generator-motor, and the pump are arranged on the same shaft that the rotational speed is determining and is a common parameter. Therefore, the rotational speed has to be selected in order to meet the requirements of all three machines. The limiting quantities in many cases are the mechanical forces, mostly the centrifugal forces acting on the rotating components, mainly the generator rotor. In case a Pelton turbine is applied the centrifugal forces acting on the Pelton runner may also be a limiting factor. In many cases this also limits the rotational speed of the pump.

Depending on the head, flow, and speed the pump will be designed as a single or multi-stage pump (see Figure 14.6). For the pump a high rotational speed would be an advantage, as the diameter would be smaller and thus the cost for the machine would be lower. There are, however, limitations, as already outlined. Nowadays for lower heads of up to 300 m the pump would be designed as single stage, for heads higher than 800 m the pump favorably will be designed as multi-stage. In the range between 300 m and 800 m single- or multi-stage solutions may be possible. In multi-stage pumps the total head will be distributed to multiple stages with a runner in each stage (see Figure 14.7). Between the runner stages the return channel will guide the flow from the runner outlet to the next runner inlet. Submergence of the pump will also impact the selection as this determines the cavitation behavior at the inlet of the pump runner. High head radial pumps have to be designed free of cavitation in order to avoid cavitation damage of the components. For this reason the tail water level needs to be

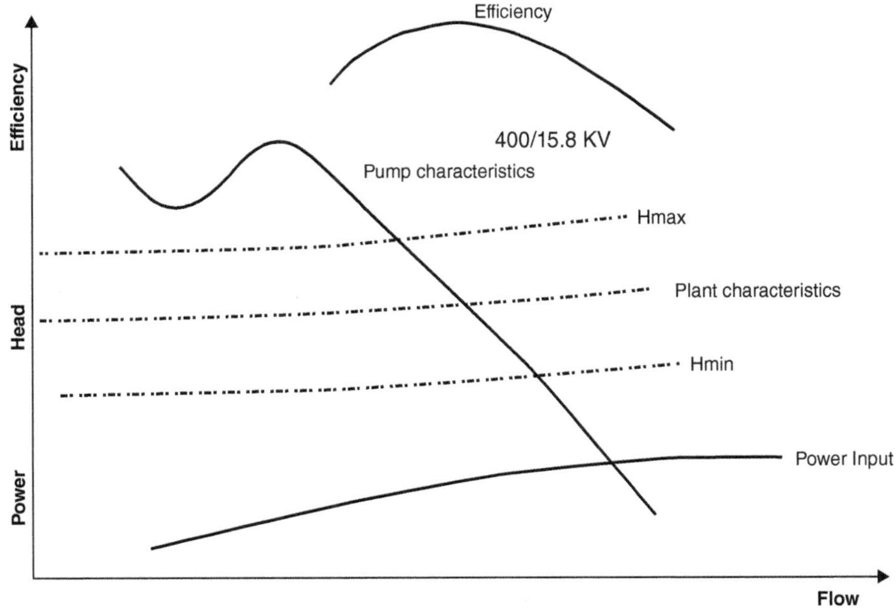

Figure 14.6 Typical characteristics of plant and storage pump.

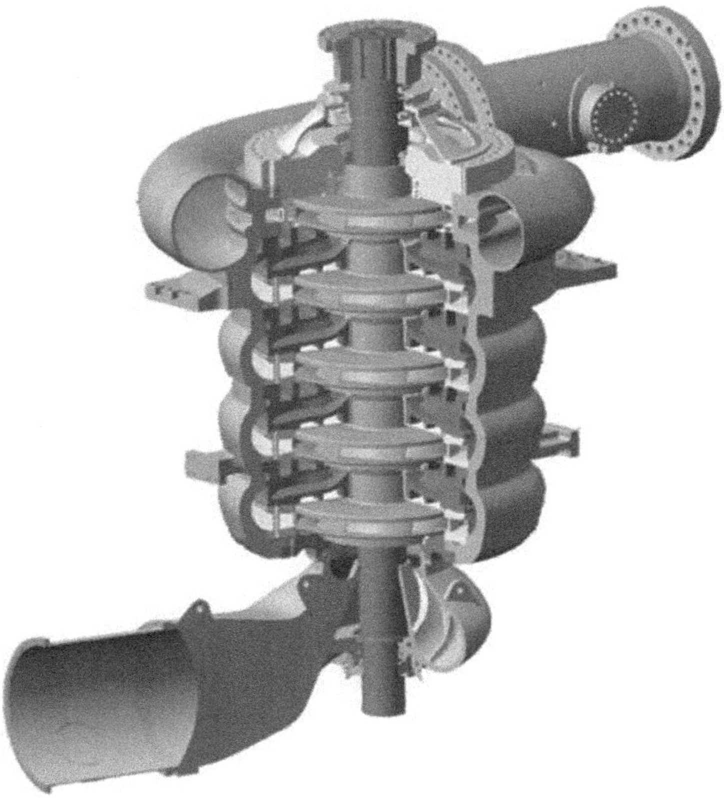

Figure 14.7 3D view of the five-stage pump Veytaux II (FMHL+), Alpiq, Switzerland (Source: Voith Hydro GmbH & Co. KG).

significantly higher than the pump inlet level, thus a certain submergence of the pump is required.

For single-stage pumps the increasing diameter and higher rotational speed will lead to increasing disk friction and leakage losses and increased noise. Multi-stage units will have additional losses due to the flow in the return channel but have a smaller diameter. Overall, a balance between the technical features, operational behavior, and the operational and maintenance cost has to be achieved.

14.1.4 Advantages of Ternary Units and Comparison to Pump Turbines

In the development of pumped storage power plants various arrangements of pump, generator-motor and turbine have been realized. Nowadays, mainly ternary units or reversible pump turbines are applied with both solutions showing advantages for specific applications.

For the ternary units the investments are higher than for pump turbines. The costs for the civil structure as well as for the machinery equipment are higher for ternary units. Further the costs for operation and maintenance are higher for ternary units as more machines have to be maintained.

The advantages of the ternary units are the better performance of the entire set. As for each machine the pump and the turbine may be optimized for its particular operation the efficiency of such units will be higher than for pump turbines. Even though nowadays pump turbines are highly developed the efficiency will be slightly lower due to the combined operation of the machines.

With respect to high flexibility in operation and load changes ternary units also offer considerable advantages. Due to the fact that for ternary units pump and turbine are rotating in the same direction no change of rotation is necessary where the mode of operation from pump to turbine and vice versa is required.

Further the pump is normally started filled with water. The power input required for such a start up is supplied either by the generator-motor or the turbine. A suitable coupling device for such a start up and synchronization procedure, like a hydraulic torque converter, is required, in order to keep the power input within a reasonable range. Further no time is needed for de-aeration of the pump as the pump is filled with water all the time.

Pump turbines frequently are started filled with air in order to reduce the power consumption during the start up procedure. After synchronization the air has to exhaust from the pump turbine during refilling with water. This procedure of aeration, exhausting, and refilling with water takes a certain time, which is not needed for ternary units.

14.1.4.1 Transition Times of Ternary Units

With respect to transition times the ternary units show a clear advantage compared to reversible pump turbines especially in pump start (see Figure 14.8). As for ternary units no change of direction of rotation and usually no filling with water are needed for pump start up, the transition times can be up considerably shorter than for comparable pump turbines. For pump turbines the transition times are in the magnitude of minutes whereas for ternary units they are in the magnitude of seconds. This brings a considerably advantage for ternary units in flexibility and short transition times during operation changes. The transition times may vary in detail, depending on the waterway system, the transient behavior and type of machines in the ternary unit, typical values are given in Figure 14.8.

14.1.5 Examples of Pumped Storage Plants with Ternary Units

14.1.5.1 Haeusling

The pumped storage plant Haeusling in Austria owned by Verbund (www.verbund.at) utilizes the energy between the reservoir Zillergruendl (1850 MASL) realized with a double curvature arch dam and the Stillup reservoir (1108 MASL) and was commissioned in 1986. This pumped storage power plant consists of two sets with a Francis

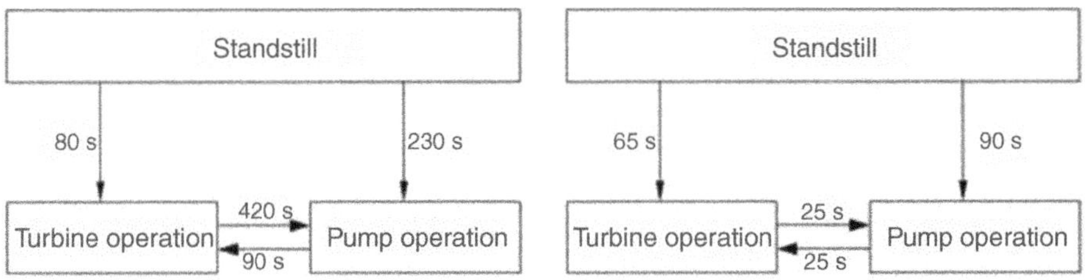

Figure 14.8 Typical transition time, left reversible pump turbine, right ternary unit with Pelton turbine and torque converter.

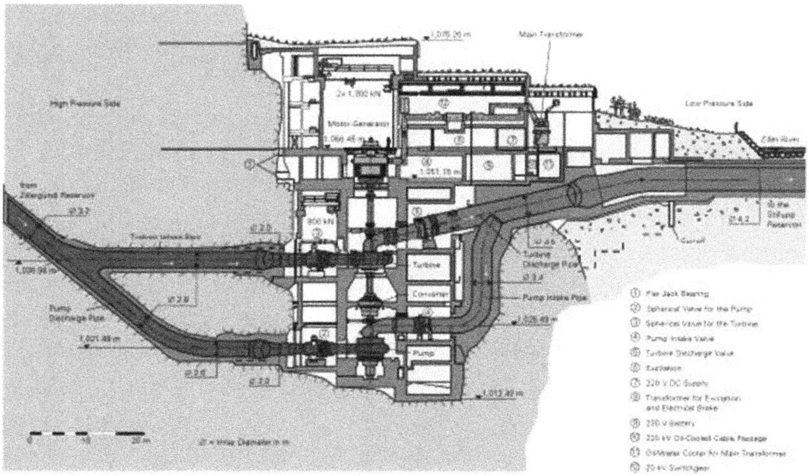

Figure 14.9 Pumped storage plant Haeusling, Verbund Austria (Source: Verbund).

turbine, motor generator, hydraulic torque converter, and a two stage pump, as shown in Figure 14.9. The powerhouse is designed in 40 m deep shaft construction and is built as a free-standing reinforced concrete cylinder with a diameter of 32.8 m.

For the Haeusling power plant a classical conception with two vertical ternary units of 180 MW each has been selected (see Table 14.1). In the Haeusling plant the generator is arranged on top followed by the fixed coupled turbine. Between the turbine and pump at the bottom the hydraulic torque converter is arranged in order to ensure quick pump start up [2–4]. The pump is designed as a single flow, two stage pump. With a maximum power output of P = 180 MW and a head of H = 734 m the Francis turbine at the time still held the world record regarding head. For the Francis turbine as well as for the pump the submergence was adequate, thus the machines could be erected contrary to orthodoxy with the draft tube on top [2, 3]. Thus the tailrace tunnel could be arranged more favorably. As an additional small disadvantage, for each dewatering and filling of the turbine the draft tube also has to be emptied, leading to a longer transition time from turbine mode to synchronous condenser operation due to bigger drainage volume [2, 3]. With adequate dimensioning of the draining and filling equipment the impact on the transition times could be minimized and become acceptable.

Based on the experience of the Rosshag [4] power plant, where small distortions of the power house caused troubles on the machine bearings, it was decided to found the power house completely on solid rock and not to embed the hydraulic machines into

Table 14.1 Main data of the pumped storage plant Haeusling.

Main data Haeusling	
Speed	600 rpm
Static Head	620–734 m
Turbine max. power	180 MW
Pump max. power	180 MW

concrete. In this way it is possible to connect the machines on both sides to the conduits, via pressure compensated expansion sleeves, thus the shaft alignment can be monitored and re-adjusted [2–3]. The primary shaft from generator to turbine and torque converter is equipped with five guide bearings and the thrust bearing at the upper end of the generator. The secondary shaft from torque converter to pump is equipped with two guide bearings and the thrust bearing at the lower end.

14.1.5.2 Kopswerk II

In 2008 Vorarlberger Illwerke in Austria commissioned the three units at the pumped storage plant Kopswerk II (Table 14.2) [5]. The pumped storage power plant Kopswerk II has been designed to regulate the grid in turbine as well as in pump operation (Figure 14.10). The power plant uses the head between the Kops reservoir (1800 MASL) and the Partenen-Rifa reservoir (1000 MASL).

The three units are located in an underground powerhouse, which is accessible via an access tunnel. The impressive power house is one of the largest rock caverns in the world regarding the ratio height to width which was determined by the vertical arrangement of the three units. The volume of rock excavated was approximately 113 000 – m^3, width approximately 30.5 m, height approximately 60.5 m, and length approximately 88 m. The cavern has to withstand high forces acting from any side and sophisticated mathematical simulations of the geological conditions together with measurements were necessary.

The three vertical units consist of a six nozzle Pelton turbine, a generator-motor, a torque converter, and a three stage pump, with a total height of approximately 38 m. In order to meet the market demands, the criteria for selection and design of the units were: the ability for regulation between -450 MW and 450 MW; a minimum generation of 3 × 150 MW; and short transition times. Various machine concepts of pump-turbines, ternary units with Francis or Pelton turbines were evaluated. Finally the concept was chosen with the Pelton turbine on top of the generator-motor.

The six jet Pelton turbine had been chosen due to the better operation behavior in part load and the ability to operate at very low part load. Due to the fact that the setting level of the Pelton turbine is below the lowest tail water level, the Pelton turbine and the tail water chamber have to be pressurized. Detailed investigations were necessary to evaluate the amount of dissolved and undissolved air and the effects on the pump operation and the elasticity in the system.

In order to achieve the ability for power regulation in pump operation too, the units were designed to operate in hydraulic short-circuit. For fixed speed, normally the pump input power is fixed for operation at a certain head. With operation at hydraulic short-circuit the pump is consuming energy according to the operation head.

Table 14.2 Main data Kopswerk II.

Main data Kopswerk II	
Speed	500 rpm
Static Head	723–818 m
Turbine max. power	190 MW
Pump max. power	180 MW

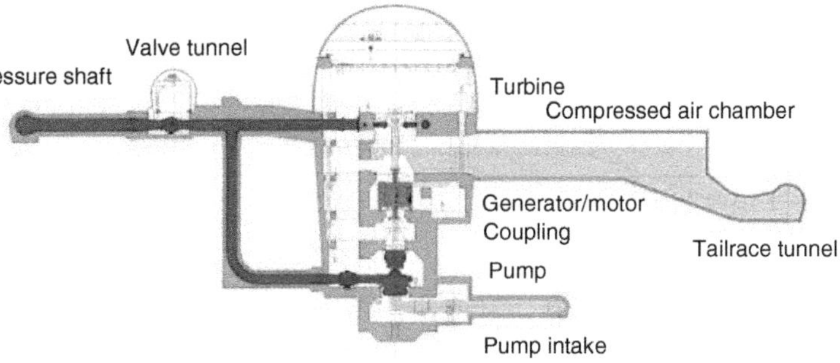

Figure 14.10 Cross section of cavern and machine arrangement of Kopswerk II pumped storage plant (Source: Illwerke).

Simultaneously energy is generated by the turbine operation. In total the energy taken out of the grid is the difference between energy consumption of the pump and energy production of the turbine. The flow of the pump partly the turbine and partly goes upstream to the head water reservoir. In this way a variable surplus of power in the grid can be compensated by controllable total input power.

14.1.5.3 Obervermuntwerk II

One of the latest installations of ternary units is the pumped storage plant Obervermuntwerk II realized by Illwerke in Austria [5–7] and commissioned in 2017, see Figure 14.11. At the very beginning of the project the target was to increase the energy storage capacity of the group. The increase of windpower and photovoltaics in Germany introduced a change in the requirements of Obervermuntwerk II (Table 14.3). Fast reaction time and operational flexibility became more valuable and thus the higher investment in ternary units compared to pump-turbines could be justified.

In order to meet the requirements of the energy market two sets of highly flexible, rapidly and fully controllable machines with Francis turbines and pumps were installed [6–7]. In the two horizontal sets a single stage pump is connected to the generator-motor by a hydraulic torque converter. The hydraulic torque converter enables acceleration and synchronization of the water filled pump within shortest time.

The Francis turbine is connected to the generator-motor by a curved tooth coupling, which is switchable during rotation. A new development of this curved tooth coupling enables connection of the two gear rings even when they are rotating at slightly different speeds, thus the coupling time can be reduced for turbine operation. By means of the switchable coupling the turbine runner can be de-coupled during pump mode and

Table 14.3 Main data Obervermuntwerk II.

Main data Obervermuntwerk II	
Speed	428 rpm
Static Head	243–311 m
Turbine max. power	180 MW
Pump max. power	180 MW

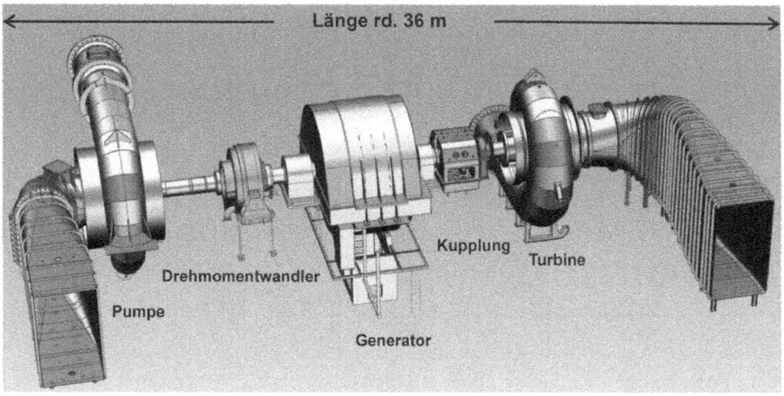

Figure 14.11 Machine arrangement of Obervermuntwerk II (Source: Voith, Andritz Hydro.)

phase shifting mode, thus ventilation losses can be avoided. This brings huge benefits especially for turbine condenser operation as the turbine remains filled with water and no aeration and refilling is necessary. In case of aeration also degassing of the water has to be ensured, which would lead to longer channels and thus to increased civil engineering costs.

In order to meet the flexibility in operation in nearly the full range of -100% - +100% of load, the operation of hydraulic short-circuit was planned. Further an advanced design concept enables operation of the Francis turbines down to low part load without aeration. As the hydraulic short-circuit is planned, aeration of the flow in the Francis turbines at low load is not desirable. In case of aeration the air injected at the outlet of the Francis turbine could enter the pump and cause disturbance in the pump runner during hydraulic short-circuit.

The single stage pump is equipped with adjustable guide vanes in order to reduce power input during pump start up. The horizontal arrangement of the units has been chosen in order to optimize the geometry of the powerhouse. The more balanced proportions of the compact rock cavity compared to a high and slim one contribute to the optimization of the civil engineering cost. The horizontal arrangement also offers better accessibility to the machines, even though the spiral casing is embedded in concrete.

The ternary units require higher investments and are more complex in operation and maintenance compared to pump turbines. This is however compensated for by the higher efficiency in both pump and turbine operation and the improved flexibility and faster transition times.

The variation of power and the potential control band of power of the ternary units of Obervermuntwerk II is shown in Figure 14.12. In a wide range of net heads the full range of power between 100% generation and 100% pump can be covered. At heads lower than 250 m a small gap arises because pump power is increasing with decreasing head whereas the maximum turbine output is decreasing [6].

14.1.5.4 Veytaux II

A further modern installation of ternary units is realized in the pumped storage plant Veytaux II by FMHL in Switzerland [8–9] and commissioned in 2017, see Figure 14.13. The owner company FMHL (Forces Motrices Hongrin-Léman SA) belongs to the shareholders Romande Energie SA, Alpiq Suisse SA, Group-E SA, and City of

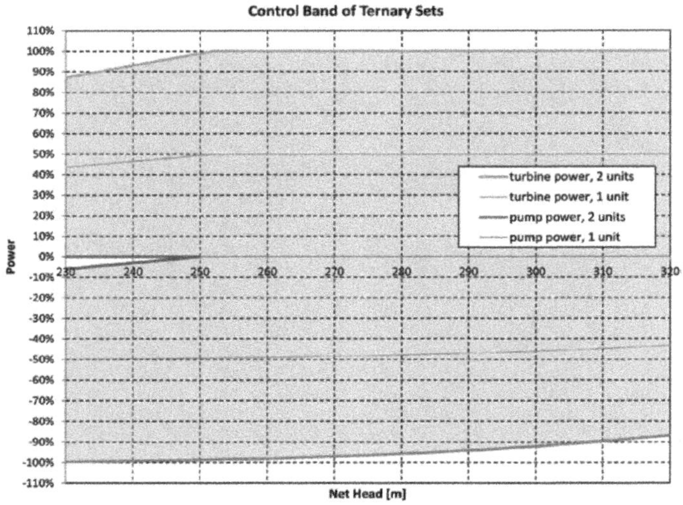

Figure 14.12 Control band of ternary units Oberwermuntwerk II, Illwerke (Source: Illwerke).

Lausanne. Alpiq Suisse SA, as owner representative, is the company in charge of supervising the study and implementing the new FMHL + project also called Veytaux II. The existing pumped storage plant Veytaux I was extended by the installation of Veytaux II. In Veytaux I four ternary units consisting of double runner Pelton turbines with two nozzles each and a power output of 60 MW and five stage pumps are installed. With the extension by another two ternary sets of 120 MW each in Veytaux II, the total power output of the plant has been doubled to 480 MW. The increased flexibility is enabling peak energy generation and will contribute to the growing demand on balancing energy in western Switzerland.

In Veytaux II the head difference between the existing reservoir Hongrin (1255 MASL) and the Lake Geneva is utilized (Table 14.4). The reservoir Hongrin contains a volume of 52 Mm^3 of water, the dam is realized as a double arch dam with a maximum height of 123 m. The powerhouse is built as an underground cavern. The two ternary sets consiststing of a generator-motor, a Pelton turbine with five nozzles, and a five stage pump at the bottom (see Figure 14.13). The pump is connected to the generator-motor by a shifting clutch, therefore it only may be connected in standstill and the transition time is a little longer than with a hydraulic torque converter.

The new pumped storage plant Veytaux II has been erected parallel to the existing scheme Veytaux I, while it was in operation. A particular challenge during the

Table 14.4 Main data Veytaux II.

Main data Veytaux II	
Speed	500 rpm
Static Head	879–884 m
Turbine max. power	120 MW
Pump max. power	120 MW

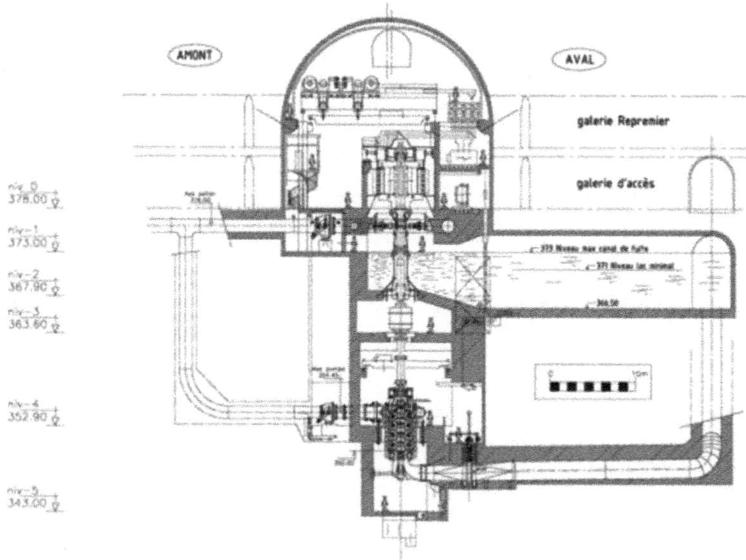

Figure 14.13 Cross section of cavern and machine arrangement of Veytaux II pumped storage plant, FMHL (Source: copy FMHL).

construction stage was to prevent harm to the operation of the existing scheme and to minimize the outage time during the connection of the waterways.

Further special attention had to be given to the selection of the machine arrangement. It was decided that the existing pressure shaft should be used, therefore, special care had to be taken to the transient behavior and overpressure of the units in order to maintain safety.

References

[1] Matt, P. (2016). Die Kraftwerksanlage Obervermunt II Und Deren Flexibilität. *VGB PowerTech Journal* 3.

[2] Grein, H., Mayr, W., and Bernhard, E. (1988). Layout and design of the Haeusling Francis Turbines operating under world record head. In: *IAHR Symposium*, Trondheim.

[3] Mayr, W. (1987). *Gedanken Zur Maschinellen Ausrüstung, Vertrauen in Die Kraft Des Wassers*. 40 Jahre Tauernkraftwerke AG, Koska, Wien.

[4] www.verbund.at (accessed 21 December 2021).

[5] www.illwerkevkw.at (accessed 21 December 2021).

[6] Meusburger, P. and Goekler, G. (2014). Obervermuntwerk II Pumped Storage Plant of Vorarlberger Illwerke AG – A Tailor-made, Customized Conception for the Changing Electricity Market. *Hydro 2014*, Cernobbio, Italy.

[7] Meusburger, P. and Werle, L. *The Hydraulic Machinery of Obervermuntwerk II*, xxxx.

[8] www.alpiq.ch (accessed 21 December 2021).

[9] www.fmhl.ch (accessed 21 December 2021).

15

Hydro-Mechanical Equipment

Claudia Pollak-Reibenwein

Department Penstock and Gates, ANDRITZ HYDRO GmbH, Lunzerstrasse 78, 4030 Linz, Austria

15.1 Steel-lined Pressure Conduits

15.1.1 Introduction

Water conducting systems – regardless whether for turbine operation or pumped turbine operation – comprise different structures chapters 13, 14 and 16. In turbine operation direction the water is transported from the free surface water level of the upper reservoir via a partly steel lined water conducting system to the lower reservoir. On its way downstream the water passes an intake structure followed by low-pressure conduits and tunnels. Before entering an inclined and usually high pressurized steel penstock (or vertical steel-lined or concreted pressure shaft) a surge tank (or surge chamber) can be arranged in the case of a rather long pressure conduit to minimize water hammer by limiting the rise and fall of pressure (e.g., caused by sudden valve closure) within a penstock. If two or more units are fed from one penstock the steel liner is split into branches before entering the power station. This system is called manifold. After leaving the power station the water is conducted via low-pressure conduits or tailrace tunnels being steel-lined or concrete lined to the lower reservoir and exits the water conveyance system by an outflow structure. Within the water conducting system several control and shut-off devices are installed for safe plant operation (see Section 15.2).

This chapter mainly deals with underground and embedded steel-lined sections of the penstock system and its hydro-mechanical facilities made of steel. The main difference in the design of steel structures for pumped storage pressure conduits versus steel liners for conventional turbine operation is the additional consideration of fatigue aspects comparable to the design of mechanical parts.

15.1.2 General Layout of Pumped Storage Pressure Conduits

A general schematic representation of a pumped storage pressure conduit is given in Figure 15.1. Following the water flow in turbine operation mode the water is passes through the following sections.

- Service Chambers (SC)

Service chambers with hydraulic operating devices for opening and closing reasons are normally positioned at the intake behind the upper reservoir, after the surge tank but

Advances in Energy Storage: Latest Developments from R&D to the Market, First Edition.
Edited by Andreas Hauer.
© 2022 John Wiley & Sons, Ltd. Published 2022 by John Wiley & Sons, Ltd.

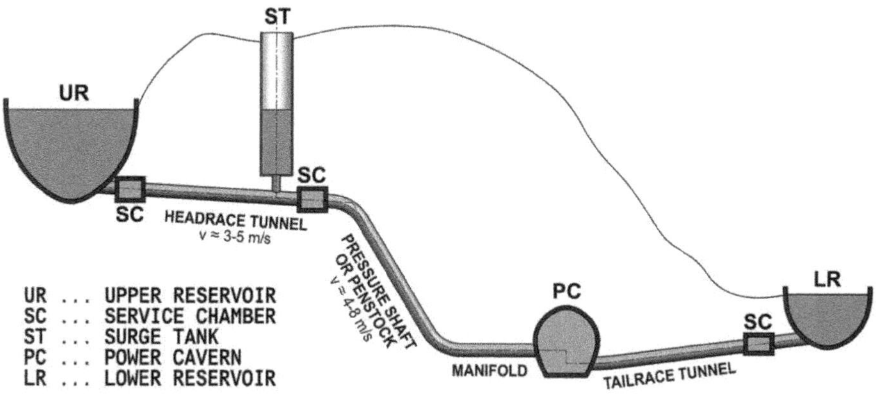

Figure 15.1 General layout of pumped storage pressure conduits.

before the pressure shaft or penstock, in front of the power units (within the Power Cavern) and finally, concerning pumped storage plants, near the outlet to the lower reservoir.

- Tunnels

Tunnels are defined as underground passageways, commonly in rock. These are usually identified as headrace and tailraice tunnels. Depending on the maximum head and the surround properties they can be steel-lined or unlined (concrete/shotcrete lining).

- Surge Tank (ST)

A surge tank or surge chamber provides plant stability by limiting the pressure rise and fall within a long penstock, helping regulate the flow and improve turbine speed regulation. Its location may be before entering the penstock and/or behind the downstream service chamber.

- Steel Liners and Penstocks

Steel liners and penstocks are designed to resist structural instability caused by high internal pressures (including water hammer) during operation and external pressures during dewatering or maintenance in empty condition. Another purpose is avoiding minimum water leakage and providing for maximum hydraulic performance by minimizing the total head loss with respect to minimizing the annual cost of a penstock. For further reading on this see [1].

The steel liner shells are typically made of large round steel cross-sections of welded steel plates.

Special attention has to be paid to welded joints with regard to fatigue assessment and careful testing should be executed (ultrasonic and radiographic methods). For corrosion protection and tightness reasons steel liners are usually coated on the inner wet surfaces.

Steel liners and penstocks are not only single straight pipeline pieces. They are equipped with additional special components:

Stiffener or Buckling Rings

Stiffener or buckling rings are commonly attached to the exterior surface of the steel-liner shell in order to avoid buckling in case of decisive external pressure loading.

Anchor or Thrust Rings

The longitudinal forces resulting from static and dynamic fluid action on the closed pressure conduit have to be absorbed or balanced to maintain integrity of the system. Theses forces occur at changes of size or at changes in the direction of flow. Axial forces resulting from constrained contraction or expansion due to thermal changes or pulling forces due to valves being closed (hydraulic thrust) cause additional longitudinal forces. If these forces cannot be adequately transferred to the surrounding concrete, anchor or thrust rings are usually located on the pipe branches directly before entering the power house to securely introduce these axial forces into the backfill concrete with adequate safety against punching shear of the concrete. Anchor rings are in most cases simple ring flanges welded to the outside surface of the pipe. According to [2] the bearing capacity is not equally split between several serially installed thrust rings – the first thrust ring is loaded by roughly 50% of the total axial force, which has to be considered in the design.

Bends, Elbows, and Reducers

Changes of penstock alignment ask for curved pipe sections called bends or elbows. The bends are commonly fabricated from mitered pipe segments that are joined by full penetration welds. Increased stress concentrations of hoop tension stress occur along the inside edge at the intersection of two segments. These additional circumferential stresses can be analytically calculated according to [3] considering the surrounding concrete of elastically embedded bends. The miter squares should not exceed 8°. Reducer pipes are usually provided for very long penstocks, where the diameter of the pipes has to be reduced as the total head on the pipes increase. The reducer pipe is a frustum of a cone and its length should keep the angle of convergence between 5 and 10°.

Manholes

Manholes are installed to provide access to the pipe interior for inspection, maintenance, and repair by personnel and equipment. Usually they are situated close to valves and close to low points that might need to be drained for inspection reasons. The typical diameter of these shell openings is 600 mm, mainly executed as cylindrical or ellipsoidal access openings. Due to the huge opening, the penstock shell has to be reinforced by collars or saddle like plates welded around. Curved fairing plates are installed to cover the opening if high turbulence flow is to be expected. A flat cover plate is fitted to the nozzle flange by bolts. Regarding the pretension of bolted connections special attention should be drawn to an appropriate design, which means considering concentrically clamped components, preventing gaps in the interface and avoiding eccentric loading conditions – for more information on that refer to e.g., [4]. A sealing gasket (commonly O-ring seal type) between nozzle flange and manhole cover plate is provided to prevent leakage.

Grouting Plugs

Grouting plugs are used both for rock consolidation and for grouting between backfill concrete and the steel liner. The design of grouting plugs and especially the execution of the welded connection of the plug to the shell mainly affects fatigue assessment. Recent results regarding fatigue analysis of grouting plugs can be found in [5].

Closure or Pressure Heads

Closure heads are commonly used to block off the end of a pipeline for the purpose of hydrostatic pressure testing (weld stress relief, tightness testing) or as a permanent closure. These domed ends can be manufactured as ellipsoidal, torispherical, hemispherical, or cylindrical covers. In the case of permanent covers access can be provided by manholes located on the cover. Their position impacts stress assessment – for determination of stress increasing factors refer to e.g., [6]. For these structures partial plastic deformation is allowed though limited to a certain percentage (less than 1% acc. to AD 2000-Merkblatt B3).

Several more appurtenances and components along the penstock span are for example, air vents and valves, expansion joints, flow measurement devices, and dewatering drains – for further reading on these components refer to e.g., [7].

- Bifurcations and Manifolds

For multiple hydroelectric pump turbines the main penstock may be branched into two or more linings. If two units are fed by one penstock the branching is named bifurcation or wye branch. Wye branches can be symmetric or asymmetric. In case of three units the branching is called trifurcation. A so-called manifold consists of several successive bifurcations and one branch outlet as depicted in Figure 15.2.

The wye branches have to be carefully designed with respect to the reduction of hydraulic losses for the hydro power facilities. Advanced stress analysis of these structures by means of finite element methods is state-of-the-art. The overall system's load carrying behavior has to be taken into consideration when setting up a model of a branching system. The steel shell of bifurcations is typically reinforced by externally welded ring girders or internally arranged sickle plates due to the interruption of the hoop action.

- Transitions

Steel lined transitions are usually installed at the entrance and exit of a gate housing where the cylindrical cross-section of the lining has to be transferred to a square or

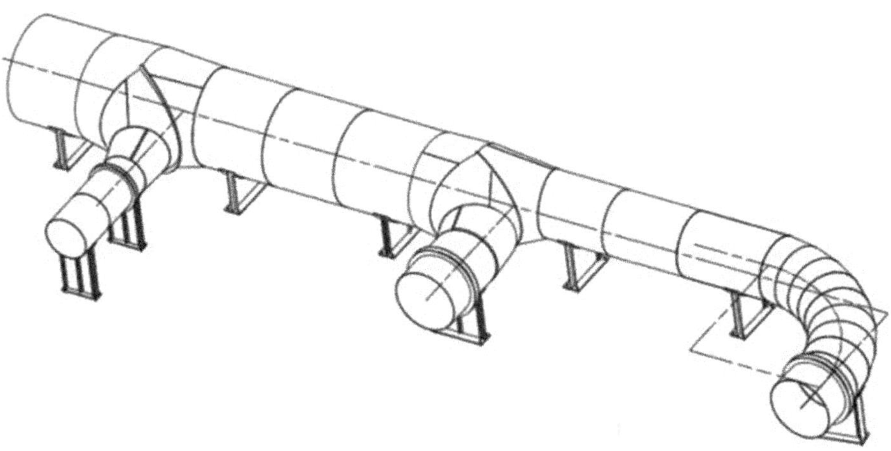

Figure 15.2 Example of a typical manifold consisting of two bifurcations and one branch outlet.

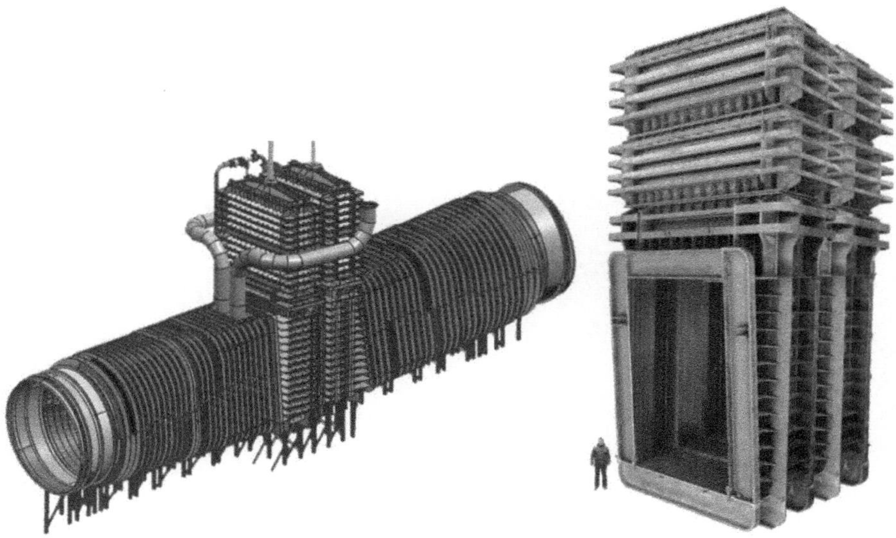

Figure 15.3 Left: Model of downstream transition and gate housing of Nant de Drance pumped storage HPP; Right: Photomontage of Nant de Drance gate housing.

rectangular cross-section because of the square- or rectangular-shaped shut-off devices. A tremendous example for this is the transition and gate housing of Nant de Drance as depicted in Figure 15.3. Due to huge stress concentrations in the corner regions the whole steel structure is commonly stiffened by frames or accomplished with anchor studs – both welded to the exterior shell surface.

15.1.2.1 Recent Layout Examples of Pumped Storage Penstocks in Europe's Alpine Region

In the past years tendencies in Europe's Alpine region have led to the renewal of existing large hydro power plants to pumped storage power plants and/or the development of new pumped storage power plants. This development has been favored by technologies such as hydraulic short-circuit operation, deeper drawdown storage levels due to improved venting systems and surge tank avoidance due to short pressure conduits. Figures 15.4 to 15.6 present some major examples of recent pumped storage projects in Europe.

15.1.3 Loading Conditions and Main Analytical Approaches

15.1.3.1 Typical Pressure Lines for Penstock Design

The basic input data for the calculation of penstocks including appurtenances for safe operation are given by the pressure lines for internal and external water pressure along the penstock axis as depicted in Figure 15.7. The internal pressures are measured by the distance between the pipe centerline and the elevation of the static level. Due to sudden shut-off or steady speed regulation of turbines elastic waves are generated and travel up- and downstream along the closed conduit causing increases and decreases in pressure. These pressure changes are referred to as water hammer or transient pressure and ought to be considered in the design pressure by superposing the maximum transient pressure proportionally over the developed length of the penstock axis with the static head.

NANT DE DRANCE PUMPED STORAGE HPP COMMISSIONING 2018		
MAX. DESIGN PRESSURE	INTERNAL DIAMETER	MATERIAL
6 N/mm²	DN3000 ÷ DN7700	S355ML, S690QL, Alform700M

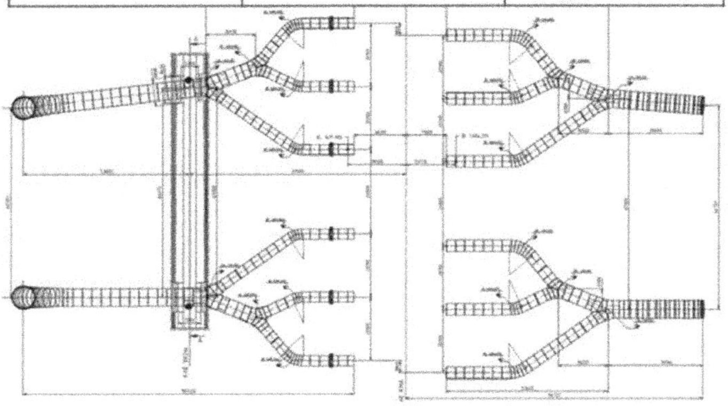

Figure 15.4 Nant de Drance Pumped Storage HPP (Switzerland) – main data and general layout.

LIMBERG II PUMPED STORAGE HPP COMMISSIONING 2011		
MAX. DESIGN PRESSURE	INTERNAL DIAMETER	MATERIAL
6,5 N/mm²	DN2200 ÷ DN5700	S355ML, S460ML, S500ML, S550ML, S690QL

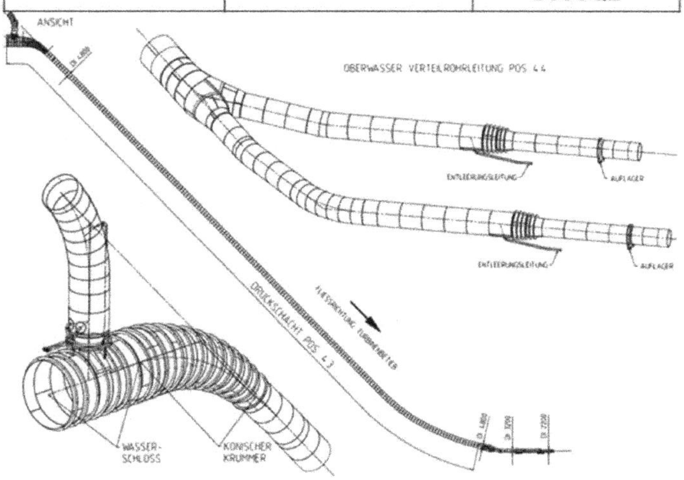

Figure 15.5 Limberg II Pumped Storage HPP (Austria) – main data and general layout.

KOPS II PUMPED STORAGE HPP COMMISSIONING 2008		
MAX. DESIGN PRESSURE	INTERNAL DIAMETER	MATERIAL
11 N/mm²	DN1500 ÷ DN7000	S285ML, S355ML, S460ML, Ald500ML1, Ald550ML1, Ald700QL1

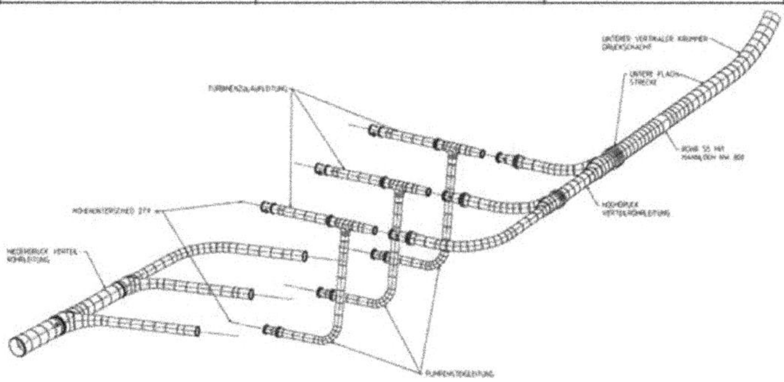

Figure 15.6 Kops II Pumped Storage HPP (Austria) – main data and general layout.

The external pressure line represents the measured difference between the elevation of the seepage or groundwater level and the pipe centerline.

For fatigue assessment usually an equivalent constant amplitude stress range, which results in the same fatigue life as for the design spectrum is determined and distributed over the developed length of the steel liner axis.

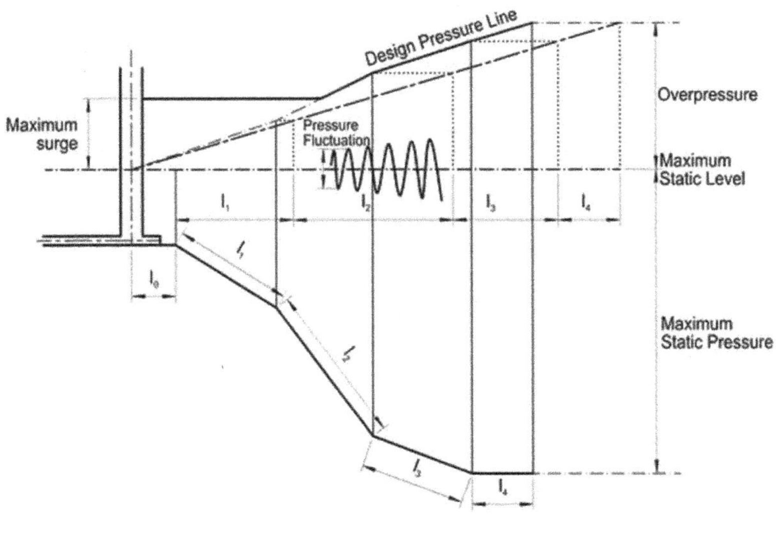

Figure 15.7 Typical design pressure lines (modified and reproduced with permission from C.E.C.T., 1984 © SNCT Publications [8]).

15.1.3.2 Minimum Shell Thickness

Formulas for minimum shell thickness values should guarantee for a limitation of possible deformations during fabrication, transport, and installation and can be obtained from different standards or are specified by the contractor. A common formula for the minimum shell thickness of steel penstocks is given in Equation (15.1).

$$t_{\min} = \frac{D + 800}{400} \tag{15.1}$$

with: D nominal diameter of steel liner in [mm]

Usually a corrosion allowance in the range of 0–2 mm depending on the applied corrosion protection system and lifetime is specified by the contractor. This value has to be considered for the design calculations.

15.1.3.3 Design Loads

- Internal Pressure Design

The required thickness of a cylindrical thin-walled pipe shell subjected to internal pressure is obtained by limiting the circumferential or membrane hoop stress to a certain stress level, which depends on the applied safety concept. In general the wall thickness of a self-supporting smooth and straight cylindrical pipe shell can be determined by Equation (15.2).

$$t = \frac{p_{\mathrm{i}} \cdot D}{2 \cdot \sigma_{\mathrm{all}}} \tag{15.2}$$

with: t required wall thickness to resist internal design pressure (without considering value for corrosion allowance or undercut), [mm]

p internal design pressure, [N/mm^2]

σ_{all} allowable stress (depending on applied safety concept), [N/mm^2]

With respect to an economic design (which means keeping the wall thicknesses within reasonable limits) the collaboration or interaction of the surrounding rock continuum can be taken into account if rock overburden and rock conditions are favorable. In view of possible scatter of rock characteristics along the pressure conduit profile an appropriate safety against yielding of steel should be met (common practice is to limit the pipe stresses under uniform bedding conditions to 60% of the material's yield strength). A detailed method for the determination of the ratio of pressure carried by the steel liner to the total internal pressure is given for example in [9] and depends on the stiffness ratio of lining and rock mass. If rock collaboration is considered then additionally an exceptional load case covering the fictive rock failure has to be applied. In this case the working stress of the fully unsupported pipe may not exceed 90% of the material's yield strength.

- External Pressure Design

The cylindrical pipe shell of a steel liner subjected to external pressure (groundwater, seepage water, grouting pressure) has to be calculated against collapse due to buckling. In the case of embedded steel liners the surrounding backfill concrete limits the radial deformation outwards and thus increases the shell's buckling resistance. The installation of stiffener rings also leads to enhanced buckling resistance. Another alternative against buckling is pressure relief valves (refer to [10]). The most common analytical formulae for the determination of the critical external buckling pressure under embedded conditions are presented by [11, 12], and [13]. These formulae are based on the assumption that the buckling of a cylindrical pipe occurs in a single lobe. The critical buckling pressure is determined by supposing a radial gap between the steel liner and the backfill surround due to concrete shrinkage and temperature difference in reference to the material temperature during assembling of the liner. Values for the gap or clearance between steel liner and backfill surround are given in [8] and mainly depend on rock properties. The buckling resistance of cylindrical pipe shells rapidly decreases under the presence of imperfections. Imperfections can be caused during fabrication, manufacture, and assembling/concreting at site and cannot be precluded. These imperfections have to be considered in the calculation and are referred to as pipe ovality. According to C.E.C.T. the ovality tolerance of an erected pipe shall not exceed 1% of the radius which is expressed in Equation (15.3).

$$(D_{max} - D_{min}) \leq 0.04 \cdot R \tag{15.3}$$

with: D_{max} maximum measured diameter, [mm],
D_{min} minimum measured diameter, [mm]
R theoretical internal radius, [mm]

If the buckling stability is investigated by means of finite element methods then accurate simulation of shell imperfection is of major importance (e.g., scaling and combination of different characteristic eigen values). More information on collapse analysis of shell structures with finite element methods is given in [14].

As for ring stiffened steel liners the buckling resistance of the tube between the external stiffener rings has to be assessed separately. The pipe portion between the stiffeners might buckle prior to the portion "shell plus stiffener." The tube's critical radial deformation is inwards without support of the surround backfill concrete.

Clearly summarized critical external buckling pressure formulae for smooth and ring stiffened pipes are given in ASCE manual No. 79 [7].

- Loads due to Thermal Effects

Both the steel liner and its appurtenances might be subjected to thermal loads in service condition (filled pipe) and/or in dewatered condition (empty pipe). For the calculation the temperature differential related to the temperature at the end of the penstock assembly is of interest here.

- Fatigue Loads

In contrast to conventional turbine operated steel liners and penstocks, pumped storage pressure conduits are exposed to not negligible pressure fluctuations, which have to be considered in a separate fatigue load case. The most critical components are usually welded or bolted connections. As for the calculation of these structural details fatigue resistance classes are provided by several standards and guidelines such as IIW [15–17] or [18]. Additionally, weld S-N curves can be found in [19]. Special emphasis should be given to the design of welds – for example, ground welds increase fatigue life.

- Seismic Loads

Additional hydrodynamic pressures caused by ground shaking during an earthquake event have to be considered for all underground structures even if earthquake load combination is not the governing design load case – for more information especially for penstock design refer to [20].

- Loads due to Transport and Installation

Transport, handling, method of installation, concreting, and grouting result in different load cases and support reactions. It has to be guaranteed that the pipes can resist these loads and meet specified tolerances. Concreting and grouting pressures do not act uniformly on the pipe circumference and have to be analyzed with respect to deformations and buckling stability. Additional measures like internal supports (crosses, rings) and securing against buoyancy (anchors, external supports) may be necessary to keep deformations and stresses within specified limits.

15.1.3.4 Main Stresses

The main general primary membrane stresses for thin-walled cylindrical pipes within a closed water conduit are summarized below in Equations (15.4) to (15.7).

- Circumferential or hoop stress:

$$\sigma_{\mathrm{R}} = p_{\mathrm{i}} \cdot \frac{R}{t} \tag{15.4}$$

- Longitudinal stress due to Poisson's effect:

$$\sigma_{\mathrm{Poi}} = \mathrm{v} \cdot \sigma_{\mathrm{R}} \tag{15.5}$$

- Longitudinal stress due to temperature:

$$\sigma_{\mathrm{T}} = \alpha_{\mathrm{t}} \cdot \mathrm{E} \cdot \Delta T \tag{15.6}$$

- Longitudinal stress due to pressure heads or closed valves:

$$\sigma_{\mathrm{L}} = \frac{p_{\mathrm{i}} \cdot R^2 \cdot \pi}{2 \cdot R \cdot \pi \cdot t} = 0.5 \cdot \frac{p_{\mathrm{i}} \cdot R}{t} = 0.5 \cdot \sigma_{\mathrm{R}} \tag{15.7}$$

with:

$\mathrm{v} = 0.3$	Poisson's ratio for steel
$\alpha_{\mathrm{t}} = 1.2 \cdot 10^{-5}$	coefficient of thermal expansion for steel, [1/K]
$\mathrm{E} \approx 210000 \mathrm{N/mm^2}$	modulus of elasticity for steel
ΔT	temperature differential, [K]

Additional primary local membrane and bending stresses occur for instance at the intersection of pipe segments, in the vicinity of ring and nozzle attachments, at openings, and have to be evaluated separately. Besides those mentioned, secondary membrane plus bending stresses and peak or notch stresses are relevant especially for stress (and fatigue) evaluation based on finite element methods. A detailed overview of all stress components and examples of stress classification can be found in [7].

15.1.4 Safety Concepts and Application of Standards

Regarding the design of steel penstocks several international standards and guidelines with their own integral safety concepts are available throughout the world. The most common European standard for the design of steel penstocks is [8, 18] and EN 1993 (a common steel construction code, parts 1–6 especially for shells, parts 1–8 for connections, parts 1–9 for fatigue) [16, 21, 22]. English speaking countries prefer applying ASME BPVC [19] and ASCE manual No. 79 [7] for steel penstocks. Besides those mentioned the Japanese technical standard for gates and penstocks [23] offers a lot of information concerning penstock (and gates) design.

The safety concepts within these guidelines may be different – therefore it is recommended to remain consistent in applying the same standard or guideline for all penstock components to be analyzed within a project. C.E.C.T. for example, indicates global safety factors or so-called dimensioning factors for different loading scenarios and different penstock conditions (e.g., exposed, buried, and buried and concreted) but only for general membrane or primary stress assessment derived by application of analytical formulae. Its safety concept is based on limiting the yield stress (meaning the apparent elastic limit guaranteed for a certain material) as illustrated in Equation (15.8).

$$\sigma_{eq} \le \frac{Re}{C'} \tag{15.8}$$

with: σ_{eq} equivalent stress (acc. to von-Mises criterion) for each point considered (actual stress), [N/mm^2]

Re minimum yield stress, [N/mm^2]

C' dimensioning factor (e.g., 1.5 for cylindrical pipes of a buried and concreted penstock and the occurrence of permanent forces)

The safety concept of ASME BPVC, ASCE manual No. 79, and EN13445-3 is based on an allowable stress intensity S. The determination of this basic allowable stress value is given in Equation (15.9) and applies for nominal design stress.

$$S = \mathrm{MIN}\left(\frac{2}{3}\cdot S_Y; \frac{S_T}{2.4}\right) \tag{15.9}$$

with: S_Y minimum specified yield strength, [N/mm^2]

S_T minimum specified tensile strength, [N/mm^2]

For secondary and peak stress assessment these standards clearly indicate stress categories and limits of stress as indicated in Equation (15.10) as an example for the local primary membrane stress (P_L) at for instance the junction of a cylindrical to a conical section.

$$P_L \le 1.5 \cdot K \cdot S \tag{15.10}$$

with: P_L local primary membrane stress, [N/mm^2]

K increase factor (e.g. K = 1 for normal operating condition)

S see Equation (15.9) [N/mm^2]

For two- or three-dimensional states of stress the equivalent stress according to the maximum shear stress theory and/or the maximum distortion energy has to be determined [19] recommends applying von-Mises criterion, [7] accepts both von-Mises and Tresca criteria).

In contrast to the above concepts EN 1993-1-6 (strength and stability of shells) [21] is based on a semi-probabilistic approach with partial safety factors considered both on material or resistance side and on action side. The partial safety factors for action depend on the one hand on the type of load (e.g., 1.35 for dead loads, 1.5 for live loads) and on the other hand on the load combination by introducing additional combination factors. On the resistance side a material dependent partial safety factor of e.g., 1.1 has to be applied. For more information on applicable partial safety factors refer to [24]. Thus the allowable stress value for primary stresses can be determined according to Equation (15.11) in a very simplified way.

$$\frac{R_{\mathrm{k}}}{\gamma_{\mathrm{M}}} \geq E_{\mathrm{d}} \tag{15.11}$$

with: E_{d} design value of action effects (including partial factors for actions and combination factors), [N/mm^2]

R_{k} characteristic value of the resistance, [N/mm^2]

γ_{M} material dependent partial factor on the resistance side

15.1.5 Aspects of Material Choice

The application of high-strength steels (characterized by a yield strength over 600 N/mm^2) for highly pressurized conduits allows for reduced wall thicknesses and results in lighter components and easier transport and handling conditions. Another advantage is less volume for weld seams and hence shorter installation time. But the challenge is its sensitive welding behavior – especially potential hydrogen-induced cold cracking. According to [25] thermomechanically treated steel grades are less prone to the effect of cold cracking whereas quenched and tempered steel grades should be welded with caution due to their susceptibility to cold cracking moving from the weld's heat-affected zone into the weld deposit. Hydrogen-induced cold cracks in high-strength weld metals are usually initiated perpendicular to the weld direction.

15.2 Typical Control and Shut-Off Devices for Pumped Storage Plants

15.2.1 General Arrangement of Control and Shut-Off Devices

The typical arrangement of control and shut-off devices is depicted in Figure 15.8. The penstock protection valves are located in the valve chamber on the upstream side of the pressure conduit. The main inlet valve is positioned in front of the turbine in the power station. Finally the maintenance gate or valve on the downstream side is either arranged in the power station or in a separate valve chamber.

15.2.2 Gates and their Main Applications

Gates for pumped storage plants are usually submerged and installed within the pressure conduit. This circumstance leads to the fact that discharge often occurs under high pressures and therefore only gates characterized by a rigid structure and

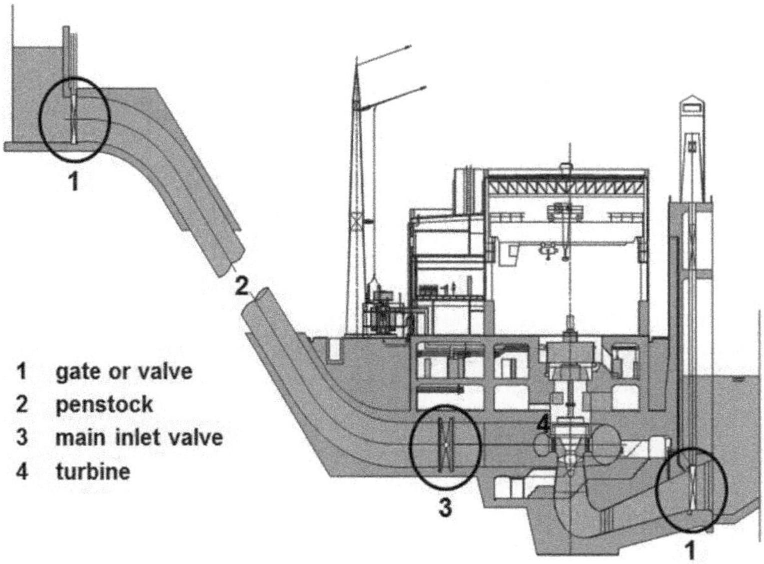

Figure 15.8 General arrangement of gates and valves.

unfailing operation should be applied. [26] distinguishes between intake gates, control and guard gates, emergency closure gates, and maintenance gates.

TSGP [23] summarizes the design conditions for hydraulic gates as follows:

- Safety against predictable loads,
- Sufficient water tightness,
- Easy and reliable operation,
- High durability,
- No harmful vibrations during operation, AND
- Easy maintenance.

15.2.2.1 Intake Gates

Intake gates can be sealed both on the upstream and on the downstream side. Upstream sealing gates are typically installed shortly behind the intake in a shaft. If intake gates operate as flow control gates or have to be shut against flow in an emergency they should be equipped with an oil hydraulic servo-motor for hoisting reasons. Intake gates that do not have to shut against flow can be rope suspended and lowered or hoisted by a conventional winch. It may be necessary to provide aeration pipes. Further types of intake gates are slide gates, radial gates, cylinder gates, and caterpillar gates.

15.2.2.2 Control and Guard Gates

Control and guard gates can be designed as vertical-lift, roller, or slide gates. They are servo-motor operated.

15.2.2.3 Emergency Closure Gates and Maintenance Gates

In general a complete operational group includes two identical gates – one behind the other – which are retracted into steel-lined gate slots (refer to Figure 15.3) The difference between an emergency closure gate and a maintenance gate is that the former one closes against flow and allows for crack open (filling of penstock) whereas the latter one commonly closes and opens under balanced conditions.

15.2.3 Valves and their Main Applications

Valves fulfill various important tasks within hydro power plants. They are safe closure and shut-off devices in pressure conduits and provide tight sealing in waterways for maintenance purposes on hydraulic machinery. As safety elements they shut off the maximum flow in the minimum possible time by respecting the maximum admissible pressure in the penstock. At maintenance works, valves are used as a sealing plug of the penstock. The typical design range of butterfly and spherical valves is depicted in Figure 15.9 and is dependent on the design head and the diameter.

Spherical valves are mainly used in high head applications (e.g., HPP Gerlos II, Austria, 1 spherical valve with a nominal diameter of 1800 mm and a design pressure of 72 bar) whereas butterfly valves are selected for the low to medium head range (e.g., HPP Nant de Drance – 2 butterfly valves with a nominal diameter of 6000 mm and a design pressure of 10 bar).

Both valves allow flow in both directions and fit perfectly for the use in the waterways of reversible pump turbines. The main components can be made either from castings or forged or welded structures based on size and pressure rating criteria.

15.2.3.1 Spherical Valves

This type of valve is characterized by minimum head loss due to complete free through-flow. Spherical valves are usually installed as high-pressure turbine inlet and pump valves and are equipped with automatically controlled service and maintenance sealing. The control system is operated by oil-hydraulic and the water pressure system. The water for the valve control system can be taken from the upstream side. Oil pressure for the control system is provided by the turbine governor oil-hydraulic system.

15.2.3.2 Butterfly Valves

Butterfly valves are equipped with either a lens type or a lattice door (through-flow) type valve disk. The lattice door types are nowadays commonly installed. Butterfly valves are mainly used as safety valves, turbine inlet valves, and pump inlet valves

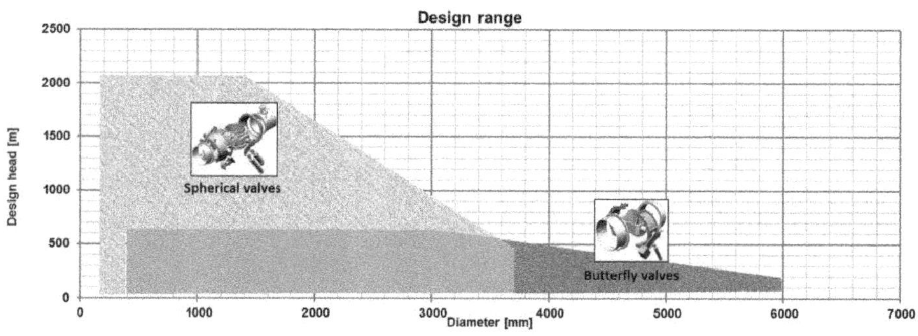

Figure 15.9 Design range of shut-off valves; Courtesy of ANDRITZ HYDRO GmbH.

for low to medium design pressures. They are operated by oil hydraulic systems for opening and closing or by closing weight and hydraulic pressure for opening. For turbine inlet valves oil pressure can be taken from the turbine governor hydraulic oil system. The sealing system is made of flexible, adjustable rubber/metal type to reduce leakage to a minimum. The lattice door type valves may also be equipped with a maintenance seal.

References

[1] Singhal, M.K. and Arun, K. (2015). Optimum design of penstock for hydro projects. *International Journal of Energy and Power Engineering* 4 (4): 216–226.

[2] Ofner, R. and Greiner, R. (2007). Beanspruchung von Schubringen für Druckrohrleitungen. *Stahlbau*, 76 Jahrgang, Heft 10, 739–746.

[3] Lechner, A., Greiner, R., and Krueger, K. (2013). Stress analysis in rock-embedded miter bends – New derivation and extension of existing formulae for the use in penstocks. *High Strength Steels for Hydropower Plants. Design Concepts – Pressure Conduits. Proceedings of the 3rd International Conference*, (ed. H. Cerjak, N. Enzinger, R. Greiner, et al.), 227–240. Graz.

[4] VDI 2230 Part 1: Systematic calculation of highly stressed bolted joints. Joints with one cylindrical bolt. Verein Deutscher Ingenieure, November 2015.

[5] Greiner, R. and Lechner, A. (2013). Fatigue strength of high strength steel-linings with different types of grouting openings. *High Strength Steels for Hydropower Plants. Design Concepts – Pressure Conduits. Proceedings of the 3rd International Conference* (ed. H. Cerjak, N. Enzinger, R. Greiner, et al.), 241–250. Graz.

[6] AD 2000-Merkblatt B3. (2011). Gewölbte Böden unter innerem und äußerem Überdruck. Verband der TÜV e. V. Berlin.

[7] ASCE Manuals and Reports on Engineering Practice No.79. (2012). *Steel Penstocks*, 2nd ed. Reston, VA: American Society of Civil Engineers.

[8] C.E.C.T. *Recommendations for the Design, Manufacture and Erection of Steel Penstocks of Welded Construction for Hydro Electric Installations.* Paris: SNCT Publications, Comité Européen de la Chaudronnerie et de la Tolerie, edition 1979, including modifications 1984.

[9] Seeber, G. (1999). *Druckstollen und Druckschächte: Bemessung-Konstruktion-Ausführung*. Stuttgart: Enke.

[10] Schleiss, A.J. and Manso, P.A. (2012). Design of pressure relief valves for protection of steel-lined pressure shafts and tunnels against buckling during emptying. *Rock Mechanics and Rock Engineering* 45 (1): 11–20.

[11] Amstutz, E. (1969). Das Einbeulen von Schacht- und Stollenpanzerungen. *Schweizerische Bauzeitung* 87 Jahrgang: 541–549.

[12] Jacobsen, S. (1974). Buckling of circular rings and cylindrical tubes restrained against radial displacement under external pressure. *International Water Power and Dam Construction* 26 (12): 400–407.

[13] Montel, R. (1960). Formule semi-empirique pour la determination de la pression extérieure limite d'instabilité des conduits métalliques lisses noyées dans du béton. *La Houille Blanche* 15 (5): 560–568.

[14] Schmidt, H. (2009). Stabilität stählerner Schalentragwerke. In: *Stahlbaukalender 2009*, (ed. U. Kuhlmann), 11. Jahrgang, 530–612. Berlin: Ernst & Sohn.

[15] Hobbacher, A. (2014). *Recommendations for Fatigue Design of Welded Joints and Components*. IIW – International Institute of Welding, doc. XIII-2460-13/XV-1440-13, Paris.

[16] EN 1993-1-9: Design of steel structures – Part 1-9: Fatigue; German version EN 1993-1-9:2005 + AC:2009, European Committee for Standardization (CEN), Brussels.

[17] FKM-Guideline: Analytical Strength Assessment of Components in Mechanical Engineering; 5th rev edn., 2003, English version; Forschungskuratorium Maschinenbau (FKM), Frankfurt/Main.

[18] DIN EN 13445-3, Unfired pressure vessels – Part 3: Design, German version EN 13445-3:2014, only on CD-ROM, European Committee for Standardization (CEN), Brussels.

[19] ASME BPVC (Boiler & Pressure Vessel Code). (2015). *Section VIII Rules of Construction for Pressure Vessels*. Division 2 *Alternative Rules*. New York: The American Society of Mechanical Engineers.

[20] Wieland, M. (2005). Hydrodynamic pressures. *International Journal Water Power and Dam Construction (Oct.)*: 18–24.

[21] EN 1993-1-6: Design of steel structures – Part 1–6: Strength and stability of shell structures (including Corrigendum AC:2009); English translation of DIN EN 1993-1-6: 2010-12, European Committee for Standardization (CEN), Brussels.

[22] EN 1993-1-8: Design of steel structures – Part 1–8: Design of joints; German version EN 1993-1-8:2005 + AC:2009, European Committee for Standardization (CEN), Brussels.

[23] TSGP Technical Standards for Gates and Penstocks. Hydraulic Gate & Penstock Association, 1986, Tokyo.

[24] EN 1990: Basis of structural design; German version EN 1990:2002 + A1:2005 + A1:2005/AC:2010, European Committee for Standardization, Brussels.

[25] Cerjak, H., Enzinger, N., and Pudar, M. (2013). Development, experiences and qualification of steel grades for hydro power conduits. *High Strength Steels for Hydropower Plants. Design Concepts – Pressure Conduits. Proceedings of the 3rd International Conference* (ed. H. Cerjak, N. Enzinger, and R. Greiner), 1–5. Graz.

[26] Lewin, J. (1995). *Hydraulic Gates and Valves in Free Surface Flow and Submerged Outlets*. London: Thomas Telford Publications.

16

Pumped Storage Machines – Hydraulic Short-circuit Operation

Thomas Gaal and Manfred Sallaberger

16.1 Hydraulic Short-circuit Operation

16.1.1 Introduction

Hydraulic short-circuit allows the regulation of storage pumps in pumped storage power plants.

The flexibility in operation of pumped storage plants may be restricted by the missing availability of pump input power. The power output of hydraulic turbines can be varied from part load to full load (depending on the turbine type this may vary from 50 to 100% load for a Francis turbine and from 10 to 100 % for a Pelton turbine). For storage pumps at synchronous speed, the input power is fixed at a certain pump head and cannot be changed. Thus, the pump can only be operated if the required input power is supplied by the grid.

In hydraulic short-circuit both pump and turbine are operated at the same time. The storage pump operates at the input power requested at the respective head. The power output of the turbine may be varied within the normal operation range of the turbine. Thus the net power consumption taken from the grid is the difference between the pump input and turbine output power. During this operation part of the flow of the pump may pass the turbine and part may flow to the upper reservoir. By hydraulic short-circuit operation the power varies between 100% turbine output and the storage pump input power at a certain head (there may be a gap left in the range where the turbine cannot be operated).

As early as the 1950s hydraulic short-circuit operation had been implemented in hydro power plants. This operation mode was used mainly for grid regulation purposes. In the following decades the operation mode became less important due to a higher overall capacity of the grid as well as regulation services (i.e., frequency response) for which the grid were not paid.

In recent years the importance of fast regulation in the grid such as frequency response made a contribution to the revival of hydraulic short-circuit operation. Some units went into operation especially designed for this operation mode, as well as plants being converted for it.

Advances in Energy Storage: Latest Developments from R&D to the Market, First Edition.
Edited by Andreas Hauer.
© 2022 John Wiley & Sons, Ltd. Published 2022 by John Wiley & Sons, Ltd.

16.1.2 Regulation of Hydro Turbines and Storage Pumps

Single stage turbines and single stage pump turbines can be operated down to the minimal permitted load. The minimal load depends on the type of turbine and specific design of the unit (i.e., specific speed). Francis or pump turbines can be operated down to minimal 30% up to 50% of the full load capacity. Pelton turbine – especially in vertical and multiple jets arrangements – can be operated down to 10%, or even a 5% load (see Figure 16.1).

In comparison to hydro turbines, storage pumps have a defined working point, which is the intersection (working) point between the storage pump and the hydraulic system characteristic (see Figure 16.2). There are two main concepts of changing the working point of the storage pump to enable its regulation.

I) Variable speed: Changing the speed of the storage pump allows change at the working point at high efficiency. Small changes in speed have a big impact on the power consumption since speed (n) and power consumption (P) are related to $P \sim n^3$.

For reversible variable pump-turbines the speed in turbine and pump modes can be different (see Figure 16.3). Benefits include high efficiencies both in pump and turbine operation and safety against cavitation.

II) Hydraulic short-circuit operation.

16.1.3 Example of Hydraulic Short-circuit

In the following example the pumped storage plant consists of a power plant equipped with two Francis-turbine generating units with a power output of approx. 35 MW each and a storage pump plant equipped with two storage pump units of

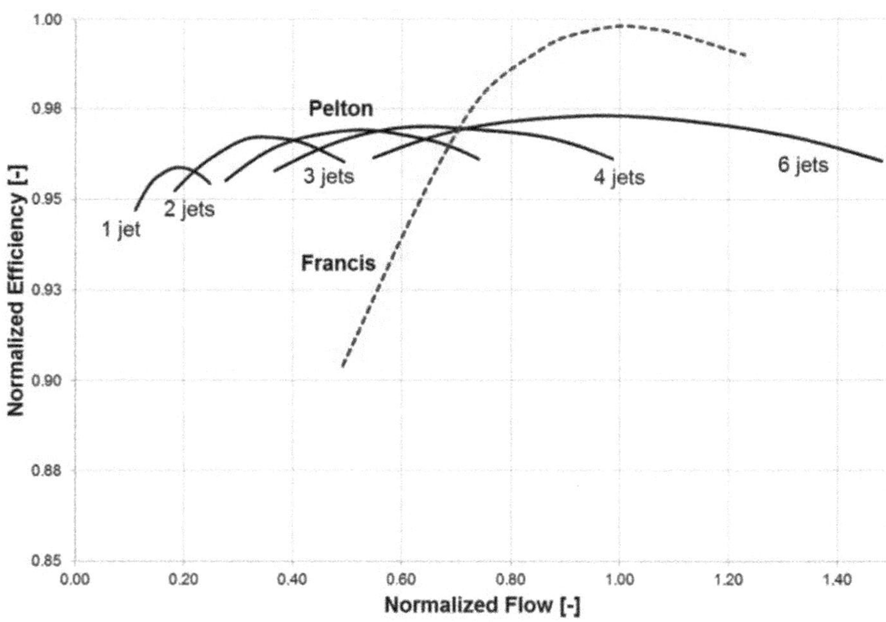

Figure 16.1 Full- and part load flexibility of Francis and Pelton Turbines. Source: Axpo or Andritz

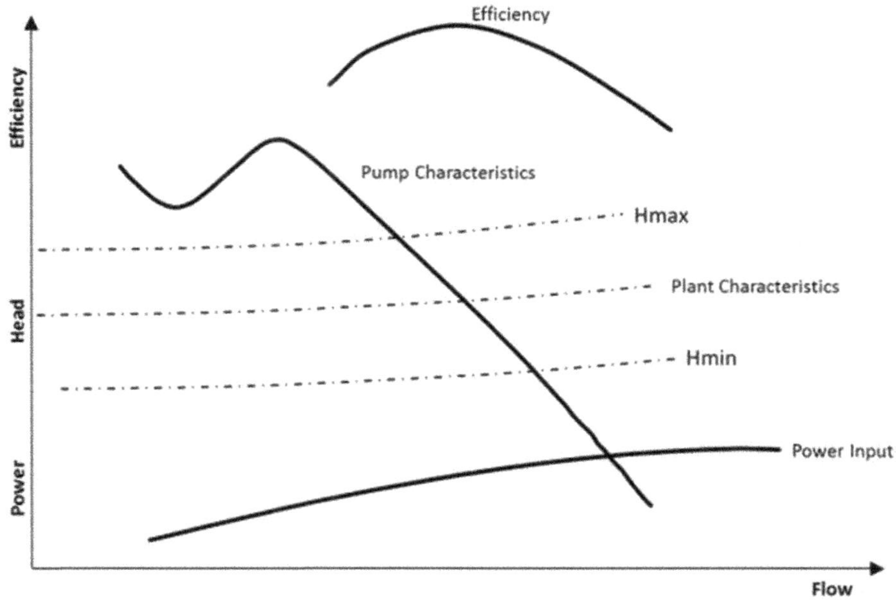

Figure 16.2 Characteristic and intersection point of pump and hydraulic system. Source: Axpo or Andritz

approximately. 20 MW each, which were initially not designed for hydraulic short-circuit operation (see Figure 16.4).

All machines, turbines as well as storage pumps, are electrically connected to the same grid feed-in point.

Figure 16.5, shows the possible operation mode using hydraulic short-circuit in this example. On the horizontal axis the power input/output is displayed, and on the

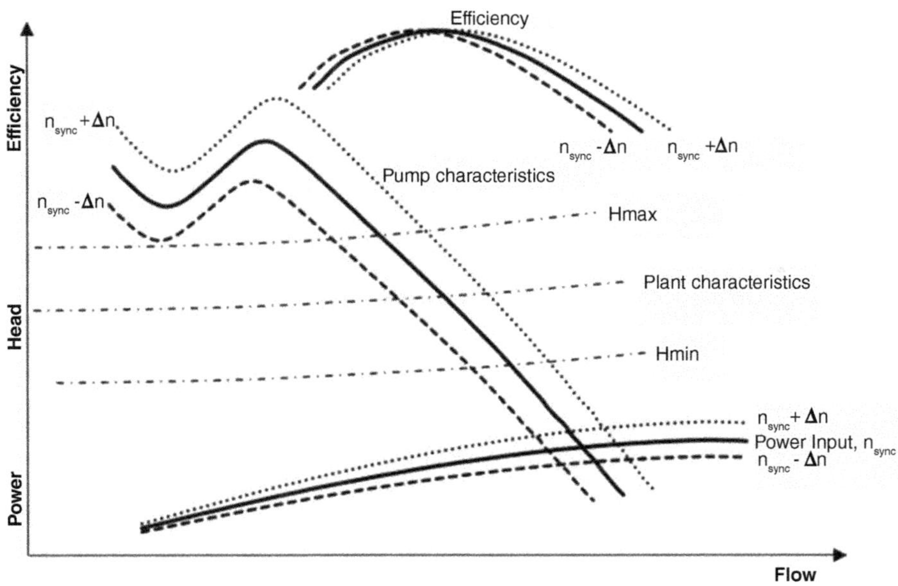

Figure 16.3 Store pump characteristic, reversible pump-turbine with variable speed. Source: Axpo or Andritz

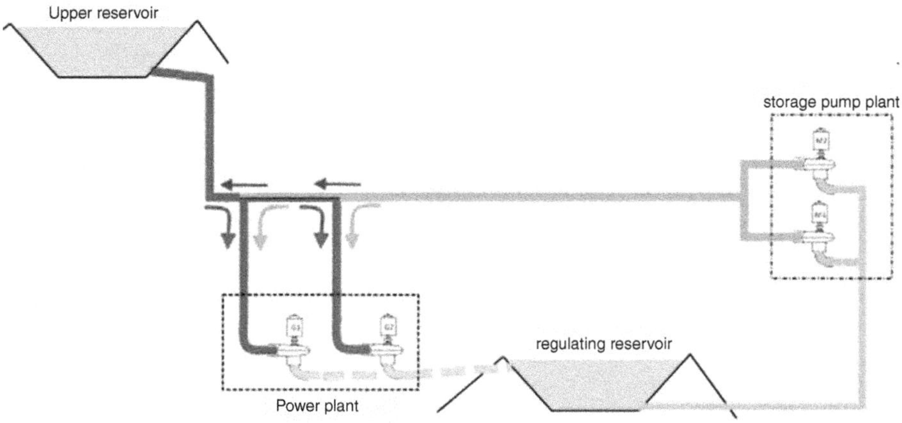

Figure 16.4 Pumped storage plant equipped with two turbines and two pumps in different plants. Source: Axpo or Andritz

vertical axis the efficiency. Since the speed of the storage pumps cannot be changed the pumps can only be operated a respective head (-20 and -40 MW power consumption) at their efficiency at this working point.

The turbines have an operation range from full load down to part load operation as low as 20 MW. Using both turbines a load range from 70 MW down to 20 MW can be covered with a small gap from 35 MW to 40 MW.

With hydraulic short-circuit the regulation range can be extended significantly from 70 MW production down to -20 MW consumption. Some operation points, i.e., 10 MW or 30 MW power production can be achieved by using different set ups of pump(s) and turbine(s) in operation. Looking at the overall efficiency, as well as minimizing the total operation hours of all machines, it is clear which combination of pump(s) and turbine(s) will be the chosen operation mode at the requested operation point.

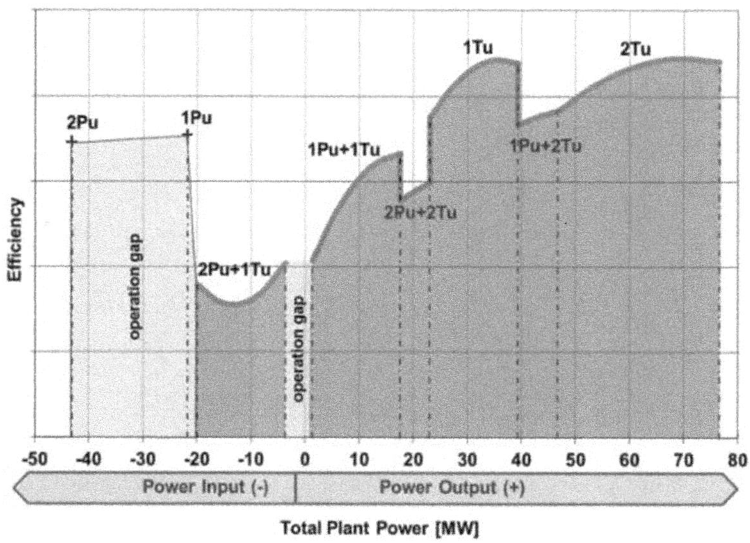

Figure 16.5 Operation characteristic of pump(s), turbine(s) at hydraulic short-circuit operation. Source: Axpo or Andritz

The load gap between -5 to -10 MW cannot be covered since the minimal part load capacity of the Francis turbines is restricted.

16.1.4 Purpose and Efficiency

In the past, the power consumption of large storage pumps was in some cases a challenge to the grid, today mainly economic reasons are the goal for hydraulic short-circuit operation. Possible benefits for hydraulic short-circuit operation are:

i) Frequency response: Fast load changes due to fast and easy load changing ability of the turbine. Frequency response services might be paid by the national grid operator.
ii) Grid services can be provided, with no water being lost in the hydraulic plant system, i.e., control of production/consumption in a regulation zone.

The efficiency of short-circuit operation depends of the different plant concepts as well as the efficiency of the involved components. Figure 16.6 shows the different energy losses of the components of a ternary unit in short-circuit operation. The most important – since it is the largest – is the losses of the hydraulic machines, the electrical machines and the hydraulic system have a smaller impact on efficiency.

Plant set-ups with single shaft arrangements have the advantage of reduced losses and a reduced number of machines involved in the cycle (motor generator). Efficiencies in the range from 75 to 83% can be achieved.

16.1.5 Different Power Plant Concepts

Looking at the variety of hydro power plants – specifically built or converted to operate a short-circuit – two general concepts were realized:

- **Single shaft arrangement – ternary units**: Storage pump and turbine are mounted on a single shaft
- **Multi-shaft arrangement:** Pump(s), turbine(s), or pump-turbine(s) have no common shaft, hydraulic machines are connected through hydraulic system. Therefore the generating and pump units can be installed in the same or different power houses.

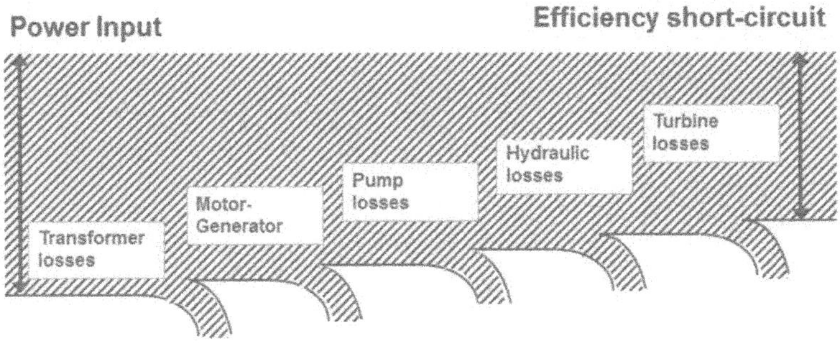

Figure 16.6 Simplified diagram of the efficiency of short-circuit operation of a ternary unit. Source: Axpo or Andritz

On both concepts various different set-ups are in operation. Sections 16.1.6 and 16.1.7 highlight a few without aiming at a comprehensive list.

16.1.6 Hydraulic Short-circuit with Ternary Units

Plant set-ups with horizontal and vertical shaft arrangements were built. In recent years vertical shaft arrangements have been more common. At high head most units consist of multiple stage storage pumps in combination with Francis or Pelton turbines, depending on the water head and requirements of operation flexibility. As shown in Section 16.1.2 a combination of a storage pump with a Pelton turbine has a higher operation flexibility since the turbine can be operated down to very low load.

These plants, especially the hydraulic system between the storage pump and the turbine are designed for hydraulic short-circuit operation. The hydraulic losses between the storage pump and turbine are minimized as well as the hydraulic layout being optimized to achieve optimal flow velocity distribution at the inlet of the turbine. This is in order to enable long term short-circuit operation.

Kopswerk II, in Austria, commissioned in 2007 is a remarkable example of a power plant compromising high head ternary units. The power plant consists of the upper reservoir, "Speicher Kops" with a capacity of 42 M m^3 and smaller lower reservoir, "Ausgleichsbecken Rifa." The power cavern contains three ternary units with a total power output of 525 MW at turbine operation.

The vertical shaft of the ternary units consists of a 175 MW Pelton turbine on the top, a 200 MVA synchronous motor generator in the middle, a 200 MVA hydraulic converter (for start-up of the pump), and a 150 MW storage pump at the bottom (see Figure 16.7).

Remarkable in this concept is the motor generator, which is placed below the Pelton turbine and the Pelton turbine operates at overpressure to atmosphere. The overpressure Pelton turbine allows the turbine to be to positioned at a lower installation level, which reduces the length of the overall shaft and therefore the height of the cavern.

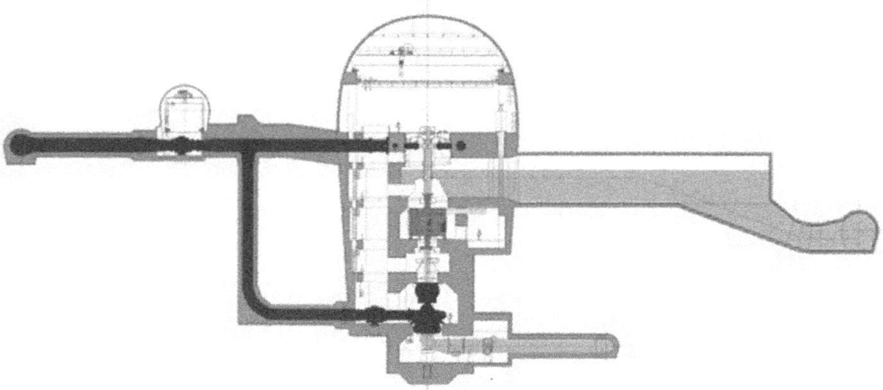

Figure 16.7 Cross section of the cavern of pump storage power plant Kopswerk II (Reproduced with permission from Vorarlberger Illwerke AG, © illwerke vkw).

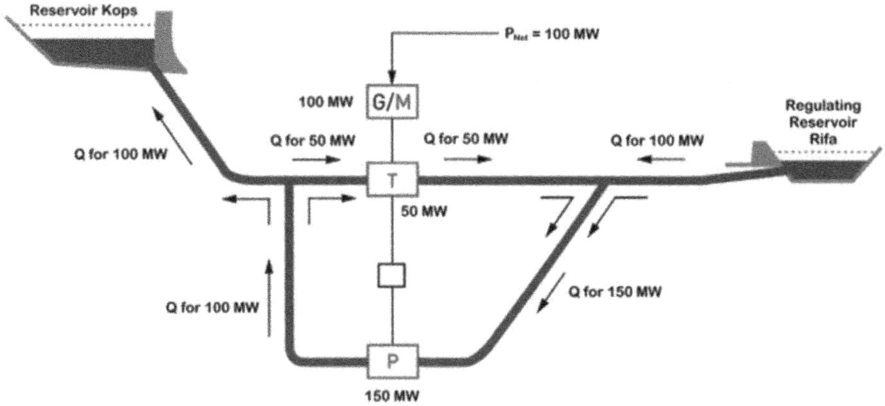

Figure 16.8 Plant- set up of Kopswerk II (Reproduced with permission from Vorarlberger Illwerke AG, © illwerke vkw).

Figure 16.8 shows one unit of Kopswerk II operating at 100 MW pump load. The storage pump operates at a certain head given by the reservoir "Speicher Kops." Since the requested power consumption is 100 MW, a flow equal to 50 MW will be diverted from the pump outlet, bifurcation to the Pelton turbine. The motor generator supplies the energy difference between the 150 MW pump load consumption and the 50 MW turbine generation.

16.1.7 Hydraulic Short-circuit with Multi-shaft Arrangements

The pump storage and the run-of-river power plants of Kraftwerke Linth Limmern AG in Switzerland were built and commissioned in the 1960s. The pump storage consists of an upper reservoir "Limmernboden" with a usable capacity of 87 M m^3 and the lower reservoir, "Becken Tierfehd" with 0.57 M m^3. The pumped storage power plant is designed as a cavern compromising in total five Pelton units (three units of the Limmern stage, two units of the Hintersand stage) and two 17 MW storage pump units Hintersand. The storage pumps were mainly used to pump water in the reservoir Limmernboden during snow melt in the spring (see Figure 16.9).

In order to achieve higher operation flexibility a four stage Francis pump-turbine was commissioned in 2007. In addition the lower reservoir was doubled in capacity by building another reservoir.

The reversible multi-stage pump-turbine with a water head of approximately 1000 m has a synchronous motor generator and a load capacity of 140 MW in pump as well as turbine operation. Due to the multi-stage layout of the pump-turbine, operation is restricted to a defined working point in both pump and turbine mode (cross section see Figure 16.10).

A few years after commercial operation started the first technical feasibility studies into hydraulic short-circuit were made in order to increase the operation flexibility. Therefore the pump-turbine was operated in pump mode and at the same time one of the Limmern turbines was operated. Both plants were electrically connected to the same grid feed-in point.

Step by step, short-circuit operations were implemented. Today, and on request, all the hydraulic machines can be operated simultaneously and this allows maximum

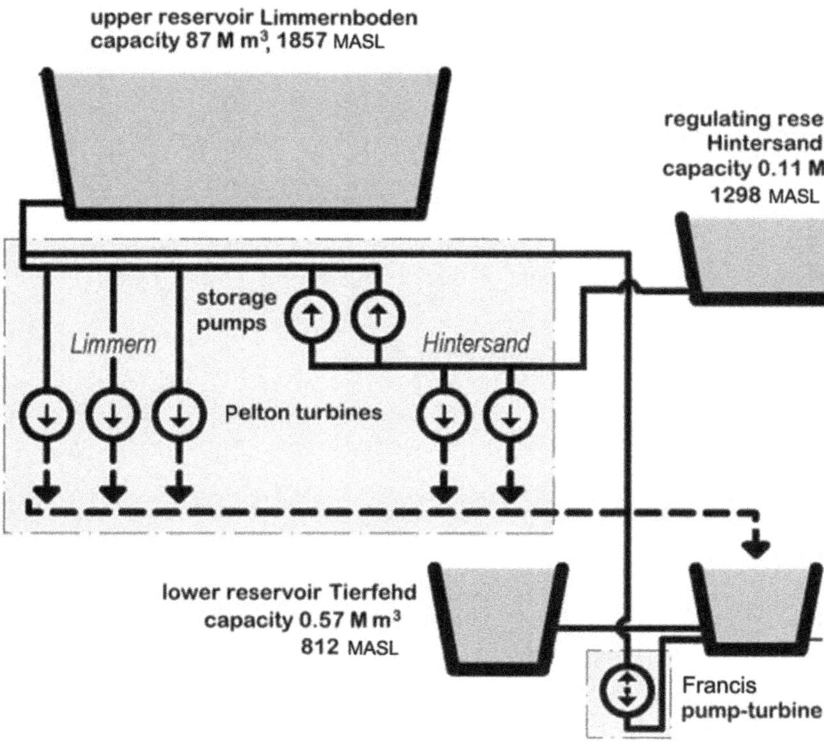

Figure 16.9 Storage plant set-up, Kraftwerke Linth-Limmern, cavern Tierfehd (Reproduced with permission from Axpo Power AG, 2016).

operation flexibility in short-circuit operation and allows the water to be kept in the pumped storage scheme's system.

16.1.8 Comparison of Concepts

Both concepts have technical, commercial, and other advantages as well as disadvantages, therefore both concepts in various set-ups can be found. Some plants are specifically built for hydraulic short-circuit but more were converted using this operation mode. Some of the benefits of both concepts are listed but it is not intended to be comprehensive.

Benefits of single shaft arrangement:

- Only one motor generator is needed – reduced costs and electrical losses. Depending on the electrical set-up only one transformer is necessary.
- Length of hydraulic system for water circulation to the turbine is shorter which leads to reduced hydraulic losses.
- In general higher overall efficiency.

Benefits of multi-shaft arrangement:

- Boundary conditions for shaft arrangement are set by the minimum required suction pressure of the storage pump i.e., for a Pelton turbine for the minimal required distant from runner to maximal tailwater level.

Figure 16.10 Reversible Francis four stage pump-turbine, Kraftwerke Linth-Limmern (Reproduced with permission from Andritz hydro AG, 2016).

- At Kopswerk II, the overall shaft length is due to these boundary conditions approximately 40 m in height, which leads to an impressive cavern size, but also to significant costs.
- Higher redundancy of the equipment due to separate units (turbine and generator, storage pump and motor), with impact on the costs.
- Higher flexibility in den design, layout of the power plant. The storage pump and power plant can be designed and optimized individually.

Some of the pumped storage power plants with multi-shaft design even have underground caverns as well as surface power houses. Therefore the range of concepts that can be found is much wider.

16.1.9 Implementation Hydraulic Short-Circuit in Existing Plants

In order to extend the flexibility in operation in existing pumped storage power plants the concept of hydraulic short-circuit is applied. This kind of operation can be applied at both, single-shaft as well as multi-shaft arrangements. Technical clarifications ensure safe and stable operation of exiting installations in this new mode of operation.

The conversion of existing pumped storage plants for hydraulic short-circuit operation has been applied successfully in a number of installations. An example of a conversion is shown in Section 16.1.7 for Kraftwerke Linth-Limmern AG, in Switzerland.

Due to the fact that, in most cases, the existing power plants were not built for hydraulic short-circuit operation, technical as well as commercial investigations are required.

16.1.9.1 Steps to Realization

A first basic technical clarification will give information about the feasibility of short-circuit operation. Simulations of cost of operation and the potential revenues will demonstrate the economic profit.

When first economical and technical feasibility were proven, detailed technical investigations and calculations as well as comprehensive economical simulations follow.

The technical investigations are complemented by site tests and measurements to prove safe and stable operation in the new mode. Regular inspection over a longer period of time ensures reliable long term operation of the plant.

16.1.9.2 Operational Range

Detailed technical clarifications of the operation range of the pump and turbine has to be investigated as a first step. As part of the flow will pass the circuit from pump to the turbine and part will be pumped to the upper reservoir the losses in the hydraulic circuit will be changed compared to the original operation set-up. Therefore the head for pump and turbine will change. This has to be investigated in the operational range of pump and turbine with respect to the admissible boundaries for continuous operation.

16.1.9.3 Losses and Flow Quality

The hydraulic system and especially the bifurcations of the originally planned pumped storage plant were not designed for hydraulic short-circuit operation. Therefore the flow will pass some bifurcations in the opposite way than they were originally designe for. This new direction of flow may lead to increased losses, pulsation, and increased turbulence within the flow.

For evaluation of the flow in the pipe system modern methods of flow simulation may be applied. By simulation of the unsteady, turbulent flow the pressure and velocity field and turbulence quantities can be achieved. For technical applications the turbulent flow is described by the time averaged Navier-Stokes equations. Various turbulence models for solution of these equations are available in literature and commercial programs.

As a result the pressure and velocity field in the entire system will be available. The losses of total energy in the entire system and in parts of it can be evaluated. Visualization of the flow by streamlines demonstrates the qualitative flow pattern is shown in Figure 16.11.

Due to the operation outside the original design specification of the bifurcations areas of low pressure may occur, which may lead to cavitation. Depending on the operation and flow rate such areas can be detected.

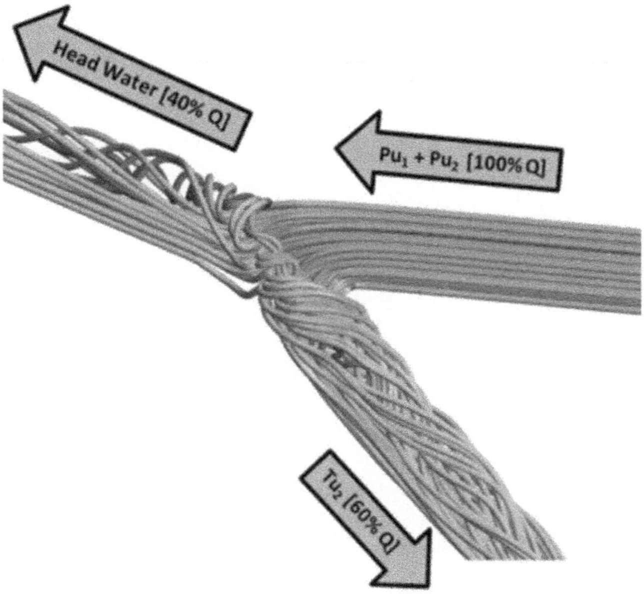

Figure 16.11 CFD flow simulation in bifurcation (Reproduced with permission from Andritz Hydro AG, 2016).

At the bifurcation distortion of the velocity profile an increase of turbulence may occur. These distorted velocity profile and flow with increased turbulence will enter the turbine or pump-turbine. The flow simulation will help to evaluate the disturbance of the flow. If necessary flow simulation for the turbine can be applied with these flow profiles as inlet boundary condition (see Figure 16.12).

Pelton turbines in comparison with Francis turbines are more sensitive in respect to changes of flow profile. Changes in flow profile might influence the jet quality in a negative way and can therefore lead to long term problems.

16.1.9.4 Transient Analysis

For the original operation of the plant the characteristics and parameters like energy losses, pressure rise, or opening and closing rates are well known. In order to ensure safe operation of the power plant during hydraulic short-circuit a transient analysis has to be performed. Opening and closing rates, speed rise, and pressure rise have to be verified and critical operations have to be detected. Operation points that are critical for draining or spillover of the surge chamber have to be investigated. The impact of the new operation on the run-away speed has to be verified.

16.1.9.5 Test Operation and Site Measurements

Test operation in hydraulic short-circuit will be supported by detailed technical investigations. By gradual application of the short-circuit operation the behavior of the entire system will be identified. Measurements of pressure pulsation and vibration in sensitive areas around the bifurcation or evaluation of the losses in certain areas will provide a solid technical base. Acoustic measurements around bifurcations can give indications about the flow behavior (see Figure 16.13).

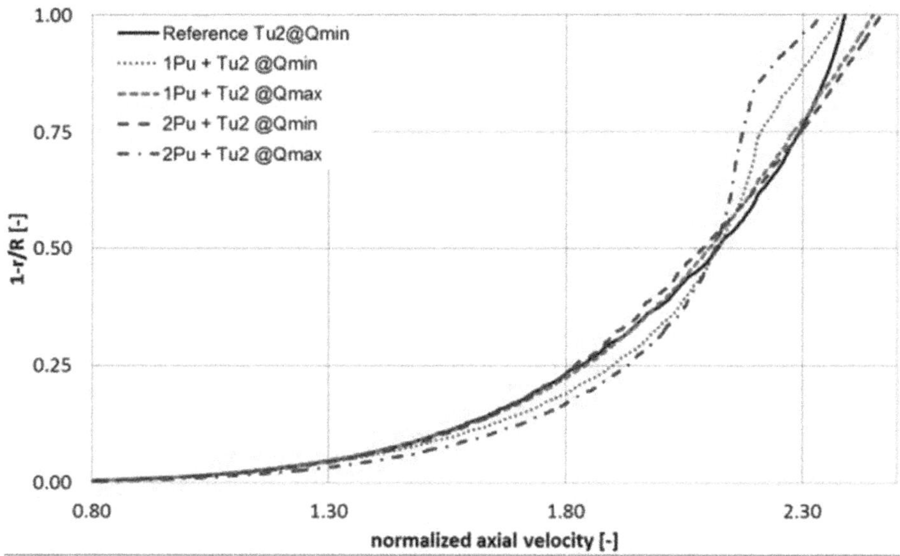

Figure 16.12 CFD flow simulation, quantitative comparison normal operation with short-circuit at inlet of turbine for different load points (Reproduced with permission from Andritz Hydro AG, 2016).

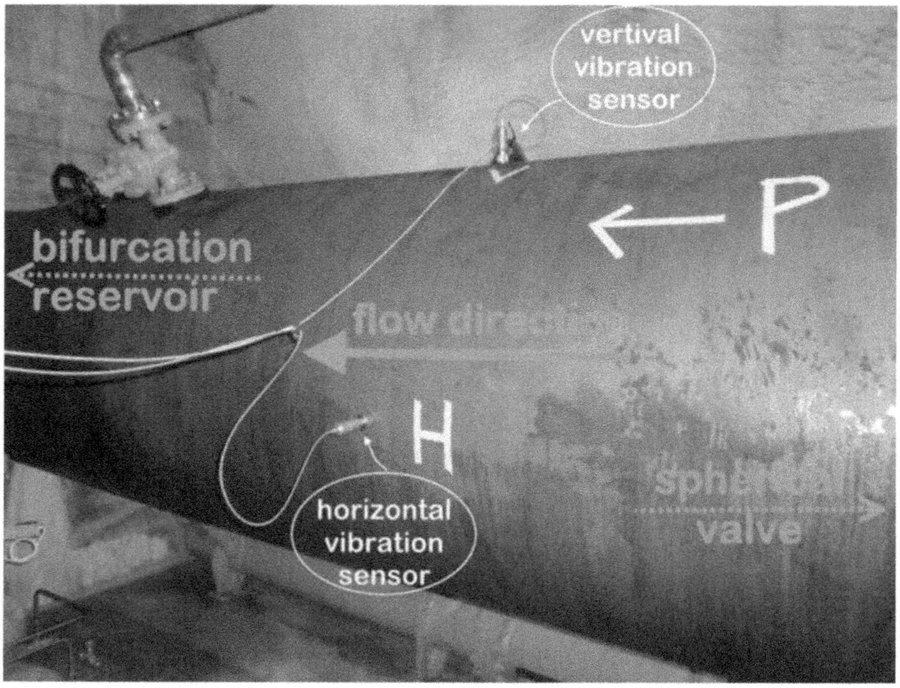

Figure 16.13 On site measurement of vibration and pulsation near the bifurcation.

Special care should be applied when water used in the short-circuit system contains hard particles, like sediments. Erosion in the hydraulic machines as well as in the hydraulic system (i.e., bifurcation) can occur. In order to avoid damage to the turbine's hard surface a coating might be applied to the most exposed parts, such as the wicked gate, runner, and labyrinth seals.

Part III

Mechanical Energy Storage, Compressed Air Energy Storage, and Flywheels

17

Compressed Air Energy Storage

Are the Market and Technical Knowledge Ready?

Pierre Bérest,[1] *Benoît Brouard,*[2] *Louis Londe,*[3] *and Arnaud Réveillère*[3]

[1] *LMS, Ecole Polytechnique, route de Saclay, 91128 Palaiseau, France*
[2] *Brouard Consulting SAS, 101 rue du Temple, 75003 Paris, France*
[3] *Geostock, 2 Rue des Martinets, 92500 Rueil-Malmaison, France*

17.1 Introduction

17.1.1 Need for Electricity Storage

The balance between the supply and demand of electricity, necessary for the proper functioning of electrical grids, is becoming increasingly fragile. The increased use of intermittent renewable energy sources, mainly wind, is creating additional sources of fragility. Weather hazards cause fluctuations in production, which are independent of consumption. New situations have to be handled: overproduction of electricity during off-peak hours, and means of production that cannot be relied upon during peak periods. The increased volatility in the price of electricity is one indicator of these tensions. In France, on the 8 February 2012, a peak of 102 GW in consumption brought the market price up to 2000€/MWh. In contrast, Germany experienced in 2010 periods of negative pricing as low as -500 €/MWh, caused by a surplus in wind turbine production [1].

Faced with these tensions, different solutions can be implemented: the use of a peaking power plant, a better interconnection of the electrical grid, management of the demand, and energy storage. More than 100 GW of electricity storage exists today globally, 99% of which is currently pumped-storage hydroelectricity (PSH). But development of additional PSH capacity is constrained by local opposition and limited suitable areas. With a 290 MW and a 110 MW plant currently operating since 1978 and 1991, respectively, Compressed Air Energy Storage (CAES) is one of the largest among other techniques currently used for storing large quantities of electricity. In a CAES, the surplus power (generated by renewable sources typically) is utilized to compress air in an underground cavern at a high pressure (5 MPa to 7 MPa is typical). When power demand is high, the stored air is released; natural gas is used to heat the air, which fires a turbine generator to produce electricity. Although hard rock caverns could be foreseen, the two existing CAES and most projects consider salt caverns for air storage. Indeed, when a salt cavern is available, it is the most attractive solution.

17.1.2 Isothermal Compressed Air Energy Storage and Adiabatic Compressed Air Energy Storage

Compression of air generates a temperature increase; the air is warmer after compression. Expansion generates a temperature decrease. If no extra heat is added, the air

Advances in Energy Storage: Latest Developments from R&D to the Market, First Edition.
Edited by Andreas Hauer.
© 2022 John Wiley & Sons, Ltd. Published 2022 by John Wiley & Sons, Ltd.

will be much colder after expansion. If the heat generated during compression can be stored and used during expansion, the efficiency of the storage improves considerably. There are two main ways in which a CAES system can deal with the heat:

- **Adiabatic Compressed Air Energy Storage (A-CAES)**

During storage, Adiabatic Compressed Air Energy Storage (A-CAES) stores the heat produced by compression and uses it for warming up the pressurized air prior to going through the expander during unloading (Figure 17.1).

The theoretical efficiency of A-CAES approaches 100% when insulation is perfect, but in practice round trip efficiency of 70% is to be expected. Heat can be stored in a solid such as concrete or stone, or in a fluid such as water or hot oil (up to 300°C) or molten salt solutions (600°C).

- **Non-adiabatic or Isothermal Compressed Air Energy Storage (I-CAES)**

In the present chapter, "Isothermal Compressed Air Energy Storage (I-CAES)" (Figure 17.2), refers to the case also called "non-adiabatic" or "diabatic" CAES, in which much of the compression heat is dissipated through intercoolers (thus approaching isothermal compression) into the atmosphere as waste, essentially wasting, the energy used to perform the work of compression.

Upon unloading air from storage, the pressure of this compressed air is one indicator of the amount of stored energy that remains available, which is actually potential mechanical work. Temperature is another indicator. To be able to perform this mechanical work, the air must be reheated prior to expansion in the turbine to power a generator. This reheating is typically accomplished by using a natural gas fired burner.

Figure 17.1 Adiabatic CAES principle. Source: Compressed Air Storage Caverns – CAES, Geostock Entrepose.

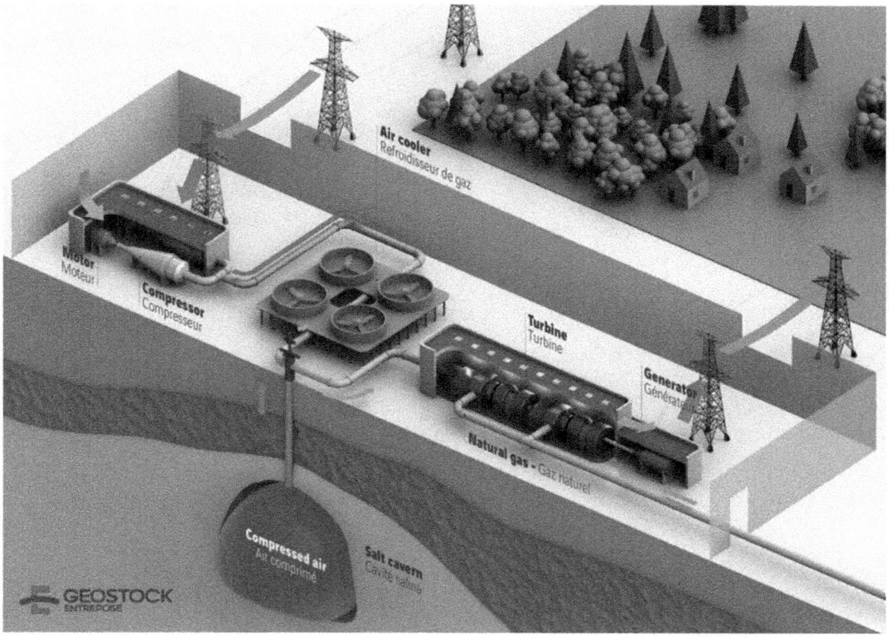

Figure 17.2 Isothermal CAES principle. Source: Compressed Air Storage Caverns – CAES, Geostock Entrepose.

According to [2], a "Standard" I-CAES plant requires 0.7 kWh of electricity and 1.1 kWh (3800 Btu) of high heating value (HHV) of gas for each kWh of energy output, corresponding to an energy recovery efficiency of about 55%.

The present chapter identifies the main historical developments (Section 17.2), discusses new challenges raised by air storage in salt caverns (Section 17.3), reviews recent projects (Section 17.4), and proposes a business case on a typical CAES project to provide some insights to the recent developments (Section 17.5).

17.2 Historical Developments

17.2.1 Huntorf, Germany

The Huntorf plant (Figure 17.3) is located in north Germany, and was commissioned in 1978 as the world's first CAES plant. Its main characteristics according to [3] are the following:

Cavern characteristics	
• Caverns depth (casing shoe)	~645 m/GL (~2100 ft/GL)
• Caverns geometric volume (in 2 caverns)	310 000 m^3 (2 Mbbl)
• Caverns pressure range	4.3–7 MPa (620–1000 psi)
Power system	
• Expander output power delivered to the grid	290 MW
• Compressor power	60 MW
• Unloading duration	4 h
• Ramp-up time	10 min

The cycle efficiency of the plant is reported as about 42%.

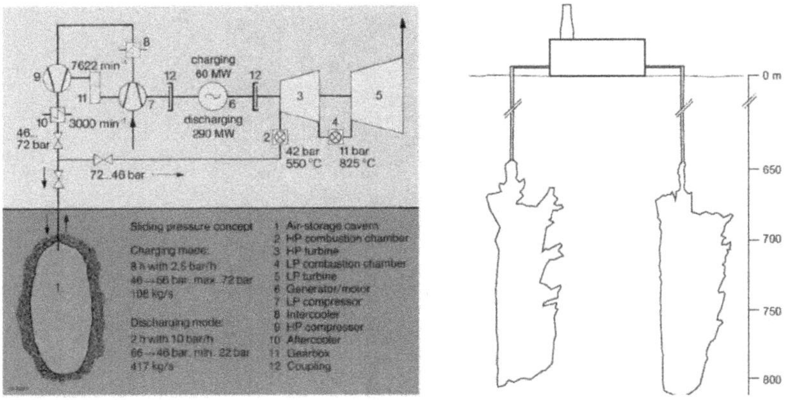

Figure 17.3 Left: Schematic layout of the Huntorf plant. From [4]. Right: The two caverns and the plant at the same scale from [3].

17.2.2 McIntosh

In 1991, a 110 MW plant was constructed in McIntosh, Alabama (Figure 17.4), and improved upon the Huntorf design by incorporating a recuperator (air-to-air heat exchanger) to preheat air from the cavern using exhaust heat from the turbines.

Its main characteristics are (based on [5–6]):

Cavern characteristics	
• Cavern depth (casing shoe)	442 m/GL (1450 ft/GL)
• Cavern geometric volume (in 2 caverns)	580 000 m^3 (3.6 Mbbl)
• Cavern pressure range	4.5–7.4 MPa (650–1000 psi)
Power system	
• Expander output power delivered to the grid	110 MW
• Compressor power	49 MW
• Unloading duration	26 h
• Ramp-up time	15 min

Thanks to the heat recuperator, the fuel consumption is reduced by 22–25% and the cycle efficiency reaches 54%.

17.2.3 Other Large-scale Projects

Several other large-scale projects were developed. Among others, a feasibility study of a 300 MW plant in two caverns at 545 and 880 m/GL was conducted in 1993 in Israel.

17.3 Challenges Raised by Air Storage in Salt Caverns

17.3.1 Introduction

Technologies required to build and operate a CAES are available and proven. In this section, salt caverns are of special interest.

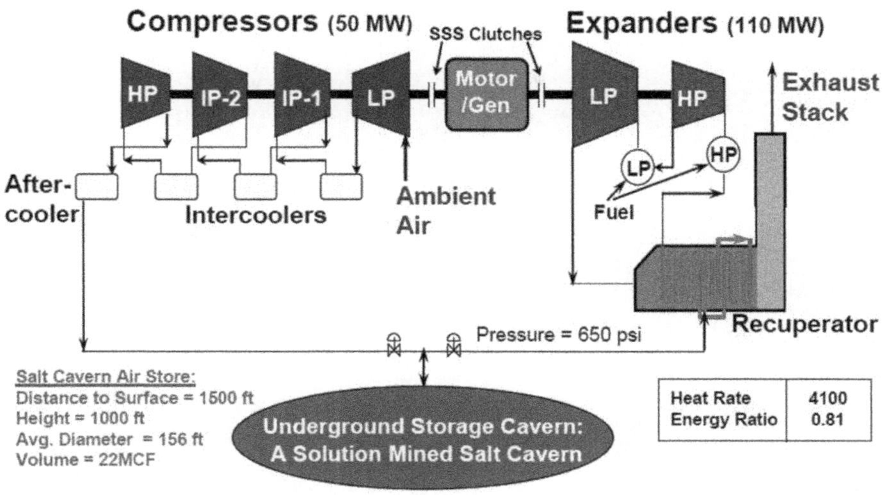

Figure 17.4 Schematic layout of the McIntosh CAES plant, including the heat recuperator.

Several thousands of salt caverns are currently used worldwide for storing gaseous, liquid or liquefied hydrocarbons, air or hydrogen. These caverns are created through solution-mining: a well is drilled to the salt formation; it is cased and cemented to the formation; a smaller tube is set inside the casing, like a straw in a bottle (Figure 17.5). Soft water is injected through the central tube; water leaches out the salt and the formed brine is withdrawn from the cavern through the annular space between the central string and the casing. After one year or so, a large cavern is created. Its depth ranges from 200 m to 2000 m and its volume ranges from 10 000 m^3 to 1 0000 00 m^3. When solution-mining is completed, the cavern is filled with saturated brine. In the case of gas storage caverns, gas then is injected through the annular space and brine is withdrawn through the central tubing ("de-brining"). A small amount of brine is left in the cavern bottom, and gas in the cavern is wet. The amount of gas that can be injected or withdrawn from the cavern depends on the gas pressure in the cavern. Minimum and maximum gas pressures typically are 25% and 80% of the geostatic pressure (i.e. the weight of the overburden, which is proportional to cavern depth; it is 22 MPa at a 1000-m depth). Some of these caverns have been operating for more than 50 years. A lot of experience has been gained. This technology can be considered as mature. Salt caverns have experienced a relatively small number of incidents [7] and can be considered as the safest way of storing large amounts of hydrocarbons [8]. Safety concerns such as tightness or consequences of a blow-out are much less in an air storage cavern.

However, CAES raises a couple of new problems that demand specific answers. CAES can be operated on a daily cycle and experience large pressure and temperature swings, in sharp contrast with natural gas storage caverns, which are often operated on a yearly basis (gas is stored in summer and withdrawn in winter). In

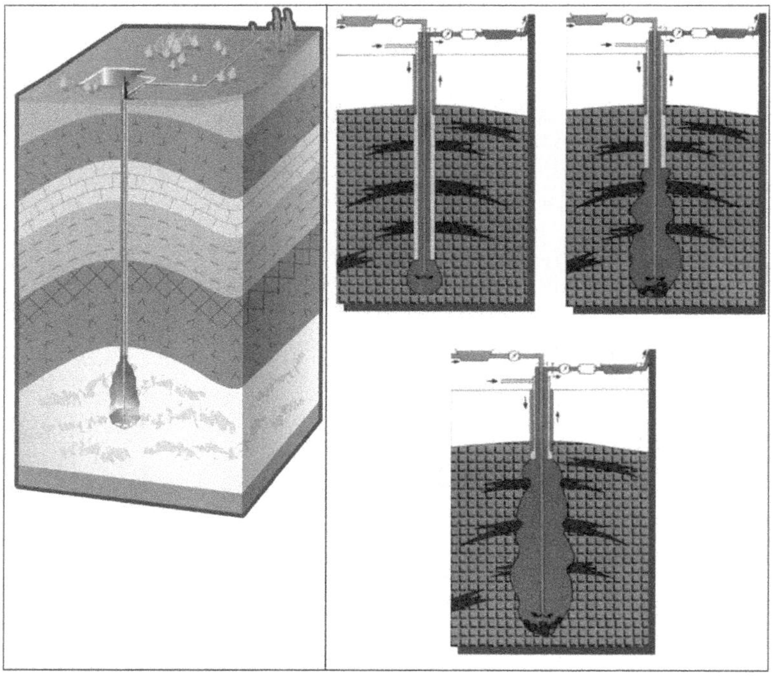

Figure 17.5 Caverns in a salt layer (left) and solution-mining of a salt cavern (right).

addition, wet air can be more aggressive than natural gas for the casing that links the cavern to ground level. These issues are discussed in this chapter.

17.3.2 Thermomechanical Behavior of Salt Caverns

Large pressure and temperature swings are generated when gas is withdrawn or injected from a gas cavern (Figure 17.6). Withdrawal (when demand is maximum) is easier and faster than injection (during the night). Turbines in which high-pressure gas from the cavern is injected are most efficient when injection pressure does not vary too much. A typical cavern pressure range is P = 5–7 MPa. In other words, a typical withdrawal rate is 0.5 MPa/hour. Injection rates are slower. The energy balance can be written as Equation (17.1)

$$m\dot{h} = v\dot{P} + Q + <\dot{m}>(h^{inj} - h) \tag{17.1}$$

Where m is the gas mass, $<\dot{m}> = \dot{m}$ when gas is injected ($\dot{m} > 0$) and $<\dot{m}> = 0$ when gas is withdrawn ($\dot{m} < 0$); v is the gas volume per unit of mass; $h = h(P,T)$ is the gas enthalpy, Q the heat flux from the rock mass. When gas is injected ($<\dot{m}> = \dot{m}$), gas pressure is continuous at the cavern neck; the gas temperature is not, and a jump in enthalpy must be considered at the cavern entrance, or $h^{inj} - h = C_P(T^{inj} - T)$. C_P is gas heat capacity when gas pressure is kept constant and $\gamma = C_p / C_v$ is the adiabaticity index; when the gas is perfect, $h = C_PT$ and $Pv = rT$; in addition, during a gas

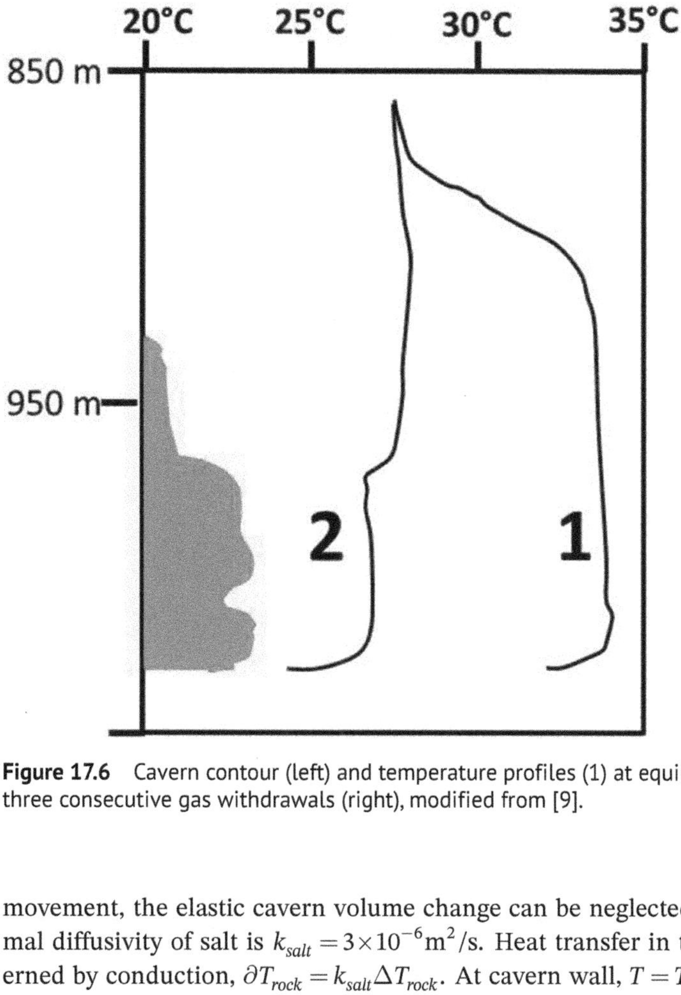

Figure 17.6 Cavern contour (left) and temperature profiles (1) at equilibrium and (2) after three consecutive gas withdrawals (right), modified from [9].

movement, the elastic cavern volume change can be neglected, $mv = V_0$. The thermal diffusivity of salt is $k_{salt} = 3\times10^{-6}\text{m}^2/\text{s}$. Heat transfer in the rock mass is governed by conduction, $\partial T_{rock} = k_{salt}\Delta T_{rock}$. At cavern wall, $T = T_{rock}$ is assumed. Only fast injection/withdrawal are considered. The depth of penetration of gas temperature changes in the rock mass after a time δt is $\sqrt{k_{salt}\delta t}$, approximately, or $\sqrt{k_{salt}\delta t} = 0.5$ m when $\delta t = 1$ day. In other words, at such a time scale, cavern surface can be considered as the sum of small flat surfaces whose total area equals the actual area of the cavern walls, or Σ, as was noted by [3], and [10]. The energy equation during withdrawal can be rewritten as in (17.2).:

$$\frac{\gamma}{\gamma-1}\frac{P\dot{T}}{T} - \dot{P} = -\frac{\Sigma K_{salt}}{V_0}\int_0^t \dot{T}_{rock}(\tau)\Big/\sqrt{\pi k_{salt}(t-\tau)}d\tau \tag{17.2}$$

(When injection stops, $\dot{m} = 0$, and this equation holds again when setting $PV_0 = m_0 rT$.) Two constants appear in this equation, $\gamma = 1.4$ (for air) and $\Sigma K_{salt}/V\sqrt{\pi k_{salt}}$. In fact, $K_{salt}/\sqrt{\pi k_{salt}} \simeq 1800\ \text{W.s}^{1/2}\text{m}^{-2}$ does not vary much from one site to the other. Solution to Equation (17.2) mainly depends on the initial conditions P_0 and T_0 and on the ratio V/Σ between the volume and the surface of the cavern: for a given pressure drop rate, the temperature drop rate is larger when the initial pressure P_0 is smaller and when

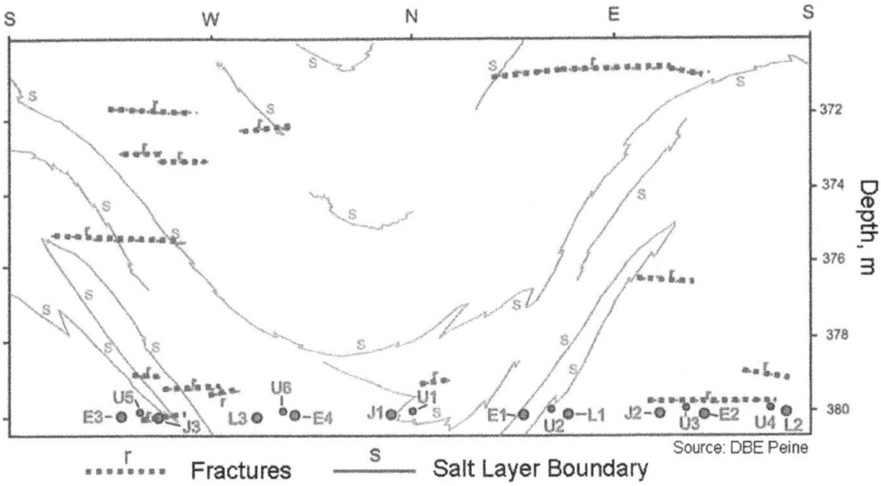

Figure 17.7 Thermal fractures at the wall of a ventilation shaft [22].

the ratio V/Σ is larger. Pressure drops are fast in a CAES and Equation (17.2) can be simplified as (17.3).

$$\frac{\delta P}{\delta T}=\frac{4}{3}\frac{\Sigma}{V}\frac{K}{\sqrt{\pi k}}\sqrt{\delta t}+\frac{\gamma}{\gamma-1}\frac{P_0}{T_0}\simeq 2.4\times 10^{-3}\frac{\Sigma}{V}(\mathrm{m})\sqrt{\delta t(\mathrm{s})}+3.5\frac{P_0(\mathrm{MPa})}{T_0(\mathrm{K})} \qquad (17.3)$$

For instance, $P_0 = 7$ MPa, $T_0 = 300$ K, $\delta t = 4$ hours, $\delta P = 5 - 7 = -2$ MPa, $\Sigma / V = 15$ m, leads to $\delta T = -20$°C. Such a temperature drop can have significant structural consequences, which were studied for instance by [11–19].

Consider for instance the case of an elongated cylindrical cavern. Both the pressure drop (by δP) and the temperature drop (by δT) generate additional tangential stresses at cavern wall, $\delta\sigma_{rr}(a,t)=\delta P_c$, $\delta\sigma_{\varphi\varphi}(a,t)=-\delta\Theta(a,t)+\delta P_c$, $\delta\sigma_{zz}(a,t)=-\delta\Theta(a,t)$ [20], where $\Theta(r,t)=E\alpha\left(T_{salt}(r,t)-T_\infty\right)/(1-\nu)$; E and α are the Young modulus and thermal expansion coefficient of salt, respectively; T_∞ is the virgin rock mass temperature at cavern depth. The change in vertical stress ($\delta\sigma_{zz}$) is especially significant. (Salt viscoplasticity tends to lessen the effects of a rapid thermal change; neglecting viscoplasticity is on the safe side.)

For instance, [21], and [22], discussed the onset of fractures in an intake air shaft in a salt formation (Figure 17.7). "During the cold season, temperatures in the shaft decreased by 20°C ... within a time period of 80 days... horizontal and vertical fractures were detected by routine inspections in the shaft. These fractures had an average spacing of about 2.8 m. The fracture aperture amounted up to several mm" (p. 365). This example is somewhat similar to the case of a CAES whose pressure drops from 7 MPa to 5 MPa in 4 hours: the temperature drop is – 20°C, typically; and the additional thermal stress is +20 MPa, much larger, in most cases, than the preexisting compressive stress: fractures are created in the cavern wall and there is a risk that cavern tightness is lost.

In fact, this risk must not be overestimated. Thermal fractures are perpendicular to the cavern wall. In a cylindrical well, fractures are mostly horizontal. At the beginning of the cooling process, when the depth of penetration of temperature changes is shallow, stresses are tensile in a thin zone at the cavern wall, and many thermal microfractures appear in the cavern wall. When cold temperatures penetrate deeper into the rock mass, propagation of many fractures stops, and only a small number continue to develop. However, in a rapidly cycled cavern, the cold front is not given time enough to penetrate deep into the rock mass. Fracture depth approximately equals the depth of the zone in which tensile stresses are larger than rock tensile strength (assuming that this strength is zero is on the safe side). During a $\delta t = 4$-hour long withdrawal, the depth of penetration of temperature changes (which also is the depth of penetration of additional tensile stresses) is $\sqrt{k_{salt}\delta t} = 0.2$ m; fractures length is small and damage can only be observed in a thin "skin" in the cavern wall.

17.3.3 Materials

As presented in Section 17.3.1, Humid air will be discharged from the cavern to the surface, exposing the completion casings to corrosion it flows through (Figure 17.8). Exposure to a corrosive fluid is not a classical problem in the salt cavern storage industry, in which stored products are essentially non-corrosive hydrocarbons. Usually, a standard carbon steel (e.g. K-55 grade) is used for the liner in contact with the stored product. Such carbon steel rusts readily when exposed to the combination of air and moisture. The resulting iron oxide surface layer (the rust) is porous and fragile, and tends to flake and fall away exposing the underlying steel to further attack, creating a pretty uniform corrosion. Therefore carbon steel cannot be used for the air liner (cf. Figure 17.8) in that

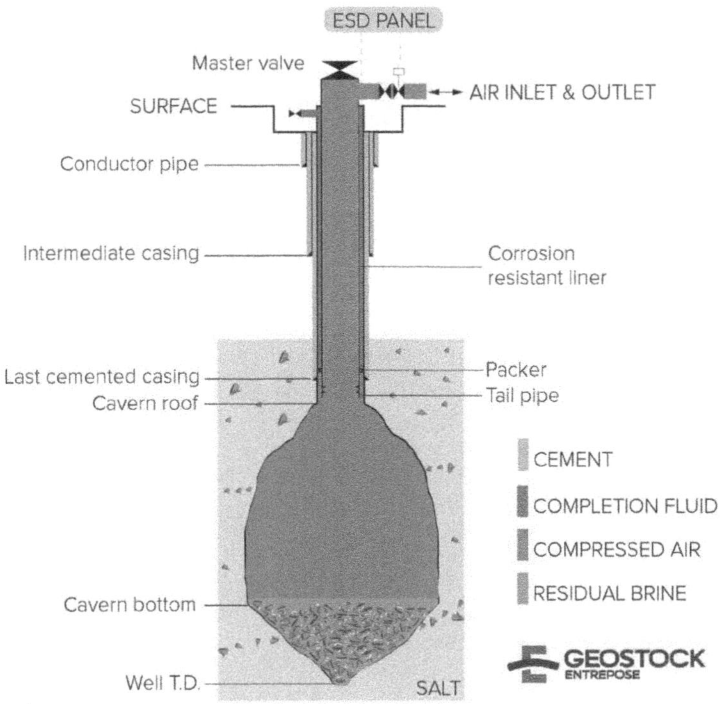

Figure 17.8 Typical well completion that can be imagined for CAES operation. Note that alternative designs could be found, for instance with no packer, due to the very large well diameters that could be considered. Source: Geostock Entrepose

would be contact with air and moisture. There are essentially two corrosion-resistant casing materials that can be used instead: stainless steel or fiber glass.

17.3.3.1 Stainless Steel Option

Stainless steel is a steel alloy with a minimum of 10.5% chromium in mass content. When exposed to a wet environment, it does not experience uniform corrosion, opposite to carbon steel, as it contains sufficient chromium to undergo passivation, spontaneously forming a microscopically thin inert surface film of chromium oxide by reaction with the oxygen in air. This passive film prevents further corrosion by blocking oxygen diffusion to the steel surface and thus preventing corrosion from spreading into the bulk of the metal. This film is self-repairing if it is scratched. If uniform corrosion is prevented, localized corrosion can occur in a number of ways, e.g. pitting corrosion. Such localized attack is most common in the presence of chloride ions, and is difficult to predict. Pitting corrosion could lead to local leakage, which might be acceptable for CAES as long as it remains limited.

17.3.3.2 Fiber Glass Option

Another corrosion-resistant option is the use of fiber glass casings, which are not subject to corrosion. Such casings are used globally for observation wells, and in some cases for highly corrosive injection or production wells that can be encountered in the oil and gas or geothermal energy industries. One of its limitations is its mechanical properties in terms of resistance to collapse, burst, and tension. But when used as a production liner, the requirements for these three design constraints are quite easily matched, and the resistance to corrosion is the most important design issue.

The Huntorf CAES plant uses fiber glass casings as the corrosion-resistant material in contact with the hot wet air. [3], mentions that, in Huntorf, a casing had to be replaced as it failed after 20 years of operation (Figure 17.9). Such a work implies keeping the cavern at atmospheric pressure for the duration of the operation.

17.4 (Selected) Recent Projects

The number of projects that have recently been under development tend to show a renewed interest in CAES technologies. Figure 17.10 and the following bullet list, briefly present a selection of the projects since 2010. Geostock has been involved in more than half of these projects.

- SACRE: French research project studying the technical and economic feasibility of adiabatic compressed air energy storage in salt (feasibility studies were done for several French locations) and mined caverns. It included heat storage in ceramics.[1]
- SEARCH: French research project focused on Advanced Adiabatic CAES, including the study of lined mined caverns and thermal energy storage pilots in rocks or ceramics (mock-up built and visible in Grenoble, France).[2]

1 http://www.agence-nationale-recherche.fr/?Project=ANR-10-STKE-0006. Accessed 28 December 2021.

2 http://www.agence-nationale-recherche.fr/Colloques/Energies2012/presentations/Search.pdf Accessed 28 December 2021. The Thermal energy storage pilots can still be visited at CEA Liten in Grenoble, as the authors did in 2017.

Figure 17.9 Fractured fiber glass casing from the Huntorf plant. From [3].

- RICAS 2020: European research project aiming at evaluating performances and feasibility of Advanced Adiabatic Compressed Air Energy Storage (AA-CAES).[3]
- DEMETAIR. Concept of small volume adiabatic CAES in the lined base of a shaft.
- SUSTAIN-X: Concept of small volume isothermal CAES in a lined mined cavern or well. The compression and expansion are to be done through a mechanical crankshaft drive train (motor) and the heat is extracted using a foam.
- NYSEG Seneca. Project of 130 to 180 MW I-CAES near Seneca lake, NY, US.[4]

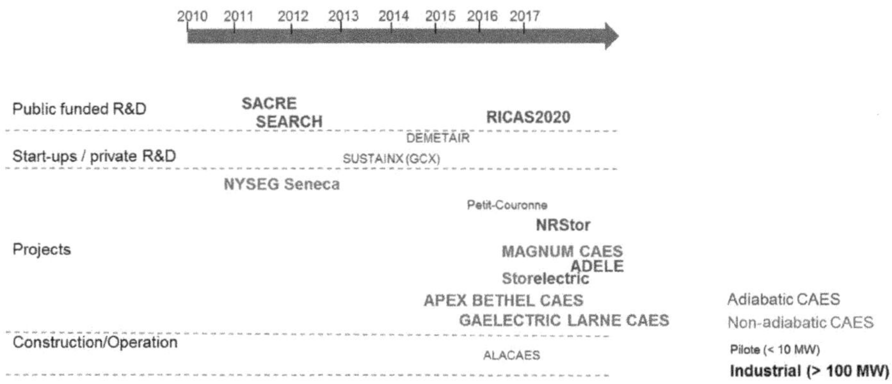

Figure 17.10 Indicative schedule of selected CAES projects after 2010.

3 www.ricas2020.eu. Accessed 28 December 2021.

4 https://www.smartgrid.gov/files/NETL-Final-Report-9-6-12.pdf. Accessed 29 August 2017.

- Petit-Couronne: Study of the conversion of a mined cavern having stored LPG into adiabatic CAES in France [23].
- NRStor: Adiabatic CAES projects in Ontario, Canada.[5]
- MAGNUM CAES: Project of a 160 MW isothermal CAES plant in Utah, US.[6]
- ADELE Adiabatic CAES. Project of an up to 200 MW A-CAES in Germany.[7]
- Storelectric: Project of either a conversion of existing gas-fired CCGT into I-CAES, or into an A-CAES using a thermal energy storage in the UK.[8]
- APEX Bethel Energy Center: Project of a 317 MW isothermal CAES facility in Texas, US.[9]
- GAELECTRIC. Projects of a 330 MW isothermal CAES in Larne (Northern Ireland, UK) and addtional projects in the UK and the Netherlands.[10]
- ALACAES: In 2016, started a pilot Adiabatic CAES in an abandonned tunnel of the Swiss Alps. With a target of a 72% round trip efficiency. The patented Thermal Energy Storage is constructed using rocks within the tunnel.[11]

17.5 Business Case

17.5.1 Various Possible Sources of Revenue

Electricity storage has the flexibility to operate in several ways: within the energy market, by trading energy to gain from arbitrage, and in ancillary markets, offering reserve, power quality, and reliability services. It can also be integrated with existing infrastructure: generators such as wind farms (to reduce balancing costs, time-shift delivery, or manage constraints); demand centers (to reduce network service charges); or networks (deferring costly upgrades to transmission and distribution systems).

These various possible services are usually offered to the grid operator, or the company that has the responsibility to balance production and demand. The revenues that can be obtained from offering these services are very regulation dependent, and therefore very country specific. It is not the intention here to review the many local markets in detail.

However, the revenues gained from providing one of these many services should come from one of the following two sources:

- **Capacity markets.** The storage operator contracts “available power” with the grid operator (or equivalent): during certain time slots, it has to be able to produce power (in MW), within a specified ramp-up time, without necessarily doing it.

5 http://www.nrstor.com/news/ieso-selects-nrstor-to-deliver-fuel-free-compressed-air-energy-storage-capacity-services. Accessed 29 August 2017.

6 http://www.westernenergyhub.com. Accessed 29 August 2017.

7 https://www.rwe.com/web/cms/mediablob/en/391748/data/364260/1/rwe-power-ag/innovations/Brochure-ADELE.pdf. Accessed 29 August 2017.

8 http://storelectric.com/how-caes-works.html. Accessed 29 August 2017.

9 http://www.apexcaes.com/project. Accessed 29 August 2017.

10 http://www.gaelectric.ie/energy-storage-projects/project-caes-larne-ni. Accessed 29 August 2017.

11 https://alacaes.com/technology/pilot-plant. Accessed 29 August 2017. Site visit may by possible, as the authors in 2016.

- **Energy markets.** The storage operator sells the energy unloaded from the storage, and is paid for the energy (in MWh) provided to the network.

In financial terms, the capacity market represents a "call option": the market pays an option premium to the CAES owner for the right to purchase energy at a specified price for a specified period of time. It may or may not exercise that right but the CAES owner keeps the premium. In the following we will show that the cases where the market exercises its right, either 50 or 350 days a year, can lead to a different technological choice.

17.5.2 Number of Days in Operation

The fact that the storage operator does not necessarily have to produce electricity implies that it does not necessarily have to operate every day. The number of operations per year may also vary over the years, as is the case of the Huntorf plant: while it was used mainly on a daily basis during its first two years, it was only used about 50 to 60 times a year after 1995 (Figure 17.11).

This evolution in number of cycles per year may be related to an evolution in the service offered by the plant, as described in 2001: "The power station is typically used today as a minute reserve: medium load power stations (coal) take 3–4 hours to generate full capacity before they can provide short-term power – the intervening time is preferentially covered by CAES plant" [3].

This example shows that the service the plant delivers can influence the number of cycles per year. More generally, it illustrates the fact that the number of cycles per year and the service a storage plant targets can influence the technological choice of the CAES solution (adiabatic or not). In the following, we will test this assumption on a Business Case applied to both A-CAES and I-CAES for two different cycles:

- 350 cycles per year case; and
- 50 cycles per year case.

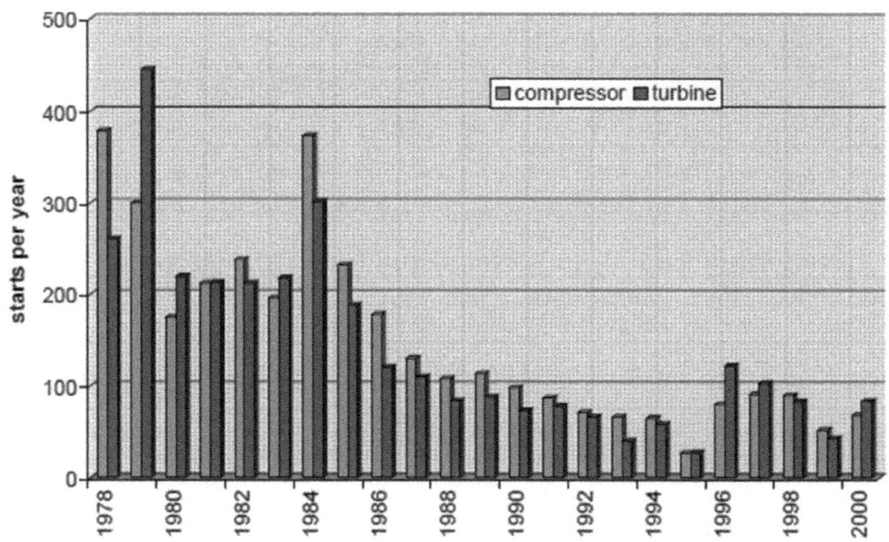

Figure 17.11 Cycles per year at Huntorf plant. From [3].

17.5.3 Cost Structure of Isothermal and Adiabatic CAES

The main characteristics of the typical plant we consider are the following.

Cavern characteristics	
• Cavern depth (casing shoe)	1000 m (3281 ft)
• Cavern geometric volume	400 000 m^3 (2.5 Mbbl)
• Operation pressure difference	2 MPa (290 psi)
Power system	
• Expander output power delivered to the grid	250 MW
• Compressor power	100 MW
• Unloading duration	4 h
Economic hypotheses	
• Natural gas (CAES case only)	18 €/MWh (5.2 €/MMBtu)
• Inflation for O&M and fuel	3% per year
• Discount rate	7%

The capital expenditure (CAPEX) of such a "typical" storage plant is estimated to be 215 M€ for the Adiabatic-CAES and 136 M€ for the Non-adiabatic-CAES. Costs breakdown are presented in Figure 17.12.

17.5.4 (Simple) Revenue Models of Both Plants

A detailed business case would require a detailed market analysis which is not the focus of the present chapter. In Figures 17.13, 17.14, and 17.15, the business case of a 100 MW CAES plant in Western Germany from [23] is extrapolated.

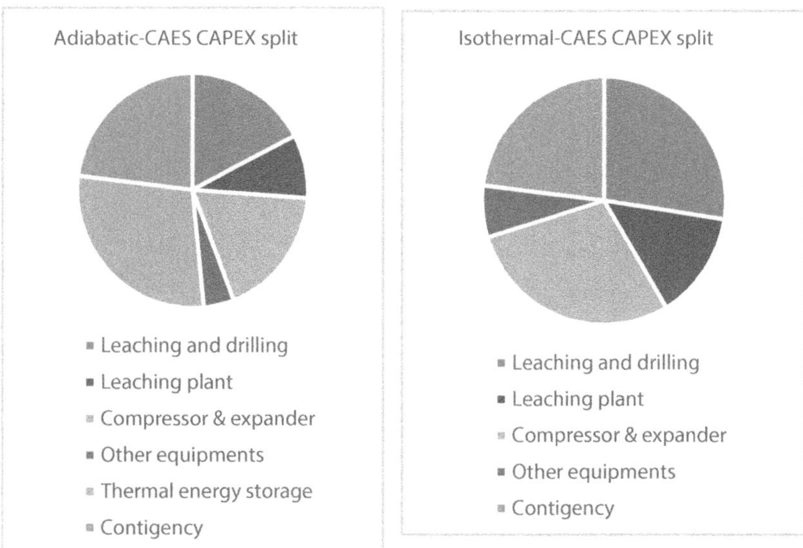

Figure 17.12 Cost breakdown in net present value.

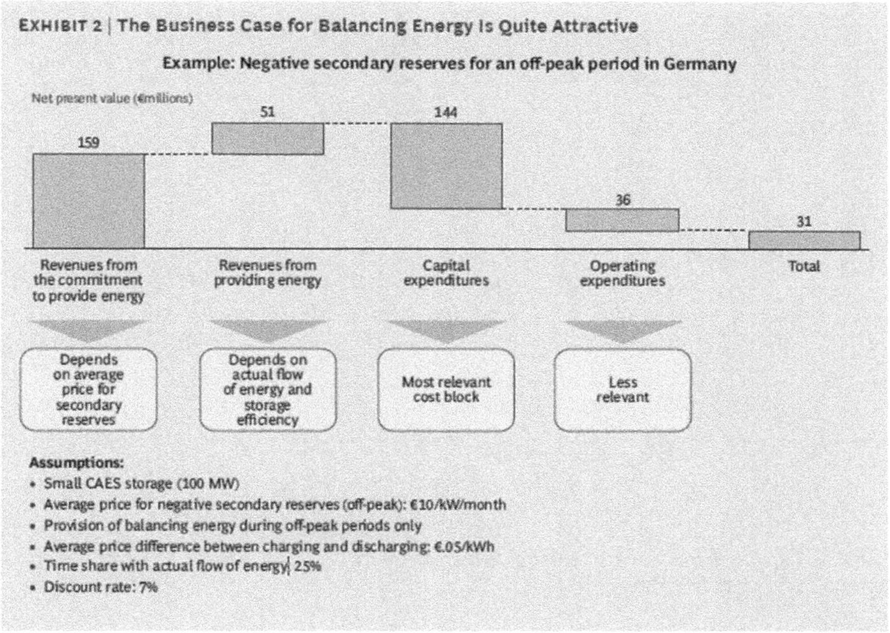

Figure 17.13 Summary of [23] business case to provide the various revenue sources. Source: BCG analysis.

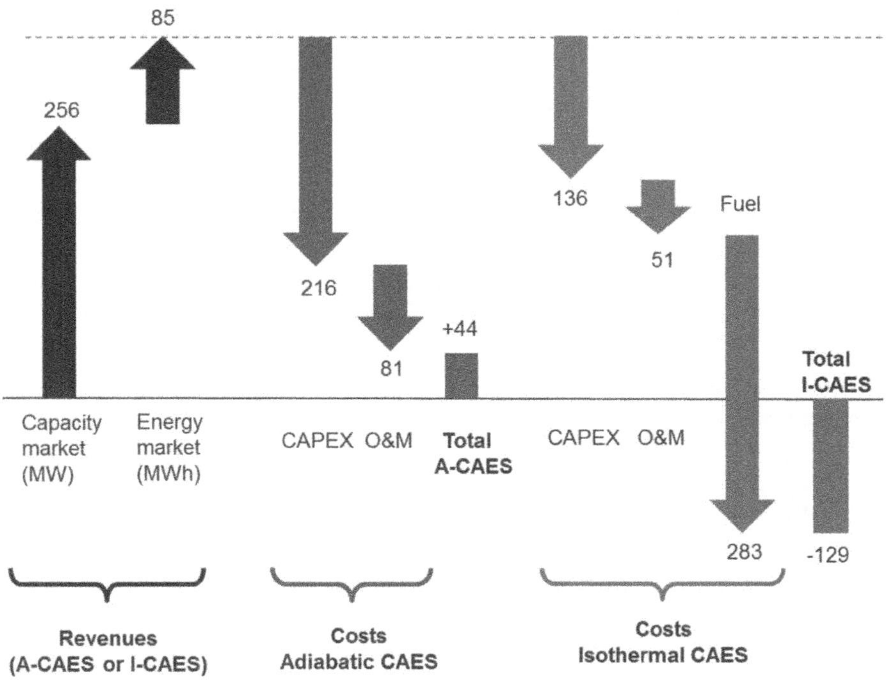

Figure 17.14 Revenues and costs net present value (M€) for A-CAES or I-CAES plants operating 350 cycles per year.

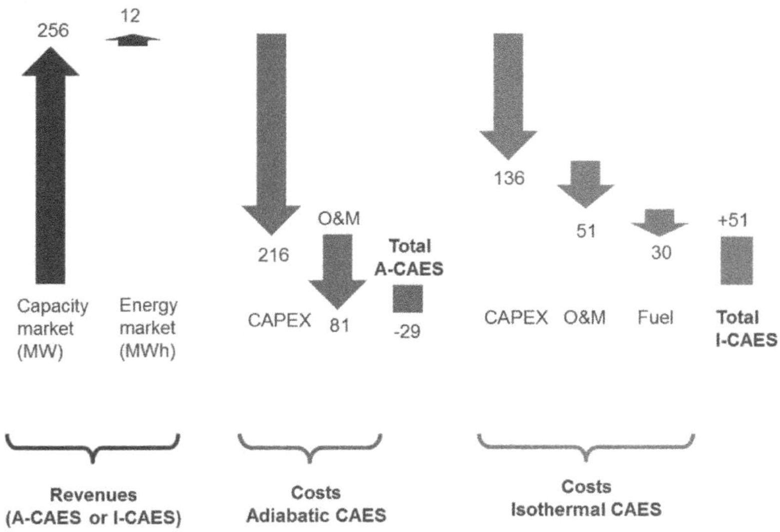

Figure 17.15 Revenues and costs net present value (M€) for A-CAES or I-CAES plants operating 50 cycles per year.

From this business case, we most note the following elements:

- 76% of the revenue comes from the capacity market,
- 24% from the energy market,
- 15% of profit for the base case (Adiabatic-CAES operating 350 days per year) is in net present value.

Using these assumptions for the revenues and the costs, it is found that when operating 350 days per year, the Adiabatic-CAES total present value is positive (+44 M €), whereas the Isothermal-CAES is not (−129 M€).

When operating 50 cycles per year, the Adiabatic-CAES total present value is negative (−29 M€), whereas the Isothermal-CAES is not (+51 M€).

In these conditions, the Isothermal-CAES that does not have to bear the investment of the Thermal Energy Storage, is more competitive when considering a limited number of operations per year (50). On the contrary, for a nearly daily use of the storage, the fuel consumption during unloading induces an increased fuel cost that makes Adiabatic-CAES more interesting.

17.6 Conclusion

The number of projects of both Adiabatic and Non-adiabatic CAES have been burgeoning over the past few years. A large part of the needed technology is currently available; challenges are raised by frequent pressure and temperature swings and the selection of materials able to withstand rapid flows of wet air; however, solutions are available. While the research or pilot projects (<10 MW power output) are mostly targeting Adiabatic-CAES, industrial projects (>100 MW) are mostly considering Non-adiabatic storage, similar to the two historical examples of Huntorf and McIntosh discussed previously.

A simple business case of a "typical" plant suggests that one explanation lies in the revenue sources and the cost structure of both techniques. Revenues for storage can come from the capacity market (commitment to be able to put electricity on the grid, without necessarily doing it) or the energy market (producing electricity). Market conditions are country specific, but tend to offer the largest share of revenue through the capacity market.

This case illustrates the fact that while Adiabatic-CAES is seductive from an engineering point of view (70% energetic efficiency), Non-adiabatic- (or Isothermal-) CAES may be better suited to the current market conditions, in which revenue sources arise mainly for the Capacity market – the commitment to provide energy without necessarily doing it. Non-adiabatic-CAES also benefits from being relatively simple, made from relatively standard equipment (an electric compressor, a – slightly specific – gas expander and a – slightly specific – salt cavern), which secures its feasibility and minimizes the initial investment. It moreover benefits from historically low prices of natural gas and from the still prevailing low cost of CO_2 emissions.

Acknowledgments

Part of this work has been funded by the French National Research Agency (ANR) as part of the FluidStory project, focused on an alternative storage technique: the EMO (Electrolysis–Methanation–Oxycombustion). CAES is used as a comparison point.

References

[1] Enea consulting. (March 2012). *Energy storage. Issues, technical solutions and development opportunities.*

[2] Schainker, R.B. (2011). Advanced Compressed Air Energy Storage (CAES) demonstration projects. *Proc. EPRI Renewable Energy Council* (6 April).

[3] Crotogino, F., Mohmeyer, K.-U., and Scharf, R. (2001). Huntorf CAES: More than 20 years of successful operation. *Proc. SMRI Spring Meeting*, 351–362. Orlando, FL.

[4] Brown Boveri & Cie (BBC). *Huntorf air storage gas turbine power plant.* Energy Supply, Publication# D GK 90202.

[5] Goodson, J.O. (1992). *History of first U.S. Compressed Air Energy Storage (CAES) plant (110 MW – 26h)* – Volume 1: *Early CAES development.* EPRI Report TR-101751.

[6] Pollak, R. (1994). *History of First U.S. Compressed Air Energy Storage (CAES) plant (110 MW – 26h)* – Volume 2: *Construction.* EPRI Report TR-101751v2.

[7] Réveillère, A. and Cracowski, P. (2016). Réhabilitation de cavités minées de Petit-Couronne en stockage d'énergie renouvelable. *Technical Sessions of CFMR/CFGI,* Paris, 03/11/2016 [in French].

[8] Evans, D.J. (2007). *An appraisal of underground gas storage technologies and incidents, for the development of risk assessment methodology.* Report by the British Geological Survey to the Health & Safety Executive, BGS Open File Report Series, OR/07/023, 2 volumes.

[9] Klafki, M., Wagler, T., Grosswig, S., and Kneer, A. (2003). Long-term downhole fibre optic temperature measurements and CFD modeling for investigation of different gas operating modes. *Proc. SMRI Fall Meeting*, 180–189. Chester, UK.

[10] Krieter, M. (2011). Influence of gas cavern's surface area on thermodynamic behaviour and operation. *Proc. SMRI Fall Meeting*, 179–184. York, UK.

[11] Bauer, S. and Sobolik, S. (2009). Pressure cycling in compressed air and natural gas storage in salt: Tracking stress states and cavern closure using 3-D Finite Element Code. *Proc. SMRI Spring Meeting*, 129. Krakow, Poland.

[12] Düsterloh, U. and Lux, K.H. (2010). Some geomechanical aspects of compressed air energy storage (CAES) in salt caverns. *Proc. SMRI Fall Meeting*, 15–27. Leipzig, Germany.

[13] Brouard, B., Frangi, A., and Bérest, P. (2011). Mechanical stability of a cavern submitted to high frequency cycles. *Proc. SMRI Spring Meeting*, 99–116. Galveston, TX.

[14] Pellizzaro, C., Bergeret, G., Leadbetter, A., and Charnavel, Y. (2011). Thermomechanical behavior of Stublach gas storage caverns. *Proc. SMRI Fall Meeting*, 161–178. York, UK.

[15] Rokahr, R., Staudtmeister, K., and Zapf, D. (2011). Rock mechanical design for a planned gas cavern field in the Preesall Project Area, Lancashire, UK. *Proc. SMRI Fall Meeting*, 189–204. York, UK.

[16] Staudtmeister, K., Zapf, D., and Leuger, B. (2011). The influence of different loading scenarios on the thermo-mechanical behavior of a gas storage cavern. *Proc. SMRI Spring Meeting*, 83–97. Galveston, TX.

[17] Bérest, P., Djakeun-Djizanne, H., et al. (2012). Rapid depressurizations: Can they lead to irreversible damage? *Proc. SMRI Spring Meeting*, 63–86. Regina, Canada.

[18] Bérest, P., Sicsic, P., and Brouard, B. (2016). Thermomechanical effects of a depressurization in a CAES. *Symp. Am. Rock Mech. Ass.* Houston, TX.

[19] Passaris, E., Yfantis, G., and Slingsby, J. (2017). Coupled thermo-mechanical analysis of a deep gas storage salt cavern in East Yorkshire, UK. *Proc. SMRI Fall Meeting*, 6–19. Münster, Germany.

[20] Boley, B.A. and Weiner, J.H. (1997). *Theory of thermal stresses*. Mineola, NY: Dover Publications Inc.

[21] Wallner, M. and Eickemeier, R. (2001). Subsidence and fractures caused by thermo-mechanical effects. *Proc. SMRI Spring Meeting*, 363–371. Orlando, FL.

[22] Zapf, D., Staudtmeister, K., and Rokahr, R.B. (2012). Analysis of thermal induced fractures in salt. *Proc. SMRI Spring Meeting*, 47–62. Regina, Saskatchewan.

[23] BCG. (2011). Revisiting energy storage. There is a business case. https://www.bcg.com/documents/file72092.pdf

[24] Réveillère, A., Bérest, P., Evans, D.J., et al. (2018). *Past salt caverns incidents database Part 1: Leakage, overfilling and blow-out*. SMRI Research Report RR2017-2.

18

The Geology, Historical Background, and Developments in CAES

David J. Evans

Retired – formerly British Geological Survey, Keyworth, Nottingham, NG12 5GG, England.

18.1 Introduction

Compressed air energy storage (CAES) is an essentially simple concept and could apply to many systems storing compressed air. Here the focus is on those systems used in support of electricity production and generation. CAES and pumped hydroelectric storage (PHS) represent the only mechanical energy storage technologies currently capable of storing large quantities of electrical energy to provide load-levelling and deliver many hours of grid-scale power output and support. Large quantities of electrical energy generated during low demand (off peak) are converted through mechanical work into significant amounts of potential energy (more precisely exergy) of high-pressure gas (air), which is injected and stored deep underground for later release to meet higher demand (peak-load) periods. It is not a new concept, with assessments, technologies and patents dating back to the 1930s [1].

The technologies behind current and future electrical energy storages based on compressed air, are largely those developed and proven for the long-term, safe storage of fuels (notably compressed natural gas), the origins of which can be traced back to 1915 (outlined later). A number of underground (geological) storage options are available, including man-made voids (salt and rock caverns, abandoned mines, and lined rock caverns) and porous rock storage (depleted hydrocarbon fields and aquifers). Geological storages offer important advantages over surface storages including: high storage capacity arising from large volumes and high operating pressures (up to 200 MPa); protection against external influences (including natural catastrophes, during periods of war and acts of terrorism); a low surface footprint plus low environmental impacts; long operating lifetimes; and low costs (relative to storage capacity) [2–3].

During the 1970s the potential of CAES to compensate for fluctuating renewable energy supplies was considered [4–5]. However, most early studies were for technical and plant designs related to load-levelling and fuel-saving applications, linked with coal and nuclear baseload power plants [6]. Many local and national governments now face requirements to reduce carbon emissions and with utility companies see CAES as facilitating development of renewable energy technologies and the integration of an increasingly higher proportion of inherently intermittent (electricity) supply, to provide grid-scale support and maintain grid stability. Although a few test plants in various storage types have operated briefly or were and are under appraisal, until recently the only commercially operational CAES plants have been hosted in solution-mined salt caverns at (Figure 18.1, Table 18.1) Huntorf (Germany)

Advances in Energy Storage: Latest Developments from R&D to the Market, First Edition.
Edited by Andreas Hauer.
© 2022 John Wiley & Sons, Ltd. Published 2022 by John Wiley & Sons, Ltd.

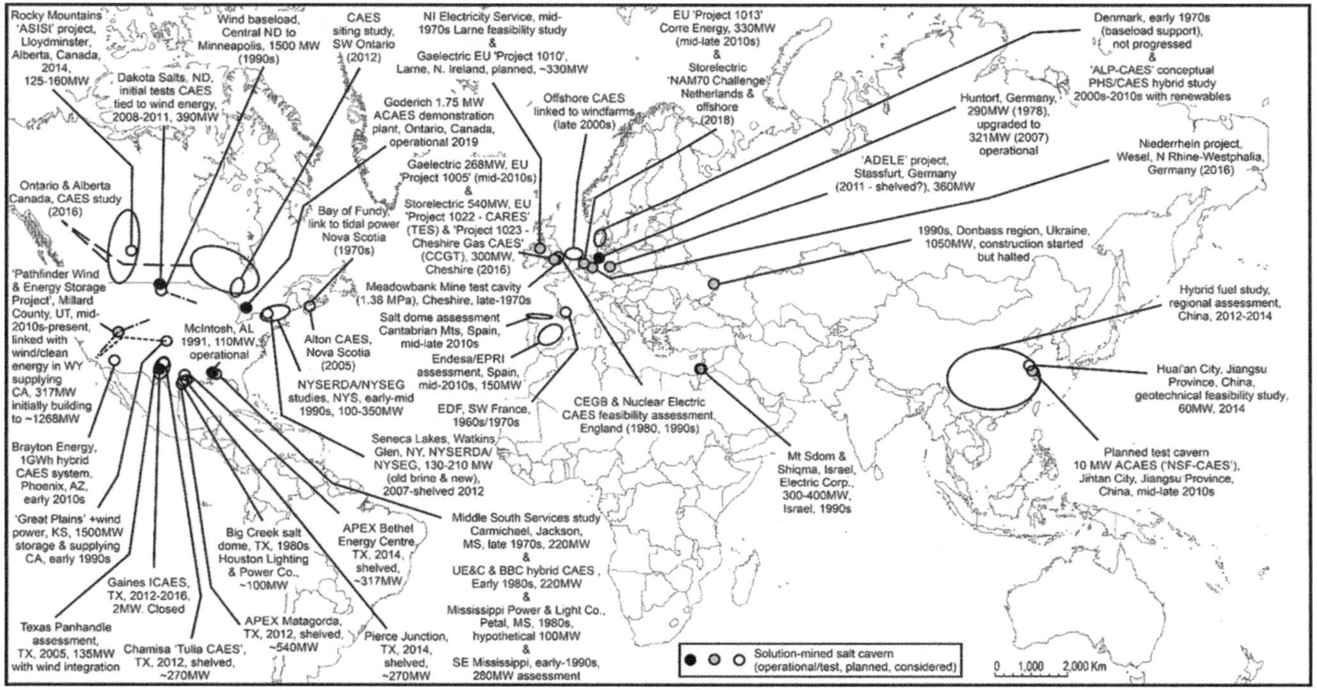

Figure 18.1 Map illustrating locations, status, and general characteristics of principal sites or locations considered for CAES in solution-mined salt caverns (refer [8] and references therein).

Table 18.1 Compressed air or CAES projects and/or assessments in solution-mined salt caverns (refer [8] and references therein).

Facility/project name	Type	Date	Location	Rocksalt type	Power capacity (MW)	Volume (m^3)	Storage depth, top (m)	Pressure range (MPa)	Status
McIntosh conceptual study	Diabatic	Early 1970s	Alabama, US	Domal	800	2 680 000 (4 cavs. x 670,000)	~460	5.0–8.0	Concept design. Informed 110MW plant built later (1991)
Denmark assessment	Diabatic	Early 1970s	Denmark	Domal					Interest, but no further details
Bay of Fundy + tidal power	Diabatic	1970s	Nova Scotia, Canada	Domal?					Early studies. Not progressed
Northern Ireland Electric Service CAES study	Diabatic	Mid-1970s	Larne, Northern Ireland	Bedded					Feasibility of CAES cavities in Larne area. Not progressed
Kansas Utility System assessment (Black and Veatch)	Diabatic?	Mid-late 1970s	Kansas, US	Bedded	150–200	708 000–1 420	~259–915 000		Early regional CAES assessment for main storage types. Not progressed
Huntorf (NKW, now EON Kraftwerk)	Diabatic	1978	N. Germany	Domal	290 upgraded to 321	310 000 (140 000 and 170 000)	650	4.3–7.5	Operational. 2 CAES caverns, 1 natural gas cavern
Carmichael Dome study (Middle South Services, DoE/EPRI)	Diabatic +recuperator	Late-1970s	Mississippi, US	Domal	220	1.39 million (total)	≤1060	5.0–7.0	Joint Doe/EPRI funded techno-economic feasibility assessments of 3 storage types. Salt cavern (2) led by MSS. 47 sites, Hazlehurst/Prothro/ Carmichael domes preferred
Mississippi Power and Light Co., Petal salt dome	Diabatic	Late-1970s–early-1980s	Hattiesburg, MS, US	Domal	~100	~720 000 (total)			Evaluation of 2 propane storage caverns. Not progressed

(Continued)

Table 18.1 (Continued)

Facility/project name	Type	Date	Location	Rocksalt type	Power capacity (MW)	Volume (m^3)	Storage depth, top (m)	Pressure range (MPa)	Status
CEGB feasibility	Diabatic	1980s	England, UK	Bedded		~570 00	~450	Up to 7.0	Feasibility study for isobaric and isochoric mode. Not progressed
UE&C and BBC hybrid CAES concept	Diabatic/ Adiabatic	Early 1980s	Carmichael, MS, US	Domal	220			5.4–8.31	Engineering design and economic study. Not progressed
Hybrid CAES concept	Adiabatic	Early 1980s	Mississippi US	Domal	220			5.4–8.31	Engineering design and economic study. Not progressed
Houston Lighting and Power Co., Big Creek salt dome	Diabatic	1980s	Fort Bend County, TX, US	Domal	~100				Evaluation of virgin site. Not progressed
McIntosh (PowerSouth)	Diabatic +recuperator	1991	Alabama, US	Domal	110	566 000	457	4.5–7.6	Operational (1991–present day). 1 CAES cavern
SE Mississippi study (Southern Company Services, APB Power, EPRI)	Diabatic	Early 1990s	Mississippi US	Domal	280				Technical, economic assessment only for salt domes in SE Mississippi
Central North Dakota wind energy system	Diabatic	Early 1990s	North Dakota, US	Bedded	1500, 20–250 hours storage	4 790 000			Wind energy potential and Great Plains of America supplying Minneapolis 800 km to SE. Not progressed
"Great Plains" Kansas wind/CAES	Diabatic	Early 1990s	Western KS, US	Bedded	1500, 50 hours				Assessing wind-generated electricity southern California

Nuclear Electric assessment	Diabatic	1990s	England, UK	Bedded	~300	~740 000–1.7 million	~450–500	5.0–7.4	Feasibility study for isobaric and isochoric mode. Not progressed
Donbass	Diabatic	1989–1991	Russia-Ukraine border	Bedded?	1050	~1.98 million			Construction began. Abandoned (collapse Soviet Union)
Shiqma (Israel Electric Corp.)	Diabatic	Mid-1990s	Israel	Bedded	300–400	29 000	1200	13–20	Assessment for up to 6 CAES caverns. Not progressed
Mt Sdom (Israel Electric Corp.)	Diabatic	Mid-1990s	Israel	Bedded	300–507	400 000	500	7.5 max.	Assessment for 2 CAES caverns. Not progressed
EPRI and NYSERDA New York State feasibility study	Diabatic	Early-mid 1990s	New York State, US	Silurian, bedded	100–350		180–915		Assessment of LPG, brine, and new caverns. Not progressed
Texas Panhandle study (SECO/Ridge Energy)	Diabatic	Early-2000s	Texas, US	Bedded	~135	807 030 (total)			Wind-CAES assessment. Up to 6 CAES caverns. Not progressed
Alton CAES (Alton Natural Gas/Landis Energy)	Diabatic	Mid-2000s	Nova Scotia, Canada	Bedded	200				Salt caverns for natural gas storage and CAES (4) supporting renewables. Dropped?
Offshore storage and renewable energy study	Diabatic	Late 2000s–mid 2010s	Southern North Sea, UK sector	Domal					Number of technical/economic feasibility studies. Not started
Finger Lakes and Silver Springs regions (Energy East Corporation, NYSEG and EPRI). ARRA funds	Diabatic	2007–2009	Seneca Lakes, New York State, US	Silurian, bedded	~180 (150–360)	~127 000	~610	2.76–11.5	Brine cavern at 1st site "shovel ready." FEED. Dropped (2010)
		2010–2012 (ARRA)			130–210	~150 000 (each)	~730	7.9–10.3	2nd site, up to 3 new caverns. FEED studies. Cancelled 2012 (cost)

(*Continued*)

Table 18.1 (Continued)

Facility/project name	Type	Date	Location	Rocksalt type	Power capacity (MW)	Volume (m^3)	Storage depth, top (m)	Pressure range (MPa)	Status
North Dakota (Sirius Exploration and EPRI)	Diabatic	2008–2011	Burke Country, ND, US	potash	390, 30 hours	~117 000 (1 Cav)	~1950		Feasibility study for CAES tied to ND wind energy. Shelved?
Gaines (Conoco Philips, General Compression)	Isothermal	2012–2016	Texas, US	Bedded	2	~42 131	745	~7	Pilot plant, 1 cavern, + wind generation, operational from 2012, closed 2016.
APEX Bethel Energy Centre (Tennessee Colony)	Diabatic	Early mid-2010s	Texas, US	Domal	317 (270–540)		~1140		Fully permitted and construction-ready for 1 cavern, supporting ERCOT renewable electricity. Stalled?
APEX Matagorda Energy Centre (Ridge Energy SS)	Diabatic	Mid-2010s	Texas, US	Domal	317–540		~1140	13.1–19.5	Advanced planning, 2 caverns permitted, supporting ERCOT renewable electricity. Dropped (post-2012)
Chamisa "Tulia CAES plant"	Diabatic	Mid-2010s	Texas, US	Bedded	270	116 620 (each)	~710	10.0–13.0	Advanced planning for 5–7 caverns, supporting ERCOT renewable electricity. Some permits obtained. Dropped?
Pierce Junction (Ridge Energy SS)	Diabatic	Mid-2010s	Texas, US	Domal	270				Assessment. Dropped?
"Pathfinder Wind and Energy Storage Project" (Haddington Ventures)	Diabatic	Mid-2010s	Delta, Millard County, UT, US	Bedded/ domal?	~300 (2x150) - ~1268	~5.1 million (4x ~1.161 mcm caverns)			Up to 4 CAES caverns assessed Central UT, linked to 2100 MW wind farm in SE WY supplying electricity to CA. Phase I 317MW, Phase II 1200MW. Progressing?

Brayton Energy hybrid CAES system	Hybrid adiabatic?	2010–present?	Phoenix, AZ, US	Bedded salt	~100 (~1 GWh)		>300	Up to 2.5	Assessing 1 cavern system using solar energy to heat compressed air
Alberta Saskatchewan Intertie Storage Project (ASISt) + wind (Rocky Mountain Power)	Diabatic	Mid-2010s	Alberta/ Saskatchewan, Canada	Bedded	160				60 hour (9600 MWh), 2 cavern wind-CAES project. Planning
Advanced 2nd Generation CAES and ACAES Plants (US DoE funds)	Adiabatic	Early 2010s	United States	Bedded/ domal	180 310 and 450				US DoE stimulus funds for 2nd generation. CAES to support Renewable Energy and Smart Grid
"ADELE" (RWE Power led consortium)	Adiabatic	2007-mid-2010s	Staßfurt, Germany	Domal	~360	~358 674	~400	~5.0–7.0	Techno-economic feasibility study. Project indefinitely suspended 2016 – economics
Endesa assessment (Endesa S.A. and EPRI)	Diabatic/ adiabatic	Mid-2010s	Spain	Bedded	150				Feasibility study on plant design and storage options – 2–3 suitable salt formations identified. Progressing?
EU "Project 1010" Islandmagee (Gaelectric)	Diabatic	Mid-2010s	N Ireland, UK	Bedded	~330	~480 000 (each)	~1500	9–12 min., 19.3 max.?	Advanced planning, 2 CAES caverns supporting renewable energy. Applications and EA report. Ceased?
EU "Project 1005" (Gaelectric)	Diabatic	Mid-2010s	Cheshire, UK	Bedded	268		~500		Feasibility and planning. Ceased?

(*Continued*)

Table 18.1 (Continued)

Facility/project name	Type	Date	Location	Rocksalt type	Power capacity (MW)	Volume (m^3)	Storage depth, top (m)	Pressure range (MPa)	Status
EU "Project 1013" (Corre Energy Storage BV)	Diabatic	Mid-2010s	Zuidwending, Netherlands	Domal?	330				Feasibility and planning for CAES and offshore wind. Delayed
EU "Project 1022 – CARES" (Storelectric)	Adiabatic	Mid-2010s	Cheshire, UK	Bedded	540 (total)		~500		Test (40 MW) and full plant (540 MW). + renewables. Planning
EU "Project 1023 – Cheshire Gas CAES" (Storelectric)	Diabatic	Mid-2010s	Cheshire, UK	Bedded	300		~500		Advanced planning retrofittable CCGT CAES plant technology storing locally generated renewable electricity. Stalled?
Energy Storage Niederrhein project	Diabatic	Mid-2010s	SE of Wesel, NW Germany	Domal			~930		Advanced design and planning stage for up to 3 CAES caverns. Stalled?
"ALP-CAES" conceptual PHS/ CAES hybrid study	Diabatic	Mid-2010s	Denmark	Domal/ bedded					Air compressed by brine pumped into cavern. +renewables. Concept
Hybrid CAES fuel system (North China Electric Power University)	Diabatic	2012–2014	N China	Bedded	~410	~400 000		6.5–7.5	System performance and techno-economic study of novel hybrid-fuel CAES with coal, supporting renewables
Huai'an City project (Sci. Foundation of China [973] Program)	Diabatic	2014	Jiangsu Province, China	Bedded	60	Min. 246 000	~1330–1470	9–12.5 max.	Geotechnical feasibility study for CAES and renewables/grid support.

"Three North Region" CAES + wind and solar (National Key R&D Program of China, Chongqing University)	Diabatic	Late 2010s	N and NE China	Bedded					Early-stage regional assessment of CAES supporting renewables/grid. Salt caverns 3rd alternative storage (Pingdingshan)
Feicheng A-CAES (Chinese Acad. Sciences, Energy Storage Alliance, Gezhouba Utd Alliance) + wind/solar	Adiabatic	2019	Feicheng, Shandong Province, China	Bedded	10 initial (1250 final)				Ground breaking ceremony Nov. 23 2019 for initial 10MW advanced CAES system. Construction 2–5 years, ultimately 1250 MW and 7500 MWh
"Salt Cave Compressed Air Energy Storage National Test and Demonstration Project" (Chinese Academy Sciences) + wind/solar	Adiabatic	Mid-late 2010s	Jintan, Jiangsu Province, China	Bedded	60 initial (300 MWh)	253,00		9.5 max.	Initial geotechnical feasibility study non-supplementary fuel (NSF-CAES). 1st phase of construction started Sept. 2020
CAES assessment N Spain (Cátedra TECFUTURA [SACYR corporation and Madrid Polytechnic Uni.])	Diabatic?	Mid-late 2010s	Cantabrian Mts, N Spain	Domal	10–100		400–800		Small-scale CAES assessment for renewable energy storage in salt domes.

(*Continued*)

Table 18.1 (Continued)

Facility/project name	Type	Date	Location	Rocksalt type	Power capacity (MW)	Volume (m^3)	Storage depth, top (m)	Pressure range (MPa)	Status
University of Waterloo and industrial partners study	Diabatic/ adiabatic	2016	Nova Scotia, Ontario and Alberta, Canada	Domal/ bedded					Study of CAES in salt caverns and renewables, optimizing economic value to microgrids.
Austria study (Austrian Research Promotion Agency)	Adiabatic	2017	Upper Austria	Bedded/ domal	10–50			Max. 1.9–12	Feasibility study of potential for brine and new caverns + renewables. Sand in TES.
NAM70 Challenge (Storelectric)	Diabatic?	2018	Netherlands and S. North Sea	Domal?					Study re-purposing oil and gas infrastructure + offshore wind
Goderich Mine (NRStor-led consortium, incl. Hydrostor)	Adiabatic	2019	Ontario, Canada	Bedded	1.75 (10+ MWh)		~550		Commissioned Nov. 2019 and world's 1st commercial ACAES plant. 1 former brine cavern

and McIntosh, Alabama (US). A small third such salt-cavern-hosted plant was commissioned in 2019 at Goderich (Canada). Drawing upon [7] and [8] this chapter reviews briefly the development of CAES, the various storage options, important projects assessing its deployment and some key issues to consider.

18.2 Operational Modes – Diabatic, Adiabatic, Isothermal (Heat), Isochoric, and Isobaric (Pressure) Operations

During large-scale underground (geological) storage of gas of all types, including air, significant changes in pressure and temperature are involved, with different modes of operation possible.

In terms of high-pressure air storages, current realistic systems operate in two modes:

- "isochoric" (constant volume) – involves pressure changes and a minimum volume of air ("cushion gas") must remain in the store, either trapped in rock pores or in a cavern to ensure its stability. The overwhelming majority of gaseous hydrocarbon cavern storages and the three commercial CAES plants using solution-mined caverns work on this principle.
- "isobaric" (constant pressure) – water-compensated operation, which for mined caverns involves the construction of water reservoirs, generally at surface, to pump water in (during emptying) and out (during filling) of the storage. Aquifer storage over the long term exhibits water compensation, but for short-term cycling reflects a more constant volume-variable pressure system [9]. Compressors and expanders work between very similar pressures, which engineering-wise is attractive for CAES, helping simplify some design features and with the majority of the stored gas/air present available for withdrawal.

In both modes withdrawn gases or air will require drying, as even in isochoric salt caverns, the stored gas absorbs water from brine remaining in the cavern sump areas. Water compensated operation also leads to problems with, for example, the "champagne effect" (air dissolved into water during high pressure storage is released on depressurization), or the unwanted and potentially unrecognized further dissolution and enlargement of salt caverns that can lead to major accidents, e.g. [10].

Compressing air to 7–10 MPa or more, increases its temperature to > 600°C, well above that with which subsurface storages can cope and requiring the air to be cooled down to well below 100°C before injection to storage [3]. Depending on how the heat generated during compression is treated, and how the retrieved air is reheated, establishes the basic CAES system classifications and determines their efficiencies. Of the operational and many emerging, hybrid technologies, three main modes exist.

- "conventional" diabatic CAES (DCAES) – the heat of compression is not captured and stored but through exchangers, is vented to the atmosphere, resulting in the need for fossil fuel (gas) to reheat the air during withdrawal and prior to going through the expander/turbine unit. Round trip efficiencies for Huntorf are around 42%, whilst the addition of a recuperator as at McIntosh, improves efficiencies to ~ 54% (see later).

- So called "second generation" adiabatic or advanced adiabatic CAES (A- or AACAES) – compression to storage pressures and removal of the heat of compression (and storage in a thermal energy storage [TES] system) is in a single stage with the stored heat used during the discharging process to reheat the air prior to going through the expander/turbine unit. No natural gas is required during the generation process and predictions are for ~ 70% efficiencies [11–12]. ACAES was considered as long ago as the early 1970s by the CEGB and EPRI [13], and again by RWE Power (Germany) between 2009 and 2013 (see later), but it was only in late 2019 that the world's first such plant was commissioned, at Goderich in Ontario, with round trip efficiencies > 60% (see later).
- Near-isothermal CAES (ICAES) – involves the subdivision of compression and decompression phases into several stages, each stage associated with only minor temperature increases, as heat is continually removed from the air during the compression cycle (interstage cooling) and continuously added during expansion. A small demonstration plant operated briefly between 2012 and 2016 at Gaines (Texas, US), with round trip efficiencies of ~ 75% (see later) [14].

18.3 Brief Review of the Historical Origins of CAES – How It All Began and Where It Is Now

This section provides a general overview of compressed air use and storage, covering the more important exemplars of CAES assessments and its development (see [7, 8]). More detail for differing storage types follows in the relevant part of section 18.4 Use of compressed air may be traced back over 4000 years, and throughout history it has been an important driver of technological progress [15]. Large-scale compressed air energy systems, based on tanks for storage, have been in wide use since the late nineteenth century, with various cities having developed systems where delivered power was used in homes, industry, and transport (e.g. trams). Patents applying to storage of compressed air in general date back to the turn of the twentieth century [1].

Amongst the first underground unlined caverns in hard rocks trialled for the storage of air (power) were in Swedish mines during the 1800s, employing schemes where water from a surface lake descended a shaft into a cavern, trapping and pressuring the air, which was then used in the mine to power drilling, etc. [16]. The Ragged Chute compressed air plant in Northern Ontario, Canada (Figure 18.2), built in 1910 at the peak of the Cobalt silver boom, operated in a similar fashion and with only two brief maintenance interludes, ran continuously for just over 70 years, until destroyed by fire in the 1980s [17–18]. The first compressed air storage scheme using a chamber (800 m^3) was constructed in the Striberg Mine (Sweden) to power mining operations (Figure 18.2), storing air at 0.7 MPa and which operated from 1910 until at least 1979 [19]. From 1936 small, unlined caverns have operated as air "reservoirs" at pressures of up to 0.8 MPa, storing air as feedstock to drilling equipment and to support electricity supply in Finnish mine operations, including the Pyhäsalmi zinc mine (Figure 18.2). In other countries compressed air was used to power mine trains, pumps, pneumatic drills, and hammers soon thereafter (Figure 18.2) [1, 19–21]. Today, rock caverns are constructed to act as compressed air cushion surge chambers attached to hydropower plants for use during power generation at, e.g. Fosdalen and Rausand in Norway (Figure 18.2) [2].

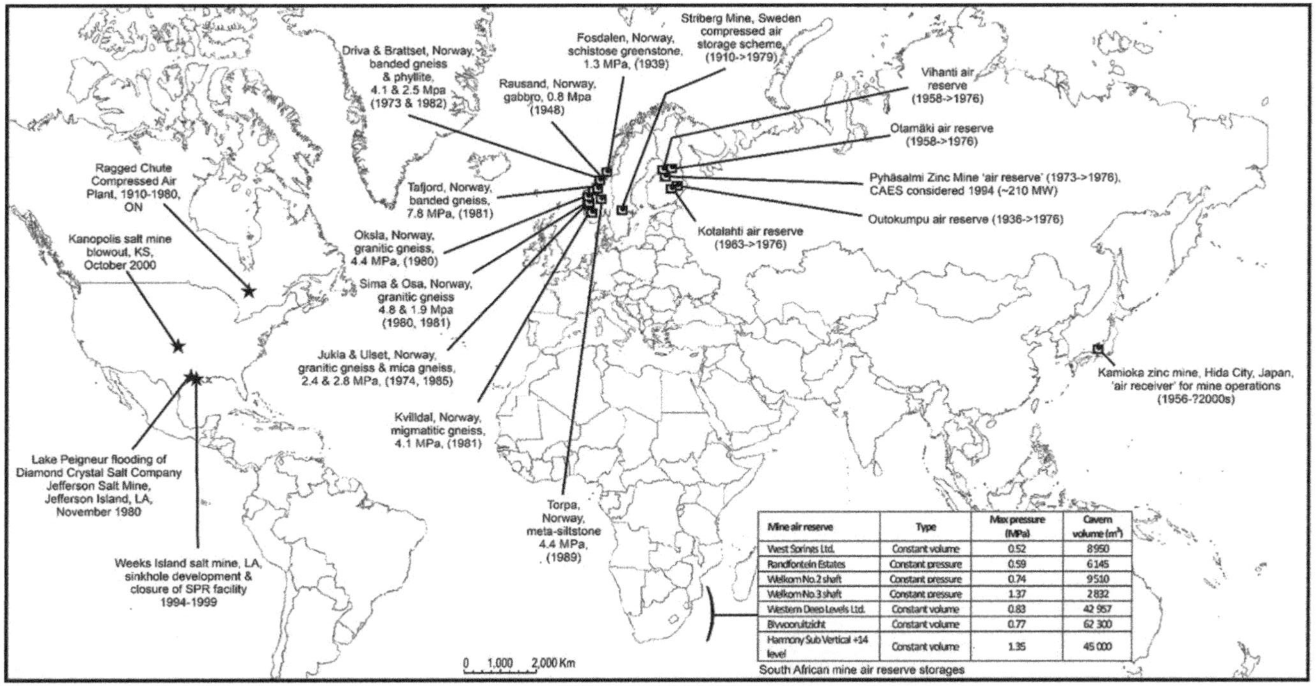

Mine air reserve	Type	Max pressure (MPa)	Cavern volume (m³)
West Springs Ltd.	Constant volume	0.52	8950
Randfontein Estates	Constant pressure	0.59	6145
Welkom No.2 shaft	Constant pressure	0.74	9510
Welkom No.3 shaft	Constant pressure	1.37	2832
Western Deep Levels Ltd.	Constant volume	0.83	42 957
Blyvooruitzicht	Constant volume	0.77	62 300
Harmony Sub Vertical +14 level	Constant volume	1.35	45 000

Figure 18.2 Map illustrating the locations of other places referred to in the text, including sites using unused mine areas for small-scale compressed air energy storage ("air reservoirs") to provide additional mine power such as in Scandinavia [20] and South Africa [21], or air surge chambers at hydroelectric power stations in Norway [2]. Refer also to [8] and references therein.

The first conceptual design work for CAES is thought to have been by the Public Service Gas and Electric Company in the late 1930s [22]. References to electrical energy storage by means of compressed air storage for utility application were included in patents issued to Gay in the US (1948) and to the turbine manufacturer Stal Laval in Sweden and the UK (1952) [1]. Circa 1951 discussions around the potential applications to US utilities of compressed air storage coupled with gas turbines took place [23], whilst in the Federal Republic of Germany, a patent for the storage of electrical energy via compressed air was issued in 1956 to Djordjevitch [24]. Between the 1950s and mid-1960s, however, little progress was made, due in large part to the lack of necessity for grid connected energy storage [7]. Djordjevitch's concept was taken by the Geological Survey of the Federal Republic of Germany in the late 1950s–early 1960s, with a recommendation that abandoned salt and potash mines represented the most technically and economically feasible geological storages for compressed air [24]. During the 1960s, the Central Electricity Generating Board (CEGB) considered the introduction of CAES into the electricity grid in England and Wales on several occasions, but concluded it was economically unattractive [1].

Not until the late 1960s and early 1970s did further developments in CAES occur when drivers such as the introduction of baseload generation from nuclear power and large lignite or coal-fired power plants, and OPEC's huge oil price rises in 1973 provided an economic case to store inexpensive off-peak power from baseload generation capacities and transfer it to peak-load hours [1, 25]. The construction of PHS plants took up much of this capacity, although its reliance on suitable topography limits it to regions with significant relief and often in environmentally sensitive areas. Various countries assessed CAES options, including in France, where EDF commissioned studies into aquifer and salt cavern storage (see later) [1] and in Northern Ireland in the early 1970s. In 1969, the need for storage capacity in northern Germany triggered the decision by the utility Nordwest-deutsche Kraftwerke (NKW) to develop the world's first conventional DCAES facility at Huntorf, construction of which commenced in 1974, with commissioning in 1978 (Figure 18.1, Table 18.1) [26]. Thereabouts suitable salt domes are present within which solution-mined salt caverns for storing large volumes of compressed natural gas were already operational, providing the potential for large amounts of compressed air storage [3]. CAES plants also offered the rapid "black start" capability required by the north German grid [26]. Today, Huntorf primarily provides the power grid with minute reserves, balancing-out shortfalls between day-ahead-forecasts for wind power and actual production. Of historical interest, the original term for the new CAES technology developed by NKW was "Air Storage System Energy Transfer" (ASSET).

The potential of CAES was thus realized and further patent applications, reviews, and assessments of large-scale CAES potential appeared [27], with Sweden, Finland, Luxembourg, former Yugoslavia, France, the US, Germany, and CEGB in the UK having shown considerable interest (Tables 18.1–18.6) [1, 28–29]. The US government and utilities showed significant interest in CAES in all geological storage options, with a number of assessments during the 1970s to early 1980s, including a broad research and predemonstration program for CAES development coordinated by the Pacific Northwest National Laboratory (PNNL) and funded by the US Department of Energy (DoE) and Electric Power Research Institute (EPRI) [1, 25]. Parallel design, siting, and cost studies, based on the three primary candidate reservoir types (caverns in bedded and domal salt, excavated caverns in hard rock environments [water-compensation or not] and porous rocks), were established to evaluate the potential for grid-scale CAES in many configurations in California, Kansas, Illinois, Indiana, Maryland, and the New England states (Tables 18.1–18.5) [1, 25, 29]. The late 1970s California Energy Commission (CEC) and US Department of Energy (DoE) backed a CAES feasibility study in California [30] that was wide-ranging, assessing porous rock (depleted field and aquifer) and existing and new mined hard rock cavern options. Studies have also included the feasibility of using storage caverns excavated by nuclear explosives [31]. The US aims, together with the CEGB in the UK [1, 29] were to minimize, or eliminate, the use of oil or gas during generation by assessing so-called "second-generation" advanced CAES concepts, including a "hybrid" ACAES system involving a single stage thermal energy storage (TES) with additional gas firing and ICAES. ACAES was seen as the most suitable and promising technology [29].

Other countries showing early interest in CAES included Israel, South Africa, and Japan, with expectations then, and at times since, high that CAES would gain utility acceptance [7, 32]. However, in the US in particular, there was not the take-up or development expected for both DCAES and second generation ACAES or ICAES technologies. Despite advanced designs for DCAES plants with heat recuperation, such as the Soyland Power Cooperative Inc. 220 MW plant in 1982, further developments for advanced designs were linked to successful implementation of DCAES, which did not occur until 1991 with McIntosh [7], following which market and economic drivers changed unfavourably.

Although discussed as early as the mid-1970s and early-1980s [1, 5], the application of CAES to compensate for fluctuating renewable energies was not a major consideration at that time. However, increasing penetration of intermittent renewable energies such as wind and photovoltaics into electricity supply systems, linked to national commitments to reduce CO_2 emissions, and the need for the balancing of supply to the grid, has driven renewed interest in CAES early in the twenty-first century. Many countries see CAES as a key enabling technology for the integration of such intermittent renewable resources [33–34]. Considerable research into CAES of all types has recently been funded by the US government and EPRI through, e.g. the American Recovery and Reinvestment Act of 2009 (ARRA), and in Europe through Horizon 2020 and the EU Trans-European Networks for Energy (TEN-E) Infrastructure Regulation (347/2013 EU), e.g. [8]. The latter aims to promote "Projects of Common Interest" that support future energy supply and security. Between 2000 and 2018 many projects around the world have been considered, including in the US, Canada, the UK, Austria, Denmark, Netherlands, Italy, Egypt, China, and South Korea (Figures 18.1, 18.3–18.5, Tables 18.1–18.6). EU-funded Framework Programmes and the French SEARCH and SACRE projects have investigated aspects of CAES, ACAES, and grid modelling. In none of these countries has the construction of commercial plants commenced. However, trials have been taking place (see section 18.4), including General Compression's Gaines ICAES pilot (now closed [14]), ALACAES's Pollegio-Loderio tunnel ACAES project, and the RICAS2020 Mine project (both ongoing) [35–36]. Thus, the first quarter of the twenty-first century may become regarded as the period of resumption of R&D on CAES technology, with its deployment on a wider scale and with greater optimism [7].

Table 18.2 Compressed air or CAES projects and/or assessments in depleted fields (see [8] and references therein) mcm: 1 million cubic metres.

Facility/project name	Type	Date	Location	Rock type	Power capacity (MW)	Volume (mcm)	Storage depth (m)	Pressure range (MPa)	Status
CEC feasibility study (Calif. Energy Commission and US DoE)	Diabatic	Early-mid 1970s	California, US	Sandstone	200–250x3 (600–750 total)	~785	670–750	~6.9	Early Acres American Inc. site feasibility and design study. 10 sites identified, one selected for study
			Princeton, Glen County, CA						Feasibility and conceptual design of preferred location. Not progressed
TVA (Tennessee Valley Authority)	Diabatic	Early 1990s	Tennessee, US	Sandstone	600–1200				Geotechnical assessment of potential. Not progressed
EPRI and NYSERDA New York State feasibility study	Diabatic	Early 1990s	New York State, US	Various	≥100		457–1525		Part of general CAES feasibility studies for state. Not progressed
Big Springs fields (NPPD)	Diabatic	Early-2010s	Nebraska, US	Dakota Sandstone	100–300	~845.5			Feasibility studies and modelling. Stalled?
Huntley field (Gaelectric)	Diabatic	Early-2010s	Montana, US	Sandstone	270				Feasibility studies and modelling of CAES to support planned wind development. Stalled?
East Keith field (Gaelectric)	Diabatic	Mid 2010s	Montana, US	Sandstone	270				Feasibility studies and modelling of CAES to support planned wind development. Stalled?
Dunnigan Hills (PG&E and EPRI)	Diabatic	2009–2010	Yolo County, California, US	Cretaceous sandstones	110		~750	~5.1	Geotechnical assessment of potential 110 MW, 37-hour CAES plant. Reservoir ok, but rejected

King Island field	PG&E (EPRI, ARRA 2009 and CEC funding)	Diabatic	Mid-2010s onwards	San Joaquin County, California, US	Mokelumne River Fm (Capital Sst.)	300	>113	~1450	Up to 18.6	Feasibility studies, EA and technical reports completed 2018. Preferred site with wells, reservoir model and air injection/pressure tests. Stalled? – costs alt. energy storage technologies?
East Island field		Diabatic	Mid-2010s onwards	San Joaquin County, California, US	Mokelumne River Fm (Capital Sst.)	300	>113	~1450	Up to 18.6	Alternative site investigated, cores retrieved. Not progressed
Thornton field	PG&E/ EPRI/ Univ. Calgary	Diabatic	Mid-2010s onwards	San Joaquin County, California, US	Mokelumne River Fm (Capital Sst.)	300 (aiding studies)	>113	~1000		Studies into gas/air mixes for oxidation and combustion/fire risk
"SMUD" CAES studies + wind integration (Sacramento Municipal Utility District)		Diabatic	Early-2010s	California, US	Sandstone	135	~343.5	~1890	~18.6	North Sherman field evaluated. Rejected
						~268				Assessment of potential and feasibility in State. 1 field identified. Stalled?
Tri-State (G&T – EPRI)		Diabatic	Mid-2010s	E. Colorado, US	Various					Feasibility study
Southern California (Edison/ Hydrodynamics)		Diabatic	2010s	California, US	Sandstone					CAES siting study and evaluation of potential structures

(Continued)

Table 18.2 (Continued)

Facility/project name	Type	Date	Location	Rock type	Power capacity (MW)	Volume (mcm)	Storage depth (m)	Pressure range (MPa)	Status
Pecos Valley project (Montoya), Austin Energy	Diabatic	2010s	West Texas, US	Various					Evaluation by hydrodynamics of 726 oil and gas-bearing structures
Alberta (Hydrodynamics for Altalink)	Diabatic	2010s	Alberta, Canada	Sandstone					Evaluation of structures, potential sites identified
Pittsfield region (Hydro dynamics for Nicor Ventures)	Diabatic	2010s	Pittsfield, Illinois, US	Sandstone					Part of regional study
Portland Wind Farm support (Monash University)	Diabatic	2011	Onshore Victoria, Australia	Sandstone?	80–240 (2 x 20–80MW turbines)				Assessment of potential and feasibility for CAES and/or PHS to support onshore wind generation
"Three North Region" CAES + wind and solar (National Key R&D Program of China, Chongqing University)	Diabatic	Late-2010s	N and NE China	Sandstones			Up to 1000		Early-stage regional assessment of CAES supporting renewables/grid. Depleted fields represent alternative storage type
Conversion NGS sites to CAES study (Mississippi State Uni, Southern Company Services)	Diabatic?	2019	Louisiana, Arkansas, US	Sandstone	180–800				Techno-economic feasibility study of re-using uneconomic NGS facility (James Sand) for CAES on Louisiana–Arkansas border

18.4 Overview of Underground (Geological) Storage Options

Clearly, critical to the success of every CAES project is a (large) repository for the safe and secure storage of air at very high pressures, which typically represents the most important and expensive component of the system. Whilst investigations into many of the CAES technologies date back to the late 1960s, the last 10–15 years has seen significant research into novel CAES operating systems. A number of different operational modes and above ground, non-geological storage options show promise. Although such systems are attractive because no underground geological storage strata or structure is required, thereby permitting the siting of CAES plants almost anywhere, they require further assessment and testing, and are generally more expensive than traditional CAES. Schemes not considered here offering readily available storages of small volume include above-ground, or shallowly buried steel vessels or pipes [37–38], energy bags secured to the seabed [39], wind turbines linked with energy storage in supporting legs [40], or those in which power is converted directly from the rotor by means of gas/air compression within the rotor blades [41].

Sections 18.4.1 to 18.4.3 provide an overview of the main storage options for high pressure gases (including air) in deep underground geological formations (salt caverns, porous rock, mined rock caverns, and lined rock caverns) and a general review of CAES projects through time, highlighting significant projects including recent developments and some key considerations for each storage type.

18.4.1 Solution-mined Salt Caverns

Halite/rocksalt ("salt"; NaCl) possesses favourable physical properties for creating large, safe, secure, high-pressure storages, being visco-plastic (self-healing), gas tight, and inert with respect to liquid and gases, including oxygen. These properties permit the construction (solution-mining) and operation of very large, unlined caverns that, correctly operated, are not only stable but suffer no pressure losses over extended periods (months). Solution-mining involves the controlled injection of water via cased boreholes into salt deposits deep underground (~300–2000 m), dissolving the salt to create a cavity, volumes of which vary from a few tens of thousand cubic metres (m^3), to greater than 1 000 000 m^3. [3] Together with high operating pressures (up to ~ 200 bar), this enables the storage of very large volumes of gas. In comparison to most other geological options, salt caverns, offer highly flexible, inexpensive storage [42] and for natural gas storages in 2014, ranked third globally, accounting for 6% of facilities [3].

The first uses of large solution-mined salt caverns ("cavities") for the storage of liquid hydrocarbons and natural gas represented major advances in technologies for bulk energy storage (see [43] and references). First conceived in Canada during the early 1940s, storage of liquid hydrocarbons in solution-mined salt caverns has been successfully employed in both former brine production caverns and newly solution-mined caverns since the early 1950s in the US and Europe. The Northern Gas Board and ICI first stored "Town Gas" in brine caverns on Teesside (England) in 1959, (see [7]) whilst a former brine cavern in Michigan (US) was converted to natural gas storage in 1961 [44]. The first storage with caverns specifically solution-mined for gas storage were

Table 18.3 Compressed air or CAES projects and/or assessments in aquifers (see [8] and references therein).

Facility/project name	Type	Date	Location	Rock type	Power capacity (MW)	Volume (mcm)	Storage depth (m)	Pressure range (MPa)	Status
Pays de Bray (EDF)	Diabatic	Early-mid 1970s	France	Various					Feasibility studies. Not progressed
Brookville Dome (National Academy of Science study)	Diabatic	Early 1970s	Illinois, US	Sandstone	150 units		~205		Brookville Dome considered for a conceptual CAES design
CEC feasibility study (Calif. Energy Commission and US DoE)	Diabatic + recuperator	Early-mid 1970s	California, US	Sandstone	200–250 x 3 (600–750 total)		~670	6.9	Acres American Inc. aquifer studies, 3 identified. Not progressed
Brookville storage site (General Electric, US Energy R&D Admin)	Diabatic + recuperator	Mid-1970s	Illinois, US	Ironton and Galesville sandstone	600 (10 hour cycle)	671	~205	1.17 – ~1.96	Conceptual design, techno-economic requirements study. Not progressed
Calcutta-Carbon field (Strata Power of Indiana)	Diabatic	Early-mid-1970s	Indiana, US	Vuggy reef dolomites	150	~850	~380	~3.2	Detailed feasibility studies and "gas testing." Not progressed
Aquifer CAES development program (Public Service Company of Indiana, DoE/EPRI)	Diabatic + recuperator	Mid-late 1970s	Illinois, Indiana, Kentuky, and Tennessee, US	Sandstone	Up to 1120			1.4–7.0	Joint Doe/EPRI funded techno-economic feasibility assessments of 3 storage types. Aquifer led by PSCI
Kansas Utility System assessment (Black and Veatch)	Diabatic	Mid-late 1970s	Kansas, US	Sst/ carbonate	150–200	>8 500 000	2 290 915	Up to 5.52	Early regional CAES assessment for main storage types. Not progressed

Hume (URS Corporation/ Hydrodynamics)	Diabatic	Late 1970s	Illinois, US	M. Devonian Carbonate	200–250	498.4	~200	~2.3	Assessment and conceptual designs and economic analyses. Not progressed
Brocton (URS Corporation/ Hydrodynamics)	Diabatic	Late 1970s	Illinois, US	M. Devonian Carbonate	200–250		~200	~2.3	Assessment and conceptual designs and economic analyses. Not progressed
Pittsfield aquifer test (DoE/EPRI)	Diabatic	Late 1970s–1984	Illinois, US	St Peter Sandstone	25 (test)		~200	~2	Extensive tests, injection of air bubble early 1980s, PNL led. Not progressed
Sesta injection test (ENEL)	Diabatic	1987–1991	Sesta, Italy	Fractured, "porous rock"	25			~2.3 (original)	Demonstration plant. Closed
Israel Electric Corporation study	Diabatic	Mid-1980s-Early 1990s	Israel	Fractured aquifer?	200–400 (3 x 100 units?)				Detailed analysis and feasibility assessments. Not progressed
Gippsland Basin assessment (for State Elec Comm. Victoria)	Diabatic + recuperator	Early 1990s	New South Wales, Aust.	Sandstone	100–300		~600	~6.3	Assessment and modelling of aquifer. Not progressed
EPRI and NYSERDA New York State feasibility study	Diabatic	Early 1990s	New York State, US	Various			457–1525		Part of general CAES feasibility studies for state. Not progressed

(*Continued*)

Table 18.3 (Continued)

Facility/project name	Type	Date	Location	Rock type	Power capacity (MW)	Volume (mcm)	Storage depth (m)	Pressure range (MPa)	Status
Iowa Stored Energy Park (IESP), Dallas Centre + wind	Diabatic	Early 2000s–2011	Iowa, US	Mt Simon Sandstone	270		~880	~9	Linked to wind. Extensive tests over 8 years. 3 wells +lab studies. Dropped
Vincent-Eagle Grove Structure (Hydrodynamics/ Prairie Wind)	Diabatic	Early 2000s	Iowa, US	Sandstone	100–200				Feasibility studies to support wind generation. Not progressed
NYSERDA and NYSEG CAES assessment	Diabatic	Late 2000s	New York State, US	Various					NYSERDA commissioned feasibility study in State
Bakersfield (PG&E with EPRI, ARRA 2009 and CEC funding)	Diabatic	Early 2010s	Kern County, CA, US	Sandstone	300				Feasibility studies and modelling. Stopped, interest switching to N California
"SMUD" CAES (Sacramento Municipal Utility District)	Diabatic	Mid 2010s	California, US	Sandstone	~268				Assessment of potential and feasibility in the State. Site identified. Stalled?
Columbia Hills (BPA and PNNL)	Diabatic	Early 2010s	Washington State, US	Fractured basalts	207		>3000	~12.5	Advanced assessment and planning
Yakima Canyon, Selah (BPA and PNNL)	Adiabatic	Early 2010s	Washington State, US	Fractured basalts	83		~4250	~49.6	TES and geothermal energy. Advanced assessment and planning

UK offshore saline aquifer potential (research)	Diabatic and adiabatic	2018	UK North Sea	Sandstones	4–12 (per well)		260–4000	Up to 25.0	PhD research into offshore aquifer potential
Faludden reservoir for CAES on Gotland (onshore). Uppsala/ Lund Univ. and Swedish Survey	Diabatic?	2018/2019	Onshore Sweden	Sandstones	~70	0.6–9.8	~270–410	2.9–4.3	Regional aquifer assessment on island to support wind generation. Est. capacity 130–1300 MWh

Table 18.4 Compressed air or CAES projects and/or assessments in abandoned mines (see [8] and references therein).

Facility/project name	Type	Date	Location	Rock type	Power capacity (MW)	Volume (m^3)	Storage depth (m)	Pressure range (MPa)	Status
CEC feasibility study (Calif. Energy Commission and US DoE)	Diabatic	Early-mid 1970s	California, US	Hard rock	200–250 x 3 (600–750 total)		~700		Acres American Inc. hard rock cavern appraisal. 12 sites identified, none found suitable
Mt Hope mine (Jersey Central Power and Light Co)	Diabatic	Early-mid-1970s	New Jersey, US	Iron ore	190–1000	764, 555 (initial)	~825	~4 Isobaric or isochoric	Initial test plant with planned 1000 MW. Existing and new caverns. Not progressed
Kamioka CAES test (Kamioka Mining and Smelting Co. Ltd.)	Diabatic	Late 1980s–early 1990s	Kamioka mine, Japan	Zinc mine			~520		Underground “air receiver” for operations (water-compensated). “Cavern” leakage testing for CAES
Norton Mine, Ohio (Ohio Edison Co.)	Diabatic	1980s	Norton, Cleveland, OH, US	Limestone	~100	8 500 000	~686		Assessment, for 10 hr charging, 10 hr discharging cycle. Not progressed
Pyhäsalmi mine (Technical Research Centre of Finland)	Diabatic	Early 1990s	Finland	Zinc mine, Precamb-rian crystalline rocks	34 (150–210 MWh)	~30 000–100 000	400–800	Isobaric or isochoric	Pre-feasibility, techno-economic studies. Not progressed
Old Grace Mine (CAES Corp.)	Diabatic	Early 1990s	Philadelphia, US	Iron ore mine, limestone host	110	~76 455	~270		EPRI grant applied for. Water-compensated. Not progressed

New York State assessment (Seneca Lakes Mine)	Diabatic	1990s	New York State, US	Salt mine	150–300	~2 000 000	~625		Part of general CAES feasibility studies for state. Not progressed
MSc CAES research study	Diabatic	Mid-1990s	South Africa	Various	200–1000	Up to 1 000 000			Old mine potential – gold mines most favourable. Not progressed
Norton Mine (Haddington Ventures FirstEnergy Generation Corp)	Diabatic	2001–2013	Norton, Cleveland, Ohio, US	Hard rock (limestone)	Up to 2700 (units of 300)	9 600 000	~722	5.5–11	Permits, advanced design and testing. Air leakage. Project cancelled (2013)
ESKOM	Diabatic/ adiabatic	1990s–2000s	South Africa	Various mines					Considered economic benefits of CAES for integrated energy plans
"Shen Lang" studies	Diabatic	Early-2000s	China	Coal mines					Method(s) developed. Patents obtained (2004). Few details
Energy Storage and Power + wind (PSEG Holdings)	Diabatic	Mid-2000s	New Jersey and New York states, US	Salt and coalmines					Theoretical concept of old mine potential plus wind. Planned?
NYSERDA and NYSEG CAES assessment	Diabatic	Late 2000s	New York State, US	Various mines (talc, zinc, granite, Fe-ore)	>100		700–1150		NYSERDA commissioned feasibility study

(Continued)

Table 18.4 (Continued)

Facility/project name	Type	Date	Location	Rock type	Power capacity (MW)	Volume (m^3)	Storage depth (m)	Pressure range (MPa)	Status
Clear Creek Power Wind and CAES (Clear Creek Power)	Diabatic	2007	Georgetown, Colorado, US	Gold/silver mine	50–1000 (525–10,000 MWh)	~100 000		3.5—8.3	One or 5 windfarms linked to one 50MW or 1000MW storage. Not progressed
Lyons Salt Mine (Empire, BSC Holding Inc., Sega Inc.)	Diabatic	2009	Kansas, US	Salt mine			~305		Feasibility of storing wind energy, small pressure test. Not progressed
Minnesota mine study, + wind (Uni. Minnesota)	Diabatic	Early 2010s	Minnesota, US	Iron ore	100				Feasibility study and assessments of mines. Pending?
"Three North Region" CAES + wind and solar (National Key R&D Program of China, Chongqing University, Chinese Academy of Science)	Diabatic	Late-2010s	N and NE China	Coal mines	~10 initial?		Up to 1000		Early-stage regional assessment of hybrid, clean CAES supporting renewables/grid. Re-using old coalmine workings preferred storage option
Angas ACAES Project (Hydrostor and ARENA [Australian Renewable Energy Agency])	Adiabatic	2019–2021	Adelaide, S. Australia	Zinc mine	5 (10 MWh)		~240	(water compensated)	Commercial demonstration facility, funded from renewable energy agency. Water compensated and due 2020/21

Table 18.5 Compressed air or CAES projects and/or assessments in mined, unlined rock caverns (see [8] and references therein).

Facility/ project name	Type	Date	Location	Rock type	Power capacity (MW)	Volume (m^3)	Storage depth (m)	Pressure range (MPa)	Status
Battelle assessment	Diabatic	Early 1970s	US	Various	174	6 116 440 (6 caverns)	~183	1.01	Preliminary study for cavities created by nuclear explosives. Not progressed
Glan (Vattenfall/ Stal Laval)	Diabatic	Early 1970s	Eastern Sweden	Metamorphic (gneiss)	220–230	~137 000	~435	4.35 max.	Advanced plans, including trial borings (1972–1973), water compensated. Not progressed
Växjö (Sydsvenska Kraftgrupp A.B./Stal Laval)	Diabatic	Early 1970s	S Sweden	Metamorphic (gneiss)	240	250 000	250	2.5 max.	Advanced plans, water-compensated storage. Not progressed
"Airstore" (Vattenfall/ Stal Laval)	Diabatic	Early-mid 1970s	Sweden	Metamorphic (gneiss)	232	~291 500	250	2.6	Advanced plans, series linked tunnels. Water compensated. Not progressed
"SITRA" subsurface power plants study	Diabatic	Early 1970s	Finland	Various hard rock (granite/ gneiss)	~200	120 000–540 000	100–600	1.5–6.0 (4.1)	Feasibility study of water compensated scheme to replace pumped hydro capacity
Ffestiniog test (CEGB)		Early-1970s	North Wales, UK	Metamorphic hard rock					Preliminary pressurization tests, small unlined gallery

(*Continued*)

Table 18.5 (Continued)

Facility/ project name	Type	Date	Location	Rock type	Power capacity (MW)	Volume (m^3)	Storage depth (m)	Pressure range (MPa)	Status
National Science Foundation (NSF) study	Diabatic	Early-mid 1970s	Illinois, Indiana, Ohio, and southern MI	Trenton-Black River carbonate	280	212 253	~490		Regional assessments. Not progressed
			Between Boston and Washington	Hard rock (granite and gneiss)	280	212 253	~490		Regional assessments. Not progressed
CEC feasibility study (Calif. Energy Commission and US DoE)	Diabatic	Early-mid 1970s	California, US	Hard rock – Lst or igneous	252.5 (x 3)	385 000	690	6.7	Early hard rock site identification (13) feasibility studies, conceptual water compensated designs. Not progressed
			Shultz Mountain, CA	Hard rock	250 x 3 (750 total)		~700		Feasibility and conceptual water-compensated design of preferred location. Not progressed
Kansas Utility System assessment (Black and Veatch)	Diabatic	Mid-late 1970s	Kansas, US	Hard rock (carb. granite, gneiss)	150–200	700,000–1 420 000	259–915	Up to 6.76	Early regional CAES assessment for main storage types Not progressed
Vianden (Société Electrique de L'Our and RWE)	Diabatic	Early-mid 1970s	Luxembourg	Devonian slates	285	100 000	500	~5	Advanced design and planning, water compensated, pumped storage scheme. Not progressed

Bremm test (Société Electrique de L'Our and RWE)		Early-mid 1970s	Mosel, Germany	"Clay slate"		6500			Test cavern for planned pumped storage scheme
Yugoslavia	Diabatic + recuperator	1970s	Former Yugoslavia	Hard rock	260				Preliminary water-compensated design concepts with heat recuperation. Not progressed
Meadowbank Mine test (I.C.I. Ltd.)		Late-1970s	Cheshire, UK	Salt mine – bedded halites		~8.53	140	Up to 1.38	Compressed air test of cavity for gas storage development
EDF CAES plant	Diabatic	1970s–early 1980s	Bretagne, France	Hard rock (granite)	250–300	300 000		2.5	Advanced planning and site characterization with test boreholes. Water compensated. Not progressed
Mt Hope mine (Jersey Central Power and Light Co)	Diabatic	Mid-1970s	New Jersey, US	Iron ore	190–1000	764 555 (initial)	~825	~4	Initial 190MW test plant with planned 1000 MW. Existing and new workings. Not progressed

(Continued)

Table 18.5 (Continued)

Facility/ project name	Type	Date	Location	Rock type	Power capacity (MW)	Volume (m^3)	Storage depth (m)	Pressure range (MPa)	Status
Potomac Electric Power Co. feasibility study (DoE/ EPRI fund)	Diabatic + recuperator	Early-mid 1970s	Maryland, US	Metamorphic (Sykesville Gneiss)	924	690 000	Up to 1500	4.8–7.6 (av. 6.9)	Joint DoE/EPRI funded techno-economic feasibility assessments of 3 storage types. URC led by PEPCO
Sykesville storage site (General Electric, US Energy R&D Admin)	Diabatic + recuperator	Early-mid1970s-	Maryland, US	Metamorphic (Sykesville Gneiss)	600 (units of 150)	1 690 000	~220 (overburden)		Conceptual techno-economic requirements, water-compensated design study. Not progressed
Jefferson Island salt mine (LSU and US DoE)	Diabatic	Late 1970s–early 1980s	Louisiana, US	Salt mine – salt dome			~400		Instrumented holes drilled in mine for CAES tests on Gulf Coast salt domes
Soyland CAES study (Soyland Power Cooperative, Inc.)	Diabatic + recuperator	Early 1980s	Illinois, US	Hard rock (Cambrian dolomite)	220		~700		To support coal-fired plant. Water compensated. Permits and contracts signed, but not progressed – power station deferred in 1982
San Diego hard rock (San Diego Gas and Electric Co.)	Diabatic	1980s	California, US	Granite	~100				Assessment, for 10 hr charging, 10 hr discharging cycle. Not progressed

EPRI and NYSERDA New York State feasibility study	Diabatic	Early 1990s	New York State, US	Dolomites or igneous	150	990 000	305–1525		Part of general water compensated CAES feasibility studies for State. Not progressed
Upper Midwest studies	Diabatic	1990s	U Midwest, US	Hard rock (granite)	220–880	290 000	450–540	4.5–5.5	Northern States Power regional water-compensated design assessment. Not progressed
		Early 2010s			Up to 1080	~114 000 (x8 caverns)	607	3.0–6.0	US DoE Wind Energy Program. Planned?
Central Research Institute of Electric Power Industry (CRIEPI)	Diabatic	Late 1980s–early 1990s	Kongo, Japan Kyushu, Japan	Granite Conglomerate	~500			4.0–8.0	Feasibility and site investigations – borehole pressure tests for two rock types
Suez + large-scale wind (University Nottingham)	Adiabatic	Mid-2010s	Suez, Egypt	Hard rock	Up to 580 (various plants)	1 248 000		4.5–7.5	Assessment of 3 large-scale energy storage system supporting wind and TES. Also LRC? Planned

(*Continued*)

Table 18.5 (Continued)

Facility/ project name	Type	Date	Location	Rock type	Power capacity (MW)	Volume (m^3)	Storage depth (m)	Pressure range (MPa)	Status
Datang CAES (Datang Power Company)	Diabatic	Mid-2010s	Datang, Inner Mongolia, China	Sandstone	300	900 000	500	5.0–8.0	Few details. Terminated – problems of air tight seals?
Pollegio-Loderio tunnel (ALACAES)	Adiabatic	2015–present	Ticino, South Switzerland	Metamorphic (Leventina Gneiss)	0.5	~1945	>450 (overburden)	7.0 max.	Small demonstration project. Operational
DEMETAIR small-scale CAES	Adiabatic	2016	Europe and US	Unspecified					Feasibility analysis and assessment of small-scale CAES using drilled shaft, possibly lined
Petit-Couronne (Geostock)	Adiabatic	2017	Northern France	Chalk		~62 500	~150		Assessment to convert LPG storage caverns to CAES. Planning?

commissioned in 1970 at Eminence, Mississippi [45]. Related to the liberalization of energy markets requiring flexible storages with frequent cycles for gas trading, the construction of salt cavern gas storages has grown considerably in recent years. Increasing installation of wind power and photovoltaics facilities, with their inherent intermittencies, suggests likely continued growth in demand.

Thick, deeply buried (>250 m) massively bedded or domal halite deposits exist in many countries [46] and provide the host to the only three operational CAES facilities that prove CAES to be a commercially viable, mature, and cost-competitive technology to meet grid-scale storage requirements. The Huntorf plant, currently operated by E.ON Kraftwerke GmbH (Figure 18.1, Table 18.1), remains operational over 40 years after commissioning, employing two air caverns 150 m in height, with a third cavern storing gas required to reheat the air during generation [26]. Initially rated at 290 MW output capacity, upgrades in 2006 raised this to ~ 321 MW, achieved within 6–9 minutes. Commissioned in 1991, the DCAES plant at McIntosh Alabama was, and remains, the world's second operational facility. Operated by PowerSouth Energy Cooperative, it utilizes one 263 m high, cigar-shaped cavern [47] and has an output of 110 MW, available within 14 minutes. It differs from Huntorf in that a recuperator captures waste heat from the turbine power plant to heat the withdrawn air, raising plant efficiency [48]. In late 2019, the Goderich facility (Ontario, Canada) became the world's third salt cavern-hosted and first commercially operational, utility-scale ACAES plant, with a 2.2 MW charge rating, providing 1.75 MW of peak power output and 10 + MWh of storage capacity [49–51].

Not unsurprisingly therefore, since the 1970s many assessments and feasibility studies on the potential for CAES in salt caverns have been undertaken around the world, based on DCAES designs similar to Huntorf, including Canada, Denmark, the UK, and France (Figure 18.1, Table 18.1). Most have remained unrealized, but a number are noteworthy. One of the earliest CAES assessments in the 1970s linked tidal power to salt cavern storage in the Bay of Fundy area, Nova Scotia [7]. Other assessments were made for storage in the Hazlehurst (SW Mississippi), Prothro (Louisiana), and Carmichael (Mississippi) salt domes, with designs and construction costs for two caverns calculated [52]. Amongst the earliest engineering and economic ACAES assessments undertaken, sponsored by EPRI, was in the early 1980s for a hybrid CAES cycle in the Carmichael salt dome (Mississippi) [25]. Two other sites in salt domes were evaluated during the early to mid 1980s, one in Texas and a second considered two former propane caverns at Petal, Mississippi [53].

Interest in salt cavern-hosted CAES continued through the 1980s into the 1990s, with most studies in the US (Figure 18.1, Table 18.1), but including in halites of Tertiary age in Israel [54]. Under the Soviet Union regime, construction started on a large (1050 MW) plant in the Donbass region near the current Ukraine/Russian border, but was abandoned following the collapse of the Soviet Union in 1991 [55]. A number of CAES proposals and assessments for salt cavern storages have emerged since the early 2000s (Figure 18.1, Table 18.1). Most are in the US and Europe, but they include projects in China and Canada [7].

Various proposed projects in the US have involved CAES linked to wind generation and grid support. In the early 1990s, an assessment was made of a wind-energy base-load system for the Great Plains of America, involving a wind farm and 1500 MW CAES plant located in central North Dakota and supplying the major demand centre

Table 18.6 Compressed air or CAES projects and/or assessments in mined lined rock caverns (LRCs; see [8] and references therein).

Facility/project name	Type	Date	Location	Host rock type	Power capacity (MW)	Volume (m^3)	Storage depth (m)	Pressure range (MPa)	Status
Sodegaura test (CRIEPI)	Diabatic	Late 1980s–early 1990s	Tokyo Bay, Sodegaura, Japan	Soft rock		~100 000	600–640	~6.0	Rock mechanical, geo-mechanical and economic assessments. Water compensated. Not progressed
Roda Stern laboratories		Late 1980s–1990s	Göteburg, Sweden	Hard rock (gneiss)		~336	60	Up to 8.0	Compressed air tests in gas storage LRC test facility
"Urban CAES tank"	Diabatic	Late 1990s	Japan	NA				2.0	Small scale test "concrete tank" in 200m deep water filled shaft. Closed
CAES-G/T Pilot Plant (New Energy Found, METI)	Diabatic	Mid 1990s–early 2000s	Kamisunagawa-cho, Hokkaido, Japan	Coalmine tunnels		78.5	450		Initial in situ tests
CAES-G/T Pilot Plant (New Energy Found, METI)	Diabatic	Mid 1990s–early 2000s	Kamisunagawa-cho, Hokkaido, Japan	Coalmine tunnels	2	~1600	450	4.0–8.0	Small pilot facility after initial in situ tests. Closed
CAES-G/T Pilot Plant (New Energy Found, METI)	Diabatic	Early 2000s	Kamisunagawa-cho, Hokkaido, Japan	Coalmine tunnels	35	30 000	450	~4.0–8.0	Planned pilot facility. Never started
LRC feasibility study (Israel Inst. of Technology)	Diabatic	Early 2000s	Israel	Hard rock	~17.5 (per cavern)	~11 300 (each unit)	>100	1.0–8.0	Feasibility and energy loss study of concrete-lined single/multiple tunnels

Kamioka (ANGAS project)		Mid 2000s	Kamioka mine, Japan	Zinc mine		240	400	0–20	Small LRC test for larger natural gas storage. Tested hydraulically and with air
SEARCH (CAES) –National Research Agency (ANR)	Adiabatic	2009–2013	France	Granite					Study of ACAES heat storage media in an LRC
Jongson pilot test (Korea Inst. Geosci. and Min Res, Kunsan Nat Uni.)	Diabatic	Early 2010s	Jongson, South. Korea	Limestone mine			100	5.0–8.0	Caverns excavated and numerous tests commenced
CAES plant study (China Datang Corp. Ren. Power Co., Ltd., Tongji Uni.)	Diabatic	Mid-2010s	Inner Mongolia, China	Granite	100	100 000		5.0–10.0	Feasibility and modelling studies. Planned
DEMETAIR small-scale CAES	Adiabatic	2016	Europe and the US	Unspecified					Feasibility analysis and assessment of small-scale CAES using drilled shaft, possibly lined
RICAS2020 Mine Project (Consortium led by Montan-universität Leoben)	Adiabatic	2017	Eisenerz, Austria	Hard rock iron ore (Limestone/ volcanic)	5 100	15 000 250 000	~220	3.0–3.6 4.8–6.7	EU Horizon 2020 funded feasibility and small test (operational) leading to full-scale test plant (planned)

of Minneapolis 800 km to the SE [56]. Other projects include one in the early 2000s in the Texas Panhandle area [57] and at least two for wind generation in the Great Plains area, linked with salt cavern storage and supplying electricity to California (Figure 18.1); in the early 2000s, an economic and technical feasibility was undertaken for the supply of wind-generated electricity in western Kansas to the Los Angeles area, with CAES in Kansas [58], whilst a second in 2014 was to develop large CAES caverns in Central Utah linked to a 2100 MW wind farm in SE Wyoming [59]. Elsewhere in the US, other projects, some having attracted ARRA and EPRI funding, have reached advanced design and permitting stages (e.g. Seneca Lakes; Table 18.1). The 270 MW "Tulia CAES project" to support the ERCOT (Electric Reliability Council of Texas) power grid, which uses renewables generated electricity, was planned in Swisher County, Texas, involving construction of five to seven caverns in Permian bedded halites [60]. Additionally, two other projects in Texas, also to support ERCOT had, by the mid-2010s, progressed to gaining permits for storage caverns in salt domes at Markham in Matagorda County [61] and Bethel in Anderson County [62].

Unfortunately, none of these projects have been progressed, with Seneca Lakes cancelled. Two notable projects exist however. Between late 2012 and 2016, Texas Dispatchable Wind LLC (a subsidiary of General Compression) commissioned and ran a small-scale, near-isothermal 2 MW CAES demonstration plant using a reconditioned former liquid hydrocarbon storage cavern for air storage, at Gaines in Texas [14, 63]. Although it is not believed to be currently operating, results should prove valuable in progressing the technology. Already alluded to earlier, in November 2019 the small Goderich ACAES plant in Ontario (Canada) commenced operation as the first commercially operational, utility-scale ACAES facility. Press reports regarding the storage are somewhat muddled, with some suggesting abandoned tunnels and caverns in a salt mine [64], however, the facility uses a former Compass Minerals' brine cavern [65], operating at depths of around 550 m [64].

Elsewhere, CAES projects using salt caverns have been under consideration in Germany, Austria, Spain, Denmark, the Netherlands, and the UK (Figure 18.1, Table 18.1). The RWE led "ADELE" project was to develop at Stassfurt (Germany) the first ACAES plant [66], feasibility studies for which started in 2007 with construction planned in 2016. The TES system was progressed [12] but the 360 MW project appears halted indefinitely. In the UK, Storelectric and Gaelectric were progressing projects that have received EU-funding as part of developing energy infrastructure (Table 18.1). Most advanced was Gaelectric's application to construct a 200–300 MW plant at Larne in Northern Ireland (EU "Project 1010"). With caverns at depths in excess of 1500 m [67], this would be the deepest CAES facility constructed. However, the collapse of the Gaelectric group of companies in 2018 means the future of the various projects are unclear, with the exception of that in the Netherlands, owned subsequently by Corre Energy BV [68]. Storelectric Limited was assessing various CAES proposals including retro-fittable CCGT CAES plant technology (EU "Project 1023") and a TES system (EU "Project 1022 – 'CARES'") with development in two stages: a 40 MW test facility followed by a 540 MW plant [69]. A general assessment of the UK's exergy storage potential through CAES in salt caverns constructed in massively bedded halite deposits was undertaken during the EPSRC-funded IMAGES project. The process of cavern location and volume calculations was outlined in [70] whilst [71] present the development of a methodology and tool to calculate the energy (exergy) storage capacity (in MWh) of a salt cavern-based CAES system, using a gas storage site example in eastern England. Exergy storage estimates for the main UK bedded halites using the tool suggest CAES could make a significant contribution towards the annual UK electricity demand of ~300 TWh in 2018 [72].

An assessment of the realizable ACAES plant potential between 10–50 MW in Upper Austria using salt caverns (mostly old and newly solution-mined, but also dry-mined voids) has been undertaken. A TES system using sand would be used, with overall storage potential calculated at > 4 GWh by 2030 [73]. Also investigated in Denmark was the potential of an adiabatic liquid piston CAES (ALP-CAES) system, re-using brine caverns [74]. Development of CAES in Canada has been investigated with the Alton Project (2005, Nova Scotia) [75] and the ~ 160 MW Alberta Saskatchewan Intertie Storage project, using one or two caverns constructed in bedded Middle Devonian halites and linked to wind-generated electricity [76]. Meanwhile, in China, the successful development of gas storage caverns in the bedded Tertiary halites of the Jiangsu Province, eastern China has raised interest in CAES [77–78]. Both former brine and newly constructed caverns are under consideration in the Jintan and Huai'an salt mining areas, with a geotechnical feasibility analysis for a 60 MW AACAES plant in the Huai'an City area completed.

18.4.1.1 Key Considerations for CAES

Salt caverns offer a well-understood, flexible, responsive, and reliable storage option, with few disadvantages. Until recently, CAES caverns have been limited by the pressure range of compressors and turbines, restricting CAES to depths between ~ 500 and ~ 1300 m depth [79]. Improvements in compressor technology have permitted consideration of CAES to depths of around 1500 m [80].

Storage of natural gas in salt caverns developed mainly for seasonal storage purposes, involving a small number of pressure cycles with low-moderate gas-production rates (or rates of pressure change – typically 0.8–2 MPa/day) each year [81–82]. Storage is between permitted maximum and minimum cavern operational pressures, determined from the geomechanical properties of the halite derived from extensive laboratory testing of lithologies, storage depth, and storage cycle numbers. Liberalized gas markets increasingly require gas storage caverns to operate under higher-frequency cycles and typically may be emptied in 10 days and refilled in 30 days or less [83]. Some cavern gas storages in the UK and Europe now operate on weekly, or even daily cycles. CAES cycles are more frequent than even high-frequency cycled gas storage caverns (typically one cycle per day over extended periods) and are designed to deliver their full power capacity over a very short time period. Both types of high-frequency cycling involve higher "gas" production rate/pressure rate changes (0.5–1.5 MPa/hr), which leads to extreme, repeated, and rapid mechanical and thermal loading of the cavern walls/roof [26, 82]. The permissible operational pressure range is necessarily smaller for such storages, with maximum rates for cavern filling and emptying governed by the maximum flow rates in the boreholes and the maximum allowable cavern pressure reduction rates.

CAES involves injecting hot or warm air and salt creep behaviour is a function of load and temperature, which increases with depth. The response of halite/rock salt to stress is strongly temperature dependent with elevated temperatures being the most important cause of salt cavern and dry salt mine opening closures [24]: for CAES, an upper temperature limit of 80°C was recommended, although maximum air temperatures could be higher during injection periods. Cavern air temperature at Huntorf is ~ 50°C [84], whilst at McIntosh, the maximum air temperature, at least during early operations, reached a maximum of ~ 55°C [85]. For the ADELE project, cavern air temperatures of 33–47°C were anticipated [66].

Regarding environmental concerns, the handling of waste water (brine) generated as a by-product during the construction phase of a gas or CAES cavern may be an issue, being dealt with in a number of ways: either it is supplied to the chemicals industry, as has been the case during construction of gas storage caverns in Cheshire, or where no market exists, is disposed of either by injection into a saline aquifer as in some US storages, or as with the gas storage caverns in eastern England, offshore via pipelines. If water-compensated systems are considered, then care is required that cycling of the compensating brine does not lead to further dissolution of the salt formation, metal corrosion, and undetected enlargement of the cavern, affecting stability, performance, and safety [10]. A saturated salt water is required, which generally requires a large reservoir at the surface that may also bring potential environmental risk/hazards.

18.4.2 Porous Rock – Depleted Hydrocarbon Fields and Saline Aquifers

Porous sedimentary rocks such as sandstones contain microscopic voids or pores (matrix porosity), which can contain water and, in some cases, hydrocarbon fluids or gases. If the pores are connected, then the rock is permeable permitting flow of fluids or gases. Natural gas is stored by injection (through boreholes cased and sealed into the overlying cap rock), into porous rocks, but flow/production rates are limited by the resistance of the pore matrix. Even with good porosity/permeability, drilling of additional wells is generally required to increase injection and withdrawal rates. Porosity and permeability may also be present through carbonate rocks affected by dissolution, or fracture systems in low porosity rocks, including crystalline rocks, producing hydrocarbon fields or aquifers and offering CAES potential.

18.4.2.1 Depleted Hydrocarbon Fields

The principal of underground gas storage started with an experiment at a depleted gasfield in Welland County, Ontario (Canada) in 1915, followed in 1916 by the first long-term gas storage facility in the depleted Zoar gasfield, Buffalo, New York, US [86], which remains operational today. Depleted hydrocarbon fields provide the largest gas storages, both volumetrically and numerically, such that as at 2014, worldwide they accounted for around 81% of the stored natural gas volume [3].

For gas storage using depleted fields, several advantages exist; the reservoir and trap have proven capabilities to retain oil, gases, or condensates over many millions of years, and as the initial exploration and subsequent production activities mean the reservoir and trapping structure is usually well defined, lower investment costs exist. It may also be possible to reuse existing infrastructure such as wells or pipelines for storage operations. However, not all depleted fields are suitable for storage and certain reservoir properties are required, including its depth (usually between 500 m and 2500 m) and adequate porosity and permeability. Storage cycles are generally longer for porous rock storages, which cannot be filled/emptied at the same rates as salt caverns, or other mined voids. Consequently, the primary use of depleted oil and gas reservoirs is for the seasonal storage of natural gas. They thus offer less flexible storages compared to salt or mined caverns.

Partly as a consequence of the above, no development of CAES in depleted oil or gas fields exists to date. However, a number of assessments and feasibility studies have been undertaken (Figure 18.3, Table 18.2). Between 2012 and 2015, depleted field studies were under investigation for CAES by KPC&L, NPPD, NYPA, NYSERDA, PG&E, SMUD, Tri-state, and Xcel, with support from NETL and EPRI [37, 87]. Significant among these were the PG&E-led investigations in California, particularly an advanced 300 MW, 10-h CAES plant using a depleted gas reservoir, which received significant funding from the ARRA 2009 Energy Smart Grid Demonstrations Program (specifically for advanced CAES demonstration projects) [37], California Public Utilities Commission (CPUC), and the California Energy Commission (CEC), with the hope a commercial plant could on-line by 2020–2021. An early PG&E and EPRI interest thought to have good potential and capable of supporting a 110 MW, 37-hour CAES plant, was the Dunnigan Hills depleted gasfield, NW of Sacramento, northern California [88]. Interest, however, lay to the south and a tie-up with the 4500 MW Tehachapi Wind Project in southern California with both depleted field and aquifer storage was considered [88]. Over 120 sites were considered initially, reduced to six candidates, from which three top sites in late 2012 would be chosen for drilling, coring, and reservoir modelling work [89]. This included the Bakersfield saline aquifer site (see later). By February 2013 all permits and approvals required to begin site preparation, borehole drilling, and core retrieval at two sites were obtained: at King Island and East Island, San Joaquin County [90–91], with a third site anticipated [92]. The first phase of the feasibility study was completed in June 2013 with the collection of core samples at the two sites, which were sent for laboratory analysis to determine permeability, porosity and sample mineralogy [90–91]. Drilling depths were 1370 m to 1520 m. Core results proved good permeability, porosity, and other geologic characteristics suitable for CAES. King Island was selected as the top site for an Air Injection Test (AIT), with a reservoir model constructed based on 3D seismic reflection data and the engineering, procurement, and site preparation completed for the AIT involving a series of pressure tests [91]. Reports suggest that 10 MW of compression equipment was installed at King Island with compression testing planned for early 2015 [87]. The current status is unclear.

Other projects involving depleted fields for which extensive feasibility studies and planning had been completed have included (Figure 18.3, Table 18.2): in support of wind generation Portland, Victoria, Australia [92], Nebraska Public Power District (NPPD) using the Big Springs gas fields in western Dakota [38], and the Sacramento Municipal Utility District (SMUD) assessment for sites in Sacramento, California [93]. NPPD had acquired storage rights in 2011 and undertaken feasibility and modelling studies of the reservoir by early 2012, which suggested the site could support a CAES plant [38]. Legal control was obtained for at least one site by the end of 2012 and NPPD had planned to drill and obtain core, and perform air injection and cycling tests of up to ~ 85 Mm^3 (3 Bf^3) at the site during 2014 [93–94]. No further details are available. SMUD were interested in a very large, long-duration CAES storage

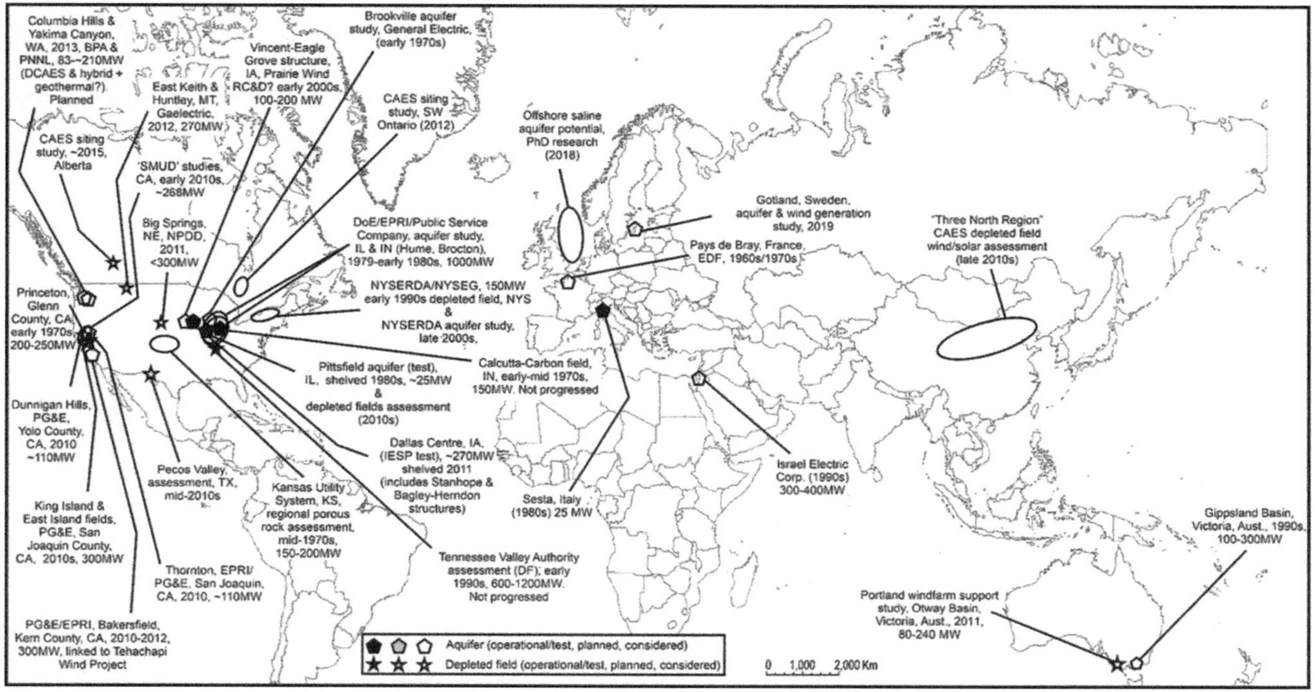

Figure 18.3 Map illustrating locations, status, and general characteristics of principal sites or locations considered for CAES in porous rock (see [8] and references therein).

(months rather than hours) for seasonal operation and had evaluated over 30 potential sites in its California service area [38]. The North Sherman Island depleted gas field was investigated and rejected, with problems encountered including the high hydrostatic pressure due to its depth of 1890 m, permeability of only 270 mD that would have necessitated a large number of wells per CAES plant and faulting within the reservoir [94]. By late 2012 two sites were identified; a depleted field and an aquifer above a depleted field (Tables 18.2 and 18.3) [95]. Initial geophysical investigations and modelling were complete for both sites in 2013, with SMUD working towards injection well drilling, building of a test bubble, and completion of cyclic testing by Q1 2015 [96]. No further details are readily available.

Key considerations for CAES. Considerable technical challenges exist when considering depleted fields for CAES. Most importantly residual hydrocarbons remain in the depleted reservoir presenting safety issues [97], particularly regarding the potential formation of ignitable air-gas mixtures. EPRI with PG&E and the University of Calgary investigated air-gas mixing when storing and withdrawing air from a depleted gas reservoir, when it could ignite at surface [38, 88]. Downhole, reservoir permeability is too low for a flame to propagate: even if a spark were somehow introduced underground, the flow of oxygen through the porosity would not be fast enough to fuel a flame, which would travel only a few millimetres before self-extinguishing. During formation and operation of the air bubble, compressed air is expected to displace most of the remaining natural gas to the periphery [93]. Repeated cycling of air through the reservoir would flush out remaining gas to present a lessening risk, with air-gas mixing thus not expected to be a substantial problem. However, the possibility of a flammable air-gas mixture reaching a well bore and thence the surface presents a safety concern requiring further study, requiring well-bore designs and equipment to ensure it cannot occur.

Porosity and permeability are critical properties in gas storage and maybe more so for CAES with air (physical energy storage) having a lower volumetric energy density (2.4–3 kWh/m^3) [79] and thus value and suitability for long-term storage applications compared with natural gas (chemical energy and ~ 1100 kWh/m^3). Typical porosities in reservoir rocks range from 10% to 30%, with at least 15% required for CAES, whilst permeabilities maybe be as high as 2500 mD in some reservoirs, with > 250 mD generally required for CAES [94]. Reservoir performance and mechanical properties of porous sedimentary rocks may be variably affected by interstitial fluids with increased pressures and temperatures. Together with possible chemical interactions, temperature effects may include increased effective stresses, decreased load-bearing capacity and increased fluid-rock interactions involving mineral dissolution, hydrolytic weakening and cementation, all of which may adversely affect reservoir and system performance. Reaction between the injected air (oxygen) and mineral constituents (e.g. pyrite) of the reservoir rock can lead to oxygen depletion and also the potential for bacterial/micro-organism growth, leading to deterioration in reservoir performance.

The effects of production and repeated pressure cycling on the reservoir and overlying caprock succession in terms of containment, performance, and hazards require consideration. During production, reservoir pressures decline with a loss of support to and potential collapse of the reservoir, causing porosity/permeability reduction affecting not only performance but producing subsidence and fracturing of the caprock/overburden succession. Safety concerns may arise with repressuring during storage, opening fractures and leading to escape of either native gas and/or air from storage into overlying strata and potentially to the surface.

18.4.2.2 Saline Aquifers

Porous and permeable water-bearing strata (saline aquifers) offer potential for gas storage, providing suitable trapping structure, adequate reservoir porosity/permeability and caprock exist and can be proved. The major difference to depleted hydrocarbon reservoirs is that the pore space in aquifers does not now, or in the majority of cases has never, contained hydrocarbons. The suitability of an aquifer formation/trap for storage must first be verified by undertaking extensive and generally expensive exploration, reservoir testing, and site characterization activities. The injection of the storage gas at a pressure higher than the hydrostatic head at discovery, displaces a volume of water equal to the volume of gas both laterally and into the deeper-lying areas of the aquifer, creating a gas cap. Otherwise, the properties and operation of an aquifer gas storage are broadly analogous to those of a storage in a converted depleted hydrocarbon field.

The first illustration of injection into essentially an aquifer ("water sand") was in 1930 during experiments at the Muldraugh field in Kentucky [86]. In 1941 injection of ~ 424,753 m^3 of gas into a Pennsylvanian aquifer took place at New Harmony in Illinois [98], and in 1946 development of the first major aquifer gas storage at the Doe Run field, Kentucky [86]. Worldwide, natural gas storage in aquifers is a now well-established and standard technology, especially in the US, France, and Germany. As of 2014 aquifer storage was second only to depleted field storages, providing around 13% of global storage capacity [3].

Early assessments of CAES considered aquifers (Figure 18.3, Table 18.3), which follows the same principal as natural gas storage: injection of compressed air creates an "air bubble" within the aquifer, with the minimum pressure of the aquifer generally required to exceed the pressure required at the gas turbine inlet. Amongst the earliest aquifer CAES studies was the development program, jointly sponsored by the US DoE and EPRI and led by Public Service Company of Indiana (PSCI) during the 1970s in which CAES in Illinois and Indiana, where aquifer gas storages were already established in both sandstone and carbonate reef reservoirs, was assessed [99, 100]. Fourteen potential sites were reduced to seven for consideration of a 1000-MW CAES facility [25]. The Calcutta-Carbon field (Indiana) with a reefal limestone reservoir, identified originally for gas storage and with reservoir and injection tests completed, was made available for CAES evaluation and development [99], but was not progressed. During the 1980s extended injection tests were conducted on sandstones in the Pittsfield "dome" (Illinois) [25, 88]. Selected from 11 candidate sites and with DOE funding and led by Pacific Northwest Laboratory (PNL), a comprehensive investigation into the practical feasibility for CAES in aquifers was undertaken between 1982 and mid-1984 [25]. A 25 MW (2 MPa/290 psi/20 bar) mini-CAES test facility undertook injection and partial withdrawal of air over several years, providing operational data on aquifer performance in terms of air bubble growth, reservoir stability, water coning, particle production, geochemical reaction, and thermal storage. A key finding was considerable oxygen depletion following storage due to the presence of pyrites in the reservoir. The test facility finally closed in 1985.

Interest in aquifer storage continued through the 1990s, including in Israel [54] and the Gippsland Basin in SE Australia [101], the 2000s and into the 2010s (Figure 18.3, Table 18.3). Most significant was the Iowa Stored Energy Park (ISEP) project [102], which investigated a 270 MW CAES plant linked to wind generation. Investigations ceased in 2011 after detailed geological investigations determined the aquifer identified could not support the planned capacity with oxygen depletion a likely problem. Nevertheless, the project left a huge legacy in terms of the volume of data for future reference. Other aquifer sites assessed in the early 2010s have been those by PG&E under ARRA-funding ("Bakersfield" prospect) [103] and by SMUD [95] in their California service territories. Nothing was progressed.

Two feasibility studies in Europe highlight current interest to support offshore wind energy generation. Estimates for UK offshore aquifer storage suggest the resource could offer considerable inter-seasonal storage capacity, equivalent to ~ 160% of the UK's electricity consumption for early 2017 [104]. A CAES feasibility study of the Middle Cambrian Faludden sandstone reservoir in several structures was undertaken on Gotland, Sweden (Figure 18.3, Table 18.3) [105]. Estimates for energy storage capacity, based upon a plant with a power output of 70 MW for 18.5 h, suggest that industrial scale CAES is possible, with energy storage capacities up to 1300 MWh.

Fundamental questions remain regarding the hydrological and energetic performance of CAES in aquifers. This is despite aquifer performance for CAES having been examined in numerous studies, including a large US government research and testing program over 30 years ago, more recent testing of an aquifer site in Iowa, and modelling studies, all suggesting aquifer CAES is feasible [106, 107].

Key considerations for CAES. Unlike depleted fields, an aquifer site has no proven capability for retention of hydrocarbons, nor have it or the caprock been pressurized to the levels likely to occur during storage and reached when conditioning the reservoir – the aquifer contains only water and to begin displacing the pore water and replacing it with natural gas or air, high injection pressures are required. Once injected, a certain amount of stored gas (air) is physically unrecoverable, being trapped in the pore spaces ("cushion gas"). Since aquifers are naturally full of water, cushion gas requirements may be as high as 80% of the total "gas" volume. Water coning could occur during storage operations, whereby water infiltrates the near-wellbore region and is produced instead of air. This all results in less flexibility in injecting and withdrawing the stored gases. Construction of associated above and below ground infrastructure may be required and consequently the effort and costs of developing aquifer sites are generally higher than the two other main storage options.

Generally, aquifer storage relies on a strong water drive whereby regional hydraulic gradients produce a flow of water in the aquifer, which provides pressure support to the storage with minimal pressure drop at the wellbore. If an aquifer is isolated or compartmentalized by faults, pressure drop may be rapid and the drive lost.

CAES air-stream flow rates mean that particles of nearly any size arising from reservoir damage or generated within a few feet of the wellbore, could be picked up and carried in the airflow to the CAES turbomachinery, damaging turbine blades and other sensitive equipment [6]. As with depleted fields, consideration of the consequences, with any potential deleterious effects on aquifer reservoir properties arising, that the introduction of air (oxygen) has on the mineral constituents of the reservoir rock at elevated temperatures and pressures is required. Aquifer tests to date (Pittsfield and Iowa), found real risks of oxygen depletion, growth of micro-organisms, with loss of porosity and permeability exist. Furthermore, wellbore desaturation and reversible air-water interface movements provide the potential for geochemical reactions and the generation, transport, and deposition of fine particulate matter in the reservoir and near wellbore regions, reducing performance.

18.4.2.3 Fractured Reservoirs

Though less common than "typical" matrix porosity reservoir types, reservoirs relying on naturally fractured rocks possess porosity and permeability, producing commercial hydrocarbon fields and aquifers around the world. Fractures can be of many types and the predictability, performance and containment of such systems is more difficult to ascertain. The reservoirs are more difficult to develop as such and developing them as gas storages is rare and it is even rarer that they are considered for CAES. However, examples exist, with fractured "hard rock aquifers" having been considered in Israel during the 1980s–1990s [42]. At Sesta (Italy), ENEL conducted a small 25 MW R&D CAES test program between 1987 and 1991 in a reportedly fractured aquifer that had previously hosted a carbon dioxide "bubble" [42, 108]. Though successful, operations at a nearby geothermal field caused problems with the storage horizon operations with the project terminated in December 1991.

More recently, CAES has been considered in the Pacific Northwest, east of the Cascade Mountain Range in eastern Washington and Oregon states, where flood basalts of the thick and often deeply buried Columbia River Basalt Group (CRBG) host the depleted Rattlesnake Hills gasfield [109, 110]. The CRBG comprises numerous lava flows each of which comprises various lithologies, with a rubbly, fractured or brecciated top and base. Together they form "interflow units," which locally form aquifers and are connected by various fracture systems in the denser flow interiors. The occurrences of gas in the basalts led to interest in developing gas storage facilities and more recently, consideration of a first of its kind CAES plant [110]. Storages would develop several interflow units as "multiple-stacked" injection reservoirs. Bonneville Power Administration (BPA) together with the Pacific Northwest National Laboratory (PNNL) evaluated the technical and economic feasibility of developing potential sites, with two identified: Columba Hills and Yakima Canyon (Washington). Both projects have progressed through conceptual design stages, with one combining an 83 MW adiabatic CAES plant (TES using salts) and geothermal energy to pre-heat the withdrawn air (Figure 18.3, Table 18.3). Their current status is unclear.

Key considerations for CAES. Basalts represent chemically reactive reservoir rocks and compressed air injection could potentially induce precipitation of iron oxyhydroxides and carbonate minerals, degrading reservoir permeability with time [110]. Watering out could be a problem if fracture porosity provides faster migration routes in some regions.

18.4.3 Abandoned Mines – Salt and Non-salt

In many areas of the world, the extraction of natural resources such as coal, salt, metals/ores, and limestone has left significant underground voids (workings), which depending upon local geological conditions and depth, may be utilized for the storage of liquids (e.g. water and hydrocarbons) and gases. Containment is generally reliant upon an impermeable rock mass and hydrostatic pressure. This section briefly summarizes assessments, tests, and pilot schemes for CAES undertaken in abandoned mines areas (Figure 18.4, Table 18.4). In some mines, small existing or newly mined tunnels have provided test sites for mined unlined rock (URC) and lined rock cavern (LRC) storage, summaries of which follow in Sections 18.4.5 and 18.4.6.

Few examples exist for natural gas storage in abandoned mines and the gas in such resources was considered. The first patent for an unlined cavern storage was in 1939, but it was not until after the successful conversion of an abandoned feldspar mine to an unlined oil storage at Härsbacka outside Stockholm in Sweden in 1948 (and which continued into at least the mid-1990s), that the technology was considered for gas storage [111, 112]. The first storage of natural gas in abandoned mines took place at the flooded Leyden coalmine in Colorado (US) in 1961, which operated successfully until its closure due to leakage in 2001 [10, 113]. Other abandoned coalmines used to store gas for short periods were at Anderlues and Peronnes in Belgium [114].

The storage option is costly and generally only considered in regions lacking alternative geological storage. Criteria controlling the prospects and development of an abandoned mine storage are the homogeneity of the rock mass and whether it is a competent, tight (impervious) and stable rock type over the entire area of the mine to be pressurized (see later). Consequently, very few examples of abandoned mines used for gas storage exist. However, the successful construction and operation of the Leyden mine as a gas storage facility demonstrates both the technical feasibility and economic viability of converting abandoned mines to gas storage [9].

Given the success of solution-mined salt cavern storage (refer earlier), gas (including air) storage in former dry salt mines appears to represent a promising option. However, only one such example, opened in 1971 at Burggraf-Bernsdorf in Germany, is currently operational as a small natural gas "buffer" storage. Around 5 Mm^3 of gas is stored at a depth of ~ 580 m and at pressures of 1.2 to 3.6 MPa [115]. CAES has not been widely considered in worked-out areas of salt mines. In the 1980s the Weeks Island salt mine hosted some tests (see Section 18.4.4) and in 2010, the Lyons Salt Mine, Kansas, was linked to storing wind energy as compressed air in Permian age salts (Table 18.4) [116]. Some testing was undertaken but nothing was progressed. Safety aspects and some important considerations associated with dry salt mines are discussed later.

Other than tests, there has been no use of former mines in non-salt rocks for CAES. However, worked areas of the Kamioka zinc mine (Japan) were used as hydraulically compensated "airstores" from 1956 with some early CAES tests reportedly undertaken [117]. Investigations and assessments have appeared periodically since the mid-1970s (Figure 18.4, Table 18.4), including at the Mt Hope mine in New Jersey (US), in gold mines in South Africa and at the Pyhäsalmi zinc mine in Finland, where air storage would use either water curtains or not, with both isochoric or isobaric storage mode considered [118]. Interest in CAES in abandoned mine space continued into the twenty-first century, with a number of projects in the US, including in New York State [119] and in Minnesota during the mid-2010s. In Minnesota, the economic and technical aspects of constructing CAES plants for both large- and small-scale CAES systems, making use of existing underground mine workings of Cuyuna, Mesabi, and Vermilion iron ranges, were reviewed [120]. The study also considered

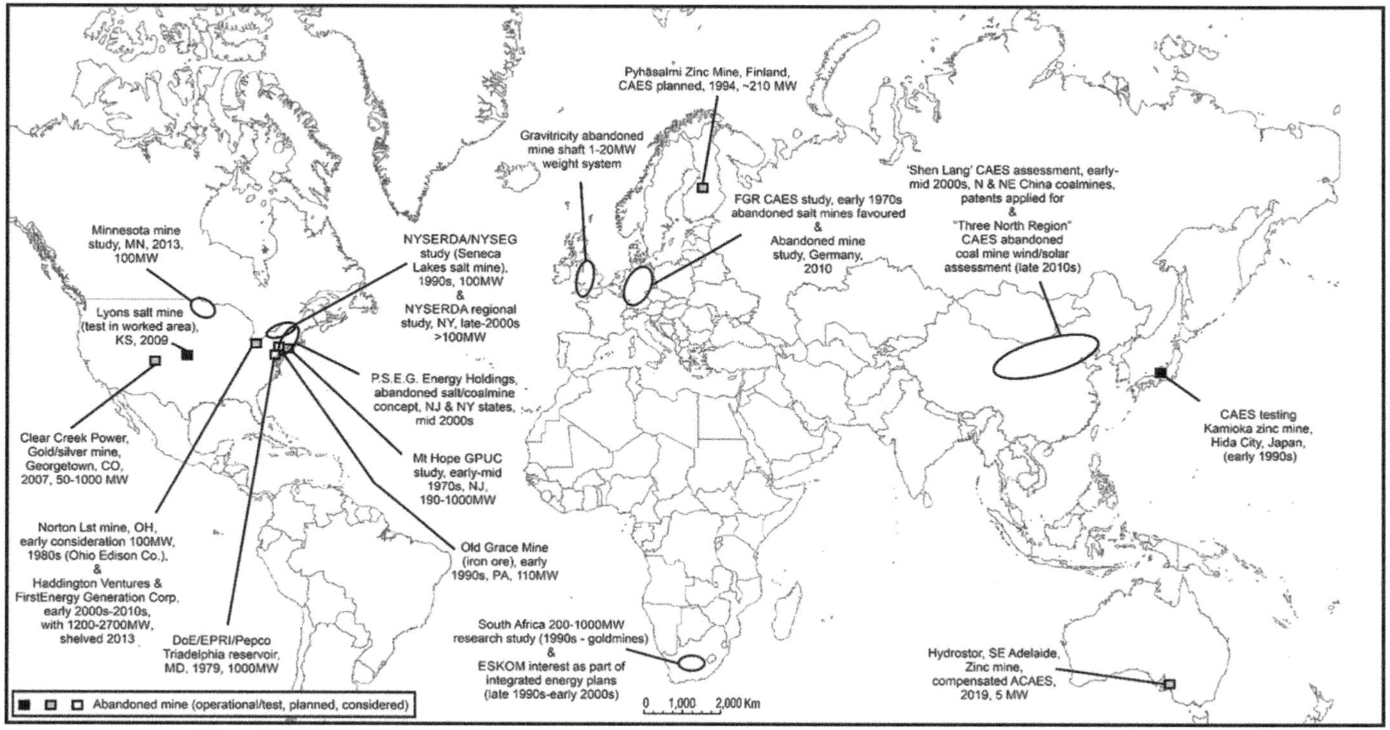

Figure 18.4 Map illustrating the locations, status, and general characteristics of principal sites or regions considered for CAES in abandoned mines, both halite and non-halite rock types (see [8] and references therein).

excavating a mine for the combined purposes of extracting valuable ore and creating a CAES cavern, with a number of challenges identified. The use of former mines for CAES was assessed in Germany in 2010 [121]. The most significant CAES project proposed in such storage was a 2700 MW CAES plant in a former limestone mine at Norton, Ohio (US). First considered in the 1980s [53], most recent interests started in 2000 [122], with investigations having continued through until 2013, when the project was suspended [123]. It now appears abandoned, although the developer has an agreement with the city of Norton until 2025.

Economic and feasibility studies were undertaken into the potential for an up-to 1000 MW CAES facility with 10 000 MWh of energy capacity co-located with wind generation in Colorado [124]. One option was a 50 MW CAES turbine with ~ 525 MWh storage capacity, linked with Clear Creek Power LLC's wind turbine project under development in the mountains near Georgetown, Colorado. Use of a large abandoned gold and silver mine in heavily fractured hard rock, owned by the wind developer and located beneath the wind site, was considered but the project has not been progressed.

More recently, Hydrostor Australia Pty Ltd, part-funded by the Australian Renewable Energy Agency (ARENA), announced construction of a 5 MW/10 MWh commercial demonstration ACAES facility with the conversion of a brownfield site, including a decommissioned Angas Zinc Mine into a clean energy hub at Strathalbyn, 60 km southeast of Adelaide, Australia [125]. Due for commissioning in 2020/2021, the project plans to use the existing mine as the air-storage cavern, operating under constant pressure, using hydrostatic head from a water column and an above ground reservoir. The heat of compression will be stored and used during withdrawal.

In 2017, Gravitricity announced plans for a novel storage system involving large cylindrical weights suspended in disused mine shafts from 150 m to 1500 m deep [126]. Raising or lowering the weight absorbs or generates electrical power, delivering 1–25 MW. The aim was for a 250 kW concept demonstrator in 2019 with the first full-scale prototype by 2020 at a disused mine in the UK.

18.4.3.1 Key Considerations for CAES

Immediate physical limitations of using abandoned mines are that mines inevitably follow the mineral or commodity of value, as a result of which, void shapes are often irregular and not particularly suited to CAES or gas storage operations. Additionally, the pre-determined mine location in relation to demand may prove uneconomical for an underground storage plant [111]. Similar to depleted fields former coal mines represent a hazard from mixing of the air with coalbed-derived methane and the resulting fire/explosion risk. Mining often involves or causes subsidence, collapse, and fracturing in the overburden, creating potential migration paths away from the storage. Additionally, pressure-tight seals between the storage and the access shafts, or other tunnels are required. Critical to any storage in abandoned mines is verifying the tightness of the entire area of the mine exposed to storage pressures, together with suitable seals, both overlying and between the storage and the access shafts. Use of old mined areas for high-pressure storage in close proximity to any remaining workings would require careful risk assessment, especially so if operated as a water-compensated scheme. Equally, nearby mining activity could affect adversely a CAES project through blasting, fracturing, and possible leakage. Water ingress may lead to dissolution and precipitation of minerals, perhaps exacerbated by the presence of air (oxygen) during CAES operations.

Some problems encountered in former salt mines are, for safety reasons, worthy of note. Sections of the Weeks Island mine in Louisiana (US) were converted and used as part of the US Strategic Petroleum Reserve (SPR) between 1980 and its final closure at the end of 1999 [127]. Closure resulted from the development between 1991 and early 1995 of two sinkholes directly over the edges of the SPR oil storage chambers, which caused seepage of water into the storage. Leaks of groundwater into other salt mines in Louisiana and elsewhere had previously led to flooding and eventual abandonment [128]. Perhaps the most catastrophic event was that of an oilwell inadvertently intersecting an area of the Jefferson Salt Mine 400 m below the shallow Lake Peigneur, Louisiana (US) in November 1980 [129]. Lake water leaked through the borehole into the underground mine, causing calamitous sinkhole development, draining of the entire lake and flooding of the mine. The dramatic blowout event at the Kanopolis salt mine in Kansas, US [130], although not storage-related, highlights the problems of increased air pressures underground and the need to identify and seal old shafts. Groundwaters seeped down and into the mine, pressuring up air trapped in the mine space, which eventually blew out through a concealed shaft. Finally, hydrostatic tests for CAES in the Lyons Salt mine, Kansas, US (Figure 18.2), involved the isolation of a tunnel section and pressuring it up with brine. Although there was no measurable pressure drop inside, the bulkhead was found to have moved forwards 1.9 cm (¾″), pressing into the halite formation [115].

18.4.4 Mined Voids (URC) – Unlined New Rock Caverns in Halite and Non-halite Rocks

Although abandoned dry salt mines have been considered for storage, few if any examples exist for newly, dry-mined URCs in salt bodies being considered for gas storage purposes, due partly to the ease and low costs of solution-mining salt caverns. Areas of salt mines have been used as small test facilities. In the late 1970s, a small instrumented test cavity was constructed 140 m underground in the Northwich Halite at the Meadowbank rocksalt mine, Winsford, Cheshire, UK to investigate the mechanical response of rocksalt under conditions for operating gas storage caverns at similar depths (Figure 18.5, Table 18.5) [131]. In-situ pressure tests up to 1.38 MPa were undertaken in the cavity using a compressed air system. Elsewhere, as part of the US DoE assessment of CAES in Gulf Coast salt domes in the late 1970s–early 1980s, a series of instrumented holes were drilled in salt pillars of the working Jefferson Island salt mine, Louisiana, US [132].

By contrast, purposely constructed, conventionally mined, URCs in non-halite rocks have long been employed for both liquid and gaseous hydrocarbon storage. Compared with solution-mined caverns and porous storages, construction of mined, unlined non-halite rock caverns (URCs) comes with higher costs and higher levels of risk and accident due to excavation methods (drilling, blasting, and rock removal). Storages comprise either single caverns or a series of shafts, ramps, tunnels, and drifts excavated out of competent rock masses. To improve hydrostatic pressure and containment, construction of water curtains of various designs increases storage pressures, but also costs. Such storage is generally therefore only considered in regions lacking alternative cheaper storage options.

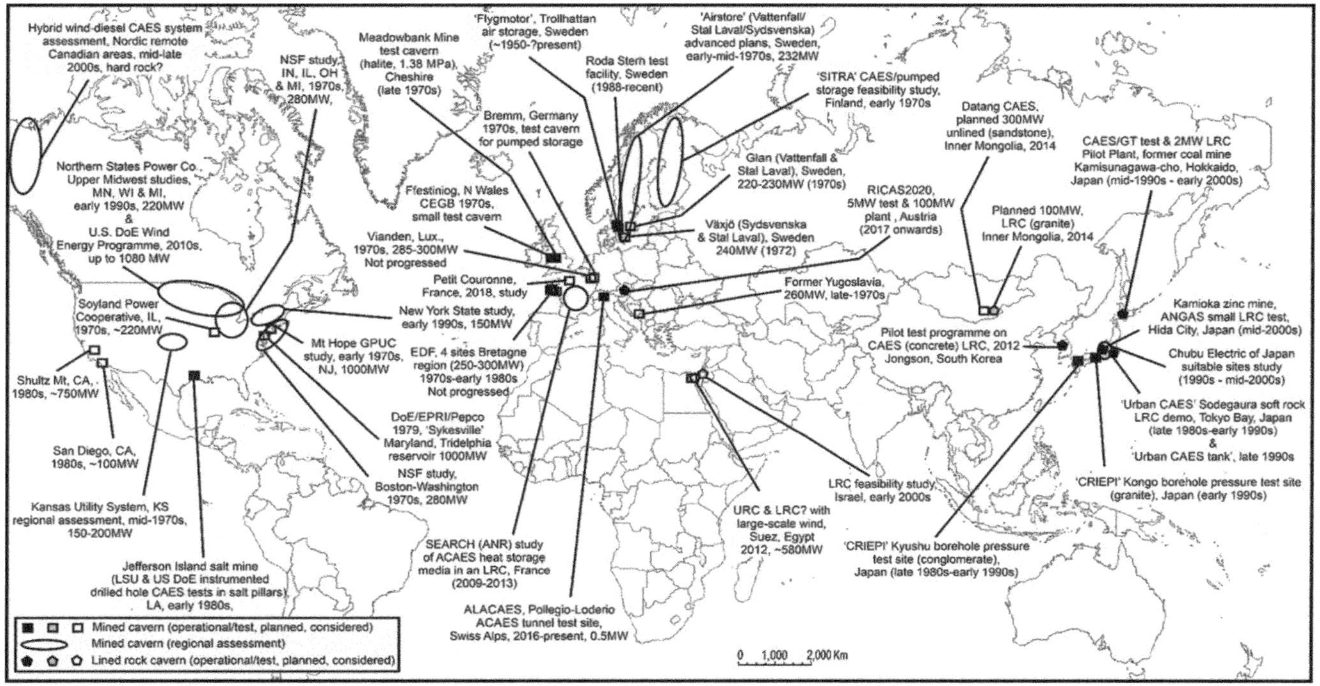

Figure 18.5 Map illustrating the locations, status, and general characteristics of principal sites or regions considered for CAES in unlined mined rock caverns/voids (ULRCs) and lined rock caverns (LRCs; see [8] and references therein).

The first test proving secure storage and containment of petroleum-derived liquid (gasoline) was in 1951 using a small (30 m^3) cavern outside Stockholm [112]. The first small (3000 m^3) URC purposely commissioned for LPG storage was constructed in a shale formation in Texas (US) in 1950, with the first such crude oil storages commissioned in 1965 and 1967 in Sweden and Norway respectively [111, 112]. Many more examples have followed in the US and Europe, particularly in high-grade metamorphic rocks in Norway [2] and Finland [133], but also elsewhere in the world. Caverns are constructed in shales, limestones, and dolomites in the US [112], and in chalks and limestones in France [134, 135] and England [136]. Around the world examples are found in Zimbabwe (granite) [137], Australia (sandstone) [138], Korea (metasediments and gneiss) [139], India (gneiss) [139], and Singapore (siltstones/sandstones) [140]. The world's first URC natural gas storage facility was opened at Háje in the Czech Republic in 1998 with a total tunnel length of ~ 45 km. It is constructed in granodoirite at a depth of ~ 1000 m, operating between 2 and 9.5 MPa [141]. There is no requirement for a water curtain due to the very low permeability of the host rock, hydrostatic pressure, and grouting in areas affected by fractures. Interest has even been shown for bulk hydrogen storage in URCs [142].

Despite the successes of URCs for liquid and gaseous hydrocarbon storage and previously noted small rock cavern compressed air storage schemes operating in association with mines and PHS plants (Figure 18.2), they have not been deployed for CAES purposes. However, a number of feasibility studies and assessments have been undertaken (Figure 18.5, Table 18.5). Supported by the Swedish firm Stal-Laval, a CAES plant utilizing a URC was first investigated in the late 1940s but cavern construction costs were considered too high [143–145]. The early 1950s saw interest in pyramid-shaped caverns excavated for use in isobaric storage schemes [23]. Stal-Laval continued to support various CAES assessments of URCs during the late 1960s and early 1970s in a project often referred to as "Airstore," with two sites of note, both hydrostatically compensated [1, 146, 147]: at Glan in eastern Sweden and Växjö, southern Sweden. In the UK, the Central Electricity Generating Board (CEGB) considered introducing CAES to the grid in England and Wales in the 1960s, conducting preliminary pressurization tests of an unlined, test gallery at the Ffestiniog PHS site in North Wales [1]. It was not progressed due to the existing PHS plant and lack of suitable rock type locations.

The 1970s to early 1980s saw a number of regional assessments and technical feasibility studies into water-compensated URC storage undertaken in the US (Figure 18.5, Table 18.5). Areas considered included Illinois, California, Kansas, New York State, Maryland, New Jersey, and Washington, DC [1, 148–152]. One significant 1970s DoE/EPRI-funded study, led by the Potomac Electric Power Company (PEPCO), was the assessment of a CAES plant with conventionally mined, water-compensated, hard-rock URC at a depth up to 1500 m [150]. A site was chosen in the uniform and massive Sykesville gneiss near the Triadelphia Reservoir and Sykesville, Maryland for study (Figure 18.5), although nothing resulted. Another project involved Soyland Power Cooperative Inc. and a planned 220 MW DCAES plant using a water-compensated URC and heat recuperation to support a coal-fired power station in Illinois [25]. Plans were advanced and contracts signed, however, construction of the coal-fired plant was deferred in 1982 due to a lesser demand for electricity, impacting directly Soyland's plans to construct the CAES plant [153].

Various other countries/utilities considered CAES in URCs at this time, including (Figure 18.5, Table 18.5) [154–156]: a 260 MW plant in Yugoslavia, Electricité de France (EDF) with construction of a 250–300 MW CAES plant in the Bretagne region of northern France using a number of caverns in a granite operating a hydraulically compensated system, at Vianden on the Luxembourg-German border with a ~ 300 MW plant in support of a large hydroelectric plant using a hydraulically compensated URC in Devonian slates. At the latter, a small test cavern was excavated at Bremm, near Mosel [155], although the main plant was never constructed. The Roda Stern Rock Laboratory (RSRL) was built near Gothenburg, Sweden during the late 1980s to assist development of natural gas storage in URCs. A number of caverns were constructed in Precambrian metamorphic gneisses at a depth of ~ 60 m and a series of tests in Cavern A during 1990–1992, including ascertaining leakage rates, were pertinent to the development of CAES [157].

A lack of halite deposits and other suitable geological options forced Japan to consider CAES in both URC and lined rock caverns (LRCs) from the late 1980s (Figure 18.5, Tables 18.5 and 18.6), the latter covered in Section 18.4.5. Assessments started with the drilling of boreholes ~ 400 m deep into granitic rocks of Cretaceous age at Kongu 30 km south of Osaka, in which pressure tests were conducted to determine gas tightness [158–160]. A 600 m deep borehole similarly tested Palaeogene sedimentary strata on Kyushu, SW Japan [161]. The late 1980s also saw tests on an unlined cavern, sealed by concrete plugs at a depth of ~ 530 m at the Kamioka mine in west central Japan, with few other details available [117]. Despite these initial studies, nothing was developed.

URC storage is once again under consideration in various countries (Table 18.5). Sandia National Laboratories (Sandia) led a feasibility study into a 1080 MW CAES plant in hard (granitic/granodioritic) rock for the US upper Midwest in support of wind energy [162], whilst a ~ 300 MW ACAES plant, using a 1.25 mcm URC in support of large-scale grid integration of wind energy in the Suez area of Egypt, was under consideration mid-2010s [163]. A 300 MW CAES facility is reportedly planned, using a cavern excavated in sandstone at a depth of ~ 500 m near Datang, Inner Mongolia, China [164]. Conversion of LPG chalk storage caverns at Petit-Couronne, northern France into a daily ACAES plant has also been investigated, with two options considered: CAES alone, or CAES coupled to a pumped energy transfer system [135]. The ALACAES project, re-purposing part of a disused unlined tunnel ~ 450 m deep in the Swiss Alps, is a 0.5 MW ACAES demonstration plant incorporating a TES within the storage chamber [36]. Construction commenced in June 2014, with commissioning and first testing having begun in 2015, during which the main plugs suffered significant leakage. The problem was rectified and extended tests of the storage and TES were ongoing [165].

18.4.4.1 Key Considerations for CAES

URC storages require significant investment both in terms of the excavation of large voids in often hard rocks and also the costs of ensuring cavern stability, gas/air containment and infrastructure. URC costs are around \$30/kWh, compared with \$1/kWh for salt caverns, or \$0.1/kWh for porous rock storage [42]. Costs for URCs may therefore be prohibitive and further increased if there are containment or stability issues arising from major joint sets that could lead to inflow of water, or rockfalls during construction, or under cyclic pressuring and which requires detailed site and geological characterization studies. Special grouting or rock bolting maybe necessary to stabilize areas of the cavern walls or roof. Additionally, and as alluded to, the installation and maintenance of water curtains may be required to enhance containment, increasing costs further.

If operated in water-compensation mode (isobaric) then the champagne effect may be problematical and lead to problems with machinery and safety, etc. Oxygen dissolved in water during storage comes out of solution at shallower depths and lower pressures.

Disposal of the mined waste requires consideration. If not for mineral extraction or readily used in construction, additional costs could be incurred with transport, dumping, grading, shaping, vegetation, and stabilization of the waste piles [162].

A cautionary note is provided by a water-compensated rock cavern air storage system at the then Volvo Flygmotor facility at Trollhättan in Sweden in 1960. The system was overcharged causing a large air bubble to enter the water shaft, blowing the fluid up to surface and destroying the building on top [147].

18.4.5 Lined Rock Caverns (LRC)

Modern LRCs represent a variation on the URC storage type, whereby conventionally mined caverns are lined with an artificial, gas-tight barrier. This generally comprises concrete and stainless-steel sheet, but possibly also some form of rubber lining. However, they represent the most expensive storage type, but may offer the only alternatives in areas lacking other geological options, regions with more local distributed generation, or where competition for space exists.

Obvious similarities exist between LRCs and early steel tanks built in the shallow subsurface in basalts on Hawaii and granites at Trondheim (Norway) during and after World War II [166–168]. Excavated both above and below ground water level, construction was with welded steel plate and concrete infill to the cavern wall. Further underground oil storage "tanks" lined with steel plates were built east of Trondheim in 1960–1962. However, there was little further development of LRCs until the period 1985 to 1992 when Swedish energy companies investigated the mechanical behaviour of different cavern linings and their tightness to gas [169]. The technology was proven for natural gas storage following the construction and testing (to 50 MPa) of three differing caverns in a pilot plant at Grängesberg in south-central Sweden between 1988 and 1993 [170]. This led to the construction and commissioning in 2004 of a demonstration plant at Skallen near Halmstad in SW Sweden, using a vertical, cylindrical cavern (52 m high and 36 m diameter) of 40 000 m^3 in crystalline gneiss at a depth of 115 m. During storage operations, maximum pressures are between 15–30 MPa and at no time is the gas in contact with the surrounding rock mass, being contained in the LRC, gas pipes etc. [169]. Still in service, the cavern lining comprises an inner steel liner, providing a gas-tight seal, with a concrete layer and steel reinforcement mesh between the steel liner and the rock wall. A drainage system is inbuilt to take groundwater away and avoid over-pressuring of the steel liner at low storage pressures.

Over the years, assessments and tests of CAES LRC potential in a number of countries lacking other suitable rock types have been undertaken, most notably in Japan and Korea (Figure 18.5, Table 18.6). A program known as "Urban CAES" in the mid-late 1980s assessed both hard and soft rock LRCs in Japan [171, 172]. Soft sediment LRC focussed on the area of Sodegaura near Tokyo Bay, with many designs and excavation methods assessed. The preferred designs were vertically enlongate "spheres" at a depth of ~ 600 m, accessed by a shaft and operated isobarically. The concept was not progressed. A variant of the "Urban CAES" LRC concept was part of a successful short-term test, which involved lowering a reinforced concrete tank to the bottom of a water-filled shaft, 1.5 m in diameter and 200 m deep. Air at up to 2 MPa was pumped into the tank and the effects of injection–withdrawal on tank and water/air behaviour observed [173].

During planning of a 500 MW gas turbine plant supported by a 100 000 m^3 cavern, several in situ tests were undertaken from the early 1990s in worked-out areas of a coal mine at Kamisunagawa-cho, Hokkaido, northern Japan [174, 175]. The excavated caverns forming the test and pilot study were at around 450 m depth and accessed via existing coal mine tunnels. Initially, tests took place successfully over a pressure range of 4–8 MPa on a small experimental cylindrical LRC (16 x 2.5 m, 78.5 m^3) comprising concrete segments and butyl rubber lining. Commencing 1998, construction of a larger cylindrical LRC (57 x 6 m, 1611 m^3) began, supporting a 2 MW, 4-hour storage pilot CAES gas turbine (CAES-G/T) plant for load-levelling purposes. The cavern, lined by reinforced concrete segments with concrete back fill and a butyl-rubber lining, was completed in 2000 and the facility commissioned in June 2001. Pilot plant operations were successful and although there were plans to develop a 35-MW, 8-hour CAES plant and research on rubber linings continued [176], nothing was developed. During studies developing an LRC natural gas storage, the ANGAS project successfully tested a small LRC using compressed air at the Kamioka Mine in Japan [177].

In the early 2010s, investigations into LRC storage commenced at Jongson in Korea, with tests of two pilot caverns (PC1 and PC2), constructed in a working limestone mine above the water table at ~ 100 m [178]. The cylindrical caverns were concrete-lined, with an inner diameter of 5 m and 10 m and 20 m in length. They had differing plug constructions (tapered and wedge shaped) and the linings tested were of two types: PC1 comprised 500 mm butyl rubber sheets with iron rods also embedded in the concrete lining for reinforcement purposes, PC2 comprised 300 mm steel plates with no iron rods. The caverns were pressured up to a maximum of 5 MPa to test the stability of the plug types, with a taper-shaped plug found to be the more stable. Thermodynamic and geomechanical modelling of LRC performances were also undertaken to assess air tightness and energy balance for underground LRCs at relatively shallow depth, where operational pressures are much higher than both ambient fluid pressure and in situ stress [179–182]. Geomechanical modelling indicates high tensile stresses may develop in the lining leading to radial cracks and air leakage. This would be minor and quickly detected by pressure monitoring outside the concrete lining [182].

A 100 MW CAES plant using an LRC in granite is planned in the Inner Mongolian region (China), (Figure 18.5, Table 18.6) [183]. The "cavern" will comprise horizontal tunnels of 5 m radius, installed with a 0.5 m thick concrete lining and a 3 mm thick butyl rubber sheet inner seal. In Austria, the EU-funded RICAS2020 project is investigating AACAES potential, using an LRC constructed in an area of a worked limestone-hosted iron ore mine at Eisenerz in Austria, with the TES independent of the geology [35]. A small-scale 5 MW pilot LRC facility is being built, with a full-scale, 100 MW test power plant facility planned (Table 18.6).

18.4.5.1 Key Considerations for CAES

Although the costs of LRC are the highest in comparison to other geological storage options, LRC has a major advantage of deployment in many more situations, offering "siting flexibility." Although tunnels are routinely driven through "soft rock" terrains, the construction and storage of materials in caverns excavated in soft rocks, even with linings, is rare. This is due in large part to the weakness of the rocks themselves. As with salt caverns, no reactions between the oxygen and the components of the LRC are expected, especially if the cavern interior is moisture-free.

Key geotechnical and structural factors to consider in LRC storage include: uplift failure of the overlying rock up to the surface, particularly for shallow caverns (main principle of the LRC concept is load from the high gas pressure inside the cavern is taken by the surrounding rock mass, with the lining only to ensure the cavern is gas tight); failure and loss of tightness of the sealing membrane; and shearing of the plug sealing the cavern [184]. Loss of tightness of the cavity decreases not only the efficiency of the system, but may also lead to stability issues, as high air pressures within the overlying rock mass increase uplift/blowout risk.

Leakage may occur through an initial defect in the inner containment liner, such as imperfect welds and construction joints, the thin steel shell, or the lining may fail through fatigue induced by expansion or buckling during cyclical pressurization. When considering the success of natural gas storage in LRCs, it should be remembered the frequency of storage cycles is lower than those in CAES, where daily/hourly cycles are required. This imparts differing geomechanical and thermal stresses on the cavern walls and lining requiring careful consideration when increasing cyclicity.

18.5 Summary

The underground (geological) storage of large volumes of gases, including air, has been a proven technology for many decades. To date, the only successful commercially operational CAES facilities have been diabatic plants constructed using salt caverns in Germany (1978), the US (1991), and in 2019, the world's first ACAES plant in Canada. Increasing reliance on renewable energy sources has again raised the profile of CAES and there are now many studies assessing the potential deployment of CAES for grid-scale support in all storage types. A pilot ICAES facility using a salt cavern operated for around 4 years in Texas, but closed in 2016. A number of tests have attempted to demonstrate aquifer storage for CAES, but have so far found conditions unfavourable due mainly to problems of oxygen loss and reservoir performance. A test adiabatic plant commenced operation in 2015 using an unlined tunnel in the Swiss Alps and there are advanced plans for ACAES testing of a lined tunnel in a worked-out part of an Austrian mine. The first quarter of the twenty-first century may, in the future, become regarded as the turning point when research into and the development of CAES technology resumed, ultimately leading to its routine deployment within the electrical energy supply sector.

Acknowledgments

Financial support from the Engineering and Physical Science Research Council (EPSRC) funded Grand Challenge [Energy Storage] IMAGES project (EP/K002228) is gratefully acknowledged. The author also thanks Fritz Crotogino, Gregor Schneider, Jon Busby, and colleagues in the EPSRC-funded IMAGES project for numerous discussions, which helped set the framework for and improved the manuscript. John Williams kindly assisted with minor revisions to Figure 18.5. Permission to publish is granted by the Director, British Geological Survey.

References

[1] General Electric. (1976). Economic and technical feasibility of CAES. General Electric Co. Final Report Prepared for the United States Energy Research and Development Administration Office of Conservation, Contract E (11-1)-2559, 399 pp. http://www.osti.gov/scitech/servlets/purl/7195898.

[2] Grøv, E. (2006). Storage of hydrocarbon products in unlined rock caverns. Norwegian Tunnelling Society Publication No. 15, 19–28.

[3] Crotogino, F., Schneider, G.-S., and Evans, D.J. (2017). Renewable energy storage in geological formations. *Proceedings Institute of Mechanical Engineering, Part A: Journal Power Energy* 232: 100–114.

[4] Eldridge, F.R. (1976). *Wind Energy Conversion Systems Using Compressed Air Storage*. Virginia: McLean.

[5] Glendenning, I. (1981). Compressed air storage. *Physics in Technology* 12: 103–110.

[6] Succar, S. and Williams, R. (2008). Compressed air energy storage: Theory, resources, and applications for wind power. Technical report Princeton Environmental Institute, 81 pp.

[7] Budt, M., Wolf, D., Span, R., and Yan, J. (2016). A review on compressed air energy storage: Basic principles, past milestones and recent developments. *Applied Energy* 170: 250–268.

[8] Evans, D.J. (2016). Review of underground fuel storage options, CAES plants and assessments – Locations and operational parameters. British Geological Survey, Commercial in Confidence Commissioned Report, CR/16/198, 462 pp (revised Feb. 2019).

[9] Doherty, T.J. (1981). Air storage systems. *Proceedings of International Conference on Seasonal Thermal Energy Storage and Compressed Air Energy Storage, CONF-811066*, vol. 2, 455–468.

[10] Evans, D.J. (2009). A review of underground fuel storage problems and putting risk into perspective with other areas of the energy supply chain. In: *Underground Gas Storage: Worldwide Experiences and Future Development in the UK and Europe*, vol. 313 (ed. D.J. Evans and R.A. Chadwick). Geological Society of London Special Publication. The Geological Society, London, Special Publications, 313, 173–216. DOI: 10.1144/SP313.12 ©The Geological Society of London.

[11] Zunft, S., Cristoph, J., Koller, M., and Bullough, C. (2006). Adiabatic compressed air energy storage for the grid integration of wind power. *Sixth International Workshop on Large-Scale Integration of Wind Power and Transmission Networks for Offshore Windfarms* (26–28 October 2006). Delft, the Netherlands. 6 pp. https://core.ac.uk/download/pdf/11128191.pdf.

[12] Zunft, S., Dreissigacker, V., Bieber, M., et al. (2017). Electricity storage with adiabatic compressed air energy storage: Results of the BMWi-project ADELE-ING. *International ETG Congress* (28–29 November 2017), 448–452: Bonn, Germany.

[13] Glendenning, I. (1976). Long-term prospects for compressed air storage. *Applied Energy* 2: 39–56.

[14] Marcus, D. (2011). Fuel-free geologic compressed air energy storage from renewable power. General Compression Task #1 Deliverable Report for New Technology Implementation Grant Program 582-11-13126-3225. October 24, 2011, 9 pp. https://www.tceq.texas.gov/assets/public/implementation/air/terp/ntig/prog_rpts/GC_Task1.pdf.

[15] Smil, V. (1994). *Energy in World History*. Boulder, CO: Westview Press, 300 pp.

[16] Morfeldt, C.-O. and Lindblom, U.E. (1976). Unlined rock caverns for compressed air storage. *Proceedings of the Workshop on Compressed Air Energy Storage System*. Airlie House, VA (18–19 December 1975). ERDA-76-124, 174–198.

[17] Legget, R.F. (1979). 75-year old underground compressed air plant still in use in Canada. *Underground Space* 4: 29–31.

[18] Tollinsky, N. (2013). Prof aims to revive Ragged Chute technology. Technology, online article, 13 August 2013. http://www.sudburyminingsolutions.com/prof-aims-to-revive-ragged-chute-technology.html.

[19] Hocking, G. (1979). Parametric cyclic thermal and pressure analysis of underground openings in crystalline rock. *Proceedings Compressed Air Energy Storage Symposium*, Pacific Grove, CA (15–18 May 1978), vol. 2, 609–634.

[20] Matikainen, R. (1977). Structure and operation of underground hydraulic compressed air reservoirs in Finland. *Storage in Excavated Rock Caverns: Rockstore 77, Proceedings of the First International Symposium* (ed. M. Bergman), (5–8 September), vol. 3, 705–710. Stockholm: Pergamon Press.

[21] Luke, R. (1996). Compressed air storage for electricity generation in South Africa. A dissertation submitted to the Faculty of Engineering, University of Cape Town in partial fulfilment of the requirements for the degree of Master in Engineering. Cape Town, 206 pp.

[22] Kalhammer, F.R. and Schneider, T.R. (1976). Energy storage. *Annual Review of Energy* 1: 311–343.

[23] Anonymous. (1951). Off-peak air storage for peaking gas turbines. Electrical World, 15 January, 64–65.

[24] Allen, R.D., Doherty, T.J., and Fossum, A.F. (1982). Geotechnical factors and guidelines for storage of compressed air in solution mined salt cavities. Report prepared by Pacific Northwest Laboratory for Battelle and the U.S. Department of Energy under contract DE-AC06-76RLO 1830. Report number PNL-4242, 105 pp.

[25] Allen, R.D., Doherty, T.J., and Kannberg, L.D. (1985). Summary of selected compressed air energy storage sites. Report prepared by Pacific Northwest Laboratory for Battelle and the U.S. Department of Energy under contract DE-AC06-76RLO 1830. Report nos. PNL-5091, 112 pp.

[26] Crotogino, F., Mohmeyer, K.-U., and Scharf, R. (2001). Huntorf CAES: More than 20 years of successful operation. *SMRI Spring Meeting*, Orlando, FL (15–18 April 2001), 7 pp.

[27] Kannberg, L.D. and Loscutoff, W.V. (1981). Compressed air energy storage technology status. *Paper presented at the American Society of Mechanical Engineers*

Winter Annual Meeting (15–20 November 1981). Washington, DC, Pacific Northwest Laboratory, Richland, Washington, Report number PNL-SA-9375.

[28] Lang, W.F. (1974). Underground energy storage for electrical power generation. *Proceedings of Fourth Symposium on Salt*, Houston, TX (8–12 April 1973). Northern Ohio Geological Society, Cleveland, Ohio, 1974, 323–327.

[29] Glendenning, I. (1979) Advanced compressed air storage – An appraisal. *Proceedings of the 1978 Compressed Air Energy Storage Symposium, CONF-780599*, National Technical Information Service, Springfield, VA, vol. 2, 687–761.

[30] Hobson, M.J., Heath, E.G., Giramonti, A.J., and Adent, W.A. (1979). Feasibility of CAES in California. *Proceedings of the Compressed Air Energy Storage Symposium, CONF-780599*, National Technical Information Service, Springfield, VA, vol. 1, 179–212.

[31] Fryer, B.C. (1973). Air storage peaking power plants – Utilizing modified industrial gas turbines and cavities created with nuclear explosives. Report by Battelle Pacific Northwest Laboratories Richland, Washington 99352, Report Number BNWL-1748, 37pp. http://www.iaea.org/inis/collection/NCLCollectionStore/_Public/04/082/4082760.pdf.

[32] Boyd, D.W., Buckley, O.E., and Clark, C.E., Jr. (1983). Assessment of market potential of compressed air energy storage systems. *Journal of Energy* 7: 549–556.

[33] Mason, J., Fthenakis, V., Zweibel, K., et al. (2008). Coupling PV and CAES power plants to transform intermittent PV electricity into a dispatchable electricity source. *Progress in Photovoltaics: Research and Applications* 16: 649–668.

[34] Lund, H. and Salgi, G. (2009). The role of compressed air energy storage (CAES) in future sustainable energy systems. *Energy Conversion and Management* 50: 1172–1179.

[35] Perillo, G. (2016). Report on material qualification including guidelines for material selection/development; Deliverable D6.1. Report for RICAS2020 Design Study for European Underground Infrastructure related to Advanced Adiabatic Compressed Air Energy Storage. 67 pp. http://www.ricas2020.eu/media/deliverable-d4-1-numerical-simulations.

[36] Zanganeh, G., Calisesi, Y., and Brüniger, R. (2016). Demonstration of the ability of caverns for compressed air storage with thermal energy recuperation. ALACAES Final Report to Swiss Federal Office of Energy, 28 November 2016, 98 pp. https://www.aramis.admin.ch/Default.aspx?DocumentID=35281andLoad=true.

[37] EPRI. (2010). Compressed air energy storage newsletter. Electric Power Research Institute (EPRI), April, 2010, 7 pp.

[38] EPRI. (2012). Compressed air energy storage newsletter. Electric Power Research Institute (EPRI), April, 2012, 8 pp.

[39] Pimm, A., Garvey, S.D., and de Jong, M. (2011). Design and testing of Energy Bags for underwater compressed air energy storage. *Energy* 66: 496–508.

[40] Li, P.Y., Loth, E., Simon, T.W., et al. (2011). Compressed air energy storage for offshore wind turbines. International Fluid Power Exhibition (IFPE), Las Vegas, NV, March, 2011, 7 pp.

[41] Garvey, S.D. (2015). Integrating energy storage with renewable energy generation. *Wind Engineering* 39: 129–140.

[42] EPRI. (2003). EPRI-DOE handbook of energy storage for transmission and distribution applications: Electricity energy storage technology options. Electric

Power Research Institute (EPRI), Palo Alto, CA, and the U.S. Department of Energy (DoE), Washington, DC. EPRI Report, 516 pp.

[43] Thoms, R.L. and Gehle, R.M. (2000). A brief history of salt cavern use (keynote paper). *Proceedings 8th World Salt Symposium* (ed. R.M. Geertman), I, 207–214. Elsevier.

[44] Vance, T.B. (1963). Southeastern Michigan gas company's high pressure gas holder in a salt strata. *International Salt Symposium*, Cleveland, OH (April 1962), 597–603.

[45] Allen, K. (1972). Eminence dome – natural gas storage in salt comes of age. *Journal of Petroleum Technology* 24: 1299–1301.

[46] Gilhaus, A., Crotogino, F., Albes, D., et al. (2006). Compilation and evaluation of bedded salt deposit and bedded salt cavern characteristics important to successful cavern sealing and abandonment. SMRI Research Report No. 2006-2-SMRI, Clarks Summit (PA), US, 2006.

[47] Goodson, J.O. (1992). History of first U.S. Compressed Air Energy Storage (CAES) plant (110 MW 26 h). Volume 1: Early CAES development. Electric Power Research Institute (EPRI) Final Report, April 1994, TR-101751-V2, 202 pp.

[48] Serata, S. and Mehta, B. (1993). Design and stability of salt caverns for compressed air energy storage (CAES). *Proceedings 7th Symposium on Salt*, vol. 1, 395–402. Amsterdam: Elsevier.

[49] NRStor. (2016). Compressed air energy storage demonstration project. NRStor Inc online newsletter, November 2016, 2 pp. https://wise.uwaterloo.ca/calendar/compressed_air_energy_storage_demonstration_project

[50] Peretick, K. (2019). Energy storage in the real world: Practical applications and considerations. *Michigan Agency for Energy "Plugging Into Storage – Energy Storage Symposium,"* held 25 March, 2019. https://www.michigan.gov/documents/energy/Energy_Storage_Session_4-Energy_Storage_in_the_Real_World_Peretick_652341_7.pdf.

[51] Hydrostor. (2019). Hydrostor and NRStor announce completion of world's first commercial advanced-CAES facility. *Globe Newswire* online article, November 25, 2019 09:10 ET. https://www.globenewswire.com/news-release/2019/11/25/1952039/0/en/Hydrostor-and-NRStor-Announce-Completion-of-World-s-First-Commercial-Advanced-CAES-Facility.html (accessed November 2020).

[52] Mitchell, A.I. (1979). Compressed air storage. Solution Mining Research Institute (SMRI), Fall Meeting, 15 October 1994, Toronto, Canada, 8 pp.

[53] Mehta, B.R. and Spencer, D. (1988). Siting compressed air energy plants. *Tunnelling and Underground Space Technology* 3 (3): 295–299.

[54] Brokman, S., Weiner, D., and Olek, S. (1994). Salt caverns for compressed air energy storage in Israel. Solution Mining Research Institute (SMRI) Fall Meeting 1994, Hannover, Germany, 19 pp.

[55] Donadei, S. and Schneider, G.-S. (2016). Compressed air energy storage in underground formations. In: *Storing Energy, with Special Reference to Renewable Energy Sources* (ed. T.M. Letcher), pp. 113–133. Cambridge, MA, Elsevier Inc.

[56] Cavallo, A.J. (1995). High-capacity factor wind energy systems. *Journal of Solar Engineering* 117: 137–143.

[57] Ridge Energy. (2005). The economic impact of CAES on wind in TX, OK, and NM storage and grid services. Report by Ridge Energy Storage and Grid Services L.P. to Texas State Energy Conservation Office, Final Report, 27 June 2005, 114 pp. http://www.ridgeenergystorage.com/re_wind_projects-compressed2005.pdf.

[58] Cavallo, A.J. and Keck, M.B. (1995). Cost effective seasonal storage of wind energy. In: *Wind Energy SED*, vol. 16 (ed. W.D. Musial, S.M. Hock, and D.E. Berg), 119–125. American Society of Mechanical Engineers (ASME), energy sources technology conference and exhibition, Houston, TX (United States), 29 Jan–1 Feb 1995.

[59] EPRI. (2015) Compressed air energy storage newsletter. Electric Power Research Institute (EPRI), August, 2015, 10 pp.

[60] Chamisa. (2013). Overview of the Tulia CAES bulk electric storage project. http://www.slideshare.net/ptpblog/compressed-air-wind-energy-storage.

[61] CH2MHILL. (2012). Prevention of significant deterioration greenhouse gas permit application Matagorda energy center, Matagorda County, Texas. Report prepared for Prepared for APEX Matagorda Energy Center, LLC., Submitted to U.S. Environmental Protection Agency Region 6, November 2012, 68 pp.

[62] Overton, T.W. (2014). The year energy storage hit its stride. Online *PowerMag* article posted 1 May 2014. http://www.powermag.com/the-year-energy-storage-hit-its-stride/?pagenum=6.

[63] Lyman, R.J. (2014). Fuel-free geologic compressed air energy storage from renewable power. General Compression Task #7 Deliverable Report for New Technology Implementation Grant Program 582-11-13126-3225. March 25, 2014, Resubmitted April 30, 2014, 20 pp. http://tceq.com/assets/public/implementation/air/terp/ntig/prog_rpts/2014_GC_Task7.pdf.

[64] Bueckert, K. (2017). Compressed air energy storehouse approved for old Goderich salt cave. *CBC News* online article, posted 11 May, 2017 3:34 PM ET, Last Updated: May 12, 2017.

[65] Butler, C. (2019). How an old Goderich salt mine could one day save you money on your hydro bill. *CBC News* online article, posted: 24 November 2019 7:00 AM ET, last updated: 25 November 2019. https://www.cbc.ca/news/canada/london/goderich-ontario-compressed-air-energy-storage-1.5369478.

[66] Banach, A. and Klafki, M. (2012). Adiabatic CAES project – ADELE: Cavern aspects. *SMRI Fall Technical Conference* (1–2 October 2012). Bremen, Germany. 11 pp.

[67] Haughey, C. (2015). Larne CAES: A project update. Gaelectric energy storage: The missing link. Gaelectric, 2015, Company newsletter available online. http://www.gaelectric.ie/wpx/wp-content/uploads/2015/09/Gaelectric-Supplement-June-2015.pdf.

[68] EU. (2018). Project 1013 – CAES Zuidwending, NL: Compressed air energy storage. https://tyndp.entsoe.eu/tyndp2018/projects/storage_projects/1013.

[69] Storelectric. (2018). Monitoring of PCIs 2018 – Electricity. Progress report for Projects of Common Interest (PCIs) pursuant to article 5 (4) of regulation (EU) No 347/2013. Report submitted to the Agency for the Cooperation of Energy Regulators (ACER), 2nd March 2018, 23 pp.

[70] Parkes, D., Evans, D.J., Williamson, P., and Williams, J.D.O. (2018). Estimating available salt volume for potential CAES development: A case study using the Northwich Halite of the Cheshire Basin. *Journal of Energy Storage* 18: 50–61.

[71] He, W., Luo, X., Evans, D., et al. (2017). Exergy storage of compressed air in cavern and cavern volume estimation of the large-scale compressed air energy storage system. *Applied Energy* 208: 745–757.

[72] Evans, D.. Parkes, D., Dooner, M., et al. (2021). Salt Cavern Exergy Storage Capacity Potential of UK Massively Bedded Halites, Using Compressed Air Energy Storage (CAES). *Applied Science*, 11, 4728. https://doi.org/10.3390/app11114728

[73] Hämmerle, M., Haider, M., Willinger, R., et al. (2017). Saline cavern adiabatic compressed air energy storage using sand as heat storage material. *Journal of Sustainable Development of Energy, Water and Environment Systems* 5: 32–45.

[74] Pedersen, A.S., Elmegaard, B., Christensen, C.H., et al. (2013). Status and recommendations for RD&D on energy storage technologies in a Danish context. Report developed for the Danish energy research programmes and is published jointly by the Danish Energy Authority (EUDP and Green Labs DK), Energinet.dk (ForskEL and ForskVE), the Danish Council for Strategic Research and the Danish Energy Association (ELFORSK). February 2014, 136 pp. https://www.energinet.dk/SiteCollectionDocuments/Danske%20dokumenter/Forskning%20-%20PSO-projekter/RDD%20Energy%20storage_ex%20app.pdf.

[75] Alton. (2010). Compressed air energy storage. Alton Natural Gas Storage LP presentation given at Nova Scotia Energy Research and Development meeting 27 May 2010. https://novascotia.ca/nse/ea/Alton/Section1-4.pdf

[76] Renne, K. (2015) Salt caverns in eastern Alberta created to store energy. *Journal of Commerce* (online), January 29, 2015, 4 pp. https://canada.constructconnect.com/joc/news/Technology/2015/1/Salt-caverns-in-eastern-Alberta-created-to-store-energy-1005431W.

[77] Ma, H.L., Yang, C.H., Liu, J.F., and Chen, J.W. (2013). The influence of cyclic loading on deformation of rock salt. *3rd ISRM Symposium on Rock Characterisation, Modelling and Engineering Design Methods* (eds. Xia-Ting Feng, John A. Hudson, and Fei Tan), SINOROCK 2013, 18–20 June 2013, 63–68. Shanghai, China: Taylor and Francis.

[78] Zhang, G., Li, Y., Daemen, J.J.K., et al. (2014). Geotechnical feasibility analysis of Compressed Air Energy Storage (CAES) in bedded salt formations: A case study in Huai'an City, China. *Rock Mechanics and Rock Engineering* 48: 1–17.

[79] Kepplinger, J. and Donadei, S. (2011). Large-scale energy storage in geological formations. *6th Dubrovnik Conference on Sustainable Development of Energy, Water and Environment Systems*, Dubrovnik, 13 pp.

[80] Dresser Rand. (2013). Dresser-rand awarded Compressed Air Energy Storage (CAES) project. Dresser Rand Press Release, 1 July, 2013. http://investor.dresser-rand.com/releasedetail.cfm?ReleaseID=774720.

[81] Gilhaus, A. (2007). Natural gas storage in salt caverns – Present status, developments and future trends in Europe. *SMRI Spring Conference* (29 April–2 May, 2007). Basel, Switzerland. 19 pp.

[82] Bérest, P., Brouard, B., Djakeun-Djizanne, H., and Hévin, G. (2011). Thermomechanical aspects of high frequency cycling in salt storage caverns. *International Gas Union Research Conference 2011* (October 2011). Seoul, South Korea. 23 pp.

[83] Spreckels, H. and Crotogino, F. (2002). Salt caverns for peak shaving – Reservoirs for seasonal balance? New market requirements and appropriate storage solutions. SMRI Fall 2002 Meeting, 6–9 October 2002, Bad Ischl, Austria, 15 pp.

[84] Brown Boveri. (1994). Huntorf air storage gas turbine power plant. BBC Brown Boveri publication Energy Supply, Number D GK 90202 E, 14 pp. http://www.solarplan.org/Research/BBC_Huntorf_engl.pdf.

[85] Mehta, B.R. (1991). First U.S. compressed-air energy storage plant status. Solution Mining Research Institute, Fall Meeting, Las Vegas, NV, 28 October 1991, 35 pp.

[86] Grow, G.C. (1976). The storage of natural gas. *American Chemical Society* 19: 99–134.

[87] EPRI. (2015). Compressed air energy storage newsletter. Electric Power Research Institute (EPRI), January 2015, 6 pp.

[88] EPRI. (2011). Compressed air energy storage newsletter. Electric Power Research Institute (EPRI), October 2011, 10 pp.

[89] EPRI. (2012). Compressed air energy storage newsletter. Electric Power Research Institute (EPRI), January 2012, 9 pp.

[90] NETL. (2014). Final environmental assessment for the Pacific Gas and Electric Company (PG&E) Compressed Air Energy Storage (CAES) compression testing phase project, San Joaquin County, California. National Energy Technology Laboratory (NETL), U.S. Department of Energy report, May 2014, DOE/EA-1752, 74 pp. https://www.netl.doe.gov/File%20Library/Library/Environmental%20Assessments/PG-E_CAES_Concurrence_Final-EA_04-30-2014.pdf.

[91] Bender, D., Byrne, R., and Borneo, D. (2015). ARRA energy storage demonstration projects: Lessons learned and recommendations: A study for the DOE energy storage systems program. Report by Sandia National Laboratories, June 2015, SAND2015-5242, 64 pp.

[92] Hessami, M.-A. and Bowly, D.R. (2011). Economic feasibility and optimisation of an energy storage system for Portland Wind Farm (Victoria, Australia). *Applied Energy* 88: 2755–2763.

[93] EPRI. (2013). Compressed air energy storage newsletter. Electric Power Research Institute (EPRI), April 2013, 8 pp.

[94] EPRI. (2012). Compressed air energy storage newsletter. Electric Power Research Institute (EPRI), October 2012, 8 pp.

[95] EPRI. (2013) Compressed air energy storage newsletter. Electric Power Research Institute (EPRI), January 2013, 8 pp.

[96] EPRI. (2013). Compressed air energy storage newsletter. Electric Power Research Institute (EPRI), October 2013, 8 pp.

[97] Grubelich, M.C., Bauer, S.J., and Cooper, P.W. (2011). Potential hazards of compressed air energy storage in depleted natural gas reservoirs. Sandia National Laboratories, Sandia Report, number SAND2011-5930, 23 pp.

[98] Buschbach, T.C. and Bond, D.C. (1967). Underground storage of natural gas in Illinois – 1967. Illinois Petroleum 86, Illinois State Geological Survey, Department of Registration and Education, Urbana, IL, 60 pp.

[99] Lang, W.J. (1976). Compressed air storage considerations for reef-type aquifer structures in Indiana. *Proceedings of the Workshop on Compressed Air Energy Storage System*, Airlie House, VA (18–19 December 1975). ERDA-76-124, 258–277.

[100] Skjei, R.E., King, M.J., and Stewart, D.H. (1980) Energy Storage in Aquifers. In: *Subsurface Space (3 volumes), Volume 2 – Environmental Protection, Low Cost Storage Energy Savings. International Society for Rock Mechanics (ISRM) International Symposium – Rockstore 80* (ed. M. Bergman), (23–27 June), pp. 721–729. Stockholm, Sweden: Pergamon Press Ltd.

[101] McKinley, T., Kirkland, J., and Wilson, D. (1992). Evaluation of Gippsland Basin onshore structures for compressed air energy storage. *Gippsland Basin Symposium*, Melbourne (22–23 June 1992), 289–298.

[102] Schulte, R.H., Critelli, N., Jr, Holst, K., and Huff, G. (2012). Lessons from Iowa: Development of a 270 megawatt compressed air energy storage project in midwest independent system operator. A study for the DOE energy storage systems program. Report prepared by Sandia National Laboratories, Sandia Report, Report Number: SAND2012-0388, 97 pp.

[103] Sandia. (2010). ARRA energy storage demonstrations. https://www.sandia.gov/ess-ssl/docs/ARRA_StorDemos_10-6-10.pdf.

[104] Mouli-Castillo, J., Wilkinson, M., Mignard, D., et al. (2019). Inter-seasonal compressed air energy storage using saline aquifers. *Nature Energy* 4: 131–139.

[105] Sopher, D., Juhlin, C., Levendal, T., et al. (2019). Evaluation of the subsurface compressed air energy storage (CAES) potential on Gotland, Sweden. *Environmental Earth Sciences* 78: 197–213.

[106] Denholm, P. and Holloway, T. (2005). Improved accounting of emissions from utility energy storage system operation. *Environmental Science and Technology* 39: 9016–9022.

[107] Oldenburg, C. and Pan, L. (2013). Porous media compressed-air energy storage (PM-CAES): Theory and simulation of the coupled wellbore–reservoir system. *Transport in Porous Media* 97: 201–221.

[108] Dinelli, G., Lozza, G., and Macchi, E. (1988). A feasibility study of CAES plants for peak load generation. Proceedings 23rd Intersociety Energy Conversion Engineering Conf. (IECEC), Denver, CO. American Society of Mechanical Engineers (ASME), pp. 417–424.

[109] Reidel, S.P., Johnson, V.G., and Spane, F.A. (2002). Natural gas storage in basalt aquifers of the Columbia Basin, Pacific Northwest USA: A guide to site characterization. Report prepared for the U.S. Department of Energy by Pacific Northwest National Laboratory (PNNL) Richland, Washington 99352, Report Number PNNL-13962, August 2002, 277 pp. http://www.pnl.gov/main/publications/external/technical_reports/PNNL-13962.pdf.

[110] McGrail, B.P., Cabe, J.E., Davidson, C.L., et al. (2013). Techno-economic performance evaluation of compressed air energy storage in the Pacific Northwest. Report prepared for US Department of Energy and Bonneville Power Administration under Intergovernmental Contract #54177 by Pacific Northwest National Laboratory (PNNL), Richland, Washington 99352, Report Number PNNL-22235, February 2013, 140 pp.

[111] Wettlegren, G. (1979). Underground oil storage in disused mines. *Underground Space* 3: 9–17.

[112] Lindblom, U.E. (1994). History and present status of hydrocarbon storage in excavated rock caverns. *Rock Mechanics in Petroleum Engineering*. Proceedings, Eurock 94, Delft, Netherlands, SPE/ISRM International Conference, 29–31 August 1994. SPE/ISRM Rock Mechanics, Eurock'94, 1994 Balkema, Rotterdam, Delft, The Netherlands, pp. 663–670.

[113] Schultz, K. (1998). Gas storage at the abandoned Leyden coal mine near Denver, Colorado. EPA report, contract 68-W5-0018, prepared by Raven Ridge Resources Inc., 12 pp.

[114] Piessens, K. and Dusar, M. (2004). CO_2 sequestration in abandoned coalmines in Belgium. *Geologica Belgica* 7: 165–180.

[115] SBN. (2013). Schachtbau Nordhausen. Schachtbau Nordhausen (SBN) company brochure, 36 pp. http://www.schachtbau.de/export/sites/www.schachtbau.de/pdf/prospekte/SBN-Brochure_Web_en.pdf.

[116] SEGA. (2010). Sega Inc. Investigates Compressed Air Energy Storage (CAES) in salt mine. Online article published 24 April 2010. http://www.segainc.com/04/sega-inc-investigates-compressed-air-energy-storage-caes-in-salt-mine.

[117] Saito, S. (1991). Mining methods and the use of inner space at Kamioka mine. *Rock Mechanics in Japan* 4: 63–66.

[118] Väätäinen, A., Särkkä, P., Sipilä, K., and Wistbacka, M. (1995). Compressed air energy storage in mine: Prefeasibility study. *8th ISRM Congress* (25–29 September). Tokyo, Japan. International Society for Rock Mechanics and Rock Engineering, 485–489.

[119] NYSEG. (2009). Compressed air energy storage Engineering and economic study. New York State Electric and Gas (NYSEG) final report prepared for the New York State Energy Research and Development Authority (NYSERDA), No. 10-09, December 2009, 250 pp.

[120] Fosnacht, D.R., Wilson, E.J., Marr, J.D., et al. (2015). Compressed air energy storage (CAES) in northern Minnesota using underground mine workings and above ground features. Natural Resources Research Institute, University of Minnesota Duluth, Technical Report NRRI/TR-2015/54, 478 pp.

[121] Beck, H.P. and Franz, O.T. (2010). Energy storage in abandoned mines – a method to stabilise the German power grid. In: *Underground Storage of CO_2 and Energy* (ed. M.Z. Hou, H. Xie, and J.S. Yoon), 261–269. Proceedings of Second Sino-German Energy Conference, China, July 2010 CRC Press, Taylor & Francis Group.

[122] Sandia. (2001). Sandia assists with mine assessment: Solution to some of country's energy woes might be little more than hot air. Sandia National Laboratories news release, 24 April 2001. https://www.sandia.gov/media/NewsRel/NR2001/norton.htm.

[123] Funk, J. (2013). FirstEnergy postpones project to generate electricity with compressed air. Online article, posted 5 July 2013 at 1:44 PM. https://www.cleveland.com/business/index.ssf/2013/07/firstenergy_postpones_project.html.

[124] Moutoux, R. and Barnes, F. (2007). Wind integrated compressed air energy storage in Colorado. http://www.colorado.edu/engineering/energystorage/files/EESAT2007/EESAT_Wind_Integrated_CAES_Paper.pdf.

[125] ARENA. (2019). South Australian zinc mine to be converted into Australia's first compressed air facility for renewable energy storage. Australian Renewable Energy Agency (ARENA) Media Release, 8 February 2019, 2 pp. https://arena.gov.au/assets/2019/02/ARENA-Media-Release_Hydrostor-compressed-air-pilot-project-at-South-Australian-disused-mine-080219.pdf.

[126] Kingsland, P. (2017). Could air be the next big battery breakthrough? Power-Technology online article, 23 July 2017. https://www.power-technology.com/features/featurecould-air-be-the-next-big-battery-breakthrough-5864457.

[127] Molecke, M.A. (2000). Final state of the Strategic Petroleum Reserve (SPR) Weeks Island Mine. Sandia National Laboratories Report, SAND2000-0400, 63 pp.

[128] Coates, G.K., Lee, C.A., McClain, W.C., and Senseny, P.E. (1981). Failure of man-made cavities in salt and surface subsidence due to mining. Sandia National Laboratories Report SAND81-7145, 136 pp.

[129] Johnson, K.S. (2005). Salt dissolution and subsidence or collapse caused by human activities. In: *Humans as Geologic Agents: Boulder, Colorado, Geological Society of America Reviews in Engineering Geology*, vol. XVI (ed. J. Ehlen, W.E. Haneberg, and

R.A. Larson), 101–110. The Geological Society of America. https://doi.org/10.1130/2005.4016(09)

[130] van Sambeek, L. (2009). Natural compressed air storage: A catastrophe at a Kansas salt mine. *Proceedings 9th International Symposium on Salt*, vol. 1, 621–632. Beijing: Zuoliang Shaed.

[131] Passaris, E.K.S. (1979). The rheological behaviour of rock salt as determined in an in-situ pressurised test cavity. *Proceedings 4th Congress International Society of Rock Mechanics*, Montreux, 1, 257–264.

[132] Thoms, R.L. and Gehle, R.M. (1994). The Jefferson Island Mine flooding revisited. Solution Mining Research Institute (SMRI), Spring Meeting, Houston, TX, 24-27 April, 20pp.

[133] Johansson, S. and Lahtinen, R. (1976). Oil storage in rock caverns in Finland. *Tunnelling 76*, 13: 41–58.

[134] Maury, V. (1977). An example of underground storage in soft rock (chalk). *Storage in Excavated Rock Caverns. Rockstore 77, Proceedings of the First International Symposium* (ed. M. Bergman), Stockholm (5–8 September, 1977), 681–689.

[135] Réveillère, A. and Londe, L. (2017). Compressed air energy storage: A new beginning? *Solution Mining Research Institute (SMRI) Fall 2017 Technical Conference Münster*, Germany (25–26 September 2017), 12 pp.

[136] Trotter, J.T., Thompson, D.M.T., and Paterson, T.J.M. (1985). First mined underground storage in Great Britain. *Tunnelling'85: Proceedings of the Fourth International Symposium* (ed. M.J. Jones), Brighton, England (10–15 March 1985). Institution of Mining and Metallurgy, London, 3–12.

[137] Sturk, R. and Stille, H. (1995). Design and excavation of rock caverns for fuel storage – a case study from Zimbabwe. *Tunnels and Deep Space* 10: 193–201.

[138] De Ambrosis, L.P. and Kotze, G.P. (2004). Stress induced roof collapses during construction of the Sydney LPG storage cavern. *Proceedings of the 9th Australia New Zealand Conference on Geomehcanics*, Auckland (8–11 February 2004), 1, 159–165.

[139] You, T. and Amantini, E. (2010). Forty-five years of geotechnical engineering feedback in underground caverns. Geostock paper available online, 4 pp. http://www.geostockgroup.com/uploads/tx_templavoila/2010_Forty_five_years_01.pdf (accessed 17 November 2015).

[140] Zhao, J., Ng, W.L., Cai, J.G., et al. (2004). Feasibility of underground hydrocarbon storage caverns at Jurong Island. *Tunnelling and Underground Space Technology* 19: 364.

[141] Lu, M. (2010). Rock engineering problems related to underground hydrocarbon Storage. *Journal of Rock Mechanics and Geotechnical Engineering* 2: 289–297.

[142] Lindblom, U.E. (1985). A conceptual design for compressed hydrogen storage in mined caverns. *International Journal of Hydrogen Energy* 10: 667–675.

[143] Berg, N. and Norén, D. (1969). Compressed air power plants – Some views on underground compressed air storage. *Svenska Kraftverksföreningens Publikationer* 536 (1969: 6): 87–111.

[144] Olsson, E.K.A. (1970). Air storage power plant. *Mechanical Engineering* 92 (11): 20–24.

[145] Robertsson, H. (1975). Stal-Laval's activities in the field of compressed air storage. *Proceedings of the Workshop on Compressed Air Energy Storage System*. Airlie House, Virginia, 18 and 19 December, 1975. ERDA-76-124, 73–88.

[146] Jeffs, E. (1973). *Air storage: The alternative peak power source*. Energy International – Miller Freeman Publications, 23–25.

[147] Pellin, J. (1980). Actual C.A.E.S. technology in Europe. In: *Energy Storage: A Vital Element in Mankind's Quest for Survival and Progress*, (ed. J. Silverman), pp. 472–484. *Proceedings*. 1st International Assembly: Papers, Pergamon Press.

[148] Haydock, J.L. and Milne, I.A. (1976). Siting and excavation of cavities in hard rock for compressed air storage. *Proceedings of the Workshop on Compressed Air Energy Storage System*. Airlie House, Virginia, 18 and 19 December 1975. ERDA-76-124, 199–227.

[149] Loane, E.S. and Mitchell, W.S. (1976). Mt Hope studies of compressed air storage. *Proceedings of the Workshop on Compressed Air Energy Storage System*. Airlie House, Virginia, 18 and 19 December 1975. ERDA-76-124, 238–257.

[150] Willett, C. and Curtis, R.H. (1980). Hard rock caverns for underground pumped hydroelectric and compressed air energy storage. *International Society for Rock Mechanics and Rock Engineering, ISRM International Symposium – Rockstore 80*, Stockholm, Sweden (23–27 June 1980), 779–788.

[151] Allen, R.D., Doherty, T.J., and Fossum, A.F. (1982). Geotechnical factors and guidelines for storage of compressed air in excavated hard rock caverns. Report prepared by Pacific Northwest Laboratory for Battelle and the U.S. Department of Energy under contract DE-AC06-76RLO 1830. Report number PNL-4180, 163 pp.

[152] Swenson, E. and Potashnik, B. (1994). Evaluation of benefits and identification of sites for a CAES plant in New York State. Report prepared by Energy Storage and Power Consultants for the Electric Power Research Institute, Final Report, August 1994, 218 pp.

[153] Berk, B., Hurwitch, J.W., and Schuller, C.R. (1983). CAES and UPH technologies: An uncertain future. *Underground Space* 7: 342–346.

[154] Mailhe, P., Comes, G., and Perami, R. (1977). Geological and geotechnical process for the siting of a hydropneumatic pumped storage plant in Britanny (France). *Storage in Excavated Rock Caverns: Rockstore 77, Proceedings of the First International Symposium* (ed. M. Bergman), (5–8 September), vol. 3, 495–500. Stockholm: Pergamon Press.

[155] Wittke, W., Pierau, B., and Schetelig, K. (1977). Planning of a compressed-air pumped storage scheme at Vianden/Luxembourg. *Rockstore 77, Proceedings of the First International Symposium*, Stockholm (5–8 September 1977), 367–376.

[156] Lehmann, J. (1981) Air storage gas turbine power plants a major distribution for energy. *International conference on Energy Storage*, Brighton, UK, 29 April–1 May 1981, Paper R3, 327–336.

[157] Lindblom, U.E. and Söder, C.O. (1992). Pilot tests of compressed air containment in a rock cavern. *Second International Conference on Compressed Air Energy Storage*, 8.3-3–8.3-18.

[158] Hayashi, M. (1991) Rock mechanics of compressed air energy storage. *Rock Mechanics in Japan: Volume VI, Japanese Committee for ISRM (JCISRM)* (ed. H. Koide et al.), 50–57.

[159] Nakagawa, K., Shidahara, T., and Ohyama, T. (1991). Field tests on air tightness of rock for compressed air storage. *Proceedings 7th International Society for Rock Mechanics (ISRM) Congress*, A. A. Balkemea, vol. l, 135–138.

[160] Shidahara, T., Oyama, T., and Nakagawa, K. (1993). The hydrogeology of granitic rocks in deep boreholes used for compressed air storage. *Engineering Geology* 35: 207–213.

[161] Shidahara, T., Oyama, T., Nakagawa, K., et al. (2000). Geotechnical evaluation of a conglomerate for compressed air energy storage: The influence of the sedimentary cycle and filling minerals in the rock matrix. *Engineering Geology* 56: 125–135.

[162] Bauer, S.J., Gaither, K.N., Webb, S.W., and Nelson, C. (2012). Compressed air energy storage in hard rock feasibility study. SANDIA Report, prepared by Sandia National Laboratories Albuquerque, New Mexico 87185 and Livermore, California 94550. Report Number SAND2012-0540, February 2012, 139 pp.

[163] Ramadan, O., Omer, S., Jradi, M., et al. (2015). Analysis of compressed air energy storage for large-scale wind energy in Suez, Egypt. *International Journal of Low-Carbon Technologies* 11: 476–488.

[164] Zhuang, X., Huang, R., Liang, C., and Rabczuk, T. (2014). A coupled thermo-hydro-mechanical model of jointed hard rock for compressed air energy storage. *Mathematical Problems in Engineering* 2014: 11 pp.

[165] Geissbühler, L., Becattini, V., Zanaganeh, G., et al. (2018). Pilot-scale demonstration of advanced adiabatic compressed air energy storage, part 1: Plant description and tests with sensible thermal-energy storage. *Journal of Energy Storage* 17: 129–139.

[166] Morfeldt, C.-O. (1983). Storage of petroleum products in man-made caverns in Sweden. *Bulletin of the International Association Engineering Geology* 28: 17–30.

[167] Haug, S.M. and Broch, E. (2007). 1. Storage of oil and gas in rock caverns: History and development. Underground constructions for the Norwegian oil and gas industry. Publication of the Norsk Forening For Fjellsprengningsteknikk (The Norwegian Tunnelling Society: NFF), Publication No. 16, p. 9.

[168] EPA. (2017). What is the red hill bulk fuel storage facility? United States Environmental Protection Agency (EPA). https://www.epa.gov/red-hill/what-red-hill-bulk-fuel-storage-facility.

[169] Tengborg, P., Johansson, J., and Durup, J.G. (2014). Storage of highly compressed gases in underground lined rock caverns – more than 10 years of experience. *Proceedings of the World Tunnel Congress 2014 – Tunnels for a better Life*, Foz do Iguaçu, Brazil, 7 pp.

[170] Glamheden, R. and Curtis, P. (2006). Excavation of a cavern for high-pressure storage of natural gas. *Tunnelling and Underground Space Technology* 21: 56–67.

[171] Hayashi, M. (1989). Submerged geo-spacing for compressed air energy storage in deep soft rock: Provisional study of feasibility related to urban siting in future. In: *Storage of Gases in Rock Caverns* (ed. J. Olsen and B. Nilsen), 377–387. Rotterdam: Balkema.

[172] Hayashi, M. (1990). Feasibility study of submerged geo-spacing in deep soft rock aiming at urban CAES. *Proceedings Japan Society of Civil Engineers* 418 (III-13): 17–44.

[173] Yamashita, Y. and Takano, K. (1998). Practical tests for a new urban CAES tank. Obayashi Corporation Technical Research Institute, No.57, 1998, 73–76. https://www.obayashi.co.jp/technology/shoho/057/1998_057_13.pdf.

[174] Ishihata, T. (1997). Underground compressed air energy storage facility for CAES-GT power plant utilising an airtight lining. *International Society for Rock Mechanics* 5: 17–21.

[175] Takahashi, K., Yasuda, T., Endoh, M., and Kurosaki, M. (2002). Application of dynamic simulation to CAES G/T control system development. *Proceedings of American Society of Mechanical Engineers (ASME) Turbo Expo 2002: Power for Land, Sea, and Air Volume 2, Turbo Expo 2002, Parts A and B*, Amsterdam, The Netherlands (3–6 June 2002). Paper No. GT2002-30036, 153–161. http://proceedings.asmedigitalcollection.asme.org/proceeding.aspx?articleid=1570685.

[176] Terashita, F., Takagi, S., Kohjiya, S., and Naito, Y. (2005). Airtight butyl rubber under high pressures in the storage tank of CAES-G/T system power plant. *Journal of Applied Polymer Science* 95: 173–177.

[177] Okuno, T., Wakabayashi, N., Niimi, K., et al. (2009). Advanced natural gas storage system and verification tests of lined rock cavern – ANGAS project in Japan. *International Journal of the Japanese Committee of Rock Mechanics* 5: 95–102.

[178] Song, W.K., Ryu, D.W., and Lee, Y.K. (2012). Stability analysis of concrete plugs in a pilot cavern for compressed air energy storage. In: *Harmonising Rock Engineering and the Environment* (ed. Q. Qian and Y. Zhou), 1813–1816. London: Taylor and Francis Group. http://wenku.baidu.com/view/b4cd97ef102de2bd960588f9.html.

[179] Kim, H.M., Rutqvist, J., and Choi, B.H. (2012). Feasibility analysis of underground compressed air energy storage in lined rock caverns using the TOUGH-FLAC simulator. *TOUGH Symposium 2012*, Berkeley, CA (17–19 September 2012), 464–472.

[180] Kim, H.M., Rutqvist, J., Jeong, J.-H., et al. (2012). Characterizing excavation damaged zone and stability of pressurized lined rock caverns for underground compressed air energy storage. *Rock Mechanics and Rock Engineering* 46 (5): 1113–1124.

[181] Kim, H.M., Rutqvist, J., Ryu, D.W., et al. (2012). Exploring the concept of compressed air energy storage (CAES) in lined rock caverns at shallow depth: A modelling study of air tightness and energy balance. *Applied Energy* 92: 653–667.

[182] Rutqvist, J., Synn, J.-H., Kim, H.-M., et al. (2012). Modelling of coupled thermodynamic and geomechanical performance of underground compressed air energy storage in lined rock caverns. *International Journal of Rock Mechanics and Mining Sciences* 52: 71–81.

[183] Zhou, S.-W., Xia, -C.-C., Du, S.-D., et al. (2015). An analytical solution for mechanical responses induced by temperature and air pressure in a lined rock cavern for underground compressed air energy storage. *Rock Mechanics and Rock Engineering* 48: 749–770.

[184] Perazzelli, P. and Anagnostou, G. (2016). Design issues for compressed air energy storage in sealed underground cavities. *Journal of Rock Mechanics and Geotechnical Engineering* 8: 314–328.

19

Compressed Air Energy Storage in Aquifer and Depleted Gas Storage Reservoirs

Michael J. King, and George Moridis

The Hydrodynamics Group LLC Edmonds Washington USA

19.1 Introduction

Compressed air energy storage (CAES) is a process for storing and delivering energy as electricity. A CAES facility consists of an electric generation system and an energy storage system (Figure 19.1). Off-peak electricity at night is stored as air pressure in a geological storage vessel. During intermediate and peak demand periods, the compressed air is released from the pressurized energy storage system, heated by the combustion of natural gas, and used to drive high-efficiency turbines to produce electricity.

CAES technology requires the matching of operating pressure and flow rate of a geological air storage vessel to the pressure and flow requirements of traditional turbo-generator equipment. The current approach to developing a CAES facility is to find a geological air storage vessel with the depth, volume, and hydraulic characteristics that match the pressure and air mass flow requirements of commercially available turbo-generator equipment, rather than modifying the design of electric generating equipment for every potential air storage structure. Major mechanical components of the CAES system are the motor-compressor train(s) and the CAES turbine-generator train (Figure 19.1).

The energy storage system stores potential energy in the form of pressure; it also stores the fluid mass required by the turbine and the oxygen needed for combustion in the turbine. The stored energy and mass are used to turn the electricity-generating turbine. Key elements of the energy storage system are the air storage vessel and air injection and withdrawal wells (Figure 19.1). Because of the high pressure and large air mass required to operate the turbo-machinery, earth-based storage vessels are the only viable air storage option. Earth-based structures suitable for service as air storage vessels include (1) aquifer geologic structures, (2) depleted natural gas reservoirs, (3) solution mined salt caverns, and (4) abandoned mines. The focus of this chapter is the design analysis of both an aquifer geological structure and a depleted gas reservoir to match the CAES turbo-machinery's operating mass flow rate and pressure requirements.

19.2 History of CAES Development

The concept of CAES was developed in the early 1970s. Drs D.L. Katz and E.R. Lady at the University of Michigan published one of the original works on CAES in 1976,

Advances in Energy Storage: Latest Developments from R&D to the Market, First Edition.
Edited by Andreas Hauer.
© 2022 John Wiley & Sons, Ltd. Published 2022 by John Wiley & Sons, Ltd.

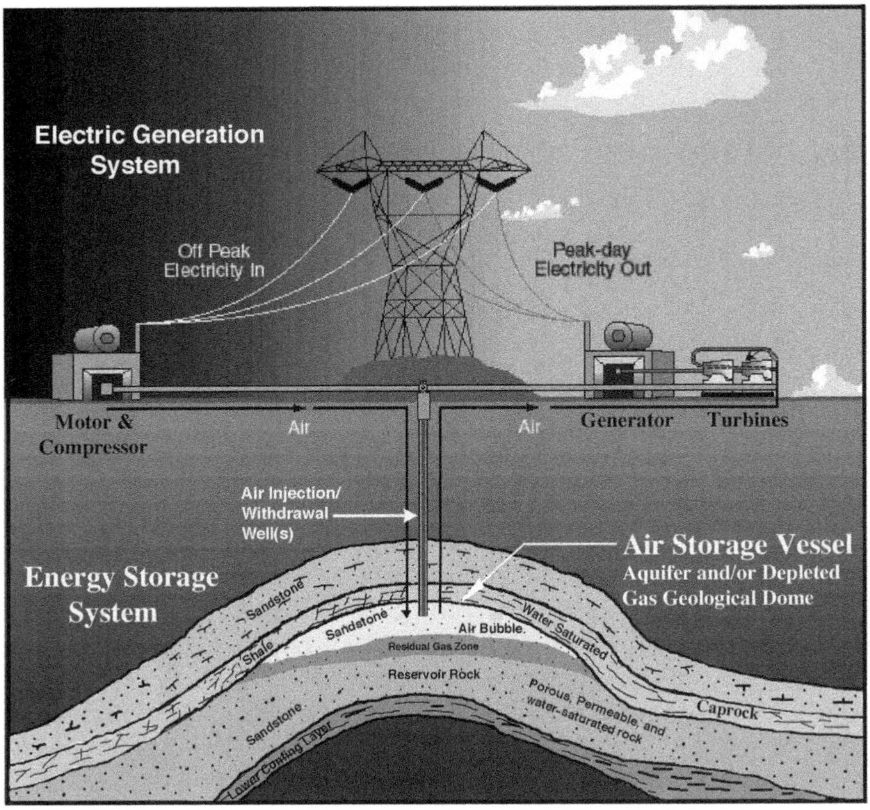

Figure 19.1 Elements of a CAES storage system.

entitled *Compressed Air Storage for Electric Power Generation* [1]. In their book, they applied natural gas storage principles to CAES.

The first commercial CAES facility was successfully constructed in Huntorf, Germany in 1978 [2]. The original Huntorf CAES plant was 290 megawatts (MW), and was operated by E.N. Kraftwerke utilizing a solution-mined salt cavity. Subsequently, the McIntosh 110 MW CAES power plant was constructed in 1991 for Alabama Electric Cooperative (AEC) at McIntosh, Alabama, US. Like the Huntorf CAES facility, AEC used a solution-mined salt cavity as the air storage vessel.

The US Department of Energy and the Electric Power Research Institute (EPRI) conducted an air injection field experiment at the Pittsfield structure in Illinois in 1982 to investigate the use of an aquifer air storage system. The experiment demonstrated the technical feasibility of air storage in an aquifer (water-bearing) geological structure. The results of this research were published by EPRI [3] in a report available to EPRI member utilities. Pacific Gas and Electric conducted an air injection test in a depleted gas field in California from 2015 to 2016. This test proved the viability of CAES in a depleted gas field.

A number of CAES siting and development projects have been conducted since the late 1980s. CAES projects in California, Iowa, Montana, Nebraska, Texas, and Alberta, Canada were going to use aquifer and depleted gas field storage vessels. A CAES project in Norton, Ohio was going to use an abandoned limestone mine. Active CAES projects under development include the use of solution-mined salt caverns in Texas,

Utah, and Larne in Northern Ireland. M.J. King and M.J. McGill published an article that provides an overview of CAES technology and the market in the 2009 *Encyclopedia of Energy Engineering Technology* [4] .

19.3 Power Train Requirements

As previously stated, CAES technology requires the matching of operating pressure and flow rates of a natural air storage vessel to the pressure and flow requirements of the traditional turbo-generator equipment. A limited number of electric power plant manufacturing companies have invested in the design of CAES power plant facilities. Dresser-Rand, a Siemens Business, designed and constructed the McIntosh CAES power plant and is developing CAES power plant designs for multiple CAES projects. Dresser-Rand's CAES turbo-machinery design requirements for their 135 MW and 160 MW high pressure CAES power plant units are used for the energy storage system design analysis discussed in this chapter.

The two key criteria for the operation of turbo-generator machinery are air mass flow rate and flowing air pressure. An air mass flow rate of 400 pounds per second (#/sec) is needed to operate the 135 MW and 160 MW Dresser-Rand CAES turbo-generator units. A flowing air pressure of 920 pounds per square inch absolute (psia) is required for the 135 MW unit, and a pressure of 2150 psia is needed for the 160 MW unit.

19.3.1 Required Air Mass Flow Rate

The design of the required air flow rate for the storage reservoir is defined in millions of standard cubic feet per day (MMscf/d) versus an air mass flow rate defined in #/sec. The conversion of 400 #/sec to MMscf/d is as follows:

1. **Required Air Flow Rate (*M*) (one pound mass (lbm) = one pound (lb) = 0.45359 kilogram (kg))**
   Minimum Turbine Inlet Air Flow Rate = 9600 lbs/MW/hour (hr) or 400 #/sec.
   400 #/sec = 181.44 kg/sec.
   Total Daily Flow Rate = 181.44 kg/sec x 16 hr run time/day x 3600 seconds (sec).
   $M = 10.45 \times 10^6$ kg/16 hr run time
   Or
   Minimum Turbine Inlet Air Flow Rate: 400 lbm/sec. = 1,440,000 lbm/hr.
   = 181.44 kg/sec = 653 kg/hr
   Total Daily Air Mass Flow Rate (16 hrs) = 23.04×10^6 lbm/day
   = 10.45×10^6 kg/day
   Gas Density (ρ_{air})
   a. Assume Ideal Gas Law Applies
      $\rho_{air} = P \times 28.97/RT$ Where: ρ_{air} = density of air, kg/meters^3 (m^3)
      P = pressure in Pascal (Pa)
      R = air constant (8314)
      T = temperature (Absolute-Kelvin (K))
   b. Pressure, P
      Assume P = 830 psig (pounds per square in gauge) or 844.7 psia or 57.46 atmospheres = 5.82×10^6 Pa

c. Temperature, T

Assume $T = 29.11^\circ$ Centigrade (C) = 302.25° K or 110° Fahrenheit (F)

Then $\rho_{air} = P \times 28.97/RT = (5.82 \times 10^6 \times 28.9)/(8314 \times 302.25)$

$\rho_{air} = 66.93\ kg/m^3$

2. **Required Air Volume (V) and Reservoir Air Flow Rate**

   $V = M/\rho_{air} = 10.45 \times 10^6$ kg/16 hr run time/66.93 kg/m^3

   $V = 1.56 \times 10^5\ m^3$ or 5.51×10^6 feet^3 (ft^3)

   Or 5.51 MMscf/16 hr run time or 8.27 MMscf/day flow rate (24 hours)

3. **To convert the volumetric rate Q_V in MMscf/d (air production units) to the mass rate Q_M in kg/second (sec) (units used by the compressor):**

   Multiply Q_V by the following factors:

   (1) 1/86, 400 (conversion from per-day to per-sec)

   (2) 0.0283 (conversion from ft^3 to m^3)

   (3) 1.1857 (the density of air at standard conditions)

   Thus, the cumulative conversion factor is:

   0.0283*1.1857/86 400 = 3.884e-7

4. **Required Flow Rate in equivalent airflow rate at standard cubic feet units is:**

   $Q_V = Q_M/3.884e\text{-}7$

   $Q_V = 181.44$ kg/sec/3.884e-7

   $Q_V = 467$ MMscf/d or approximately 0.5 billion cubic feet (Bcf) per day

An approximate 0.5 Bcf per day air flow rate (rate for a single CAES power train unit) from an aquifer or depleted gas reservoir is significant relative to historical natural gas storage production rates from high deliverability gas fields. A number of CAES projects intend to use either two 135 MW or two 160 MW in a single air storage vessel. In those cases, the required airflow rate will double to 1 Bcf per day. To support a 0.5 Bcf per day airflow rate, an air storage bubble containing several Bcf of air and multiple air injection/withdrawal wells will be required.

19.3.2 Required Flowing Air Pressure

Dresser-Rand's CAES power plant equipment can operate over a range of operating pressures to the turbine inlet for power outputs from 60% to 100% gross power output. As stated earlier, the minimum required air pressure at the turbine for 100% gross power output is 920 psia for the 135 MW unit, and 2150 psia for the 160 MW unit.

19.4 How Does a CAES Energy Storage System Work? Matching the Storage System to CAES Power Train Requirements

CAES systems in depleted gas storage reservoirs are feasible because we know that natural gas has been contained in deep geological structures for millions of years. We also know it is feasible in aquifer storage structures due to more than 70 years of experience using aquifer geologic structures and depleted gas storage reservoirs for seasonal natural gas storage service. However, the challenge lies in the matching of the energy storage system to the CAES power train requirements defined earlier.

The goals of any air storage vessel selection and characterization study are to confirm the stability of the vessel at the anticipated air storage pressures, and to determine (and quantify, if possible) the potential for air migration from the vessel. The design and operation of an underground air storage system is first based on the concept of multiple barriers to air migration from the structure. The stored air pressures must not compromise the stability or integrity of the geological vessel by creating potential pathways (fractures) for air migration, and/or the loss of storage space due to collapse or chemical reactions. Design is also based on reservoir engineering hydraulic flow principles.

The key elements of an air storage facility are (1) a geological containment structure, (2) an air medium (geological formation), (3) a system of injection and withdrawal wells, and (4) surface compression. Figure 19.2 provides a geological cross-section through the crest of a CAES aquifer structure. The CAES aquifer structure system consists of a porous medium (e.g. sandstone or porous carbonate rocks) and an anticlinal/dome geological structure that forms the air containment vessel (Figure 19.2). The stored air is contained within the confined limits or "air spill point" of the dome structure. The air storage vessel is developed by injecting air through a system of wells into the aquifer geological structure to create an air bubble to support the air mass flow rate and pressure required to operate the turbo-machinery.

The geological structure is confined by a relatively impermeable and water saturated caprock that is typically a shale or carbonate type rock (Figure 19.2). The air storage system is further confined by approximately 3000 ft of rock overburden.

The hydraulic barriers of the CAES aquifer storage system are a function of (1) the lithostatic rock pressures within the geological structure rocks, and (2) the hydrostatic

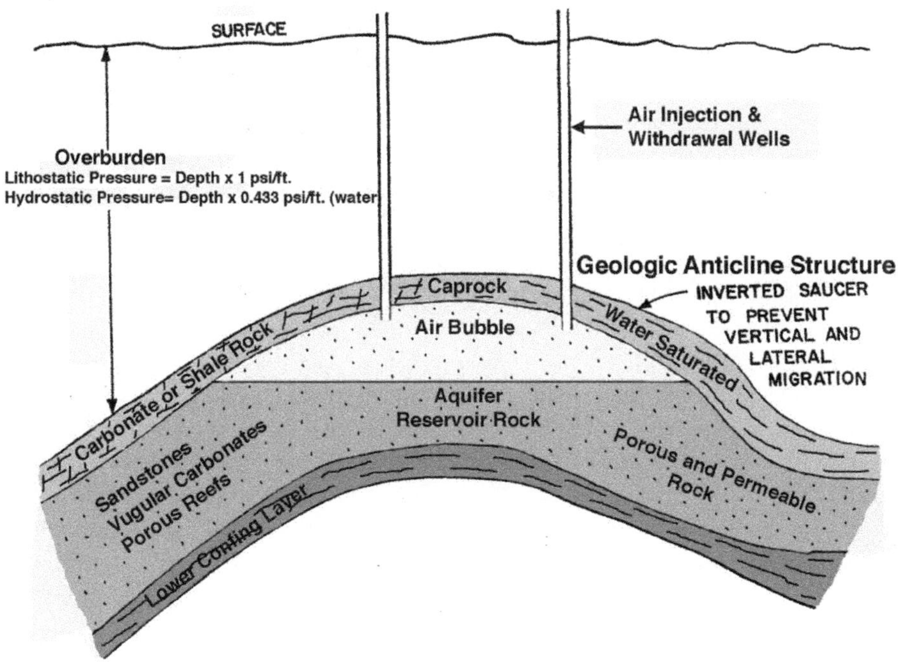

Figure 19.2 Sketch of a typical CAES aquifer geologic structure.

fluid pressures within both the caprock and permeable air storage rock medium. The stored air bubble is contained within the storage system by limiting the air pressure in the bubble to between 0.65 to 0.8% less than the lithostatic pressure within the caprock. Air is typically stored in the storage system at pressures that cycle above and below the hydrostatic pressure for relatively equal periods. The hydrostatic pressure provides a driving force within the aquifer that counters the migration of air laterally into the regional aquifer system. This force acts like a compensating pressure against the air bubble pressure. This condition is referred to as a "water-drive" aquifer storage system. Thus, the net volume of natural gas stored is relatively constant. Operating pressures at the surface are designed to respect this range of aquifer-operating pressures. The hydraulic properties of the caprock medium also provide a primary barrier to vertical air migration due to capillary pressures within the rock.

Underground air storage bubbles contain what is known as "base air" or *cushion air*. This is the volume of air that must remain in the storage facility to provide the required pressurization to extract the air. In the normal operation of the storage facility, this cushion air remains underground. The storage system works most efficiently if the pressure differences between the injection and the production cycles do not vary widely. This leads to a need to maintain a large volume of air bubble in the storage vessel, and maintaining a more or less constant pressure keeps operating costs lower. The air that is produced during the electric power generation period is called the *working air* volume.

19.4.1 CAES in a Depleted Gas Reservoir

Of the types of oil-and gas-bearing geological fields, only depleted natural gas fields are an option for CAES development. The presence of oil in the reservoir would damage the turbine blades in the CAES power generation equipment. Also, the potential for oxygen loss in the stored air and the potential of in situ combustion in the residual oil in the reservoir is an unacceptable risk. Depleted gas reservoirs are those formations that have already been tapped for a majority of their recoverable natural gas. This leaves space in an underground formation geologically capable of holding air. Depleted reservoirs are particularly attractive for CAES use because their geological characteristics are already well known.

Natural gas is found in both water-drive aquifer structures (Figure 19.3) and, geological stratigraphic or fault traps. Figure 19.3 shows a geological cross-section through the crest of a CAES water-drive depleted gas aquifer type reservoir. A CAES water-drive depleted gas reservoir behaves like any aquifer gas storage system, described earlier, with the only difference being the presence of residual gas. The CAES water-drive depleted gas reservoir requires the displacement of the residual gas in the reservoir from the working gas volume to prevent the potential of combustion of the gas in the turbo-machinery, as illustrated in Figure 19.3. It is also necessary to displace the residual gas from the working-air volume to prevent oxidation of the stored air into the turbo-machinery. Note that there is an air and residual natural gas mixing zone at the extremities of the air bubble at depth and at the outer boundaries of the reservoir (Figure 19.3).

Depletion gas traps are described as "volumetric" type gas reservoirs. As the natural gas was produced from the field the reservoir pressure continued to drop. There is no

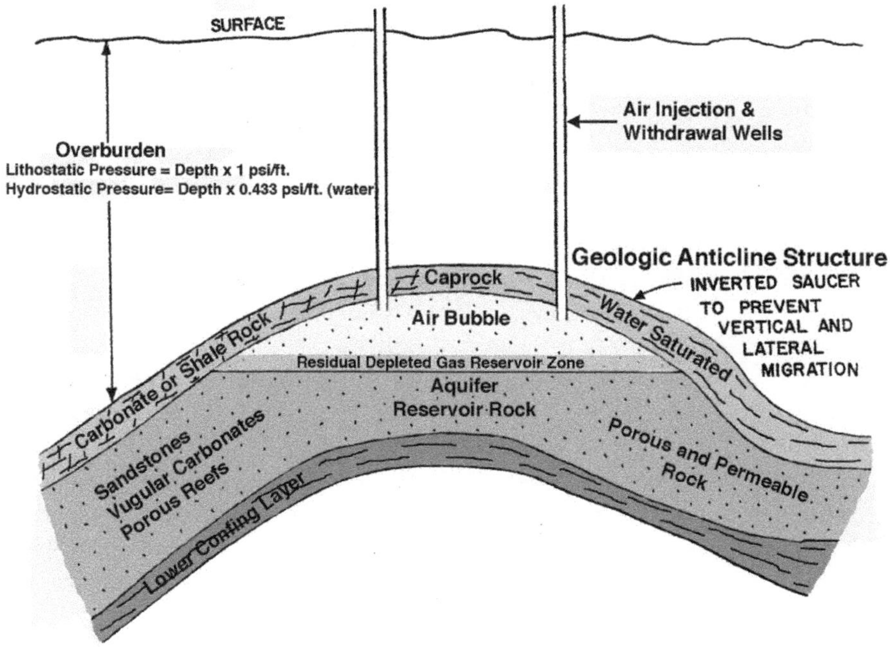

Figure 19.3 Sketch of a typical CAES depleted gas aquifer type reservoir.

water-drive in a volumetric type gas field. Gas production is stopped when an uneconomically low reservoir pressure is achieved. The use of a volumetric type depleted gas reservoirs for CAES provides a reliable type of CAES storage system, but it requires the replacement of essentially all of the produced gas with air to restore the reservoir to the original discovery pressure.

19.4.2 Matching the CAES Storage System to the Turbo-machinery

The matching of the CAES storage system to the turbo-machinery involves: (1) the determination of the required reservoir volume, (2) the design of the air storage pressure cycle, and (3) the reservoir air flow rates (deliverability) relative to the number of air injection/withdrawal wells. The specific design conditions for these parameters are provided.

19.4.2.1 Required Reservoir Air Storage Volume

The air storage potential within geological aquifer structures and depleted gas reservoirs is significant. These geological structures typically have surface areas of several square miles. In addition, the thickness of the air storage porous media (geological formation) can be several hundred feet thick. The available pore space within the porous air storage media is typically on the order of 25% to 30% of the available storage volume within the containment geological structure.

The development of the air storage bubble in the aquifer structure or reservoir involves the injection of air into the formation through wells to build up pressure. In this sense, the underground formation becomes a pressurized natural air container. As the pressure in the storage facility increases to that which is required by the turbo-machinery, air can be extracted more readily.

Working air is the volume of air in the storage reservoir that can be extracted during the normal operation of the storage facility. This is the air that is being stored and withdrawn to operate the turbo-machinery. The capacity of storage facilities normally refers to their working-air capacity. At the beginning of a withdrawal cycle, the pressure inside the storage facility is at its highest; meaning working air can be withdrawn at a high rate. As the volume of air inside the storage facility drops, pressure (and thus deliverability) in the storage facility also decreases.

The air storage capacities of selected CAES candidate aquifer structures, or reservoirs have been in the order of 10 Bcf to > 30 Bcf. These air storage volumes exceed the working-air volume of 0.5 to 1 Bcf per day by 10 to 30 times that which is needed to operate the CAES turbo-machinery.

19.4.2.2 Required Reservoir Air Pressure Cycling

The minimum required air pressure values at the storage system wellhead(s) are the minimum pressures at the turbine plus any pipeline friction pressure loss from the wellhead to the turbine inlet. The minimum air bubble pressure in the reservoir is the wellhead pressure plus any well casing pressure friction losses.

The pressure regime within a porous media storage system is a function of the depth to the top of the geological storage structure that defines the hydrostatic pressures of aquifer fluids within the porous storage media and caprock materials. Essentially all rock below a given depth under the ground surface contains groundwater. This includes aquifer type oil and gas reservoirs at various stages of depletion. The pressure gradient of these groundwater fluids results in a linear relationship between depth and hydrostatic pressure (Figure 19.4). The pressure gradient of reservoir fluids is specific

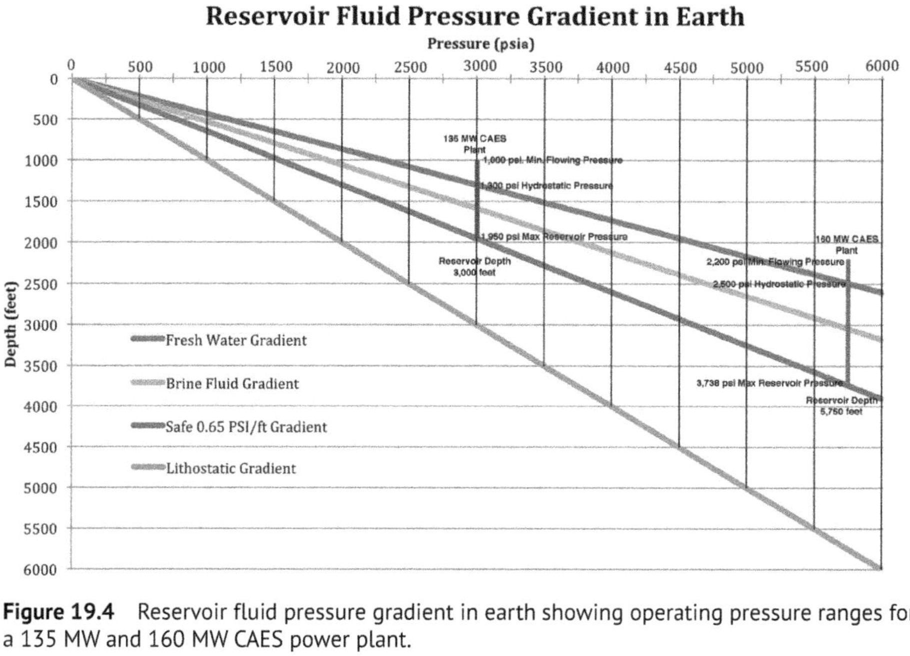

Figure 19.4 Reservoir fluid pressure gradient in earth showing operating pressure ranges for a 135 MW and 160 MW CAES power plant.

to the type of fluid, such as fresh water, and brine. Figure 19.4 illustrates the pressure gradient with depth for water and brine aquifer fluids.

Air storage projects work most efficiently if the pressure differences between the injection and the production cycles do not vary greatly (Figure 19.5). Withdrawal and injection of only a fraction of the total air in the aquifer structure or reservoir reduces the pressure difference between the withdrawal and injection cycles. This principle leads to maintaining a large volume of air in the aquifer structure or reservoir and maintaining a more-or-less constant pressure.

Both aquifer structures and reservoirs also have pressure constraints that should not be exceeded (Figures 19.4 and 19.5). The maximum design pressure is defined by some percentage of the lithostatic rock pressure that prevents hydraulic fracturing of the air storage system rocks that might cause the system to leak. The lithostatic rock pressure is approximately 1.0 psi per foot of depth (Figure 19.4). Safe air storage pressure to prevent hydraulic rock fracturing ranges from 0.6 to 0.8 of the lithostatic pressure. State regulatory agencies have set the maximum air storage pressures between 0.6 to 0.65 of the lithostatic pressure in aquifer structures and depleted gas reservoirs (Figure 19.4). The minimum design pressure is set to maintain the required volume of cushion air and an adequate pressure differential at the wellhead to support air deliverability.

19.4.2.3 Reservoir Air Flow Rates – Deliverability

Geologically, aquifer and depleted reservoir formations must have relatively high permeability and porosity. The porosity of the formation determines the amount of air that it may hold, while its permeability determines the rate at which air flows through the formation, which in turn determines the rate of injection and withdrawal of working air.

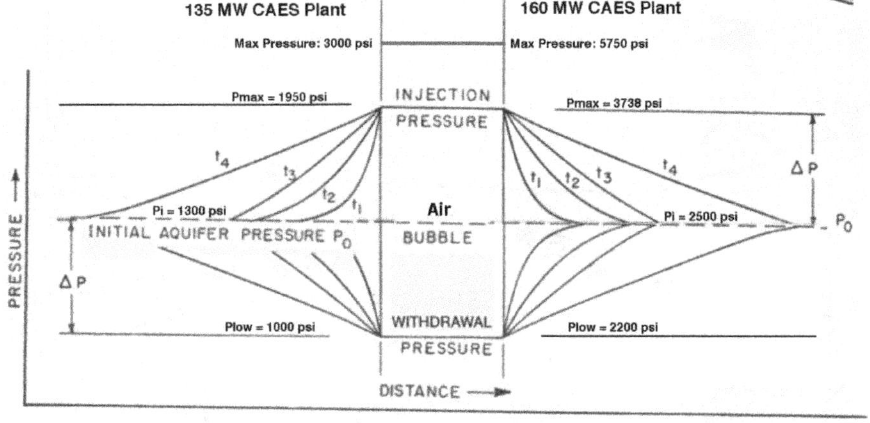

Figure 19.5 Diagram showing distribution of air pressure in a CAES storage system.

Deliverability is the rate at which air can be injected and withdrawn from storage. There are many factors that affect the deliverability (injection into and production rate from individual air wells within the air storage system), as illustrated and defined in Equation (19.1) (Figure 19.6) [1, 5].

$$Q = \frac{P_f^2 - P_w^2 \times K \times h}{\left(3.418 \times 10^7\right) \times T \times z \times u \times \left(\ln r_f / r_w\right)} \tag{19.1}$$

Where:
Q = production rate (MMcf/day)
P_f = Average reservoir pressure (psia)
P_w = Flowing bottom hole pressure (psia)
$T = 460° + °F$ = Rankine
z = compressibility factor
u = gas viscosity, cp
K = permeability, milliDarcy's (mD)
h = reservoir thickness, ft.
r_f = radius of influence, ft.
r_w = well radius, ft.

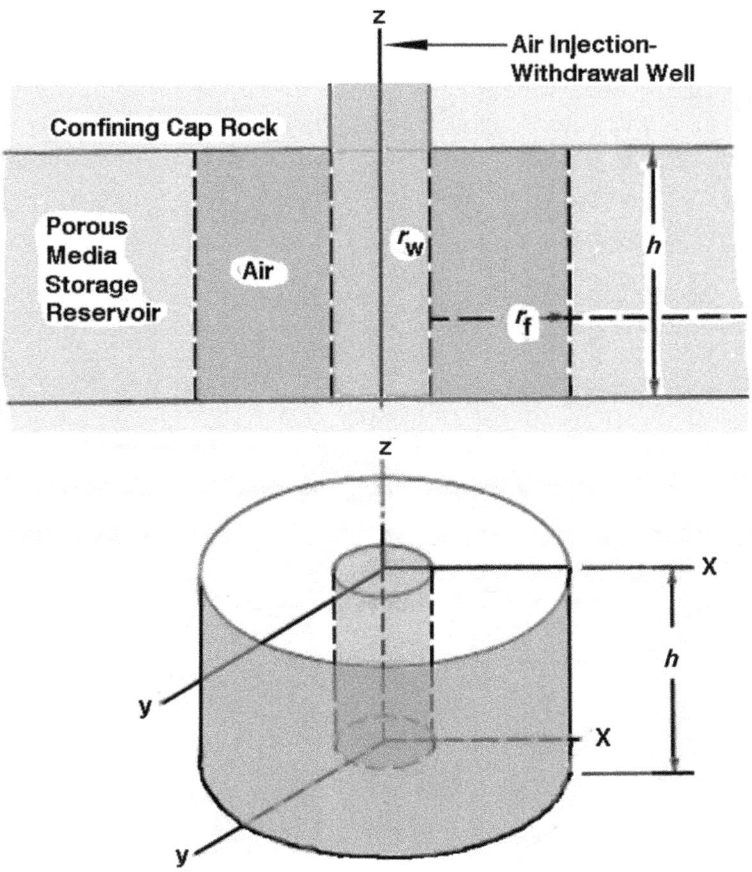

Figure 19.6 Diagram illustrating Equation (19.1) parameters.

Note: $P_f - P_w$ represents the flowing delta pressure: the pressure drop between the air storage vessel pressure and the flowing pressure in the air production wells.

Beyond the delta pressure available to produce air from the storage reservoir formation, the permeability (K) and thickness (h) of the air bubble within the formation are the most important. These two flow parameters are important attributes in selecting a CAES aquifer structure and depleted gas reservoir storage system. Acceptable reservoir hydraulic properties and delta pressures to support well deliverability of 0.5 Bcf per day are: $K > 600$ mD, $h > 30$ feet, and a delta pressure of < 400 psia.

19.5 Advantages and Disadvantages of CAES in Aquifer Structures and Depleted Gas Reservoirs

CAES development and operation of both aquifer geological structures and depleted gas reservoirs have advantages and disadvantages.

19.5.1 Advantages and Disadvantages of CAES in Aquifer Structures

Numerous natural gas storage fields have demonstrated the effectiveness of CAES in an aquifer structure since the 1950s. Many of these operating gas storage fields meet the design criteria for a CAES geological storage system discussed earlier. The results of the Pittsfield CAES field test program demonstrated the technical feasibility of CAES in an aquifer structure. However, CAES in an aquifer storage media is problematic during the candidate CAES site selection and development process.

Unlike proven gas production reservoirs, limited geological and hydraulic data is available on aquifer storage structures. Because of the economic value of natural gas, funds are expended to develop a gas reservoir using the results of geophysical surveys, exploration drilling, and core/laboratory testing of reservoir rocks. Some aquifer structures have been investigated for groundwater production, but for the most part remain a geological unknown. As a result, economically risky exploratory seismic surveying and drilling is necessary to confirm the geometry of the confining geological structure and air storage hydraulic rock properties. The verification of the integrity of the caprock in an aquifer storage structure can only be confirmed for the purposes of the project development by the injection of a test volume of air, often at considerable cost. Many aquifer storage structures tested or developed for natural gas fields failed because of gas leakage through the caprock during injection testing, or leaked natural gas after creation of the gas storage bubble.

An additional potential issue with CAES in an aquifer structure is the loss of oxygen content in the stored air. The Pittsfield field air storage test showed a potential of oxygen loss in the stored air due to the presence of iron pyrite minerals in the host storage media. This loss of oxygen was supported by theoretical analysis of potential chemical reactions between air and groundwater, and air and the host rock.

19.5.2 Advantage of CAES in Depleted Gas Reservoirs

As previously mentioned, depleted gas fields have contained natural gas for millions of years and their geological characteristics are already well known. The geological framework of the depleted gas reservoir is known at depth because of the numerous exploration and gas production wells that penetrated the field. Well completion records that may be available include well construction data, geophysical well logs, and possible core laboratory test data. Many of these fields were further characterized by two-dimensional and three-dimensional seismic reflection surveys over the areal extent of the field.

Oil and gas production companies typically collect extensive gas production well data. Data available from depleted gas reservoir by well are daily gas production volumes, static and flowing gas pressures, daily water production volumes, and/or drill stem test or controlled well flow test data. Reservoir fluid chemistry may also be available. This data is essential for determining reservoir hydraulic properties at various locations in the depleted gas reservoirs. The analysis of this data using a numerical CAES reservoir-modeling analysis is necessary to design the CAES storage system.

19.5.3 Disadvantages of CAES in Depleted Gas Reservoirs

One of the key issues of air storage in a depleted reservoir is the mixing of air with residual natural gas in the storage reservoir. The potential for spontaneous combustion in the depleted reservoir, the combustibility of the air/natural gas mix in the turbine, and detrimental chemical reactions, such as oxygen depletion in the stored air have been evaluated in laboratory experiments, air injection reservoir field tests, and with air and natural gas mixing reservoir simulation models.

Laboratory testing involved the circulation of air and methane through reservoir core samples at reservoir pressure conditions for an extended period [6]. The results of these laboratory tests showed essentially no oxidation of the air because of the low temperature of the air/gas mixture. This was confirmed from theoretical analysis of potential air and methane reactions. Laboratory and air injection testing during the US DoE and EPRI Pittsfield field experiment show a potential loss of oxygen content in the stored air because of chemical reactions between air and various ferrous minerals in the porous media, such as iron pyrite. The resolution of this issue is to select a CAES storage reservoir media with no ferrous minerals.

A potential for a combustible air/gas mix exists during the initial injection of air into an air injection/withdrawal wellbore. Prior to the initial injection of air into a wellbore, the casing is typically filled with natural gas. During the initial injection of air into the wellbore there will be a short period of time when the concentration of natural gas to air is within the combustible range of 4% to 8%. The simple solution to this problem is to fill the wellbore with nitrogen to displace any residual natural gas out of the wellbore into the reservoir.

The issue of measurable concentrations of methane in the produced air going to the turbo-machinery has been evaluated using reservoir chemical mixing models. The solution to this issue is to effectively displace any residual methane to the extremities of the reservoir out of the working gas portion of the air bubble (Figure 19.7). This requires: (1) a close spacing of air injection/withdrawal wells, and (2) stage development of the air bubble (intermittent air injection) to allow the buoyancy of the air to displace both the residual gas and groundwater from the system.

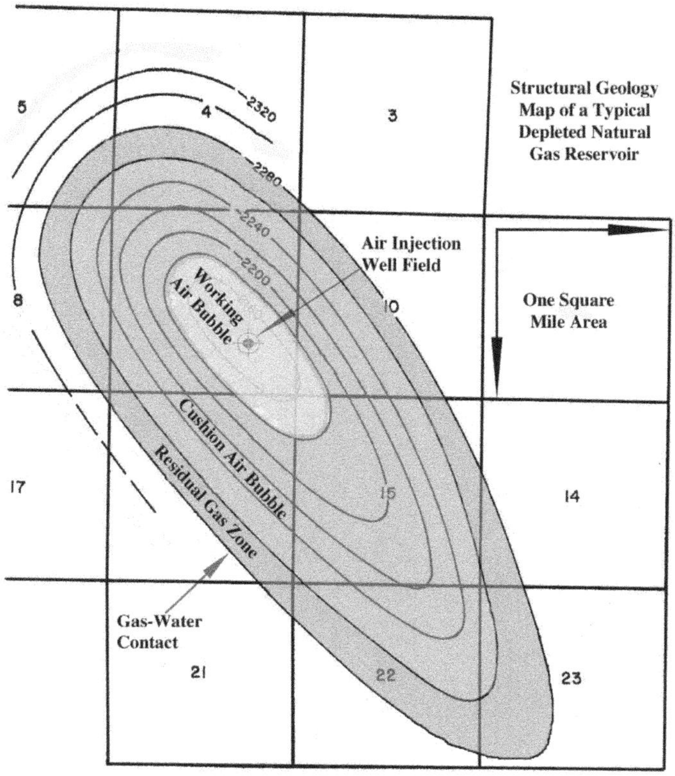

Figure 19.7 Sketch of distribution of air and residual natural gas in a CAES depleted gas storage reservoir.

19.6 CAES Storage System Design Tools, Development, and Operation

The design, development, and operation of any CAES storage system represent a process that is too complex to be accomplished using simple reservoir engineering analytical tools. Our primary CAES storage system design tool is a reservoir simulation model program.

The development of a CAES reservoir simulation model involves the preparation of a digitized geological framework model using a model grid system. Reservoir properties are then imported into the model grid system to represent the aerial distribution of hydraulic rock properties. In the case of depleted gas storage systems, historical gas production data is entered into the model for each well. The modeling code is then run in an attempt to simulate the history of individual natural gas production by well. Individual well reservoir properties are adjusted in the model until an acceptable match with the gas production history data is acheived. The final reservoir model will then be a realistic representation of the gas production history of the field and indicate reservoir conditions at the time of abandonment. The resulting model is then used to simulate the development of the air bubble and the cycling of air during CAES operations. The model is used as a tool to design the number of required air injection/withdrawal wells to support the turbo-machinery, and the required well spacing. The model is also used to determine the required working and cushion air storage volumes to support the CAES power plant.

A CAES reservoir simulation model is also an effective tool to design an aquifer air storage system. The primary difference is the lack of natural gas production history data to develop an accurate determination of aquifer properties over the entire storage structure. The results of an air injection test program will be necessary to provide the required hydraulic rock property data.

A number of commerical numerical reservoir simulation modeling programs are available to simulate CAES reservoir development and operating conditions. Most reservoir-modeling programs are finite difference codes that model the interaction of reservoir flow conditions between cells within a grid pattern over the boundaries of the reservoir. Each cell is assigned various reservoir engineering properties to represent changes in geological and rock properties over the extent of the reservoir. Reservoir-modeling programs are divided in to “Black Oil” and multi-phase flow codes. Black oil reservoir-modeling codes do not consider changes in the composition of the hydrocarbons within the reservoir. Multi-phase flow codes consider the thermodynamic and chemical interactions of reservoir fluids. These reservoir - modeling programs are divided into open source and commerical software. Some of the most well known reservoir-modeling programs are:

Open Source Programs

1. BOAST (Black Oil Applied Simulation Tool)
2. MRST (MATLAB Reservoir Simulation Toolbox)
3. OPM (Open Porous Media)
4. TOUGH2 Family of Codes (Transport of Unsaturated Groundwater and Heat: United States Department of Energy Source Code)

Commerical Source Programs

1. CMG Suite (IMEX, GEM, and STARS Computer Modeling Group)
2. ECLIPSE (SIS, formerly know as GeoQuest and Schlumberge)
3. ExcSim (Microsoft Excel-based program)
4. Nexus (Landmark Graphics a Haliburton product)
5. ResAssure (by ResAssure)
6. tNavigator (Rock flow Dynamics)
7. ReservoirGrail (by GrailQuest)

The TOUGH2, as well as TOUGH2 + Air (its most modern realization) is one of the most effective CAES reservoir-modeling tools because of its ability to model complex interactions of multi-phase reservoir fluids over a range of storage temperatures. This is a family of multi-dimensional, multi-component models that are designed for the simulation of the coupled transport of fluids and heat in porous and/or fractured geologic media. Since its first release in 1987 by a team of researchers at the Lawrence Berkeley National Laboratory (a US Department of Energy laboratory) the program has been fully verified against actual reservoir field data. Verification details of the TOUGH2 code can be found in a 1996 textbook [7].

The TOUGH family of codes have been enriched by a large number of modules that describe a variety of important Equations-Of-State (EOS, involving various mass components). The appropriate EOS that best simulate CAES reservoir conditions is called GasH2O. The GasH2O EOS simulates gas mixture that includes air (described as a mixture of two components, O_2 and N_2) and a gaseous alkane with the properties of the residual natural gas in the depleted reservoir. This EOS includes the complete thermodynamics of the system with practically no modification. It provides a full description of the thermo-physical properties of the various phases and components (e.g. density, viscosity, enthalpy), and can handle the various processes encountered during a CAES operation: flow through porous media, relative permeability and capillary pressure issues, phase changes (e.g. water evaporation during injection in the vicinity of the wellbore), system and fluid compressibility, gas dissolution in the reservoir water, and temperature changes during injection and production operations.

The simulation of potential loss of oxygen of the stored air is accomplished by use of the GasH2O EOS coupled with the TOUGHREACT EOS (another member of the TOUGH family). Such coupling is seamless (as it involves members of the same family of codes). A separate analysis is conducted to determine if the composition of the gases and the conditions at any point in the reservoir and in the wellbore at any time during the CAES operations can reach an unsafe level of a methane/air mix that can lead to a potential explosion. It is important to note that earlier work on the subject has shown this to be a practically impossible situation under reservoir conditions within a CAES storage system.

19.7 Summary

CAES is a proven technology to store bulk energy by converting off-peak generated electricity into compressed air, and then using this compressed air to generate electricity during peak power demand periods. A CAES facility consists of an electric generation system and an energy storage system. The electric generation system utilizes commercially available turbo-electric generation systems as demonstrated by the successful operation of the Huntoff Germany and McIntosh Alabama CAES facilities. Both of these facilities utilized solution-mined salt caverns to store the required pressurized air supply.

Aquifer geological structures and depleted gas storage reservoirs are viable compressed air energy storage systems. These porous media CAES storage systems have been storing natural gas successfully for over a 60-year period, demonstrating the technical viability of these two types of energy storage systems. Air injection testing in the Pittsfield aquifer storage field and in a California depleted gas field further demonstrated the technical viability of these CAES storage systems.

CAES requires the matching of the air storage properties of the earth-based geological storage structure to the mass flow rate and pressure requirements to operate the electric turbo-machinery system. Dresser-Rand's (a current manufacturer of CAES power plant equipment) turbo-machinery requirements are:

- 400 #/sec Mass Air Flow Rate per Unit
- 135 MW Unit – 920 psia
- 160 MW Unit – 2150 psia

These air mass flow rates and pressure requirements define the siting and operating criteria for the porous media CAES geological structure. Key site selection and design parameters are:

1. size and depth of the air storage reservoir,
2. air storage capacity of the storage reservoir, and
3. potential air pressure and flow rates from wells.

Numerous candidate aquifer and depleted gas storage fields have been identified in the US and Canada that fully support CAES utilization.

The candidate site selection and development of a CAES aquifer structure is inhibited by the limited amount of geological data necessary to confirm the geometry of the air storage structure and data on the hydraulic properties of the storage formation, unlike a depleted gas reservoir. Exploration seismic surveys over the extent of the geological structure and exploration corehole drilling are necessary to confirm the merits of a structure for CAES development. A field air injection test program is then necessary to confirm the air storage potential of the candidate structure. The cost of the exploration and field-testing program versus the potential of air leakage from the storage vessel must be considered.

The issues associated with the mixing of stored air and residual natural gas in a depleted gas storage reservoir have been extensively evaluated using air field injection testing, laboratory air and methane mixing tests, and numerical reservoir simulation modeling. The results of these tests and modeling indicate that a combustible concentration of methane can be prevented by proper development of the original air storage bubble. The potential of spontaneous combustion is essentially non-existent. The potential of depletion of oxygen in the stored air due to oxidation reaction is a non-issue because of the low stored air temperature regime. A potential for oxygen depletion due to the presence of ferrous minerals in the reservoir storage media must be considered. Selection of a CAES structure absent of ferrous minerals will help resolve this air oxidation issue.

Reservoir simulation modeling is an essential tool to the design and operation of a CAES storage system. Numerous open source and commercial reservoir-modeling programs are available for simulating CAES reservoir conditions. Of these, the TOUGH2 suite of model codes is one of the most effective in simulating the complex fluid mixing and thermodynamics/pressure condition in a CAES porous media system. The EOS TOUGH2 + Air, GasH2O, and TOUGHREACT modules have been successfully used to simulate CAES development and operation of several CAES candidate aquifer and depleted gas reservoirs.

References

[1] Katz, D.L. and Lady, E.R. (1976). *Compressed Air Storage for Electric Power Generation*. Ulrich's Books, Inc., Ann Arbor, MI (Eq. 3–19).

[2] Crotogino, F., Mohmeyer, K., and Scharf, R. (2001). Huntorf CAES: *More than 20 years of successful operation*. *Proceedings of the SMRI Spring Meeting*. Orlando, FL (15–18 April).

[3] EPRI (Electric Power Research Institute). (1990). *Compressed-Air Energy Storage: Pittsfield Aquifer Field Test, Test Data: Engineering Analysis and Evaluation*. EPRI GS-6688, prepared by ANR Storage Company, February.

[4] King, M.J. and McGill, M.J. (2009). Compressed air energy storage. In: *Encyclopedia of Energy Engineering, Volume 1:* 1, 1–14. New York, NY: Taylor & Francis Group, Editor: Barney Capehart.

[5] Katz, D.L. (1971). *Monitoring Gas Storage Reservoirs*, paper SPE 3287 presented at the 1971 SPE Midwest Oil and Gas Industry Symposium, Chicago, IL., April 1–2.

[6] Pruss, K., Simmons, A., Wu, W.S., and Moridis, G. (1996). *TOUGH2 software qualification*. Lawrence Berkeley National Laboratory Publication 38383, February.

[7] Wood, J. (2014). *Air-methane mixing experiment final report*. Southwest Research Institute Private Consultant Report to Sacramento Municipal Utility, 14 April.

20

Open Accumulator Isothermal Compressed Air Energy Storage (OA-ICAES) System

Perry Y. Li, Eric Loth, Chao (Chris) Qin, Terrence W. Simon, and James D. Van de Ven

20.1 Introduction

Despite their abundance, fossil fuels, such as coal, oil, and natural gas, will produce greenhouse gases and will exacerbate the state of global warming and climate change. Displacing fossil fuels by clean and renewable energy is necessary to reverse this trend. Although renewable resources such as wind or solar energy are quite abundant, they have a significant drawback – they are not available on-demand and their availabilities are often unpredictable. For example, wind speed varies from hour to hour and from day to day with higher intensity wind usually occurring at night; solar energy is available only during the day and when clouds are not in the way. There is often a mismatch between when these renewable energies are available and when demand is high.

The variable and stochastic nature of these renewable energies pose a significant challenge in incorporating a large amount of these resources into the electrical grid. The challenge lies in that when the renewable resources become unavailable, other energy resources that can be called upon quickly are needed. Currently, natural gas power plants, with sufficient capacity to compensate for variability of the renewables, are used for this purpose. Besides having a carbon dioxide (a greenhouse gas) footprint, these natural gas "peaker" plants are expensive to build, maintain, and operate. Their use increases the overall cost and diminishes the environment benefits of renewable energies. To an extent, a smart electrical grid that connects a broad portfolio of generators and demands across a wide geographic domain can mitigate the need for "peaker" plants by compensating for the variability, provided that adequate transmission capacity and a responsive market exist. Alternatively, energy storage can be used in conjunction with wind and solar to store excess energy that is to be regenerated when demand is high. This allows renewable energies to be predictable and available on-demand, eliminating the need for "peaker" plants.

We focus now on wind energy. If an energy storage system is configured to store energy locally at the wind turbine/wind farm, and prior to generating electricity, the electrical components, including the generator, power electronics, and interconnects, can be downsized for the mean power instead of peak power. This is significant since the capacity factor of a typical zone-six wind turbine is only 47% [1] (with many other turbines having a capacity factor below 40%), meaning that downsizing by 50% is possible. The savings in capital expense (CAPEX) is even more significant for off-shore

Advances in Energy Storage: Latest Developments from R&D to the Market, First Edition.
Edited by Andreas Hauer.
© 2022 John Wiley & Sons, Ltd. Published 2022 by John Wiley & Sons, Ltd.

wind installations since electrical collection and transmission costs are four times the cost of on-shore deployment, accounting for 15% of the total cost (versus 6% for on-shore). The cost saving for off-shore collection and transmission alone would be 20% [2, 3]. Since energy price can vary by an order of magnitude during a hot summer day, it is much more useful (and profitable) to generate electricity according to demand rather than according to wind availability, as is currently practiced. The amount of wind energy that can be captured will also be increased as it is no longer necessary to reserve for control, through curtailment, a portion (typically 5%) of the available power. The cut-off wind speed (to protect the turbine) for maximum power tracking can also be increased to the limit of the mechanical structure or the limit of the Power-Take-Off (PTO), instead of that of the electrical generator.

Storing the electrical energy locally in batteries is an option, except for the cost, need for AC/DC/AC conversion, low power density, weight, limitation in charge/discharge cycles, end-of-life environmental impact, and low efficiency in cold climates. Also the wind turbines would require a full-sized generator and power electronics to accommodate the available power. Pumped hydro storage (PHS) and conventional compressed air energy storage (CAES) systems are economical utility-scale methods (several times less expensive than batteries [4]) but since they require a large reservoir or an underground cavern, they depend greatly on geographies. Existing conventional CAES systems, such as the plants in McIntosh, US or Huntdorf, Germany, use excess electricity to compress air (up to 80 bar [8 MPa]) into a cavern, and the energy is recuperated by burning a mixture of the natural gas with the compressed air. The requirement for an additional hydrocarbon fuel and the low efficiency of the storage-regneration cycle itself (estimated at 29–36% [5]) diminish the attractiveness of the conventional CAES processes.

CAES has a competitive energy and power density, especially if operated at high pressure. If the compressed air pressure is raised to 350 bar (35 MPa), the energy is compressed and extracted in a near-isothermal manner, and if open accumulator architecture (see Figure 20.1) is used, as much as 170MJ (47 kWh) can be stored per m^3 of volume. For a PHS to achieve an equivalent energy density, the reservoir would need a 17 km elevation! Gases at 350 bar (35 MPa), and even higher, pressures are routinely stored in steel or composite hydraulic accumulators and scuba tanks. It is expected that with larger vessels, improved designs, and economy of scale, the storage vessels can be quite cost effective. Use of engineered pressure vessels allows CAES to be sited anywhere without geological constraints. To store an average power of 3 MW (for a peak 5 MW turbine) continuously for 8, a volume of 500 m^3 is needed. Such a vessel is compatible with a turbine of this scale (100 m tall and a blade span of 60 m). The storage vessel can even be integrated into the structure of the turbine tower, or used as ballast.

In this chapter, we describe research for a novel CAES concept for wind turbines (or other renewable energy sources with mechanical output) that was first proposed in [6, 7] for storing energy locally prior to electricity generation. In this system, as shown in Figure 20.1, compressed air is stored in high-pressure (200–350 bar [20–35 MPa]) vessels; the compressor/expander used to store and extract energy operates nearly isothermally so that it is efficient and a variable hydraulic drive, instead of a mechanical gearbox, is used for power transmission. This improves the reliability of the transmission system and allows the generator and storage system to be housed down-tower,

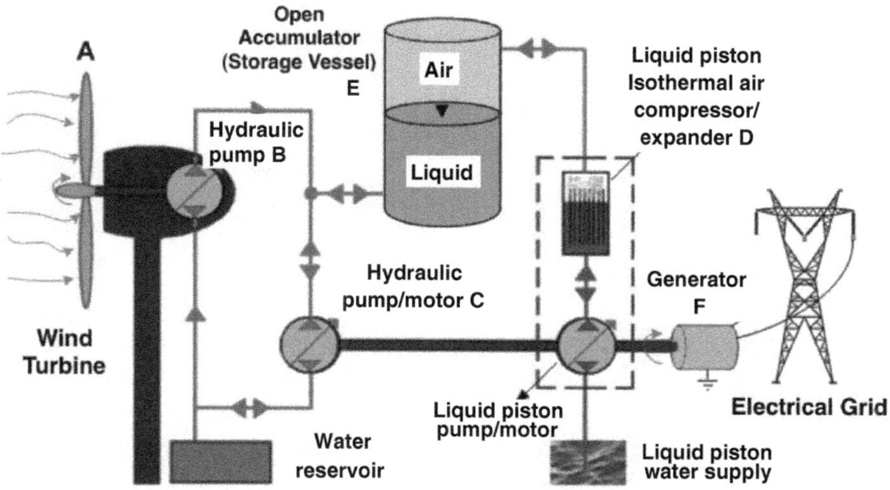

Figure 20.1 Open Accumulator Isothermal Compressed Air Energy Storage System OA-ICAES system for a wind tubine.

thus reducing construction and repair costs. In addition, a cost-effective, fixed-speed induction generator can be used instead of the combination of a permanent magnet synchronous generator and power electronics for frequency and voltage conversion. Since the system can be turned on and off in less than 0.1s, it can be used to provide lucrative ancillary services such as grid frequency regulation. It is well suited for this application because of its cost effectiveness for utility scaled deployment. Target storage and capacity costs, in 2015, were in the order of \$150/(kWH) and \$1000/(kW) respectively.

The two major challenges for realizing this concept are: (i) the compressor/expanders are generally not very efficient or powerful and (ii) the pressure in the storage vessel reduces as compressed air in the storage vessel depletes, making it difficult for the air compressor/expander to maintain either its efficiency or power at all energy levels. The first challenge is overcome by developing a near-isothermal air compressor/expander with enhanced heat transfer. This is achieved using a liquid piston compressor/expander [8] in conjunction with porous media inserts [9] and injected droplet sprays [10]. The latter challenge, which leads to low energy density (hence requiring large storage vessels), is overcome by deploying the *open accumulator* configuration with a dual-chamber storage vessel for both liquid and compressed air, such that energy can be stored/retrieved hydraulically and pneumatically [11]. By coordinating the hydraulic and pneumatic paths, the pressure can be maintained regardless of the energy content. Because hydraulic components are more power-dense, and can be turned on/off quickly, they can be used for peak transient power while the air compressor/expander can be downsized for steady power. An appropriate controller that coordinates these two power paths is essential for the simultaneous pressure regulation, tracking of the desired generator power, and maximizing wind power capture in the presence of supply or demand power variations. The efficiency and performance of the CAES system would depend significantly on the design of the controller.

The rest of this chapter is organized as follows. Section 20.2 describes the overall system architecture. Section 20.3 presents the liquid piston isothermal compressor/expander. Section 20.4 describes the spray cooling concept. Section 20.5 summarizes the control system concepts. Sections 20.6 and 20.7 contain discussion and concluding remarks.

20.2 Open Accumulator Isothermal Compressed Air Energy Storage (OA-ICAES) System Architecture

The proposed CAES with the open accumulator architecture [11] is shown in Figure 20.1. A variable displacement hydraulic pump (B) attached to the wind turbine rotor (A) in the nacelle converts wind power to hydraulic power. At down-tower, a variable displacement hydraulic pump/motor (C), a near-isothermal liquid piston air compressor/expander (D), and a fixed speed induction generator (F) are connected in tandem on a common shaft. They are powered by the pump (B) and exchange power with the storage vessel (E) with both liquid and compressed air at the same pressure. This allows energy to be stored in or extracted from (E) either hydraulically (as in a conventional hydraulic accumulator) or pneumatically (as in a conventional air receiver). In both cases, energy is stored in the compressed air. By coordinating the hydraulic and pneumatic power paths, the pressure in the storage vessel (E) can be controlled independently of the energy content, unlike a conventional closed hydraulic accumulator with only a hydraulic port or a compressed air receiver with only a pneumatic port. For example, as compressed air is being released from (E), some liquid can be added to reduce the compressed air volume to maintain the pressure. The hydraulic fluid is preferably water (for cost and environmental reasons) so the hydraulic components (B, C, E) should also be compatible with water.

In conventional CAES systems, such as the ones in McIntosh, AL, US or in Huntorf, Germany, the storage volume contains compressed air only, typically up to 70–80 bar (7–8 MPa). Pressure varies greatly as energy is stored or depleted. When the pressure is low, greater airflow is needed for the same power. This limits the compressor/expander power, or makes the low-pressure compressed air unusable. The large pressure variation also makes it difficult to optimize components to operate efficiently throughout the pressure range. For this reason, the McIntosh CAES plant limits pressure within a 50–78 bar (5–7.8 MPa) range. Since the compressed air that remains at the lower pressure is not used, the energy density is low[1] (16.5 MJm^{-3} or 4.6 $kWhm^{-3}$). If the remaining compressed air at 50 bar (5 MPa) is allowed to expand to ambient pressure, the energy density can be increased by 65%.

In the open accumulator concept [11], the storage vessel holds both liquid and air. Notice from Figure 20.1 that energy can be added/subtracted to/from the storage vessel pneumatically via the compressor/expander or hydraulically via the hydraulic pump/motors. The hydraulic power path stores energy by compressing the air already in the vessel with the addition of liquid similar to a hydraulic accumulator. This path is power-dense, but the amount of energy that can be stored this way is relatively low

1 Assuming energy is extracted isothermally.

as the maximum pressure limit is reached quickly. It is well suited for high-power transient events such as wind gusts or periods of sudden generation power demand. The pneumatic power path stores energy by putting more compressed air into the vessel. This path is energy-dense, but the power density is much lower than the hydraulic path, so it is more suited for steady power. The hydraulic path is 20–30 times more power-dense than the pneumatic power path, whereas the pneumatic path is 20 times more energy-dense than the hydraulic path [11]. This dual power path property offers several benefits:

1) **Increase energy density:** By controlling the liquid volume, pressure can be maintained regardless of the amount of compressed air stored. This makes all the compressed air usable at the system's nominal pressure, thus realizing the full energy capacity. At 210 bar (21 MPa), the energy density is 91.4 MJm^{-3} (25.4 $kWhm^{-3}$), a 5.5-fold increase. This increase makes it economically feasible to use pressure vessels instead of underground caverns to store the compressed air in grid-scale applications.
2) **Downsizing compressor/expander:** Since hydraulic pump/motors are cheaper and more compact than similarly powered pneumatic compressor/expanders, one can reduce the cost and size by utilizing the hydraulic power path for large transient peaks (e.g. from wind gust) and downsizing the air compressor/expander for steady power.
3) **Optimizing efficiency:** With pressure maintained in a narrow range, components need only be optimized over this range. Furthermore, the redundancy afforded by the dual power path provides an extra degree of freedom to optimize system efficiency while storing/releasing the desired amount of energy and maintaining pressure within a reasonable range. Simulation (see Section 20.5) shows that an efficiency increase of 4% is possible.
4) **Rapid response:** Hydraulic actuation can usually be achieved in 10–100 ms. This allows the storage system to respond rapidly to sudden changes in power input or power demand. This rapid response can be utilized to provide ancillary services for the electrical grid, such as frequency regulation or grid stabilization.

The near-isothermal liquid piston air compressor/expander is a critical component of the system as it is responsible for transforming mechanical energy into stored energy, and vice versa. Thus, for the system to be economical, it must be both efficient and power-dense. Sections 20.3 and 20.4 describe the various techniques being used to achieve these characteristics.

20.3 Liquid Piston Isothermal Compressor/Expander

Making the air compressor/expander efficient and power-dense is challenging because of thermal effects. For example, if compression/expansion by 350-fold were done isentropically, the result would be extreme temperature variations about the ambient temperature (+1200 K/−250 K), which the materials cannot normally tolerate. Even if this is possible, the compressed air, which has been heated during compression, is expected to stay within the storage vessel for hours before reuse. During this time, it will cool to the ambient temperature. Figure 20.2 illustrates the relationship between the pressure-volume curve and the work input and output. The work input W_{in} is the area under the P–V curve above the initial pressure P_0 for compression from P_0 to rP_0, including the isobaric ejection of the compressed air into the reservoir at rP_0. The output work is the area under the expansion P–V curve from rP_0 to P_0 including the initial isobaric charging at rP_0. To be efficient, both the compression and expansion P–V curves must be close to each other. They must also be close to the isothermal P–V curve at ambient temperature T_0 if the heat sink and source are also at T_0. Conventionally, a high pressure gas compressor/expander consists of multiple stages with inter-cooling/warming. This creates a zig-zag pressure–volume curve consisting of successive adiabatic compression/expansion and constant volume cooling/warming segments. As the number of stages increases, compression and expansion approach the most efficient isothermal processes, at the temperature of the environment.

The minimum amount of work input and the maximum amount of work output are, therefore, the input work or output work when the process is isothermal at the ambient temperature. We refer to this as the storage energy (Equation (20.1)),

$$E_{store} := W_{in}(\zeta_{iso}) = W_{out}(\zeta_{iso}) \tag{20.1}$$

where ζ_{iso} denotes the isothermal process. The input and output efficiencies are defined in Equation (20.2).

$$\eta_{in} := \frac{E_{store}}{W_{in}}; \ \eta_{out} := \frac{W_{out}}{E_{store}} \tag{20.2}$$

Since, from the first law of thermodynamics,

$$dE = -PdV + Qdt$$

where $E(P, V)$ is the internal energy of the air being compressed/expanded, PdV is the differential compression work, $Q(h, A, \Delta T)$ is the heat transfer (rate), which is a function of the heat transfer coefficient h, the heat transfer area A, and the temperature

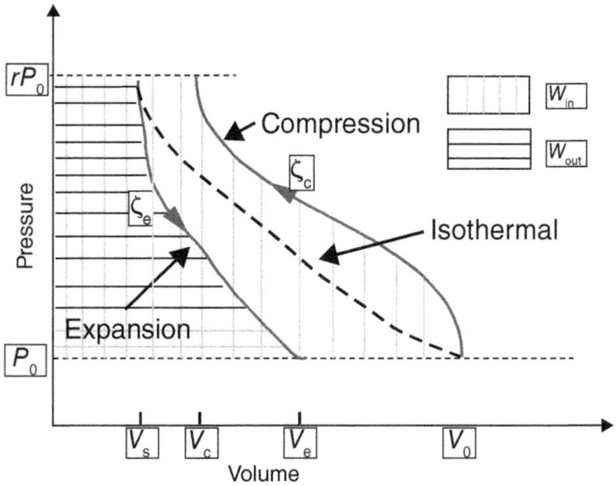

Figure 20.2 Pressure–volume plot illustrating work input (W_{in}) and output (W_{out}). The compressed air is assumed, conservatively, to return to ambient temperature inside the storage. [47]

difference between the air and the heat transfer surfaces ΔT, the time it takes to trace a pressure-volume (P–V) curve is given in Equation (20.3).

$$t = \int_{\zeta_{\text{eject/charge}}} dt = \int \frac{dE(P,V) + PdV}{Q(h, A, \Delta T)} \tag{20.3}$$

The integration is taken over the path of the compression/expansion process. From this, we see that the input and output powers, defined as in Equation (20.4),

$$Power_{\text{in}} := \frac{E_{\text{store}}}{t_{\text{c}}}; \; Power_{\text{out}} := \frac{W_{\text{out}}}{t_{\text{e}}}$$

(t_{c}, t_{e} are the times for the compression and expansion processes) are inversely proportional to process time and depend on the heat transfer rate Q in the compressor/expander, or in the inter-cooling/warming processes. Note that the *P–V* curves determine the efficiency of the process and the heat that must be transferred. A P–V curve that requires less heat to be transferred (and hence less time) tends to deviate more and more from the isothermal curve and the process becomes less and less efficient. Hence, there is an inherent trade-off between efficiency and power of a compressor/expander. For this reason, typical pneumatic compressors are either inefficient, low power or both.

To mitigate this trade-off in order to increase both efficiency and power density, the air compressor/expander is enhanced with heat transfer using a liquid piston concept [8]. A schematic is shown in Figure 20.3 that consists of an air compression/expansion chamber filled with porous media and a hydraulic pump/motor used to actuate a liquid piston. The porous material is used to increase the heat transfer surface area [9, 11–13]. A liquid (water) piston can flow through the porous material, providing a tight seal for

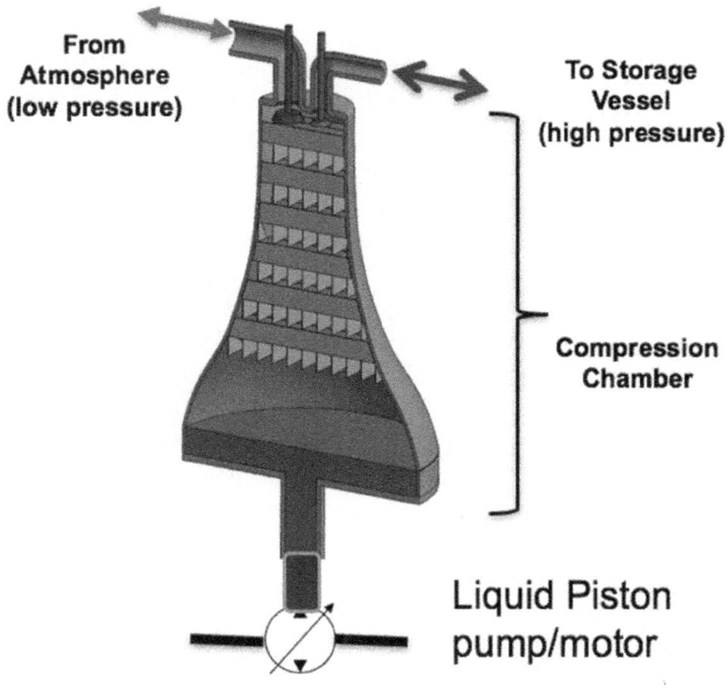

Figure 20.3 Schematic of a near-isothermal liquid piston compressor/expander.

the compressed air and serves as a heat sink and heat source for the porous media. The liquid piston is also advantageous in displacing and ejecting compressed air, and eliminating dead volumes that decrease system efficiency. The liquid piston pump/motor, which pumps water into the chamber during compression and receives water during expansion, serves as the power-take-off (PTO) interface between the compression/expansion process and the mechanical domain.

When storing energy pneumatically, the liquid piston pump/motor pumps water into the compression/expansion chamber, compressing the air within it. Thermal energy of compression is transferred to the porous material and to the water to maintain a near-isothermal operation. When the chamber pressure exceeds that of the storage vessel, a valve is opened and the compressed air is ejected and stored in the pressure vessel. The chamber is then refreshed by releasing the water and filling it with atmospheric air for the next cycle. When discharging energy, the compressed air is initially released into the expansion chamber. The valve is then closed. As the air expands, the liquid piston retreats, the liquid piston pump/motor is motored and work is derived. Thermal energy is supplied from the porous material to maintain the temperature of the expanding air.

Section 20.3.1 will discuss the porous media modeling and design. Section 20.3.2 presents how compressor/expander performance is enhanced by optimizing the motion trajectory of the liquid piston, porous media distribution and chamber shape [14–18]. Section 20.3.3 presents a new efficient liquid piston pump/motor design. Section 20.4 describes how tiny water droplets [10, 19] enhance heat transfer in this application.

20.3.1 Porous Media Heat Exchanger Modeling and Design

Effective regenerative heat exchangers require close thermal communication between the heat exchanger and the air being compressed and expanded. This requires (i) a large surface area per unit volume to effect that heat transfer with a small temperature difference, (ii) maintaining a temperature difference as heat transfer proceeds, and (iii) a heat exchange material that has a high thermal capacity relative to that of air. In addition, a low pressure drop is also desirable. Heat exchanger geometries that offer large ratios of heat-transfer area to volume include beds of beads, honeycombs, a series of parallel small tubes, screens or fiber beds, open-cell foam, and parallel plates.

Figure 20.4 shows two styles of heat exchangers: honeycomb and open-cell foam. Honeycomb regenerators that have short flow-direction sections to allow some mixing between layers and reinitiation of thermal boundary layers at the beginning of each segment are more effective. Open-cell foams tend to have higher pressure drop than the honeycomb, but offers good lateral mixing. They are also more inclined to trap the air or water at the interface between the two passes through the matrix. A geometry that our group has found most effective is the interrupted plate regenerator, such as shown in Figure 20.5. It has the repeating thermal boundary layer feature, offers effective mixing between plate layers and, if important, allows effective draining of liquid and minimal trapping of gas when liquid is present. The interrupted plate regenerator matrices with multiple layers can be fabricated with 3-D printing using ABS plastic or metal powder [20, 21].

To analyze the effectiveness of the heat exchanger regenerator matrix that consists of a matrix of solid webs with air in the interstitial spaces, a porous medium modeling

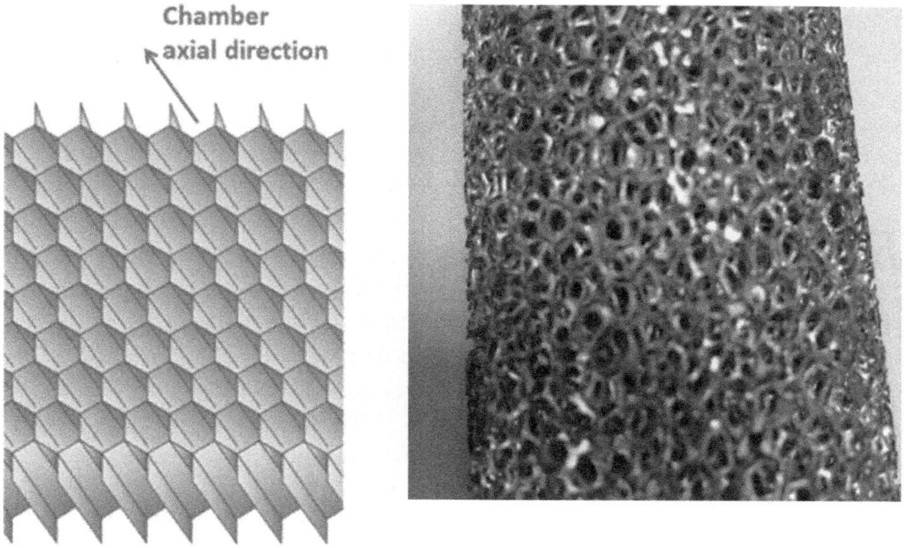

Figure 20.4 Two styles of porous media heat exchangers: honeycomb and open-cell foam.

approach is used. Modeling is done by averaging over the fine features of the flow using such terms as porosity (the interstitial volume as a fraction of the total volume), permeability, and inertial coefficients (for computing the losses due to flow forces within the porous region). Porosity can be determined by the geometry of the medium, permeability and inertial coefficients are either measured or computed via detailed flow calculation on a segment of the porous medium, called the Representative Elementary Volume (REV). One thermal equation is needed for the solid region and one for the air region. They are also averaged over the REV. Modeling of the processes within the REV involves determining the heat transfer between the solid and the air using the area-per-unit volume and a heat transfer coefficient per unit volume, h_V. The area is computed from the geometry and h_V is determined by measurement or by detailed computation within the REV.

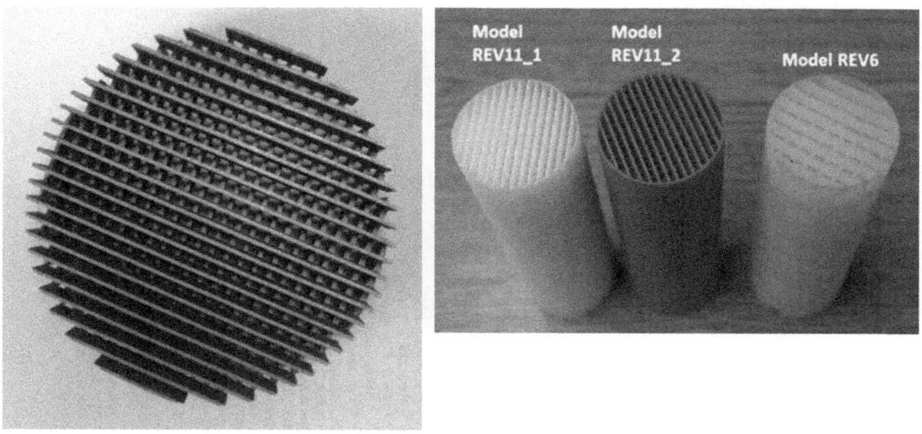

Figure 20.5 Interrupted plate heat exchanger fabricated using ABS plastic or metal powder by 3-D printing [20, 22].

Important to the CAES analysis is the compression work in the air energy equation. Dissipation is usually too small to include. Conduction within the air and within the solid are computed with models from the literature or are measured or computed in the REV calculations. They are usually not very important terms relative to the terms for heat transfer between air and solid. If eddies are present in the interstitial spaces, mixing (transport) by dispersion is included. Modeling terms for dispersion can be measured or computed in the REV calculations. With these modeling parameters, the flow and temperature fields, averaged over the pore regions, can be computed. The overall porous medium model is then used to compute the air temperature and pressure rises due to compression and the thermal energy storage in the solid part of the porous material. Once developed, the models are used to determine the performance of the regenerator, as affected by: (i) material, (ii) geometry, (iii) how the porous medium is distributed within the compression space [23, 24], (iv) the shape of the compressor vessel, and (v) the pumped flow rate vs time during compression [17, 25, 26]; and generally, the design and operation of the compressor.

Figure 20.6 shows the REV geometry used to compute the pore-scale flow and heat transfer modeling terms as well as the flow and solid temperature fields. Modeling terms are extracted for use in the porous medium model. An example is the heat transfer coefficient in Figure 20.7 for various conditions. It is developed from many REV simulations. One can see in these heat transfer coefficient results a transition, with an increasing Reynolds number, from laminar-like flow to turbulent-like flow. Also computed from the REV simulations are models for pressure drop, dispersion, and dissipation to be used in the porous medium model. A comparison of computed efficiencies for the plastic and metal interrupted plate and honeycomb regenerators is given in [22]. When there is no porous medium in the compressor, strong 3-D flow features develop, as computed by 3-D analysis [27]. The heat exchange matrix suppresses the motion of these 3-D flow features.

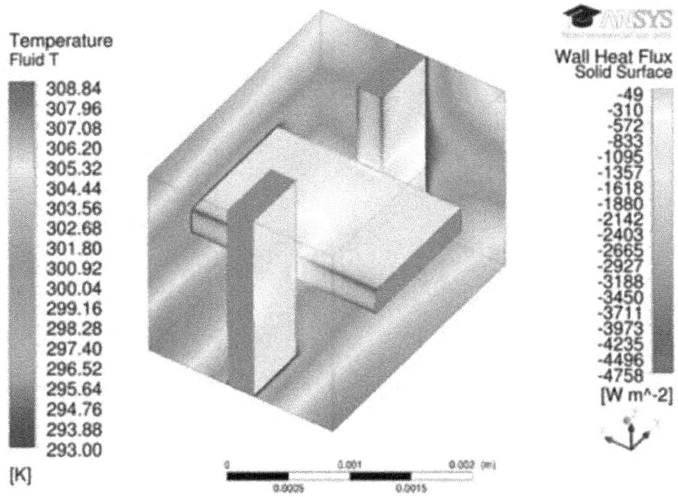

Figure 20.6 Temperature and wall heat flux in a unit cell (REV) of an interrupted plate heat exchanger (from [47]).

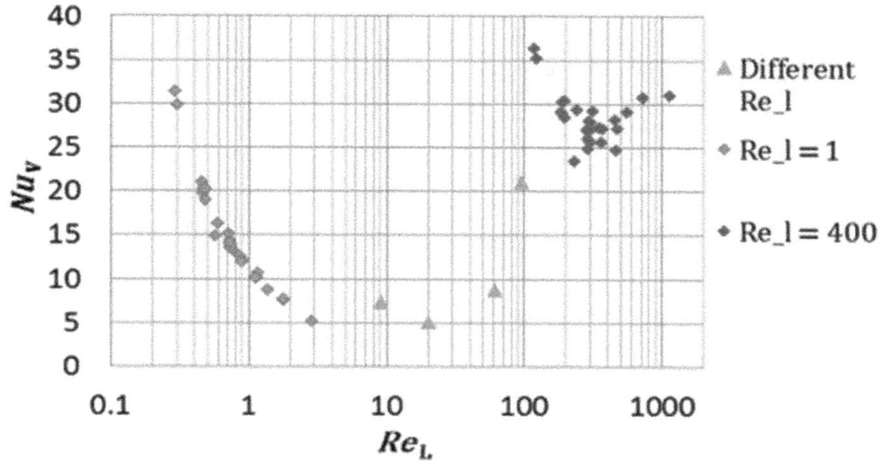

Figure 20.7 Dimensionless volume heat transfer coefficient (Nusselt number) $Nu_V = h_{av} L^2/k$ vs. Reynolds number for the interrupted plate heat exchanger. The different points represent different geometries, as given by different l values [20].

Recently, it was discovered that matrices with plates that are not aligned with the flow but are at an angle of attack give increased heat transfer coefficients with only minor pressure loss penalties. This suggests that a tilted matrix, interrupted plate design as shown in Figure 20.8 will increase heat exchanger performance [28].

Verification of the computed values is essential. In the case of the interrupted plate heat exchanger, computation with the REV model is generally reliable, though such features as flow instabilities leading to mixing and transition to turbulence may not be accurately computed. But in more complex geometries, like bead beds or open-cell porous foams [9], the REV computations are not so reliable, in isolation, and experiments for verification are needed. Detailed data for pore-level processes, like heat transfer coefficients, are difficult to extract but comparisons of gas space volume and bulk temperatures during compression, measured and computed, can be made. Figure 20.9 shows such a comparison.

The efficacy of using a liquid piston with porous media to improve the efficiency/power density trade-off has been demonstrated in both low-pressure (1–10 bar [0.1–1 MPa]) and high-pressure (7–210 bar [0.7–21 MPa]) compression and expansion

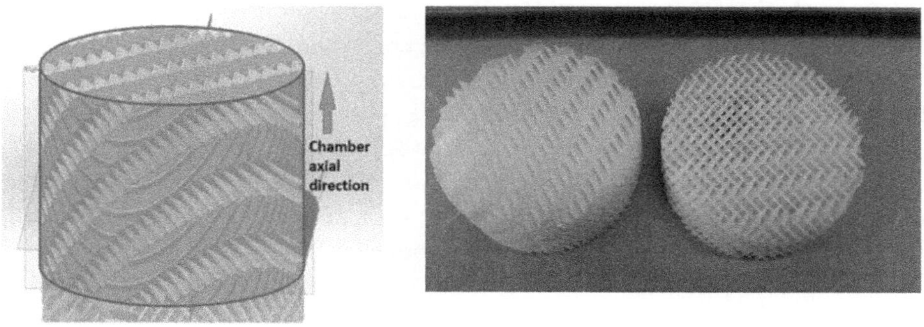

Figure 20.8 Tilted matrix, interrupted plate heat exchanger.

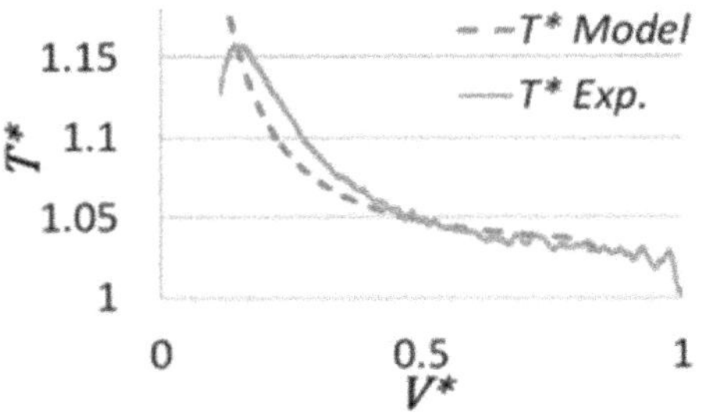

Figure 20.9 Comparison of measured and experimental (from a zero-D model) temperatures. Dimensionless temperature: T/T_{init}; dimensionless air volume: V / V_{init} (from [20]).

in bench top experiments [12–13, 29]. With constant velocity compression/expansion profiles, adding an interrupted plate (2.5 mm plate distance) heat exchanger (Figure 20.6) uniformly in space leads to an over 10 times increase in power density at 90% efficiency (Figure 20.10). The increase means that for the same power capability, the compressor/expander size can be made one-tenth the original size (and proportionately cheaper). It was confirmed, as hypothesized, that the increase in heat transfer area is the dominant contributor to the improvement. A study of the placement of the porous medium in the chamber revealed that the porous media is most useful when it is at the top of the chamber where the high pressure air resides. This suggests that, with a given amount of porous medium solid volume, its distribution in the chamber can have an important effect on performance and should be optimized (see Section 20.3.2).

20.3.2 Optimization of Compression/Expansion Trajectory, Porous Medium Distribution, and Chamber Shape

As shown in Figure 20.2, the choice of the P–V curve determines the work and efficiency of the process. Given heat transfer coefficient (h) and surface area (A), this choice also determines the process time and hence, power. It has been shown through theoretical [14–16] and numerical [17] analyses as well as in experiments [25, 30], that an optimal P–V curve can significantly increase power density for the same efficiency (3–5 times improvement over linear or sinusoidal trajectories). A computationally efficient method has been developed using Dynamic Programming to determine the solution to this Pareto optimal control problem under general constraints and heat transfer correlations [18]. The optimal compression/expansion process typically consists of a sequence of fast-slow-fast segments. In the simple case when the product hA of the heat transfer coefficient, h, and surface area, A, is a constant and there is no constraint on the piston speed, the optimal trajectory consists of an adiabatic segment (infinitely fast), an isothermal segment at an elevated temperature (slow), followed by a final adiabatic segement (fast).

The compression/expansion trajectory has also been optimized in combination with the distribution of the porous media, and the shape of the compression/expansion chamber [18]. Here, the total porosity of the porous media is given but how it is distributed is optimized. The chamber shape is optimized with the constraint on the volume and the maximum length. Table 20.1 shows that the power density of the compressor/expander can be increased by 21 times over the case with uniform chamber shape,

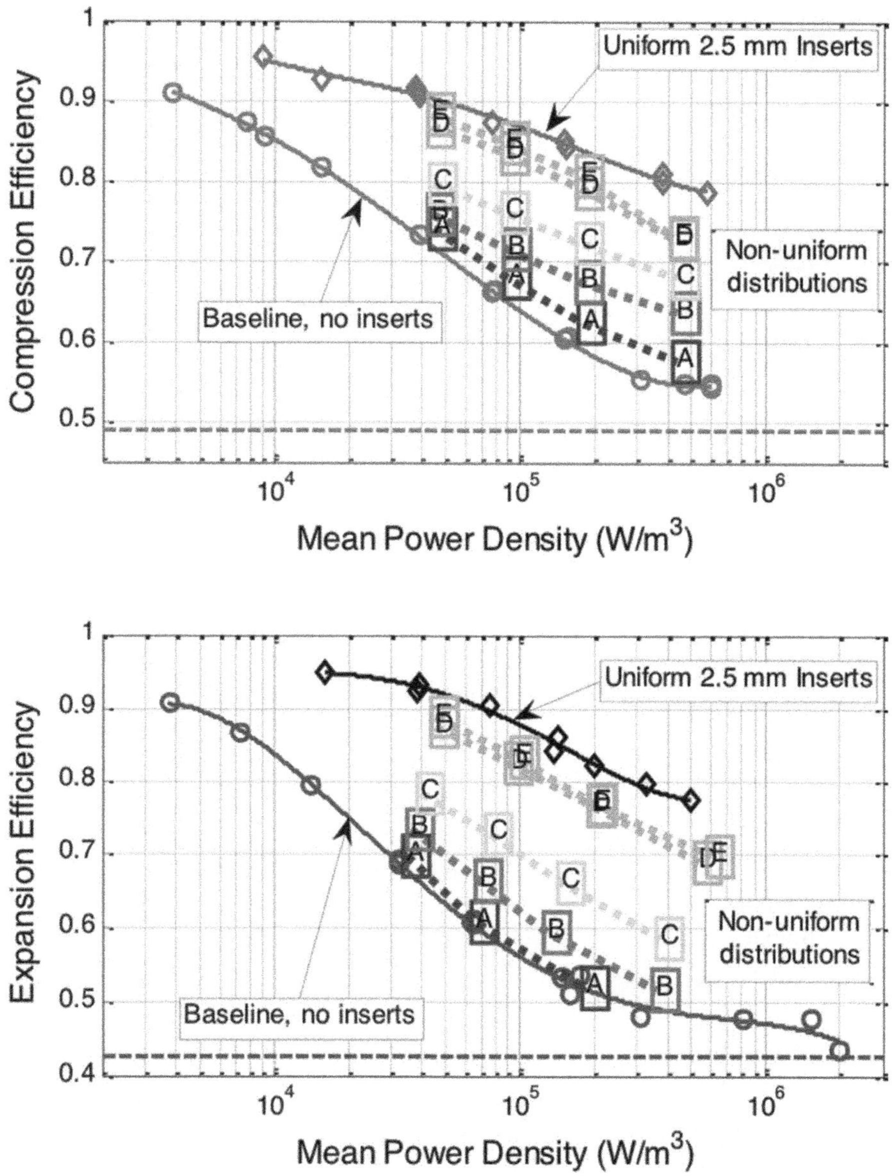

Figure 20.10 Efficiency vs. power-density with and without porous media. Top: Compression; Bottom: Expansion. A, B, C, D, E indicate different porous medium distributions (from [12]).

uniform porous medium distribution and uniform piston speed while maintaining overall efficiency of 92%. This leads to a second stage (7–210 bar [0.7–21 MPa]) compression/expansion time of 1.5s and a power density of 1.47 MW/m^3 (does not include the intake portion of the cycle, which is not limited by heat transfer). Figure 20.11 shows the optimal compression trajectory, porous medium distribution, and chamber shape. The combined effect of introducing a porous medium and of optimizing the trajectory, distribution of porous medium, and chamber shape is that power density has

Table 20.1 Effect of optimization of porosity distribution, chamber shape and compression trajectory (From [18]).

Cases	Porosity	Flow Rate	Shape	Efficiency (%)	Compression Time (s)	Power Density (kW/m^3)
1	Uniform	Constant (43cc/s)	Uniform	92	33	71.2
2	Uniform	Optimal	Uniform	92	10.8	217.3
3	Optimal	Constant (149cc/s)	Uniform	92	9.6	245.6
4	Optimal	Optimal	Uniform	92	3.5	669.3
5	Optimal	Optimal	Optimal	92	1.6	1470

been increased by over 200 times, i.e. for a given power density and desired efficiency, the size of the compressor/expander chamber size has been reduced by 200 times!

An issue with implementing the optimal trajectory is that it requires a large liquid piston pump/motor to accommodate the fast segment of the trajectory, but it must also operate at low displacement for the slow segment. A possible approach to mitigating this issue is with the use of an intensifier as proposed in [31].

20.3.3 Efficient Power-take-off via an Adjustable Linkage Liquid Piston Pump/Motor

The discussions in Sections 20.3.1 and 20.3.2 have focused on the gas compression/ expansion processes that take place in the compression/expansion chamber. The hydraulic (water) pump/motor that provides the liquid piston flow must also be efficient and power-dense. This pump/motor is coupled to the hydraulic pump/motor and the synchronous AC generator. To meet the large variations in flow rate required for the optimal liquid piston trajectory, while operating at the constant speed of the synchronous AC generator, the pump/motor displacement must be variable. The

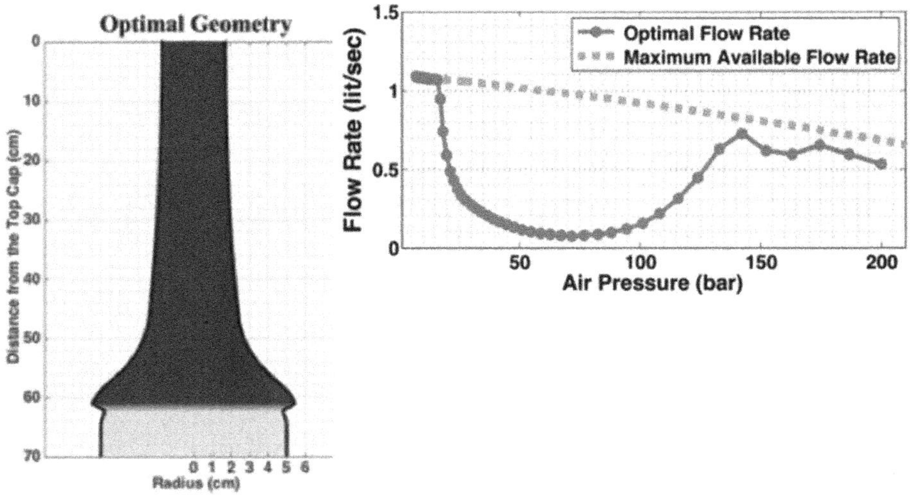

Figure 20.11 Sample optimal compression (liquid piston flow rate) trajectory (left), porous media distribution, and compression chamber shape (right). Blue color indicates porous media; yellow indicates empty.

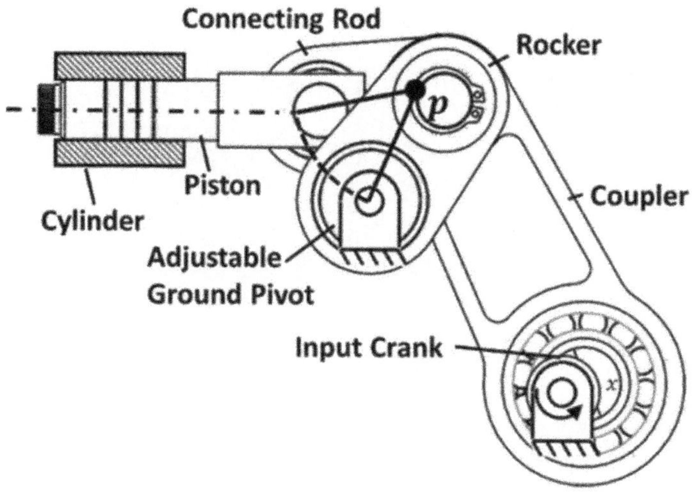

Figure 20.12 Diagram of the adjustable six-bar linkage of the variable displacement linkage pump.

optimal trajectory requires that this pump/motor operates at low displacement and low pressure for a large portion of the liquid piston stroke, which is a region of low efficiency for most variable displacement pumps. Furthermore, most variable displacement pump architectures use the working fluid to lubricate the pump mechanism, creating tribological, corrosion, and leakage challenges when pumping water. For these reasons, a new water hydraulic pump/motor architecture was developed specifically for controlling the liquid piston flow rate in the OA-ICAES system.

The new water pump architecture, the Variable Displacement Linkage Pump (VDLP), uses an adjustable six-bar linkage to drive the pistons. As seen in Figure 20.12, the adjustable linkage is driven by an input crank that causes reciprocation of the piston through the kinematic constraints of the coupler link, connecting rod, and rocker link. By moving the adjustable ground pivot of the rocker link, the stroke length of the piston is varied. As the rocker link and connecting rod are the same length, when the adjustable ground pivot becomes co-linear with the axis of the piston, there is no translation of the piston, resulting in zero pump displacement. In contrast to most variable displacement pump architectures where the unswept volume in the cylinder increases as the displacement decreases, the piston in the VDLP reaches the same top-dead-center position regardless of the linkage displacement. This minimizes the unswept fluid volume and the associated compressible energy loss. There are four kinematic configurations that achieve both constant top-dead-center and reach zero displacement, each with trade-offs regarding piston stroke, linkage footprint area, and transmission angles [32, 33].

An important metric of any pump is the flow ripple, the variation in the flow rate through a cycle. Excessive flow ripple is detrimental in the liquid piston compressor as it increases the acceleration of the liquid piston, which can lead to water–air interface instability at operating frequencies lower than possible with a smooth flow rate. The first way the flow ripple is minimized in the VDLP is through the use of multiple cylinders. The focus for the ICAES work has been on a three-cylinder, inline pump design, where the crank throws are phase shifted 120 degrees

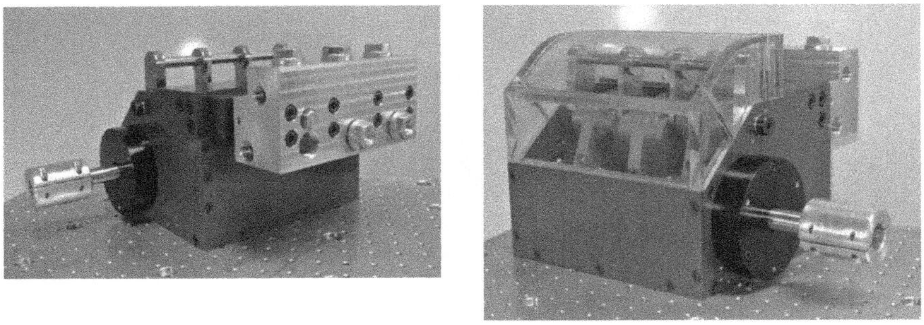

Figure 20.13 Three-cylinder inline VDLP.

(Figure 20.13) although other configurations, such as radial pistons have also been explored. Second, the design of the link lengths provides some control over the displacement trajectory of the pistons throughout the stroke. Designing pump kinematics with consideration for the addition of the flow rate from the multiple cylinders, including effects due to fluid compressibility, allows the flow ripple to be significantly decreased.

The VDLP uses all pin joints, in contrast to the spherical and planar hydrostatic bearings found in axial piston and other pump architectures, which are the largest source of energy loss [34]. Rolling element bearings are used throughout the linkage, including the main and crank pins on the split crankshaft. These bearings are all splash lubricated. The rotation of most of the joints in the linkage decreases as the displacement of the piston decreases, causing the mechanical energy loss to scale with output power. The low friction and scaling energy loss enable the adjustable linkage pump to have good mechanical efficiency across a wide displacement range.

The VDLP can operate with water and other non-lubricating fluids by preventing the pumped fluid from entering the pump case. The ceramic pistons use a high-pressure v-packing seal. Behind the high pressure seal is a drain port and a low-pressure wiper seal. The piston is supported by a crosshead bearing, which is splash lubricated from the pump linkage. The crosshead bearing not only centers the piston in the seal, but also reacts to radial loads generated by the linkage. The positive piston seal and low unswept volume result in good volumetric efficiency.

To maximize the efficiency of the VDLP for the ICAES system, the pump parameters are optimized for the compression/expansion cycle. For the optimization, a mathematical model was constructed of the dynamic behavior of the pump and the primary energy loss mechanisms [33, 35]. The dynamic model accounts for the fluid inertia in the inlet and lines, the dynamics of the inlet and outlet valves, the pressure dynamics in the cylinder, the linkage dynamics, and the fluid dynamics in the manifold. The primary energy loss mechanisms are the friction in the rolling-element bearings, viscous friction and leakage of the piston, throttling of flow across the inlet and outlet valves, and shaft seal friction. The model was validated on three different generations of prototypes, demonstrating good agreement in terms of dynamic behavior and efficiency [36, 37]. For the optimized pump geometry, the model predicts an efficiency of over 90% over all displacements above 10%, which is a significant improvement over typical axial piston pumps, as seen in Figure 20.14.

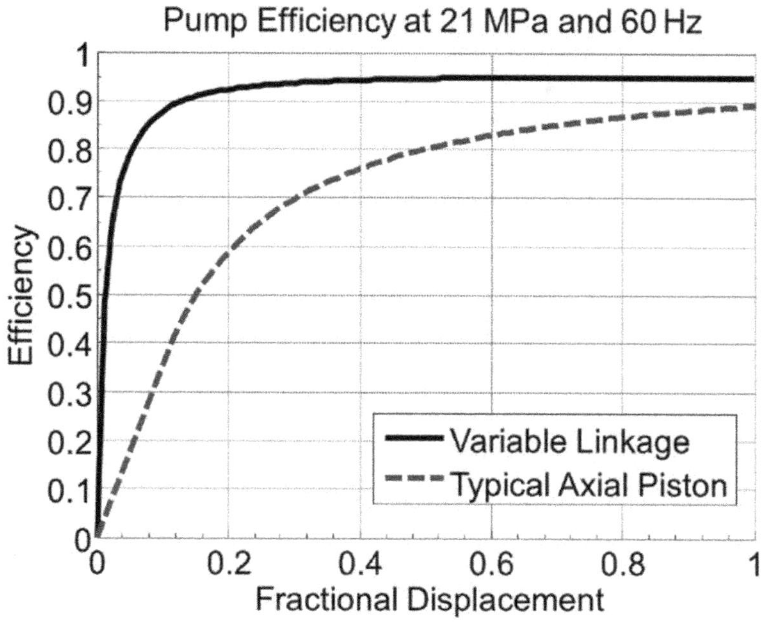

Figure 20.14 Predicted efficiency of the optimized VDLP at 21 MPa and 30 Hz.

20.4 Using Water Droplet Spray to Enhance Heat Transfer

The liquid piston approach above has a significant disadvantage when compressing from, or expanding air to, atmospheric pressure. A large liquid piston pump/motor is needed to provide the liquid flow rate equivalent to the air volume flow rate at atmospheric pressure. Such high-flow, low-pressure operating conditions are ill suited for hydraulic pump/motors. Moreover, advantages of sealing, eliminating dead volume, and optimizing compression/expansion are also less essential at low pressure. Thus, the liquid piston approach will be used for the high pressure stage, starting from 10 bar (1 MPa). This reduces the required liquid flow rate and pump/motor size by 10 times!

For the low pressure ($\sim$1–0 bar [0.1–1 MPa]) stage, a mechanical piston approach, running at high frequency (20 Hz), similar to a crank-shaft engine can be used. For heat transfer, minute water droplets are injected directly into the chamber [38]. Water has a high heat capacity so a small amount ($\sim$1/1000 by volume) of water, if properly distributed in the air space, can keep the process of air compression/expansion nearly isothermal. With smaller sized droplets, the total surface area for heat transfer increases dramatically for a given mass fraction of water. The heat transfer coefficient also increases as droplet size decreases due to the shrinking of the boundary layer. For 20 μm, 50 μm, and 100 μm droplets, the heat transfer coefficients are of the order 6000, 5000, and 3500 $Wm^{-2}K^{-1}$ respectively! Similar to optimizing the compression/expansion trajectory, for a fixed amount of water to be sprayed, the trajectory of water sprays can also be optimized to maximize performance [19].

A schematic of the direct injection spray system is shown in Figure 20.15. The compression chamber is oriented vertically to allow a liquid piston to be used in

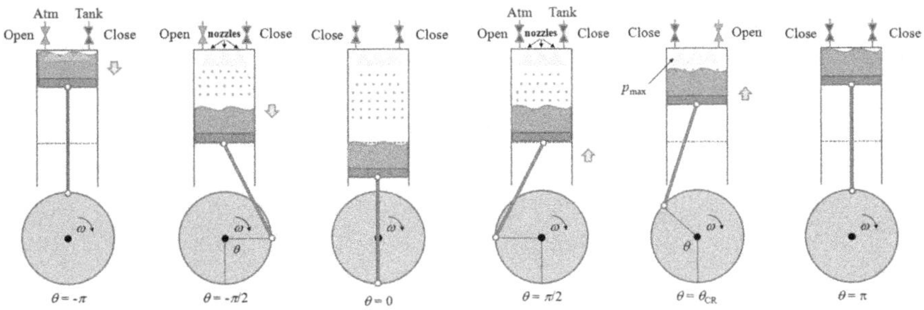

Figure 20.15 Concept of direct injection as a function of stroke angle.

conjunction with spray. As the piston descends, direct injection initiates water spray with air suction simultaneously, so water droplets have filled the chamber when the suction stroke finishes. As the piston ascends, and the compression stroke continues, water spray injection continues to keep the chamber full of droplets, thereby maximizing heat transfer. When the air pressure reaches the prescribed value of compression, the valve is opened to the next compression stage or to the accumulator tank, and the compressed air is pushed out of the compression chamber during the rest of the compression stroke. During expansion, the same amount of heat transfer has to be supplied by droplets at ambient temperature again, to ensure expansion at a constant temperature. This way, the power used to run the compressor during the energy storage stage can ideally be completely recovered during the energy regeneration stage. Such an isothermal process will have an ideal efficiency of 100%. Note that the energy consumed by spraying is very small [10].

To determine the amount of water employed for cooling, the total spray mass is considered. The amount of water injected is defined by using total discharged spray mass in a single compression cycle. This injected mass can be expressed non-dimensionally in terms of the injected mass loading, defined as the ratio of water spray mass discharged (m_{spray}) to the mass of dry air drawn into the piston chamber (m_a), i.e. $ML = m_{spray}/m_a$.

The injected mass loading (ML) is based on the liquid flow rate of a given nozzle, the injection period, and the compressed gas mass dependent on the intake pressure and the initial compression chamber volume. In addition, one may define the transient mass loading (ML_t) based on the instantaneous spray mass aloft in the gas. This parameter is initially zero and increases during the spray injection period. If the droplet mass is never lost, once injected, $ML_t = ML$ at the end of spray injection. However, droplet losses caused by hitting the chamber wall or piston surface will result in $ML_t < ML$. These droplet losses may be counteracted by internal vortices that can help keep the droplets aloft in the gas for a longer time.

Theoretically, to determine overall efficiency, one may characterize the pressure rise as bounded between the isothermal and the adiabatic processes. During compression, one may define the polytropic index (n) as the relationship between pressure increase and volume decrease:

$$\frac{P_2}{P_1} = \left(\frac{V_1}{V_2}\right)^n$$

This index represents the overall performance of the compression with the lowest value being ideal. In the case of isothermal compression (where work is used exclusively to increase pressure and no work is lost to internal energy rise), the temperature is constant. As a result, the pressure is linearly proportional to density (and inversely proportional to volume) during compression so that n is unity. This condition will result if there is high heat transfer between the droplets and the gas. In the case of adiabatic compression, this index will equal the gas specific heat ratio ($\gamma_g = 1.4$ for air). This condition will result in the least amount of work directed at pressure rise since some work is instead used to create a temperature rise. For CAES, this work employed for temperature rise is considered to be wasted, since it is lost to the ambient conditions when the compressed air is stored for longer periods.

The polytrophic index for the case of isothermal compression (minimum work requirement) can be obtained by considering infinitely small drops so that the heat transfer is infinitely fast. In this limit, the polytropic index can be expressed in terms of the liquid spray mass loading and the specific heats of the droplets (c_d) and of the gas (c_p) as in Equation (20.5).

$$n = \gamma_g \frac{1 + ML(c_d / c_p)}{1 + \gamma_g ML(c_d / c_p)} \tag{20.5}$$

This expression represents the smallest possible polytropic index for a given droplet mass loading. When the mass loading becomes infinite, the polytropic index trends to 1, which indicates isothermal compression. Thus, theoretically, discharging a very large amount of liquid in small droplets (to promote fast heat transfer) helps to achieve near-isothermal compression. However, practically, it is difficult for a single nozzle to produce large mass quantities while ensuring small droplets.

If the pressure ratio and volume ratio is known for a certain compression process, the average polytropic index can be calculated. However, the polytropic index for the actual compression process will vary with time depending on the instantaneous heat transfer rate and the transient mass loading. The entire compression process can be discretized into a number of sub-processes, where polytropic indexes can be assumed to be constants for each discrete time interval.

The test conditions investigated are consistent for a single compression/expansion chamber with a geometry typical of conventional piston cylinders. In particular, a cylinder compression chamber with an inner diameter of 15 cm, a length of 30 cm, and a pressure ratio (initial value to final value) of 10 are employed. A single pressure-swirl nozzle orifice is located in the center of the piston top wall. For spray discharge, pressure-swirl nozzles were found to be most effective for liquid piston compression. Herein, the spray-cooling concept is investigated to achieve nearly isothermal compression for a three-stage compression system. The compression ratios for the three stages are 10 : 1, 7 : 1 , and 5 : 1, based on ratios that are reasonable to achieve in terms of sealing and mechanical operation, and leading to a final pressure of 350 bar (35 MPa). For fabrication simplicity, it can be desirable to set the dimensions of the chambers in all three stages to be equal. Three different cases are considered to examine the operational performance by varying the intake/initial pressure and the spray mass injected. Case One is a first-stage cylinder with an initial air pressure of one atmosphere and the same nozzle conditions as [10], which corresponds to a total injected mass loading of 0.12. Case Two is also a first-stage cylinder but the spray flow rate is adjusted to achieve the maximum experimental flow rate, which allows droplets with the same mean diameter as Case One, resulting in an injected mass loading of 1.6. Case Three uses the same spray conditions as Case Two but for a second-stage cylinder with an initial air pressure of 9.6 atmospheres, which resulted in an injected mass loading of 0.185 (due to the higher density and mass of air in the cylinder). Therefore, comparing Cases One and Two allows the effect of mass loading to be understood for a fixed initial pressure, while comparing Cases Two and Three allows the effect of initial pressure to be understood for a fixed injected liquid mass. In addition, comparing Cases One and Three allows the effect of initial pressure to be understood for a similar mass loading.

In the 1-D simulation [38], gas and liquid phase ordinary differential equations are integrated in time with a finite difference approach, while the 2-D simulation [39] employs an Eulerian numerical approach for the gas phase and the Lagrangian approach for the liquid phase. To describe the influence of droplet heat transfer on compression thermodynamics, P–V curves are shown in Figure 20.16 (left). In order to make the first- and second-stage curves be comparable, a pressure ratio is used for the vertical axis. The results show that injected mass loading is a primary indicator for compression performance for both 1-D and 2-D simulations. This is evident by comparing cases with similar mass loadings but different spray conditions, whose P–V curves are close to each other. For large mass loading, both 1-D and 2-D models show a substantial improvement tending toward isothermal conditions as the ML increases. Shown in Figure 20.16 (right), all of the cases are compared with the theoretical value, which assumes infinitely small drop sizes (for a given mass loading) so that droplet heat transfer is immediate and assumes that the average mass loading is based on the total drop mass injected minus the mass of drops hitting the piston surface. For the $ML = 1.6$ case, the efficiency is significantly higher than 90% for both 1-D and 2-D results. The efficiency for this case is somewhat lower than the theoretical result, owing to droplet losses and the thermal inertia of finite droplet diameters. In the lower mass loading cases, the 2-D simulations actually predict higher efficiency than theory, which can be explained by an over-prediction of droplet deposition on the wall surface. This is because of the vortex formed in the chamber, which tends

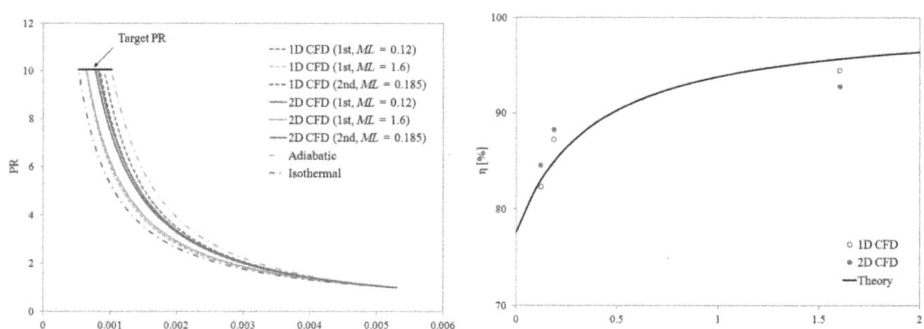

Figure 20.16 Left: Pressure ratio vs. piston volume during compression. Right: Compression efficiency for different injected mass loading (*ML*) values. 1-D and 2-D CFD indicate 1-D and 2-D computational procedures.

to prevent droplet losses. Despite these differences, it is interesting to note that the overall compression efficiency can be reasonably estimated by theory, i.e. is primarily a function of mass loading for present conditions.

20.5 Systems and Control

Control systems are needed for the proposed OA-ICAES in Figure 20.1 for proper operation. They are needed at the plant operation level, and at the supervisory control level.

The plant level controller in [40] uses the control inputs of: (i) the displacement of the pump in the nacelle, (ii) the displacement of the down-tower pump/motor, and (iii) the displacement of the near-isothermal air compressor/expander. The primary objectives are (a) to optimize the wind energy capture above and below the cut-off wind speeds (the so-called regions 2 and 3), and (b) to fulfill the power demand (at the desired 60 Hz frequency without using power electronics). Secondarily, the control system can also choose to maintain the open accumulator pressure close to the nominal value (210 bar [21 MPa]). Maximum wind energy capture requires controlling the wind turbine speed according to the optimal tip speed ratio. By controlling the slip speed of the common shaft relative to the synchronous speed of the generator, desired power/current output is satisfied. Excess or deficit power is made up by the energy storage either with the power-dense hydraulic power path for transient power or the energy-dense pneumatic path for steady power.

As mentioned earlier, the open accumulator has two power paths – a hydraulic path and a pneumatic path, to store energy in or regenerate energy from the storage. The hydraulic path is power-dense whereas the pneumatic path is energy-dense. The control system in [18] uses the hydraulic power path to accommodate high frequency, transient power from the wind or from the demand; and the pneumatic path is used for low-frequency steady power. An interesting consequence of this approach is that the open accumulator absorbs and filters the high-frequency wind gust directly without needing to actuate either the hydraulic or the pneumatic power paths. Figure 20.17 shows this result as well as the ability of the controller to optimize wind energy capture and to meet power demand. Notice that pressure variation is <1 bar even as the accumulator is absorbing the high-frequency wind power variation.

A supervisory controller determines when the system should store or generate energy. In [41], a supervisory controller, which does so in order to optimize the total revenue, is developed. This controller takes into account the components' efficiency characteristics and, currently, assumes that the wind and electricity variations are known in advance. Not surprisingly, the control tends to store energy when the electricity price is low, and generate energy when the electricity price is high. Figure 20.18 shows a sample simulated control results using historical wind and electricity price data. Here, the system is allowed to store energy from the grid when the electricity price is low. As expected the system generates energy when the electricity price is high and stores energy when it is low. Table 20.2 shows a sample comparison of the revenue and other operating statistics for the wind turbine in a one-week summer period under various scenarios.

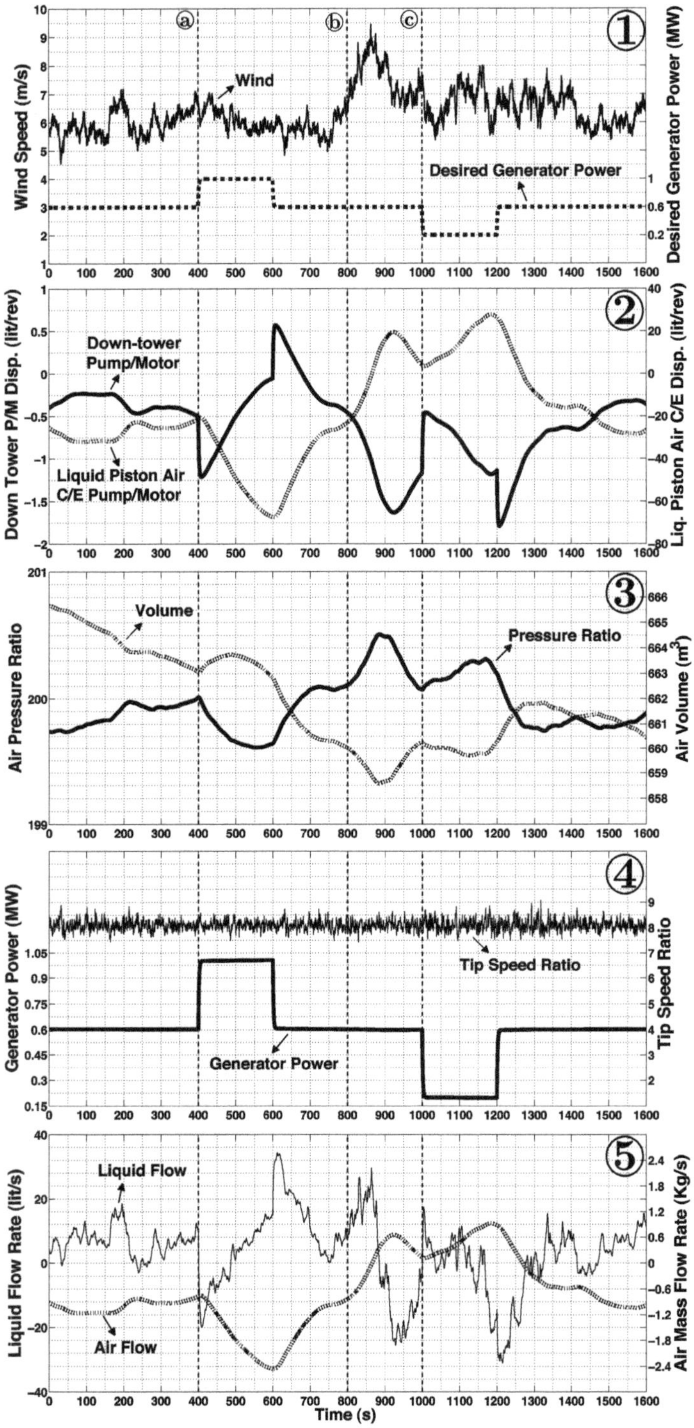

Figure 20.17 Sample results with a stochastic wind speed profile and step changes in power demand (taken from [7]).

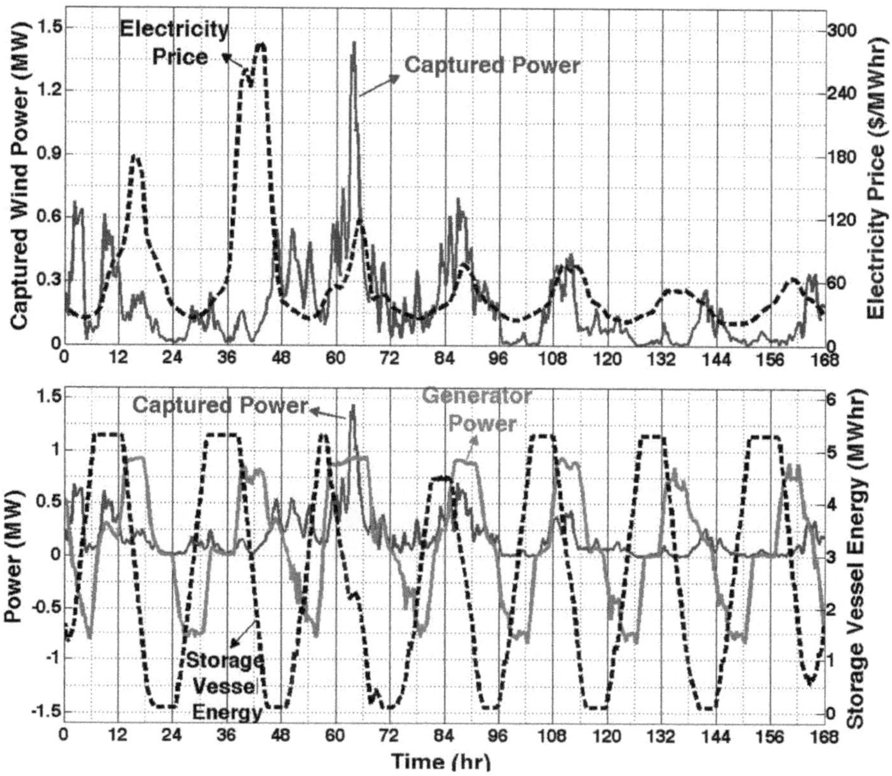

Figure 20.18 Sample supervisory control results for a two-week period in July.

For a typical summer week (when the price variation is large), the storage system increases the revenue by nearly 140% when the system can buy and sell electricity to/from the grid despite the lower efficiency of the hydraulic components compared to the gearbox system. It is also interesting to note that allowing the accumulator

Table 20.2 Energy captured, efficiency, and revenues for a 1.5 MW wind turbine over two 7-day periods July, 2012.

	Wind captured (MWh)	Electricity sold (MWh)	Efficiency (%)	Revenue ($)/% increase over case 1
Conventional wind turbine	30.1	25.7	85.5	1570
Hydrostatic wind turbine w/no storage	30.1	20.7	68.5	1268/-19.3
Proposed system w/3.6 MWh storage (sell, constant pressure)	30.1	18.7	62	2068/ + 31.7
Proposed system w/3.6 MWh storage (sell/buy, constant pressure)	30.1	36.3	64.6	2801/ + 78.3
Proposed system w/3.6 MWh storage (sell, varying pressure)	30.1	20.3	66.8	2353.8/ + 49.9
Proposed system w/3.6 MWh storage (sell/buy, varying pressure)	30.1	49.7	68.4	3737/ + 137.9

pressure to vary rather than be tightly regulated at the nominal pressure, tends to increase revenue by 10–15% and efficiency by up to 4%. Notice that the revenue increase due to price arbitrage depends significantly on the price variation. During winter periods, the revenue increase is much less significant (~14%) [7, 40].

The supervisory controller has provided some important insight into both control and design. For example, we learn that it is advantageous for the open accumulator pressure to remain low (with limits) to increase system efficiency. Also, by simulating the system with different accumulator volumes and compressor/expander power capabilities, one can trade-off between size of the storage and the power capability to gain the same revenue.

20.6 Discussion

In this chapter, we have reviewed the concept of a novel compressed air energy storage (CAES) system that is free of fossil fuel, not tied to geological restrictions, and is well suited when a power source in the mechanical domain is available. For this reason, we anticipate that wind turbines, especially off-shore wind turbines will be ideal applications. However, as shown in the supervisory control results in Section 20.5, the system can also be profitably used for storing excess electricity. In this function, the system serves as a generic storage node on the grid, such as at a substation.

It has often been said that electricity price arbitrage is the least valuable attribute of storage. As an operator, the ancillary control services for the grid, such as frequency regulation or grid stabilization can be much more profitable. Such services are now provided by the so called spinning reserves, which are basically rotating inertias that can respond instantaneously to provide or absorb power mechanically. The proposed OA-ICAES system is well positioned to provide these for two reasons: (i) it preserves the synchronous machine in the system that has a physical rotating inertia, and (ii) extra energy can be provided or absorbed by the storage system through the actuation of the hydraulic power path. Hydraulic actuation typically has high control bandwidth since only a small mass needs to be actuated to control displacements. These physical attributes, together with a control system that can tightly regulate the speed of the synchronous machine, can create a virtual spinning reserve that is as fast as a physical inertia but can absorb/provide a nearly unlimited amount of energy. By controlling the synchronous machine speed, the local frequency at the storage node can be regulated. An intriguing possibility is to see if more can be done by coordinating with the rest of the grid, to help stabilize the rest of grid. More research is needed regarding these possibilities.

The performance of the near-isothermal compressor/expander is critical for the AO-ICAES system. In this chapter, we focused mostly on the heat transfer challenges since they limit the trade-off between power density and efficiency. However, there are other aspects that also need attention. We briefly discuss two of these.

One aspect is the mass transfer processes between the water and air in the system, since air and water share the same volume, both inside the compression/expansion chamber (with either liquid piston or water spray concepts) and also in the storage vessel. Inside the compressor, having air dissolved into the water or trapped by water films prevents that compressed air from being stored in the storage tank. Thus, it represents work that went into the air compressor without the benefit of storing the resulting compressed air. A computation of this process can be found in [41]. On the other hand, evaporation of water and condensation of steam from and to tiny droplets inside the compression/expansion chamber may enhance heat transfer and hence the performance of the compressor/expander. Preliminary results indicate that for high efficiency operation when the compressed air temperature is kept low, the effect of such phase change is not significant [42]. Inside the storage vessel of the open accumulator, high pressure air can also dissolve into the water as governed by Henry's Law. At 200 bar (20 MPa), 3.7 g of air is dissolved in 1 kg of water. This contains 1.5 kJ of energy or 7.7% of the hydraulic energy. To recuperate this energy, care is needed in the design and operation of the hydraulic motor connected to the liquid port of the storage vessel. The energy in the dissolved compressed air can be retrieved by controlling the valve timing of the hydraulic motor such that the air is released from solution and then allowed to fully expand. Recent work on so called digital hydraulics using active valvings would be applicable to this end [43, 44].

Another aspect that needs attention is the propensity for air and water to be trapped inside the porous matrix. During compression, trapped air contributes to increased dead volume and lost work since the compressed air bubbles consume work but do not end up in the storage vessel. Similarly, during expansion, trapped air initially at low pressure is compressed by compressed air from storage, thus reducing work output and efficiency. Incomplete water drainage, on the other hand, reduces the volume for fresh air intake, thus reducing the effective displacement of the compressor. Water that remains can likely bridge the porous medium cells trapping air within the matrix. Without treatment, water hold-up in the ABS interrupted plate heat exchangers can be as much as 80% of the dry weight of the exchangers. To solve this problem, processes for applying durable nano-textured coatings that make the surfaces super-hydrophobic have been developed [45]. With such coatings, water hold-up is decreased by eightfold at ambient pressure. Further work is needed to ensure proper water drainage and to prevent air trapping during high pressure operation in the compressor/expander.

The CAES system described in this chapter is based upon near-isothermal compression/expansion. In our analysis, a conservative assumption is made that the compressed air returns to ambient temperature in the storage before it is expanded for regeneration. If the storage vessels can be kept warm, either through insulation or by heating (e.g. via solar or waste heat), energy output will increase. Taken to the extreme, when the compressed air can be reheated before expansion, to the temperature at the exit of the compression phase, the process resembles the approaches of advanced adiabatic CAES [46] and thermal energy storage (TES). It should be noted, however, the design trade-offs between the isothermal and adiabatic regimes are quite different. For example, for effective TES, a high temperature must be attained and heat should not escape during the compression process itself. This is in contrast to isothermal CAES where temperature should be kept low and heat transfer during compression is desirable.

20.7 Conclusions

In this chapter, an open accumulator isothermal compressed air energy storage system (OA-ICAES) has been described. The open accumulator architecture offers high energy-density, high transient power-density, and fast response; whereas the isothermal compressor/expander offers improved efficiency and power density. It can be coupled with a wind-turbine to provide predictable and on-demand renewable energy, and to regulate grid frequency. The system can be sited anywhere (unlike pumped hydro) and does not need fossil fuel (unlike existing, conventional CAES). Its competitive advantages over various electrical batteries are significantly lower cost, higher reliability, higher power density, greater longevity, and unlimited charge/recharge cycles.

To realize this concept, the isothermal compressor/expander with augmented heat transfer capability (via the liquid piston with porous media and spray cooling/warming concepts) and efficient power take-off (PTO) was developed. Optimization of the trajectory, porous medium distribution, and compressor/expander chamber shape have a dramatic effect on the efficiency–power-density trade-off. Control systems that operate at the plant and supervisory level have also been developed to enable the best system performance.

By offering a cost-effective and efficient grid-scale energy storage concept, larger amounts of renewable resources can be better and more reliably integrated into the electrical grid.

Acknowledgment

This work is supported by the National Science Foundation under Grant EFRI 1038294 and the Institute for Renewable Energy and Environments (IREE) at the University of Minnesota under Grant RM-0027-11. Many post-doctoral and graduate student researchers contributed greatly to this work, in particular, Dr Mohsen Saadat, Dr Farzad Shirazi, Mr Jacob Wieberdink, Dr Shawn Willheim, Mr Bo Yan, and Dr Chao Zhang.

References

[1] Spera, D. (2009). *Wind Turbine Technology – Fundamental Concepts of Wind Turbing Engineering*, 2e. New York: ASME Press.

[2] Green, J., Bowen, A., Fingersh, L., and Wan, Y. (2007). Electrical collection and transmission for offshore wind power. National Renewable Energy Laboratory (NREL), CO.

[3] Musial, W. and Butterfield, S. (2004). Future for offshore wind energy in the United States. National Renewable Energy Laboratory (NREL), CO.

[4] Carnegie, R., Gothan, D., Nderitu, D., and Preckel, P. (2013). *Utility scale energy storage systems – Benefits, applications and technologies*. State Utility Forecasting Group.

[5] Elmegaard, B. and Brix, W. (2011). Efficiency of compressed air energy storage. *The 24th International Conference on Efficiency, Cost, Optimization, Simulation and Environmental Impact of Energy Systems.*

[6] Li, P., Loth, E., Simon, T., et al. (2011). Compressed air energy storage for offshore wind turbines. *Presented at the IFPE (International Fluid Power Exposition)* (22–26 March 2011). Las Vegas, NV.

[7] Saadat, M. and Li, P.Y. (January 2015). Modeling and control of an open accumulator compressed air energy storage (CAES) system for wind turbines. *Applied Energy* 137: 603–616.

[8] Van de Ven, J.D. and Li, P.Y. (October 2009). Liquid piston gas compression. *Applied Energy* 86 (10): 2183–2191.

[9] Zhang, C., Wieberdink, J., Shirazi, F., et al. (2013). Numerical investigation of metal-foam filled liquid piston compressor using a two-energy equation formulation based on experimentally validated models. IMECE2013-63854, *2013 ASME-IMECE conference*, San Diego, CA.

[10] Qin, C., Loth, E., Li, P.Y., et al. (2014). Spray cooling concept for wind based compressed air energy storage. *Journal of Renewable and Sustainable Energy* 6 (4): 0423125.

[11] Li, P.Y., Van de Ven, J.D., and Sancken, C. (2007). Open accumulator concept for compact fluid power energy storage. *Proceedings of the 2007 ASME-IMECE*, Seattle, WA.

[12] Wieberdink, J. (2014). Increasing efficiency and power density of a liquid piston air compressor/expander with porous media heat transfer elements. MS Thesis, Department of Mechanical Engineering, University of Minnesota.

[13] Wieberdink, J., Li, P.Y., Simon, T.W., and Van de Ven, J.D. (2018). Effects of porous media insert on the efficiency and power density of a high pressure (210 bar) liquid piston air compressor/expander – an experimental study. *Applied Energy* 212: 1025–1037.

[14] Sancken, C. and Li, P.Y. (2009). Optimal efficiency-power relationship for an air motor-compressor in an energy storage and regeneration system. *ASME Dynamic Systems Control Conference / Bath Fluid Power and Motion Control Symposium.* Hollywood, CA. Vol. 48937, pp. 403–410.

[15] Rice, A.T. and Li, P.Y. (2011). Optimal efficiency-power tradeoff for an air motor/compressor with volume varying heat transfer capabilities. *ASME Dynamic Systems and Control Conference*, Arlington, VA. Vol. 54754, pp. 145–152.

[16] Rice, A.T., Li, P.Y., and Sancken, C.J. (09 October 2017). Optimal efficiency-power tradeoff for an air compressor/expander. *ASME Journal of Dynamic Systems, Measurement and Control* 140 (2): 021011.

[17] Saadat, M., Li, P.Y., and Simon, T.W. (2012). Optimal trajectories for a liquid piston compressor/expander in a compressed air energy storage system with consideration of heat transfer and friction. *American Control Conference*, Montreal, Canada. pp. 1800–1805.

[18] Saadat, M. and Li, P.Y. (2015). Combined optimal design and control of a near-isothermal air compressor/expander for a CAES sysem for wind turbine. *ASME Dynamic Systems and Control Conference*, Columbus, OH. ASME Paper No. DSCC2015-9957, 10.

[19] Saadat, M., Shirazi, F., and Li, P.Y. (July 2013). Optimal trajectories of water spray for a liquid piston air compressor. *ASME 2013 Summer Heat Transfer Conference HT2013-17611*, Minneapolis, MN.

[20] Zhang, C., Shirazi, F., Yan, B., et al. (2013). Design of an interrupted-plate heat exchanger used in a liquid-piston compression chamber for compressed air energy storage. HT2013-17484, *ASME Summer Heat Transfer Conference*, Minneapolis, MN.

[21] Zhang, C., Simon, T.W., Li, P.Y., and Van de Ven, J. (2015). Numerical modeling of three-dimensional heat transfer and fluid flow through interrupted plates using unit cell scale. *Journal of Porous Media* 6 (2): 145–158, and *Proceedings of the 15th International Heat Transfer Conference, IHTC-15*, August 10–15, 2014, Kyoto, Japan.

[22] Zhang, C., Simon, T.W., Wieberdink, J., et al. (2014). Numerical analysis of heat exchangers used in a liquid piston compressor using a one-dimensional model with an embedded two-dimensional sub-model. IMECE2014-38567, *ASME 2014 International Mechanical Engineering Congress and Exposition* (14–20 November). Montreal, Canada.

[23] Zhang, C., Simon, T.W., and Li, P. (2013). Optimization of the axial porosity distribution of porous inserts in a liquid piston gas compressor using a one-dimensional formulation. IMECE2013-63862, *2013 ASME-IMECE Conference*, San Diego, CA.

[24] Zhang, C., Yan, B., Wieberdink, J., et al. (2013). Thermal analysis of a compressor for application to compressed air energy storage. IWHT2013-001, *Proceedings of IWHT2013 2nd International Workshop on Heat Transfer Advances for Energy Conservation and Pollution Control* (18–21 October 2013). Xi'an, China.

[25] Shirazi, F., Saadat, M., Yan, B., et al. (2013). Optimal control experimentation of compression trajectories for a liquid Piston Air Compressor. *ASME 2013 Summer Heat Transfer Conference*, HT2013-17613, Minneapolis, MN.

[26] Shirazi, F., Saadat, M., Yan, B., et al. (2013). Iterative optimal and adaptive control of a near isothermal liquid piston air compressor in a compressed air energy storage system. 2013 *American Control Conference*, Washington, DC. pp. 2934–2939.

[27] Zhang, C., Simon, T.W., and Li, P.Y. (2012). Storage power and efficiency analysis based on CFD for air compressors used for compressed air energy storage. IMECE2012-88985, Houston, TX.

[28] Zhang, C. (2015). CFD simulations and thermal design for application to compressed air energy storage. PhD Thesis, Mechanical Engineering Department, University of Minnesota.

[29] Yan, B., Wieberdink, J., Shirazi, F., et al. (2015). Experimental study of heat transfer enhancement in a liquid piston compressor/expander using porous media inserts. *Applied Energy* 154: 40–50.

[30] Saadat, M., Srivatsa, A., Li, P.Y., and Simon, T.W. (2016). Air compression performance improvement via trajecotry optimization – experimental validation. *2016 ASME Dynamic Systems and Control Conference*, Minneapolis, MN, #9825.

[31] Li, P.Y. and Saadat, M. (2015). An approach to reduce the flow requirement for a liquid piston air compressor/expander in a compressed air energy storage (CAES) system. *2nd Offshore Energy and Storage Symposium (OSES)*, 1–3 July, Edinburgh, UK.

[32] Wilhelm, S. and Van de Ven, J. (2011). Synthesis of a variable displacement linkage for a hydraulic transformer. *ASME International Design Engineering Technical Conference*, Washington, DC.

[33] Wilhelm, S. and Van de Ven, J.D. (2013). Design and testing of an adjustable linkage for a variable displacement pump. *Journal of Mechanisms and Robotics* 5 (4): 041008.

[34] Ivantysyn, J. and Ivantysynova, M. (2001). *Hydrostatic Pumps and Motors: Principles, Design, Performance, Modelling, Analysis, Control and Testing*. New Dehli, India: Tech Books International.

[35] Wilhelm, S. and Van de Ven, J.D. (2014). Design of a variable displacement triplex pump. *National Conference on Fluid Power, International Fluid Power Exposition*, Las Vegas, NV.

[36] Wilhelm, S. and Van de Ven, J.D. (2014). Efficiency testing of an adjustable linkage triplex pump. *ASME/BATH 2014 Symposium on Fluid Power & Motion Control*, FPMC2014-7856. Bath, UK.

[37] Wilhelm, S. and Van de Ven, J.D. (2015). Adjustable linkage pump: Efficiency modeling and experimental validation. *ASME Journal of Mechanisms and Robotics* 7 (3):031013. https://doi.org/10.1115/1.4028293

[38] Qin, C. and Loth, E. (2014). Liquid piston compression efficiency with droplet heat transfer. *Applied Energy* 114: 539–550

[39] Qin, C. and Loth, E. (2016). Simulation of spray direct injection for compressed air energy storage. *Applied Thermal Engineering* 95: 24–34

[40] Saadat, M. and Li, P.Y. (2014). Revenue maximization of electricity generation for a wind turbine integrated with a compressed air energy storage system. *Proceedings of the 2014 American Control Conference*, Portland, OR, pp. 1560–1565.

[41] Huang, L., Simon, T., Li, P.Y., and Hummelt, E.J. (2015). Analysis of dissolved gas in the application of liquid piston gas compression. *1st Thermal and Fluid Engineering Summer Conference, TFESC* (9–12 August 2015). New York City, NY: TFESC-12869.

[42] Srivatsa, A. and Li, P.Y. (2016). The effect of moisture on the efficiency and power density of a liquid piston air compressor/expander. *2016 ASME Dynamic Systems and Control Conference*, Paper 9957, Minneapolis, MN.

[43] Linjama, M. (2011). Digital fluid power – state of the art. *Twelfth Scandinavian International Conference on Fluid Power*, Tampere, Finland.

[44] Ehsan, M., Rampen, W., and Salter, S. (2000). Modeling of digital-displacement pump-motors and their application as hydraulic drives for nonuniform loads. *ASME Journal of Dynamic Systems Measurement and Control* 122: 210–215.

[45] de Samaniego Steta, F. (2010). Modeling of an Advanced Adiabatic Compressed Air Energy Storage (AA-CAES) unit and an optimal model-based operation strategy for its integration into power markets. MS Thesis, ETH, Zurich.

[46] Milionis, A., Dang, K., Prato, M., and Loth, E. (2015). Liquid repellent nanocomposites obtained from one-step water-based spray. *Journal of Materials Chemistry A* 3 (24) 12880–12889.

[47] Zhang, C., Li, P. Y., Van de Ven, J. D., & Simon, T. W. (2017). Oscillatory Flow Regenerative Heat Exchangers for Application to Energy Storage Systems. *Proceeding of Second Thermal and Fluids Engineering Conference.*

Part IV

Chemical Energy Storage

21

Hydrogen (or Syngas) Generation – Solar Thermal

Jonathan Scheffe, Dylan McCord, and Diego Gordon

University of Florida
Department of Mechanical and Aerospace Engineering Gainesville, FL 32611

21.1 Introduction

Solar energy is emitted from the sun in the form of electromagnetic radiation that closely approximates the spectral distribution of a blackbody at 5777 K, as described by Planck's Law [1]. Due to the inverse square law, the total solar irradiance decreases as the radiation travels from the sun to the earth's atmosphere, where it is approximately 1367 W m^{-2} (G_{on}) [1].[1] As the solar radiation penetrates earth's atmosphere, it is attenuated through light scattering and absorption, further decreasing the total irradiance (G_n) [1].[2] The resultant solar spectral irradiance for air mass ratios (m)of 1 and 1.5, as well as extraterrestrial and that of a blackbody at 5777 K are shown in Figure 21.1 [2].[3] As seen for $m = 1$ (gray) and 1.5 (gray) solar irradiance decreases for small wavelengths due to scattering and at higher wavelengths in selected bands due to absorption, primarily by H_2O, CO_2, and O_3. The resultant total irradiation for $m = 1$ and 1.5 are 934 and 814.9 W m^{-2}, respectively. As seen, the majority (48%) of the solar radiation falls in the visible range (380 nm < λ < 780 nm), followed by 46% in the infrared (λ > 780 nm) and 6% in the ultraviolet (λ < 380 nm). The sun's energy may be considered as discrete packets of quantized energy, or photons, whose energy (E_{photon}) is dependent on its wavelength (λ), or frequency (ν), as described mathematically by:

$$E_{photon} = h\nu = hc/\lambda \tag{21.1}$$

where h is Planck's constant (6.6256×10^{-34} J s or 4.1357×10^{-15} eV s) and c is the speed of light specific to the medium through which it is traveling.

Because of earth's rotation on its axis and the varying declination of the sun with respect to earth's equator, the spectral irradiance varies with time of day and year for a fixed position on earth's surface [1]. Therefore, as solar energy conversion technologies become cheaper (such as photovoltaics), there is a growing demand to store this energy in a form where it can be used on demand rather than relying on traditional fossil technologies when the sun isn't shining. The conversion of solar energy into

1 G_{on} is the irradiance on a surface normal to propagation outside earth's atmosphere.
2 G_n irradiance on a surface normal to propagation inside earth's atmosphere.
3 m is a ratio of the atmosphere mass through which radiation passes to that of the atmosphere mass through which it passes when the sun is at its zenith.

Advances in Energy Storage: Latest Developments from R&D to the Market, First Edition.
Edited by Andreas Hauer.
© 2022 John Wiley & Sons, Ltd. Published 2022 by John Wiley & Sons, Ltd.

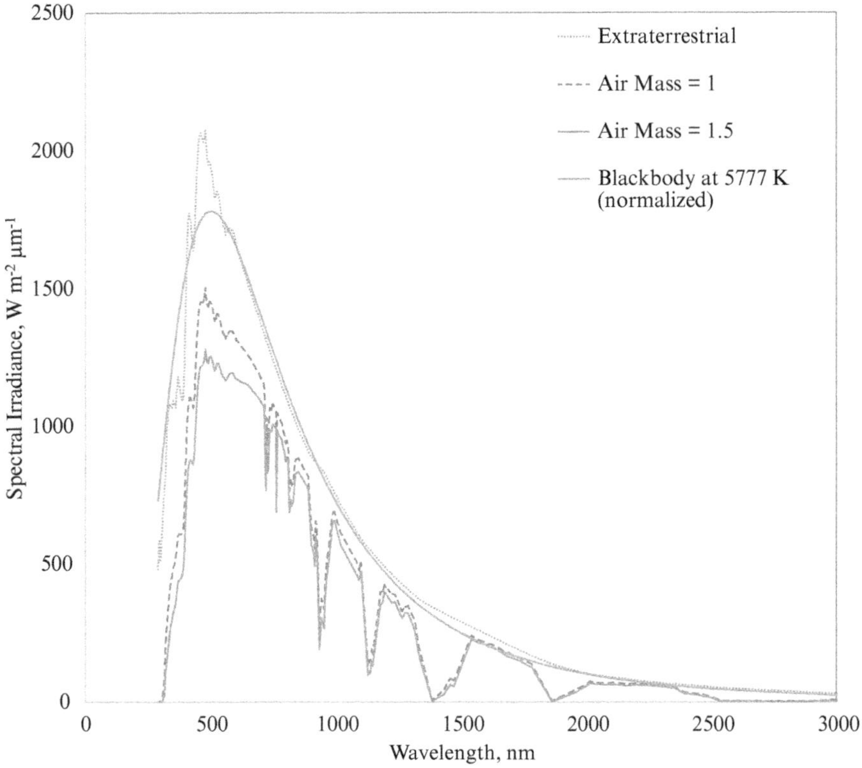

Figure 21.1 Solar spectral irradiance versus wavelength for extraterrestrial radiation, $m = 1$ and $m = 1.5$. For reference the normalized spectral distribution of a blackbody at 5777 K is included. For air mass calculations, $\alpha = 0.66$ and $\beta = 0.085$.

chemical means (i.e. hydrogen or syngas),[4] hereby referred to as solar fuels, is one pathway among several (e.g. thermal, batteries,[5] pumped hydro, compressed air) that has the potential to be cost effective and efficient, while at the same time capable of converting solar energy into transportable fuels such as hydrogen, gasoline, diesel, and kerosene, several of which are compatible with our energy infrastructure as it stands today.

21.1.1 Storage of Solar Energy in Chemical Bonds

Motivating chemical energy storage is the fact that compared to batteries, the gravimetric energy density of fuels such as diesel are 100 times greater [3]. Further, fuels such as diesel, gasoline, and kerosene are readily integrated with our current energy infrastructure and offer the possibility of long term, seasonal storage. The conversion of solar energy into fuels may involve either the direct (photons or thermal energy) or

4 Synthesis gas (syngas) is a mixture of H_2 and CO, usually with a H_2/CO ratio from 1/1 to 3/1.
5 While batteries do constitute a form of chemical energy storage, they are more formally an electrochemical energy storage pathway and will not be discussed in this chapter.

indirect (electrons from an alternative storage technology or stored thermal)[6] utilization of solar energy to drive an endothermic chemical reaction while the sun is plentiful. This can be described easily using the example of water dissociation into H_2 and O_2, as shown below.

$$H_2O(l) \leftrightarrow H_2(g) + 0.5O_2(g)\left(\Delta h = 285.8 \text{ kJ mol}^{-1}\right) \tag{21.2}$$

In the forward direction the reaction is not spontaneous and requires energy to proceed. This energy is equal to the enthalpy of reaction (Δh) and for water dissociation is 285.8 kJ per mol H_2O or 2.962 eV at 298.15 K [4]. When reversing the reaction, the reaction enthalpy may be released to provide heat and drive a thermodynamic cycle such as a fuel cell [5] or heat engine [6].

The conversion of sunlight to fuels (H_2 itself being one) may be accomplished through a variety of pathways as indicated in Figure 21.2, some of which even occur in nature such as photosynthesis.[7] In all scenarios, the primary energy source is photons from the sun and the primary feedstocks to be considered here are H_2O and/or CO_2. Only these feedstocks are considered as they are the primary byproducts of high energy density liquid fuel or H_2 combustion and constitute a convenient form of energy storage that has the potential to operate in a closed loop cycle (i.e. no net emissions) if coupled with sequestration.

Photons ("Energy Input," far left side of Figure 21.2) may be used to drive the process of interest either through its conversion to a free electron or through absorption and the generation of heat ("Converted To," middle of Figure 21.2). These energy

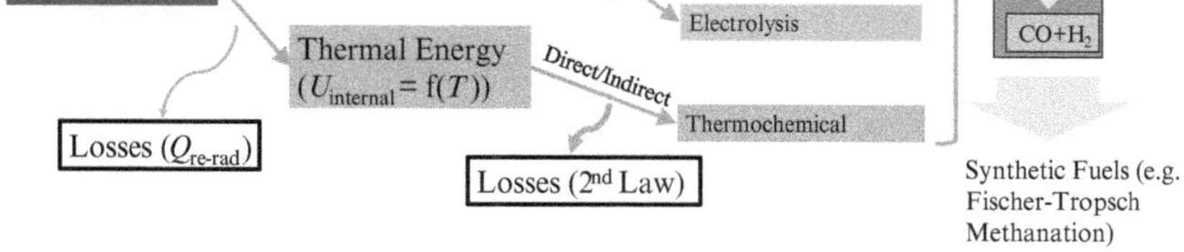

Figure 21.2 Potential pathways to convert solar energy to combustible fuels. The two primary pathways involve conversion of photons to free electrons or their absorption as thermal energy. The processes shown here are not exhaustive.

6 This is similar to the working principle of concentrated solar power plants coupled with thermal storage. There, a working fluid such as molten salt is heated during the day and stored in a thermal reservoir. The thermal energy from the reservoir is then extracted on demand to run a power plant 24/7 [1].

7 Photosynthesis is essentially the conversion on photons, water, and carbon dioxide to carbohydrates. The reversible reaction, which releases energy can easily be seen by burning plant matter which produces heat.

forms may then be used to drive the endothermic reduction of H_2O and/or CO_2 via a variety of processes ("Processes," right side of Figure 21.2) to produce either H_2 and or synthesis gas ($CO + H_2$), the building blocks of synthetic fuels.[8] For many catalytic pathways that convert H_2O and or CO_2 to fuels this is a simplified explanation; rather than produce synthesis gas the hydrocarbon fuel is directly converted [7]. For example, consider the photo-assisted catalytic conversion of H_2O and CO_2 into methane or other hydrocarbon fuels [8].

21.1.1.1 Electron Driven Processes

The conversion of photons to free electrons may be achieved in a semiconductor by ejecting an electron from the semiconductor's valence band to the conduction band; this occurs if the photon energy is greater than the semiconductor's bandgap energy (E_G). For example, E_G of Silicon (Si), a ubiquitous semiconductor used in the solar industry, is 1.12 eV [9], meaning that photons with wavelengths equal to or less than 1.104 μm can eject electrons to the conduction band with exactly this energy; photons with larger wavelengths are either absorbed as heat, reflected, or transmitted. The excess energy associated with smaller wavelength photons is also primarily released as heat.[9] The ejected electrons may then be used to drive a variety of processes, either directly or indirectly, such as those highlighted in Figure 21.2. Direct usage refers to applications in which electrons are ejected and utilized within the same process, such as photoelectrochemical production of H_2 [10], where the semiconductor and electrode exist as a single photoelectrode. Indirect usage refers to the coupling of two independently operating processes, such as photovoltaic energy production to generate electrical energy, followed by electrolysis of water.

21.1.1.2 Thermally Driven Processes

Photons may also be absorbed thermally in a medium to generate heat, which may then be used for a variety of applications (a straightforward and ubiquitous usage is solar driven hot water heaters).

However, in this chapter we are concerned with applications in which the heat is used to drive chemical reactions; hereby referred to as thermochemical reactions. Thermochemical reactions may refer to simple reversible chemical reactions in which the reaction is driven in the forward direction when energy is available, then reversed to release heat on demand (e.g. $Co_3O_4\,(s) \leftrightarrow 3CoO\,(s) + 0.5O_2\,(g)$) [11], or reactions in which the solar derived heat is converted to combustible fuels such as H_2 or synthesis gas. The conversion to fuels is typically a more energetic process than the former and requires higher temperatures (usually between 1000 and 2000 K) because the nature of gases from which they are derived; namely H_2O or CO_2 which are inherently extremely stable molecules. Because the entire solar spectrum is capable of being utilized if photons are absorbed as heat,[10] and the reactions operate at elevated temperatures, these

8 Synthesis gas is the basic building block of a range of synthetically derived fuels (e.g. diesel, kerosene) via catalytic pathways such as Fischer-Tropsch synthesis or methanation.

9 For most processes that utilize electrons, the production of heat is usually not desirable.

10 Either using materials that have high absorptivity over the entire solar spectrum, such as Pt black, or through the design of a blackbody enclosure which ensures that there are enough internal reflections that incident photons are eventually absorbed.

processes offer a thermodynamically attractive pathway to solar fuel production [12]. A more detailed discussion follows in Section 21.2.

21.1.2 Solar Concentration and Absorption as Heat

In order to achieve temperatures suitable to drive thermochemical reactions, sunlight is usually concentrated,[11] through a variety of possible mechanisms, to increase the radiative flux on an absorbing surface. Usually 3-D point focused concentration (as opposed to lower concentration line focus 2-D technologies), such as a parabolic dish, heliostat field or Fresnel concentrator, is required to achieve the concentration ratios necessary (e.g. > 500 suns) [1, 13–14] for driving these high temperature processes. An exemplary demonstration of 3-D optical concentration shown in 2-D is shown in Figure 21.3 using a parabolic shape. Here, incident photons with irradiance G over and area A_1, are reflected to a focal plane with irradiance G_{in} and area A_2. The reflected photons are not reflected to a single point because of imperfections in the mirror surface and a non-zero solid angle formed between the earth and sun (i.e. sun rays are not perfectly parallel). Assuming perfect reflectivity and no spillage (radiation that does

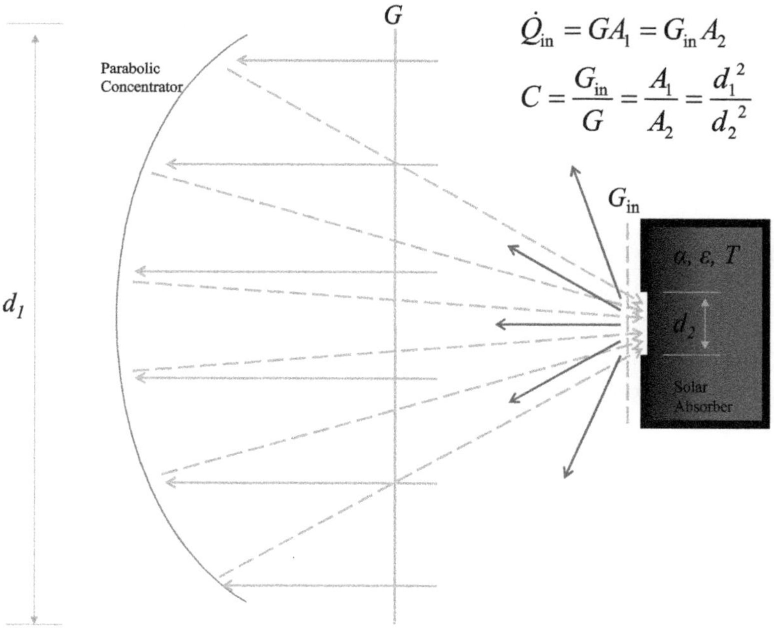

Figure 21.3 Schematic depicticing solar concentration in three dimensions using a 2-D parabolic shape. Radiation (solid lines) is incident normal to the projected area of the parabolic surface, where it is reflected towards its focal point (dashed lines). The concentrated radiation is then absorbed in a solar absorber which re-radiates to its surroundings at a rate proportional to its temperature to the fourth power.

11 Concentration (C) refers to the use of reflecting or refracting optics to concentrate sunlight beyond the typical value of 1 kW m^{-2}, or 1 sun, to values as high as 10 000 kW m^{-2}, or 10 000 suns.

not fall within the aperture), then the total power ($\dot{Q}_{in}$) at each plane (shown in dashed and solid vertical lines) should be equivalent according to the first law of thermodynamics, and thus $GA_1 = G_{in}A_2$. The ratio of G_{in}/G, or A_1/A_2 in the case of perfect reflectivity, is referred to as the optical concentration ratio (C) and is a measure of the increase in radiative flux as a result of concentration. Any reflectivity less than unity or spillage losses will cause the optical concentration to decrease in a proportional manner. Refer to [15] for a general discussion and for a more detailed discussion refer to a textbook on geometric optics [16–17].

Following concentration and entrance through the aperture, the thermal radiation is absorbed in a cavity or receiver as heat, some of which then re-radiates back to the surroundings (shown as dark arrows). This re-radiation is referred to as radiative losses ($\dot{Q}_{rad}$) that inherently occur as temperature increases. Overall, $\dot{Q}_{rad}$ is proportional to the emissivity (ε) of the absorbing cavity (for a blackbody $\varepsilon = 1$), the Stephan-Boltzmann constant ($\sigma = 5.67 \times 10^{-8}$ W m^{-2} K^{-4}), the aperture area through which radiation passes, in this case A_2, and its temperature to the fourth power, shown in Equation (21.3).

$$\dot{Q}_{rad} = \varepsilon\sigma A\left(T^4 - T_{surroundings}{}^4\right) \approx \varepsilon\sigma AT^4 \tag{21.3}$$

The absorption efficiency (η_{abs}) of solar irradiance to heat is defined as the amount of radiation absorbed ($\dot{Q}_{abs}$) divided by $\dot{Q}_{in}$. Thus, mathematically η_{abs} decreases with temperature to the fourth power and increases with increasing concentration ratio, as shown in Equation (21.4).

$$\eta_{abs} = \frac{\dot{Q}_{abs}}{\dot{Q}_{in}} = \frac{\dot{Q}_{in} - \dot{Q}_{rad}}{\dot{Q}_{in}} = 1 - \frac{\varepsilon\sigma T^4}{GC} \tag{21.4}$$

It is typical for solar cavity receivers to approach the absorptivity of a blackbody ($\alpha = \varepsilon \approx 1$) because of the large number of internal reflections. η_{abs} versus temperature, plotted for concentration ratios from 1000 to 10 000 suns, are shown in Figure 21.4. They can be seen in the top solid curves that begin at 1 and decrease to zero at elevated temperatures. As seen, η_{abs} increases with increasing concentration ratio for a fixed temperature because the area through which re-radiation may occur is decreased (i.e. aperture area has to decrease in order for C to increase).

Ultimately, it is the aim of solar thermochemical fuel processes to convert the net heat that is absorbed, to work (in this case the work is considered to be H_2 or synthesis gas). In a seminal *Science* publication in 1977 by Fletcher and Moen [18], this is shown to be limited by the product of the Carnot efficiency (η_{Carnot}) and η_{abs}, where η_{Carnot} is defined as:

$$\eta_{Carnot} = 1 - \frac{T_{low}}{T} \tag{21.5}$$

Here the low temperature (T_{low}) is the temperature at which the fuel is eventually combusted and the high temperature (T) is the temperature at which solar energy is used to drive the chemical reaction of interest (e.g. $H_2O \rightarrow H_2 + 0.5O_2$). The product of η_{abs} and η_{Carnot} is therefore the maximum efficiency (η_{max}) at which one could convert solar energy to fuels [18]. η_{abs}, η_{Carnot}, and $\eta_{Carnot}\eta_{abs}$ are all plotted in Figure 21.4 and as expected, there is a peak efficiency for a given concentration ratio where the efficiency is maximized (gray points). In general the temperature where peak efficiency occurs increases with increasing concentration ratio. Importantly, this shows that for

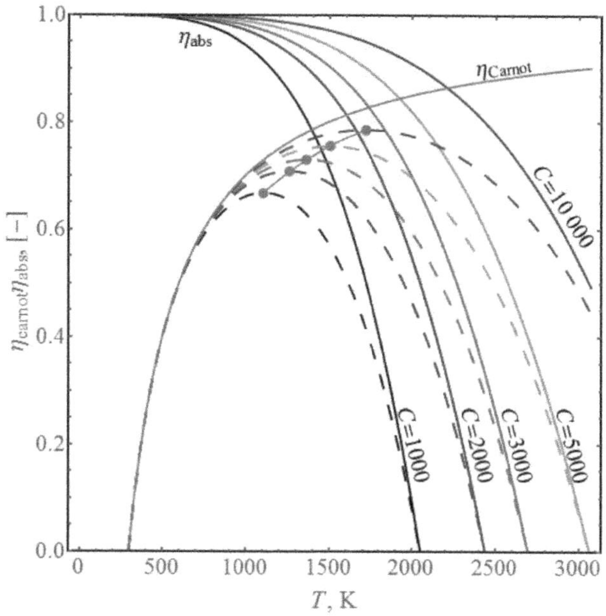

Figure 21.4 Dashed lines represent the maximum theoretical solar to fuel efficiency via thermochemical pathways. These lines are the product of the absorption efficiency, which decreases with temperature to the fourth power, and Carnot efficieny which increases with increasing operating temperature. Insprired by Fletcher and Moen, *Science* 1977 [18].

realistically high concentration ratios[12] (e.g. 500–5000 suns), maximum solar to fuel conversion efficiencies are greater than 70%. In terms of the economics of solar fuel processes, the efficiency is the principle metric that dictates the economic viability, or levelized cost of energy (LCOE), of the process because it directly scales with the amount of capital equipment required [14].

21.2 Solar Thermochemical Processes

There are several approaches to convert absorbed solar energy to fuels via thermally driven, or "thermochemical" pathways. The primary pathways through which solar thermochemical conversion to H_2 or synthesis gas can be achieved are either fuel reforming processes (e.g. steam methane reforming) where a fuel and H_2O (and sometimes CO_2) are used as feedstocks, or thermolysis/redox cycles in which no fuel precursor is required and only H_2O or CO_2 are used as feedstocks. These are both indicated on the left and right hand sides in Figure 21.5, respectively. A detailed description of these processes, especially with respect to their thermodynamics, will be the remaining focus of this chapter.

12 Realistically in the sense that when the concentration ratio approaches extreme values (e.g. 10 000 suns) the spillage is usually so high because of imperfect optics that it is not practical for most applications.

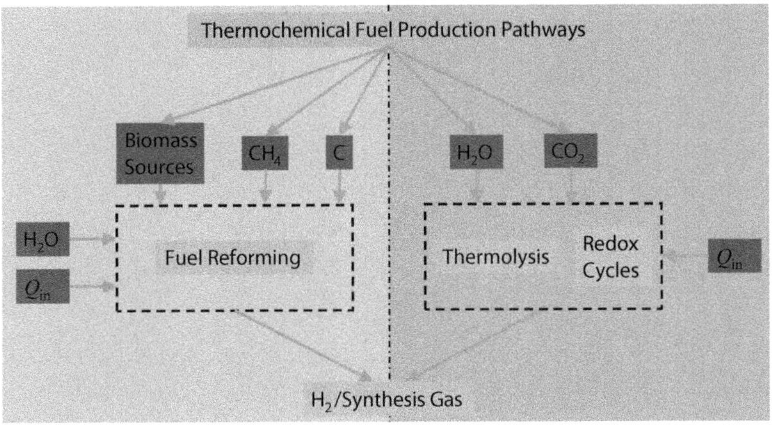

Figure 21.5 Thermochemical fuel production pathways that utilze concentrated sunlight to drive an endothermic resulting in H_2 or synthesis gas. On the left fuel reforming cycles drive an endothermic reforming reaction and on the right are pathways that only utilize H_2O or CO_2 as net feedstocks.

21.2.1 Solar Reforming Processes

For most fuel reforming process that are non-solar, some of the carbonaceous feedstock is combusted (e.g. CH_4 (g) + $2O_2$ (g) → $2H_2O$ (g) + CO_2 (g)) to provide the heat necessary heat to drive an endothermic reforming reaction (CH_4 (g) + H_2O (g) → $3H_2$ (g) + CO (g)). As a result, the net heating value of the products is less than the primary feedstock. For example, consider steam methane reforming, which is the most established method to produce industrial H_2. Here, the net steam reforming reaction may be expressed by:

$$H_2O(l) + CH_4(g) \rightarrow 3H_2(g) + CO(g) \tag{21.6}$$

This reaction is the net reaction of the following three formation reactions

$$H_2O(l) \rightarrow H_2(g) + 0.5O_2(g), -\Delta H^\circ_{f,H_2O} = 285.83\text{ kJ @}298.15\text{K} \tag{21.7}$$

$$CH_4(g) \rightarrow C(s) + 2H_2(g), -\Delta H^\circ_{f,CH_4} = 74.873\text{ kJ @ }298.15\text{ K} \tag{21.8}$$

$$0.5O_2(g) + C(s) \rightarrow CO(g), \Delta H^\circ_{f,CO} = -110.527\text{ kJ @}298.15\text{K} \tag{21.9}$$

where $\Delta H^\circ_{f,i}$ is the standard formation enthalpy of species i at standard conditions (298.15 K). Thus, the net reaction enthalpy of the steam reforming reaction ($-\Delta H^\circ_{f,H_2O}$ + $-\Delta H^\circ_{f,CH_4}$ + $\Delta H^\circ_{f,CO}$) is 250.2 kJ at 298.15 K. To drive this endothermic reaction, some of the methane feedstock is combusted according to the following reaction:

$$2O_2(g) + CH_4(g) \rightarrow 2H_2O(l) + CO_2(g) \tag{21.10}$$

which is a product of the following formation reactions.

$$CH_4(g) \rightarrow C(s) + 2H_2(g), -\Delta H^\circ_{f,CH_4} = 74.873\text{ kJ @}298.15\text{K} \tag{21.11}$$

$$O_2(g) + C(s) \rightarrow CO_2(g), \Delta H^\circ_{f,CO_2} = -393.522\text{ kJ @}298.15\text{K} \tag{21.12}$$

$$O_2(g) + 2H_2(s) \rightarrow 2H_2O(l), 2\Delta H^\circ_{f,H_2O} = -571.66\text{ kJ @}298.15\text{K} \tag{21.13}$$

Thus, the net reaction enthalpy $\left(-\Delta H^{\circ}{}_{f,CH_4} + \Delta H^{\circ}{}_{f,CO_2} + 2\Delta H^{\circ}{}_{f,H_2O}\right)$ is -890.3 kJ at 298.15 K. Therefore, for every mole of CH_4 that is reformed (21.6) that requires 250.2 kJ, 0.28 moles must be combusted (21.10) to provide the required process heat, at a minimum (250.2 kJ/890.3 kJ = 0.28). In reality it is substantially greater because of inefficiencies and because the reaction enthalpy decreases at higher temperatures (e.g. for methane combustion ΔH° = -800 kJ at 900 K) [4]. For a more thorough description of reaction enthalpies and their temperature dependence refer to *Introduction to Chemical Engineering Thermodynamics* 7th Edition, specifically Chapter 4 entitled Heat Effects [19], or another appropriate thermodynamics or chemistry textbook.

In solar-driven fuel reforming processes, solar energy is used as process heat to drive the endothermic reforming reaction rather than combustion of the feedstock [20]. As a result the feedstock is "upgraded" because the heating value of the products are greater than the primary feedstock, and solar energy is effectively stored in chemical form. A wide variety of solar reforming processes have been proposed in the literature, and compared to the non-fuel reforming processes on the right side of Figure 21.5, these typically occur at relatively moderate temperatures (e.g. < 1000°C). Therefore, they are practically more straightforward because demands on reactor and construction materials are not as strict. As a result, these technologies are relatively mature and several concepts have been tested at the prototype reactor scale. For example, reactors have been developed and tested for the reforming of natural gas [21], methanol [22], activated carbon [23], coal [23], coke [23–24], beech char [25], biochar [26], bagasse [27], corn stover, Kentucky bluegrass [28], and a wide variety of industrial waste products [29]. It is important to note that solar reforming processes are not carbon neutral unless the carbon comes from a source that is continuously replenished by CO_2 from the atmosphere (e.g. algae, corn stover, bagasse, etc.).

To date, the highest reported efficiency for solar driven steam methane reforming is 69% by Zheng et al. from Pacific Northwest National Laboratory (PNNL) [30]. Here, the efficiency is defined as the change in the heating value between the products (syngas) and reactants (methane) divided by the solar power input. A photograph of their reactor, which they call solar thermochemical advanced reactor system (STARS) can be seen in Figure 21.6 on the left. The concept combines an aperture where concentrated sunlight enters that is coupled with a catalytic methane reforming system based on micro and

Figure 21.6 On the left is the STARS solar methane reforming reactor with the aperture for sunlight on its right hand side. On the left is the STARS systems mounted to a tracking parabolic concentrator for prototype testing [28].

meso channel reactors and heat exchangers. The micro and meso channel heat exchanger/reactor system was developed independently of any intended solar applications over several decades and recently adopted to meet the demands of concentrated solar applications. An example of the system mounted to a parabolic dish is shown in Figure 21.6 on the right. The upper operating temperature is of the order of 840°C, moderate by gas splitting solar thermochemical standards, and this technology nicely demonstrates the potential maturity, robustness, and feasibility of solar driven reforming while operating under realistic conditions. Current barriers to commercialization likely lie in the overall cost of the reactor system, integration with syngas storage because of the inherently transient nature of the sun, and startup/shutdown associated with nighttime and cloud cover. These are issues that all concentrated solar syngas systems must consider.

An example of a prototype aerosol reactor, developed by Kruesi et al. and used for the reformation of bagasse (sugar cane residue) particles [27] is shown in Figure 21.7. Here, solid bagasse particles and H_2O (g) enter a tubular reactor that is housed within an insulated cavity type solar receiver. Concentrated radiation enters through the receiver aperture (right side), is further concentrated in a secondary concentrator (here a compound parabolic concentrator, or CPC).[13] Initially the bagasse is rapidly pyrolyzed in the pyrolysis zone, where the bagasse is converted to char and undesired secondary products such as ash. Following pyrolysis it is gasified, or reacted with H_2O (g), to form synthesis gas. Because the gasification reaction occurs relatively slowly, this aerosol reactor incorporated a porous structure to increase the residence time of the char particles in the hot zone. Depending on the reactant composition and structure, a variety of reactor configurations have been proposed, but their discussion is outside of the scope of this chapter. Please refer to the references [20–29] for a more detailed description of several concepts.

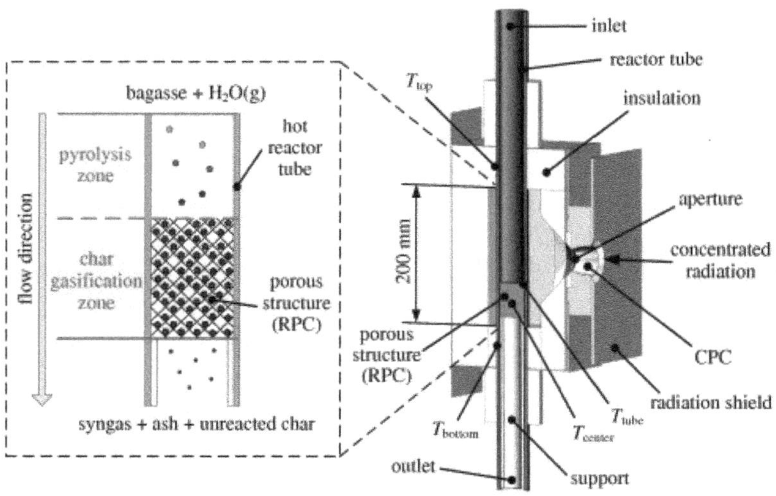

Figure 21.7 From Kruesi et al. Solar driven bagasse reactor to produce synthesis gas [27].

13 CPC is short for Compound Parabolic Concentrator, a nonimaging concentrating device that increases the geometric concentration ratio (A_{in}/A_{out}). Refer to Winston's text entitled *Nonimaging Optics* [16].

21.2.2 Solar Driven Thermolysis

Conceptually, the simplest path to H_2 or synthesis gas is the direct thermolysis of H_2O or CO_2. Additionally, these pathways are inherently carbon neutral, as opposed to fuel reforming processes, because their net inputs are simply the byproducts of fuel combustion. Their net chemical reactions are shown below.

$$H_2O(l) \rightarrow H_2(g) + 0.5O_2(g), -\Delta H^\circ_{f,H_2O} = 285.83\ \text{kJ @}298.15\text{K}^4 \tag{21.14}$$

$$CO_2(g) \rightarrow CO(g) + 0.5O_2(g), \Delta H^\circ_{CO_2} = 283.0\ \text{kJ @}298.15\text{K}^4 \tag{21.15}$$

Where ΔH°_{CO2} is the enthalpy change of CO_2 thermolysis referenced to standard state. This reaction enthalpy change is determined by recognizing that it is the sum of the following two formation reactions ($\Delta H^\circ_{CO2} = -\Delta H^\circ_{f,CO2} + \Delta H^\circ_{f,CO}$) shown in Equations (21.16 and 21.17).

$$CO_2(g) \rightarrow C(g) + O_2(g), -\Delta H^\circ_{f,CO_2} = 393.522\ \text{kJ @}298.15\text{K} \tag{21.16}$$

$$C(g) + 0.5O_2(g) \rightarrow CO(g), \Delta H^\circ_{f,CO} = -110.527\ \text{kJ @}298.15\text{K} \tag{21.17}$$

While conceptually simple, these reactions are difficult to achieve in practice because of their extreme temperatures (usually $\geq$ 2273 K) and the favorability of the reverse reactions as the gaseous products cool. Thus, to prevent product recombination, thermolysis reactions are usually conducted in a high temperature oxygen conducting ceramic, such as ZrO_2, where the outside of the ceramic is kept at a low oxygen partial pressure [31–32]. As the reaction progresses, the produced O_2 diffuses across the solid membrane and away from either H_2/CO, thereby preventing recombination. Other techniques such as rapid quenching of the product gases have been utilized but with limited success [31]. Usually thermolysis reaction temperatures are limited by the temperature stability of the oxygen conducting ceramic (usually less than 2273 K) or other reactor construction materials and are therefore conducted well below where they are thermodynamically favorable at ambient pressure $\left(\Delta G_{rxn} >> 0 \text{ at } 2273\text{ K}\right)$. As a result, reaction extents are usually very small or a large amount of mechanical work (e.g. vacuum pumping) is required. In both instances, the efficiency is hindered; in the former, an excess of thermal energy is supplied that may only be recovered through efficient, high temperature heat exchange, or the latter because of the second law losses associated with producing the work. Further, even if materials were stable enough to operate where the reactions are more thermodynamically favorable, the re-radiation losses would be so large that they would have a severe and negative effect on the efficiency (c.f. Figure 21.4).

21.2.3 Thermochemical Redox Cycles – Background

Rather than using electrical work, another possibility to decrease the operating temperature of H_2O or CO_2 thermolysis while still utilizing only thermal energy is through the implementation of a redox cycle where the thermolysis reaction is split into two or more reactions, where one of them has a larger entropy change than the net entropy change of the reaction [33–34]. In general, as the number of reactions increases, the upper operating temperature decreases but the complexity associated with the process increases [33]. Therefore, for practical reasons, most interest in H_2O/CO_2 redox splitting cycles in recent years has been focused on two-step cycles.

Most two-step redox cycles drive a high temperature endothermic reduction of a solid metal oxide (MeO) in the first step (solar driven step); temperatures are typically greater than 1000°C and dependent on the metal oxide used.[14] Following reduction, the reduced metal oxide (Me) is then subsequently oxidized in an exothermic reaction at lower temperatures by H_2O or CO_2 to re-oxidize it produce gaseous H_2 or CO. The net reaction is simply the dissociation of H_2O or CO_2 [35]. The two-step MeO based redox cycle for H_2O splitting and the net reaction is shown below.

$$\mathrm{MeO} \xrightarrow{\Delta H_{\mathrm{red}}} \mathrm{Me} + 0.5\mathrm{O_2(g)} \quad \text{(Reduction Step 1)} \tag{21.18}$$

$$\mathrm{Me} + \mathrm{H_2O(g)} \xrightarrow{\Delta H_{\mathrm{ox}}} \mathrm{MeO} + \mathrm{H_2(g)} \quad \text{(Oxidation Step 2)} \tag{21.19}$$

$$\mathrm{H_2O(g)} \xrightarrow{\Delta H_{\mathrm{H2O}}} \mathrm{H_2(g)} + 0.5\mathrm{O_2(g)} \quad \text{(Net Reaction)} \tag{21.20}$$

The reason that these cycles result in a lower operating temperature than the direct thermolysis of H_2O (or CO_2) is that the first step takes advantage of the fact that the entropy change (ΔS) required to reduce the oxide is greater than for water dissociation (but also enthalpy is greater). For example, refer to the left subplot of Figure 21.8 that shows the Gibbs free energy change (ΔG) for the reduction reaction of a hypothetical metal oxide MeO alongside the H_2O thermolysis reaction. The equilibrium of both reactions is dictated by ΔG, where less than 0 indicates the reaction is more favorable in forward direction and greater than zero more favorable in the reverse direction. $\Delta G = \Delta H - T\Delta S$, and as seen here the entropy change is greater (steeper slope) for the oxide reduction, resulting in $\Delta G = 0$ at a lower temperature, meaning the reaction is more likely to proceed in the forward direction compared to thermolysis.

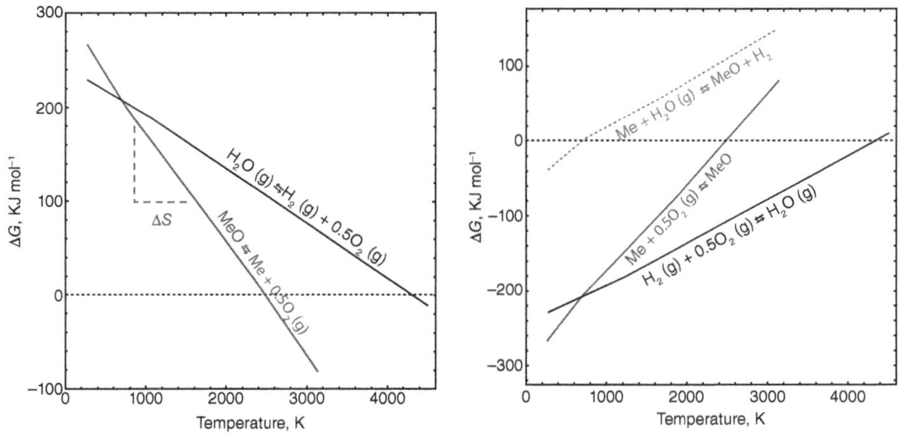

Figure 21.8 Left) Gibbs free energy change versus temperature of the water thermolysis reaction and metal oxide reduction reaction. (Right) Gibbs free energy change versus temperature of the metal oxide oxidation reaction with steam (dashed line), oxygen (gray line) and H_2 oxidation (black line).

14 Details of the metal oxides used will be discussed in subsequent sections.

The oxidation of the reduced oxide with H_2O is the sum of the following two reactions.

$$\mathrm{Me} + 0.5\mathrm{O_2}(\mathrm{g}) \rightarrow \mathrm{MeO} \tag{21.21}$$

$$\mathrm{H_2O}(\mathrm{g}) \rightarrow \mathrm{H_2}(\mathrm{g}) + 0.5\mathrm{O_2}(\mathrm{g}) \tag{21.22}$$

$$\mathrm{Me} + \mathrm{H_2O}(\mathrm{g}) \rightarrow \mathrm{MeO} + \mathrm{H_2}(\mathrm{g}) \qquad \text{(Net Reaction)} \tag{21.23}$$

The Gibbs free energy changes of reactions (21.21), (21.22), and (21.23) are shown on the right subplot of Figure 21.8. As seen, Equations (21.21), (21.22) are simply the reverse of the oxide reduction and water thermolysis reactions and thus the oxide oxidation has a larger entropy change. The net reaction (21.23) (dashed line) is simply the difference in ΔG between the two reactions. As seen, it is most negative at the lowest temperatures, indicating that oxidation is thermodynamically most favorable at the lowest temperature possible.[15] Effectively at these low temperatures the metal has a higher affinity for oxygen than H_2 does, as indicated by the more negative Gibbs at the lowest temperatures. However, as temperature increases, the reverse is true and oxidation becomes less favorable. Usually the oxidation temperature is conducted at the highest temperature possible where thermodynamics is not hindered because of improved kinetics (i.e. more rapid reaction rates) and higher efficiencies because of smaller temperature swings between reduction and oxidation reactions that require a heat input. Descriptions of thermochemical redox cycles detailing this and the state of the art metal oxides used are reviewed in the ensuing sections (c.f. Sections 21.2.5, 21.2.7, 21.2.10), but first the basics of reaction equilibrium will be discussed in order to better facilitate understanding.

21.2.4 Reaction Equilibrium

A basic understanding of the fundamental equations of chemical reaction equilibria and Gibbs free energy data may be used to appreciate the thermodynamic favorability of thermochemical reactions (or any chemical reaction). According to [19], the general symbolic representation of a reversible chemical reaction may be expressed as:

$$aA + bB + \ldots \rightleftarrows cC + dD + \ldots \tag{21.24}$$

Where A,B,C and D represent chemical formulas and a,b,c and d represent stoichiometric coefficients (ν). The equilibrium constant (K) of a reaction is defined as

$$K = \frac{a_C^c a_D^d \ldots}{a_A^a a_B^b \ldots} \tag{21.25}$$

Where a_i is the activity of species i. For an ideal gas where fugacity (f_i) is equal to the pressure (P) and the standard state pressure is 1 atm ($f_i^\circ = P = 1$ atm), $a_i = x_i = p_i$.[16] x_i is the molar fraction of species i and p_i refers to the partial pressure

15 It is important to note that the enthalpy change of the metal oxide reduction must be larger than the enthalpy change of water dissociation, or else there will be NO temperatures where the oxide oxidation is thermodynamically favorable.
16 This comes from the fact that $a_i = f_i / f_i^\circ$ and $f_i = x_i P = p_i$.

of species i. K is related to the standard Gibbs free energy change of a reaction ($\Delta G°$) according to:

$$\Delta G° = -RT \ln K \tag{21.26}$$

where R is the ideal gas constant and T is temperature. Therefore, the equilibrium constant and standard Gibbs free energy change of reaction (21.14) assuming gas ideality at 1 atm total pressure is the following.

$$K_{H_2O} = \frac{p_{H_2} p_{O_2}^{1/2}}{p_{H_2O}} = \frac{x_{H_2} x_{O_2}^{1/2}}{x_{H_2O}} \tag{21.27}$$

$$\Delta G°_{H_2O} = -RT \ln K_{H_2O} \tag{21.28}$$

Similar equations could be derived for reaction (21.15). When $\Delta G° > 0$ or $K < 1$, the reaction is more favorable in the reverse direction (e.g. denominator in (21.27) is greater than the numerator) and when $\Delta G° < 0$ or $K > 1$, the reaction is more favorable in the forward direction (e.g. numerator in (21.27) is greater than the denominator). Thus, K and $\Delta G°$ are an indicator of the thermodynamic favorability for the reaction to occur. It should be recognized that:

$$\Delta G°_{H_2O} = -\Delta G°_{f,H_2O} \text{ and } K_{H_2O} = 1/K_{f,H_2O} \tag{21.29}$$

and

$$\Delta G°_{CO_2} = -\Delta G°_{f,CO_2} + \Delta G°_{f,CO} \text{ and } K_{CO_2} = K_{f,CO} / K_{f,CO_2} \tag{21.30}$$

where $\Delta G°_{f,i}$ is the standard Gibbs free energy of formation of species i. $\Delta G°$ for H_2O and CO_2 thermolysis reactions (21.14) and (21.15) as a function of temperature, obtained from equilibrium data in *NIST JANAF Thermochemical Tables* [4], as shown in the left subplot of Figure 21.9. $\Delta G°$ is very large and does not equal 0 ($K = 1$) until 4310 K and 3339 K for H_2O and CO_2, respectively. These high temperatures make qualitative sense because these are known to be extremely stable molecules.

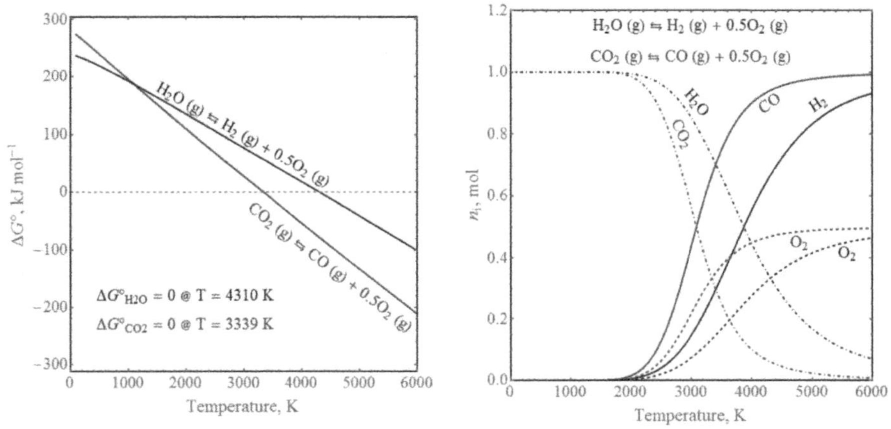

Figure 21.9 (Left) $\Delta G°$ shown versus temperature for H_2O and CO_2 thermolysis reactions. Data extracted from *NIST JANAF Thermochemical Tables*. (Right) Equilibrium yields of H_2O and CO_2 thermolysis reactions assuming only H_2, CO, and O_2 may be produced.

Equilibrium yields may be calculated in a straightforward manner with knowledge of either ΔG° or K. For this, the reaction coordinate (ε) is useful; ε effectively characterizes the extent to which a reaction has occurred. ε may be related to moles of species i at equilibrium (n_i) through the following relationship [19]:

$$\pm\varepsilon = \frac{\Delta n_i}{v_i} = \frac{n_i - n_{i,0}}{v_i} \tag{21.31}$$

where $n_{i,0}$ is the initial number of moles present of species i and v_i is the stoichiometric coefficient of species i. A positive sign is attributed for product species and negative sign for reactant species. x_i is related to n_i and the total number of moles in the system, n_{total} by the following equation:

$$x_i = \frac{n_i}{n_{total}} = \frac{n_i}{\sum n_i} \tag{21.32}$$

Thus, the equilibrium constant (or ΔG°) may be related to the reaction coordinate and the initial number of moles of each species in the system. If the equilibrium constant (or ΔG°) is known as a function of temperature, then equilibrium yields of each species may then be determined, or vice versa. For example, consider a system in which 1 mole of H_2O (g) is initially present ($n_{H_2O,i} = 1$) and knowledge of the final equilibrium yields of H_2O, H_2, and O_2 at a given temperature and pressure are desired. Combining (21.27), (21.31), and (21.32) the following two equations can be derived.

$$K_{H_2O} = \frac{x_{H_2} x_{O_2}^{1/2}}{x_{H_2O}} = \frac{\left(\dfrac{\varepsilon}{n_{total}}\right)\left(\dfrac{1/2\varepsilon}{n_{total}}\right)^{1/2}}{\left(\dfrac{1-\varepsilon}{n_{total}}\right)} \tag{21.33}$$

$$n_{total} = n_{H_2} + n_{O_2} + n_{H_2O} = \varepsilon + 1/2\varepsilon + 1 - \varepsilon = 1 + 1/2\varepsilon \tag{21.34}$$

From [4], at 2000 K, K_{H_2O} ($1/K_{f,H_2O}$) of reaction (21.14) is equal to 0.00029 which is $\ll 1$; thus we expect mole fractions of products to be low. Solving for equilibrium yields using Equations (21.33) and (21.34), we confirm that product yields are low; $\varepsilon = 0.0055$, $x_{H_2} = 0.0055$, $x_{O_2} = 0.00274$ and $x_{H_2O} = 0.9912$. At higher temperatures, for example near where $K_{H_2O} = 1$ ($T = 4300$ K), we see that conversion is much higher; $\varepsilon = 0.663$, $x_{H_2} = 0.498$, $x_{O_2} = 0.25$ and $x_{H_2O} = 0.253$. In Figure 21.9 (right subplot), we show molar yields of all product species versus temperature for H_2O and CO_2 thermolysis reactions. Here, we have assumed that the only products that may be formed are H_2, CO, and O_2[17]. As seen, the decomposition of CO_2 is expected to occur at more moderate temperatures than H_2O and is largely the result of the larger entropy change (ΔS°) associated with decomposition which dictates the slope of the ΔG° curve versus temperature ($\Delta G^\circ = \Delta H^\circ - T\Delta S^\circ$). For more details regarding chemical reaction equilibria (especially multi-reaction equilibria) and determination of equilibrium yields refer to [19] or another suitable thermodynamics textbook.

17 Practically, other products may form at higher temperatures, such as OH, O_3, etc. This calculation has limited the potential products only for simplicity in order to demonstrate the thermodynamic impact of Gibbs and equilibrium on the favorability of the gas splitting reactions.

21.2.4.1 A Note on Thermolysis versus Electrolysis

For any chemical reaction, the total energy necessary to drive the reaction is equal to the standard reaction enthalpy (ΔH°) which is usually only slightly dependent on temperature. Thus, when the reaction is driven at $\Delta G^\circ = 0$, all of the energy supplied to the system is supplied as heat ($\Delta H^\circ = T\Delta S^\circ = Q$). However, if operation at lower temperatures is desired where ΔG° is positive and the reaction is not spontaneous, work (e.g. electrical, W_{el}, or mechanical work, W_{pv}) may be supplied to the system in addition to thermal energy ($\Delta H^\circ = Q + W = Q + \Delta G^\circ$) to shift the reaction equilibrium to the right side. Electrolysis of H_2O is perhaps the most ubiquitous and straightforward means of H_2O splitting and may be performed at room temperature where no thermal energy is supplied $\Delta H^\circ \sim \Delta G^\circ \sim W_{el}$, or elevated temperatures (High Temperature Electrolysis) where a mixture of thermal energy and electrical work is supplied. Because electrical work is usually derived from a thermal energy source, it is thermodynamically most attractive to use as much thermal energy as possible to drive the splitting of H_2O or CO_2. However, the elevated temperatures at which H_2O and CO_2 thermolysis are thermodynamically most attractive are prohibitively high (as discussed previously).

21.2.5 Iron Oxide Based Thermochemical Redox Cycles

The first realistic MeO cycle proposed was an iron oxide based redox cycle by Nakamura et al. in 1977 [36]. The cycle is based on the stoichiometric reduction of magnetite (Fe_3O_4) to hematite (FeO), where Fe changes from the 3+ to 2+ oxidation state. Thus, the first step is the endothermic and solar driven reduction of magnetite to hematite and the subsequent release of gaseous oxygen, and the second step is the re-oxidation of hematite with H_2O (or CO_2) to produce H_2 (or CO).

Reduction:

$$Fe_3O_4 \xrightarrow{\Delta H_{Fe3O4}} 3FeO + 0.5O_2(g) \tag{21.35}$$

Oxidation:

$$3FeO + H_2O\,(g) \xrightarrow{\Delta H_{ox}} Fe_3O_4 + H_2(g) \tag{21.36}$$

21.2.5.1 Thermodynamics of Iron Oxide Based Redox Cycles

While conceptually simple, the thermodynamics of these oxide systems are relatively complex, with a variety of different crystallographic and intermediate oxidation states existing between Fe_3O_4/FeO depending on the operating temperature and pressure [37–42]. However, for simplification purposes we will demonstrate the redox cycle here based only on the stoichiometric reaction, where the thermodynamics can be relatively easily described. The reduction of Fe_3O_4 may be seen as the sum of the reverse formation reaction of Fe_3O_4 and the formation reaction of FeO, as seen below, where the thermodynamics of each formation reaction are well documented [4].

$$Fe_3O_4 \rightarrow 3Fe + 2O_2(g), -\Delta H^\circ_{f,Fe_3O_4} = 1120.89\ kJ\ @\ 298.15\ K^4 \tag{21.37}$$

$$3Fe + 3/2O_2(g) \rightarrow 3FeO,\ 3\Delta H^\circ_{f,FeO} = -816.13\ kJ@298.15\ K^4 \tag{21.38}$$

Using the reaction coordinate method we can predict the equilibrium reduction extent of magnetite at a given temperature and pressure. To do so requires knowledge of either the Gibbs free energy or equilibrium constant of the reaction, which may then be related to the reaction coordinate by combining Equations (21.25), (21.26), (21.31) and (21.32) and knowing that $p_i = x_i P$, where P is the absolute pressure. Thus, we can show that:

$$K_{F_e3O_4} = \exp\left(\frac{-\Delta G^\circ_{Fe_3O_4}}{RT}\right) = \frac{a_{FeO}{}^3\, p_{O_2}^{1/2}}{a_{Fe_3O_4}} = \left(\frac{1/2\varepsilon}{n_{total}} P\right)^{1/2} \tag{21.39}$$

This assumes unity activities of FeO and Fe_3O_4. $\Delta G^\circ_{Fe_3O_4}$ may be determined by summing the Gibbs formation free energies of Fe_3O_4 and FeO ($\Delta G^\circ{}_{f,Fe_3O_4}$ and $\Delta G^\circ{}_{f,FeO}$). If knowledge of Gibbs formation free energies are not tabulated directly (as is common in *NIST Chemistry WebBook*) then these values may be determined from knowledge of the free energy function (FEF) [43] which mathematically is:

$$FEF(298\text{ K}) = \frac{G^\circ(T) - H^\circ(298\text{ K})}{T} = \frac{H^\circ(T) - H^\circ(298\text{ K})}{T} - S^\circ(T) \tag{21.40}$$

The Gibbs formation free energy is then related to FEF through the following relationship, for the reaction $v_1 E_1 + v_2 E_2 + \ldots = B$:

$$\frac{\Delta G_{f,B}(T)}{T} = FEF_B(298\text{ K}) - \sum_i v_i \left[FEF_{E_i}(298\text{ K})\right] + \frac{\Delta H_{f,B}(298\text{ K})}{T} \tag{21.41}$$

In Figure 21.10 we show the gas phase of oxygen resulting from reduction of Fe_3O_4 to FeO as a function of temperature and two different total pressures. All thermodynamic data is derived from *NIST Chemistry WebBook* using polynomial fits but tabulated data is also available in [4]. For these calculations a total pressure of 1 atm was assumed for the right curve, 1×10^{-4} atm for the left curve. At 1 atm, gas phase oxygen does not

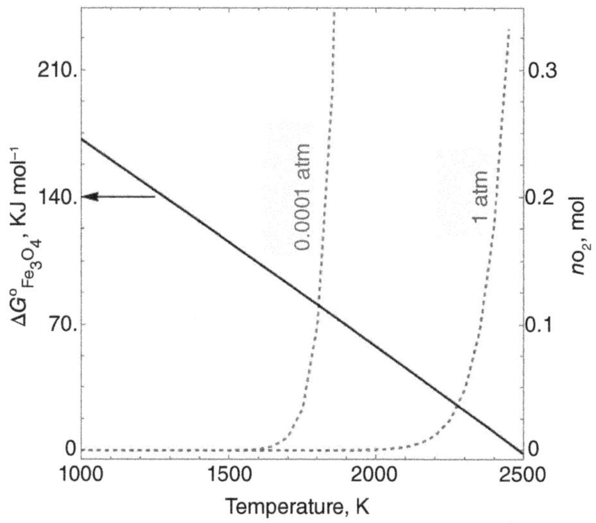

Figure 21.10 Standard Gibbs free energy change (left axis) for the reaction $Fe_3O_4 \rightarrow 3FeO + 0.5O_2$. Also shown is the equilibrium gas phase O_2 on the right axis.

increase appreciably until $\Delta G^{\circ}_{Fe_3O_4}$ approaches zero, as expected. For example, at 2000 K where $\Delta G^{\circ}_{Fe_3O_4} = 57.8\ \text{kJ mol}^{-1}$, n_{O_2} equals 5×10^{-4} moles, but at 2450 K where $\Delta G^{\circ}_{Fe_3O_4}$ approaches zero (4.2 kJ mol^{-1}), n_{O_2} equals 0.33 moles. While this temperature is high, in practice it is alleviated by increasing the total number of gas phase species in the system (e.g. inert sweep gas) or decreasing the total pressure, as evidenced in the left curve where the reaction goes to completion near 1850 K when operating at 1×10^{-4} atm. This is especially important for the iron oxide system because of the fact that FeO melts at 1650 K, well below the decomposition temperature of Fe_3O_4 at ambient pressure. In fact, this low melting temperature is one of the reasons that this process has not been realized successfully in prototype solar reactors like other candidate materials such as CeO_2 (discussed later).

The oxidation reaction is simply the reverse of the reduction formation reactions summed with the thermolysis of H_2O. The reactions are shown below:

$$3\text{Fe} + 2\text{O}_2(\text{g}) \rightarrow \text{Fe}_3\text{O}_4, \Delta H^{\circ}_{\text{f,Fe}_3\text{O}_4} = -1120.90\ \text{kJ @ 298.15 K}^4 \quad (21.42)$$

$$3\text{FeO} \rightarrow 1.5\text{O}_2(\text{g}) + 3\text{Fe}, -3\Delta H^{\circ}_{\text{f,FeO}} = 816.12\ \text{kJ @ 298.15 K}^4 \quad (21.43)$$

$$\text{H}_2\text{O(l)} \rightarrow 0.5\text{O}_2(\text{g}) + \text{H}_2(\text{g}), -\Delta H^{\circ}_{\text{f,H}_2\text{O}} = 285.83\ \text{kJ @ 298.15 K}^4 \quad (21.44)$$

$$3\text{FeO} + \text{H}_2\text{O (l)} \rightarrow \text{Fe}_3\text{O}_4 + \text{H}_2(\text{g}),\ \Delta H_{\text{GS}}\ \text{(Net Reaction)}$$

where the subscript GS in the bottom reaction (Net Reaction) enthalpy stands for Gas Splitting. To determine equilibrium compositions we may again turn to the reaction coordinate methodology and show that $\varepsilon = -\left(n_{H_2O} - n_{H_2O,i}\right)$ and $\varepsilon = n_{H_2}$, where it is assumed that there is no H_2 initially present in the system and the initial amount of H_2O is $n_{H_2O,i}$. This assumes that all of the reactions above are in equilibrium (oxygen partial pressure of each reaction is equivalent) and thus there is no net production of gas phase oxygen. Thus, knowing either the formation equilibrium constants or Gibbs free energy formation energies, equilibrium H_2 product yields may be determined according to Equations (21.45) and (21.46). Comparable calculations could be performed using CO_2 instead of H_2O [44].

$$K_{GS} = K_{Fe_3O_4} \frac{1}{K_{FeO}} \frac{1}{K_{H_2O}} = \frac{x_{H_2}}{x_{H_2O}} = \frac{\varepsilon}{n_{H_2O,i} - \varepsilon} \quad (21.45)$$

$$\Delta G_{GS} = \Delta G_{f,Fe_3O_4} - \Delta G_{f,FeO} - \Delta G_{f,H_2O} = -RT \ln \frac{x_{H_2}}{x_{H_2O}}$$

$$= -RT \ln \frac{\varepsilon}{n_{H_2O,i} - \varepsilon} \quad (21.46)$$

Shown in Figure 21.11 is the predicted equilibrium molar composition of H_2 per mol FeO and the Gibbs free energy change of the H_2O gas splitting reaction (ΔG_{GS}) versus oxidation temperature. As seen, ΔG_{GS} is negative for the lowest temperatures indicating that the oxidation reaction is thermodynamically favorable in this regime ($K_{GS} \gg 1$) but near 675 K ΔG_{GS} is equal to zero. As seen, for the lower temperature range the equilibrium H_2 yield approaches 1 (1 is indicitive of complete conversion) but at higher temperatures this decreases relatively quickly. If the amount of initial H_2O moles is increased from 1 to 10 however, the expected equilibrium H_2 yields are increased at higher temperatures. This doesn't come without a penalty however because all of the

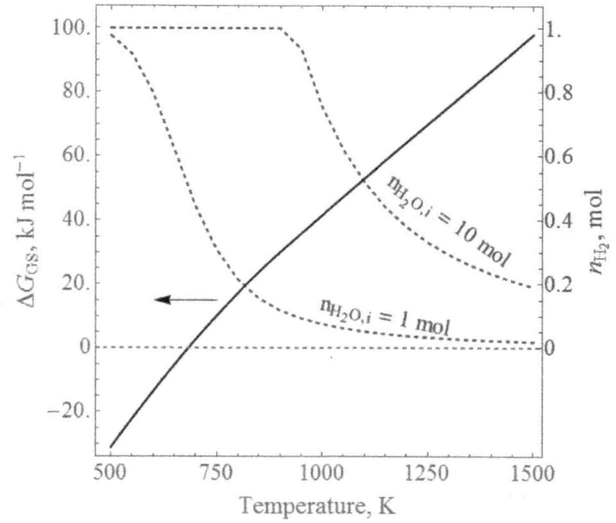

Figure 21.11 Equilibrium H_2 yields (dashed lines) resulting from H_2O dissociation as a function of oxidation temperature, alongside ΔG_{GS} (solid curve), for the reaction $3FeO + H_2O(l) \rightarrow Fe_3O_4 + H_2(g)$. H_2 yields are shown for initial amounts of H_2O equal to 1 and 10 moles.

initial H_2O must be heated to the oxidation temperature of interest and the cost benefit of doing so relative to the gain in H_2 yields should be considered. This is discussed extensively in several thermodynamic analyses where solar-to-fuel conversion efficiencies are calculated over a wide range of operating conditions [45–48].

The trend of ΔG_{GS} with respect to temperature should be expected for all thermochemical redox materials; namely oxidation with H_2O or CO_2 is more thermodynamically favorable at low temperatures. In this favorable regime, ΔG_{GS} is negative which implies that the Gibbs free energy of oxidation of the reduced oxide with oxygen $\left(3FeO+0.5O_2 \rightarrow Fe_3O_4\right)$ is more favorable than oxidizing H_2 with oxygen $\left(H_2+0.5O_2 \rightarrow H_2O\right)$; effectively the reduced oxide wants the oxygen bound to the H_2O molecule more than H_2 does, and as a result the H_2O molecule is dissociated.

21.2.5.2 Practical Considerations of Iron Oxide Based Redox Cycles

One of the major issues associated with this cycle is the relatively low melting temperature of FeO, which occurs at 1650 K, below the temperature where the reduction reaction is conducted. As a result, severe sintering has been reported which typically limits the practicality of this process using pure magnetite [49–51]. To circumvent some of these issues, the materials are usually mixed with transition metals to form spinels (e.g. $NiFe_2O_4$, $CoFe_2O_4$) [42, 52–53], which are deposited on stabilizing supports [50, 54], or dissolved in ZrO_2 based oxides [55]. The incorporation of transition metals is known to reduce the reduction temperature, and are thus less likely to sinter upon formation of FeO. The dissolution into ZrO_2 based supports has been shown to enhance the reaction kinetics that are mostly bulk transport limited. This is because the oxygen exchange through ZrO_2 is much faster than Fe_3O_4 [55].

Although thermodynamics dictates that the GS reaction is most favorable at the lowest temperature possible, there are other practical reasons why the reaction is usually driven at temperatures closer to 1000 K and sometimes only a few hundred degrees cooler than the reduction temperature (a typical reduction temperature range is on the order of 1673 to 1873 K depending on the material used, operating pressure, sweep gas flowrate, reduction extent, etc.). The primary reason has to do with chemical kinetics and solid state diffusion, which dictate the rate that a reaction will occur. For example, consider from experience the fact that metallic iron is known to be relatively stable in air at ambient conditions. However, the oxidation of iron with O_2 to form hematite (Fe_2O_3) is thermodynamically predicted at these conditions. The reason it does not oxidize readily (at least in the bulk) is because the reaction and diffusion rates are so slow at ambient temperature. Only at higher temperatures, where chemical kinetics and cation diffusion through the bulk are rapid enough, does it begin to noticeably oxidize.

Another reason that oxidation temperatures are elevated has to do with the fact that redox cycles are operated between two temperature reservoirs, the hot reduction reactor and the cooler oxidation reactor. The cooler the oxidation reaction occurs, the more sensible heat will have to be added to the system to bring the reactants back to the reduction temperature. Here again there is a tradeoff between minimizing the temperature difference between reduction and oxidation temperatures, in order to decrease sensible heating requirements, and maximizing oxidation yields. In the case of both deposition on and dissolution of the oxide in supports, one must also consider the energy penalty associated with heating the unreactive mass from the oxidation temperature to the reduction temperature. This has a negative impact on the overall efficiency and is widely considered not to be a practical solution to obtain high solar to fuel conversion efficiencies. To date, the CR5 is the only known solar reactor that has utilized iron oxide based materials and has been tested at the 16 kWth scale [56]. This reactor is unique compared to most in the sense that it has alternating, rotating rings of reacting material that effectively serve as a solid-solid heat exchanger with the goal of boosting efficiency.

In addition to the morphological changes that occur as a result of sintering, oxidation proceeds via an outwardly growing "shell" which can also present practical problems related to morphological variability with time. This is because oxidation proceeds via diffusion of cation species to the surface, rather than bulk oxygen, where they are oxidized.

21.2.6 Other Redox Cycles

There are a number of metal oxide redox pairs proposed in the literature that fall into one of two distinct classes of redox cycles – stoichiometric and nonstoichiometric. Stoichiometric refers to the fact that the reduction of the oxide proceeds primarily from one distinct phase to another (e.g. $Fe_3O_4 \rightarrow 3FeO$); sometimes the transition is from one crystallographic phase to another or sometimes a complete change in state of matter, from solid to vapor or liquid. As stated earlier, iron oxide was the first stoichiometric metal oxide based cycle proposed but there have been several adaptations to this system through the introduction of other cations that are substituted for iron, so-call ferrites such as $CoFe_2O_4$ [53–54] and $NiFe_2O_4$ [52]. In addition to iron oxide based cycles, ZnO [57–58] and to a lesser extent SnO_2 [59] cycles that are based on the reduction of the oxide to a volatile metal (e.g ZnO (s) $\rightarrow$ Zn (g) + 0.5O_2 (g)) have been investigated.

The difficulty in these cycles lies in the production of the gaseous vapor products (Zn (g) or SnO (g)) that want to rapidly recombine with O_2 when the gas mixture is cooled. Thus, fairly elaborate quenching mechanisms or gas separation techniques must be employed to avoid recombination, similar to direct thermolysis of H_2O or CO_2. For a thorough discussion of stoichiometric cycles there are a host of review articles that the reader is referred to [12, 60–64].

21.2.7 Ceria-based Nonstoichiometric Redox Cycles

More recently, research efforts have been focused on nonstoichiometric redox cycles in which there exist a spectrum of oxide oxidation states without undergoing a crystallographic or state of matter phase transition. Perhaps most understood and studied is the nonstoichiometric oxide ceria, or $CeO_{2\text{-}\delta}$, where δ refers to the nonstoichiometry of the oxide and is directly proportional to the average oxidation state of Ce in the oxide. Chueh [67]. first proposed the use of this nonstoichiometric oxide cycle in which $CeO_{2\text{-}\delta}$ releases O_2 (g) during the reduction reaction that is proportional to its change in nonstoichiometry, or $\Delta\delta$[18] [65]. Thus, the reduction occurs according to the following reaction:

Reduction:

$$CeO_{2-\delta i} \overset{\Delta H_{rxn}}{\rightarrow} CeO_{2-\delta f} + \frac{\delta_f - \delta_i}{2} O_2(g) \tag{21.47}$$

Where δ_i is and δ_f are the initial and final nonstoichiometries prior to and following reduction. The oxidation reaction to split H_2O or CO_2 proceeds as expected according to

Oxidation:

$$CeO_{2-\delta f} + (\delta_f - \delta_i)H_2O \overset{\Delta H_{GS}}{\rightarrow} CeO_{2-\delta i} + (\delta_f - \delta_i)H_2 \tag{21.48}$$

21.2.7.1 Thermodynamics of Ceria-based Nonstoiciometric Redox Cycles

The nonstoichiometry of ceria is strongly dependent on temperature and oxygen partial pressure (p_{O_2}) and has been studied in detail by several authors, and is documented nicely by [66]. The advantage of ceria-based systems compared to other stoichiometric systems such as iron oxide is that nonstoichiometric ceria, $CeO_{2\text{-}\delta}$, is extremely stable, both morphologically and crystallographically, over a wide temperature range and therefore is not as affected by sintering and subsequent deactivation [65, 67–68]. Further, kinetics are extremely rapid compared to the iron oxide system due to the high ambipolar oxygen diffusion rates in the bulk, which is dictated by its high ionic and electronic conductivities [65]. Because oxygen is the primary species that is transported in the bulk, growing scales also are not formed like in the iron-oxide-based systems. Oxygen nonstoichiometry of ceria as a function of temperature and oxygen partial pressure, as measured by [66] in a thermogravimetric analyzer, are shown in Figure 21.12. As seen, δ increases with increasing temperature and decreasing oxygen partial pressure.

---

18 It should be noted that Flamant et al. proposed the use of a ceria based cycle prior to Chueh et al. [67] but in that work they proposed reducing all the way to the Ce_2O_3 phase.

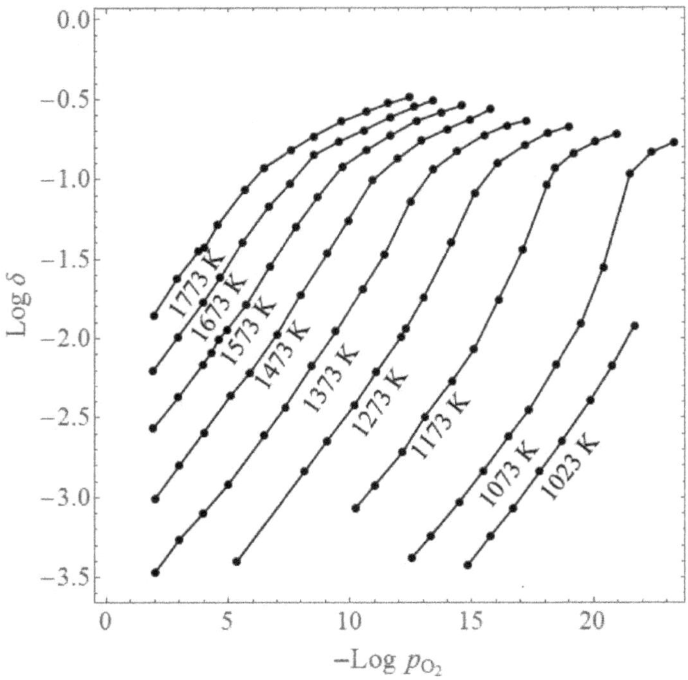

Figure 21.12 Ceria ($CeO_{2-\delta}$) oxygen nonstoichiometry shown as a function of p_{O_2} for temperatures between 1023 K to 1773 K. Data are taken from [63, 66].

21.2.7.2 Ceria Reduction

Thermodynamic data of nonstoichiometric systems such as ceria are not as readily tabulated like that for iron oxide and other stoichiometric reactions. However, the data may be easily extracted from measured nonstoichiometry data such as that shown in Figure 21.12. For this, an infinitesimally small change in nonstoichiometry is considered, as shown below:

$$\lim_{\delta_f - \delta_i = 0} \frac{1}{\delta_f - \delta_i} CeO_{2-\delta_i} \rightarrow \frac{1}{\delta_f - \delta_i} CeO_{2-\delta_f} + \frac{1}{2} O_2(g) \tag{21.49}$$

Assuming that the activity of the solid reactants and products in the above reaction are the same, it can be shown that

$$\Delta g_O(\delta,T) = -RT \ln p_{O_2}(\delta,T)^{1/2} \tag{21.50}$$

where Δg_o is the partial molar Gibbs free energy which is a function of both nonstoichiometry and temperature [65–66, 69]. Δg_o may be related to the partial molar enthalpy (Δh_o) and partial molar entropy (Δs_o) through the well-known relation

$$\Delta g_O(\delta,T) = \Delta h_O(\delta) - T\Delta s_O(\delta) \tag{21.51}$$

And upon rearrangement of Equations (21.50) and (21.51) it can be shown that

$$\ln p_{O_2}(\delta,T)^{1/2} = -\left(\frac{1}{T}\right)\frac{\Delta h_O(\delta)}{R} + \frac{\Delta s_O(\delta)}{R} \tag{21.52}$$

Thus, the slope of a plot of ln $p_{O_2}^{1/2}$ versus inverse temperature (at a constant nonstoichiometry) is proportional to Δh_o and the intercept is proportional to Δs_o.

When acquiring data like that in Figure 21.12, it is important to obtain enough data points at different temperatures and oxygen partial pressures at a fixed delta (effectively a horizontal cross-section through the plot) to give confidence to the fitted data (e.g., more than two points). Results for Δh_o and Δs_o using experimental nonstoicheometry data obtained from Panlener et al. [66] are shown in Figure 21.13. As seen, enthalpy decreases with increasing nonstoichiometry from about 480 to 400 kJ mol^{-1} as δ increases from 0 to 0.2. Entropy also decreases with increasing nonstoichiometry but even more strongly than enthalpy; it decreases from about 300 J mol^{-1} K^{-1} to 150 J mol^{-1} K^{-1} as δ increases from 0 to 0.2. Once Δh_o and Δs_o are known as a function of nonstoichiometry, it is then possible to calculate predicted oxygen partial pressures or nonstoichiometries for reduction conditions by using one of the two as an input alongside temperature using Equation (21.52).

Example Question 1:
Given ceria at 1773 K in an environment with an oxygen partial pressure of 10^{-5} atm, calculate the oxygen non-stoichiometry (δ) at equilibrium using an iterative, guess and check, approach.

Solution 1:
The question states that a sample of ceria has a temperature of 1773 K and exists in an environment with an oxygen partial pressure (p_{O_2}). The oxygen non-stoichiometry of ceria can be found as a function of these two properties by relating the two well-known equations for partial molar Gibbs free energy, Equations (21.50) and (21.51).

$$\Delta g_o(\delta,T) = -RT\ln p_{O_2}(\delta,T)^{1/2} \tag{21.50}$$

$$\Delta g_o(\delta,T) = \Delta h_o(\delta) - T\Delta s_o(\delta) \tag{21.51}$$

Equating these two relationships and dividing by T, we arrive at Equation (21.52).

$$\ln p_{O_2}(\delta,T)^{1/2} = -\left(\frac{1}{T}\right)\frac{\Delta h_o(\delta)}{R} + \frac{\Delta s_o(\delta)}{R} \tag{21.52}$$

Here, the left side of the equation can be directly calculated, and values for temperature (T) and the universal gas constant (R) can be plugged in to reach an easily iterated equation.

$$-5.756 = -\left(\frac{1}{1773\text{K}}\right)\frac{\Delta h_o(\delta)}{8.314\dfrac{\text{J}}{\text{mol}\cdot\text{K}}} + \frac{\Delta s_o(\delta)}{8.314\dfrac{\text{J}}{\text{mol}\cdot\text{K}}}$$

Using an initial guess of the oxygen non-stoichiometry (δ), refer to data obtained from Panlener et al. in Figure 21.13 to retrieve enthalpy and entropy values. Then calculate the right-hand side (RHS) of the equation and compare to the left-hand side (LHS).

Initial guess will be taken as: δ = 0.05

(RHS) = -5.399

Second guess will be δ = 0.055

RHS = -5.706

Continuing iterations until RHS = LHS will converge on the exact answer of δ = 0.0589

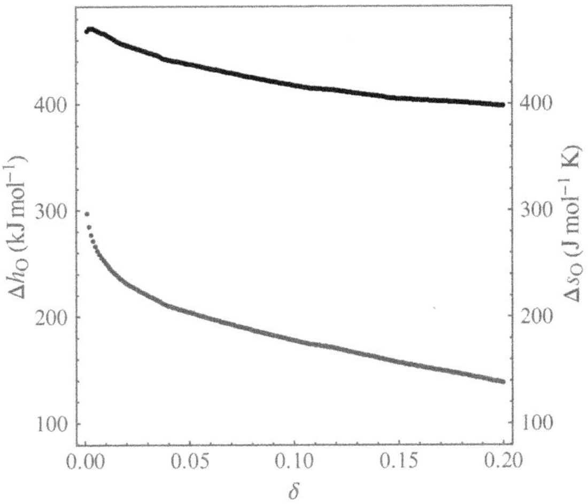

Figure 21.13 Partial molar enthalpy (Δh_o) and entropy (Δs_o) plotted as a function of δ in $CeO_{2-\delta}$. Nonstoichiometry equilibrium data obtained from [63, 66].

Note: In order to obtain accurate values of enthalpy and entropy, it is often useful to fit a polynomial curve to the data found in Figure 21.13 and use your favorite mathematical software to calculate the respective values based on your guess of oxygen non-stoichiometry.

Example Question 2:
Given ceria at 1773 K in an environment with an oxygen partial pressure of 10^{-5} atm, calculate the oxygen non-stoichiometry (δ) at equilibrium by employing a minimization technique using your favorite mathematical software.

Solution 2:
Note: This solution will employ MATLAB to solve for oxygen non-stoichiometry (δ)

The question provides that a sample of ceria has a temperature of 1773 K and exists in an environment of oxygen partial pressure (Po_2). The oxygen non-stoichiometry of ceria can be found as a function of these two properties by relating the two well-known equations for partial molar Gibbs free energy, Equations (21.50) and (21.51), and arriving at Equation (21.52) using the same method as seen in Solution 1.

$$\ln p_{O_2}(\delta,T)^{1/2} = -\left(\frac{1}{T}\right)\frac{\Delta h_o(\delta)}{R} + \frac{\Delta s_o(\delta)}{R} \tag{21.52}$$

In this solution, the MATLAB function "fzero" will be used to search for a non-stoichiometry that will satisfy the following version of Equation (21.52), where all terms have been rearranged to equal zero.

$$\ln p_{O_2}(\delta,T)^{1/2} + \left(\frac{1}{T}\right)\frac{\Delta h_o(\delta)}{R} - \frac{\Delta s_o(\delta)}{R} = 0$$

Before using the fzero solver, functions for changes in enthalpy and entropy as a function of oxygen non-stoichiometry should be defined. This can be done in several ways including fitting a polynomial to existing data or using existing data with an interpolating function. In the blocks of code seen in this solution, these functions are dH(x) and dS(x). For example, see example Question 4

The fzero function takes an input of a function (f) and an initial guess (x) in the format fzero(f,x)

An example of code employing this technique is seen below, along with the MATLAB output of the answer, delta = 0.0589.

```
>> T=1773; Po2=1e-05; R=8.314;
>> f=@(x)log(Po2^0.5)+((1/T)*(dH(x)/R))-(dS(x)/R);
>> delta=fzero(f,0.05)
delta = 0.0589020711130001
```

Note: Using simple coding loops in MATLAB or another software, this minimization technique can be employed across a range of conditions to reproduce the data seen in Figure 21.12.

Example Question 3:
Compute change in partial molar entropy and change in partial molar enthalpy as a function of oxygen non-stoichiometry for ceria.

Solution3:
In order to calculate the change in entropy and change in enthalpy of ceria at a given condition, one must recall Equation (21.52).

$$\ln p_{O_2}\left(\delta,T\right)^{1/2} = -\left(\frac{1}{T}\right)\frac{\Delta h_o(\delta)}{R} + \frac{\Delta s_o(\delta)}{R} \tag{21.52}$$

Comparing this equation to slope-intercept form, $y = mx + b$, it can be seen that by plotting the natural log of the square root of oxygen partial pressure on the vertical axis and the reciprocal of temperature on the horizontal axis, the slope of the resulting plot will be proportional to the change in enthalpy and the vertical axis intercept will be proportional to the change in entropy, at a fixed nonstoichiometry.

Using data such as that presented in Figure 21.12 (obtained experimentally), oxygen partial pressure values for each isotherm at a range of constant delta values should be tabulated. This will result in a data set that can be easily converted to represent $lnp_{O_2}(\delta,T)^{1/2}$. For example, for δ = 0.01 the data looks like the following:

$\ln p_{O_2}\left(\delta,T\right)^{1/2}$	-3.256	-5.313	-7.884	-10.570	-13.827	-17.598	-21.997	-24.569
1/T	5.98E-4	6.37E-4	6.79E-4	7.28E-4	7.86E-4	8.52E-4	9.32E-4	9.78E-4

Note that data is not measured for the highest temperature at 1773 K for this nonstoichiometry.

Plotting the $\ln p_{O_2}\left(\delta,T\right)^{1/2}$ against $\left(\frac{1}{T}\right)$ should be linear with a negative slope for each constant delta value, as shown below.

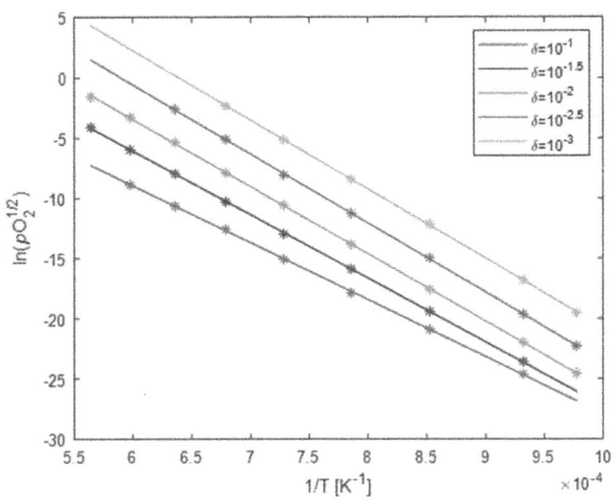

Calculating the slope of each line will result in $\frac{\Delta h_o(\delta_i)}{R}$ where δ_i is one constant delta value. The respective y-intercepts will correspond to $\frac{\Delta s_o(\delta_i)}{R}$. Scaling these values by the universal gas constant (R), in this case equal to 8.314 kJ mol^{-1} K^{-1} and plotting them against a range of deltas will result in the plot seen in Figure 21.13 that shows partial molar enthalpy and partial molar entropy as a function of oxygen non-stoichiometry. See https://faculty.eng.ufl.edu/jonathan-scheffe/solar-energy-educational-resources/ for full solution and csv data

21.2.7.3 Ceria Oxidation with H_2O (or CO_2)

During oxidation with H_2O, the following reactions occur.

$$\frac{1}{\delta_f - \delta_i}CeO_{2\text{-}\delta_f} + \frac{1}{2}O_2(g) \rightarrow \frac{1}{\delta_f - \delta_i}CeO_{2\text{-}\delta_i}, -\Delta g_O(T,\delta) \tag{21.53}$$

$$H_2O(l) \rightarrow \tfrac{1}{2}O_2(g) + H_2(g) \tag{21.54}$$

This results in the net oxidation reaction proceeding according to

$$\frac{1}{\delta_f - \delta_i}CeO_{2\text{-}\delta_f} + H_2O(l) \rightarrow \frac{1}{\delta_f - \delta_i}CeO_{2\text{-}\delta_i} + H_2(g), \Delta G_{ox} \tag{21.55}$$

A qualitative indicator of oxidation thermodynamic favorability may be seen by comparing the magnitudes of -Δg_O (Equation (21.53)) and $\Delta G^\circ_{f,H_2O}$ (reverse of Equation (21.54)). By doing so, the thermodynamic favorability of either nonstoichiometric ceria or H_2 to consume gaseous O_2 may be discerned. These quantities are plotted in Figure 21.14 in dotted lines for δ_i = 0.001, 0.01 and 0.1. Also plotted in dashed lines are the Gibbs free energy change (ΔG_{ox}) of the net oxidation reaction, Equation (21.55). As seen, the slope of each of the -Δg_O lines is greater than $\Delta G^\circ_{f,H_2O}$ which is one criteria necessary for thermochemical redox cycles, as discussed in the section above entitled "Thermochemical Redox Cycles." The temperature where -Δg_O crosses $\Delta G^\circ_{f,H_2O}$ (a second criteria, resulting from a larger enthalpy change) increases with increasing δ_i, which indicates that the reaction from δ_f to that specific nonstoichiometry is possible at increasingly higher temperatures. Put another way, to oxidize to lower δ_i where oxidation conversions are greater ($\delta_f - \delta_i$), the temperature should be as low as possible. For δ_i decreasing from 0.1 to 0.001, the temperatures that this intersects $\Delta G^\circ_{f,H2O}$ are 1386 K, 1115 K and 912 K, respectively. As seen in the dashed curves, ΔG_{ox} is negative below each of these respective temperatures, indicating their favorability. Also note that nearly complete oxidation

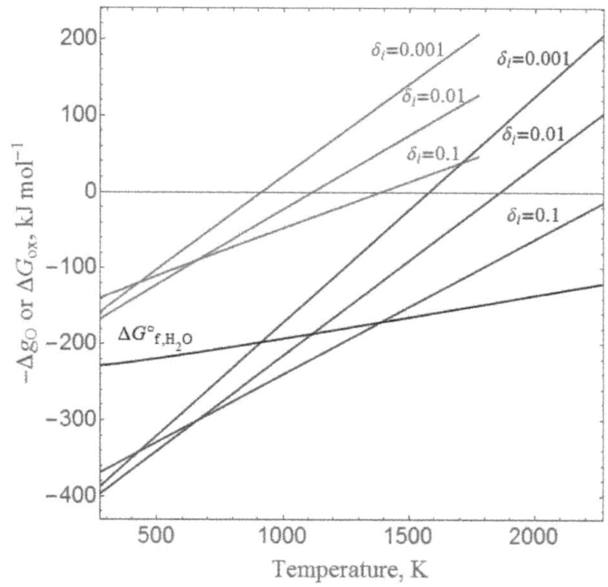

Figure 21.14 Partial molar Gibbs free energy change for the ceria oxidation reaction, $-\Delta g_O$ (dotted), Gibbs free energy change of the water formation reaction, $\Delta G^\circ_{f,H2O}$ (whole line) and Gibbs free energy change for the oxidation of ceria with H_2O, ΔG_{ox} (dashed), all as a function of temperature. δ_i refers to the nonstoichiometry in ceria, $CeO_{2-\delta i}$, following oxidation.

to 0.001 is favorable below 912 K, compared to ~ 675 K for FeO oxidation to Fe_3O_4 (c.f. Figure 21.11), meaning that reduced ceria should oxidize more readily at higher temperatures, which is beneficial from a kinetic and efficiency perspective.

The approach used to determine oxidation yields is similar to that discussed prior for iron oxide, but instead of oxide formation energies, partial molar properties are used. It is assumed that reactions (21.53) and (21.54) above are in equilibrium and thus the oxygen partial pressures are equal. Therefore, combining Equations (21.50) and (21.27) we can show that

$$\Delta g_O(\delta_i, T) = -RT \ln \frac{K_{H_2O} p_{H_2O}}{p_{H_2}} \tag{21.56}$$

Assigning reaction coordinates as discussed prior we can relate the partial pressures to moles of each species at equilibrium to show the following.

$$\Delta g_O(\delta_i, T) = -RT \ln \frac{K_{H_2O}\left(n_{H_2O,i} - \varepsilon\right)}{\varepsilon} \tag{21.57}$$

where

$$\varepsilon = \delta_f - \delta_i \tag{21.58}$$

Therefore, the reaction coordinate can be solved iteratively using Equation (21.57). Results for H_2 yields are shown in Figure 21.15 as a function of temperature. Here, the total pressure was kept constant at 1 atm, δ_f was fixed at 0.1 (following reduction), and $n_{H_2O,i}$ was set to δ_f, $10\delta_f$ and $100\delta_f$. When $n_{H_2O,i} = \delta_f$, oxidation is nearly 100% complete (mol $H_2 = \delta_f$) for temperatures lower than 800 K, reaches 90% completion at about 1019 K and then steadily decreases with increasing temperature; only 10% completion is reached at 1596 K. For larger amounts of initial H_2O in the system, oxidation extents are greater at higher temperatures, as one would expect

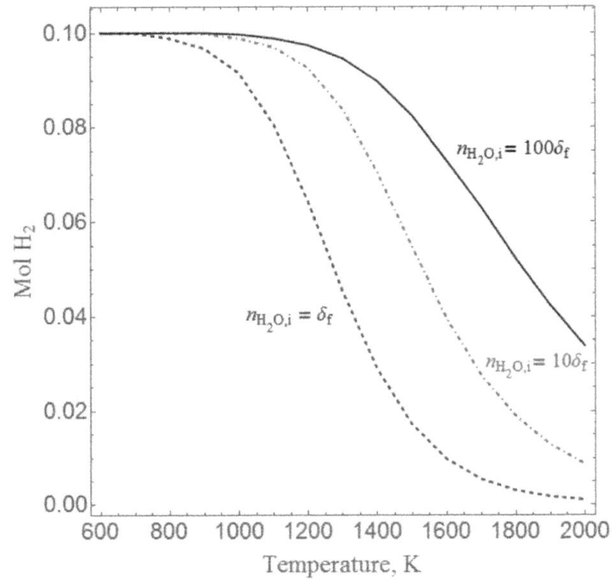

Figure 21.15 H_2 produced as a result of oxidation of $CeO_{2-\delta f}$ with H_2O. Total pressure is kept constant at 1 atm and δ_f is set 0.1. The same general trends hold for different values of δ_f.

from Le Chatlier's principle. For example, when $n_{H_2O,i} = 100\delta_f$, oxidation is nearly 100% complete at 1200 K and lower. 90% completion is achieved at 1396 K and 10% at 2529 K (outside scale of figure). While at first glance this may appear to be a convenient way to increase oxidation yields, the conversion of H_2O to H_2 should be considered because of sensible heating requirements necessary to bring the excess H_2O to the oxidation temperature. Here, conversion (α) is defined as the moles of water converted to H_2 divided by the total amount of water introduced into the system, shown mathematically as.

$$\alpha = \frac{n_{H_2O,i} - n_{H_2O}}{n_{H_2O,i}} \tag{21.59}$$

When $n_{H_2O,i} = \delta_f$, conversion is the same as oxidation completion described prior. However, when $n_{H_2O,i} = 100\delta_f$, conversion will always be ≪ 100% because the amount of H_2O introduced into the system is much larger than the total number of moles that may react to produce H_2. Thus, at the conditions considered prior (1200 K, 1396 K, and 2529 K), conversion is only 1%, 0.9%, and 0.1%. The importance of conversion becomes apparent when comparing the magnitudes of enthalpies of heating H_2O to the oxidation temperature of interest relative to the higher heating value (HHV) of H_2. For example, consider oxidation at 1396 K where $n_{H_2O,i} = 100\delta_f$, and conversion is equal to 0.9%. This implies that 10 mol of H_2O must be heated and 0.09 mol of H_2 is produced. The enthalpy required to heat 10 mol of H_2O from ambient conditions to 1396 K is 870.3 kJ, while the HHV is only 25.7 kJ.

If the goal is to dissociate CO_2 rather than H_2O, similar calculations can be performed using the reaction coordinate methodology. For CO_2 splitting it is important to consider that C(s) may be thermodynamically predicted rather than CO(g). Nevertheless, this has not been observed experimentally and it is typical to only consider the $CO_2 \rightarrow CO + ½\, O_2$ reaction.

Example Question 4:
Given data arrays of $\Delta h_o(\text{kJ}/\text{mol})$ and $\Delta s_o(\text{J}/\text{mol}*\text{K})$ versus δ for Ceria, fit a third degree polynomial function suitable for describing Δh_o and Δs_o as a function of δ using the *curve*fit_ function from Python's *scipy.optimize* library.

Solution 4:
Note: For this we will use Python and import tabulated data from a *.csv* file.

```
In [1]: #Import necessary libraries and data arrays from files.
        import numpy as np
        import pandas as pd
        delta_h = pd.read_csv('Change_in_EnthalpyVSDelta.csv',
        names=['delta','$\Delta h_o$ (kJ/mol)'])
        delta_s = pd.read_csv('Change_in_EntropyVSDelta.csv',
        names=['delta','$\Delta s_o$ (J/mol*K)'])
        delta_h.head()

Out[1]:
```

	delta	Δh_o (kJ / mol)
0	0.006319	472.083
1	0.011813	465.863
2	0.016758	461.339
3	0.021703	455.685
4	0.026374	452.292

Plot extracted data by slicing the data array in different columns. The first column represents δ and the second column represents either Δh_o or Δs_o

```
In [2]: #Import necessary libraries and plot raw data for
        better visualization
        import matplotlib.pyplot as plt
        plt.plot(delta_h.iloc[:,0], delta_h.iloc[:,1], 'bo',
        markersize=3, label='$\Delta h_o$ data')
        plt.plot(delta_s.iloc[:,0], delta_s.iloc[:,1], 'gs',
        markersize=3, label='$\Delta s_o$ data')
        plt.xlabel('$\delta$')
        plt.ylabel('$\Delta h_o$'+' (kJ $mol^{-1})$'+', '+'$\
        Delta s_o$'+' (J $mol^{-1}K^{-1})$')
        plt.legend
        loc='best')
        plt.ylim(0,500)
        plt.title
        ('Enthalpy and
        Entropy versus
        Delta')
        plt.show()
```

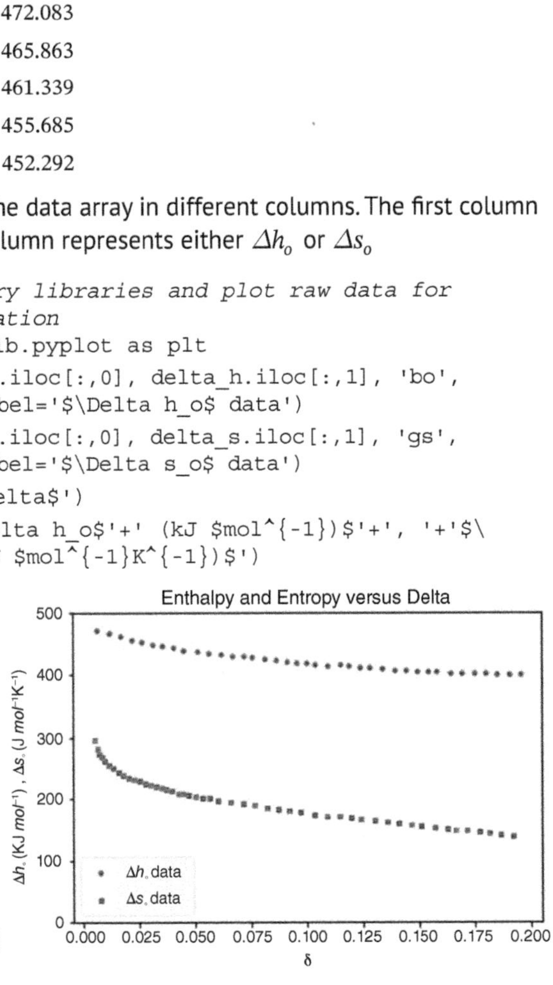

Define a polynomial function *poly* dependent on delta (**d**) and the different coefficients of the polynomial (**c1,c2,c3,c4**)

```
In [3]: def poly(d,c1,c2,c3,c4):
          return c1*d**3+c2*d**2+c3*d+c4
In [4]: #Import necessary libraries and calculate the polynomial
        coefficients
        from scipy.optimize import curve_fit
        coeff1, pcov = curve_fit(poly, delta_h.iloc[:,0],
        delta_h.iloc[:,1])
        coeff2, pcov = curve_fit(poly, delta_s.iloc[:,0],
        delta_s.iloc[:,1])
        print('Coefficients for delta_h : ', coeff1)
        print('Coefficients for delta_s : ', coeff2)
        Coefficients for delta_h : [-8899.81954683 4542.24540679
        -943.45358723 475.59544598]
        Coefficients for delta_s : [-48944.72003456
        17642.77453637 -2368.87714186 284.14274668]
```

Plot curve fit polynomial function *poly* with the obtained coefficients along with the previously plotted raw data to show comparison

```
In [5]: d_fit = np.arange(0,0.2,0.001) #Define new array for d
        plt.plot(d_fit, poly(d_fit, *coeff1), 'r', label='$\Delta
        h_o$ curve fit')
        plt.plot(d_fit, poly(d_fit, *coeff2), 'k', label='$\Delta
        s_o$ curve fit')
        plt.plot(delta_h.iloc[:,0], delta_h.iloc[:,1], 'bo',
        markersize=3, label='$\Delta h_o$ data')
        plt.plot(delta_s.iloc[:,0], delta_s.iloc[:,1], 'gs',
        markersize=3, label='$\Delta s_o$ data')
        plt.xlabel('$\delta$')
        plt.ylabel('$\Delta h_o$'+' (kJ $mol^{-1})$'+',
        '+'$\Delta s_o$'+' (J $mol^{-1}K^{-1})$')
        plt.legend(loc='best', ncol=2)
        plt.ylim(0,500)
        plt.title('Enthalpy and Entropy versus Delta with Curve Fit')
        plt.show()
```

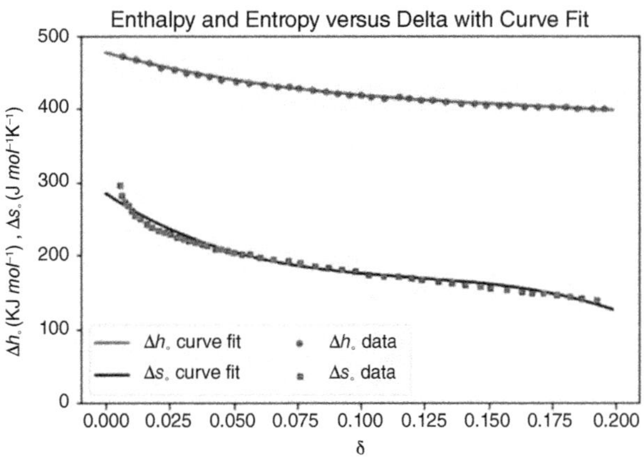

Solution: Δh_o and Δs_o can now be computed as a function of δ and the curve fit coefficients found, *coeff1* and *coeff2* respectively. See https://faculty.eng.ufl.edu/jonathan-scheffe/solar-energy-educational-resources/ for full solution and csv data

Example Question 5:
Given a temperature of operation $T = 1273$ K for ceria and an initial amount of moles of water $n_{\mathrm{H_2O,i}} = \delta_f = 0.1$. Solve for δ_i using an iterative approach.

Background: The partial molar Gibbs free energy for Ceria Δg_o, which is a function of both nonstoichiometry and temperature, is expressed as two equations below:

$$\Delta g_o(\delta,T) = -RT\ln p_{O_2}(\delta,T)^{1/2} \tag{21.50}$$

$$\Delta g_o(\delta,T) = \Delta h_o(\delta) - T\Delta s_o(\delta) \tag{21.51}$$

Also, the oxygen partial pressure p_{O_2} can be obtained from the reaction equilibrium analysis of the dissociation of H_2O. From (21.33) we recognize that the reaction coordinate ε is equal to $\delta_\mathrm{f} - \delta_\mathrm{i}$,

$$K_{H_2O} = \frac{\left(\frac{\in}{n_\mathrm{total}}\right) * p_{\mathrm{O_2}}^{1/2}}{\left(\frac{n_{\mathrm{H_2O,i}} - \in}{n_\mathrm{total}}\right)} = \frac{(\delta_\mathrm{f} - \delta_\mathrm{i}) * p_{\mathrm{O_2}}^{1/2}}{n_{\mathrm{H_2O,i}} - (\delta_\mathrm{f} - \delta_\mathrm{i})}$$

$$p_{\mathrm{O_2}}^{1/2} = \frac{n_{\mathrm{H_2O,i}} - (\delta_\mathrm{f} - \delta_\mathrm{i})}{(\delta_\mathrm{f} - \delta_\mathrm{i}) * K_{\mathrm{f,H_2O}}}$$

Solution 5:
Note: For this we will use Python, and import data for the equilibrium constant of formation of water $K_{\mathrm{f,H_2O}}$ from the NIST-JANAF website. A curve fitting procedure needs to be applied, similar to Example Question 4.

```
In [6]: #Import Equilibrium Constant Data for formation of
        Water, 1bar, (1,g)
        Kf_H2O = pd.read_csv('H2O_Equilibrium_Constant.csv',
        names=['Temperature (K)','log $K_f$'])
        Kf_H2O.head()

Out[6]:
```

	Temperature (K)	log K_f
0	280.00	44.796
1	298.15	41.546
2	300.00	41.237
3	320.00	38.131
4	340.00	35.396

The data has a linear relationship when plotted versus $1/T$ on the x-axis.

```
In [7]: #Plot raw data for better visualization
        plt.plot(1/(Kf_H2O.iloc[:,0]), Kf_H2O.iloc[:,1], 'bo',
        markersize=3, label='$K_f$ data')
        plt.xlabel('$1/T$'+' (K)')
        plt.ylabel('log '+'$K_f$')
        plt.title('Equilibrium Constant for Formation of Water')
        plt.legend()
        plt.show()
```

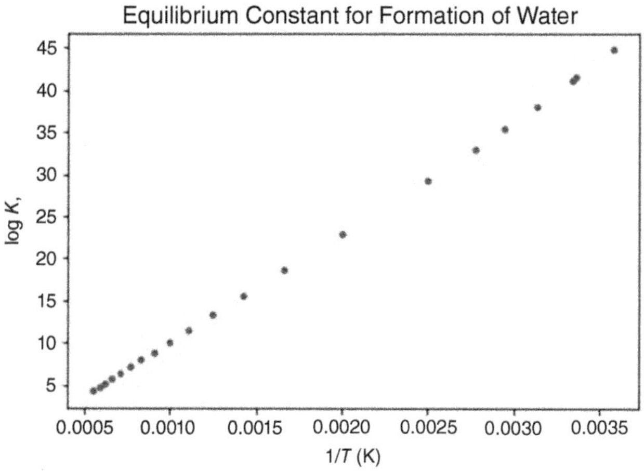

Define a function named *linear* dependent on the **(x)**, the slope **(m)** and the y-intercept **(b)**. Calculate the *linear* function coefficients. Then, plot the curve fit function *linear* with the obtained coefficients along with the previously plotted raw data to show comparison.

```
In [8]: def linear(x, m, b):
            return m*x+b

In [9]: #Calculate the linear function coefficients in the same
        way as Example Question 4
        coeff3, pcov = curve_fit(linear, 1/(Kf_H2O.iloc[:,0]),
        Kf_H2O.iloc[:,1])
        print('Coefficients for Kf : ', coeff3)
        Coefficients for Kf : [ 1.32377167e+04 -3.21602829e+00]

In [10]: x_fit = np.arange(0,0.0035,0.00005) #Define new array for x
        plt.plot(1/(Kf_H2O.iloc[:,0]), Kf_H2O.iloc[:,1], 'bo',
        markersize=3, label='$K_f$ data')
        plt.plot(x_fit, linear(x_fit, *coeff3), 'r', label='$K_f$
        curve fit')
        plt.xlabel('$1/T$'+' (K)')
        plt.ylabel('log '+'$K_f$')
        plt.title('Equilibrium Constant for Formation of Water
        with Curve
        Fit')
        plt.legend()
        plt.show()
```

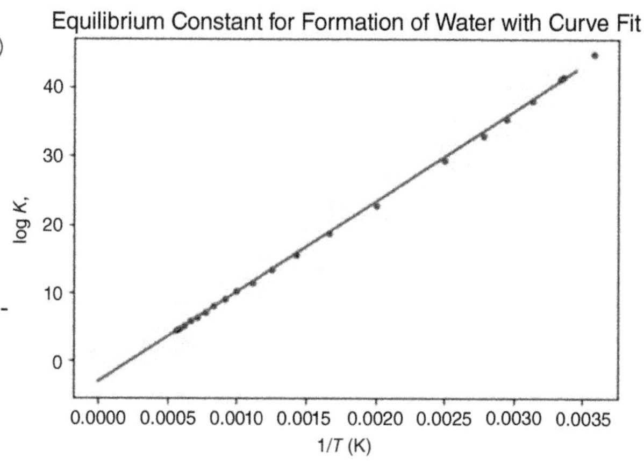

Define functions for the partial molar Gibbs free energy as a function of both nonstoichiometry and temperature, make sure to replace the expression for oxygen partial pressure into (21.50). δ_i is represented as d.

```
In [11]: #Define variables and fixed parameters
         R = 0.008314 #(kJ/mol K)
         T = 1273 #(K)
         d_f = 0.1 #(moles)
         n_H2O = df #(moles)

In [12]: def Gibbs1(d):
             return -R*T*np.log((n_H2O-(d_f-d))/((d_f-d)*10**
         (linear((1/T), *coeff3))))
         def Gibbs2(d):
             return poly(d, *coeff1)-poly(d, *coeff2)*T/1000
```

Trial and error approach. Iteratively input values for delta such that $0.0<\delta_i<\delta_f$. Modify values in both functions until the result is approximately the same. If we substract both functions the value should get close to zero when the function values are the same.

```
In [13]: #Try d=0.08
         attempt = Gibbs1(0.08)-Gibbs2(0.08)
         print(attempt)
         -31.955065691423272

In [14]: #Try d=0.09
         attempt = Gibbs1(0.09)-Gibbs2(0.09)
         print(attempt)
         -42.389362187631036

In [15]: #Try d=0.06
         attempt = Gibbs1(0.06)-Gibbs2(0.06)
         print(attempt)
         -14.49090419841059

In [16]: #Try d=0.045
         attempt = Gibbs1(0.045)-Gibbs2(0.045)
         print(attempt)
         1.4560637768568085
```

Solution: $\delta_i=0.045$ approximately. See https://faculty.eng.ufl.edu/jonathan-scheffe/solar-energy-educational-resources/ for full solution and csv data

Example Question 6:
Given a temperature of operation $T=1273\mathrm{K}$ for Ceria and an initial amount of moles of water $n_{\mathrm{H_2O,i}}=\delta_{\mathrm{f}}=0.1$. Solve for δ_{i} and the H_2 yield. Create a solver/minimizer using the *minimize* function from the *scipy.optimize* library.

Solution 6:
Note: For this we will use Python, and remember that $H_2=\delta_{\mathrm{f}}-\delta_{\mathrm{i}}$, which is the difference between the final and initial nonstoichiometries.

Define the function *objective*, which is the absolute value of the difference between both partial molar Gibbs free energy expressions, in that way we guarantee the result gets as close possible to zero. δ_i is represented as d.

```
In [17]: #Import necessary libraries for minimization procedure
         and define objective function to minimize
         from scipy.optimize import minimize
         def objective(d):
            return np.abs(poly(d,*coeff1)-
         poly(d,*coeff2)*T/1000+(R*T*np.log((n_H2O-(d_f-d))/
         ((d_f-d)*10**(linear((1/T),*coeff3))))))
```

The minimization method we are using, **(SLSQP)**, requires a set of initial guesses and boundaries for every value. Define those parameters, then minimize the *objective* function.

```
In [18]: initial_guess = 0.01
         solution = minimize(objective, initial_guess,
         method='SLSQP', tol=1e-10,
         options={'disp': False}, bounds=[(0,0.2)])
         print('Value of delta_i is: %.4f' %solution.x[0])
         Value of delta_i is: 0.0462

In [19]: print('H_2 yield is: %.4f' %(df-solution.x[0]))
         H_2 yield is: 0.0538
```

Solution: $\delta_i = 0.0462$ and $H_2 = 0.0538$. See https://faculty.eng.ufl.edu/jonathan-scheffe/solar-energy-educational-resources/ for full solution and csv data

Example Question 7:
Given six different temperatures of reduction $T_H = 1573, 1673, 1773, 1873, 1973, 2073\,\mathrm{K}$, temperature of oxidation $800\,K < T_L < 1200\,K$, and oxygen partial pressure $p_{O_2} = 10^{-5}$ atm for Ceria. Compute H_2 yield to reproduce data from [44, 67] Figure 21.17a .

Solution 7:
Note: For this we will use Python, and remember that $n_{H_2O,i} = \delta_f$ (initial moles of water are equal to the nonstoichiometry after reduction).

Perform minimization procedure to find the nonstoichiometry after reduction **(d_f)** knowing the relationship in between p_{O_2}, T, and δ given by (21.52):

$$\ln p_{O_2}(\delta, T)^{1/2} = -\left(\frac{1}{T}\right)\frac{\Delta h_o(\delta)}{R} + \frac{\Delta s_o(\delta)}{R} \qquad (21.52)$$

Define function *objective_red* depending only on delta **(d)** similar to Example Question 6. However, the minimization procedure will repeat using a *for loop* for every temperature of reduction **(T_H)**. Then, the values of delta **(d)** will be stored in the array **(d_f)** for later use.

```
In [20]: #Define variables and fixed parameters
         T_H = np.array([2073, 1973, 1873, 1773, 1673, 1573]) #(K)
         p_O2 = 1e-5 #(atm)
         d_f = np.array([])

In [21]: for i in range(0, T_H.size):
           def objective_red(d):
             return  np.abs(np.log(p_O2**0.5)+(poly(d,*coeff1)/
         (T_H[i]*R))-
         (poly(d,*coeff2)/(1000*R)))
         initial_guess_red = 0.01
         #Minimize function objective_red
```

```
        sol_red = minimize(objective_red, initial_guess_red, method=
        'SLSQP', tol=1e-10, options={'disp': False}, bounds=[(0,0.2)])
        d_f = np.append(d_f, sol_red.x[0])
```

Perform minimization procedure to find nonstoichiometry after oxidation **(d_i)** similar to Example Question 6. Define function *objective_ox* depending only on delta **(d)**. The minimization procedure will repeat using a *for loop* for different temperatures of oxidation in the range $800\,K < T_L < 1200\,K$. Then the values of delta **(d)** will be stored in the **(H_2)** array, remembering that $H_2 = \delta_f - \delta_i$. See https://faculty.eng.ufl.edu/jonathan-scheffe/solar-energy-educational-resources/ for full solution and csv data

```
In [22]: #Define variables and fixed parameters
         n_H2O = d_f #(mol)
         T_L = np.arange(800,1200,10) #(K)

In [23]:for ii in range(0, d_f.size-1):
         H_2 = np.array([])
         for i in range(0, T_L.size):
         def objective_ox(d):
         return np.abs(poly(d,*coeff1)-poly(d,*coeff2)*T_L[i]/1000+
        (R*T_L[i]*np.log((n_H2O[ii]-(d_f[ii]-d))/((d_f[ii]-d)
        *10**(linear((1/T_L[i]),*coeff3))))))initial_guess_ox = 0.01

        #Minimize function objective_ox

        sol_ox = minimize(objective_ox, initial_guess_ox,
        method='SLSQP', tol=1e-10, options={'disp': False},
        bounds=[(0,0.2)])
        H_2 = np.append(H_2, d_f[ii]-sol_ox.x[0])

        #For every T_H plot H_2 productivity as a function of T_L
        plt.plot(T_L, H_2, 'r')
        plt.plot(T_L, np.repeat(d_f[ii], 40), 'b--', linewidth=1)
        plt.xlabel('Oxidation Tempertature, '+' $T_L (K)$')
        plt.ylim(0, 0.2) plt.ylabel('$H_2$'+' Productivity')

        #Plot image with labels and text to reproduce original image
        as accurately as possible
        plt.text(810, 0.182, '$T_H=2073$ K')
        plt.text(810, 0.16, '$1973$ K')
        plt.text(810, 0.105, '$1873$ K')
        plt.text(810, 0.065, '$1773$ K')
        plt.text(810, 0.042, '$1673$ K')
        plt.text(810, 0.024, '$1573$ K')
        plt.plot(T_L,
        np.repeat(d_f[5],
        40), 'b--',
        linewidth=1)
        plt.title('$H_2$
        Productivity
        versus Oxidation
        Temperature')
        plt.show()
```

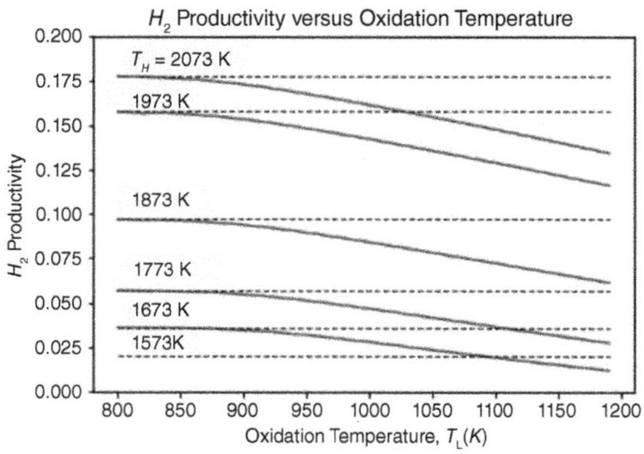

21.2.8 Efficiency of Redox Cycles

The solar to fuel conversion efficiency ($\eta_{\text{solar-to-fuel}}$) of thermochemical redox cycles is strongly tied to the temperature difference between reduction (T_{red}) and oxidation (T_{ox}) steps and the oxidation conversion at T_{ox} ($n_{H2}/n_{H2O,i}$). The former is related to the fact that the metal oxide must be heated each cycle from T_{ox} to T_{red} and the latter is related to the sensible energy required to heat the H_2O or CO_2 from ambient conditions to T_{ox}. A schematic of the cycle indicating the most pertinent heat transfer and mass flows is shown in Figure 21.16. Here, the material considered is nonstoichiometric ceria but the analysis may be applied to any two step redox cycle. An energy balance for this system yields.

$$\begin{aligned}
&Q_{\text{solar},1} + Q_{\text{solar},2} + \left(h_{H_2O}|_{T_{\text{OX}}} - h_{H_2O}|_{T\text{amb}}\right)n_{H_2O,i} = Q_{\text{re-rad, sensible}} \\
&+Q_{\text{re-rad, red}} + Q_{\text{rejection}} + Q_{\text{oxidation}} + \left(h_{O_2}|_{T_{\text{RED}}} - h_{O_2}|_{T\text{amb}}\right)n_{O_2} \\
&+\left(h_{H_2O}|_{T_{\text{OX}}} - h_{H_2O}|_{T\text{amb}}\right)n_{H_2O} + \left(h_{H_2}|_{T_{\text{OX}}} - h_{H_2}|_{T\text{amb}}\right)n_{H_2} \\
&+\text{HHV}_{H_2}n_{H_2}
\end{aligned} \tag{21.60}$$

$Q_{\text{solar},1}$ is the solar energy required to heat the oxide from T_{ox} to T_{red}, $Q_{\text{solar},2}$ is the solar energy needed to reduce the oxide, $Q_{\text{re-rad, sensible}}$ and $Q_{\text{re-rad, red}}$ are the radiative losses during sensible heating and reduction, $Q_{\text{rejection}}$ is the energy rejected from the system to cool the particles from T_{red} to T_{ox}, $Q_{\text{oxidation}}$ is the energy rejected from the system during exothermic oxidation, h_i is the specific enthalpy of species i at the specified temperature and n_i is the number of moles of species i. HHV_{H_2} is the higher heating value of H_2 and included in the energy balance assuming that the H_2 produced may be

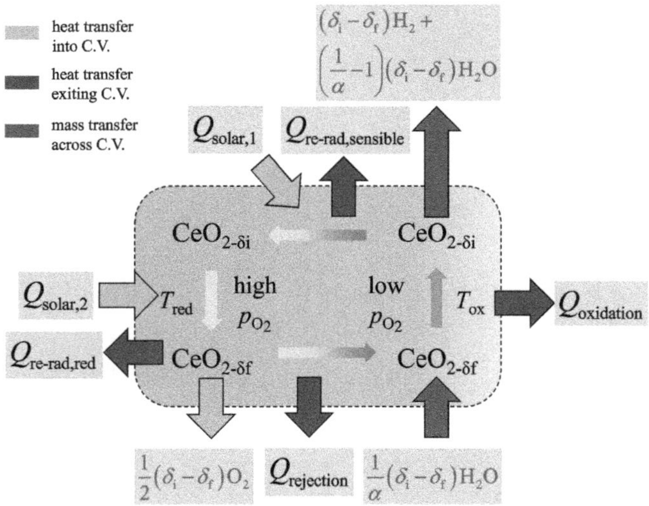

Figure 21.16 Heat and mass flows that accompany a ceria based thermochemical cycle; the analysis is conceptually similar for any two step thermochemical cycle. The major energy components required to drive the cycle are the result of the energy required to heat the oxide (Q_{sensible}), reduce it (Q_{red}), offset the radiative losses ($Q_{\text{re-rad,sensible}}$ and $Q_{\text{re-rad,red}}$) and heat the steam (or CO_2).

combusted to produce heat. $Q_{solar,1}$ is related to the sensible energy stored in the oxide, $Q_{sensible}$, through the following relationships.

$$Q_{sensible} = Q_{solar,1} - Q_{rerad,sensible} \tag{21.61}$$

$$Q_{sensible} = n_{CeO_2} \int_{T_{OX}}^{T_{RED}} c_p dT \tag{21.62}$$

For ceria, specific heat data are available from several resources as a function of temperature and nonstoichiometry and extrapolation is applied for high temperatures where data are not available [70]. $Q_{solar,2}$ is related to the enthalpy change of the oxide during reduction, Q_{red}, which is related to the partial molar enthalpy of ceria and should be integrated over the change in nonstoichiometry, as shown below.

$$Q_{red} = Q_{solar,2} - Q_{rerad,red} \tag{21.63}$$

$$Q_{red} = n_{CeO_2} \int_{\delta_f}^{\delta_i} \Delta h_o d\delta \tag{21.64}$$

The addition of solar energy to the system via an aperture means that the radiative losses $Q_{re\text{-}rad,sensible}$ and $Q_{re\text{-}rad,red}$ must be accounted for. These terms may be calculated assuming a blackbody cavity emitting radiation to the surroundings that are at ambient temperature. Because these terms do not necessarily occur at constant temperature, the losses should be integrated over the temperatures and times during which the heat transfer occurs. However, to simplify things it is usually assumed that the radiative losses all occur at the highest temperature, T_{red}, in which case they become equal to,

$$\left(Q_{re\text{-}rad,\ sensible} + Q_{re\text{-}rad,\ red}\right) = Q_{rad} = \left(1 - \eta_{abs}\right)\left(Q_{solar,1} + Q_{solar,2}\right) \tag{21.65}$$

where Q_{rad} is the total radiative heat transfer losses and η_{abs} was defined prior in Equation (21.4). Therefore, knowing the equilibrium yield of H_2 which is a function of the initial reduction extent (which is a function of T_{red} and p_{O_2}), $n_{H_2O,i}$ and T_{red}, the solar-to-fuel conversion efficiency ($\eta_{solar\text{-}to\text{-}fuel}$) can be calculated as a function of each of these parameters according to the standard efficiency definition in Equation (21.66).

$$\eta_{solar\text{-}to\text{-}fuel} = \frac{W_{out}}{Q_{in}} = \frac{n_{H_2} HHV_{H_2}}{Q_{solar,1} + Q_{solar,2} + \left(h_{H_2O T_{OX}} \middle| - h_{H_2O}\middle|_{Tamb}\right) n_{H_2O,i}} \tag{21.66}$$

It should be noted that energy inputs can be decreased with heat exchangers that take advantage of the hot gases exiting the system and heat rejection from solids that are recirculated within the system boundary, for example when cooling or during oxidation. For example, solid–solid heat exchangers that recover heat as particles are cooling from T_{red} to T_{ox} such as Sandia's CR5 and circulating particle reactor operate [56, 71], or gas–gas heat exchangers that take advantage of the hot gases exiting the system as in the University of Minnesota's solar reactor [72]. There are numerous research papers that deal with this topic and investigate the impact that heat exchange, operation temperatures, and pressures have on the overall theoretical efficiency [45–46, 48, 71, 73–75]. In general, depending on the operating conditions and degree of heat recovery, $\eta_{solar\text{-}to\text{-}fuel}$ can vary between only a few percent (e.g. during isothermal operation of reduction and oxidation steps and no heat recuperation) to 30% when using ceria. It should be noted that these values do not account for any spillage of solar radiation and typically heat losses except for radiation. Further, the energy required to obtain the oxygen partial pressures in the range of 10^{-3} to 10^{-5} atm during the reduction step are not included here but should be for a more thorough analysis; depending on the pressures assumed these may not be insignificant values. For example, it is typical to use an inert sweep gas such as Ar, which requires energy to produce, or vacuum pumping which requires mechanical work. The highest reported experimental efficiency is 5.25 % as reported by Marxer et al. in 2017 using ceria and vacuum pumping during reduction [76]. They note that vacuum pumping has the added benefit of limiting the heat transfer through the sides of their reactor. Seasonal variations regarding DNI and its impact on total fuel yields should also be considered when evaluating the overall economic cost of such a system. The economics are strongly tied to the solar-to-fuel conversion efficiency and have been discussed in several works [46, 77–78]. This is because the size of the concentrating infrastructure is the largest cost contributor and this size directly scales with efficiency. For example, a thermochemical cycle that is twice as efficient as another will require half the concentrating area to produce the same amount of fuel (assuming equal concentration ratios and spillage) and thus the concentrating equipment cost will be roughly cut in half.

Chueh et al. discuss how the efficiency of a ceria based thermochemical cycle varies as a function of T_{red} and T_{ox} assuming that only a limited amount of H_2O is introduced into the system to drive oxidation; namely $n_{H_2O,i} = \delta_f$. A summary of their results is shown in Figure 21.17. As seen, the efficiency largely scales with the amount of H_2 produced and the temperature swing required to produce it. For example, focusing on subplot a of Figure 21.17, the most H_2 is produced where ceria has been most reduced (i.e. at the highest reduction temperature of 2073 K). Looking at the subsequent subplot in Figure 21.17b, it is clear that efficiency largely scales with these yields, but there are optimal oxidation temperatures where the efficiency is maximized. This is because for the highest oxidation temperatures, H_2 yields begin to decrease, and at the lowest temperatures, the temperature swing and thus heating demands of the solid increase. This can be seen in Figure 21.17d where all of the enthalpy terms per mol of H_2 are tabulated versus reduction temperature. As seen, for the lowest reduction temperatures, the largest contributor is the enthalpy required to heat ceria. As the reduction temperature increases along with increasing yields of H_2, this contribution decreases, resulting in overall higher efficiencies.

21.2.9 Experimental Demonstration of the Ceria Based Cycle

There have been a number of prototype solar reactors proposed and developed for a number of metal oxide redox pairs other than ceria [56, 79–85]. In recent years however, the overwhelming majority of reactors have been focused on ceria based redox cycles because of its aforementioned favorable properties. The reactors vary from particle based systems [47, 86] that propose circulating particles through separate reduction and oxidation reactors, to batch type systems [67–68, 72, 76, 87] where ceria is stationary and the operating conditions within the reactor are varied with time to drive either reduction or oxidation.

Because of the simplicity of batch type systems, these have been especially attractive and to date have set the record for solar thermochemical splitting efficiencies, starting

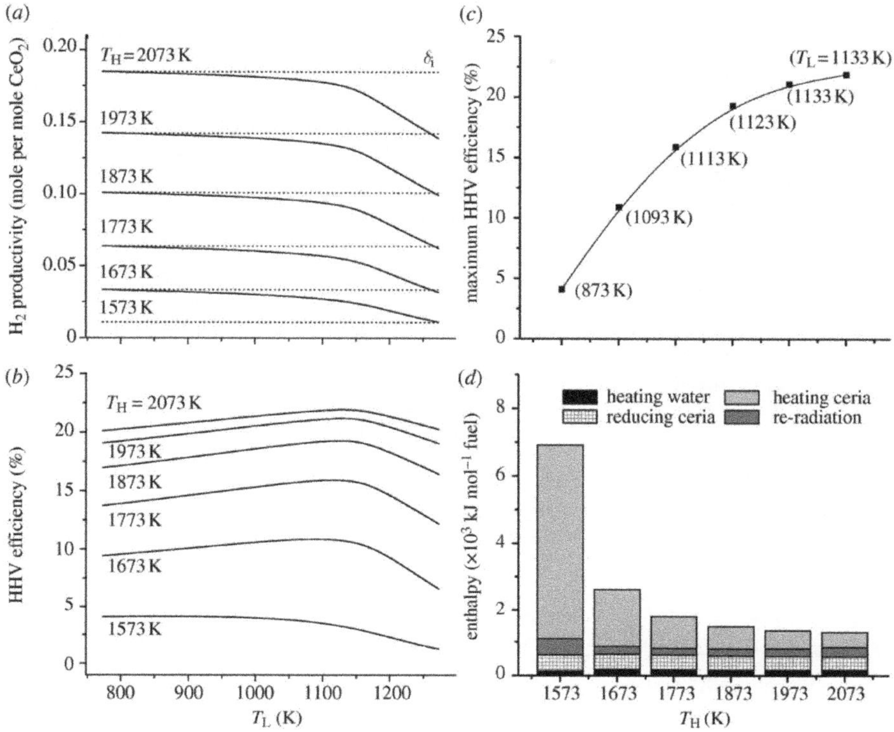

Figure 21.17 From [67]. a) H_2 productivity versus oxidation temperature (here shown as T_L), for fixed reduction temperatures, T_H. As seen for low T_L, ceria is almost fully oxidized and $n_{H2} = \delta_i$ (here δ_i is the nonstoichiometry after reduction, or δ_i as we have referred to it). b) $\eta_{solar\text{-}to\text{-}fuel}$ versus T_H and T_L. c) Maximum efficiency versus T_H. d) Energy flows related to heating H_2O, heating ceria, reduction of ceria and re-radiation. For all calculations, $pO_2 = 10^{-5}$ atm during reduction and for oxidation $n_{H_2O,i} = \delta_i$.

in 2010 when Chueh et al. demonstrated the dissociation of CO_2 and H_2O in a cavity type receiver using porous ceria bricks housed inside. Here they demonstrated efficiencies of less than 1%. Importantly, they recognized that the low efficiencies were resultant of the fact that their reaction rates during reduction where solar energy is supplied were limited by heat transfer [67]. Following this, Furler et al. aimed to improve heat transfer rates by producing reticulated porous ceramic (RPC) ceria parts that act as more effective volumetric absorbers of radiation because of their large pore sizes; they demonstrated CO_2 splitting efficiencies of 1.73% [68]. While this represented a dramatic improvement because of improved heat transfer, the authors recognized that oxidation reaction rates were kinetically limited and much slower compared to the prior Chueh et al. work as a result of the large length scales (and specific surface area) inherent to RPCs. Therefore, RPCs by the same authors were then developed with dual scale pore sizes; large pores for efficient and volumetric absorption of solar radiation where heat transfer is limiting and small pores to increase the specific surface area and hence oxidation rates, where kinetics are rate limiting [87].

Recently Marxer et al. demonstrated CO_2 splitting efficiencies with this dual scale RPC of 5.25% [76]. Also impressively, they demonstrated 500 consecutive redox cycles and 100% selectivity of CO_2 to CO with 83% molar conversion. A schematic and

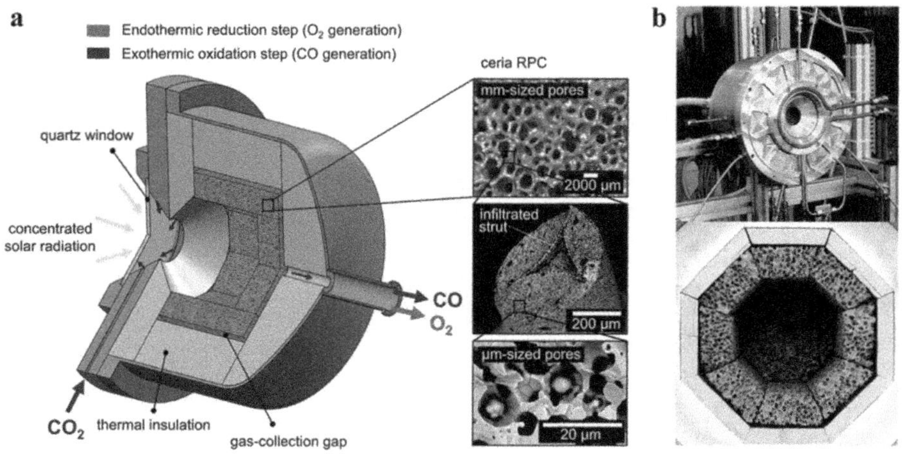

Figure 21.18 Schematic of ceria based cavity reactor developed by [76]. This system utilizes a dual scale RPC that serves as a good solar absorber and enhances heat transfer during reduction and has high specific surface area to enhance oxidation reaction rates.

photograph of the reactor that was utilized, along with the RPC is shown in Figure 21.18. As seen, the RPC is contained within the cavity receiver and is enclosed with a quartz window to separate it from the ambient environment. The reactor is operated by first subjecting the ceria RPC to concentrated solar radiation at either ambient pressure with a sweep gas or reduced pressure (10 mbar) with vacuum pumping, resulting in oxygen evolution. Following this, the temperature is cooled by removing the concentrated solar radiation (in this case by turning off the solar simulator that was utilized but there are practical ways to do this on a real solar furnace) and introducing CO_2 into the system to produce CO.

Experimental results of the aforementioned reactor are shown in Figure 21.19. The left hand side of the figure shows the reduction reaction and plots the temperature and the corresponding oxygen evolution (solid lines correspond to vacuum operation and dashed to ambient pressure). As seen, when the similar power is increased at time = 0 min, the temperature begins to increase to a peak of 1500°C, alongside the oxygen which scales with temperature. Here, the reaction is not driven to equilibrium, as indicated by the rapid decrease in oxygen production rates because of a decrease in power input. This is related to the fact that optimal efficiency does not usually occur if reaction extents are driven to equilibrium because of diminishing yields with time, even while the power input to maintain the temperature remains roughly constant. Once the temperature decreases to about 1100°C, CO_2 is injected into the reactor and a large increase in CO production is seen, followed by a slower decline. Once completed, the cycle is re-initiated by stopping the delivery of CO_2 and increasing the power input again. From these results, it can be seen that the reduction and subsequent oxidation yields are greater during vacuum operation, and as a result efficiencies were greatest when operating the cycle below ambient pressure. This was primarily a result of the fact that the reduction rates were enhanced with vacuum operation and heat losses were decreased across the insulation to the outside of the cavity because of decreased convective losses. This record thermochemical efficiency of 5.25% is competitive with benchmark's set by PV combined with CO_2 electrolysis [88].

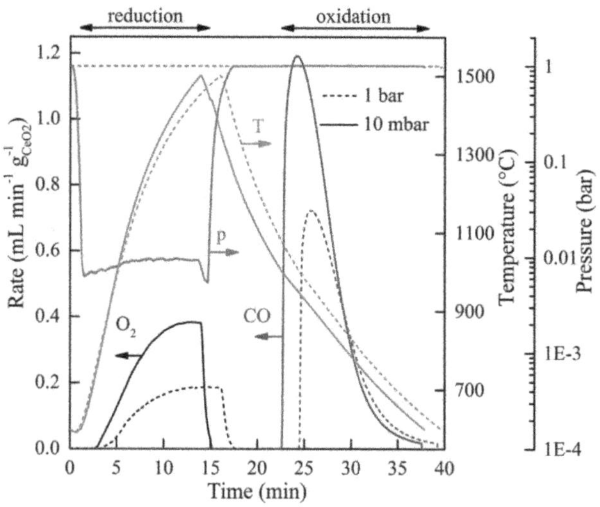

Figure 21.19 Experimental results during operation of Marxer et al.'s [76] solar reactor. On the left is reduction that is initiated by introducing concentrated radiation into the cavity. As a result the temperature increases and oxygen production rates increase. This is followed by oxidation of ceria with CO_2 to produce CO that is initiated by decreasing the solar flux incident into the reactor, causing the temperature to decrease.

21.2.10 Emerging Redox Materials

In recent years, there have been a host of new oxides that have shown promise when it comes to improvement over ceria. The major driving force for these materials comes from the fact that in order for ceria to maximize its efficiency, temperatures on the order of 1500°C and higher should be used. This is related to the fact that the reduction extent at lower temperatures is relatively small at the pressures typically used, which puts an upper limit on the H_2 yields and thus efficiency. When operating at temperatures of 1500°C or higher, the robustness of ceria regarding sintering, its reactivity with other reactor components, sublimation, and the stability of reactor components all come into play. When operating at an average temperature of 1500°C, locations where the radiation is incident can be even hundreds of degrees hotter and suffer most severely. Further, the demands on concentrating optics to achieve the concentration ratios necessary with limited losses due to spillage have not been developed, at least for heliostat driven tower based systems: these usually operate at temperatures lower than 1000°C. In general, there have been a host of materials proposed, but initially they were all based on various forms of doped ceria using Zr^{4+}, Hf^{4+}, and others [89–90]. In general the 4+ dopants were all shown to reduce at more moderate temperatures than ceria and in general increase reduction extent with increasing degree of doping. For lower valence dopants however, this effect was suppressed [91]. In 2013, two classes of perovskites were proposed based on La-Sr-Mn [92] and La-Sr-Mn-Al [93] systems. The perovskites, of the form $ABO_{3-\delta}$, were also shown to be more easily reduced than ceria and McDaniel et al. demonstrated fuel productivity on the order of 5 times ceria (albeit for low reduction temperatures of 1350°C where ceria is not readily reduced) [93]. From there, a host of related perovskite compositions have been proposed, including La-Ca-Mn [69, 94],

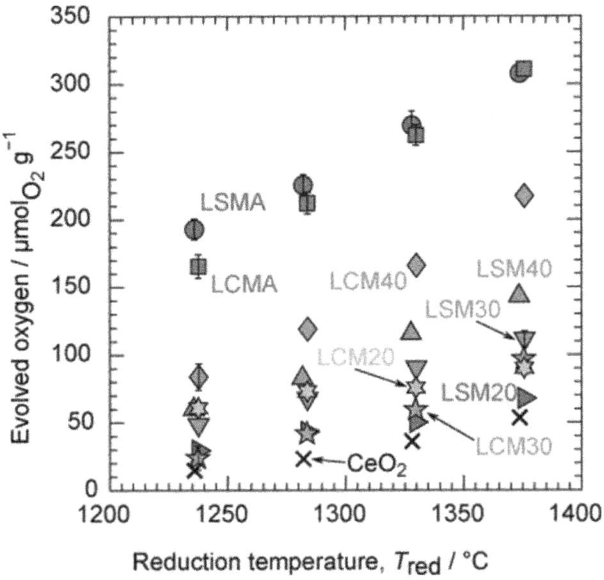

Figure 21.20 Measured amounts of evolved oxygen from thermogravimetric analysis. From [69]. The letters in the acronym stand for the cations (e.g. L = La) and the number represents the number of moles (times 100) of second cation doped on the A-site.

La-Ca-Mn-Al [69, 94], Y-Sr-Mn [95], and La-Sr-Cr-Co [96], all of which have shown the ability to reduce more readily than ceria. More recently papers have been published promoting materials based on $BaCe_{0.25}Mn_{0.75}O_{3-\delta}$ and poly cation oxides of the family $(FeMgCoNi)O_x$ ($x \approx 1.2$). Cooper et al. nicely compare the oxygen yields of a select number of $La_{1-x}(Sr,Ca)_xMn_{1-y}Al_yO_3$ perovskites with ceria, as shown in Figure 21.20 [69]. In general, for both Ca^{2+} and Sr^{2+} doped LSM and LCM perovskites, evolved oxygen increases in increasing A-site doping concentration (except for LCM20 and LCM30) and for 20, 30, and 40 mol-% A-site dopant concentrations they all reduce more readily than pure ceria. When Al^{3+} is doped on the B-site (i.e. LSMA and LCMA where A-site Sr^{2+} and Ca^{2+} concentrations are 40 mol-% and B-site Al^{3+} concentrations are 40 mol-%), reduction extents are even greater. Overwhelmingly, there is a strong correlation between the reduction enthalpy and the oxygen evolution at a given reduction temperature; namely, as the reduction enthalpy decreases, so too does the reduction temperature. This makes sense given that the driving force for reduction, or Gibbs free energy change of the reaction, is directly related. These trends are discussed in detail in a recent review paper by [64].

While all of these materials can more easily reduce than ceria, reduction ability is only half of the story. The materials also must have an affinity for oxidation with H_2O, which is also tied to the reduction enthalpy. The lower it is, the closer it becomes to the enthalpy required to split water, and eventually at the point where its magnitude is lower, the cycle is no longer tenable. Such a scenario can be imagined with a figure similar to Figure 21.8 but with a lower enthalpy, in which case

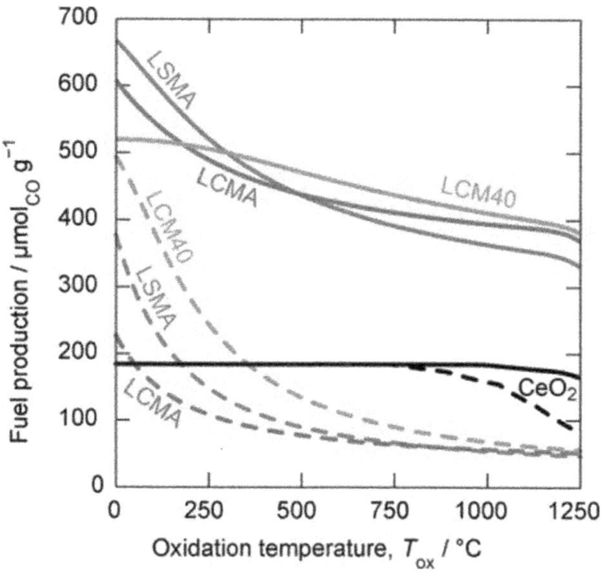

Figure 21.21 From Cooper et al. Predicted equilibrium yields from thermodynamic calculations, all samples have first been subjected to reduction conditions at 1400°C and $p_{O_2} = 10^{-5}$ atm. The reduction extents are $\delta_f = 0.15, 0.13, 0.11$, and 0.03 for LSMA, LCMA, LCM40, and CeO_2, respectively. Following this, reduced samples are then oxidized with either 1 mol CO_2 per mol oxide (dashed lines) or 100 mol CO_2 per mole oxide (solid lines).

the curve indicating oxidation favorability would always be greater than zero. The implications of this on fuel yields can be observed in Figure 21.21 from Cooper et al. [69]. In this figure, which that shows predicted equilibrium yields from thermodynamic calculations, all samples have first been subjected to reduction conditions at 1400°C and $p_{O_2} = 10^{-5}$ atm. The reduction extents are $\delta_f = 0.15, 0.13, 0.11$, and 0.03 for LSMA, LCMA, LCM40, and CeO_2, respectively. Following this, reduced samples are then oxidized with either 1 mol CO_2 per mol oxide (dashed lines) or 100 mol CO_2 per mole oxide (solid lines). Although all of the perovskites reduce substantially more than ceria, the oxidation yields when oxidant amounts are low are lower than ceria under most conditions (high temperature). For example, consider the dashed LCM40 sample that is initially reduced to $\delta_f = 0.11$ versus ceria that is reduced to $\delta_f = 0.03$. The fuel yields are lower than ceria until about 400°C where oxidation becomes more favorable and the oxidation conversions increase; at this temperature and lower, the kinetics would likely be to slow for the cycle to be tenable. When the oxidant yields are increased by a factor of 100, the fuel yields for all perovskites are higher than ceria, but at the expense of excess heating requirements which will hinder the efficiency. These results highlight the importance of considering oxidation favorability as well as reduction extent when evaluating candidate redox materials. Ultimately, only a thermodynamic analysis that couples reaction equilibrium of both reduction and oxidation steps with required heat and mass flows can predict the potential suitability of materials to be used in thermochemical redox cycles.

References

[1] Duffie, J. and Beckman, W. (2013). *Solar Engineering of Thermal Processes.* Hoboken, New Jersey: John Wiley & Sons, Inc.

[2] Mecherikunnel, A.T. and Richmond, J.C. (1980). *Spectral Distribution of Solar Radiation*, GODDARD SPACE FLIGHT CENTER, NASA Technical Memorandum 82021 https://ntrs.nasa.gov/api/citations/19810016493/downloads/19810016493.pdf

[3] Chu, S., Cui, Y., and Liu, N. (2016). *Nature Materials* 16: 16.

[4] NIST-JANAF. (2013). *NIST-JANAF Thermochemical Tables.*

[5] Larminie, J., Dicks, A., and McDonald, M.S. (2003). *Fuel Cell Systems Explained.* Chichester, UK: J. Wiley.

[6] White, C.M., Steeper, R.R., and Lutz, A.E. (2006). *International Journal of Hydrogen Energy* 31: 1292–1305.

[7] Lewis, N.S. (2016). Research opportunities to advance solar energy utilization. *Science* 351 (6271).

[8] Roy, S.C., Varghese, O.K., Paulose, M., et al. (2010). *ACS Nano* 4: 1259–1278.

[9] Honsberg, C. and Bowden, S. PVEDUCATION.ORG.

[10] Van de Krol, R. and Grätzel, M. (2012). *Photoelectrochemical Hydrogen Production.* Vol 90. New York: Springer.

[11] Alonso, E., Gallo, A., Pérez-Rábago, C., et al. (2016). *Thermodynamic Study of CuO/Cu2O and Co3O4/CoO Redox Pairs for Solar Energy Thermochemical Storage*, Vol. 1734, 050004. Melville, NY: AIP Publishing LLC); https://doi.org/10.1063/1.4949102

[12] Scheffe, J.R. and Steinfeld, A. (2014). *Materials Today* 17: 341–348.

[13] Steinfeld, A. and Meier, A. (2004). *Encyclopedia of Energy* 5: 623–637.

[14] Romero, M. and Steinfeld, A. (2012). *Energy & Environmental Science* 5: 9234–9245.

[15] Steinfeld, A. and Palumbo, R. (2001). *Encyclopedia of Physical Science and Technology* 15: 237–256.

[16] Winston, R., Miñano, J.C., and Benitez, P.G. (2005). *Nonimaging Optics*. London: Academic Press.

[17] Chaves, J. (2017). *Introduction to Nonimaging Optics*. Boca Raton: CRC Press.

[18] Fletcher, E.A. and Moen, R.L. (1977). *Science* 197: 1050.

[19] Smith, J.M., Van Ness, H.C., and Abbott, M.M. (1996). *Sat* 18: 1–3.

[20] Piatkowski, N., Wieckert, C., Weimer, A.W., et al. (2011). *Energy & Environmental Science* 4: 73–82.

[21] Dahl, J.K., Weimer, A.W., Lewandowski, A., et al. (2004). *Industrial & Engineering Chemistry Research* 43: 5489–5495.

[22] Liu, Q., Jin, H., Hong, H., et al. (2011). *International Journal of Energy Research* 35: 52–60.

[23] Gregg, D.W., Taylor, R.W., Campbell, J.H., et al. (1980). *Solar Energy* 25: 353–364.

[24] Z'Graggen, A., Haueter, P., Maag, G., et al. (2007). *International Journal of Hydrogen Energy* 32: 992–996.

[25] Kruesi, M., Jovanovic, Z.R., Haselbacher, A., et al. (2015). *AIChE Journal* 61: 867–879.

[26] Melchior, T., Perkins, C., Lichty, P., et al. (2009). *Chemical Engineering and Processing: Process Intensification* 48: 1279–1287.

[27] Kruesi, M., Jovanovic, Z.R., and Steinfeld, A. (2014). *Fuel* 117 *Part A*: 680–687.

[28] Lichty, P., Perkins, C., Woodruff, B., et al. (2010). *Journal of Solar Energy Engineering* 132: 011012-011012-011017.

[29] Piatkowski, N., Wieckert, C., and Steinfeld, A. (2009). *Fuel Processing Technology* 90: 360–366.

[30] Zheng, R., Diver, R., Caldwell, D., et al. (2015). *Energy Procedia* 69: 1192–1200.

[31] Lede, J., Lapicque, F., Villermaux, J., et al. (1982). *International Journal of Hydrogen Energy* 7: 939–950.

[32] Traynor, A.J. and Jensen, R.J. (2002). *Industrial & Engineering Chemistry Research* 41: 1935–1939.

[33] Glandt, E.D. and Myers, A.L. (1976). *Industrial & Engineering Chemistry Process Design and Development* 15: 100–108.

[34] Funk, J.E. and Reinstrom, R.M. (1966). *Industrial & Engineering Chemistry Process Design and Development* 5: 336–342.

[35] Bilgen, E. and Bilgen, C. (1982). *International Journal of Hydrogen Energy* 7: 637–644.

[36] Nakamura, T. (1977). *Solar Energy* 19: 467–475.

[37] Darken, L. and Gurry, R.W. (1945). *Journal of the American Chemical Society* 67: 1398–1412.

[38] Bransky, I. and Hed, A.Z. (1968). *Journal of the American Ceramic Society* 51: 231-231.

[39] Salmon, O.N. (1961). *The Journal of Physical Chemistry* 65: 550–556.

[40] Nowotny, J. and Rekas, M. (1989). *Journal of the American Ceramic Society* 72: 1221–1228.

[41] Pelton, A.D., Schmalzried, H., and Sticher, J. (1979). *Berichte der Bunsengesellschaft für Physikalische Chemie* 83: 241–252.

[42] Allendorf, M.D., Diver, R.B., Siegel, N.P., et al. (2008). *Energy & Fuels* 22: 4115–4124.

[43] Jacobson, N. (2001). *Journal of Chemical Education* 78: 814.

[44] Chueh, W.C. and Haile, S.M. (2010). *Philosophical Transactions of the Royal Society of London A: Mathematical, Physical and Engineering Sciences* 368: 3269–3294.

[45] Lapp, J., Davidson, J.H., and Lipiński, W. (2012). *Energy* 37: 591–600.

[46] Siegel, N.P., Miller, J.E., Ermanoski, I., et al. (2013). *Industrial & Engineering Chemistry Research* 52: 3276–3286.

[47] Ermanoski, I., Miller, J.E., and Allendorf, M.D. (2014). *Physical Chemistry Chemical Physics* 16: 8418–8427.

[48] Lin, M. and Haussener, S. (2015). *Energy* 88: 667–679.

[49] Steinfeld, A., Sanders, S., and Palumbo, R. (1999). *Solar Energy* 65: 43–53.

[50] Kodama, T., Nakamuro, Y., and Mizuno, T. (2006). *Journal of Solar Energy Engineering* 128: 3–7.

[51] Charvin, P., Abanades, S., Flamant, G., et al. (2007). *Energy* 32: 1124–1133.

[52] Kodama, T., Gokon, N., and Yamamoto, R. (2008). *Solar Energy* 82: 73–79.

[53] Scheffe, J.R., McDaniel, A.H., Allendorf, M.D., et al. (2013). *Energy & Environmental Science* 6: 963–973.

[54] Miller, J.E., Allendorf, M.D., Diver, R.B., et al. (2008). *Journal of Materials Science* 43: 4714–4728.

[55] Coker, E.N., Ambrosini, A., Rodriguez, M.A., et al. (2011). *Journal of Materials Chemistry* 21: 10767–10776.

[56] Diver, R.B., Miller, J.E., Siegel, N.P. et al. (2010). January. Testing of a CR5 solar thermochemical heat engine prototype. *Energy Sustainability* 43956: 97–104
[57] Loutzenhiser, P.G., Meier, A., and Steinfeld, A. (2010). Review of the two-step H_2O/CO_2-splitting solar thermochemical cycle based on Zn/ZnO redox reactions. *Materials* 3 (11): 4922–4938.
[58] Stamatiou, A., Loutzenhiser, P.G., and Steinfeld, A. (2010). *Chemistry of Materials* 22: 851–859.
[59] Abanades, S., Charvin, P., Lemont, F., et al. (2008). *International Journal of Hydrogen Energy* 33: 6021–6030.
[60] Perkins, C. and Weimer, A.W. (2004). *International Journal of Hydrogen Energy* 29: 1587–1599.
[61] Steinfeld, A. (2005). *Solar Energy* 78: 603–615.
[62] Kodama, T. and Gokon, N. (2007). *Chemical Reviews* 107: 4048–4077.
[63] Miller, J.E., McDaniel, A.H., and Allendorf, M.D. (2014). *Advanced Energy Materials* 4: 1300469–n/a.
[64] Carrillo, R.J. and Scheffe, J.R. (2017). Advances and trends in redox materials for solar thermochemical fuel production. *Solar Energy* 156 (2017): 3–20.
[65] Chueh, W.C. and Haile, S.M. (2009). *ChemSusChem* 2: 735–739.
[66] Panlener, R.J., Blumenthal, R.N., and Garnier, J.E. (1975). *Journal of Physics and Chemistry of Solids* 36: 1213–1222.
[67] Chueh, W.C., Falter, C., Abbott, M., et al. (2010). *Science* 330: 1797–1801.
[68] Furler, P., Scheffe, J., Gorbar, M., et al. (2012). *Energy & Fuels* 26: 7051–7059.
[69] Cooper, T., Scheffe, J.R., Galvez, M.E., et al. (2015). *Energy Technology* 3: 1130–1142.
[70] Riess, I., Ricken, M., and Noëlting, J. (1985). *Journal of Solid State Chemistry* 57: 314–322.
[71] Ermanoski, I., Grobbel, J., Singh, A., et al. (2016, May). Design and construction of a cascading pressure reactor prototype for solar-thermochemical hydrogen production. *AIP Conference Proceedings* 1734 (1): 120001). Melville, NY: AIP Publishing LLC.
[72] Hathaway, B.J., Bala Chandran, R., Gladen, A.C., et al. (2016). *Energy & Fuels* 30: 6654–6661.
[73] Bader, R., Venstrom, L.J., Davidson, J.H., et al. (2013). *Energy & Fuels* 27: 5533–5544.
[74] Venstrom, L.J., De Smith, R.M., Hao, Y., et al. (2014). *Energy & Fuels* 28: 2732–2742.
[75] Carrillo, R.J. and Scheffe, J.R. (2019). *Energy & Fuels* 33: 12871–12884.
[76] Marxer, D., Furler, P., Takacs, M., et al. (2017). *Energy & Environmental Science* 10: 1142–1149.
[77] Kim, J., Henao, C.A., Johnson, T.A., et al. (2011). *Energy & Environmental Science* 4: 3122–3132.
[78] Stechel, E.B. and Miller, J.E. (2013). *Journal of CO2 Utilization* 1: 28–36.
[79] Gokon, N., Takahashi, S., Yamamoto, H., et al. (2008). *International Journal of Hydrogen Energy* 33: 2189–2199.
[80] Kaneko, H., Kodama, T., Gokon, N., et al. (2004). *Solar Energy* 76: 317–322.
[81] Chambon, M., Abanades, S., and Flamant, G. (2011). *AIChE Journal* 57: 2264–2273.
[82] Roeb, M., Neises, M., Säck, J.-P., et al. (2009). *International Journal of Hydrogen Energy* 34: 4537–4545.

[83] Kaneko, H., Miura, T., Fuse, A., et al. (2007). *Energy & Fuels* 21: 2287–2293.

[84] Koepf, E., Villasmil, W., and Meier, A. (2016). *Applied Energy* 165: 1004–1023.

[85] Säck, J.P., Breuer, S., Cotelli, P., et al. (2016). *Solar Energy* 135: 232–241.

[86] Welte, M., Barhoumi, R., Zbinden, A., et al. (2016). *Industrial & Engineering Chemistry Research* 55: 10618–10625.

[87] Furler, P., Scheffe, J., Marxer, D., et al. (2014). *Physical Chemistry Chemical Physics* 16: 10503–10511.

[88] Schreier, M., Curvat, L., Giordano, F., et al. (2015). Efficient photosynthesis of carbon monoxide from CO_2 using perovskite photovoltaics. *Nature Communications* 6 (1): 1–6.

[89] Scheffe, J.R., Jacot, R., Patzke, G.R., et al. (2013). *The Journal of Physical Chemistry C* 117: 24104–24114.

[90] Takacs, M., Scheffe, J.R., and Steinfeld, A. (2015). *Physical Chemistry Chemical Physics* 17: 7813–7822.

[91] Scheffe, J.R. and Steinfeld, A. (2012). *Energy & Fuels* 26: 1928–1936.

[92] Scheffe, J.R., Weibel, D., and Steinfeld, A. (2013). *Energy & Fuels* 27: 4250–4257.

[93] McDaniel, A.H., Miller, E.C., Arifin, D., et al. (2013). *Energy & Environmental Science* 6: 2424–2428.

[94] Takacs, M., Hoes, M., Caduff, M., et al. (2016). *Acta Materialia* 103: 700–710.

[95] Nair, M.M. and Abanades, S. (2016). *ChemistrySelect* 1: 4449–4457.

[96] Bork, A.H., Kubicek, M., Struzik, M., et al. (2015). *Journal of Materials Chemistry A* 3: 15546–15557.

22

Power-to-Liquids - Conversion of CO_2 and Renewable H_2 to Methanol

Robin J. White

Sustainable Catalytic Materials Group, Division Hydrogen Technologies, Fraunhofer Institute for Solar Energy Systems ISE, Germany Materials Research & Technology Department, Luxembourg Institute of Science and Technology, Luxembourg

22.1 Introduction

We are living in and indeed experiencing the consequences of the "*Anthropocene*" era. This era has been essentially created and influenced by the activities of human society and has led to increased atmospheric CO_2 (and other greenhouse gas (GHG)) concentration and associated increased average global temperature and changes in climatic cycles. It is the generally accepted opinion in the scientific community that this results from fossil fuel combustion (e.g. to supply heat and electricity) resulting in increasing emissions of GHGs during the industrialization of the last century [1]. The rapid and frankly unprecedented advancement of technology and its associated energy demand of our society in this time frame have likewise supported population growth, improvements in the quality of life standards, and a resulting increase in life expectancy. This has been most prevalent in western or First world industrialized nations. Through the exploitation of natural resources and reserves, increasingly focused over the last century on non-renewable fossil fuel reserves (i.e. oil, natural gas, coal), GHG emissions arising from their combustion (e.g. in electricity generation), are having a direct impact on climate cycling and stability. It could also be argued that the geographical location of fossil fuels also strongly influences international policy and decision making. Independently, it is important to highlight that without a stable climatic system, continuous improvements in quality of life, access to food and water, etc. will become increasingly challenging. Concurrently, global population will continue to increase (ca. 10 billion by 2050), with supply and indeed demand for affordable and reliable energy thus ever more important if overall global living standards are to follow population growth (e.g. primary energy consumption globally was 538.68 EJ ($E = 10^{18}$) in 2012) [2]. Therefore how does society, at both national and international levels, meet the demands of this population size for energy, water, food, and material needs? Given agreements such as COP21-23 [3], this has to be achieved whilst reducing GHG emissions to limit in a best case scenario the overall global average temperature increase to <2°C, ideally ≤1.5°C. It is perhaps evident to the reader that this will be only achieved through an ever increasing unity between traditionally separated sectors (e.g. the chemical and energy production sectors) and likewise a rapid and high level adoption of sustainable technologies as supported by national and international drivers and legislation. This will hopefully result in a significant

Advances in Energy Storage: Latest Developments from R&D to the Market, First Edition.
Edited by Andreas Hauer.
© 2022 John Wiley & Sons, Ltd. Published 2022 by John Wiley & Sons, Ltd.

socio-economic change (e.g. the initial principles of the German "*Energiewende*") [4], leading to the establishment of (e.g. trans-national) renewable energy-based economies.

Currently fossil hydrocarbons are the main sources of feedstocks and fuels for the modern chemical, energy, and transport sectors [5–8]. Fossil hydrocarbons are near perfect energy carriers with regard to their volumetric energy density and physical state, but as a consequence of their finite reserves, global distribution, and associated emissions, these can only be considered as primary energy sources only in the short term. Derived from the products of natural photosynthesis, hydrocarbons (e.g. crude oil, coal, and natural gas) are stored sunlight (photonic energy) and sequestered carbon (i.e. as produced via photo-, bio-, and geo-chemical processes over millennia). As a consequence of their combustion at a rate at which natural cycles cannot compete (and in turn replace), the overall net concentration of CO_2 is increasing in the "free" atmospheric form as opposed to the sequestered or stored form.

Natural photosynthesis based on the (bio-)catalytic conversion of CO_2 and H_2O (as powered by solar energy) produces (oxy)hydrocarbons (e.g. glucose, cellulose, lignin, starch, etc.) and O_2. These products are supplemented in the overall biosphere by secondary or tertiary metabolic products (e.g. from herbivores). On time scales not appreciable to our society (i.e. millennia), the resulting biomass provides the basis for the geo-production of the fossil fuels used so extensively over the last 50–100 years by our society. Whilst some reserves (e.g. natural gas) are relatively plentiful (but currently economically unattractive), there is a continuing debate concerning the remaining fossil fuel reserves – i.e. should we continue to extract or "*leave them in the ground*" [9–12]. Therefore there is an ever increasing focus and a continued desire to instigate sustainable socio-economic and environmentally friendlier "*Circular Economies*" based on renewables, biomass, CO_2, and H_2O, with the ambition to supply energy carriers, materials, and chemicals with as minimal an environmental burden as possible (Scheme 22.1) [13–15].

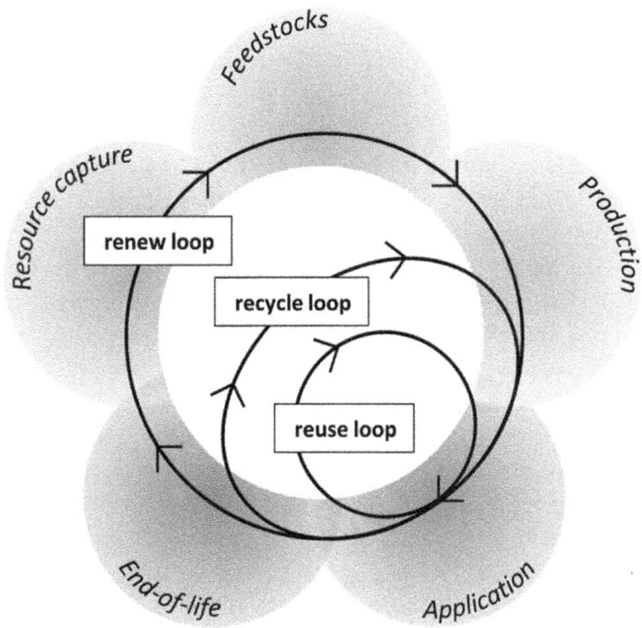

Scheme 22.1 The chemical product cradle-to-cradle life cycle stages in a circular economy as proposed by Clark et al. (Reproduced from [15] with permission from The Royal Society of Chemistry.)

The aim of this chapter is to provide an overview on such economies with a particular reference to the simplest of the alcohols, methanol (CH_3OH). As will be discussed, this alcohol can be synthesized from H_2 (e.g. as supplied from H_2O splitting) and CO_2 (e.g. as capture from industrial waste point sources) powered by renewable energy (RE). This approach has obvious analogies to the production of sugars in natural photosynthesis (i.e. from a H_2O and CO_2 basis). Ultimately, this approach will be integrated with different conventional and evolving sectors (e.g. the biorefinery, steel manufacture, biogas plants, etc.; Scheme 22.2) [16]. The elaboration of such an approach to the production of platform chemicals, fuels, or energy carriers in this manner has been referred to in the literature as "*Artificial Photosynthesis*" [17], or "*Technical Photosynthesis*" [18], although currently within the European Union in countries such as Germany or Denmark, it is commonly referred to as a "*Power-to-X*" process where X = a gas, liquid, chemical, etc. (e.g. *Power-to-Methanol*) [19], where RE powers the production of a wide range of sustainable energy carriers, fuels, chemicals, and materials (and not just alcohols) [20–22]. Methanol then acts as a platform for a range of other fuels or energy carriers including Oxymethylene dimethyl ethers (OME) which are currently receiving attention as potential diesel fuels or fuel additives [23].

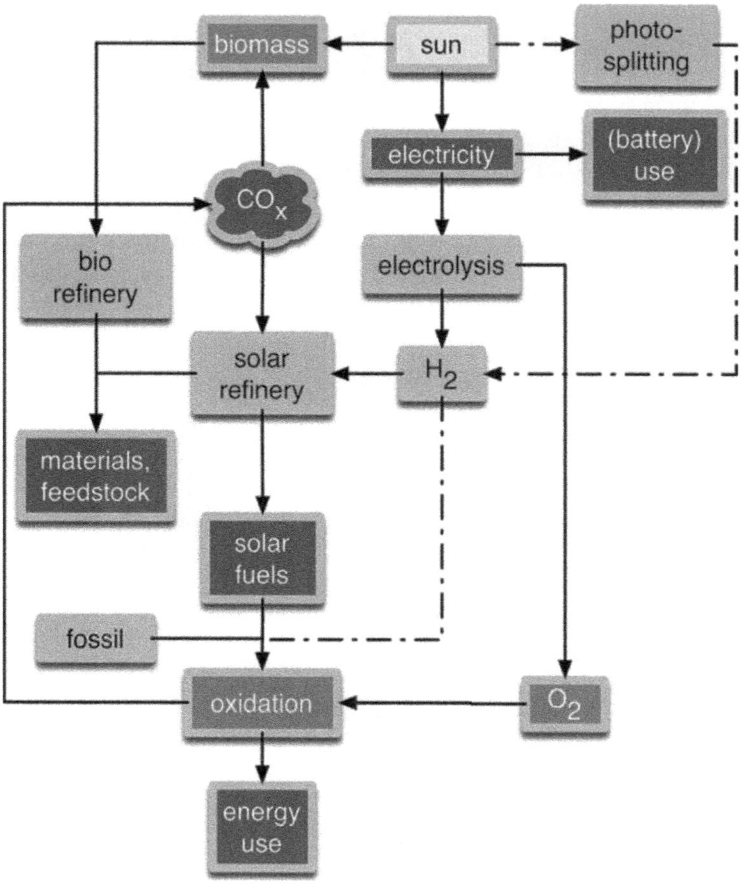

Scheme 22.2 A generalized design of a future sustainable energy/chemical/fuel system (as proposed by Prof. R. Schlögl). (*Dashed lines indicate elements that may contribute in later development stages of the energy system. The light gray boxes indicate elements that are not yet existing and/or fit for world scale applications*). (Reproduced with permission from [16].)

To support and establish *Power-to-X* schemes, it is envisioned that a variety of sustainable chemical/energy systems will be required based on carbon-neutral energy supplies (e.g. solar, wind, tidal, etc.), whilst the utilization of one sector's waste (e.g. CO_2 rich exhaust gases) as the "carbon" source will be critical to improve overall ecological footprint minimization [14, 24]. Extensive process synergies and integration will have to be identified, often involving industrial sectors that have not worked in close proximity in the past (e.g. steel production and chemical plants). Conversely, there is also the challenge of "*economies of scale*" employed in classical industrial plant considerations (e.g. Mega-Methanol plants). Coupling for example of gases arising from small to medium scale municipal waste remediation activities (e.g. biogas plants as sources of CO_2 operating at the 500 kW_{el}), may require the adoption of small scale chemical production facilities (e.g. 50 $Kt(CH_3OH)$/annum) which might (e.g. without support) in the short term render limited economic competitiveness. This is a very complex area and multiple scenarios (Figure 22.1) can be envisioned leading to different potentials, and likewise capital and operating expenditure for "*Power-to-X*" plants [24]. There are also other important factors to consider in the design of such plants that are beyond the scope of this chapter (e.g. non-continuous operation, regulations, taxation structures, etc.).

Power-to-Liquid (PtL) and specifically *Power-to-Methanol* plants will utilize CO_2 or CO_X which if sourced from (e.g. high concentration) industrial waste points (e.g. steel, cement, biogas production facilities) may enable a rapid reduction in the rate at which CO_2 emissions enter the atmosphere. An example schematic of such an approach is presented (Scheme 22.3). These approaches will contribute to the desire to instigate "*Carbon Capture and Utilization*" (CCU) strategies, whilst delivering ecological and (ideally given current market structure) economic benefits (e.g. through emissions trading, carbon credits, etc.) [25–36]. Using CO_2 as a C_1 feedstock (e.g. in fuel or chemical synthesis), these approaches have already been demonstrated to result in reduced GHG emissions (e.g. vs conventional production) [14, 37–39]. As a reference, the production of conventional fossil-derived methanol results in production of 0.50 to

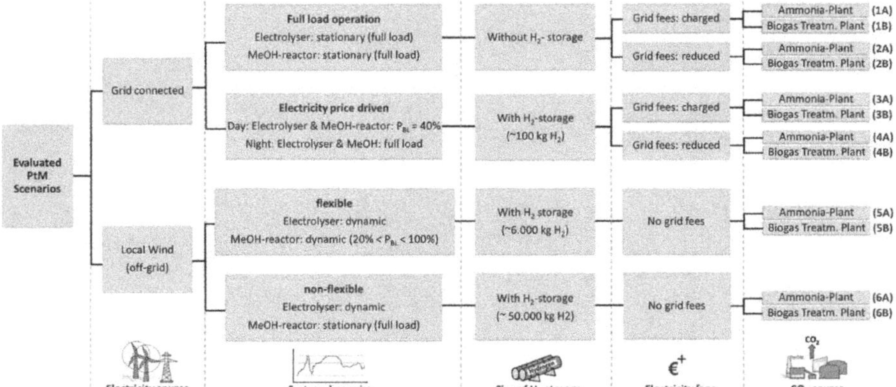

Figure 22.1 Overview of the six different "Power-to-Methanol" scenarios (1–6) evaluated in the study from White et al.; each differing in the sourced CO_2 – from (A) an ammonia-plant or (B) a biogas upgrading plant. (Reproduced with permission from [24].)

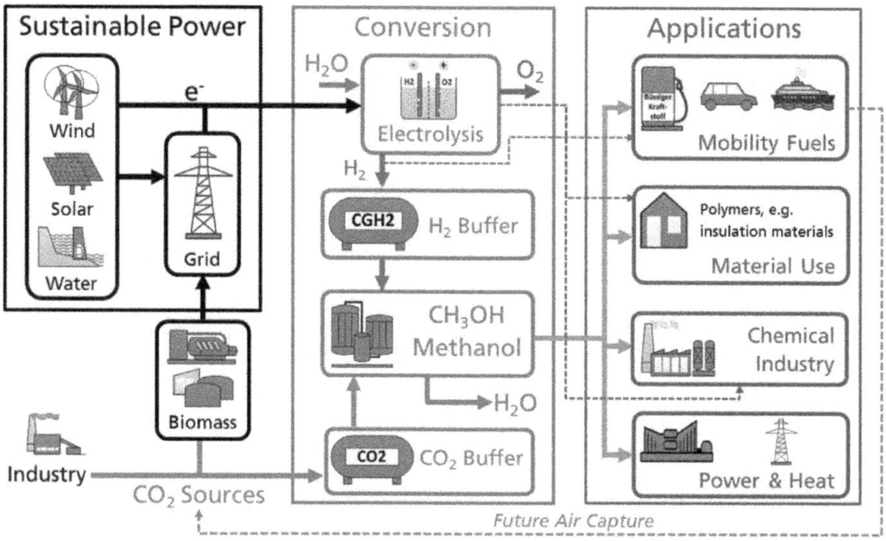

Scheme 22.3 A potential "Power-to-Liquid" concept for the production of methanol based on the conversion of captured CO_2 (e.g. as sourced from industry point sources) and H_2 (as supplied by renewable energy driven water splitting; e.g. through electrolysis), with potential application indicated in a general manner. To create circularity in the scheme, it will be necessary in the medium-to-long term to implement air capture systems and recycling back to e.g. fuels. (*NB: As examples, H_2 can also be used in the mobility sector, whilst the O_2 by-product from water splitting could also potentially find applications in material production or the chemical industry*).

0.77 t(CO_{2eq})/t(MeOH), including direct specific CO_2 emissions of 0.24 t(CO_{2dir})/t(MeOH) [40–41].

Concerning GHG emissions at the global level, CO_2 emitted from fossil fuels and industry was recorded at ca. 37 billion metric tons in 2017. It is expected, based on "*business as normal*" trends of the last decade that emissions will continue to increase to >45 Gt/a by 2050 [42–44]. Therefore a significant amount of "carbon" could be potentially "captured" and utilized in PtL schemes. Whilst legislation is under consideration to further encourage adoption of CCU-related chemistry and reduce GHG emissions, at the time of writing the capture of industrial CO_2 and its utilization in chemical processes is not included in the EU Emission Trading System (EU ETS[1]) and, for this reason, does not lead to CO_2 Certificate savings (i.e. not as for Carbon Capture and Storage (CCS)) [45]. The associated capture technology (e.g. chemical or physical capture, gas cleaning, etc.) needed for CCS will also be required for CCU, and it is hoped that the ETS will soon include CO_2 reuse technologies. Once appropriate legal (e.g. EU Directives) and economic frameworks are employed then the utilization of CO_2 waste streams will hopefully become more attractive to industry, thus transforming CO_2 from a commercial and ecological liability to an industrial carbon source [46–47]. Going forward a combination of "CO_2 avoidance cost" together with detailed

1 Annex I of the EU Directive 2003/87/EC underlines that "under the EU ETS rules for any other transfer" than in the case of long-term geological storage "of CO_2 out of the installation, no subtraction of CO_2 from the installation's emissions is allowed".

Life Cycle Assessment (LCA) should be used as key-criteria for the assessment and selection of different routes to fuels and chemicals from PtL [48]. Positively, this is will be supported by a rapidly growing and intended expansion of installed RE capacity (e.g. in Germany and the EU in general) over the next few decades. The growth of RE will facilitate a further discussion of PtX/CCU technologies, particularly as the RE price per ct€ per kWh continues to drop.

Sequestering or storing CO_2 via its transformation into new molecules can be achieved via chemical reactions each with their own energetic hurdles to overcome (e.g. a poly(carbonate) – a comparatively low energy process) [33, 49]. When one (or more) of the C-O bonds is "broken" as is mechanistically the case for CO, CH_3OH, or hydrocarbon synthesis, chemical reduction (e.g. with H_2) is required – and thus a comparatively energy consumptive process. The energy to power these conversions should therefore come directly or indirectly from RE, if net emission targets are to be met. The rapidly growing contribution of RE to electrical grids must be (e.g. in periods of excess) supported by the appropriate technology and catalysis (e.g. electrolysis) to split H_2O to produce H_2, in turn used for the energetic reduction of C–O bonds in CO_2 (and by extension for the production of energy carrier molecules).

22.2 Methanol Synthesis

The production of CH_3OH at the industrial level has been conducted for well over a hundred years. Therefore, there is a significant amount of know-how available regarding catalysis, process, and plant development. Methanol is currently produced at the industrial scale in large scale plants, with the largest known operations running at ca. 3000 t/day. As driven by increasing demand and technology/plant improvements, larger scale operations are under consideration capable of producing >5000 t/day. The largest known plant currently in operation is the Mega-Methanol plant operated by Methanex® in Trinidad & Tobago, with a production capacity of ca. 2.5 Mt/annum.

Regarding the synthesis of CH_3OH in the "classical" (syngas) manner, a number of points are worth discussing as they relate to synthesis strategies based on CO_2 and H_2 feeds. The synthesis from syngas (i.e. mixture of CO_2, CO, and H_2) is based on three reaction equilibria (Eq. 22.1–22.3). The exact syngas ratio is conventionally defined by the carbon source employed (e.g. syngas from coal gasification requires more conditioning via sweetening or water gas shift chemistry than syngas derived from CH_4 reforming). Methanol production in China is predominantly based on coal gasification, whilst "*Gas-to-Liquid*" plants (i.e. steam methane reforming based) are commonly operated in the Middle East. As a consequence of the "carbon" source (e.g. coal vs methane), the resulting methanol products have differing ecological footprints [50]. Despite growing coal-to-methanol capacities in China, 55% of the total installed production capacity is based on natural gas steam reforming [51].

On a pure stoichiometric reaction basis, the synthesis of methanol (i.e. based on syngas) proceeds through the following. It is generally accepted that the direct hydrogenation of CO occurs (Eq. (22.1)) whilst depending on the exact reaction conditions the water-gas shift (WGS) reaction can also proceed (Eq. (22.2), leading to a higher CO_2 content which can also undergo direct hydrogenation of CO_2 (Eq. (22.3)). It is perhaps obvious based on these reactions that the WGS reaction should ideally be avoided as it consumes the supplied H_2 [52].

$$CO + 2H_2 \rightleftharpoons CH_3OH \qquad \Delta H_R^0 = -91\,\text{kJ}\,\text{mol}^{-1} \tag{22.1}$$

$$CO + H_2O \rightleftharpoons CO_2 + H_2 \qquad \Delta H_R^0 = -41\,\text{kJ}\,\text{mol}^{-1} \tag{22.2}$$

$$CO_2 + 3H_2 \rightleftharpoons CH_3OH + H_2O \qquad \Delta H_R^0 = -50\,\text{kJ}\,\text{mol}^{-1} \tag{22.3}$$

As a product, liquid raw methanol (i.e. a mix of methanol, H_2O, and unreacted, dissolved gases), is produced. In a traditional production plant, this product stream is condensed after the main reactor, and typically removed through a gas/liquid separator. The unreacted gases that were previously dissolved can then be recycled. To avoid problems with inert compound accumulation, a small purge is often conducted – the scale of which is dependent on the amount of recycling. Pure methanol is ultimately produced via a purification step typically conducted using distillation.

The syngas quality/composition plays a critical role in methanol synthesis process design. The composition can in principle be tailored to define the molar fraction ratios of CO_2 (x_{CO2}) and CO (x_{CO}). The composition, as determined by the so-called "*carbon oxide ratio*" (COR) is an important consideration as the respective reaction enthalpies Eqs. (22.1) and (22.3) dictate that an overall methanol synthesis reactor temperature rise occurs with a decreasing COR (and vice versa). COR is defined by the following equation [53–55]:

$$COR = \frac{x_{CO_2}}{x_{CO_2} + x_{CO}} \tag{22.4}$$

It is important to note that the H_2O content in the methanol synthesis reactor has been found to encourage catalyst deactivation and decreases activity (e.g. via active site "flooding") [53, 56–58]. As a consequence it is considered desirable for conventional methanol synthesis to keep the COR as low as possible. This is an important consideration regarding the direct hydrogenation of CO_2, as inherently more H_2O will be produced. Another important consequence of operating methanol synthesis with increasing CO_2 content in the feed is the impact on the overall reaction kinetic and equilibrium yield. With increasing CO_2 content, the reaction kinetic is known to decelerate and thus reduce equilibrium yield [59–60]. Based on Eqs. 22.1–22.3, it is clear that the hydrogenation of CO and CO_2 requires different molar amounts of H_2. In this respect, the stoichiometric number (SN) is defined by the molar fractions of H_2 (x_{H2}), CO and CO_2 in the feed gas [61]:

$$SN = \frac{x_{H_2} - x_{CO_2}}{x_{CO} + x_{CO_2}} \tag{22.5}$$

It is known that a stoichiometric conversion SN = 2 is preferred, however at the industrial scale, a slightly increased H_2 content is applied [62–63]. The SN can be controlled via addition of CO or CO_2. A SN < 2 is not recommended as a H_2 shortage in the syngas makeup leads to undesired by-product formation [64–65].

Adding to the complexity of the syngas makeup chemistry, inert gases (e.g. N_2, CH_4, Ar, or He) are also often found (i.e. non-participatory components). They act as diluents, reducing the respective partial pressures but also assist to counteract reactor temperature rise (e.g. as a consequence of heat capacity). Regarding recycling, inert components in the syngas can accumulate in "looping" (i.e. recycling of unreacted components), which must be compensated through either through purging or changing the recycling ratio (e.g. in turn negatively impacting on process economics) [66–68]. Alongside the inert gases, a number of impurities in the syngas may also be found depending on the carbon source, which in turn impacts on process design (e.g. gas sweetening). Typically impurities (e.g. H_2S) deactivate the catalyst, which is usually based on Cu/ZnO (e.g. Zn has a very strong affinity for sulfur with the favoured formation of ZnS leading to deactivation) [56, 69–70]. As a consequence, conventional methanol synthesis plants incorporate additional technology (and indeed process steps) (e.g. upstream gas cleaning, complex downstream distillation, etc.) to either clean the syngas/product. This in turns has additional investment and operational costs.

22.3 Catalysts for Methanol Synthesis

Regarding catalysts, the material of choice is Al_2O_3 (e.g. 5–10 mol.%) supported Cu nanoparticles (NPs), with ZnO NPs as a "chemical promoter" (denoted hereon as Cu/ZnO-cat) [71]. This catalyst functions well with conventional syngas under classical methanol synthesis conditions (T = 513–533 K; p = 50–100 bar) [72]. Positively, with the increased interest in the direct hydrogenation of captured CO_2; the same Cu/ZnO-based catalysts are also applicable for this direct hydrogenation [73]. Approaching methanol synthesis in this manner brings the additional challenge of H_2O handling and the greater (e.g. stoichiometric) consumption of H_2. At the process level, the continuous removal of H_2O through membranes or indeed through distillation is an attractive proposition and acts to shift reaction equilibrium in favour of CH_3OH production (e.g. as demonstrated in the CAMERE process) [74]. Reaction thermodynamics indicate that a shift to lower temperatures (e.g. ca. 200°C) would favor overall methanol productivity [75] and likewise reduce the recycling loop demand (e.g. and therefore a reduction in operational costs) [24]. However it should be noted, at these temperatures, H_2O is relatively strongly bound or "sticky" to the surface of the catalyst. In this regard, previous reports including that of Liu et al. propose that strong water adsorption effectively blocks the active site at the catalyst surface [76]. Further research has led to the suggestion that chemisorbed water is in fact not the "blocking" species but rather it is either chemisorbed as carbonate or formate (i.e. from the reaction of H_2O and CO_2) [60]. The impact of exposing Cu NPs to H_2O leads to "deactivation"; for example via blocking of hydrogen adsorption sites, and influencing the morphology/oxidative state of the Cu NPs (e.g. in turn influencing active site composition/catalyst activity).

Regarding a deeper understanding of the Cu/ZnO system and active site composition, during the last decade remarkable advances (as underpinned by increasing access to high resolution microscopy and complimentary computational methods), have been made. This has led to an evolving discussion, and the proposition that a mixture of high crystal lattice strain (e.g. in Cu NPs at the interface with ZnO NPs), as

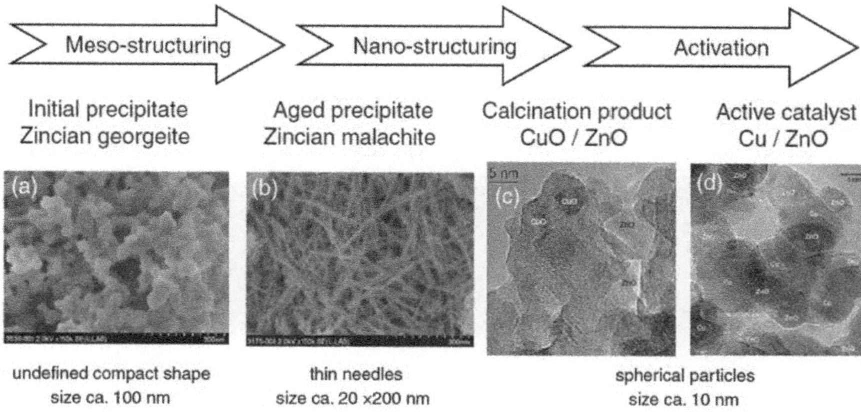

Figure 22.2 Electron microscopy images of the materials obtained by Behrens at different stages in the course of methanol synthesis (i.e. $Cu/ZnO/Al_2O_3$) catalyst preparation. (Reproduced with permission from [79].)

influenced by a high number of crystal defects (e.g. edge sites) leads to higher methanol synthesis activity [77–78]. In this regard, the works of Behrens, Schlögl, and Chorkendorff, Nørskov, et al. are to be noted [79–84]. Following on from earlier reports [85–88], Behrens et al. have extensively investigated different chemical promoter/support chemistry (e.g. MgO, Al_2O_3,) [89–90], and preparation procedures (e.g. nanostructuring, co-precipitation, ageing, etc.) [91], on the activity of binary Cu/ZnO-based model systems (Figure 22.2).

The complex interplay between these parameters and their impact on catalytic activity in CH_3OH synthesis is often referred to as "*Chemical Memory*" [92]. Combining characterization of the Cu/ZnO cat by high resolution microscopy with high powered computation modelling (DFT) indicates that steps at the Cu NP surface generate sites required to provide high activity in the hydrogenation reaction (Figure 22.2(c)) [82]. It has been reported that these steps are stabilized by bulk crystal defects (e.g. stacking faults or twin boundaries) terminating at the NP surface. In the literature, it is suggested that activity enhancement is the result of stronger binding of reaction intermediates at "stepped sites" and the associated lower energy barrier. As a consequence, it is thus theorized that $Zn^{\delta+}$ interfaces with these Cu steps, which results in a dynamic strong metal-support interaction (SMSI), with the partial coverage of the Cu NP by a ZnO layer. These discussions have led to the suggestion that a further substitution of Cu for Zn at the step sites will result in a further enhancement of intermediate state binding and in turn catalytic activity.

Characterization of the nano-interplay and roles of the different components of the Cu/ZnO-cat (e.g. under applied reaction conditions) represents a very active area of research. Some recent reports suggest that the active phase corresponds to a CuZn alloy as formed during the reaction conditions, via the partial reduction of the ZnO phase [94–95]. As further elaboration of the reports of Behrens et al., the analysis of industrial catalysts have been reported to contain ZnO_X-overlayers on Cu NPs [96]. More recent experimental results and analysis have also demonstrated that ZnO_X NPs on a Cu(111) surface generates higher catalytic activity (i.e. as compared to the conventional Cu/ZnO-cat [97–101]. These reports and associated data lead to the

Figure 22.3 TEM micrographs of a) a Fritz-Haber-Insitute (FHI) methanol synthesis catalyst (calcined) and b–d) FHI-methanol synthesis catalyst (reduced). c) HRTEM image of a Cu NPs with defects (arrowheads indicate twin formation). d) HAADF–STEM image represent Cu and Zn, respectively. (Reproduced with permission from [93].)

strong assertion that the sites responsible for the synthesis of methanol may indeed correspond to a ZnO phase or a somewhat unique ZnO-Cu nano-interface [84, 102]. In this context, a ZnO_X/Cu(111) system was proposed and explored in previous works [97, 103], whilst a more recent report highlights the potential of a Cu-supported $Zn_6(OH)_7$ as a possible catalyst [98]. This system has been further investigated from a computational standpoint by Reichenbach et al. in [104]. This work highlights further the potential and importance of such structures in further describing methanol synthesis catalysts, whilst exploring global minima structures, and the broad independence from explicit size and composition of the reduced ZnO/Cu inverse catalyst on theoretical catalytic performance. Reichenbach et al. [104] highlight in particular the suitability of Zn_XO_YH/Cu(111) and ZnO-nanoribbon/Cu(111) as models for the industrial ZnO_X/Cu catalyst and show that the reaction proceeds via the formate pathway, thus avoiding CO intermediates.

The presented results of Reichenbach et al. indicate that the formation of methanol could be directed by varying the oxide type (e.g. from ZnO_X to CeO_X), as suggested previously by Behrens [105]. Reichenbach et al. reported that the initial CO_2 binding

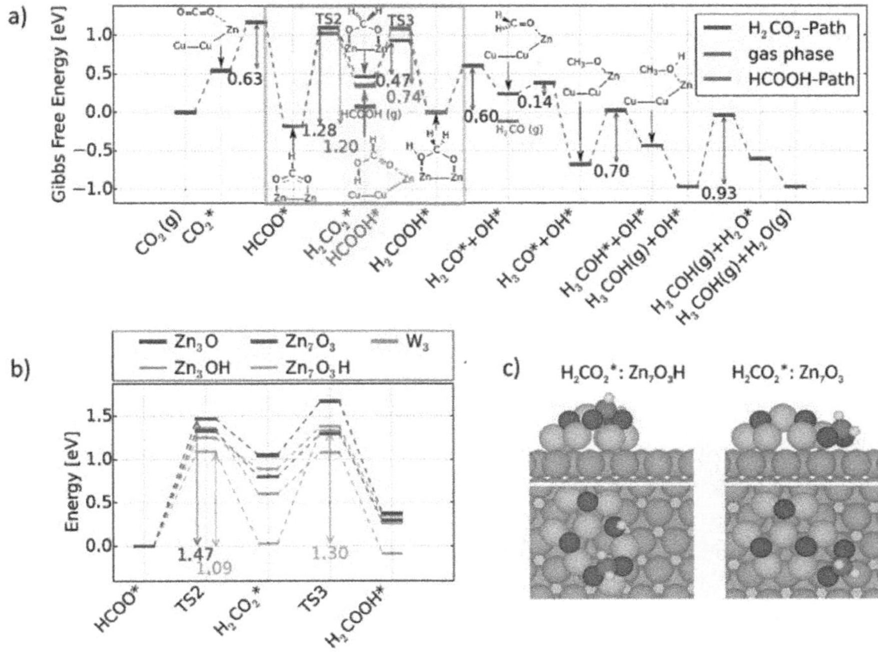

Figure 22.4 a) Reaction pathway of CO_2 hydrogenation to methanol on the Zn_3OH/Cu(111)-system and the Gibbs free energy diagram calculated at T = 500 K and partial pressures of 40; 10; 1; 1, and 1 bars of H_2, CO_2, methanol, CO, and H_2O, respectively as reported by Reichenbach et al. (*Indicated energies are relative to CO_2 (g), six hours*/Cu(111) far from ZnO and the bare cluster Zn_3OH/Cu(111). Gas-phase species are marked by (g), while molecules marked with * are adsorbed. In addition, energy levels of transition states are denoted with the corresponding forward barrier*). Only the metal atoms participating in bonds with the reactants are indicated. (b) Electronic energies of the rate limiting barriers enclosed in the rectangle in (a) for different model structures. (c) Structure of H_2CO_2* intermediates on Zn_7O_3H and Zn_7O_3. (Reproduced with permission from [104].)

energy as well as the rate determining barriers (Figure 22.4; HCOO ➔ H_2CO_2) are very similar for various reduced ZnO structures supported on Cu(111). This mechanism occurs independently of changes in the form and size of reduced ZnO and suggests that the reaction path is generally the same on any under coordinated Zn sites in ZnO/Cu catalysts (i.e. ZnO in contact with metallic Cu). Reichenbach et al. calculated that the entropy gain in the gas-phase causes all stable species (CO_2, CO, HCOOH, H_2CO, H_3COH, and H_2O) to desorb. This is important to note as it inhibits poisoning through water accumulation, although desorption itself simultaneously drops the methanol formation rate (i.e. as it competes with HCOOH ➔ H_2COOH and H_2CO ➔ H_3CO reactions). Therefore, it is proposed that further optimization of the reaction temperature assists the reaction in this respect, but notably hinders overcoming of the reaction energy barriers.

These critical insights into catalyst structure open opportunities for fundamental manipulation and direction of active site chemistry, importantly based on highly abundant metals (e.g. Cu, Zn). These reports highlight the potential for further design and optimization that will underpin further large scale and increasingly sustainable production of methanol. The work of the aforementioned researchers highlights further the intricate relationship between a high Cu surface area, defects/step/edge site chemistry and a dynamic SMSI to generate catalysts with optimal activity. Regarding future methanol synthesis, catalytic technology is already available at the commercial level, reducing the requirement for significant investment in new catalyst development although investments will be required to develop infrastructure/hardware to deal with a presumed increase in worldwide production to meet different sector demands (e.g. consider volume of liquid fuel market).

22.4 Transitioning to Sustainable Methanol Production

The use of traditional feedstocks in methanol production has been extensively reviewed [63, 68, 106–109]. Regarding the development of methanol production based on the use of sustainable precursors, the use of (municipal) waste and biomass (i.e. via gasification to syngas) is often considered (e.g. "*Waste-to-Chemicals*", Netherlands; Scheme 22.4) [110]), and may be suitable in specific locations (e.g. close to large refuse collection areas). However, given its chemical heterogeneity (and indeed the desire to reduce overall waste production), gasification can only been seen going forward as a minor source for the production of syngas and in turn methanol. The synthesis of methanol based on the hydrogenation of CO_2 therefore has potential advantages. Inherently, CO_2 as captured from waste streams (e.g. industrial plant exhausts) can be supplied in a relatively pure manner. This in turn increases catalyst lifetime, simplifies base process design, and produces an inherently purer methanol product (e.g. important if the alcohol is to be later used as a fuel). Therefore, in the context of "*sector coupling*," the carbon/CO_2 source for the synthesis of methanol has an impact on the potential economic viability of the overall process. Likewise, this gives an opportunity for CO_x

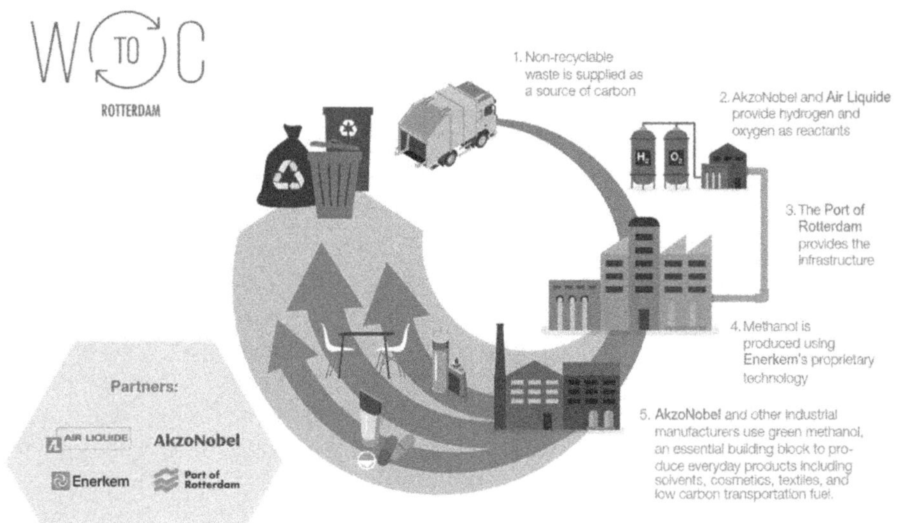

Scheme 22.4 The "*Waste-to-Chemicals*" project concept under introduction in the Port of Rotterdam, Netherlands involving the indicated partners. (*Source*: Enerkem® corporate website – [111].)

emitters to transform a waste gas from a liability to a potential carbon feedstock for fuel and chemical synthesis.

A potential benefit is the transformation of waste into a value added product and the overall mitigation of greenhouse gas emissions. The (bio-)anerobic digestion of municipal or industrial organic (e.g. food) wastes to produce both CH_4 and CO_2 (as a typically unwanted side product) through the further installation of biogas plants, could provide a source of two relevant energy carriers, namely bio-natural gas (e.g. fed to the existing natural gas network) and a liquid fuel/platform chemical (e.g. methanol, which could also be "dropped" into existing liquid fuel provision infrastructure). However, like syngas derived from biomass gasification, most syngas mixes originating from these sources typically contain organic and inorganic impurities, making comprehensive feed gas cleaning necessary. In China, methanol predominantly is produced from syngas derived from the gasification of abundant coal supplies. In this context, coal has the benefit of being inexpensive (e.g. as compared to natural gas), however its continued use will inhibit the transition to RE-based energy and chemical sectors, whilst its extraction and indeed gasification are environmentally unsound (e.g. vast H_2O demand, significantly higher CO_2 emissions arising from gas conditioning and elevated investment costs compared to a NG-based synthesis process) [64, 107, 112], and thus will not be considered further in this chapter.

In the book "*Beyond Oil and Gas: The Methanol Economy*®", Olah, Goeppert, and Prakash outline a new energy/chemical economy based around the capture of CO_2 and its conversion to CH_3OH (Figure 22.5) [64]. This book, the first in the English language world to propose the concept, strongly advocates the use of CH_3OH as a base from which to access future chemical, fuel, and material needs. Earlier, a similar concept was put forward (originally in German) by Asinger [63] and [113]. It will be well known to the chemist, that this simplest of alcohols, under standard, ambient conditions, CH_3OH is a liquid and therefore can potentially be viewed as a practical and

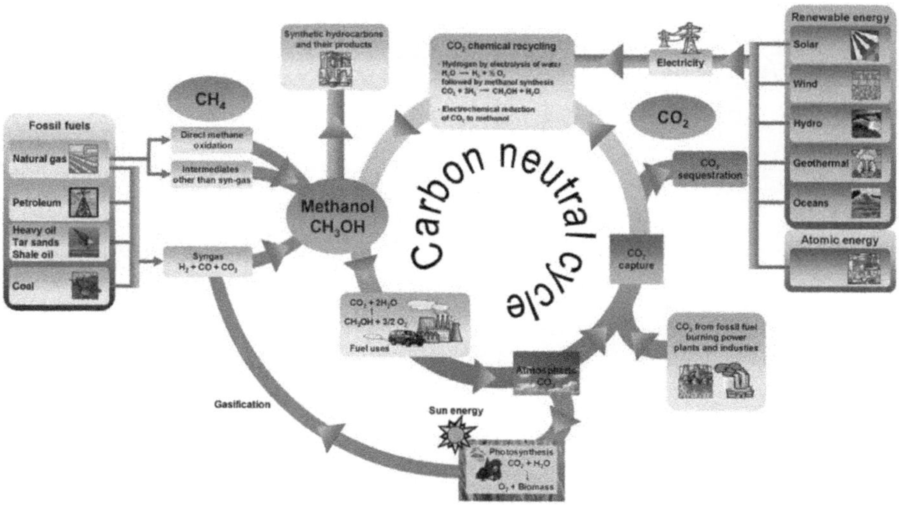

Figure 22.5 An Anthropogenic Carbon Cycle based on CH_3OH cycling, as proposed by [114, 115]. (Reproduced with permission from [114–115].)

safer alternative to (heavily compressed) H_2 (e.g. in terms of energy storage density, transport fuel, etc.).

CH_3OH can be stored, transported, and "*dropped in*" in to the liquid fuel infrastructure, with minimal modifications (*compare with the infrastructural demands of 700 bar(H_2)*). Irrespective of production (e.g. renewable *vs.* non-renewable), the efficient storage, transfer, and distribution of H_2 (e.g. in the large scales needed to support mobility), faces a number of challenges, although in countries like Germany or Japan this is being pursued (e.g. see https://h2.live/en). Under ambient conditions, gaseous H_2 is a limiting energy carrier as a result of a low volumetric/weight energy relationship. Although offering benefits compared to standard Li-ion battery technology, at 15 °C and at a pressure of 1 bar H_2 (g) has a density of

Table 22.1 Properties of a range of different energy carriers/fuels.

Energy Carrier/Fuel	Energy Density (volume)/ MJ/L	Specific Energy (weight)/ MJ/kg	Energy (volume/ weight) ratio	Density/kg/m^3	Standard enthalpy of combustion/ kJ/mol
Gasoline fuel (e.g. C_8H_{18})	33	52–39	0.74	630–830	−47.3 (kJ g^{-1})
Diesel (e.g. $C_{10}H_{20}$ to $C_{15}H_{28}$)	36	43	0.84	840	−44.8 (kJ g^{-1})
Methanol (H_3COH)	17.9	20	0.90	791	−726.0
Dimethyl Ether (H_3C-O-CH_3)	19	28	0.68	666	−1460.0
Oxymethylene Ethers (OME)* (H_3CO-$(CH_2O)_n$-CH_3)	–	19.4 (LHV)	–	1070	–
Ethanol (CH_3CH_2OH)	21	27	0.78	789	−1371.0
Jet Fuel (e.g. kerosene/$C_{>12}H_{>26}$)	37.6	43.8	0.86	820	−4620.0
Liquid Petroleum Gas (Propane/ $CH_3CH_2CH_3$)	25 (LHV)	46 (LHV) 48–50 (HHV)	0.54	540	−2202.0
Methane (CH_4)	0.037	55.6	6.7×10^{-4}	0.67 (NTP) 465 (LNG)	−890.3
Hydrogen (H_2)	0.01 (NTP) 4.5 (*l*) (@ 690 bar)	123	8.1×10^{-5} (0.037)	0.09 (NTP)	−286.0
Ammonia (NH_3)	14.1	32.8	0.43	0.682 (at 240 K/1 bar) 0.603 (at 298 K/10 bar)	−382.0

LNG = Liquified Natural Gas; NTP = Normal Temperature and Pressure; HHV = Higher Heating Value; LHV = Lower Heating Value; *OMEs = H_3CO–$(CH_2O)_n$–CH_3 where n = 3–5.

0.09 kg/m or in terms of energy density = 143 MJ/kg (Table 22.1; Figure 22.2). Compressing H_2 (g) improves energy density properties (e.g. 5.6 MJ/L (at 700 bar and 15 °C) or 4.5 kg/m at 690 bar) or cooled (e.g. t < −200 °C), but requires an "energy" input and therefore the overall net energy cannot compete with traditional hydrocarbon liquid fuels [116]. Progress is being made regarding material development (e.g. metal organic frameworks, hydrides), which could be viable storage media facilitating the reversible storage of large amounts of H_2 [117–118]. Regardless of these challenges, H_2 will play an extremely important role in any future chemical/energy/fuel provision system. Likewise, it is foreseen that methanol (e.g. as H_2 carrier) will play also a significant role (e.g. in unison with an expected roll out of electrolysers as powered from grids rich in RE).

Regarding fuels and energy carriers, methanol is known to be an excellent fuel for internal combustion engines (ICE) with a high octane rating (Octane No. (ON): ≥100; conventional petrol = 95–100). For a comparison of different fuel options see Table 22.1). Likewise a comparison between various different energy storage options is also provided (Figure 22.6) [116].

Liquid fuels are easily handled within the current fuel distribution network and in many cases the infrastructure for one liquid fuel is suitable for distribution of another (or at least with minor modifications). Regarding the use of methanol in mobility, modern ICEs require slight modification to run on this C1 alcohol [119]. Additionally, dehydration chemistry (e.g. based on zeolite (H-ZSM-5) catalysis) can be used to syntheize dimethyl ether (CH_3OCH_3; DME (g)), a high cetane diesel substitute (Cetane No.: ≥55) [120]. CH_3OH can also be used as a basis to produce higher hydrocarbon fuels for example based on "*Methanol-to-Gasoline*" technology developed by Exxon-Mobil in the mid-1980s [121]. Likewise, sources of sustainable syngas can also be

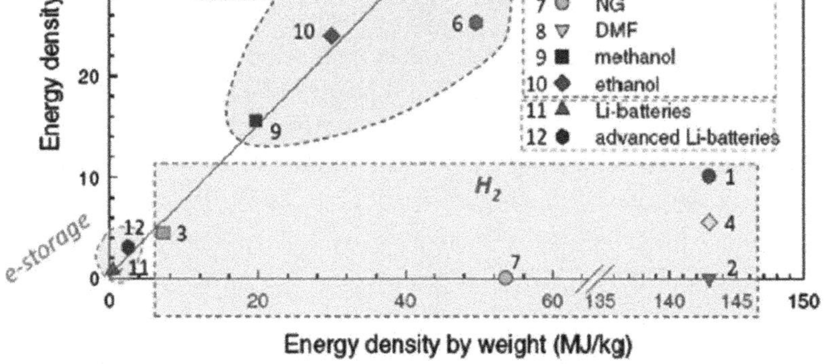

Figure 22.6 Energy density per unit weight (MJ/kg) *vs.* per unit volume (MJ/L) for a range potential storage media. (Reproduced with permission from [116].)

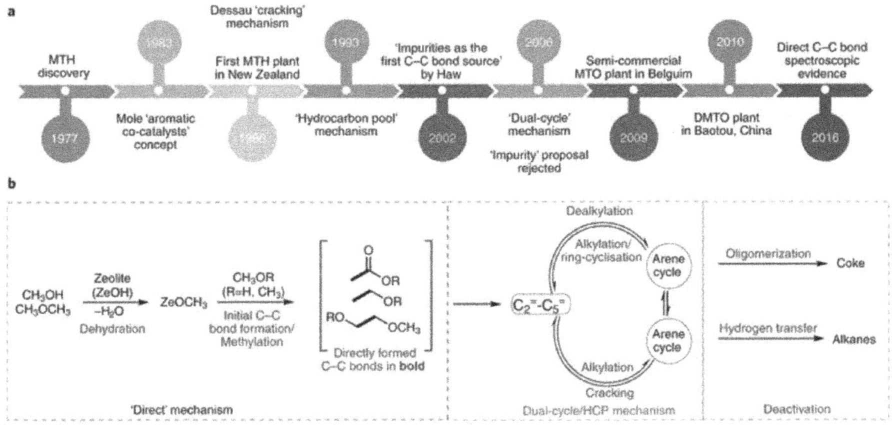

Scheme 22.5 (a) A chronology depiction over ca. 40 years of efforts to mechanistically describe the initial induction and steady state periods for "*Methanol-to-Hydrocarbons*" (MTH) synthesis (as produced by [129]). (b) A simplified MTH mechanism demonstrating that the aromatic cycle leads to light olefin formation, but also as a scaffold for coke formation; thus, olefins formation is preferred over the olefins cycle (Acronyms: MTO, methanol-to-olefin; DMTO, dimethylether-to-olefin; ZeOH, zeolite acid sites). (Reproduced with permission from [129].)

used in the production of a range of hydrocarbons (e.g. waxes, jet fuel, etc.) via well-known and industrially established Fischer-Tropsch synthesis. The production of fuels in this manner will be necessary to satisfy sectors such as shipping and aviation. These topics have been extensively discussed and reviewed elsewhere [122–130]. A recent review by Yarulina et al. provides an insightful description of the development of this technology and associated mechanistic discussions over the last 40 years (Scheme 22.5) [129].

The chemical possibilities of methanol in the chemical industry are relatively well known and is currently a large volume (ca. 90–100 Mt/a) commodity chemical. Going forward as the chemical industry transitions to increasingly sustainable production pathways, this simple alcohol will continue to be an important platform [63, 131]. Currently, the CH_3OH industry is very dynamic. The main feedstock for production globally however is still fossil fuels (i.e. natural gas or coals) with associated emissions of CO_2. In 2010, the annual global demand for CH_3OH (and derivatives; e.g. DME) was ca. 45.6×10^6 t (where t refers to metric tonnes) and was expected to grow to $>90 \times 10^6$ t by the end of the decade (excluding ethylene and propylene production). This growth is driven by the increased demand for cleaner chemistry (e.g. synthetic fuels, MtO, biodiesel) and increased regulations designed to encourage the adoption of renewable fuels and chemicals (e.g. the renewable energy directive (2009/28/EC)) [132]. However, it must be noted that if a Methanol Economy is to be fully implemented then global production must be dramatically increased.

Established (industrial) chemistry can be employed to convert CH_3OH into a variety of important industrial derivatives (e.g. ethylene, propylene – based on analogous technology to MTH technology) [130] – and subsequently a significant majority of our consumer plastics (e.g. poly(propylene)). Likewise, established chemistry and catalysis convert methanol into other important commodity chemicals such as acetic acid or formaldehyde (e.g. via the FORMOX® process) [133]. The production of formaldehyde will be a critical building block in the production of OMEs, which if based on captured CO_2 and renewable H_2, provide an option to generate synthetic diesel substitutes [23].

22.5 Elaboration of a Methanol Economy

To further elaborate, the concept of Olah, Asinger, and colleagues will to a degree require certain new chemical technologies. For example, if atmospheric CO_2 levels are to be reduced (e.g. via efficient membrane flue gas separations), this will require capture technologies from point emission sources (e.g. cement factories) and ultimately, if a true circular economy is to be established, then it must be captured from the air in an economically and ecologically sound manner – a significant challenge given $[CO_2]$ = ca. 400 ppm Regarding these technologies, the previously published reviews [134–136], and are worth visiting for a more in depth discussion of these technologies. New approaches to CO_2 use as a carbon platform are also being developed and ultimately will represent a significant mechanism for longer "CO_2" storage/energy storage [32, 34, 137]. It is these approaches and indeed the continuing integration of RE with "chemistry," that opens opportunities for the chemical and energy industries to "green" their public image. The same approach will provide a degree of "future proofing" as fossil reserves become increasingly scarce/economically unviable. Likewise to capture technologies, new chemical technologies (e.g. photo(electro)chemical water splitting, CO_2 capture materials) can continue to be developed until techno-economic/market conditions are appropriate for their large scale introduction.

With regard to CO_2 mitigation, and its recycling to CH_3OH (higher fuels and other chemicals), inputs of H_2 and/or heat/electrical energy are required. It is therefore necessary to ask which technologies and chemistries are best suited at the present time (and in the future) to achieve this goal in the most efficient manner possible. Briefly, in such a scheme there would be five general steps constituting a general process for sustainable CH_3OH synthesis based on waste CO_2 and RE utilization:

i. CO_2 capture/concentration/purification
ii. H_2 and/or syngas production
iii. Methanol synthesis
iv. Methanol purification (if required)

and, if required,

v. (a) conversion of methanol (e.g. to DME, OMEs, olefins, and other fuels/chemicals) *and/or* (b) recovery of "stored" electrical energy (e.g. reforming, Direct Methanol Fuel Cells (DMFC), or dehydrogenation to H_2) for mobility/power generation.

A wide variety of scenarios has been proposed and predominantly focus on optimizing H_2 or syngas production for a desired end product (e.g. CH_3OH *vs* hydrocarbon fuels), with electrical or heat energy being supplied ideally from (e.g. renewable) CO_2-free sources [138]. This concept can be generalized as multicomponent flow sheet(s) depicting the key processes and also the potential to integrate associated recycling of energy (e.g. exothermic CH_3OH synthesis) and emissions (Figure 22.5) [24].

Methanol production based on this concept (Figure 22.7) can be divided into six primary steps based on the following considerations:

i. Where does the electricity come from? (e.g. generated from renewables on-site or purchased from the grid).
ii. Sourcing of H_2 – e.g. obtained via H_2O electrolysis or industrial streams providing high concentrations of H_2 (e.g. coke oven gas from steel mills). H_2 should not be generated (in-)directly from fossil resources (e.g. steam reforming of CH_4 and H_2O electrolysis powered by the electricity grid mix). Regarding overall economics, a by-product of H_2O electrolysis is high purity O_2, which could potentially be sold for further value.
iii. From where is CO_2 sourced? For example, is it captured from industrial processes, biogas, or ambient air? (Please note, different sources will require different purification strategies.)
iv. Feed (H_2 and CO_2) compression and the associated energy demand (this will be source dependent).
v. Is a temporary storage of H_2 required? If so, the size will be related to the fluctuation of the electricity production and the range of the system dynamics (e.g. when operating an electrolyser).
vi. How is the overall process operated? In continuous or dynamic operation? This is an important consideration and may impact on technology choice and indeed component/catalyst lifetime.

For the synthesis of methanol via such a process it can be assumed at the current state of industrial technology that the catalytic conversion of H_2 and CO_2 is based on established reactors, catalysis, etc. Methanol synthesis is followed by a purification step (e.g. based on distillation). Approaching the simulation and optimization of this approach makes a number of assumptions (e.g. relating to setting variable or fixed

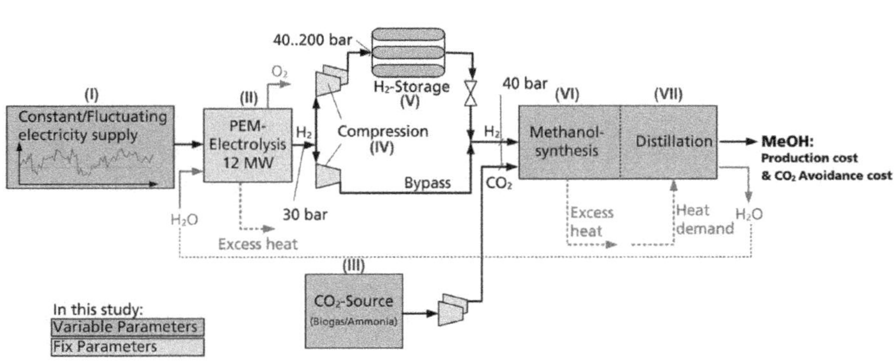

Figure 22.7 Production of methanol based on renewable energy, water electrolysis, and recycled carbon dioxide, indicating potential variable and fixed parameters as reported and analysed by [24]. (Reproduced with permission from [24].)

parameters in Figure 22.7; fixed, such as polymer electrolyte membrane electrolysis, compression and pressure levels, and distillation; variables, such as power supply, CO_2 source, H_2 storage, and MeOH synthesis). In the previous report [24], the evaluation of the economic feasibility demonstrated that methanol production in this manner strongly depends on the electricity and H_2 production cost, CO_2 cost (also reflecting any introduced carbon taxes), specific electrolyser cost, and the possible dynamics of the methanol reactor (having an impact on the necessity and size of H_2 storage). Indirect parameters such as carbon taxes could positively encourage adoption of PtL methanol and as such can be considered desirable.

It was noted in [24] that MeOH production (denoted as eMeOH) in Germany (under the reported conditions) is currently not economically competitive (Figure 22.8). The authors highlighted that investment and operational costs are currently inhibiting and thus eMeOH production price exceeds expected revenues. The authors proposed that a high degree of capacity utilization is advantageous and favourable for low eMeOH production cost. Therefore maximized plant operation (e.g. in terms of hours) is also highlighted as an important factor in achieving economic feasibility. A sensitivity analysis in this report demonstrated that a decrease in the specific investment costs of the electrolysis unit leads to a major economic efficiency improvement (target value: 440 € kW^{-1}), whilst H_2 storage costs (e.g. for a wind driven production scenario) was also shown to have a strong impact on overall production costs. The possibility of underground H_2 storage in salt caverns (e.g. in northern Germany) was highlighted as a potential alternative, whilst in the long-term it was proposed that a more dynamic operation of MeOH synthesis could reduce H_2 storage demands. As can be observed in this study and indeed throughout the literature, the cost of H_2 production from renewables remains the key bottleneck (at least from a chemistry/process point of view) in the further elaboration of methanol production in this manner.

Regarding the ecological impact of methanol production based on such schemes, Qin et al. calculated a CO_2-equivalent emission of 2.97 $t_{CO2,eq}/t_{CH3OH}$ based on a

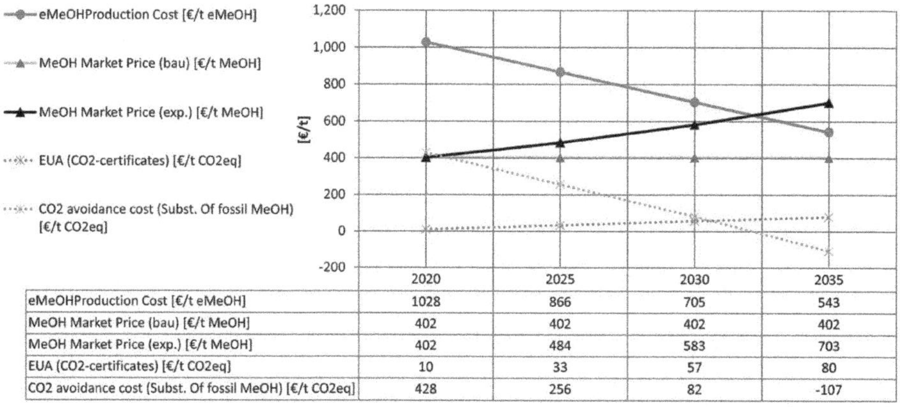

	2020	2025	2030	2035
eMeOHProduction Cost [€/t eMeOH]	1028	866	705	543
MeOH Market Price (bau) [€/t MeOH]	402	402	402	402
MeOH Market Price (exp.) [€/t MeOH]	402	484	583	703
EUA (CO2-certificates) [€/t CO2eq]	10	33	57	80
CO2 avoidance cost (Subst. Of fossil MeOH) [€/t CO2eq]	428	256	82	-107

Figure 22.8 Estimation for the future development of green methanol production cost and CO_2 avoidance cost in the case of fossil methanol substitution as reported in [24]. (Reproduced with permission from [24].)

cradle-to-gate life cycle assessment (LCA) of a coal-based methanol production plant in China [139]. For a NG-based process the according CO_2-equivalent amounts to 0.85 $t_{CO_2,eq}/t_{CH_3OH}$ (http://dechema.de/Low_carbon_chemical_industry.html). In comparison, the direct hydrogenation of captured CO_2 to produce methanol has been reported to have a CO_2-equivalent of −0.67 $t_{CO_2,eq}/t_{CH_3OH}$. This indicates possible carbon emission savings of 1.52 $t_{CO_2,eq}/t_{CH_3OH}$ showing the sensitivity of methanol pricing towards political regulations such as taxation and trading systems [140] (http://dechema.de/Low_carbon_chemical_industry.html) [38].

22.5.1 Sourcing Carbon for the "Circular" Production of Methanol

The debate regarding the use and implementation of sustainable feedstocks and CO_2-neutral processes for methanol synthesis arrived in the scientific community in the early 1990s via contributions from both academic and industrial research groups [141–146]. The following discussion highlights two types of alternative feedstocks: CO_2 point sources (e.g. biogas plants) and carbon oxide rich industrial off-gases from existing industrial processes (e.g. cement production). In this context, off-gases arising from steel production have significant potential, with particular respect to sector coupling and establishment of circular economies [140]. Demonstration facilities at the small and industrial scale have led to enhanced knowledge concerning the advantages and drawbacks of CO_2-based methanol synthesis [74, 143, 146–151]. Industrially relevant experience has been obtained regarding the advantage of operating methanol synthesis with CO_2-rich syngas, leading to a lower heat duty emerging from CO_2 hydrogenation [141, 146, 152–154]. Another advantage is the lower catalyst selectivity towards hydrocarbon side-products (e.g. ketones), in turn reducing product purification costs [57, 143, 148, 154–156]. Regarding drawbacks, CO_2-based methanol synthesis requires an overall larger H_2 supply (i.e. due to the formation of H_2O), which in turns reduces the loop hydrogen efficiency (LHE), equilibrium conversion, and space time yield [143, 148, 157]. Some reported processes look to directly hydrogenate CO_2 to methanol [148], whilst other consider CO_2-based process concepts whereby the feed gas is conditioned by shifting CO_2 to CO via reverse WGS chemistry (e.g. the so-called CAMERE process) [74, 145, 158]. Even under optimized process conditions, the disadvantage of an elevated H_2 consumption (and cost) persists [63]. This highlights further the necessity of an economic and sustainable H_2 supply to support future methanol (and derivatives) synthesis [64, 149].

22.5.2 Example 1 – Steel Mill Gases

Compared to CCU schemes, typically based on CO_2-rich "make-up gas" (MUG), a feed stream with elevated CO content typically results in an increased external loop hydrogen efficiency (LHE_{ext}). In this regard, waste gases from steel mills (e.g. blast furnaces) have the potential to provide large volume streams containing H_2, CO, and CO_2 (http://dechema.de/Low_carbon_chemical_industry.html). The steel industry accounts for ca. 6.7% of all global CO_2 emissions (1.9 $t_{CO_2,eq}/t_{Steel}$ are emitted on average through steel production) [159–160]. As such, a number of research initiatives aim to utilize these waste gases (e.g. for synthetic purposes) and thus decrease the overall carbon footprint [161–164].

The compositions of three main steel mill gases are shown in Figure 22.8 (i.e. blast furnace gas (BFG), coke oven gas (COG), and basic oxygen furnace gas (BOFG)) [165]. With a fraction of ca. 87% of the total gas volume, BFG is the predominant potential syngas source [165]. However, it has an inert gas (N_2) fraction of ca. 50% [164, 166–167]. BOFG and COG have smaller inert gas content, but are only available in comparatively small volumes. The COG alternative can be viewed as a potential H_2 source [168–170], that could be transformed to conventional syngas through reforming technology [169, 171–175]. Notably BOFG and BFG are both H_2 light (BFG: SN ≈ −0.40 and BOFG: SN ≈ −0.15) and will require additional H_2 input to generate the required SN. Alternatively, the removal of CO_2 from BFG could be conducted to generate a secondary pure CO_2 stream which could in turn be used to produce chemicals [176].

The European Union in particular is leading research initiatives concerning the reduction and recycling of CO_2 from major industrial processes as supported by legislative frameworks (e.g. as supported by national initiatives) [178]. Such initiatives create new challenges for energy-intensive industries and indeed promote integration

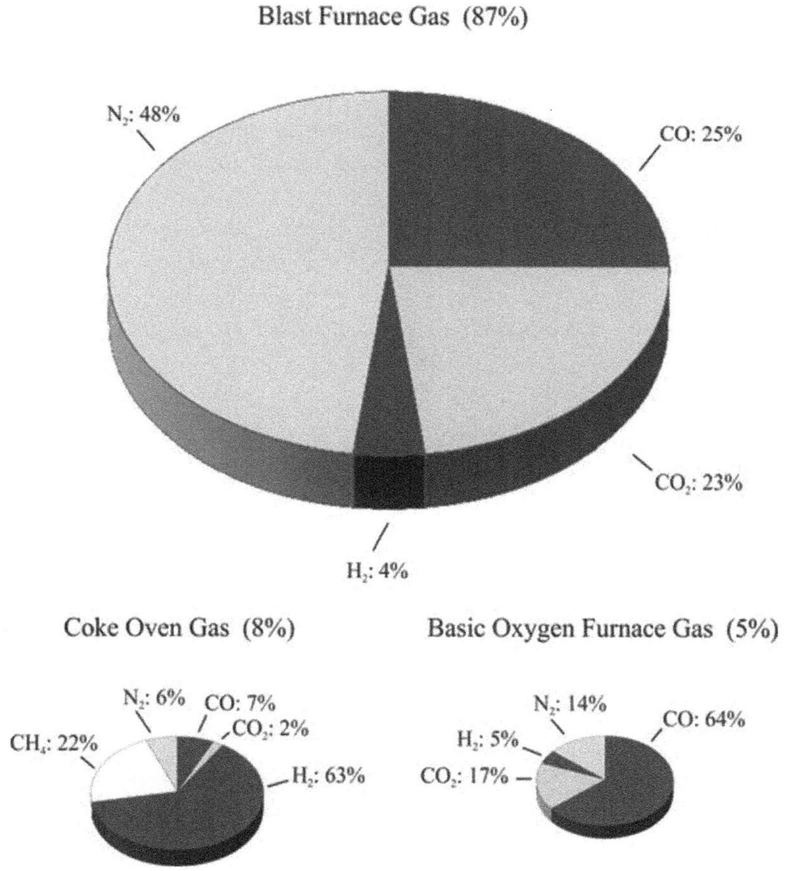

Figure 22.9 Gas composition of steel mill gas streams according to [165]; (Values in bracket give fraction of the total volumetric gas stream emitted from the steel mill.) (Reproduced with permission from [177].)

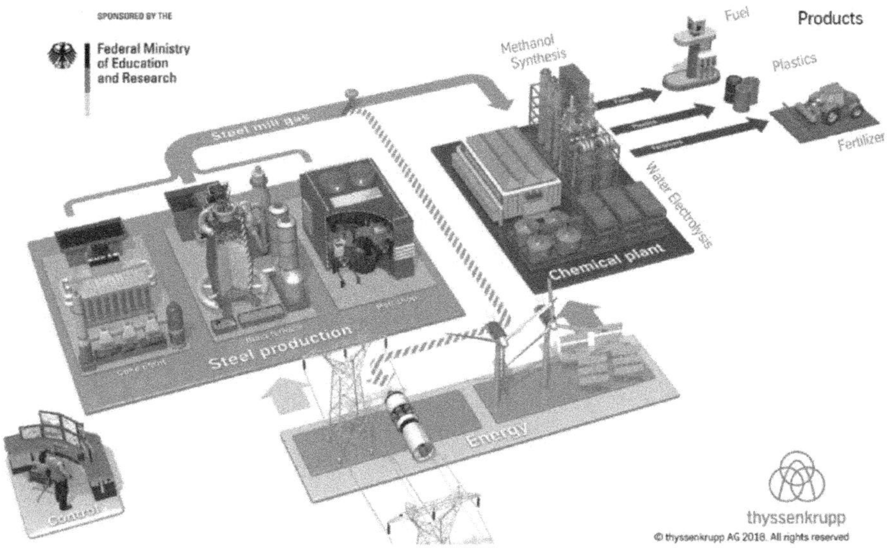

Scheme 22.6 Systematic overview of the "*Carbon2Chem*®" concept. (*Source*: Thyssenkrupp® - Reproduced from Website - [181].)

and sector coupling. One project in this context through the EU-Horizon2020 "SPIRE" program is "FReSMe," which is the follow-on of previously funded "MefCO2" and "STEPWISE" projects [161, 179–180]. It focuses on CO_2 capture and methanol synthesis at a Swedish steel mill. A comparable project funded at EU member national level, as introduced by the German Ministry for Education and Research (BMBF), is "*Carbon2Chem*®" (Scheme 22.6) [164]. In this project, steel mill waste gases (e.g. top oven wastes) are to be used in the production of a variety of platform chemicals, fuels, and polymers. In one sub-project methanol will produced, whilst in others this approach is taken a step further and considers the production of higher hydrocarbons, ammonia, urea, and polymers. Depending on the required precursor gas mix (e.g. top gases are carbon oxide rich), renewable energy will be used to provide H_2 via water electrolysis to generate the correct syngas stoichiometry.

The transnational EU-INTERREG project "*Carbon2Value*" looks to establish value chains arising from converting steel mill waste gases, with partners from Belgium, UK, France, and the Netherlands. It focuses on the production of ethanol and synthetic naptha [183]. The H2020-funded "*Steelanol*" project focuses on the development of fermentation processes from these gases (e.g. to produce bio-ethanol) [162–163].

22.5.3 Example 2 – Carbon Recycling International (CRI)®

The company Carbon Recycling International® (denoted hereon as CRI) founded near Reijavik, Iceland in 2006, is perhaps the most well-known commercial example of a PtL company. CRI exploits geographical conditions in Iceland to combine geothermal (and in principle industrial) CO_2 gases sources with H_2 produced using RE-powered (low temperature) PEM electrolysis to produce methanol (available on the market as Vulcanol®) [184]. The first commercial plant at the CRI site came online in 2011 – named as the *George Olah Renewable Methanol Plant* (Figure 22.10). This demonstration facility as realized at the near industrial scale, also considers the option of performing reverse WGS chemistry in order to increase the CO content of the MUG – opening flexibility regarding the syngas feed [182]. This plant operates at a 5 × 10^6 L/annum production scale. CRI indicate the possibility to reclaim 4.5 × 10^3 t(CO_2) per year from the atmosphere. In recent years, the company has received significant investments (e.g. Methanex®; \$5 x 10^6 (US)), demonstrating the expectation of sustainable CH_3OH in future markets. The company is looking to grow and the intention is to continue to grow production capability (a 50 × 10^6 L/annum facility is coming online). CRI has designed its production system in modules which in principle can be exported to any desired location. The company has recently started exporting to the European market, where the sustainable CH_3OH is intended to be blended with conventional petrol. More recently, a joint venture (JV) was founded named "CRI Ji Xin Environment Protection Technology Co. Ltd. (CRIJX)" in 2017 (Shanghai, China). This JV incorporates CRI, the Chinese automotive company, Geely Holdings, and the Zixin Environmental Protection Technology Company. The objective of the JV is to promote and invest in the green methanol industry with a particular focus on the Chinese market. This will be done based on CRI's proprietary technology (i.e. at a 50 000 t/annum methanol production capacity). Through the MefCO2 project, CRI with project partners has inaugurated an "Emissions-to-Liquids" plant at a power plant in Niederaußem, Germany. Here methanol will be synthesised from waste CO_2 emissions with H_2 produced by electrolysis (from the company Hydrogenics®), coupled to one of the CRIs process modules [183].

Other commercial ventures in this field, include SunFire GmbH (Germany) [185], who look to convert renewable electricity to liquid hydrocarbons (via high temperature electrolysis and FT synthesis). In the United States, Ceramatec Ltd. were developing solid oxide electrolysers (SOEC) to produce syngas from the co-electrolysis of H_2O/CO_2 powered by abundant wind electricity in the state of Wyoming. Both SunFire GmbH [185] and Ceramatec Ltd. were the first commercial entities to develop and introduce SOEC technology to the market for sustainable fuel synthesis. Other commercial ventures have considered alternative sources of syngas for CH_3OH or DME production. For example, BioMCM (and the SuperMethanol/GtM concept project) have considered waste glycerol (as sourced from bio-diesel production) gasification as a route to

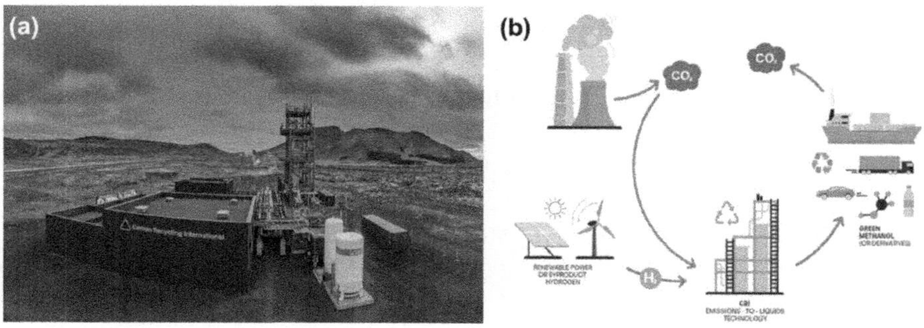

Figure 22.10 (a) The George Olah Renewable Methanol plant at the HS Orka Svartsengi installation near Rejkavik, Iceland and (b) The Methanol (tradename "Vulcanol®") synthesis approach as proposed by Carbon Recycling International® based on industrial CO_2 hydrogenation using H_2 produced from H_2O electrolysis. (*Source*: [181].)

inexpensive syngas [186]. A plant was in operation until recently in Delfzijl, Netherlands based on this concept.

22.6 Conclusion and Summary

If CH_3OH is produced using RE, then electrical energy stored in this liquid can be converted to kinetic energy (e.g. the internal combustion engine (ICE)) or viewed as a chemical flow battery (at a very large scale) whereby electrical energy is recovered using emerging technologies such as direct alcohol fuel cells (e.g. direct methanol fuel cells). Concepts such as Power-to-Liquids, the Methanol Economy, etc. would then represent a closed (e.g. low carbon emitting) energy/chemical provision loop. The sustainability credentials of such schemes will only be of interest long term and at the scales necessary to reach national and international emissions targets provided the CO_2 emitted (e.g. from mobility) are ultimately captured again and recycled. Herein lies the necessity of air capture, unless large scale rewilding or reforestation programmes are introduced – which seems challenging given an ever growing global population. The implementation of such schemes attempts to reverse the "carbon" consumption of the last century via an anthropogenic carbon cycle, analogous to the natural carbon cycle [114–115]. This means that the capture and recycling of CO_2 will provide the basis for the synthesis of future CH_3OH supplies, while fossil-based CH_3OH is phased out.

Regarding scale up and obviously in the context of delivery mobility fuels, future large-scale "green" methanol plants as coupled with existing chemical or heavy industry, would have advantages: instant access to low-cost grid electricity, availability of (highly concentrated) CO_2 streams (requiring purification) or unused H_2 in off-gases, if necessary access to high temperature process heat/steam and last but not least, the vast cost-reducing effect of economies of scale. In this context, it could be envisioned to bring together a number of sectors together at one site. An example as proposed by ThyssenKrupp® is shown in Scheme 22.7, whereby carbon oxide emissions and hydrogen (as supplied from RE powered water splitting) lay the basis for the production of a wide range of energy carriers, chemicals, and fuels for the mobility, chemical, and agricultural sectors. It will only be via such sector coupling that national and international emission targets will be reached. Methanol as sourced from PtL technologies will be an important pillar in integrated-energy concepts within (e.g. European) markets of electricity, heat, and mobility. A slow integration of RE into the mobility sector and chemical industries will only impede climate protection programmes [187].

As mentioned earlier, current market conditions (e.g. natural gas price) could be potentially inhibiting regarding the roll out of PtL methanol production. To support and encourage the adoption of such technologies, certain policy instruments will be required. In this context, amendment of general policy frameworks and taxation systems (e.g. special grid fees for energy storage and grid-supporting technologies, and an improved EU ETS) would be a good start. Likewise, a taxation structure to generate CO_2 Certificates and market conditions suitable to promote a business case are necessary. The EU is leading on these approaches with the proposed "double counting" of second generation biofuels based on the iLUC Directive 2015/1513/EC [189]

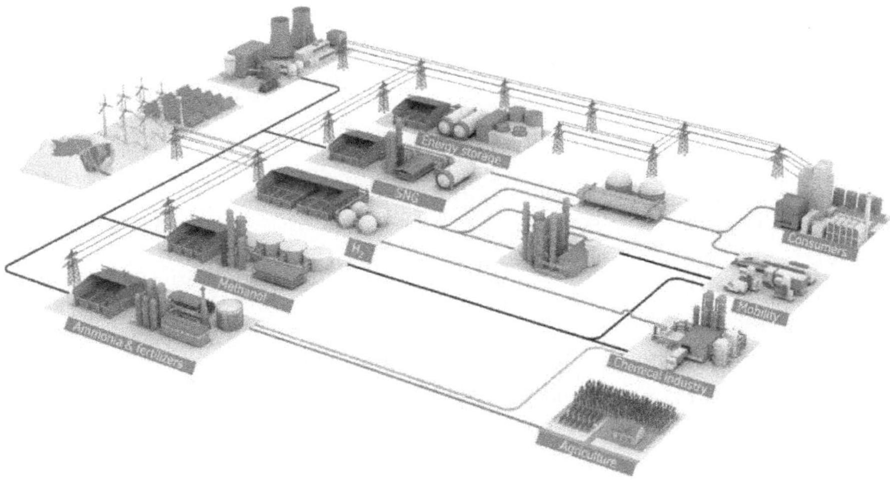

Scheme 22.7 "*Future Hydrogen Value Chains*" as proposed by ThyssenKrupp® - the coupling of waste gases from one sector with the energy sector through the production of H_2 (as sourced from H_2O splitting) to produce a range of energy carriers, chemicals, and fuels for the mobility, chemical, and agricultural sectors. (*Source*: [188].)

of the RE directive RED (2009/28/EC) [190] a good start regarding inclusion of CO_2-based fuels.

Considering a more global context, the development of PtL/CCU technologies in regions with higher solar irradiance (e.g. north Africa, Australia, Chile, etc.) and/or reliable wind power (e.g. Chile, Peru, South Africa, or Scandinavia) potentially already indicate an opportunity to provide very low RE costs. An RE electricity cost of ca. 2.5 ct€ per kWh_{el} (reflecting recent price bids for PV electricity) [191] and a higher degree of capacity utilization due to a less fluctuating RE potential makes the production of PtL methanol in the near future economically viable (e.g. ~500€ t/(MeOH)). Regarding ecological footprint, fuels and chemicals produced in these regions should ideally be used locally and not simply exported to larger markets. Further installation of pilot plants and associated operational experiences, coupled with expected economies of scale, will lay a path to improved PtL economics, even in countries where RE electricity costs are currently prohibitive. Therefore the challenge to chemists and chemical engineers is the efficient integration of RE with efficient, selective, and cost effective chemistry to provide the basis for chemical, materials, and fuel production. But like the coupling of the energy and chemistry sectors, efforts here alone will not suffice and "coupling" of many branches of the natural and social sciences as well as industry (as promoted by ever increasingly stringent legislation) will be required to achieve sustainable, circular economies.

References

[1] Zalasiewicz, J., Williams, M., Steffen, W., et al. (2010). *Environmental Science & Technology* 44 (7): 2228.
[2] http://www.eia.gov.
[3] Balibar, S. (2017). *Comptes Rendus Physique* 18 (7–8): 479.
[4] Jacobs, D. (2012). *Renewable Energy Law and Policy Review* 3 (4): 223.
[5] Sadorsky, P. (1999). *Energy Economics* 21 (5): 449.
[6] Fleisch, T.H., Sills, R.A., and Briscoe, M.D. (2002). *Journal of Natural Gas Chemistry* 11 (1–2): 1.
[7] Armaroli, V.B.N. (2007). *Angewandte Chemie International Edition* 46 (1–2): 52.
[8] Khatib, H. (2014). *Energy Policy* 64: 71.
[9] http://www.newstatesman.com/sci-tech/2012/10/its-not-safe-leave-fossil-fuels-ground.
[10] Adelman, M.A., Houghton, J.C., Kaufman, G., et al. (1983). *Energy Resources in an Uncertain Future*. Cambridge, MA: Ballinger.
[11] Hoffert, M.I., Caldeira, K., Benford, G., et al. (2002). *Science* 298 (5595): 981.
[12] Hall, C., Tharakan, P., Hallock, J., et al. (2003). *Nature* 426 (6964): 318.
[13] Kalmykova, Y., Sadagopan, M., and Rosado, L. (2018). *Resources, Conservation and Recycling* 135: 190.
[14] Sternberg, A. and Bardow, A. (2015). *Energy Environ. Sci.* 8 (2): 389.
[15] Clark, J.H., Farmer, T.J., Herrero-Davila, L., et al. (2016). *Green Chemistry* 18 (14): 3914.
[16] Schlögl, R. (2016). *Topics in Catalysis* 59 (8-9): 772.
[17] Whang, D.R. and Apaydin, D.H. (2018). *ChemPhotoChem* 2 (3): 148.
[18] Haas, T., Krause, R., Weber, R., et al. (2018). *Nature Catalysis* 1 (1): 32.
[19] Blondelle, J., *European Framework for Power-to-X*. https://ec.europa.eu/jrc/sites/default/files/BlondelleDGRTD.pdf
[20] Kourkoumpas, D.S., Papadimou, E., Atsonios, K., et al. (2016). *International Journal of Hydrogen Energy* 41 (38): 16674.
[21] Ghaib, K. and Ben-Fares, F.-Z. (2018). *Renewable and Sustainable Energy Reviews* 81: 433.
[22] Blanco, H. and Faaij, A. (2018). *Renewable and Sustainable Energy Reviews* 81: 1049.
[23] Ouda, M., Yarce, G., White, R.J., et al. (2017). *React. Chem. Eng.* 2 (1): 50.
[24] Hank, C., Gelpke, S., Schnabl, A., et al. (2018). *Sustainable Energy & Fuels* 2: 1244.
[25] Leitner, W. (1996). *Coordination Chemistry Reviews* 153: 257.
[26] Mazzotti, M. (2005). *Intergovernmental Panel on Climate Change - Report - Carbon Dioxide Capture and Storage: Ch. 7*. UK: Cambridge University Press.
[27] Sakakura, T., Choi, J.-C., and Yasuda, H. (2007). *Chemical Reviews* 107 (6): 2365.
[28] Aresta, M. (ed.). (2010). *Carbon Dioxide as Chemical Feedstock*. Germany: Weinheim.
[29] MacDowell, N., Florin, N., Buchard, A., et al. (2010). *Energy Environ. Sci.* 3 (11): 1645.
[30] Cokoja, M., Bruckmeier, C., Rieger, B., et al. (2011). *Angewandte Chemie International Edition* 50 (37): 8510.

[31] Quadrelli, E.A., Centi, G., Duplan, J.-L., et al. (2011). *ChemSusChem* 4 (9): 1194.

[32] Peters, M., Köhler, B., Kuckshinrichs, W., et al. (2011). *ChemSusChem* 4 (9): 1216–1240.

[33] Darensbourg, D.J. and Wilson, S.J. (2012). *Green Chem.* 14 (10): 2665.

[34] Markewitz, P., Kuckshinrichs, W., Leitner, W., et al. (2012). *Energy Environ. Sci.* 5 (6): 7281.

[35] Centi, G., Quadrelli, E.A., and Perathoner, S. (2013). *Energy Environ. Sci.* 6 (6): 1711.

[36] Yang, H., Zhang, C., Gao, P., et al. (2017). *Catal. Sci. Technol.* 7 (20): 4580.

[37] Romero, M. and Steinfeld, A. (2012). *Energy Environ. Sci.* 5 (11): 9234.

[38] Snoeckx, R. and Bogaerts, A. (2017). *Chemical Society Reviews* 46 (19): 5805.

[39] Marxer, D., Furler, P., Takacs, M., et al. (2017). *Energy Environ. Sci.* 10 (5): 1142.

[40] International Maritime Organization. *DNV GL.*

[41] Ecoinvent Centre. (2017). *Eecoinvent LCA-database V3.3.* Ecoinvent Centre – Swiss Centre for Life Cycle Inventories.

[42] PBL Netherlands Environmental Assessment Agency. (2013). *Trends in Global CO2 Emissions – 2013 Report.* Netherlands: The Hague.

[43] Boden, T., Marland, G., and Andres, R. (2012). *Global, Regional, and National Fossil Fuel CO2 Emissions: Carbon Dioxide Information Analysis Centre, Oak Ridge.* Tennessee, US.

[44] Marchal, V., Dellink, R., Van Vuuren, D., et al. (2001). *OECD Environmental Outlook to 2050: Climate Change, Joint Report.*

[45] European Commission (ed.). (2003). *Directive 2003/87/EC: Scheme for Greenhouse Gas Emission Allowance Trading (EU ETS).*

[46] Mikkelsen, M., Jørgensen, M., and Krebs, F.C. (2010). *Energy Environ. Sci.* 3 (1): 43.

[47] Patil, Y.P., Tambade, P.J., Jagtap, S.R., et al. (2010). *Front. Chem. Eng. China* 4 (2): 213.

[48] Zimmermann, A.W. and Schomäcker, R. (2017). *Energy Technol.* 5 (6): 850.

[49] Darensbourg, D.J. (2007). *Chemical Reviews* 107 (6): 2388.

[50] Li, J., Ma, X., Liu, H., et al. (2018). *Journal of Cleaner Production* 185: 299.

[51] http://news.ihsmarkit.com/press-release/country-industry-forecasting-media/global-methanol-demand-growth-driven-methanol-olefi.

[52] Skrzypek, J., Lachowska, M., Grzesik, M., et al. (1995). *The Chemical Engineering Journal and the Biochemical Engineering Journal* 58 (2): 101.

[53] Sun, J.T., Metcalfe, I.S., and Sahibzada, M. (1999). *Ind. Eng. Chem. Res.* 38 (10): 3868.

[54] Liu, G. (1984). *Journal of Catalysis* 90 (1): 139.

[55] Chanchlani, K.G. (1992). *Journal of Catalysis* 136 (1): 59.

[56] Kung, H.H. (1992). *Catalysis Today* 11 (4): 443.

[57] Ladebeck, J. (1993). *Hydrocarbon Processing; (United States)* 72 (3): 89.

[58] Wu, J., Saito, M., Takeuchi, M., et al. (2001). *Applied Catalysis A: General* 218 (1-2): 235.

[59] Behrens, M. (2015). *Recyclable Catalysis* 2 (1): 3343.

[60] Sahibzada, M., Metcalfe, I.S., and Chadwick, D. (1998). *Journal of Catalysis* 174 (2): 111.

[61] Lange, J.-P. (2001). *Catalysis Today* 64 (1-2): 3.
[62] Balthasar, W., Müller, D., and Wagner, U. (2016). *Verfahren zur Synthese von Methanol* (EP Patent 000003205622A1).
[63] Bertau, M. (ed.). (2014). *Methanol: The Basic Chemical and Energy Feedstock of the Future: Asinger's Vision Today*. Berlin, Heidelberg: Springer.
[64] Olah, G.A., Goeppert, A., and Prakash, G.K.S. (2009). *Beyond Oil and Gas - the Methanol Economy*. Wiley-VCH.
[65] Riaz, A., Zahedi, G., and Klemeš, J.J. (2013). *Journal of Cleaner Production* 57: 19.
[66] Rostrup-Nielsen, J.R. (2002). *Catalysis Today* 71 (3-4): 243.
[67] Dybkjær, I. and Christensen, T.S. (2001). *In Natural Gas Conversion VI*, p 435. Elsevier.
[68] Hu, Y.H. and Ruckenstein, E. (2004). *Advances in Catalysis* 48: 297.
[69] Twigg, M.V. (2003). *Topics in Catalysis* 22 (3/4): 191–203.
[70] Liu, X.-M., Lu, G.Q., Yan, Z.-F., et al. (2003). *Ind. Eng. Chem. Res.* 42 (25): 6518.
[71] French, S.A., Sokol, A.A., Bromley, S.T., et al. (2003). *Topics in Catalysis* 24 (1-4): 161.
[72] Bart, J.C.J. and Sneeden, R.P.A. (1987). *Catalysis Today* 2 (1): 1.
[73] Strunk, J., Kähler, K., Xia, X., et al. (2009). *Surface Science* 603 (10–12): 1776.
[74] Joo, O.-S., Jung, K.-D., and Yonsoo, J. (2004). *Carbon Dioxide Utilization for Global Sustainability, Proceedings of 7th International Conference on Carbon Dioxide Utilization*, 67. Elsevier.
[75] Kunkes, E. and Behrens, M. (2012). In *Chemical Energy Storage* (ed R. Schlögl and D. Gruyter), 413. Berlin.
[76] Liu, G., Willcox, D., Garland, M., et al. (1985). *Journal of Catalysis* 96 (1): 251.
[77] Kasatkin, I., Kurr, P., Kniep, B., et al. (2007). *Angewandte Chemie International Edition* 46 (38): 7324.
[78] Kurr, P., Kasatkin, I., Girgsdies, F., et al. (2008). *Applied Catalysis A: General* 348 (2): 153.
[79] Behrens, M. (2009). *Journal of Catalysis* 267 (1): 24.
[80] Studt, F., Sharafutdinov, I., Abild-Pedersen, F., et al. (2014). *Nature Chemistry* 6 (4): 320.
[81] Kuld, S., Thorhauge, M., Falsig, H., et al. (2016). *Science* 352 (6288): 969.
[82] Behrens, M., Studt, F., Kasatkin, I., et al. (2012). *Science* 336 (6083): 893.
[83] Fichtl, M.B., Schlereth, D., Jacobsen, N.,et al. (2015). *Applied Catalysis A: General* 502: 262.
[84] Studt, F., Behrens, M., Kunkes, E.L., et al. (2015). *ChemCatChem* 7 (7): 1105.
[85] Klier, K. (1982a). *Journal of Catalysis* 74 (2): 343.
[86] Klier, K. (1982b). *Advances in Catalysis* 31: 243.
[87] Chinchen, G.C., Denny, P.J., Parker, D.G., et al. (1984). *ACS Division of Fuel Chemistry* 29 (5): 178.
[88] Hansen, P.L., Wagner, J.B., Helveg, S., et al. (2002). *Science* 295: 2053.
[89] Zander, S., Kunkes, E.L., Schuster, M.E., et al. (2013). *Angewandte Chemie International Edition* 52 (25): 6536.
[90] Behrens, M., Zander, S., Kurr, P., et al. (2013). *Journal of the American Chemical Society* 135 (16): 6061.
[91] Behrens, M., Kissner, S., Girgsdies, F., et al. (2011). *Chemical Communications* 47 (6): 1701.

[92] Bems, B., Schur, M., Dassenoy, A., et al. (2003). *Chemistry - A European Journal* 9 (9): 2039.

[93] Schumann, J., Lunkenbein, T., Tarasov, A., et al. (2014). *ChemCatChem* 6 (10): 2889.

[94] Kuld, S., Conradsen, C., Moses, P.G., et al. (2014). *Angewandte Chemie International Edition* 126 (23): 6051.

[95] Li, M.M.-J., Zeng, Z., Liao, F., et al. (2016). *Journal of Catalysis* 343: 157.

[96] Lunkenbein, T., Schumann, J., Behrens, M., et al. (2015). *Angewandte Chemie International Edition* 54 (15): 4544.

[97] Senanayake, S.D., Ramírez, P.J., Waluyo, I., et al. (2016). *Journal of Physical Chemistry C* 120 (3): 1778.

[98] Kattel, S., Ramirez, P.J., Chen, J.G., et al. (2017). *Science* 355: 1296.

[99] Martinez-suarez, L., Frenzel, J., and Marx, D. (2014). *Physical Chemistry Chemical Physics* 16: 26119.

[100] Martínez-Suárez, L., Siemer, N., Frenzel, J., et al. (2015). *ACS Catalysis* 5 (7): 4201.

[101] Larmier, K., Liao, W.-C., Tada, S., et al. (2017). *Angewandte Chemie International Edition* 129: 2358.

[102] d'Alnoncourt, R.N., Xia, X., Strunk, J., et al. (2006). *Physical Chemistry Chemical Physics* 8: 1525.

[103] Fujitani, T., Nakamura, I., Uchijima, T., et al. (1997). *Surface Science* 383: 285.

[104] Reichenbach, T., Mondal, K., Jäger, M., et al. (2018). *Journal of Catalysis* 360: 168.

[105] Behrens, M. (2014). *Angewandte Chemie International Edition* 53: 12022.

[106] Farina, G.L. (1992). *Hydrocarbon Processing* 71 (3): 77.

[107] Rostrup-Nielsen, J.R., Sehested, J., and Nørskov, J.K. (2002). *Advances in Catalysis* 47: 65.

[108] Muradov, N. and Vezirolu, T. (2005). *International Journal of Hydrogen Energy* 30 (3): 225.

[109] Higman, C. and Van Der Burgt, M. (2011). *Gasification*. Burlington: Elsevier Science.

[110] https://www.portofrotterdam.com/de/geschaeftsmoeglichkeiten/hafen-der-zukunft/energiewende/kreislaufwirtschaft/waste-to-chemicals (last accessed October 2018).

[111] https://enerkem.com/facilities/projects-and-partnerships/rotterdam-en.

[112] Pieterman, M. (2001). *The historical development of innovative and energy efficient synthesis gas production technology*, Report.

[113] Asinger, F. (1986). *Methanol – Chemie- Und Energierohstoff* (ed F. Asinger, S.B. Heidelberg, and H. Berlin), 1.

[114] Olah, G.A. (2005). *Angewandte Chemie International Edition* 44 (18): 2636.

[115] Olah, G.A., Prakash, G.K.S., and Goeppert, A. (2011). *Journal of the American Chemical Society* 133 (33): 12881.

[116] Centi, G. and Perathoner, S. (2011). *Greenhouse Gases: Science and Technology* 1 (1): 21.

[117] Li, S.-L. and Xu, Q. (2013). *Energy Environ. Sci.* 6 (6): 1656.

[118] Zhao, D., Yuan, D., and Zhou, H.-C. (2008). *Energy Environ. Sci.* 1 (2): 222.

[119] Zhen, X. and Wang, Y. (2015). *Renewable and Sustainable Energy Reviews* 52: 477.

[120] Arcoumanis, C., Bae, C., Crookes, R., et al. (2008). *Fuel* 87 (7): 1014.

[121] Stöcker, M. (1999). *Microporous and Mesoporous Materials* 29 (1-2): 3.
[122] Lanzafame, P., Centi, G., and Perathoner, S. (2014). *Chemical Society Reviews* 43 (22): 7562.
[123] Klerk, A.D. (2011). *Energy Environ. Sci.* 4 (4): 1177.
[124] Bond, J.Q., Upadhye, A.A., Olcay, H., et al. (2014). *Energy Environ. Sci.* 7 (4): 1500.
[125] Rye, L., Blakey, S., and Wilson, C.W. (2010). *Energy Environ. Sci.* 3 (1): 17.
[126] Dorner, R.W., Hardy, D.R., Williams, F.W., et al. (2010). *Energy Environ. Sci.* 3 (7): 884.
[127] Baliban, R.C., Elia, J.A., and Floudas, C.A. (2013). *Energy Environ. Sci.* 6 (1): 267.
[128] Nicodème, T., Berchem, T., Jacquet, N., et al. (2018). *Renewable and Sustainable Energy Reviews* 88: 151.
[129] Yarulina, I., Chowdhury, A.D., Meirer, F., et al. (2018). *Nature Catalysis* 1 (6): 398.
[130] Olsbye, U., Svelle, S., Bjørgen, M., et al. (2012). *Angewandte Chemie International Edition* 51 (24): 5810.
[131] Bertau, M., Offermanns, H., Menges, G., et al. (2010). *Chemie Ingenieur Technik* 82 (12): 2055.
[132] European Parliament. (2009). *Directive 2009/28/EC of the European Parliament and of the Council of 23 April 2009 on the promotion of the use of energy from renewable sources and amending and subsequently repealing Directives 2001/77/EC and 2003/30/EC (Text with EEA relevance): RED.*
[133] Ochoa, J.V. and Cavani, F., eds., *CHAPTER 8. Gas-Phase Oxidation of Alcohols: Innovation in Industrial Technologies and Recent Developments.* DOI:10.1039/9781782621652-00203 Corpus ID: 101996967
[134] Bui, M., Adjiman, C.S., Bardow, A., et al. (2018). *Energy & Enviromental Science* 11: 1062.
[135] Bhattacharyya, D. and Miller, D.C. (2017). *Current Opinion in Chemical Engineering* 17: 78.
[136] Koytsoumpa, E.I., Bergins, C., and Kakaras, E. (2018). *The Journal of Supercritical Fluids* 132: 3.
[137] Aresta, M. and Dibenedetto, A. (2007). *Dalton Transactions* 28: 2975. DOI: 10.1039/b700658f.
[138] Ridjan, I., Vad Mathiesen, B., Connolly, D., et al. (2013). *Energy* 57: 76.
[139] Qin, Z., Zhai, G., Wu, X., et al. (2016). *Energy Conversion and Management* 124: 168.
[140] http://dechema.de/Low_carbon_chemical_industry.html.
[141] Arakawa, H., Dubois, J.-L., and Sayama, K. (1992). *Energy Conversion and Management* 33 (5–8): 521.
[142] Fujiwara, M., Ando, H., Tanaka, M., et al. (1994). *BCSJ* 67 (2): 546.
[143] Ushikoshi, K., Mori, K., Kubota, T., et al. (2000). *Appl. Organometal. Chem.* 14 (12): 819.
[144] Lee, J.S., Moon, K.I., Lee, S.H., et al. (1995). *Catalysis Letters* 34 (1-2): 93.
[145] Joo, O.-S., Jung, K.-D., Moon, I., et al. (1999). *Ind. Eng. Chem. Res.* 38 (5): 1808.
[146] Goehna, H. and Koenig, P. (1994). *Chemtech* 24 (6): 36.
[147] Saito, M., Takeuchi, M., Watanabe, T., et al. (1997). *Energy Conversion and Management* 38: 403–408.
[148] Pontzen, F., Liebner, W., Gronemann, V., et al. (2011). *Catalysis Today* 171 (1): 242.
[149] Centi, G. and Perathoner, S. (2009). *Catalysis Today* 148 (3-4): 191.

[150] Doss, B., Ramos, C., and Atkins, S. (2009). *Energy Fuels* 23 (9): 4647.

[151] https://www.thyssenkrupp-industrial-solutions.com/media/download_1/methanol_plants_eng.pdf.

[152] Lee, J.S., Lee, K.-Y., Lee, S.Y., and Kim, Y.G. (1993). *Journal of Catalysis* 144 (2): 414.

[153] Rahimpour, M.R. (2008). *Fuel Processing Technology* 89 (5): 556.

[154] Goehna, H. and Koenig, P. (1994b). *Verfahren zur Erzeugung von Methanol* (DE Patent 000004416425A1).

[155] Saito, M., Takeuchi, M., Fujitani, T., et al. (2000). *Appl. Organometal. Chem.* 14 (12): 763.

[156] Zhang, Y., Fei, J., Yu, Y., et al. (2006). *Energy Conversion and Management* 47 (18–19): 3360.

[157] Schlögl, R. (ed.). (2012). *Chemical Energy Storage*. Berlin: De Gruyter.

[158] Joo, O.-S. (2000). In *Energy and Fuels* (ed American Chemical Society), 686–689.

[159] Mousa, E.A., Ahmed, H.M., Viswanathan, N.N., and Larsson, M. (2016). In *Ironmaking and Steelmaking Processes: Greenhouse Emissions, Control, and Reduction* (ed P. Cavaliere), 173. Springer International Publishing.

[160] Cavaliere, P. and Silvello, A. (2016). In *Ironmaking and Steelmaking Processes: Greenhouse Emissions, Control, and Reduction* (ed P. Cavaliere), 151. Springer International Publishing, Cham.

[161] https://www.sintef.no/globalassets/project/tccs-9/presentasjoner/d5/7%9720170614-fresme-h2020-project%97dissemination.pdf.

[162] http://www.steelanol.eu/en.

[163] https://www.carbon2value.be/en.

[164] Oles, M., Lüke, W., Kleinschmidt, R., et al. (2018). *Chemie Ingenieur Technik* 90 (1-2): 169.

[165] Iosif, A.-M., Hanrot, F., Birat, J.-P., et al. (2010). *International Journal of Life Cycle Assessment* 15 (3): 304.

[166] Harp, G., Tran, K.-C., Bergins, C., et al. (2015). *Proceedings of METEC* 2.

[167] Lundgren, J., Ekbom, T., Hulteberg, C., et al. (2013). *Applied Energy* 112: 431.

[168] Bukhtiyarova, M., Lunkenbein, T., Kähler, K., et al. (2017). *Catalysis Letters* 147 (2): 416.

[169] Bermúdez, J.M., Arenillas, A., Luque, R., et al. (2013a). *Fuel Processing Technology* 110: 150.

[170] Joseck, F., Wang, M., and Wu, Y. (2008). *International Journal of Hydrogen Energy* 33 (4): 1445.

[171] Bermúdez, J.M., Ferrera-Lorenzo, N., Luque, S., et al. (2013b). *Fuel Processing Technology* 115: 215.

[172] Guo, J., Hou, Z., Gao, J., et al. (2008). *Energy Fuels* 22 (3): 1444.

[173] Shen, J., Wang, Z.-Z., Yang, H.-W., et al. (2007). *Energy Fuels* 21 (6): 3588.

[174] Bermúdez, J.M., Fidalgo, B., Arenillas, A., et al. (2010). *Fuel* 89 (10): 2897.

[175] Bermúdez, J.M., Arenillas, A., and Menéndez, J.A. (2012). *Chemical Engineering Science* 82: 95.

[176] Varone, A. and Ferrari, M. (2015). *Renewable and Sustainable Energy Reviews* 45: 207.

[177] Nestler, F., Krueger, M., Full, J., et al. (2018). *Chemie Ingenieur Technik* 90 (10): 1409.

[178] https://www.bmwi.de/Redaktion/DE/Downloads/E/energiekonzept-2010.pdf?__blob=publicationFile&v=3.

[179] https://www.spire2030.eu/what/walking-the-spire-roadmap/spire-vision.

[180] http://www.mefco2.eu/pdf/mefco2-teaser.pdf.

[181] http://www.circulary.eu/project/carbon2chem.

[182] Ingolfsson, O., Jonsson, F.R., Shulenberger, A., et al. (2007). *Process for producing liquid fuel from carbon dioxide and water* (WO Patent 002007108014A1).

[183] http://www.carbonrecycling.is/news/2018/11/1/cri-co2-to-methanol-plant-erected-in-germany-ck6nx.

[184] http://www.carbonrecycling.is.

[185] https://www.sunfire.de/de.

[186] http://www.biomcn.eu.

[187] Norman, G., Sandau, F., Zimmermann, B., et al. ed. (2014). *Geschäftsmodell Energiewende: Eine Antwort auf das "Die-Kosten-der-Energiewende" - Argument.*

[188] https://www.thyssenkrupp-uhde-chlorine-engineers.com/en/products/water-electrolysis-hydrogen-production/power-to-gas.

[189] European Commission (ed.). (2015). *Quality of Petrol and Diesel Fuels: Directive 2015/1513/EC.*

[190] European Commission (ed.). (2009). *Promotion of the Use of Energy from Renewable: Directive 2009/28/EC.*

[191] https://www.bloomberg.com/news/articles/2016-08-19/solar-sells-in-chile-for-cheapest-ever-at-half-the-price-of-coal.

23

Hydrogenation Energy Recovery – Small Molecule Liquid Organic Hydrogen Carriers and Catalytic Dehydrogenation

Jong-Hoo Choi,[1] Dominic van der Waals,[1] Thomas Zell,[2] Robert Langer,[3,4] and Martin H. G. Prechtl[1,5]

[1] *Universität zu Köln, Department für Chemie, Greinstraße 6, D-50939 Köln, Germany*
[2] *Adama Makhteshim Ltd., PO Box 60 Industrial Zone, Beer Sheva, 8410001, Israel.*
[3] *Philipps-Universität Marburg, Hans-Meerwein Straße, D-35043 Marburg, Germany*
[4] *Martin-Luther-Universität Halle, Kurt-Mothes Straße 2, D-06120 Halle, Germany*
[5] *Instituto Superior Técnico, University of Lisbon, Av. Rovisco Pais 1, 1049-001 Lisboa, Portugal*

23.1 Introduction

23.1.1 The Arguments for LOHCs

It is apparent that a method for the safe, efficient, and inexpensive storage of hydrogen (H_2) is required to fulfil the needs of a hydrogen economy in cases when the hydrogen which is produced cannot be used directly on the production site. This problem can vary from very large scale applications, such as protecting against the intermittence of wholesale energy supply and demand throughout national networks, to smaller storage requirements, e.g. H_2 storage solutions of Hydrogen-Fuel-Cell Vehicles (HFCVs) [1]. For potential mobile applications, the volume-efficient storage of H_2 by compression is not favored as it requires very high pressure or very low temperature [2]. Under these circumstances, the use of heavy storage tanks might be a risk to safe transportation. [2] In this context, the concept of chemical hydrogen storage, which implies that the hydrogen is bound in a chemical compound, with improved safety, is a promising approach for mobile applications. In combination with an on board catalytic system, hydrogen can be evolved from the chemical storage material on demand.

Notably, throughout the literature there is no unified definition of the term LOHC. In this work, the term is used for all H_2 carrier compounds, which are liquid in the hydrogenated state, but not necessarily in their dehydrogenated form. This includes compounds, which are in the dehydrogenated state gaseous under standard conditions (e.g. carbon dioxide).

According to the US Office of Energy Efficiency and Renewable Energy, certain benchmarks for efficient mobile hydrogen storages systems are set including chemical and physical properties such as the available H_2 content, decomposition temperature, regenerability, and gas purity. Potential compounds should have a H_2 content of at least 5.5 wt.%, a low decomposition temperature in the presence of a catalyst, fast H_2 release capability as well as a fast recharge process [3]. Six compound classes are listed which could potentially comply with at least one of these conditions along with their temperature range in which a literature-known catalyst can operate effectively to generate hydrogen from the substrates:

Advances in Energy Storage: Latest Developments from R&D to the Market, First Edition.
Edited by Andreas Hauer.
© 2022 John Wiley & Sons, Ltd. Published 2022 by John Wiley & Sons, Ltd.

(i) Amino boranes ($\leq$ 19.5 wt%(H_2); H_2-release at > 130°C)
(ii) Hydrocarbons (6.1 – 7.2 wt%(H_2); H_2-release at 250 – 300°C),
(iii) Metal hydrides (< 8.4 wt%(H_2); H_2-release at 25°C (hydrolytic) and 200°C (thermolytic))
(iv) Formic acid (4.4 wt.%(H_2); H_2-release at ambient temperature),
(v) Methanol (denoted hereon as MeOH) (12.6 wt.%(H_2); H_2-release at 65 – 95°C)

and

(vi) Water (11.0 wt.%(H_2); H_2-release at 1000°C or via electrolysis).

Amine boranes (i) have high H_2 content and are highly reactive towards organometallic catalysts in hydrogen generating reactions and therefore seem very promising candidates as H_2 storage compounds. Unfortunately, the regeneration of the dehydrogenated species is highly problematic. For hydrocarbons (ii) and metal hydrides (iii), the regeneration of the dehydrogenated materials also remains a major drawback; additionally, the relatively high temperatures for catalytic H_2 release reactions are required. Water (vi) is an excellent, clean hydrogen source, but it is ruled out for mobile applications, due to excessive energy demand required to split the water molecule [2, 4]. Looking at the remaining substrates (iv) and (v), formic acid and MeOH are interesting candidates due to their low catalytic decomposition temperature. These substrates are rather inexpensive, readily available in bulk amounts, and have fair theoretical hydrogen storage capacities (wt.%(H_2)) [5–7]. Taking these points into account, the use of all of these substrates carry advantages and disadvantages. In addition to those substrates, compounds such as ethanolamine or ethanol are also highly interesting for catalytic hydrogen generation. These compounds can be dehydrogenated easily, and their regeneration via hydrogenation is more practical than for the C_1–based molecules. Molecules with multiple -OH groups (e.g. polyols) have also been considered as high potential LOHCs. Throughout this chapter, the potential use of C_1 (and related) LOHCs, such as MeOH, formic acid, formaldehyde/methane diol, ethanol. etc. will be discussed.

23.1.2 An Overview of High Potential C_1 Molecules as LOHCs

In terms of the dehydrogenation of LOHCs in general and single-carbon (C_1) molecules in particular, they present many beneficial physical and economic properties. One of the significant challenges impeding the adoption of these substrates is the ability to release the high levels of "*chemically stored*" hydrogen in an efficient manner, using inexpensive available catalysts, importantly without concomitant production of potential fuel cell catalyst poisons (e.g. carbon monoxide (CO)). Recent investigations by numerous research groups have led to steady and important progress in this field. The use of C_1 molecules as hydrogen fuel sources include many economic, environmental, and practical benefits; Table 23.1 presents the key substrates that have so far been investigated and will be covered in this work. Furthermore, we will also provide a somewhat simplified summary of the key strengths and weaknesses associated with the use of each one of the substrates.

It is worth highlighting the ready availability of all of these C_1 substrates all of which are produced industrially on a mega tonne scale. MeOH, one of the most important

Table 23.1 Potential small liquid organic hydrogen fuel sources – C_1 molecules.

C1 substrate	wt.% H_2	Benefits	Weaknesses	Notes
MeOH (Aqueous)	12.6 (12.1)	High hydrogen content, high volumetric density, inexpensive & readily available, generally safe to transport.	Toxic particularly to humans. First dehydrogenation step is endergonic.	Depending on conditions dehydrogenation can result in byproducts which potentially limit hydrogen production or viable applicability.
Formaldehyde	6.7	Polymeric *para*-formaldehyde (or also the cyclic trioxane) offers an increased potential density. Formaldehyde reforming is exergonic.	Hazardous in gaseous form. Potential for poisoning of fuel cells.	For hydrogen storage applications, formaldehyde is generally considered as a precursor for methanediol.
Methanediol	8.4	High gravimetric wt.%(H_2) with many of the same storage and application benefits as MeOH.	Dilution with water required to avoid paraformaldehyde formation.	Formed readily in equilibrium through the addition of water to formaldehyde.
Formic acid	4.4	Can produce clean gas for fuel cell usage upon catalytic decomposition. Inexpensive and readily available.	Lowest gravimetric wt.%(H_2) of the C_1 class.	

platform chemicals (ca. 110 Mton per annum), is produced from synthesis gas (a mixture of CO, CO_2 and H_2), itself available from both fossil fuels (e.g. steam methane reforming) and biomass (e.g. via gasification). One major industrial application of MeOH is in the production of formaldehyde for use in plastics manufacture (e.g. via the FORMOX® process). Formic acid which in addition to having numerous biological roles, can be produced from MeOH via methyl formate formation and subsequent hydrolysis.

23.1.3 The Ideal Concept Based on C_1 Substrates

An idealized concept for a hydrogen storage system based on C1 substrates (MeOH, formaldehyde/methanediol, and formic acid) is described in Figure 23.1. This schematic drawing is closely related to the so-called MeOH economy model [5] In this context, the hydrogenation of CO_2, for example to formic acid, is the key step of the hydrogen storage cycle, (Figure 23.1). In accordance with a sustainable and economic method, the production of hydrogen itself plays a major role as well. On the global scale, the vast majority of hydrogen however is still produced through steam reforming from hydrocarbon sources (e.g. steam methane reforming), [8], although

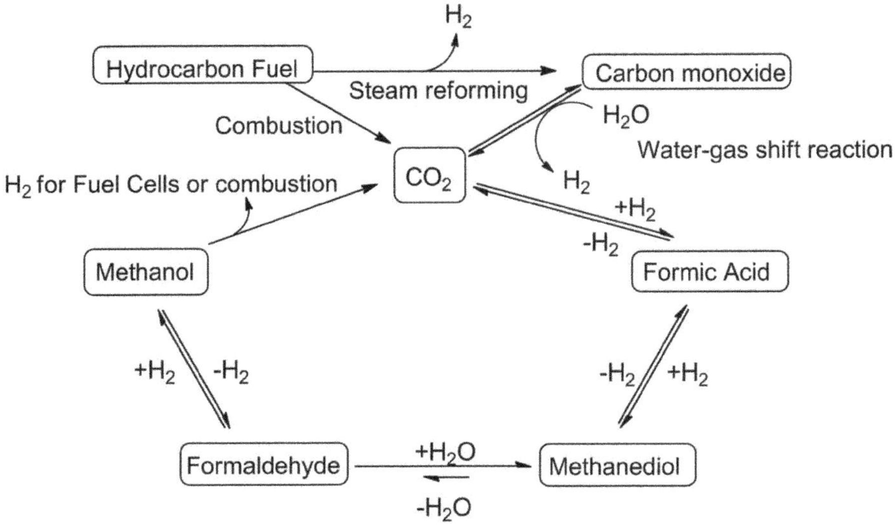

Figure 23.1 C_1 substrates with potential hydrogen storage molecules.

this method has an significant environmental impact. Despite the relatively high costs of hydrogen production, recent future-orientated developments have been made. The production of hydrogen can be carried out on a large scale in both renewable and carbon-neutral processes, for example based on renewable energy powered electrolysis of water (as discussed elsewhere in this book) or the production of hydrogen from biomass gasification [9]. The utilization of carbon monoxide to produce formaldehyde and MeOH on an industrial scale is ruled out in this idealized scenario, as carbon monoxide is typically obtained from fossil fuels. In turn, the great challenge of hydrogen storage involving C_1 substrates relies on the technical ability to effectively and sustainably utilize carbon dioxide (CO_2) resulting in a carbon neutral cycle.

Whereas the reductive transformation of CO_2 to other C_1 hydrogen fuel sources is commonly conducted over heterogeneous catalysts, homogeneous catalyst-based approaches to form these substrates have recently received significant interest [10–13]. The reduction of CO_2 to formates and formic acid, the first step of the processes, has been well explored; high turnover numbers (TONs) for the production of formic acid have been reported using catalysts based on metals such as iridium [14] and ruthenium [15]. It has also been shown that the addition of MeOH allows for the production of further equivalents of MeOH via the formation and reduction of methyl formate [16]. Similar methodologies applying the formation of a reactive intermediate such as esters and carbamates from CO_2 with subsequent reduction to generate MeOH have been applied by the groups of [17] and [18].

This chapter focuses on recent progress in the field of homogeneously catalyzed hydrogen production from the C_1 molecules mentioned in Table 23.1 as well as other small organic hydrogen storage materials (e.g. ethanol, amino-alcohols, and glycerol). Particular attention will be given to the seminal advances made, potential methodologies, technologies opened up, and the future outlook for the individual C_1 sources as well as the field of H_2 generation from small liquid organics as a whole.

23.2 Methanol (CH_3OH)

23.2.1 Homogeneous Catalytic Dehydrogenation

MeOH can be considered one of the most promising LOHCs. The significant benefit for using MeOH pertains to its high hydrogen density, low price, and ready availability. In the means of mobility, MeOH (b.p.: 64.7°C) can be stored relatively safely compared to pure hydrogen-gas [5]. Production capacity of MeOH worldwide is currently at a 110 Mtonne per annum scale; [19] (http://www.methanol.org/the-methanol-industry) whilst the vast majority of this capacity is related to MeOH production via synthesis gas (e.g. based on steam methane reforming), there is a growing interest for MeOH production from renewable sources such as synthesis gas produced from biomass and waste sources. Furthermore, methane as a feedstock for methanol production is an abundant natural-gas, which does not cause deforestation and monoculture plantations like for the production of ethanol [20]. In a CO_2-neutral pathway, carbon dioxide can additionally be converted into synthesis gas (by water-gas-shift chemistry), and subsequently into MeOH, using hydrogen produced from the electrolysis of water using renewable energy sources creating a carbon neutral route to MeOH production [5, 21]. The direct hydrogenation of CO_2 to MeOH, already a highly relevant topic in academia, [15, 21] is also receiving increasing commercial interest (e.g. Carbon Recycling International, Iceland).

Whilst the gravimetric weight of hydrogen in MeOH is high compared to other hydrogen storage materials; the quantity of H_2 released during dehydrogenation is dependent on the mechanism of dehydrogenation. This mechanism depends, in turn, on the reagents, reaction conditions, and the dehydrogenation catalyst employed. Different methods to release hydrogen from MeOH are listed below along with the maximum theoretical hydrogen storage capacity. The maximum theoretical hydrogen storage capacity always relates to the stoichiometry of the hydrogen generating reaction and neglects additional compounds and additives which contribute to the overall weight of the system. This includes additives such as solvents, bases, (pre) and (co) catalysts, and also the weight of the reaction vessel of the reactor. For aqueous methanol dehydrogenation, for example, the 12.1 wt.%(H_2) are calculated by the net reaction of one equiv. MeOH with one equiv. H_2O to result in the formation of three equiv. H_2 and one equiv. of CO_2.

1. MeOH dehydrogenation generating methyl formate (6.3 wt.%(H_2))
2. Aqueous phase MeOH reforming (12.1 wt.%(H_2))
3. Chemo enzymatic MeOH reforming (8.4 wt.%(H_2))

Figure 23.2 shows a general scheme for catalytic MeOH dehydrogenation. Seminal work that has contributed to the study of each of the routes as well as the related advantages and disadvantages of each pathway will be discussed in the following sections. Route 1 and 2 are dehydrogenation pathways, which are established in homogeneous catalysis, while Route 3 is still an explorative pathway in academia combining bio-chemistry with organometallic chemistry.

23.2.1.1 Methanol Dehydrogenation Generating Methyl Formate (Route 1)

The production of hydrogen from MeOH has been known for many years; upon addition of an appropriate dehydrogenation catalyst to MeOH the formation of formaldehyde can occur with the concomitant release of one equivalent of molecular hydrogen

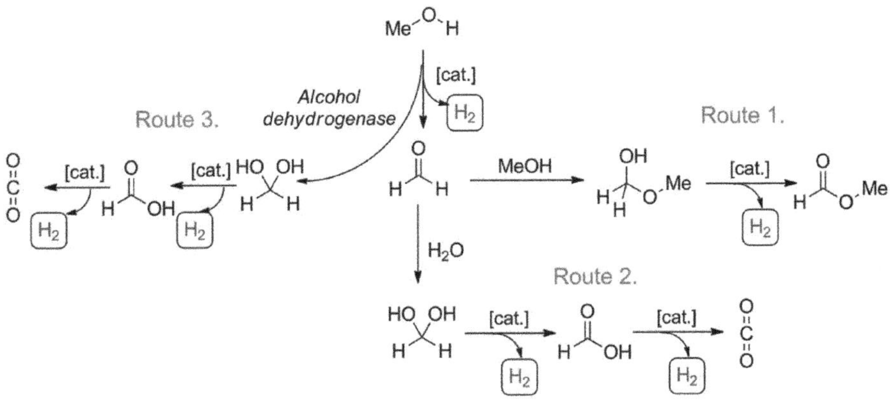

Figure 23.2 Potential routes for methanol dehydrogenation (for simple depiction and clarity equilibrium arrows have not been shown).

from the catalyst per MeOH (Figure 23.3). Once the electrophilic formaldehyde has been produced it is able to undergo nucleophilic addition of an additional molecule of MeOH, reversibly generating a hemiacetal. Once formed, the hemiacetal can undergo either further nucleophilic attack generating dimethoxymethane or, alternatively, undergo catalytic dehydrogenation, releasing another equivalent of hydrogen to form methyl formate.

Clearly, these routes do not release the maximum amount of hydrogen available from MeOH. In practice, the hydrogen storage content is commonly in a range of 3.2–6.3 wt.%(H_2) depending on the product ratio of dimethoxymethane and methyl formate formed.

One of the earliest reported homogeneous catalytic methodologies for the dehydrogenation of MeOH in this way was published in 1985 [22]. In this study, a ruthenium catalyst was used for the production of methyl formate from MeOH concomitant with the release of hydrogen to reduce ketones such as cyclohexanone or methyl vinyl ketone to their corresponding alcohols. This work was based on previous studies, which employed ruthenium, rhodium, osmium, and iridium catalysts for the transfer of hydrogen from MeOH to acceptor substrates [23]. Different catalytic activities were observed for the different catalysts. It was reported that a ruthenium phosphine precatalyst [$Ru(PPh_3)_3Cl_2$] was the most effective for the release of hydrogen from MeOH with a TON of 65 (with respect to H_2 production), in a typical reaction set-up of

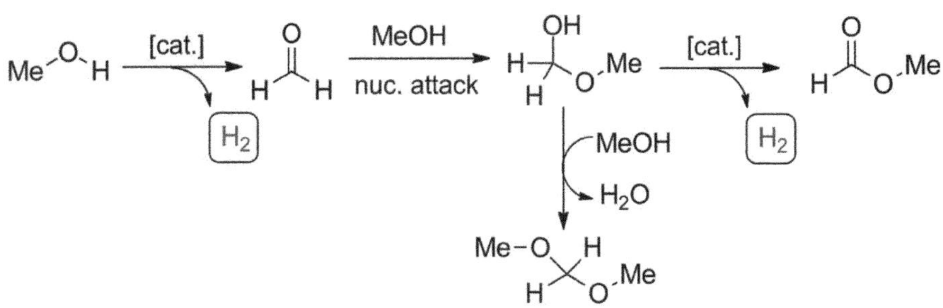

Figure 23.3 Mechanism for dehydrogenative ester and acetal formation from methanol (for simple depiction and clarity equilibrium arrows have not been shown).

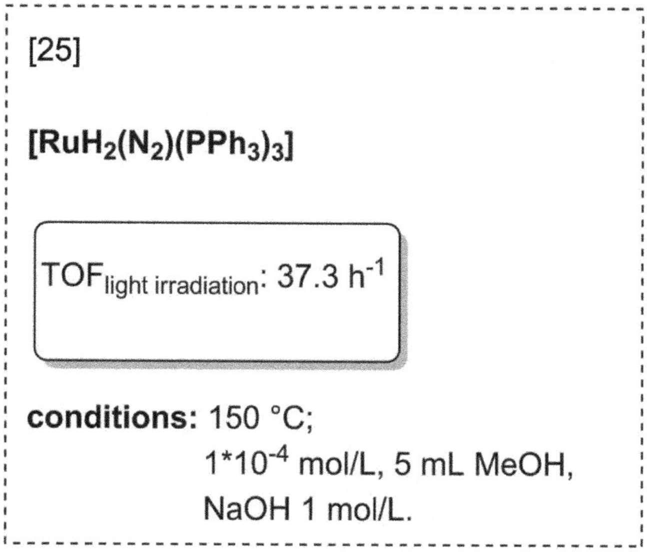

Figure 23.4 Selected catalytic systems for the decomposition of methanol to hydrogen [25].

250 mmol MeOH, 0.08 mmol catalyst at 150°C over 18 h under argon, whereby a mixture of major products was obtained, including methane, carbon dioxide or carbon monoxide. Importantly, only trace levels of other gaseous products were reported. Similarly, ruthenium-hydride phosphine catalysts [$RuH_2(PPh_3)_4$], along with a selected number of rhodium examples, were employed by [24] for the formation of a range of ester products from alcoholic substrates with concomitant hydrogen release [24]. The possibility of the activation of ruthenium hydride precatalysts by irradiation has been illustrated in early work by Cole-Hamilton, which showed that the reaction rate of hydrogen release from alcohols including MeOH could be increased by up to a factor of 10 with the irradiation of a [$RuH_2(N_2)(PPh_3)_3$] precatalyst containing reaction mixture (Figures 23.4 and 23.5) [25]. Under the reaction conditions, 150°C, $1*10^{-4}$

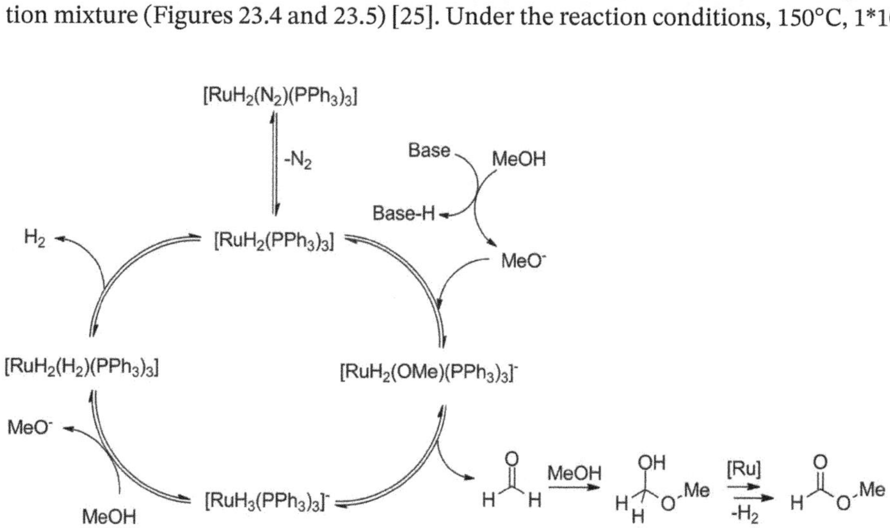

Figure 23.5 Mechanism for ruthenium hydride catalysed dehydrogenation of methanol under basic conditions.

mol/L catalyst concentration, 5 mL system, effective TOF of 6.4 h^{-1} were reached. Under irradiation a photochemical acceleration of the ruthenium catalyzed reaction took place resulting in a TOF of 37.3 h^{-1}. For other alcohols such as EtOH, PrOH, or BuOH TOF over 200 h^{-1} were reached and the highest TOFs was observed for ethane-1,2-diol to be over 1000 h^{-1}. As a drawback in this system, high loadings of base are required (1 mol/L) to generate alkoxide anions to be coordinated to the ruthenium catalyst.

Mechanistic studies of the reaction were conducted using *in vitro* [26] and *in silico* methodologies [27] with the generalized proposed mechanism depicted in Figure 23.5 showing a typical ruthenium catalyzed hydrogenation/dehydrogenation network in presence of base. More recently, [28], [29], and [30] have further developed this approach with respect to optimized reaction conditions and expansion to hydrogen evolution from a wider variety of alcohols.

23.2.1.2 Aqueous Phase Dehydrogenation of MeOH (Route 2)

The liberation of hydrogen from methanol in the presence of water is one of the most promising approaches for methanol dehydrogenation. This is primarily due to the high gravimetric density of hydrogen which can be released. In this reaction, formally complete oxidation of the methanol carbon atom to carbon dioxide takes place, with concomitant release of three equivalents of hydrogen for each equivalent of MeOH in presence of water. The aqueous dehydrogenation of MeOH mechanism has the benefit of producing highly pure hydrogen with respect to the absence of potential fuel cell poisons (e.g. carbon monoxide). The generalized mechanism is based on the initial dehydrogenation of MeOH to yield formaldehyde as shown (Figure 23.6); this electrophilic molecule is then hydrated to generate methanediol which can subsequently be dehydrogenated to formic acid and eventually to carbon dioxide.

An early report by Cole-Hamilton regarding the production of hydrogen from aqueous alcohols using a rhodium catalyst is noteworthy, although the equivalents of hydrogen produced are not as high as in state-of-the-art systems nowadays, and the extension of the methodology towards MeOH dehydrogenation was not possible [26]. One of the recent highlights on the development of the catalytic homogeneous aqueous MeOH dehydrogenation was reported in 2013 [31–34]. In study [34] the authors described the first organometallic catalyst, defined at the molecular scale, for the dehydrogenation of MeOH. This catalyst was found to yield more than one equivalent of hydrogen per MeOH. The system described makes use of aliphatic PNP pincer ruthenium complexes and is based on previous work in the field. Particularly for the ruthenium catalyst **A** (Figure 23.7) used at mild temperatures between 65–95°C the addition of an excess of base (KOH) to a defined ratio of MeOH to water is necessary for its high activity to dehydrogenate MeOH completely to CO_2 without any undesired by-products such as the catalyst poison CO. Despite the high base content of 8.0 M KOH, with

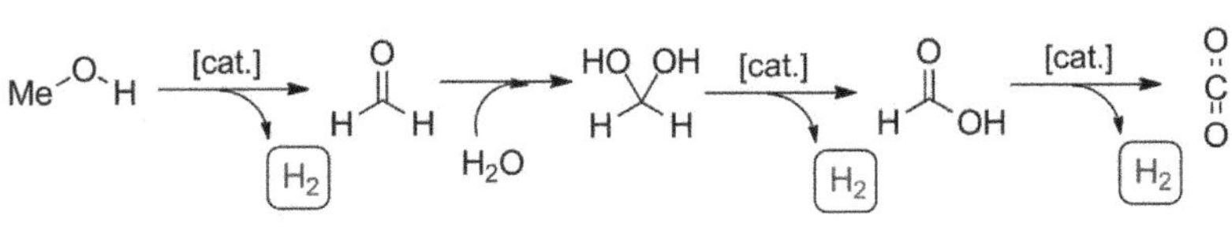

Figure 23.6 Mechanism for the aqueous dehydrogenation of methanol.

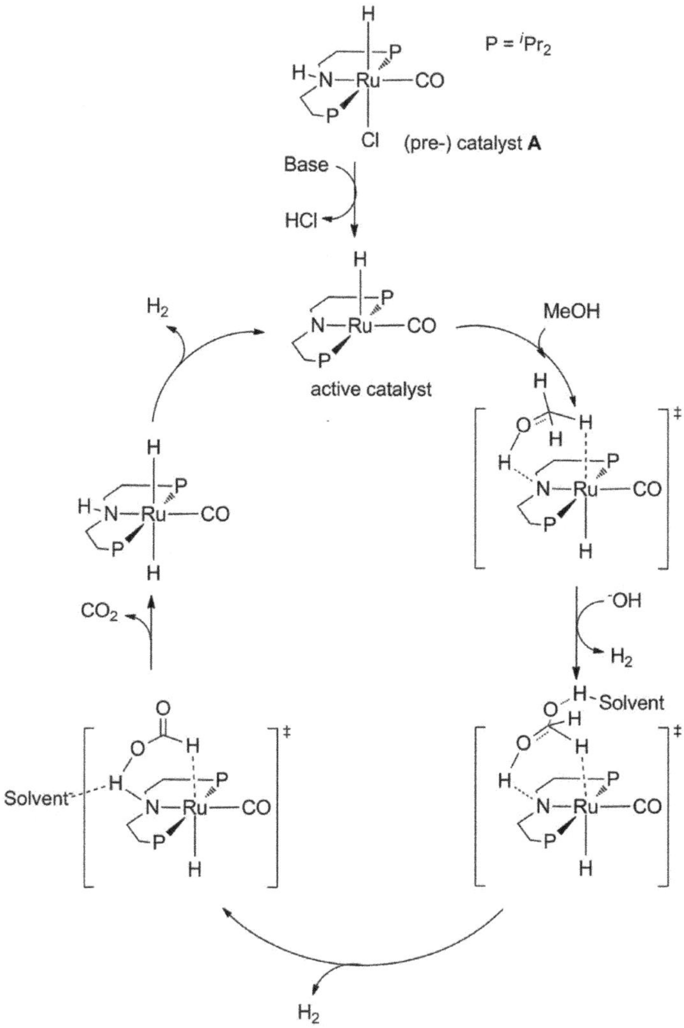

Figure 23.7 Proposed catalytic cycle for methanol dehydrogenation in basic aqueous medium using an aliphatic PNP ruthenium pincer complex [33].

a maximum observed TON of > 350 000 after 24 days "*online*" and up to a TOF of 4700 h^{-1} and yields up to 59%, this method (10–40 mL MeOH/H_2O, 9:1) is currently the most effective system in homogeneous catalysis. Their studies have shown that an excess of base accelerates the reaction, supporting the generation of anionic intermediates and by-products such as formaldehyde, formic acid, and CO_2 (Figure 23.7). The reaction undergoes a typical pathway of heterolytic activation of methanol in the presence of pincer complex providing a cooperative ligand backbone which acts as internal base, hence proton acceptor (nitrogen), followed by subsequent β-hydride elimination of the coordinated substrate. In the course of this reaction three equivalents of hydrogen can be extruded in addition to CO_2 formation.

In terms of pincer ligands, the ruthenium PNN complex (catalyst **B**, Figure 23.8) developed by [35], which is one of the most active dehydrogenation catalysts, is also

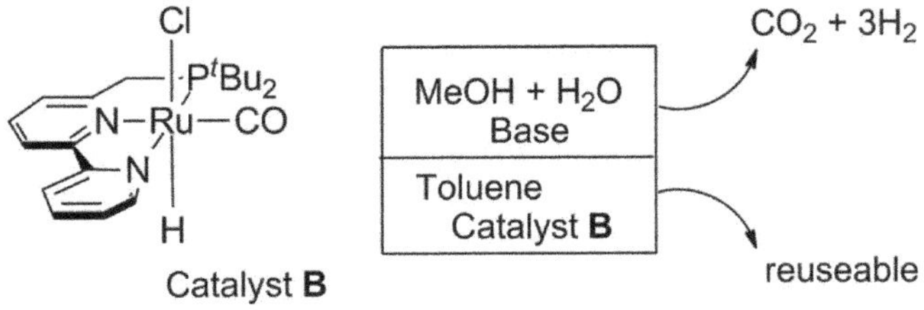

Figure 23.8 Catalyst B for the dehydrogenation of methanol in a two-phase catalytic system [35].

noteworthy in this context [35]. In this system, the catalyst dissolved in toluene serves as a "*stationary*" solvent, while MeOH, water as well as base can be added. A H_2 yield of 77%, average TOF of 43 h^{-1} after 9 days was reached at 100°C with a catalyst loading of 0.025 mol% in 2 mL toluene, 2 mL water and 20 mmol MeOH and two equivalents of KOH as base. Without any significant decrease of the catalytic activity, within 27 days, a total TON of around 29 000 was reached. Despite the decrease of the gravimetric capacity of methanol in the system with the involvement of a stationary organic solvent, the catalyst shows to be very robust as the catalyst containing solvent can be reused and even stored for several months (Figure 23.8) [35].

This methodology of base-supported MeOH reforming in aqueous medium was further expanded in early 2013.[36]. The authors report the use of a ruthenium complex (catalyst **C**, Figure 23.9) employing a non-innocent chelating trop-diazadiene ligand system (trop$_2$dad = 1,4-bis(5H-dibenzo[a,d]cyclohepten-5-yl)-1,4-diazabuta-1,3-diene; Figure 23.9) to cleanly release the maximum volume of hydrogen from a mixture of 2 mmol MeOH and 2 mmol of NEt_3 in 1 mL water, and 0.5 mL THF, in which the catalyst was dissolved before. For the initial conversion of MeOH to CO_2, a

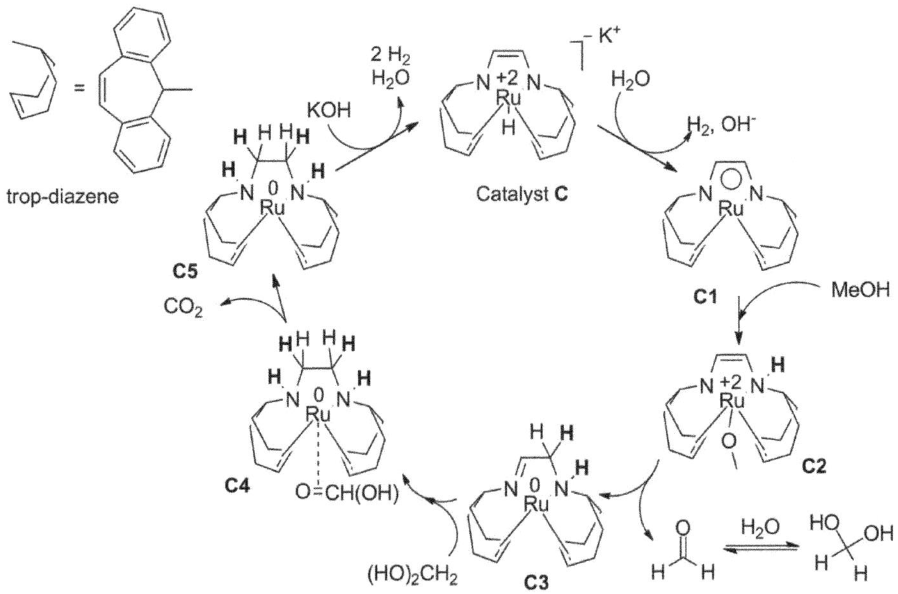

Figure 23.9 Mechanism for the catalytic aqueous dehydrogenation of methanol by catalyst C [36].

TOF of 16 h^{-1} was reached over 10 h, with 0.5 mol% catalyst loading at 90°C. A simplified proposed catalytic mechanism for the dehydrogenation highlighting the active role of the ligand system is provided in Figure 23.8 [36]. The key-steps in this catalyst system utilize the cooperative ligand system, which temporarily stores the molecular hydrogen, while the oxidation state of the ruthenium metal switches between 0 and + II in the following step, formic acid is dehydrogenated by the catalyst with the evolution of CO_2 and H_2. While the direct conversion of MeOH to CO_2 is limited to low TOFs, the authors reported the direct conversion of formic acid (1M in dioxane) leading to 24 000 h^{-1} with 0.01 mol% catalyst loading at 90°C within several minutes.

Both of these mechanisms require the use of a base for catalyst activation, as well as to bind the carbon dioxide, which is produced. The use of a [Ru-MACHO-BH] catalytic system in coordination with a [$Ru(H)_2(dppe)_2$] catalyst has recently been reported. This system does not need additional base which has many practical and environmental benefits. The system works by promoting the dehydrogenation of formic acid, an intermediate in the dehydrogenation of MeOH which can deactivate the initial catalyst [37].

The use of other transition metals as alternatives to ruthenium, was also investigated and a bi-catalytic aqueous dehydrogenation of MeOH, based on iridium PNP catalysts and an iron tetraphosphine catalyst were presented. Depression of gas production was seen for the iridium catalyst whilst no H_2 was seen in the iron complex. Progress has recently been made to expand the scope of metal complexes which can be applied for aqueous MeOH dehydrogenation by a substantial amount of research groups. [38] recently reported on the use of iridium N-heterocyclic carbene (NHC) catalysts for the dehydrogenation of MeOH in the presence of hydroxide and water nucleophiles. Very low levels of carbon monoxide gas (< 8 ppm), were produced during the dehydrogenation; a key benefit for potential fuel cell applications.

The use of an ionic iridium systems for the dehydrogenation of alcohols have also been described [39] which could be extended to aqueous systems [40]. Optimization of the system led to the possibility of the MeOH dehydrogenation using an iridium catalyst, which activity is pH dependant [41]. For this reaction, a mechanism was proposed by the authors (Figure 23.10). The catalyst **D** can be employed at various pH-ranges under mild temperatures in H_2O/MeOH mixtures (1:4). This offers the advantage, that the high loadings of base, which are typical for the reported systems described earlier, can be reduced to 0.046 mmol*L^{-1} NaOH. In a set-up of 20 mmol MeOH, 80 mmol H_2O, only 0.1 mmol NaOH and 0.02 mmol catalyst were needed to obtain 64% of H_2. Under these conditions a TON of 10 510 was reached after 150 h (TOF of 70 h^{-1}).

Initially the methanol reacts in a ligand exchange reaction with the hydroxo ligand under formation of the corresponding methoxy complex (D1), this is followed by a β-hydride elimination resulting in a hydride complex (D2). In the next step the carbonyl of the cooperative ligand acts as proton acceptor, resulting in the neutral complex (D3). This complex rapidly loses molecular hydrogen by recombination of the protic and hydridic hydrogen stored in the complex.Whilst the catalytic systems discussed so far have many positive advantages with regards to catalytic durability and high efficiency; the desire to transition away from the use of platinum group metals (PGMs) is of growing concern across catalysis due to both rising cost of raw materials and long term supply limitations. For this reason, the use of earth abundant metals to replace or complement PGMs has been researched. In this case,

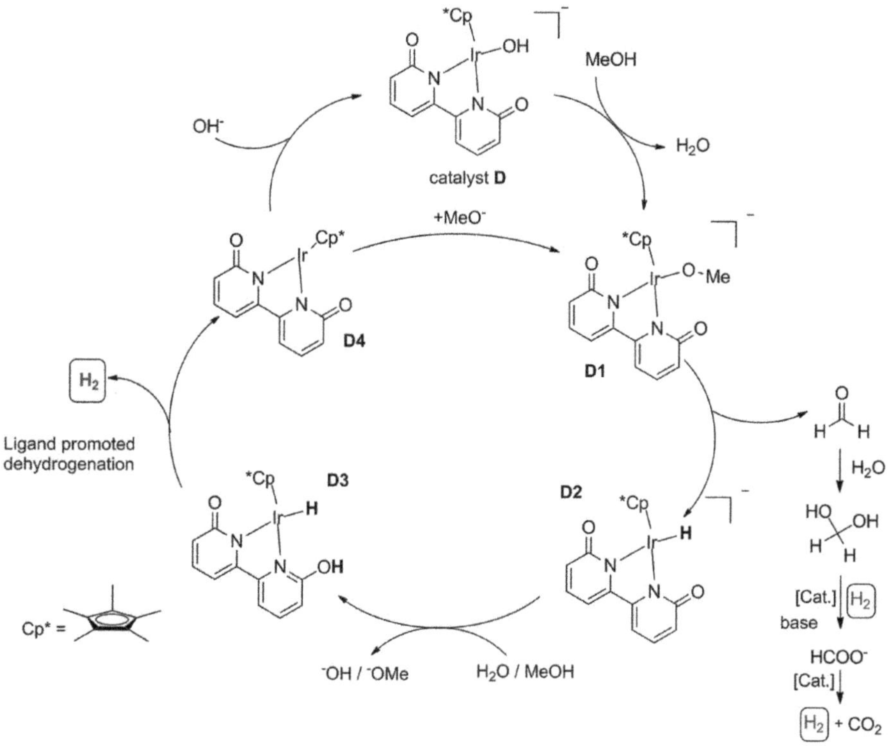

Figure 23.10 Ligand promoted iridium catalysed aqueous dehydrogenation of methanol [41].

an iron PNP pincer complex system (catalyst **E**, Figure 23.11) has been reported by [42]. Notably, this catalyst is similar ruthenium PNP complex reported in their earlier work, and able to dehydrogenate MeOH with TON in the range of 10 000 after 46 h. [42]. In a different approach by [43], an aliphatic iron PNP catalyst (catalyst **F**, Figure 23.11) was employed along with a Lewis Acid, such as $LiBF_4$, as a supporting co-catalyst and ethyl acetate as solvent, giving a TON of 51 000 within 94 h. [43]. Importantly, it was shown that the production of hydrogen was almost entirely CO free under mild conditions, based on this earth abundant metal catalyst.

23.2.1.3 Chemoenzymatic Reforming of Methanol (Route 3)

Organometallic catalysts have been effective tools for MeOH dehydrogenation. However, Route 3, Figure 23.2, describes a different approach which combines two systems for MeOH dehydrogenation, namely the combination of an enzyme system with an organometallic system. As a proof of principle, the focus of this section is directed towards the theoretical possibility rather than at high TONs or TOFs.

This concept mimics nature's way of converting MeOH as performed by enzymes [44–45]. Throughout evolution, microorganisms have found various ways to survive harsh surroundings (e.g. high temperature or pressure, sunlight or oxygen deficient conditions, etc.). In this context, some organisms have developed

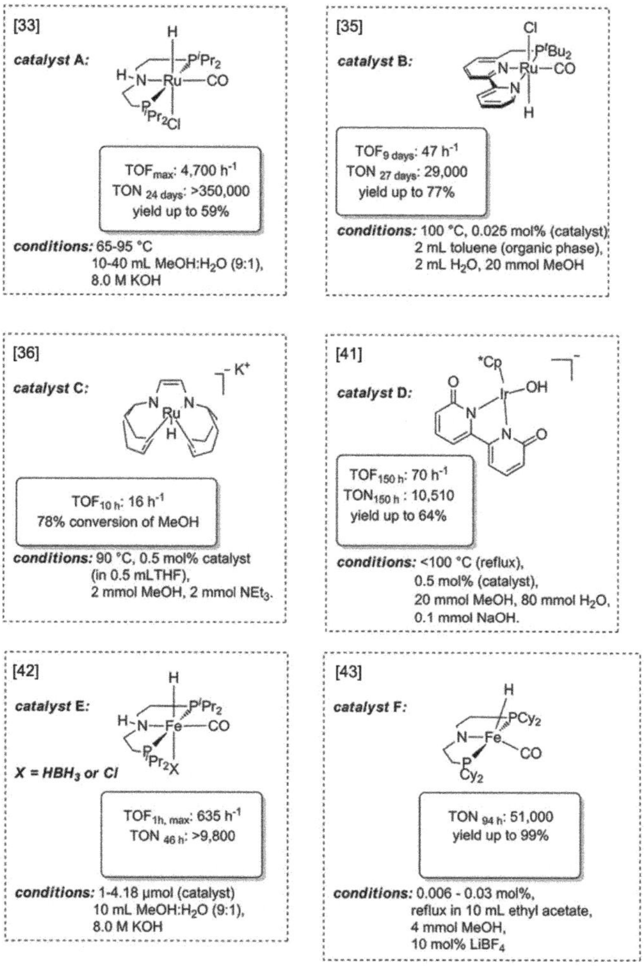

Figure 23.11 Selected catalytic systems for the dehydrogenation of methanol to hydrogen and carbon dioxide.

metabolic pathways based on methane or MeOH and their conversion via well-defined low temperature pathways [46–47]. It is noteworthy that the cell metabolism process found in microorganisms, such as in aerobic methylotrophic yeasts and bacteria, rely on formaldehyde formed through oxidase/catalase-catalysed formal oxygenation of MeOH. The generated formaldehyde hydrate is at the same time a source of reduction equivalents, such as NAD(P)H, and a feedstock for the formation of carbohydrates [48]. Followed by the dehydrogenation of formaldehyde via nucleophilic attack by co-factors such as glutathione, S-(hydroxymethyl)-glutathione (GSH) is formed [49]. The hemithioacetal GSH reacts with the GSH dependent dehydrogenase and forces a hydride transfer turning NAD^+ to NADH and forming S-formylglutathione. S-formylglutathione can also undergo a hydrolysis to form formic acid [50–51].

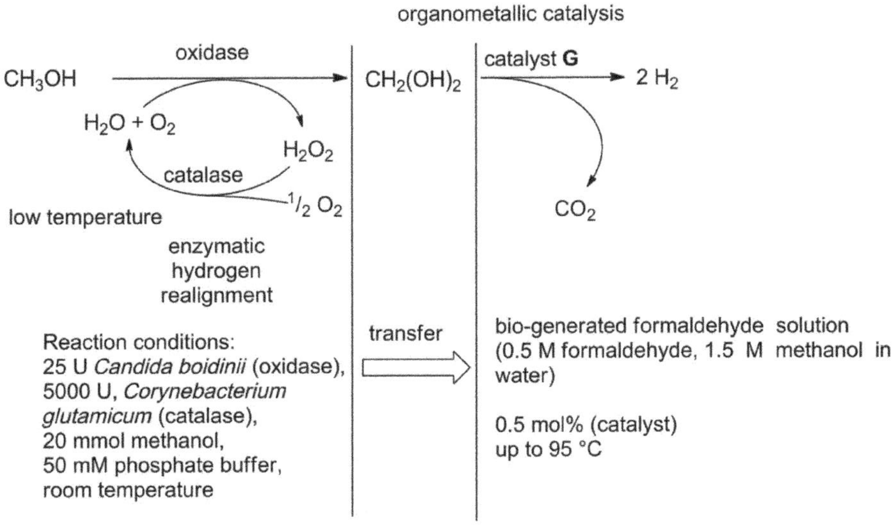

Figure 23.12 Concept of combining enzymatic and organometallic catalytic systems employing the ruthenium arene complex catalyst [Ru(p-cymene)Cl_2] G.

Inspired by nature, the concept of combining nature's way to transform MeOH with an organometallic catalyst, is able to effectively dehydrogenate formaldehyde/methanediol to hydrogen and carbon dioxide (Figure 23.12).

Seminal work has been conducted in 2015 by [44–45]. In these studies, an artificial set-up was built, which allows identified enzymes to transform MeOH to formaldehyde, respectively methanediol in aqueous medium, with a following dehydrogenation step by an air- and water-stable dimeric ruthenium arene complex at low temperatures. Among the enzymes, *Canadida boidinii* proved to be the most promising candidate, showing high activity as well as robustness by evolving methanediol concentrations up to 0.53 M after 24 h. While the employed ruthenium catalyst lacks the ability to dehydrogenate MeOH at lower temperatures, they have been reported to be active to decompose methanediol even at ambient temperature [52]. The catalyst **G**, a dimeric ruthenium arene complex for the dehydrogenation of formaldehyde/methanediol is shown in Figure 23.13. Compared to the organometallic complexes

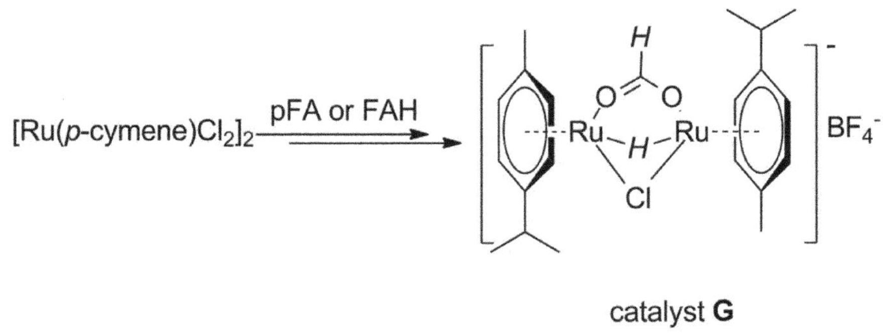

Figure 23.13 Activation of a dimeric ruthenium arene complex with paraformaldehyde (pFA) or formaldehyde hydrate (FAH).

discussed in the previous section, the employed complex system has the advantages of high robustness towards air and water, and does not require further additives (e.g. a base or Lewis acid).

Combining these two disciplines, the first MeOH reforming at room temperature was achieved. Investigating daily examples in Nature for the future, artificial bio set-ups can be optimized, while specific enzymes can be obtained via fermentation. This concept opens up new opportunities for the application of LOHCs in mobile applications, most importantly under low operation temperatures and without the necessity of excess base.

23.3 Formaldehyde/Methanediol (CH_2O/CH_2OHOH)

Following enzyme supported methanol reforming, the potential of aqueous formaldehyde as an LOHC will be discussed. In the current literature, the use of formaldehyde as the primary hydrogen source/carrier seems neglected so far. Formaldehyde is well-known for its broad application as a C1 building block in chemical synthesis, for example in resin materials or disinfectant preparation. With an annual production of ca. 41 Mtonne in 2012, formaldehyde is one of the most important chemicals [3]. While MeOH or formic acid are defined molecules under normal conditions, formaldehyde, if not gaseous, typically forms the solid product paraformaldehyde, based on 8–100 formaldehyde units. However, in aqueous solution (commonly referred to as Formalin), formaldehyde reacts with water to form its aldehyde hydrate, referred to as methanediol (Figure 23.14).

Theoretically, the water molecule is "*absorbed*" by the formaldehyde, as a result, the methanediol intermediate is solely present [53]. As already implied in the aqueous MeOH dehydrogenation, the interaction with water is the key point, which makes formaldehyde, and respectively Paraformaldehyde, highly interesting as a potential chemical hydrogen carrier. While formaldehyde itself has a weight efficiency of 6.7 wt.%(H_2), methanediol has an increased weight efficiency of 8.4 wt.%(H_2), both of which between that of its molecule analogues MeOH (12.6 wt.%(H_2)) and formic acid (4.4 wt.%(H_2)). This subchapter discusses the possible homogeneously catalysed dehydrogenation method of methanediol rather than of formaldehyde.

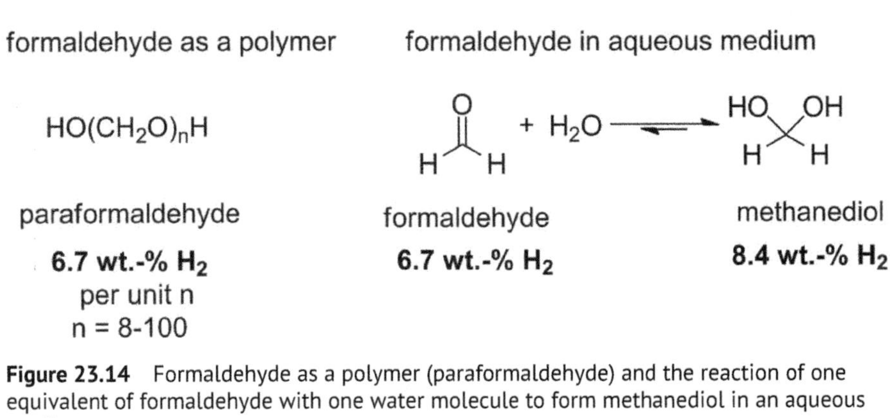

Figure 23.14 Formaldehyde as a polymer (paraformaldehyde) and the reaction of one equivalent of formaldehyde with one water molecule to form methanediol in an aqueous medium.

In the context of this discussion, a simple but a fundamental question arises: *what are the main arguments for using formaldehyde as a H_2-source, when it can readily be generated from MeOH?*

As a matter of fact, in terms of theoretical weight efficiency and accessibility, there are no obvious advantages of using formaldehyde over MeOH, since MeOH has a higher weight efficiency and it serves industrially as the main formaldehyde precursor. The argument for using formaldehyde relies on different aspects:

(i) The interaction with water to methanediol increases the potential H_2 content.
(ii) The aqueous formaldehyde solution is non-flammable.
(iii) Formaldehyde decomposition is exergonic in contrast to methanol decomposition

Therefore, a viable catalytic system can take advantage of the mentioned aspects, provided that a clean decomposition of methanediol can be assured without the production of unwanted by-products (e.g. carbon monoxide). Furthermore, an enzyme supported production of formaldehyde (as discussed earlier) can also be considered advantageous. A desirable pathway to the dehydrogenation of formaldehyde via methanediol in two initial steps is depicted in Figure 23.15. Firstly, the dehydrogenation of formaldehyde or paraformaldyde takes place via methanediol to formic acid proceeds followed by the dehydrogenation of formic acid to carbon dioxide. In regard to the involvement of H_2O in the dehydrogenation steps, an indirect water splitting process has been proposed. The low-temperature chemoenzymatic methanol reforming via aqueous methanediol has the advantage for potential mobile low-temperature devices that high-temperature formaldehyde production from methanol (via the FORMOX-process) is not required. Thus, a less energy-consuming all over reaction can be achieved.

With a theoretical maximum of 8.4 wt.%(H_2) in methanediol, the application of the catalytic system with catalyst **G** used with technical formaldehyde solution (37 wt.%) contains ca. 5%, which is still higher than formic acid, but lower than the desired 8.4 wt.%(H_2). After filing a patent (2013), in 2014 [52] reported for their system that catalyst loadings can be as low as 0.1 mol% and catalysts operates best at 95°C. Notably, also at lower temperatures, decent catalytic activities have been observed. Moreover, this system is not directly dependent on pH, despite correlation to the substrates, and has been found to tolerate a wide pH range (over 2 – 9), giving typical conversion of ca. 75% within 1.5 h. After reaching this conversion the system can simply be recharged with water and paraformaldehyde [52].

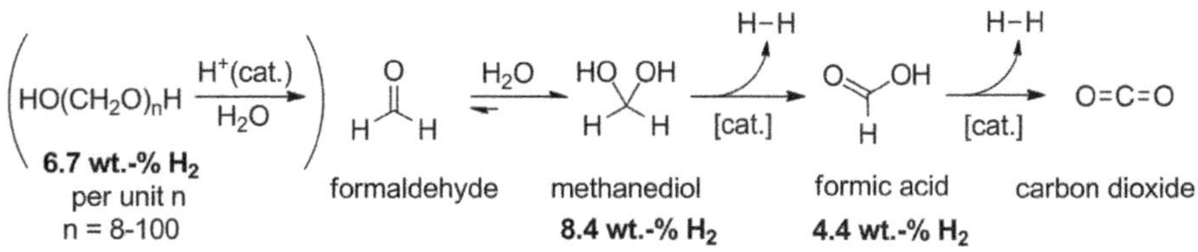

Figure 23.15 Catalytic dehydrogenation of methanediol derived from paraformaldehyde. In two steps methanediol can be dehydrogenated to carbon dioxide with formic acid as intermediate.

23.4 Formic Acid (HCO_2H)

23.4.1 General Aspects, Thermodynamics, and Reversibility

Over the last few decades the dehydrogenation of formic acid has been a focus of interest for many research groups [54–59]. Formic acid has a hydrogen content of 4.4 wt.%(H_2). This may be lower than for other small LOHCs, but for formic acid, reversible hydrogen storage is possible and has been reported, see earlier. Pure formic acid is corrosive, but has a rather low volatility and toxicity as compared to other LOHCs, and is non-flammable under standard conditions. In principle, formic acid can be catalytically decomposed in a dehydrogenation/decarboxylation pathway to generate CO_2 and H_2 or in a dehydration/decarbonylation pathway to produce CO and H_2O (Figure 23.16) [55]. Both reactions are thermodynamically favored under standard conditions, with $\Delta G^0 = -32.9\ kJ\ mol^{-1}$ and $\Delta G^0 = -20.7\ kJ\ mol^{-1}$, respectively.

However, the dehydration pathway poses a problem for hydrogen storage applications employing fuel cells. The presence of small amounts of CO in the H_2/CO_2 gas mixtures has a detrimental effect on fuel cell performance as CO is fuel cell catalyst poison [54]. Minimizing CO content to ppm or even ppb levels is therefore considered crucial for these applications and achieving this goal requires highly selective dehydrogenation catalysts.

A reversible hydrogen storage cycle based on the dehydrogenation of formic acid and the hydrogenation of CO_2 is depicted in Figure 23.17). In this cycle the hydrogen liberation step corresponds to the decomposition of formic acid (or a formic acid derivative, e.g. formic acid–tertiary amine adduct) to H_2 and CO_2 (or CO_2 derivative, e.g. carbon dioxide–tertiary amine adduct). The hydrogen produced in this way can then be directly used in fuel cells or in traditional combustion engines with water being the sole by-product of the energy cycle. CO_2 (or the CO_2 derivative, e.g. carbon dioxide–tertiary amine adduct) can be hydrogenated in a second chemical reaction to reproduce formic acid (or a formic acid derivative, e.g. formic acid–tertiary amine adduct). This reaction can in principle also be conducted off-site. This means that the CO_2 liberated to the atmosphere and/or off-site atmospheric CO_2 can be recycled in formic acid synthesis – *a potentially CO_2-neutral cycle necessitates that the greenhouse gas CO_2 is efficiently recycled.* In other words, for running this system CO_2 neutral, the C_1 source for the formic acid production needs to be CO_2. This principle can be compared to a rechargeable battery. In the "hydrogen storage battery," where hydrogen is the energy

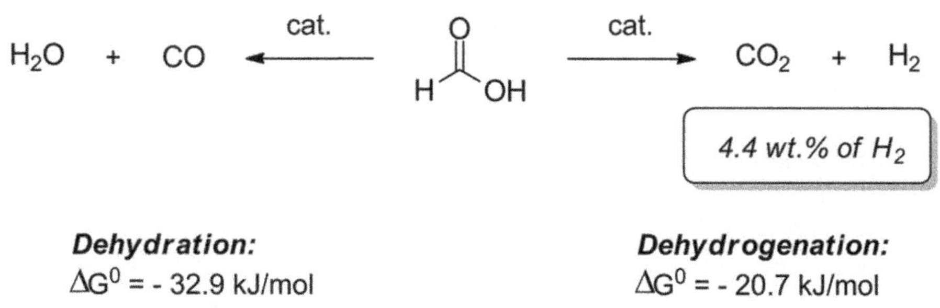

Figure 23.16 Dehydration and dehydrogenation of formic acid [55].

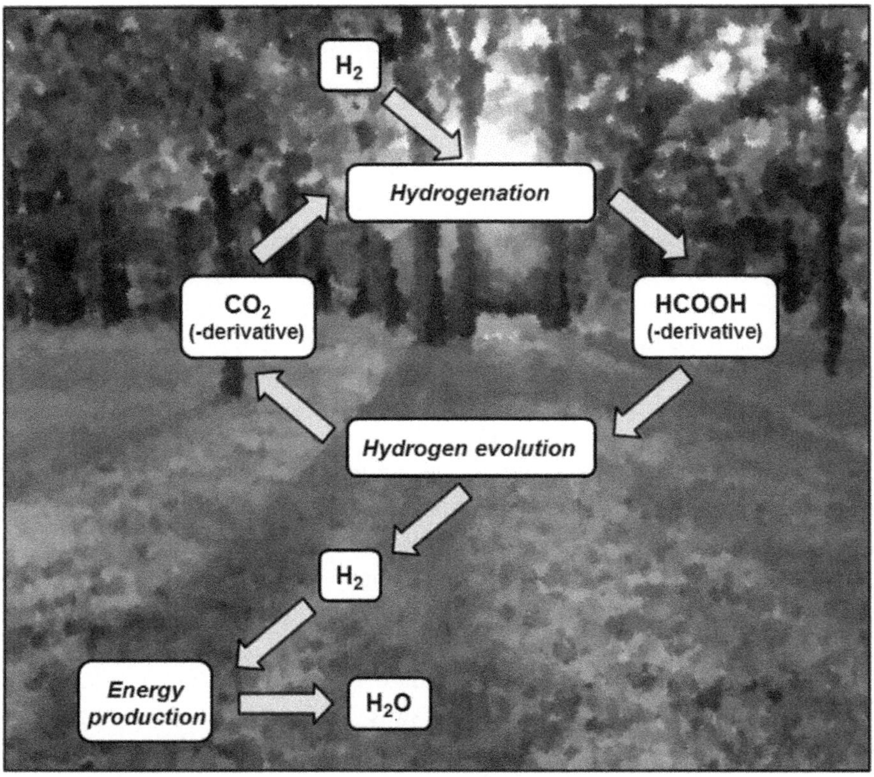

Figure 23.17 A hydrogen storage cycle based on CO_2 hydrogenation and formic acid dehydrogenation.

carrier, the hydrogenation step to obtain formic acid corresponds to the charging and the dehydrogenation step is the equivalent to discharging of the system with energy/hydrogen.

At the industrial scale, ca. 800 000 tonne/year of formic acid are produced currently via the reaction of MeOH with carbon monoxide and a strong base [60]. Formic acid and formic acid-amine mixtures can in principle also be obtained from CO_2 by catalytic hydrogenation [60–66], or by electrochemical [67–68], or by photocatalytic reduction [69]. Notably, the hydrogenation reaction of CO_2 to formic acid is exothermic under standard conditions, but highly entropically unfavored and therefore endergonic ($\Delta G^0 = 33$ KJ mol^{-1}: Figure 23.18). However, homogenous catalytic hydrogenation can indeed be achieved by influencing the thermodynamic parameters (e.g. via the addition of base or through solvation). An attractive approach is the addition of amines. Owing to the strongly exothermic deprotonation reaction of the formic acid with amines yielding ammonium formate, the process turns exergonic. Using this effect, [63] developed a two-phase process for the homogenously catalysed hydrogenation of CO_2 indirectly to formic acid via the corresponding ammonium formate complex [Hex_3NH][HCOO]. This ammonium formate is immiscible in trihexylamine ($NHex_3$) and forms, in an excess of the amine, a second phase consisting of $NHex_3$. In their study, [$Ru(H)_2(PEt_3)_4$] was used as precatalyst and in the process the catalyst is recovered by extraction to the $NHex_3$ phase. After the hydrogenation reaction a $NHex_3$

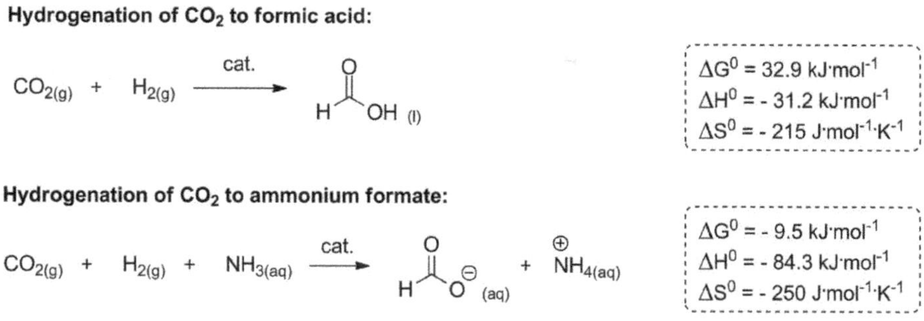

Figure 23.18 Hydrogenation reactions of CO_2 to formic acid and ammonium formate [63].

phase containing the catalyst and a diol phase containing the [Hex_3NH][HCOO] adduct was obtained. The diol/product phase consisting of [Hex_3NH][HCOO] and a diol (e.g. 2-methyl-1,3-propanediole) is separated by phase separation. From this phase formic acid is then obtained by fractional distillation in which the formic acid-amine adduct is thermally split into formic acid and $NHex_3$. Both the diol as well as the $NHex_3$ can be reused for the hydrogenation reaction.

The hydrogenation of CO_2 in water or DMSO is also possible without the addition of any base or buffer [60]. In fact, [60] recently reported on formic acid production under acidic conditions in water and DMSO as solvents. In water, the free enthalpy is due to the equilibria between $CO_2/H_2CO_3/HCO_3^-$, $\Delta G^0 = -4$ kJ mol^{-1} and therefore significantly favored over the hydrogenation in the absence of solvent to give liquid formic acid. Using 2.74 mM [$RuCl_2(PTA)_4$] as precatalyst in DMSO at 50°C and 100 bar of H_2/CO_2 mixture (1:1 partial pressures) up to 1.93 M (ca. 15 vol.%) solution of formic acid in DMSO were obtained after 120 h. In addition, it should also be mentioned that hydrogenation reactions of supercritical CO_2 ($scCO_2$) have been investigated by various research groups [64]. In this context, [70] have presented a continuous process using $scCO_2$ as mobile phase and where both the catalyst and the base are immobilized in one single process unit. The formic acid can then be isolated by decompression of the $scCO_2$. Although this may not be an economical way for the synthesis of formic acid from CO_2, simply because of the high energy costs of generating $scCO_2$, this system shows high catalytic activities. Further work may be directed towards the extraction of formic acid from the immobilized phase in order to minimize the amount of the $scCO_2$ undergoing evaporation.

The use of polymer electrolyte membrane-based direct formic acid fuel cells (DFAFC) is an alternative to the decomposition of formic acid to CO_2 and H_2 [71]. DFAFCs are proton exchange membrane fuel cells, where the formic acid is fed directly into the fuel cell. Fuel cells with air cathodes, which both actively and passively breathe have been developed. In the fuel cells formic acid is converted to carbon dioxide, water, and energy. The oxygen reduction at the cathode (e.g. a supported Pt-based catalyst) is a four-electron process with a potential of $E_0 = 1.23$ V (vs. SHE), and the direct oxidation of formic acid (usually over supported or unsupported Pd- or Pt-based catalyst materials) is a two-electron process with a potential of $E_0 \sim -0.25$ V (vs. SHE). The reactions at the cathode and anode and the overall reaction are provided in Figure 23.19. Major drawbacks of this technology, however, are the intrinsically high

Anode reaction:	$HCOOH \longrightarrow CO_2 + 2\,H^{\oplus} + 2\,e^{\ominus}$	$E^0 = \sim -0.25$ V (vs. SHE)
Cathode reaction:	$0.5\,O_2 + 2\,H^{\oplus} + 2\,e^{\ominus} \longrightarrow H_2O$	$E^0 = 1.23$ V (vs. SHE)
Net reaction:	$HCOOH + 0.5\,O_2 \longrightarrow CO_2 + H_2O$	OCV = ~ 1.48 V

Figure 23.19 Reactions taking place in direct formic acid fuel cells (DFAFC).

sensitivity of the fuel cells to poisoning by CO and the crossover of formic acid through the fuel cell membrane (e.g. leading to deactivation).

23.4.2 Mechanistic Considerations

The catalytic dehydrogenation of formic acid requires the transfer of a proton and a hydride to a metal-based catalyst, in order to facilitate the liberation of H_2 from the substrate. Due to the comparably high acidity of this substrate different mechanistic scenarios are viable. Besides formic acid, other substrates (e.g. alcohols) can serve as hydrogen storage molecules. Different mechanisms, described in the literature for their catalytic dehydrogenation, will be discussed briefly in the following section (Figure 23.20).

For most substrates the catalytic dehydrogenation requires the protonation of a coordinated hydrido ligand in the catalyst (**I**), yielding a dihydrogen-coordinated complex (**II**). In general, those dihydrogen complexes can either be real catalytic intermediates (local minimum on the reaction coordinate) or transition states. Depending on the acidity of the substrate the direct protonation of the metal hydride by the substrate can be facile (e.g. via $\mathbf{TS_f}$); alternatively, a ligand site containing a protic functional group can facilitate this reaction step. In this latter case two different mechanisms are known. On the one hand the hydride can either be directly protonated by transfer of an acidic proton from the acidic ligand site (via $\mathbf{TS_e}$), alternatively the functional group facilitates pre-coordination of the substrate, via hydrogen bonding, followed by protonation of the hydrido ligand (via $\mathbf{TS_d}$). Liberation of H_2 results in the formation of an

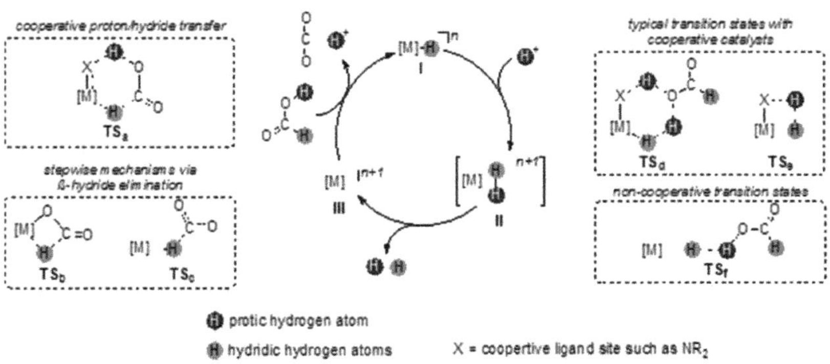

Figure 23.20 Generalized elementary steps in acceptor-less dehydrogenations of formic acid to carbon dioxide.

unsaturated species (**III**) which has a "*vacant*" coordination site. Again, several mechanistic pathways have been observed for the hydride transfer. For cooperative catalysts, which contain a functional group capable of reversibly accepting and releasing protons, the concerted transfer of a proton and a hydride via the transition state $\mathbf{TS_a}$ is possible. For "*non-functionalized*" complexes the most common reaction pathway requires a vacant coordination site in the *cis*-position to the coordinated, deprotonated substrate in order to proceed in a ß-hydride elimination step ($\mathbf{TS_b}$). A rather rare event is the dissociation of the deprotonated substrate, followed by hydride elimination step without precoordination or a vacant coordination site in *cis*-position ($\mathbf{TS_c}$). Alternatively, for formate ligands, a strictly intramolecular ß-hydride elimination process is also known. [72] have found that the ß-hydride elimination from the formate ligand of the iron pincer complex [(tBu-PNP)Fe(H)(CO)(η^1-OOCH)] proceeds through a non-classical intramolecular mechanism. According to DFT calculations the intramolecular pathway is energetically favored in this case by 29 kJ · mol^{-1} for this complex over the dissociation of the ligand.

In general, the dehydrogenation of formic acid does not require a functionalized transition metal catalyst and can be efficiently performed over "simple" transition metal phosphine complexes. As formic acid is rather acidic, the formation of dihydrogen intermediates (**II**) can easily proceed via direct protonation of the metal hydride intermediate **I**. Furthermore, hydride transfer from formates can take place without pre-coordination or a vacant coordination site (via $\mathbf{TS_c}$). Nevertheless, several highly active transition metal catalysts containing chelating cooperative ligands have been reported, which are likely to react via the indicated cooperative pathways ($\mathbf{TS_d}$, $\mathbf{TS_e}$ and $\mathbf{TS_a}$).

23.4.3 Homogeneous Catalysts for Formic Acid Dehydrogenation

Various homogenous [54–59] and heterogeneous [54, 59] catalysts for formic acid dehydrogenation have been described and reviewed comprehensively [54–55]. A summary of some selected homogenous catalysts is provided in Figure 23.21. Interestingly, besides noble metal based catalysts, iron catalysts have also shown very high activities for this dehydrogenation reaction [58, 71–73]. In 2011, the research groups of [74] published work regarding in situ generated iron catalyst **H** for the selective dehydrogenation of formic acid. This catalyst is composed of $[Fe(H_2O)_6](BF_4)_2$ and a four-fold excess of the tetradentate ligand PP_3 (PP_3 = tris[(2-diphenylphosphino)ethyl]phosphine). Compared to other non-noble metal catalysts this catalyst system showed unprecedented catalytic activity and catalyst stability at that time. High Turnover Frequency (TOF_{max} = 9425 h^{-1}) and Turnover Numbers (TON_{max} = 92 417) were obtained in propylene carbonate at 80°C. Notably no further additives, except the excess of ligand, were needed to achieve this catalytic activity. Shortly afterwards in 2013, [72] reported on the application of the molecularly defined PNP iron pincer complex **I** as a selective and efficient formic acid dehydrogenation catalyst. For efficient catalytic performances this catalytic system requires the presence of tertiary amine additives. Catalytic dehydrogenation reactions using **I** were conducted at 40°C, and gave TON_{max} of 100 000 and a TOF_{max} of 836 h^{-1}. In 2014, [75] described the application of a similar PNP iron pincer complex **J** as a catalyst for hydrogen liberation from formic acid. Similarly to complex **I**, a high catalytic activity was obtained in reactions

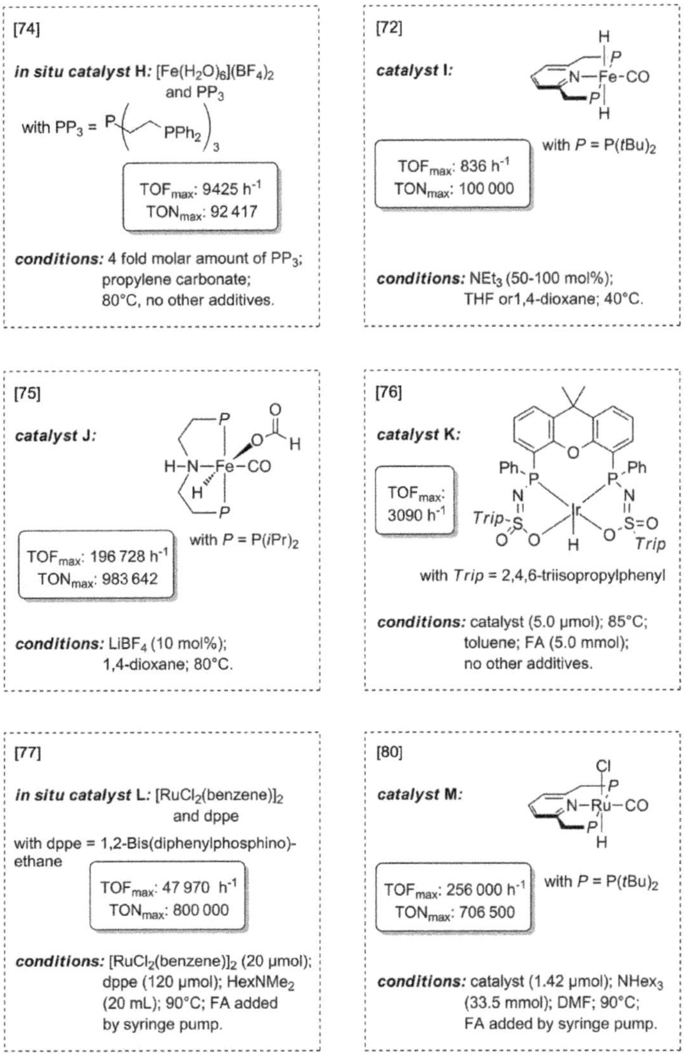

Figure 23.21 Selected examples of homogenous FA dehydrogenation catalysts [72, 74–77, 80].

using triethyl amine (50 mol%) as an additive. Additionally, the authors found not only the presence of triethyl amine is beneficial for the catalysis, but also that Lewis acids facilitate the catalysis. After optimization of the reaction conditions, $TOF_{max} = 197\,728\ h^{-1}$ and $TON_{max} = 983\,642$ were observed in the presence of 10 mol% $LiBF_4$ at 80°C in 1,4-dioxane. Ir catalyst **K** was reported by [76] in 2015. This catalyst is capable of dehydrogenating formic acid at high initial rates. A TOF_{max} for the hydrogen liberation of 3 090 h^{-1} was observed in toluene at 85°C without further additives. Unfortunately, no information was given in their publication on the long term stability of the catalytic system.

In 2012 [77] reported on the efficient dehydrogenation of formic acid catalysed by the *in situ* generated Ru catalyst **L**. This catalyst is composed of the dimeric precursor $[RuCl_2(benzene)]_2$ in combination with a six-fold molar excess of dppe (=

1,2-bis(diphenylphosphino)ethane). When used in $HexNMe_2$ as solvent TOF_{max} = 47 970 h^{-1} and TON_{max} = 800 000 were achieved when formic acid was added by a dosing pump at 90°C. This was at the time the highest TON observed for formic acid dehydrogenation. However, maybe more importantly, this work provided the proof of the feasibility of reversible hydrogen storage using this system. In parallel to the formic acid dehydrogenation the CO_2 hydrogenation by Ru catalysts in the presence of amines was also studied. The authors also demonstrated that such reactions can in fact be used as hydrogen storage systems. Up to 8 consecutive cycles of the hydrogenation/dehydrogenation reactions were successfully conducted using $[RuCl_2(dppm)_2]$ as catalyst in DMF (20 mL) in the presence of NEt_3 (2 mL). Formic acid formation by hydrogenation of carbon dioxide was performed in a commercial autoclave (100 mL volume) using a mixture of 60 bar of a H_2 and CO_2 (30 bar each). The reaction was performed until an overall pressure of ca. 45 bar was reached. Under the given conditions the hydrogenation reached this criterion overnight at room temperature. Dehydrogenation of the formic acid amine adduct was then observed upon releasing the pressure from the system. The amount of gas evolved was measured during 5 h stirring at room temperature. In the first seven cycles, 1620 to 2170 mL of gas mixture was generated in this manner. In comparison, when decomposing under identical conditions 1.67 mL formic acid (44.3 mmol) a gas volume of 2105 mL was obtained. To compensate for the small amounts of NEt_3 evaporating during the gas release 0.1 mL NEt_3 were added to each cycle.

Similar to the work in [77], in 2014 [80] reported the application of the [78, 79] type Ru-PNP pincer catalyst **M** for the reversible hydrogenation of CO_2 in presence of DBU (=1,8-diazabicyclo[5.4.0]undec-7-ene) and subsequent decomposition of the formic acid amine adduct [80]. The highest catalyst stability and activity (TON_{max} = 706 500 and TOF_{max} = 256 000 h^{-1}) for the formic acid decomposition was observed using 1.42 μmol **M** in the presence of 33.5 mmol $NHex_3$ in DMF at 90°C. The hydrogen storage experiments were performed under similar conditions using DBU instead of $NHex_3$. The hydrogenation step was conducted at different overall pressures of 1:1 mixtures of CO_2 and H_2 (5 bar or 40 bar) at 65°C in order to demonstrate the high activity of the system even at low pressure. Acid to amine ratios of 1.6 and 1.1 were obtained for the high and low pressure hydrogenation procedures, respectively. The dehydrogenation was then observed upon releasing the pressure from the system. The hydrogen release reaction was performed at 90°C. Both the hydrogenation (3 h) and the dehydrogenation (1 h) steps proceeded fairly fast under these conditions. Notably, in 2014 [81] also reported on reversible hydrogen storage based on formic acid-DBU adduct decomposition with a Ru complex with a tetradentate PNNP ligand. Subsequent cyclic hydrogenation/dehydrogenation experiments were performed with a catalyst loading of 0.075 mol% and both hydrogenation as well as the dehydrogenation were performed at 100°C. For charging the system, which corresponds to the hydrogenation reaction of CO_2, 70 bar of H_2 was added, together with dry ice. During the heating of the system a pressure of ca. 140 bar was built up. The hydrogenation was then performed until a pressure of ca. 100 bar was reached. Next, the hydrogen release was performed under ambient pressure.

23.5 Other Alcohols, Diols, and Amino Alcohols

Besides hydrogen storage by MeOH, see earlier, other alcohols are, in principle, also suitable for reversible hydrogen storage applications. Homogenous [31–32, 34] and heterogeneous [82] catalytic systems for hydrogen liberation from different alcohols in the absence of any hydrogen acceptors have been developed in the last decades and reviewed comprehensively in the recent literature. Depending on the alcohols used, different dehydrogenated species (aldehydes, ketones, and/or carboxylic acid derivatives) are obtained by the dehydrogenation process. The outcome of the dehydrogenation reaction can be influenced by the choice of catalyst and the reaction conditions.

23.5.1 General Aspects, Thermodynamics, and Reversibility

The dehydrogenation of primary alcohols and derivatives thereof, such as diols and amino alcohols, offers several advantages as a hydrogen storage compound, for example in comparison to formic acid. The utilization of alcohols with low carbon content would in principle allow storing hydrogen with comparably high gravimetric capacities. Table 23.2 summarizes known hydrogen liberation reactions and their corresponding capacities. The highest capacity among this series of 12.1 wt.%(H_2) can be achieved using MeOH/water mixtures (Entry 1), but this reaction suffers from the same problems as formic acid: the reaction products are gaseous and require further treatment to capture the formed CO_2 in order to provide a reversible storage system. Furthermore, for the reversible reaction, namely the hydrogenation of carbon dioxide to MeOH, only a few examples are described in the literature and it remains questionable at the moment, whether this catalytic hydrogenation proceeds in aqueous media. The heavier analogue ethanol exhibits a significantly reduced gravimetric capacity of 4.4 wt.%(H_2) (Entry 2), which is comparable to formic acid. Nonetheless, ethanol

Table 23.2 Hydrogen liberation from small molecule LOHCs and their gravimetric capacities.

Entry	Liberation reaction		Gravimetric capacity/wt.%(H_2)
1	MeOH + H_2O ⟶ CO_2 + 3 H_2		12.1 (12.5)[a]
2	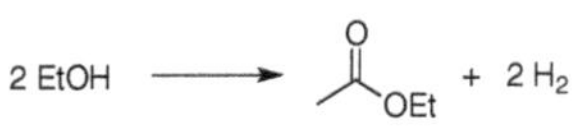		4.38
3	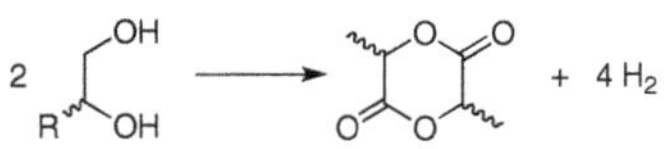	R = H	6.49
		R = CH_3	5.30
4	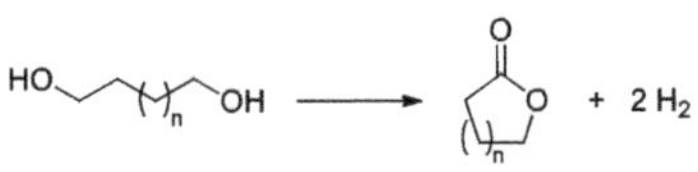	n = 1	5.30
		n = 2	4.48
5	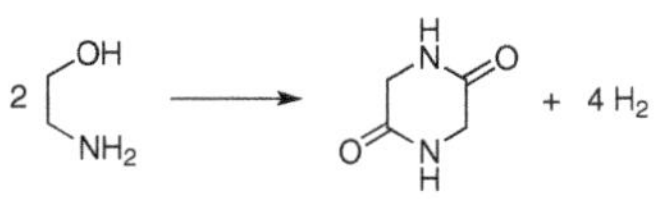		6.60
6	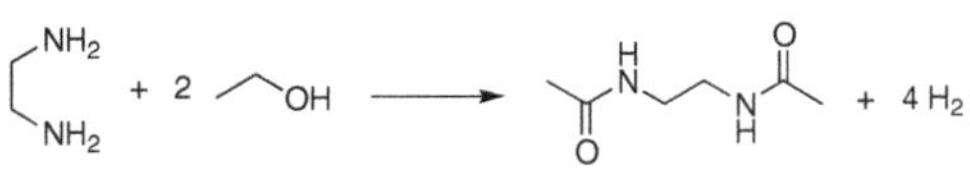		5.30

[a] *Gravimetric capacity based on MeOH/H2O with the capacity based on MeOH in parenthesis.*

dehydrogenation usually yields ethyl acetate, whose hydrogenation can be achieved with certain types of homogeneous catalysts. More attractive hydrogen storage in terms of capacity are diols (Entries 3–4), whose dehydrogenation yields different products depending on the substitution pattern. 1,2-diols, such as ethylene or propylene glycol, are efficiently dehydrogenated to cyclic dilactides (Entry 3), resulting in capacities of up to 6.49 wt.%(H_2). The dehydrogenation of 1,3- and 1,4-diols selectively leads to the formation of lactones (Entry 4), with H_2-capacities of up to 5.3 wt.%(H_2). Importantly, the reverse reaction, the hydrogenation of lactones and lactides, is efficiently catalysed by cooperative catalysts. Using ethanolamine or mixtures of ethanol and diamines as LOHCs has enormous potential for hydrogen storage applications, due to the high capacity and the practical advantage that hydrogen is the only gaseous reaction product. Accordingly, up to 6.60 wt.%(H_2) can be liberated from 2-aminoethanol, yielding a cyclic diglycide as reaction product (Entry 5). Slightly lower capacities can be realized when ethanol is dehydrogenated in the presence of 1,2-diamino-ethane (Entry 6).

Besides the theoretical and practically achievable capacities of suitable LOHCs, the thermodynamics of the corresponding hydrogenation and dehydrogenation reactions must be considered. A major drawback of formic acid as a storage compound is that the reaction for hydrogen storage, namely the hydrogenation of CO_2 to formic acid, requires additives to become thermodynamically favorable. For dehydrogenation, no additive is required, but due to the endothermic nature of these reactions higher temperature and/or removal of the released hydrogen is necessary to drive the reaction to high conversions.

Based on experimentally derived enthalpies ($\Delta_F H°$) and entropies ($\Delta_F S°$) for formation of simple molecules it is possible to obtain thermodynamic data for the dehydrogenation reactions (Figure 23.22).[83] (http://webbook.nist.gov/chemistry) With a reaction enthalpy of 75.9 kJ mol^{-1} the formation of ethyl acetate from ethanol is highly endothermic. Consequently, this entropy-driven reaction becomes thermodynamically favorable only at elevated reaction temperature. Assuming a temperature independence of the thermodynamic functions (which is not the case in reality, but may serve here as simplification) it is possible to calculate the equilibrium temperature ($\Delta_R G = 0$) of 383.9 K. This indicates that the dehydrogenation of ethanol is thermodynamically uphill under the typical reaction conditions (80–90°C external heating temperature). Accordingly, the hydrogen in these reactions has to be removed from the reaction mixture in order to drive the reaction towards the product side (Le Châtelier's principle). This situation changes for diols, which usually have significantly higher boiling points and whose dehydrogenation reactions are expected to have similar reaction enthalpies and entropies.

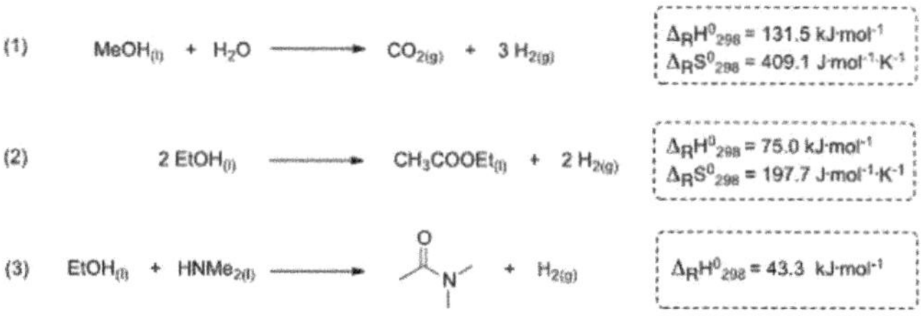

Figure 23.22 Thermodynamic data for selected reactions [83] (http://webbook.nist.gov/chemistry).

The dehydrogenative coupling reactions of primary alcohols with amines are also highly attractive as potential hydrogen storage systems (Table 23.2). As experimental data for the thermodynamics of these reactions are rare, the hypothetical formation of acetamide from ethanol and dimethylamine is used here to evaluate the thermodynamics of those reactions. A closer look at the available data reveals that the amide forming dehydrogenation reaction seems to be much less endothermic than the previously discussed ester forming reaction.

For the realization of a practical and viable hydrogen storage system the gravimetric storage capacity has to be maximized, highlighting a very important boundary condition for hydrogen storage based on LOHCs and homogeneous catalysts: the reactions have to work under neat conditions in the absence of an additional solvent and/or other additives in substantial amounts.

23.5.2 Mechanistic Considerations

In general, the same mechanistic scenarios as for formic acid (ß-hydride elimination and hydrogen liberation) are applicable for alcohol dehydrogenation as well. A striking difference between alcohols and formic acid is the lower acidity of the former, which usually is not sufficient for a direct protonation of hydride intermediates. Accordingly, to overcome this limitation acid co-catalysts were used in early examples of these catalytic reactions. More recently cooperative transition metal complexes, which facilitate the formation of dihydrogen intermediates by ligand-assisted proton transfer, were successfully applied as catalysts. Another mechanistic difference arises from the fact that the alcohol dehydrogenation reactions, discussed earlier, include two dehydrogenation steps (Figure 23.23) [78]. The first step is the formation of an aldehyde **I** that is assumed to be in equilibrium with the corresponding hemiacetal **II** in the case of an ester-forming reaction. In the case of an amide-forming reaction, the aldehyde intermediate reacts with an amine to generate the corresponding hemiaminal **IV**. A second equivalent of hydrogen is eliminated from the hemiacetal **II**, yielding the corresponding ester **III**. For some catalysts it has been demonstrated that pathways including a metal mediated *Tshchenko*-type oxidation of the aldehyde **I** are viable

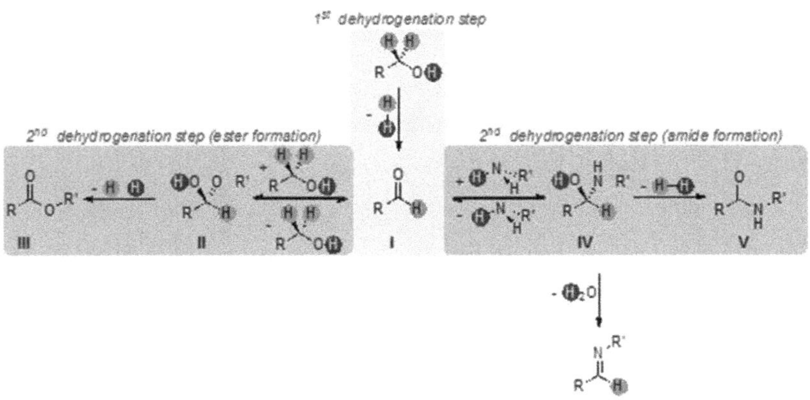

Figure 23.23 Reaction pathways in the dehydrogenative coupling of alcohols, including possible side reactions.

[84]. In the presence of amines the formation of hemiaminals **IV** is assumed, from which a second equivalent of hydrogen is eliminated to give the corresponding amide **V**. It is important to note that for the attractive amide formation reaction the dehydration of the hemiaminal to result in the formation of imines is a competing side reaction. The selectivity in those reactions strongly depends on the nature of the catalyst and on the reaction conditions. [79].

23.5.3 Homogeneous Catalysts for Hydrogen Liberation from Alcohols, Diols, and Amino Alcohols

In 1975, [85] reported on the catalytic dehydrogenation of various alcohols catalysed by the Ru complex [$Ru(OCOCF_3)(CO)(PPh_3)_2$] **A** (Figure 23.24). The dehydrogenation reactions were conducted in neat alcohol under reflux in the presence of the catalyst (4.5×10^{-3} molar) and an excess of trifluoroacetic acid. It was found that the catalyst is most active when 12 equivalents of trifluoroacetic acid with respect to **N** are added. Under these conditions primary and secondary alcohols were dehydrogenated to give aldehydes and ketones, respectively. The highest TOF of 2400 h^{-1} was observed for benzyl alcohol, and notably MeOH did not react under these conditions. A mechanism based on displacement of one of the trifluoroacetate ligands by the corresponding alcohol under formation of trifluoroacetic acid and the corresponding alkoxide complex **O** was proposed (Figure 23.24). **O** undergoes rapid β-hydride elimination to give the corresponding aldehyde and hydride complex **P**. (Pre)catalyst **N** is then recovered by the reaction of **P** with one equivalent of trifluoroacetic acid under liberation of H_2. In terms of a potential application as hydrogen storage reaction ethanol has with 4.34 wt.%(H_2), the highest hydrogen content from the examples described in this study.

Other catalysts for the selective acceptor-less dehydrogenation of alcohols to aldehydes or ketones were later reported (see selected examples in Figure 23.25). [86] demonstrated in 1982 that substitution of the triphenylphosphine ligands by the chelating ligand dppe (=1,2-Bis(diphenylphosphino)ethane) in precatalyst **N** results in an enhanced catalytic activity for the dehydrogenation of cyclohexanol. Similar to the

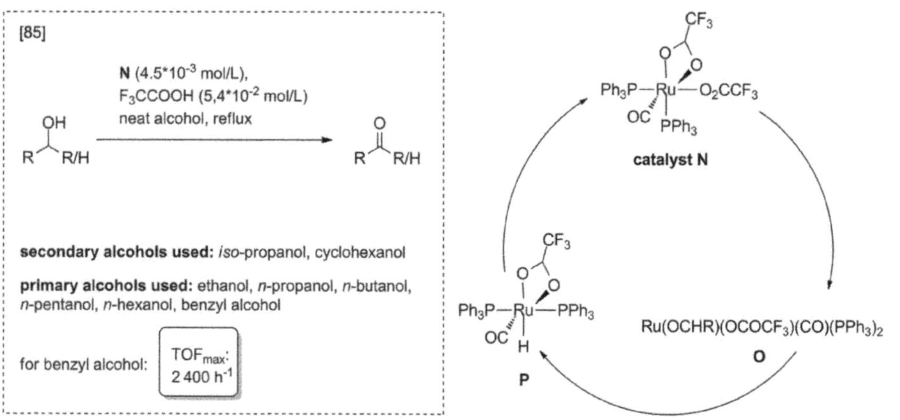

Figure 23.24 Dehydrogenation of primary and secondary alcohols to aldehydes and ketones, respectively [85].

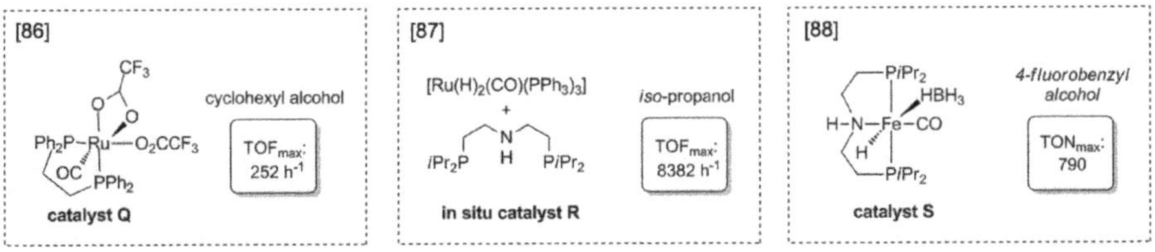

Figure 23.25 Selected examples of homogeneous catalysts for the dehydrogenation of primary and secondary alcohols [86, 88–89].

protocol for the dehydrogenation of alcohols by **N** precatalyst **Q** was also applied in the presence of additional acid. In 2011 [87] reported a screening study of different *in situ* generated catalysts for dehydrogenation of *iso*-propanol. For these screening experiments different Ru-precursors were combined with tridentate PNP ligands. These potentially bifunctional catalysts [78–79] were applied in the absence of additional acid or base. The most effective catalyst **Q**, was obtained based on catalyst loading of 4.0 ppm, generating a TOF of 8382 h^{-1} (measured over the first 2 h of the reaction). Besides Ru-based catalysts other metals, such as Fe, have also been shown to selectively catalyse the dehydrogenation of alcohols to ketones. In 2014, [88] found that complex **S**, bearing the same ligand as *in situ* catalyst **R**, is capable of catalysing dehydrogenation of secondary alcohols. Depending on the substrate, TONs ranging from 64 (for cyclohexanol) to 790 (for 4-fluorobenzyl alcohol) were reported. Interestingly, substrates with primary and secondary alcohol groups gave selective dehydrogenation of the secondary alcohol. On the other hand the diols 1,2-Benzenedimethanol and 1,5-pentanediol resulted in the formation of the corresponding lactones.

More important in terms of capacity is the dehydrogenation of primary alcohol derivatives to esters and amides. The reports about transition metal catalysed acceptorless dehydrogenation reactions of primary alcohols dramatically increased in recent years. In the following paragraph the most active catalysts are highlighted and their common properties are summarized accordingly.

When moving from MeOH (for MeOH dehydrogenation see Section 23.2) to longer chain analogues, examples of transition metal catalysts for the dehydrogenation of primary alcohols to esters and hydrogen are more common and can be achieved with various types of catalyst. The dehydrogenation of ethanol to ethyl acetate, however, remains challenging, due to the low boiling points of the reactants and products. Under these conditions only few catalysts operate efficiently with high TOFs (Figure 23.26). The most

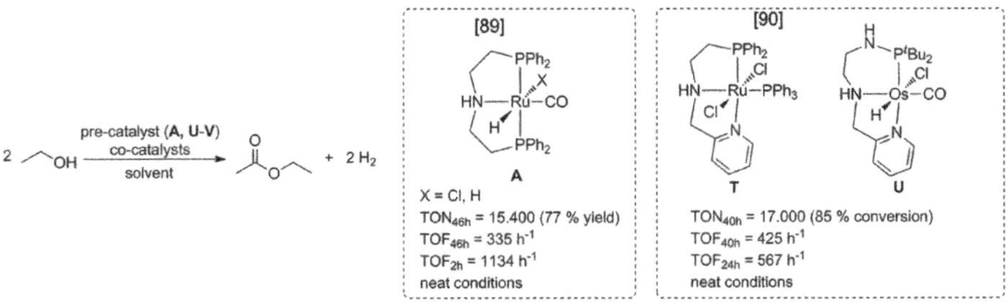

Figure 23.26 Selected examples for homogeneous catalysts for the dehydrogenation of ethanol to ethyl acetate and hydrogen.

active catalysts among those are cooperative ruthenium PNP-pincer-type complexes: complex **A** that shows excellent activity in the dehydrogenation of MeOH/water mixtures is also an efficient catalyst for the liberation of hydrogen from ethanol. Within 46 h, catalyst **A** produces ethyl acetate and hydrogen with a TON of 15 400, high TOFs and an overall yield of 77% [89]. In parallel, [90] developed ruthenium and osmium catalysts involving PNN-type pincer ligands with a central amine group, believed to be essential as a cooperative ligand site that reversibly provides protons and thereby participates in transition states [84, 90]. The ruthenium complex **T** shows slightly higher productivities and activities than **A**, and works efficiently under neat conditions, too. The corresponding osmium complexes (e.g. **U**) were found to be slightly less active than complex **T**.

Considering the theoretical gravimetric capacities of longer chain alcohols, their catalytic dehydrogenation to esters is not attractive for hydrogen storage and will thus not be discussed here.

Hydrogen liberation from longer chain diols offers a high potential. 1,4-butanediol for example exhibits a high gravimetric capacity of 4.4 wt.%(H_2) for the dehydrogenation to γ-butyrolactone. Ruthenium complex **V** is capable of catalysing this reaction (Figure 23.27) [30]. This ruthenium(II) complex with ethylene diamine as potentially cooperative ligand produces γ-butyrolactone at 205°C reaction temperature with a TON of 17 000 at more than 99% yield. A second example that is worth mentioning and operates in refluxing 1,4-butanediol is the iridium complex **W**, producing γ-butyrolactone in high yield with high TONs and TOFs [91].

All examples discussed so far in this chapter relied on the oxidation of the alcohol via formation of an additional C-O-bond with associated hydrogen liberation. The formation of amides, which includes the formation a new C-N-bond, is a highly attractive approach for hydrogen storage, due to the high H_2 storage capacities that can be achieved (Entries 5 and 6, Table 23.2). However, dehydrogenative coupling reactions of alcohols with amines to amides are rare and were reported for the first time by [92] in 2007. Very recently attempts have been made to use this reaction for hydrogen storage applications (Figure 23.28) [93]. The ruthenium PNN-type Pincer complexes **B** and **X** were used to liberate hydrogen from ethanolamine, leading to the formation of glycine anhydride and a mixture of linear peptides [94]. Although the maximum TON and TOF achieved in this transformation are significantly lower than for previous examples, this is a remarkable H_2 storage system offering high potential for further development. The dehydrogenative coupling of ethanol with ethylene diamine yields

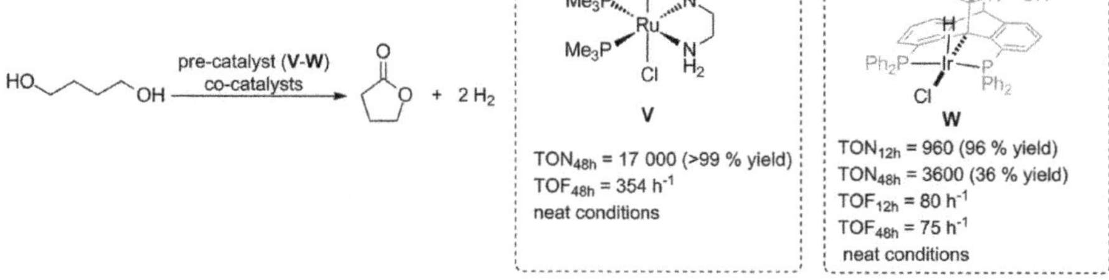

Figure 23.27 Selected examples for homogeneous catalysts for the dehydrogenation of 1,4-butanediol to γ-butyrolactone and hydrogen.

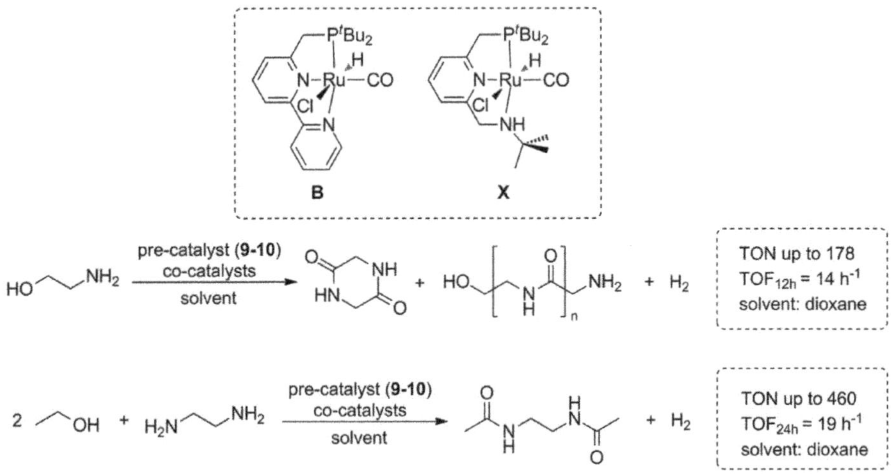

Figure 23.28 Selected examples for homogeneous catalysts for the dehydrogenation of aminoethanol and ethanol/ethylene diamine mixtures.

N,N'-diacetylethylenediamine and a series of incomplete dehydrogenation products, including N-(2-aminoethyl)acetamide and N-ethylideneethane-1,2-diamine [93]. In addition to this selectivity issue, the activity and productivity of **B** and **X** were found to be comparably low in these reactions. An additional challenge, namely the development of protocols for performing these reactions under neat conditions still remains.

23.6 Summary and Outlook

MeOH is one of the most attractive and viable hydrogen storage compounds due to its high gravimetric hydrogen density, low price and high availability. Although the partial conversion of MeOH and other alcohols through dehydrogenative coupling to generate esters has been well investigated, this type of reaction does not liberate the maximum amount H_2 possible. Therefore, this type of reaction seems on the first sight less attractive for hydrogen storage systems as compared to dehydrogenation of MeOH/water mixtures.

The dehydrogenation reaction of methanol/water mixtures releases significantly higher gravimetric H_2 content, offering a genuine opportunity for widespread hydrogen storage applications. The possibility of homogeneous aqueous MeOH dehydrogenation was achieved by the application of pincer complexes, which have potentially cooperative ligand systems. These catalysts enable the activation of the substrates and subsequent molecular hydrogen release. Additionally room-temperature methanol reforming has been demonstrated using a chemo-enzymatic concept which has potentially some advantages for mobile applications where low-temperature energy conversion processes is preferred.

The future for the application of MeOH as a hydrogen storage solution remains a challenge. The improvement of the activity and catalyst durability of the catalytic systems are a desirable long-term goal, which is needed to match the requirements of real-world applications in the next decades. The current state-of-the-art of fundamental research is still far away from meeting those requirements in terms of robustness and weight efficiencies for stationary and mobile applications. The ability to produce hydrogen under mild conditions and with minimal carbon monoxide contaminants is however a considerable benefit of homogeneous catalytic methodologies. Formaldehyde and paraformaldehyde can be considered as high potential LOHCs; especially with the employment of a robust and effective homogeneous catalytic system and it has been shown that the dehydrogenation process can be repeated by simply charging the reactor vessel with the substrate. The combination of an enzymatic catalytic system to transform MeOH to formaldehyde in the first step even allows for reforming at room temperature. This implies that formaldehyde can be produced in situ at low temperature from methanol without the requirement of energy-intensive formaldehyde production processes from methanol (FORMOX process). In consequence the endergonic methanol oxidation process can be performed near room temperature rather than at elevated temperature (>400°C). This is an attractive aspect at least for mobile applications rather than for stationary industrial applications. In this case, it is worth investigating possible interplays between bio- and organometallic chemistry.

Further work to extend the range of eco-sustainable and inexpensive metals from which molecularly defined catalysts can be prepared will be required to improve the practical application and large scale commercialization of the methodologies along with attempts to circumvent the use of sophisticated and expensive phosphine ligands to reduce the cost of catalysis. It is envisioned that a greater use of bio-catalytic reactivity will be applied in future studies due to the high TOFs that enzymatic catalytic centres can attain coupled with the mild reaction conditions they require.

Clearly, formic acid is the hydrogen storage compound with the lowest H_2 storage capacity among the C1 based storage compounds discussed here. However, a broad variety of catalysts have been found to efficiently catalyse the dehydrogenation of formic acid to CO_2 and H_2 under various conditions. A significant amount of mechanistic investigations on different systems has been reported. This knowledge may also provide a valuable basis for the development of related hydrogen storage systems. Importantly, for formic acid it has been demonstrated that hydrogen can be stored reversibly.

References

[1] Felderhoff, M., Weidenthaler, C., von Helmolt, R., et al. (2007). Hydrogen storage: The remaining scientific and technological challenges. *Physical Chemistry Chemical Physics* 9: 2643–2653.

[2] Eberle, U., Felderhoff, M., and Schuth, F. (2009). Chemical and physical solutions for hydrogen storage. *Angewandte Chemie International Edition* 48: 6608–6630.

[3] Merchant-Research-Consulting-Ltd. (2016). Formaldehyde: 2016 world market outlook and forecast up to 2020.

[4] Sahler, S., Konnerth, H., Knoblauch, N., et al. (2013). Hydrogen storage in amine boranes: Ionic liquid supported thermal dehydrogenation of ethylene diamine bisborane. *International Journal of Hydrogen Energy* 38: 3283–3290.

[5] Olah, G.A. (2005). Beyond oil and gas: The methanol economy. *Angewandte Chemie International Edition* 44: 2636–2639.
[6] Fellay, C., Dyson, P.J., and Laurenczy, G. (2008). A viable hydrogen-storage system based on selective formic acid decomposition with a ruthenium catalyst. *Angewandte Chemie International Edition* 47: 3966–3968.
[7] Loges, B., Boddien, A., Junge, H., et al. (2008). Controlled generation of hydrogen from formic acid amine adducts at room temperature and application in H2/O2 fuel cells. *Angewandte Chemie International Edition* 47: 3962–3965.
[8] Ogden, J. (1999). Prospects for building a hydrogen energy infrastructure. *Annual Review of Energy and the Environment* 24: 227–279.
[9] Dincer, I. (2012). Green methods for hydrogen production. *International Journal of Hydrogen Energy* 37: 1954–1971.
[10] Jessop, P.G., Joó, F., and Tai, C.-C. (2004). Recent advances in the homogeneous hydrogenation of carbon dioxide. *Coordination Chemistry Reviews* 248: 2425–2442.
[11] Enthaler, S. (2008). Carbon dioxide—The hydrogen-storage material of the future? *ChemSusChem* 1: 801–804.
[12] Enthaler, S., von Langermann, J., and Schmidt, T. (2010). Carbon dioxide and formic acid – The couple for environmental-friendly hydrogen storage? *Energy & Environmental Science* 3: 1207.
[13] Wang, W., Wang, S., Ma, X., et al. (2011). Recent advances in catalytic hydrogenation of carbon dioxide. *Chemical Society Reviews* 40: 3703–3727.
[14] Tanaka, R., Yamashita, M., and Nozaki, K. (2009). Catalytic hydrogenation of carbon dioxide using Ir(III)-pincer complexes. *Journal of American Chemistry* 131: 14168–14169.
[15] Jessop, P., Ikariya, T., and Noyori, R. (1994). Homogeneous catalytic hydrogenation of supercritical carbon dioxide. *Nature* 368: 231–233.
[16] Huff, C.A. and Sanford, M.S. (2013). Catalytic CO2 Hydrogenation to formate by a ruthenium pincer complex. *ACS Catalysis* 3: 2412–2416.
[17] Wesselbaum, S., vom Stein, T., Klankermayer, J., et al. (2012). Hydrogenation of carbon dioxide to methanol by using a homogeneous ruthenium-phosphine catalyst. *Angewandte Chemie International Edition* 51: 7499–7502.
[18] Balaraman, E., Gunanathan, C., Zhang, J., et al. (2011). Efficient hydrogenation of organic carbonates, carbamates and formates indicates alternative routes to methanol based on CO2 and CO. *Nature Chemistry* 3: 609–614.
[19] http://www.methanol.org/the-methanol-industry. Methanol Institute, Bruxelles (EU). Accessed 3 August 2021.
[20] Andrade de Sá, S., Palmer, C., and Engel, S. (2012). Ethanol production, food and forests. *Environmental and Resource Economics* 51: 1–21.
[21] Kothandaraman, J., Goeppert, A., Czaun, M., et al. (2016). Conversion of CO2 from air into methanol using a polyamine and a homogeneous ruthenium catalyst. *Journal of the American Chemical Society* 138: 778–781.
[22] Smith, T., Aplin, R., and Maitlis, P. (1985). The ruthenium-catalysed conversion of mathanol into methyl formate. *Journal of Organometallic Chemistry* 291: 13–14.
[23] Smith, T. and Maitlis, P. (1985). Methanol as a hydrogen donor in reactions homogeneously catalysed by ruthenium and rhodium complexes. *Journal of Organometallic Chemistry* 289: 385–395.

[24] Murahashi, S., Naota, T., Ito, K., et al. (1987). Ruthenium-catalyzed oxidative transformation of alcohols and aldehydes to esters and lactones. *Journal of Organometallic Chemistry* 52: 4319–4327.

[25] Morton, D. and Cole-Hamilton, D.J. (1988). Molecular hydrogen complexes in catalysis: Highly efficient hydrogen production from alcoholic substrates catalysed by ruthenium complexes. *Journal of the Chemical Society, Chemical Communications*: 1154–1156, doi.org/10.1039/C39880001154

[26] Morton, D. and Cole-Hamilton, D. 1987. Rapid thermal hydrogen production from alcohols catalysed by [Rh(2,2'-bipyridyl)2]Cl. *Journal of the Chemical Society, Chemical Communications*: 248–249 doi.org/10.1039/C39870000248

[27] Sieffert, N. and Buhl, M. (2010). Hydrogen generation from alcohols catalyzed by ruthenium-triphenylphosphine complexes: Multiple reaction pathways. *Journal of American Chemistry* 132: 8056–8070.

[28] Adair, G.R.A. and Williams, J.M.J. (2005). Oxidant-free oxidation: Ruthenium catalysed dehydrogenation of alcohols. *Tetrahedron Letters* 46: 8233–8235.

[29] Ligthart, G.B.W.L., Meijer, R.H., Donners, M.P.J., et al. (2003). Highly sustainable catalytic dehydrogenation of alcohols with evolution of hydrogen gas. *Tetrahedron Letters* 44: 1507–1509.

[30] Zhao, J. and Hartwig, J.F. (2005). Acceptorless, neat, ruthenium-catalyzed dehydrogenative cyclization of diols to lactones. *Organometallics* 24: 2441–2446.

[31] Johnson, T.C., Morris, D.J., and Wills, M. (2010). Hydrogen generation from formic acid and alcohols using homogeneous catalysts. *Chemical Society Reviews* 39: 81–88.

[32] Dalebrook, A.F., Gan, W., Grasemann, M., et al. (2013). Hydrogen storage: Beyond conventional methods. *Chemical Communications (Camb)* 49: 8735–8751.

[33] Nielsen, M., Alberico, E., Baumann, W., et al. (2013). Low-temperature aqueous-phase methanol dehydrogenation to hydrogen and carbon dioxide. *Nature* 495: 85–89.

[34] Trincado, M., Banerjee, D., and Grützmacher, H. (2014). Molecular catalysts for hydrogen production from alcohols. *Energy & Environmental Science* 7: 2464–2503.

[35] Hu, P., Diskin-posner, Y., Ben-david, Y., et al. (2014). Reusable homogeneous catalytic system for hydrogen production from methanol and water. *ACS Catalysis* 4: 2649–2652.

[36] Rodriguez-Lugo, R.E., Trincado, M., Vogt, M., et al. (2013). A homogeneous transition metal complex for clean hydrogen production from methanol-water mixtures. *Nature Chemistry* 5: 342–347.

[37] Monney, A., Barsch, E., Sponholz, P., et al. (2014). Base-free hydrogen generation from methanol using a bi-catalytic system. *Chemical Communications (Camb)* 50: 707–709.

[38] Campos, J., Sharninghausen, L.S., Manas, M.G., et al. (2015). Methanol dehydrogenation by iridium N-heterocyclic carbene complexes. *Inorganic Chemistry* 54: 5079–5084.

[39] Fujita, K.-I., Yoshida, T., Imori, Y., et al. (2011). Dehydrogenative oxidation of primary and secondary alcohols catalyzed by a Cp*Ir complex having a functional C,N-chelate ligand. *Organic Letters* 13: 2278–2281.

[40] Kawahara, R., Fujita, K., and Yamaguchi, R. (2012). Dehydrogenative oxidation of alcohols in aqueous media using water-soluble and reusable Cp*Ir catalysts bearing a functional bipyridine ligand. *Journal of American Chemistry* 134: 3643–3646.

[41] Fujita, K., Kawahara, R., Aikawa, T., et al. (2015). Hydrogen production from a methanol-water solution catalyzed by an anionic iridium complex bearing a functional bipyridonate ligand under weakly basic conditions. *Angewandte Chemie International Edition* 54: 9057–9060.

[42] Alberico, E., Sponholz, P., Cordes, C., et al. (2013). Selective hydrogen production from methanol with a defined iron pincer catalyst under mild conditions. *Angewandte Chemie International Edition* 52: 14162–14166.

[43] Bielinski, E.A., Förster, M., Zhang, Y., et al. (2015). Base-free methanol dehydrogenation using a pincer-supported iron compound and lewis acid co-catalyst. *ACS Catalysis* 5: 2404–2415.

[44] Heim, L.E., Thiel, D., Gedig, C., et al. (2015). Bioinduced room-temperature methanol reforming. *Angewandte Chemie International Edition* 54: 10308–10312.

[45] Heim, L.E., Thiel, D., Gedig, C., et al. (2015). Bioinduzierte methanolreformierung. *Angewandte Chemie* 127: 10447–10451.

[46] Clarke, P.H. (1983). *The Biochemistry of Methylotrophs* by C. Anthony Academic Press; London, New York, 1982 xvi + 432 pages. *FEBS Letters* 160: 303.

[47] Chistoserdova, L., Kalyuzhnaya, M.G., and Lidstrom, M.E. (2009). The expanding world of methylotrophic metabolism. *Annual Review of Microbiology* 63: 477–499.

[48] Cereghino, J.L. and Cregg, J.M. (2000). Heterologous protein expression in the methylotrophic yeast Pichia pastoris. *FEMS Microbiology Reviews* 24: 45–66.

[49] Vorholt, J.A. (2002). Cofactor-dependent pathways of formaldehyde oxidation in methylotrophic bacteria. *Archives of Microbiology* 178: 239–249.

[50] Yang, Z.-N., Bosron, W.F., and Hurley, T.D. (1997). Structure of human $\chi\chi$ alcohol dehydrogenase: A glutathione-dependent formaldehyde dehydrogenase1. *Journal of Molecular Biology* 265: 330–343.

[51] Sanghani, P.C., Robinson, H., Bosron, W.F., et al. (2002). Human glutathione-dependent formaldehyde dehydrogenase. Structures of apo, binary, and inhibitory ternary complexes. *Biochemistry* 41: 10778–10786.

[52] Heim, L.E., Schlorer, N.E., Choi, J.H., et al. (2014). Selective and mild hydrogen production using water and formaldehyde. *Nature Communications* 5: 3621.

[53] Schore, K.P.C.V.A.N.E. (2005). *Organic Chemistry*. Weinheim: Wiley-VCH.

[54] Loges, B., Boddien, A., Gärtner, F., et al. (2010). Catalytic generation of hydrogen from formic acid and its derivatives: Useful hydrogen storage materials. *Topics in Catalysis* 53: 902–914.

[55] Grasemann, M. and Laurenczy, G. (2012). Formic acid as a hydrogen source – Recent developments and future trends. *Energy & Environmental Science* 5: 8171–8181.

[56] Zell, T. and Langer, R. (2015). Iron-catalyzed hydrogenation and dehydrogenation reactions with relevance to reversible hydrogen storage applications. *Recyclable Catalysis* 2: 87–109.

[57] Zell, T. and Milstein, D. (2015). Hydrogenation and dehydrogenation iron pincer catalysts capable of metal-ligand cooperation by aromatization/dearomatization. *Accounts of Chemical Research* 48: 1979–1994.

[58] Mellmann, D., Sponholz, P., Junge, H., et al. (2016). Formic acid as a hydrogen storage material – Development of homogeneous catalysts for selective hydrogen release. *Chemical Society Reviews* 45: 3954–3988.

[59] Singh, A.K., Singh, S., et al. (2016). Hydrogen energy future with formic acid: A renewable chemical hydrogen storage system. *Catalysis Science & Technology* 6: 12–40.

[60] Moret, S., Dyson, P.J., and Laurenczy, G. (2014). Direct synthesis of formic acid from carbon dioxide by hydrogenation in acidic media. *Nature Communications* 5: 4017.

[61] Jessop, P.G. (2006). Homogeneous hydrogenation of carbon dioxide. In: *The Handbook of Homogeneous Hydrogenation*, (ed. J.G. de Vries and C.J. Elsevier), pp. 489–511. Wiley-VCH Verlag GmbH.

[62] Federsel, C., Jackstell, R., and Beller, M. (2010). State-of-the-art catalysts for hydrogenation of carbon dioxide. *Angewandte Chemie International Edition* 49: 6254–6257.

[63] Schaub, T. and Paciello, R.A. (2011). A process for the synthesis of formic acid by CO2 hydrogenation: Thermodynamic aspects and the role of CO. *Angewandte Chemie International Edition* 50: 7278–7282.

[64] Appel, A.M., Bercaw, J.E., Bocarsly, A.B., et al. (2013). Frontiers, opportunities, and challenges in biochemical and chemical catalysis of CO2 fixation. *Chemical Reviews* 113: 6621–6658.

[65] Wang, W.-H., Himeda, Y., Muckerman, J.T., et al. (2015). CO2 hydrogenation to formate and methanol as an alternative to photo- and electrochemical CO2 Reduction. *Chemical Reviews* 115: 12936–12973.

[66] Klankermayer, J., Wesselbaum, S., Beydoun, K., et al. (2016). Selective catalytic synthesis using the combination of carbon dioxide and hydrogen: Catalytic chess at the interface of energy and chemistry. *Angewandte Chemie International Edition* 55: 7296–7343.

[67] Costentin, C., Robert, M., and Saveant, J.-M. (2013). Catalysis of the electrochemical reduction of carbon dioxide. *Chemical Society Reviews* 42: 2423–2436.

[68] Taheri, A. and Berben, L.A. (2016). Making C-H bonds with CO2: Production of formate by molecular electrocatalysts. *Chemical Communications* 52: 1768–1777.

[69] Corma, A. and Garcia, H. (2013). Photocatalytic reduction of CO2 for fuel production: Possibilities and challenges. *Journal of Catalysis* 308: 168–175.

[70] Wesselbaum, S., Hintermair, U., and Leitner, W. (2012). Continuous-flow hydrogenation of carbon dioxide to pure formic acid using an integrated scCO2 process with immobilized catalyst and base. *Angewandte Chemie International Edition* 51: 8585–8588.

[71] Yu, X. and Pickup, P.G. (2008). Recent advances in direct formic acid fuel cells (DFAFC). *Journal of Power Sources* 182: 124–132.

[72] Zell, T., Butschke, B., Ben-david, Y., et al. (2013). Efficient hydrogen liberation from formic acid catalyzed by a well-defined iron pincer complex under mild conditions. *Chemistry – A European Journal* 19: 8068–8072.

[73] Enthaler, S. and Loges, B. (2012). The rise of the iron age in hydrogen evolution? *ChemCatChem* 4: 323–325.

[74] Boddien, A., Mellmann, D., Gartner, F., et al. (2011). Efficient dehydrogenation of formic acid using an iron catalyst. *Science* 333: 1733–1736.

[75] Bielinski, E.A., Lagaditis, P.O., Zhang, Y., et al. (2014). Lewis acid-assisted formic acid dehydrogenation using a pincer-supported iron catalyst. *Journal of the American Chemical Society* 136: 10234–10237.

[76] Oldenhof, S., Lutz, M., de Bruin, B., et al. (2015). Dehydrogenation of formic acid by Ir-bisMETAMORPhos complexes: Experimental and computational insight into the role of a cooperative ligand. *Chemical Science*. 6: 1027–1034

[77] Boddien, A., Federsel, C., Sponholz, P., et al. (2012). Towards the development of a hydrogen battery. *Energy & Environmental Science* 5: 8907–8911.

[78] Gunanathan, C. and Milstein, D. (2011). Metal–ligand cooperation by aromatization–dearomatization: A new paradigm in bond activation and "green" catalysis. *Accounts of Chemical Research* 44: 588–602.

[79] Gunanathan, C. and Milstein, D. (2013). Applications of acceptorless dehydrogenation and related transformations in chemical synthesis. *Science* 341: 1229712.

[80] Filonenko, G.A., van Putten, R., Schulpen, E.N., et al. (2014). Highly efficient reversible hydrogenation of carbon dioxide to formates using a ruthenium PNP-pincer catalyst. *ChemCatChem* 6: 1526–1530.

[81] Hsu, S.F., Rommel, S., Eversfield, P., et al. (2014). A rechargeable hydrogen battery based on Ru catalysis. *Angewandte Chemie International Edition* 53: 7074–7078.

[82] Scott, M.S. and Idriss, H. (2010). Heterogeneous catalysis for hydrogen production. In: *Handbook of Green Chemistry* (ed. Paul T. Anastas), chapter 10, (ed.Robert H. Crabtree), pp. 223–246. Wiley-VCH Verlag GmbH & Co. KGaA.

[83] http://webbook.nist.gov/chemistry. NIST Webbook Chemistry, *National Institute of Standards and Technology, US Department of Commerce*, Washington, DC: (USA). Accessed 3 August 2021.

[84] Gusev, D.G. (2016). Dehydrogenative coupling of ethanol and ester hydrogenation catalyzed by pincer-type YNP complexes. *ACS Catalysis* 6: 6967–6981.

[85] Dobscn, A. and Robinson, S.D. (1975). Catalytic dehydrogenation of primary and secondary alcohols by Ru(OCOCF3)2(CO)(PPh3)2. *Journal of Organometallic Chemistry* 87: C52–C53.

[86] Jung, C.W. and Garrou, P.E. (1982). Dehydrogenation of alcohols and hydrogenation of aldehydes using homogeneous ruthenium catalysts. *Organometallics* 1: 658–666.

[87] Nielsen, M., Kammer, A., Cozzula, D., et al. (2011). Efficient hydrogen production from alcohols under mild reaction conditions. *Angewandte Chemie International Edition* 50: 9593–9597.

[88] Chakraborty, S., Lagaditis, P.O., Förster, M., et al. (2014). Well-defined iron catalysts for the acceptorless reversible dehydrogenation-hydrogenation of alcohols and ketones. *ACS Catalysis* 4: 3994–4003.

[89] Nielsen, M., Junge, H., Kammer, A., et al. (2012). Towards a green process for bulk-scale synthesis of ethyl acetate: Efficient acceptorless dehydrogenation of ethanol. *Angewandte Chemie International Edition* 51: 5711–5713.

[90] Spasyuk, D. and Gusev, D.G. (2012). Acceptorless dehydrogenative coupling of ethanol and hydrogenation of esters and imines. *Organometallics* 31: 5239–5242.

[91] Musa, S., Shaposhnikov, I., Cohen, S., et al. (2011). Ligand–metal cooperation in PCP pincer complexes: Rational design and catalytic activity in acceptorless dehydrogenation of alcohols. *Angewandte Chemie International Edition* 50: 3533–3537.

[92] Gunanathan, C., Ben-david, Y., and Milstein, D. (2007). Direct synthesis of amides from alcohols and amines with liberation of H2. *Science* 317: 790–792.

[93] Hu, P., Ben-David, Y., and Milstein, D. (2016). Rechargeable hydrogen storage system based on the dehydrogenative coupling of ethylenediamine with ethanol. *Angewandte Chemie International Edition* 55: 1061–1064.

[94] Hu, P., Fogler, E., Diskin-Posner, Y., et al. (2015). A novel liquid organic hydrogen carrier system based on catalytic peptide formation and hydrogenation. *Nature Communications* 6: 6859.

24

Hydrogen Energy Recovery – H_2- Based Fuel Cells

Nada Zamel and Ulf Groos

Fraunhofer Institute for Solar Energy Systems, ISE Heidnehofstrasse 2, 79110, Freiburg, Germany

24.1 Introduction

Increasing demand for energy, energy security, and the need to minimize the impact on the environment related to energy are the major drivers for the research and development of alternative energy technologies. The dominance of a single energy resource also shakes the security of access to and distribution of energy. Hence, diversity in energy sources is crucial in meeting today's (and indeed tomorrow's) energy demands. Enter the Hydrogen economy where H_2 is used as an energy carrier. Not only are the methods of H_2 production, storage, and distribution of importance, but also its recovery and use are crucial for the development of the economy. In this regard, hydrogen-based fuel cells are considered to be a promising technology for this energy conversion. These electrochemical cells are devices that convert the chemical energy of H_2 efficiently into electrical energy. Applications of these fuel cells span a wide spectrum with varying electrical power capacity, such as transportation, portable power, backup power, and distributed generation. In this regard, if the H_2 can be supplied by sustainable means (e.g. electro-, photo-chemical water splitting), then the development and implementation of such technology can contribute to Greenhouse Gas emission reduction in the aforementioned applications. To fulfill such applications, a vast range of fuel cell types exist, with the main common aspect being their structure consisting of an anode (negative side) and a cathode (positive side) with an electrolyte separating the two electrodes. The classification of fuel cells is often made in consideration to their electrolyte. Some of the most common H_2-based fuel cells are listed in Table 24.1.

24.1.1 Market Shares

According to the Fuel Cell Review, in 2020, the shipment of polymer electrolyte membrane fuel cell (PEMFC) systems continued to be in the lead (Figure 24.1). This lead in shipment was attributed to the adaptability of the integration of this technology within a wide range of applications, specifically in transport, stationary and portable fields. This contrasts with other fuel cell technologies, which are application specific, such as Molten Carbonate FC (MCFC), Phosphoric Acid FC (PAFC), solid Oxide FC (SOFC), Alkaline FC (AFC) and Direct Methanol FC (DMFC). The forecasted increase in PEM

Advances in Energy Storage: Latest Developments from R&D to the Market, First Edition.
Edited by Andreas Hauer.
© 2022 John Wiley & Sons, Ltd. Published 2022 by John Wiley & Sons, Ltd.

Table 24.1 Fuel cell by type, its electrolyte, applications, advantages and disadvantages.

Fuel Cell Type	Electrolyte	Application Fields	Advantages	Disadvantages
Polymer Electrolyte Membrane	Perfluorosulfonic acid	backup power, portable power, distributed generation and transportation	solid electrolyte reduces corrosion and electrolyte management problems, low temperature and quick startup	expensive catalysts, sensitive to fuel impurities, low temperature waste heat
Alkaline	aqueous solution of potassium hydroxide soaked in a matrix	military, space	fast cathode reaction leads to high performance, low cost components	sensitive to CO_2 in fuel and air, electrolyte management issues
Phosphoric Acid	phosphoric acid soaked in a matrix	distributed generation	higher temperature enables combined heat and power, increased tolerance to fuel impurities	Pt catalyst, long startup time, low current and power
Molten Carbonate	solution of lithium, sodium and/or potassium carbonates, soaked in a matrix	electric utility, distributed generation	high efficiency, fuel flexibility, catalyst flexibility, suitable for combined heat and power	high temperature
Solid Oxide	Yttria stabilized zirconia	auxiliary power, electric utility, distributed generation	high efficiency, fuel flexibility, catalyst flexibility, solid electrolyte, suitable for combined heat and power and CHHP, hybrid/GT cycle	high temperature corrosion and breakdown of cell components, high temperature operation requires long startup time and limits

fuel cell sales in 2020 was mainly led by passenger cares and micro-combined heat and power (CHP) shipments. SOFCs remain second in terms of unit numbers, while PAFC are the second largest fuel cell type shipped by capacity. The growth in SOFC units comes largely from an overall rise in Ene-Farm (CHP systems) shipments. The shipment of PAFCs is led by Doosan Fuel Cell and with smaller stacks from Fuji Electric. The numbers of MCFCs and AFCs shipped remain very small. DMFCs remain the mainstay of portable systems, where the high stack costs are offset by the ease of refueling and long runtimes.

In this chapter, discussion will primarily focus on PEMFCs, covering the principles of the Oxygen Reduction Reaction (ORR), and the Hydrogen Oxidation Reaction, HOR. Furthermore, a summary of some of the pressing research and development questions is also provided along with a brief introduction to characterization techniques often utilized to answer these questions.

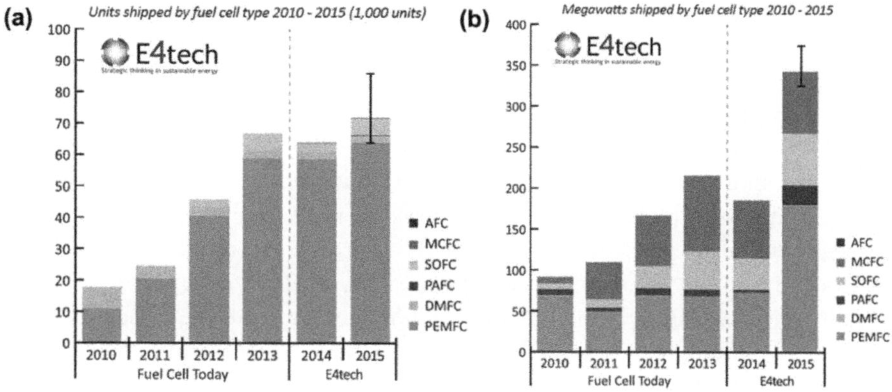

Figure 24.1 The different types of Fuel Cell technologies in terms of (a) Units shipped per year and (b) Megawatts per year [1].

24.2 Polymer Electrolyte Membrane Fuel Cells

PEMFCs, as illustrated in Figure 24.2, convert the chemical energy of H_2 efficiently into useful electrical energy according to the following reaction:

$$\frac{1}{2}O_2 + H_2 \rightarrow H_2O + Heat + elecrial\ energy \tag{24.1}$$

A PEMFC is composed of an anode electrode, where H_2 is supplied as the fuel and a cathode electrode, where oxygen or air is supplied as the oxidant. The porous anode and cathode electrodes are composed of a gas diffusion layer (GDL), a catalyst layer (CL) and a microporous layer (MPL) sandwiched between the GDL and CL. The catalyst

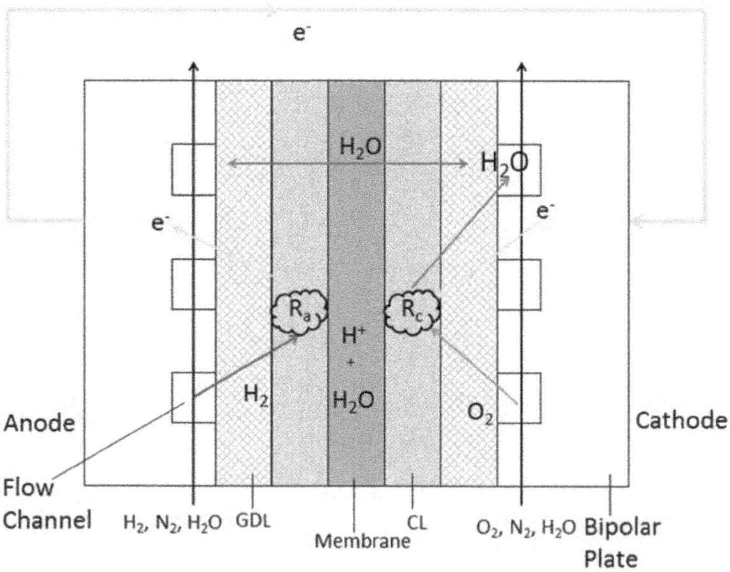

Figure 24.2 Schematic and operation principles of a Polymer Electrolyte Membrane Fuel Cell.

layer acts as the heart of the cell, over which the half-cell reactions occur. These components come together to form the complete cell, which are normally "stacked" in multiple to produce a fuel cell system (i.e. composed of multiple cells and associated system elements regarding fuel supply, heat management, monitoring, etc.) (Figure 24.3).

In order for the electrochemical reactions in the PEMFC to proceed with minimum losses, continuous pathways for proton, reactants/products, and electron transport must be established. These pathways are dictated by the catalyst layer's heterogeneous structure. During operation, H_2 is provided through the flow channels grooved into the so-called anode bipolar plate, while air is provided through the cathode flow channels in the cathode bipolar plate. At the anode catalyst layer, the H_2 undergoes an oxidation process where it is split into protons and electrons. The HOR occurs on the platinum site in three elementary reaction steps as given below:

$$H_2 + 2Pt \leftrightarrow 2H - Pt \qquad \textit{Tafel Reaction} \tag{24.2}$$

$$Pt - H \leftrightarrow H^+ + e^- + Pt \qquad \textit{Volmer Reaction} \tag{24.3}$$

$$H_2 + Pt \leftrightarrow H - Pt + H^+ + e^- \qquad \textit{Heyrovsky Reaction} \tag{24.4}$$

$$H_2 \leftrightarrow 2H^+ + 2e^- \qquad \textit{Overall HOR} \tag{24.5}$$

In the Tafel-Volmer pathway, the dissociative adsorption of a H_2 molecule is followed by two separate one electron oxidations of the H adatoms. In the Heyrovsky-Volmer pathway, a one electron oxidation occurs simultaneously with chemisorption, followed by another one electron oxidation of the H adatom [2–4].

At the cathode catalyst layer, the oxygen is said to be reduced to H_2O according to:

$$\frac{1}{2}O_2 + 2H^+ + 2e^- \rightarrow H_2O \qquad \textit{ORR} \tag{24.6}$$

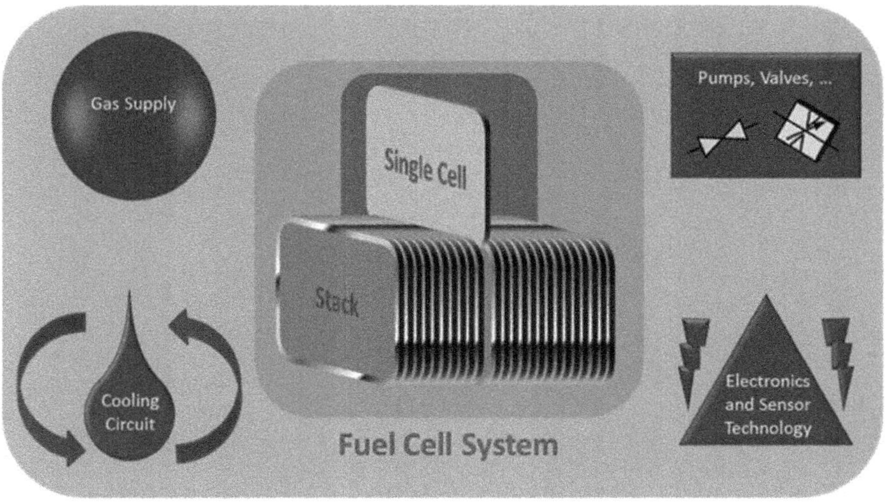

Figure 24.3 A fuel cell system consists of the stack and the auxiliary units. The stack is made up of individual cells that are connected to each other in series – Fraunhofer ISE©.

Much debate surrounds the actual mechanisms in which the ORR proceeds. Hence, in the literature various mechanisms have been proposed based on respective theoretical and experimental investigations. Two models have been proposed, the associative and dissociative models. In the associative model, the charge transfer to the adsorbed oxygen molecule, with or without simultaneous proton transfer is considered to be the rate determining step as suggested by [5–7].

$$M + O_2 \rightarrow M - O_2 \tag{24.7}$$

$$M - O_2 + H^+ + e^- \rightarrow M - O_2H \tag{24.8}$$

$$M - O_2H + 3H^+ + 3e^- \rightarrow 2H_2O + M \tag{24.9}$$

where M represents a metal such as Pt and Equation (24.8) is the rate determining step.

Norskøv et al. examined the thermodynamic properties of both the associative and dissociative mechanism [8–9]. Their results illustrated that both mechanisms are possible and are dependent on the electrode potential. At high electrode potential, the activation energy barrier for oxygen dissociation will increase; thus, the associative mechanism is dominant. Under typical reaction conditions, the O coverage was assumed to be appreciable. The reaction rate was suggested to be controlled by one of the electron-proton transfer steps. The authors suggested that the dissociative mechanism follows the three reaction steps given below:

$$\frac{1}{2}O_2 + * \rightarrow O^* \tag{24.10}$$

$$O^* + H^+ + e^- \rightarrow HO^* \tag{24.11}$$

$$HO^* + H^+ + e^- \rightarrow H_2O + * \tag{24.12}$$

where * denotes a site on the catalyst surface.

The performance of the cell is directly dependent on the overall reaction rate and is usually expressed in terms of a polarization curve, which is the relationship between the cell voltage *vs.* current density (Figure 24.4).

The performance of a PEMFC is characterized by three different regions; activation polarization region, ohmic polarization region and concentration polarization region. The activation polarization region occurs due to sluggish electro-catalysis at high voltages. At this region, an activation energy is required to drive the reaction. Cell voltage is high, but current density is very low; hence, the power density is low and this region is not very useful for practical FC operation. The second region is associated with ohmic losses across the membrane and the electrodes. The third region is associated with high current densities. At this region, transport limitations are very high and affect the overall performance of the cell. Further, when simulating transport phenomena in the cell, the distribution of species becomes very sensitive to the transport coefficients.

24.2.1 Structure of Polymer Electrolyte Membrane Fuel Cells

24.2.1.1 The Gas Diffusion Layer

The GDL is situated between the bipolar plates and the catalyst layers. It provides a flow path for reactant supply and product removal; hence, the use of an electrically

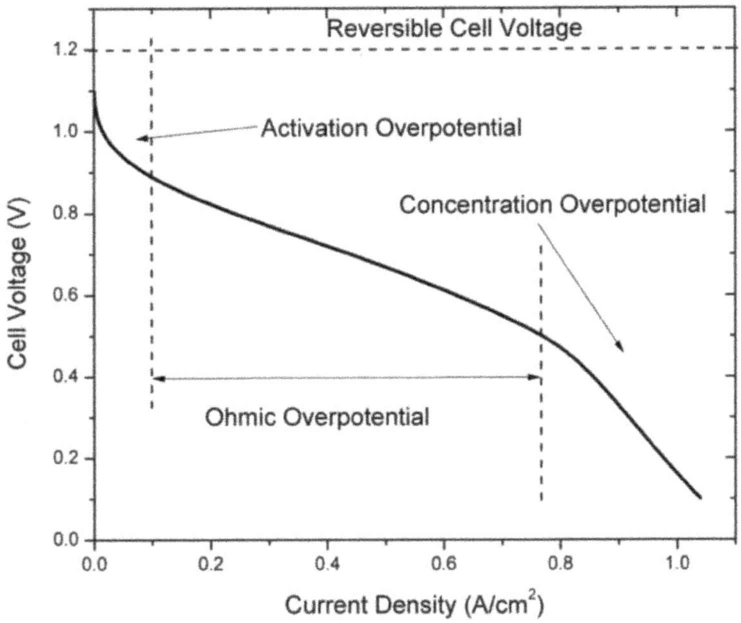

Figure 24.4 A demonstrative polarization curve illustrating the three types of over-potential.

conductive porous material is essential. The void region is needed to provide a path for the free diffusion of gaseous species and the removal of the reactant products (in gas and liquid phases), while the solid is used as an electrical and thermal conductor. The conductive material of choice in the design of the GDL is carbon and carbon paper, which is made hydrophobic by adding polytetrafluorethylene (PTFE), is commonly used. Carbon paper is best described as a carbon/carbon composite since after heat treatment it consists of carbon fibers, of around 7 μm diameter, held together by a carbon matrix [10]. The carbon matrix that holds the fibers together is usually referred to as a binder and is 5–15 % of the total weight of the final product. The manufacturing process of carbon paper consists of various stages of heat treatment (carbonization and graphitization). While the fibers possess graphite-like properties, the resin-based portion (mostly the binder) of the composite does not graphitize and remains as amorphous carbon [10].

Due to the manufacturing process, carbon paper is highly anisotropic since carbonization and graphitization are typically achieved by stacking many sheets or layers of carbon fibers in a horizontal or vertical batch furnace. The structure of carbon paper can be analyzed using SEM images as well as the pore size distribution. SEM image analysis illustrates the face and cross-sectional views of commercial TORAY carbon paper (Figure 24.4). From this figure, it is easily seen that the carbon paper is highly anisotropic and its PTFE content significantly affects its structure. With low concentrations of PTFE, the PTFE is concentrated around the surface of the layer. An increase in concentration results in the penetration of the PTFE deeper into the bulk of the material to the point it looks very "globular." Further, it is apparent that the addition of PTFE changes the structure of the binder. In the presence of PTFE, the binder is no longer smooth and takes on a shape of a web-like porous material. The face views of the carbon paper from scanning electron microscopy (SEM) image analysis (Figures 24.5a and 24.5b) indicate

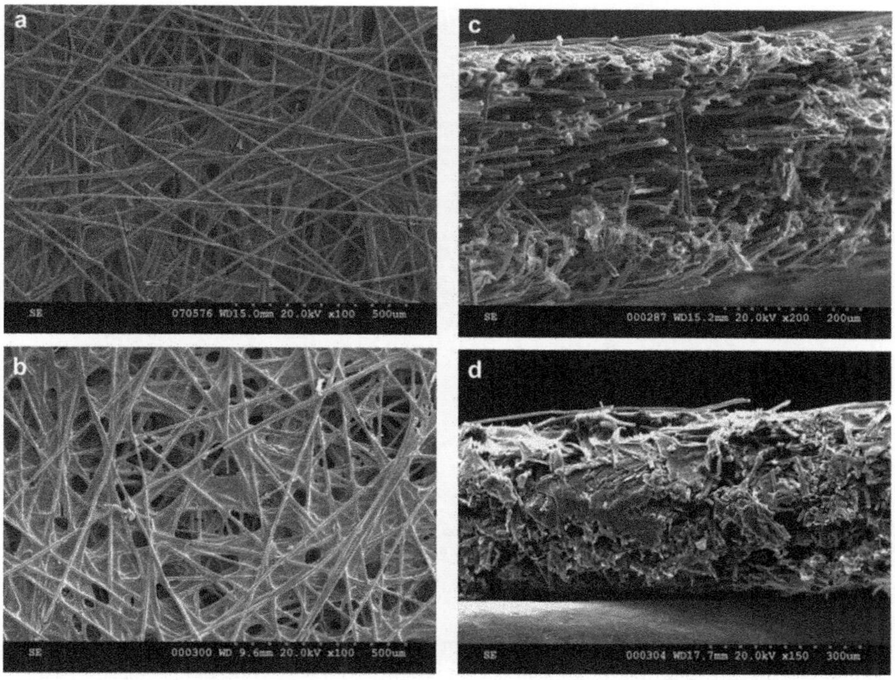

Figure 24.5 SEM images of TORAY-TPGH-120, a commonly used carbon paper as the gas diffusion media in PEM fuel cells; the different images illustrate the effect of PTFE loading on the carbon paper structure: (a) face view with 0% PTFE, (b) face view with 50% PTEF, (c) cross-sectional view with 20% PTFE, and (d) cross-sectional view with 50% PTFE. The face views a and b show the carbon paper structure in the in-plane direction (also called xy plane), the cross-sectional view c and d show the structure in the through-plane direction (also called xz plane); the z axis is defined as the direction pointing through the thickness of the carbon paper. The coordinate system xyz is defined for convenience of description throughout [11]. Reproduced with permission from Elsevier.

that the carbon paper structure in the in-plane direction (also called xy plane for convenience of description) is composed of randomly arranged carbon fibers; and it is often considered direction independent macroscopically so that the x and y coordinate can be oriented arbitrarily. The cross-sectional views (Figures 24.5c and 24.5d) show the structure in the through-plane direction (also called the xz plane) with the z axis being the direction pointing through the thickness of the carbon paper. Clearly, the through-plane structure is made of circular cross sections of the carbon fibers and is significantly affected by the PTFE loading; and the depth of the PTFE penetration into the interior of the carbon paper increases with the PTFE loading.

The effect of the PTFE content on the overall structure of the carbon paper GDL can also be measured by its macropore distribution. In this regard, [11] have applied standard porosimetry to measure this effect (Figure 24.6). The overall macropore volume of the carbon paper sample decreases with the increase in Teflon. At a macropore radius $r \leq 3000$ nm, the effect of Teflon on the overall pore volume is minimal. This implies that PTFE does not penetrate macropores with a radius < 3 μm. At $3000 \leq r \leq 5000$ nm, a transition region is measured and is independent of the Teflon

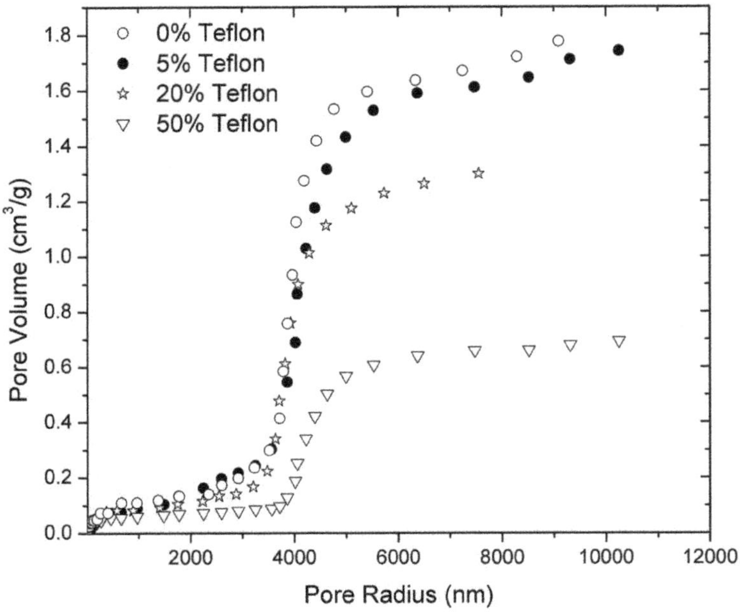

Figure 24.6 Effect of Teflon loading on the pore volume [11], the pore volume is based on the total weight of the carbon paper sample including Teflon and is in grams. Reproduced with permission from Elsevier.

loading. Finally, the macropore volume reaches a plateau that is linear with the PTFE loading.

Typically, a microporous layer (MPL) is introduced between the GDL and the catalyst layer to decrease transport resistances between the two layers. The MPL is composed of carbon agglomerates with a pore size between those of the GDL and the catalyst layer (Figure 24.7). The MPL consists of a powdery mixture of carbon black and PTFE particles and is spread onto the GDL then sintered so that the PTFE can bind the powder together. It is proposed that the MPL positively enhances the performance of the cell due in part to three main reasons [12]:

(i) reduction of ohmic losses mainly due to enhanced hydration of the electrolyte membrane and the reduction of the overall resistance between the GDL and CL;
(ii) reduction of the likelihood of flooding by facilitating the removal of liquid H_2O through the anode;

and

(iii) increase of the chemical and mechanical stability of the CL and the membrane.

24.2.1.2 The Catalyst Layer

The importance of the catalyst layer to PEMFC performance is dictated by its function and its location between the GDL and the electrolyte membrane. As the area where the electrochemical half-cell reactions occur, it must provide continuous pathways for various species, primarily:

(i) a continuous path for efficient transport of protons,

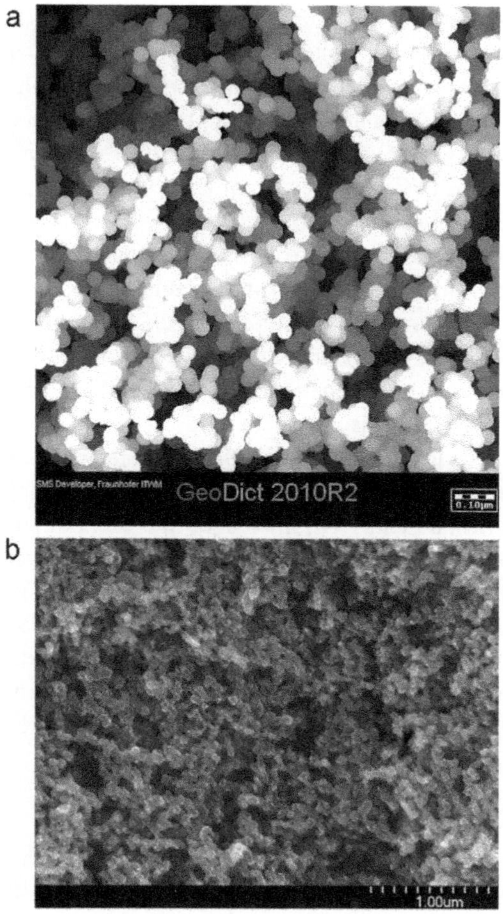

Figure 24.7 (a) Cross-section of the reconstructed microporous layer [13] reproduced with permission from Elsevier; (b) SEM image of a microporous layer [14] reproduced with permission from Elsevier.

(ii) a continuous pore network for the transport of reactants/products and for efficient H_2O removal,

and

(iii) continuous passage for the conduction of electrons between the catalyst layer and the current collection.

In turn, the catalyst layer's structure is heterogeneous and complex and its optimization is required for the increase of its activity and the performance of the cell. This is further complicated by the requirement for its stability and durability over the lifetime of the cell and under various operating conditions.

The preparation method of the catalyst layer has a direct effect on the structure of the layer and hence directly affects the power density of the cell. This connection is highlighted through the history of catalyst layer preparation. In 1967, Niedrach used PTFE as the binder to hold fine powder metal catalyst particles together [15]. However, proton conduction was lacking within these PTFE-bonded catalyst layers as Raistrick later showed [16]. A method of impregnating Nafion® solution through the surface of a finished catalyst layer was, hence, developed. This was a breakthrough in that it highlighted the importance of a 3-phase interface. The main breakthrough in catalyst layer structure design came when Wilson and Gottesfeld eliminated the use of PTFE all together by using recast Nafion® ionomer as the binder and significantly reduced the amount of Pt loadings [17–18]. Subsequently, Pt-black, i.e. unsupported Pt, was replaced with Pt supported on carbon to reduce the metal content. This state-of-the-art catalyst layer is typically formed by the dispersion of an ink comprising a catalyst deposited onto a support (e.g. Pt/C), binder (e.g. perfluorinated sulfonic acid (PFSA) ionomer) and a dispersing solvent. The structure of the layer is, hence, shaped within the ink by the agglomeration of the carbon particles and the aggregation of the ionomer. The dispersion medium governs the ink's properties, such as aggregation size of the catalyst/ionomer particles, viscosity, rate of solidification, and ultimately the physical and mass transport properties of the catalyst layer [19]. The ionomer used in the preparation of the catalyst layer is required to be a proton conductor. Any imbalance in the amount of the ionomer can result in transport losses; too much results in flooding the catalyst layer (slow diffusion of gases) and too little results in poor ionic conductivity and reduced active sites. Understanding how the ionomer distributes within the catalyst layer is important in optimization studies and a number of literature reports indicate this is directly related to aggregation of the ionomer solution [19–20]. The durability and stability of the catalyst layer is hence governed by its catalyst utilization, rate of support degradation (e.g. carbon corrosion), and ionomer stability [21–22].

24.2.1.3 The Electrolyte Membrane

The electrolyte membrane is located between the catalyst layers of the anode and the cathode. It acts as an ion conductor, so that ions can migrate from the anode catalyst layer to the cathode catalyst layer. To conduct the protons efficiently, the electrolyte membrane must be well hydrated without risking flooding of the porous electrodes. It is also an electron insulator, so that electrons are forced to migrate through the external circuit; thus, providing electric power output. Finally, it must be impermeable to the reactants to prevent their crossover from the anode to the cathode and vice versa; thus, controlling the electrochemical reaction to proceed as expected. On the other hand, the reactant gases, H_2 and O_2, in the pore region of the catalyst layers, have to be dissolved in the hydrated ionomer membrane covering the supported catalyst particles, and then diffuse to the catalyst surface for the electrochemical reactions to occur. This implies that it might be desirable to use an ionomer in the catalyst layer that has a high protonic conductivity and high diffusion coefficient and to use another membrane as the electrolyte that has a high protonic conductivity, but a low diffusion coefficient to avoid reactant gas crossover. However, this approach creates an interface between the catalyst layer and the membrane electrolyte, which not only increases the interfacial resistance to proton transport between the layers, but also increases the likelihood of the delamination and local cracks between the layers due to the thermal and mechanical cycling during the operation of the cell. Consequently, for typical PEMFC applications, the same material is used as ionomer and electrolyte membrane.

In this context, Nafion® is the most widely used as the electrolyte membrane in PEMFCs. Chemically, Nafion is a sulfonated tetrafluoroethylene-based fluoropolymer-copolymer developed by the Dow group. The sulfonate side chains, which make up the structure of Nafion, ensure its stability and high performance. As chemical degradation of Nafion is an issue, alternative materials have been investigated. A detailed review of membranes for use in PEMFCs is given by [23], where 60 alternatives to Nafion were identified and categorized based on their performance (e.g. ill-suited or suitable), and technology readiness (e.g. pending more research), for use in automotive PEMFCs. The rejection of 46 materials in this overview was based on 13 reasons; mainly, a lack of FC Data (Conceptual), expense, fire risk, low durability, poor performance, H_2O soluble, shrinkage (low flexibility), low conductivity, low thermal stability, low stability, depolymerization, or no longer in production.

Another type of membrane that has gained much interest lately is the short-side chain (SSC) and low equivalent weight PFSA. The short-side chain polymer was developed by Dow Chemicals [24–25], but its production was discontinued due to the high cost of its synthesis. However, its production was later commenced by Solvay Solexis under the commercial name of Hyflon Ion, which was then later renamed as Aquivion^{TM}. It is often referred to as the short-side chain (SSC) polymer in literature as its main feature is a shorter pendant side-chain that carries the functional ion-exchange group. A comparison between Nafion and Aquivion membranes reveals that SSC polymers possess a much higher crystallinity at a given equivalent weight and a higher glass transition temperature [26]. Thus, an SSC membrane is promising for high temperatures and low humidity operation [27]. The enhanced crystallinity also allows for the decrease of the equivalent weight of the polymer without the risk of membrane dissolution or excessive swelling. In turn, a higher charge carrier concentration can be achieved, increasing the proton conductivity of the polymer [28].

In general, when choosing a new membrane material type, two considerations are important [29–31]:

(i) durability and stability

and

(ii) compatibility with the ionomer in the catalyst layer.

The resistance of the membrane electrode assembly (MEA) is affected by the "connectedness" of the catalyst layer and the membrane; specifically, the ionomer and the membrane. Delamination of catalyst layers prepared with Nafion ionomer from s-Polyether ether ketone (PEEK) membranes was observed after some operational hours [32]; hence, affecting the long-term stability of the cell. However, the use of s-PEEK as the ionomer in the catalyst layer and as the electrolyte membrane showed enhanced adhesion between the CL and membrane.

24.3 Topics of Research

24.3.1 Contamination

Due to the various components comprising a PEMFC, contamination and degradation sources are abundant (Figure 24.8). The sources of various contaminants

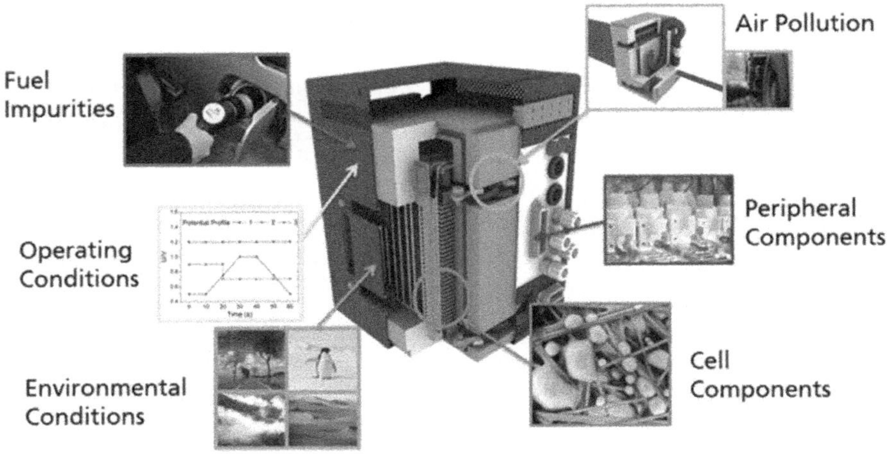

Figure 24.8 Sources of degradation in PEM fuel cell system – Fraunhofer ISE©.

originate from the fuel (H_2), oxidant (air) and component materials used in the PEMFC system. Contamination of the fuel and oxidant feed results in catalytic contamination (and in turn deactivation). Impurities in the air feed appear to be many and vary significantly depending on the air quality; however NO_x and SO_x seem to be the dominant air contaminant species. The presence of these species even at very low concentrations can result in severe losses in the performance of the cathode. At the cathode, these impurities originate from atmospheric air pollutants. Their concentration may vary from one location to the other; however, their effect is similar due to the sensitivity of the catalyst towards these contaminants. Similarly, impurities in the fuel feed, (e.g. CO, CO_2, H_2S, N_3H, and CH_4), result in severe degradation of the anode performance. Due to the low operating temperature of PEMFCs and the technical and practical issues associated with the economic method of H_2 production via steam reforming of hydrocarbon fuels (mostly natural methane gas), contamination of the H_2 fuel stream by these species is inevitable. The use of high purity H_2 as the fuel for PEMFC is associated with many limitations from infrastructure to refueling processes to on-board and off-board storage issues. The infrastructure to deliver H_2 is almost none existent, the refueling process is very slow, and the on-board and off-board storage of H_2 is very heavy. Hence, hydrocarbon fuels are usually used. To decrease the overall concentration of impurities, cleaning and purification is implemented after the reforming process, resulting in an extra cost element in production. However, impurities even at very low concentrations can still hinder the cell performance.

Cations, such as Fe^{3+}, Ni^{2+}, Cu^{2+}, and Cr^{3+} from the material of the bipolar plate are known also to hinder the performance of the PEMFC tremendously. These ions can attack the structure of the electrolyte membrane resulting in its chemical and mechanical degradation. Other contamination sources originate from sealing gaskets, coolants, and compressor oils. They act as impurities, which can block void regions in the porous components of the cell.

Table 24.2 H_2 fuel quality specification: Current ISO (ISO 14687-2:2012) and SAE directory for transportation H_2.

Species	Maximum Concentration (μmol/mol) (ppm)
Water	5
Total hydrocarbons	2
Oxygen	5
Helium	300
Nitrogen/Argon	100
Carbon dioxide	2
Carbon monoxide	0.2
Total sulfur compounds	0.004
Formaldehyde	0.01
Formic acid	0.2
Ammonia	0.1
Total halogenated compounds	0.05
Maximum particulates concentration	1.0 mg/kg

24.3.1.1 Contamination of the Anode Feed

The issue of anode feed contamination can stem from various sources. First, the use of pure H_2 as the fuel is associated with many limitations from infrastructure to refueling process to on-board and off-board storage issues. Second, with the low operating temperatures of PEM fuel cells, the use of an impure H_2 feed can result in contamination of the catalyst sites and consequently the reduction of the performance of the cell overtime. Third, the constant demand to lower the catalyst loading in the catalyst layer leads to stricter specifications for the purity of H_2. This is specifically of an issue for transportation applications. Hence, the quality of the H_2 fuel must adhere to strict specifications (i.e. Table 24.2).

In the literature, the presence of CO, CO_2, and H_2S are often investigated. These three contaminants are catalytic contaminants as they cover the available catalyst sites reducing the available active area for the HOR to proceed. In the presence of these contaminants, the mechanism in which the HOR proceeds is altered.

With the presence of CO in the fuel stream, the monoxide adsorbs at the platinum sites according to [33]:

$$\mathrm{CO} + \mathrm{Pt} \leftrightarrow \mathrm{CO} - \mathrm{Pt} \tag{24.13}$$

The adsorbed CO can then be electro-oxidized at higher electrode potentials via the "reactant pair" mechanism [33]:

$$\mathrm{CO} - \mathrm{Pt} + \mathrm{H_2O} \leftrightarrow \mathrm{Pt} + \mathrm{CO_2} + 2\mathrm{H}^+ + 2\mathrm{e}^- \tag{24.14}$$

Equations (24.13) and (24.14) illustrate the mechanism of CO poisoning [34]. The CO molecule chemisorbs on the platinum reaction sites creating an obstacle for the H_2 molecules to reach the reaction sites. This is especially true since the CO-Pt bond is much stronger than that of the H-Pt bond [35]. The removal of CO via electro-oxidation once the CO is adsorbed is very minimal since the electro-oxidation of CO on Pt occurs within a potential range of 0.6–0.9 V. With fewer available sites for the H_2 electro-oxidation, this potential range is difficult to achieve.

Similarly to CO poisoning, the presence of CO_2 in the H_2 feed can lead to contamination of the catalyst layer. Studies with the aim to understand the mechanism in which CO_2 affects the performance of the cell have shown that CO_2 encourages the formation of CO via the reverse water gas shift reaction (RWGS). [36] reported that CO_2 is electro-reduced by platinum hydrides at low potentials following Equation (24.15):

$$CO_2 + 2Pt - H \rightarrow Pt - CO + H_2O + Pt \tag{24.15}$$

The Pt – CO bond created by the reduction of carbon dioxide affects H_2 electro-oxidation in the same manner as the Pt – CO formed via CO_2 reduction. Although, the reduction of CO_2 polarizes H_2 electro-oxidation, others studies have been performed to investigate the role of water-gas shift chemistry:

$$CO + H_2O \leftrightarrow CO_2 + H_2 \qquad \Delta H^0_{298} = -41.1 kJmol^{-1} \tag{24.16}$$

The above reaction is often used for purification of H_2 produced from reforming hydrocarbons and alcohols (e.g. steam methane reforming to produce syngas mixtures or H_2). For further information in regards to this cleaning process, the reader is encouraged to visit [37–44].

Sulfur is another contaminant that must be avoided in the fuel stream. At low temperatures, a trace amount of sulfur can result in sever losses in PEMFC performance. This loss occurs due to the adsorption of sulfur compounds at the catalytic sites, thereby reducing the number of available active sites to participate in the HOR. The kinetics of H_2S poisoning have been investigated previously by [45], who proposed that there are two adsorbed forms of S on the platinum catalyst, one strongly and the other weakly bound. [46] proposed that the surface reactions followed by H_2S follow:

$$H_2S + Pt \leftrightarrow Pt - S + H_2 \tag{24.17}$$

$$Pt - H + H_2S \leftrightarrow Pt - S + \frac{3}{2}H_2 \tag{24.18}$$

The formation of platinum sulfide (Pt-S) can either follow Equations (24.17) and (24.18) or could also be formed electrochemically as suggested by [47] according to Equation (24.19):

$$H_2S + Pt \leftrightarrow Pt - S + 2H^+ + 2e^- \tag{24.19}$$

Furthermore, [48] proposed that platinum disulfide (Pt-S_2) could also be formed according to the reaction Equation (24.20):

$$H_2S + Pt - S \leftrightarrow Pt - S_2 + 2H^+ + 2e^- \tag{24.20}$$

The presence of liquid H_2O at the anode catalyst layer can also take part in the electrochemical reaction. The adsorbed sulfur can oxidize via reaction with H_2O. This mechanism (Equations (24.21) and (24.22)) was proposed by [49] and confirmed by [47] and [50]:

$$Pt - S + 3H_2O \leftrightarrow SO_3 + 6H^+ + 6e^- + Pt \quad (24.21)$$

$$Pt - S + 4H_2O \leftrightarrow SO_4^{2-} + 8H^+ + 6e^- + Pt \quad (24.22)$$

24.3.1.2 Contamination of the Cathode Feed

Similarly to the contamination of the anode feed, at the cathode, contaminants in the air stream coupled with the low operating temperature of PEMFCs can result in the poisoning of the catalyst. Some of the most common impurities are SO_x, NO_x, O_3, and volatile organic compounds. These contaminates arise from the use of atmospheric air as the oxidant of choice and can vary from one geographical location to the next, depending on the air quality of the region (e.g. urban vs. rural).

The presence of NO_2 in the air stream was investigated by [51] shedding light on the mechanism in which this contaminant affects the cell. They concluded that following repetitive poisoning of the cell by NO_2, a full recovery, once neat air was provided, was no longer attainable. Using cyclic voltammetry analysis, they concluded that NO_2 is not a catalyst surface contaminant, but rather the ionomer and/or the catalyst-ionomer interface could be affected by the exposure. It was hypothesized in the report of [51] that NO_2 is electrochemically reduced at the cathode, competing with the oxygen for Pt sites according to Equation (24.23):

$$NO_2 + 8H^+ + 7e^- \rightarrow NH_4^+ + 2H_2O \quad (24.23)$$

Further analysis of the dosage concentration by [51] also showed that the rate of poisoning is not a strong function of the concentration.

Another interesting analysis of the effects of NO_x on the performance of the cell was performed by [52]. In this work, it was concluded that a full recovery is achievable after exposure to a NO_x mixture. Keeping in mind the findings in [20], [52] proposed different mechanisms to better understand the decrease in performance due to exposure to NO_x. It was observed that NO can readily react with oxygen to form NO_2 in the presence of Pt catalysis (Equation (24.24)):

$$2NO + O_2 \rightarrow 2NO_2 \quad (24.24)$$

Comparing the concentration of NO and NO_2 in the air at the outlet of the cathode and at the end of the same manifold without a fuel cell, they found that the concentration of the NO_x is almost identical. They suggested that the reaction of nitrogen dioxide with H_2O to form nitrous acid can explain this negligible change in concentration. The reaction follows:

$$2NO_2 + H_2O \rightarrow HNO_3 + HNO_2 \quad (24.25)$$

In the presence of oxygen, however, the relatively unstable nitrous acid can be converted to nitric acid as follows:

$$4HNO_2 + 2O_2 \rightarrow 4HNO_3 \quad (24.26)$$

Nitric acid can easily dissociate in wet conditions to release protons; hence, the concentration of protons at the cathode will increase, increasing the cathode over-potential.

Similarly to the anode, sulfur compounds adsorb onto the platinum sites preventing the oxygen from being reduced to H_2O. [53–55] reported a significant loss in Pt mass activity with the presence of sulfur in the oxidant supply. They further argued that in the presence of sulfur, the reaction pathway of the ORR on the Pt/Vulcan catalyst changed from a four electron to a two electron process. [56] measured the polarization curve of a PEM fuel cell supplied with air containing 5 ppm H_2S or SO_2. They showed that the contamination of the cathode by H_2S and SO_2 followed the same trend and overlap; implying that sulfur is the determinant species. Further, they measured an increase in F^- release rate at the exhaust when the cell was supplied with S-containing gases. Consequently, it was concluded that the poisoning of the cathode catalyst by S-containing compounds inhibits the ORR ability, increases the amount of production of H_2O_2 and causes progression of the decomposition of the electrolyte membrane components of ionomers. This is also in agreement with the findings of [55].

The harmful effect of SO_2 in the oxidant stream was also reported by [57]. The authors here investigated the effect of a mixture of contaminants in the air feed. They showed that once sulfur was introduced performance was dramatically decreased. In the presence of NO_2, this reduction, however, was less. They suggested that as the gas mixture was added, the adsorption reaction of NO_2 took place first and the adsorption of SO_2 was suppressed on activated carbon.

24.3.1.3 Contamination/Degradation of the Electrolyte Membrane

Chemical degradation of the electrolyte membrane is often attributed to the oxidizing power of H_2O_2, which is formed via the two-electron ORR pathway (Equation (24.27)). In the presence of metal ions, H_2O_2 allows for the formation of radicals, which can attack the electrolyte membrane. The two-electron pathway has been investigated in various studies [58–59] and is given as:

$$O_2 + 2H^+ + 2e^- \rightarrow H_2O_2 \tag{24.27}$$

On the anode side, H_2O_2 does not form during the reduction reaction at the anode, but rather due to oxygen crossover from the cathode [60–65]. Its formation follows Equations (24.28) and (24.29):

$$H^+ + O_2 \rightarrow \bullet OOH \tag{24.28}$$

$$\bullet OOH + H^+ \rightarrow H_2O_2 \tag{24.29}$$

Metal ions, such as Fe^{2+}, Cu^{2+}, Mg^{2+}, and others, can arise from the manufacturing process or evolve during operation from the end plates [66–68]. In a study by [66], a fuel cell with stainless steel plates was found to degrade significantly after 960 hours of continuous operation. Upon replacing these stainless steel plates with iron-free, aluminum alloy plates, little or no degradation was observed over a further 1200 hours.

24.3.2 Water Management

Water management plays a significant role in the design of the components of the cell as well as its operation strategies. This stems from the necessity to increase the proton conduction properties of the electrolyte membrane and ionomer, at the same time avoiding flooding in the porous electrodes. For acceptable membrane proton conductivity, the membrane must be well hydrated without the risk of swelling due to overhydration. The membrane hydration is often achieved by the use of humidified reactant gases, where the inlet relative humidity is often carefully chosen depending on the load on which the cell operates. This intricate relationship between the origins of H_2O in the cell is of importance in order to avoid flooding of the pores in the catalyst layer and the gas diffusion layer. In the event of flooding, the pores available for the diffusion of reactants would be severely decreased, resulting in a decrease in performance. Many review papers have been written on this topic from various perspectives; the reader is referred to the references [69–73].

Appropriate H_2O management is crucial to achieve the highest performance in the FC and guarantee the stability and durability of its components over the lifetime of the cell. Some of the most widely used management strategies are:

- *Optimum operational conditions* – including inlet humidity of the feed streams, cell temperature, operating current density, back pressure, air and fuel stoichiometry, etc. Many studies can be found in literature where the effect of these conditions have been investigated; Examples are found in references [74–76].
- *Cell system design* – this management strategy relies on optimization of the overall system design, including the humidification system and the flow field design.
- *MEA material and structure design* – under such a strategy, the design of the electrolyte membrane, the catalyst layer, the gas diffusion layer, and the microporous layer is of interest. The choice of material and the structure of each component will directly affect H_2O transport in the cell. The interaction between each of these components is specifically of interest and hence, much thought is put into the interfaces between layers. Some exemplary literature covering this strategy are to be found in references [77–80].

24.3.3 Degradation

Degradation in PEMFCs is thought of in terms of their reliability, stability, and durability [81]. These terms are used to define the ability of the cell/stack to operate and serve their function under pre-defined conditions, to age in such a way that the irreversible decrease in performance does not affect the target performance and to recover power lost during continuous operation. It is, hence, necessary to quantify this definition from a long term operation perspective, specifically focusing on the loss of voltage over time. To do so, long term operation analysis is used, which is, however, expensive and time consuming. Hence, the use of accelerated stress tests to investigate the effect of various stressors on the performance of the cell is preferred.

A good summary on degradation mechanisms and mitigation strategies is given in reference [81]. The major failure modes in PEMFCs are attributed to their components, the membrane, the catalyst layer, the gas diffusion layer, and the bipolar plate. Degradation of the electrolyte membrane originates from various sources, including mechanical, thermal, and chemical stressors. Relative humidity cycling leads to a cycling in the amount of H_2O adsorbed by the membrane and eventually creates mechanical stress that leads to pinhole formation, perforations, cracks, and tears. Such mechanical failure can occur randomly, but causes catastrophic failures [82]. Thermal degradation of the polymer electrolyte can occur due to a local increase in temperature, normally referred to as a hot spot. For optimal operation, temperatures between 60–80°C are preferred, above this range breakdown of the PFSA membrane may occur. Detailed mechanisms of the breakdown of the PFSA structure under high temperatures are given in references [83–84]. The chemical degradation of the membrane occurs due to crossover of reactants from anode to cathode and vice versa. The exothermic reaction between H_2 and O_2 potentially leads to the formation of pinholes in the membrane. Furthermore, the crossover of O_2 from cathode to anode can lead to the formation of peroxide and hydro-peroxide radicals, which attack the chemical composition of the membrane and ionomer. These types of degradation were investigated in references [60, 85] and are discussed in more detail in the review article [86].

Degradation of the catalyst layer is often defined based on the stability and durability of the catalyst and its support. The degradation of the catalyst is attributed to changes in the (nano) particle size and its distribution during operation. Specifically, three mechanisms describing the structural change of the catalyst have been proposed in literature. The first phenomenon is known as Ostwald ripening [87]. Under this mechanism the small catalyst clusters dissolve in the ionomer phase and redeposit on the surface of large particles; hence, leading to particle growth. The second mechanism is described in terms of the agglomeration of catalyst particles on the support [88]. The third mechanism is attributed to the minimization of the cluster's Gibbs free energy [89]. A consensus on which phenomenon drives the clustering of the catalyst particles is yet to be formed. However, it is agreed that this change in particle size can lead to a decrease in catalyst activity. The second degradation mechanism that results in the degradation of the catalyst layer is attributed to the support. Carbon-based materials are the most widely used support in PEMFCs. However, the operation conditions of the FC can lead to its corrosion. Consequently, the loss of support is inevitable and results in the loss of the electrochemical active surface area and the reduction of catalyst layer durability and stability. In reference [90], the authors summarize alternative catalysts and supports that are under investigation to minimize the degradation of the catalyst layer.

The loss of stability and durability of the gas diffusion layer and its microporous layer is attributed to the loss of hydrophobicity and changes in its electrical and thermal properties as well as its mechanical stability. This form of degradation has been investigated by various studies (e.g. [91–92].

Finally, bipolar plate degradation plays an important role in determining the lifetime of the PEMFC. The material from which the bipolar plates are manufactured must have high corrosion and chemical resistance, low density, and high electrical and thermal conductivity. Hence, graphite composites are often used. However, due to their high manufacturing costs along with their low durability, under shock and vibration, metals are favored as the material choice. In general, noble metals possess high corrosion resistance; however, they are expensive and their use might not be sustainable long term. Hence, stainless steel is the material of choice for the mass manufacturing of metallic bipolar plates. Due to the environment in which these plates are used, they are prone to corrosion, which is harmful for the rest of the cell's components and leads to the decrease in the plate's electrical conductivity and mechanical stability [67, 93–94]. Hence, to avoid the negative consequences of corrosion, the development of coatings is crucial.

24.3.4 Estimation of Material Properties

Numerical studies are often utilized to understand the transport within PEMFCs. These studies focus on modeling the multi-phase transport of reactant and product species, moment, heat (energy), electrons, and protons in the components of the cell. These intertwined phenomena govern the operation and design and ultimately the performance of the cell. Modeling these mechanisms is often carried out from a macro-homogeneous perspective, where the actual structure of the component is ignored. Thus, the transport processes are determined with their respective effective properties. These properties include the effective diffusion coefficient for mass transfer, effective thermal conductivity for heat transfer, effective electronic conductivity for electron transfer, effective protonic conductivity for proton transfer, intrinsic and relative permeability as well as capillary pressure for fluid flow, etc. Therefore, accurate determination of these properties is crucial. In the past few years, there has been much focus on determining these properties, one from the perspective of model accuracy and two from the perspective of defining material specifications for future production. A detailed overview of research activities on the field as well as a summary of how to measure these properties is discussed in reference [95].

24.4 Characterization Techniques

Various characterization techniques, ex- and in-situ, have been employed in the investigation of the performance of PEMFCs. Basic analysis of the various components' structure involves gas sorption or Hg porosimetry analysis to determine the surface area, pore size, and distribution, SEM analysis to evaluate agglomerate sizes in the CL and to analyze the pore structure of the GDL/MPL, Transmission Electron Microscopy (TEM) to analysis Pt nanoparticle and ionomer distribution in the CL. Other ex-situ analysis involves the determination of the various parameters of the individual components, such as diffusion coefficient, thermal conductivity, electrical conductivity, capillary pressure, and permeability. In situ analyses involve the use of electrochemical characterization techniques to investigate the kinetics and mass transport resistances within the GDL/MPL, CL, and membrane. Mainly, this involves the use of cyclic voltammetry (CV) to determine the electrochemical surface area of the catalyst, linear sweep voltammetry (LSV) to determine gas crossover, electrochemical impedance spectroscopy (EIS) to determine protonic conductivity, polarization curve analysis under high oxygen stoichiometry to determine the ORR kinetics, and under load to evaluate the mass transport resistance. The following section provides a brief overview of the various techniques used to evaluate the performance of the cell.

24.4.1 Electrochemical Techniques

24.4.1.1 Polarization Curves

Fitting the polarization curve to the general Equation (24.30) can be used to estimate various parameters pertaining to the contribution of each loss mechanism.

$$E = E_r + b \log i^\circ - b \log i - R_u i \tag{24.30}$$

where E_r is the reversible potential for the electrode, b is the Tafel slope, i° is the exchange current density for the ORR in the Pt/C catalyst and R_u represents the total contributions of the polarization components in the linear region of the polarization curve, which include the electron transfer resistance, the resistance of the electrolyte and the mass transfer resistance due to the gas crossing through the diffusion and catalytic layers.

24.4.1.2 Cyclic Voltammetry

Cyclic voltammetry (CV) is utilized to estimate the electrochemical active surface area (ECSA) and the double layer capacitance (C_{DL}). As the name of the technique suggests, the potential of the fuel cell is swept back and forth between two set voltage limits and the current is recorded. The potential scanning rate is linear with time and in most studies is in the range of 1 to 1000 mV s^{-1} [95]. To record a CV, the cell is often fed with hydrated H_2 at one electrode, which acts as both the counter and reference electrode. The other side of the cell, the working electrode, is flushed with an inert gas, often N_2. The cycle is repeated until no change in the CV shape is observed.

Although cyclic voltammetry, Figure 24.9, is a well-established technique to measure the electrochemical active surface area of an electrode, one must be aware of the artifacts that can arise due to the measurement methodology; please refer to reference [96] for more details.

24.4.1.3 Linear Sweep Voltammetry

Linear sweep voltammetry (LSV) is used to determine the fuel crossover rate. Similarly to CV, LSV is run using an inert gas, e.g. nitrogen, on the cathode, while H_2 is supplied

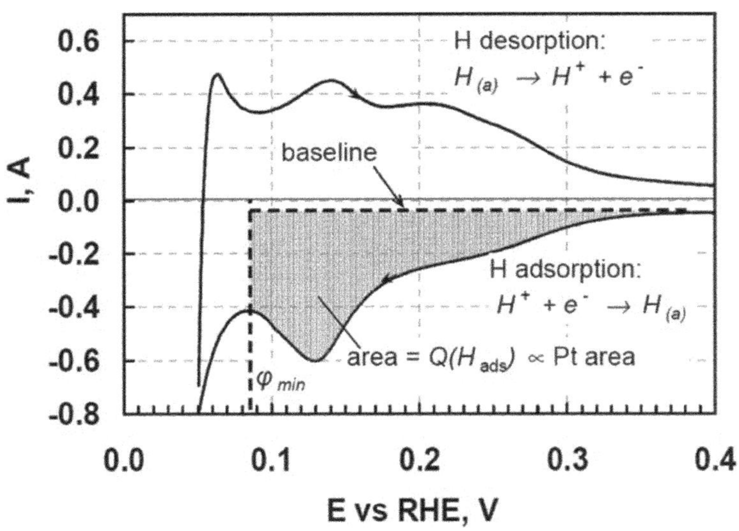

Figure 24.9 Schematic voltammogram illustrating how the total charge for atomic hydrogen adsorption is determined by integrating below the baseline drawn from the double layer capacitance region at $E \approx 0.4$ V and the cathodic CV curve down to the lowest potential minimum in cathodic current (φ_{min}) [96]. Reproduced by permission of The Electrochemical Society.

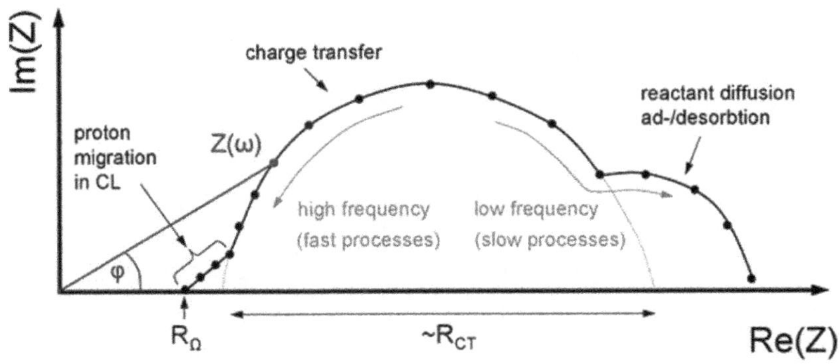

Figure 24.10 A typical electrochemical impedance spectrum of a PEM fuel cell – Fraunhofer ISE©.

to the fuel cell anode. The potential of the cathode is swept by means of a linear potential scan to potentials at which any H_2 (g) present at the cathode is instantaneously oxidized.

24.4.1.4 Electrochemical Impedance Spectroscopy

Electrochemical impedance spectroscopy (EIS) is measured based on the linear response theory, which implies that the system response to an interfering signal is linear; refer to Figure 24.10. To ensure this linearity holds, it is important to select a small amplitude of interfering signal, $\hat{\imath}$. As a rule of thumb, 2–5% of the DC signal value is often used as the perturbing signal in galvanostatic mode and $\leq$ 10 mV in potentiostatic mode [97]. The cell must reach a steady state before an EIS measurement can be carried out.

The membrane resistance can be obtained using EIS by running the cell with H_2/Air. The membrane resistance + contact and all electrical resistances are often referred to as the high frequency resistance as it is the intercept at the high frequency of the real impedance part of the Nyquist plot. The HFR was used by [98] to investigate the effect of cell potential on the hydration along the channel length of a segmented fuel cell.

24.4.2 Physical Techniques

24.4.2.1 Imaging Techniques

Imaging of various components of the cell at various stages of manufacturing is used to investigate the structural effects on cell performance. In Figure 24.11, some illustrations of how these techniques can be used to investigate the CL are given. These techniques include:

i. Cryogenic TEM can be used to study the interaction between carbon black and Nafion ionomer in the catalyst ink.
ii. Identical location TEM can be used to study the degradation process and its effect on the structure of the component.
iii. Focused ion beam (FIB)/SEM has been used to create 3-dimensional reconstruction of the catalyst layer.

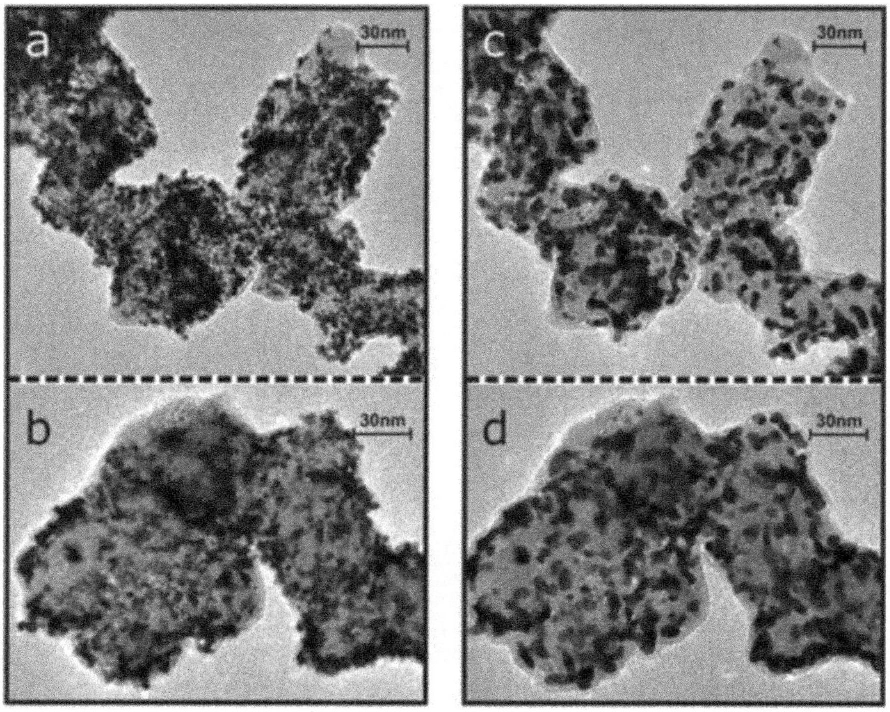

(I)

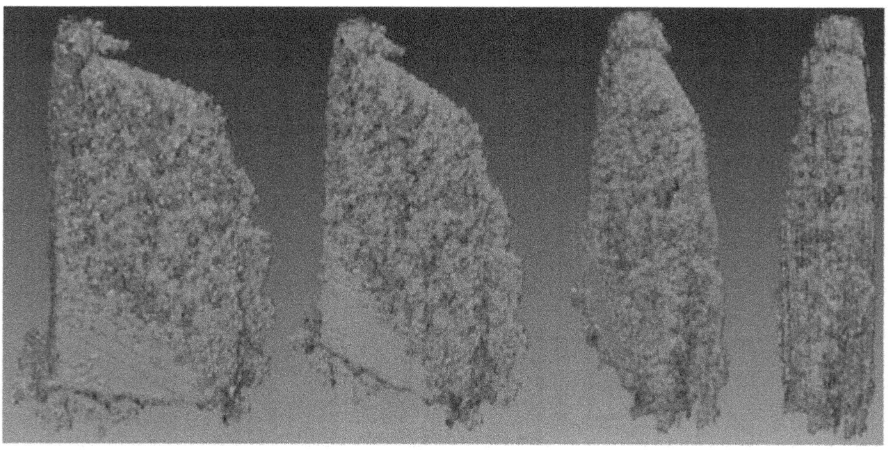

(II)

Figure 24.11 (I) Identical Location-TEM of two representative areas of catalyst II before (a and b) and after (c and d) 2 h accelerated degradation treatment consisting of potential cycling between 0.4 and 1.4 V_{RHE} at 1 V s^{-1} Reproduced from Ref. [99] with permission from The Royal Society of Chemistry.; (II) Sample A: Simultaneous presentation of 3D chemical imaging of the carbon support and ionomer components [100]. Reproduced by permission of The Electrochemical Society; (III) (a) A cube of 340 nm side length of the SEMt reconstruction. (b) The reconstructed Pt volume information in the same spatial area as the SEMt reconstruction. (c) The superposition of (a) and (b) and visualizes the location of Pt catalyst particle volume and surface information within the porous carbon matrix. [101] Reproduced with permission from Elsevier.

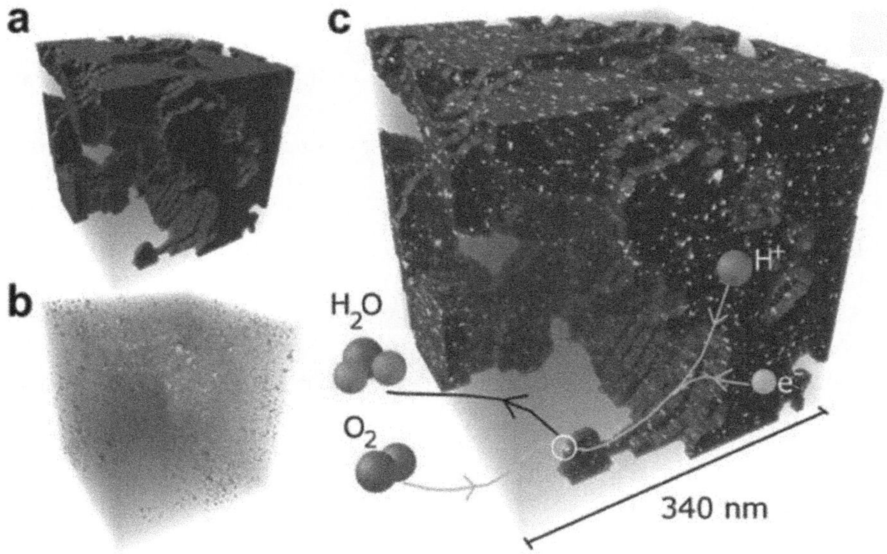

(III)

Figure 24.11 (Continued)

iv. Soft X-ray radiography can be used to visualize liquid H_2O in the cathode catalyst layer.
v. Scanning Transmission Soft X-ray Spectrotomography can be used to show the differences between materials in a component, such as interaction between the carbon-support/catalyst and ionomer in the catalyst layer.
vi. SEM can be used to shed light onto the structure of the gas diffusion layer, whilst coupled analytics (e.g. Energy-dispersive X-ray spectroscopy (EDX)) can be used to generate information regarding elemental content and distribution (e.g. via mapping).

24.4.2.2 Material Properties

Techniques to measure the material properties, gas diffusion coefficient, thermal, electrical and ionic conductivity, and permeability, of the gas diffusion layer, catalyst layer, and the electrolyte membrane have been summarized in references [90, 92, and 95].

24.4.2.3 Other Characterization Techniques

i. Dynamic Vapor Sorption (DVS) is a gravimetric technique for investigating H_2O sorption/desorption on particulate materials.
ii. Method of standard porosimetry is another method to characterize the pore size distribution, the porosity, and the capillary pressure of the CL, MPL, and GDL.
iii. ICP (inductively coupled plasma), can be used not only to characterize the material composition mainly associated with contamination, but also to measure the Pt loading in the CL.
iv. X-ray diffraction can be used to determine crystalline phases present in the catalyst layer.

v. Ultra small angle x-ray scattering (USAXS) has been used to investigate the size and micro-geometry of aggregates in the catalyst ink.
vi. Grazing incidence small angle x-ray scattering (GISAXS) is used to investigate the surface morphology of ionomer films.
vii. Current sensing atomic force microscopy (AFM) is used for measuring the local conductivity of ionomer film samples.
viii. An environmental scanning electron microscope (ESEM) is used to evaluate the impact of phase change of H_2O by cycling the H_2O inside the catalyst layer between liquid-vapor-ice.
ix. Electron spin resonance (ESR) spectroscopy is a means for estimating the number of unpaired electrons in a structure. In the case of the catalyst layer, this type of technique is used to investigate the interaction between the catalyst and its support.

24.5 Conclusions

In this chapter, an overview of H_2-based fuel cells with a focus on PEMFC was put forward. We chose to concentrate on the technology of PEMFC due to its market share and the potential of their growth over the coming years. The focus of the chapter was on discussing the fundamentals of PEMFC and the pressing technical challenges associated with their commercialization. Specifically, we summarized the intermediate steps in the oxygen reduction reaction and hydrogen reduction reaction. These two reactions are governed by the intertwined transport phenomena in the cell and are fundamental in determining the performance and durability of the cell. Hence, attention was particularly given to four research topics of interest, namely, contamination, degradation, water management, and material development. Available characterization techniques were also summarized.

References

[1] E4tech. The Fuel Cell Industry Review 2020. www.FuelCellIndustryReview.com.

[2] Tafel, J. (1905). Polarization with cathodic hydrogen liberation. *Zeitschrift fur Physikalische Chemie* 50: 641–712.

[3] Erdey-Grúz, T. and Volmer, M. (1930). Theory of hydrogen overpotential. *Zeitschrift fur Physikalische Chemie* 150: 203–213.

[4] Heyrovsky, J. (1927). A theory of overpotential. *Recueil des Travaux Chimiques des Pays-Bas* 46: 582.

[5] Damjanovic, A. and Brusic, V. (1967). Electrode kinetics of oxygen reduction on oxide free platinum electrodes. *Electrochimica Acta* 12 (6): 615–628.

[6] Damjanovic, A. and Genshaw, M.A. (1970). Dependence of the kinetics of oxygen dissolution at Pt on the conditions for adsorption of reaction. *Electrochimica Acta* 15 (7): 1281–1283.

[7] Sepa, D.B., Vojnovic, M.V., Vracar, L.M., et al. (1987). Different views regarding the kinetics and mechanisms of oxygen reduction at Pt and Pd electrodes. *Electrochimica Acta* 32 (1): 129–134.

[8] Norskøv, J.K., Rossmeisl, J., Logadottir, A., et al. (2004). Origin of the overpotential for oxygen reduction at a fuel cell cathode. *Journal of Physical Chemistry B* 108 (46): 17886–17892.

[9] Karlberg, G.S., Rossmeisl, J., and Norskøv, J.K. (2007). Estimations of electric field effects on the oxygen reduction reaction based on the density functional theory. *Physical Chemistry Chemical Physics* 9 (37): 5158–5161.

[10] Mathias, M., Roth, J., Fleming, J., et al. (2003). *Diffusion Media Materials and Characterization, Handbook of Fuel Cells-Fundamentals, Technology and Applications*, Vol. 3. John Wiley and Sons.

[11] Zamel, N., Litovsky, E., Shakhshir, S., et al. (2011). Measurement of the in-plane thermal conductivity of carbon paper diffusion media in the temperature range of -20 to +120°C. *Applied Energy* 88 (9): 3042–3050.

[12] Weber, A.Z. and Newman, J. (2005). Effects of microporous layers in polymer electrolyte fuel cells. *Journal of the Electrochemical Society* 152 (4): A677–A688.

[13] Zamel, N., Becker, J., and Wiegmann, A. (2012). Estimating the thermal conductivity and diffusion coefficient of the microporous layer of polymer electrolyte membrane fuel cells. *Journal of Power Sources* 207: 70–80.

[14] Chun, J.H., Park, K.T., Jo, D.H., et al. (2010). Determination of the pore size distribution of micro porous layer in PEMFC using pore forming agents under various drying conditions. *International Journal of Hydrogen Energy* 35 (20): 11148–11153.

[15] Niedrach, US Patent 3,297,484 (1967).

[16] Raistrick, L.D. (1986). Diaphragms, separators, and ion exchange membranes, PV86-13. In: *The Electrochemical Society Proceedings Series* (ed. J.W. van Zee, R.E. White, K. Kinoshita, and H.S. Burney), 172. Pennington, NJ. Electrochemical Society.

[17] Wilson, M.S. and Gottesfeld, S. (1992). Thin film catalyst layers for polymer electrolyte fuel cell electrodes. *Journal of Applied Electrochemistry* 22 (1): 1–7.

[18] Wilson, M.S. and Gottesfeld, S. (1992). High performance catalyzed membranes of ultra-low Pt loadings for polymer electrolyte fuel cells. *Journal of the Electrochemical Society* 139 (2): L28–L30.

[19] Holdcroft, S. (2014). Fuel cell catalyst layers: A polymer science perspective. *Chemistry of Materials* 26 (1): 381–393.

[20] Malek, K., Eikerling, M., Wang, Q.P., et al. (2007). Self-organization in catalyst layers of polymer electrolyte fuel cells. *Journal of Physical Chemistry C* 111 (36): 13627–13634.

[21] Xia, Z., Wang, Q., Eikerling, M., et al. (2008). Effectiveness factor of Pt utilization in cathode catalyst layer of polymer electrolyte fuel cells. *Canadian Journal of Chemistry* 86 (7): 657–667.

[22] Debe, M.K. (2003). Novel catalysts, catalysts support and catalysts coated membrane methods. In: *Handbook of Fuel Cells – Fundamentals Technology and Applications* (ed. W. Vielstich, A. Lamm, and H.A. Gasteiger), ch 45. John Wiley & Sons.

[23] Mehta, V. and Cooper, J.S. (2003). Review and analysis of PEM fuel cell design and manufacturing. *Journal of Power Sources* 114 (1): 32–53.

[24] Prater, K. (1990). The renaissance of the solid polymer fuel cell. *Journal of Power Sources* 29 (1–2): 239–250.

[25] Eisman, G.A. (1990). The application of Dow Chemical's perfluorinated membranes in proton-exchange membrane fuel cells. *Journal of Power Sources* 29 (3–4): 389–398.

[26] Ghielmi, A., Vaccarono, P., Troglia, C., et al. (2005). Proton exchange membranes based on the short-side-chain perfluorinated ionomer. *Journal of Power Sources* 145 (2): 108–115.

[27] Gebert, M. (2010). Aquivion™ – The short-side-chain and low-EW PFSA for next generation PEFCs expands production and utilization. Proceedings in: 18th World Hydrogen Energy Conference 2010, Essen, Germany (16–21 May 2010).

[28] Kreuer, K.D., Schuster, M., Obliers, B., et al. (2008). Short-side-chain proton conducting perfluorosulfonic acid ionomers: Why they perform better in PEM fuel cells. *Journal of Power Sources* 178 (2): 499–509.

[29] von Kraemer, S., Puchner, M., Jannasch, P., et al. (2006). Gas diffusion electrodes and membrane electrode assemblies based on sulfonated polysulfone for high temperature PEMFC. *Journal of the Electrochemical Society* 153 (11): A2077–A2084.

[30] Kerres, J., Zhang, W., Jörissen, L., et al. (2002). Application of different types of polyaryl-blend-membranes in DMFC. *Journal of New Materials for Electrochemical Systems* 5: 97–107.

[31] Kim, Y.S., Sumner, M.J., Harrison, W.L., et al. (2004). Direct methanol fuel cell performance of disulfonated poly(arylene ether benzonitrile) copolymers. *Journal of the Electrochemical Society* 151 (12): A2150–A2156.

[32] Jung, H.Y., Cho, K.Y., Sung, K.A., et al. (2006). The effect of sulfonated poly(ether ether ketone) as an electrode binder for direct methanol fuel cell (DMFC). *Journal of Power Sources* 163 (1): 56–59.

[33] Gilman, S. (1964). The mechanism of electrochemical oxidation of carbon monoxide and methanol on platinum II: The "reactant pair" mechanism for electrochemical oxidation of carbon monoxide and methanol. *Journal of the Electrochemical Society* 68 (1): 70–80.

[34] Vogel, W., Lundquist, J., Ross, P., et al. (1975). Reaction pathways and poisons-II: The rate controlling step for electrochemical oxidation of hydrogen on Pt in acid and poisoning of the reaction by CO. *Electrochimica Acta* 20 (1): 79–93.

[35] Baschuk, J.J. and Li, X. (2001). Carbon monoxide poisoning of proton exchange membrane fuel cells. *International Journal of Energy Research* 25 (8): 695–715.

[36] Giner, J. (1963). Electrochemical reduction of CO_2 on platinum electrodes in acid solutions. *Electrochimica Acta* 8 (11): 857–865.

[37] Giunta, P., Amadeo, N., and Laborde, M. (2006). Simulation of a low temperature water gas shift reactor using the heterogeneous model/application to a PEM fuel cell. *Journal of Power Sources* 156 (2): 489–496.

[38] Radhakrishnan, R., Willigan, R.R., Dardas, Z., et al. (2006). Water gas shift activity of noble metals supported on ceria-zirconia oxides. *AIChE Journal* 52 (5): 1888–1894.

[39] Batista, M.S., Santiago, E.I., Assaf, E.M., et al. (2005). Evaluation of water-gas shift and CO methanation processes for purification of reformate gases and the coupling to a PEM fuel cell system. *Journal of Power Sources* 145 (1): 50–54.

[40] Germani, G., Alphonse, P., Courty, M., et al. (2005). Platinum/ceria/alumina catalysts on microstructures for carbon monoxide conversion. *Catalysis Today* 110 (1–2): 114–120.

[41] Radhakrishnan, R., Willigan, R.R., Dardas, Z., et al. (2006). Water gas shift activity and kinetics of Pt/Ru catalysts supported on ceria-zirconia oxides. *Applied Catalysis B: Environmental* 66 (1–2): 23–28.

[42] de Bruijn, F.A., Papageorgopoulos, D.C., Sitters, E.F., et al. (2002). The influence of carbon dioxide on PEM fuel cell anodes. *Journal of Power Sources* 110 (1): 117–124.

[43] Janssen, G.J.M. (2004). Modeling study of CO_2 poisoning on PEMFC anodes. *Journal of Power Sources* 136 (1): 45–54.

[44] Minutillo, M. and Perna, A. (2008). Behaviour modeling of a PEMFC operating on diluted hydrogen feed. *International Journal of Energy Research* 32 (14): 1297–1308.

[45] Contractor, A. and Lal, H. (1979). Two forms of chemisorbed sulfur on platinum and related studies. *Journal of Electroanalytical Chemistry* 96 (2): 175–181.

[46] Mathieu, M. and Primet, M. (1984). Sulfurization and regeneration of platinum. *Applied Catalysis* 9 (3): 361–370.

[47] Mohtadi, R., Lee, W., Cowan, S., et al. (2003). Effects of hydrogen sulfide on the performance of a PEMFC. *Electrochemistry and Solid State Letters* 6 (12): A272–A274.

[48] Najdeker, E. and Bishop, E. (1973). The formation and behavior of platinum sulfide on platinum electrodes. *Journal of Electroanalytical Chemistry* 41 (1): 79–87.

[49] Loučka, T. (1973). The formation and reduction of an oxide layer on a platinum electrode partially covered with adsorbed sulfur. *Journal of Electroanalytical Chemistry* 44 (2): 221–227.

[50] Wang, Y., Yan, H., and Wang, E. (2001). The electrochemical oxidation and the quantitative determination of hydrogen sulfide on a solid polymer electrolyte-based system. *Journal of Electroanalytical Chemistry* 497 (1–2): 163–167.

[51] Mohtadi, R., Lee, W.-K., and van Zee, J.W. (2004). Assessing durability of cathodes exposed to common air impurities. *Journal of Power Sources* 138 (1–2): 216–225.

[52] Yang, D., Ma, J., Xu, L., et al. (2006). The effect of nitrogen oxides in air on the performance of proton exchange membrane fuel cell. *Electrochimica Acta* 51: 4039–4044.

[53] Garsany, Y., Baturina, O.A., Pietron, J.J., et al. (2006). Impact of sulfur dioxide on the performance of the PEMFC cathodes. *ECS Transactions* 3 (1): 685–694.

[54] Garsany, Y., Baturina, O.A., and Swider-Lyons, K.E. (2007). Impact of sulfur dioxide on the oxygen reduction reaction at Pt/Vulcan carbon electrocatalysts. *Journal of the Electrochemical Society* 154 (7): B670–B675.

[55] Garsany, Y., Baturina, O.A., and Swider-Lyons, K.E. (2009). Oxygen reduction reaction kinetics of SO_2 contaminated Pt_3Co and Pt/Vulcan carbon electrocatalysts. *Journal of the Electrochemical Society* 156 (7): B848–B855.

[56] Imamura, D. and Hashimasa, Y. (2007). Effect of sulfur containing compounds on fuel cell performance. *ECS Transactions* 11 (1): 853–862.

[57] Jing, F., Hou, M., Shi, W., et al. (2007). The effect of ambient contamination on PEMFC performance. *Journal of Power Sources* 166 (1): 172–176.

[58] Sethuraman, V.A., Weidner, J.W., Haung, A.T., et al. (2008). Durability of perfluorosulfonic acid and hydrocarbon membranes: Effect of humidity and temperature. *Journal of the Electrochemical Society* 155 (2): B119–B124.

[59] Antoine, O. and Durand, R. (2000). RRDE study of oxygen reduction on Pt nanoparticles inside Nafion®: H_2O_2 production in PEMFC cathode conditions. *Journal of Applied Electrochemistry* 30 (7): 839–844.

[60] Endoh, E., Terazono, S., Widjaja, H., et al. (2004). Degradation study of MEA for PEMFCs under low humidity conditions. *Electrochemical and Solid-State Letters* 7 (7): A209–A211.

[61] Xie, J., Wood III, D.L., Wayne, D.M., et al. (2005). Durability of PEFCs at high humidity conditions. *Journal of the Electrochemical Society* 152 (1): A104–A113.

[62] Amphlett, J., Baumert, R., Mann, R., et al. (1995). Performance modeling of the Ballard Mark IV solid polymer electrolyte fuel cell: I. Mechanistic model development. *Journal of the Electrochemical Society* 142 (1): 1–8.

[63] Aoki, M., Uchida, H., and Watanabe, M. (2005). Novel evaluation method for degradation rate of polymer electrolytes in fuel cells. *Electrochemistry Communications* 7 (12): 1434–1438.

[64] Merlo, L., Ghielmi, A., Cirillo, L., et al. (2007). Resistance to peroxide degradation of Hyflon® ion membranes. *Journal of Power Sources* 171 (1): 140–147.

[65] Inaba, M., Kinumoto, T., Kiriake, M., et al. (2006). Gas crossover and membrane degradation in polymer electrolyte fuel cells. *Electrochimica Acta* 51 (26): 5746–5753.

[66] Pozio, A., Silva, R.F., De Francesco, M., et al. (2003). Nafion® degradation in PEFCs from end plate iron contamination. *Electrochimica Acta* 48 (11): 1543–1549.

[67] Wind, J., Späh, R., Kaiser, W., et al. (2002). Metallic bipolar plates for PEM fuel cells. *Journal of Power Sources* 105 (2): 256–260.

[68] Makkus, R.C., Janssen, A.H.H., de Bruijn, F.A., et al. (2000). Use of stainless steel for cost competitive bipolar plates in the SPFC. *Journal of Power Sources* 86 (1–2): 274–282.

[69] Ji, M. and Wei, Z. (2009). A review of water management in polymer electrolyte membrane fuel cells. *Energies* 2 (4): 1057–1106.

[70] Bazylak, A. (2009). Liquid water visualization in PEM fuel cells: A review. *International Journal of Hydrogen Energy* 34 (9): 3845–3857.

[71] Li, H., Tang, Y., Wang, Z., et al. (2008). A review of water flooding issues in the proton exchange membrane fuel cell. *Journal of Power Sources* 178 (1): 103–117.

[72] Dai, W., Wang, H., Yuan, X.-X., et al. (2009). A review on water balance in the membrane electrode assembly of proton exchange membrane fuel cells. *International Journal of Hydrogen Energy* 34 (23): 9461–9478.

[73] Jiao, K. and Li, X. (2011). Water transport in polymer electrolyte membrane fuel cells. *Progress in Energy and Combustion Science* 37 (1): 221–291.

[74] Natarajan, D. and Nguyen, T.V. (2005). Current distribution in PEM fuel cells. Part 1: Oxygen and fuel flow rate effects. *AIChE Journal* 51 (9): 2587–2598.

[75] He, W.S., Lin, G.Y., and Nguyen, T.V. (2003). Diagnostic tool to detect electrode flooding in proton-exchange-membrane fuel cells. *AIChE Jorunal* 49 (12): 3221–3228.

[76] Voss, H.H., Wilkinson, D.P., Pickup, P.G., et al. (1995). Anode water removal: A water management and diagnostic technique for solid polymer fuel cells. *Electrochimica Acta* 40 (3): 321–328.

[77] Watanabe, M., Satoh, Y., and Shimura, C. (1993). Management of the water content in polymer electrolyte membranes with porous fiber wicks. *Journal of the Electrochemical Society* 140 (11): 3190–3193.

[78] Wang, E.D., Shi, P.F., and Du, C.Y. (2008). Treatment and characterization of gas diffusion layers by sucrose carbonization for PEMFC applications. *Electrochemical Communication* 10 (4): 555–558.

[79] Jiao, K. and Zhou, B. (2007). Innovative gas diffusion layers and their water removal characteristics in PEM fuel cell cathode. *Journal of Power Sources* 169 (2): 296–314.

[80] Okada, T., Wakayama, N.I., Wang, L.B., et al. (2003). The effect of magnetic field on the oxygen reduction reaction and its application in polymer electrolyte fuel cells. *Electrochimica Acta* 48 (5): 531–539.

[81] Wu, J., Yuan, X.Z., Martin, J.J., et al. (2008). A review of PEM fuel cell durability: Degradation mechanisms and mitigation strategies. *Journal of Power Sources* 184 (1): 104–119.

[82] Huange, X., Solasi, R., Zou, Y., et al. (2006). Mechanical endurance of polymer electrolyte membrane and PEM fuel cell durability. *Journal of Polymer Science, Part B: Polymer Physics* 44 (16): 2346:2357.

[83] Wilkie, C.A., Thomsen, J.R., and Mittleman, M.L. (1991). Interaction of poly(methyl methacrylate) and nafions. *Journal of Applied Polymer Science* 42 (4): 901–909.

[84] Samms, S.R., Wasmus, S., and Savinell, R.F. (1996). Thermal stability of Nafion® in simulated fuel cell environments. *Journal of the Electrochemical Society* 143 (5): 1498–1504.

[85] Büchi, F.N., Gupta, B., Haas, O., et al. (1995). Study of radiation-grafted FEP-G-polystyrene membranes as polymer electrolytes in fuel cells. *Electrochimica Acta* 40 (3): 345–353.

[86] Zamel, N. and Li, X. (2011). Effect of contaminants on polymer electrolyte membrane fuel cells. *Progress in Energy and Combustion Science* 37 (3): 292–329.

[87] Watanabe, M., Mizukami, T., Tsurumi, K., et al. (1994). Activity and stability of ordered and disordered Co-Pt alloys for phosphoric acid fuel cells. *Journal of the Electrochemical Society* 141 (10): 2659–2668.

[88] Zhai, Y., Zhang, H., Xing, D., et al. (2007). The stability of Pt/C catalyst in H_2PO_4/PBI PEMFC during high temperature life test. *Journal of Power Sources* 164 (1): 126–133.

[89] Ascarelli, P., Contini, V., and Giorgi, R. (2002). Formation process of nanocrystalline materials from x-ray diffraction profile analysis: Application to platinum catalysts. *Journal of Applied Physics* 91 (7): 4556–4561.

[90] Zamel, N. (2016). The catalyst layer and its dimensionality – A look into its ingredients and how to characterize their effects. *Journal of Power Sources* 309: 141–159.

[91] St-Pierre, J. and Jia, N. (2002). Successful demonstration of Ballard PEMFCS for space shuttle applications. *Journal of New Materials for Electrochemical Systems* 5 (4): 263–271.

[92] Lee, C. and Merida, W. (2007). Gas diffusion layer durability under steady-state and freezing conditions. *Journal of Power Sources* 164 (1): 141–153.

[93] Barbir, F. (2005). *PEM Fuel Cell: Theory and Practice*. New York: Elsevier Academic Press, 99–133.

[94] Ma, L., Warthesen, S., and Shores, D.A. (2000). Evaluation of materials for bipolar plates in PEMFCs. *Journal of New Materials for Electrochemical Systems* 3 (3): 221–228.

[95] Qi, Z. (2008). Electrochemical methods for catalyst activity evaluation. In: *PEM Fuel Cell Electrocatalysts and Catalyst Layers: Fundamentals and Applications* (ed. J. Zhang). London: Springer.

[96] Carter, R.N., Kocha, S.S., Wagner, F., et al. (2007). Artifacts in measuring electrode catalyst area of fuel cells through cyclic voltammetry. *ECS Transactions* 11 (1): 403–410.

[97] Yuan, X.Z., Song, C., Wang, H., et al. (2010). *Electrochemical Impedance Spectroscopy in PEM Fuel Cells: Fundamentals and Applications*. London: Springer.

[98] Gerteisen, D., Zamel, N., Sadeler, C., et al. (2012). Effect of operation conditions on current density distribution and high frequency resistance in a segmented PEM fuel cell. *International Journal of Hydrogen Energy* 37 (9): 7736–7744.

[99] Hartl, K., Hanzlik, M., and Arenz, M. (2011). IL-TEM investigations on the degradation mechanism of Pt/C electrocatalysts with different carbon supports. *Energy and Environmental Science* 4: 234–238.

[100] Berejnov, V., Susac, D., Stumper, J., et al. (2012). 3D chemical mapping of PEM fuel cell cathodes by scanning transmission soft X-ray spectrotomography. *ECS Transactions* 50 (2): 361–368.

[101] Thiele, S., Fürstenhaupt, T., Banham, D., et al. (2013). Multiscale tomography of nanoporous carbon-supported noble metal catalyst layers. *Journal of Power Sources* 228: 185–192.

Part V

Thermal Energy Storage

25

Thermal Energy Storage – An Introduction

Andreas Hauer and Eberhard Laevemann*

ZAE Bayern, Bavarian Center for Applied Energy Resaerch, Division: Energy Storage, Walther-Meissner-Str. 6, 85748 Garching, Germany

25.1 Introduction

25.1.1 Relevance of Thermal Energy Storage

In the public discussion on energy storage in the context of the transformation of our energy system towards an increasing share of renewable energies the focus is on electrical storage, mostly batteries. But looking at the final energy demand of most developed countries, the demand for thermal energy is more than 50%. In Germany the demand for heat and cold in the building and industrial sector was 56% in 2014 [1].

Also the installed storage capacity of thermal energy storage is much higher compared to electrical storage technologies. In Germany about 7 GW of pumped hydro plants are available. This corresponds to about 4 TWh/a assuming 600 operating hours per year. Pumped hydro stands for more than 95% of all installed storage capacity in the electricity sector. In comparison, about 22 GW thermal energy storages are installed and contribute to about 29 TWh/a (assuming about 1500 operating hours per year) [2].

25.1.2 Fields of Application

Basically energy storage can contribute to the integration of renewable energies and to the increase in energy efficiency in order to support a future sustainable energy system.

Thermal energy storage technologies are best suited for the integration of renewable heat from solar thermal systems in the building sector (heating and domestic hot water) and for the supply of process heat for industry. Today, with a higher share of renewable electricity from PV and wind in the grid, power-to-heat applications require storage capacities for a better utilization of the variable regenerative sources. The same holds for demand side management measures, where thermal energy storage can be a tool to keep up production.

Thermal energy storage solutions can help in improving energy efficiency in the building sector e.g. by increasing the thermal capacity of the building itself thus leveling out the day/night temperature differences. Also industrial processes can be optimized by the utilization of waste heat via thermal storage devices. Electricity production by combined heat and power systems can be made independent of the actual heat demand by thermal storage installations. Even in the mobility sector thermal storage can support the air-conditioning of electric vehicles.

Advances in Energy Storage: Latest Developments from R&D to the Market, First Edition.
Edited by Andreas Hauer.
© 2022 John Wiley & Sons, Ltd. Published 2022 by John Wiley & Sons, Ltd.

This variety of application fields can be served by a number of different thermal energy storage technologies, with a broad spectrum of different storage properties.

25.2 Characteristic Parameters of Thermal Energy Storage

25.2.1 Prologue: What Is Thermal Energy?

Before listing the characteristic parameters of thermal energy storages, one particularity of thermal energy in comparison to other forms of energy needs mentioning. Thermal energy – heat and cold – is process-dependent, not a state variable. Other than state variables like temperature, pressure, or entropy, process-dependent variables are not defined for a certain state but for a certain process. In other words they do not allow for true statements in the form of: "Medium A contains amount X of heat (or cold)." Only statements like "Medium A has the thermodynamic potential to transfer amount X of heat (or cold) to a different, defined medium" can be true. Therefore, also the storage capacity, the amount of heat/cold stored in a given volume or mass of storage medium, is a process variable and not a material property.

Consequently, the amount of heat stored is only properly defined for given reference conditions.

25.2.1.1 Fundamental Steps for the Storage of Thermal Energy

A storage process consists of a minimum of three steps (Figure 25.1):

1. Charging
2. Storage
3. Discharging

A thermal energy storage is charged with, stores, and supplies heat/cold energy. Since heat is a process variable, all these steps need to be taken into account, when quantifying the storage of thermal energy.

25.2.1.2 Energy Conservation, Exergy, and Anergy

The principle of conservation of energy renders the term "energy consumption" technically incorrect, meaning it makes no sense to speak of "energy storage" since energy is conserved anyway.

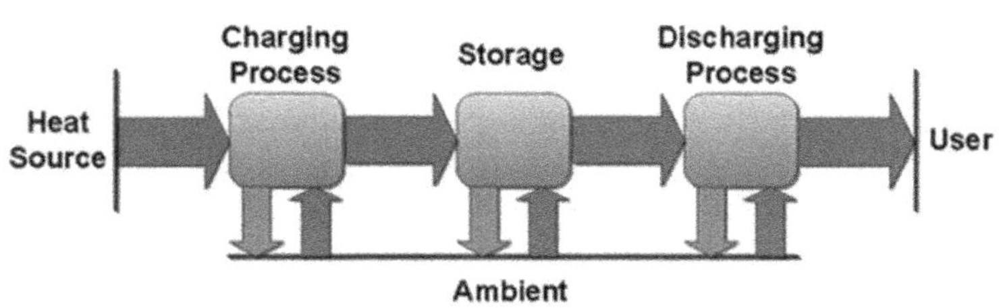

Figure 25.1 Thermal energy storage is a combination of several processes in which the storage is charged, thermal energy is stored, and, finally, supplied to a consumer. (ZAE Bayern).

Nevertheless, different forms of thermal energy are more or less “usable” than others. Take, for example, a pot of boiling water which slowly cools down to ambient temperature. Once said has been reached, the cooled water can no longer be employed as a heat source. The energy, however, has been conserved. It has only been transformed into a less “usable” form.

To describe this connection, the concept of exergy and anergy was introduced. Figure 25.2 shows equations splitting a certain quantity of heat ΔQ into exergy and anergy. Exergy can be transformed into work. Anergy cannot be utilized any further and has to be discharged to the surroundings after exergy has been separated. Exergy describes the quality of the heat stored. Its definition always depends on a reference case defined by the current surrounding conditions.

In Figure 25.2, the ambient temperature is T_a. More generally formulated, the chemical potential of the surroundings is the reference condition. The exergy of the stored energy, with regard to the temporal and local ambient conditions of the consumer, can be considered a quality measure for said energy. The amount of energy readily available to the user equals this value.

Figure 25.3 shows how the shares of exergy and anergy vary during charging, storage, and discharging. Although the energy is conserved, the exergy (depicted as light grey blocks) disappears and is given off to the surroundings as anergy. The temperature of the stored heat decreases to the ambient temperature level over the process.

Figure 25.4 shows that the heat’s temperature does not necessarily have to decrease during the process. E.g. a heat pump can be used to raise the amount of heat stored by raising the share of exergy. This can increase the storage capacity.

25.2.2 Storage Capacity

The storage capacity is the maximum amount of energy made available to the consumer by a storage under defined process conditions. This value, as aforementioned, is highly dependent on a reference case.

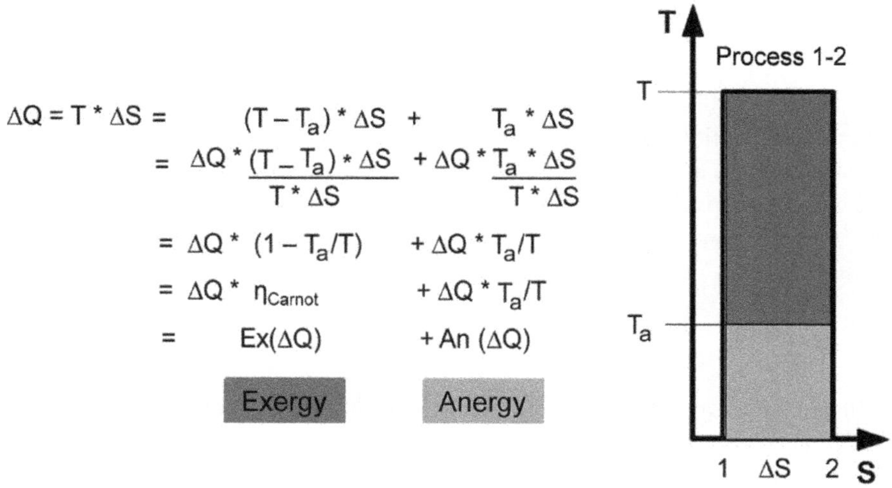

Figure 25.2 The concept of exergy and anergy in reference conditions (ambient), with ΔQ representing heat, T representing temperature, ΔS representing the difference in entropy and η_{Carnot} representing the Carnot efficiency value. (ZAE Bayern).

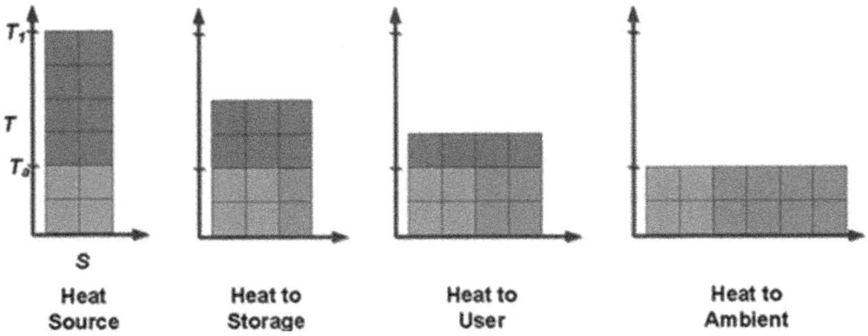

Figure 25.3 When heat is stored and consumed it is converted into anergy. Energy, being the sum of both terms, is conserved. (ZAE Bayern).

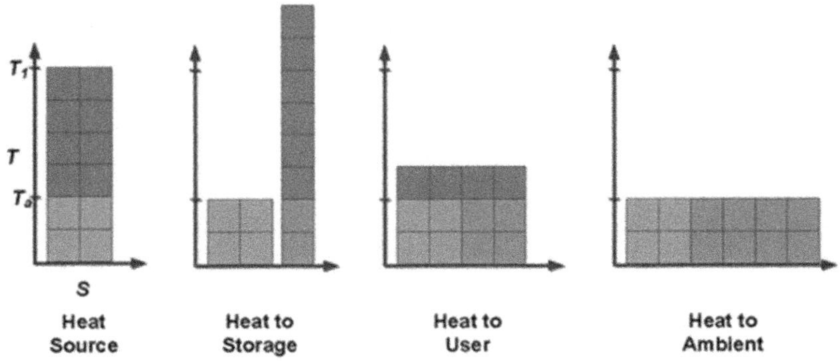

Figure 25.4 When using heat pumps the temperature of the energy stored can be above the temperature of the heat supplied by the source. (ZAE Bayern).

The storage capacity depends on the physical or chemical effect employed. If a storage medium is heated, its capacity is proportional to the temperature deviation from the reference value. If the heating furthermore triggers a phase change, the enthalpy of melting or evaporation adds to the capacity. For chemical heat storage reactions, the enthalpy of reaction is critical. This, in turn, is highly dependent on the ambient conditions.

Storage capacity can be given as an absolute value in J or kWh. Usually, however, it is set as a specific, volumetric (with regard to volume), or gravimetric (with regard to mass) value. In the literature, specific storage capacity is usually given with regard to the storage medium alone. For particular applications, however, regard to the entire storage component, including e.g. insulation or even the entire storage system, makes more sense. For economic evaluation, the storage capacity needs to be put in proportion to storage cost.

25.2.3 Thermal Power

Charging and discharging power indicate how quickly a storage can load and unload a specified amount of heat. The value is, again, dependent on process parameters and not a state variable of the storage.

Power is energy over time [J/s]. This value is crucial for the optimization of a storage for the charging and discharging profile of its current application. The requirements for charging and discharging can vary widely. For example, solar thermal water heating can be performed at low power levels. For discharging, however, the full amount of power needed to, for example, fill a bathtub needs to be available at once.

The thermal power – contrary to the storage capacity – does not depend on the physical or chemical storage process. The crucial parameter here is the heat transfer and also, for chemical reactions, the mass transfer. Besides the driving slope, defined by the total temperature difference, the technical implementation of, for example, a heat exchanger or a reactor with a specified exchange surface is important. The power of a storage is likely to be given as a nominal value for design parameters. The actual power then depends on the present process parameters.

The charging/discharging power of a storage can (analogous to its storage capacity) be given as absolute values (in W) or in reference to the storage's volume or mass (in W/m^3 or W/kg). For an economic view, the storage's performance needs to be put in proportion to the cost.

25.2.4 Storage Efficiency

The storage efficiency describes the ratio between energy taken up during charging and energy supplied during discharging. Generally, the amount of useful heat tends to be smaller than the amount fed into the storage due to the storage's losses during the storage period. Also, the amount of exergy stored decreases in proportion with temperature.

The storage efficiency is highly dependent on the conditions of the storage process and the system it is integrated into. Especially storage duration and temperature level, in combination with the storage's thermal insulation (except in some chemical storage reactions) and the ambient temperature determine its loss and, therefore, its storage efficiency. Therefore, besides the relation between energy levels, the relation between charging and discharging exergy levels can be defined as the "exergetic" storage efficiency. Herein, the loss of "quality" (= temperature) of stored heat is the decisive factor. Comparing different storage technologies is difficult, since all boundary conditions have to be taken into full account for doing so.

The degree of utilization is dimensionless and commonly expressed as a percentage.

For a number of chemical reactions, the storage efficiency is determined differently than for sensible or latent storages.

25.2.5 Storage Cycles

The term "storage cycle" denominates the time elapsed between the charging and discharging of a storage. The sum of charging, storage, and discharging time (and, possibly, another period before the next charging step) equals the duration of one storage cycle. A "full" cycle describes the maximum amount of thermal energy charged and discharged to and from the storage.

The actual storage cycles are dependent on the heat source's (supply) and consumer's (demand) temporal profiles, as well as on the storage capacity and charging and discharging power of the storage. Storage cycle length for thermal energy storages varies from minutes to as much as one year. The storage of solar heat from summer to winter, for example, requires the longest storage cycles for thermal energy storage applications. In this application, charging and storage can begin as early in the year as spring, while discharging can begin in autumn and last all through winter. The cycle, from "fully discharged" (at the end of winter) over "fully charged" (at the end of summer) back to "fully discharged," elapses one year in this application. Under realistic conditions, whatsoever, there is usually an overlap of seasonal and short-term storage (by "non-full" cycles), resulting in more than one single storage cycle per year (more likely 1.2–1.5 cycles).

If a storage cycle includes irreversible processes that compromise the storage's capacity, the possible number of storage cycles is limited. The stability of thermal storages is determined by their degradation over a specified number of cycles.

Running a large number of annual cycles has a positive effect on a storage's economic efficiency. Under such conditions, the system has a high energy throughput, frequently provides stored energy, and, therefore, can quickly amortize.

25.3 The Physical Storage Principle – Sensible, Latent, and Thermochemical

The processes the storage of thermal energy is based on can be characterized thermodynamically. The different types of storage – sensible, latent, and thermochemical – thereby, are subject to different dependencies. While for sensible heat storage, a change in heat content equals a temperature change of the storage medium, for latent heat storage, this process is accompanied by a phase change, i.e. a change in physical structure. Thermochemical storages use a chemical reaction or an ad- or absorption process (based on physical forces). For these, the enthalpy of reaction determines the energetic throughput.

The transition between physical and chemical structural change is an aspect of the binding forces among the elements and, therefore, of its strength. Theoretically, this transition is fluent. Weak bonds, up to and including the hydrogen bond, are usually called physical bonds.

Thermal energy can be stored in different ways. Subsequently, the terms "sensible" and "latent" heat as well as thermochemical storage will be introduced [3].

25.3.1 Sensible Storage of Thermal Energy

For sensible storage a storage medium is heated up or cooled down. Due to its high specific heat capacity, non-toxicity, and low price the medium most commonly used is water. Small storages are used as buffer storages in solar thermal systems (for water heating) for storage over several days or weeks. Large storages (of up to several thousand m^3) are designed for seasonal storage of solar heat, usually coupled with a district heating network. Existing solar district heat projects with large seasonal heat storages in Germany, Sweden, Canada, and other countries were designed to cover about more than half of the overall heat demand for room and water heating with solar energy. Heat and cold can also be stored underground. For example, thermal energy can be extracted from the ground during winter with a heat pump and cooled down to about

5°C. During summer, on the other hand, the "cold" from the ground can be used directly for room cooling.

Solids can particularly be used for sensible heat storages at very high temperatures (< 500°C). Rock, sand, and concrete storages are the subjects of current research in these temperature ranges.

25.3.1.1 Specific Heat Capacity

Decisive for a storage's overall heat capacity is, foremost, its specific heat capacity. The heat absorbed therefore is

$$Q_{Sens} = \int_{T_1}^{T_2} C_P dT = \int_{T_1}^{T_2} \left(\frac{\partial S}{\partial T} \right)_P dT = \bar{T} \Delta S_{12}$$

whereby the absorbed heat Q equals the product of the specific heat capacity c_P and the increase in temperature from T_1 to T_2. If this specific heat capacity is replaced with the change of entropy per temperature, the heat equals the product of average temperature $\bar{T}$ and enthalpy increase ΔS. This proves that a material's potential capacity for entropy is linked to its storage capacity and a physical quality of the storage material.

25.3.2 Latent Heat Storages

In addition to the change in temperature, latent heat storages employ a phase change of the storage medium. Due to its sufficiently high enthalpy and sufficiently small effect on the medium's volume, the change usually employed is melting and solidification between the liquid and solid state. Hence, these storages allow for significantly larger amounts of thermal energy to be stored at small temperature differences than in sensible storages. This is advantageous, among others, in cold storage. So-called Phase Change Materials, PCMs, with a melting point of 20–25°C integrated into a building's structure can help maintain a comfortable room temperature and prevent overheating (Figure 25.5). The high storage capacity helps to keep the storage geometry compact.

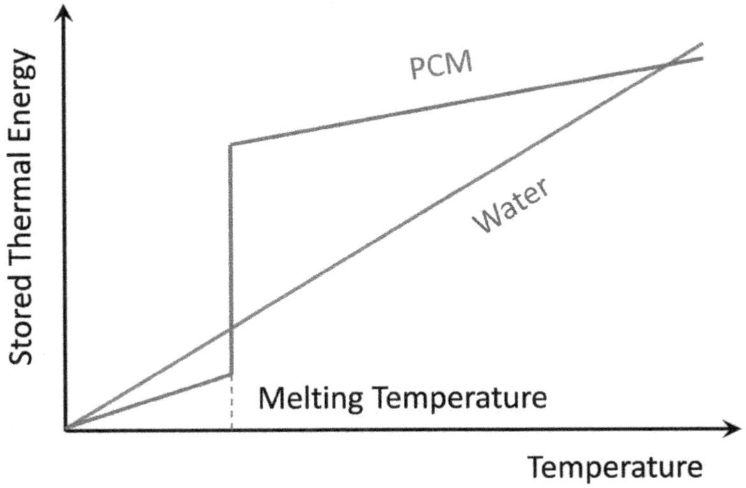

Figure 25.5 Dependency between stored heat and temperature for sensible and latent storage.

Suitable storage materials for a wide temperature range of about −40°C to far exceeding 1000°C are known.

So-called Phase Change Slurries (PCS) are interesting, especially for the transportation of stored energy, since they remain pumpable when solid. In slurries, the PCM is suspended in a carrier fluid which remains liquid or the material is only partly frozen with solid crystals floating in the remaining liquid. Ice slurries for storage and distribution of heat/cold in buildings or industrial facilities are the most common application of these PCS.

25.3.2.1 Heat of Fusion

For a latent heat storage, the stored heat is

$$Q_{\text{Lat}} = \int_{T_1}^{T_2} c_p dT + \Delta H_{\text{is}} = \bar{T}\Delta S_{12} + \text{T}_{\text{is}}\Delta S_{\text{is}} \approx \bar{T}(\Delta S_{12} + \Delta S_{\text{is}}) > Q_{\text{sens}}$$

whereby ΔH_{ls} is the enthalpy of fusion for the phase change between solid and liquid. During this phase change, which happens at temperature T_{ls}, the change of entropy ΔS_{ls} happens. Therefore, the stored heat Q_{Lat} of a latent heat storage is usually larger than the stored heat of a sensible storage Q_{Sens} if the rise in temperature conducted includes a phase change at all.

In theory it would be possible to utilize the phase change between liquid and solid. In this case much higher enthalpies would be involved. For example the heat of evaporation of water is about 10 times higher than the heat of fusion. However it remains almost impossible to deal with the very large volume change during evaporation in any technical application concerning design and construction of the storage so far.

25.3.3 Thermochemical Storage Processes

Another means of storing thermal energy is via reversible chemical reactions. Such systems have a high energy storage density and can adjust their temperature level to changing requirements during charging and discharging. The most thorough research in this area has been conducted on ad- and absorption processes. Open sorption storages are presently being evaluated in terms of their suitability for the utilization of industrial waste heat. Especially in combination with industrial drying processes, there is potential for efficient and economically interesting systems. The possibility of – in addition to storage – transforming heat into cold is put to use in applications such as solar air conditioning in buildings.

The term thermochemical heat storage is somewhat fuzzy since it refers to a range of storage systems not solely based on chemical reactions. Indeed, sorption storages are more likely to be “physical” storages. Nevertheless, this book prefers the common term.

25.3.3.1 Enthalpy of Reaction

Thermochemical heat storage processes, such as the adsorption of water vapour on solid adsorbents, can be understood as reversible chemical reactions. The stored energy equals the specific enthalpy of the underlying reaction. The ideal reaction scheme would be the thermal, e.g. forced by the heat of absorption (endothermal), reversible dissociation of a condensed (i.e. solid or liquid) compound AB into a condensed reaction product A and a gaseous product B:

$$AB \leftrightarrow A + B_{(g)}$$

B is preferred gaseous for two reasons, since a gaseous phase can easily be separated from the condensed phases A and AB [4]. Thereby, loss-free long-time storage can be achieved through separation of the reactants and prevention of the reverse reaction. The gas formation reaction in the direction AB → A + B, i. e. during charging of the storage, causes a sharp rise in entropy. This increases the storage's capacity for heat absorption.

The gaseous component can be stored separately after the dissociation process. This, however, calls for large storage volumes, significantly reducing storage density.

In order to avoid the necessity of incorporating the entropy into the thermochemical storage, the gaseous phase can be condensed in a closed adsorption system if the operating conditions allow this. Such a storage is shown schematically in Figure 25.6. The gaseous phase exits the absorber and is condensed in the condenser. Waste heat Q′ and entropy S′ exit the system during condensation. For determination of the energy storage density, the amount of stored heat can now be put in relation to the volume of the adsorbent plus the volume of the condensed component B.

For the discharging process, the arrows in Figure 25.6 reverse their direction. During the reverse reaction, heat from the surroundings is used to vaporize the condensed component B. Therefore, the surroundings are deprived of heat: they are cooled. In the reactor, gaseous component B reacts with component A and the reaction heat is released. When discharging such storage systems, heat can be supplied as well as cold.

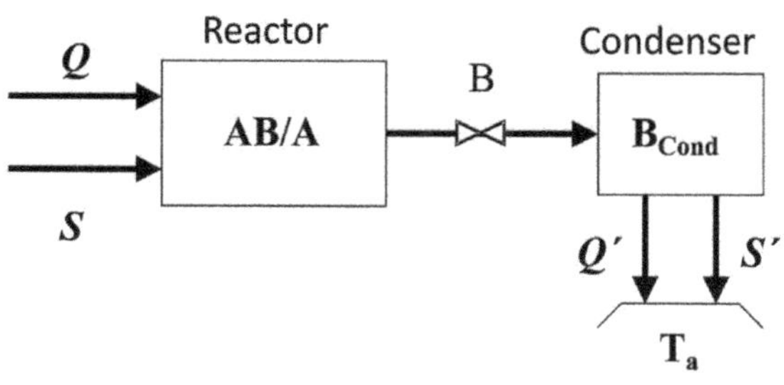

Figure 25.6 Closed heat storage via thermochemical dissociation of AB and subsequent condensation of gaseous component B.

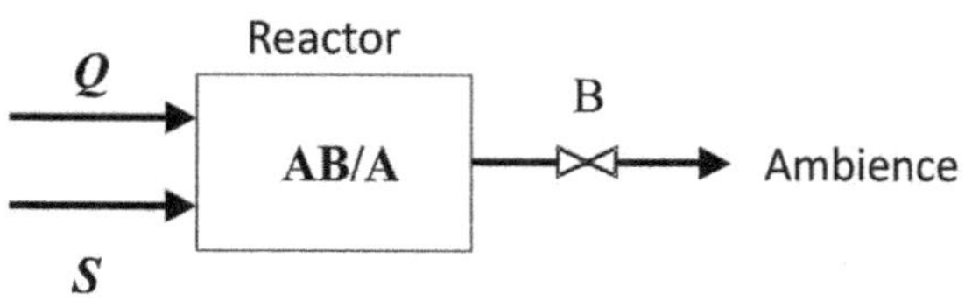

Figure 25.7 Open thermochemical heat storage.

Open adsorption systems provide another option: If the gaseous component B is a natural constituent of the atmosphere, it can be stored temporarily in the surroundings without adding to the storage volume (cf. Figure 25.7). In this case, the entropy exits the storage system with the substance flow of the gaseous component B. The attainable energy storage density is then defined as the heat available from the storage per volume (or mass) of the condensed component A only. In thermochemical storages, same is the adsorbent. This advantage makes open adsorption systems using water vapour very interesting for heat storage. However, component B needs to be available in an adequate concentration during discharging.

25.4 Design of a Thermal Energy Storage and Integration into an Energy System

25.4.1 From the Storage Material to the System

Thermal storages consist of several components: the storage material itself, installations which facilitate transport of thermal energy into and out of the material, and, finally, a container holding the material–likely to be insulated. The storage is integrated into an energy system from which it is fed with and into which it feeds energy. Figure 25.8 shows the storage material, the component, and the system schematically [3].

25.4.1.1 Storage Material

The storage material takes in thermal energy. In doing so, its physical state changes. A warm water storage, for example, will experience a rise in temperature during charging.

For determining the storage's volumetric capacity, the density of the storage material evidently plays a decisive role.

For thermal storages utilizing a phase change or reversible chemical reactions, other material characteristics, such as melting point, melting enthalpy, and heat of reaction need to be taken into account.

Other important factors for the characterization of storage materials, regarding a system's charging and discharging power, are thermal conductivity and, where applicable, viscosity. For reasons of practicality, the material furthermore needs sufficient physical integrity under working conditions and consistent material properties.

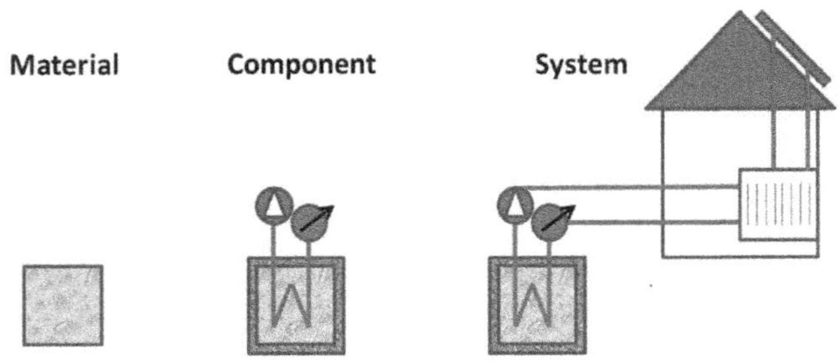

Figure 25.8 Schematic representation of storage material, storage component, and storage system (ZAE Bayern).

For practical applications, the material cost is another important factor. It can depend on the manufacturing process as well as on the quantity produced. The former can pose a problem when looking for new, suitable materials. Different properties, such as toxicity, recyclability, or availability, can also be decisive factors for practical use.

Materials research in the field of thermal energy storage is a highly active branch of research. New material classes are constantly examined and tested for their usability in storages.

25.4.1.2 Storage Component

When making a storage out of a storage material, it has to be made sure that the necessary amount of thermal energy can be transported into and out of the storage over the desired time span. If the thermal conductivity of the storage material alone is insufficient for this task, it requires the integration of heat exchanging structures into the material. Such enhancement of charging and discharging power is necessary for many practical applications. Exceptions from the norm are cases in which the storage medium is also the heat carrier or even the utilized medium (as in hot water storages).

The component, made up the storage material and some form of heat exchanger, is commonly placed inside of a thermally insulated container. This container has a double functionality: on the one hand, it encloses and protects the storage material and the other installations, on the other hand, it minimizes thermal losses and helps to pass the greatest possible share of exergy on to the user.

The boundaries of the term storage component are fuzzy. In some applications further elements such as pumps, volume flow gauges, valves, or the like are also considered parts of the storage component.

The storage component, consisting of storage material, heat exchanger, and container, can in its entirety be characterized quantitatively by its storage capacity, which is smaller for the entire component than for the storage material alone, or by its thermal power output.

25.4.1.3 Storage System

The storage component now needs to be integrated into a storage system. It interacts strongly with this system. Only in this interplay of storage component and system can the characteristic variables storage efficiency and storage cycle be properly defined. The system includes the energy source used to charge the storage; it determines the period of time over which it is stored, and which requirements the user has for the storage. In particular, the system determines the range of temperature available or necessary.

Only when integrated into a system can the economic efficiency of a storage become visible and the heat supply quantified.

The system is defined as including all the components necessary for the full storage cycle: source, storage, and sink. The design of a system needs to correlate to a particular problem such as: how to heat a building using solar heat? In this scenario, the building, equipped with a solar collector, warm water storage, and heating, is the system. The storage needs to be compatible with the temporal profiles of heat supply and heating demand.

A larger system can be chosen for different problems like, for example: How can the entire German energy supply be switched to renewables? In this case, the system includes the entire German energy grid with all of its sources and consumers.

Calculating the storage demand and finding the (technically and economically) best way to cover it for a system like that are, of course, extremely complex problems, which are currently the subject of intensive research.

25.4.1.4 Interplay of the Different Levels

The materials are developed into components which are fitted to a system. This system is then supposed to cover the storage demand. Simultaneously, the material needs to be fitted to the component in the system. Therefore, the material developer needs to have a clear vision of the final system to be able to develop an adequate material.

In summary, the transitions between the levels (material, component, and system) need to be clearly defined in order to avoid misconceptions, misinterpretations, and misdevelopments of a technical and economic nature.

25.5 Thermal Energy Storage Classification

The wide variety of thermal energy storage types and possible applications make it difficult to maintain an overview of the field. Therefore, this chapter introduces a classification of storage types in their respective applications. This classification tool allows for a concise characterization and categorization of thermal storages.

25.5.1 Which Demands Should the Storage Meet in Application?

Thermal energy storages can be classified according to the qualities demanded from them in practical applications. Criteria are the application itself, the demanded discharging temperature, or specific qualities, such as capacity or thermal power output.

25.5.1.1 Classification According to Area of Application

Most thermal storages are integrated into a system in order **to raise its efficiency** or **to integrate fluctuating energy sources** generally aiming, on a higher level, at a reduction of primary energy demand and CO_2 emission.

An increase in efficiency is usually achieved by storing heat/cold, which cannot be used at the time, for use at a later time or in a different place. A typical example is the utilization of industrial waste heat. For the integration of renewables, tasks such as the reduction of peak loads or supply at low availability are in focus.

Thermal energy storages can be classified, according to their physical location in an energy system, as **central** (installed near the energy source) or **decentral** (installed near the consumer). Hence, a molten salt storage in a solar thermal plant is central, whereas a latent heat storage for storing cold in an office building is decentral (or "distributed").

Most applications of thermal energy storages call for **stationary** systems, installed in a permanent location. This is equally true for central and decentral storages. A second option are **mobile** heat storages which facilitate the transport of surplus energy to a potential consumer. Possible realizations of this concept are the transport of the entire storage container or pumping of the storage medium (e.g. PCM-Slurries) over long distances.

Particularly high demands have to be met by storages built to provide energy independently from the grid. These storages need to meet the requirements laid down by the consumer in terms of capacity and energy provision.

25.5.1.2 Classification According to Temperature Level

Figure 25.9 depicts different temperature levels for possible applications of thermal energy storages.

Temperature levels are commonly divided into three sectors.

- Cold
- Heat
- High temperature

The cold sector includes freezing (temperatures below −18°C) and cooling (normally 5–18°C). Also, for cold a distinction can be made between cooling of goods, e.g. foods (normally at 5–18°C), and cooling of rooms, i. e. air-conditioning (normally between 6 and 18°C).

The heat sector can also be subdivided: space heating temperatures usually range between 25°C (surface heating) and 90°C (radiators). For hot water supply, for hygienic reasons, temperatures exceeding 65°C need to be temporarily possible. Storages in heating systems usually require low power output, while storages for service water supply require very high power output.

The high temperature storage sector is very heterogeneous and bundles all applications beyond the heat sector. It includes process heat for the food industry (e.g. for sterilization) at temperatures just above 100°C as well as heat supply for solar thermal plants at over 400°C.

25.5.1.3 Classification According to Storage Capacity, Thermal Output, and Storage Time

Different applications pose different requirements for storage properties, storage capacity, charging and discharging capacity, as well as for storage time.

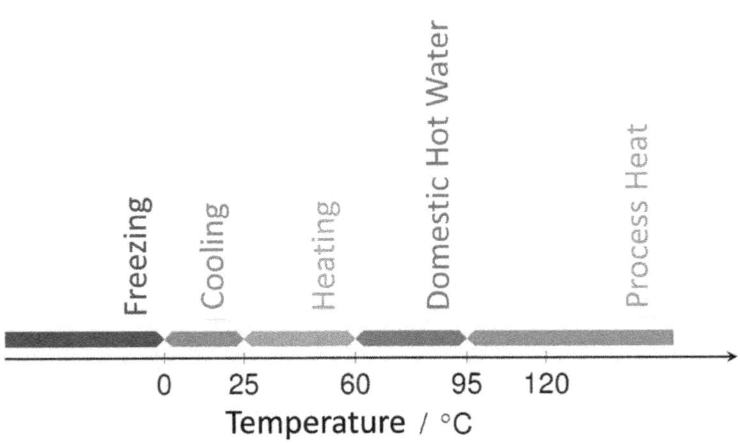

Figure 25.9 Rough classification of temperature levels referring to typical applications.

Storages with **high storage capacity/energy density** are most important in areas where space is scarce or large amounts of energy need to be stored. Mobile storages in general make for a good example of such since large volume or high weight are clearly disadvantageous for them.

Storages with **high thermal power output** are needed for areas such as hot water supply. Certain heat storage technologies, such as latent heat storages, are less suitable for such tasks in principle. The high thermal output levels necessary would have to be achieved by addition of complex heat exchanging components, while in conventional hot water storages, where the supplied hot water is, at the same time, the storage medium, high thermal output is much easier to achieve.

Another possible distinction is between **long- and short-term storages**. Long-term storages have storage cycles of several weeks up to one year (seasonal storage). Thermal short-term storages store heat over time spans ranging between hours and days. The two categories differ vastly in which storage technologies are suitable for them in, both, technological and economic respect.

As a general rule, long-term storages need to be capable of storing large amounts of energy at low charging and discharging power. Short-term storages, on the other hand, need high power output but store small amounts of heat.

25.6 Conclusions

Since thermal energy stands for more than 50% of the final energy demand in Europe, thermal energy storages can substantially support Europe's heat and cold supply. Applications like power-to-heat and power-to-cold combined with storage technologies are able to open the heat and cold market to the electricity sector and provide a continous and economic availability.

The different thermal energy storage technologies – sensible, latent, and thermochemical – can cover the wide range of applications with their unique storage properties and will play an important role within our future energy systems. However it remains important to choose the most suitable technology for the actual application. Therefore the complete system – energy source, storage, and final consumer – and its temporal supply and demand fluctuations have to be taken into account.

Acknowledgments

Many of the definitions and expressions used in the text were developed from discussions within the energy storage programme of the International Energy Agency, IEA, which is officially called Technology Cooperation Palatform "Energy Conservation through Energy Storage, ECES." A very similar discussion within the German Energy Storage Association, Bundesverband Energiespeicher BVES, also helped to clarify many of the elements in this topic. We want to thank all colleagues involved in this process for their valuable contributions.

References

[1] Arbeitsgemeinschaft Energiebilanzen (AGEB), Bundesverband der energie- und Wasserwirtschaft BDEW. http://www.bmwi.de/DE/Themen/Energie/Energiedaten-und-analysen/Energiedaten/energiegewinnung-energieverbrauch.html

[2] Stawiarski, K.-H. (2013). *Das Potential der Wärmepumpe Markttrends, Wohnkomfort, Umwelt- und Klimaschutz*. Berlin: Bundesverband Wärmepumpe (BWP) e.V., ISH 2013. https://docplayer.org/29043151-Das-potential-der-waermepumpe-markttrends-wohnkomfort-umwelt-und-klimaschutz-karl-heinz-stawiarski-bundesverband-waermepumpe-bwp-e-v.html

[3] Hauer, A., Hiebler, S., and Reuß, M. (2012). *Wärmespeicher*. BINE Fachbuch, Fully revised 5th edition. Stuttgart: Fraunhofer IRB Verlag. ISBN 978-3-8167-8366-4

[4] Sizmann, R. (1989). Speicherung thermischer Energie – Eine Übersicht. Technical report. Stuttgart: BMFT Statusseminar Thermische Energispeicherung.

26

New Phase Change Materials for Latent Heat Storage

Elena Palomo del Barrio[1] *and Fouzia Achchaq*[2]

[1] *CIC energiGUNE, Albert Einstein 48, Miñano, Álava 01510, SP*
[2] *Institut de Mécanique et d'Ingénierie (UMR CNRS 5295), Université de Bordeaux, CNRS, Talence, FR*

26.1 Introduction

In the framework of thermal energy storage (TES), the term phase change materials (PCMs) refers to a class of materials that exploit phase transition phenomena for storing energy. Compared to sensible heat storage, the storage technologies based on PCMs (latent heat storage technologies) are particularly attractive for applications where energy has to be stored/delivered over a narrow temperature range or/and when compactness is a premium. Indeed, PCMs are capable of absorbing or releasing a great amount of energy in a form of latent heat during transitions at nearly constant temperature. They enable compact TES systems with volumetric storage capacity five to ten times greater than that of sensible heat storage systems.

The study of PCMs, pioneered by [1], did not received much attention until the late 1970s, when PCMs started to be applied on a larger scale for buildings and solar energy systems. Most early studies of latent heat storage focused on inorganic salt hydrates because of their high energy storage density, relatively high thermal conductivity, and low cost. However, they have some obvious disadvantages such as supercooling, cycling instability, and corrosion. In an effort to avoid some of the problems inherent to salt hydrates, the research interest turned toward the study of numerous organic substances and their mixtures as novel PCMs, as well as to the enhancement of heat capacity, thermal stability, thermal conductivity, and durability of composites enclosing PCMs. A large number of PCMs, covering a wide range of working temperatures, are available today. Materials that exhibit phase change below 15°C are used in cooling applications and the food industry, while materials above 90°C can be used for absorption refrigeration, industrial waste heat recovery, solar thermal power generation, and in the aerospace industry. PCMs with phase change around 18–70°C are suitable for thermal comfort in textiles and buildings as well as in domestic hot water production and space heating.

The research on PCMs has been increasing tremendously over the last 25 years. Abhat published one of the earliest reviews on PCMs [2] that summarized the studies carried out on materials with phase change in the temperature range from 0°C to 120°C. Since then, more than 190 review articles have been published. They represent a great contribution to the collection and briefing of research studies associated to latent heat storage and PCMs, which also provide assistance for further developments and applications.

Advances in Energy Storage: Latest Developments from R&D to the Market, First Edition.
Edited by Andreas Hauer.
© 2022 John Wiley & Sons, Ltd. Published 2022 by John Wiley & Sons, Ltd.

This chapter provides a brief overview of PCMs for TES, covering extensively studied PCMs as well as emerging materials. It focuses on materials with phase change below 200°C because the corresponding TES applications have greater market potential than those at higher temperatures. Besides, there is a large number of possible PCMs groups that can be used below 200°C, whereas only a few PCMs categories apply at higher temperatures. The chapter is structured as follows. In the second section, the fundamentals of PCMs for TES are briefly recalled. PCMs are then classified according to their type of phase change and their chemical nature. The criteria generally adopted for selecting a PCM are also presented and discussed. Extensively investigated PCMs (i.e. salt hydrates, paraffin waxes, fatty acids and their derivatives, polyethylene glycol, and polyalcohols) are shortly introduced in Section 26.3.1, including a discussion about their respective advantages and drawbacks. Section 26.3.2 deals with other promising or emerging PCMs such as new renewable-based PCMs, PCMs with high potential for achieving compact TES and soli-solid PCMs. For improving the PCMs performance or/and promoting the easy handling and integration into the applications, many research groups develop encapsulated PCMs and composite PCM. An overview of such developments is provided in the fourth section. The commercial status of PCMs is presented in Section 26.5 and Section 26.6 discusses future development directions in the field. Most of the cited papers are recent review articles where the references to the original works can be found.

26.2 Fundamentals, Materials, Groups, and Properties

26.2.1 Fundamentals

Three states of matter (often referred as phases) are observable in everyday life, namely solid, liquid, and gas. At well defined conditions of temperature and pressure, transitions among them take place. On heating, a crystalline solid becomes liquid at the so-called melting temperature, a liquid vaporizes at the boiling point corresponding to the working pressure, and a solid transforms directly into vapor at the sublimation temperature, given that the pressure is lower than the triple point of the substance. In the reverse direction (cooling), a liquid may solidify and a gas can condense either as a liquid or a solid. Other possible phase changes are those in which one crystalline form transforms into another one (solid–solid transitions) or those in which a liquid becomes a glass (non-crystalline, amorphous solid).

All these transformations (except liquid–glass) are first order transitions, which are characterized by discontinuous changes in entropy and volume. There is, therefore, a latent heat (the system absorbs or gives energy to its environment) associated to the transition. Moreover, for a single component system, they take place at interrelated constant pressure and temperature. The molar (or gravimetric) energy involved in the transformation (phase transition latent heat) corresponds to the difference in molar (or gravimetric) enthalpy between the final and the initial phases. Heat must be supplied when a more condensed phase transforms into a more disordered one, and it is released in the reverse way. At constant pressure, the enthalpy change (Δh_{tr}) at the transition temperature (T_{tr}) relates to the entropy change (Δs_{tr}) as:

$$\Delta h_{\mathrm{tr}} = T_{\mathrm{tr}} \Delta s_{\mathrm{tr}} \tag{26.1}$$

It is well-known that the entropy is closely related to the molecular structure, decreasing with increasing particle (atoms, ions, molecules) packing and structural ordering of the phase. Therefore, Δs_{tr} is generally higher in solid–gas than in liquid–gas transformations, which in turn usually exhibit higher entropy change than solid–liquid transformations. Solid–solid transformations are generally those with the lowest Δs_{tr} values. Accordingly, the following rule generally applies:

$$\Delta h_{\mathrm{solid-gas}} > \Delta h_{\mathrm{liquid-gas}} > \Delta h_{\mathrm{solid-liquid}} > \Delta h_{\mathrm{solid-solid}} \tag{26.2}$$

In a perspective trying to maximize the energy density of the storage system, the phase transitions with higher enthalpy change should be preferred. However, the volume change accompanying solid–gas and liquid–gas transformations is very large and would require complex technology. Significantly smaller volume changes occur, usually ca. 10% or less, with solid–liquid and solid–solid transformations. This makes them economically and practically attractive as materials for TES systems despite their smaller heat of phase transition. Numerous PCMs with a solid–liquid transition show relatively high latent heat of fusion, whereas only a limited number of PCMs with solid–solid transition have sufficient latent heat for TES applications. Therefore, solid–liquid PCMs are the most widely used PCMs so far. However, employing solid–solid PCMs can avoid the leakage problem that characterizes solid–liquid PCMs at temperatures above the melting point. Consequently, when providing sufficient heat of phase transition, solid–solid PCMs can be an attractive alternative to solid–liquid PCMs.

The melting point (T_{m}) as well as the latent heat of fusion (Δh_{m}) provide a measure of the energy required to disrupt the crystal lattice of a compound. They are hence determined by the strength of the crystal lattice, which, in turn, is controlled primarily by three factors: intermolecular forces, molecular symmetry, and the conformational degrees of freedom of a molecule. Therefore, the melting temperature and the latent heat of fusion tend to correlate so that Δh_{m} is usually larger for PCMs with higher T_{m} values. Besides, chemical bonds can be classified as follows in decreasing order of strength: ionic, covalent, metallic, hydrogen, and van-der-Waals. Accordingly, melting salts involves breaking ionic bonds and the corresponding melting point and latent heat of fusion are usually higher than those of n-alkanes series, where the intermolecular forces holding the molecules together are Van-der-Waals forces.

The energy stored when heating a solid single component from temperature T_1 to T_2 ($T_1 < T_{\mathrm{m}} < T_2$) can be written as:

$$Q = \int_{T_1}^{T_{\mathrm{m}}} mc_{\mathrm{p}}(T)dT + m\Delta h_{\mathrm{m}} + \int_{T_{\mathrm{m}}}^{T_2} mc_{\mathrm{p}}(T)dT \tag{26.3}$$

where m represents the mass, c_{p} is the specific heat at constant pressure, Δh_{m} is the latent heat of fusion and T_{m} is the melting point. The two integral terms in this equation represent the sensible heat stored by heating the solid compound from T_1 to T_{m}, then the liquid from T_{m} to T_2. The amount of energy stored by latent heat is given by $m\Delta h_{\mathrm{m}}$. A similar equation applies for solid–solid transitions.

Under thermodynamic equilibrium conditions, the crystallization temperature equals the melting point and the energy delivered on cooling from T_2 to T_1 is equal to $-Q$ (Equation (26.3)). However, unlike melting, freezing is an out-of-equilibrium process in which metastable states can be observed. Indeed, a chemical compound could remain liquid at temperatures far below the melting point. This phenomenon is known as undercooling (also referred as to supercooling or subcooling) and was pointed out and explained by Willard Gibbs in the nineteenth century. Two steps can be distinguished in crystallization, namely nucleation and subsequent crystal growth. During the former one, crystalline clusters – the precursors of bulk crystal formation – form and disintegrate continuously by thermodynamic fluctuations. Occasionally a cluster may form that is large enough to continue growing. The smallest crystalline cluster that can grow to become a bulk crystal is known as the critical nucleus. Nucleation induced in the bulk of the liquid in the absence of solid surfaces is called homogeneous nucleation. If a solid surface – whether a container wall, an impurity, or a pre-existent crystal – is involved, the process is called heterogeneous nucleation. Both, homogeneous and heterogeneous nucleation are thermally activated processes which are usually quantified by the number of nuclei surpassing the nucleation barrier per unit volume/area per unit time (shortly, nucleation rate). PCMs with poor nucleating ability (low nucleation rate) result in supercooling of the liquid, which makes a TES system difficult to control and impedes the withdrawal of heat at the desired time. Moreover, the heat of the exothermic transition may be smaller than the stored heat because part of this is used by the supercooled PCM to reach the equilibrium temperature. To avoid or minimize supercooling, nucleating agents are added, or small amounts of crystals are retained in the system to act as nucleation sites, which allow reducing the nucleation energy barrier.

In addition to single component systems, PCMs mixtures can also be of great interest for TES. Indeed, they often allow to substantially enlarge the temperature range of application of a PCM group and offer the possibility of adjusting the melting point to have the best match for the application's requirements. Therefore, numerous multicomponent systems (mostly binary and ternary systems) encompassing either organic or inorganic or even organic and inorganic chemical compounds have been studied in order to discover suitable PCM blends. Multicomponent systems with at least one of the following features are of particular interest:

- Phase diagrams with solid-state miscibility at all proportions (isomorphous sytem) or over a wide composition range. As shown in Figure 26.1, this kind of system provides an easy way to adjust the melting point of the PCMs. Depending on the chosen composition the melting temperature of the mixture can be smoothly varied within the temperature interval determined by the melting points of the single components. The latent heat of the mixture is also in between those of the single components, even though its value is often lower than that predicted by the theory of mixtures. Typical examples of isomorphous systems with a narrow temperature window between solidus and liquidus lines are those of alkanes, benzene compounds and polyalcohols.
- Phase diagrams with eutectic points (eutectic systems). Many multicomponent systems form eutectic mixtures, which are homogeneous solid mixtures of the single

components forming a joint super-lattice at defined composition. The eutectic forms on cooling by solids co-precipitation from the liquid. This reaction is an invariant reaction taking place at constant temperature. As shown in Figure 26.1, the eutectic temperature is the lowest possible melting temperature over all of the mixing ratios. A eutectic mixture hence provides a lower melting point than either of its constituents while still melting congruently, but generally with the cost of significant latent heat reduction. The use of out-of-eutectic compositions is not recommended for TES applications because they melt incongruently, thus leading to density differences, phase separation, and sedimentation in containers causing serious technical problems.

26.2.2 PCMs Classification and Criteria for Selection

Over the last 40 years different classes of materials have been considered as PCMs for heat storage applications. They can be divided into three main groups, based on the temperature ranges over which the phase transition takes place: (i) low temperature PCMs – with phase transition temperature below 15°C, (ii) mid temperature PCMs, the most popular – with phase transition temperatures in the range 15–90°C, and (iii) high temperature PCMs with a phase transition above 90°C. PCMs can also be classified according to their type of phase transition and their chemical nature. Figure 26.2 shows the classification proposed by [3] as well as the main groups of PCMs considered so far.

The literature survey carried out by [4] covers more than 700 individual PCMs, including blends. As shown in Figure 26.3, over half of the studied materials have transition temperatures below 100°C, and only salts and metallic alloys have significant number of materials above 200°C. It can also be seen that there is a large number of possible material categories for the lower temperatures TES applications, while the medium and high temperatures show a significant drop-off in the number of possible

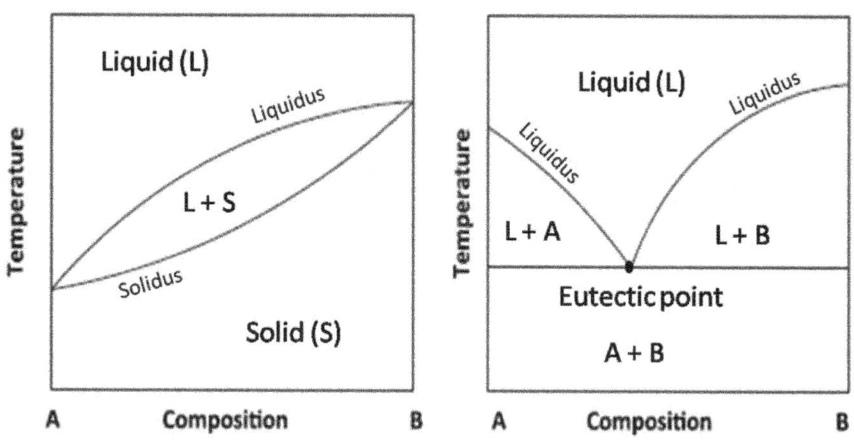

Figure 26.1 Schematic A-B binary phase diagrams (temperature-composition map): isomorphous system (left side); simple eutectic system (right side).

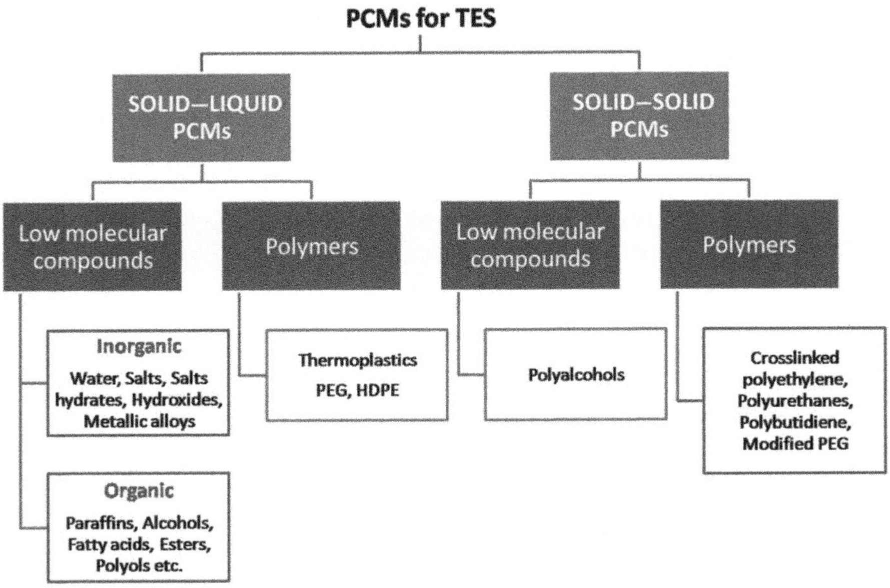

Figure 26.2 Classification of solid–liquid and solid–solid PCMs and main material groups.

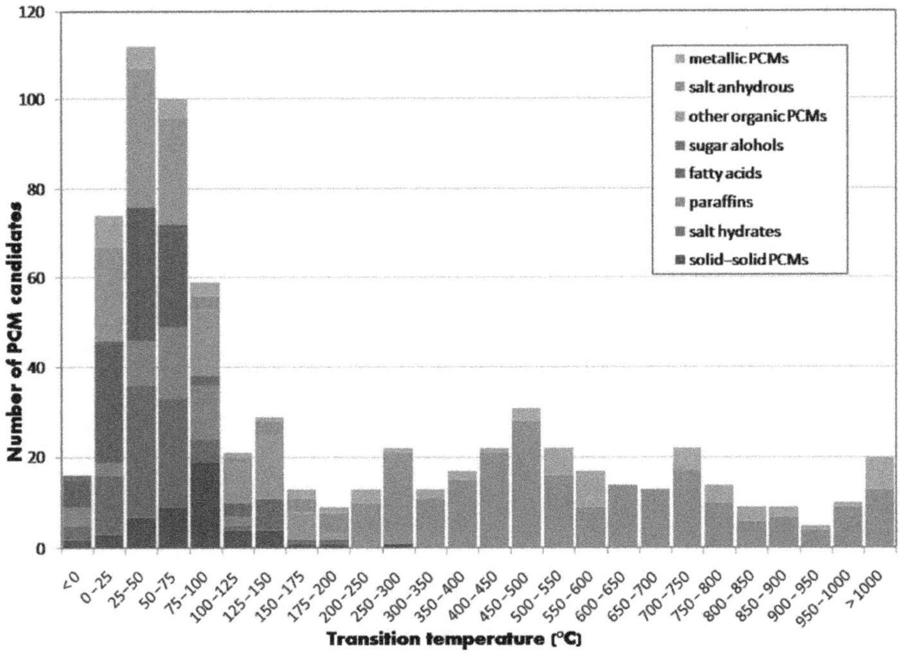

Figure 26.3 Transition temperature (°C) distribution of PCM candidates, sorted by material category [4].

types of materials. The phase transition latent heat typically ranges from 50 $J.g^{-1}$ to 350 $J.g^{-1}$ for PCMs with melting point below 200°C, whereas it takes values from 50 $J.g^{-1}$ to 1000 $J.g^{-1}$ (and beyond) for materials with higher melting points.

The ideal PCM should meet a number of criteria related to the desired thermophysical, kinetic, chemical, economic, and environmental properties [2]. The main characteristics for a good PCM are summarized in Table 26.1. In practice, those criteria are not fully met by most PCMs. However, recent progress in the design of novel materials (see Section 26.4) has opened new possibilities for enhanced performance with extended lifetime.

In addition to those characteristics, proper design of a PCM-based TES system requires sufficient knowledge of material properties. At a minimum, a properly formulated heat transfer model for PCMs behavior simulation would require melting/transition temperature (T_m or T_{tr}), latent heat of fusion (Δh_m) or transition (Δh_{tr}), specific heat (c_p), density (ρ), and thermal conductivity (k). The latter three properties should be known for both solid and liquid state. Moreover, their temperature dependency should also be specified, as they can often change significantly over large temperature ranges and around the melting point. Another property required to enable modeling

Table 26.1 Criteria for the selection of a PCM.

Thermal properties	• Transition temperature in the desired operating range • High latent heat of phase transition per unit mass, to minimize the amount of material required for a given storage capacity • High phase transition latent heat per unit volume, to achieve TES systems as compact as possible • High specific heat, to provide additional sensible heat storage • High thermal conductivity of both phases, to avoid over-sizing of the heat exchangers
Physical properties	• Small volume change on phase transformation and low vapor pressure at the operating temperature, to reduce the complexity of the storage vessel and that of the heat exchanger • Congruent melting, to avoid phase separation and sedimentation in containers causing serious technical problems • High density, so that a smaller container volume holds the storage material
Kinetic properties	• No or little subcooling during freezing, to have almost equal temperature working conditions both in charge and in discharge modes • High nucleation rate and crystal growth rate, to achieve controlled and fast heat delivery
Chemical properties	• Long-term chemical stability • Completely reversible freeze/melt cycle • Compatibility with the construction materials • No corrosion influence on the construction materials • Non-toxic, non-flammable and non-explosive to ensure safety
Economic & environmental properties	• Readily available in large quantities • Inexpensive or with affordable cost • Easy recycling and treatment • Good environmental performances based on Life Cycle Assessment

of convective effects is viscosity (ν), which is also dependent on temperature. Depending on the specific PCM considered, the nucleation rate and the crystal growth velocity should also be given. Finally, for PCM mixtures other than eutectics, the phase diagram of the corresponding multicomponent system is required as well.

It is worth noting that of the 704 materials indicated in the Figure 26.3, only 89 included data for T_m, Δh_m, c_p, ρ and k, while over 400 of the others only include T_m and Δh_m. As a result, while the figures suggest a broad range of available materials, individual efforts to make use of any particular material will likely also require an investment in additional material characterization.

26.3 Currently Used and Emerging Phase Change Materials

26.3.1 Extensively Investigated PCMs

Many different types of PCMs have been considered for their use in TES systems (see Figure 26.2). However, only a few groups have been extensively studied. Among solid–liquid inorganic PCMs, the most widely used (besides water) are salt hydrates. Organic solid–liquid PCMs such as paraffin waxes, fatty acids and their derivatives, and poly(ethylene glycol) have also been studied by numerous research groups. Regarding solid-solid PCMs, polyalcohols are the most analyzed so far. A short description of all these PCM groups is provided in this section, including a discussion about their respective advantages and drawbacks.

26.3.1.1 Salt Hydrates and Paraffin Waxes

Salt hydrates and paraffin waxes are the most notable group of PCMs extensively studied for their use in TES, which currently dominate the PCMs market:

- *Salt hydrates.* A salt hydrate is an inorganic salt containing n moles of water molecules that form a typical solid of general formula $AB.nH_2O$. The bonds are usually ion–dipole bonds or hydrogen bonds and the water molecules are located and oriented in the structure in a well-defined manner. The solid–liquid transition of salt hydrates is actually dehydration/hydration of a salt, a process very similar to melting/freezing in thermodynamics. At the transition temperature, a hydrated crystal breaks up into a lower hydrate that contains $m < n$ moles of water or into the anhydrous form of the salt and water. The most readily available salt hydrates are $CaCl_2.6H_2O$ and $Na_2SO_4.10H_2O$ although many other salt hydrates (including eutectic mixtures) have also been studied as PCMs for TES [2, 5–6].
- *Alkanes and paraffin waxes.* Straight chain n-alkanes (also referred as to pure paraffins) are saturated hydrocarbons with chemical formula C_nH_{2n+2}. The intermolecular forces holding n-alkane molecules together are van-der-Waals forces. Therefore, both the melting point and the latent heat of fusion increase progressively with the carbon chain length. Alkanes covering a range from C_1 to C_{100} with a melting temperature range of about −183°C to 115°C have been studied in-depth. Besides, it is shown that many binary alkane systems make solid-state solution with very narrow temperature window (2–5°C) between solidus and liquidus lines. As already mentioned, this property allows a blend to be tailored to a particular melting range while

broadening the temperature range covered by alkanes. However, due to cost considerations, only technical graded paraffin waxes are used in practice. A paraffin wax is a blend of n-alkanes (typically > 75 wt%) and other hydrocarbon molecules (typically < 25 wt%). They are produced at a large scale as by-product in crude oil refining and, therefore, they are relatively cheap. However, they show significantly reduced latent heats relative to the alkane series.

Practically all available original experimental data on the key physical properties (melting temperatures, latent heat of fusion, specific heat, density, thermal conductivity, thermal diffusivity, and viscosity) of these PCMs are summarized in the review articles by [2, 5–10]. As shown in Figure 26.4, salt hydrates and paraffins are available over a broad temperature range and show comparable values of latent heat per unit mass. However, the density of salt hydrates (ca. 1440–1960 $kg.m^{-3}$ in liquid) is much higher than that of paraffins (ca. 725–846 $kg.m^{-3}$ in liquid) and, therefore, the former provide much higher values of latent heat per unit volume. A comparative analysis between salt hydrates and paraffins as PCMs for TES, with indication of merits and demerits of each relative to the other, is provided in Table 26.2.

26.3.1.2 Fatty Acids and their Derivatives

The growing interest in the use of fats and oils for sustainable chemistry has led to a growing importance of PCMs prepared from them, such as fatty acids and their derivatives. Fatty acids are some of the primary components of naturally occurring vegetable

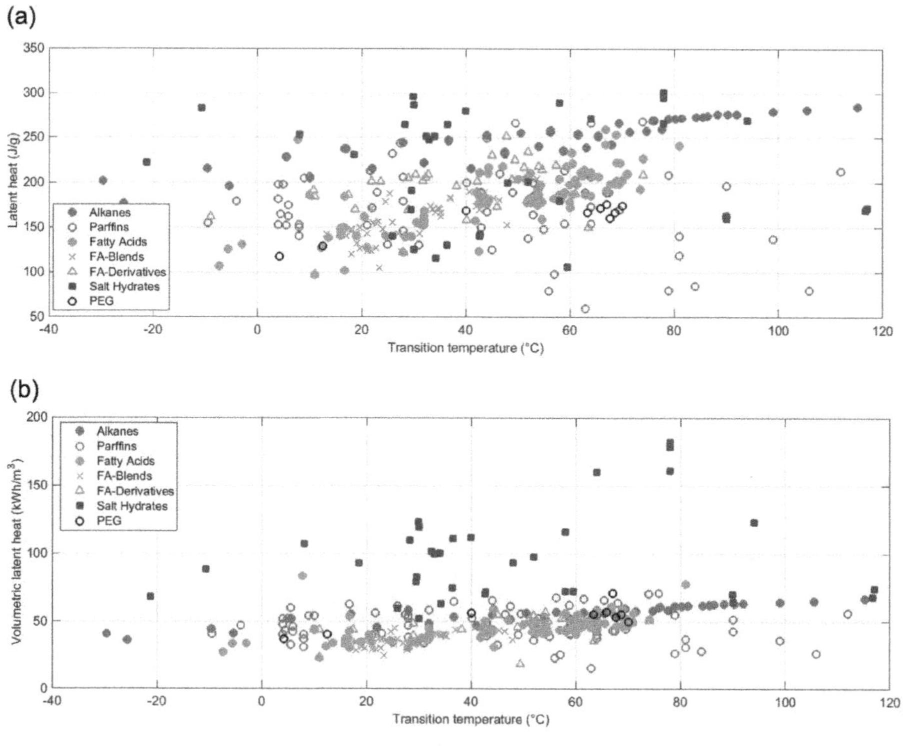

Figure 26.4 Covered transition temperature ranges and corresponding phase change latent heat for extensively studied PCMs: (a) latent heat per unit mass; and (b) latent heat per unit volume.

Table 26.2 Advantages and disadvantages of salt hydrates and paraffins as PCMs for TES.

Paraffins	Salt hydrates
Merits	
• Available in a large temperature range with easy tailoring of the melting point • Congruent melting and good nucleation properties, with no or little supercooling • Chemically inert and stable below 500°C • Safe, reliable and non-corrosive	• Higher latent heat of fusion per unit volume and comparatively high thermal conductivity (typically between 0.4 and 0.8 $W.m^{-1}.K^{-1}$) • Compatibility with plastics • Many salt hydrates are sufficiently inexpensive for the use in TES
Demerits	
• Low density which results in lower volumetric energy density • Low thermal conductivity (typically less than 0.25 $W.m^{-1}.K^{-1}$) • Non compatibility with plastic containers • Slow oxidation when exposed to oxygen, which requires leak-proof containers • Higher cost than salts hydrates. Only technical graded paraffin waxes have affordable cost for TES • Moderately flammable	• Many salt hydrates melt incongruently resulting in poor cycling stability. To avoid or minimize this problem, the addition of gelling or thickening agents is usually required [11] • Most salt hydrates have poor nucleating properties resulting in significant supercooling. The use of nucleation agents is hence necessary to reduce supercooling effects [7]. • Corrosive effects on metallic elements of the TES system

and animal oils, and are characterized by the general formula $CH_3(CH_2)_{2n}COOH$. The natural rather than synthetic source of these materials reduces concerns about long term material availability relative to petroleum based paraffins. Fatty acids with number of carbon atoms in the molecule ranging from 8 to 22 (e.g. caprilic, capric, lauric, mystitic, palmitic, stearic acids) proved to have suitable phase change characteristics, such as congruent melting and little undercooling, with demonstrated long term stability [12]. They share many of the characteristics of paraffins, including low density (840–900 $kg.m^{-3}$), low thermal conductivity (0.14–0.17 $W.m^{-1}.K^{-1}$) and significant expansion on melting (10–15%). However, they typically have lower latent heats than paraffins in similar melting ranges while having higher costs (2–3 times higher). In addition, while generally nontoxic some fatty acids can be mildly corrosive complicating container design. Covered temperature ranges (−7–80°C) and corresponding latent heat of fusion (98–259 $J.g^{-1}$) values are displayed in Figure 26.4.

To adjust the melting temperature of fatty acids, binary fatty acid systems have also been investigated. It is proven that when fatty acids differ by six or more carbon atoms, a simple eutectic forms. As shown in Figure 26.4, studied fatty acid eutectics cover a temperature range from 13°C to 53°C with latent heat of fusion of about 126–200 $J.g^{-1}$. Derivative products of fatty acids, including alcohols and esters with other compounds, have been studied as well. Among them, fatty acid esters (prepared by esterification reactions of fatty acids with alcohols) have attracted much attention because of their ability to overcome some undesired properties of fatty acids, such as bad odor, corrosion, and high sublimation rate in the melt form. On the whole, compared to fatty acids, fatty acid derivatives have lower melting temperature and provide slightly less latent heat (Figure 26.4).

26.3.1.3 Polyethylene Glycol (PEG)

Polyethylene glycol (PEG), also referred to as polyoxyethylene (POE) or polyethylene oxide (PEO), is an important semi-crystalline polymer composed of linear dimethyl ether chains with hydroxyl ending groups, $HO\text{-}CH_2\text{-}(CH_2\text{-}O\text{-}CH_2)_n\text{-}CH_2\text{-}OH$, and accordingly possess dual feature of water solubility and organic solubility. PEGs with molar mass ranging from 400 to 1000000 $g.mol^{-1}$ have been studied as PCMs for TES, covering a temperature range from 3.2 to 68.7°C with latent heats of fusion of about 91.4–187.9 $J.g^{-1}$ (see Figure 26.4). Their thermophysical properties and main features as PCMs are summarized in [3, 9] and [13]. PEGs have the following desirable characteristics: congruent melting, relatively high heat of fusion, low and moderate melting temperature intervals, and low vapor pressure when melted. Moreover, they are chemically and thermally stable, non-flammable, biodegradable, non-toxic, non-corrosive, and inexpensive. In addition, the melting temperatures and the latent heat values of PEGs increase gradually as the average molar weight (MW) of the PEG chains increase. Therefore, the melting point and heat capacity of a PEG system can be easily adjusted by selecting or mixing PEGs with different MWs. Against these desirable characteristics, there is a considerable supercooling effect (ca. 15–35°C depending on the MW) due to the difficulty of crystallization of long PEG chains during cooling. Thus, preventing the supercooling of PEGs besides the improvement of their thermal conductivity (~0.3 $W.m^{-1}.K^{-1}$) is an important issue for their future use in heat storage systems.

26.3.1.4 Polyalcohols and Polyalcohol Derivatives

Alcohol and amine derivatives of 2,2-dimethylpropane (Neopentane, NP, C(CH3)4) are the most analyzed solid–solid PCMs with practical potential for TES applications. These are sometimes referred as "plastic crystals" and undergo solid–solid phase transition from a layered or chained low temperature structure (tetragonal, monoclinic, etc.) to a high temperature homogeneous face centered cubic crystals (plastic phase). The polyalcohol (polyol) series of these compounds include pentaerythritol (PE), pentaglycerine (PG), and neopentylglycol (NPG). The amine (NH2)-substituted derivatives of polyols include 2-amino-2-methyl-1,3-propanediol (AMPL), and tris(hydroxyl methyl)aminomethane (TAM). The solid–solid phase transition in these organic compounds takes place at temperatures between 48°C and 188°C and it is characterized by an unusually large enthalpy of transition, ranging from 110 $J.g^{-1}$ to 290 $J.g^{-1}$ [13], because of the reversible breaking of nearest neighbor hydrogen bonds in the molecular crystals at the transition temperature. They also offer small volumetric change, no leakage, small erosion to the device, long lifespan, and do not have segregation or phase separation problems. Binary and ternary mixtures of these compounds have also been studied to design new solid–solid PCMs for storage applications at lower temperatures. It is shown that the isomorphism between the crystalline structure of some of the studied polyalcohols makes the existence of solid solutions possible at all proportions or over wide composition range. The transition temperature as well as the corresponding latent heat are hence easily adjustable between those of the pure constituents. The PE–PG system has the greatest latent heat values (140–290 $J.g^{-1}$ depending on the composition) and covers transition temperatures from 81°C to 188°C. As for

PEG, supercooling is the major issue of polyalcohols, requiring the development of appropriate nucleation agents to control it.

26.3.2 Emerging PCMs

While the majority of the investigations of organic PCMs have focused on the previously described categories, numerous other materials covered in the literature have shown the potential for utility as PCMs, including alcohols, ketones, phenols, amines, and others. However, in spite of the tremendous R&D efforts carried out over the last 25 years, the cost of latent heat storage technologies is still quite high (10–50 €/kWh) and the compactness is often insufficient to cope with large-scale applications. Different R&D strategies aiming at reducing the investment costs can be envisaged: (i) *Reducing the cost of the storage material*. For example, the cost of paraffin products is 2900–3500 €/ton, whereas the PCM cost should be less than 1–2 €/kg to achieve cost-effective storage solutions. This aim can be reached only if technical grade materials produced with a minimum of technological processing are used. (ii) *Increasing the TES system compactness*. Reducing the volume needed for a given thermal storage capacity allows scaling-down the cost of containers and heat exchangers. The use of PCMs with very high volumetric latent heat values should notably contribute to this end. (iii) *Simplifying the storage technology*. As already mentioned, employing solid–solid PCMs can avoid the problems of PCM leakage and volume expansion that characterize solid–liquid PCMs. Using solid–solid PCMs (provided they have high latent heat of transition) should hence contribute to simplify the TES system. For instance, solid–solid PCMs do not require separate packing which would inevitably increase the cost of the system. Moreover, they can be shaped so that the heat exchanger can be avoided provided that there is compatibility between the solid–solid PCM and the heat transfer fluid.

This section briefly describes a number of emerging PCMs which offer attractive prospects for either cost reduction or compactness increase or both, such as low cost bio-based PCMs, PCMs with unusual high latent heat values per unit volume, and new solid–solid PCMs. Additionally, most of them are renewable-based PCMs and, therefore, potentially valuable bio-based alternative to paraffin waxes and salt hydrates.

26.3.2.1 Low Cost, Renewable-based Materials

Natural waxes made from underused and renewable feedstocks, such as soybean oils, coconut oils, palm oils, and low-value animal fats and greases, are a relatively new class of materials termed as bio-based PCMs in the literature [9, 13]. They contain a large number of identified chemical (> 97% C_{16} and C_{18} fatty acids) as well as unidentified ones, which are all environmentally safe and non-toxic. They possess storage properties comparable to those of paraffins. They have good thermal stability for thousands of melting and freezing cycles, with no risk of oxidation, because they are fully hydrogenated. Also, they have latent heat values similar to conventional paraffins. They are expected to undergo small volume changes, and have no supercooling during freezing. They have melting congruency, and they can also be manufactured such that the melting point can be varied between −22.7°C and 78.3°C. Due to these advantages, together with their very low cost, bio-based PCMs are considered as promising PCMs for TES applications.

26.3.2.2 PCMs with Significant Potential for Compactness Increase

Compared to the organic materials previously described, salts hydrates are characterized by very high volumetric latent heat values (see Figures 26.4 and 26.5) which result from relatively high gravimetric latent heats (100–300 $J.g^{-1}$) combined with high density values (ca. 1440–1960 $kg.m^{-3}$ in liquid). This feature makes salt hydrates excellent PCMs candidates for achieving compact TES. However, they have some obvious disadvantages such as supercooling, cycling instability, and corrosion. Although uncommon, some organic PCMs can show storage performances comparable to those of salt hydrates while avoiding partially their drawbacks. In particular, a number of sugar alcohols and carboxylic acids exhibit unusual high volumetric latent heats and, therefore, they could be promising alternative to achieve compact TES.

Sugar alcohols, also referred as to polyols or polyalcohols, are a sub-class of carbohydrates whose defining characteristic is the occurrence of an alcohol group (>CH-OH) in place of the carbonyl group (>C = O) in the aldose and ketose moieties of mono-, di-, oligo, and polysaccharides. Some polyols are found naturally in various fruits and vegetables. However, most of them are produced by chemical reduction of carbohydrates. They have received less attention than other organic PCMs in the TES literature because of the higher transition temperature of single sugar alcohols (90–200°C). However, some of them (e.g. xylitol, erythritol, D-mannitol, dulcitol) have volumetric latent heat as much as twice that of other organic materials (see Figure 26.5) due to the hydrogen bonds in the molecular crystals. Recently, straight chain monosaccharide polyols with the general formula $C_nH_{2(n+1)}O_n$ (n = 4–6), as well as their eutectic mixtures, have been studied in depth for industrial TES applications and solar energy seasonal storage [14]. On the whole, compared to other organic PCMs, polyols are bio-sourced materials, cover a temperature range outside that of paraffins and fatty acids (70–200°C), provide unusual high volumetric energy density (up to 160 $kWh.m^{-3}$), have relatively high thermal conductivity (0.3–1.3 $W.m^{-1}.K^{-1}$ in solid) and are non-toxic, non-flammable, and compatible with metals and plastics. They also have affordable cost (1–3 €/kg for most of them). However, preventing supercooling and improving thermal endurance still are important issues for the future use of polyols in TES systems.

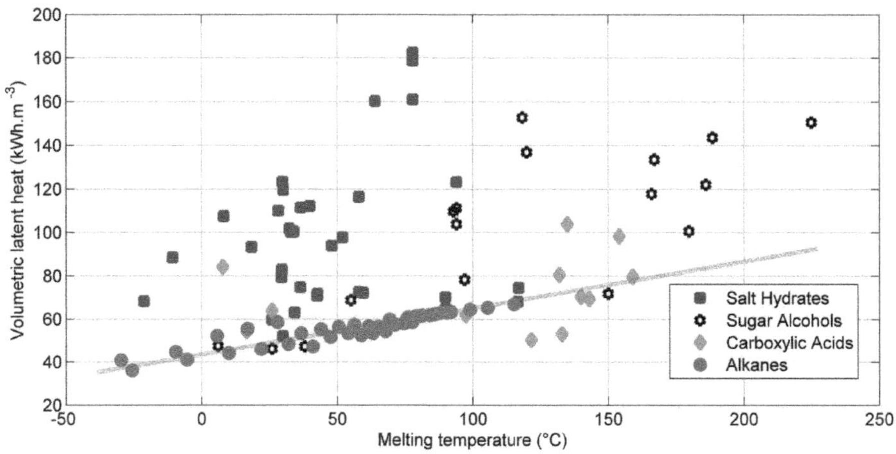

Figure 26.5 Latent heat values for unit volume vs. melting temperature for different PCMs groups.

Carboxylic acid based materials [4] also have high volumetric latent heat values (60–104 kWh.m^{-3}) at temperatures ranging from 8°C to 160°C (Figure 26.5). The gravimetric latent heat values are between 200 and 350 J.g^{-1} and the density is about 1200–1500 kg.m^{-3}. Moreover, they have no supercooling during freezing. On the contrary, their thermal conductivity is low (< 0.2 W.m^{-1}.K^{-1}). Most of them have good thermal stability under nitrogen atmosphere up to ca. 50°C above the melting point. Although they are found in nature, both as free acids or salts, industrial processes are employed for manufacturing them.

Looking for a *cooperative effect in PCMs blends* could also be an efficient way for increasing volumetric energy density. For instance, synergetic action of the components was found in different binary polyether-fatty acid blends [3], with latent heats 15–30% higher than for pure fatty acid and polyether. The origin of this effect probably arises from the formation and decay of hydrogen bonding during the phase transition process.

26.3.2.3 Solid–solid PCMs

Polymeric materials are beginning to play a growing role as solid–solid PCMs. They offer the scope for making significant chemical and physical modification which may facilitate the production of specific end products with "tailor made" energy storage properties. Most promising polymeric PCMs so far are [3]: (i) *Crosslinked polyethylene.* These copolymers are obtained by chemical graft of PEG onto a cellulose-derivative skeleton. The phase transition temperature of these solid–solid PCMs could be adjusted from room temperature to 50°C by changing the content and molecular weight of the PEG side chains. Novel biodegradable blends such as PEG/potato starch, PEG/agarose and PEG/chitosan are also being investigated as solid–solid (or form-stable) PCMs. (ii) *Polyurethanes.* Polyurethanes constitute a class of versatile macromolecular materials that derive most of their useful properties from the incompatibility of the hard segments, made from diisocyanate and a chain extender, and soft segments usually composed of long-chain diols. Specific interactions between the phases via hydrogen bonding, connected with phase intermixing and reorganization processes, influence the microdomain structure and properties of these promising macromolecular candidates for energy storage applications. Polyurethanes containing PEG segments have been synthesized by several research groups and proven to show solid–solid transitions at suitable temperatures (19–60°C) with high transition enthalpies ranging from 100 to 179 J.g^{-1}.

Other materials have also received attention for solid–solid TES applications [4]. A series of organometallic compounds called *layered perovskites*, represented as $(n\text{-}C_nH_{2n+1}NH_3)_2MX_4$ where M is a metal atom, X is a halogen, and n varies from 10 to 16, have been characterized in several reports for X = Cl. The transition temperature range covered by these materials is 33–160°C, with latent heats of transition ranging from 60 J.g^{-1} to 154 J.g^{-1}. However, the solid-state transition takes place over a temperature range of ca. 10–50°C. Finally, solid–solid transitions of interest for TES have also been observed in a number of *dialkyl ammonium salts* of the form $C_{2n}H_{4n}XNH_4$, with X being a salt anion. They exhibit solid–solid phase transitions from −2°C to 190°C and transition enthalpies up to 186 kJ/kg.

26.4 Approaches to Improve PCMs' Properties

Compared to the ideal PCM described in Table 26.1, many of the studied PCMs exhibit one or several undesirable properties. Table 26.3 summarizes frequently encountered problems with indication of the concerned material groups. To avoid or to minimize such problems, several approaches have been proposed. For instance, adding a nucleating agent with a crystal structure similar to that of the energy storage material reduces the amount of subcooling [7]. In addition, uniformly distributing the nucleating seed agent throughout the phase-change media helps increase the rate of crystallization in PCMs with slow crystallization. Moreover, segregation and sedimentation problems appearing in PCMs with incongruent melting could be avoided by adding gelling or thickening agents such as cellulose derivatives, silica gel, sepiolite, diatomaceous earth, and cross-linked polymeric hydrogels [11]. However, using composite PCMs is the

Table 26.3 Frequently encountered problems, consequences and concerned PCM groups.

Problem	Consequences	Concerned materials
Low thermal conductivity	Over-sized heat exchangers Increased investment cost	All organic PCMs (excepted polyols)
High volume change on phase transformation (ca. 10% or higher)	Increases the complexity of the storage vessel and that of the heat exchanger, reducing effective energy density and increasing investment cost	Most liquid–solid PCMs
Incongruent melting	Phase separation and sedimentation in containers causing serious technical problems. Requires specific storage system design, which reduces effective energy density and increases cost	Salt hydrates
Poor nucleation ability	Significant subcooling during freezing. If acceptable, it makes it difficult to control the system discharge and reduces recovered heat	Salt hydrates PGE, Polyols Polyhydric alcohols
Poor long-term stability	Short life-time. Unacceptable if there is no means of prevention, as for instance using leak-proof containers with inert gas atmospheres	Sugar alcohols Carboxylic acids
Incompatibility with construction materials (corrosion or other)	Reduces the choice of construction materials and/or requires specific protection, thus leading to higher investment cost	Salt hydrates Paraffins
Flammability	Limits the applicability, especially in buildings. Requires the incorporation of flame retardant agents to the PCMs	Paraffins In a lesser extent, fatty acids and their derivatives
Non-renewable materials, poor environmental performances	Resources depletion and other negative effects that jeopardize the sustainability	Paraffins Salt hydrates

approach attracting the greatest interest because of the ability of these novel materials to face most of the problems encountered in solid–liquid PCMs. Therefore, R&D in this area has been increasing tremendously over the last 20 years. A large number of composite materials have been proposed and investigated, covering micro-encapsulated PCMs, shape-stabilized materials, and others types of composites. Review articles published by [3, 15] and [13] summarize the studies carried out on composite PCMs as well as the different processing techniques used. This chapter gives a short overview of the state-of-the-art with focus on advantages and drawbacks analysis.

26.4.1 Composite PCMs Classification and Short Description

The term composite PCM refers to a class of materials composed of a supporting material which encloses the PCM. Their development was primarily motivated by the problem of low thermal conductivity of most organic PCMs and the problem of liquid leakage that affects all solid–liquid PCMs. However, composite PCMs can also provide many other advantages such as protection from the external environment, local management of volume changes, lower flammability, phase separation and sedimentation problem reduction, and easier handling and integration into the application. The most widely studied composite PCMs so far are micro-encapsulated PCMs and the so called shape-stabilized PCMs. However, composite materials overcoming efficiently the thermal conductivity problem although useless to prevent liquid leakage have also been proposed and will be described as well.

Micro-encapsulation is the process of coating individual particles or droplets with a film to produce capsules at micrometer to milliliter size (<1000 μm), known as a microcapsule, microparticles, or microspheres. Capsules at nanometer sizes (<1000 nm) are known as nanocapsules, nanoparticles, or nano spheres. Micro- or nano-encapsulated PCMs (μEPCMs) are composed of two main parts (Figure 26.6): PCMs as a core with polymer, inorganic, or hybrid material as a shell. The shell will act as a container to protect PCMs from the external environment and to avoid liquid leakage. Micro- or nano capsules could exist in several shapes such as spherical, tubular, oval, or irregular shape. PCMs with a melting temperature ranging from −10°C to 80°C can be turned into micro- or nanocapsules. μEPCMs can be prepared using either physico-mechanical, chemical,

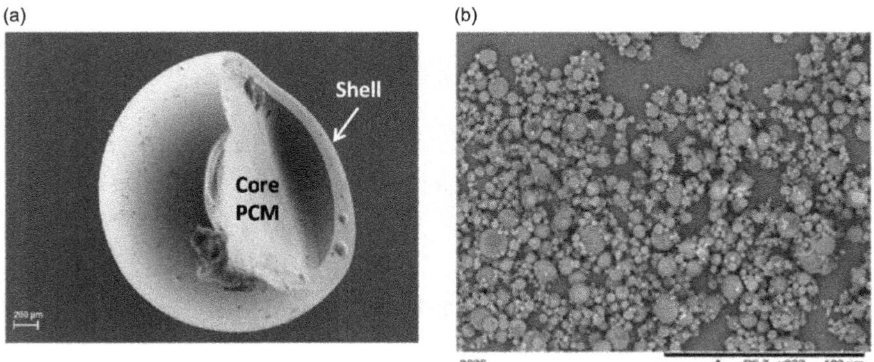

Figure 26.6 SEM images of μEPCMs: (a) Xylitol encapsulated with organic-inorganic hybrid polymer shell (SAM.SSA, FP7 2012-2015 – courtesy of Fraunhofer ISC); (b) d-mannitol/PMMA micro-capsules (SAM.SSA, FP7 2012-2015 – courtesy of Aidico).

or physico chemical methods depending on the physico chemical properties of the PCMs, the shell materials to be used and the preferred size to be obtained [13, 15]. In-situ polymerization (interfacial, emulsion, or suspension polymerization) is the most often used method for micro- or nano-encapsulation of organic PCMs. The most analyzed core materials are paraffins and fatty acids, whereas urea-formaldehyde resin, melamine-formaldehyde resin, and polyurethanes are usually selected as shell materials. An appropriate choice of the shell material and processing method yields encapsulation efficiency values from 60 to 90% and PCM mass proportion up to 80%. The encapsulation of salts hydrates cannot be achieved by using organic polymers as shell material because of the chemical incompatibility with salt hydrates. On the contrary, salts hydrates are properly encapsulated by using either organic-inorganic hybrid polymers (ORMOCER®s) or inorganic compounds (SiO_2, TiO_2) as shell materials. In the former case, a co-extrusion technique combined with UV-curing is used to encapsulate the PCM, whereas physico-chemical sol-gel method is employed in the latter case. For TES applications, μEPCMs are not used as received but incorporated into other supporting materials. For instance, PCM wall or PCM wallboards can be achieved by incorporating μEPCMs into construction materials such as gypsum, cement, concrete, wood, etc. The incorporated μEPCMs increase the thermal mass of the wall so indoor temperature swings are reduced and, therefore, the thermal comfort in the building is increased. Also, the μEPCMs can be disperse in a fluid (mostly water) to form a pumpable slurry (PCM slurry), that can be used as an energy transport and storage medium.

Shape-stabilized phase change materials (SPCMs), also referred as to form-stable PCMs, can be defined as a new class of materials that have the ability to keep the shape of PCMs as in a solid state even when the temperature of PCMs is over the melting point (liquid state). SPCMs can be classified into two major groups, depending on the nature of the supporting material (Figure 26.7):

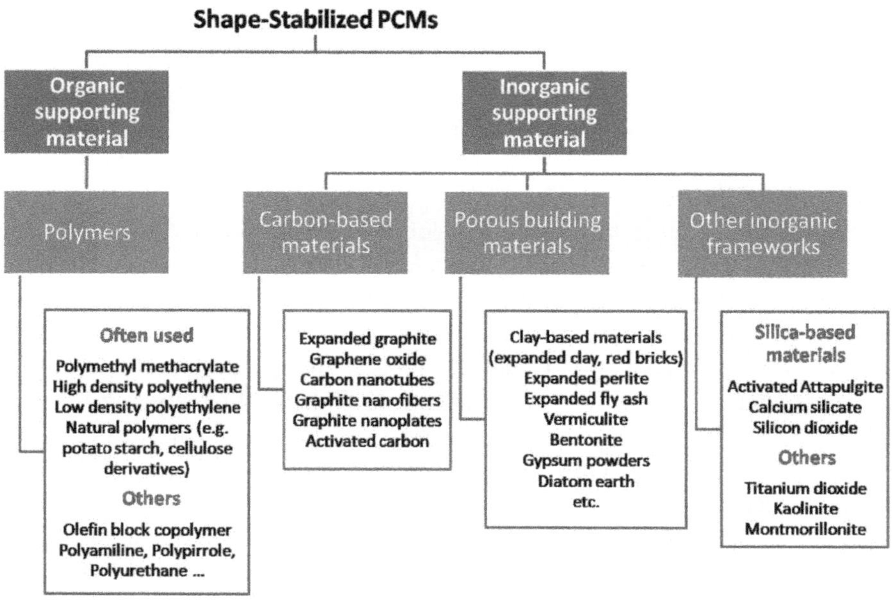

Figure 26.7 SPCMs classification according to the nature of the supporting material and short list of investigated supporting materials.

- *SPCM with supporting organic material based on polymers.* These SPCMs are produced by preparing a liquid mixture of PCM and supporting material then cooling below the glass transition temperature of the supporting material, until the mixture becomes solid. PCMs are hence encapsulated into the polymeric structure. As shown in Figure 26.7, many different types of polymers can be used as supporting material. An appropriate selection of this one allows PCM mass proportion up to 80% without leakage. Moreover, flame retardant agents as well as conductive nano-fillers can be easily added to reduce PCMs' flammability and to increase the thermal conductivity of the PCMs, respectively.
- *SPCM with supporting inorganic porous materials.* Most of the SPCM of this group are produced by incorporating liquid PCM into the porous structure of the supporting material by vacuum assisted infiltration method. However, SPCMs using SiO_2 or TiO_2 based frameworks are obtained by sol-gel method. Whatever may be the production method, the PCM is encapsulated into the porous structure so that the leakage problem is avoided by capillary forces and eventually by hydrogen bonding. As shown in Figure 26.7, many types of nano- and meso-porous materials have been studied. Most of them have high specific surface area and open porosity, excellent thermo chemical stability and wide availability. Moreover, carbon-based supporting materials have good thermal conductivity. The characteristics of the porous structure, specially their open porosity and specific surface area, determine the amount of PCM that can be held by the supporting material. PCM mass proportion values ranging from 35 to 95% have been reported in the literature. The highest values (80–95 %wt) correspond to carbon-based supporting materials. The size of the pores is also an important factor. Indeed, if the pores are too small, Gibbs effects can be observed. In general, the melting point is depressed and the latent heat is decreased (even suppressed). However, in some cases, if a strong attractive interaction occurs between the liquid and pore surface, the melting point is shift toward higher temperatures and the latent heat could be increased.

Most of the work carried out on SPCMs has been done by using paraffins, fatty acids, and their derivatives and PEG.

Other types of composite materials have been developed with the specific objective of increasing the thermal conductivity of organic PCMs [16–17]. Although these composite materials are unable to overcome liquid seepage, some of them can be highly efficient in enhancing heat transfer. Combining PCMs of low thermal conductivity with highly conductive materials is a logical solution to come up with new material with enhanced effective thermal conductivity. Placement of metal fillers, foams, and wools into the PCM have been investigated and practiced for some time. Most of the published papers deal with the dispersion of metal and metal oxide nanoparticles in organic PCM. However, the PCMs composites prepared using such type of heat transfer promoters often show poor stability because of the large difference between the density of the solid inclusions and that of the PCM. Compared with the aforementioned fillers, carbon-based nanostructures (i.e. nanotubes, nanofibers, nanoplates) have smaller density difference to the organic matrix, which make them have better stability in the PCM. Besides, carbon-based nanostructures, particularly carbon nanotubes (CNTs), exhibit far greater enhancement of thermal conductivity in comparison with metallic and metal oxide nanoparticles. However, the experimental results are still much closer to the lower bound conductivity model than to the upper rule of mixing because the thermal conductivity improvement is limited by interface thermal resistance between the fillers. Using conductive porous structures, such as metallic or carbon foams, as hosts for the PCM is an alternative route for heat transfer enhancement that allows overcoming such a drawback. Graphite foams saturated with PCMs were demonstrated to provide tailorable thermal conductivity from few W/m/K up to several tens of W/m/K with, obviously, decreasing open porosities with the conductivity. The thermal conductivity enhancement ratio (per volume fraction of carbon) typically ranges from 2 to 5 in such materials. Ultrathin graphite foams (UGF) with volume fractions as low as 0.8–1.2 and thermal conductivity from 1.5 W/m/K to 2.8 W/m/K, achieving PCM thermal conductivity enhancement ratios as high as 10–15, have been recently proposed [18]. It was also demonstrated that the thermal conductivity of UGF/PCM composites can be increased further (up to 4.1 W/m/K) by growing long CNT networks directly from the graphite struts of UGFs into the pore space [19]. Contrary to SPCM with supporting carbon-based materials, the pores in most graphitic foams are too big for the capillary forces to avoid the liquid seepage.

26.4.2 Advantages and Drawbacks Analysis

As already mentioned, μEPCMs and SPCMs allow overcoming the problem of liquid leakage that affects all solid–liquid PCMs while providing easier materials handling and integration into the application. Besides these important advantages, many other benefits can be obtained by using these types of composite materials. They are shortly described in Table 26.4, where a qualitative evaluation of μEPCMs and SPCMs is provided.

In spite of the aforementioned advantages, μEPCMs and SPCMs have some drawbacks such as increased cost of the storage material, loss of the energy storage capacity and supercooling amplification. Also, the thermal conductivity enhancement ratios achieved are very low compared to those obtained by composite materials based on metallic or graphitic foams.

Compared to μEPCMs, SPCMs have several important advantages:

- Encapsulation technology using polymers into micro- or nanosized particles involve high preparation cost (ca. 50% of the end-material cost). On the contrary, SPCMs need a simple preparation step and, therefore, it is expected that they result in much cheaper end-materials.
- Contrary to SPCMs, most of the developed encapsulation methods were not able to improve the thermal conductivity of organic PCMs. Indeed, the limitation imposed to heat transfer by the thermal resistance of the shell usually counterbalances the positive effect of the large surface to volume ratio of the capsules as well as that of conductive nano-filler eventually added to the core.
- SPCMs also could solve the problems associated with micro- or nano-encapsulation technology, such as the effect of different parameters and type of preparation methods. It is worth noting that the preparation method of SCPMs is only dependent on the organic or inorganic nature of the supporting material, but independent of the polymer or the inorganic framework used. In addition, melted PCM leakage when a capsule shell is damaged is another problem of μEPCMs that can be avoided by using SPCMs.

Therefore, increasing attention is given to the development of shape-stabilized PCMs. However, μEPCMs remain an essential class of composite PCMs because they have better mixing properties with construction materials than SPCMs. In addition, PCM slurries cannot be formed by using SPCMs.

Table 26.4 Frequently encountered problems and benefits obtained by using μEPCMs or SPCMs.

Problem	Benefits obtained by using μEPCMs or SPCMs
Low thermal conductivity	SPCMs, especially those with supporting carbon-based material, lead to significant increase of the apparent thermal conductivity of low conductive PCMs. On the contrary, most of the developed μEPCMs (especially micro-sized) are not able to improve it.
High volume change on phase transformation	Most of the developed μEPCMs and SPCMs allow managing the volume change of the PCMs during phase transitions at a local scale (capsule scale or pore scale), thus simplifying the storage system design. However, the shell mechanical strength could be an issue in the μEPCMs.
Incongruent melting	Both types of composites prevent that the solid phase precipitates out of the liquid during the non-congruent melting, thereby separating macroscopically the phases. Indeed, phase separation is restricted to the microscopic distances determined by the size of the micro- or nano-capsules or by the size of the pores in supporting porous materials.
Poor nucleation ability	Although some times the supercooling can be reduced because the supporting material provides appropriate sites for heterogeneous nucleation, it is generally amplified because of the small size of capsules/pores and the discontinuity of the PCM phase. This is especially true for nano-encapsulated PCMs.
Poor long-term stability	In μEPCMs, the shell acts as a container that protects the PCM from external environment, thus contributing to the durability of the PCM. Obviously, the shell must be compatible with the enclosed PCM as well as leak-proff. In a lesser extent, SPCMs also can contribute to extend the PCMs lifetime although they don't warrant tightness.
Incompatibility with construction materials (corrosion or other)	As the PCM in μEPCMs and SPCMs is protected from the external environment, the latter is protected from PCMs. Indeed, contact between the PCM and the construction materials is completely avoided (μEPCMs) or significantly reduced (SPCMs).
Flammability	The incorporation of flame retardant agents can be easily achieved in both types of composites. Moreover, many inorganic supporting materials have flame retardant properties.

Concerning the different classes of SPCMs, those using organic polymers as supporting material can be easily shaped and prepared with desired dimensions, whereas this is more complicate for those based on inorganic frameworks. On the contrary, the latter can be easily incorporated into building materials. Among SPCMs with inorganic supports, those using carbon-based materials are attracting increasing attention because generally leading to superior performances. Indeed, their large surface area and low density is expected to enhance the shape stabilization capability which finally will minimize the loss of energy storage capacity. Moreover, carbon-based supports have good thermal conductivity, thus enhancing the heat transfer ability of the PCM without additives. In addition, carbon based materials are flexible materials because they can be used to encapsulate almost all types of PCMs due to their excellent chemical stability. Also, synergetic effects in which the latent heat of the PCM is increased have been observed in paraffin, fatty acid, and PEG based materials due to the strong attractive interaction that occurs between the liquid PCM and the carbon surface.

26.5 Commercial Status

PCMs offer tremendous potential to fulfill the growing energy needs for cooling and heating applications across various industries, including construction, commercial refrigeration, textiles, transportation packaging for temperature-sensitive products, several solar energy based systems, electronics, and biomedical materials. The use of different PCM technologies is expected to improve the global efforts to conserve energy in the wake of fast depleting fossil fuels. However, effective integration of PCMs in end-products remains a challenging task. The increasing demand for energy-saving and ecologically friendly systems is driving the growth of the global PCM market, which was estimated to grow from $460 million in 2013 to $1.15 billion by 2018, at a compound annual growth rate of 20.1% for the same period. The highest growth is observed in Europe (ca. 42%), where the majority of PCM producers are established, but large markets also exist in North America. Building and construction industries presently are the largest application market of organic PCMs due to the globally increasing requirement for cooling buildings, which in turn has arisen in consequence of the shift from heavy thermal mass design to lightweight architecture. This application contributed 22% to the global PCM market revenues in 2009.

Many international companies such as BASF, Climator, Cristopia, EPS Ltd., Mitsubishi Chemical Corporation, Rubitherm GmbH, TEAP, and Witco market PCMs and PCM-based products. Currently, there are more than 50 different PCMs commercially available, most of which are based on modified compositions of salt hydrates, paraffins, or eutectic salt water solutions. The paraffin-based PCM market commands the largest share of the overall PCM market in terms of value, while salt hydrate-based PCMs lead the market in terms of volume. Apart from encapsulated PCMs, companies are also marketing products such as PCM wallboards, PCM-polymer or PCM-silica dry composite powders for use in a wide range of applications including building and construction, HVAC (heating, ventilation, and air conditioning), textiles, fixed refrigeration, and thermal energy storage. The extensive lists of the commercially available PCMs and PCM containing products have been published by e.g. [5] and [10] in their review articles. The price of PCM products varies from 0.5 to 10 €/kg depending on the type of PCM, the purity grade, the additives needed (e.g. viscosity modifiers, nucleating agents, other stabilizers) and the processing method used to prepare end-material (e.g. packaging, micro-encapsulation).

26.6 Future Development Directions

Thermal energy storage using PCMs is a dynamically growing research area and the interest in this research field can be illustrated by the number of research papers published in the last two decades. Since 2003, an exponential growth in the number of publications can be seen, with more than 10 000 papers published in 2014.

Future trends in the area of PCMs will most likely be related to the use of nano- and bio-based materials:

- Nanomaterials such as micro-encapsulated PCMs or shape-stabilized PCMs offer vast possibilities in the design of novel advanced composites with far superior

properties to those of traditional materials. If properly designed and manufactured they may exhibit remarkable advantages and provide a new stimulus for the development of more efficient PCMs. In recent years, a lot of composite PCMs have been reported, prepared by different processing techniques. However, the influence of the physicochemical properties of the supporting material, as well as that of the size of the capsules or the characteristics of the porous structures used, on the performances of the composite PCM still are misunderstood. Besides, further research is needed to reduce preparation cost, especially for micro-encapsulated PCMs, and to minimize the loss of energy storage capacity due to the presence of the supporting materials.
- Bio-based materials made, at least partially, of renewably resources are an alternative to petroleum-based chemicals. With the aim of extending the renewable-based PCMs beyond fatty acids and their derivatives, new PCM groups have been recently produced and/or investigated. Natural waxes made from underused and renewable feedstocks are of special interest because of their very low cost. Although the research in this area has been carried out, more in-depth exploration is still required.

A part from these obvious trends, further R&D addressing compactness and storage system simplification are also needed:

- Because of the focus on gravimetric latent heat, until recently organic PCMs with unusual high volumetric latent heat such as sugar alcohols and carboxylic acids have received limited attention. Besides, metals have really not been considered as yet in the PCM literature. This is despite their generally high volumetric latent heat which, in combination with their high thermal conductivity, could result in much higher system energy density with proper design. Also, synergetic effects leading to significantly increased latent heat have been observed in different PCMs blends and SPCMs, likely related with hydrogen bonding or strong attractive interactions between the liquid and pore surface. Part of the future R&D efforts should be devoted to the production of new materials with high potential for ultra-compact TES.
- Solid–solid PCMs have also received only limited attention, and those that have been studied generally have low-to-moderate specific and volumetric latent heat. The lack of a liquid phase, however, gives them a number of advantages over traditional PCMs, such as elimination of liquid seepage and void space for volume change and freedom to use additives or multi-constituent blends without concern for liquid phase segregation. While there has been more study into heterogeneous form-stable blends of PCMs with non-melting materials, further research into homogenous solid-state PCMs could produce materials with useful system level performance.
- Most of the approaches proposed to increase the thermal conductivity of organic PCMs provide very low enhancement ratios per unit volume fraction of the conductive phase added. Ultrathin graphite foams with open porosity as high as 98.8–99.2% and thermal conductivity up to 2.8 W/m/K, achieving PCM thermal conductivity enhancement ratios of 10–15, have been recently proposed. However, the composite PCMs obtained by infiltration of these foams are not able to avoid the leakage problem. Further research would be required to produce new foams with very high porosity and thermal conductivity, while having cellular structure and much small pores.

Most of the approaches used to reduce supercooling when needed are based on the use of nucleating agents with a crystal structure similar to that of the energy storage material. Therefore, the researchers have to do the repetitive work of searching adapted nucleating agents to ensure the usability of each PCM that exhibits unacceptable supercooling. The use of external fields such as electrical or magnetic fields has yet not been considered in the PCM literature despite the potential that these approaches could offer to activate and to control the crystallization of different PCM groups, such as polyols or salt hydrates.

Focusing on the primary PCM function of energy storage, many reviews limit material data to melting temperature and gravimetric latent heat, which is insufficient for modeling or providing a proper material comparison. Both volumetric and specific latent heat values are needed for a minimum material comparison, and thermal conductivity and specific heat are additionally required for transient thermal modeling. It has also been observed that very few studies are available in which temperature dependence thermo-physical properties such as conductivity, density, heat capacity, and viscosity of prospective PCMs are measured. Moreover, there are no international technical standard methods available to determine the thermal properties of PCMs. Therefore, the researchers have to do repetitive work to ensure the thermo-physical properties of new materials and there is often a large variability in the results. In addition to reliable PCM data bases and standardized measurement techniques, a general standard for testing PCMs cyclability and thermal endurance is also missing.

References

[1] Telkes, M. and Raymond, E. (1949). Storing solar heat in chemicals – A report on the Dover house. *Heating and Ventilating* 46: 80–86.

[2] Abhat, A. (1983). Low temperature latent heat thermal energy storage: Heat storage materials. *Solar Energy* 30: 313–332.

[3] Pielichowska, K. and Pielichowski, K. (2014). Phase change materials for thermal energy storage. *Progress in Materials Science* 65: 67–123.

[4] Jankowski, N.R. and McCluskey, E.P. (2014). A review of phase change materials for vehicle component thermal buffering. *Applied Energy* 113: 1525–1561.

[5] Zalba, B., Marin, J.M., Cabeza, L.F., et al. (2003). Review on thermal energy storage with phase change: Materials, heat transfer analysis and applications. *Applied Thermal Engineering* 23: 251–283.

[6] Kenisarin, M. and Mahkamov, K. (2007). Solar energy storage using phase change materials. *Renewable and Sustainable Energy Reviews* 11: 1913–1965.

[7] Sharma, A., Tyagi, V.V., Chen, C.R., et al. (2009). Review on thermal energy storage with phase change materials and applications. *Renewable and Sustainable Energy Reviews* 13: 318–345.

[8] Kenisarin, M.M. (2014). Themophysical properties of some organic phase change materials for latent heat storage. A review. *Solar Energy* 107: 553–575.

[9] Sharma, R.K., Gamesan, P., Tyagi, V.V., et al. (2015). Developments in organic solid-liquid phase change materials and their applications in thermal energy storage. *Energy Conversion and Management* 92: 193–228.

[10] Kenisarin, M.M. and Mahkamov, K. (2016). Salt hydrates as latent heat storage materials: Thermophysical properties and costs. *Solar Energy Materials and Solar Cells* 145: 255–285.

[11] Cabeza, L.F., Svensson, G., Hiebler, S., et al. (2003). Thermal performance of sodium acetate trihydrate thickened with different materials as phase change energy storage material. *Applied Thermal Engineering* 23: 1697–1704.

[12] Yuan, Y., Zhang, N., Tao, W., et al. (2014). Fatty acids as phase change materials: A review. *Renewable and Sustainable Energy Reviews* 29: 482–498.

[13] Sarier, N. and Onder, E. (2012). Organic phase change materials and their textile applications: An overview. *Thermochimica Acta* 540: 7–60.

[14] Palomo del Barrio, E., Cadoret, R., Daranlot, J., et al. (2016). New sugar alcohols mixtures for long-term thermal energy storage applications at temperatures between 70 °C and 100 °C. *Solar Energy Materials and Solar Cells* 155: 454–468.

[15] Khadiran, T., Hussein, M.Z., Zainal, Z., et al. (2015). Encapsulation techniques for organic phase change materials as thermal energy storage medium: A review. *Solar Energy Materials and Solar Cells* 143: 78–98.

[16] Fan, L. and Khodadadi, J.M. (2011). Thermal conductivity enhancement of phase change materials for thermal energy storage: A review. *Renewable and Sustainable Energy Reviews* 15: 24–46.

[17] Khodadadi, J.M., Fan, L., and Babaei, H. (2013). Thermal conductivity enhancement of nanostructure-based colloidal suspensions utilized as phase change materials for thermal energy storage: A review. *Renewable and Sustainable Energy Reviews* 24: 418–444.

[18] Ji, H., Sellan, D.P., Pettes, M.T., et al. (2014). Enhanced thermal conductivity of phase change materials with ultrathin-graphite foams for thermal energy storage. *Energy & Environmental Science* 7: 1185–1192.

[19] Kholmanov, I., Kim, J., Ou, E., et al. (2015). Continuous carbon nanotube-ultrathin graphite hybrid foams for increased thermal conductivity and supressed subcooling in composite phase change materials. *ACS Nano* 9: 11699–11707.

27

Sorption Material Developments for TES Applications

Alenka Ristić

Department of Inorganic Chemistry and Technology
National Institute of Chemistry
Hajdrihova 19, SI-1001 Ljubljana, Slovenia

27.1 Introduction

Development of a new generation of thermal energy storage systems that would enable efficient utilization of low-temperature solar energy or waste heat/combined heat and power (CHP) heat for heating and cooling applications is the focus of recent R&D activities world-wide. Presently, sensible heat storage (SHS) is the most extended way of storing thermal energy; however, this method presents large limitations, like the large storage size due to the storage density of the heat storage media usually required or the temperature swing created from the sensible addition and extraction of energy, to name only two of them. For example, water tank storage is limited to store heat for a very short-term (hours) of relatively small facilities. Novel techniques are needed when the objective is to store a large energy amount captured from the sun in summer to use during winter (called seasonal heat storage), or when large sources of industrial waste heat, are available, or when it is required to store heat captured on sunny days in winter (short-term storage).

The two new storage concepts that have been intensively studied recently to overcome the above limitations are: latent heat storage (LHS), which uses the heat of liquefaction of phase change materials (PCM) for storage, and thermochemical heat storage (TCS), which is based on reversible chemical or physical sorption of gases in solids. TCS is considered as the only storage concept with a potential for long-term, also seasonal (several months), thermal energy storage. The amount of energy stored per unit volume in TCM is, in theory and in practice, higher (several hundreds of kWh m^{-3}) than that of sensible heat in water and other media and there are no heat losses over a long time, even without any insulation. Furthermore, it is considered superior to LHS technology, which suffers from segregation of PCM material in the system upon melting, super cooling, corrosion, etc. high costs, and safety issues. TCS technology as well shows disadvantages, like degradation, corrosiveness, and the high costs of adsorbents, and complex process technology.

27.1.1 Thermochemical Heat Storage (TCS)

Thermochemical heat storage can utilize the reversible solid:gas adsorption, which can be divided into chemical adsorption (chemisorption) or physical adsorption

Advances in Energy Storage: Latest Developments from R&D to the Market, First Edition.
Edited by Andreas Hauer.
© 2022 John Wiley & Sons, Ltd. Published 2022 by John Wiley & Sons, Ltd.

(physisorption) [1]. Physical adsorption is a phenomenon when the molecules are weakly bound to the material surfaces, while chemical adsorption is due to the strong (e.g. covalent) bonds of the molecules in the compounds. Different working pairs (adsorbent/adsorbate) considered mainly employ water as the adsorbate due to its environmentally friendly properties and high evaporation heat. Other adsorbates like ammonia, methanol, and ethanol can be used as well [2]. Adsorbents or sorption materials can be divided into (i) porous materials, where physical adsorption occurs, like zeolites, silica gels, aluminophosphates, metal organic frameworks (MOFs), (ii) solids based on chemical adsorption such as salt hydrates like chlorides, nitrates, sulfates, and (iii) materials based on chemical and physical adsorption like composites of salts integrated into porous matrix. Water sorption storage usually consists of two processes. Under the influence of a heat supply, water is desorbed from the material (an endothermic phenomenon referred to as the charging or activation of material). When water is adsorbed on adsorbent, the heat of adsorption is released (an exothermic phenomenon referred to as a material's discharge or deactivation). Figure 27.1 shows a schematic presentation of adsorption (discharging) and desorption (charging) processes on the porous solid.

Energy can therefore be stored with negligible thermal loss because thermal energy is not kept in either sensible or latent form, but as a chemical potential. The TCS has the potential to make extensive use of a solar thermal energy and low temperature residual heat from industry, thus leading to a low carbon energy society. In this chapter, porous sorption materials based on physical sorption (physisorption) of water will be considered only.

27.1.2 Basic Criteria for Suitable Sorption Materials

Sorption materials play a crucial role in the development of sorption storage systems. The suitable materials should meet the following basic criteria [2–5], which depend

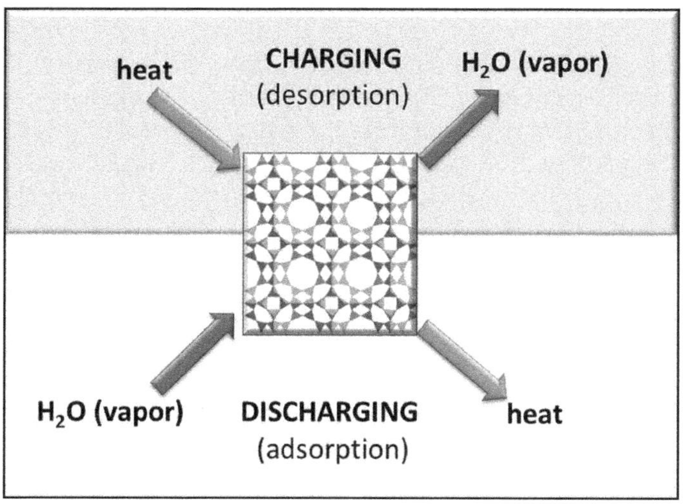

Figure 27.1 Schematic presentation of adsorption (discharging) and desorption (charging) process on adsorbent.

also on the storage time as indicated (short term storage or daily storage/long term or seasonal storage):

1. Energy densities above 200 kWh m^{-3} at activation temperatures below 140 °C and an usable temperature lift (between evaporator and adsorber) of at least 25 K.
2. High sorption capacity (uptake of adsorbate).
3. Low regeneration (charging) temperature (desorption of water below 140 °C to follow e.g. the solar collector operating temperature).
4. Good thermal and chemical stability under working conditions like temperature (T) and pressure (p) for at least 50 cycles of adsorption/desorption (seasonal storage) or at least 3000 cycles (daily storage).
5. Adequate kinetics of the sorption-desorption stages and adequate thermal conductivity in order to match the loading/unloading power characteristics in final applications (important for daily storage).
6. Environmental safety, non-toxicity, low global warming potential.
7. Non-corrosiveness of materials.
8. Low to moderate cost of materials production (<5 € kg^{-1} for daily storage, < 2 € kg^{1} for seasonal storage).
9. Low energy intensity of materials production (especially important for energetic amortization of seasonal storage).
10. Suitability for shaping of the material.

At the moment, no material currently available could satisfy all the listed requirements. Therefore, development of sorption materials for TES remains the challenge to be solved. Another challenge is the suitability of the sorption material at the system level according to the basic criteria. The adsorption system can be divided into open and closed systems. An open system exchanges mass and energy with the environment and operates at ambient pressure, while a closed system exchange only energy with the environment, and are normally evacuated [2].

27.1.3 Determination of Sorption Properties

A lot of attention has been devoted to the synthesis of inorganic porous water adsorbents due to the needs of materials with high sorption capacity during the last few decades. The water sorption behavior of a porous sorbent, one of the important characteristics of any water adsorbent depends on many parameters such as the structure and the chemical composition of the porous material (e.g. Si : Al ratio for zeolites), the presence of charged species, and hydration level which can be best represented in the form of adsorption isotherms. According to the International Union of Pure and Applied Chemistry (IUPAC) classification of water sorption isotherms are divided in seven types (Figure 27.2, [6]).

Sorption materials that show high affinity to water are very hydrophilic materials and exhibit a Type I isotherm. These materials represent very fast reaction saturation at low relative pressure ($p\ p_0^{-1}$) which is followed by consistent adsorption over a wide range of relative pressure due to the water saturation in the pores. Hydrophilic materials exhibit sorption isotherms Type II or IV, where a considerably high sorption capacity of water at low relative pressure and moderate relative pressure is measured. Additionally, the rare Type VI step-like isotherm is also characteristic of hydrophilic

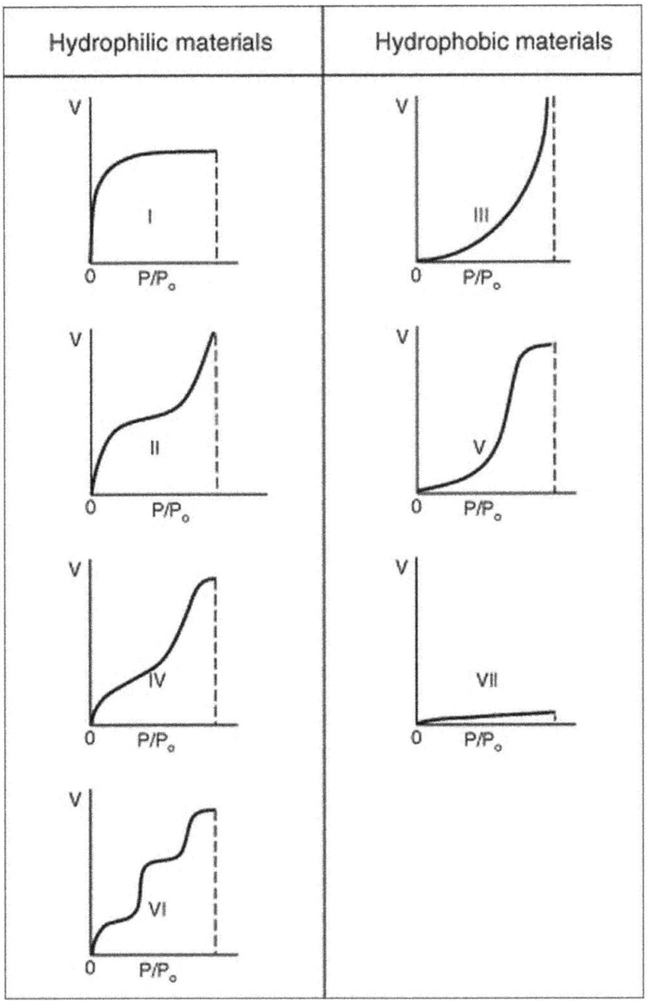

Figure 27.2 Adsorption isotherms classification according to IUPAC: Type I – very hydrophilic material, Type II – hydrophilic material, Type III – hydrophobic/ low hydrophilic material, Type IV- hydrophilic material, Type V- hydrophobic/low hydrophilic material with weak sorbent-water interactions, Type VI- hydrophilic material with multiple sorbent-water interactions and stepwise sorption, Type VII – very hydrophobic material (reprinted from publication *Microporous and Mesoporous Materials*, 114, Eng-Poh Ng, Svetlana Mintova, Nanoporous materials with enhanced hydrophilicity and high water sorption capacity, 1-26, 2008, with permission from Elsevier).

sorbent having stepwise sorption. In contrast, sorption isotherms Types III and V describe hydrophobic or low hydrophilic materials with low sorption at low relative pressure and sometimes moderate sorption at the middle relative pressure, and suddenly high water sorption at relative pressure in close proximity to 1. For highly hydrophobic sorbents, isotherm Type VII is attributed, which gives a low sorption of water throughout the entire relative pressure range.

27.2 Sorption Materials

The great progress in material science offers innovative sorption materials besides traditional ones for thermochemical heat storage.

27.2.1 Traditional Sorption Materials

Traditional sorption materials like highly hydrophilic zeolites and weak hydrophilic silica gels with water vapor systems have been one of the most studied and tested adsorbent/adsorbate pairs for application in thermochemical heat storage.

27.2.1.1 Zeolites

27.2.1.1.1 Structural and Sorption Properties

Zeolites are microporous crystalline alkali or alkali-earth aluminosilicate materials with a three dimensional framework, containing uniformly distributed channels and/or cavities with a pore dimension range of 0.3–0.8 nm. Frameworks consist of SiO_4 and AlO_4 tetrahedra with loosely bound metals as charge-compensating cations (e.g. extra-framework cations), which can be readily exchanged. Different zeolites structures can be formed via connecting the SiO_4 and AlO_4 tetrahedra in networks. 255 (up to date) different zeolite framework types are known [7] and some of them are found in nature (about 40 types). Depending on the structure type, the zeolite framework composition can be varied to different extents. Thus the diversity of zeolitic materials is extremely large, from both structural and chemical points of view. Zeolites are highly hydrophilic sorbents and the hydrophilicity of zeolites is directly dependent on their Al content, i.e. the Si : Al ratio of the zeolite framework, framework structure type, and on the exchangeable (extra-framework) cation sites [6]. Zeolites with a low Si : Al ratio show mainly a Type I water sorption isotherm. The high affinity to water at low partial pressure influences on the temperature of dehydration. High temperatures of dehydration (200–300 °C) are the consequence of the strong interaction between the charged frameworks and water molecules [8]. The extra-framework cations (alkali- or alkali-earth metals), which balance the negative charge of the zeolite frameworks, and their distinctive interactions with the water molecules determine the sorption ability of the zeolites. During the adsorption process, water molecules interact with these cations and form aqua-complexes [9]. Water sorption capacity depends on the type of extra-framework cations and degree of ion-exchange [10] and decreases dramatically from lithium to cesium due to the decrease in the free volume of the large cavities of zeolites as a result of the increased cation diameter [11]. Jänchen et al. have systematically studied the effect of cations on the water sorption uptake [12]. Zeolite A and Zeolites X and Y were ion-exchanged with different cations and they found that the water uptake is dependent and proportional to the cation loadings. It was revealed that zeolites exchanged with smaller cations (e.g. Mg) have higher water sorption capacity. This was explained by the replacement of two Na^+ ions (0.102 nm) by one smaller Mg^{2+} (0.072 nm), thus enhancing the water sorption through enlarging the pore volume of the zeolite [13]. Changing chemical composition and surface chemistry of the zeolites by preparation, calcination, or post-synthesis treatment influences the enhancement of the hydrophilic properties. Thus, the combination of pore characteristics, structural arrangement, and chemical composition determines the overall properties of the zeolite and its storage application. The hydrophilic character of zeolites can be tailored by post-synthesis modifications. The chemical composition can be changed by demetallation [14] (e.g. removal of framework forming elements) and ion exchange. Medium hydrophilic or hydrophobic-hydrophilic materials (exhibit S-shape of water isotherm) can be prepared by dealumination (e.g. removal of framework aluminium), a method for chemical and structural modification of zeolites. Lutz prepared Zeolite Y by steaming at temperatures higher than 300 °C [15]. The DeAluminated Zeolite Y (denoted DAY) contained a higher Si : Al ratio (Si : Al = 7) and lower amount of charge compensating cations resulting in a lower concentration of charges and adsorption sites, and a decrease of the water sorption capacity was observed. On the other hand, chemical treatment with H_4EDTA or HCl increased water sorption capacity of the binder-free Zeolite NaY [8]. Dealumination prepared by steaming or chemical treatment controlled the hydrophilic behavior of NaY and contributed to defined low desorption temperatures [8, 16]. Dealumination has another effect on the Zeolite Y, namely a secondary pore system can be formed, thus preparing

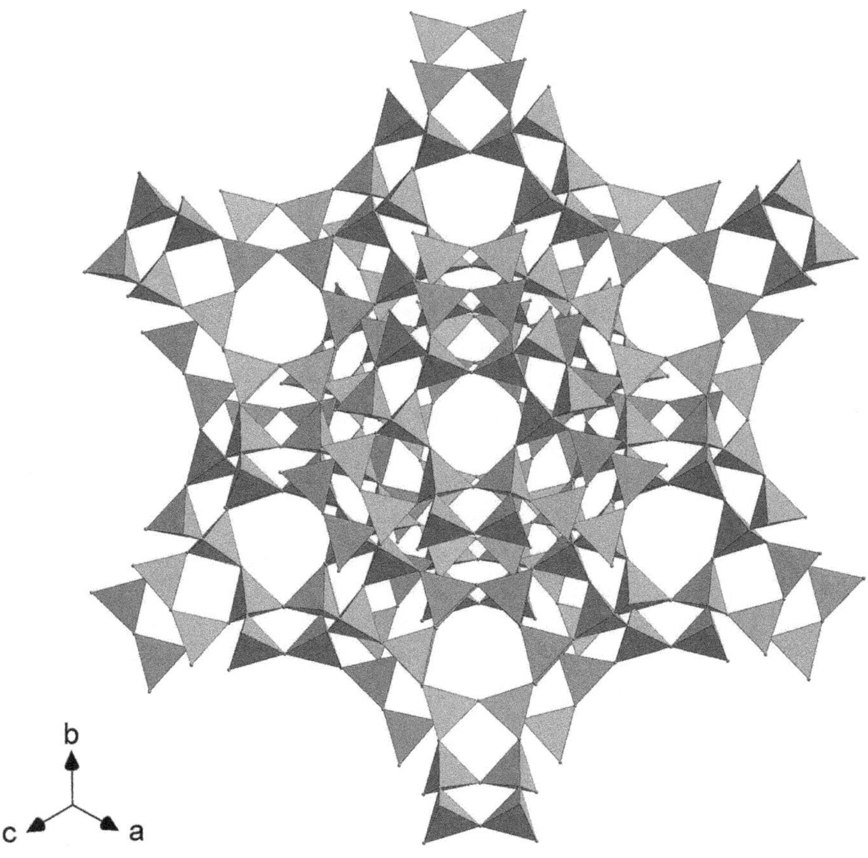

Figure 27.3 Framework structure of Zeolite X.

hierarchical zeolites [17], which can promote mass transport and increase water sorption capacity [8]. Differently from modifications by cations exchange or variation of the Al : Si ratio by dealumination, an alternative way is the functionalization of zeolites by chemisorption of organosilanes, which resulted in an increased hydrophobicity that corresponded to a change from the typical Type I isotherm of microporous hydrophilic materials toward the characteristic S-shaped (Type V) isotherm [18]. Mainly commercially available Zeolites A, Y, and X were investigated for TES. The framework of Zeolite A is constructed of sodalite cages which are connected with double 4 rings into network, forming cavities with pore diameters of 1.14 nm and pore diameter entrance of 0.42 nm, while sodalite cages in Zeolite X or Y are connected with double 6 rings forming larger pore diameter cavities of 1.3 nm and pore entrance of 0.74 nm. The crystal structure of Zeolite X is shown in Figure 27.3.

27.2.1.1.2 Storage Properties of Commercial Zeolites

Investigations were performed on powdered and shape formed (grains, pellets, extrudates) zeolites containing weak hydrophilic binders (ca 20%), which decreased water sorption capacity compared to equivalent powdered zeolites. Inorganic binders or binding agents are usually clays, silicates, and alumina, which are mixed with powdered zeolites in order to prepare shape forms. Depending on operating conditions (desorption temperature: 180 °C, adsorption: 20 °C/2.3 kPa) in closed system commercial Zeolite X possessed energy storage densities of 110 kWh m^{-3}, respectively, while the exchanged zeolites showed higher storage densities from 128 to 160 kWh m^{-3} [12]. Recently new opportunities such as manufacturing of granulated binder-free (e.g. the intermediate binder – metakaolin was transformed into zeolite) zeolites (A, X, and Y) [19] and the modifications of chemical composition have enabled tailoring the hydrophilic character of storage materials. Water sorption capacity of binder-free zeolites has been determined to be higher comparing to zeolites containing binders by 14% for Zeolite A and 16% for Zeolite X [8]. These binderless zeolites show sufficient hydrothermal stability, higher specific energies, and the potential for a better performance density of the storage. Both open and closed storage tests have shown comparable adsorption capacities and specific energies for the binder-free zeolites [8, 20]. The experimental investigations have shown that binderless Zeolite X is characterized by a high water uptake and very fast reaction kinetics. The high water uptake is required for a high energy storage density whereas high reaction kinetics are essential for a high thermal power output during adsorption [21]. The process of the discharging performance of an open thermal adsorption storage was investigated by tests in a 1.5 L lab-storage with Zeolite 4A beads and honeycomb structures both with and without binder. The binder containing materials, 4A beads, and 4A-honeycombs, have the lowest adsorption capacities due to the remaining binder in the structures. Differently, beads and honeycombs of the similar dimension and geometry as the former bodies but without binder show, as expected, higher adsorption capacities. The 4A-honeycombs show 82% of the adsorption capacity of the binderless zeolite honeycombs (4ABF-honeycombs) (0.269 g g^{-1}) because of the 20% of the binder in the 4A-honeycombs. The binderless zeolite honeycombs showed the best adsorption dynamics [22]. The EU COMTES project (AEE – Institute for Sustainable Technologies, Austria) with a prototype cylindrical vessel (closed system, charging temperature: 180 °C) containing about 164 kg of binderless Zeolite X grains with 30 wt% of water adsorption capacity showed a volumetric storage density of the zeolite bulk of 113 kWh m^{-3} [23].

Figure 27.4 Mobile heat storage system at ZAE Bayern.

This is 71% of the theoretical maximal storage density, mainly due to the relatively high residual water vapor content of the zeolite. The binderless Zeolite X was used in SolSpaces project in an open system (charging temperature of 180 °C at a water vapor partial pressure of 1 kPa and adsorption temperature of 20 °C; solar thermal space heating) at IGTE Stuttgart, which shows storage density of 163 kWh m^{-3} [24], while the mobile storage system (Figure 27.4) for utilization of waste heat and working with a packed bed of Zeolite 13X with binder (14 t) in an open system (charging temperature of 250 °C and discharging temperature of 60 °C and 0.09 kg kg-1 humidity) achieved energy density of 214 kWh m^{-3} [25].

27.2.1.1.3 Natural Zeolites and Storage Properties

Compared to synthetic zeolites, natural zeolites are less expensive and easy to access. Most naturally occurring zeolites (e.g. those found in volcanic sedimentary rocks like chabazite, clinoptilolite, faujasite, mordenite, phillipsite) are hydrophilic due to their relatively low Si : Al molar ratio. Among them, clinoptilolite is the most abundant and it is commonly considered a low-cost adsorbent. Chabazite-rich tuff was investigated for thermal energy storage 37 years ago [26]. Natural chabazite from Nova Scotia possesses higher energy density than NaLSX (Na Low Silica zeolite X) with Si : Al = 1 [27]. Recently, hydrophilic properties of clinoptilolite-tuff from Greece and Serbia were investigated for TES [28]. The results showed lower adsorption water capacities (0.14–0.16 g g^{-1}) than the adsorption water capacities of synthetic zeolites and storage densities of 90 Wh kg^{-1} (20 °C/RH of 79%) for both materials.

27.2.1.2 Silica Gels

27.2.1.2.1 Structural and Sorption Properties

Silica gels with the general formula $SiO_2 \cdot xH_2O$ are porous compounds with high affinity to water vapor and large water sorption capacity at low humidity, cheap, and can be easily regenerated (150 °C) [29]. Figure 27.5 shows the TEM picture of the

Figure 27.5 TEM photograph of silica gel.

silica gel. This class of materials has a medium energy density and a limited temperature lift [30].

Basically, the selectivity of silica gel towards water is based on its pore structure and the distribution of silanol (Si–OH) groups [6]. The amount of silanol groups in the silica gels is a key parameter that determines the amount of adsorbed water and degree of surface modification. A high water sorption capacity of the silica gel can be achieved through reducing the grain size, whereby the small size of the substances leads to the enhancement of the surface area. The pore size of silica gels plays a significant role in water sorption processes. It is known that adsorbents with smaller pore diameter have higher active sites per surface area than those having larger pore diameter. Another approach to improve sorption capacity of the silica gel is the incorporation of transition metals during the synthesis process. A modified silica gel containing iron shows three times higher water uptake at the same conditions [31]. Mainly, commercially produced silica gels, such as Fuji types A and RD, Davison gel types 125 and 127 (Grace), and KSMG were studied for heat transformation applications in the past. These materials had pore diameters from 2 to 3.5 nm and surface areas from 400 to 700 m^2g^{-1} [32].

27.2.1.2.1 Storage Properties

On demonstration scale, a thermochemical storage based on microporous silica gel/water in a closed system was developed for solar space heating in a single-family house within the MODESTORE project [33]. In the pilot plant system an energy storage density of approximately 50 kWh m^{-3} (charging temperature of 88 °C at water vapor pressure of 1.9–3.5 kPa), [2] was achieved.

27.2.2 Innovative Sorption Materials

New material classes such as microporous aluminophosphates ($AlPO_4$), metal organic frameworks (MOFs) and composite materials of hygroscopic salts incorporated into the porous carrier material are the subject of the intensive research.

27.2.2.1 Aluminophosphates

27.2.2.1.1 Structural and Sorption Properties

Recent investigations of microporous aluminophosphates (APO-n or $AlPO_4$-n; n = structure type) and their modified analogues (SAPO-n and MeAPO-n; S = Si, Me = transition metal) revealed their advantages over zeolites concerning charging temperature and storage density. The framework of aluminophosphates, consisting of AlO_4 tetrahedra, and AlO_5 or AlO_6 polyhedra, and PO_4 tetrahedra, are neutral unlike the zeolites which have extra-framework cations and negatively charged framework, therefore they have no ion-exchange capacity or acidic catalytic properties. The characteristics of aluminophosphate molecular sieves include a univariant framework composition with Al : P = 1, a high degree of structural diversity and a wide range of pore sizes and volumes. Their surface selectivity is mildly hydrophilic. These materials exhibit excellent thermal (up to 1000 °C) and hydrothermal stability up to 600 °C (steam) [34]. Aluminophosphates are prepared by hydrothermal, solvothermal, or ionothermal treatments of aluminophosphate gels or solutions containing organic templates or structure-directing agents, which have to be removed after the synthesis in order to prepare microporous materials. Some aluminophosphate adsorbents exhibit Type V water isotherms due to interactions of water to sorbent and water to water through hydrogen bonding [6]. To date, the emphasis in heat storage research has been on the aluminophosphates with incorporated silicon or metal cations, where the presence of dopants enhances the water sorption capacity because structural defects enable better interactions of polar water molecules with the surface. However, dislodgement of silicon or metal cations from the framework sites and framework structure degradation, which is quite common after a few water desorption/sorption cycles, lowers the performance of these aluminophosphate materials.

27.2.2.1.2 Storage Properties

The most studied adsorbent is SAPO-34 [27, 35–36] (also known as AQSOA FAM Z02), a commercially applied adsorbent, besides $AlPO_4$-5 (AQSOA FAM Z05), FeAPO-5 (AQSOA FAM Z01), and $AlPO_4$-18 [30, 37–39]. Frazzica and Brancato verified hydrothermal stability under severe working conditions, shelf and cycle conditions, of the commercial AQSOA-Z02 adsorbent. The shelf test was performed for 21 days at 90 °C and 7.3 kPa of water vapor pressure. 1550 cycles between 90 °C and 40 °C at 7.3 kPa of water vapor pressure were performed as the cycle test. The structural characterization of the adsorbent confirmed its stability under the given conditions [40]. SAPO-34 and $AlPO_4$-18 are found to have energy densities of 171 Wh kg^{-1} and 215 Wh kg^{-1}, respectively at discharging temperature of 40 °C and at the water vapor pressure of 5.6 kPa and charging temperature of 95 °C and at the water vapor pressure of 1.2 kPa [41]. On the other hand, a research of Brancato and Frazzica, $AlPO_4$-18, FAPO-34, and SAPO-34 achieved energy storage capacities of 139 Wh kg^{-1}, 118 Wh kg^{-1}, and 94 Wh kg^{-1}, respectively, at

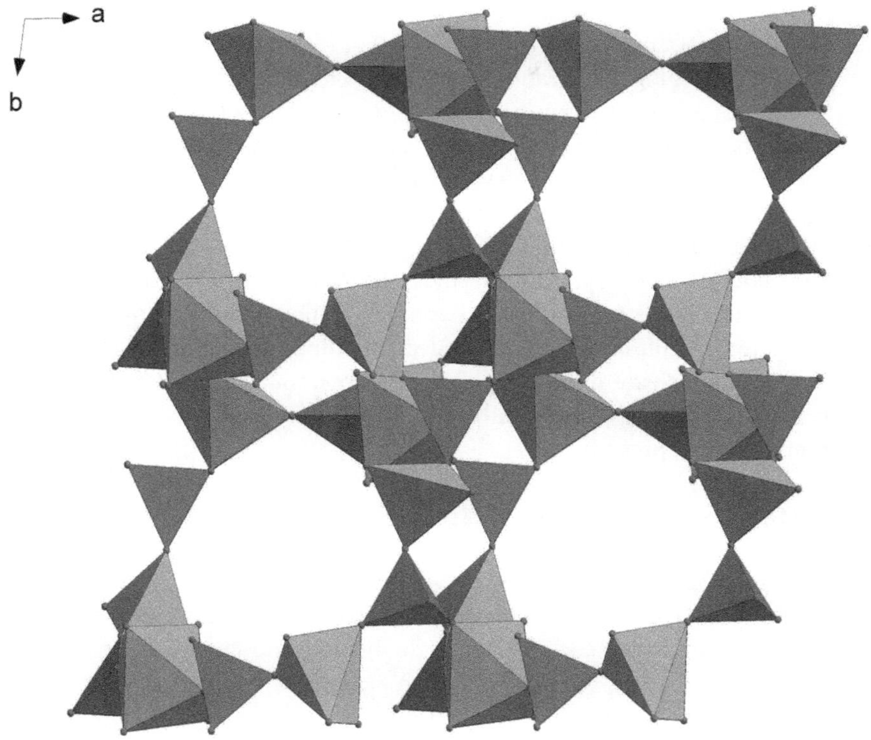

Figure 27.6 Dehydrated aluminophosphate APO-Tric with chabazite structure.

charging temperature of 90 °C and at the water vapor pressure of 1.2 kPa and discharging temperature of 35 °C and at the water vapor pressure of 4.2 kPa [39]. A study of Ristić et al. enabled to clearly reveal the structure-property relationship in three aluminophosphates with S-shaped water isotherm (SAPO-34, $AlPO_4$-18, and APO-Tric). Specifically APO-Tric (Tric is a short name for triclinic crystal symmetry of chabazite structure) material has shown high water uptake of 0.318 g g^{-1} and cycling stability (decrease of water uptake for 3.5% after 40 cycles of adsorption/desorption between temperatures of 140 and 40 °C at the water vapor pressure of 5.6 kPa) [42]. Energy storage capacity of APO-Tric at the charging temperature of 100 °C and the discharging temperature of 25 °C at water vapor pressure 1.2 kPa, was determined to be (320 Wh kg^{-1}) 472 kWh m^{-3} [43]. Examination of the crystal structures of hydrated APO-Tric (e.g. hydrated $AlPO_4$-34) shows that the driving force for the determined water sorption process is the formation of highly ordered water clusters in the pores, which is enabled by rapid and reversible changes in the Al coordination and optimal pore diameters [44]. The ease with which changes in the Al coordination can occur in APO-Tric is related to the use of the fluoride route in the synthesis. The crystal structure of dehydrated APO-Tric, which is shown in Figure 27.6, is similar to the structure of SAPO-34 material.

The water can be desorbed at 95 °C, which is another important characteristic for its utilization. At even lower temperature, at 60 °C, all water can be removed from the microporous aluminophosphate with LTA topology ($AlPO_4$-LTA), which can be easily achieved using most common flat plate solar collectors. It possesses the same structure like well-known water adsorbent Zeolite A. The material adsorbs water in a very narrow relative-pressure range (0.10–0.15), exhibits the highest water uptake (0.420 g g^{-1}) and energy storage capacity of (373 Wh kg^{-1}) 527 kWh m^{-3} (charging temperature of 60 °C and discharging temperature of 25 °C at water vapor pressure of 1.2 kPa). Furthermore, this water adsorbent outperforms (e.g. energy storage density) all other aluminophosphates as well as metal-organic porous materials tested to the present [43]. It shows also remarkable cycling stability (decrease of water uptake less than 2% after 40 cycles of adsorption/desorption between temperatures of 150 and 25 °C and relative humidity of 0.3) [43]. In $AlPO_4$-LTA, energy is predominantly stored in the hydrogen-bonded clusters of water molecules within the pores, showing on one-step hydration mechanism. Two drawbacks restrict the industrial implementation of aluminophosphates, organic templates are needed for their production and the estimated production costs, which are above acceptable cost for seasonal and mid-term heat storage systems, but could be economically feasible in some other heat transformation applications (e.g. heat pumps).

27.2.2.2 Metal-Organic Frameworks (MOFs)

27.2.2.2.1 Structural Properties

Metal-organic framework (MOF) structures are another class of very hydrophilic adsorbents and they have been intensively studied for the use in adsorption heat

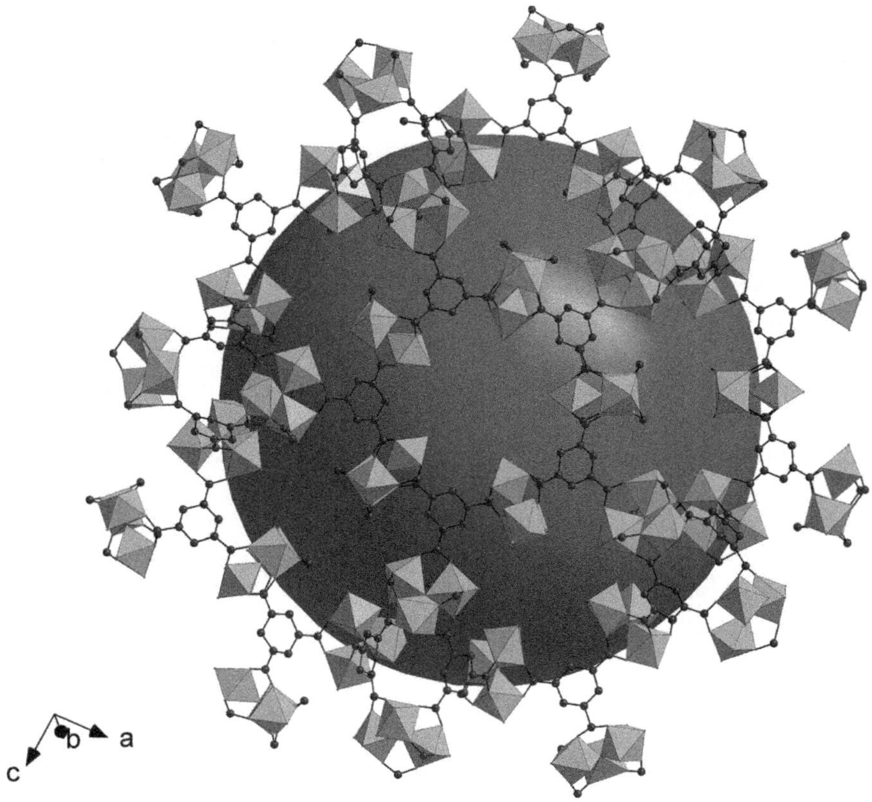

Figure 27.7 MIL-100(Fe) as water adsorbent is a suitable candidate for thermally driven heat pumps.

transformation processes (heat pumps). The isotherms of these compounds show S-shape, which is advantageous for this application, as a large loading lift can be achieved within a narrow range of relative pressure. MOFs are inorganic-organic hybrid materials based on metal ions or metal ion clusters as nodes, which are linked by organic, at least bidentic ligands. They are crystalline open porous materials with a one, two, or three-dimensional framework like zeolites and aluminophosphates. Figure 27.7 shows a three-dimensional framework of MIL-100(Fe), which is a large pore iron(III) trimesate. MIL-100(Fe) is composed of inorganic building blocks representing octahedrally coordinated Fe(III) centres arranged in trimeric clusters which are further connected through trimesate ligands into a three-dimensional (zeotype) framework.

27.2.2.2.2 Sorption and Storage Properties

Compared to traditional adsorbents (zeolites, silica gels) used in heat pump applications, MOFs exhibit a much richer variety in composition, pore structure, and topology. The cluster/linker concept allows tuning of pore structures and chemical functionality over a wide range. These materials possess unique features such as huge surface area, large pore volume, and an unprecedented geometric, chemical, and physicochemical variability, which are due to their tunable composition. A comprehensive overview on the available structures and possible heat transformation applications can be found in extensive reviews [4, 45–46]. With regard to the application in adsorption heat transformation processes, the main focus lies on water adsorption properties and storage capacities, which are strongly connected to the specific pore volume [45]. With regard to the application, the adsorption step of several MOFs occurs at a high relative pressure ($pp_0^{-1} > 0.35$), which reduces the useful loading lift within the possible cycle. A crucial point for the applications is the hydrothermal instability of most MOFs [47]. Review papers from 2014 and 2015 considered most of the limitations of MOFs for heat pump applications, by pointing out the most severe one, which is low thermal or hydrothermal stability [46]. Several compounds show a large degradation within a few cycles. On the other hand, CAU-10(Al)-H, MIL-100(Fe) (Figure 27.7), MIL-101(Cr), MOF-801(Zr), MOF-841(Zr), Al-fumarate, porous Zn-carboxylate exhibit good hydrothermal stability [4, 38, 48]. MIL-100(Fe) was found to have maximal water sorption capacity of 0.400 g g^{-1} and energy sorption capacity of 267 Wh kg^{-1} at adsorption temperature at 40 °C/5.6 kPa and desorption temperature at 90 °C/5.6 kPa [45]. The water adsorption characteristics of CAU-10-H show high hydrophilicity with a steep (0.340 g g^{-1}) and characteristic S-shaped adsorption isotherm already at a low relative humidity around $p\ p_0^{-1} = 0.2$. With regard to possible applications in sorptive heat transformation processes, hydrothermal stability was tested by short- and long-term cycle tests. No degradation of crystallinity was observed even after 700 adsorption/desorption cycles [49]. Higher water adsorption capacity (0.360 g g^{-1}) compared to CAU-10-H was determined for MOF-801. Krajnc et al. determined lower energy storage capacity of 323 Wh kg^{-1} (514 kWh m^{-3}) for this adsorbent in the same working conditions as for $AlPO_4$-LTA (charging temperature of 60 °C and discharging temperature at 25 °C at water vapor pressure 0.5 kPa) [43]. Recently, comparatively high heat of adsorption of NH_2-UiO-66 and, especially, NH_2-MIL-125 were determined offering the possibility for low-temperature driven applications, e.g. domestic heating [50]. However, the production costs of MOFs are in the range of aluminophosphates.

27.2.2.3 Two-component Sorbents of Porous Matrix and Salt Hydrates

27.2.2.3.1 Structural Properties

The most versatile class of sorbents for thermal energy storage are two-components or composites, which have been prepared for the enhancement of thermo-physical properties of sorption materials, like water sorption capacity, heat and mass transfer on one side, and on the other hand to avoid or diminish the deliquescence, swelling, and agglomeration of salt hydrates during sorption/desorption cycles. Thus the advantages of the pure porous material which act as a matrix and hygroscopic salt hydrates (chlorides, bromides, sulfates, and nitrates of alkali and alkali-earth metals) are combined. These sorbents have the advantage to tailor sorption properties by varying the salt chemical nature and content, porous structure, and to present a low desorption temperature, a low price, and a simple production method [51]. Water sorption behavior of the composite materials is a combination of three mechanisms: adsorption by the host matrix, chemical reaction between water and salt resulting in formation of hydrates and absorption by the salt aqueous solution in the pores. The role of the porous matrix is important since the matrix adsorbs an adsorbate and serves as a dispersion medium, which forms a required salt particle size and a high salt surface area, affects the state and properties of the salt, and provides heat supply through the solid [52]. The salt interacts with water to increase the sorption capacity, which is greater than water sorption capacity of the porous matrix. Salt hydrates are also very attractive thermochemical storage materials, concerning high storage densities, relatively low cost, but suffer from very slow sorption kinetics, low thermal conductivity and severe materials degradation. A drawback of hygroscopic chlorides is the formation of solution during hydration and the corrosion of the reactor. Due to the deliquescence, swelling, and agglomeration of salt hydrates after some sorption/desorption cycles, the reaction is impossible to continue. An overview on the salt hydrates and composite sorbents as thermochemical storage materials can be found in reviews [1–2, 53–55]. The most studied salt hydrates in the composites in the last years have been LiCl, $CaCl_2$, $MgCl_2$, $MgSO_4$, $Al_2(SO_4)_3$, $LiNO_3$, and lately $SrBr_2$. Studies on water sorption properties of the composites revealed that the formation of salt hydrates in the dispersed state happened at a lower relative pressure of water vapor with respect to the bulk material. This makes a composite containing a dispersed salt a much more efficient sorbent than a common bulk salt. The main disadvantage of the composite sorbents can be the leakage of the hygroscopic salt from the pores, resulting in a degradation of the adsorption properties in some extreme cases. Several solutions [52] have been proposed to avoid these risks by a proper selection of salt content in the composite, binding the salt cations to the matrix surface and covering the external surface of the composite grains with a hydrophobic coating, which is only permeable for vapor but not for liquid.

Silica gels have been the most used porous matrices for the preparation of two-component sorbents in the past. Composites known as Selective Water Sorbents (SWS) containing mesoporous silica gel KSK or vermiculite (clay) and salts $CaCl_2$ and/or $LiNO_3$ were prepared more than 25 years ago [56]. Since 1996, different microporous and mesoporous materials have been used as matrices for TES water sorbents like binder-free CaX zeolite [57], mesoporous ordered silicates (MCM-41, SBA-15) [58], mesostructured iron silicate (FeKIL-2) [31], silica-alumina [59], silica gel [60], clays [61], MOFs [62–63]. Multi-Wall Carbon Nano-Tube (MWCNT) [64], activated alumina [65], and silicone foam; [66] were used for the preparation of the composite sorbents.

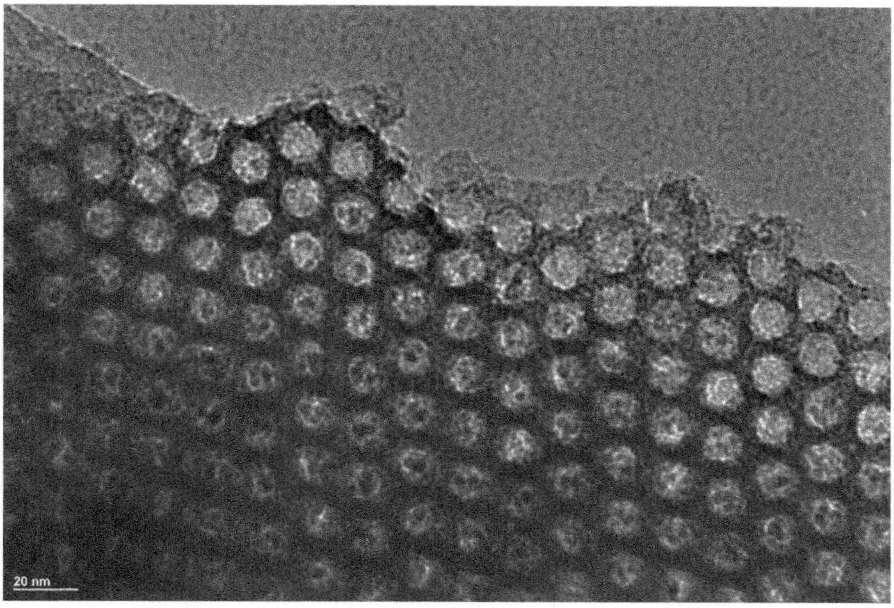

Figure 27.8 TEM photograph of hexagonal pore arrangement in SBA-15.

Mesoporous ordered silicates (MCM-41, SBA-15, PHTS) synthesized by long-chain surfactant molecules were used as matrices in several studies [58, 67–69]. These materials have different pore sizes (2–15 nm), large surface areas (> 500 m^2g^{-1}), excellent thermal stability and hexagonal pore arrangement. The hexagonal pore arrangement can be observed in Figure 27.8, which shows a TEM photograph of parallel cylindrical pores of uniform size in SBA-15 matrix. Recently, it has been found out that synthesis procedure and post-synthesis thermal treatment (calcination) have a great influence on water sorption behavior of SBA-15, which correlates mainly with the amount and type of silanol groups on the silica surface [70]. An increased uptake of water over a broad range of relative pressure can be obtained for SBA-15 material with high micropore volume and high surface polarity, which can be tailored by calcination, while lower temperature of aging during synthesis procedure was beneficial for higher water uptake at low relative pressure [69].

27.2.2.3.2 Sorption and Storage Properties

These mesoporous silicates (SBA-15, MCM-41, PHTS) exhibit water isotherms of Type V according to the IUPAC classification, characteristic of mesoporous materials having a hydrophobic surface, thus presenting small amounts of adsorbed water at lower relative pressures [6] The maximal water sorption capacity of mesoporous silicates is higher than of zeolites. The water sorption capacities of the composites with MCM-41, SBA-15, and PHTS [67, 69, 71] with $CaCl_2$ were considerably higher than for the composite obtained from mesoporous silica gel KSK. MCM-41 containing 7 wt% of $Al_2(SO_4)_3$ indicated higher storage capacity (170 Wh kg^{-1}) than the composite with SBA-15 matrix and the same amount of salt hydrate (93 Wh kg^{-1}, at 20 °C/30% RH) [58]. The composite of mesoporous alumina-silica (AS) and 24 wt% of calcium chloride possesses heat storage capacity of 205 Wh kg^{-1} (at 20 °C/28% RH) [59], while the composite of mesostructured iron silicate and 7 wt% of the same salt showed heat storage capacity of 156 Wh kg^{-1}. The working conditions were a maximum desorption temperature of 150 °C at 5.6 kPa condenser pressures and a very low minimum adsorption temperature of 25 °C at 1.2 kPa evaporator pressures. The latter matrix contains large mesopores from 4 to 29 nm, large pore volume, and high surface area together with additional adsorption sites due to the incorporated iron into silica. The repeated hydrothermal stability is another important value with regard to a cyclic adsorption process for heat storage applications. The composite $CaCl_2$-FeKIL-2 (7 wt% $CaCl_2$) showed good hydrothermal stability (3.6% loss of water uptake capacity) after 20 cycles of sorption/desorption between temperatures of 150 °C and 40 °C and at water vapor pressure of 5.6 kPa. After the cycling test no leakage of the salt hydrate from the composites was detected [31] [42]. The energy storage capacity of the confined $CaCl_2$ (43 wt%) in a silica gel equaled 300 Wh kg^{-1}, considering adsorption temperature of 30 °C, and desorption temperatures of 80 °C and adsorption pressure of 1.2 kPa [60], while higher heat storage capacity was achieved for LiCl/MWCNT under conditions (adsorption temperature at 35 °C/1.2 kPa and desorption temperature at 75 °C/4.2 kPa), which reached 470 W h kg^{-1} [64].

Another group of materials which are also used as hydrophilic matrices are clays (for example: attapulgite and vermiculite) which are natural silicate minerals with layered structures, having pore sizes from less than 2 nm and up to 50 nm and large pore volumes > 2.4 cm^3g^{-1}. The energy storage capacity of the composite, containing attapulgite and a salt mixture of 20 wt% $MgSO_4$ and 80 wt% $MgCl_2$, was determined to be 442 Wh kg^{-1} by calorimetry, (at 30 °C/85% RH) and desorption temperature of 130 °C [72]. Vermiculite and expanded vermiculite were used for the preparation of the composites with several salts, such as LiCl, $LiNO_3$, $MgSO_4$, $CaCl_2$, LiBr, $Ca(NO_3)_2$, K_2CO_3, and $SrBr_2$, which can achieve energy storage capacities from 100 to 640 Wh kg^{-1}, depending on different working conditions and the content and type of the confined salt [61, 73, 74–76].

Zeolites NaA [77], NaX [78], NaY, HY, mordenite [79], and CaX [57] have been used as matrices and combined with 15 wt% of salt hydrates (LiBr, $MgSO_4$, $MgCl_2$, and $CaCl_2$) with high theoretical storage densities [54]. More information about salt hydrates as thermochemical sorbents can be found in reviews [1–2, 55, 80]. Mainly, two preparation procedures were used: wet impregnation and incipient wetness impregnation. During these procedures non-controlled ion exchange of sodium cations occurs together with the formation of sodium chloride or sulfate or bromide [57]. This effect is usually ignored in the papers about two-component sorbents containing zeolites and hygroscopic salt hydrates. The heat storage density of the composite CaX with 15 wt% $CaCl_2$, which was determined on a lab-scale level (desorption at 110 °C and water vapor pressure of 0.02kPa; adsorption at 30 °C and water vapor pressure: 0.5–3.3 kPa) reached values up to 260 kWh m^{-3} [57]. On the demonstration level, a process design for a long-term thermochemical energy storage integrated into a solar thermal combi system was developed for the composite material of zeolite and salt at IGTE [81]. The regeneration of material was performed at a temperature of 130 °C and ambient humidity. Composite materials of salt impregnated zeolites exhibit very good reaction kinetic and specific thermal energy storage density, and mechanical and thermal stability (low thermal degradation). Also the measurement of thermochemical storage material based on impregnated natural zeolite clinoptilolite with LiCl/$MgSO_4$

and $MgCl_2/MgSO_4$ (with maximum desorption temperature of 110 °C, adsorption temperature: 25° C, RH of 95%) in a rotating heat drum showed good controllability of the process, sufficient high temperature lifts for room heating applications, maximum reactor temperatures in the range of 50 °C, up to 5 kWh of released heat for one batch containing around 70 kg [82].

Only a few MOF-based composites (MIL-100(Fe), MIL-101(Cr), UiO-66) with $CaCl_2$, LiCl, and SrBr2 have been reported so far. As host matrices, MOFs with large pore volumes, large specific surface areas, suitable pore diameters, and robust pore structure with good chemical and thermal stability were selected. The composite of MIL-101(Cr) with 62 wt% $CaCl_2$ salt loading achieved the energy storage capacity of 450 Wh kg^{-1} in the cycle conditions of 30 °C and 1.25 kPa for adsorption and 80 °C and 1.25 kPa for desorption [62]. On the orther hand, the same MOF containing 63 wt% $SrBr_2$ under the same conditons reached an energy storage capacity of 375 Wh kg^{-1} [63].

27.3 Future Developments

Resent developments on sorption materials for TES have been presented. A significant progress in sorption materials was achieved from zeolites containing binders to binder-free zeolites, from inorganic porous solids (aluminophosphates) to inorganic-organic frameworks (MOFs), and from two-component sorbents containing silica gels to composites with ordered mesoporous solids (like SBA-15) or disordered mesostructured iron silicate (FeKIL-2) as well as natural zeolites (clinoptilolite). These progresses were focused on the development of new materials in order to increase water sorption capacity and consequently energy storage density. To achieve an efficient and economically feasible TES system, further research on materials is needed by developing new sorbents with higher storage density, low charging temperature, good cyclic ability, and low price. New approaches and combinations of synthesis methods and post-synthesis modifications of solid sorbents offer an extensive range for tailoring specific properties, for example strong hydrophilic to medium hydrophilic properties of porous sorption materials or pore architecture by introducing larger pores into porous solids for preparing hierarchical adsorbents in order to enhance thermal energy storage and other heat transformation applications. New types of composites can be developed by using natural porous materials or hierarchical adsorbents or porous polymers as matrices with the appropriate content of the confined hygroscopic salt hydrates or porous adsorbents. The future of new sorption materials for TES applications is bright.

Acknowledgments

The author thanks ZAE Bayern (Federal German Ministry of Economic Affairs and Energy, project code 03ESP138A) and Slovenian Research Agency (research program P1-0021-0104 and project L1-7665). Dr. Matjaž Mazaj (National Institute of Chemistry, Slovenia) is acknowledged for providing TEM pictures and crystal structures of porous solids.

References

[1] Yu, N., Wang, R.Z., and Wang, L.W. (2013). Sorption thermal storage for solar energy. *Progress in Energy and Combustion Science* 39 (5): 489–514.

[2] Scapino, L., Zondag, H.A., Van Bael, J., et al. (2017). Sorption heat storage for long-term low-temperature applications: A review on the advancements at material and prototype scale. *Applied Energy* 190: 920–948.

[3] Tatsidjodoung, P., Le Pierrs, N., and Luo, L. (2013). A review of potential materials for thermal energy storage in building applications. *Renewable and Sustainable Energy Reviews* 18: 327–349.

[4] Hastürk, E., Ernst, S.J., and Janiak, C. (2019). Recent advances in adsorption heat transformation focusing on the development of adsorbent materials. *Current Opinion in Chemical Engineering* 24: 26–36.

[5] Wu, H., Salles, F., and Zajac, J. (2019). A critical review of solid materials for low-temperature thermochemical storage of solar energy based on solid-vapour adsorption in view of space heating uses. *Molecules* 24 (5): 945.

[6] Ng, E.P. and Mintova, S. (2008). Nanoporous materials with enhanced hydrophilicity and high water sorption capacity. *Microporous and Mesoporous Materials* 114 (1–3): 1–26.

[7] Database of zeolite structures. http://www.iza-structure.org/databases.

[8] Ristić, A., Fischer, F., Hauer, A., et al. (2018). Improved performance of binder-free zeolite Y for low-temperature sorption heat storage. *Journal of Materials Chemistry A* 6 (24): 11521–11530.

[9] Hunger, B., Klepel, O., Kirschhock, C., et al. (1999). Interaction of water with alkali-metal cation-exchanged X type zeolites: A temperature-programmed desorption (TPD) and X-ray diffraction study. *Langmuir* 15 (18): 5937–5941.

[10] Beta, I.A., Böhlig, H., and Hunger, B. (2004). Structure of adsorption complexes of water in zeolites of different types studied by infrared spectroscopy and inelastic neutron scattering. *Physical Chemistry Chemical Physics* 6 (8): 1975–1981.

[11] Yang, S. and Navrotsky, A. (2000). Energetics of formation and hydration of ion-exchanged zeolite Y. *Microporous and Mesoporous Materials* 37 (1-2): 175–186.

[12] Jänchen, J., Ackermann, D., Stach, H., et al. (2004). Studies of the water adsorption on zeolites and modified mesoporous materials for seasonal storage of solar heat. *Solar Energy* 76 (1-3): 339–344.

[13] Stach, H., Mugele, J., Jänchen, J., et al. (2005). Influence of cycle temperatures on the thermochemical heat storage densities in the systems water/microporous and water/mesoporous adsorbents, *Adsorption*, 11: 393-404.

[14] Valtchev, V., Majano, G., Mintova, S., et al. (2013). Tailored crystalline microporous materials by post-synthesis modification. *Chemical Society Reviews* 42 (1): 263–290.

[15] Lutz, W. (2014). Zeolite Y: Synthesis, modification, and properties – A case revisited. *Advances in Materials Science and Engineering* 2014: 20p.

[16] Herzog, T.H., Jänchen, J., Kontogeorgopoulos, E.M., et al. (2014). Steamed zeolites for heat pump applications and solar driven thermal adsorption storage. *Energy Procedia* 48: 380–383.

[17] Verboekend, D., Vilé, G., and Pérez-Ramírez, J. (2012). Hierarchical Y and USY zeolites designed by post-synthetic strategies. *Advanced Functional Materials* 22 (5): 916–928.

[18] Bonaccorsi, L., Bruzzaniti, P., Calabrese, L., et al. (2016). Organosilanes functionalization of alumino-silica zeolites for water adsorption applications. *Microporous and Mesoporous Materials* 234: 113–119.

[19] Schumann, K., Unger, B., Brandt, A., et al. (2012). Investigation on the pore structure of binderless zeolite 13X shapes. *Microporous and Mesoporous Materials* 154: 119–123.

[20] Jänchen, J., Schumann, K., Thrun, E., et al. (2012). Preparation, hydrothermal stability and thermal adsorption storage properties of binderless zeolite beads. *International Journal of Low-Carbon Technologies* 7 (4): 275–279.

[21] Mette, B., Kerskes, H., Drück, H., et al. (2014). Experimental and numerical investigations on the water vapor adsorption isotherms and kinetics of binderless zeolite 13X. *International Journal of Heat and Mass Transfer* 71: 555–561.

[22] Jänchen, J., Herzog, T.H., Gleichmann, K., et al. (2015). Performance of an open thermal adsorption storage system with Linde type A zeolites: Beads versus honeycombs. *Microporous and Mesoporous Materials* 207: 179–184.

[23] van Helden, W., Wagner, W., Schubert, V., et al. (2014). Solid sorption in the COMTES project: First experimental assessment. Gleisdorf Solar, Gleisdorf, Austria, 26–27 June 2014.

[24] Weber, R., Asenbeck, S., Kerskes, H., et al. (2016). SolSpaces – Testing and performance analysis of a segmented sorption store for solar thermal space heating. *Energy Procedia* 91: 250–258.

[25] Krönauer, A., Lävemann, E., Brückner, S., et al. (2015). Mobile sorption heat storage in industrial waste heat recovery. *Energy Procedia* 73: 272–280.

[26] Aiello, R., Nastro, A., and Colella, C. (1984). Solar energy storage through water adsorption-desorption cycles in zeolitic tuffs. *Thermochimica Acta* 79 (C): 271–278.

[27] Jänchen, J., Stach, H., and Hellwig, U. (2008). Water sorption in faujasite- and chabazite type zeolites of varying lattice composition for heat storage applications. *Studies in Surface Science and Catalysis* 174A: 599–602.

[28] Jänchen, J., Herzog, T.H., and Thrun, E. (2015). Natural zeolites in thermal adsorption storage and building materials for solar energy utilization in houses, SASEC2015. *Third Southern African Solar Energy Conference* (11–13 May 2015). South Africa: Kruger National Park.

[29] Dawoud, B. and Aristov, Y. (2003). Experimental study on the kinetics of water vapor sorption on selective water sorbents silica gel and alumina under typical operation conditions of sorption heat pumps. *International Journal of Heat and Mass Transfer* 46: 273–281.

[30] Henninger, S.K., Jeremias, F., Kummer, H., et al. (2012). Novel sorption materials for solar heating and cooling. *Energy Procedia* 30: 279–288.

[31] Ristić, A., Maučec, D., Henninger, S.K., et al. (2012). New two-component water sorbent $CaCl_2$-FeKIL2 for solar thermal energy storage. *Microporous and Mesoporous Materials* 164: 266–272.

[32] Aristov, Y.I. and Gordeeva, L.G. (2009). "Salt in a porous matrix" adsorbents: Design of the phase composition and sorption properties. *Kinetics and Catalysis* 50: 72–79.

[33] Jähnig, D., Hausner, R., Wagner, W., et al. (2006). Thermo-chemical storage for solar space heating in a single-family house. *Proceedings of ECOSTOCK conference, Richard Stockton College*, New Jersey.

[34] Flanigen, E.M., Broach, R.W., and Wilson, S.T. (2010). Introduction. In: *Zeolites in Industrial Separation and Catalysis,* (ed. Santi Kulprathipanja), pp. 1–26. Weinheim: WILEY-VCH Verlag GmbH & Co. KGaA.

[35] Jänchen, J. and Stach, H. (2014). Shaping adsorption properties of nano-porous molecular sieves for solar thermal energy storage and heat pump applications. *Solar Energy* 104: 16–18.

[36] Palomba, V., Vasta, S., and Freni, A. (2017). Experimental testing of AQSOA FAM Z02/water adsorption system for heat and cold storage. *Applied Thermal Engineering* 124: 967–974.

[37] Kakiuchi, H., Shimooka, S., Iwade, M., et al. (2005). Water vapor adsorbent FAM-Z02 and its applicability to adsorption heat pump. *Kagaku Kogaku Ronbunshu* 31 (4): 273–277.

[38] Henninger, S.K., Ernst, S.J., Gordeeva, L., et al. (2017). New materials for adsorption heat transformation and storage. *Renewable Energy* 110: 59–68.

[39] Brancato, V. and Frazzica, A. (2018). Characterisation and comparative analysis of zeotype water adsorbents for heat transformation applications. *Solar Energy Materials and Solar Cells* 180: 91–102.

[40] Frazzica, A. and Brancato, V. (2018). Verification of hydrothermal stability of adsorbent materials for thermal energy storage. *International Journal of Energy Research* 43 (12): 6161–6170.

[41] Henninger, S.K., Schmidt, F.P., and Henning, H.-M. (2010). Water adsorption characteristics of novel materials for heat transformation applications. *Applied Thermal Engineering* 30 (13): 1692–1702.

[42] Ristić, A., Logar, N.Z., Henninger, S.K., et al. (2012). The performance of small-pore microporous aluminophosphates in low-temperature solar energy storage: The structure-property relationship. *Advanced Functional Materials* 22 (9): 1952–1957.

[43] Krajnc, A., Varlec, J., Mazaj, M., et al. (2017). Superior performance of microporous aluminophosphate with LTA topology in solar-energy storage and heat reallocation. *Advanced Energy Materials* 7, 201601815-1-201601815-8.

[44] Varlec, J., Krajnc, A., Mazaj, M., et al. (2016). Dehydration of AlPO4-34 studied by variable-temperature NMR, XRD and first-principles calculations. *New Journal of Chemistry* 40 (5): 4178–4186.

[45] Henninger, S.K., Jeremias, F., Kummer, H., et al. (2012). MOFs for use in adsorption heat pump processes. *European Journal of Inorganic Chemistry* 16: 2625–2634.

[46] De Lange, M.F., Verouden, K.J.F.M., Vlugt, T.J.H., et al. (2015). Adsorption-driven heat pumps: The potential of metal-organic frameworks. *Chemical Reviews* 115 (22): 12205–12250.

[47] Burtch, N.C., Jasuja, H., and Walton, K.S. (2014). Water stability and adsorption in metal-organic frameworks. *Chemical Reviews* 114 (20): 10575–10612.

[48] Birsa Čelič, T., Mazaj, M., Guillou, N., et al. (2013). Study of hydrothermal stability and water sorption characteristics of 3-dimensional Zn-based trimesate. *The Journal of Physical Chemistry C* 117 (8): 14608–14617.

[49] Fröhlich, D., Henninger, S.K., and Janiak, C. (2014). Multicycle water vapor stability of microporous breathing MOF aluminium isophthalate CAU-10-H. *Dalton Transactions* 43 (41): 15300–15304.

[50] Jeremias, F., Lozan, V., Henninger, S.K., et al. (2013). Programming MOFs for water sorption: Amino-functionalized MIL-125 and UiO-66 for heat transformation and heat storage applications. *Dalton Transactions* 42 (45): 15967–15973.

[51] Aristov, Y.I. (2013). Challenging offers of material science for adsorption heat transformation: A review. *Applied Thermal Engineering* 50 (2): 1610–1618.

[52] Gordeeva, L.G. and Aristov, Y.I. (2012). Composites "salt inside porous matrix" for adsorption heat transformation: A current state-of-the-art and new trends. *International Journal of Low-Carbon Technologies* 7 (4): 288–302.

[53] Aristov, Y.I. (2014). Concept of adsorbent optimal for adsorptive cooling/heating. *Thermal Engineering* 72 (2): 166–175.

[54] N'Tsoukpoe, K.E., Schmidt, T., Rammelberg, H.U., et al. (2014). A systematic multi-step screening of numerous salt hydrates for low temperature thermochemical energy storage. *Applied Energy* 124: 1–16.

[55] Aydin, D., Casey, S.P., and Riffat, S. (2015). The latest advancements on thermochemical heat storage systems. *Renewable and Sustainable Energy Reviews* 41: 356–367.

[56] Aristov, Y.I., Tokarev, M.M., Cacciola, G., et al. (1996). Selective water sorbents for multiple applications, 1. $CaCl_2$ confined in mesopores of silica gel: Sorption properties. *Reaction Kinetics and Catalysis Letters* 59 (2): 325–333.

[57] Nonnen, T., Beckert, S., Gleichmann, K., et al. (2016). Testing of a thermochemical long-term heat storage system based on a zeolite-salt composite. *Chemical Engineering & Technology* 88 (3): 363–371.

[58] Jabbari-Hichri, A., Bennici, S., and Auroux, A. (2016). Effect of aluminum sulfate addition on the thermal storage performance of mesoporous SBA-15 and MCM-41 materials. *Solar Energy Materials and Solar Cells* 149: 232–241.

[59] Jabbari-Hichri, A., Bennici, S., and Auroux, A. (2015). Enhancing the heat storage density of silica–alumina by addition of hygroscopic salts ($CaCl_2$, $Ba(OH)_2$, and $LiNO_3$). *Solar Energy Materials and Solar Cells* 140: 351–360.

[60] Courbon, E., D'Ans, P., Permyakova, A., et al. (2017). A new composite sorbent based on SrBr2 and silica gel for solar energy storage application with high energy storage density and stability. *Applied Energy* 190: 1184–1194.

[61] Casey, S.P., Elvins, J., Riffat, S., et al. (2014). Salt impregnated desiccant matrices for "open" thermochemical energy storage – Selection, synthesis and characterisation of candidate materials. *Energy and Buildings* 84: 412–425.

[62] Permyakova, A., Wang, S.J., Courbon, E., et al. (2017). Design of salt-metal organic framework composites for seasonal heat storage applications. *Journal of Materials Chemistry A* 5: 12889–12898.

[63] D'Ans, P., Courbon, E., Permyakova, A., et al. (2019). A new strontium bromide MOF composite with improved performance for solar energy storage application. *Journal of Energy Storage* 25: 100881.

[64] Grekova, A., Gordeeva, L., and Aristov, Y. (2016). Composite sorbents "Li/Ca halogenides inside multi-wall carbon nano-tubes" for thermal energy storage. *Solar Energy Materials and Solar Cells* 155: 176–183.

[65] Zhang, Y.N., Wang, R.Z., and Li, T.X. (2018). Thermochemical characterizations of high-stable activated alumina/LiCl composites with multistage sorption process for thermal storage. *Energy* 156: 240–249.

[66] Calabrese, L., Brancato, V., Palomba, V., et al. (2019). Magnesium sulphate-silicone foam composites for thermochemical energy storage: Assessment of dehydration behaviour and mechanical stability. *Solar Energy Materials and Solar Cells* 200: 109992.

[67] Ponomarenko, I.V., Glaznev, I.S., Gubar, A.V., et al. (2010). Synthesis and water sorption properties of a new composite "$CaCl_2$ confined into SBA-15 pores." *Microporous and Mesoporous Materials* 129 (1-2): 243–250.

[68] Jänchen, J. and Stach, H. (2012). Adsorption properties of porous materials for solar thermal energy storage and heat pump applications. *Energy Procedia*: 30: 289–293.

[69] Ristić, A. and Logar, N.Z. (2019). New composite water sorbents $CaCl_2$–PHTS for low-temperature sorption heat storage: Determination of structural properties. *Nanomaterials* 9 (1): 1–16.

[70] Maaz, S., Rose, M., and Palkovits, R. (2016). Systematic investigation of the pore structure and surface properties of SBA-15 by water vapor physisorption. *Microporous and Mesoporous Materials* 220: 183–187.

[71] Ristić, A. and Henninger, S.K. (2014). Sorption composite materials for solar thermal energy storage. *Energy Procedia* 48: 977–981.

[72] Posern, K. and Kaps, C. (2010). Calorimetric studies of thermochemical heat storage materials based on mixtures of $MgSO_4$ and $MgCl_2$. *Thermochimica Acta* 502 (1–2): 73–76.

[73] Grekova, A.D., Gordeeva, L.G., and Aristov, Y.I. (2017). Composite "LiCl/vermiculite" as advanced water sorbent for thermal energy storage. *Applied Thermal Engineering* 124: 1401–1408.

[74] Sutton, R.J., Jewell, E., Elvins, J., et al. (2018). Characterising the discharge cycle of $CaCl_2$ and $LiNO_3$ hydrated salts within a vermiculite composite scaffold for thermochemical storage. *Energy and Buildings* 162: 109–120.

[75] Brancato, V., Gordeeva, L.G., Sapienza, A., et al. (2019). Experimental characterization of the LiCl/vermiculite composite for sorption heat storage applications. *International Journal of Refrigeration* 105: 92–100.

[76] Shkatulov, A.I., Houben, J., Fischer, H., et al. (2020). Stabilization of K_2CO_3 in vermiculite for thermochemical energy storage. *Renewable Energy* 150: 990–1000.

[77] Mette, B., Kerskes, H., Drück, H., et al. (2013). New highly efficient regeneration process for thermochemical energy storage. *Applied Energy* 109: 352–359.

[78] Hongois, S., Kuznik, F., Stevens, P., et al. (2011). Development and characterisation of a new $MgSO_4$-zeolite composite for long-term thermal energy storage. *Solar Energy Materials and Solar Cells* 95 (7): 1831–1837.

[79] Whiting, G.T., Grondin, D., Stosic, D., et al. (2014). Zeolite-$MgCl_2$ composites as potential long-term heat storage materials: Influence of zeolite properties on heats of water sorption. *Solar Energy Materials and Solar Cells* 128: 289–295.

[80] N'Tsoukpoe, K.E., Rammelberg, H.U., Lele, A.F., et al. (2015). A review on the use of calcium chloride in applied thermal engineering. *Applied Thermal Engineering* 75: 513–531.

[81] Mette, B., Kerskes, H., and Drück, H. (2012). Concepts of long-term thermochemical energy storage for solar thermal applications – Selected examples. *Energy Procedia* 30: 321–330.

[82] Zettl, B., Englmair, G., and Somitsch, W. (2015). An open sorption heat storage concept and materials for building heat supply. *Energy Procedia* 73: 297–304.

28

Vacuum Super Insulated Thermal Storage Systems for Buildings and Industrial Applications

Thomas Beikircher and Matthias Rottmann

Bavarian Center for Applied Energy Research (ZAE Bayern), Divison Energy Storage, Walther-Meißner-Str. 6, 85748 Garching, Germany

28.1 Introduction

28.1.1 Thermal Storage Efficiency, Thermal Losses, and Temperature Decay

Efficient thermal energy storage becomes important with continuing substitution of fossil fuels by regenerative energies, which are commonly not continuously available. Thermal storage compensates the temporal difference between heat demand and supply. Furthermore, there is a large CO_2 reduction potential in the industry, if today's waste heat, often at medium and high temperature levels, is stored with minimum losses for later (re-)use. The efficiency of a thermal storage is defined as the ratio between discharged heat E_{out} and charged heat E_{in}, where the difference between both is the thermal loss power through the storage surface integrated over the storage time. Additionally, thermal losses at the heat exchanger during charging or discharging can occur.

The storage loss power $\dot{Q}_{loss}$(W) can be calculated if thermal loss coefficient U (W m^{-2} K^{-1}), surface area A (m^2) of the storage, and difference between the spatial mean storage temperature T (K) and the ambient temperature T_a (K) are known. In a one-dimensional approximation, it is assumed that the heat loss is strictly vertical to the storage surface, which is usually sufficiently accurate. When pipe losses are also neglected, U is given by:

$$U = \lambda / d \tag{28.1}$$

where d (m) is the thickness of the storage insulation, and λ (W m^{-1} K^{-1}) is its thermal conductivity. For $\dot{Q}_{loss}$:

$$\dot{Q}_{loss} = \frac{\lambda}{d} A(T - T_a) \tag{28.2}$$

While being exact only for planar geometries, this equation is also a good approximation for cylindrical and spherical storages, if the inner storage diameter is significantly larger than insulation thickness d.

Advances in Energy Storage: Latest Developments from R&D to the Market, First Edition.
Edited by Andreas Hauer.
© 2022 John Wiley & Sons, Ltd. Published 2022 by John Wiley & Sons, Ltd.

During the storage process, T decreases because the stored thermal energy E (J) is reduced (minus sign) by $\dot{Q}_{loss}$:

$$-\frac{dE}{dt} = -\rho V c_p \frac{dT}{dt} = \dot{Q}_{loss} = \frac{\lambda}{d} A\,(T - T_a) \tag{28.3}$$

Solving for T yields an exponential temperature decay over the storage time t, starting from the initial spatial mean storage temperature T_0 and depending on storage volume V, heat capacity of the storage medium c_p, as well as on λ, A, d and T_a:

$$T(t) = (T_0 - T_a)\cdot \exp\left(-\frac{\lambda A}{d\rho V c_p} t\right) + T_a = (T_0 - T_a)\cdot \exp\left(-\frac{\ln 2}{t_{1/2}} t\right) + T_a \tag{28.4}$$

The half-life period $t_{1/2}$ of the storage temperature, i.e. the time which passes until the original temperature difference between storage and ambience has been halved, is directly proportional to d/λ. As a consequence of Equation (28.4), the most important parameter for efficient heat storage – besides a small surface-to-volume ratio A/V – is the storage insulation, where a high ratio between thickness and thermal conductivity is favorable.

28.1.2 Conventional Storage Insulation

Domestic thermal storages for room heating and hot water supply up to 100°C have typical volumes between 0.2 and 1 m^3, and are usually insulated with 10 cm mineral wool, EPP, EPS, Melamine, or PUR. The thermal conductivity of these materials is between 0.03 and 0.05 W m^{-1} K^{-1}. Medium-sized (up to 1000 m^3) district heat storages are often insulated with mineral wool. Large storages (10 000 m^3 and more), which are used, e.g. in solar district heating [1], are insulated with typically 1 m of granular bulk materials like expanded perlite, expanded clay, glass granules, or foam glass gravel [2–3], where thermal conductivities range between 0.05 and 0.1 W m^{-1} K^{-1}. Since granular materials are self-adapting to arbitrary geometries, they have advantages when insulating big and individually shaped large storages. Moreover, these materials are relatively resistant against humidity, and thus more appropriate than mineral wool in the prevalent case of a buried storage.

All storage insulation materials (fibres, foams, or hollow grains in the range of mm down to μm) have a filigree structure, resulting in low solid body conduction, and show small voids filled with air. Therefore, convection is suppressed, because the buoyancy forces do not exceed friction caused by adhesion of molecules at the large inner surface. Radiation is at least partly reduced by extinction (i.e. absorption and scattering of thermal radiation) within the insulation, but gas heat conduction is fully present, resulting in typical thermal conductivities between 0.03 and 0.05 W m^{-1} K^{-1} at ambient conditions. The replacement of air during the foaming process of, e.g. PUR with heavier inert gases like pentane reduces the gas heat conduction from 0.026 W m^{-1} K^{-1} (for air at 25°C) down to values of 0.015 W m^{-1} K^{-1}, which yields total thermal conductivities below 0.03 W m^{-1} K^{-1}. Due to thermally driven diffusion of air back into the pores, this advantage vanishes quickly after exposing the insulation to higher storage temperatures. Also improper installation, humidity [3], and shrinking effects caused by aging often deteriorate conventional insulations after a short time.

The currently best non-evacuated insulation material is silica aerogel with a thermal conductivity of 0.013 W m^{-1} K^{-1} at 25°C [4]. Due to the aerogel's nanopores, which are in the same dimension as the mean free path of air at ambient conditions (60 nm), air heat conduction is reduced by a significant fraction. In addition, radiative transfer is reduced effectively by opacifying additives [5]. Opacified silica aerogel has excellent insulation properties, but is currently too expensive (> 2500 € m^{-3}) for technical application in heat storages. A similar material is fumed silica, which is also nanoporous, but not as expensive (approx. 1000 € m^{-3}) as aerogel.

28.1.3 Development of Vacuum Super Insulated Storages

The extremely low thermal conductivities of evacuated expanded perlite at −200°C (down to 0.001 W m^{-1} K^{-1}) have been successfully used to insulate liquefied gas storages since the 1950s. These 20-fold better insulation properties in comparison to conventional insulation materials have been the motivation for the development of perlite-based medium and high temperature heat storages with highest efficiency at ZAE Bayern, starting in 2009: First, vacuum super insulation (VSI) with low-density ($\approx$ 50 kg m^{-3}) expanded perlite powder, as used in cryogenic applications, has been applied for thermal insulation of hot water storages. Expanded perlite is a highly porous powder of natural origin, which is non-flammable and non-toxic. A main advantage compared to microporous materials (aerogels, fumed silica) is the low price (ca. 50–250 € m^{-3} depending on density between around 50 and 200 kg/m^3), which allows the fabrication of highly efficient, yet economical storages. Meanwhile, these storages are commercially available, and are also sold for industrial applications up to 160°C. At these medium temperatures, thermal conductivities between 0.007 and 0.01 W m^{-1} K^{-1} are commercially realized. Combined with an insulation thickness of 20 cm, the half-life period $t_{1/2}$ (compare Equation (28.4)) of a VSI storage is thus 3- to 10-times longer than for conventionally insulated storages (see also Figure 28.4). For hot water storages within solar thermal systems, this enables seasonal storage of the solar heated water.

Due to the great success at medium temperatures up to 160°C, VSI technology with expanded perlite has now been considered also for high-temperature applications up to 800°C on the hot side. Using high density expanded perlite ($\approx$ 180 kg m^{-3} at about 250 €/m^3), and by adding SiC as opacifier to increase absorption and scattering of infrared radiation, ultra-low thermal conductivities below 0.02 W m^{-1} K^{-1} have been recently realized at ZAE Bayern [6–7]. An alternative VSI powder especially for high-temperature applications is fumed silica, which is already used in vacuum insulation panels (VIPs) at temperatures between −70 and 100°C.

This chapter will present the physical fundamentals of VSI and its technical application in heat storages up to 160°C, which today is commercially realized using low-density expanded perlite. In addition, the potential of high-temperature VSI storages up to 800°C on the basis of high-density expanded perlite is discussed, which requires both different heat storage media and adjustments of the insulation material, mainly by adding opacifiers. Furthermore, promising applications apart from thermal storages are presented, some of which may also require the usage of alternative VSI powders, e.g. fumed silica.

28.2 VSI with Expanded Perlite for Highly Efficient and Economical Thermal Storages

28.2.1 Fundamentals of VSI Storages

The best thermal insulation with thermal conductivities below 0.01 W m^{-1} K^{-1}, and even as low as 0.003 W m^{-1} K^{-1} at ambient conditions, can be realized if vacuum is applied to an insulation material in order to completely suppress air heat conduction [5]. This technique is generally referred to as vacuum super insulation (VSI). In contrast to vacuum insulation (VI), which is known from Thermos® flasks or Dewar vessels, vacuum super insulation also reduces radiative transfer between the hot and cold surfaces, inversely proportional to their distance, by filling the interspace with structured material. Hence, above a certain insulation thickness, VSI is always superior to VI, where heat flow is independent of insulation thickness. On the other hand, VI is the best insulation method for small separations between low emitting metallic surfaces. There are two different variants how VSI can be realized technically: In a multilayer insulation (MLI), parallel metallic foils with low thermal emissivity ($\varepsilon < 0.05$) are used as reflective radiation shields. Evacuated (high vacuum, pressure below 0.001 mbar) spacers with low solid thermal conductivity keep the single foils at sufficient distance from each other to avoid thermal bridges by direct metal contact. For a more detailed overview of MLI, see e.g. [8]. In the second, more economical VSI variant, evacuated powders are used, and the reduction of radiative transfer happens by absorption and scattering at the single powder particles. Infrared opacifiers, which are more effective than the base VSI powders with respect to absorption and scattering – at least in certain wavelength intervals, where extinction of base powder is low – may be added especially at high temperatures. For a complete suppression of gas conduction within the evacuated powder insulation, fine vacuum (0.01–1 mbar) or even rough vacuum (1–10 mbar) is sufficient, which is another advantage compared to MLI. The required vacuum pressure depends on the (effective) pore diameter of the powder, since the confinement of air in the pores limits the mean free path of the air molecules, which is proportional to gas heat conduction. Solid conduction within the powder is greatly reduced by the small contact areas between the single grains, which represent large thermal resistances. In principle, every conventional insulation material with an open-pore structure can be evacuated to obtain a vacuum super insulation (VSI). Here, the material either has to mechanically withstand the atmospheric pressure of 100 kN m^{-2} by itself, or it must be enveloped by a load-bearing structure.

To realize a powder VSI for thermal storages, two load-bearing cylindrical or spherical steel tanks are placed inside each other and connected with special fixings, which can be suspensions or supports with a thermal resistance as high as possible to avoid additional heat losses via thermal bridges [8]. Subsequently, the annulus between both containers is filled with VSI powder material and evacuated. This principle, as used in cryogenic tanks for the storage of ultra-cold liquid gases, is shown in Figure 28.1. For maximum cost effectiveness, and for store temperatures below 200°C, low-density expanded perlite is the ideal VSI powder due to its low price of 50 to 100 € m^{-3}.

Perlite is a natural volcanic glass (obsidian) and consists mainly of SiO_2 (70%) and Al_2O_3 (15%). To produce an efficient insulation material, it is technically treated in an expansion process at temperatures between 800 and 1000°C, where it expands to more

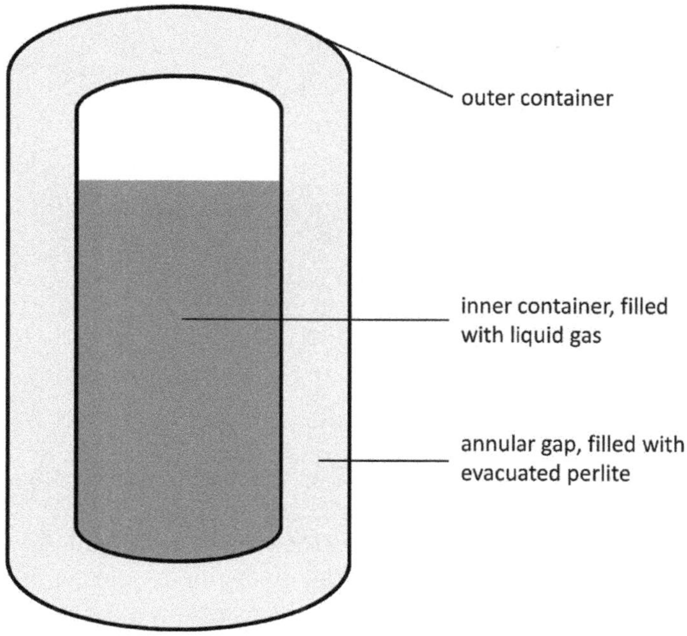

Figure 28.1 Sketch of a typical cryogenic storage for liquid gases with expanded perlite as insulating powder.

than 20 times its original volume. Due to the high temperature, the raw material softens and chemically bound water evaporates, which results in the formation of internal pores. The expanded low-density perlite "Technoperl C 1.5" of the Austrian manufacturer Stauss Perlite, which is used in cryogenic applications, and which has also been applied for long-term hot water storages (see later), has a density of 55 kg m^{-3} ex factory, a porosity of 0.95–0.98, and a maximum water content of 2% (mass fraction). After filling the storage annulus, the material may compact itself to higher densities up to 90–95 kg m^{-3}. The structure of perlite can be seen in Figure 28.2.

28.2.2 VSI Heat Storages for Temperatures up to 160°C

28.2.2.1 Market Review and State of the Art

Using evacuated, expanded low-density perlite (e.g. Technoperl C 1,5), ZAE Bayern has developed a high-performance and economical heat storage for temperatures up to 160°C in cooperation with the German manufacturer Hummelsberger GmbH in a publicly granted research project (Federal Ministry for the Environment, Nature Conservation and Nuclear Safety, grant number 0325964A), [9–10].

Meanwhile, the industrial partner of the completed research project, Bavarian manufacturer Hummelsberger GmbH in Mühldorf, and recently also the Bavarian producer Sirch GmbH in Kaufbeuren, sell commercial VSI tanks based on the project results. For hot water storage at temperatures up to 110°C (pressurized water at 3 bar), Hummelsberger is now producing containers with volumes between 2 and 40 m^3. Originally, the heat storages had been designed for applications within domestic solar

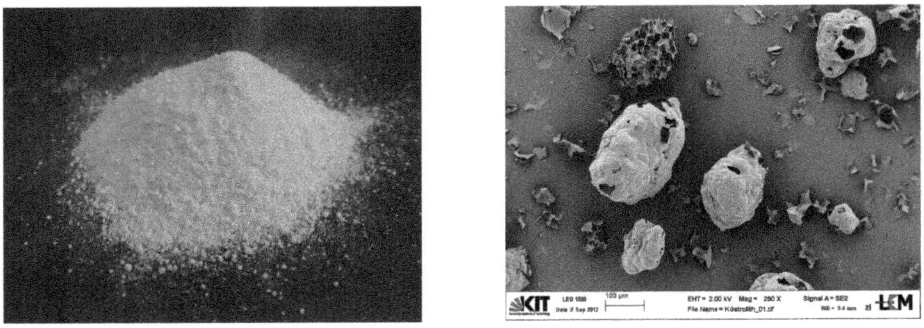

Figure 28.2 (left) Expanded perlite "Technoperl C 1.5" (Stauss Perlite); (right) SEM picture of single grains of expanded perlite (picture by KIT, Karlsruhe).

thermal systems. Therefore, the Hummelsberger VSI storage includes a patented stratification unit (see Figure 28.5) to optimally charge the storage with water at varying temperatures coming from the solar thermal system. The stratification device prevents mixing of water layers to reach high storage efficiencies and is self-regulating, i.e. driven only by density differences of water layers at different temperatures. The actual functionality of the stratification unit has been experimentally attested by ZAE Bayern. There are also some other storage tanks on the market that claim to reach effective stratification through their devices. Tests at ZAE Bayern revealed that this is rarely the case. For industrial applications, e.g. the storage of waste heat, Hummelsberger GmbH also sells VSI storages for temperatures up to 160°C.

The heat transport in the evacuated (0.01 mbar) perlite insulation consists mainly of two transport mechanisms, solid heat conduction and thermal radiation, leading to a total thermal conductivity of around 0.008 W m^{-1} K^{-1} ($T = 110°C$, $\rho = 90$ kg m^{-3}). Natural air convection is completely and gas heat conduction nearly fully suppressed (< 0.001 W m^{-1} K^{-1} at $T < 110°C$) due to the small pores and the fine vacuum applied. Solid heat conduction is small (about 0.005 W m^{-1} K^{-1} up to 110°C) due to the high contact resistances between the filigree perlite grains, see Figure 28.2 (right). The radiation thermal conductivity typically amounts to 0.003 W m^{-1} K^{-1} or less below 110°C storage temperature. These values have been measured at ZAE Bayern in the laboratory. If the vacuum is reduced or deteriorates with time, gas heat conduction increases, and total thermal conductivity as a function of vacuum pressure can be read from Figure 28.3.

For the application in hot water storages up to 160°C, low-density expanded perlite has the advantage that it is relatively cheap (50–100 € m^{-3}), which allows the economical realization of super insulated storages. In comparison to other insulation materials, it is also beneficial that expanded perlite is non-toxic and non-flammable. Furthermore, a long-time deterioration of the insulating properties due to moisture intake is impossible because of the vacuum-tight steel envelope.

The low thermal conductivity of the evacuated perlite in the storage envelope leads to a very low heat loss of a VSI storage in comparison to a (otherwise equally designed) conventionally insulated tank, see Figure 28.4. Here, the exponential temperature decay according to Equation (28.4) is shown for a storage volume of 16.5 m^3, a strictly

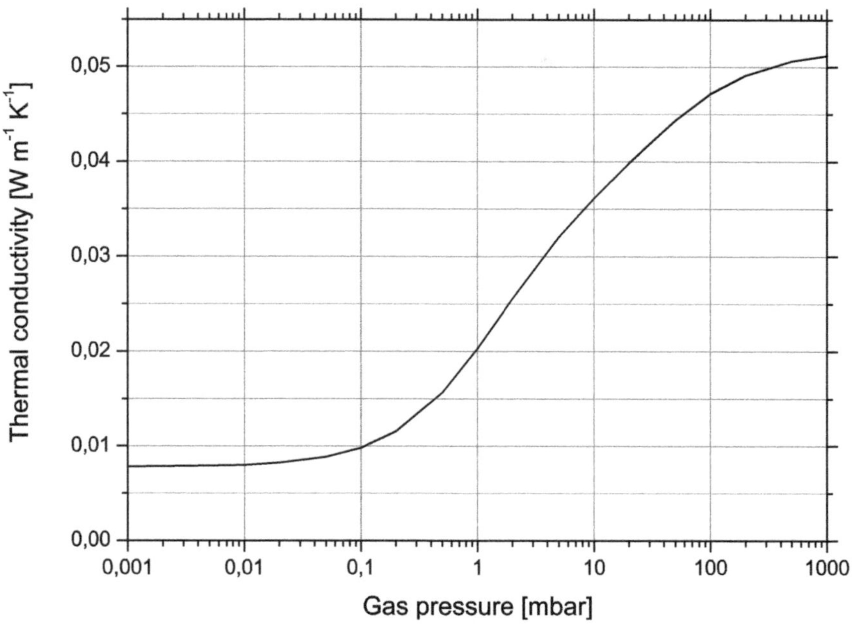

Figure 28.3 Experimentally determined thermal conductivity of expanded perlite "Technoperl C 1.5" (Stauss Perlite), varied with gas pressure at 110°C storage temperature.

cylindrical geometry and an insulation thickness of 20 cm. A similar real prototype VSI hot water storage that is shown in Figure 28.5 experimentally corroborated the predictions from Figure 28.4. For the calculations, an ambient temperature of 15°C, a radius of 1 m (1.2 m) and a height of 5.25 m (5.45 m) for the inner (outer) tank is assumed, leading to an effective mean radius and height of an imaginary insulation cylinder of 1.1 m, and 5.35 m respectively. As Figure 28.4 shows, the half-life of the storage temperature, i.e. the time when the difference between storage temperature and ambient temperature has reached half of its initial the value, is 10 months (305 days) for the VSI storage tank, and only three months (98 days) for a conventionally (PUR) insulated tank in new condition and with optimal insulation ($\lambda = 0.025$ W m^{-1} K^{-1}). It can also be seen, that the VSI tank is a real seasonal storage: Even after half a year, hot water at 115°C produced in summer (e.g. by vacuum tube solar collectors) can be discharged at a high temperature level of about 81°C in winter, where it is directly usable for both room heating and service water (e.g. for showering) without conventional auxiliary heaters. In comparison, the temperature level in the conventional storage has degraded to a value of 43°C after half a year, and the heat from summer has been nearly completely lost. In addition, the thermal losses of the conventional storage will heat up the building in an uncontrolled way, especially during summer and autumn.

28.2.2.2 Application Within Solar Thermal Systems

The first VSI storage prototype developed in 2010 was originally designed unpressurized for a solar-assisted thermal application in single- and multi-family houses up to 100°C, see Figure 28.5.

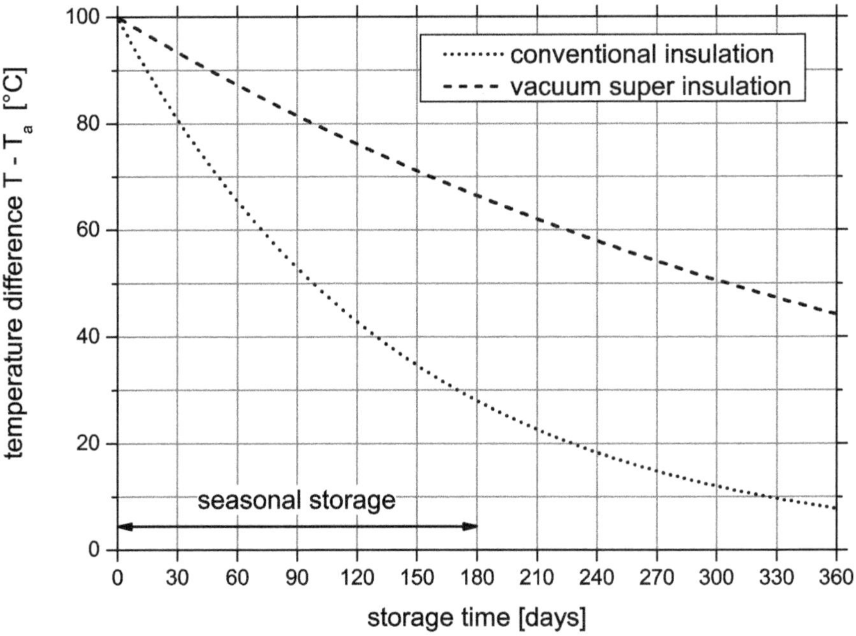

Figure 28.4 Temperature decrease in two identically constructed storage tanks with 16.5 m^3 volume and 200 mm insulation thickness. The VSI storage (measured, dashed line) is compared to the best-case scenario of a conventional tank insulated with PUR foam (calculated, dotted line). It takes 10 months to reach half of the initial temperature difference for the VSI tank, but only three months for the PUR insulated tank.

Prices for such VSI storages are higher than for tanks with standard insulation and depend on size and insulation quality. For small volumes (< 5 m^3), the VSI variant is two to three times more expensive, depending on the applied vacuum and the insulation thickness. However, for large sizes (> 20 m^3), VSI technology is only 1.5 times more expensive and shows significantly lower heat losses, three to five times lower than in conventional storages.

A simulation of the solar fraction for a solar thermal system has been made for various house standards, solar collector areas, and different storage tank sizes and storage insulations [11]. Hereby, the solar fraction can be compared for conventional and VSI storage. The result for a low-energy house with an energy demand of 50 kWh m^{-2} a^{-1} with four inhabitants can be seen in Figure 28.6. Although an increase of the solar fraction f_{sav} is achieved by the VSI tank (see Figure 28.6), especially for f_{sav} values close to 1 (solar houses), the thermal benefit is not the main reason for the use of a VSI tank in a domestic solar thermal system where only temperatures below 100°C occur. More important are its long-term stability against humidity and corrosion, and the possibility to place the VSI tank directly in the building without heating up the rooms in summer. Alternatively, since VSI containers are weatherproof due to their steel tank-in-tank design, they can be placed outside or even buried underground to increase living space.

To use the VSI storage technology also in large storage tanks (e.g. for solar district heating) with sizes larger than 500 m^3 [1], a new construction concept needs to be realized, since the tank-in-tank design is economically not feasible anymore.

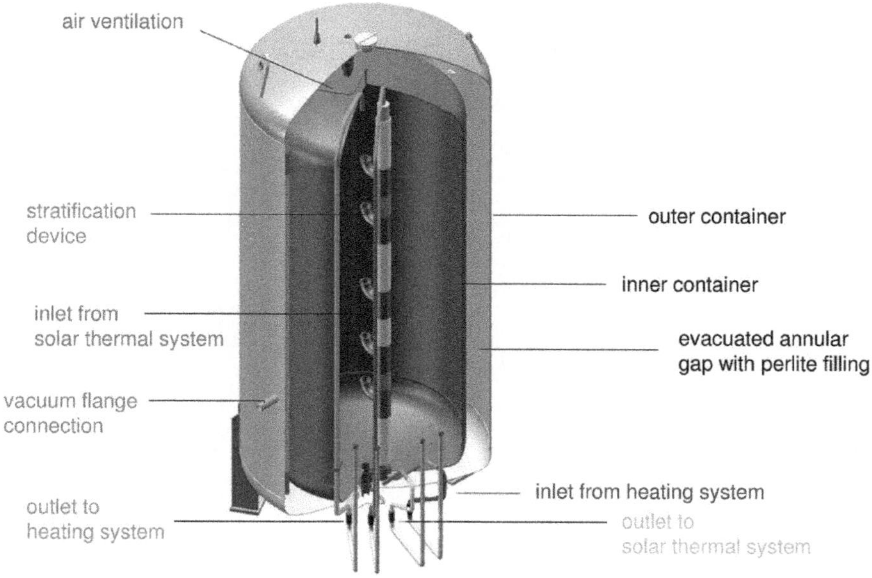

Figure 28.5 Setup of the VSI storage tank for solar thermal applications with central stratification unit.

Prefabricated VSI panels combined with concrete plates that form the tank are under investigation. For lower temperatures, vacuum insulation panels (VIPs) with a load-bearing core of fumed silica have been successfully developed at the ZAE Bayern [9].

28.2.2.3 Physical Description of Heat Transport in Expanded Perlite

The contributions of the single heat transport mechanisms to the total thermal conductivity of evacuated expanded perlite at temperatures up to 110°C have already been briefly discussed. The largest contribution to the effective thermal conductivity $\lambda_{eff} = 0.008$ W m^{-1} K^{-1} is caused by solid conduction λ_s which amounts to 0.005 W m^{-1} K^{-1}. Since gas conduction is suppressed in vacuum, the remaining 0.003 W m^{-1} K^{-1} are due to radiation. With higher storage temperatures, the radiative thermal conductivity increases significantly, since it is proportional to the third power of the mean insulation temperature (see Equation (28.6)). Consequently, at higher temperatures special attention has to be paid in order to reduce radiative transport, e.g. by higher density ρ or higher extinction e^*. In contrast, the other heat transport mechanisms show only small, i.e. at the most linear, changes with temperature, and hence contribute only little to the absolute value of λ_{eff} at higher temperatures.

Altogether, the effective thermal conductivity can be split into four addends, which are associated with the different heat transport mechanisms: radiative heat transport (λ_r) solid conduction (λ_s), gas conduction (λ_g) and coupling effects (λ_c) [19]:

$$\lambda_{eff} = \lambda_r + \lambda_s + \lambda_g + \lambda_c \tag{28.5}$$

Here, the coupling term describes all interference effects between the single heat transport mechanisms, which are individually discussed in the following.

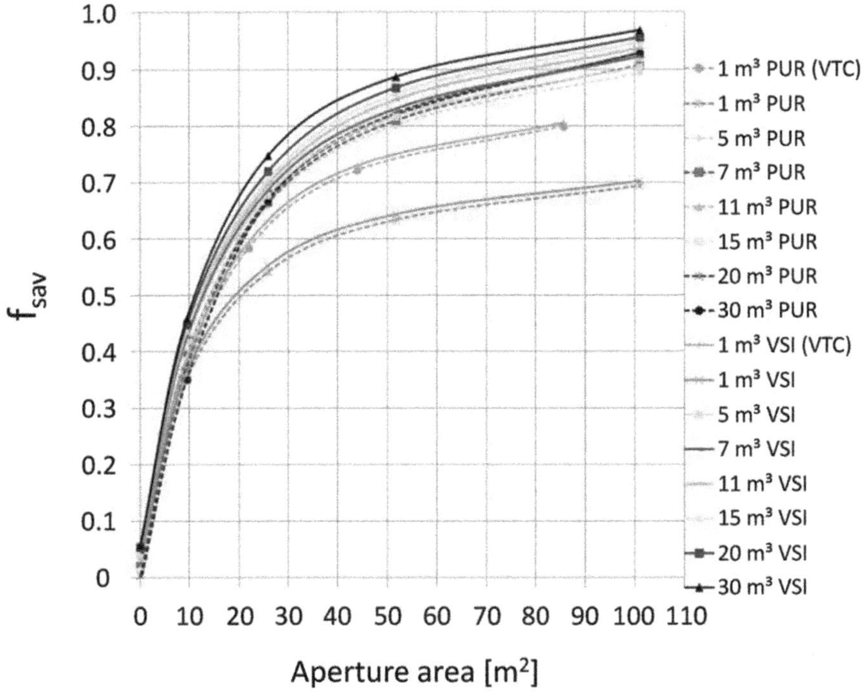

Figure 28.6 Solar fraction f_{sav}, i.e. the amount of energy provided by the solar thermal system divided by the total energy demand, for a single-family low-energy house (150 m^2 living area, four inhabitants, 50 kWh m^{-2} a^{-1} heating demand, domestic hot water temperature 60°C) as function of the applied solar collector aperture area in m^2. VSI and conventionally (PUR) insulated storage tanks of different sizes are compared. All simulations use flat plate collectors, apart from the ones marked with VTC (vacuum tube collector).

Radiative thermal conductivity: The heat transport by radiation within evacuated perlite powder – as in general for optically thick insulations – can be treated by a diffusion model, which leads to a radiative thermal conductivity λ_r [5]:

$$\lambda_r = \frac{16\, n^2 \sigma T_r^3}{3\, \rho e^*(T_r)} \tag{28.6}$$

where σ is the Stefan-Boltzmann constant, n is the refractive index (which is close to 1 for highly porous materials), and ρ is the powder density (bulk density). It can be seen that λ_r grows with T_r^3, where T_r is a mean radiation temperature, calculated from the storage temperature T and the ambient temperature T_a, which are the temperatures at the hot and cold side of the perlite insulation:

$$T_r = \sqrt[3]{\frac{1}{4}(T^2 + T_a^2)(T + T_a)} \tag{28.7}$$

Roughly speaking, T_r is a kind of mean insulation temperature, which can also be approximated by the arithmetic mean for low temperature differences $T - T_a$.

The quantity e^* in Equation (28.6) is the mass-specific extinction coefficient, which is a measure for the reduction of radiative transport caused by absorption and scattering within the powder insulation, normalized to the powder density. In general, absorption and scattering are wavelength-dependent, and therefore, the spectral mass-specific extinction coefficient e^*_λ must be experimentally determined, e.g. from Fourier transform infrared (FTIR) spectroscopy. e^*_λ describes radiation extinction as a function of wavelength, and is shown in Figure 28.7 (left) for two different expanded perlites. Additionally, dimensionless albedo a is plotted, which is defined as the share of scattering in total extinction and characterizes the type of extinction (between a limit of pure scattering if $a = 1$ and a limit of pure absorption if $a = 0$). To obtain total mass-specific extinction coefficient e*, which has to be inserted into Equation (28.6), a weighted mean value (Rosseland mean) has to be calculated by integrating over the whole wavelength spectrum [12]. Here, the contributions of e^*_λ at the different wavelengths are weighted according to their fraction of intensity in the Planck spectrum. Since the spectral distribution of the Planck spectrum varies with temperature, also e* becomes temperature-dependent. In Figure 28.7 (right), the Rosseland mean extinction coefficient e* has been calculated from the measured spectral values for two different expanded perlites, and is shown as a function of mean radiation temperature between 20°C and 800°C. Typical values are between 35 and 55 $m^2\ kg^{-1}$. Higher values for e* (typically > 100 $m^2\ kg^{-1}$) can be achieved by adding absorbing or scattering opacifiers like silicon carbide SiC, iron oxide Fe_3O_4 or carbon soot [13].

The diffusion approximation for radiative transport as applied here is valid, if the insulating material is optically thick, which means that the geometric thickness of the insulation material is at least 15-times larger than the mean free path of infrared photons [14]. With a mean specific extinction coefficient e* around 0.04 $m^2\ g^{-1}$ (0.045 $m^2\ g^{-1}$), see Figure 28.7, and bulk densities ρ between 55 $kg\ m^{-3}$ and 90 $kg\ m^{-3}$ (180 $kg\ m^{-3}$), the total extinction coefficient $E = e^*\ \rho$ [15] lies between 2200 m^{-1} and 3600 m^{-1} (8100 m^{-1}) for low-density (high-density) expanded perlite, respectively. Since E is the inverse of the photon mean free path l_f [15], and thus l_f ranges from 0.1 to 0.5 mm, an insulation thickness of 1 cm is already sufficient to fulfill the requirement of optical thickness for all expanded perlites.

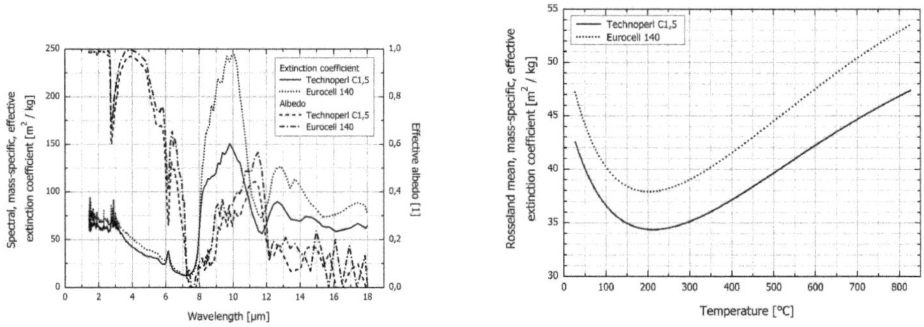

Figure 28.7 (left) Experimental spectral mass-specific extinction coefficient e*λ and albedo of low density perlite "Technoperl C 1.5" and high density perlite "Eurocell 140" (both by Stauss Perlite) as determined by FTIR spectroscopy at the ZAE Bayern and plotted against wavelength (1.4–1.8 µm). (right) Mass-specific total extinction coefficient e* of "Technoperl C 1.5" and "Eurocell 140" versus radiation temperature T_r (20–800°C).

Heat transport by gas conduction: The thermal conductivity λ_g of the gas inside the pores as a function of pressure and temperature as shown in Figure 28.3 can be described by the relation:

$$\lambda_g = \frac{\lambda_0(T_m)}{1+\frac{p_{1/2}}{p}} \tag{28.8}$$

where $\lambda_0(T)$ denotes the thermal conductivity of air at normal pressure, and T_m is the mean insulation temperature. For perlite C1.5 of Stauss perlite, the experimentally determined, characteristic half-value pressure $p_{1/2}$ with air as pore gas is 1.38 mbar, while p is the actual vacuum pressure in the insulation [9]. By evacuation below 0.1 mbar, gas heat conduction becomes negligible in practice.

Heat transport by solid body conduction: From experiments at ZAE Bayern, a linear correlation between λ_s and bulk density ρ (in kg m^{-3}) has been obtained for low-density expanded perlite Technoperl C1.5 [16]:

$$\lambda_s = [0.0709 \cdot \rho - 1 \cdot 45] \cdot 10^{-3} \frac{W}{mK} \tag{28.9}$$

At mean temperatures below 150°C, a temperature dependency of solid conduction was not observable within the measurement accuracy. At higher temperatures, expanded perlite is expected to show a $\lambda_s(T)$ behavior similar to silica glass [17], since usually, the solid thermal conductivity of porous materials follows the temperature dependency of the massive, non-porous solid [18].

Equation (28.9) is only valid if the material is compressed by its own weight or by vibrations, which cause the single perlite grains to rearrange themselves in a closer packing. This is the standard situation in today's VSI storages consisting of two concentric steel cylinders, since the outer container absorbs the atmospheric pressure completely. If the load-bearing envelope is missing, and perlite is compressed by mechanic load instead, $\lambda_s(\rho)$ shows considerably higher values. A possible explanation is, that the material is damaged by high mechanical forces acting on the filigree structures, which shortens the correlated thermal resistances. Accordingly, care has to be taken when expanded perlite is to be used without load-bearing envelope, as for example in a VIP. For the two different compression methods (with and without mechanical load), the solid thermal conductivity of perlite Technoperl C1.5 as a function of density is shown in in Figure 28.8, which has been drawn from measurements at ZAE Bayern at mean sample temperatures between 20 and 150°C.

Since the solid and radiative thermal conductivities depend on bulk density in opposing ways, an optimum density exists, where the sum of both thermal conductivities is minimized. According to Equations (28.6) and (28.9), solid body thermal conductivity is directly proportional to density, whereas radiative thermal conductivity is indirectly proportional to ρ. Figure 28.9 shows the sum $\lambda_r + \lambda_s$ for low-density expanded perlite and different mean temperatures T_r up to 450 K, which corresponds to about 300°C storage temperature and 20°C ambient temperature. It can be seen, that the optimum density is approximately 55 kg m^{-3} for lower temperatures (i.e. close to the ex-factory density of Technoperl C1,5), and around 100 kg m^{-3} for T_r = 450 K (i.e. close to the maximum density that can be achieved by self-rearrangement of perlite grains,

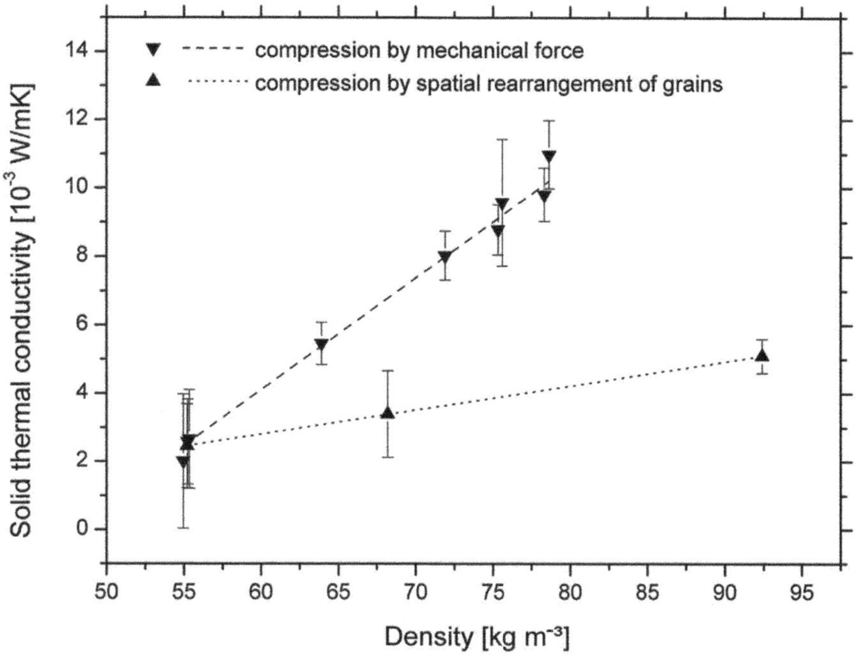

Figure 28.8 Measured solid thermal conductivity as a function of density of Technoperl C1,5, varied either by spatial self-rearrangement or by mechanical compression. The uncompressed density ex-factory amounts to 55 kg m^{-3}. During the measurements, the mean sample temperature has been varied between 20 and 150°C. However, a temperature dependency of the solid thermal conductivity has not been observable within the measurement accuracy.

compare Section 2.1 and Figure 28.8). For even higher temperatures, according to Equation (28.6), λ_r dominates over λ_s and high-density (180 kg m^{-3}) expanded perlite is significantly better suited, because $\lambda_r \sim \frac{1}{\rho}$, even if λ_s increases with ρ. Recently, thermal conductivity measurements have been done in a guarded hot plate apparatus at ZAE Bayern, showing that pure evacuated expanded perlite is suitable as efficient store insulation up to hot side temperatures of 200°C with low-density and up to 400°C with high-density perlite [7]. At even higher storage temperatures up to 800°C, high-density perlite with opacifier should be applied for best insulation values [6, 19].

Heat transport by coupling effects: The coupling thermal conductivity λ_c includes all interference effects, on the one hand between radiative transport and conduction in general, and on the other hand between solid body and gas conduction. Due to the sufficiently high optical thickness (see earlier), radiation and conduction mechanisms decouple completely, and the simple addition of the related thermal conductivities is valid [20]. What remains is the coupling between solid and gas heat conduction: In the presence of gas molecules, the contact resistances between perlite grains can be shorted by gas conduction, which leads to an increase of the total thermal conductivity. This kind of coupling can be physically described as a serial connection of thermal resistances. For evacuated low-density perlite at $T < 100°C$, the coupling between solid and gas heat conduction has been measured and analytically modeled [9]. Very recently,

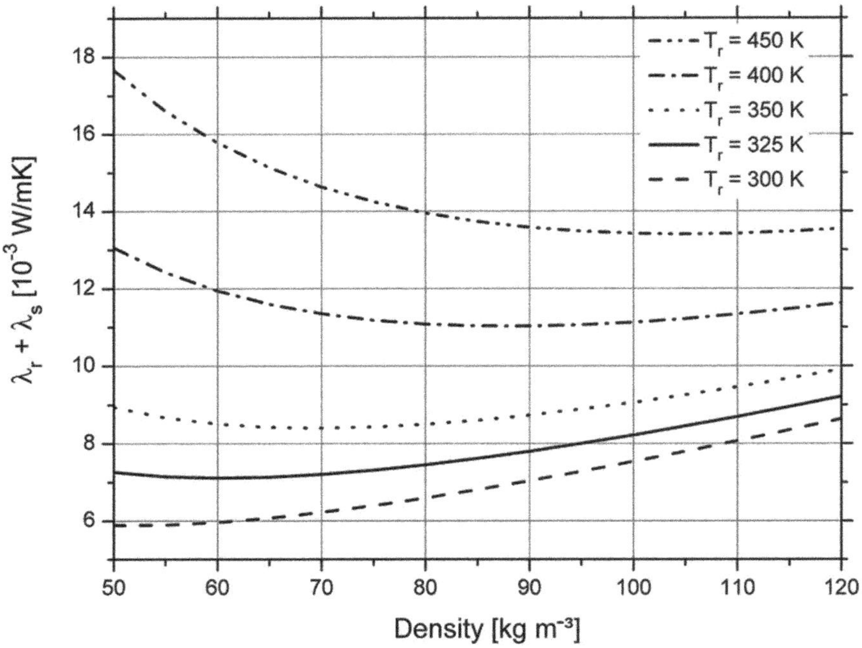

Figure 28.9 Sum of radiative and solid thermal conductivity as a function of density according to Equations (28.6) and (28.9) for low density perlite Technoperl C1.5 at different mean radiation temperatures.

also coupling in high-density perlite has been measured and successfully modeled for arbitrary vacuum pressure and up to 800°C by [6]. However, since λ_s becomes noticeable only at vacuum pressures above 0.1 mbar, the coupling effect is negligible in the VSI application with a typical operating pressure of 0.05 mbar and below.

28.2.3 VSI Storage with Expanded Perlite at Medium and High Temperatures

Naturally, the exceptional insulation properties of VSI become even more beneficial at higher temperatures. Therefore, the VSI storage concept should advantageously be applied to industrial processes or power plants (solar and conventional) up to 800°C. Note, that at such high temperatures, alternative envelope materials have to be found, because the cheap standard steel as used in today's commercial VSI storages does not withstand temperatures higher than 250°C.

From Equation (28.6) it is clear, that at high temperatures, only high-density perlite is suitable to realize a highly insulating storage envelope. Very recently at ZAE Bayern, experiments with Eurocell 140 (by Stauss Perlite, Austria) have been done in guarded hot plate and hot wire apparatus as well as in FTIR spectrometer to experimentally characterize thermal conductivities (total, radiative, and solid) between 20°C and 800°C and for vacuum pressures between 0.001 mbar and 10^3 mbar [6–7, 19]. Preparing a mixture of 40 wt% opacifying SiC (Silcar P8 by ESK-SIC, Frechen) and 60 wt% Eurocell 140 with a resulting density of 263 kg m^{-3}, an ultra-low thermal conductivity of 0.013 W m^{-1} K^{-1} at 400°C mean temperature has been measured. This already corresponds to a storage application with 675°C hot side insulated against ambient temperature. Even at 600°C mean temperature, total λ was below 0.02 W m^{-1} K^{-1}.

Figure 28.10 compares, depending on mean temperature, thermal conductivities of today's best conventional insulation materials and new super insulation based on pure and opacified high-density perlite Eurocell 140 in evacuated state and additionally at ambient pressure: With the exception of fumed silica boards that are very expensive, the new perlite-based high temperature vacuum super insulation shows much better insulation values than today's insulating materials. The improvement in thermal conductivity amounts to a factor of 3 to 5 compared to conventional non-evacuated insulations at high temperatures. The thermal conductivities of commercial high-temperature insulation materials presented in Figure 28.10 (calcium silicate, alkaline earth silicate wool, opacified fumed silica board) have been drawn from the technical data sheets of selected manufacturers [21–23].

Also with respect to cost effectiveness, opacified perlite powder has great potential for high-temperature applications: The price of expanded perlite mixed with opacifier is estimated to be about 300 € m^{-3}. Adding the costs for a vacuum-tight envelope, which are in the order of 1000 € m^{-3}, high-temperature VSI with expanded perlite is still more economical than opacified microporous silica boards, which typically cost 2500 € m^{-3} and more, but need no evacuation due to nanoporosity. Conversely, the prices for conventional non-evacuated high-temperature insulation materials (calcium silicate boards, mineral wool, silica fiber mats) range from 400 to around 1000 € m^{-3}. Thus, VSI with opacified perlite is in general more expensive than these materials, but its insulation efficiency is far better (compare Figure 28.10).

But even without vacuum, opacified (but up to 400°C also pure) high-density expanded perlite shows good insulation values, comparable to commercial high temperature insulation materials. Here, the recently identified optimum mixture of expanded high density perlite and SiC, with expected prices of about 300 €/m^3 is even cheaper than competing materials.

The experimental results reported in this chapter have been obtained during a publicly granted research project (2018–2022) to investigate thermal conductivity of non-opacified and opacified expanded perlite powder at high temperatures at ZAE Bayern. This research has been supported by the German Federal Ministry for Economic Affairs and Energy (BMWi), based on a resolution of the German Bundestag [grant number 03ET1639A].

28.3 Storage Media for Medium and High Temperatures

Thermal storages currently available on the market almost exclusively store sensible heat in the form of hot water. This limits the temperature to a maximum of 100°C (non-pressurized), respectively 150°C (pressurized, 6 bar). For higher storage temperatures, a short overview of suitable storage media is given:

- Pressurized Steam: Typical storage temperatures are 200°C at 16 bar pressure, corresponding to T_m = 110°C (ambient temperature 20°C), where an evacuated VSI-storage with pure perlite is highly efficient and three to five times better insulating than a conventional storage.
- Thermo-oil: Maximum application temperatures are up to 400°C. Since the oil is rather expensive, the combination with cheap stones is advantageous (rock bed

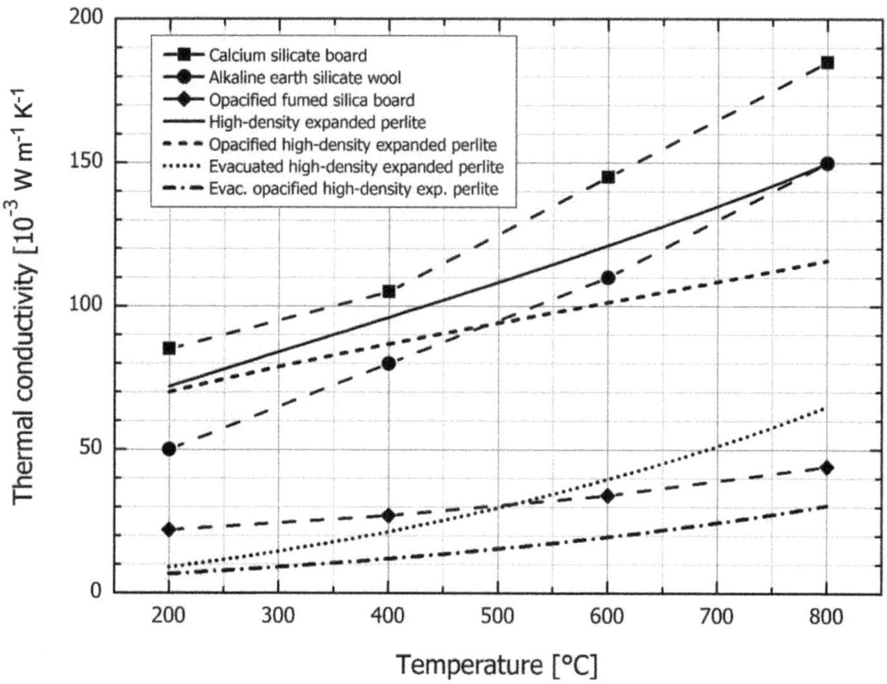

Figure 28.10 Thermal conductivity as a function of mean insulation temperature for commercial high-temperature insulation materials and vacuum super insulation with expanded perlite, both with and without opacificier. The values for the commercial materials have been obtained from data sheets of manufacturers [21–23]. The values for high-density expanded perlite have been determined from a calculation model, which is based on experimental data and reproduces the measurements with a deviation of 5% or less [6].

immersed in thermo-oil). Because the volumetric heat capacity of stone bulk is higher than that of thermo-oil, the storage volume can be reduced by about 20%, which leads to a further cost reduction. In comparison to a pure thermo-oil storage tank, the expansion vessel can be smaller, and a good temperature stratification is possible at high power transmission. Furthermore, thermo-oil is often used to discharge waste heat in industrial processes, so a direct integration of a thermo-oil storage is possible. In a currently running, federally granted research project at ZAE Bayern (grant number 03ESP370B, 2013–2018, German Federal Ministry for Economic Affairs and Energy), the waste heat of a melting furnace is stored at 250°C for later re-use, and the technical and economical potential of VSI for this application is being investigated.

- Phase change materials (PCM): So far, there is no storage with PCM for medium temperatures available on the market. Nevertheless, there is a lot of ongoing research on this topic. PCMs are especially suitable when loading and unloading of the storage is done by condensation and evaporation. Here, storage systems which can absorb and release heat within a small temperature range are necessary. As storage material, molten salts on the basis of nitrate and nitrite with different melting points are used, e. g. $NaNO_3$ or KNO_3, with storage operating temperatures of 308 and 334°C, respectively. At these temperature levels, a highly insulating storage envelope is important and can be realized by an evacuated perlite VSI. Blending the perlite with opacifiers is not strictly necessary, because T_m amounts to about 160–180°C, yielding total thermal conductivities λ_{eff} below 0.02 W m^{-1} K^{-1} (see Figure 28.10), which corresponds to an approximately 2.5-times better insulation than in conventional storages.

- Solid materials: Solid storage media are used in energy intensive branches like glass or steel production. Regenerators for industrial furnaces are installed to preheat combustion air with the heat from exhaust gas, and use ceramics as storage material. Concrete storages are under investigation for CSP plants [24]. In domestic night storage heaters, stones with a high portion of maghemite are used for temperatures up to 650°C. In this temperature range ($T_m > 300$°C), opacified perlite is obligatory to be significantly better than conventional insulation.

28.4 VSI and VSI Storages in Industrial Applications

28.4.1 Applications

To reduce CO_2 emissions worldwide, especially energy intensive industries with process temperatures above 300°C must become more energy efficient. This can be achieved by realizing energy saving heat production processes but also by using more efficient thermal insulation, or by storing intermediate heat in highly efficient storages for later re-use. As an example, the following industrial high-temperature applications up to 800°C could be energetically improved using VSI with (opacified) expanded perlite:

- Insulation of transport containers for molten metals, especially aluminum.
- Insulation of intermediate thermal storages for waste heat.
- Insulation of pipings.
- Insulation of furnaces, process chambers, feeders for glass furnaces, etc.

At still higher temperatures up to 1000°C, VSI with fumed silica powder is possible as an alternative. Fumed silica has the additional advantage of being load-bearing, and shows a similar price like high-density perlite (200–300 € m^{-3}). Also mixtures between fumed silica and expanded perlite are possible. Currently, ZAE Bayern is investigating these possibilities together with industrial companies in Germany, and applications are intended to be realized in future research projects.

28.4.2 Economic Considerations

In this section, the economic efficiency of a VSI storage tank is shown for two exemplary applications. First, a comparison between a conventional storage insulated with mineral wool and a VSI storage with expanded perlite is made for the case of a 120 m^3 (6 × 20 m^3) outdoor storage with storage temperatures up to 250°C at a foundry (German research project 03ESP370B, 2013–2018, Federal Ministry for Economic Affairs and Energy): The conventional storage envelope with 200 mm mineral wool and a galvanized steel sheet for weather protection costs about 7000 €, whereas the 200 mm VSI envelope costs 13 000 €. However, since leaking hot thermo-oil in contact with mineral wool is flammable, the conventional storage has to be double-walled to ensure plant safety. This causes further costs of 3000 €, totaling 10 000 € for the conventional storage. Since the VSI storage is already double-walled, no additional costs occur for fire protection. The annually saved thermal energy due to reduced thermal losses of the VSI storage (5-times better thermal insulation) amounts to 8800 kWh, corresponding to 330 € at an expected energy price of 3.8 cents per kWh. Thus, after nine years, the additional investment of 3000 € for the VSI-storage has paid off as a consequence of the better thermal performance.

In the above example, the amortization period is rather high, which is caused by the fact that the storage is loaded and unloaded 200 times per year in this particular industrial application. Since storage times are consequently very low, the VSI cannot unfold its full potential. If both higher application temperatures (up to 800°C) and higher storage intervals are considered, the amortization period reduces to typical values between two and three years.

Independent calculations for the German breweries were done in [25]. For the calculation, a 12 m^3 VSI storage with 200 mm perlite insulation and a storage temperature of 82°C in outdoor installation is compared to a standard tank with mineral wool insulation. The yearly heat saving amounts to 16 000 kWh at extra investment costs of 10 000 €. Hence, the static amortization time is six years compared to 100 mm standard mineral wool insulation and 7.6 years to a conventional tank with 200 mm mineral wool. If these numbers are projected to the whole brewery industry, possible energetic savings of 44 000 MWh per year corresponding to 16 000 t CO_2 can be achieved, if conventional systems are replaced by VSI systems.

28.4.3 Energetic Amortization Time

Finally, the energetic amortization time of a perlite based VSI storage has been calculated for the example of the above-mentioned foundry [10]. The cumulative energy demand for the production of the four main components (steel, expanded perlite, rock, and thermo-oil) of six 20 m^3 storage tanks amounts to 720 MWh. The primary energy saving with 200 cycles per year totals 1750 MWh. Hence, the energy amortization time is only five months, which is an excellent result and shows the high potential of VSI storages in industrial applications.

28.5 Conclusions

Perlite-based VSI heat storages show excellent insulation values (three to five times better compared to fabric-new conventionally insulated storages) and are humidity insensitive. Expanded perlite is low-cost (50–300 € m^{-3}, depending on density and opacifying additives), non-toxic, and can withstand temperatures up to 800°C. A moderate vacuum of 0.1 mbar, which can be easily realized technically, is sufficient to essentially suppress gas heat conduction. Perlite VSI can be advantageously used for seasonal storage in solar energy applications, where it reduces the solar collector area that is required to obtain a certain solar fraction. In industry, the substitution of conventional storages by perlite-based VSI storages efficiently reduces CO_2-emission. For industry applications at 300–800°C storage temperature, the monetary amortization period typically amounts to some years, while the energetic amortization time is below half a year. The addition of infrared opacifiers is necessary to keep the advantage of three times better insulation efficiency compared to standard insulation materials also at high temperatures. All aspects show the great economic and environmental potential of VSI storages also for medium and high temperatures. Apart from storage applications, VSI with expanded perlite can also be used as a thermal insulation for various industrial applications at high temperatures.

References

[1] Beikircher, T. and Demharter, M. (2013). Heat transport in evacuated perlite powders for super-insulated long-term storages up to 300°C. *Journal of Heat Transfer* 135 (5): 051301. (11 pages).

[2] http://www.bine.info/fileadmin/content/Publikationen/Englische_Infos/projekt_0211_engl_internetx.pdf (accessed 15 April 2017).

[3] http://www.bine.info/fileadmin/content/Presse/Projektinfos_2014/PM_14_2014/ProjektInfo_1414_engl_internetx.pdf (accessed 15 April 2017).

[4] Borst, W. and Fricke, J. (2014). *Essentials of Energy Technology*. Weinheim: JW-VCH. ISBN: 978-3-527-33416-2, 2014.

[5] Caps, R., Trunzer, A., Büttner, D., et al. (1984). Spectral transmission and reflection properties of high temperature insulation materials. *International Journal of Heat and Mass Transfer* 27 (10): 1865–1872.

[6] Caps, R. and Fricke, J. (2000). Thermal conductivity of opacified powder filler materials for vacuum insulations. *International Journal of Thermophysics* 21 (2): 445–452.

[7] Calsitherm Silikatbaustoffe GmbH: Datenblatt Silcal / Microcal. http://www.calsitherm.de/pdf/Silcal_MC_DE_20120816_DR.pdf (accessed 30 May 2016).

[8] https://belglas.files.wordpress.com/2016/03/demharter_master_thesis_2011.pdf

[9] Fricke, J., Schwab, H., and Heinemann, U. (2006). Vacuum insulation panels – Exciting thermal properties and most challenging applications. *International Journal of Thermophysics* 27 (4): 1123–1139.

[10] Beikircher, T., Reuß, M., and Streib, G. (2015). Vacuum super insulated heat storage up to 400°C. *International Renewable Energy Storage Conference, IRES, Düsseldorf, 2015*. https://www.researchgate.net/publication/283449078_VACUUM_SUPER_INSULATED_HEAT_STORAGE_UP_TO_400_C

[11] Kaganer, M.G. (1969). Thermal insulation in cryogenic engineering. Israel Program for Scientific Translations, Jerusalem.

[12] Laing, D. and Zunft, S. (2015). Using concrete and other solid storage media in thermal energy storage (TES) systems. In: *Advances in Thermal Energy Storage Systems: Methods and Applications* (ed. L.F. Cabeza): 65–86. Cambridge: Elsevier.

[13] Lu, X., Wang, P., Arduini-Schuster, M.C., et al. (1992). Thermal transport in organic and opacified silica monolithic aerogels. *Journal of Non-Crystalline Solids* 145: 207–210.

[14] Morgan Advanced Materials PLC: Data sheet WDS Ultra, Stand. (2015). http://www.morganthermalceramics.com/sites/default/files/datasheets/wds_ultra_data_sheet_english_mi_0.pdf (accessed 30 May 2016).

[15] Ochs, F., Heidemann, W., and Müller-Steinhagen, H. (2008). Effective thermal conductivity of moistened insulation materials as a function of temperature. *International Journal of Heat and Mass Transfer* 51: 539–552.

[16] Ochs, F., Heidemann, W., and Müller-Steinhagen, H. (2009). Performance of large-scale seasonal thermal energy storage. *ASME Journal of Solar Energy Engineering* special Issue "Energy Storage," 131 (11): 0410051–0410057.

[17] Reiss, H. (2013). Superisolierungen. In: *VDI-Wärmeatlas* (ed. Verein Deutscher Ingenieure), 1153–1217. Berlin: Springer.

[18] Rettelbach, T. (1996). *Der Wärmetransport in evakuierten Pulvern bei Temperaturen zwischen 10 K und 275 K.* Dissertation, Universität Würzburg.

[19] Rottmann, M. (2020). *Untersuchung des Wärmetransports in expandiertem Perlit zur Entwicklung einer Vakuumpulverisolation für Hochtemperatur-Anwendungen bis 800 °C.* Dissertation, Technical University Munich accepted as dissertation by physics department at 20.1.2021, https://mediatum.ub.tum.de/doc/1559836/1559836.pdf

[20] Rottmann, M., Beikircher, T., and Ebert, H. (2020). Thermal conductivity of evacuated expanded perlite measured with guarded-hot-plate and transient-hot-wire method at temperatures between 295 K and 1073 K. *International Journal of Thermal Sciences* 152: 106338. doi: 10.1016/j.ijthermalsci.2020.106338.

[21] Rottmann, M., Beikircher, T., Ebert, H., et al. (2021). Thermal conductivity and extinction coefficient of opacified expanded perlite for vacuum super insulation up to 1073 K. *International Journal of Thermal Sciences*, Vol 163, May, 106813. https://doi.org/10.1016/j.ijthermalsci.2020.106813

[22] Siegel, R. and Howell, J.R. (1972). *Thermal Radiation Heat Transfer*. Tokyo: McGraw Hill.

[23] Stolz, F. (2013). Wie in der Thermoskanne. Brauindustrie 4.

[24] Touloukian, Y.S. (ed.) (1970). *Thermophysical Properties of Matter; Volume 2; Thermal Conductivity; Non-metallic Solids*. New York: IFI/Plenum.

[25] Unifrax Ltd.: data sheet "Insulfrax LT Blanket", Stand: 04.06.2013 http://www.unifrax.eu.com/web/Audit.nsf/ByUNID/8BB4E86EC95B65EA85257FBD006353D6/$File/Insulfrax%20LT%20Blanket%20EN.pdf (accessed 24 May 2016).

29

Heat Transfer Enhancement for Latent Heat Storage Components

Jaume Gasia[1], Laia Miró[1], Alvaro de Gracia[1], and Luisa F. Cabeza[1,]*

[1] *GREiA research group, Universitat de Lleida, Pere de Cabrera s/n, 25001, Lleida, Spain*

29.1 Introduction

Latent heat thermal energy storage (LHTES) is known to be a promising thermal energy storage (TES) technique because of its high energy storage density and the nearly isothermal melting/solidification processes at the phase change transition temperature of the TES material, from now on named phase change material (PCM).

During phase change, two simultaneous heat transfer processes are known to occur. On the one hand, between the heat transfer fluid (HTF) and the PCM, and on the other hand, between regions of PCM which are under different thermal gradients. These processes are carried out until the PCM has totally undergone phase change. During the melting process, the HTF transfers heat to the PCM, which is solid at the beginning of the process. Then, the melting front starts to increase from the heat transfer surface, the solid layer decreases and convection becomes the dominant heat transfer mechanism. On the contrary, during the solidification process, the HTF absorbs heat from the PCM, which is liquid at the beginning of the process. A solid layer is created around the heat transfer surface and conduction becomes the dominant heat transfer mechanism.

The heat sources that supply the thermal energy, but especially the final users that consume this energy, require the melting and solidification processes to be done within a specific period of time and under specific operation conditions [1]. Generally, the low thermal conductivity values are found in the current cost-effective PCM in both the solid and liquid state (fatty acids: ~0.15 W/m·K; paraffins: ~0.20 W/m·K; salts: ~0.6 W/m·K [2]). Moreover, the use of these materials in such processes presents low values of natural convection coefficients on the liquid region of the PCM if compared to forced convection, which limits the heat transfer rates during these processes. In order to overcome this drawback, the implementation of heat transfer enhancement techniques has been studied. These heat transfer enhancement techniques are divided depending on whether they are focused on enhancing the heat transfer between the HTF and the PCM, or focused on enhancing the heat transfer within the PCM itself. The first ones consist of adding extended surfaces to the heat transfer surface, such as fins or heat pipes, moving the PCM while undergoing phase change, or placing different PCM in series. The second ones are focused on combining the PCM with highly conductive materials such as nanoparticles, graphite, and metal foams.

Advances in Energy Storage: Latest Developments from R&D to the Market, First Edition.
Edited by Andreas Hauer.
© 2022 John Wiley & Sons, Ltd. Published 2022 by John Wiley & Sons, Ltd.

The decision of selecting the previously defined techniques must be carefully taken depending on the requirements of the system (type of process, temperature range, etc.). Hence, in the present chapter, the most studied LHTES systems and heat transfer enhancement procedures for both improving the heat transfer between the HTF and the PCM, and inside the PCM itself, are presented.

29.2 Heat Transfer Enhancement Techniques

29.2.1 Heat Transfer Enhancement between the HTF and the PCM

The HTF is the responsible for carrying thermal energy from the heat source to the LHTES system, by means of the PCM, and from the LHTES systems to the final users. Hence, a study of heat transfer between the HTF and the PCM is required, which is usually hindered by the thermophysical properties of the PCM.

Three main enhancements techniques have been identified to increase the heat transfer between the HTF and the PCM: the addition of extended surfaces, forcing the movement of the PCM while undergoing phase change, and the multiple PCM configuration (Figure 29.1).

Table 29.1 shows the main characteristics of the heat transfer enhancement techniques between the HTF and the PCM. The main advantages and disadvantages, as well as the enhancements related to the different techniques and the main materials used are shown.

29.2.1.1 Addition of Extended Surfaces

This enhancement technique provides an increase of the heat transfer rate of the process as a result of increasing the heat transfer surface with a material that has a much higher thermal conductivity than the PCM. Hence, with this technique the distance between the remote PCM regions and the heat transfer surface is decreased and the thermal resistance between both materials is reduced. However, two main drawbacks are associated with this technique: on the one hand, the increase of costs due to the

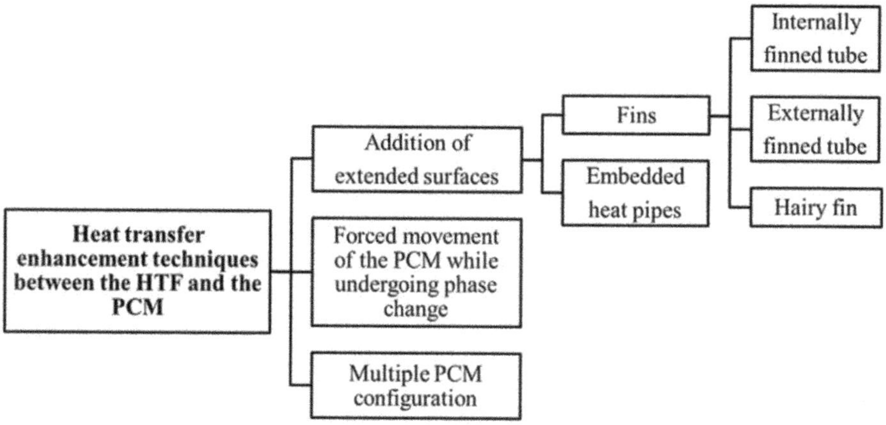

Figure 29.1 Classification of the heat transfer enhancement techniques focused on enhancing the heat transfer between HTF and PCM.

addition of the extended surfaces material and, on the other hand, the decrease of the packing factor of the TES system due to the increase of the volume occupied by the extended surfaces. This factor represents the percentage of volume occupied by the PCM per volume of the storage container.

Two different enhancement techniques have been identified regarding this technique: the addition of fins and the addition of heat pipes on the HTF heat transfer surface.

Table 29.1 Main characteristics of the heat transfer enhancement techniques between the HTF and the PCM.

	Enhancement technique			
	Addition of extended surfaces		**Forced movement of the PCM while undergoing phase change**	**Multiple PCM configuration**
	Fins	**Embedded heat pipes**		
Advantages	- Increase of heat transfer surface		- Control of the heat transfer - Same packing factor - Avoid phase segregation	- More uniform heat transfer rate - Higher outlet HTF temperature - Exergy increase - Same packing factor
Disadvantages	- Decrease of the packing factor - Increase of costs	- Decrease of the packing factor - Increase of costs	- Requirement of an external force - Increase of costs	- Increase of costs
Enhancement	Thermal conductivity: depending on number of fins, fin thickness, fin length, space between fins, HTF tube and fins geometry, and fin material	Thermal conductivity: depending on heat pipes orientation, HTF flow rate, module length, condenser section tube length, evaporator section tube length, and the vapor core radius	Effectiveness and phase change process time: depending on the velocity of both the PCM and HTF	Energy rate: depending on the selection of the PCM
Materials used	- Aluminium - Stainless steel - Carbon steel - Graphite	- Stainless steel - Carbon steel	---	---

29.2.1.1.1 Fins

The study of adding fins as an enhancement technique has been focused on two different points of view: first, how they are embedded into the heat transfer surface and second, how the modification of the main geometrical parameters and the fins' material influence the melting and solidification processes.

Regarding the way the fins are embedded into the heat transfer surface, three main geometries are identified: internally finned tubes, externally finned tubes, and hairy fins. The effect of adding internal longitudinal fins was numerically studied and compared under the same boundary conditions with a plain tube by [3] (Figure 29.2). It was found that the melting process was significantly accelerated by adding internal fins because of the increase of the Nusselt number and the direct conduction of heat to the PCM. Moreover, the melting volume fraction (MVF) was also increased. This increase was proportional to the increase of the internal fin thickness and height, as well as to the number of fins, since a higher heat transfer surface was created for the convection mechanism.

Several authors have numerically studied the addition of external fins to HTF tubes in melting and solidification processes, focusing on the influence of the number of fins, the fin thickness, the height or length of the fin, the fin spacing, the material of the fin, and the shape of the fins (Figure 29.2).

The modification of the number of fins and their length or width has influence on the heat transfer rate and on the phase change processes time (Figures 29.3a and 29.3c)

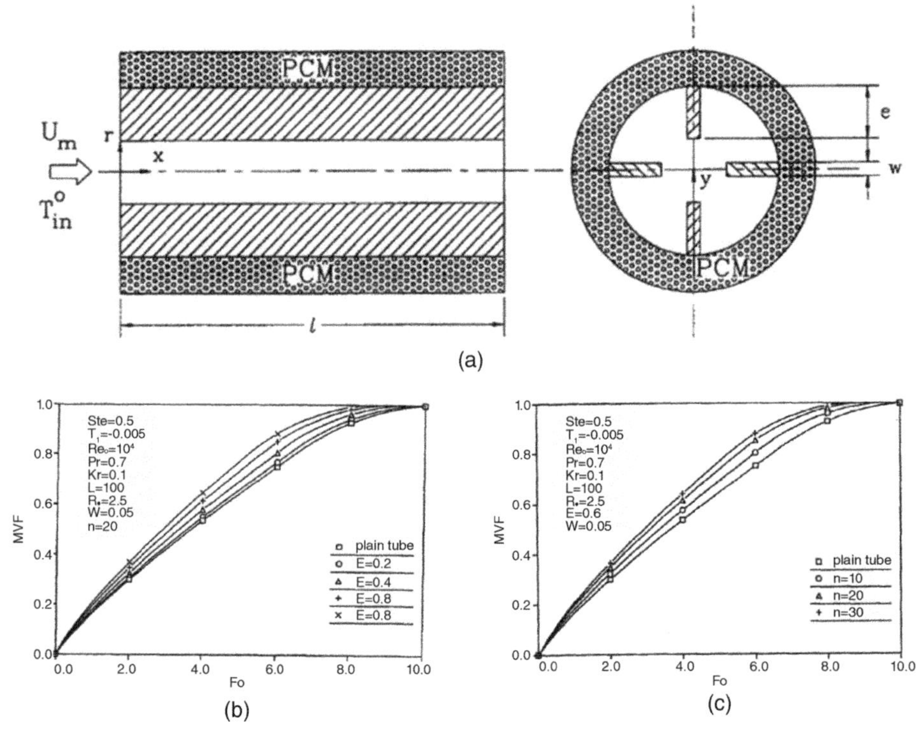

Figure 29.2 (a) Longitudinal (left) and radial (right) section of an internally finned tube; (b) Effect of the fin height on the MVF; (c) Effect of the fin number on the MVF [3].

[4–5]. By increasing these parameters, the heat transfer surface between the HTF and the PCM is increased. As a consequence, a larger amount of PCM is in contact with a surface at a different temperature and therefore a higher heat transfer between the PCM and the HTF. Moreover, the number of fins has a much higher influence in the initial stages of the processes while this influence decreases as the processes continue because of the thermal resistance of the PCM. However, when the number of fins or the fin length is increased, the volume occupied by the fins within the TES system increases and therefore the volume of PCM is reduced. On the one hand, this fact implies a reduction of the heat capacity of the system and on the other hand, shorter phase change processes periods.

Regarding the fin thickness, it influences mainly the phase change process time (Figures 29.3b and 29.3f) [5–7]. By increasing the thickness of the fins, a reduction on the phase change process time can be obtained because a larger volume within the TES system is occupied by the fins and therefore a smaller volume of PCM is present in the system. However, there is a maximum fin thickness, and beyond this optimum value the improvement on the phase change process time is not significant and only a decrease on the heat capacity of the system is obtained [5]. Moreover, the fact of modifying the fins' thickness practically has no effect on the heat transfer rate since the heat transfer surface between the HTF and the PCM remains almost constant. Hence, this modification shows worse improvement if compared to modifying the number of fins, so a thin fin is desired for a better performance.

Furthermore, varying the space between fins is closely linked with the modification of the number of fins since the TES system volume is constant in all the cases. If the distance between fins is large, a smaller number of fins can be place at the same space and therefore lower heat transfer rates and slower phase change processes are obtained (Figure 29.3d).

In addition, the variation of the fin geometry was numerically studied by [8]. They compared the effect of four different configurations (a dimpled tube, a cone-fined tube, a helically-fined tube and a reference smooth tube) on the phase change process time and heat transfer rate (Figure 29.3e). Regarding the phase change process time, the three different tubes configurations showed improvements up to 30% because of the increase of the heat transfer surface and a reduction of the PCM volume. Regarding the heat transfer rates, enhancements up to 50% during the PCM latent stage was obtained.

The material of the fins influences the phase change process time and the heat transfer rate (Figure 29.3f) [7]. By modifying the material of the fins (Table 29.2), the thermal conductivity of the heat transfer surface is increased and therefore the thermal resistance between the HTF and the PCM is reduced. As a consequence, the fins reach closer temperatures to the HTF ones, creating larger temperature gradients between the PCM and the HTF, which increase the heat transfer rate and reduce the phase change process times.

These studies indicate that the number of fins, fins' length, and fins' material have much more significant effect on the heat transfer rate and phase change processes time than the thickness and shape of the fins.

The addition of fins at higher scales has been experimentally addressed by [6]. They performed a study comparing the behavior of a sample of 150 kg of PCM (hydroquinone) under the same boundary conditions in two identical shell-and-tube storage tanks with the

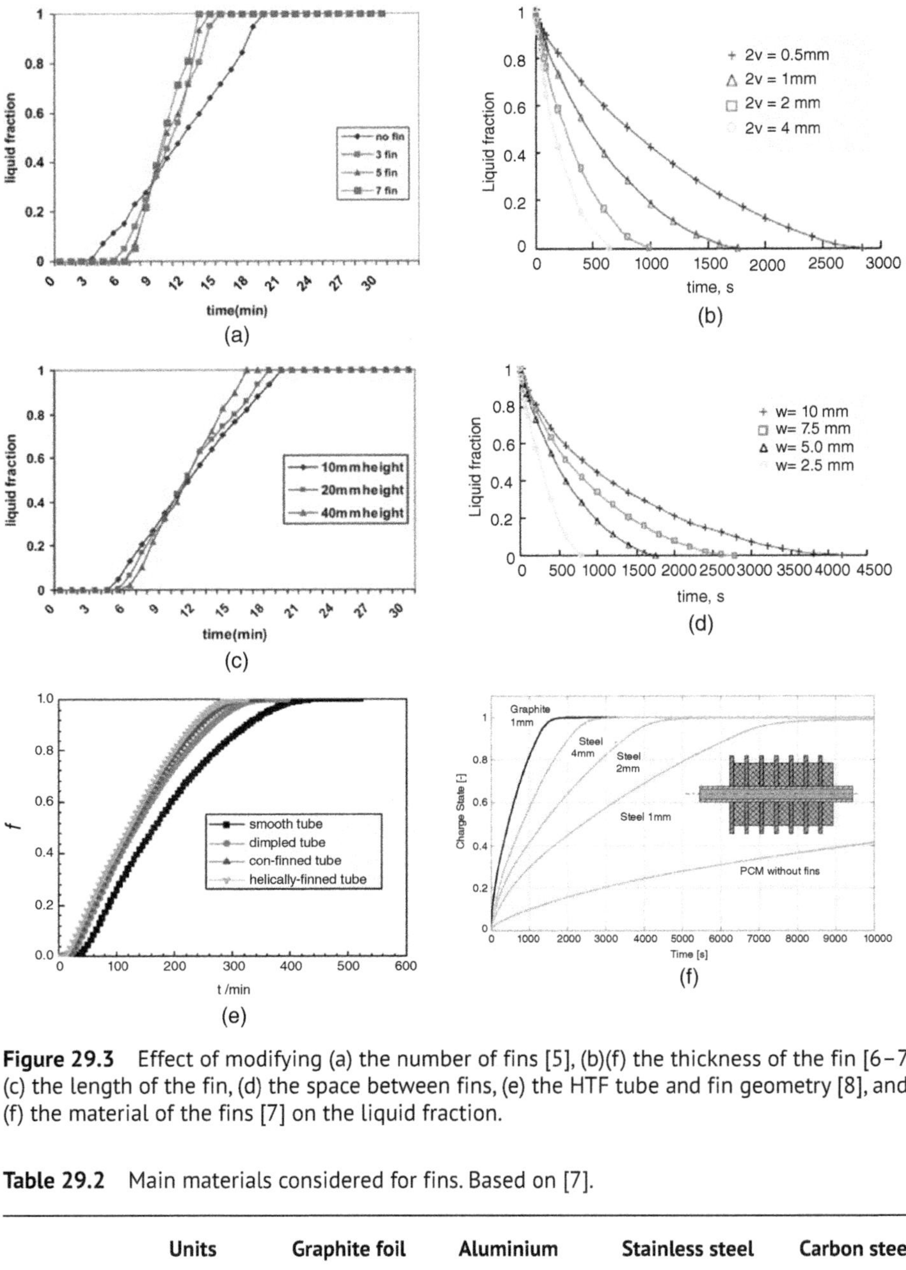

Figure 29.3 Effect of modifying (a) the number of fins [5], (b)(f) the thickness of the fin [6–7], (c) the length of the fin, (d) the space between fins, (e) the HTF tube and fin geometry [8], and (f) the material of the fins [7] on the liquid fraction.

Table 29.2 Main materials considered for fins. Based on [7].

	Units	Graphite foil	Aluminium	Stainless steel	Carbon steel
Thermal conductivity	W/m·K	150	200	20	30
Density	kg/m^3	1000	2700	7800	7800
Cost	USD/m^3	10 000	7000	20 000	15 000

difference that one of them incorporated 196 squared fins on the tubes bundle (Figure 29.4). Results showed that the fact of increasing the heat transfer surface by ~300% reduced the PCM volume by ~10%, and therefore the melting and solidification periods were

Figure 29.4 (a) TES system without fins and (b) TES system with squared fins used [6].

decreased. Moreover, it was found that, in terms of heat transfer rates, the addition is especially recommended for partial melting and solidification processes.

Finally, the effect of addition of hairy fins has been numerically studied by [10]. This configuration consists of a primary rectangular fin with a large number of slender secondary rods attached on its surface (Figure 29.5). The enhancement in heat transfer because of the hairy fin was found to increase when the secondary rods diameter decreases, since they become more flexible and can easily swim near the primary fin surface which increases the convection coefficient.

29.2.1.1.2 Embedded Heat Pipes (or Thermosyphons)

The embedded heat pipes enhancement technique, also referred in the literature as thermosyphons, consists of several small closed tubes attached to the main HTF pipe. These tubes are filled with a working fluid which undergoes a liquid–vapor phase change (Figure 29.6). Hence, in this system various heat transfer processes are involved. On the one hand, evaporation and condensation of the heat pipe working

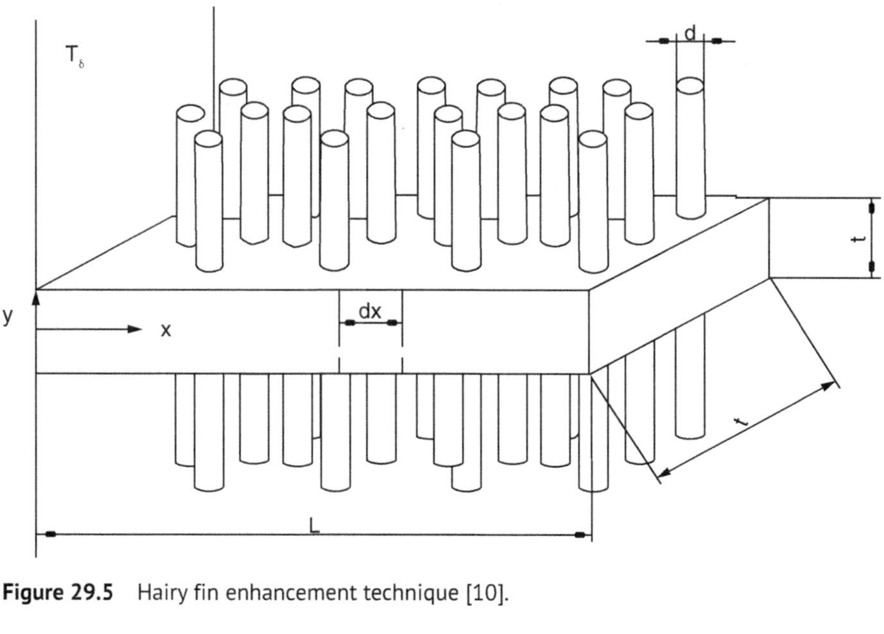

Figure 29.5 Hairy fin enhancement technique [10].

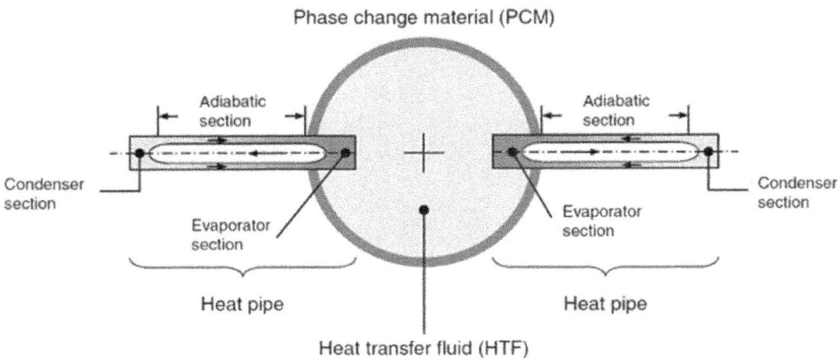

Figure 29.6 Embedded heat pipes working principle [11].

fluid, and on the other hand melting and solidification of the PCM. So far, this technique has only been numerically addressed.

Two numerical models were developed by [11] and [12] to analyze the heat transfer in a high temperature LHTES unit with two different arrangements depending on the location of the PCM and the HTF in the system, as seen in Figure 29.7. In Module 1, the HTF internally flowed within the HTF tubes and the PCM surrounded them. In Module 2, the PCM was embedded in cylindrical tubes and the HTF externally flowed transverse to the PCM tubes. The orientation of the heat pipes was found to play a minor role in Module 1, while in Module 2 the orientation was an important factor in determining the thermal response of the system. Both modules were found to have higher efficiency and a faster melting rate if compared to a system without this enhancement technique. The thermal behavior enhancement was found to be dependent on the heat pipes length and radius, and on if it is a melting or a solidification process [12]. Besides, [13] conceptually studied, from an economic point of view, the incorporation of heat pipes' in a LHTES system, using LiCl/KCl (44/56 wt%) as PCM, for a 50 MW_e CSP plant and compared the results with a two-tank molten TES system. They estimated the capital costs for the storage container, the PCM, the heat pipes, and the internal architecture of the storage container based upon size and mass, and real data from published studies. First results showed the economic competitiveness of such system, with a potential reduction of 15% in the capital costs.

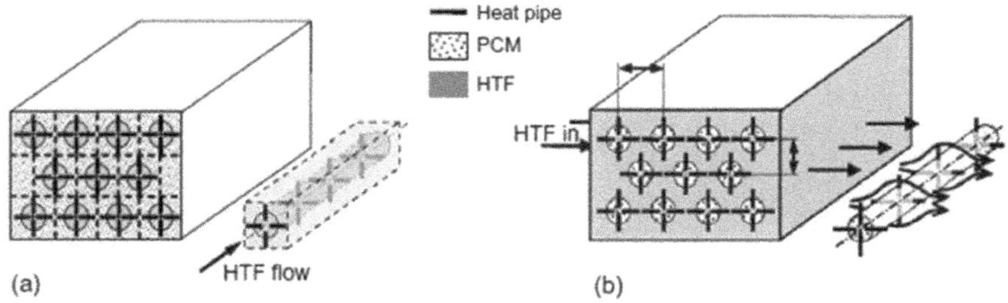

Figure 29.7 Heat pipes modules [11]: (a) Module 1: The HTF internally flows within the HTF tubes and the PCM surrounds them; (b) Module 2: The PCM is embedded in cylindrical tubes and the HTF externally flows transverse to the tube axis.

29.2.1.2 Forced Movement of the PCM while Undergoing Phase Change

A new technique focused on enhancing the heat transfer between the HTF and the PCM has been recently proposed. It consists of actively moving, with an external force (pumps or other transport mechanisms), the PCM while it is undergoing phase change (both melting and solidification). As a result, during the melting process the thermal resistance is reduced due to the fact that the overall heat transfer coefficient is increased because of the forced convection dominance, and during the solidification process the thermal resistance is reduced because the solid layer around the heat transfer surface is continuously removed.

This enhancement technique has three main advantages. First, the heat transfer rate between the PCM and the HTF, as well as the melting and solidification processes periods, can be controlled by adjusting the flow rate of both the PCM and HTF. Second, the PCM phase segregation is prevented because the PCM is continuously moving. And third, the volume of PCM is maintained constant within the TES system (same packing factor) since no external elements (like fins or heat pipes) are introduced. On the other hand, the main disadvantage of this enhancement technique is that an external electrical energy source is required to create the forced movement of the PCM, with its associated energy consumption.

Currently, there are three different concepts dealing with the forced movement of the PCM while undergoing phase change, which are still in the early stages and therefore little information is available: dynamic melting, screw heat exchanger (SHE), and PCM flux. [14] and [15] presented the dynamic melting concept. Its working principle is based on creating melted PCM paths around the HTF tubes, which are afterwards recirculated with an external pump, creating a continuous flow of liquid PCM. They numerically and experimentally evaluated this enhancement technique during the melting process of different PCMs in two shell-and-tube heat exchangers (Figures 29.8a and 29.8b). Results showed an improvement on the effectiveness and melting process time which strongly depended on the flow rates of both the HTF and PCM. It was observed that much higher enhancements were obtained when the PCM flow rate was equal or higher than the HTF flow rate. [16] presented the prototype design of a screw heat exchanger (SHE), whose working principle consists of transporting the PCM along an helicoidally heat transfer surface during both melting and solidification processes. The phase change takes place during this transport as a consequence of the heat transfer with the HTF, which flows through the hollow shaft, screws, and trough (Figure 29.8c). So far there are not numerical results showing the performance of this system. Finally, [17] presented the PCM flux concept and numerically investigated the enhancements in terms of phase change process time and heat transfer rates. This technique consists of moving, parallel to a heat source, macro encapsulated PCM containers with a transport mechanism (Figure 29.8d). The containers are separated from the energy source with an intermediate fluid to ensure the thermal conduction. Preliminary results showed nearly constant power over time, which can be controlled by modifying the PCM velocity.

29.2.1.3 Multiple PCM Configuration

The multiple PCM configuration, also referred in the literature as cascaded or multistage configuration, is a heat transfer enhancement technique which consists of two or more LHTES systems arranged in series in a decreasing order of the melting

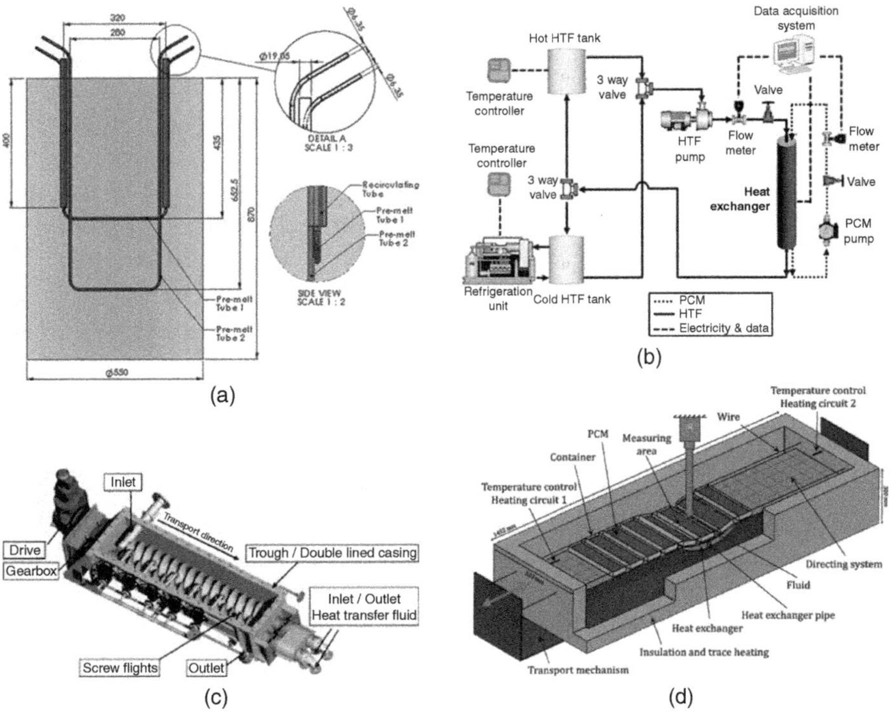

Figure 29.8 Prototypes of the forced movement of the PCM while undergoing phase change enhancement technique: (a) [14], (b) [18], (c) [16] and (d) [17].

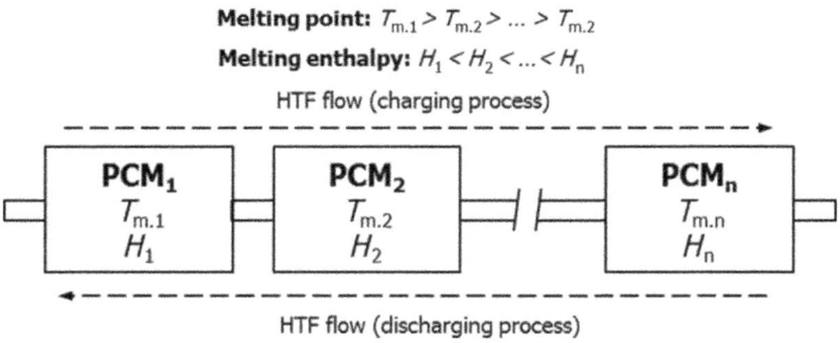

Figure 29.9 Working principle of a multiple PCM configuration [19].

temperatures and in an increasing order of the heat storage capacities of their PCM during the melting process and the other way around during the solidification process, as observed in the Figure 29.9 [19]. This technique presents two main advantages if compared to a single PCM configuration. First, a more constant heat transfer rate between the HTF and the PCM during longer periods. Despite the fact that the HTF temperature decreases/increases during the melting/solidification processes in the direction of the HTF flow, the temperature difference between the HTF and the PCM is kept almost constant. And second, an increase of the exergy efficiency, since more energy from/to the HTF can be absorbed/released. On the other hand, the main disadvantage is that a bad selection of PCMs not only can increase the economic costs but also can reduce the performance of the LHTES system.

This technique has been mainly performed with numerical study cases and very few cases have been experimentally carried out [19–20]. Regarding the experimental studies, [18] evaluated the multiple PCM configuration with two PCM at pilot plant scale (> 100 kg) and compared the effectiveness and the temperature profiles of both the PCMs and the HTF to single PCM configurations with both materials separately. The selected PCM were d-mannitol (T_m = 167 °C and H = 316 kJ/kg) and hydroquinone (T_m = 172.4 °C and H = 258 kJ/kg). Results showed an enhancement in the effectiveness of almost 20% and a higher uniformity on the outlet HTF temperature. [20] experimentally evaluated the multiple PCM configuration with three PCM: KNO_3 (T_m = 306 °C and H = 172 kJ/kg), 95.5 wt% KNO_3 + 4.5 wt% KCl (T_m = 320 °C and H = 74 kJ/kg) and $NaNO_3$ (T_m = 335 °C and H = 95 kJ/kg) and observed a reduction of the HTF outlet temperature variations and an energy rate enhancement.

29.2.2 Heat Transfer Enhancement Within the PCM

The heat transfer enhancement techniques within the PCM are focused on increasing the thermal conductivity of the PCM and therefore decreasing the melting and solidification processes time as a result of combining highly conductive materials with the selected PCM. Three main enhancements techniques have been identified: combination of PCM with nanoparticle, PCM and graphite composites, and the combination of PCM with metal foams and porous materials (Figure 29.10).

Table 29.3 shows the main characteristics of the heat transfer enhancement techniques within the PCM. The main advantages and disadvantages as well as the enhancements related to the different techniques, the main materials used, and their characteristics are also shown in this table.

29.2.2.1 Combination of PCM with Metal Foams and Porous Materials

This section includes the combination of metallic and highly porous materials (also referred in the literature as metals, porous metals, foamed metal, or metallic foam) with PCM. [21] reviewed the research progress of this enhancement technique and studied the effect of introducing metal foams or porous materials on the conduction, convection, and phase change process of the composite. They observed that these composites have higher effective thermal conductivity and thus they could improve the uniformity of the temperature distribution and the phase change process time, especially during solidification. However, the presence of the porous structures of the

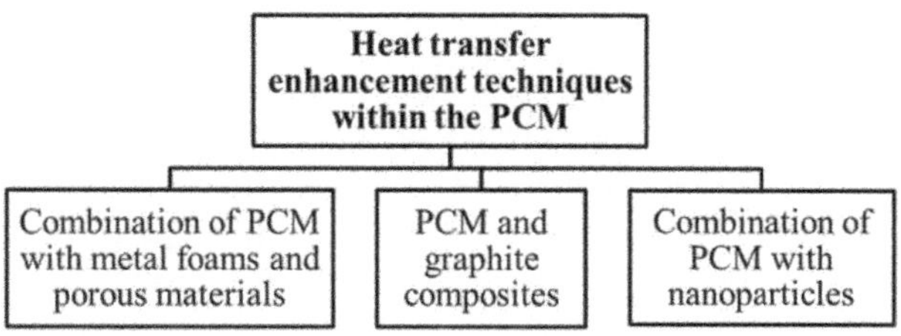

Figure 29.10 Classification of the enhancement techniques within the PCM.

Table 29.3 Main characteristics of the heat transfer enhancement techniques within the PCM.

	Enhancement technique		
	Combination of PCM with nanoparticles	**PCM and graphite composites**	**Combination of PCM with metal foams and porous materials**
Advantages	Increase of the effective thermal conductivity		
Disadvantages	- Suppression of the natural convection in liquid state - Increase of costs - Decrease of the packing factor	- Reduction of latent heat - Increase of costs - Possibility of PCM leakages - Decrease of the packing factor	- Difficulty on the synthesis - Difficulty on controlling the size and dimensions of the nanoparticles - Settlement of the nanoparticles - Increase of costs - Decrease of the packing factor
Enhancement	Thermal conductivity: depending on the material, porosity and pore density	Thermal conductivity: depending on the combination technique, type of graphite, concentration, porosity, and working temperature	Thermal conductivity: depending on the nanoparticles material
Materials used	- Cu - Steel - Al - Ni	- Natural graphite - Expanded graphite - Matrix	- SiO_2 - Cu - ZrO_2 - MWCNT/SWCNT* - TiO_2 - Graphene - Al_2O_3
Concentration (C), size (S), pore density (PD), porosity (P)	PD: 0–40 PPI P: 70–95%	C: 0–30 wt% S: 0–1 mm	C: 0–20 wt% S: 0–100 nm

*MWCNT: Multi-walled carbon nanotubes; SWCNT: Single-walled carbon nanotubes;

metal foam reduces drastically the natural convection and limits the enhancement during the melting process. [22] observed that the enhancement of the heat transfer coefficient was reduced from a factor of 28.1 times in the solid state by conduction to a factor of 3.1 times in the liquid state by the combination of natural convection and conduction. Hence, one has to consider the effects of the metal foam material, the metal foam porosity, the metal foam pore density (also named cell size), and the PCM state on the heat transfer (liquid or solid), when working with metal foams and porous materials composites. Otherwise, one can obtain unexpected values regarding the energy that can be stored in the composite and phase change process time.

The modification of the material of the metal foam has influence basically on the solidification process time. The higher the thermal conductivity of the metal foam, the lower the thermal resistance within the composite and therefore the better the heat transfer process enhancement. During the melting process, the modification of the thermal conductivity has practically no effect since convection is the main heat transfer mechanism. [23] experimentally and numerically investigated the thermal enhancement of an eutectic mixture by combining it with copper ($\lambda = 398$ W/m·K) and nickel metal foams ($\lambda = 91.4$ W/m·K) (same porosity and pore density) and observed that in the solidification process, where conduction is the dominant mechanism, the presence of copper and nickel foams enhanced the process by 28.8% and 19.3%, respectively.

The modification of the metal foam porosity, as a result of changing the equivalent diameter or the cross-section shape of fibres, has influence on the heat transfer rate, the phase change processes time, and the energy stored [22]. By decreasing the porosity, the volume of metal foam is increased and therefore the volume of PCM which can be included into the TES system is decreased. Hence, the total energy stored by the composite is reduced. Moreover, a lower porosity of the metal foam also implies that a larger surface of high thermal conductivity material contacts directly with the PCM, reducing the phase change process time and increasing the heat transfer rate (Table 29.4).

The modification of the pore density, as a result of changing the size of the pores, has influence on the phase change processes time and the heat transfer rate [22]. By increasing the pore density, the natural convection of the liquid PCM is decreased and therefore a slower melting process and lower heat flux values are obtained. Moreover, when the pore density is low, the influence of the porosity is not relevant since the natural convection of the liquid PCM is strong enough [24].

29.2.2.2 PCM and Graphite Composites

The combination of graphite in the form of powder or matrix with the PCM is the most widely studied heat transfer enhancement technique within the PCM because of its optimal thermophysical properties (high thermal conductivity and low density) and the chemical resistance of graphite to corrosive environments [1].

The combination of PCM and graphite is done following different techniques regarding the process stage of the graphite, as Figure 29.11 shows. First, natural graphite (NG) can be directly combined with the PCM with mechanical dispersion. Second, the

Table 29.4 Influence of the porosity and the pore density on the melting process time, solidification process time, and heat flux. Based on [22].

Porosity [%]	Pore density [PPI]	Melting time [s]	Solidification time [s]	Heat flux [W/m^2]
Pure PCM		739	5406	3793
95	5	256	466	8892
95	10	221	419	8410
95	20	233	412	5913
95	30	286	420	4024
90	5	179	262	11 966
90	10	152	224	10 840
90	20	169	215	7347
90	30	214	213	6093

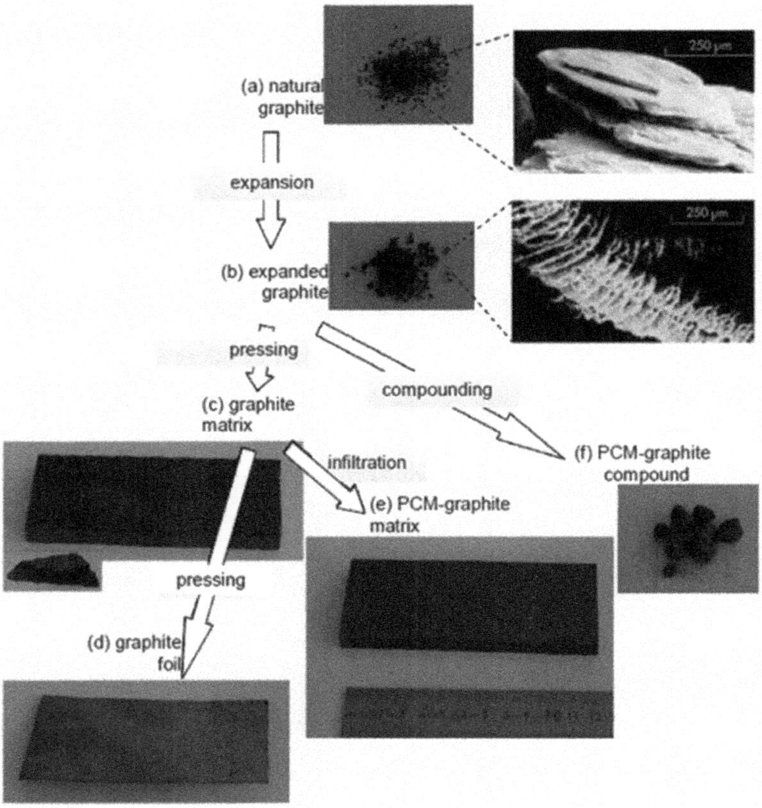

Figure 29.11 Processing of graphite from natural graphite to different products [13].

NG can be expanded in order to obtain expanded graphite (EG), which can be either mechanically dispersed into the PCM or mixed with the PCM to be further compressed at room temperature. In both, dispersion and compression techniques, the type of graphite, its concentration, distribution, and orientation should be taken into account to determine the desired enhancement. Finally, the expanded graphite can be pressed to obtain a graphite matrix, which is used to be impregnated or infiltrated with PCM. In this technique, the porosity and capillarity forces between the matrix and the PCM should be considered in order to avoid PCM leakages and a large reduction of the natural convection when the PCM is in liquid state. Among the impregnation, infiltration, compression, and dispersion techniques, the latter provides the worst enhancement since the graphite connection chains, which are the responsible for enhancing the thermal conductivity, are either not formed because only isolated microstructures are created, or are the weakest. On the contrary, the other three techniques allow the creation of interconnected chains or a completed network of graphite, which highly improve the effective thermal conductivity of the composite. However, the reduction of the latent heat, which is proportional to the concentration of graphite and porosity, is the biggest disadvantage of the PCM and graphite composites enhancement technique.

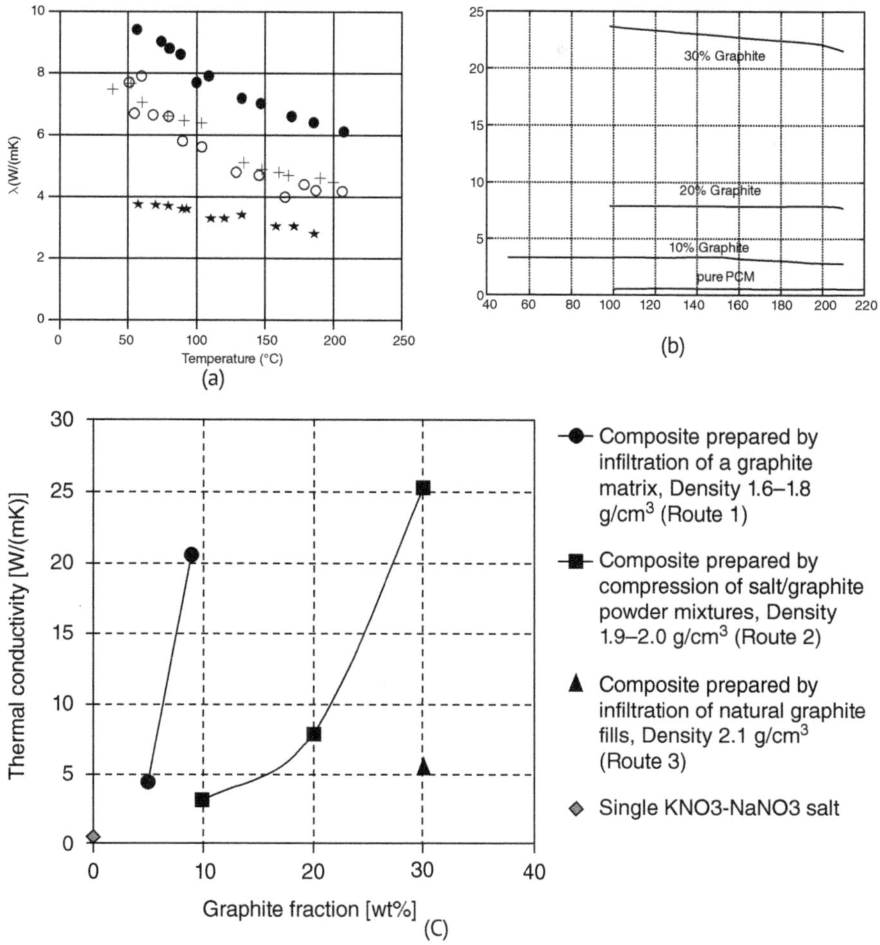

Figure 29.12 Influence on the thermal conductivity of (a) the temperature and type of graphite ((★) Graphite flakes I, (✚) Graphite flakes II, (o) Natural graphite, (•) Expanded graphite), (b) the graphite concentration and (c) the combination technique (Infiltration and compression) [18].

Several experimental studies have been carried out regarding this enhancement technique. Figure 29.12 shows some results regarding the influence of the temperature, the type of graphite, the graphite concentration, and the combination technique on the thermal conductivity of an eutectic mixture of KNO_3 and $NaNO_3$, which turns to be one of the most recurrent enhanced materials due to its applications in real facilities. Results showed that regarding temperature, the higher the temperature the lower the thermal conductivity; regarding the type of graphite, EG shows a better performance; regarding the enhancement technique, compression showed much more improvement, demonstrating what was theoretically stated; and regarding the concentration, the higher the concentration the higher enhancement.

29.2.2.3 Combination of PCM with Nanoparticles

The combination of PCM with nano-sized particles (nanoparticles, nanotubes, and nanofibers) is an enhancement technique that has gained relevance in the last few years. However, a lack of information about their behavior is still found. The combination of PCM with nanoparticles can be performed whether in liquid state (nanofluid), where a free-form structure is created, or in solid (nanocomposite), where a fixed-form structure is created. For a better understanding of this section, both concepts will be presented as nanomaterials.

The advantages of nanomaterials are: higher effective thermal conductivity and higher specific surface, good stability on the dispersion of the nanoparticles, liquid-like properties of the nanoparticles when the nanomaterial is melted, and the reduction of the subcooling effect. All these advantages lead to a higher thermal storage in reduced spaces and a reduction of the melting and solidification process length. Moreover, the user can adjust the nanomaterial properties depending on the material of the nanoparticles, their concentration, size, and geometry. The disadvantages of their use are the difficulty on their synthesis, the difficulty on controlling the size and dimensions of the nanoparticles, the settling of the nanoparticles after some cycles as a result of the density differences, and the toxicity of some of the nanomaterials proposed up to now. Therefore, the study of behavior of the nanomaterials should be carefully addressed for each specific application [25]. It has been observed in the literature that exact predictions of this enhancement technique are difficult to calculate, despite the fact that both the nanoparticles and PCM thermophysical properties are known, due to the current unpredictability on how the nanoparticles cluster or aggregate between them, the molecular connection between the PCM and the nanoparticles, the Brownian motion of the nanoparticles, and the variation of the convection in liquid state.

[18] and [25] reviewed and summarized the different studies concerning nanomaterials at high and low temperatures, respectively. The main tested PCM found in literature concerning low temperatures are paraffin, palmitic acids, and water-like materials, and concerning high temperature are carbonates, nitrates, and cements. The main nanoparticles used as enhancers are silica (SiO_2), titania (TiO_2), single-walled carbon nanotubes (SWCNT), multi-walled carbon nanotubes (MWCNT), grapheme, copper, and alumina (Al_2O_3) in a concentration between 0 and 20 wt% with dimensions up to 50 nm. Results from laboratory analyses showed very different and unrelated results in terms of effective thermal conductivity, latent heat, and melting temperature depending on the material (type of material, concentration, size) and temperature evaluated. The reason lies in the lack of normalization on the techniques used to evaluate the thermophysical properties and the lack of knowledge on the atomic connections between the PCM and the nanoparticles.

29.3 Technology Development and Commercial Status

One of the techniques used to assess the maturity of evolving technologies during their development and early operations and to compare technologies is the technology readiness level (TRL) [26]. This scale classifies the technology development and

Table 29.5 Technology readiness levels summary. Based on [26].

Level	Definition
TRL 1	Basic principles observed
TRL 2	Technology concept formulated
TRL 3	Experimental proof of concept at laboratory scale
TRL 4	Technology validated in small scale
TRL 5	Development at real scale
TRL 6	System prototype demonstration in simulated operational environment
TRL 7	System prototype demonstration in real operational environment
TRL 8	System complete and qualified, first commercial prototype
TRL 9	Commercial application

commercial status in nine levels, from the basic research until its commercial application (Table 29.5). This scale classifies the maturity of a technology in nine levels, from the basic research until its commercial application (Table 29.6).

The TRL of the different enhancement techniques presented in this chapter are shown in Table 29.6. It can be observed that all the techniques, except the addition of heat pipes and the combination of PCM with nanoparticles, have already been tested at small scale and estimations of the potential enhancement for real applications have already been done. Moreover, the addition of fins already has commercial applications.

Table 29.6 Technology readiness levels of the different heat transfer enhancement techniques.

	Heat transfer enhancement between the HTF and the PCM				Heat transfer enhancement within the PCM		
	Addition of extended surfaces		Forced movement of the PCM while undergoing phase change	Multiple PCM configuration	Combination of PCM with metal foams and porous materials	PCM and graphite composites	Combination of PCM with nanoparticles
	Fins	Heat Pipes					
TRL 1							
TRL 2		X					
TRL 3							X
TRL 4			X		X	X	
TRL 5							
TRL 6				X			
TRL 7							
TRL 8							
TRL 9	X						

Acknowledgments

The work is partially funded by the Spanish government (ENE2015-64117-C5-1-R and ENE2015-64117-C5-3-R). This project has received funding from the European Commission Seventh Framework Programme (FP/2007–2013) under Grant agreement N°PIRSES-GA-2013-610692 (INNOSTORAGE) and from the European Union's Horizon 2020 research and innovation programme under grant agreement No 657466 (INPATH-TES). The authors would like to thank the Catalan Government for the quality accreditation given to their research group GREA (2014 SGR 123). Jaume Gasia would like to thank the Departament d'Universitats, Recerca i Societat de la Informació de la Generalitat de Catalunya for his research fellowship (2016FI_B 00047). Laia Miró would like to thank the Spanish Government for her research fellowship (BES-2012-051861). Alvaro de Gracia would like to thank Ministerio de Economia y Competitividad de España for Grant Juan de la Cierva FJCI-2014-19940.

References

[1] Cabeza, L.F., Castell, A., Barreneche, C., et al. (2011). Materials used as PCM in thermal energy storage in buildings: A review. *Renewable and Sustainable Energy Reviews* 15 (3): 1675–1695.

[2] Chen, J., Yang, D., Jiang, J., et al. (2014). Research progress of Phase Change Materials (PCMs) embedded with metal foam (a review). *Procedia Materials Science* 4: 389–394.

[3] European Commission. (2014). *Horizon 2020 – WORK Programme 2014 – 2015. General annexes revised.* [online] https://ec.europa.eu/research/participants/portal/doc/call/h2020/common/1617621-part_19_general_annexes_v.2.0_en.pdf (accessed 4 April 2016).

[4] Gasia, J., Miró, L., and Cabeza, L.F. (2016). Materials and system requirements of high temperature thermal energy storage systems: A review. Part 2: Thermal conductivity enhancement techniques. *Renewable and Sustainable Energy Reviews* 60 (July): 1584–1601 https://doi.org/10.1016/j.rser.2016.03.019

[5] Gasia, J., Tay, N.H.S., Belusko, M., et al. (2016). Experimental investigation of the effect of dynamic melting in a cylindrical shell-and-tube heat exchanger using water as PCM. *Applied Energy* 185 (1): 136–145. https://doi.org/10.1016/j.apenergy.2016.10.042.

[6] Gil, A., Oró, E., Miró, L., et al. (2014). Experimental analysis of hydroquinone used as phase change material (PCM) to be applied in solar cooling refrigeration. *International Journal of Refrigeration* 39: 95–103.

[7] Guo, C. and Zhang, W. (2008). Numerical simulation and parametric study on new type of high temperature latent heat thermal energy storage system. *Energy Conversion and Management* 49 (5): 919–927.

[8] Hosseinizadeh, S., Tan, F., and Moosania, S. (2011). Experimental and numerical studies on performance of PCM-based heat sink with different configurations of internal fins. *Applied Thermal Engineering* 31 (17–18): 3827–3838.

[9] Khaled, A. (2007). Heat transfer enhancement in hairy fin systems. *Applied Thermal Engineering* 27 (1): 250–257.

[10] Khodadadi, J., Fan, L., and Babaei, H. (2013). Thermal conductivity enhancement of nanostructure-based colloidal suspensions utilized as phase change materials for thermal energy storage: A review. *Renewable and Sustainable Energy Reviews* 24: 418–444.

[11] Laing, D., Bauer, T., Steinmann, W.D., et al. (2009). Advanced high temperature latent heat storage system – Design and test results. *The 11th International Conference on Thermal Energy Storage – Effstock*, Stockholm.

[12] Li, Z. and Wu, Z. (2014). Numerical study on the thermal behavior of phase change materials (PCMs) embedded in porous metal matrix. *Solar Energy* 99: 172–184.

[13] Mehling, H. and Cabeza, L. (2008). *Heat and Cold Storage with PCM*. Berlin: Springer.

[14] Michels, H. and Pitz-Paal, R. (2007). Cascaded latent heat storage for parabolic trough solar power plants. *Solar Energy* 81 (6): 829–837.

[15] Nithyanandam, K. and Pitchumani, R. (2011). Analysis and optimization of a latent thermal energy storage system with embedded heat pipes. *International Journal of Heat and Mass Transfer* 54 (21–22): 4596–4610.

[16] Peiró, G., Gasia, J., Miró, L., et al. (2015). Experimental evaluation at pilot plant scale of multiple PCMs (cascaded) vs. single PCM configuration for thermal energy storage. *Renewable Energy* 83: 729–736.

[17] Pointner, H., Steinmann, W., and Eck, M. (2014). Introduction of the PCM flux concept for latent heat storage. *Energy Procedia* 57: 643–652.

[18] Robak, C., Bergman, T., and Faghri, A. (2011). Economic evaluation of latent heat thermal energy storage using embedded thermosyphons for concentrating solar power applications. *Solar Energy* 85 (10): 2461–2473.

[19] Seeniraj, R., Velraj, R., and Lakshmi Narasimhan, N. (2002). Thermal analysis of a finned-tube LHTS module for a solar dynamic power system. *Heat and Mass Transfer* 38 (4–5): 409–417.

[20] Shabgard, H., Bergman, T., Sharifi, N., et al. (2010). High temperature latent heat thermal energy storage using heat pipes. *International Journal of Heat and Mass Transfer* 53 (15–16): 2979–2988.

[21] Tao, Y., He, Y., and Qu, Z. (2012). Numerical study on performance of molten salt phase change thermal energy storage system with enhanced tubes. *Solar Energy* 86 (5): 1155–1163.

[22] Tay, N., Bruno, F., and Belusko, M. (2013). Experimental investigation of dynamic melting in a tube-in-tank PCM system. *Applied Energy* 104: 137–148.

[23] Yang, J., Yang, L., Xu, C., et al. (2015). Numerical analysis on thermal behavior of solid–liquid phase change within copper foam with varying porosity. *International Journal of Heat and Mass Transfer* 84: 1008–1018.

[24] Zhang, P., Xiao, X., Meng, Z., et al. (2015). Heat transfer characteristics of a molten-salt thermal energy storage unit with and without heat transfer enhancement. *Applied Energy* 137: 758–772.

[25] Zhang, Y. and Faghri, A. (1996). Heat transfer enhancement in latent heat thermal energy storage system by using the internally finned tube. *International Journal of Heat and Mass Transfer* 39 (15): 3165–3173.

[26] Zipf, V., Neuhäuser, A., Willert, D., et al. (2013). High temperature latent heat storage with a screw heat exchanger: Design of prototype. *Applied Energy* 109: 462–469.

30

Reactor Design for Thermochemical Energy Storage Systems

Wim Van Helden

AEE INTEC, Feldgasse 19, 8200 Gleisdorf, Austria

30.1 Requirements for TCM Reactors

The functionality of thermal storage devices in general is to store heat at a certain quality for a given time. The quality of the heat stored is either determined by the temperature level in the case of sensible or latent heat storage or by the exergy of the heat stored in case of thermochemical storage.

The storage process can be subdivided into three phases: the charging, the storage period, and the discharging. As the thermochemical process is characterized by a reaction between two substances, for instance, the sorbate and the sorbent, the process involves both heat and mass transfer and is characterized by two temperatures at charging and two temperatures at discharging. In recent publications, these four temperatures were proposed to serve as anchors for a proper comparison of different thermal storage concepts for a given application. In theory, the thermochemical storage reactions can take place between two or more components of any physical phase. In this chapter, we will only look at solid–vapor and liquid–vapor reactions, and limit the vapor to water vapor, although reactions with other vapors and the same solids are also studied in practice, for instance with vapors of ethanol, methanol, or ammonia.

A division of TCM systems is into open and closed systems. In the first, there is no open connection to the ambient and only heat is exchanged between the system and its environment. The second has an open connection to the ambient, with which mass and possibly also heat is exchanged. See Figure 30.1.

This chapter first describes the physical processes taking place at the charging and discharging phases. These processes determine the design challenges for a thermochemical reactor. In the second part the solutions or solution directions to these design challenges will be given along a subdivision of reactor types and a number of examples from recent research and development projects.

30.2 Charging and Discharging Processes in TCM Reactors

Thermochemical storage reactions are composed of a number of physical processes. In the phase of discharging the storage, two substances have to be brought into contact with each other on molecular or nano-pores scale. For the combination

Advances in Energy Storage: Latest Developments from R&D to the Market, First Edition.
Edited by Andreas Hauer.
© 2022 John Wiley & Sons, Ltd. Published 2022 by John Wiley & Sons, Ltd.

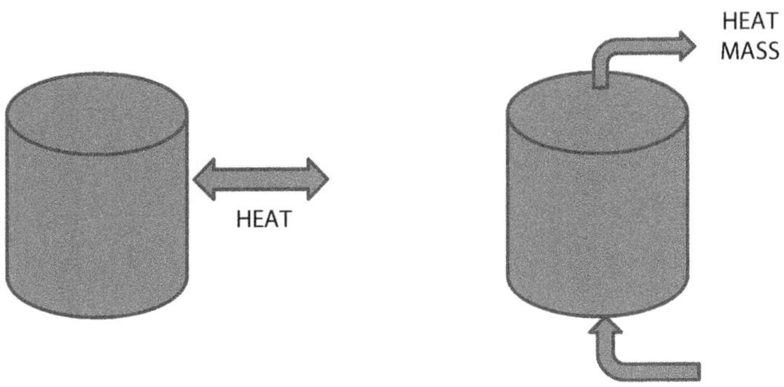

Figure 30.1 Schematic representation of closed systems (left) and open systems (right).

liquid–vapor this means surface interaction, for solid–vapor it means mass transfer of the vapor into the nano-pores of the material. When the two come into contact, thermal energy is generated. This energy has to be transported out of the material to a heat exchanger, through the solid or the liquid and through the vapor phase and then taken up by the heat exchanger. In the charging process, the path is taken along the reverse direction, starting with heat transfer from heat exchanger to the nano-pores level, producing the vapor phase and transportation of this vapor out of the pores to the condenser. The design challenge is to minimize both the heat transfer and mass transfer losses. In Section 30.2.1 these transport mechanisms will be studied in more detail.

30.2.1 Heat Transfer

30.2.1.1 Heat Transfer in the Storage Material

Most thermochemical storage materials are bad to very bad heat conductors. The heat conductivity of the materials is composed of the intrinsic material heat conductivity and the porosity of the material. The first cannot be influenced, while the second can be improved by compacting the storage material, reducing also the possibilities for mass transfer in the compacted elements. Another way that is being investigated to increase the thermal conductivity is by adding quantities of a well conducting material, like metals or carbon. Although the materials' properties improvement is not directly connected to the design of TCM reactors, the material elemental shape does have influence on the vapor transport in the bulk, on the heat transfer between adjacent material elements, and on the contact between element and the heat transfer component. Therefore the elemental shape of the storage material is a design parameter of the reactor.

The material element shape determines also the fill factor of the material in the reactor volume. For instance, if the material elements are perfect spheres with identical diameter stacked in hexagonal way, there will be void space in the reactor volume of 29%. This can be reduced somewhat further by using smaller spheres to fill the voids partially, but at the same time the vapor flow will have more difficulties in permeating through the bulk.

30.2.1.2 Heat Transfer in the Vapor

During charging and discharging in a closed system, the heat transfer in the vapor is almost completely determined by the vapor transport, the so-called convective heat transfer. During the storage period, with the vapor still in the reactor, the heat transfer is determined by the conductivity of the vapor/TCM ensemble.

For thermochemical reactions at temperatures below 200°C the water vapor pressures are low, typically 5 kPa and lower. In a storage reactor with only the active vapor present, pressures will be also low and vapor transport is determined by the low absolute pressure and consequently also by small pressure differences. The flow and mass transfer consequences of this are discussed further later.

In open thermochemical storage systems, the vapor is mixed with air or another carrier gas. These systems generally have atmospheric or higher pressures but the partial pressure of the vapor is low. With the carrier gas the convection can be increased and the transport of the vapor to the storage material made more effective through the reactor geometry. For an optimal thermochemical reaction, a sufficiently long contact time between vapor and storage material should be guaranteed. In practice, this is mostly done in fluidized bed reactors or in moving bed reactors, which will be described later.

30.2.1.3 Heat Transfer in the Heat Exchanger

In an ideal case, the heat exchanger geometry should be such that all individual storage material elements have direct contact to the heat exchanger. This, however, would lead to a large amount of heat exchanger material used, to possible pressure losses in the vapor flow and to reduced heat exchanger efficiency because of the high heat capacity. An optimum has to be found between the average distance from material elements to the heat exchanger surface. The heat exchanger material has to be compatible with the TCM. Salt hydrates can be corrosive and therefore the traditional metals used for heat exchangers should be coated in cases of some salts. An alternative is to use polymeric materials for the heat exchangers, with lower efficiencies as a consequence. The highest performance is given with copper or aluminium heat exchangers, those being also the materials most often applied in practical applications of heat exchangers, like in industrial processes, automotive, and refrigeration.

From a geometrical point of view, standard shapes for heat exchangers are used, next to tailor-made shapes. Most common are the fin and tube heat exchangers for solid thermochemical materials and the plate heat exchanger or the tube bundle heat exchanger for liquid TCM. Fin and tube heat exchangers are widely applied for air-to-liquid cases, with fin spacing in the order of a few millimetres or less. Heat exchanger mass production technologies are not yet fit for the larger fin spacing that is required for thermochemical storage applications.

30.2.2 Mass Transfer

30.2.2.1 Kinetics in the Storage Material

Within the thermochemical storage material, the type of material determines the speed at which enthalpy is released. Especially with salt hydrates with a series of sequential hydration steps, there can be a variation of enthalpy release as a function of time in the material leading to a variation in power to the heat exchanger. When this variation is very pronounced, the heat exchanger geometry can be altered in a way to promote smoothing of the power release over time. This process is, however, also influenced by the vapor transport in the reactor and only if a validated numerical simulation of the processes in the reactor is available can the power provision of the reactor over time be optimized with heat exchanger geometry. This is a field still in its early development.

30.2.2.2 Vapor Flow in the Bulk Material

Vapor has to be transported between the storage material and the evaporator/condenser. The vapor flow will be in pure form in case of an evacuated system and in other cases the vapor will be transported by the carrier gas. The shape of the material elements and the amount of material in the reactor and the shape of the reactor determine what the pressure drop will be. In practice, the equivalent porosity of the material is calculated to determine what the hydrodynamic pressure loss will be.

For carrier-gas vapor mixtures, the pressure drop can be compensated by the fan power and will be only critical for very dense or very large beds of material.

In the case of evacuated systems, the pressure losses have to be minimized, as the total pressure difference between the evaporator/condenser and the storage material is causing an additional temperature difference due to the different saturation pressures. For instance with water vapor, every mBar of pressure difference leads to an additional temperature difference of about 1 K. When designing the reactor, the pressure losses can be minimized by having short distances in the storage material bed and by choosing large diameter valves and vapor channels between the components in the storage system.

As already mentioned, typical pressures for closed, low to medium temperature TCM storage systems are between 1 and 5 kPa. Although the corresponding pressures are not yet high vacuum, the flow of the vapor will be hindered when other gases are present in the reactor. These other gases spring either from out gassing of the storage material or from leakages in the system. As the presence of these gases can seriously affect the performance of the storage reactor and the evaporator/condenser, they should be prevented or removed as effectively as possible.

30.2.2.3 Evaporation and Condensation

When atmospheric pressures are used in the system, standard solutions for the evaporator/condenser can be used, like a vertical plate heat exchanger or a horizontal tube bundle heat exchanger. Evaporation requires, wetting of the heat exchanger surface by film or droplets falling down by gravitation or by using hydrophilic sheets on, or as surfaces of, the heat exchanger.

At low pressures, for closed TCM storages, the vapor transport is pressure driven and much less effective as the convection driven transport. This leads to much higher demands on the efficiency of especially the evaporator [1]. As the efficiency of the complete storage system is highly dependent on the total pressure drop of the vapor between the storage material and the evaporator/condenser, the efficiency is helped when the temperature drop from heat exchanger to vapor in the evaporator is minimized. A perfect situation would be when the total surface of the heat exchanger is covered with a thin film of liquid sorbate, minimising the temperature drop over the film and maximising the vapor production rate. In practice, this complete wetting under low pressure conditions is hard to realize. Investigations are ongoing to optimize the geometry of the evaporator, in combination with methods to increase the surface wetting, like surface roughening or grooved surfaces [2].

30.2.3 Pressure

With low pressure systems, there are extra demands on the reactor vessel and other components regarding mechanical strength and vacuum tightness. The consequences of mechanical strength demands are that in practice the reactor vessel will have cylindrical shapes, in order to minimize cost with maximum strength. Vacuum tightness demands a minimal number of flange connections and good quality of other connections, like welding or brazing. Leaking in of air, together with non-condensable gases can form a serious problem in functioning of the storage. For adsorption, CO_2 in the air will be adsorbed by the material and can lead to partial blocking of the sorption material and reduction of the capacity and power of the storage.

30.3 Types of Reactors and Examples of Design Solutions

In this section a number of examples from realized or ongoing thermochemical storage reactor designs is discussed and the design challenges and solutions described.

30.3.1 Fixed Bed

In a fixed bed reactor, the storage material lies still in the reactor volume, and the vapor has to be transported through the volume either by a carrier gas like air in open systems or by pressure differences in closed evacuated systems. Due to the low thermal conductivity of most TCM the optimization of heat input and output has to be guaranteed with a proper design of the heat exchanger in the bed.

In the framework of the development line A in the COMTES project [3–4] an optimization of the heat exchanger geometry in a cylindrical vessel containing zeolite 13XBF in a closed system was performed, see Figure 30.2. Numerical simulation of the heat and mass transport in the volume, reduced to a two-dimensional problem, served to find an optimum in terms of length of the discharge period with given minimal output power.

It was found that the average distance between heat exchanger plates and the zeolite material was of more influence than the actual geometry of the heat exchanger, and that therefore the geometry could be optimized also in terms of material use and manufacturability. The star-shaped plate and tube heat exchanger was chosen for further prototyping and system experiments in the COMTES development line A.

The heat transfer from and to the active material is less restricting if the vapor is forced through the fixed bed with a carrier gas. In open systems, this carrier gas is air, as for instance in the prototype system built at the University of Lyon, France in the TESS project, see Figure 30.3.

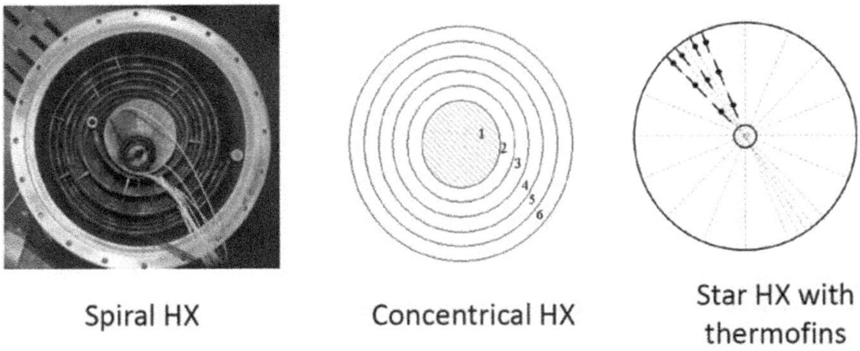

Figure 30.2 Three geometries of the heat exchanger in a cylindrical vessel. Optimization studies for the COMTES project, development line A.

The air plus vapor flow is in vertical direction through the zeolite beds. The heat in the air flow is exchanged with external air to liquid heat exchangers. The advantage of external heat exchangers is the smaller material use for the heat exchanger, while the disadvantage is the higher auxiliary energy needed for the air flow through the beds.

30.3.2 Fluidized Bed

In a fluidized bed, the solid particles are held in motion by a gas flow. The turbulence created leads to a very effective heat and mass transfer between the solid and the gas. The challenges with a fluidized bed are the transport of the solid material through the system and the auxiliary power requirements for the carrier gas flow. A comparison on reactor level between fluidized bed reactor and two other types was made in [5] in which the fluidized bed reactor was not found the best candidate. So far, no prototypes of TCM fluidized bed reactors at low temperatures are known.

Figure 30.3 Two zeolite fixed bed reactor in the TESS project, University of Lyon, France.

30.3.3 Moving Bed

With the moving bed principle, the active material is transported through the reactor, in which the vapor and heat exchange take place. The very big advantage of this principle is that the storage system capacity is uncoupled from the power of the system. The reactor can be made to fit the power requirements, while the amount of material stored in a separate bulk container determines the total capacity of the system. This leads to lower costs on both the side of the reactor and of the bulk storage. The challenge in these systems is the transport of the active material to and in the reactor.

In the SOTHERCO project, a reactor was developed in which the storage material particles are transported through mechanical vibration of the disc-shaped reactor floor, see Figure 30.4. The vibrations cause the particles to move from the centre to the edge, where they are picked up and transported to a subsequent reactor. Air with water vapor is moved upward through the perforated disc and the vapor is effectively exchanged to the storage material.

Another way to transport the storage material through the reactor is developed in the German project ENERCHEM. Here, the grains of zeolite or of a salt hydrate–zeolite composite material are slowly transported down a heat exchanger passing a cross-flow of air, see Figure 30.5. Depicted are the flows in case of discharging the storage material, with the incoming air at a given relative humidity, for instance coming from the return air of a domestic ventilation system. The rotation of a transport wheel at the bottom of the heat exchanger controls the mass flow of the storage material. It is a two-stage heat transfer reaction which showed a much better mass exchange efficiency than a single stage concept. After the second stage, the outgoing air flow immediately passes an air-liquid heat exchanger.

30.3.4 Liquid Flow

In absorption processes two liquids mix and release heat. When prior to the mixing one of the fluids is brought to the process in vapor phase, the heat of condensation is

Figure 30.4 Disc reactor developed and tested in the SOTHERCO project.

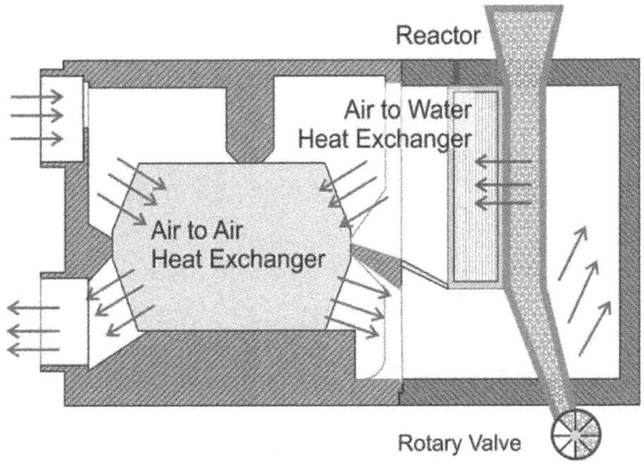

Figure 30.5 Cross section of the test set up in the ENERCHEM project, with at the right the cross-flow air to solid grains heat exchanger. The grains are slowly transported down the heat exchanger by the transport wheel at the bottom.

added to the heat generated, increasing the total storage enthalpy. This is the most commonly used principle for absorption for thermal storage, defining the design targets for a reactor for absorption: bringing together a liquid and a vapor effectively and extract the generated heat. The absorption of the vapor can only take place at the surface of the liquid. Best performance of the reactor can be achieved when the heat exchanger surface is fully covered with a thin film of liquid. This is where the technical challenge is.

In the COMTES development line B, an absorption reactor, consisting of an absorber/desorber unit, was developed [6]. It consists of a horizontal tube bundle heat exchanger with on top a unit that distributes droplets of the liquid, sodium lye, on the top tubes. The liquid then trickles down and passes all tubes underneath. The experiments showed that the droplets not evenly distributed over the surface, but formed wet channels and left large part of the tubes not wetted. Consequently, the measured performance was less than the design value. Solutions to a better wetting were found in surface roughening of the tubes and in adding surfactants to the liquid.

Another challenge for absorber/desorber design is the changing viscosity of the sodium hydroxide with changing concentration. This leads to a non-linear mass flow and thus non-linear heat transfer over the height of the heat exchanger. Solutions to this still need to be developed.

30.4 Conclusions and Outlook

The simultaneous heat and mass transfer requirements for thermochemical reactors pose serious design challenges. Heat transfer in general is difficult because of the intrinsic bad heat conductivity of thermochemical materials. And the transport of mass, in practice mostly water vapor, is limited by a number of parameters in the processes, like material element size and shape, reactor geometry, and heat exchanger possibilities.

There is a growing group of researchers working on the design and development of thermochemical reactors, like fixed bed and moving bed reactors for solid–vapor systems and horizontal tube bundles for absorption systems. First solutions to better performance of the reactors have been found, but more work is needed to arrive at better solutions that also have promising possibilities for the cost effectiveness of the process.

References

[1] Dang, B.N., Van Helden, W., and Luke, A. (October 2017). Investigation of water evaporation for closed sorption storage systems. *Energy Procedia* 135: 504–512.

[2] Dudita, M., Daguenet-Frick, X., and Gantenbein, P. (2018). Closed sorption seasonal thermal energy storage with aqueous sodium hydroxide. *Conference on Sustainable Energy*, doi: 10.1007/978-3-319-63215-5_18

[3] Fumey, B., Weber, R., Gantenbein, P., et al. (2014). Closed sorption heat storage based on aqueous sodium hydroxide. *Energy Procedia* 48: 337–346.

[4] Köll, R., Van Helden, W., Engel, G., et al. (2017). An experimental investigation of a realistic-scale seasonal solar adsorption storage system for buildings. *Solar Energy* 155: 388–397. doi: 10.1016/j.solener.2017.06.043.

[5] Engel, G., Asenbeck, S., Köll, R., et al. (in preparation). Simulation of a seasonal, solar-driven sorption storage heating system. *Energy Storage*, 13: 40-47. October 2017. doi: 10.1016/j.est.2017.06.001

[6] Zondag, H.A., Van Essen, V.M., Schuitema, R., et al. (2009). Engineering assessment of reactor designs for thermochemical storage of solar heat. *Effstock 2009 - Thermal Energy Storage for Energy Efficiency and Sustainability*, Stockholm, Sweden (14–17 June 2009).

31

Phase Change Materials in Buildings – State of the Art

Thomas Haussmann, Tabea Obergfell, and Stefan Gschwander

Fraunhofer ISE, Freiburg, Switzerland

31.1 Introduction

For years, latent heat storage systems have been researched with the aim of being able to air-condition buildings more energy-efficiently or to improve user comfort without active air conditioning. Latent heat storage materials (PCM) exploit the effect of almost isothermal latent heat storage when melting and crystallizing a material in order to keep temperatures in the building or technical application more constant. This can reduce overheating in the building in summer or prevent cooling. Phase transition solid–liquid is used typically, sometimes also solid–solid. Although the phase tranition solid–gaseaous seems to be advantegous in terms of enthalpy, it is technically very difficult to handle the increase of volume or pressure. The first steps in using PCMs in buildings were taken as early as 1948 with the Dover Sun House, where solar heat was stored with Glauber salt as a PCM in order to heat the building during the night.

In the following decades, latent heat storage in buildings was again and again the main focus of research. At the end of the last millennium, the development of PCM applications in buildings once again came into focus in order to achieve climate policy goals. In the European Union (EU), existing buildings account for about 40% of energy use. A more efficient use of energy, the increased use of renewable energies, and new developments in the field of PCM have made new development approaches for the building sector possible. PCMs are mainly used for applications in the field of thermal comfort insulation, i.e. the avoidance or reduction of overheating in buildings. A smaller part of the applications focuses on the provision of heat, which can also be provided via PCM storage tanks.

The possible uses of PCM in buildings are manifold and are not uniformly grouped in the literature. The main groups are active, passive, centralized, and decentralized applications as shown in Table 31.3. There are also different definitions for the definition of active and passive applications. We use the following definitions in this chapter.

Passive cooling uses ambient cold or heat that is stored in a PCM in order to be able to retrieve it when required. A special feature is that no cold or heat generation takes place. Energy is only used to transport the heat or cold. Night ventilation concepts are a frequently implemented system in combination with PCM. The systems for implementing the concepts are manifold, for example by simply opening windows or by complex ventilation systems. The use of ground cooling in combination with cooling ceilings is another example. However, such passive concepts are essentially dependent on the availability of natural cold or heat. If this is not sufficiently available, the

Advances in Energy Storage: Latest Developments from R&D to the Market, First Edition.
Edited by Andreas Hauer.
© 2022 John Wiley & Sons, Ltd. Published 2022 by John Wiley & Sons, Ltd.

functionality is no longer given. Energetically passive applications are therefore always the most favorable solution, since only energy must be used for the transport. However, the lack of planning possibilities for passive systems is currently preventing a further spread of passive concepts.

Active systems are based on the demand-driven generation of cold. This can lead to high loads in heating and cooling supply systems as well as in electrical supply networks, especially in refrigeration. In such applications, PCMs can provide a temporal decoupling of demand from generation by means of peak shifting. Newer approaches to sector coupling, e.g. power to heat, make targeted use of these decoupling possibilities. An additional effect made possible by peak shifiting is the more efficient generation of heat or cold by operating the systems when the boundary conditions are favorable, e.g. compression refrigeration systems at night or air heat pumps during the day. An alternative advantage of PCM in building applications is peak shaving. Storage tanks can be charged with the necessary energy over longer periods of time, meaning that systems no longer have to be designed for peak loads. These can be covered by the stored energy in the PCM. This means that such systems can be made smaller, which offers cost advantages. In addition, by accumulating over longer periods of time, energy resources can be used which typically provide only low power or whose power is cost-intensive. Especially renewable energies can be better integrated into supply systems.

The integration of PCM in applications always requires encapsulation. In the past, there were frequent attempts to impregnate materials with PCM, but here the integrated PCM led to material incompatibilities, which could also have effects on the mechanical stability, or could escape during the usage phase. Depending on the capsule size, the encapsulation technologies are divided into micro, meso, and macro capsules. While microencapsulation is a new development of the last 20 years, meso and macro capsules have been used for a longer time. A frequent disadvantage of macrocapsules is the risk of damage from the outside, as well as internal effects such as corrosion or segregation of PCM mixtures. Microcapsules, on the other hand, are located in the range of a few nanometers in size and are therefore often too small for mechanical damage, but at the same time have a very large surface area for heat exchange. Microcapsules have therefore found their way into a wide variety of building materials, especially for passive cooling, and are currently often the most stable solution over the long term, shown within the project report of the PCM Eval research project focussing on the evalution of the performance of real used building equipped with PCM products after more than a decade of operation [1]. Technical solutions for the microencapsulation of salt hydrates are either not sufficiently stable in the long term or cannot be produced economically enough at this point in time.

In addition to paraffins, the material class of salt hydrates is also predominantly used for applications in buildings. With a higher storage density and non-flammability, salt hydrates offer significant advantages over paraffins. Due to their high corrosiveness and hydrophilic properties, however, the encapsulation materials have to meet high requirements.

The use of PCM in buildings is often still considered to be close to research. However, the number of buildings equipped with PCM systems or building materials is constantly increasing, as is the number of products and systems with integrated PCM. In Germany alone, approx. 100 projects have been realized, with a large number of buildings currently unknown to the authors, so that a multiple can be assumed. Worldwide, for example, a Google Maps Overlay on the website of the RAL Gütegemeinschaft PCM [2] shows a multitude of realized objects. Here, too, only those manufacturers are considered who are also part of the quality association.

Obstacles to the spread of PCM systems in buildings are the high price, the lack of design aids, the lack of planning security, the lack of funding opportunities, and the lack of knowledge about long-term stability. In the past, a lack of design and planning aids has repeatedly led to incorrect system planning, but often also unpredictable user behavior led to the malfunction of the planned system. PCM systems are very sensitive to a precise design, since the effectiveness is only given in a small temperature range, the melting temperature range. The German standard VDI 2164 was therefore developed to provide necessary support for the most frequently implemented PCM systems. At the national level in the Netherlands, similar results are achieved with ISSO 111. Precise material properties are essential for system planning and development. For this reason the RAL Quality Association PCM GZ896 was founded in 2004. Here detailed regulations were developed for the determination of the enthalpy temperature curve, cycle stability, and thermal conductivity, and also the data presentation. A similar approach is followed in ASTM C1784 especially for the use of a HeatFlowMeter for material characterization.

31.2 Materials

The use of PCMs as thermal mass is widely discussed and it is obvious that PCMs used in the building sector should have melting and crystallization temperatures in the temperature range in which buildings are operated. They should provide a high melting enthalpy and they should offer charging and discharging performances which are in the required range of the application. The required properties can be broken down into thermophysical, chemical, economic, and environmental. Table 31.1 gives an overview of these requirements.

Table 31.1 Overview on required properties for PCMs applied to buildings.

Thermophysical properties	suitable phase transition temperatures high melting enthalpy less supercooling high specific heat high density small volume expansion during phase transition fast crystallization rate
Chemical properties	thermal stability, no degradation over time nontoxic nonflammable non corrosive
Economic porperties	cost efficient available
Environmental properties	low embodied energy high recycle potential no or minor environmental impact

Table 31.2 Pros and cons for the PCM classes.

Material Class	Advantages	Disadvantages
Organics	Large number of materials with different melting temperatures are available Reasonable melting enthalpy minor supercooling fast crystallization rate almost no segregation	Minor thermal conductivity low density flammable sufficient volume change more expensive
Inorganics	many materials available that show congruent melting thermally stable high melting enthalpy low cost high density small volume change with phase transition	corrosive supercooling change of properties caused by hydration/water content is possible during thermal cycling phase separation is possible can be toxic

For the building sector there are mainly organic and inorganic PCMs in use. Paraffins and fatty acids are the main materials that come in operation for the organic class and hydrated salts and there eutectic mixtures are the most used inorganics. A eutectic mixture shows the minimum melting temperature of two or more components and most valuable eutectics melt congruently [3]. Mixtures are possible of only organic or inorganic PCMs as well as mixtures of both classes. Both classes, the organic and inorganic, have benefits and drawbacks which are listed in Table 31.2. A high number of different organics are available so it´s easier to find materials within the required melting range. Many organic materials show a high melting enthalpy but with densities almost below 1 kg/L their storage density is lower than for salt hydrate which offer densities higher than 1 kg/L, some materials show densities up to 2 kg/L. Flammability is also a drawback if organics are used compared to hydrated salts especially if they should be integrated into buildings. Especially paraffins are very stable PCM since they almost do not react with other materials. Figures 31.1 and Figure 31.2 depict some examples of organic, inorganic, and eutectic materials which show melting temperatures appropriate for buildings.

31.2.1 Overview Encapsulation Technologies

As it is hardly possible to incorporate PCMs directly into buildings it is necessary to choose technologies that prevent the PCM from leaking out of the building component or structure in which it should be integrated. Encapsulation is the most used method, others are shape stabilization or to use solid–solid PCMs. The most used technology commercially is encapsulation. Encapsulations should be strong enough to withstand mechanical and thermal stress and they should be resistant to corrosion and other interaction with the PCM and the surrounding [4]. The encapsulation material should provide a good barrier to prevent the PCM and the environment from harmful interaction,

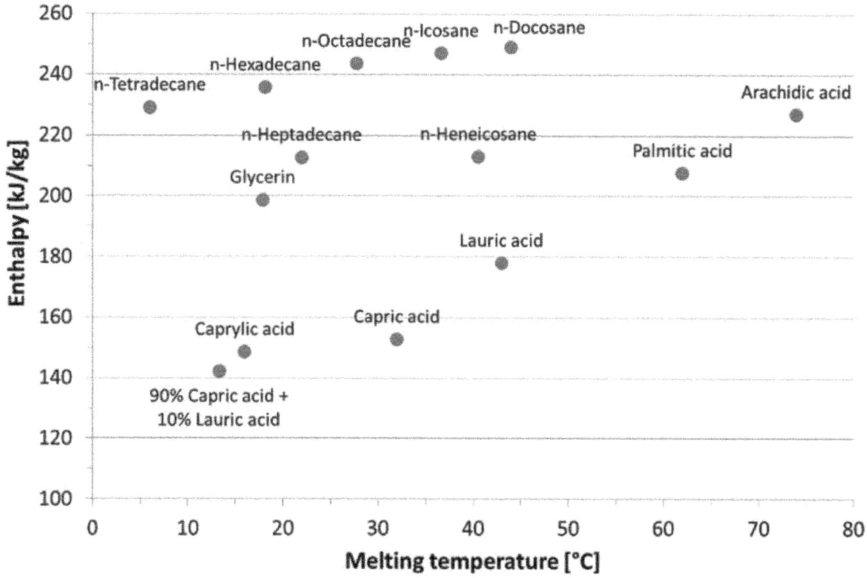

Figure 31.1 Overview on some organic PCM.

should offer a large surface for high heat transfer rate and should be easy to handle on the construction site.

The most used encapsulation technologies are macro- or microencapsulation. So far only organic PCMs are commercial available as microencapsulated PCM since the classic and most cost efficient manufacturing processes (coacervation and interfacial polymerization) are water based and are good to handle with

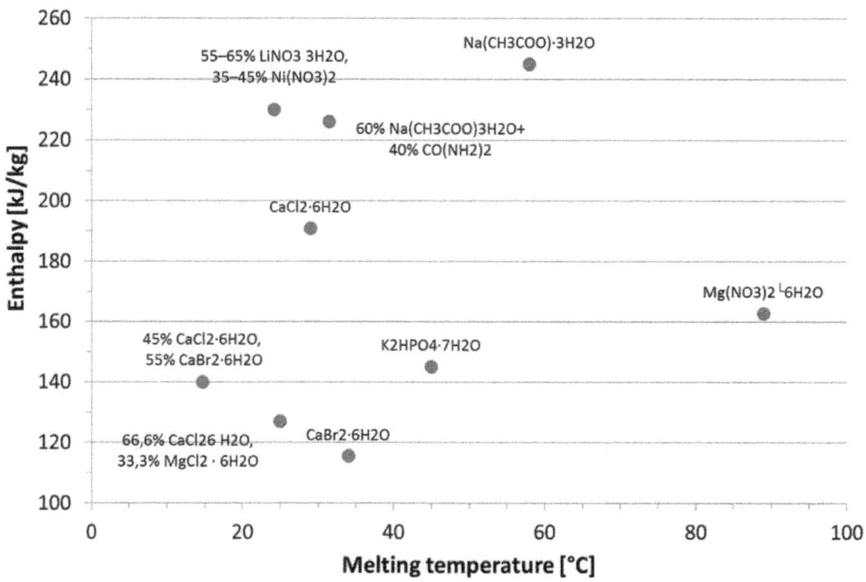

Figure 31.2 Overview on some salt hydrates and eutectics.

Table 31.3 Matrix of Applications.

	Passive	active
Decentral	Construction materials, example Section 31.3.1	Chilled ceilings, example Section 31.3.3
Central	Central Ventilation System, example Section 31.3.2	Central cold and heat storages, example Section 31.3.4 and Section 31.3.5

hydrophobic core materials. Due to this it is very difficult to microencapsulate hydrated salts using these processes. Even if this can be done successfully the wall thicknesses of the organic shell materials is very often too untight for water vapor which diffuses through the nano scaled wall. Additionally hydrated salts show a strong influence of volume on supercooling, small volumes cause massive supercooling. So there are much more challenges to microencapsulate hydrated salts than for organic materials. In the application microencapsulated PCM provides a huge surface to volume ratio which leads to a good heat transfer between the PCM and the surrounding.

Macro encapsulation is also a very common method to bring PCM into applications. Basically there are metallic and polymeric encapsulations available. Shapes can be like tubes, spheres, panels, ashlar-formed, and others. The shape of the container should provide a good surface to volume ratio to reach high heat transfer rates. Metallic encapsulations are very often used to encapsulate paraffin since it is very capable to creep through nonpolar polymeric wall materials. Additionally compared to polymers metals provide a high thermal conductivity which is positive for the heat transfer. It is also possible to embed a heat conduction structure inside the capsules to increase heat transport out of the PCM. Polymeric encapsulations are very suitable for hydrated salts as there is no issue with corrosion but the wall has to provide a barrier against water migration.

31.2.2 Overview on Products

In the last 20 years of PCM research many PCM products have been developed. However, the PCM market is subject to constant change, so that some products have been only available for a short period of time, have been withdrawn from the market, or are only available on specific customer request. Table 31.4 gives an incomplete overview of PCM products from different manufacturers.

31.3 Example of Building Integration of PCM

In this chapter we would like to present some exemplary applications which, however, can only represent a fraction of the possibilities. As already shown, we group the applications according to the matrix in Table 31.3. However, an unambiguous assignment is sometimes difficult because active and passive functions are often combined.

Table 31.4 Overview of manufacturers of PCM products ready for building applications.

Overview of manufacturers of PCM products ready for building applications								
Source	**Manufacturer**	**Product**	**Phase change Temperature(°C)**	**Latent heat Capacity(kJ/kg)**	**PCM Material**	**Type[a]**	**Building Product**	**Availability**
[A]	DuPont	EnErgain	18–24	515kJ/m^2	Paraffin wax compound	Organic	Thermal mass panels	No longer available
[A]	Knauf	Comfort board			BASF micronal	Organic	Gypsum board with microcapsules of PCM	All no longer available
		Smartboard 23	23	200kJ/m^2	BASF micronal	Organic		
		Smartboard 26	26	330kJ/m^2	BASF micronal	Organic		
[A]	RGEES	PB29p	29	33kWh/m^3		Organic	Spheres 75 mm in diameter filled with PCM	No longer available
		PB22p	22	0.1 kWh/ft^2		Inorganic	Tubes filled with PCM	No longer available
		PB24p	24	0.1 kWh/ft^2		Inorganic	Tubes filled with PCM	No longer available
		PB29p	29	0.1 kWh/ft^2		Inorganic	Tubes filled with PCM	No longer available
[A]	Phase Change Energy Solutions	BioPCM at M27	23		Bio-based	Organic	Mats filled with PCM	X
		BioPCM at M51	25		Bio-based	Organic	Mats filled with PCM	X
		BioPCM at M91	27		Bio-based	Organic	Mats filled with PCM	X
		Thermastix			Bio-based	Organic	Stick filled with PCM	X
[A]	Dorken	Delta-cool 24	24		Salt hydrate	Inorganic		No longer available
		Delta-cool 28	28		Salt hydrate	Inorganic		No longer available
[A]	Salca	K-Block	applicable from 5°C to 90°C		Salt hydrate	Inorganic	Mat filled with PCM	X
[B]		Thermupod	applicable from 5°C to 90°C		Salt hydrate	Inorganic	Coffee pad shaped plastic, filled with PCM, increasing heat storage capacity in buffer vessel	X

(*Continued*)

Table 31.4 (Continued)

Overview of manufacturers of PCM products ready for building applications

Source	Manufacturer	Product	Phase change Temperature(°C)	Latent heat Capacity(kJ/kg)	PCM Material	Type[a]	Building Product	Availability
[A]	SGL group	Ecophit GC20	22	85kJ/kg			graphite building material, optionally with PCM	No longer available
		Ecophit LC20	22	140kJ/kg				No longer available
[A]	National gypsum	ThermalCORE			BASF micronal	Organic	Gypsum board with microcapsules of PCM covered with a fibreglass mat	No longer available
[A]	H+H Deutschland BASF micronal	CelBLoc Plus			BASF micronal	Organic	Poreous Concrete block with microcapsules of PCM	No longer available
[C]	Maxit Deutschland	Maxit clima	24–26		BASF micronal	Organic	Plaster with integrated microcapsules of PCM	No longer available
[D]	Ilkazell Isoliertechink	llkatherm walls, ceilings and sails	23					X
[A]	Monodraught	Coolphase					Low energy ventilation system with PCM as thermal battery	X
[A]	Tate	EcoCore	75.2 °F	147 Btu	Vegetable bio-based	Organic	steel welded shells filled with PCM	X
[E]	Emco Bau-und klimatechink	EmcoCool GK-PCM		200kj/m^2			cooling ceiling	X

[F]	Autarkis	Crystal Beam Module					PCM induction unit	X
		Crystal Climate Cooler					PCM ceiling	X
		Crytal Climate Floor					PCM floor as thermal battery	X
[A]	Armstrong World Industries	Coolzone			BASF micronal	Organic	Chilled roofs system	No longer available
[A]	Trox	FSL-B-PCM					Supply air unit with PCM	No longer available
[A]	Sunamp	Heat batteries			Salt-based	Inorganic	thermal battery filled with PCM combined with PV and heat pump	X
[H]	Global Energy System Europe BV	THERMAVAR PCM climate floor PCM24	17–19 22–23 Applicable from −30°C to +89°C	496 Wh/m^2 396 Wh/m^2	Salt hydrate	Inorganic	HDPE-panel filled with PCM	X
		THERMAVAR PCM climate floor PCM29			Salt hydrate	Inorganic	HDPE-panel filled with PCM	X
		THERMAVAR 18			Salt hydrate	Inorganic	Aluminium ceiling panel	X
		THERMAVAR 22			Salt hydrate	Inorganic	Aluminium ceiling panel	
		GAIA Energy storage ball					High Density Polyethylene, Polypropylene or stainless steel in a spherical shape; diameter 80mm–125mm	X X
		CALDAMEG heating system			Salt hydrate or eutectic solution	Inorganic	CO_2 heat pump combined with PCM buffer	X

(Continued)

Table 31.4 (Continued)

Overview of manufacturers of PCM products ready for building applications

Source	Manufacturer	Product	Phase change Temperature(°C)	Latent heat Capacity(kJ/kg)	PCM Material	Type[a]	Building Product	Availability
[I]	Phase Change Material Products Ltd.	FlatICE					HDPE-container filled with PCM	X
		TubeICE						X
		BallICE					PCM-filled HDPE tube design	No longer available
		ICEpacks					spheres filled with PCM	X
		PCM pouches						
[J]	Lehm Orange	Clay building board			BASF Micronal	Organic	Clay board with microcapsules of PCM	No longer available
[K]	Saint-Gobain Weber	Weber.mur clima			BASF Micronal	Organic	Wall plaster containing microcapsules of PCM	No longer available
		Alba balance			BASF Micronal	Organic	Gypsum board with microcapsules of PCM	X
[L]	Enersolve	Modulo					thermal storage, optional with PCM	X
[M]	Die Latentspeicher Company	PowerTank			Paraffin	Organic	thermal battery filled with PCM	X
[N]	H.M. Heizkörper	Thermal battery			Paraffin	Organic	thermal battery filled with PCM, uses supercooling effect	X
Source	**Manufacturer**	**Product**	**Phase Change Temperature**		**Heat Storage Capacity (Wh/m^2) Tvis solid–liquid (%)**	**Materical**	**type**	**Availability**

[P]	GLASSX	GLASSX crystal	26–30		1185 0–28/4–45	Salt hydrates	Inorganic	X
		GLASSX comfort	26–30		1185 0–38/4–55	Salt hydrates	Inorganic	X
		GLASSX slims	26–30		1185 0–38/4–45	Salt hydrates	Inorganic	No longer available
		GLASSX store	26–30		1185 0–38/4–55	Salt hydrates	Inorganic	X

[a] Many manufacturers do not give a full description of the mixtures used in PCMs. Hence, type has only been limited to organic and inorganic, though some of these maybe eutectic mixtures as well.

[A] Kalnaes, S.E. and Jelle, B.P. (2015). *Phase change materials and products for building applications: A state-of-the-art review and future research opportunities. Energy and Buildings*, 94: 150–176.
[B] http://www.salcabv.nl/index.asp?CategorieID=6&Taal=EN
[C] https://www.badenova.de/mediapool/media/dokumente/unternehmensbereiche_1/stab_1/innovationsfonds/abschlussberichte/2004_4/200416abschlussberichtmaxit.pdf
[D] https://www.ilkazell.de/en/special-links/downloads.html
[E] https://www.emco-klima.com/kompetenzen/produkte/emcocool.html
[F] https://www.orangeclimate.com/products
[G] https://www.sunamp.com/residential
[H] https://www.global-e-systems.com/en/products
[I] www.pcmproducts.net
[J] www.lehmorange.de
[K] http://www.rigips.com/alba.html
[L] https://www.enersolve.de/Modulo-Thermalspeicher.html
[M] http://www.latentspeicher.com/de
[N] https://hm-heizkoerper.de/die-thermobatterie/#section-technologie
[O] https://www.glassx.ch/de/produkte/pcm-fassadenprodukte

31.3.1 Example of Passive Decentral PCM System: Office Building in Tübingen from the 1950s Modernized to a Passive-house Standard [2]

The office building in Figure 31.3 was part of a barracks areal which was listed as a historical monument. In 2003 the building was modernized to the passive-house standard. During the restructuring process the rooftop was completely replaced and converted to workspaces and the ground floor was energetically optimized.

The walls and ceilings of the first floor were equipped with PCM (Micronal) containing Smartboards (Knauf) in order to increase thermal mass of the light-weight construction. Melting range of PCM was between 26°–28°C which is not suitable to stabilize operative room temperature around 26°C. In order to ensure sufficient thermal comfort, a melting range between 24°–26°C would have been a better choice.

In addition the building is equipped with a ventilation system combined with a countercurrent heat exchanger. By an upstream brine–air heat exchanger ambient air can be chilled during summer. Mechanical night ventilation ensures an air flow rate of 4000 m^3/h (which corresponds to an air change rate of 2). The use of jet nozzles which lead incoming air directly under the ceilings improves heat transfer compared to natural night ventilation. Hence, storage mass can be de-heated faster.

Overall performance of the PCM floor is satisfying, nevertheless limitations during hot spells became obvious. [2] noticed that during measuring periods from 2005 to 2006 thermal comfort was very good despite the absence of active air cooling. During several hot weeks, only, in 2006 the PCM couldn't work anymore as it was not possible to discharge it during night due to the high room and ambient temperatures.

31.3.2 Example of Passive Central PCM System: Courtyard Building Düsseldorf

The central objective of the new construction of the District Court in Düsseldorf (completion 2009) was to provide the energy required for the air conditioning, if

Figure 31.3 Outside view of the Tübingen building with passive PCM boards.

possible via regenerative energies or waste heat utilization. Two technologies were used in the building for this purpose: First, an air–earth heat exchanger and secondly a PCM storage tank, both upstream of a conventional air-conditioning system. The air–earth heat echanger was dimensioned in such a way that the entire air volume flow can be tempered, whereas the PCM storage tank only preconditions a partial volume flow due to the high investment costs. Both technologies are connected in series. Figure 31.4 shows a schematic of the preconditioning of the supply air in winter and summer mode.

The aim was to pre-condition the supply air in summer and winter, so that the final air conditioning requires less energy. Cool night air is used as a cold source for summer cooling and waste heat at 40°C from the cooling of the IT systems, which can no longer be used elsewhere in the building directly and would be dissipated to the environment, is used as a heat source for supply air heating. Rubitherm CSM modules filled with SP 21 (Figure 31.5), a salt hydrate and paraffin mixture, was selected as the PCM material. The melting point of 21°C enables both preheating and cooling of the supply air to approx. 21°C with the same material. In the course of testing the PCM storage, was splitted in two parts with SP 21 and SP 25 to better support winter or summer mode. The CSM modules were installed horizontally in the storage tank so that the supply air can flow around them directly. The horizontal storage of the plates reduces the effects of the segregation, typical for salt hydrates, a separation of the

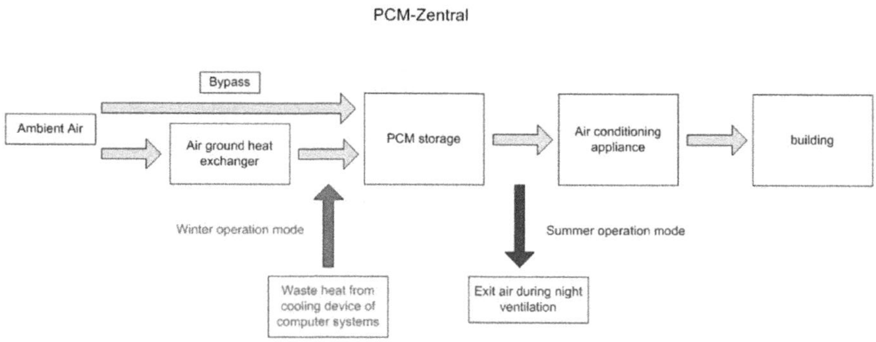

Figure 31.4 Air treatment at Düsseldorf test plant.

Figure 31.5 Left: CSM module as PCM product used in the storage, middle: string of four cubes assembled for testing at that time the plates were mounted vertically, later they are mounted horizontally, right: final installation at Düsseldorf.

individual components of the material due to differences in density. A total of 6048 CSM plates with a total weight of approx. 11 tons were installed in the whole PCM storage. For this purpose, 14 CSM plates were assembled to form a module, 18 modules each forming a cube with an account length of approx. 1x1m. Four cubes are connected in series to form a string, of which a total of six were assembled in the building (Figure 31.5). This modular concept allows a very simple adaptation of the storage capacity (total number of CSM modules), as well as the temperature profile (length of one string), to the requirements of other buildings. Additionally a simple mounting of the storage tank is possible.

Experience in operation has shown that in winter operation both technologies for pre-air conditioning can be very well combined with each other. While the air/earth heat exchanger heats the supply air up to 15–18°C almost independently of the outside temperature, a further temperature increase can be achieved via the SP25, which was implemented later, so that the supply air can be brought almost to the target temperature using regenerative energies and waste heat. In summer operation, cooling by the PCM storage tank could not be tested, as the upstream air/earth heat exchanger could already achieve complete supply air cooling to approx. 15–18°C independent of the outside temperature here as well. Even towards the end of the summer periods, no thermal exhaustion of the air/earth heat exchanger could be detected within the framework of the project.

The storage facility was operated successfully for a total of almost 10 years. During winter operation, approx. 300 kWh of thermal energy was stored in the PCM storage tank every day and fed back into the supply air during the day.

31.3.3 Example of Active Decentral PCM System: Chilled Ceilings with Integrated PCMs in Würzburg

A chilled ceiling using PCMs for passive cooling during the daytime was installed in the Energy Efficiency Center, Würzburg [5]. Macroencapsulated PCM modules from Rubitherm Technologies GmbH and Doerken GmbH & Co. KG (melting at 23°C) are integrated into water cooled cooling ceilings, which are individually optimized Plafotherm®-Boards from Lindner AG (Figure 31.6). The ceiling allows (passive) space cooling in daytime as the PCM absorbs heat while melting. During night the ceiling is regenerated by cold stored in a central storage. In this way, a reduction of the daily cooling load peaks and a shift of the cooling demand from day to night can be achieved. To guarantee the regeneration of the PCM (freezing at about 21°C) over night, cold water with a temperature between 16°C and 18°C is required to ensure the needed cooling power. If the PCM melts completely during the course of the day or if its passive cooling capacity is not sufficient, the cooling ceilings can also be actively flown through with chilled water like conventional chilled ceilings. Important for the energy-efficient operation of a PCM cooling ceiling is the adjustment of the control strategy towards a demand-optimized regeneration. In contrast to a time-based regeneration, the loading condition of the cooling ceiling must be taken into account. The monitoring of the building showed, that the PCM ceilings reduced the energy demand compared to conventional ceilings without PCM up to 13% within a one year period. At the same time, the cooling load has been shifted to the night, so that further advantages arise here due to the more efficient cold generation.

Figure 31.6 opened chilled ceiling showing the implemented PCM modules (*source*: ZAE Bayern)[5].

31.3.4 Example of Active Central Cold Storage: Office Building Stuttgart

Cold storages can be used to decouple the cold demand from the supply and thus relieve the elecricity grid by producing cold netreactively. As a high share of the total, energy is used as heat and cold, thermal storages offer the potential to provide storage capacity at comparable low costs and high efficiencies [6]. Coupled to compression chillers it is possible to use thermal storages for demand side management by operating the chiller when there is a high share of renewables in the grid or to shave peak load demands. In Germany many existing office buildings are air conditioned and cooled by using compression chillers. Only a few buildings are equipped with cold storages so their electricity demand to operate the chillers has to be covered simultaneous with the cooling demand. These systems are potentially usable for demand site management if they would be equipped with thermal storages. Due to the relatively small temperature differences of cooling networks in buildings, e.g. 12–18°C PCM storages can be beneficial. PCM storages are supposed to provide large storage capacities within narrow temperature differences and therefore require small volumes compared to chilled water storages. More challenging can be the thermal power, which is typically smaller.

In the basement of an office building a water and a PCM storage with a volume of 16 m^3 each was installed to serve as cold storage for a CHP plant. The aim was to have the CHP running depending on electricity net and not on the cold demand. Figure 31.7 shows a schematic drawing of the installation. Due to the installation in the fourth basement floor and the available space the storages have been installed in place as open non pressurized system connected to the building's pressurized hydraulic network via heat exchangers. In this building the chilled water system is used to operate

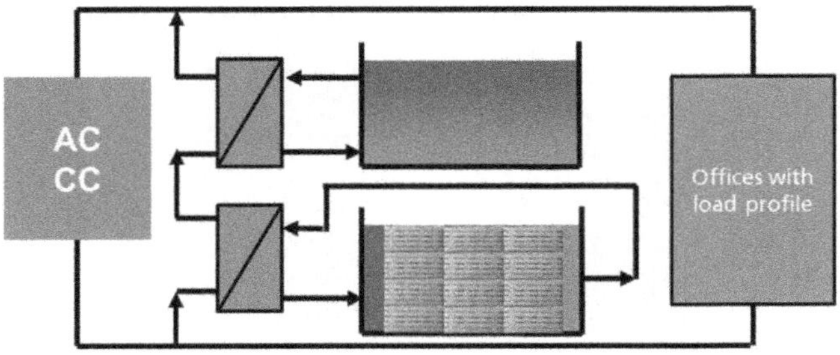

Figure 31.7 Simplified schematic drawing of the parallel installation of the storages, water at the top, PCM at the bottom. Not all possibilities to connect both storages are drawn. Arrows indicate the flow direction for charging the storages.

fan coils in the offices and in meeting rooms. The system is operated by using compression chillers (CC) with 1.6 MW and an absorption chiller (AC) with a nominal power of 0.5 MW. The temperature difference of the chilled water system is designed to be 6 K (12°C supply and 18°C return).

FlatIce provided by PCM-Products was chosen as PCM macrocapsule for the PCM storage filled with S15 PCM, a salthydrate melting at ~ 15°C. The capsules are made of HDPE with a dimension of 500 × 250 × 30 mm. Figure 31.8 shows the FlatIce system in the 16 m^3 storage vessel. The storage contains 2400 Flatice capsules providing a nominal storage capacity of 440 kWh (12–18°C). Additionally the PCM storage contains 1.4 m^3 water which provides a storage capacity of additional 9.8 kWh. The 16 m^3 water storage provides a capacity of about 110 kWh between 12 and 18°C in comparison.

The thermal power of the water storage is basically controlled by the volume flow and the temperature difference. The PCM storage can provide a higher storage capacity but due to the limited thermal power and the influence of the flow distribution over the capsules in the storage it is not possible to fully charge and discharge it. This is also the reason why it is hardly possible to use the full nominal PCM storage capacity of about 440 kWh. In the best case it was possible to reach 335 kWh at a flow rate of 15 m^3/h to charge and discharge it within 24 hours. The Flatice capsules show

Figure 31.8 PCM storage filled with FlatIce capsules.

a good stability. There was no release of PCM visible within an operation of one year. As the shape of the HDPE capsules is changing due to the phase transition of the encapsulated PCM it is hardly possible to get defined flow channels between the capsules which lead to poor flow distributions in operation decreasing the thermal power. Additionally the capsules have a thickness of about 30 mm which is too thick to reach high thermal discharging and charging power. Therefore further development of PCM capsule systems should consider to decrease the size and to provide more defined capsule shapes.

31.3.5 Example of Active Central Heat Storage: Residential Building Weberstedt

In addition to cold storages, PCM can also be used as heat storage. Compared to water storage tanks, however, the heat storage density of PCM storage tanks is only advantageous in small temperature differences. Heat sources are often available at a relatively high temperature level, even with renewable energies, e.g. solar thermal systems. However, if this temperature level cannot be reached, e.g. in spring, autumn or winter, a PCM storage tank can be advantageous as this lower available temperature level can be better utilized. In addition, heat storage at a lower temperature level offers the advantage of reduced heat losses during storage.

In a residential building in Weberstedt, a demonstration unit of a PCM heat storage system was set up, which additionally makes use of the supercooling effect to store heat loss-free over longer periods of time. The storage tank can cool down again to room temperature without triggering crystallization again. If necessary, this is triggered in a targeted manner and the heat is released again. Complete melting of the PCM is a prerequisite for reaching the supercooled melt. For this purpose, the storage tank must be heated to approx. 70–90°C.

In the test facility to serve a residential buildings heating system with heat, solar thermal collectors are used as a heat source, if required supported by a gas burner as backup. The heat is stored in the so-called Thermobattery of HM Heizkörper shown in Figure 31.9. Each cell of the Thermobattery consists of four independent cylinders, each filled with sodium acetate trihydrate (melting temperature 56°C). The subdivision of the storage tank into individual cylinders

Figure 31.9 Demonstration building with solar thermal collectors as heat source, right: Thermobattery prototype installed.

offers the advantage that the stored heat can be released according to demand. If crystallization is triggered, it can no longer be stopped, so compartimentization is advantageous here.

In the demonstration plant, the stable supercooled melt was regularly reached and storage efficiencies, stored energy divided by the released energy, of over 80% could be achieved.

31.4 Planning Boundary Conditions

Internationally, there are only a few norms and standards that support the use of PCM in buildings. Material characterization of PCM is described in the quality and testing specifications of the RAL Quality Association PCM (GZ 896). ASTM C1784 describes another method for determining material properties using a HeatFlowMeter (HFM). Based on this, VDI 2164 describes planning and design of different PCM systems in buildings and thus provides planners with the necessary basics. For the Netherlands the ISSO 111 guideline was published.

31.4.1 RAL Quality Association PCM

RAL Quality Marks are a predominantly German quality seal, which do not make any legally binding specifications, but can nevertheless be the basis for these and are mandatory for products that want to advertise with this quality mark. Well-known RAL quality marks are the classification of colors or the “Blue Angel,” a quality seal for environmentally friendly products in Germany. The RAL Quality Association PCM (Quality Mark GZ 896 [7]) was founded in 2004 by leading, predominantly German manufacturers and research institutions with the aim of improving the comparability of PCM by defining comparable material characteristics and uniform quality criteria. The quality and testing regulations have been published since 2009. Since the foundation of the RAL Quality Association, the standard has been further internationalized and, for example, English, American, Indian, Chinese, and Korean manufacturers have joined the Quality Assurance Association. Central material parameters are the enthalpy temperature curve and the cycle stability of the material. Further parameters that have been defined are thermal conductivity, segregation, and tightness of encapsulations. In addition to the specification of how these parameters are to be determined, it is also defined how the results are to be presented. Another major innovation introduced by the RAL Quality Association is the classification into PCM, PCM compound, PCM object, and PCM system. PCMs are pure materials or material mixtures that are to be used as PCMs. For technical use, these materials are often combined with other materials, e.g. graphite to increase thermal conductivity. This combination is called a PCM compound. If an object such as a plate or bag is made from a PCM or PCM-C, this is referred to as a PCM object. Finally, PCM systems are products or applications in which several parts are combined and interact with each other, e.g. cooling ceilings. This distinction is important, as the type of quality criteria and testing depends on the PCM class.

31.4.1.1 Enthalpy Temperature Curve

The specification of a melting temperature and enthalpy is not sufficient to describe the thermal properties of a PCM. Most PCMs have a melting and crystallization range that can be extended over several Kelvin. At the same time, the materials usually exhibit hysteresis between melting and crystallization. If all these parameters are not known, it is not possible to select a suitable PCM for a specific application. The RAL Quality Association therefore specifies that both the melting and the crystallization of a PCM must be specified in maximum 1 Kelvin intervals and that the specifications must cover the entire melting and crystallization process. This makes different PCMs directly comparable with each other. The determination of enthalpy is usually carried out with Differential Scanning Calorimeters (DSC). The sample is heated at a constant heating rate and cooled down again to determine the heat flow into or out of the sample. However, the temperature determination in the DSC takes place outside the crucible, so that there is always a temperature difference between the measured and real sample temperature. This difference depends on the selected heating rate, but should be as close as possible to the thermal equilibrium to determine the material properties. A heating rate test is therefore introduced to determine the instrument and sample dependent heating rate. Here the heating rate is halved until the difference in peak temperatures is below 0.2 K. The isothermal melting process means that the heat flow must be continuously adjusted to achieve a constant heating rate. More energy must be supplied or dissipated, especially in phase changes. A distinction can be made between sensitive and latent heat. For reasons of comparability of the materials, the Quality Association stipulates that the specified energy must always contain the sensitive and latent heat content.

31.4.1.2 Cycle Stability

PCMs must be able to withstand a defined number of melting and crystallization cycles undamaged in the respective application. Typical ageing effects are a cycle-dependent decrease in enthalpy, changes in the enthalpy temperature curve or defects in the encapsulation. The RAL Quality Association defines how the cycle stability can be reliably determined. In addition to cycle parameters such as temperature limits, heating rates and isothermal phases, the number of samples and damage events are also defined. Thus the enthalpy may only decrease by 10% compared to the initial value or the characteristic temperatures may only change by +- 1 K. The cycle stability is divided into classes. Depending on the application of the PCM, a certain cycle class must be proven in order to receive the quality seal of the Quality Association. The highest class A requires, e.g. 10 000 successfully completed cycles, the simplest class F can be reached from 50 cycles. It should be noted that cycle stability is only an indicator of real long-term stability, as will be described in detail in Section 31.5. In principle, however, cycling should be based on real use in order to reflect it as accurately as possible.

The further quality criteria thermal conductivity, segregation (i.e. segregation due to density differences), and tightness of the encapsulation must only be determined if necessary. Uniform specifications as to how these criteria are to be tested are still being researched in part.

31.4.2 ASTM C1784

The ASTM C1784 "Standard Test Method for Using a Heat Flow Meter Apparatus for Measuring thermal Storage Properties of Phase Change Materials and Products" is an international standard for determining thermal properties using a heat flow meter. In contrast to RAL, a specific measurement method is described here. With the exception of the public sector in the USA, however, the use of the standard is also voluntary. Heat flow meters essentially consist of two tempered metal plates between which a sample is placed. By measuring the heat flow into or out of the sample, the specific heat capacity, enthalpy, and melting range can be determined. A significant advantage over DSCs is the significantly larger sample quantity, which also makes it possible to measure inhomogeneous materials representatively and therefore complements DSC measurement options. The ASTM C1784 is largely based on the specifications of the RAL Quality Association. The same classification of the PCM types is used and the enthalpy for melting and crystallization is given in interval steps. Instead of the dynamic temperature change with constant heating rate frequently used with DSC, the gradual temperature change is used. In principle, this is also permissible for measurements according to RAL. The sample is first conditioned to a starting temperature below the melting range. This is followed by a sudden temperature change at interval height, with a subsequent isothermal phase. The isothermal phase must last at least until the entire sample is again isothermal, i.e. no heat exchange with the environment takes place. This sudden temperature change is repeated over the entire melting and crystallization range. Within the interval limits, an exact temperature assignment of the measured enthalpy is possible, but requires more measuring time compared to continuous measurement due to the additional holding phases.

31.4.3 VDI 2164

Based on the determination of the exact material properties, VDI 2164 supports planners and users of PCM systems in buildings with design and performance determination. Detailed calculation rules for dimensioning the PCM systems as well as options for determining the system parameters required for dimensioning are described. The following systems are dealt with in VDI 2164:

- passive surface heating and cooling systems, e.g. gypsum plasterboards with PCM, which are only regenerated via a ventilation concept. There is no active cold generation, but active ventilation can be integrated.
- active surface heating and cooling systems, e.g. cooling ceilings
- Decentralized ventilation systems for cooling operation, e.g. parapet units for retrofitting in existing buildings
- Central ventilation systems for heating and cooling operation, e.g. storage tanks in central ventilation systems
- Energy storage in hydraulic systems

31.4.4 ISSO 111

The ISSO 111 Phase Change Materials was designed for the ISSO series of standards applicable in the Netherlands. In the first four chapters, ISSO 111 provides an overview of PCM materials, the thermophysical relationships, their encapsulation options, PCM semi-finished products and end products. Chapter 5 describes the possible uses of PCM in buildings, in particular the passive use of PCM products and active air conditioning systems. ISSO 111 is only available in Dutch and has not been translated into English.

31.5 Long Term Experience

31.5.1 Material Stability

One reason for the PCM's niche existence is probably lack of information about long term performance of the materials, which is quite important as investment decisions for buildings are always made with a medium to long term planning horizon [1]. The topic of aging of PCM has only been addressed so far in RAL GZ 896 where requirements for tests of cycle stability are set. According to these guidelines samples are heated and cooled alternating over the melting/crystallization point for a certain number of repetitions. However, this cycle stability test does not give any information about time-dependent ageing processes. Especially not about those favored by incomplete melting/crystallization cycles [6]. In order to find out whether PCM still works after several years of operation, in a public funded project called "PCM Eval" PCM material samples were collected in as many buildings as possible. Unfortunately only a few building owners allowed us to pick a piece of their PCM walls or ceilings leading to a rather small sample size of eleven objects so far. Unfortunately, for only four of these samples reference samples or measurements were available. The samples were measured by differential scanning calorimeter (DSC) using a Setaram microDSCIIIa device to indicate the real material aging since installation. The Tian-Calvet construction principle, with heat flux sensors surrounding the crucibles containing sample and reference, enables the usage of larger samples up to 1 ml compared to standard heat flux DSCs. Determination of phase transition temperature and stored heat was done according to the German standard RAL GZ 896 Phase change materials. But due to the small available sample masses, only one sample measurement, including three heating and cooling cycles, were processed in the DSC. By simultaneously measuring a crucible containing the sample and a similar empty crucible the resulting heat flux into the sample can be determined by substracting the heat flux of the empty crucible from the sample crucible.

Results of DSC analyses of PCM after several – up to 15 – years of operation are displayed in Figure 31.10. Although only few samples could be collected, measurement results indicate that ageing processes influence functionality of paraffin based PCMs clearly. Unfortunately we were not able to test other PCM products than those containing microencapsulated paraffin; therefore this statement is only true for this material and is not necessarily true for other materials. Furthermore, reliable statements on the extent of reduction of thermal capacity require further samples. It is also important to mention: at the time of installation in these buildings microencapsulated PCM were still under development. Several improvements have been made in the meantime on both the gypsum plasterboard and the microencapsulation. But as a take-home message we'd like to mention that the functionality of paraffin based microencapsulated PCM is affected by daily melting and crystallizing processes or diffusion processes and thus lose capacity over time.

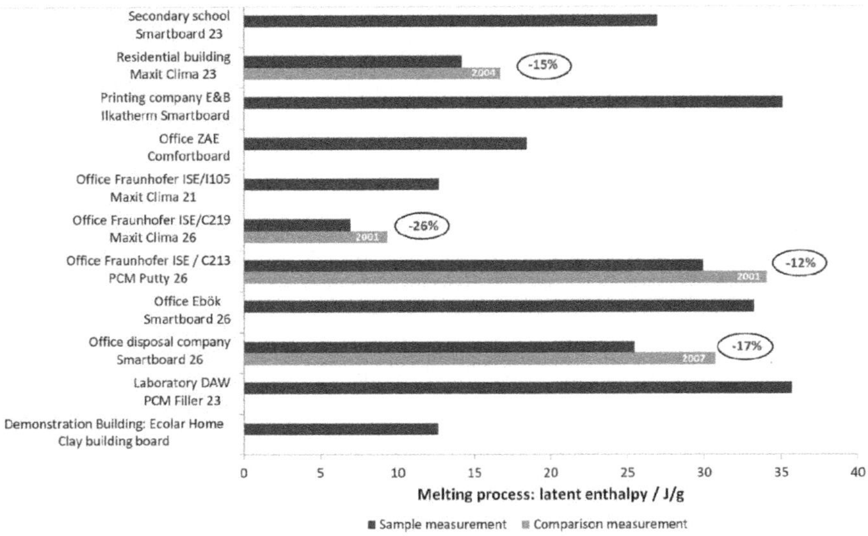

Figure 31.10 Results of DSC measurements of Micronal containing PCM product after several years of operation. All samples were collected and measured in 2018. Percent values in circles indicate reduction of latent heat capacity compared to comparison measurement, year dates (white) indicate year of comparison measurement.

31.5.2 Planning vs. Usage

Besides the objective criteria of technical functionality, the subjective sensations of the PCM buildings' users are very important for the evaluation of the PCM system functionality. In order to generate new insights concerning this topic we carried out a survey on user satisfaction and the planners' perspectives.

We worked with qualitative interviews and written questionnaires. Overall, eleven users and four planners or architects were interviewed and 16 written questionnaires were filled out by users. It is well known from literature that individual control options (opening/closing of windows, shutters, doors, etc.) strongly influence perceived thermal comfort. Our results confirmed this relationship: if users could not use windows as they wanted, they felt very restricted and unsatisfied. Some of the questioned people knew about the request of keeping windows closed, but not about the reason. This lack of knowledge causing "wrong" user behavior (from a technical perspective) made it difficult for the entire PCM system to work properly. One common case was that users opened windows although outside temperature exceeded room temperature which was – as the users stated – because they felt a strong conflict between desire for fresh air and keeping warmth outside. Also non-optimal use of blinds due to a conflict between letting light in and keeping warmth outside was very common. And finally we recorded some cases where the room occupancy was higher than formerly planned which led to higher energy inputs and therefore poorly working PCM systems.

Our surveys showed that often PCM systems do not work satisfactory – from a technical perspective as well as in respect of users' perception – which is due to planning errors and non-optimal user behavior. It can be concluded that a strict implementation of the designed system is essential. Deviations of planned internal loads, input of solar energy or night ventilation are difficult. A provision of systematic information might help users take behavior-related decisions more consciously and thus favor proper functioning of PCM systems.

References

[1] PCM Eval; https://www.tib.eu/de/suchen/id/TIBKAT:1753320283/PCM-EVAL-Evaluierung-PCM-Bestandsgeb%C3%A4ude-Langzeiterfahrungen?cHash=2a0379ad9094922c905819214390dce9

[2] Eicker, U., Seeberger, P., Fischer, H., et al. (2007). *Evaluierung eines im Passivhaus-Standard sanierten Bürogebäudes hinsichtlich des energetischen und raumklimatischen Verhaltens*. Schlussbericht.

[3] FelixRegin, A., Solanki, S.C., and Saini, J.S. (2008). Heat transfer characteristics of thermal energy storage system using PCM capsules: A review. *Renewable and Sustainable Energy Reviews* 12 (9): 2438–2458.

[4] Weismann, S., Büttner, D., Chhugani, B., et al. (2019). *Betriebsoptimierung, Monitoring, Entwicklung von interdisziplinären Steuerungs- und Regelungskonzepten und Forschungsbetrieb an innovativen Anlagen und Komponenten eines Forschungs- und Demonstrationsgebäudes, deutscher Beitrag zum IEA EBC Annex 65; MoniResearch : Schlussbericht : Laufzeit: 01.05.2014 bis 30.04.2019*. Garching: Bayerisches Zentrum für Angewandte Energieforschung. https://doi.org/10.2314/KXP:1690250003.

[5] Kenisarinn, K.M., and Mahkamov, K. (2016). Salt hydrates as latent heat storage materials: Thermo physical properties and costs. *Solar Energy Materials & Solar Cells* 145: 255–286.

[6] Smallbone, A., Jülch, V., Wardle, R. et al. (2017). Levelised cost of storage for pumped heat energy storage in comparison with other energy storage technologies. *Energy Conversion and Management* 152: 221–228.

[7] RAL Gütezeichen Phase change materials. https://www.ral-guetezeichen.de/gz-einzelansicht/?gz=gz_896 (last accessed 15 March 2019).

32

Industrial Applications of Thermal Energy Storage Systems

Viktoria Martin and Ningwei Justin Chiu

KTH-Royal Institute of Technology, Dept. of Energy Technology, Stockholm, SWEDEN.

32.1 Why Thermal Energy Storage in Industry?

On a global level, industrial energy use is in the order of 157 EJ of energy annually [1]. Effective energy management must allow for industrial energy demand to be matched to be cost-effective, and secure energy supply. With increasing environmental awareness and commitment, it is also essential to minimize the use of fossil-based energy and instead shift to locally available, renewable resources.

In a recent study, the potential of CO_2 mitigation through large-scale integration of Thermal Energy Storage (TES) in energy intensive sectors was assessed [2]. Utility plants for power generation (Combined Heat and Power (CHP), Solar), refrigeration, and the horticultural industry, are examples considered, and the analysis showed that the amount of CO_2 mitigated is significant when integrating TES. However, a comparison of options is difficult since the storage is designed in relation to each specific application, and each application impacts the overall energy system to varying extents. Nevertheless, figures published show that integrated TES leads to CO_2-mitigation ranging from several hundred to over a thousand kg per MWh energy turn over (whether heat, cold, or electricity).

The potential for resource efficiency, and CO_2 mitigation through the integration of TES in industrial scale processes come from three resulting aspects: (i) TES integration facilitates large-scale implementation of Renewable Energy Sources (RES), (ii) TES integration maximizes the opportunity for process integration, and (iii) TES integration allows for the strategic export of industrial energy surplus to external users, such as energy utilities. These aspects are further described in the following sections.

32.1.1 Maximizing the Use of Renewable Energy Sources through TES

A huge effort towards switching from fossil to renewable energy sources (RES) is under way, e.g. resulting in roadmaps such as the Energy Technology Perspectives put together by the International Energy Agency [3]. In that work, it is for example stated that electricity generation shall be de-carbonized by 2050. IPCC reports on no, or at least very few, technical barriers towards large-scale integration of RES. Instead, focus must be on a portfolio of RES (solar, wind, biomass), cogeneration of power, heat, and cold, as well as the integration of storage along various energy supply chains [4]. For industry, IPCC highlights energy efficiency, switching from fossil-fuel usage to RE-based electricity or hydrogen, and making residual energy (biomass, heat, etc.)

Advances in Energy Storage: Latest Developments from R&D to the Market, First Edition.
Edited by Andreas Hauer.
© 2022 John Wiley & Sons, Ltd. Published 2022 by John Wiley & Sons, Ltd.

available to external usage [4]. No specific mention of storage is however made in this context, even though this is a key issue when encouraging a switch from today's fuel-based systems for heating and cooling, to electricity as an energy carrier. The impact of the industrial demand on any electricity grid will be huge, and the levelling of peak hour demand essential since many RE electricity options are also fluctuating in nature. Here, although electricity is the suggested energy carrier, the industrial thermal demand side can be controlled through the integration of TES.

One example where the integration of TES immediately leads to a higher solar fraction in electrical power generation is Concentrating Solar Power (CSP). From a technical point of view, CSP with integrated TES can play an important part of an RES portfolio interacting with other fluctuating options like wind and photovoltaic (PV) [5]. This is schematically shown in Figure 32.1.

Heat for operating the CSP power block can then be provided by either the solar collector field or the TES. However, although a techno-economical optimization indeed showed the solar fraction and CO_2-mitigation to increase with TES integration, the levelized cost of electricity also increased as compared to covering the heat demand with a back-up burner [5]. This can be attributed to the relatively high cost for TES technology. Presently, attention is given to finding cost-effective, and technically robust TES alternatives for CSP applications. For a comprehensive review on thermal energy storage options for CSP, the work by [6] can be mentioned. In there, materials and technical concepts are reviewed along with how to theoretically model such systems. An alternative to the so-far most widely employed molten salt storage, is rock-bed thermocline storage presently considered by many (e.g. [7–8]). In another recent work, special focus has been given to the emerging thermochemical storage technology, concluding that the reversible reaction involving calcium carbonate is by far the most researched, whereas the manganese oxide reaction appears particularly promising for desert climate since no water is involved, but only air [9].

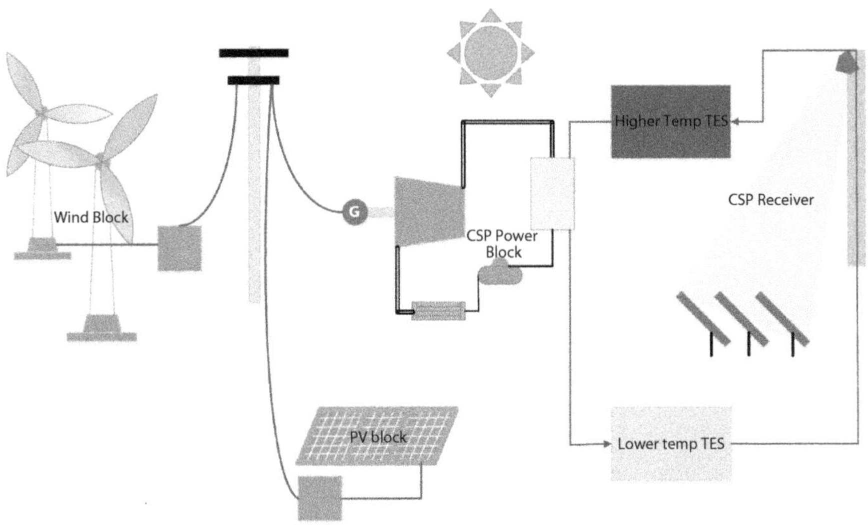

Figure 32.1 Principle RES power portfolio.

The Wind Block can also interact with TES for stable power output, to avoid the need to dump excess power, and minimize fluctuations in the grid. In Germany, for example, the large scale implementation of wind power has created problems of excess power in the system since base-load "fossil fuel" generation cannot ramp down in response to excess power in the grid [10]. On one hand, measures to alleviate this problem can be taken in the market (reduce feed-in tariffs for renewable electricity, connect to the European "supergrid"), but working with storage on the demand-side, like the built environment, is also an option. At the utility scale, excess power can be used to generate hot water for a district heating system, so-called power-to-heat. In one exemplifying study, the combination of wind power, combined heat and power, and TES was analysed for Finland [11]. One motivating factor to choose TES over, e.g. batteries are the order of magnitude difference in cost, per installed kWh, with TES being the cheaper alternative. Connecting the power system to a heating system, e.g. District Heating, opens up the opportunity to use TES and power to heat. In the national-level case study conducted, the results showed that the optimal TES size should increase from today's 17 GWh up to 100 GWh [11].

Aside from balancing renewable power generation through TES only through power-to-heat, hybrid storage solutions combining TES with battery, are very promising for the future [12]. Also, by implementing thermal cold storage chillers can operate at full capacity, regardless of cooling demand, during hours of excess electricity availability (meaning low price).

Finally, also the PV-block can benefit the power-to-heat/cold concept described for wind. In a high-share renewable energy system, the fluctuating sources will be many, and storage of energy will cost-effectively alleviate the mismatch between supply and demand.

32.1.2 Process Integration and Storage

Accomplishing energy efficiency is crucial to managing the global goals on energy sustainability. As stated in the World Energy Outlook 2015 [13], it is planned for energy efficiency measure to limit the growth in energy demand to one third until 2040, while the global economy is expected to grow by 150%. For industrial scale systems, process integration is a powerful tool for accomplishing energy efficiency through techno-economic optimization [14]. It can be done without integrating any type of energy storage, but it can be done even better with the integration of storage.

For example, industrial scale processes are often operating in batch mode due to discrepancies in supply (of process input commodities) and demand (for process product), as well as other inherent time fluctuations. The integration of TES in such processes is said to improve economics, controllability, and energy efficiency. [15] explored process integration with TES for a number of case studies on batch-type processes. Since industrial processes are complex to begin with, often with 10 + steps involving heating or cooling, adding TES can make the process even more complex by adding at least one heat exchanger and two storages per process step to recover heat. Thus, [12] present a structured and simplified way of integrating TES in an analysis of process integration. In this way, it turns out that although TES add cost and certain complexity to the overall system, the number of heat exchangers for obtaining maximum heat recovery is reduced which is beneficial.

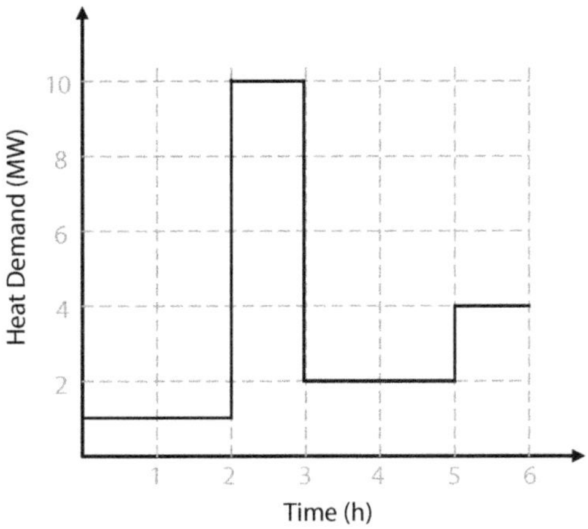

Figure 32.2 Examplifying heat demand pattern in batch process (based on [18]).

Despite the potential complexity of including TES in process integration, excluding the time variable in the overall optimization leads to sub-optimal results. An extensive review on process integration with batch processes has recently been published, including methodologies that allow for identifying opportunities for storage, optimally sizing such storage, and evaluating the dynamics in the storage like level of charge over time [16]. In this review, a simple example is outlined [17], where a utility shows a demand pattern according to Figure 32.2. As shown, there is a peak demand of 10 MW which is only needed for one hour, and in the described base-case this is supplied by two boilers sized 4 MW and 6 MW, respectively. In one scenario discussed, the 6 MW boiler is taken out of commission, and the question then is if, instead of replacing the boiler, the 4 MW boiler is enough when integrated with a heat storage?

To answer this question, depicting the situation in a "ΔQ vs time"-diagram is used for targeting an optimum size of the heat storage [18]. This is re-drawn here in Figure 32.3. As shown, the energy can be supplied over the six hour period using only the 4 MW boiler. However, a heat storage is needed to provide enough power. Here, we must then target the optimally sized storage. In the time interval through hour 2, there is 8 MWh of heat available while only 2 MWh is needed. Thus, 6 MWh can be stored. In the time interval following this surplus, between hour 2 and 3, there is a demand of 10 MWh while only 4 MWh is available, such that there is a deficit of 6 MWh. In this simple example, there is then a perfect match to the surplus in the previous interval.

Thus, with integrated TES, the flexibility of batch processes may be maintained, while still enabling process internal heat integration. The final decision on replacing the boiler in the example above, as compared to adding on a TES, will be based on complex input such as plans for future demand and process development. However, it is very likely that investment cost and technology lifetime will be deciding factors. Boiler capacity in the MW size cost approximately 0.5 M€/MW (depending on type [17]).Then, an approximate investment of 3 M€ for replacing the 6 MW boiler should be compared to the cost of, e.g. a hot water storage. The storage cost depends very much on the price of steel, but a hot water storage price of 100€/m^3 is reasonable for this example. Then, the required storage volume to hold 6 MWh of accumulated heat will depend on the operational temperature difference, ΔT, which this example gives no indication of. However, if a storage

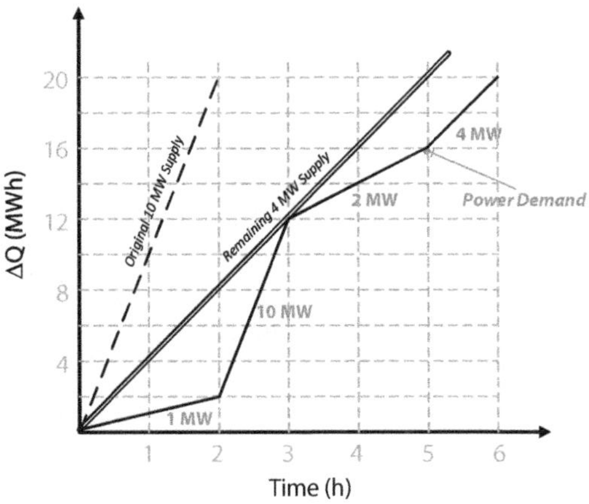

Figure 32.3 Accumulated energy demand and availability vs. time.

operational ΔT of 50°C is assumed, then approximately 100 m^3 of storage is needed, resulting in an investment cost under 100 k€. The investment in TES in this case is thus significantly cheaper than investing in a new boiler.

32.1.3 Industrial Surplus Heat as a Resource in a Sustainable Energy System

Energy use in industries represents a third of the global energy demand [3], where only part of the energy input is converted into useful end-products. The rest, in the order of 10 000 TWh/year, by the laws of thermodynamics, must be dissipated as heat in one way or the other. Naturally, the quality of this heat (the exergy content) is a deciding factor whether it is worth recovering, or not.

In a recent report to the UK department for Energy and Climate Change (UK-DECC), the potential for using industrial surplus heat was explored from many aspects [19]. In a structured way, it is described how the opportunity for recovering and using the heat is depending on factors like:

1. The characteristics of the heat sources and heat sinks.
2. The compatibility of sources and sinks with regards to, e.g. temperature, timing, and location.
3. The cost and efficiency of available recovery technologies, cost of energy, and any available economic incentives.

With regards to the characteristics, surplus heat can be carried by liquid, gas, or in solids. With the solids, the radiative heat transfer mode can be an especially interesting way of recovering high temperature heat from cooling goods. Nevertheless, TES will be an essential component to help matching source to demand with regards to timing as well as location – effective transportation of heat from industrial sites can be done utilizing TES in innovative district energy applications and will be discussed in more detail further into this chapter.

In the previously mentioned UK-DECC report, assessing the availability of surplus heat, the authors considered the potential from four perspectives: available surplus

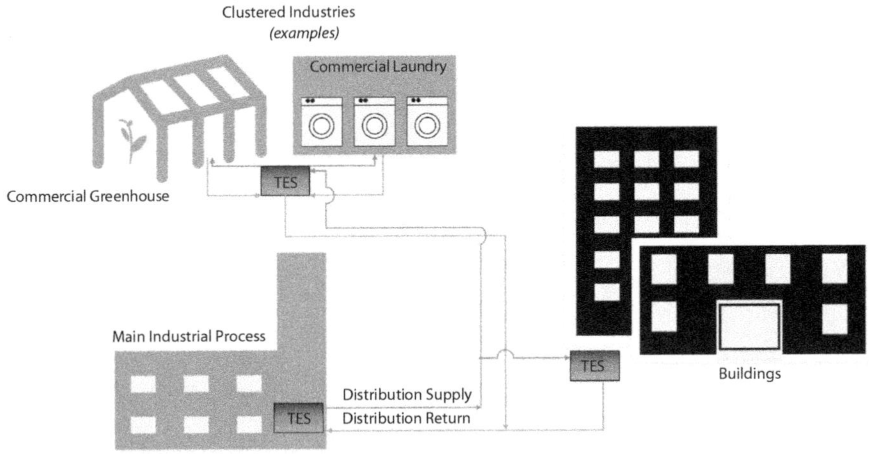

Figure 32.4 Surplus of industrial process heat servicing adjacent clustered industries, and the built environment through integration of TES.

heat, technically feasible for heat recovery, economically feasible, and commercially feasible. The figures reported for UK are then 48 TWh/year available surplus heat, out of an approximate annual industrial energy demand of close to 60 TWh. Here, 11 TWh/year was deemed technically feasible to recover, 7 TWh/year economically feasible, and 5 TWh/year commercially feasible. Commercial feasibility also considers factors like increased process complexity and risk for the core process. This study did however NOT consider any TES to help address the matching between source and sink [16].

Another example is Sweden, with approximately 4 TWh/year industrial surplus heat utilized already in the district heating networks [20], the theoretical potential is estimated to be 10 TWh/year [21]. For the EU27 as a whole, approximately 750 TWh/year is assessed for the Heat Road Map Europe [22].

Part of why there is such a large surplus not used is due to the fact that much of the heat comes out as low grade heat [23–24], and cannot be used economically to satisfy the industry's internal heating demand or generate electrical power through a thermal process like the Organic Rankine Cycle (ORC). With innovative integration of TES to match this available heat surplus, the potential for external use is significant. Hence, after maximizing internal process integration, this surplus can be a major source of energy for a region as a whole. The establishing of smart cluster networks of industries to share energy commodities, and intelligent connection to a society's heating and cooling infrastructure as a whole are key factors in enabling the use of such a surplus. TES comes in to any such successful system by allowing matching supply and demand, especially when keeping in mind the batch-type nature of many of the industrial processes. Intelligent use of industrial surplus heat is exemplified in Figure 32.4.

32.2 Integration of TES in Industrial Scale Applications

Integrating TES in industrial scale processes requires the consideration of a number of technical design parameters regarding the storage, like:

- Temperature range and desired mode of temperature operation (fixed temperature discharging vs variable temperature discharge).
- Energy density (e.g. kWh/m^3), linked to volume requirement
- Power requirement (rate of charge and discharge of the storage)

In later sections, a number of examples of energy intensive industries are discussed, with special focus on the benefits of integrating TES: district energy utilities; steel manufacturing; and the pulp and paper industry. There are of course many more industrial processes to consider, but the general conclusions from the following discussion will be similar.

32.2.1 Combined Heat and Power and District Heating

Combined Heat and Power (CHP) improves the overall energy efficiency of generating heat and power, as compared to the separate generation of these commodities. The link between heat and electricity systems are clearly described by the International Energy Agency [25], where co-generation and district heating and cooling (DHC), in combine, is described as an integrated sustainable energy network for the future. In the same report, energy storage is put forward to ensure the flexibility in regulating the electricity and heat outputs while minimizing heat losses. Also, as previously discussed a large district energy system with TES incorporated can serve as a "balancing-tool" by letting it absorb excess electricity (converting it to hot/cold water) when needed. With integrated TES, local, intermittent heat sources (e.g. solar and industrial surplus heat) can also be integrated with the co-generation and DHC system (Figure 32.5).

An increasing need for electrical power during peak hours affects the price on a deregulated market, as does the availability of electricity generated at low cost. For example, the fluctuation of electricity price managed by the Nordic and Baltic European power market Nord Pool http://www.nordpoolspot.com is illustrated in Figure 32.6.

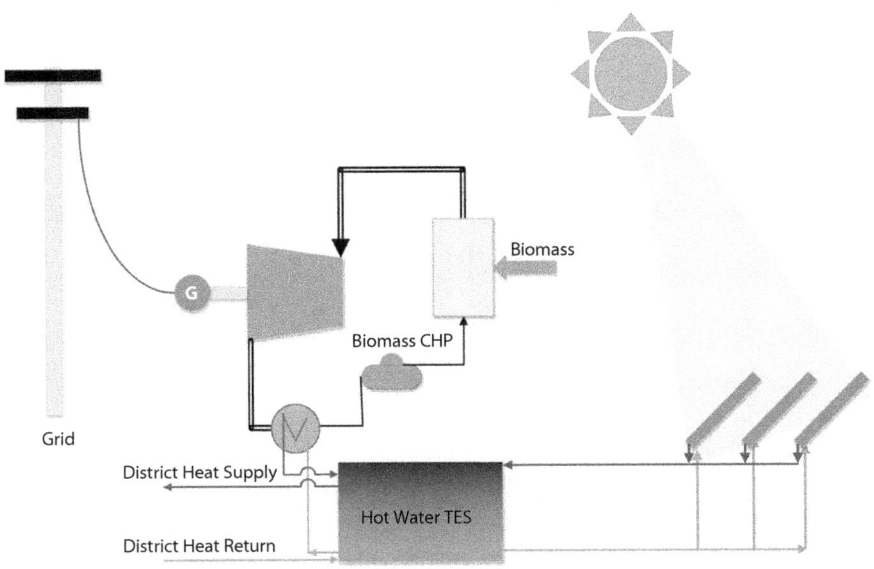

Figure 32.5 Co-generation-based district heating system interacting with intermittent solar heat.

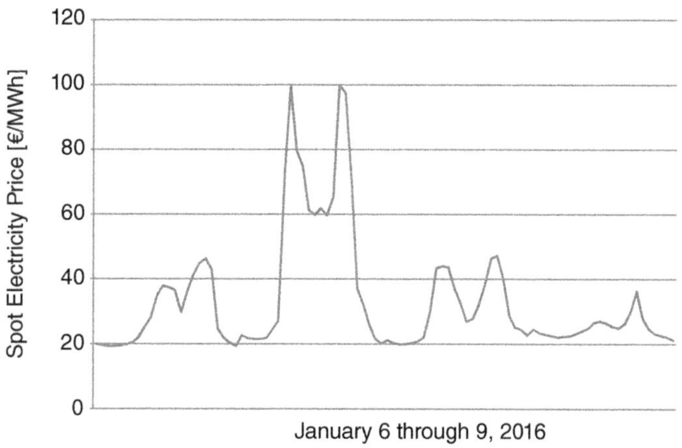

Figure 32.6 Example of variation of spot electricity price in the Swedish market, Oct through Nov 2021 [26].

Then, there are large economic incentives for de-coupling the generation of heat and power, which is regularly done in, e.g. Swedish and Danish District Heating systems linked to a CHP plant. Here, TES in terms of stratified hot water storage is the conventional technology. Aside from economically sound, such a TES integration also leads to a robust operation based on the most environmentally sound and cost-effective part of the plant (like biomass-based back-pressure steam turbine block), minimizing the use of back-up equipment such as oil-fired boilers. These storage solutions are short-term in design, however they are sufficiently large such that they would be fit for inter-seasonal storage in a smaller-scale system. With TES integration, operation can continue during short stops in generating equipment, aside from having a heat (or cold) reserve to meet sudden load changes. Also, CHP plants can temporarily increase the power generation (during hours with high demand, and high electricity price) which then results in lower heat generation such that the TES helps to still manage the heat demand [27].

To some extent, the district energy network itself can provide storage by increasing the supply temperature. However, a central storage tank is normally needed as well [24]. When steam is used as heat carrier, a sliding pressure accumulator is used, e.g. Ruth's Accumulator where steam is charged as pressurized water which flashes at a gliding pressure upon discharge. With hot water as a heat carrier, stratified hot water storage is used.

Energy storage systems in district heating networks have shown potential in contributing to user load demand leveling, in maximizing renewable energy intake, and in increasing the network capacity. As one example, Sweden has an extensive district heating network, the amount of total primary energy used for heating amounted to 62TWh in 2013 [28] where 15% is still fossil fuel-based production, and around 8% is heat pump- and electric-based heat production. When integrating TES, the utility peak-demand often covered by fossil-fuel can be lowered, which in Sweden can result in a cost-effective mitigation of 620 kt (kiloton) of CO_2 annually as shown in one study [29]. The techno-economic feasibility of TES was at the center of attention, e.g. the size of TES. Thus, the level of reduction in fossil-fuel use and CO_2-mitigation was derived

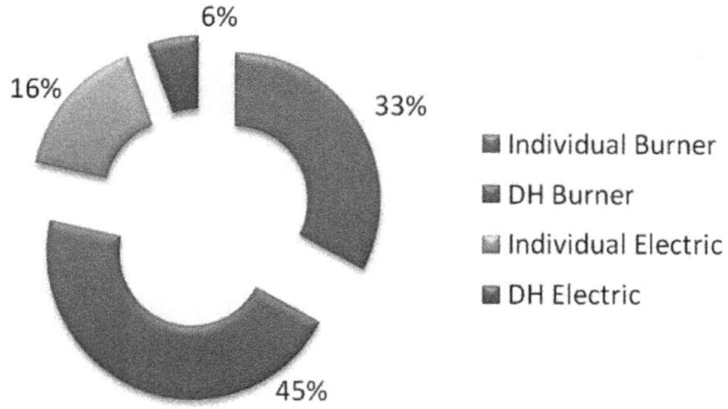

Figure 32.7 CO_2 emission reduction potential by means of heating in the Swedish context [29].

from the cost breakeven point of capital expenditure of TES systems in relation to the achieved heating generation cost reduction (GCR), and the avoided fossil fuel cost (AFFC) in a conventional system. The GCR accounted for fixed and variable operation and maintenance costs while the AFFC is proportional to the share of fuels consumed in the energy system. Considering the Swedish Energy mix for Heating, the results from this study is shown in Figure 32.7. Integrating TES with individual and district heating oil burner type of generation have the largest CO_2 emission reduction potential reaching 80% of total achievable CO_2-mitigation. Such a distribution of techno-economically feasible CO_2 emission reduction share is country specific, and depending on the country specific share of primary energy sources for heating.

32.2.2 District Cooling

With increasing high density buildings in cities, and very specific cooling demand from industrial processes, district cooling systems have become more popular. District cooling is an industrial-sized system that distributes chilled water from central sources to end users for indoor temperature and humidity control, application cooling, and industrial processes [30]. Heat sinks for district cooling networks can vary: e.g, geothermal boreholes, ambient air and surface water in lakes, rivers, or the sea. The energy sources used to generate cold may be renewable energy such as wind, solar, biomass, with conversion using electric chillers, sorption chillers, and water pumps.

The use of cold storage in district cooling is very important since cooling demand tends to fluctuate to a large extent. Integration of TES to district cooling may alleviate peak hour district cooling demand, increase network cooling capacity, and improve production efficiency which ultimately results in cost and environmental benefits. Table 32.1 summarizes examples of district cooling installations, of which several are integrated with TES.

In industrial-scale cold production, the integration of TES can significantly alleviate the peak power demand in the electrical system when electrical chillers are used. Also, TES enables maximal use of free-cooling options, and minimizes the need to start any back-up electrical chillers during peak hours.

Sweden accounts for around 1 TWh of annual district cooling (DC) supply [25]. Due to the low temperature difference between supply and return lines in DC systems (normally only 10°C), they deliver a lower amount of thermal capacity per unit pipe

Table 32.1 World-wide examples of large district cooling systems, based on [31].

Country	Project	Capacity (t)	Users	Chillers	Thermal Storage System
Japan	Yokohama MM21	57 510	Offices, commercial buildings, hotels, public facilities, dwelling units	Absorption chillers and electrical chillers	Ice storage system
Canada	Montréal City Central District heating and cooling system	2857	Convention centers, hotels, office buildings, a railway station, and a university campus	Four chillers	
Qatar	The Pearl Qatar	30 000	Residents, hotels, shopping centers, high-end retail, and lavish hospitality	26 chillers with 2600 t each and 26 chillers with 2400 t each	
Singapore	Marina Bay	260 000	Office building, hotel, convention centre, shopping, casino, theaters	Chillers	Ice storage system
UAE	Business Bay Executive Towers	32 500	High-rise office and residential towers, retail and mixed-use office complexes, and a luxury hotel	16 chillers	
US	Texas Medical Center	120 000	44 buildings of medical institutions	Chillers with VSD	Chilled water storage system
US	District cooling St. Paul	25 000	100 buildings including office buildings, government buildings, museums, banks	Electrical chillers and absorption chillers	Chilled water storage system

volume, as compared to district heating networks. Integrated TES for peak shaving in DC systems is then an option to increase the existing network capacity. One example is shown in Figure 32.8a, where by performing daily load shift, the required peak energy supply is lowered by 40% for an office building in Stockholm [32]. An economic feasibility study was performed to benchmark a PCM-based TES against: (i) stratified chilled water storage, and (ii) auxiliary chiller capacity [29]. Figure 32.8b shows the schematic of the case with 6°C supply and 16°C return temperature from/to the network and a required 8°C feed-in and 18°C feedback temperature from the end-user. When the demand is low during off-peak hours, surplus cooling capacity is used to charge the storage unit at 13°C. When the demand rises, stored cooling is used for dry cooling purpose at temperature above 14°C. The difference between storage

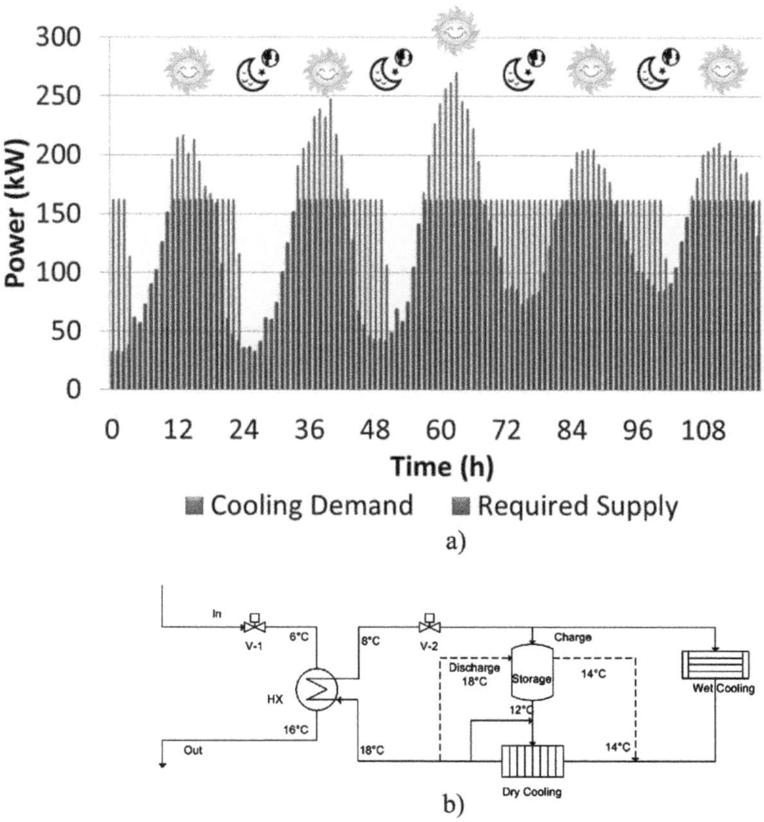

Figure 32.8 A) Load leveling with 40% peak power reduction b) schematic of tes charging (continuous line) and discharging (dashed line) [32].

temperature and extraction temperature is due to temperature jump to and from the storage system.

Currently, the cost for the user is based on the flow rate and not on the actual extracted energy, this causes penalty for the user. With this penalty, a PCM-based TES was only an economically viable option, as compared to a stratified chilled water tank, at < 5% of peak shaving. As compared to additional auxiliary chillers at the customer side, the PCM TES was economically viable in the range of 1% to 9% of peak shaving. On the other hand, if such penalty is taken away, PCM based TES would be cost effective at < 9% of peak shaving as compared to stratified chilled water tank and at 1% to 11% of peak shaving as compared to auxiliary chiller. With steadfast work on lowering the cost of the merging PCM TES technology, the feasibility is likely to improve. Finally, it is important to point out that the feasibility is highly dependent on the shape of the cooling load curve, such that if the peak demand is sharper (i.e. less time duration for high power demand, and then less energy per unit peak power) then the feasibility of TES increases.

32.2.3 Steel Industry

The steel industry is an important part of society, and is expected to grow by 50% until 2050 [33]. World-wide, the estimated turnover is 900 billion USD, and the production

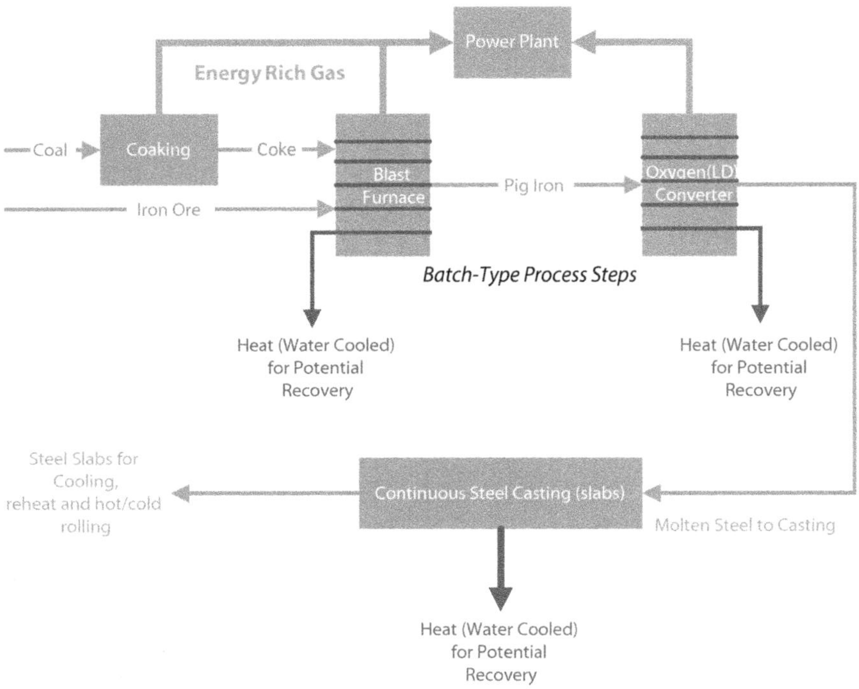

Figure 32.9 Steel making process from ore in a simplified schematic.

of crude steel in 2014 amounted to 1665 million tonnes [27]. With an average (primary steel and secondary, recycled steel) energy intensity of 20 GJ/t [34] (world-wide then order of 9000 TWh/year), the cost of energy constitute between 20 to 40% of the total cost of steel-making. Thus, it is clear that any energy efficiency measure can significantly help in making the production more cost-effective, and if TES can aid the practical realization of such measures, it must be considered.

The production of primary steel from ore is carried out by the reduction of the ore to iron in the blast furnace, conventionally by adding coke. Following this step, the so-called pig iron is then transferred to the batch-type oxygen converter (LD converter) where the iron is de-carbonized, forming steel. Following this, the steel is then cast into slabs, which in turn are treated and rolled (hot/cold) and into various products with very precise properties [35]. A simplified schematic of this process is depicted in Figure 32.9, including the major energy flows in the process in terms of energy rich gas (containing H_2, CH_4, and CO in various compositions).

As shown, the blast furnace, and oxygen converter processes are batch-type processes which means that a gas-holder storage is beneficial when recovering the energy rich gas. This is done to a large extent, to enable on-site power generation using a steam-turbine cycle. However, aside from the gas, a lot of heat is generated in the processes and this is conventionally rejected to a heat sink through a water-cooled system. The temperature in the blast-furnace and oxygen converter is high, but the heat rejection take place at a traditionally low temperature, well below 100°C, using large amounts of cooling water. With some adjustments, the heat could instead be recovered at a temperature suitable to service external heating demand through a district heating system. Then, normal stratified hot water TES is commonly integrated at the district heating side (e.g. as exemplified in Figure 32.4). For internal recovery, the technically most feasible use of excess heat would be to convert the heat into electricity, e.g. through ORC [21]. This is since the process ovens for reheat in the after-treatment of the crude steel (hot treatment and rolling) are operated on either gas or electricity. This sensitive part of the process demands the precision of such rapidly controlled ovens. For power generation of excess heat, the development of high temperature recovery processes, including suitable materials for high temperature TES, is of fundamental importance, like: the effective storage of steam; Phase Change Materials (PCM); or Sorption Processes [36]. The cost-effectiveness of such solutions will only be achieved through innovation-oriented R&D of integrated systems: manufacturing process-ORC-TES, hand in hand.

32.2.4 Pulp and Paper Industry

The pulp and paper industry is one industrial sector where each plant is highly dependent on energy efficiency measures in order to stay competitive on a global market. The complexity of the energy system in pulping processes is described in detail by [37]. The most common pulping process converting wood chips into pulp is the Kraft process, which consumes steam for e.g. delignification, evaporation of the thin liquor, and drying of the pulp (Figure 32.10). In the thermomechanical pulping process (second most common process) the fibers are mechanically separated through friction. This process requires electricity, which in the process is converted into heat.

In the pulp and paper manufacturing, heat (surplus) generated in one unit operation may be utilized either in other unit operations, or be distributed for external use. From an energy storage point of view, the opportunities to aid in energy management lie in the batch-type nature of these processes, e.g. the cooking/delignification sub-process of the Kraft process.

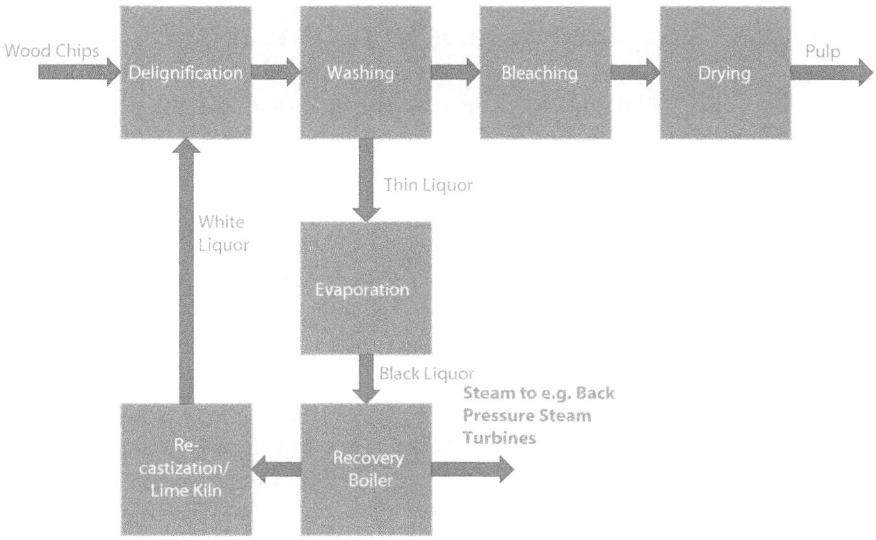

Figure 32.10 The Kraft pulping process in a simplified schematic [34].

[38] studied the influence of short term variations in a Kraft pulp mill's energy system on the opportunities for energy savings. In process integration studies considering annual averages, approximately 18% of process heat savings were identified through process integration. If instead 10 minute averages were used, short term variations were included, and the opportunity for savings was lowered slightly to 16%. One reason for this that was discussed is that if heat exchangers are sized based on annual average duties, they are not large enough to handle possible peak hours appearing in real operation (seen in 10 minute averages). Payback time for investing in larger heat exchangers ranged between one and six years, which is why it was concluded that it is not a likely solution, since the benefit in terms of energy savings is relatively small. Thermal energy storage (e.g. steam and hot water, already common in such pulp mills) is another option discussed and it was then pointed out that the many energy flows (steam, hot water) come at various temperatures, which makes it complicated to install only one TES for the purpose. However, with careful optimization of plants (or sections of plants) to be newly built or retrofitted, it is very likely that TES integration will serve as a key component for technical robustness, and cost-effectiveness.

For the Kraft pulping process, although there is already a surplus of heat, a number of measures can be made to further make heat available from the process, e.g. novel evaporation design, and recovery of heat from flue gases [39]. Also, when comparing a variety of ways to utilize the surplus heat outside the pulping process, [36] showed that the surplus can be economically extracted from the mill as increased generated electricity, district heat, and by selling bark (saving internally on the demand for bark), and lignin. These positive results were assessed without considering the integration of TES, thus assuming immediate dispatch to any electricity and district heating network. With integration of TES, like a steam accumulator, the matching of supply to the most profitable demand situation (e.g. generate electricity when the price is optimal), can further aid the economic benefits of a strategy putting process surplus energy as an important by-product for this competitive industry.

32.3 Mobile TES in Innovative Energy Distribution

One important aspect highlighted with regards to being able to use industrial surplus heat to satisfy external heat demands is the match between source and demand in location [16]. Transportation of energy is presently done over large distances by transportation of fuel like oil, with an energy density of 10 MWh/tonne, or bio-fuel with an energy density of 1 MWh/tonne. Also hot water is used as a media for transportation of thermal energy over shorter distances, like in pipe-based district heating and cooling systems. For district heating, the energy density in a water-based system is in the order of 50 kWh/tonne depending on the allowable temperature range (like 90°C supply and 45°C return temperature). With regards to heat transportation, district heating is primarily suitable for densely populated areas, which are often located far away from industry that generates surplus heat in its process. Putting the heat on rails, truck or a boat could enable increased utilization of such industrial surplus in the built environment – one solution for a sustainable energy situation in the future. However, for cost-effectiveness, the energy density of the mobile TES (M-TES) must then be much higher than what is possible with water.

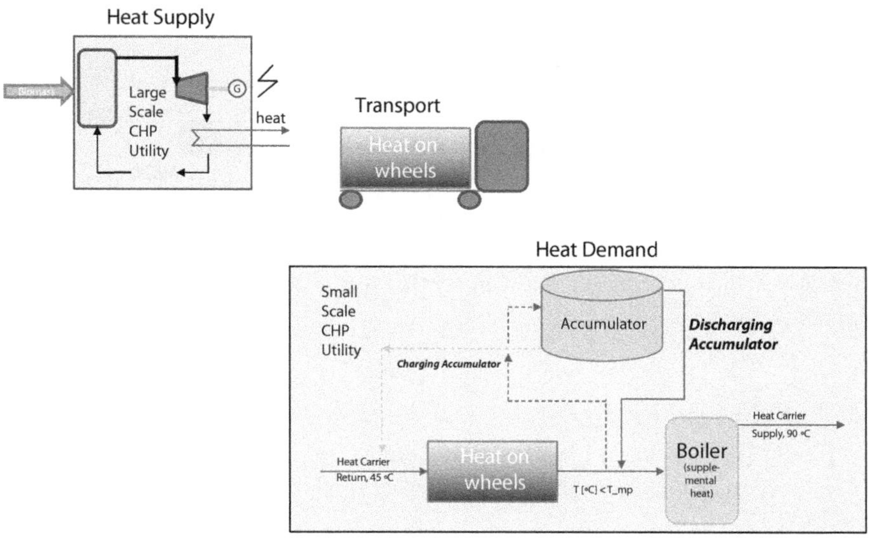

Figure 32.11 Mobile TES concept, connecting a heat source with surplus to an end user some distance away.

Recent activities in TES technology have shown enhanced technical development, and an increased potential for implementing advanced thermal energy storage using phase change materials (PCM) or chemical reaction [33]. With advanced TES, storage capacities up to 0.5 MWh/tonne are theoretically possible which makes it enticing to try to utilize these technologies for the transportation of heat and cold.

A conceptual schematic of the M-TES-concept is depicted in Figure 32.11. In this example, M-TES connects a larger-scale CHP utility to an end user some distance away.

[40] examined the economic feasibility of such conceptual M-TES based on PCM as storage material. They showed that the cost of heat was directly proportional to the transport distance, while inversely proportional to the heat demand. Also, acceptable payback times were obtained for distances in the range of 10 to 30 km. In a recent specific case study, the transportation of heat from the biomass-based CHP plant utility in Gävle, Sweden, to a smaller community Hedesunda has been studied [41]. In this study, the optimized cost of transported thermal energy amounts to 49–86 €/MWh_{th} with CO_2 emission ranging from 10 to 150 kg CO_2/MWh_{th} for maritime, rail, and road transports. In comparison with fossil fuel based direct burners, the emission may be reduced by up to 200 kg CO_2/MWh_{th}. Through a sensitivity analysis, it was shown that by reducing the PCM price by half, the levelized cost of transported energy may be further reduced by 20% to 30%. Optimized logistics and operating strategy making use of non-linear charge/discharge thermal power may cut down high cost road transportation based transported energy by 25%. Halving the transportation time on slower transportation means, such as maritime, may lead to 5% to 15% lower levelized cost. Improved design of the heat exchanger also accelerates the recharge cycle, and improves the cost benefit.

[42] proposed sorption technology based M-TES. In this concept, surplus heat from a waste incineration plant charged an open zeolite TES using hot air at 130°C. The TES

was used in a drying process 7 km from the incineration plant. Operation showed the practical functionality of the concept, including no degradation of the storage material within the test period. A cost of heat of around 70 €/MWh_{th} was forecasted.

Clearly, the attractiveness of the M-TES concept lies in transporting heat from a heat source to an end user through a non-fixed distribution system, thus providing flexibility in delivering from diverse heat sources to various end users. However, the concept needs further steps in developing, demonstrating, and commercializing. Also, industrial surplus heat is currently considered as carbon free and without cost. In reality, once the surplus heat is being used, it is likely that it will bear an equal portion of the emission and energy cost.

32.4 Concluding Remarks

A few characteristic conclusions can be made on industrial applications of TES:

- TES always introduces an additional temperature gap between heat source/sink, and the end-user.
- Short-term TES are as essential as longer term TES since they enable operational control of processes, and alleviate peak demand – thermal (heat/cold), as well as electricity demand.
- CO_2 mitigation may be achieved with peak shifting of marginal fossil fuel based electricity/thermal energy production.

Traditionally, TES is used sparingly in industrial applications, despite excellent technical function, as well as cost-benefits of integrating TES. This has likely to do with two important aspects when it comes to the "commercial" feasibility discussed before: a) TES can increase complexity to processes, and thus increase the operational risk; b) an economically feasible TES, with payback on investment of only a few years, is still an investment that need to compete with other investments closer to the core of the business itself. With increasing cost of energy, it is likely that the benefits of TES will make an investment in such concepts attractive, if not necessary. At the same time, the development of new technical solutions for storing heat and cold with high efficiency and rate of charge/discharge is under way, to meet the coming needs of storage in a sustainable energy system.

References

[1] International Energy Agency. (2020). *Tracking Industry 2020.* International Energy Agency. www.iea.org.

[2] Cabeza, L.F., Miró, L., Oró, E., et al. (2015). CO2 mitigation accounting for Thermal Energy Storage (TES) case studies. *Applied Energy* 155: 365–377.

[3] International Energy Agency. (2012). *Energy Technology Perspectives 2012 – Pathways to a Clean Energy System*. www.iea.org/etp. ISBN: 978-92-64-17488-7.

[4] Sims, R., Mercado, P., Krewitt, W., et al. (2011). Integration of renewable energy into present and future energy systems. In: *IPCC Special Report on Renewable*

Energy Sources and Climate Change Mitigation, (ed. O. Edenhofer, R. Pichs-Madruga, Y. Sokona, et al.). Cambridge, UK and New York: Cambridge University Press.

[5] Guédez, R., Spelling, J., and Laumert, B. (2014). Thermoeconomic optimization of solar thermal power plants with storage in high-penetration renewable electricity markets. *Energy Procedia* 57: 541–550.

[6] Gil, A., Medrano, M., Martorell, I., et al. (2010). State of the art on high temperature thermal energy storage for power generation. Part 1: Concepts, materials and modellization. *Renewable and Sustainable Energy Reviews* 14: 31–55.

[7] Zanganeh, G., Pedretti, A., Haselbacher, A., et al. (2015). Design of packed bed thermal energy storage systems for high-temperature industrial process heat. *Applied Energy* 137: 812–822.

[8] Mertens, N., Alobaid, F., Frigge, L., et al. (2014). Dynamic simulation of integrated rock-bed thermocline storage for concentrated solar power. *Solar Energy* 110: 830–842.

[9] Prieto, C., Cooper, P., Fernández, A.I., et al. (2016). Review of technology: Thermochemical energy storage for concentrated solar power plants. *Renewable and Sustainable Energy Reviews* 60: 909–929.

[10] Martin, R. (2016). Germany runs up against the limits of renewables. *MIT Technology Reviews*, https://www.technologyreview.com

[11] Rinne, S. and Syri, S. (2015). The possibilities of combined heat and power production balancing large amounts of wind power in Finland. *Energy* 82: 1034–1046.

[12] Energy Storage News. (2018). *Batteries, Power-to-heat Combine at AEG's German Frequency Control Project*. https://www.energy-storage.news/news/batteries-power-to-heat-combine-at-aegs-german-frequency-control-project

[13] International Energy Agency. (2015). *World Energy Outlook 2015 Executive Summary*. www.iea.org.

[14] Majozi, T. (2013). Heat integration in batch processes. In: *Handbook of Process Integration, Minimisation of Energy and Water Use, Waste and Emissions* (ed. J. Klemes). A volume in Woodhead Publishing Series in Energy. ISBN: 978-0-85709-593-0.

[15] Stoltze, S., Mikkelsen, J., Lorentzen, B., et al. (1995). Waste-heat recovery in batch processes using heat storage. *Journal of Energy Resources Technology* 117: 142–149.

[16] Klemes, J. (ed.) (2013). *Handbook of Process Integration, Minimisation of Energy and Water Use, Waste and Emissions*. A volume in Woodhead Publishing Series in Energy, ISBN: 978-0-85709-593-0.

[17] Energinet.dk. (2016). *Technology Data for Energy Plants*, ISBN:www: 978-87-7844-931-3, accessed through www.energinet.dk.

[18] International Energy Agency. (2012). *Energy Technology Perspectives 2012 – Pathways to a Clean Energy System*. www.iea.org.

[19] Element Energy LTD. (2014). *The Potential for Recovering and Using Surplus Heat from Industry*, Final Report to the UK-DECC, https://www.gov.uk/government/publications

[20] Swedish District Heating Association. (2016). *Energy Supply* (statistics, in Swedish), www.svenskfjarrvarme.se

[21] Swedish Energy Agency (2008). *Styrmedel för industriell spillvärme* (Incentives for industrial surplus heat, report in Swedish). Report no ER 2008:15, www.energimyndigheten.se

[22] Connolly, D., Mathiesen, B.V., Østergaard, P.A., et al. (2013). *Heat Roadmap Europe 2: Second Pre-Study for the EU27*. Department of Development and Planning, Aalborg University. ISBN 978-87-91404-48-1.

[23] US Department of Energy. (2008). *Waste Heat Recovery: Technology and Opportunitie in US Industry*. http://www.eere.energy.gov

[24] Li, Y., Xia, J., Fang, H., et al. (2016). Case study on industrial surplus heat of steel plants for district heating in Northern China. *Energy* 102: 397–405.

[25] International Energy Agency. (2014). *Linking Heat and Electricity Systems*. France: Paris. www.iea.org.

[26] Noord Pool Groups. http://www.nordpoolspot.com, Lysaker, Norway.

[27] Frederiksen, S. and Werner, S. (2013). *District Heating and Cooling*. Lund, Sweden: Studentlitteratur AB. ISBN 978-91-44-08530-2.

[28] Swedish Energy Agency. (2015). Energy in Sweden 2015. www.energimyndigheten.se

[29] Chiu, N.W.J. and Martin, V. (2011). TES: Climate change mitigation solution? *International Conference on Sustainable Energy Storage*, Feb 21–25, 2011, Belfast, UK.

[30] ASHRAE. (2013). *District Cooling Guide*. Georgia, USA: ASHRAE.

[31] Gang, W., Wang, S., Xiao, F., et al. (2016). District cooling systems: Technology integration, system optimization, challenges and opportunities for applications. *Renewable and Sustainable Energy Reviews* 53: 253–264.

[32] Chiu, V., Martin, V., Setterwall, F., et al. (2009). System integration of latent heat thermal energy storage systems for comfort cooling. Presented at *the 11th International Conference on Thermal Energy Storage, EFFSTOCK 2009*, Stockholm, Sweden, June 14–18, 2009.

[33] World Steel Association. (2015). *World Steel in Figures 2015*, www.worldsteel.org

[34] Hasanbeigi, A., Price, L., Chunxia, Z., et al. (2014). Comparison of iron and steel production energy use and energy intensity in China and the U.S. *Journal of Cleaner Production* 65: 108–119.

[35] Yin, R. (2011). *Metallurgical Process Engineering*. Springer. ISBN 978-3-642-13956-7 (eBook).

[36] Rommel, M., Hauer, A., and Helden, W. (2015). *Compact Thermal Energy Storage: Material Development for System Integration – Joint Task of IEA SHC Task 42 and IEA ECES Annex 29*. www.iea-shc.org

[37] Bonhivers, J.-C. and Stuart, P.R. (2013). 25. Application of process integration methodologies in the pulp and paper industry. In: *Handbook of Process Integration (PI)*, 765–798. Woodhead Publishing Limited. https://doi.org/10.1533/9780857097255.5.765

[38] Persson, J. and Berntsson, T. (2010). Influence of short-term variations on energy-saving opportunities in a pulp mill. *Journal of Cleaner Production* 18: 935–943.

[39] Jönsson, J. and Algehed, J. (2010). Pathways to a sustainable European kraft pulp industry: Trade-offs between economy and CO2 emissions for different technologies and system solutions. *Applied Thermal Engineering* 30: 2315–2325.

[40] Li, H., Wang, W., Yan, J., et al. (2013). Economic assessment of the mobilized thermal energy storage (M-TES) system for distributed heat supply. *Applied Energy* 104: 178–186.

[41] Chiu, N.W.J., Castro Flores, J., Martin, V., et al. (2016). Industrial surplus heat transportation for use in district heating. *Energy* 110: 139–147.

[42] Krönauer, A., Lävemann, E., Brückner, S., et al. (2015). Mobile sorption heat storage in industrial waste heat recovery. *Energy Procedia* 73: 272–280.

33

Economy of Thermal Energy Storage Systems in Different Applications

Christoph Rathgeber, Eberhard Lävemann, and Andreas Hauer*

Bavarian Center for Applied Energy Research (ZAE Bayern), Walther-Meißner-Str. 6, 85748 Garching, Germany

33.1 Introduction

A major barrier to the development and the more extensive utilization of thermal energy storage (TES) technologies is cost uncertainty [1]. Reliable values or estimates for both acceptable and realized costs of different TES technologies in relevant applications are sketchy. This chapter treats the economy of thermal energy storages whose main purpose is to store and deliver a certain amount of thermal energy. In this way, the energy efficiency of a process might be increased, and the usage of fossil energy carriers reduced, thereby lowering the primary energy consumption and avoiding or reducing CO_2 emissions. In order to make a rough estimate of the economic viability of such an energy storage for a specific application and user, a simple tool was developed, which consists in determining the maximum acceptable storage capacity costs via a top-down approach and the realized storage capacity costs via a bottom-up approach.

However, certain thermal energy storage systems are installed not only to store a certain amount of heat or cold, but aim at providing additional benefits. For example, several large hot water storages are being operated by German energy suppliers in their district heating networks in order to improve the flexibility of attached CHP installations and, thus, to create higher gains at the electricity sales market. In this way, the investment in the storage is motivated in optimizing the operational strategies of the overall energy system. In other cases, the main purpose of TES systems is to provide a required magnitude of thermal power, e.g. when applying a buffer storage to quickly fill a bathtub or in the case of thermal storages as backup systems in industrial processes. Further examples of thermal energy storages whose costs should be assessed not only with the developed tool are stand-alone systems, e.g. storages for solar thermal collectors in remote off-grid locations, and comfort applications, e.g. phase change materials in textiles.

33.2 Methods to Evaluate Thermal Energy Storage Economics

Bottom-up approaches are insufficient to determine reliable limits for the acceptable costs of energy storage systems. Typically, they yield a large range of cost estimates for each energy storage technology due to uncertainties in material and manufacturing

Advances in Energy Storage: Latest Developments from R&D to the Market, First Edition.
Edited by Andreas Hauer.
© 2022 John Wiley & Sons, Ltd. Published 2022 by John Wiley & Sons, Ltd.

costs, R&D costs, installation and maintenance costs. Therefore, a top-down approach was applied to determine cost limits depending on the user type and the results were compared with realized costs per installed storage capacity calculated from data of existing systems via a bottom-up approach.

33.2.1 Top-down Approach

The top-down approach assumes that the cost of energy supplied by the storage should not exceed the costs of energy from the market[1] (hereinafter referred to as *REC* = reference energy costs). Following this assumption, the maximum acceptable storage capacity costs (hereinafter referred to as SCC_{acc}) are calculated from the discount rate of storage capital *i*, the payback period of the investment *n*, the number of storage cycles N_{cycle}, and the reference energy costs *REC* [2]. To simplify the evaluation, this analysis neglects operating costs and changes in the cost of energy production over time. Detailed information about the storage technology or implementation are not required for this approach.

Using the interest rate assigned to the capital costs and the payback period, the present value annuity factor *ANF* can be calculated to determine the present value of the energy storage capital. *ANF* as a function of payback period *n* and interest rate *i* can be calculated via Equation (33.1):

$$ANF = \frac{(i+1)^n \cdot i}{(i+1)^n - 1} \tag{33.1}$$

Interest rate *i* and payback period *n* depend on the user. Three classes of users are referred to in the following discussion. In the *industry* sector, high interest rates of 10% and above and short payback periods of 5 years and below are usual. For *building* applications, moderate interest rates of 5% and longer payback periods of 15–20 years are acceptable. In addition, one might also assume a user that can tolerate even longer payback periods of 25 years and low interest rates of 1%. The latter user class has probably political or ecological reasons for the investment and is hereinafter referred to as *enthusiast*. In Figure 33.1, the annuity factor *ANF* is plotted as a function of the payback period *n* for interest rates of 10% (solid line) indicating *industry*, 5% (dashed line) indicating *building*, and 1% (dotted line) indicating *enthusiast*.

In the *industry* sector, a payback period of five years yields an *ANF* of about 0.26. Therefore, a range of *ANF* from 0.25 to 0.30 is considered as storage capacity cost annuity for industrial users. In the *building* sector, *ANF* are within 0.07–0.10, and in the case of *enthusiasts*, consequently, low *ANF* between 0.04 and 0.06 can be achieved.

The maximum acceptable storage capacity costs SCC_{acc}, calculated in € per kWh installed storage capacity (€·kWh_{cap}^{-1}), are simply the product of the substituted reference energy costs *REC*, given in € per kWh energy (€·kWh_{en}^{-1}), and the number of storage cycles per year N_{cycle} divided by the annuity factor *ANF*:

$$SCC_{acc} = \frac{REC \cdot N_{cycle}}{ANF} \tag{33.2}$$

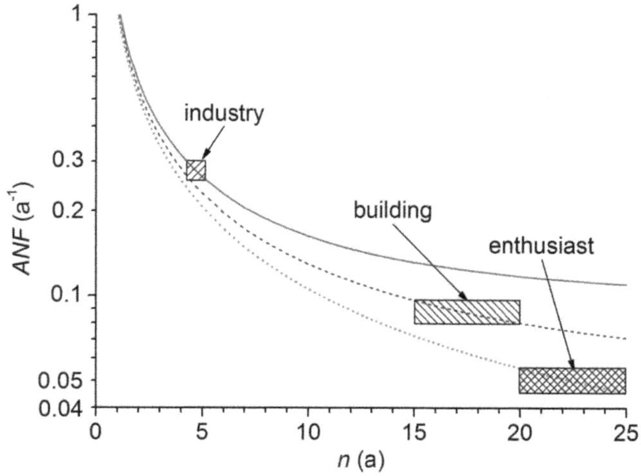

Figure 33.1 Annuity factor *ANF* as a function of payback period *n* for three user classes (industry $i = 10\%$, building $i = 5\%$, and enthusiast $i = 1\%$); framed regions indicate acceptable annuity factors for these user classes.

Equation (33.2) neglects operating costs and changes of *REC* over the payback period. Nevertheless, this analysis illustrates the relationship between acceptable storage capacity costs, the frequency of storage handling, and the costs of reference energy that is substituted by the storage system.

Similar to *ANF*, a range is considered for *REC*. As the focus of this work is to evaluate the costs of thermal energy storages, *REC* given in Table 33.1 correspond to heat or cold supply costs. Table 33.1 summarizes the economic boundary conditions of the three user classes that are taken into account in the top-down evaluation.

As an aid to orientation, expectable ranges for the costs of substituted reference energy *REC* and the storage annuity factor *ANF* are considered. In this way, a high- and a low-cost case are analyzed for each user class. The high-cost case considers the max. *REC* and the min. *ANF*, and the low-cost case the min. *REC* and the max. *ANF*, respectively. Future changes of the reference energy costs *REC* can be taken into account by adjusting the values of *REC* given in Table 33.1 appropriately, for example by considering average *REC* for the intended payback period. According to Equation (33.2), SCC_{acc} is proportional to *REC* and, hence, an increase in *REC* will cause a corresponding increase in SCC_{acc}. Operating costs should be taken into consideration if they are not negligible compared to the capital costs. Especially in the case of mobile storages, operating costs are expected to have a significant influence on the economic viability. To consider operating costs requires a modification of the top-down approach.

Table 33.1 Economic boundary conditions: costs of substituted reference energy *REC* and storage annuity factor *ANF* calculated via Equation (33.1).

User class	*REC* (€·kWh_{en}^{-1})		*ANF* (a^{-1})	
Industry	0.02	0.04	0.25	0.30
Building	0.06	0.10	0.07	0.10
Enthusiast	0.12	0.16	0.04	0.06

According to the procedure outlined earlier, the top-down approach calculates the costs per storage capacity. However, if operating costs have to be included, the costs per stored energy have to be determined, for instance on an annual basis.

33.2.2 Bottom-up Approach

The bottom-up approach focuses on the realized storage capacity costs of existing storage systems (hereinafter referred to as SCC_{real}). To investigate particular storages, a measure was developed that inquires among other technical parameters both actual and expectable investment costs *INC* of the storage. The investment costs are divided into costs of the heat storage material, costs of the storage container or reactor, and costs of the charging/discharging unit. For the bottom-up approach, sensible heat storage, latent heat storage using phase change materials (PCM), and thermochemical heat storage including sorption storage have been investigated. Besides commercially available storage systems, innovative prototypes that are the subject of ongoing research have been analyzed [3]. The realized storage capacity costs SCC_{real} are simply the investment costs *INC* divided by the installed storage capacity *SC*:

$$SCC_{real} = \frac{INC}{SC} \tag{33.3}$$

INC sums up heat storage material costs, storage container or reactor costs, and cost of charging and discharging the device. As in the case of SCC_{acc}, SCC_{real} are calculated in € per kWh installed storage capacity (€·kWh_{cap}^{-1}).

33.3 Comparison of Acceptable and Realized Storage Capacity Costs in Different TES Applications

The maximum acceptable storage capacity costs SCC_{acc} for the three user classes calculated via Equation (33.2) are plotted as a function of the annual number of storage cycles N_{cycle} in Figure 33.2. Solid lines indicate the high-cost case of each user class and dashed lines the low-cost case, respectively. A double-logarithmic scale was chosen to visualize both SCC_{acc} of long-term storages with only few cycles per year and short-term storages with several hundred cycles per year.

The results of the top-down evaluation as shown in Figure 33.2 indicate that, for a fixed cycle period N_{cycle}, SCC_{acc} depend on the user's economic environment. The low-cost case of the industry sector and the high-cost case of enthusiasts differ by a factor of about 60 in costs. Short-term storage with several hundred storage cycles per year, however, allows several hundred times higher storage costs because of the larger energy turnover.

For reasons of clarity, the comparison of SCC_{acc} (top-down approach) with SCC_{real} (bottom-up approach) is split up into four figures: long-term storages (Figure 33.3), hot-water storages up to 30 m^3 storage volume (Figure 33.4), and short-term storages (Figures 33.5 and 33.6). Sensible, PCM, and TCM storages are marked with triangles, circles, and squares, respectively. If stated by the developer/manufacturer of the storage and in order to indicate the operational readiness of a particular system,

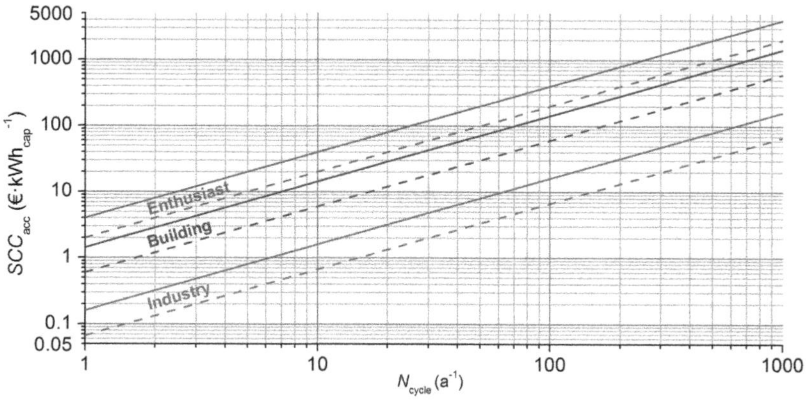

Figure 33.2 Maximum acceptable storage capacity costs SCC_{acc} for three user classes as a function of storage cycles per year N_{cycle}; enthusiast high/low case (solid/dashed line), building high/low case (solid/dashed line), and industry high/low case (solid/dashed line).

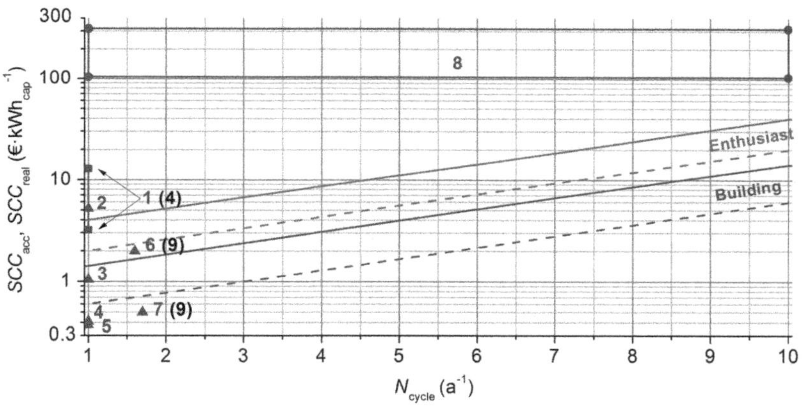

Figure 33.3 Maximum acceptable storage capacity costs (SCC_{acc}) and realized storage capacity costs (SCC_{real}) for long-term storages; numbers inside the parentheses indicate technology readiness levels (TRL) according to Horizon 2020 definition.

technology readiness levels (TRL) according to the Horizon 2020 definition[2] are indicated in parentheses along with the system number.

Seasonal TES with max. two cycles per year requires storage capacity costs below 3 €·kWh_{cap}^{-1} in the building and below 0.4 €·kWh_{cap}^{-1} in the industry sector, respectively. With respect to the storages under investigation, seasonal TES is only economical via large sensible hot water storages (cf. systems 3–7, Figure 33.3).

In the case of hot-water storage up to 30 m^3 storage volume, the building sector is usually targeted. However, these storages can become financially attractive for *industrial* applications if N_{cycle} is sufficiently high. Since these storages can be integrated in a variety of systems, exemplary ranges are indicated for N_{cycle}. The storage capacity *SC* of storages 9–13 is calculated for the maximum technically permissible temperature ranges indicated in Table 33.2. To evaluate the economics under differing application conditions, these temperature ranges have to be adjusted.

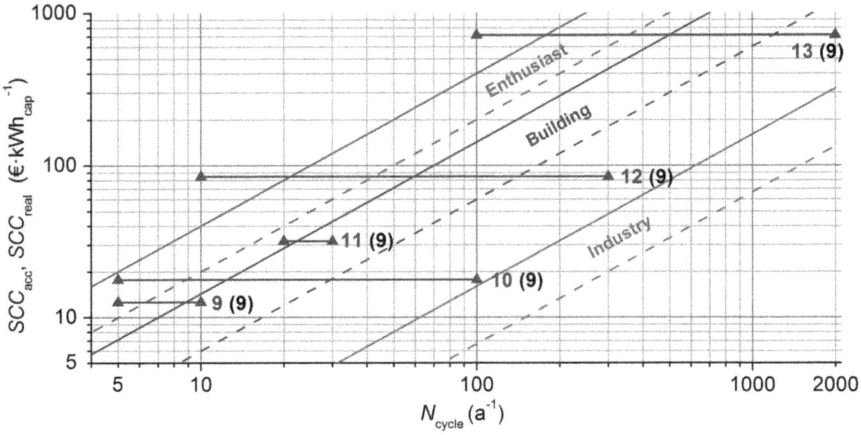

Figure 33.4 Maximum acceptable storage capacity costs (SCC_{acc}) and realized storage capacity costs (SCC_{real}) for hot water storages up to 30 m^3 storage volume; numbers inside the parentheses indicate technology readiness levels (TRL) according to Horizon 2020 definition.

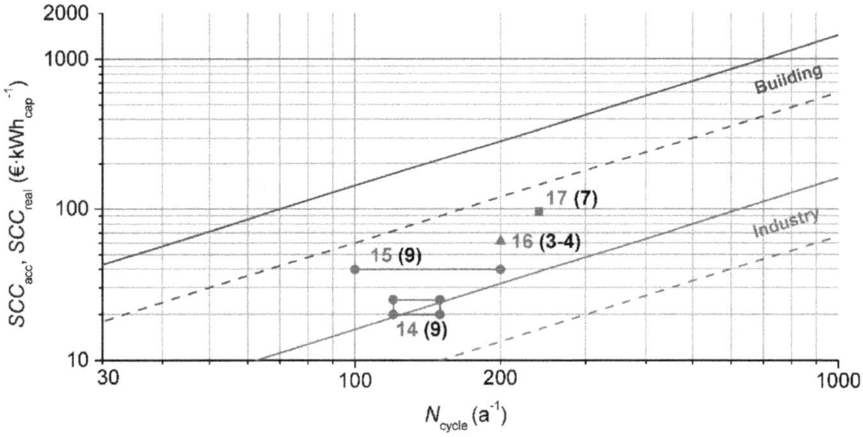

Figure 33.5 Maximum acceptable storage capacity costs (SCC_{acc}) and realized storage capacity costs (SCC_{real}) for industrial short-term storages; numbers inside the parentheses indicate technology readiness levels (TRL) according to Horizon 2020 definition.

On the other hand, in the case of short-term storages, storage systems are intended for either *industry* or *building*. Among the investigated short-term storages, systems 14–17 and 18–26 have been developed for *industry* and *building* applications, respectively.

Considering the investigated short-term storages for industrial applications (cf. Figure 33.5) it turns out that ice storages (system 14) are cost-effective, and other technologies are within reach.

In the case of the systems 1, 8, 14, 18, 20, 22, 23, and 26, cost ranges are given for SCC_{real} indicating the interval between actual costs (upper limit) and expectable costs that can be achieved in the near future (lower limit). The mobile PCM storage (system

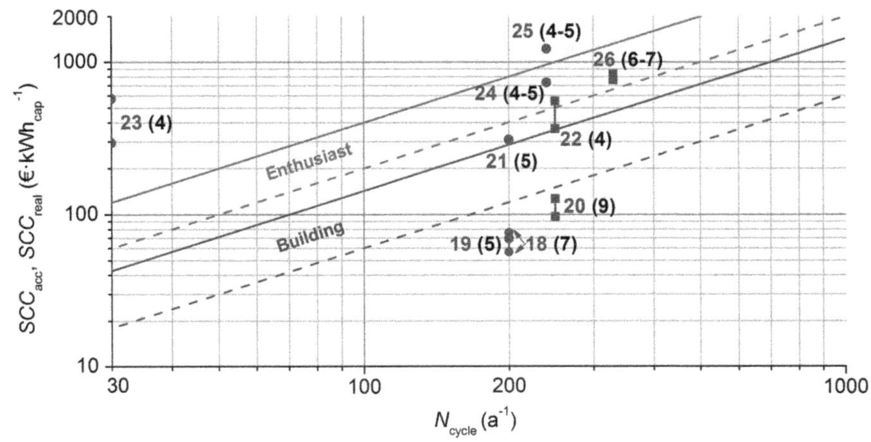

Figure 33.6 Maximum acceptable storage capacity costs (SCC_{acc}) and realized storage capacity costs (SCC_{real}) for short-term storages in buildings; numbers inside the parentheses indicate technology readiness levels (TRL) according to Horizon 2020 definition.

Table 33.2 Specifications of thermal energy storages investigated via Bottom-up approach: technology readiness levels TRL according to Horizon 2020 definition, annual number of storage cycles N_{cycle}, investment costs *INC*, installed storage capacity *SC*, and realized storage capacity costs SCC_{real}.

Storage system (Institution(s))	TRL	Ncycle (a^{-1})	*INC* (€)	*SC* (kWh_{cap})	SCC_{real} (€·kWh_{cap}^{-1})
1: NaOH sorption storage; seasonal storage for domestic applications (Empa)	4	1	8000–32 400	2500	3.2–13.0
2: Ottrupgård (1995); 1500 m^3 hot water; 35–60°C (PlanEnergi)	-	1	225 500	43 500	5.18
3: Sunstore 2 (2003); 10000 m^3 hot water; 35–90°C (PlanEnergi)	-	1	671 100	638 000	1.05
4: Sunstore 3 (2013); 60000 m^3 hot water; 10–90°C (PlanEnergi)	-	1	2 671 100	6 960 000	0.38
5: Sunstore 4 (2012); 75000 m^3 hot water; 10–90°C (PlanEnergi)	-	1	2 281 900	5 570 000	0.41
6: Ackermannbogen; 6000 m^3 hot water; 20–90°C [5] (ZAE Bayern)	9	1.6	942 400	472 400	1.99
7: Attenkirchen; hot water + borehole heat exchanger; 7000 m^3 H_2O_{eq}; 10–90°C (ZAE Bayern)	9	1.7	327 300	654 600	0.50
8: Seasonal storage based on supercooled sodium acetate trihydrate [6–7] (DTU, Univ. Graz)	-	1–10	2700–4120	13–26	104–317
9: Vacuum super insulated hot water storage; 30 m^3; 5–95°C (ZAE Bayern, Hummelsberger GmbH)	9	5–10[a]	37 888	3020	12.5
10: allSTOR VPS/3 2000/3-7; 2000 l hot water; 5–95°C (Vaillant GmbH)	9	5–100[a]	3559	202	17.6

(Continued)

Table 33.2 (Continued)

Storage system (Institution(s))	TRL	*Ncycle* (a^{-1})	*INC* (€)	*SC* (kWh_{cap})	SCC_{real} (€·kWh_{cap}^{-1})
11: Vacuum super insulated hot water storage; 5 m^3; 5–95°C (ZAE Bayern, Hummelsberger GmbH)	9	20–30[a]	15 962	504	31.7
12: actoSTOR VIH RL 500–60; 500 l hot water; 5–110°C (Vaillant GmbH)	9	10–300[a]	4953	58.7	84.4
13: actoSTOR VIH CL 20 second; 20 l potable water; 10–70°C; (Vaillant GmbH)	9	100–2000[a]	965	1.35	715
14: Ice storages based on nodules filled with water/ice; installations in Europe (Cristopia)	9	120–150	-	-	20–25
15: Mobile PCM storage (sodium acetate trihydrate); 40–90°C (Univ. Bayreuth, LaTherm)	9	100–200	99 000	2500	39.6
16: Sensible dual media storage up to 300°C; stone + heat transfer oil (ZAE Bayern, Gießerei Heunisch)	3–4	200	400 000	6500	61.5
17: Mobile sorption heat storage (2x14 t zeolite) for industrial waste heat recovery [8–9] (ZAE Bayern)	7	240	440 000	4600	95.7
18: 1 m^3 PCM storage ($CaCl_2·6H_2O$) for solar heating and cooling; 22–36°C (ZAE Bayern)	7	200	4700–6300	83	56.6–75.9
19: TubeICE; modular PCM tubes (Salt hydrate + graphite); 30–70°C (VITO)	5	200	900	13	69.2
20: Dishwasher with sorption drying; 1.5 kg zeolite; (ZAE Bayern)	9	250	29–38	0.3	96.7–127
21: PCM storage (RT58); 0.2 m^3; 30–70°C (VITO)	5	200	1850	6	308
22: LiBr sorption storage for domestic applications (ZAE Bayern)	4	250	31 000–47 000	85	365–553
23: PCM-Air heat exchanger; RT27; free-cooling [10–11] (Univ. Zaragoza)	4	30	2000–3900	6.8	294–574
24: VDSF: ventilated double skin facade + PCM (SP21); free cooling (Univ. Lleida)	4–5	240	5133	7	733
25: PCM (hydroquinone) storage for solar applications; 145–187°C (Univ. Lleida)	4–5	240	16 768	13.7	1223
26: Plate based PCM storage (RT60) for a domestic micro-CHP installation [12] (Univ. Basque Country)	6–7	330	5500–6000	7.2	764–833

[a] Storages can be integrated in a variety of systems with different N_{cycle}.

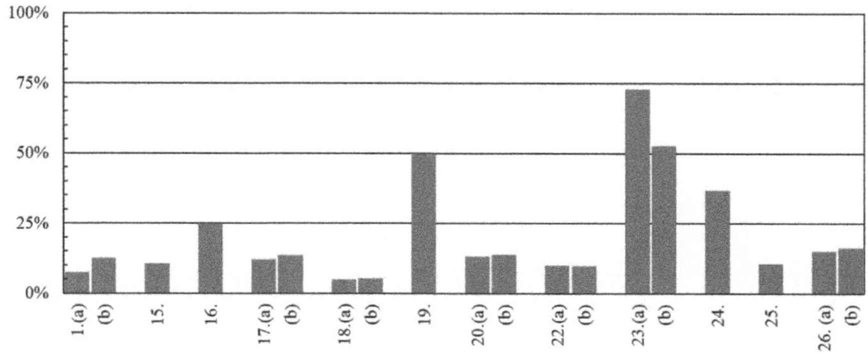

Figure 33.7 Fraction of costs of the heat storage material of the actual (a) and expectable (b) investment costs (*INC*); storage systems are numbered as in Table 33.2.

15) is intended to be operated for 100–200 cycles per year depending on the application.

INC of the large water storages 2–7 are DMC (direct material costs) of the installed systems. *INC* of the commercial water storages 9–13 are list prices [4]. In the case of the other investigated systems, *INC* correspond to DMC of prototypes or estimated DMC and, therefore, numbers are roughly rounded. In addition, most of these prototypes are subject of ongoing research and, hence, at a lower TRL with higher investment costs.

Relevant specifications of the investigated storage systems are listed in Table 33.2.

In order to identify major cost drivers and cost reduction potentials for the investigated storages, the composition of the investment costs *INC* has been analyzed. Figure 33.7 illustrates how *INC* of the thermal energy storages under investigation (except hot water storages) are divided into costs of the heat storage material itself and costs of the surrounding container or reactor including charging/discharging device. If available, both actual (a) and expectable (b) costs are given.

In the majority of cases, the costs of the container including charging/discharging device exceed the costs of the heat storage material by far. For 9 out of 12 investigated storages, the costs of the heat storage material account for 25% or less of the total *INC*. In just two cases, the costs of the heat storage material account for 50% or more of *INC*. The composition of both actual and expectable investment costs indicates the significant potential to reduce storage costs by developing cost-effective storage containers and charging/discharging devices. Therefore, rendering TES systems cost-effective requires both the identification of cost-effective heat storage materials and the employment of cost-effective storage container or reactor components.

33.4 Discussion on the Major Influencing Factors on the Economics of Thermal Energy Storage

In order to assess the comparison of maximum acceptable with realized storage capacity costs, the results for storage 17 (mobile sorption heat storage) are discussed in more detail as an example. Therefore, a sensitivity analysis of the maximum acceptable storage capacity costs SCC_{acc} has been carried out (Figure 33.8). The first solid line

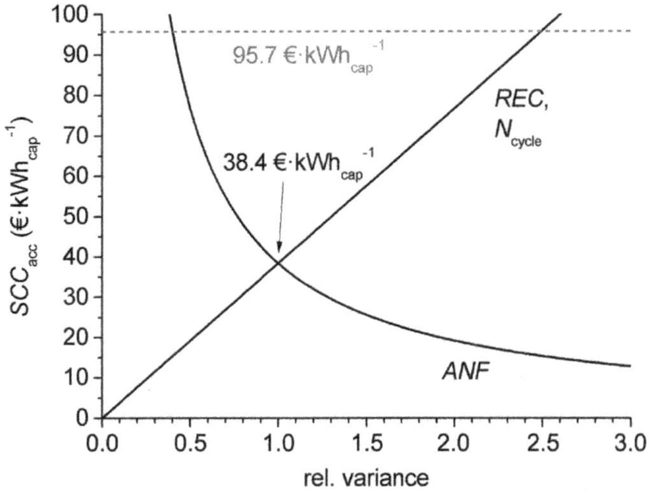

Figure 33.8 Sensitivity analysis of the maximum acceptable storage capacity costs SCC_{acc} for the Mobile sorption heat storage (system 17) calculated via Equation (33.2). Realized storage capacity costs SCC_{real} of 95.7 €·kWh_{cap}^{-1} are indicated as a horizontal dashed line.

indicates the reciprocal dependence of SCC_{acc} from the annuity factor *ANF*, and the second solid line the linear dependence of SCC_{acc} from both the reference energy costs *REC* and the number of storage cycles per year N_{cycle}.

The basic scenario (rel. variance = 1.0) corresponds to the conditions given in Table 33.2 and Figure 33.5: The upper limit of SCC_{acc} for the industry sector ($ANF = 0.25\ a^{-1}$, $REC = 0.04$ €·kWh_{en}^{-1}) considering $N_{cycle} = 240\ a^{-1}$ amounts to 38.4 €·kWh_{cap}^{-1}. However, the actually realized storage capacity costs are SCC_{real} = 95.7 €·kWh_{cap}^{-1}, indicated as a horizontal dashed line in Figure 33.8. The storage would be economically viable if $SCC_{real} \leq SCC_{acc}$. This can be achieved either by decreasing SCC_{real} or increasing SCC_{acc}. A reduction of SCC_{real} to 38.4 €·kWh_{cap}^{-1} requires either a reduction of investment costs *INC* from currently 440 000 € to approx. 176 000 € or an increase of the installed storage capacity *SC* from currently 4600 kWh to approx. 11 500 kWh. A possible increase of SCC_{acc} to 95.7 €·kWh_{cap}^{-1} is indicated with the solid lines in Figure 33.7. Either *ANF* is reduced by 40% from 0.25 a^{-1} to about 0.1 a^{-1}, or *REC* is increased by a factor of 2.5 from 0.04 €·kWh_{en}^{-1} to about 0.1 €·kWh_{en}^{-1}, or N_{cycle} is increased by a factor of 2.5 from 240 a^{-1} to about 600 a^{-1}. These changes of boundary conditions are necessary, if $SCC_{real} \leq SCC_{acc}$ would be achieved by changing only one of the conditions. However, if a combination of several conditions can be changed appropriately, the individual modifications are smaller.

33.5 Conclusions

A simple tool for the economic evaluation of thermal energy storages via a top-down and a bottom-up approach has been developed and tested on various existing storages. For a specific user class and application, this tool provides a rough estimate of the economic viability of a thermal energy storage whose main purpose is to store and deliver a certain amount of heat or cold. The top-down approach indicates some important findings in thermal energy storage economics that have often been ignored. First, for a fixed storage period, the maximum acceptable storage costs depend on the user's economic environment (e.g. *industry* or *building*) due to variances in payback period, discount rate, and costs of reference energy from the market. Second, the annual number of storage cycles has by far the largest influence on the maximum acceptable storage capacity costs and the cost effectiveness of storages. If the annual number of storage cycles is sufficiently high, all TES technologies can become economically competitive. Third, scenarios exist under which most storage technologies are economical. In this case, systems should be compared with regard to physical and technical attributes.

The bottom-up approach has been applied to analyze the costs of 26 thermal energy storages. Unlike commercial water storages, several innovative storages are the subjects of ongoing research at a lower TRL and, hence, their corresponding costs are approximately determined. The comparison of SCC_{acc} and SCC_{real} indicates that, at present, seasonal storage is only economical using large hot water storages; other technologies require at least an order of magnitude reduction in costs. That implies that the development of storage systems that allow a high annual number of storage cycles is economically favorable over seasonal storages with exactly one cycle per year because of the larger energy turnover. In addition, the bottom-up approach showed that a major fraction of the investment costs of the investigated storages are not costs of the heat storage material itself but costs of the storage container or reactor including charging/discharging unit. Therefore, R&D activities on cost-effective TES systems have to consider both cost-effective heat storage materials and cost-effective storage container or reactor components.

Acknowledgments

This study was part of IEA SHC Task 42/ECES Annex 29 “Compact Thermal Energy Storage – Material Development and System Integration” (http://task42.iea-shc.org). The work of ZAE Bayern was part of the project PC-Cools_V and supported by the German Federal Ministry for Economic Affairs and Energy under the project code 03ESP138A. The responsibility for the content of this publication is with the authors.

Notes

1 Communicated e.g. by Dr. Rainer Tamme, German Aerospace Center (DLR), in many presentations for the past 20 years.

2 Technology readiness levels (TRL) according to Horizon 2020 definition: http://ec.europa.eu/research/participants/data/ref/h2020/wp/2014_2015/annexes/h2020-wp1415-annex-g-trl_en.pdf.

References

[1] Campos-Celador, A., Diarce, G., Terés-Zubiaga, J., et al. (2013). Design of a finned plate latent heat thermal energy storage system for domestic applications. *SHC 2013 – International Conference on Solar Heating and Cooling for Buildings and Industry*, Freiburg, Germany (23–25 September 2013).

[2] Dallmayer, W., Kuckelkorn, J., Radspieler, M., et al. (2010). Begleitforschung: Solare Nahwärme am Ackermannbogen in München – SNAB. http://www.tib.uni-hannover.de/en.html

[3] Dolado, P., Lázaro, A., Marín, J.M., et al. (2011). Characterization of melting and solidification in a real scale PCM-air heat exchanger: Experimental results and empirical model. *Renewable Energy* 36: 2906–2917.

[4] Englmair, G., Dannemand, M., Johansen, J.B., et al. (2015). Testing of PCM heat storage modules with solar collectors as heat source. *SHC 2015 – International Conference on Solar Heating and Cooling for Buildings and Industry*, Istanbul, Turkey (2–4 December 2015).

[5] Hauer, A., Quinnell, J., and Lävemann, E. (2013). Energy storage technologies – Characteristics, comparison, and synergies. In: *Transition to Renewable Energy Systems* (ed. D. Stolten and V. Scherer), 557–577. Wiley-VCH.

[6] Johansen, J.B., Dannemand, M., Kong, W., et al. (2015). Practical application of compact long term heat storage. Laboratory testing of solar combi system with compact long term PCM heat storage. *SHC 2015 – International Conference on Solar Heating and Cooling for Buildings and Industry*, Istanbul, Turkey (2–4 December 2015).

[7] Krönauer, A., Lävemann, E., Brückner, S., et al. (2015). Mobile Sorptionsspeicher zur industriellen Abwärmenutzung, Grundlagen und Demonstrationsanlage (MobS II). http://www.tib.uni-hannover.de/en.html

[8] Krönauer, A., Lävemann, E., Brückner, S., et al. (2015). Mobile sorption heat storage in industrial waste heat recovery. *Proceedings of the 9th International Renewable Energy Storage Conference (IRES 2015)*, Düsseldorf, Germany (9–11 March 2015).

[9] Lázaro, A., Dolado, P., Marín, J.M., et al. (2009). PCM-air heat exchangers for free-cooling applications in buildings: Experimental results of two real-scale prototypes. *Energy Conversion and Management* 50: 439–443.

[10] Rathgeber, C., Hiebler, S., Lävemann, E., et al. (2015). IEA SHC Task 42/ECES Annex 29 – A simple tool for the economic evaluation of thermal energy storages. *SHC 2015 – International Conference on Solar Heating and Cooling for Buildings and Industry*, Istanbul, Turkey (2–4 December 2015).

[11] Rathgeber, C., Hiebler, S., Lävemann, E., et al. (2015). Economic evaluation of thermal energy storages via top-down and bottom-up approach. *Greenstock 2015 – 13th International Conference on Energy Storage*, Beijing, China.

[12] Rathgeber, C., Lävemann, E., and Hauer, A. (2015). Economic top-down evaluation of the costs of energy storages – A simple economic truth in two equations. *Journal of Energy Storage* 2: 43–46.

Part VI

Energy Storage Concepts, Regulations, and Markets

34

Energy Storage Can Stop Global Warming

Halime Ö. Paksoy

Department of Chemistry, Cukurova University, Adana, Turkey

34.1 Introduction

Energy storage is a very urgent consideration and a matter of life or death or death for human beings and the earth. The devastating results of climate change are already affecting our lives, economy, and security of our future. Rapid actions need to be taken before it is too late. Energy storage has two imperative roles in these rapid actions against climate change.

1. Building resilience for critical infrastructures against more frequent natural disasters seen as a result of climate change and other urgent global health crisis such as the recent COVID-19 pandemic.
2. Decreasing CO_2 emissions to mitigate climate change.

34.1.1 Energy Storage for Resilience of Critical Infrastructures

Energy plays a vital role in secure and continuous functioning of critical infrastructures. Any nation's security and public health relies on these infrastructures that may vary with geopolitical conditions of each country. Some examples are power plants, industrial, government facilities, information centers, health care, financial services, food and agriculture, emergency services [1]. Energy is essential to provide five main critical services to these infrastructures: communications, transportation, heating and cooling, health care facilities, and water supply. Common usage of these services leads to complex interdependencies of critical infrastructures [2]. An interruption in any of these sources generates immediate ripple effects in sectors near or far–posing high risks to communities.

Climate change consequences appear as heat waves, floods and storms that occur more often and with higher intensities each season. Meteorological disturbances trigger cascade effects in critical infrastructures. In case of a power outage, the average total cost in Germany is estimated as 430 million Euros/day [3]. A similar scenario for the US would be several tens of billions of dollars per year [4]. This harm may be avoided by investing in correct forms of storing energy. The ultimate goal is saving human lives. Energy storage can be incorporated in any of the following steps of the value chain of an energy system [4]:

- generation,
- transmittance,

Advances in Energy Storage: Latest Developments from R&D to the Market, First Edition.
Edited by Andreas Hauer.
© 2022 John Wiley & Sons, Ltd. Published 2022 by John Wiley & Sons, Ltd.

- distribution, and
- end-user.

Using energy storage alleviates losses from disruptions and enables uninterrupted and secure supplies of energy.

Climate change related natural disasters are not the only cases where energy storage has a building resilience role. For other natural disasters like earthquakes energy storage can bring relief. Shortly after the massive earthquake and tsunami disasters that destroyed four nuclear plants in Japan in 2011, a new thermal energy storage project with the dual purpose of disaster relief was introduced [5]. Vertical water storage tanks stacked on top of one another connected by flexible pistons were used for high-rise buildings in densely populated metropolitans. For a 65 m high building vertical storage tanks of 210 m^3 on each side of the building were installed. Water storage tanks were designed for peak energy demand control, load shifting, and also serving as water reservoir in case of disasters.

34.1.2 Energy Storage for Mitigating Climate Change

Energy systems around the world today rely heavily on fossil fuels. Total primary energy supply (TPES) and total final consumption (TFC) in the world have reached 13 699 Mtoe (megatonnes of oil equivalent) and 9425 Mtoe in 2014 respectively [6]. The difference of 4274 Mtoe between these values is the energy lost as waste heat during energy conversion or use. This shows that about 30% of the TPES is lost and is not recovered. In 2014 66.4% of TFC was met by fossil fuels and 66.7% of the electricity generation that accounts for 18.1% of the TFC was also met by fossil fuels [6].

The climate change threatening the earth's future is a result of our fossil fuel dependence. Fossil fuels are responsible for these emissions that have reached 32 381 $MtCO_2$ in 2014 and annual CO_2 emissions have increased by more than 50% since 1973 [6].

In 2015, the average concentration of CO_2 in the atmosphere reached to 399 ppm, 40% higher than the pre-industrial level with an average increase of 2 ppm/year in the past decade [7]. The Intergovernmental Panel on Climate Change (IPCC) states that the human influence on climate change is clear and evident from increasing greenhouse gas emissions [8]. Energy has the highest share of 68% in all human activities that produce greenhouse gases [7].

The Conferences of the Parties (COP21) Agreement of Paris in 2015 called countries to make commitments that ultimately will result in no more than 2°C, and ideally stay within 1.5°C, warming above pre-industrial levels. The global energy outlook of the IEA Energy Technology Perspectives [9] shows that to achieve the 2°C scenario, energy-related CO_2 emissions must be halved and the share of fossil fuels in primary energy should decrease to 45% by 2050. Contributions of different technologies to this scenario are given by [9]:

- end-use energy efficiency 38%,
- renewables 32%,
- carbon capture and sequestration (CCS) 12%,
- end-use fuel switching 10%,
- nuclear 7% and
- power generation efficiency and fuel switching 1%.

The highest contributions are expected to come from energy efficiency and renewables. Energy storage is inevitable in any energy system to realize these contributions.

Energy systems are transforming into more complex structures to meet these challenges. In the traditional energy system, central generation plants provide energy to end-users through, usually, electricity and gas grids. In this top-down control approach, supply and demand can be matched by balancing the production on the supply profile. The ambitious targets of increasing renewables and de-carbonization make the new energy systems too complex. In the new energy system, prosumers (producers and consumers) of renewables can begin operating with new bottom-up control logic. The gap between production and consumption of renewables requires energy management technologies involving different kinds of storage and controls [10]. Energy storage systems will provide flexible solutions to facilitate the transition to a new energy system.

34.2 Energy Storage Technologies

Energy can be stored to be retrieved at a later time, a different place and maybe at different temperature levels. This is possible based on thermodynamic laws of energy transformations between different energy forms, such as thermal, mechanical, chemical, and magnetic. Transformations can be grouped into energy storage technologies as shown in Figure 34.1. This handbook covers all types of energy storages shown in this figure: Superconducting magnetic energy storage (SMES; see Part I), Pumped Hydro Storage (PHS; see Part II), Compressed Air Energy Storage (CAES; see Part III), Electrochemical (see Part I), Hydrogen and CO_2 (see Part IV), Thermal Energy Storage (TES; see Part V).

Each of these energy storages can be used at different steps in the value chain of an energy system or jointly with another kind of storage. System power ratings vary between kWs to GWs and duration of storage may last for seconds or even several months. Figure 34.2 shows possible locations of different energy storages in the energy value chain [11].

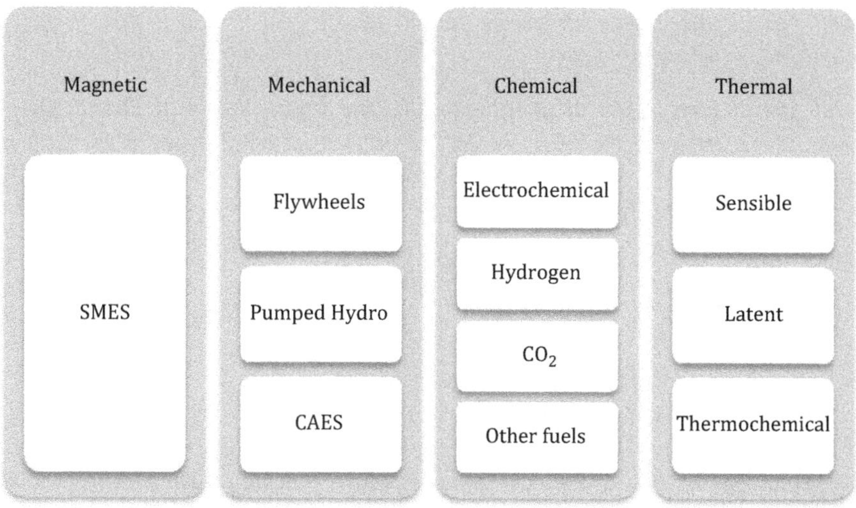

Figure 34.1 Energy transformations and energy storage technologies.

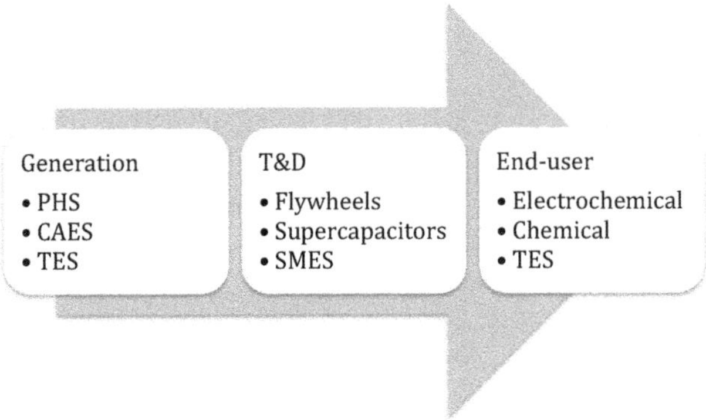

Figure 34.2 Energy storage technologies at different stages of energy value chain.

34.3 Energy Storage Systems

Energy storage presents energy savings and low carbon solutions in energy systems within a sector.

- Buildings – heating, cooling, power, water treatment for residential, service buildings, and urban energy systems.
- Industry – process heat/cold, combined heat and power.
- Power generation – renewables, hybrid systems with renewables and fossil fuels.
- Transportation – hybrid/electric vehicles, de-icing of highways, recovering brake energy in train systems.
- Agriculture – greenhouses, storage facilities, refrigeration.

There are also cross-sectoral applications of energy storage. Some examples are:

- Power ⟷ Buildings ⟷ Industry;
- Power ⟷ Buildings ⟷ Transportation;
- Power ⟷ Buildings ⟷ Agriculture; and
- Agriculture ⟷ Industry.

In such applications, surplus energy produced in one sector can be used as supply in another sector. Energy storage closes the gap between these sectors in terms of distance and time. Surplus energy from one sector can also be transformed into other energy forms before it is used in another sector. Some of these flexible concepts are called:

- power-to-heat
- power-to-*X*,
- vehicle-to grid (V2G).

The power-to-heat concept is used for power generated from renewables to be used to produce heat before it is provided to the customers. Power-to-*X* is used for concepts where *X* stands for any chemical that can be produced using renewable power. This can be H_2 that is produced by photocatalysis or by solar thermal. *X* can also be another chemical such as CO_2, which can be converted to methanol by using renewable power. In the V2G concept plug-in hybrid electric vehicles can provide storage to match supply with demand [12].

34.4 The Potentials of Energy Storage

The energy storage potential or benefits depend on the technology used and its size according to the target application. Both technical and economic aspects have to be considered. Among the parameters that influence the choice of storage technologies are: investment cost, energy or power densities, number of annual cycles, and impact on the environment. A type of storage may have a different value in a sector with varying user-demand profiles. Sometimes, storage is the only choice, where it meets a need that no other technology can provide, i.e. in mobile phones.

It is hard to estimate the overall potential of energy storage in an energy system. The potential of energy storage can be classified under the four areas shown in Figure 34.3. These groups may be inter-related and energy storage can provide benefits for more than one of these; this is called "benefits-stacking."

34.4.1 Optimizing Renewable Integration

Meeting ambitious renewable energy targets for a low-carbon future necessitates high penetration of renewables in any sector. This, in turn, requires the optimized integration of renewables. Surplus renewable energy that may be otherwise wasted at peak hours needs to be stored for later use when demand gets high. This will enable continuous use of renewables by avoiding variability and also increase the value of the investment made on renewables. Depending on the energy mix that is replaced by renewables, different levels of fossil fuels and CO_2 emissions can be saved.

In a study by [2], the potential of thermal energy storage is estimated in terms of load reduction, replaced thermal or electrical energy, and CO_2 emissions reductions in the buildings and industrial sectors. The energy systems included in buildings in this analysis were seasonal solar thermal, district heating, solar short-term storage, and passive cold storage. Within the industrial sector, combined heat and power, process heating

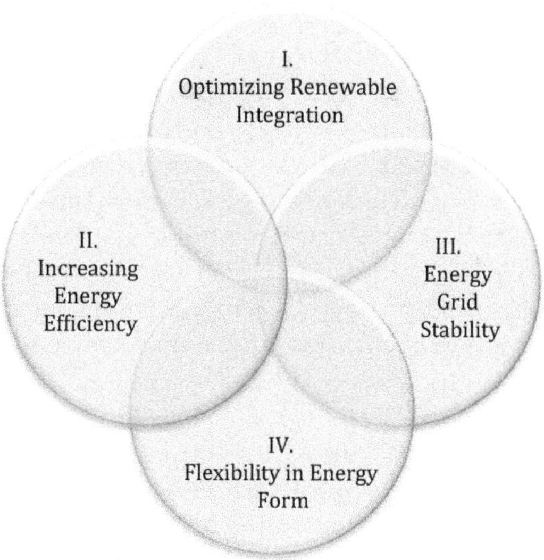

Figure 34.3 Potential of energy storage.

and cooling, and concentrated solar power plants (CSP) were analyzed. Results show that 2.6 million tonnes of CO_2 emissions can be reduced in EU-25 countries based on 2005 figures.

Optimizing the integration of renewable power generated from solar or wind into the grid can also be done by energy storage. There are different kinds of electrochemical and mechanical storage options that can allow use of renewables as continuous and reliable power sources.

Previous chapters provide detailed information on storage options. Battery storage solutions that store from the grid and feed it back to the grid require additional components like transformer, AC/DC protection equipment, inverter, and a management system. Optimizing battery capacity is important in minimizing the investment and operation cost while maximizing renewable penetration into the grid [14].

Pumped hydro storage is often used in connection to wind farms to optimize renewable power integration. At off-peak hours when wind is available and demand is low, renewable power is used to run the pumps to facilitate pump hydro storage. This enables optimized scheduling and the ability to provide operating reserves through pumped hydro storage. An important aspect to be considered for this technology is effect of climate change on water availability through the changes in rain patterns.

34.4.2 Increasing Energy Efficiency

Energy storage, depending on where it is integrated in the value chain of energy system shown in Figure 34.1, can increase energy efficiency at different levels.

On the generation level, power plants want to maximize the energy produced and the volume of energy sold. In renewable power generation plants energy storage will increase the power produced per installed renewables. Energy storage can also avoid the use of additional spinning reserves and reduce the cost of maintaining such reserves. The application of energy storage on the generation level serves both optimizing renewable integration and increasing energy efficiency.

Energy storage plays a dual role of providing both heat and electricity in photovoltaic thermal (PV-T) hybrid solar cells. The most efficient PV cells today have reached 46% efficiency [15]. These cells are gaining acceptance in concentrated photovoltaic (CSV) solar farms, but are still too expensive for flat panel rooftop or wall applications. In flat panel applications the commercial monocrystalline PVs have maximum efficiency of around 20% [16]. About 80% of the incident solar energy is wasted as heat during conversion. Part of this can be utilized to meet heat demand. PV-T uses thermal storage in PV panels to supply electricity and heat. In a commercial product such application claims to provide up to 300% more energy (www.solarwall.com). Heat removed is supplied to the HVAC (heating, ventilation, and air conditioning) system of the building in this application. The secondary benefit of using thermal energy storage is to reduce the operating temperature of the PV modules, which improves the efficiency. The power output drops by 0.65% for every 1°C increase above 25°C [17].

Thermal energy storage in CSP plants reduces the cost of electricity and increases energy performance in a cost-effective way [18, 19]. Molten salt is the most common thermal energy storage using for CSPs. In traditional power plants combined with CSP, waste heat can also be stored using thermal energy storage to increase the energy efficiency of generation and also optimizing the CSP plant [20].

For transmission and distribution level, energy storage does not directly increase the energy efficiency of the grid. By integrating energy storage for optimum consumption of renewable energy at the place of production, the load of the transmission, and distribution lines will be decreased. This will help alleviate the congestion problem in transmission lines. SMES and flywheels are used to store energy and then discharge energy from the grid in duration of seconds to minutes.

In district heating, and cooling, networks, including energy storage, can increase energy efficiency. Energy storage enables the use of renewable energy sources (solar, wind, natural heat, and cold) and waste heat and cold from the vicinity of the networks. This application of energy storage increases the capacity of the network and the amount of heat and cold that can be delivered to consumers. Using waste heat or cold sources nearby increases energy efficiency of both the supplier and the network. Underground thermal energy storage is used in district heating networks for efficient utilization of solar energy. Sweden has examples of underground thermal energy storages that provide natural heat/cold efficiently to district networks [21]. In another example from Denmark waste heat from industry is used to increase the capacity of the district heating network with underground thermal energy storages [22].

On the end-user side there are several roles energy storage can have in increasing energy efficiency: highly efficient heating and cooling, reduced cost of energy by peak shifting, and distributed energy generation opportunities.

Energy storage can also increase energy efficiency in industry. This could be done by saving primary energy used for power, heat, and cooling through integration of renewables, waste heat recovery, and inter-sectoral utilization of the waste heat. IEA ETP 2016 shows that waste heat from industry can be used as a supply to meet the demand in urban systems [16]. Transportation of waste heat from industry to other industries or urban systems can also be realized by mobile thermal energy storage systems or through pipes. Chapters 25 and 32 have more on industrial applications.

Energy saving potential of energy storage is not limited to providing heat/cold and power. Waste heat recovery in domestic appliances, thermal management of electronics, and batteries are among the applications that energy storage can increase energy efficiency. Considering the many uses of such equipments around the world, energy storage has an enormous cumulative energy management and conservation effect. Waste heat applications for dishwashers in Germany and Turkey show that energy efficiency can be increased depending on the energy storage technology.

Greenhouses are becoming more important in agriculture in food production and consuming locally produce is popular. Locally grown produce has to be transported less distance thus reducing lorries' use of fossil fuels. The increased productivity per growth area is an important factor in increasing greenhouse agriculture around the world. Greenhouses can produce efficiently in any climate – cold or hot – through the use of energy storage [23].

34.4.3 Energy Grid Stability

High penetration of renewable power as a consequence of efforts to meet renewable energy targets for low carbon future affects the stability of the energy grid. PV and wind power that are an increasing share in the energy portfolio of many countries today introduces variability and reduced production predictability in the power system. A high penetration of such renewable sources is expected to create voltage and power fluctuations that also lead to frequency fluctuations. Energy storage can provide a stable grid by regulating frequency and controlling voltage and reactive power.

Increasing the number of renewable energy prosumers feeding the grid in variable amounts and times causes problems in places like California and Germany already reaching high renewables integration targets. The humps, peaks, or even spikes occurring in the load curves can be managed with energy storage that gives optimum operation conditions for utilities. Large investments to meet peak demands can be avoided. Additional infrastructure for new transmission/distribution lines can also be reduced.

Thermal energy storage in the form of ice storage or cold-water storage is used for load leveling and peak shaving purposes in some countries and regions, especially with high cooling demand. In this application, different electricity tariffs during the day are utilized to generate cold at peak hours. Chillers are run using cheaper electricity during the night to produce ice or cold water that will be used for cooling during the day.

Managing a "green" grid with high renewable penetration has some challenges. An example can be given from California, who has an energy and environment policy initiative to produce 50% of retail electricity from renewable power by 2030. California Independent System Operator (CAISO) carried out an analysis on how the real-time electricity net demand will be affected to meet this political initiative. The results of this analysis when plotted on a net daily load curve resembles a duck shape – so called "The Duck Curve." This curve shows different regions (times of the day) that needs to be managed properly for a stable "green" grid [24].

- Early morning: when people start to wake up and demand for electricity increases with a steep ramp of 8000 MW.
- Sunrise: when a supply of solar energy is available and ramp goes down.
- Noon: when solar energy is at the maximum level for over generation.
- Sun set: when other resources are needed to meet the steep ramp of 11 000 MW going up in the first 3 hours of early evening.
- Late night: when demand starts decreasing and people are going to bed requires reduced generation.

Operating the grid under these conditions requires flexibility that can perform the following functions [24]:

- sustain downward or upward ramps,
- respond for a defined period of time,
- change ramp directions quickly,
- store energy or modify use,
- react quickly and meet expected operating levels,
- start with short notice from a zero or low-electricity operating level,
- start and stop multiple times per day, and
- accurately forecast operating capability.

The shape of "The Duck Curve" can be different depending on the time of the year and location on the earth. Energy storage can offer flexible electrical and thermal solutions to the challenges seen in this curve by:

- reacting quickly with fast responding high power alternatives;
- providing energy as long as it is needed;
- reducing demand for heating, cooling and hot water; and
- providing homogeneous indoor temperature distribution to optimize on/off times of HVAC during the day.

Insufficient flexibility in the power network can cause negative electricity pricing in liberalized electricity markets. Energy storage can be used to control or benefit from such instances. Combinations of high renewable generation and low energy consumption caused electricity prices to be negative from 7 a.m. to 5 p.m. on Sunday, 8 May 2016 in Germany. The small increase in the ratio of renewable to conventional power resulted in a hugely negative price spike [25]. Storing energy in batteries or using the energy to produce heat and/or cold and then storing it during these negative pricing periods may give favorable options for producers and consumers.

34.4.4 Flexibility in Energy Form

Energy storage by linking electricity, gas and heat/cold networks can provide flexibility in the energy forms that can be supplied to end-users.

Renewable power at times when it is abundant can be used to run heat pumps to meet heat demand or store produced heat for later use. Similarly excess renewable power can be used to run chillers/refrigerators and meet cold demand or store cold for later use.

Alternative fuels like H_2 and CO_2 that require high energy for production can be produced using renewable power. This concept is called "Power to *X*" as explained before. Electrolysis of water to produce H_2 with renewable power gives a clean fuel that can be used with a fuel cell in any system from hybrid cars to large building infrastructures. CO_2 being a greenhouse gas can be converted to CH_4 by a reforming reaction using renewable power.

34.5 Policy Frameworks

Many countries around the world are putting new policies in action to meet the demanding targets for climate change. In the core of most of these policies is increasing the proportion of renewables and energy efficiency. Energy storage is one of the key technologies that can assist fulfillment of these policies. Strong policy actions on national and local levels are needed to strengthen the position of energy storage in the energy system. Among the drivers that strongly support more use of energy storage are [11]:

- movement towards de-carbonization,
- energy security concerns,
- aging energy system infrastructure,
- decreasing cost of PVs and higher penetration into grid, and
- more emphasis on decentralized energy production.

Current policy and regulatory frameworks do not have sufficient instruments for cost-efficient deployment of energy storage. Some countries already government actions that support energy storage technologies deployment. These actions can be grouped into following types of support [11]:

- Direct mandate, market evolution (Canada, US).
- Demonstration project, performance testing (China, South Korea).
- Support of RD&D framework on energy storage, documentation (Germany).
- Direct subsidy for distributed storage (Germany).
- International Collaboration (EU, US).
- Policy framework development (EU).

The energy storage potential discussed shows that energy storage can be used anywhere in the value chain of the energy system. It can bring overlapping or separate benefits. In liberalized electricity markets, it is difficult to determine who is the owner of these and who can receive the direct payment. This makes compensation of energy storage and making investment decisions difficult. Energy storage systems do not fit into current regulatory frameworks [11].

Public–private partnerships is a model that can be employed to accelerate development and deployment of energy storage technologies. An example is Yokohama Smart City Project in Japan, where the city is collaborating with various public sectors for introducing renewable energy, managing the energy of households, buildings, and communities, and next generation transport systems (www.city.yokohoma.lg.jp). The key technology used in all of the initiatives designated as Community Energy Management System (CEMS), Home Energy Management System (HEMS), Building Energy Management System (BEMS), Factory Energy Management System (FEMS), and electric vehicles (EV) is energy storage.

34.6 Cross-cutting Aspects

Rapid implementation of energy storage starts with social awareness. Community involvement in renewables and energy efficiency targets can be as important as the policy framework. Good demonstration projects that draw attention and show the benefits of storage are strong initiators of social awareness. The success of such projects is followed by social acceptance of the technology. Joint efforts of education, governmental, and non-governmental institutes are needed to increase social awareness.

Education and training are needed to expand skilled employees, technical staff, professionals, and researchers that can develop, produce, and install energy storage technologies. There is strong link with faster development of these technologies and the R&D budget allocated to energy storage projects.

Databases can show the potential and benefits that can be achieved with energy storage. Department of Energy (DoE) has established a global energy storage database (https://www.energystorageexchange.org/projects). The database provides information on 985 electrochemical, 342 PHS, 206 TES (ice and cold water storage), 70 mechanical, and 13 hydrogen storage projects, which are all grid connected. This database does not cover all energy storage applications that are included in this handbook. There is a need for a more comprehensive database of all kinds of energy storage projects.

Standards and guidelines are needed for proper use of technologies and to avoid any mistakes. Problems seen with improper application can stop or slow down the introduction of energy storage technologies into the market.

Cybersecurity is also an important issue to consider with increasing smart technologies and the network of things that are used together with energy storage technologies. Urban energy systems that rely on communication technologies are under threat of terrorist attacks that may cause fatal malfunctioning of daily lives. Energy storage can also have an important building resilience role in such instances as explained in the Introduction.

34.7 Conclusions

Energy storage is the key technology in meeting the ambitious climate change goals. Various energy storages technologies can be applied at any level of the energy system to optimize renewable integration in more secure and efficient ways. Using energy storage increases the variety of products and services from renewable energy, which increases the competitiveness on the market and the added-value. Clear policy frameworks are needed for energy storage to take its position in the energy system. Good policy can also help financing energy storage systems and speed up market penetration. Cross-cutting aspects can be as important as policy frameworks and should be emphasized more for a successful introduction of energy storage into society.

References

[1] Andersson, O. (2015). Long-term performance of the HT_BTES in Emmaboda, Sweden, *13th International Conference on Energy Storage*, Beijing, China, (May 2015).

[2] Arce, P., Medrano, M., Gil, A., et al. (2011). Overview of thermal energy storage (TES) potential energy savings and climate change mitigation in Spain and Europe, *Applied Energy* 88(8): 2764–2774

[3] Barbour, E. (2016). A new high in renewable electricity for Germany – And a low in electricity prices! http://energystoragesense.com/author/eddie.

[4] Bender, S. (2016). The relevance of cascading effects for adapting critical infrastructures to climate change. *4th International Climate Change Adaptation Conference*, Rotterdam, The Netherlands (10–13 May 2016).

[5] Brunet, Y. (2011). *Foreword in Energy Storage* (ed. Y. Brunet). Hoboken, NJ: John Wiley & Sons Inc.

[6] California, I.S.O. (2016). What the Duck Curve tells us about managing a green grid. https://www.caiso.com/Documents/FlexibleResourcesHelpRenewables_FastFacts.pdf.

[7] Cavallaro, F. (2010). Fuzzy TOPSIS approach for assessing thermal-energy storage in concentrated solar power (CSP) systems. *Applied Energy* 87(2): 496–503.

[8] Dimroth, F., Grave, M., Beutel, P., et al. (2014). Wafer bonded four-junction GaInP/GaAs//GaInAsP/GaInAs concentrator solar cells with 44.7% efficiency. *Progress in Photovoltaics: Research and Applications* 22(3): 277–282.

[9] Fathima, H. and Palanisamy, K. (2015). Optimized sizing, selection, and economic analysis of battery energy storage for grid-connected wind-PV hybrid system. *Modelling and Simulation in Engineering* 2015, Article ID 713430: 16 pages. http://dx.doi.org/10.1155/2015/713430.

[10] Green, M., Emery, K., Hishikawa, Y., et al. (2015). Solar cell efficiency tables (Version 45). *Progress in Photovoltaics: Research and Applications* 23: 1–9.

[11] Homeland Security. (2015) Energy Sector Specific Plan. https://www.dhs.gov/publication/nipp-ssp-energy-2015.

[12] Ibamoto, T. (2011). Japan country report, *IEA ECES Executive Committee Meeting (XC71) minutes*, Istanbul, Turkey (11–12 May 2011).

[13] IEA. (2014). Technology roadmap on energy storage. http://www.iea.org/publications/freepublications/publication/technologyroadmapenergystorage.pdf.

[14] IEA. (2016). International energy agency key world energy statistics. http://www.iea.org/publications/freepublications/publication/KeyWorld2016.pdf.

[15] IEA. (2016). International energy agency CO2 emissions from fuel emissions highlights. https://www.iea.org/publications/freepublications/publication/CO2EmissionsfromFuelCombustion_Highlights_2016.pdf.

[16] IEA. (2016) International energy agency, energy technology perspectives: Towards sustainable urban energy systems (June 2016). http://www.iea.org/media/etp/etp2016/ETP2016_Webinar_ALL.pdf.

[17] IPCC. (2013). Summary for policymakers. In: *Climate Change 2013: The Physical Science Basis. Contribution of Working Group I to the Fifth Assessment Report of the Intergovernmental Panel on Climate Change* (ed. T.F. Stocker, D. Qin, G.-K. Plattner, et al.). Cambridge, UK and New York, NY: Cambridge University Press.

[18] Kuravi, S., Trahan, J., Goswami, D.Y., et al. (2013). Thermal energy storage technologies and systems for concentrating solar power plants. *Progress in Energy and Combustion Science* 39(2013): 285–319.

[19] Lund, H. and Kempton, W. (2008). Integration of renewable energy into the transport and electricity sectors through V2G. *Energy Policy* 36(9): 3478–3487.

[20] Ostergaard, P.A. and Lund, H. (2011). A renewable energy system in Frederikshavn using low-temperature geothermal energy for district heating. *Applied Energy* 88(2): 479–487.

[21] Paksoy, H.Ö. and Beyhan, B. (2015). Thermal energy storage (TES) systems for greenhouse technology. In: *Advances in Thermal Energy Storage Systems* (ed. L.F. Cabeza), 533–547. Duxford, UK: Woodhead Publishing.

[22] Radziemska, E. (2003). The effect of temperature on the power drop in crystalline silicon solar cells. *Renewable Energy* 28: 1–12.

[23] Schleicher-Tappeser, R. (2012). How renewables will change electricity markets in the next five years. *Energy Policy* 48(9): 64–75.

[24] Wang, Y., Xu, J., Chen, Z., et al. (2017). Technical and economical optimization for a typical solar hybrid coal-fired power plant in China. *Applied Thermal Engineering* 115(3): 549–557.

[25] Zimmerman, R. and Restrepo, C.E. (2009). Analyzing cascading effects within infrastructure sectors for consequence reduction, *Proceedings IEEE International Conference on Technologies for Homeland Security HST 2009*, 165–170, Waltham, MA (11 May 2009).

35

Energy Storage Participation in Electricity Markets

Tom Brijs,[1,4] *Andreas Belderbos,*[2,4] *Kris Kessels,*[3,4] *Daan Six,*[3,4] *Ronnie Belmans,*[1,4] *and Frederik Geth*[1,4]

[1] *Department of Electrical Engineering, University of Leuven (KU Leuven), Heverlee, Belgium.*
[2] *Department of Mechanical Engineering, University of Leuven (KU Leuven), Heverlee, Belgium.*
[3] *VITO, Mol, Belgium.*
[4] *EnergyVille Research Institute, Genk, Belgium.*

35.1 Introduction

Electricity, due to the laws of physics, is very much a real-time product. At any instance, consumption and generation need to be balanced, to support the stable operation of the electric power system. In a historic perspective, this meant that generation followed demand. This balancing effort used to be delivered by flexible power plants and stockpiling fuel at the supply side of the power system. Power plants typically hold significant fuel reserves, e.g. the natural gas grid with its storage capabilities for gas-fired generation, coal piles at classic thermal power stations, and nuclear fuel at nuclear power plants [1].

Due to the integration of variable renewable energy sources (RES) such as photovoltaic (PV) and wind generation, part of the flexibility at the generation side is disappearing. Conventional power plants currently experience decreasing profitability as a result of low electricity prices and a limited number of operating hours. The former is due to the close-to-zero marginal cost of RES, while the latter is due to the merit-order effect of RES. Therefore, the paradigm is shifting to find a balance between flexibility at the generation side and flexibility at the consumption side, through flexible consumption processes on the industrial, commercial, and residential level. One of the technologies to fill the remaining gap is energy storage. In addition, these technologies can be activated in neighboring regions through the electric grid, which can be considered a vehicle for flexibility.

The storage of electricity represents a combination of three functions [2]: consuming electricity, accumulating the energy in some form, and generating electricity. The grid and storage systems provide complementary services. The grid was developed historically to allow the decoupling of consumption and generation *in space*. Having multiple generation assets and customer bases connected to one system increased the simultaneity of generation and consumption and made it possible to decrease costs. Storage also allows an increase in the simultaneity of generation and consumption, but by decoupling them *in time*.

Storage is an asset in the power system that can behave as both generator and load. Storage operators can offer a range of services in the electricity market. The market may provide a number of revenue streams that offer opportunities to upgrade current

Advances in Energy Storage: Latest Developments from R&D to the Market, First Edition.
Edited by Andreas Hauer.
© 2022 John Wiley & Sons, Ltd. Published 2022 by John Wiley & Sons, Ltd.

storage assets or to build new ones. Depending on the availability of such revenue streams and the credibility of the market and its regulator, stakeholders develop business cases and make investments. Potential investors can choose from a range of storage technologies with varying degrees of maturity and techno-economic characteristics. Investment decisions must consider multiple services and markets and take an application-focused approach.

Electricity storage has demonstrated successful application in the conventional power system. Worldwide, there is more than 145 GW of installed capacity of pumped-hydro storage (PHS) in operation [3]. Although all other technologies represent less than 1% of this installed capacity in 2010 [4], this is increasing with a large number of battery energy storage systems (BESS) currently being developed. Other storage technologies improve on the limiting characteristics of PHS, e.g. the availability of significant elevation differences in short distances and energy density. Such new technologies are therefore often considered in planning studies for future application in the power system [5]. The flexibility that can be provided by storage can be parametrized through multiple metrics. Three factors largely define the flexibility of a storage system:

- rated power or power range P [W],
- ramp rate P/t [W/h] or ramp duration t [h], and
- storage capacity E [J, Wh] or energy-to-power ratio E/P [h].

The general power balance of a storage system that can both consume $P_t^c \geq 0$ and generate $P_t^d \geq 0$ electric power and store it in a buffer $E_t \geq 0$ is:

$$\underbrace{\frac{dE_t}{dt}}_{\Delta \text{Energy buffer}} = P_t = \underbrace{\underbrace{P_t^c \cdot \eta^c}_{\text{Addition}} - \underbrace{P_t^d / \eta^d}_{\text{Removal}}}_{\text{Electric origin}} + \underbrace{\underbrace{P_t^+}_{\text{Addition}} - \underbrace{P_t^-}_{\text{Removal}}}_{\text{Exogenous origin}} \tag{35.1}$$

Due to a charge efficiency η^c, only part of the consumed electric energy is converted to energy stored in the buffer during charging, while due to a discharge efficiency η^d only part of the stored energy is converted back into electric energy during discharging. Furthermore, the state of the energy buffer can change independently of the electric power flows (i.e. exogenously), through $P_t^+ \geq 0$ and $P_t^- \geq 0$.

Table 35.1 illustrates the different combinations of generators, consumers, and buffers. The distinction between electricity consumers with downstream energy storage, such as power-to-fuels and thermal energy storage, and generators with energy storage upstream from electricity generation, such as coal plants, is easily made. Some systems, e.g. nuclear plants, consume a significant amount of power in standby, however, such secondary load behavior is out of the scope of this chapter. Energy buffers with both a consumption and a generation interface to the grid are bidirectionally coupled electricity storage systems. However, the actual exploitation of such systems will determine to which degree such systems provide an electricity storage function.

The concept of storage in the context of this chapter therefore refers to systems that have the ability to buffer energy and that are bidirectionally interfaced with the electric power system. This energy buffer can be implemented based on a variety of physical principles: energy stored as thermal, chemical, electro-chemical, kinetic, or potential energy, or in the electro-magnetic field.

Table 35.1 System classification based on design.

Classification	Buffer	Equation	Example
Generator	Absent	$0 = -P_t^d / \eta^d + P_t^+$	PV
Consumer	Absent	$0 = P_t^c \cdot \eta^c - P_t^-$	Aluminum processing
Generator	Upstream	$dE_t / dt = -P_t^d / \eta^d + P_t^+$	Coal, nuclear, hydro
Consumer	Downstream	$dE_t / dt = P_t^c \cdot \eta^c - P_t^-$	P2G, thermal storage
Bidirectionally grid-coupled	Energy storage	$dE_t / dt = P_t^c \cdot \eta^c - P_t^d / \eta^d + P_t^+ - P_t^-$	BESS, PHS

This work is structured as follows. Section 35.2 discusses different storage technologies with a link to important techno-economic characteristics and application-related information. Next, Section 35.3 provides a comprehensive formulation of the techno-economic parameters by which storage plants can be characterized with a link to the operation, while Section 35.4 focuses on the services that can be provided by storage in electricity markets. Section 35.5 discusses some case studies to illustrate the interaction between techno-economic characteristics and services that storage can provide. Finally, Section 35.6 provides conclusions.

35.2 Classification of Energy Storage Options

This section describes the technological envelopes of a number of storage technologies often considered for grid integration, while providing links to the techno-economic characteristics (Section 35.3) and storage applications (Section 35.4) where appropriate. A summary and categorization of the technologies discussed in this section is given in Table 35.2.

35.2.1 Pumped-hydro Storage (PHS)

PHS represents over 99% of the worldwide installed capacity of grid-related electricity storage. This is a mature technology with relatively low energy losses. The energy capacity of PHS is a function of elevation difference and volumetric capacity of the water reservoirs. Substantial civil engineering works are required to obtain reservoirs with sizable energy storage capacity. In the industrialized world, a lot of the obvious locations may already have been used. Converting some of the current hydro power plants to PHS units is considered an attractive option to increase the available storage capacity [6]. Due to geographical requirements, both hydro plants and PHS are typically found far from population centers. Therefore, it is difficult for this technology to help resolve congestion-related issues at times of peak use of the grid. Individual PHS stations have power ratings of up to several GWs during a few hours up to a day. Improved PHS technologies are developed, to

Table 35.2 Categorization of storage technologies.

Type	Technology	Buffer	Carrier	Conversion technology
Mechanical	PHS	Gravitational field	Water	Electromotor-generator pump-turbine
	CAES	Pressure	Air	Electromotor-generator compressor-turbine
	Flywheel	Kinetic energy	Flywheel	Electromotor-generator flywheel
Electrical	Supercapacitors	Electric field	Capacitor	Power-electronic converter
	SMES	Magnetic field	Coil	Power-electronic converter
Electro-chemical	Battery	Chemical potential	Active materials	Power-electronic converter
	Flow battery	Chemical potential	Active materials	Power-electronic converter
	Fuel cell	Chemical potential	Active materials	Power-electronic converter
Chemical	Electrolysis	Hydrogen H_2	H_2O	Electro-chemical reactor
	Power-to-Methane	Methane CH_4	H_2O, CO_2	Electro-chemical and chemical reactor

increase the flexibility in terms of services and land use. On the other hand, using the sea as a reservoir or using underground reservoirs can minimize land use and lead to new suitable locations.

35.2.2 Compressed Air Energy Storage (CAES)

CAES is a next important storage technology in terms of installed capacity, however there are only two major plants, the 321 MW Huntorf plant in Germany and the 110 MW MacIntosh plant in the US. In a CAES unit, air is compressed, which also increases temperature, and energy is stored due to the increased pressure. The heat represents part of the energy converted in the process. If the compressed air cools down after pressurizing, the thermal energy is lost and the system efficiency decreases. To make up for the heat loss, a fuel, typically natural gas, is used to reheat before and during expansion of the compressed air.

Natural storage reservoirs are used as pressure reservoirs, these include salt caverns, former mines, and former natural gas sites. Consequently, the application of this technology is limited by geological features. For instance, the CAES in Huntorf, Germany, operating since 1978 uses salt caverns. CAES system ratings are in the range of hundreds of MWs for a few hours to a day. CAES is similar to PHS in terms of sizing and storage duration.

Newer CAES concepts try to improve the efficiency, negate the need of additional fuel, and the need of underground geological features. The efficiency can be improved by storing heat, for example using adiabatic CAES technology.

35.2.3 Flywheels

The energy content, stored in the mass of the rotor, depends on the rotational speed. To charge or discharge the flywheel, a variable speed electric motor with a power electronic converter is used. When the flywheel is turning it continually loses energy, because of friction-related losses.

The advantages of flywheels include high power density, long lifetime, and high efficiency when cycled frequently. The storage capacity of individual flywheels is limited, with typical capacities between 0.5 kWh and 10 kWh [7]. Therefore, numerous flywheels are installed together to increase the total capacity of an installation.

Flywheels are typically designed for seconds to a few tens of minutes of energy capacity. Consequently, they are used in applications where high power is more valuable than high energy capacity. For example, a 20 MW flywheel power plant is in operation in the US (NY state) and is used for primary frequency control. Flywheels have known limited success, with about 25 MW of flywheel capacity installed worldwide [4].

35.2.4 Supercapacitors and Superconducting Magnet Energy Storage (SMES)

Supercapacitors store energy in the electric field. Just like in electro-chemical cells, the voltage of a single basic unit is limited to a few volts. Therefore, grid application necessitates putting a number of capacitors in series. The cycle-life is very high, as is the efficiency, but the energy density is rather low.

With SMES, energy is stored in the magnetic field of a direct current flowing in a superconducting coil. To maintain superconductivity, cryogenic temperatures are required. The complexity of the cooling system, and the investment and operating cost, limit SMES to high power applications, offering MW ratings for a few seconds. The response time is low, the power ratings are very high, and although the efficiency is high the cooling system leads to additional losses. High investments are required, and if cycling is not frequent the system efficiency may decrease significantly.

35.2.5 Battery Storage

A variety of battery storage technologies have been considered for power system applications. In the recent past, sodium-sulfur was the most commonly used technology. Lithium-ion technology has caught up very quickly due to a large number of projects developed in the past few years, as a result of rapidly decreasing battery prices.

35.2.5.1 Lithium-ion

Lithium-ion refers to a class of rechargeable electro-chemical cells in which lithium ions move from the negative to the positive electrode when discharging and back when charging. Lithium-based batteries have a high electrode potential (up to 3.9 V) compared to other battery technologies (1.2 V to 2 V). Because of the high potential, even at a charge density [Ah/g] similar to other technologies, the energy density [Wh/kg] is already substantially higher. Lithium-ion batteries operate at temperatures within the normal atmospheric range.

Usually, the batteries are named after the active material of the positive electrode, e.g., lithium-cobalt-oxide (LCO), lithium-nickel-oxide (LNO), lithium-manganese-nickel-cobalt (NMC), lithium-nickel-cobalt-aluminum-oxide (NCA), lithium-iron-phosphate (LFP), and lithium-manganese-oxide spinel (LMO). Except for lithium-titanate (LTO), all technologies typically use lithium carbon LiC_6 at the negative electrode.

Different trade-offs of energy density, power density, and cost are encountered. In general though, lithium-ion batteries have a high efficiency but significant investment cost. However, prices are decreasing quickly because of economies of scale and learning effects linked to the rollout of electric vehicles.

Within the lithium-ion chemistry family, LTO and LFP chemistries are most commonly discussed for grid applications, due to their improved power, efficiency, cycle-life, and thermal stability. BESS systems with lithium-ion have been demonstrated up to a few tens of MWs with 15 minutes up to about 10 hours of rated storage capacity.

Generally, lithium-ion offers superior characteristics to the other low-temperature battery chemistries; however, it is relatively unforgiving when operating outside normal ranges of temperature, voltage, and current. The safety limits vary with the chemistry of the cells as well as the manufacturer.

35.2.5.2 Sodium-sulfur

Sodium-sulfur (NaS) is a kind of high temperature chemistry. It consists of molten salts of sodium and sulfur, materials available at low cost.

This chemistry offers a long cycle-life, high efficiency, and a good energy density. However, the heating system represents a cost and losses. NaS BESS are operated worldwide, e.g. in Japan, the US, France, and Germany. NaS is used to provide frequency regulation services on islands or to decrease variability in the output of wind power plants. Typical installed commercial systems have ratings from 1 MW up to 34 MW, with 6–7 hours of storage capacity.

Sodium-sulfur technology operates at about 290–350°C. It has low energy-specific investment.

35.2.5.3 Flow Batteries

Flow batteries use chemical components dissolved in a liquid electrolyte, which are stored in separate tanks. This separation of the active materials limits self-discharge losses, offering the potential of much longer-term storage than conventional batteries. The cycle-life and scalability are high, but the maturity is low (demonstration/pre-commercial), the energy density is low and the complexity is high because of the pumps and control systems. Typical flow battery system ratings are a few to tens of MW for a few hours.

Vanadium-vanadium (Va-Va) redox flow batteries have been widely researched, and demonstration projects have been developed. In the Va-Va chemistry, ionic solutions of vanadium, but in different ionic states, are provided to both electrodes [7]. There has been some interest in zinc-bromine flow battery systems as well and demonstration trials have been run, but only for limited power ratings of less than 500 kW [7].

35.2.6 Fuel Cells

Just like batteries, fuel cells are composed of individual electro-chemical cells consisting of two electrodes and an electrolyte. Conversely to batteries, a fuel is used. Common fuels include hydrogen, methanol, and natural gas. Generally, oxygen, taken from the atmosphere, is used as oxidizing agent. Fuel cells are categorized by the kind of electrolyte used. Common types include proton exchange membrane fuel cells, phosphoric acid fuel cells, solid oxide fuel cells, and molten carbonate fuel cells. In a unidirectional use of fuel cells, they function as a generator given the input of fuel, not as an electricity storage system. For the latter, unidirectional fuel cells can be combined with power-to-gas technology. Some fuel cell technologies can be used bidirectionally, functioning as both a generator and consumer.

35.2.7 Power-to-Gas (P2G)

P2G technology stores energy in a chemical form, by converting electricity to hydrogen and possibly to methane. P2G technology is currently in the demonstration phase with projects of several kW up to 6 MW installed electrolyzer capacity. The modular configuration of electrolyzers, composed of multiple small electrolyzer cells, allows for future P2G units ranging from several kilowatts up to a gigawatt.

Current P2G efficiencies are relatively low: the electrolysis process has an efficiency of 60–85% [8], the methanation step has an efficiency of 70–95% [9], and the reconverion of methane to electric power in open cycle of combined cycle gas turbines has an efficiency of 30–60%[10]. However, P2G may offer a high potential as storage technology due to its large energy storage capacity [11]. In the first instance, produced synthetic methane can be stored in the existing natural gas grid and storage reservoirs. Energy storage capacity could be increased greatly by taking new reservoirs such as depleted oil fields, salt caverns, and saline aquifers into operation. Hydrogen from an electrolyzer could also be directly fed into the natural gas grid, although to a limited extent. Studies report that it is possible to mix hydrogen with natural gas for volume concentrations of 1% vol. to 15% vol. [12] without modifications of the gas grid or gas consuming appliances.

Three different types of electrolyzers exist: alkaline, polymer electrolyte membrane (PEM), and solid oxide electrolyzers (SOE). Many manufactures can now produce alkaline electrolyzers capable of following the volatile generation profile of variable renewables [9]. This is due to quick start-up times in the order of seconds and almost unlimited ramping capabilities. PEM electrolyzers have dynamic characteristics similar to alkaline but have a wider operating range [13]. SOE are still in the research phase. Although their reported efficiencies are much higher than those of alkaline and PEM electrolyzers [14]) it is yet unclear whether SOE can handle dynamic operation due to the high temperatures at which they operate, which leads to high thermal inertia.

Methanation reactors are less dynamic in operation and will therefore most likely need a short-term storage (buffer) of hydrogen to have an economically optimal system configuration. The methanation reaction can take place in a catalytic or biochemical way. Many reactor designs exist for each reaction type [15]. The catalytic process is more mature and controllable, but highly sensitive to impurities in the reactant stream. Impurities form a lesser problem for biological methanation [9].

35.3 Techno-economic Energy Storage Characteristics

Individual storage plants can be characterized by a set of technical and economic characteristics. These include charge and discharge power and duration, losses and efficiency, calendar and cycle-life, volume and mass, and cost.

35.3.1 Charge and Discharge Power and Duration

The storage component of the storage system has specific technology-related limitations. For technologies such as batteries, available charge or discharge power depends on the amount of energy storage capacity installed. For others, e.g. fuel cells, these parameters are independent from the storage capacity. Charge and discharge ratings are often unequal. The information on the ability to provide power is given in one of the following normalized ways:

- C-rates [A/Ah];
- power to energy ratios [kW/kWh]; or
- charge and discharge duration [h].

C-rates are typically used to quantify the charge and discharge rates of battery cells, and are provided in manufacturers' specification sheets. For example, a 3.9 V 10 Ah Li-ion battery may have a 4 C maximum discharge rate, corresponding to a current of 40 A and a discharge duration of about 15 minutes. Charge and discharge durations, and power-to-energy ratios are typically used to describe storage subsystems and storage systems in general. Moreover, the energy storage capabilities of PHS may be provided as the number of hours of sustained operation in pump or turbine mode.

It is noted that certain technologies may have limitations on the operation. BESS, when consuming power from the grid, uses electricity to charge the batteries. When discharging, power is generated again. It is impossible for individual batteries to be simultaneously charging and discharging. Conversely, for PHS, charging and discharging at the same time is possible: a pump and a turbine interfacing the same reservoirs can operate simultaneously.

35.3.2 Losses and Efficiency

Energy efficiency of an energy conversion system is calculated as the ratio of the useful energy output to the energy input. Multiple conversion steps occur in a storage system. For example in a BESS, the grid AC power is first converted to DC, after which the batteries convert the electric energy to chemical energy. During discharging, the chemical energy is put back into the grid after is has been converted again to AC power. Finally, the roundtrip energy efficiency can be calculated.

In general, energy losses in storage systems are incurred in multiple ways:

- during charging and discharging,
- continuous self-discharge during storage,
- consumption of auxiliary systems, and
- possible maintenance cycling.

Efficiencies are generally specified for a complete cycle, as for certain technologies it is difficult to measure the energy content of the energy storage subsystem accurately, so charge and discharge losses cannot be separated. Self-discharge losses and the consumption of auxiliary systems are incurred over time. The impact of these losses on the efficiency can be assessed for a given application and time horizon through tracking of the energy balance of the system. Compared to the power rating of the system, the power of the auxiliaries is low. Nevertheless, the consumption by auxiliary systems may be substantial as it is incurred continuously.

35.3.3 Lifetime

The lifetime is the time after which the performance of a system is degraded in such a way that it can no longer perform sufficiently. For instance for batteries, the performance degradation threshold is generally defined as a reduction in rated energy capacity to 80% of its original value.

The lifetime in any application is determined either by the cycle-life in case of frequent use or by calendar life in case of infrequent use.

Calendar life is characterized independent of use. It is the expected time that the device can be used, whether or not actually operated. For instance in batteries, the calendar life depends on the occurrence of parasitic chemical reactions that progressively inhibit the function of, or destroy, the active materials.

The cycle-life takes into account the deterioration of the components due to use. The cycle-life is given as a number of cycles, sometimes called cyclability, which can occur before the performance threshold value is exceeded. For example in batteries, this is caused by the nonreversibility of the chemical reactions. Determining the cycle-life for a given depth of discharge limitation is a time and energy consuming process. The cycle-life when focusing on a certain application further depends on conditions of use such as charge and discharge rates, average state of charge, average depth of discharge of cycles, and temperature.

35.3.4 Volume and Mass

The energy density and the power density of a certain technology can be specified in terms of volume [Wh/l and W/l] or weight [Wh/kg and W/kg]. These quantities can be used to estimate the size of storage systems. Additional space use is incurred because of the need of dedicated buildings, safety installations, and heating and cooling systems. The power conversion system (inverter) also adds to the system weight and volume.

35.3.5 Cost

The cost of a storage system can be related to the dimensioning in terms of power rating, energy capacity, or both. The investment required for storage capacity can be normalized by division with the storage capacity [€/kWh]. The investment required for the power rating is typically normalized by the converter power rating [€/kW].

35.4 Energy Storage Applications

Historically, electricity storage plants have been mainly considered as an alternative for investing in peak-load generation, by charging during off-peak and discharging during peak periods [16]. However, due to the liberalization of electricity markets and the increasing integration of RES, multiple applications emerged through which storage can provide and capture value [1, 7, 17–19]. It is noted that potential applications are influenced by regulation and location (e.g. behind-the-meter or not). Although many categorizations are possible, here they are categorized in energy, grid, and reliability services.

35.4.1 Energy Services

Energy services include both arbitrage and portfolio optimization by market participants.

35.4.1.1 Arbitrage

Arbitrage can be defined as any activity in which a commodity is bought at a certain price and a similar commodity, or commodity in which the former can be converted, is sold at a higher price [20, 21]. Price spreads can be captured over time in a single market by means of storage resources, but also over multiple sequential markets, i.e. so-called multimarket arbitrage. Although the arbitrage value is directly related to the presence of these price spreads, it also depends on the price-effect of (dis)charge actions, as additional storage capacity generally reduces price spreads by increasing off-peak prices when charging as well as decreasing peak prices when discharging [21].

In the context of energy storage, single-market arbitrage is based on price differences over time: electricity is bought and stored when the price is low, typically during periods with low demand and/or high RES generation, and is sold and generated again when the price is higher, typically during periods with high demand and/or low RES generation. Single-market arbitrage can be performed at different electricity markets: in long-term unstandardized forward and standardized future markets, and the short-term day-ahead, intra-day, and real-time markets. Multimarket arbitrage allows price spreads to be captured: (i) over time in one of the considered sequential markets (i.e. single-market arbitrage), (ii) over time in several of the considered sequential markets, and (iii) in the considered sequential markets for the same delivery period by players changing their position. While for the first two options storage resources are required, the third option can also be performed by generation and consumption capacities, and by virtual bidders not operating in any physical capacity.

When physical (dis)charge actions are required to perform arbitrage, the arbitrage value is not only related to the price difference between the buy and sell price, but also to the roundtrip efficiency and effective value of the buy price. The value of the price at which energy is bought determines the cost due to energy losses.

35.4.1.2 Portfolio Optimization

Portfolio optimization by market participants is performed in different timeframes: in the investment, scheduling, and operation phase, and covers generation investment deferral, intertemporal energy shifting, and capacity firming, respectively.

At the investment stage, energy storage may decrease the need for peak power plants that are usually only operated to meet the peak demand. It can decrease the need for these power plants by storing baseload and RES generation in times of low demand and high RES generation, and by replacing peak generation in times of high demand and low RES generation. In addition, at future higher RES targets the "fuel cost" of storage, which is related to the energy losses and the price at which energy is stored, decreases. This positive correlation between the decrease in operational cost of storage and the integration of RES presents an advantage compared to conventional peak power plants. The latter faces less operational hours due to the merit-order effect of RES, and a decreasing profitability due to the lower electricity prices as a result of the close-to-zero marginal cost of RES.

At the scheduling phase, generators can maximize the value of generation by decoupling generation from grid injection, and consumers can minimize the cost of consumption by decoupling consumption from grid offtake. This is possible due to the intertemporal energy shifting ability of energy storage systems. Among others, generators can schedule the more inflexible base and mid-load generation technologies more efficiently through storage resources. This decoupling of consumption and physical withdrawal is profitable for large industrial consumers that directly participate in the wholesale markets, or have a contract with their supplier through which they are sensitive to price changes. Contrarily, this would require more dynamic tariff structures or capacity-based tariffs for residential consumers instead of the current quasi-fixed tariffs, or additional incentives for self-consumption in case of residential PV generation.

In the operation phase, market participants can use energy storage resources for capacity firming purposes. Capacity firming may include both smoothing of generation or consumption output, resulting in less volatile power profiles, and the ability to follow pre-determined output schedules to reduce imbalanced positions in real-time. The smoothing of output profiles can especially contribute to an efficient operation of conventional power plants because of their limited ramp rates, minimum up- and down-times, and minimum and maximum output level constraints. Imbalanced positions in real-time are settled by the system operator applying imbalance prices, which reflect the procurement costs of the activated reserve capacity, to these unbalanced positions.

35.4.2 Grid Services

Grid services include the provision of frequency control, voltage support, congestion management, and black-start capabilities to the transmission system operator (TSO). The TSO is responsible for the stable operation and maintenance of the transmission grid and the interconnection with neighboring regions. In the future, some of these will likely also be provided to the distribution system operator (DSO), which is responsible for the stable operation, maintenance, and development of the low- and medium-voltage grids. The provision of ancillary services to the DSO is quite innovative, as historically speaking the goal of the distribution grid was merely to transmit electric energy generated on the transmission level to consumers on the distribution level. In contrast to this rather passive character of the distribution grid and its operation, an active operation will become the new standard due to the increasing integration of distributed energy sources, demand response, and local storage systems [22].

35.4.2.1 Frequency Control

To be able to satisfy the most important condition for a stable operation of the power system, being the instantaneous balance between generation and consumption of electric energy, system operators contract reserve capacity from grid users. These reserves are activated to compensate unforeseen variations in generation and consumption. As these variations can occur on different time- and duration-scales, system operators hold different reserve categories. In the ENTSO-E[1] synchronous zone, operating reserves are categorized into Frequency Containment Reserves (FCR), Frequency Restoration Reserves (FRR), both automatic (aFRR) and manual (mFRR), and Restoration Reserves (RR). FCR are used to stabilize the frequency after a disturbance or incident in a matter of seconds, and are activated automatically. FRR aims to restore the system frequency by restoring the balance in the control zone of a TSO, thus relieving the system-wide activated FCR. FRR capacity is activated either automatically (aFRR) or manually (mFRR). Their activation is triggered by the so-called Area Control Error (ACE), calculated as the difference between the scheduled and actual power exchange of a control zone. Finally, RR can be used to progressively relieve the activated FRR or to support the FRR activation. Previously, FCR were referred to as primary reserves, aFRR were called secondary reserves, and mFRR and RR generally correspond to the concept of tertiary reserves [23].

In the context of liberalized electricity markets, these reserves are contracted through organized markets. Depending on the considered reserve product and geographical market, reserve providers are paid a reservation payment (e.g. €/MW), independent from activation, and/or an activation payment (e.g. €/MWh). If energy storage plants meet the technical requirements to provide these reserve products, they may provide operating reserves to the system operator.

35.4.2.2 Voltage Support

In order to keep the voltage level throughout the grid within the technical and contractual limits, the system operator contracts services to support the voltage. Voltage support in transmission grids is typically performed through reactive power control, e.g. using capacitor banks or synchronous machines. In the distribution grid, due to their more resistive nature, a combination of active and reactive power control can be more effective. Next to the active power dispatch capabilities inherent to energy storage, the grid interface of the storage system may also have reactive power control capabilities. Furthermore, in low-voltage grids, voltage imbalance compensation can prove valuable in supporting the voltage. In contrast to frequency control, the location largely determines whether storage systems can provide the appropriate amount of voltage support.

35.4.2.3 Congestion Management

Congestion management indicates the ability to reduce flows on congested lines. Storage systems can provide this service if they are physically located in the grid-constrained areas. The provision of congestion management is thus very location-specific. When considering a line that is often congested, the congestion issue could be

1 European Network of Transmission System Operators for Electricity.

solved by expanding the line rating, by installing generation capacity on the demand side of the congested line, or by installing storage capacity on either side of the congested line. Installing generation capacity at the demand side of the congested line would reduce flows on the line and as such relieve the issue. Storage capacity located at the demand side of the congested line could be charged during off-peak periods, and discharged during peak periods when the line would typically be constrained. Alternatively, storage capacity on the supply side of the congested line could be charged during peak periods when the line would typically be congested, and discharged when it is not constrained.

35.4.2.4 Black-start Capabilities

In the event of a black-out, the system operator has to re-activate the system step-by-step. In order to do this, the system operator calls upon capacities that can start without an external electric supply. Capacities that have the ability to do this can provide the so-called black-start service to the system operator. In the case where an energy storage plant meets the technical requirements to provide this service, and keeps a certain amount of energy stored in the buffer at all times, it can provide this black-start service. This includes an opportunity cost for the storage operator, since he cannot use all of the storage capacity to provide other services. However, in the case of PHS the effective head (elevation difference) is the lowest when the upper reservoir is almost empty, i.e. when the limit is reached of the black-start energy content. This lowers the opportunity cost for keeping a certain amount of energy stored at all times.

35.4.3 Reliability Services

Reliability services include the support of reliability on both the local and system level.

35.4.3.1 Local Level and Behind-the-meter

On the local level, consumers can use energy storage systems to either improve the quality of power used by its electric devices or to provide back-up in case of a disruption in the supply of electric energy. The former is based on the ability of storage to buffer active or reactive power imbalances between demand and supply, while the latter increases the reliability of the consumer. Storage as back-up can replace small-scale diesel generators. Storage, behind the meter, is often discussed to improve the energy-autonomy of prosumers, i.e. consumers who also operate distributed generation. As such, distributed generation, like combined-heat-and-power or PV, is not actively dispatched for power system operation but driven by external processes. Forcing these parties to be more autonomous generally mitigates the grid impact. In certain locations, like Germany, incentives have been developed, for example to invest in BESS in combination with PV.

35.4.3.2 System Level

Reliability services at the system level include the provision of firm capacity to meet peak demand, i.e. to guarantee system adequacy. The integration of variable, renewable, and decentralized energy sources is a major challenge for the operation of the power system. Their limited controllability and predictability results in an increased need for power system flexibility, while conventional power plants currently experience profitability issues due to decreased electricity prices and number of operating hours. In the long run this may impact installed generation capacities due to the lack of investment signals and lead to system adequacy and flexibility problems. Currently, several so-called "capacity mechanisms (CMs)" are discussed and implemented throughout Europe with the goal of ensuring generation adequacy. Such CMs provide alternative revenue streams, e.g. the Belgian strategic reserve mechanism, or complementarity revenue streams, e.g. the UK capacity auction mechanism, for firm capacity. Contrary to an energy-based market, where generators are remunerated for the energy they supply, e.g. in €/MWh, these mechanisms provide a remuneration that is independent of the energy output but values the availability of capacity, e.g. in €/MW.

35.4.4 Aggregation of Services

This multitude of applications makes electricity storage plants an interesting asset for a wide range of market participants. However, operating a storage plant to provide just one or a few of these services might not always result in a positive business case; profitability may require the aggregation of multiple applications while accounting for the interdependence between potential revenue streams [16, 21, 24]. Although some studies focus on the co-optimization of different storage applications [24–26], most existing work focuses on only a single application or allocates the available storage resources a priori when considering multiple applications, instead of applying a periodically performed optimization process. In addition, the sharing and operating of storage resources by different players has only been studied to a limited extent, except for the work done by [27] and [1]. They do not a priori define the applications or even the market player that the storage resources will serve at any given moment in time. The method discussed in [1] allows simultaneous inclusion of multiple resource suppliers and players competing for the right to use them, and simultaneous consideration of their offers. This can be accomplished by the development of a centralized platform where periodical auctions with sequential market clearings take place to allocate the right to use (dis)charge power capacities and energy storage capacity. Contrarily, the method discussed in [27] considers a sequential allocation to players who express their need for flexible resources at different time scales. In addition, the allocation mechanism in [1] auctions physical (dis)charge power rights and storage capacity rights, whereas the allocation in [27] is based on actual utilization profiles.

35.5 Interaction Market Opportunities and Technical Characteristics – Illustrative Case Studies

This section discusses some case studies related to the interaction between techno-economic storage parameters and services storage that are provided in the market.

35.5.1 Efficiency and Cycle-life – Arbitrage

Similar to other grid assets, there is an initial investment cost to install BESS, as well as continuous operation and maintenance costs during its lifetime. Initial costs include equipment, construction, certification, land, and engineering costs [28]. Operational costs include inspection, maintenance, replacement parts, taxation, insurance, and losses. Depending on the country-specific accounting rules, the initial cost is usually incurred over a period of months to years during which the asset is put in place, and is, in budgeting, often assigned to a specific year. Operational costs are often periodical in nature and are budgeted annually as long as the asset is operating. The lifetime of the BESS will vary, being limited by the calendar life at low cycling rates and by the cycle-life at high cycling rates.

In common variable cost models for PHS, only the energy losses are taken into account. Other variable cost components are often considered zero. For instance [29] takes the costs of the roundtrip storage efficiency into account. Such approximation is invalid for most BESS technologies, as the per-cycle opportunity cost of wear to the batteries is substantial. This cost depends on the battery cycling rate.

An arbitrage strategy can be performed with BESS to obtain positive cash flows by exploiting price differences in the electricity market. The actual strategy, considering technical limitations of BESS, can be complicated, because it may not be possible to fully charge or discharge the BESS during a single market time slot.

The variable cost of the use of batteries is illustrated in comparison with PHS in Figure 35.1. Three PHS systems with varying efficiency η_{rt} and five BESS with varying efficiency η_{rt} and specific opportunity cost of wear (depreciation) c_{depr}^{bat} are used. Three price scenarios are provided, each with a price difference of 70 €/MWh between a buy price m_{buy} and sell price m_{sell} (40–110, 5–75 or -10–60 €/MWh). A cycle representing 1 MWh of discharged energy is considered for all systems and therefore the cost of discharging is identical (- 110, -75, or -60 €). Due to efficiency losses, more energy is consumed during charging than can be generated again during discharging. For instance, to allow the $\eta_{rt} = 55\%$ PHS system to discharge 1 MWh, 1.82 MWh must be consumed first (E_{buy}), thereby obtaining a buy cost of 72.7 €/MWh based on the 40 €/MWh price. Therefore the cost of buying electricity varies as a function of both the electricity price and the storage efficiency. Next, because of the limited technical lifetime, c_{depr}^{bat} is not negligible for BESS and represents a significant expense. Finally, all costs are added together to obtain the variable cost K^{var} (positive cost indicates expense).

It is seen that a lower efficiency increases the variable cost at high buy prices and decreases the cost at negative buy prices. A higher efficiency decreases the sensitivity to the buy price. Furthermore, the specific opportunity cost of wear has a significant influence on the variable cost. However, this cost factor may come down substantially as it depends on the ratio of battery prices to cycle-life numbers, where the former are decreasing while the latter are increasing.

35.5.2 Energy-to-power Ratio – Arbitrage

The energy-to-power ratio E/P can be a limiting factor for the arbitrage revenue of a storage unit. When prices show a daily varying profile, an energy-to-power ratio of several hours is sufficient to capture all possible arbitrage revenue. However, if prices present strong monthly or seasonal variations, a limited energy-to-power ratio can cause the storage to be fully charged at times of low prices making it impossible for the storage unit to capture all possible arbitrage value. This is shown in the illustrative example provided in Figure 35.2. Storage units with a high energy-to-power ratio are

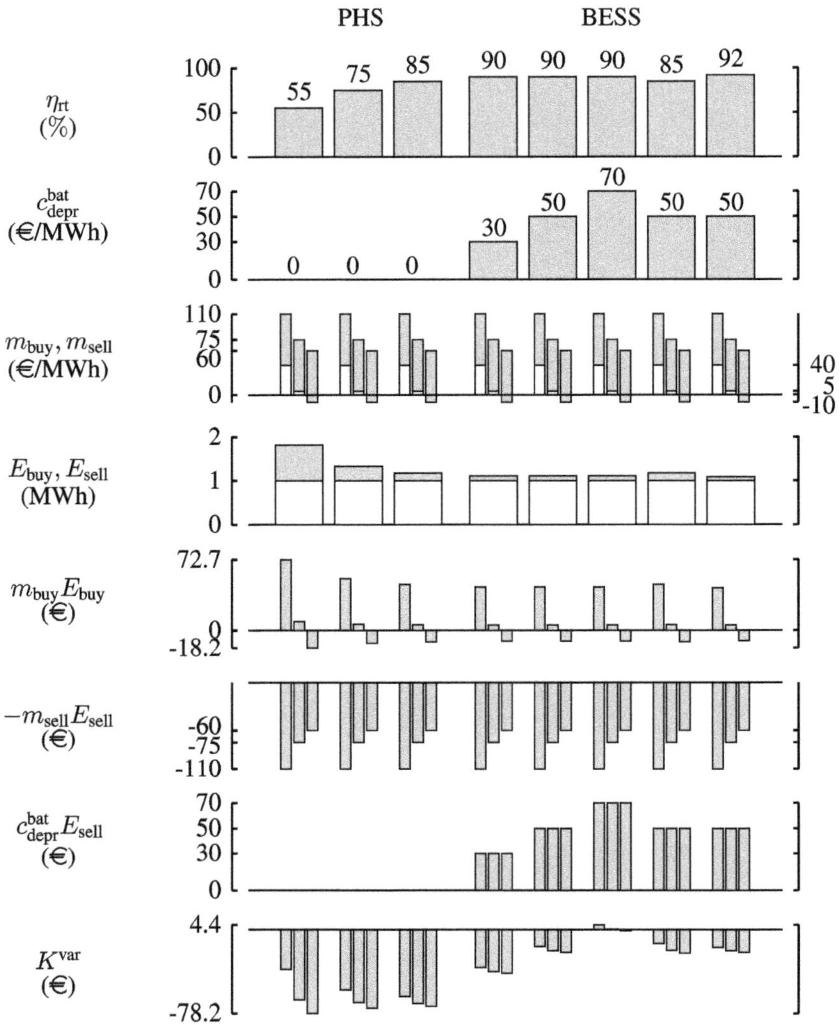

Figure 35.1 Illustration of the variable costs of storage. From top to bottom, the rows indicate: storage efficiency, specific depreciation cost, price signal, energy volume, buy cost, sell cost, depreciation cost, and variable cost [30].

capable of capturing the full arbitrage revenue for both seasonal and daily varying price profiles. They might, however, be more expensive compared to storage units with a lower energy-to-power ratio. It is therefore important to size the storage unit in accordance to the expected price profile.

35.5.3 Energy-to-power Ratio – Frequency Control

BESS use battery cells that can, within certain limits, be designed to have certain *E/P* ratios. However, mass-manufacturing, through economies of scale, favors standardization, limiting the number of cell types and chemistries available. Therefore, *E/P* ratios of cells may not be matched perfectly to all applications envisioned by system manufacturers or project developers. Nevertheless, for batteries, the *E/P* inherent in

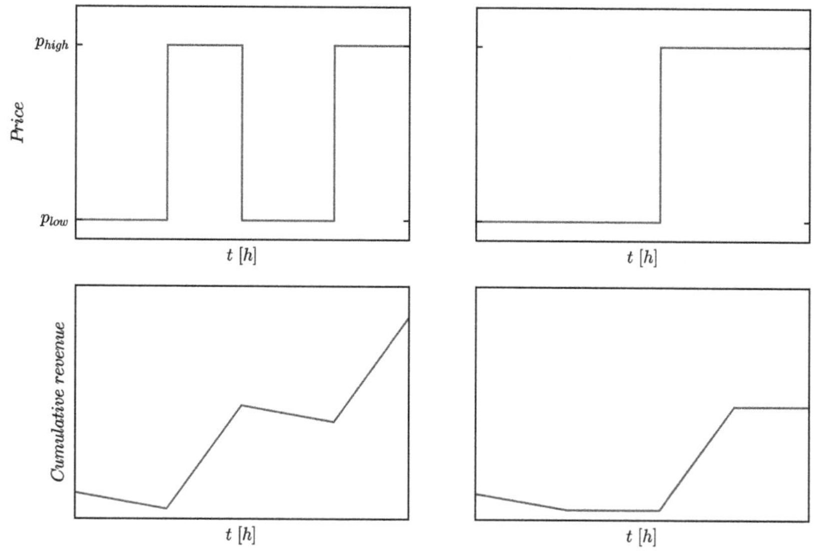

Figure 35.2 The energy-to-power ratio as a limiting factor for the arbitrage revenue.

the chemistry[2] does not force a BESS manufacturer to size the power conversion system accordingly.

The inverter can be oversized or undersized relative to the *E/P* capabilities of the batteries.[3] Oversizing the inverter power, relative to the power the batteries can provide or absorb, can be useful in case the inverter also has to deliver reactive power or unbalance compensation services. On the other hand, undersizing the inverter can be generally understood as increasing the *E/P*, as the accessible charging and/or discharging power is effectively limited by the *E/P* through power constraints on the system level.

Finally, whereas for a given system, e.g. a standardized design of a certain manufacturer, the *E/P* is known, this does not mean that the full energy storage capacity and full power exchange capabilities have to be exploited. For example in a frequency regulation application, the BESS is cycled at high power, but for very short durations, often switching rapidly from high-power charging to high-power discharging. The changes in energy content are comparatively limited. Nevertheless, the "uncycled" capacity still serves a function, i.e. limiting the extent of wear and aging, as small cycles at a medium state of charge lead to lower wear rates. Obviously, one must carefully model and validate the expected lifetime of battery cells in specific applications, as it may be challenging to extrapolate from aging results obtained on standardized test cycles.

Furthermore, in the end, developers ought to look at the smallest overall costs, not the smallest possible dimensioning of the system or the minimum investment. Battery products with an *E/P* ratio not perfectly matched (e.g. originally designed

2 It is noted that *E/P* also varies as a function of the expected duration of the peak power. For instance the short-term pulse power is higher than the sustainable value.

3 It is noted that the rated charge and discharge power may differ, therefore there is a charging *E/P* as well as a discharging *E/P*.

for EV applications), but less expensive due to economies of scale yet obtained, may overall prove to be less expensive compared to new purpose-designed systems.

35.6 Conclusions

The share of renewable energy sources in the electricity generation mix is rapidly increasing throughout Europe. Their variable nature results in an increasing need for flexibility in the system, while their close-to-zero marginal cost are pushing flexible conventional power plants out of the market. This transition increases the need for alternative sources of system flexibility. In general, four categories of flexibility sources can be employed in electricity systems: flexible supply, flexible demand, energy storage, and grid capacity. Although all flexibility sources are able to provide up- and down-ward power adjustments in order to keep the necessary balance between generation and consumption, these flexibility sources have different technical and economic characteristics. The value depends on the chosen application and obtained technical performance.

Electricity storage can be used for different applications, which can be categorized in energy, grid, and reliability services. Energy services include arbitrage and portfolio optimization of market participants. Grid services include the provision of operational reserves, congestion management, voltage support, and black-start capabilities to the TSO. Although grid services are quite a new concept for the DSO, it can be expected that electricity storage will provide ancillary services to the future distribution system as well. Reliability services include the provision of reliability on both the local and system level.

Electricity storage technologies are categorized according to the energy in which the consumed electrical energy is converted during the storage phase, being mechanical, chemical, electrical, electrochemical, and thermal energy. Worldwide, PHS has the largest installed capacity of all electricity storage technologies. At the distribution level, BESS are believed to have a significant potential because of technological innovations and reduced costs following economies of scale.

The fixed costs of storage systems are determined by the installation cost, while the variable costs consist of maintenance, cycling, and energy cost components. The energy costs depend on both the price at which electricity can be consumed, and on the roundtrip efficiency. Economic barriers to the storage business case can occur because operators may not be able to capture the full storage value in revenues due to the lack of market products compatible with storage. Additionally, the value of storage is often underestimated due to the focus on traditional applications, i.e. arbitrage between peak and off-peak times. Finally, potential revenues are missed due to operating strategies focusing on a single application, as the development of strategies considering multiple services simultaneously remains complicated.

Among others, important parameters in the regulatory framework that have the ability to impact the storage business case are grid fees, storage ownership principles, access to electricity markets and products, and grid operators' remuneration mechanisms.

References

[1] Brijs, T., Huppmann, D., Siddiqui, S., and Belmans, R. (2016). Auction-based allocation of shared electricity storage resources through physical storage rights. *Journal of Energy Storage* 7: 82–92.

[2] Ruester, S., Vasconcelos, J., He, X., et al. (2012). Electricity storage: How to facilitate its deployment and operation in the EU, Technical report. Florence, Italy: European University Institute. http://www.eui.eu/Projects/THINK/Documents/Thinktopic/THINKTopic8on-line.pdf

[3] DoE. (2016). Global energy storage database. http://www.energystorageexchange.org.

[4] *Electric Energy Storage Technology Options: A White Paper Primer on Applications, Costs, and Benefits.* (2010). Technical report 1020676. Palo Alto, CA: Electric Power Research Institute.

[5] Whittingham, M.S. (2012). History, evolution, and future status of energy storage. *Proceedings of the IEEE* 100 (Special Centennial Issue): 1518–1534.

[6] Akhil, A., Huff, G., Currier, A., et al. (2013). DOE/EPRI 2013 Electricity Storage Handbook in Collaboration with NRECA, Technical report SAND2013-5131. Albuquerque, NM and Livermore, CA: Sandia National Laboratories. http://www.sandia.gov/ess/publications/SAND2013-5131.pdf.

[7] Lacal-Arantegui, R., Fitzgerald, N., and Leahy, P. (2011). *Pumped-hydro Energy Storage: Potential for Transformation from Single Dams.* Luxembourg: Publications Office of the European Union. http://setis.ec.europa.eu/publications/jrc-setis-reports/pumped-hydro-energy-storage-potential-transformation-single-dams.

[8] Müller-Syring, G., Köppel, W., Mlaker, H., et al. (2013). Entwick-lung von modularen Konzepten zur Erzeugung, Speicherung und Einspeisung von Wasserstoff und Methan ins Erdgasnetz, Technical report, DVGW.

[9] DNV KEMA. (2013). System Analyses Power to Gas, Technical report. Groningen, The Netherlands.

[10] Belderbos, A., Delarue, E., and D'haeseleer, W. (2015). Possible role of Power-to-Gas in future energy systems. *2015 12th International Conference on the European Energy Market (EEM)*, IEEE, pp. 1–5.

[11] IEA & NEA. (2015). Projected costs of generating electricity. https://www.oecd-nea.org/jcms/pl_14756/projected-costs-of-generating-electricity-2015-edition?details=true.

[12] Iskov, H. and Rasmussen, N.B. (2013). Global screening of projects and technologies for Power-to-Gas and Bio-SNG, a reference report, Technical report. Horsholm, Denmark: Danish Gas Technology Centre.

[12] Smolinka, T., Günther, M., and Garche, J. (2011). Stand und Entwicklungspotenzial der Wasserelektrolyse zur Herstellung von Wasserstoff aus regenerativen Energien, Technical report, Frauhofer ISE.

[14] Benjaminsson, G., Benjaminsson, J., and Rudberg, R.B. (2013). Power-to-Gas – A technical review, Technical report. Malmo, Sweden.

[15] Lehner, M., Tichler, R., Steinmüller, H., and Koppe, M. (2014). *Power-to-Gas: Technology and Business Models.* Cham: Springer.

[16] Sioshansi, R. (2010). Some policy and research questions related to energy storage, Technical report. Florence, Italy: European University Institute – Robert Schuman

Centre for Advanced Studies. http://cadmus.eui.eu/bitstream/handle/1814/14394/RSCAS201058.pdf?sequence=1

[17] Denholm, P., Ela, E., Kirby, B., and Milligan, M. (2010). The role of energy storage with renewable electricity generation, Technical report, January. Golden, CO: National Renewable Energy Laboratory. http://www.nrel.gov/docs/fy10osti/47187.pdf.

[18] Eyer, J. and Corey, G. (2010). Energy storage for the electricity grid: Benefits and market potential assessment guide, Technical report, February. Albuquerque, NM and Livermore, CA: Sandia National Laboratories. http://www.sandia.gov/ess/publications/SAND2010-0815.pdf.

[19] Castillo, A. and Gayme, D.F. (2014). Grid-scale energy storage applications in renewable energy integration: A survey. *Energy Conversion and Management* 87: 885–894.

[20] Brijs, T., Geth, F., Siddiqui, S., et al. (2016). Price-based unit commitment electricity storage arbitrage with piecewise linear price-effects. *Journal of Energy Storage* 7: 52–62.

[21] Shahidehpour, M., Yamin, H., and Li, Z. (2002). *Market Operations in Electric Power Systems: Forecasting, Scheduling, and Risk Management*. New York: IEEE-Wiley.

[22] Rivero, E., Mallet, P., Stromsather, J., et al. (2015). EVOLVDSO: Assessment of the future roles of DSOs, future market architectures and regulatory frameworks for network integration of DRES. *23rd International Conference on Electricity Distribution*, Lyon, France, pp. 1–5.

[23] van Stiphout, A., De Vos, K., and Deconinck, G. (2016). The impact of operating reserves on investment planning of renewable power systems. *IEEE Transactions on Power Systems* 32 (1): 378–388.

[24] Xi, X., Sioshansi, R., and Marano, V. (2013). A stochastic dynamic programming model for co-optimization of distributed energy storage. *Energy Systems* 5 (3): 475–505.

[25] Drury, E., Denholm, P., and Sioshansi, R. (2011). The value of compressed air energy storage in energy and reserve markets. *Energy* 36 (8): 4959–4973.

[26] Moreno, R., Moreira, R., and Strbac, G. (2015). A MILP model for optimising multi-service portfolios of distributed energy storage. *Applied Energy* 137: 554–566.

[27] He, X., Delarue, E., D'haeseleer, W., and Glachant, J.-M. (2011). A novel business model for aggregating the values of electricity storage. *Energy Policy* 39 (3): 1575–1585.

[28] Willis, H.L. (1997). *Power Distribution Planning Reference Book*. New York: Marcel Dekker Inc.

[29] Schill, W.P. and Kemfert, C. (2011). Modeling strategic electricity storage: The case of pumped hydro storage in Germany. *The Energy Journal* 32 (3): 59–87.

[30] Geth, F. (2014). Battery energy storage systems and distribution grid support, PhD thesis. University of Leuven (KU Leuven), Leuven, Belgium.

36

Public Perceptions and Acceptance of Energy Storage Technologies

Per Alex Soerensen

PlanEnergi, Jyllandsgade 1, 9520 Skoerping, Denmark

36.1 Introduction

Public resistance against energy projects takes place when implementing all kinds of plants. Plants utilizing renewable energy sources such as hydropower, wind power, and biogas plants are not an exception from this and neither will energy storages be.

Some projects are met with positive public perception and acceptance and some types of projects can be met with resistance and demonstrations. This chapter tries to establish guidelines for how to gain positive perception and acceptance and avoid the opposite.

36.2 Why Resistance?

Energy storage projects can meet resistance where negative influence will occur to the sorroundings. Negative influences can stem from:

Environmental issues such as polluting emissions to water, soil, or air. This can of course be chemical emissions and particles. However, environmental issues can also stem from heating up ground water, cooling down rivers and lakes, and noise pollution.

Land use and integration in the landscapes. Some energy storage facilities might demand use of land, and land use in city areas is not only expensive but also a question of utilization of the land for residential purposes, green areas, or industrial purposes.

Outside city areas land use for purposes other than for agriculture, forestry, or recreation is always a topic for discussion and if large constructions are implemented outside industrial areas, landscape integration or integration in the surroundings will be an issue.

Economic issues can cause resistance if the project alternatives are not clear and/or if the implementation project is risky due to technical reasons or due to risky boundary conditions.

Social issues. Last but not least, resistance against projects can come from lack of information and inclusion. The feeling of powerlessness or exclusion can really cause huge frustrations, angry feelings, and resistance to projects.

Advances in Energy Storage: Latest Developments from R&D to the Market, First Edition.
Edited by Andreas Hauer.
© 2022 John Wiley & Sons, Ltd. Published 2022 by John Wiley & Sons, Ltd.

36.3 Who Will Resist?

The types of energy storages where there is a risk of a lack of public acceptance is for instance pumped hydro, CAES, the large pit heat water storages (PTES), and all kinds of underground thermal storages (ATES, BTES).

There are three main groups of stakeholders that have to be taken into account when implementing energy storage technologies where there is a risk of lack of public acceptance.

36.3.1 Politicians and Public Authorities

Politicians and public authorities will of course have to verify that relevant legislation concerning environmental, planning, and other issues is upheld. But also, politicians and civil servants are very important in securing correct information about the implementation project.

36.3.2 Public Interest Groups (NGOs)

Public interest groups can, for instance, be NGOs protecting the environment such as environment and anglers associations. But it can also be church organizations if the implementation project is near a church.

36.3.3 Groups in Direct Connection to the Projects

This group consists of neighbors to the project, land owners who have to sell or rent land to the project, customers and owners of utilities for implementing the energy storage, and companies where environmental damage can have consequences. For instance a drinking water company if there is a risk of pollution of their water sources.

36.4 Cases

In this chapter, good and bad practice is illustrated referring to implementation projects in general (not only for energy storages).

36.4.1 Borehole Storage (BTES) and Solar Thermal

A consumer-owned district heating company who already had a large ground-mounted solar thermal plant wanted to double the solar plant size and implement a pilot BTES.

The general assembly in the district heating company discussed and accepted the project. Also, the project was discussed with politicians, civil servants, and owners of the land before the general assembly and before the start of the planning procedure. A local plan procedure with a public hearing included showed no resistance to the project. To get environmental permission, a calculation of future drinking water temperature was required.

36.4.2 Pit Heat Storage (PTES) and Solar Thermal I

A consumer-owned district heating company, who already had a large ground-mounted solar thermal plant and a small PTES wanted to double the solar plant size and implement a PTES of 75 000 m^3.

The project was discussed and accepted at the general assembly in the district heating company and was discussed with politicians, civil servants, and owners of the land before a local plan procedure started. The public hearing included in the local plan procedure showed no resistance to the project.

36.4.3 Pit Heat Storage (PTES) and Solar Thermal II

A consumer-owned district heating company with natural gas fueled CHP wanted to implement a large ground-mounted solar thermal plant and a PTES covering up to 50% of the heat production. Politicians, civil servants, and environmental NGOs were informed about the project and were positive before the planning procedure started. Also, the ordinary general assembly was positive. But, during the public phase of the planning process, neighbors organized resistance. At an extraordinary general assembly, two out of three still voted for the project but the neighbors complained about the Local Plan procedure. The result was that the district heating company had to undertake an Environment Impact Assessment (EIA) where the environmental consequences of the project were explained, illustrated, and calculated in detail.

This delayed the project two years but did not stop the implementation.

36.4.4 Biogas Plant I

A private company (farmer owned) wanted to implement a biogas plant processing manure from surrounding farms within a 20 km radius. The municipality administration was informed about the project but was not active in finding an area where the plant could be built. The farmers investigated localization possibilities and found a place that also could be accepted by the municipality before the planning process started. But, as soon as the location was published, neighbors started protesting. A public information meeting attended by 200 angry people accompanied by a 4000 signature petition made the farmers give up the location.

36.4.5 Biogas Plant II

A private company (farmer owned) wanted to implement a biogas plant processing manure from surrounding farms within a 20 km radius. The municipality (administration and politicians) was informed about the project and was active in finding a location for plant implementation in cooperation with the farmers. The farmers decided to have public information meetings about the project before the planning process started and neighbors were invited to join a working group, which could follow the project and offer suggestions during the design phase. A public bus tour to a similar biogas plant was also arranged. Here, the neighbors could get an impression of the smell, traffic, and size of such a plant. Also, architects designed buildings, tanks, and integration in landscape.

Nevertheless, there was resistance from citizens in a city 2 km away from the biogas plant. The district heating company in the city was supposed to utilize heat from the biogas plant, but at the general assembly arranged by the district heating company, a proposal was made saying that the biogas plant should be placed at least 5 km from the city. This received a majority, and the farmers had to stop the project. This project is known as "The coup at Gudenaen" [1].

36.4.6 Biogas Plant III

In one region farmers had problems meeting the standards for discharge of phosphor and nitrogen. This could be solved if a biogas plant was implemented. Representatives from the farmers' association, utilities, and municipality met and drew up the idea.

The municipality found a location for the biogas plant near an incineration plant. The municipality was the land owner and the area was an industrial area in the municipal plan.

However, neighbors complained and the project developers therefore decided to improve transparency about the project. Thus, they hosted visits at existing biogas plants, hired architects to design the plant, used the newest technologies to avoid smells, and tried to reduce traffic to the plant by pumping manure.

For the municipality it was important that community embers understood that having the biogas plant would generate 1000 local jobs: an important consideration.

The plant was realized, jointly owned by farmers and by the local municipal owned supply company for electricity, heat, water, and sewage water. This project is also described in [2] as "Maabjerg BioEnergy – Transparency was the key."

36.5 Drivers for Positive Public Perceptions and Acceptance

There are some general drivers for public perception and acceptance, and there are specific drivers for the main groups mentioned in Section 36.2. The overall drivers are "What's in it for me?" credibility, and transparency.

What's in it for me is a question all stakeholders will ask themselves during a project process and the less there is a direct gain for the stakeholders' quality of life, such as cheaper energy, the more careful the project developer has to be when presenting the project. This is a big problem with biogas projects. Many stakeholders have advantages from the project but the neighbors often only have disadvantages.

Credibility or trust is another general driver [3]. If it is possible to establish a project management group where people affected by the project can find at least one person they trust, communication will be much easier. Furthermore, communication and processes have to be fair and the information has to be reliable.

Transparency is necessary to avoid frustration with decisions made. If the distribution of costs and benefits is not visible, it is a particular problem [3].

36.5.1 Drivers for Politicians and Public Authorities

The project has to be in accordance with national and municipal energy strategies and in accordance with planning, environmental, and other legal requirements.

Not all nations and municipalities have energy strategies and that can be a serious obstacle for the implementation of energy projects. On the other hand, if there has been a planning process at a national level, where different future energy scenarios have been discussed, resulting in a long-term policy backed up by a large majority in the parliament, this will be a strong implementation driver [5].

The same applies at the municipal level. For instance, several Danish municipalities have arranged planning processes involving the most important stakeholders and ended up with 100% political agreement behind a municipal strategic energy action plan.

In planning processes it is very important to show different future solutions and their consequences for environment, economy, and employment. This will result in a transparent and democratic process and a stable long-term plan. [6] favors this strategy due to experience with cases where there has only been one solution proposed.

To show that the project fulfills the legal requirements, the consequences of the project have to be described. Included in this description could be details about how the project fits into national and local energy strategies, clear visualization of landscape integration, cost and benefits (including socio-economy), local economic consequences, distribution of benefits, creation of local workplaces, processes that involve local stakeholders, and an implementation strategy.

36.5.2 Drivers for Public Interest Groups (NGOs)

Public interest groups normally have an area of interest they protect. For environmental protection organizations, it's the environmental consequences that concern them. These consequences could be discussed in co-operation with the interest groups as part of involvement of stakeholders.

36.5.3 Drivers for Groups in Direct Opposition to the Project

Groups in direct connection with the project are often afraid of project consequences that might harm them. And if the project type is unknown, that can be a problem in itself. Therefore, identifying and publicizing consequences that raise the quality of life is fundamental. And one very important part of quality of life is to be able to make own decisions on matters that influence their everyday life and if groups in direct opposition to the project are given the opportunity to have real influence this problem will be somewhat alleviated.

Thus, community-owned projects or projects where people are part of it gain public acceptance more easily; [7–9] come to the same conclusion.

The objective of involvement of groups in direct opposition to the project is to create ownership, meaning that the project is "our" project instead of "their" project.

It also means that decision processes not involving groups in direct opposition to the projects increase the risk of resistance significantly. It has to be respected that the stakeholders the are directly involved are local experts alongside other experts involved in the project.

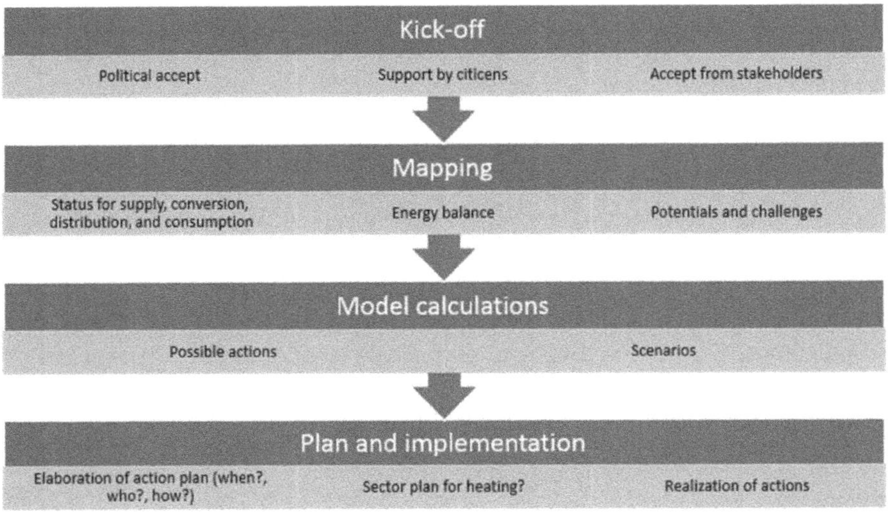

Figure 36.1 Process for elaboration of a Strategic Energy Action Plan.

36.6 Is There a Manual for Citizen Involvement?

It's impossible to construct a precise manual about how to gain positive public perception and acceptance of energy storage technology projects because projects differ and national and local conditions also differ.

However, some "rules" can be established. From previous projects it is evident that former models such as "decide-announce-defend" [10] are not adequate.

What is adequate is to have a strategic energy action plan at national and local (municipal) levels. This plan has to be elaborated in a process including the main stakeholders and must have citizen involvement. The process is illustrated in Figure 36.1.

Such processes have, in several Danish municipalities, resulted in 100% agreement from stakeholders and politicians on the content in the plan. That makes it easier to understand the reasons behind implementation projects but the design and implementation process still has to be carefully realized.

On the Danish island Samsø, several energy projects have been implemented in the period from 1996 with positive public perception and acceptance. One of the experiences from the implementation projects is that careful preparation in the first steps is a must. In the project "Implement" supported by the EU Interreg program, Samsø Energy Academy described the implementation process in "A manual on citizen involvement" [11]. The manual is described for a biogas project but is useful for other projects as well.

The steps in the manual are:

- Elaboration of baseline study including collection of information about local social habits.
- Engagement of people that know local habits and conditions.

- Identify the directly involved project stakeholders.
- Find "whats in it for me" for the involved stakeholders.
- Define objectives for involvement and a strategy for how to reach the stakeholders.
- Involve the local authorities in the project.
- The process (must be carried out from the bottom and controlled from the top).
- Communication must be clear and proactive. Communication channels must be defined.
- Objective of meetings must be clear and meetings prepared by contacting key stakeholders before the meeting and there must be discussion of "what might happen" scenarios.
- Between meetings the project should contact key stakeholders, arrange working groups, and arrange site visits to similar projects.

This method of involvement has created local ownership of all kinds of energy projects on the Island of Samsø. It is also important that there is a master plan for transition to renewable energy for the island and that this masterplan is broadly discussed and politically approved.

36.7 Perception of Acceptance of Energy Storage Technologies

Many renewable energy projects are met with public resistance because public involvement has been poor. This can be reduced or avoided for energy storage projects because environmental consequences normally don't include noise and smell. Public resistance can be further reduced by careful preparation of the projects.

The cases 3.1 and 3.2 illustrate that large thermal storages can gain positive public perception and acceptance in projects driven by consumer-owned developers if the process of information and involvement is transparent and proactive. In case 3.3 the solar thermal technology was new and caused resistance from neighbors.

The case outlined in 36.4.4 was prepared too poorly but even with a careful preparation the similar project described in 36.4.5 failed. Mainly because it is difficult to create ownership to biogas plants not owned by neighbors because they are afraid of smell and traffic.

But Biogas Plant III (36.4.6) shows that even in that type of projects, acceptance can be reached if the organization behind the project is strong enough and carefully listens to the stakeholders.

Large thermal storages have, as mentioned earlier, no problems with smell or noise. Furthermore, they can be situated in a way that fits into the landscape and adds quality to the neighborhood. Examples are: Dronninglund, where a former unattractive gravel pit was used for a PTES (see Figure 36.2); and Munich where a tank thermal energy storage (TTES) is covered with soil and is seen as a hill in the landscape (see Figure 36.3).

Figure 36.2 From gravel pit to PTES, Dronninglund, DK [9].

Figure 36.3 From TTES to a hill, Munich, DE [9].

References

[1] https://peopleandbiogas.com/en/kuppet-i-gudenaen/

[2] https://peopleandbiogas.com/en/abenhed-er-noglen-maabjerg-bioenergy/

[3] Agterbosch, S., Meertens, R.M., and Vermeulen, W.J.V. (2009). The relative importance of social and institutional conditions in the planning of wind power projects. *Renewable and Sustainable Energy Reviews* 13 (2): 395–405.

[4] IEA Wind. (January 2013). Social acceptance of wind energy projects. Expert group summary on recommended practices. http://www.socialacceptance.ch/images/rp_14_social_acceptance_final.pdf.

[5] Lund, H. (2008). *Choice Awareness and Renewable Energy Systems*. Denmark: Aalborg University.

[6] O'Hare, M., Bacow, L., and Sanderson, D. (1983). *Facility Siting and Public Opposition*. New York: Van Nostrand Reinhold Company.

[7] Olsen, B.E.. (2010). Wind energy and local acceptance: How to get beyond the Nimby Effect. *European Energy and Environmental Law Review* 19(5): 239–251.

[8] *People and Biogas. A manual on citizen involvement*. www.peopleandbiogas.com.

[9] Rogers, J.C., Simmons, E.A., Convery, I., and Wheatherall, A. (2008). Public perceptions of opportunities for community-based renewable energy projects. *Energy Policy* 36 (11): 42175–4226.

[10] Ruggiero, S., Onkila, T., and Kuittinen, V. (2014). Realizing the social acceptance of community renewable energy: A process-outcome analysis of stakeholder Influence. *Energy Research & Social Science* 4 (12): 53–63.

[11] Soerensen, P.A., Miedaner, O., and Mangold, D. (2015). *Large Sensible Hot Water Storages (Up to 100 °C) in Germany and Denmark*. Beijing: Greenstock.

[12] Warren, C.R. and McFadyen, M. (2010). Does community ownership affect public attitudes to wind energy. A case study from south-west Scotland. *Land Use Policy* 27 (2): 204–213.

37

Business Case for Energy Storage in Japan

Masaya Okumaya

37.1 Energy Consumption in Japan

Figure 38.1 shows the energy consumption in Japan, by sector and financial year (FY), from 1973 to 2012. By sector, the industrial sector has stayed flat or decreased slightly due to a continuous energy saving act following the oil crises. However, due to the improvement of living standards private sector growth, such as the residential and commercial sectors, has doubled.

The annual energy consumption total has been increasing up to the FY2007. However after that due to the effect of the economy and the Great East Japan Earthquake, the total value varies year on year. For example in FY2010 due to the recovery in the economy and the high temperature the total value grew but in FY2011 due to the earthquake the energy saving act was promoted and the value has decreased.

As of FY2012 it has grown 1.3 times compared to FY1990, since the growth of the residential/commercial sector is obvious, measures for reduction in use in these sectors should be taken.

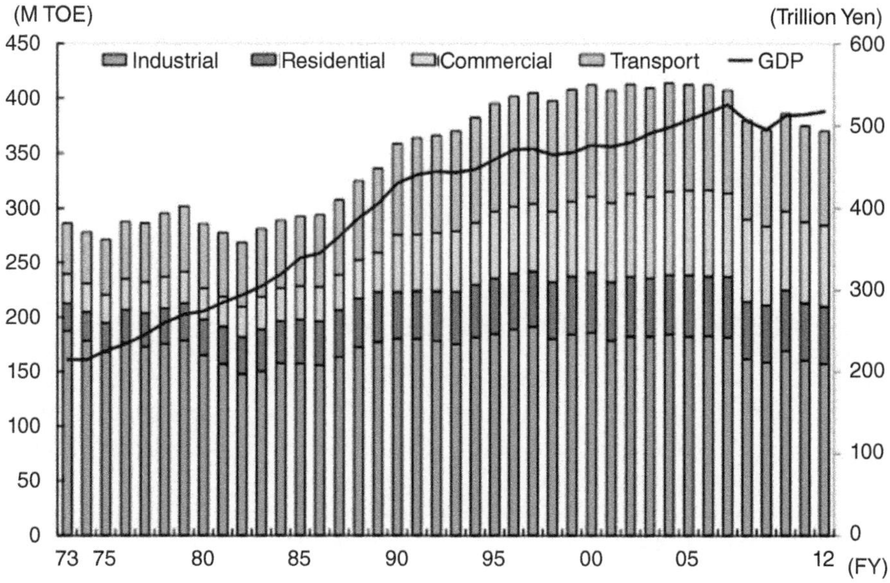

Figure 37.1 Energy consumption by sector [1] and financial year.

Advances in Energy Storage: Latest Developments from R&D to the Market, First Edition.
Edited by Andreas Hauer.
© 2022 John Wiley & Sons, Ltd. Published 2022 by John Wiley & Sons, Ltd.

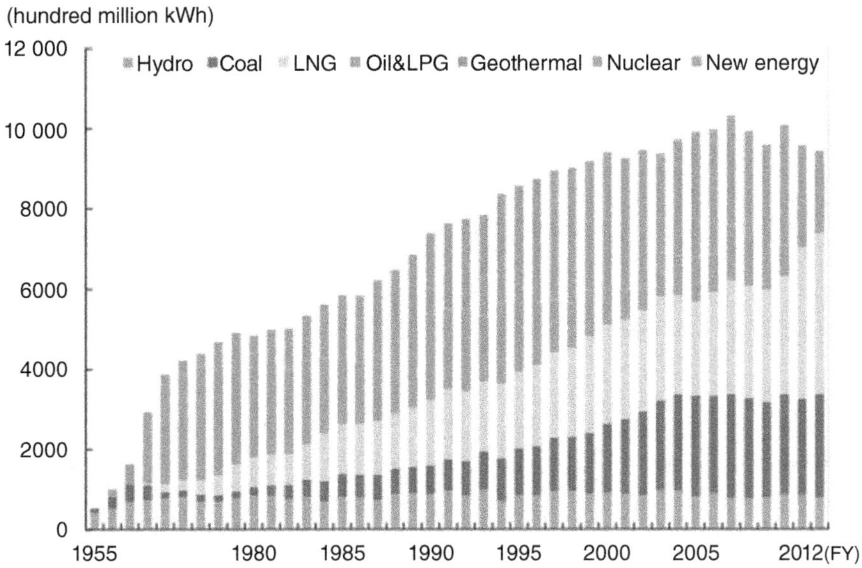

Figure 37.2 Total electric power generation by energy source [2].

37.2 Electricity Situation

37.2.1 Total Electric Power Generation by Energy Source

Figure 37.2 shows total electric power generation by energy source. In the 1970s oil and LPG accounted for a huge percentage but after the oil crisis occurred the share of nuclear and liquified natural gas (LNG) has grown rapidly up to FY2010.

After the Great East Japan Earthquake, nuclear power generation has gradually stopped and the use of oil and LNG has grown rapidly with the result that the share of fossil fuel has reached as much as 88%. Since fossil fuels have high greenhouse gas emission we need to take action. So on both the supply and demand sides highly efficient use of electricity is essential.

37.2.2 Daily Electric Demand and Annual Electric Load Factor

Figure 37.3 shows the electric load curve on the highest demand day of each year. The highest demand day appears in summer and the peak time is around 15:00 and valley is around 6:00. Due to the improvement in living standards usage has grown and the peak/valley pattern has become more obvious year by year.

Figure 37.4 shows the change of the annual electric load factor, which is the ratio of annual average power demand to peak power demand. The load factor decreased until 1995 followed by an increase in air-conditioning demand. After 1995 the load factor rose despite a load factor improvement policy including the thermal energy storage to reach 67.8% in 2011.

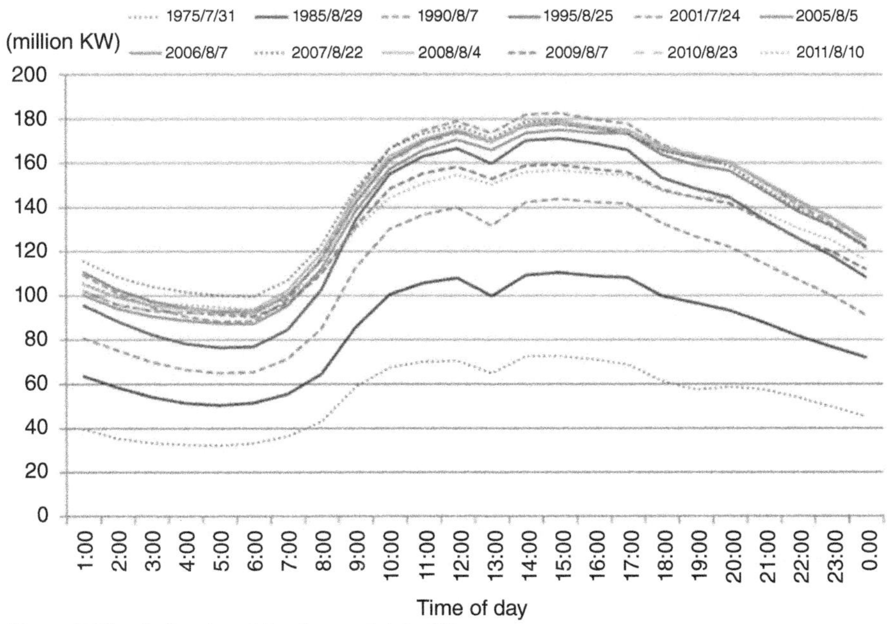

Figure 37.3 Daily electricity demand data [2].

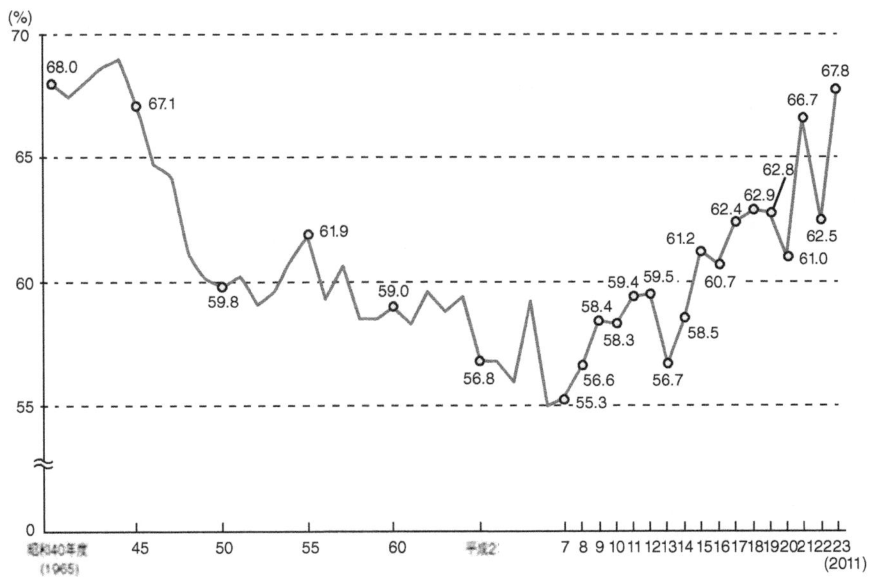

Figure 37.4 Change of the annual load factors [2].

37.3 Climate Condition and Cooling/heating Load

Figure 37.5 shows a climograph of several cities in the world and the average temperature, including several cities in Japan. Japan has a hot and humid climate conditions

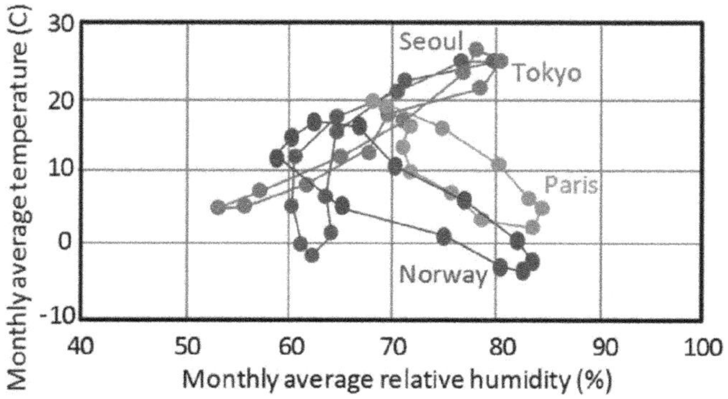

Figure 37.5 Climograph of several cities.

except in the northern part. Cooling and heating load has to be treated by heating, ventilation and air conditioning (HVAC) system, in almost all areas of Japan. Also the HVAC system has to process the latent heat in summer (dehumidification).

37.4 Situation of Thermal Energy Storage (TES) Spread

Figure 37.6 shows the changes in the quantity of thermal energy storage (stock at the end of the fiscal year). As shown in Figure 37.6 the installation thermal energy storage systems increased rapidly from 1995 to 2005 and has lead to improvement of the load factor as in Section 37.3. Ice storage, especially the package type, increased in number after 1995.

Figure 37.7 shows a change in the installation capacity of the thermal storage tanks. Capacity of water storage and central ice stanorage accounts for about 40% and 50% of the installation capacity in 2010.

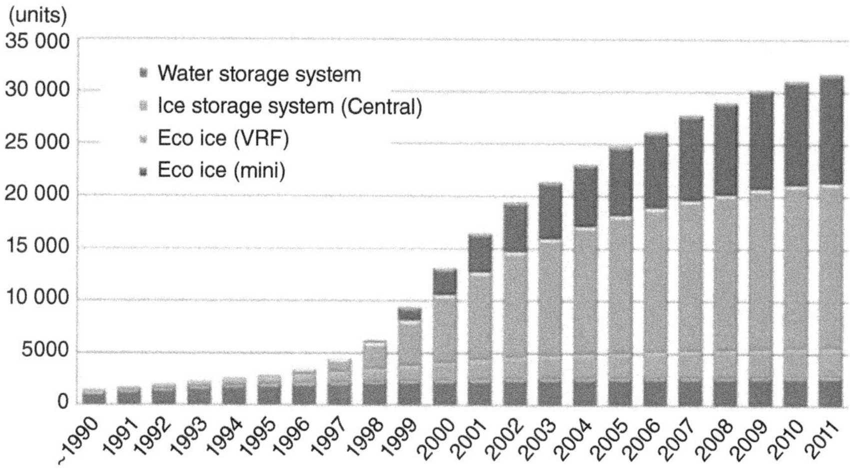

Figure 37.6 Total number of TES by each category [3].

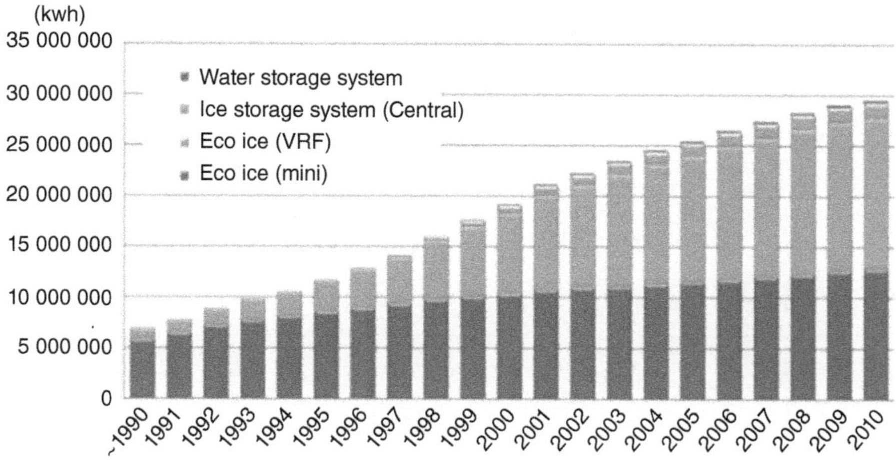

Figure 37.7 Total capacity of TES by each category.

The capacity is calculated under the assumption that:

Eco Ice/KE (Mini): regular Horse Power of outdoor unit is 6HP
Eco Ice/KE (VRF): typical Horse Power of outdoor unit is 16HP
Ice storage type/KE (Central): IPF=0.6
Water storage type: ⊿t=5 °C Capacity=Volume*deltaT*Cp

Peak shift (KW) is 1.91GW in 2011 and it is about 1% of peak electric demand (180GW in 2011).

37.5 Variation of TES

There are various mediums to store heat by TES. For TES using the temperature difference of the medium, namely sensible heat storage, there are methods to store thermal energy from water tanks installed in the building and in the concrete that provides the skeleton of the building. For TES using latent heat of the medium, there are methods using latent heat of phase change from water to ice and chemical change by temperature.

Figure 37.8 shows the history of energy storage development in Japan. The oldest actual thermal storage system was installed in a film studio in 1953. It already had the philosophy of saving energy with high load factor operation and the reduction of facility capacity. In the 1960s and 1970s, there were many studies of the thermal response of water storage tanks and mathematical and control models were adopted for actual building designs and operations. In the 1980s, many kind of thermal storage systems were developed for the reduction of the electricity demand in day-time during summer. Especially developments of the latent sensible heat materials or enlargement of storage quantity using post-and-beam were performed by many companies and universities. However, only water and ice storage systems survived to the 2010s. There were so many problems on endurance capacity, stability, manageability, economic efficiency, safety, disposability, etc.

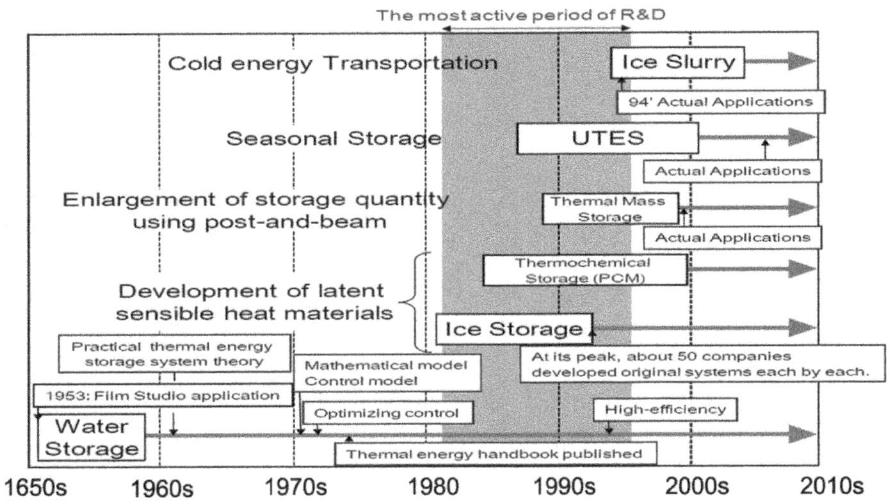

Figure 37.8 History of energy storage development in Japan.

37.6 Water Storage

One basic characteristic required of a thermal energy storage tank is that the water with different temperatures mixes as little as possible. For this purpose, there are the temperature-stratified and multi-connected mixing type tanks.

37.6.1 Temperature-Stratified Thermal Storage Tank

Stratification in chilled water storage tanks is based on the density difference of water to form horizontal layers or temperature zones. As water gets colder, it becomes denser, until it reaches 4°C. As water is cooled below this point, it becomes less dense, until it freezes. Cold water near to 4°C and up to 6°C will collect and stabilize in the lowest regions of the tank, while warmer water between 10°C and 18°C will collect in the upper regions of the tank. During the charging cycle, cold water from the chilling equipment enters the tank through diffusers at the bottom, and warm water exits the tank at the top. As the volume of chilled water increases and warm water is displaced, the thermocline rises. The total volume of water in the tank remains the same. The flow of water is reversed during discharging. The system draws chilled supply water from the low portion of the tank and sends warm return water into the top of the tank. The storage capacity of a stratified chilled water tank increases with the temperature difference between the stored chilled water and the warm water returning from the load, as in other chilled water storage methods. In addition, a larger temperature differential increases the density difference between the warm and cold water through a chilled water tank, further enhancing stratification.

37.6.2 Multi-connected Mixing Type Tank

Multi-connected mixing type are connected in series. Water in each tank, in multi-connected mixing system, is fully mixed, however as a whole it retains temperature stratification.

Generally, if there are more than 20 tanks, the mixing of water is considerably restrained and has the characteristics equal to temperature-stratified thermal storage tank.

37.7 Ice Storage

37.7.1 Background

From the beginning of the 1980s the following situation began to be utilized for example in an office building.

- Internal heat generation increased according to spread of office automation and it resulted in the increase of the cooling load.
- TES became a useful system for existing buildings for electric power load leveling.
- Reduction of fluid to transport heat was required for energy conservation.

According to above condition, reduction of the volume of storage tank was required with the storage medium of the large thermal mass.

37.7.2 Classification of Freezing and Melting Methods

Figure 37.9 shows the classification of freezing and melting method. First of all, it is classified into an indirect cooling system and a direct cooling system.

The direct cooling system makes slurry (crystal) ice with the direct injection of a refrigerant to the water tank. Tn indirect cooling system makes ice through the heat exchanger (ice maker). The indirect type is further classified into a static type, where ice stay on the surface of heat exchanger, and a dynamic type, where ice does not stay on the surface of heat exchanger.

Ice becomes thermally resistant and the ice making efficiency decreases as the thickness of ice increases. Because ice does not stay on the surface of the heat exchanger in a dynamic type, this problem can be avoided. However, it is necessary to consider the:

- energy and time to strip off ice,
- method of Stripping off ice; and
- device that prevents the ice being stripped off from forming a block of ice.

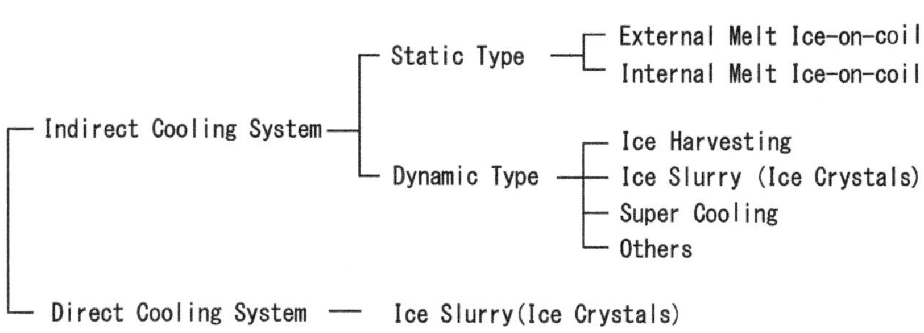

Figure 37.9 Classification of ice making and melting method.

The static type is divided into "External Melt Ice-on-coil" where ice is melted from the outside and "Internal Melt Ice-on-coil" where ice is melted from the inside. The static type is if further divided into a spiral type, a snake tube type, a cylinder multi-tube type, and a plate type, according the form of the heat exchanger. Copper, steel, and cross-linked polyethylene are used for material in the heat exchanger.

The dynamic type is divided into ice harvesting, ice slurry, and supercooling. In all the methods, fluid ice that is a mixture of liquid and thin ice plate or ice grain is made.

37.7.2.1 External Melt Ice-on-coil Storage

Figure 37.10 shows the outline of "External Melt Ice-on-coil" storage. In charging mode, cold refrigerant or glycol flows in the pipe (coil) on which ice is built. In discharging mode, warm return water flows outside the coil and melts the ice. In charging when the ice becomes thick, it becomes thermally resistant. The evaporating temperature of chiller should be low and it results in a low COP (coefficient of performance). This is reason why dynamic type ice making was developed. Bridging of ice between individual coils should be avoided because it restricts the circulation of warm water in the tank discharging mode. Therefore ice thickness control is important.

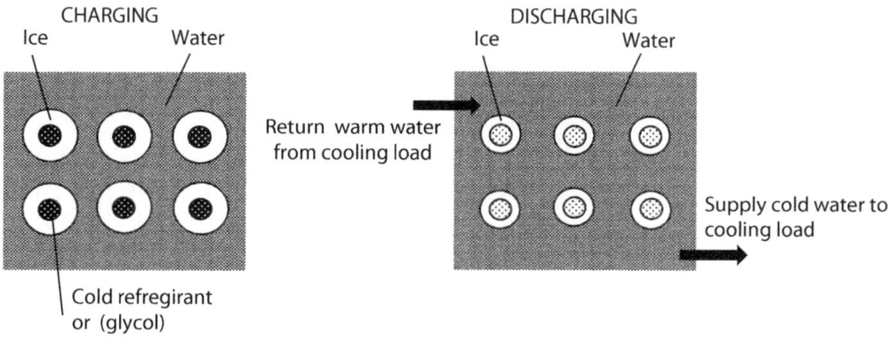

Figure 37.10 Outline of External Melt Ice-on-coil storage.

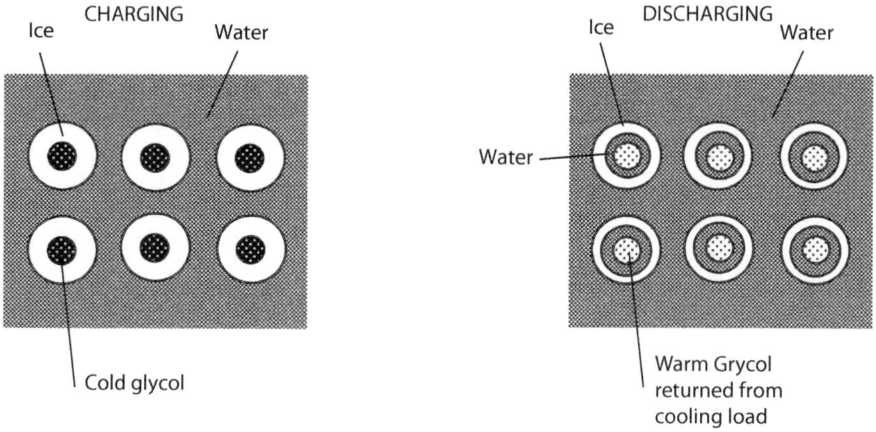

Figure 37.11 Outline of Internal Melt Ice-on-Coil storage.

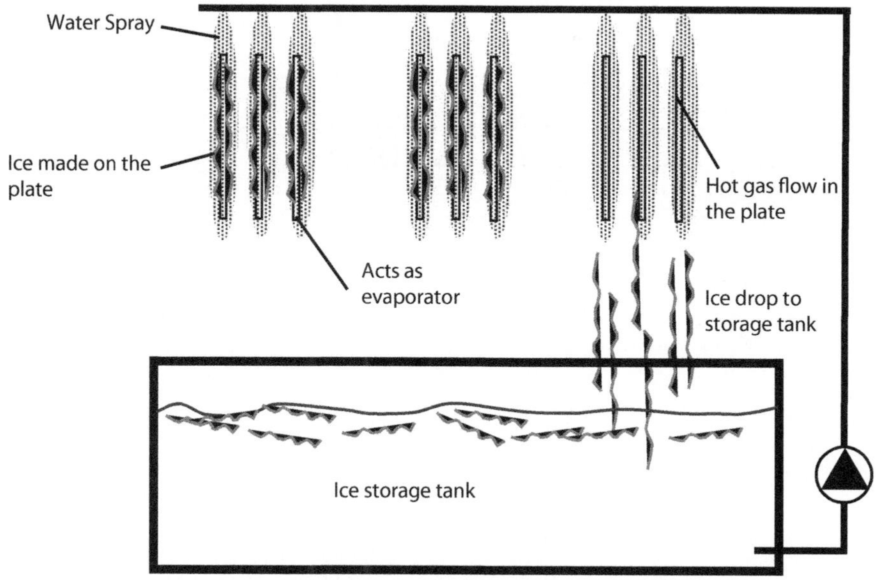

Figure 37.12 Outline of ice harvesting.

37.7.2.2 Internal Melt Ice-on-coil Storage

Figure 37.11 shows the outline of “Internal Melt Ice-on-coil” storage. In charging mode, cold glycol flows in the pipe (Coil) on which ice is built. In discharging mode, warm glycol flows in the pipe and ice is melted from the inside.

37.7.2.3 Ice Harvesting

Figure 37.12 shows the outline of “Ice Harvesting” storage. Ice is made on the surface of the flat plate heat exchanger. The flat plate heat exchangers (ice maker plate) are divided into three groups. When ice is made on certain plate, ice is harvested at another plate and is dropped into the thermal storage tank which located under the ice maker plate. The plate on which ice is made acts as an evaporator or chiller and at another plate ice is harvested with a hot gas flow through the plate. The thickness of ice is usually 10 mm to 30 mm

37.7.2.4 Super Cooling

Figure 37.13 shows the outline of “Super Cooling” storage. Water is super cooled in the heat exchanger and it is released on to the board. When super cooling is released, super-cooled water divides into 0°C water and ice.

37.7.3 Low Temperature HVAC System

Ice storage systems require low temperature outputs of the chiller. The refrigeration capacity and COP of the chiller decreases to lower the temperature of the heat media to make ice. To minimize performance decrease of the chiller, different operation modes of chiller are adopted: for charging mode (making ice) in the nighttime and discharging mode (melting ice) in daytime. Chillers can have one or two evaporators.

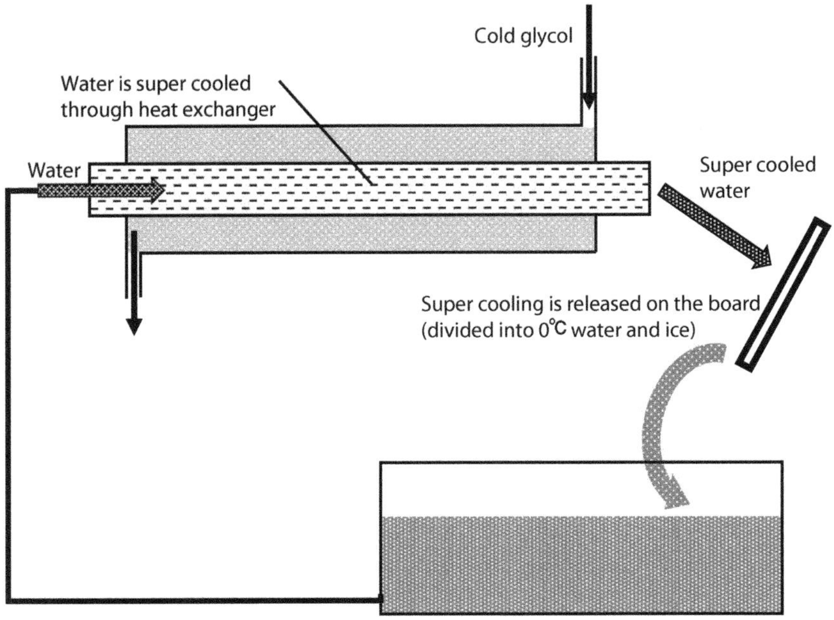

Figure 37.13 Outline of Super Cooling storage.

When the chiller has only one evaporator, it is for cooling brine. The chiller is driven at the temperature for ice making at nighttime, and it is driven at a higher brine temperature in daytime to cool the water sent to the secondary side through the heat exchanger.

In the case where the chiller has two evaporators, the evaporator for brine operates at a low temperature for charging the ice storage tank (at nighttime) and in discharging mode (daytime) the evaporator for cooling water operates at a higher temperature.

The low temperature output has to be utilized efficiently. A low temperature HVAC system is the answer to the above consideration. Low temperature HVAC systems mean large temperature differences in the system and it can reduce the flow rate of fluid (water and air) that conveys heat. Figure 37.14 shows the advantage of the low temperature HVAC system. If the supply air temperature is lowered to 10°C, the air flow rate can be

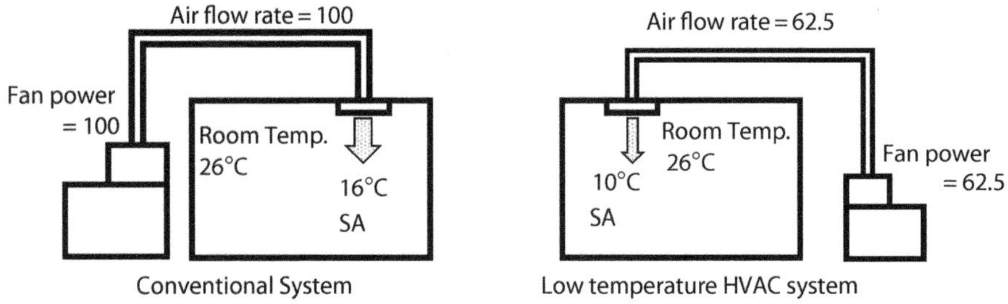

Figure 37.14 Advantage of low temperature HVAC systems.

reduced to 62.5% and it results in a reduction of fan power. In addition to energy saving, the size of duct can be reduced and the floor height of building also can be lowered. This results in the reduction of initial cost and life-cycle cost.

In the case of low temperature HVAC, there is a possibility that a cold draft is caused. Therefore it is necessary to select air diffusers carefully.

References

[1] EDMC/Directory of Energy & Economic Statistics.
[2] Infobase of "The Federation of Electric Power Companies of Japan (FEPC)."
[3] Heatpump & Thermal Storage System Data.

38

Energy Storage in the Electricity Market

Business Models and Regulatory Framework in Germany

Helena Teschner

Politics and Markets, BVES – German Energy Storage Association, Berlin

Abbreviations

BNetzA	Federal Network Agency (Bundesnetzagentur)
BSS	Battery Storage System
CHP	Combined Heat and Power (KWK)
EEG	Renewable Energies Act (Erneuerbare-Energien-Gesetz)
EEX	European Energy Exchange
ENTSO-E	European Network of Transmission System Operators for Electricity
EnWG	Energy Industry Act (Energiewirtschaftsgesetz)
EV	Electric Vehicles
FIT	Feed-in Tariff
G2V	Grid-to-Vehicle
GW	Gigawatt
KfW	Development Loan Corporation (Kreditanstalt für Wiederaufbau)
kWh	kilowatt hour
kWp	kilowatt peak
PCR	Primary Control Reserve
PTG	Power-to-gas
PTH	Power-to-heat
PTX	Power-to-X
PV	Photovoltaic
R&D	Research and Development
RES	Renewable Energy Sources
StromNEV	Electricity Grid Charges Ordinance (Strom-Netzent-geltverordnung)
TSO	Transmission System Operator
UPS	Uninterruptible Power Supply

Advances in Energy Storage: Latest Developments from R&D to the Market, First Edition.
Edited by Andreas Hauer.
© 2022 John Wiley & Sons, Ltd. Published 2022 by John Wiley & Sons, Ltd.

38.1 Introduction

In the past, the energy system consisting of the three parts generation, transport, and consumption was acting in a very predictable way. The fast roll-out of Renewable Energy Sources (RES) in an increasingly de-centralized system with fluctuating power infeed and variable consumption has promoted the importance of energy storage systems public perception drastically.

A successful energy transition can only work with flexibility options for the implementation of RES.

Given the required transition, energy storage is one of the key subjects, because it can act like a "Swiss army knife" that unfolds its different tools depending on the specific application and particular requirements (see Figure 38.1).

Pumped hydro storage has been integrated successfully in the energy system for many decades and has proven its relevance for system stability.

Battery Storage Systems (BSS) are the current center of attention. They react within milliseconds and serve many applications, e.g. by providing control energy (large-scale), by optimizing load scenarios in the commercial sector (medium-scale), but also, and especially, in residential areas (small-scale).

The variety of storage technologies is also reflected in the capability to transform electrical energy to other energy sources, e.g. to gas by "Power-to-gas" (PTG) or to heat by "Power-to-heat" (PTH). By linking the areas of electricity, gas, and heat, excess electricity can be stored for days, weeks, and months. Moreover it can be used in different sectors like transport (as fuel), industry (as raw material), and as well in the heating sector.

The linkage of sectors offers a huge potential for future energy system based on renewable energies and contributes to efficiency and to reduce carbon emissions.

However, given the current regulatory framework in Germany, those processes are mostly not, as yet, economically viable.

This chapter gives an overview of existing business models, but also refers to trends and future models like combined solutions that are enabled by digitalization.

The first section (38.2) describes the business models and distinguishes between models "before-the-meter" – like ancillary services – and the combination of technologies "behind-the-meter" for the optimization of self-consumption. Moreover, it addresses mixed business models that combine the benefits of both areas and different grid levels.

The second section (38.3) is dedicated to the German framework. Beginning with existing funding programs, the focus of this chapter lies on the analysis of the legal and regulatory barriers energy storage is confronted with and furthermore the necessary adjustments so that the different technologies can deploy their abilities in practice.

The conclusion (38.4) completes the chapter and gives an outlook on how the further roll-out and future development of energy storage technologies in Germany are expected.

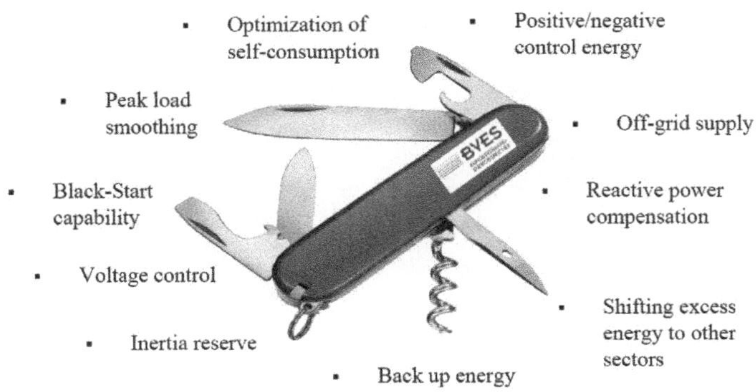

Figure 38.1 The different facets of energy storage.*Source*: BVES e.V., German Energy Storage Asscociation.

38.2 Business Models in Germany

Pumped hydro has been an integral part of the German energy supply system for several decades.

New business models exist especially for battery storage as it is ideally applicable for the provision of ancillary services like Primary Control Reserve (PCR), but as well to increase self-consumption rates in the commercial and residential sector. The trend is strikingly upwards as on top a further decline in prices is expected. Figure 38.2 shows the variety of applications that apply mainly to batteries.

Thermal energy storage like heat pumps and electrical storage heating were condemned for a long time due to their perceived excessive energy use and inefficiency. Now they appear in a new light when green electricity is used to operate them.

Thermal energy storage completes the package of self-consumption by maximizing the rate up to 90–100%. But operators of Combined Heat and Power (CHP) plants also use thermal energy storage in order to be prepared for electricity price volatility.

As shiftable additional load, Power-to-X (PTX) like PTG and PTH can help to prevent the down-regulation of RES-electricity by, additionally, giving it a beneficial use in other sectors. It contributes significantly to the decarbonization of sectors such as heat, mobility, and industry, and increases the efficiency of the whole system.

One of the business models of PTG is to benefit from cross-sectoral price differences between electricity and gas. At this moment pure arbitrage between the sectors is not worthwhile, even with a fully depreciated plant, because the current electricity price of 28.69 € cents/kWh[1] cannot overweigh the gas price of 6.74 € cents/kWh.[2]

Another hindering aspect to bringing PTG into the market is the current regulatory framework. The recent amendment of the Renewable-Energies-Act (Erneuerbare-Energien-Gesetz, EEG 2017; EEG) provides a flicker of hope as it provides a

1 Cp. graph on page 15.

2 Cp. https://1-gasvergleich.com/gaspreise.

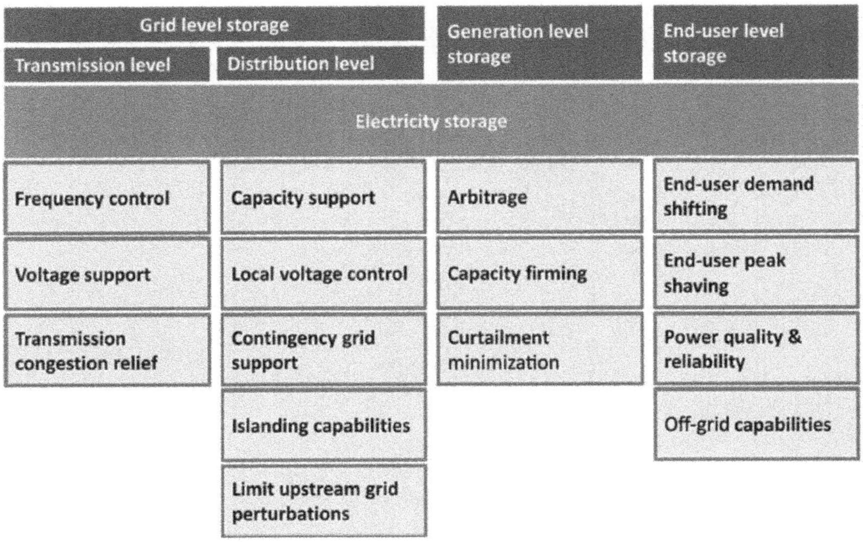

Figure 38.2 Applications for battery storage. *Source*: Policies for Storing Renewable Energy, IEA, March 2016.

remuneration mechanism for loads in cases where they are operated in a grid-supportive manner. In autumn 2016, the preparation of the statutory authorizations began that all details how PTG can benefit. In general, PTG is not that profitable yet, but only a few changes would make a lasting improvement.[3]

38.2.1 Business Models "Before-the-meter"

Energy storage devices "before-the meter" offer an extensive service potential at medium- and low-voltage level, e.g. the provision of control reserve,[4] peak load smoothing and levelling, and load buffering but also a variety of other grid-supporting services like inertia reserve, voltage control, reactive power compensation, services for redispatch, short circuit power, or black-start capability.

Energy storage "before-the meter" is a crucial element for smart grids (on islands or in rural areas) and the reduction of RES curtailment as well as the deferral of grid extensions are valuable consequences.

While most grid-supporting services currently receive no extra remuneration, as they are covered by the spinning reserve of conventional power plants, the provision of control reserve – and here especially PCR – is a predominantly profitable business case.

3 Cp. Faktenpapier Speicher, BVES and DIHK, p. 16; Dr. Klaus-Dieter Maubach, p. 2 f.

4 The transmission system operator (TSO) is in charge of maintaining the balance between electricity generation and consumption within their control areas at all times. For the performance of this task, the TSO needs different types of control reserve: primary control reserve (PCR), secondary control reserve, as well as minute reserve, which is also called tertiary control reserve. The three types differ according to their capacity and activation speed.

38.2.1.1 Provision of PCR

Since 2001, the German Transmission System Operators (TSO) have been procuring their required control reserve on an open, transparent, and non-discriminatory market for control reserve according to the guidelines of the Federal Cartel Office (Bundeskartellamt-BKartA). The procurement of primary control reserve is currently carried out at a weekly tender auction on the control reserve market with the participation of numerous bidders.[5]

In the past the control energy demand was mainly covered by conventional power plants. However, batteries can provide electricity within milliseconds and are able to deliver PCR better than any other technology.[6] The battery system is controlled automatically by the frequency of the connected grid. In the same way fluctuations of renewable energies can be compensated very effectively by reducing carbon emissions at the same time.

The weekly average price of PCR was at around 3500 € per MW in 2014 and around 3700 € per MW in 2015. On the contrary the prices dropped in 2016 for several weeks in summer to as low as 1800 €, which was influenced by different effects. Most likely this was caused for a great part by the growing number of market players, who were testing the pricing mechanism. Another big driver for the price drops were hydro electric power stations and pumped hydro storage. Once they prove a sufficient water level in their reservoir they are allowed to participate in the PCR tendering process and due to massive and lasting rainfalls this was easy to realize at a very low price level.

As the market is mainly still driven by fossil power plants, the mid-term price expectations are at around their production cost of around 1800 € to 2000 € per MW.

As a precondition for participating in the weekly PCR auctions, the offering unit has to pass a prequalification process. The process is carried out by the TSO (of the respective control area) and turned out to be a significant market entry barrier for many battery projects.

According to the European Transmission Code 2007, the necessary condition to access to the balancing market was the fulfillment of 2 × 15 minutes at full load within an hour. In 2015 the European Network of Transmission System Operators for Electricity (ENTSO-E) came up with the requirement of 2 × 30 minutes at full load specifically for technical units with limited capacity (like battery storage). This directly affects the profitability of many battery projects.

The need for this tightening is not technically justified or legally proven and contradicts the technologically open approach according to European law. The Guideline for System Operation (finalized in May 2016) includes a cost-benefit-analysis for the provision of PCR. The elaboration of a methodology as well as the analysis itself will be carried out by ENTSO-E. During this process, the requirement of 15 minutes is set, unless the TSO applies to the national regulatory authority for a different requirement between 15 and 30 minutes:

---

5 Cp. https://www.regelleistung.net/ext/?lang=en.

6 If it falls below a grid frequency of 50 Hertz, the battery feeds electricity into the grid. At a defined threshold above 50 Hertz, the battery is charging and takes excess electricity from the grid.

> Where no period has been determined [] each FCR provider shall ensure that the FCR from its FCR providing units or groups with limited energy reservoirs are able to fully activate FCR continuously for at least 15 minutes [] or for a period defined by each TSO, which shall not be greater than 30 or smaller than 15 minutes. (Article 56 (9)[7]

The criteria required to justify a different time requirement are not yet set out conclusively. Therefore, there is still no reliable basis on which investors can base their decisions.

Contrary to this ambiguity there is still a strong growth in this market segment due to falling cell prices.

38.2.1.2 Arbitrage on Electricity and Gas Markets

By decoupling the simultaneity of production and consumption, energy storage can use the price fluctuations in the electricity market in order to generate revenue. By buying electricity at a low price in times of surplus supply, "storing" it, and selling it at times of low supply at a higher price level, energy storage has a perfect business case and also supports the grid. For the past few decades pumped hydro storage has been the main source for shifting energy between day and nighttime. Such arbitrage transactions may be of interest on the energy exchange platform in Leipzig (EEX), but also on a bilateral level.

Currently those price fluctuations are small. But in the course of further expansion of renewable energy sources, the decommissioning of many conventional power plants, and the shutdown of the last nuclear power plant in Germany in 2022, a significant increase of such fluctuations and also the need for grid supporting services is to be expected.

38.2.2 Business Models "Behind-the-meter"

Through the combination of photovoltaic (PV) and a BSS, households also move their "private energy transition" forwards and become more independent from the rising electricity price at the same time. Currently there are around 35 000 storage systems installed in German households. With around 1.6 million PV systems on German roofs, the growth and retrofit potential is enormous. By 2020 the number of residential BSS is expected to grow to about 170 000.

The different optimization scenarios for the residential sector as well as for commercial applications is now explained.

38.2.2.1 Optimization of Self-consumption in Residential Areas

Owners of PV systems without a storage entity can typically only cover about 30% of their electricity requirements with self-generated solar energy. In the past, it was standard practice to sell one's excess solar power to the regional power grid operator – obtaining a guaranteed Feed-in Tariff (FIT) promoted by the EEG. The power drawn from the grid for household consumption remained standardized.

---

7 Cp. https://ec.europa.eu/energy/sites/ener/files/documents/SystemOperationGuideline%20final%28provisional%2904052016.pdf, p. 115.

Of course, these relationships originated during a time when the FIT was significantly higher than the prices for consuming electricity from the grid. Since the FIT for new PV-systems has sunk to a level much lower than the private consumer price per kilowatt hour (kWh) the situation is now worth reconsidering. Maximizing the rate of self-consumption becomes the most sensible way to benefit. The FIT is currently 12.70 € cents/kWh for new PV-installations up to 10 kilowatt peak (kWp) and 12.36 € cents/kWh for PV-installations between 10 and 40 kWp.[8]

The production costs of self-generated PV-electricity are currently between 7 and 10 € cents/kWh. Combined with a battery the costs of a self-generated kWh are currently about 30 € cents/kWh.[9] Considering the constantly increasing price for electricity – currently about 28.69 € cents/kWh[10] in the residential sector – the combined use of PV and storage is getting very close to economic viability.

Moreover, the radical decrease of lithium-ion battery cell prices is expected to continue. As a consequence, retrofitting an existing PV-system with a battery to maximize the consumption of self-generated electricity is very worthwhile when the FIT is expiring or lower than the electricity price. The potential in Germany is massive as there are about 1.6 million PV-systems installed.

Figure 38.3 shows how battery storage helps to level out generation with consumption.

The upper graph of Figure 38.3 shows the use of a PV-system without storing excess electricity – the light grey parts are unused and get fed into the grid while the red parts cannot be covered by own generation and have to be obtained from the grid.

The lower graph, of Figure 38.3, shows the same scenario, but now optimized with a battery. Only at nighttime when the battery is discharged and before the sun rises is there a requirement for a small part of electricity from the grid.

Taking into account that on average about 85% of a household's energy consumption is needed for heating and the preparation of hot water, the concept of self-consumption is completed when heat pumps and/or electrical storage heating are included. Combining a PV-system with a BSS allows the use of electricity up to 70% – instead of only 30% with no storage entity. Including the heating and hot water supply, the share of self-consumption can be maximized up to 90–100%.

Electric Vehicles (EV) are gaining in popularity and play an increasing role in these optimization models. Moreover, by means of mobile meter devices it is possible to assign electricity charged, e.g. at work to one's individual electricity bill and thus combine it with the PV generation at home.

A perfect example of the recent developments in the residential area is the project in Weinsberg.

8 Cp. BNetzA, EEG-Fördersätze für PV-Anlagen: http://www.bundesnetzagentur.de/cln_1411/DE/Sachgebiete/ElektrizitaetundGas/Unternehmen_Institutionen/ErneuerbareEnergien/Photovoltaik/DatenMeldgn_EEG-VergSaetze/DatenMeldgn_EEG-VergSaetze_node.html;jsessionid=8653A294DD4C75991A59142E7A0C68A6#doc405794bodyText4.

9 The price per installed kWh is currently on average between 1,000 € and 1,200 €.

10 Cp. graph on p. 15.

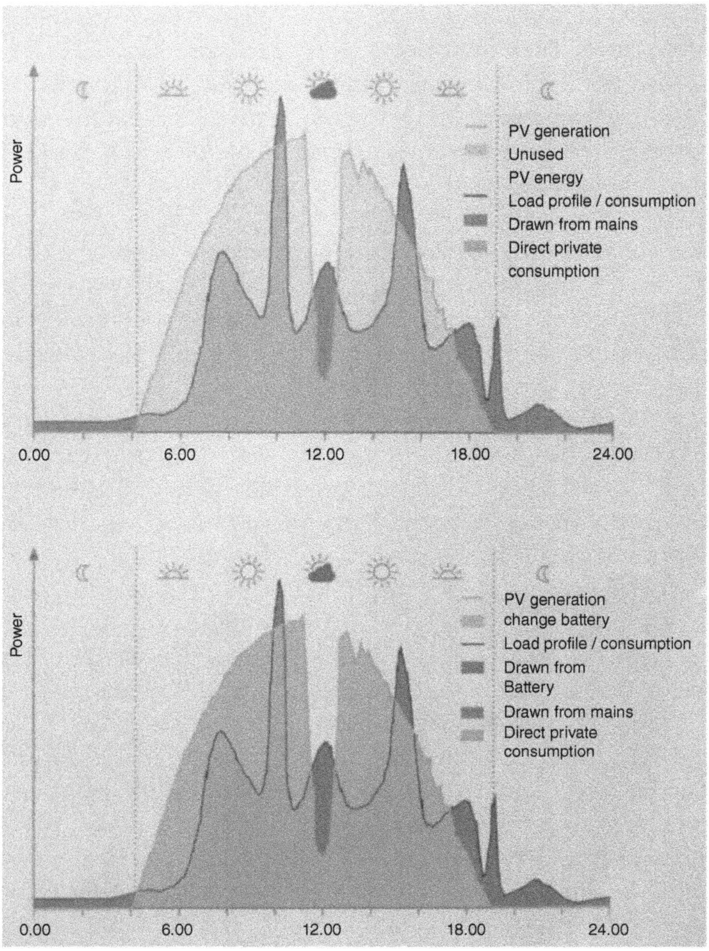

Figure 38.3 Domestic system, leveling out supply. *Source*: ads-tec GmbH, "StoraXe Home and Small Businesses."

38.2.2.1.1 The Flagship Project in Weinsberg

With the goal of providing the greatest possible degree of electrical independence, the project consortium[11] started building the model region on the outskirts of Weinsberg in the German state of Baden Württemberg.

With 23 households, 145 kW installed solar power on the roofs of the buildings, one central electrical storage system with a capacity of 150 kWh, and multiple thermal storage systems they have realized a pilot project that proves that it is possible to completely meet energy requirements with solar power (see Figure 38.4).

Solar electricity that is not immediately needed by the consumers is by means of a heat pump fed into a thermal storage unit with a capacity of 20 000 liters. In addition there are 18 smaller water tanks located at the individual buildings to cover the daily

11 The consortium consists of ads-tec GmbH, Fraunhofer ISE, Kaco new energy, Kruck und Partner, the Ministry of Environment Baden-Württemberg, and Solarworld.

demand for hot water. To assume control of all electrical and thermal components, an energy management system was created for the entire residential complex.

The pilot project impressively demonstrates that it is possible to cover almost all electricity needs and half of the heating requirements in residential areas by means of PV sources at least in those periods of the year when there is enough sun (approx. half of the year).[12]

38.2.2.2 Applications for Larger Scales – Small-and Medium-sized Enterprises and Industry

Not only households, but also larger scale applications can benefit from the use of storage technologies. Small businesses and agricultural factories can optimize self-consumption and compensate high variations in consumption and thus reduce energy costs. At the same time, storage technologies have been implemented effectively for decades to provide an uninterruptible power supply – e.g. in hospitals or at production lines that are highly sensitive to power variations.

However, for most industrial customers, the optimization of self-consumption is not yet economically viable, as they are exempt from the majority of charges and therefore have a much lower electricity subscription price. In 2015, as Figure 38.5 shows, the industrial electricity price was 15.32 € cents/kWh.

Considering the upwards tendency for industrial electricity prices on the one hand and the decreasing cost of PV and BSS on the other, the model of self-consumption might, however, also become viable in only a few years in this segment.

If at the same time the peak load from the public grid can be reduced, profitability can be reached even more quickly. Likewise, the combination with CHP accelerates the momentum of economic viability.

38.2.2.2.1 Reduction of Grid Charges by Peak-load Management and Atypical Grid Usage

For companies at medium-voltage grid level with an annual consumption of more than 100 000 kWh, grid charges are determined by the maximum annual grid load. Thus energy storage can make an important contribution to reduce grid charges by avoiding consumption peaks from the grid and the annual peak load. Ideally, the necessary power provided in times of load (consumption) peaks has been saved in times of excess energy in the grid.

According to § 19 StromNEV,[13] those companies have additionally the opportunity to reduce their grid charges up to 80% via so called atypical grid usage. The idea behind this reduction of grid charges is that less grid capacity is required (or future upgrades might be avoided) if the peak load is reduced due to the company having its individual peak load not at the same time. This relieving effect is rewarded by reduced grid charges. Storage technologies can thus help to shift the individual peak load and allow for atypical grid use.

---

12 Cp. Kaco new energy "Solutions for maximum self-consumption of solar power in residential areas", Explanation, ads-tec GmbH "StoraXe Industrial & Infrastructure", http://kaco-newenergy.com/de/lounge/modellprojekt-weinsberg.

13 Electricity Grid Charges Ordinance (Stromnetzentgeltverordnung).

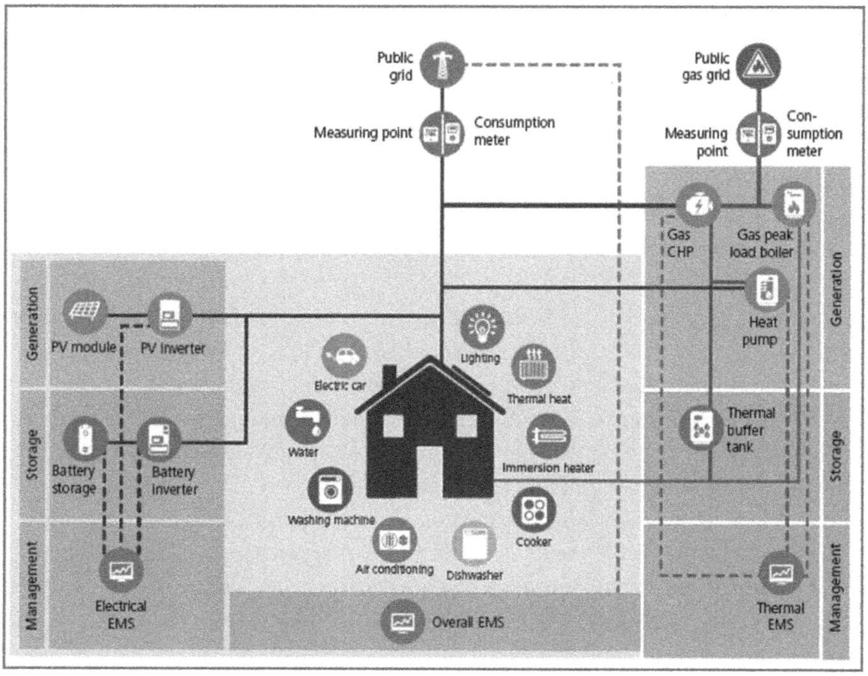

Figure 38.4 Schematic for the Weinsberg project *Source*: Kaco new energy "Solutions for maximum self-consumption of solar power in residential areas."

38.2.2.2.2 CHP

By the combined use of heat and electricity, carbon emissions are lower than when produced in two separate plants and efficiency is higher, up to 90%.

Due to the poor transportability of heat, CHP usually requires an operation near the point of heat consumption based on the time of requirement.

As a result, the co-generated electricity is "pushed" very often into the grid, regardless of whether there is a need or not. Starting the process in the early morning hours to stop the night setback of the building temperature, the electricity is generated at a time of minimum requirements.[14]

Storing the electricity for this "heat-controlled" operation usually implies higher costs than vice-versa using thermal energy storage for an "electricity-controlled" operation. In addition to the CHP bonus fee that is granted according to the EEG, the CHP operator can benefit additionally from electricity price arbitrage effects when producing in times of high demand and storing as thermal energy for when it is needed.

38.2.3 Mixed Business Models

The constant growth of self-consumption "behind-the-meter" sets new challenges for the grid operators, because not all systems act in a grid-supportive manner.

14 Cp. http://www.bafa.de/bafa/de/energie/kraft_waerme_kopplung/publikationen/flyer_kwk.pdf.

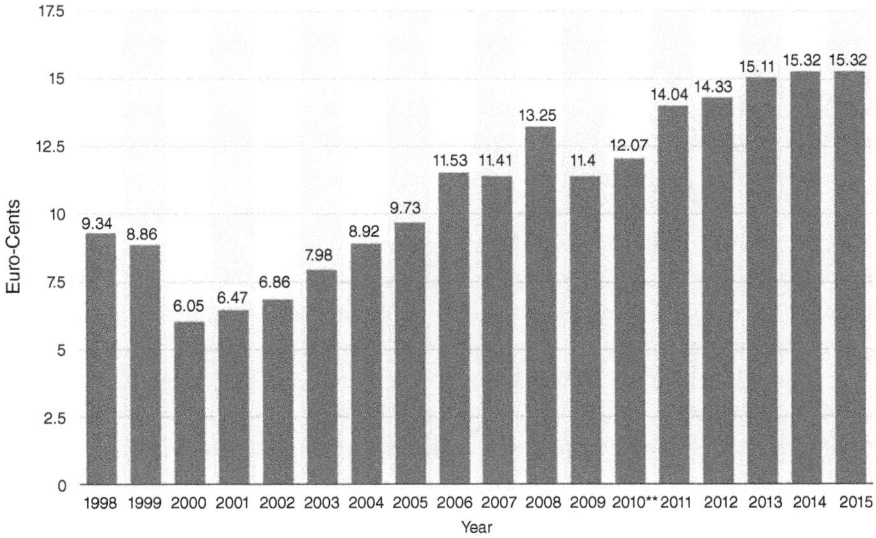

Figure 38.5 Prices by year in Euro-Cents per kWh. *Source*: http://de.statista.com/statistik/daten/studie/252029/umfrage/industriestrompreise-inkl-stromsteuer-in-deutschland.

Furthermore, it implies a continuous and serious for in clients which, is a big challenge for many local utility companies.

Mixed business models allow the sharing of different storage and generation capacities bringing them into a collaborative framework taking advantage of the complementarity of the different consumption profiles. Swarms or quarter storage concepts offer a positive effect for both consumers and the utility company. The grid is relieved and supply from conventional power plants can be increasingly reduced.

Households optimize their individual self-consumption in a centralized system and can, e.g. control with whom they wish to share their storage capacities. They participate in several market segments,[15] which allows higher returns and thus a faster amortization. In a quarter storage concept, households do not even have to invest in their own BSS, because they lease their capacity in the central storage unit.

The grid operator can better control the system and additionally benefit from grid supporting services. Furthermore, also the regional utility company benefits from such models that bring new business cases and economic value.

Another worthwhile fact worth mentioning in this context is electro-mobility as bidirectional charging (Grid-to-Vehicle (G2V) or Vehicle-to-Grid (V2G)) brings another valuable effect for grid flexibility. Germany's ambitious aim was to pass the threshold of 1 million EVs in operation by 2020. That equals a total storage capacity of 25 Gigawatts (GW), which is an enormous amount compared with 7 GW of all German pumped hydro storages.[16]

15 e.g. the provision of control energy, intraday-exchanges, bilateral Over-the-counter transactions.

16 Cp. http://www.wiwo.de/technologie/green/tech/virtuelle-speicher-elektroautos-koennen-das-stromnetz-stabilisieren/13764646.html.

Mixed business models are on the rise, but there are still some regulatory barriers that complicate the roll-out. The decisive obstacles will be addressed later in the chapter.[17]

Nevertheless, some outstanding projects have been already realized against the odds and will be explained in the Sections 38.2.3.1–38.2.3.3. They significantly underline the trend to smart grids–the smart connection of generation, grid, and consumption.

38.2.3.1 The "Electricity Bank Account" Quarter Storage

The quarter storage project "electricity bank account" of ads-tec GmbH and the utility company MVV Energie AG connects 14 households and four small businesses with a 100 kWh lithium-ion battery quarter storage. The generation side consists of 16 PV-systems and 3 CHPs (see Figure 38.6). A special measuring system controls the input and output of kWhs in the storage entity and the participants can retrieve their "kWh-balance" in the storage system on a daily basis via an online energy cloud.

This exemplary project shows how easy it is to combine the maximization of self-consumption for the homeowner and the provision of additional grid services.[18]

The main hindering factor of this business is the mandatory payment of final consumer charges. As the electricity uses the public grid on its way to the battery and back to the households, some of the fees are charged twice.[19] Due to the double tax burden (at the battery and the households), the business model is currently not economically viable and the optimization of every single household "behind-the-meter" is the more profitable model nowadays. The particular parts of the final consumer charge will be explained when the regulatory framework is discussed.

38.2.3.2 The SWARM Project

38.2.3.2.1 Provision of Control Energy

The SWARM project of Caterva GmbH and the utility company N-RGIE AG is a virtual MW-storage facility – consisting of 65 Caterva household BSS that contribute to the stabilization of the transmission power grid (see Figure 38.7).

Every participating household has a BSS with a total output of 20 kW and an energy volume of 21 kWh. The storage entities are combined to form a swarm of a virtual capacity of more than 1 MW and are tapped or refilled from a control center. If there is a need in the grid for positive or negative control energy, the battery swarm is perfectly disposed to deliver.

The SWARM was prequalified to deliver PCR in July 2015 and stands for an absolute innovation in the area of privately utilized storage systems.[20]

---

17 Cp. Faktenpapier Speicher, BVES and DIHK, p.20 f., Kaco new energy: p. 2.

18 Cp. ads-tec GbmH, http://www.ads-tec.de/en/company/press/press-release/singleview/article/inbetriebnahme-quartierspeicher-strombank-testet-neue-moeglichkeiten-zur-nutzung-erneu erba.html.

19 When feeding in to the battery all charges apart from the EEG-share and grid fees are imposed and when reaching the households the full range of final consumer charges are due.

20 Cp. Caterva GmbH, http://caterva.de/pdf/20160314_Caterva_PR_Energiemanagement_DE_final.pdf.

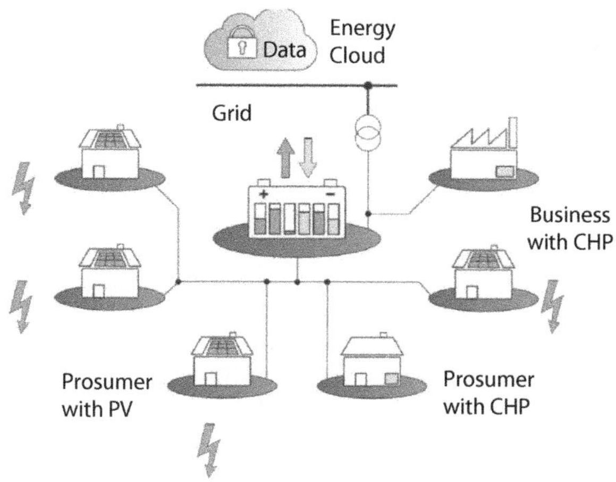

Figure 38.6 The "Electricity Bank Account" Quarter Storage. *Source*: ads-tec GmbH.

38.2.3.3 The "Economic Grid"

38.2.3.3.1 Provision of Negative Control Energy

Similar to the SWARM project (Section 38.2.3.2), this project of SENEC.IES by Deutsche Energieversorgung combines numerous household BSS into a large virtual storage entity connected via the so called "Economic Grid" (word origin: Economic and Dynamic). Besides the optimization of self-consumption the business focus is on providing negative control energy.

By receiving excess energy in times of high infeed, the "Economic Grid" contributes sustainably to grid stability and improves the efficiency of the whole system.

As negative control energy is remunerated, the resulting revenue is to used to cover the cost of running the "Economic Grid" business (e.g. staff).

The absorbed excess energy is provided for free to the customer and is delivered as electricity but also already converted to heat. In spring and autumn when there is usually a high wind infeed and thus excess electricity, the customer is covered almost exclusively by means of this approach.

Currently, the provided amount of electricity per household is about annually 800 kWh and about 2500 kWh of heat energy.[21]

38.3 Legal and Regulatory Framework – Opportunities and Barriers

38.3.1 Funding Programs

There are numerous programs that allow start-up funding for energy storage technologies on a federal level but also with a more regional focus financed directly by the

21 Cp. http://www.senec-ies.com/was-ist-economic-grid.

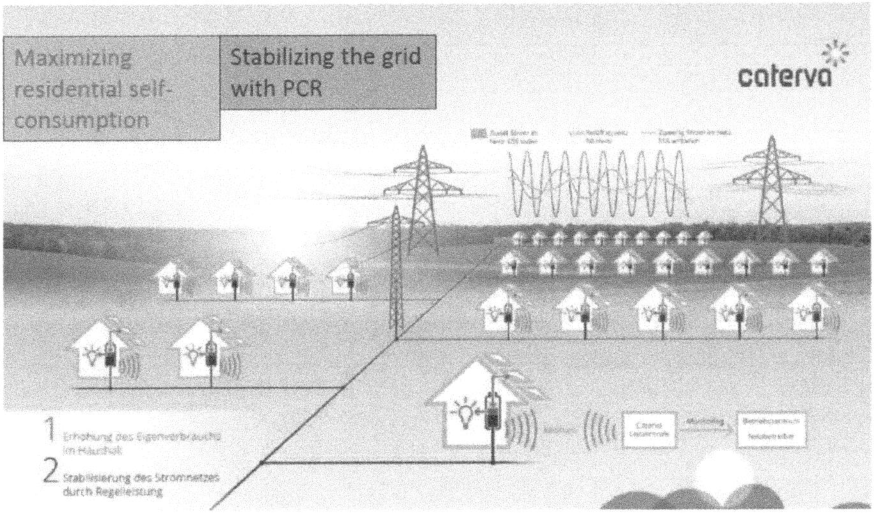

Figure 38.7 The SWARM project. *Source*: Caterva GmbH.

German federal states. The most important programs with the biggest impact will be explained here.

To enable the decisive technological breakthroughs, the Ministry of Education and Research[22] and the Ministry of Economic Affairs and Energy[23] provided 200 million € for the research initiative "Energy Storage" that is open to all technologies.[24]

In the program "Schaufenster Elektromobilität" innovative technologies and mobility concepts are being tested in four German regions. The focus is on the improvement of the charging infrastructure and the implementation of EVs in the existing (energy) infrastructure. The government supports the showcase program with a total of 180 million €.

The KfW Program Renewable Energies "Storage" of the Development Loan Corporation[25] supports the use of stationary BSS in combination with a PV-system in households up to 30 kWp (kilowatt peak) and with a low-interest loan, while the Federal Ministry of Economic Affairs and Energy covers up to 30% of the purchase costs in the form of a repayment grant. The aim of the incentive program is to stimulate the market for BSS and to support the integration of small-to medium-sized PV installations arrangements in the electricity grid. Thus, the requirement that must be met to get the grant awarded is a grid supportive operation.[26]

---

22 Bundesministerium für Bildung und Forschung (BMBF).

23 Bundesministerium für Wirtschaft und Energie (BMWi).

24 Forschungsinitiative "Energiespeicher."

25 Kreditanstalt für Wiederaufbau (KfW).

26 The maximum infeed of electricity shall not be more than 50% of the installed PV capacity at any time. This limited infeed is mandatory for the whole life of the PV system, at least for 20 years. Thus it extends even to a possible continued operation of the PV system, in case the storage entity gets decommissioned at an earlier stage.

More than 20 000 households have taken advantage of the program. The first funding period was scheduled to December 2015. Due to its great popularity and impact, the program was extended until end of 2018.[27]

The ZIM-program[28] is a funding program for SMEs with business operations in Germany. Under the program, SMEs with up to 500 employees and the research institutes they work with can be awarded grants for ambitious R&D projects focusing on the development of new products, technical services, and better production processes. ZIM is open to all technologies and sectors and a lot of projects, including energy storage, have been funded so far.[29]

38.3.2 Legal and Regulatory Barriers to Energy Storage in Germany

As already shown in the business models, energy storage is playing an increasingly important role for successful energy transition. However, at this moment there are still some legal and regulatory barriers that need to be tackled for successful market penetration.

The decisive obstacle is the fact that according to the current German framework, storage facilities are assigned to the role of consumption; in consequence, every kWh that is stored is charged with most of the final consumer fees. On top, the Federal Network Agency[30] requires the precise delineation of which electricity is used for what purpose or business. This additionally complicates the roll-out of mixed business models.[31]

Besides the intrinsic electricity price of 6.11 € cents/kWh, the following taxes and charges are currently imposed for households (see Figure 38.8):[32]

- EEG surcharge: 6.35 € cents/kWh
- Grid usage fee: 7.07 € cents/kWh
- Concession fee: 1.66 € cents/kWh
- Grid related surcharges: 0.86 € cents/kWh
- CHP, §19 StromNEV,[33] offshore liability
- Electricity tax: 2.05 € cents/kWh
- Value-added-tax: 4.58 € cents/kWh

In any case there is to be expected a further increase of these fees – in particular of the EEG surcharge and the grid usage fees–in the coming years.

In the following, the different fees and charges and how the different storage technologies are required to pay them, will be explained.

38.3.2.1 Grid Usage Fee

There are several exemptions regulated in §118 (6) of the Energy Industry Act (EnWG). Since December 2008 electrical energy storage technologies are exempted from the

---

27 Cp. https://www.kfw.de/inlandsfoerderung/Unternehmen/Energie-Umwelt/F%C3%B6rder produkte/Erneuerbare-Energien-%E2%80%93-Speicher-(275)/#1.

28 Zentrales Innovationsprogramm Mittelstand (ZIM).

29 Cp. BMWi, Infobroschüre Zentrales Innovationsprogramm Mittelstand (ZIM).

30 Bundesnetzagentur (BNetzA).

31 Cp. Faktenpapier Speicher, BVES and DIHK, p.20 f., Kaco new energy, p. 2.

32 Based on an annual consumption of 3500 kWh.

33 Cp. page 10.

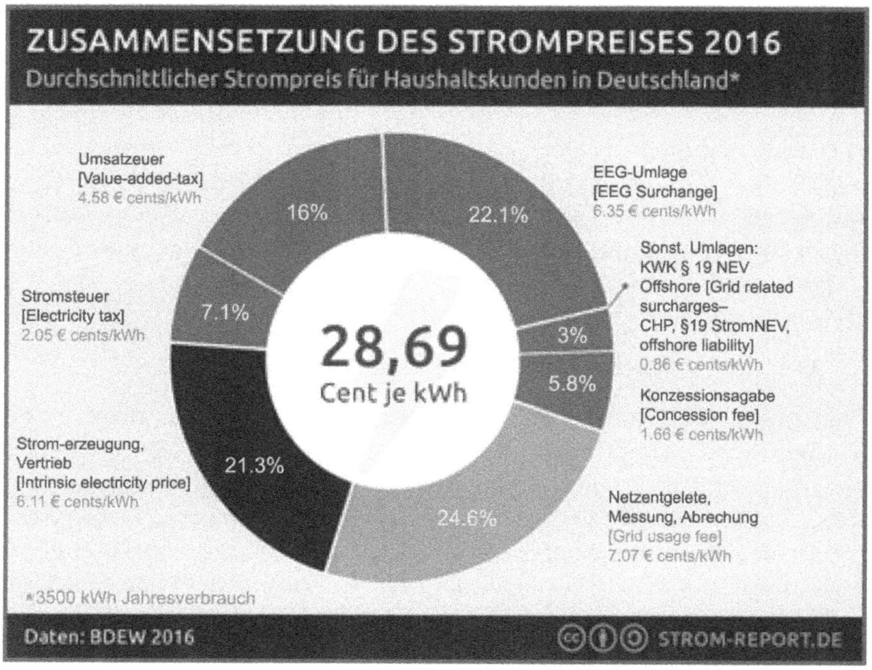

Figure 38.8 A breakdown of the cost, to the consumer, of electricity in Germany (2016). *Source*: http://strom-report.de/strompreise.

grid usage fee for 20 years (as of commissioning). This scheme applies to all systems that are connected to the grid after 4 August 2011.

Apart from this exemption, the following storage technologies are burdened with this fee:

- Pumped hydro storage: as this is the most established technology, all plants in Germany were put into operation decades ago. They can get grid usage fee relief for 10 years if they increase their efficiency. This usually requires massive investment costs, which make this option completely out of proportion to this privilege.

  With about 7 GW pumped hydro being the largest installed storage capacity in Germany and the missing viability is alarming.

  The only way of reducing their grid fees is – as described atypical grid management.
- All PTX methods (PTG, PTH) are obliged to pay the fee–unless a reconversion to electricity takes place, which is the least efficient way to use the technologies, as the electricity in this case leaves the electric sector. All other charges that apply to final consumption have to be paid.
- The exemption from grid fees and all other charges does neither apply to EV due to their mobility.

As we see according to § 118 (6) EnWG there does not even exist a level-playing field between the different storage technologies concerning the imposition with grid usage fees. Moreover, uncertainty prevails over whether the exemption from the grid usage fee should also include other grid-related charges like the concession fee, CHP, §19 StromNEV, and offshore liability surcharge.

Currently, this additional block has to be paid by all storage technologies and can be very decisive for the economic viability of storage projects.

38.3.2.2 EEG-surcharge

The EEG-surcharge is imposed for the roll-out of renewable energies according to the Renewable Energy Act (EEG).

According to EEG 2014, the payment was omitted only for grid-related storage that is used for leveling out imbalances. With the amendment (EEG 2017), all storage applications will be now exonerated[34] as soon as it is guaranteed that the share is paid by the final consumer.

At the beginning of the discussion in summer 2016, mixed business models were still not taken into consideration, because it was implied that it is impossible to separate electricity that is exempted from the fee from that which has to pay it.

According to today's state of the art, there are appropriate metering systems that are perfectly able to identify the different flows of electricity. In the end – after a long discussion – in December 2016 mixed business models were finally exonerated from the EEG-surcharge, which is a big relief for the industry and for a lot of upcoming projects.

38.3.2.3 PTX Technologies are Faced with a Huge Burden

PTX technologies can efficiently contribute to the decarbonization of sectors such as heat, mobility, and industry and increase the efficiency of the whole system. Raising the enormous potential of PTX would be a major milestone for cross-sectoral combination and also for the energy transition.

Nonetheless this ecologically sound option is currently charged with all final consumer charges when no reconversion to electricity takes place, which is the least efficient option.

The recent amendment of the EEG includes some initial stages providing a benefitting mechanism for loads operated in a grid-supportive manner. The current focus of the government lies on PTH and CHP.

PTG provides a much wider range of applications which underlines the importance of an approach that is open to all technologies.

38.3.2.4 The Absurd Situation of Charging Twice

According to the current regulatory framework some fees and charges are paid twice: in the first place by the energy storage system and then again by the final consumer. Figure 38.9 demonstrates the resulting double burden in the electricity system which has a significant impact on the economic viability and thus on the operation of storage technologies.

38.3.3 Necessary Adjustments to Pave the Way into the Market

38.3.3.1 Definition of Energy Storage as "Fourth Column of the Energy System"

Energy storage, as one basic component of a decentralized and flexible future energy system, should be enabled to deploy its multifunctional services and should

34 E.g. the optimization of self-consumption, intraday exchange or arbitrage.

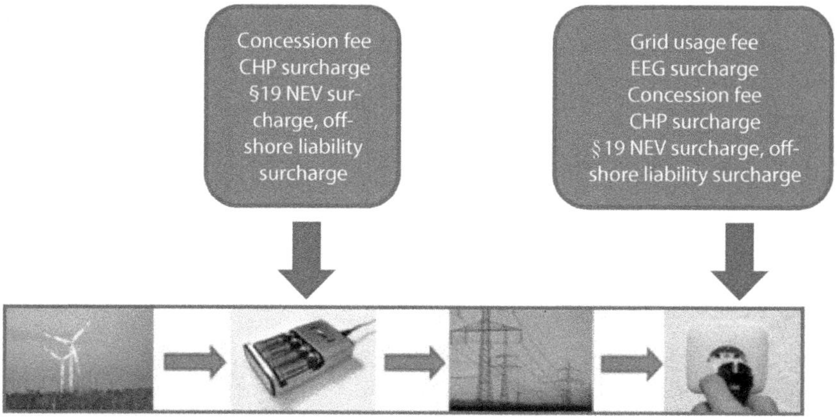

Figure 38.9 The double fee burden. *Source*: BVES+DIHK, Faktenpapier Speicher, p. 11.

therefore not be charged fees. The legal classification of energy storage by implementing it as a fourth column of the system – alongside generation, transport, and consumption – is urgently required to clarify the discussion about consumption charges.

The exoneration from fees would not imply additional costs to the system. On the contrary, it offers a huge potential of cost savings, e.g. by reducing congestion management costs. In 2015, the cost for re-dispatch in Germany came had already reached billion €. According to a broad estimate, the grid usage fee paid by pumped hydro in Germany is about 70 million € per year. Although it is an amount that massively influences their economic viability, it has quite a secondary magnitude for the whole system when comparing it to the annual sum that is expended for re-dispatch. It results particularly in a serious disadvantage for German pumped hydro plants in cross-border markets, as plants from abroad do not have to pay comparable fees.

38.3.3.2 Creating Economic Incentives by Rewarding More Services

Apart from control reserve, there are several other services that are relevant for a stable grid, such as inertia reserve, voltage control, reactive power compensation, services for re-dispatch, short circuit power, and black-start capability.

Nowadays those services are automatically covered by the spinning reserve of conventional power plants. With the increasingly fluctuating RES infeed and the subsequent decommissioning of conventional power plants there will be a stronger demand for alternative service providers. Energy storage devices can provide those services with excellent quality and also reducing carbon emissions.

As those services are not paid for individually, there is a need for change towards a market-driven return. This would have a positive effect on the whole system – because it is cost-effective and supports attempts to reduce carbon emissions.

38.3.3.3 Uniform and Transparent Standards for Prequalification

Another important topic, especially for battery storage, is the prequalification process. The prequalification process requires higher requirements from technical entities with limited capacities in order to pass. This one-sided condition for a certain technology is not legally justified or technically required, as the market of PCR is only to cover the first few seconds after an incident.

The prequalification process should be standardized in order to achieve predictability and to provide reliable legal protection. At this moment, investors cannot reliably plan their projects.

38.4 Conclusion and Outlook

In the old system where production and consumption were constant and predictable, energy storage was only used at the margin for classical arbitrage business models, mainly by pumped hydro storage shifting energy between day time and nighttime.

The energy system of the future will be based on variable renewable generation that requires flexible handling until its consumption. In this context, energy storage technologies have already gained and will continue to gain massively in importance and will become an indispensable component. Also the fact that the two biggest German utilities, E.ON and RWE, are splitting their businesses into separate parts – the old business based on fossil generation and the renewable "green" part – shows the unstoppable movement towards a flexible and increasingly de-centralized new energy system where the multivalent input of energy storage is also greatly needed.

All energy storage technologies offer different services for flexibility – also cross-sectoral. As well as having in mind the Paris-agreement (COP21), only focusing on the electricity sector is not sufficient to achieve the goal of limiting global warming to well below 2°C.[35] Carbon reductions have to be fulfilled also in other sectors such as transport and industry. P2X technologies are the perfect enabler.

Mixed business models are the current center of attention – optimizing self-consumption rates, providing control energy and improving the (distribution) grid conditions, and energy balance at the same time.

However, the actual legal and regulatory framework in Germany has not yet adapted to recent developments – so that in most contexts energy storage is assigned to the role of consumption. Even when stabilizing the grid, the stored energy gets charged with final consumption fees, which is decidedly significant in the economic viability and thus on the operation of storage technologies. Precisely, mixed business models that offer a large range of services should not be hindered but encouraged and the legal and regulatory framework has to catch up to the fast evolving new energy system.

In conclusion, energy storage plays a key role in the renewable concert of today and will do even more in the future. However, a few parameters still need to be adjusted to pave the way to a sustainable level-playing field[36] and into the market.

---

35 Cp. http://www.cop21.gouv.fr/en/parisagreement-points-that-remain-in-suspense.

36 Between the different flexibility options but also between the different storage technologies.

Further Reading

BMWi Infobroschüre Zentrales Innovationsprogramm Mittelstand (ZIM).

Deutscher Industrie und Handelskammertag (DIHK). Bundesverband Energiespeicher e.V. (BVES): Faktenpapier Energiespeicher – Rechtsrahmen. Geschäftsmodelle, Forderungen.

International Energy Agency (IEA). (March 2016). *Policies for Storing Renewable Energy*.

Kaco New Energy: Solutions for Maximum Self-consumption of Solar Power in Residential Areas.

Kreditanstalt Für Wiederaufbau (Kfw): Merkblatt KfW-Programm Erneuerbare Energien "speicher."

National Renewable Energy Laboratory (NREL), Neubauer, J. and Simpson, M. (2015). *Deployment of Behind-the-Meter Energy Storage for Demand Charge Reduction*. Golden, Co: National Renewable Energy Laboratory.

RWTH Aachen, Sauer, D.U., Lunz, B., and Magnor, D. *Marktanreizprogramm Für Dezentrale Speicher Insbesondere Für PV-Strom*. Available at: https://docplayer.org/15071946-Marktanreizprogramm-fuer-dezentrale-speicher-insbesondere-fuer-pv-strom.html.

StoraXe Home and Small Business. Leaflet. ads-tec GmbH.

StoraXe Industrial & Infrastructure. Leaflet. ads-tec GmbH.

Online Sources

BNetzA, EEG-Fördersätze Für PV-Anlagen. http://www.bundesnetzagentur.de/cln_1411/DE/Sachgebiete/ElektrizitaetundGas/Unternehmen_Institutionen/ErneuerbareEnergien/Photovoltaik/DatenMeldgn_EEG-VergSaetze/DatenMeldgn_EEG-VergSaetze_node.html;jsessionid=8653A294DD4C75991A59142E7A0C68A6#doc405794bodyText4.

Bundesamt Für Wirtschaft Und Ausfuhrkontrolle (BAFA): Kraft-Wärme-Kopplung. http://www.bafa.de/bafa/de/energie/kraft_waerme_kopplung/publikationen/flyer_kwk.pdf.

Caterva - Energiemanagement Kann Virtuelle Speicherkraftwerke Vollautomatisch Im Intraday-Handel Vermarkten. http://caterva.de/pdf/20160314_Caterva_PR_Energiemanagement_DE_final.pdf.

COP21. http://www.cop21.gouv.fr/en/parisagreement-points-that-remain-in-suspense.

Maubach, K.-D. TU Clausthal: Vorlesung Elektrizitätswirtschaft - Kapitel 6: Stromhandel. http://www.iee.tu-clausthal.de/fileadmin/downloads/Scripte/SS8819K6_25_06_2008.pdf.

EU-System Operation Guideline. https://ec.europa.eu/energy/sites/ener/files/documents/SystemOperationGuideline%20final%28provisional%2904052016.pdf.

http://de.statista.com/statistik/daten/studie/252029/umfrage/industriestrompreise-inkl-stromsteuer-in-deutschland.

http://kaco-newenergy.com/de/lounge/modellprojekt-weinsberg.

http://strom-report.de/strompreise.

http://www.ads-tec.de/en/company/press/press-release/singleview/article/inbetriebnahme-quartierspeicher-strombank-testet-neue-moeglichkeiten-zur-nutzung-erneuerba.html.

http://www.senec-ies.com/was-ist-econamic-grid.

http://www.wiwo.de/technologie/green/tech/virtuelle-speicher-elektroautos-koennen-das-stromnetz-stabilisieren/13764646.html.

https://1-gasvergleich.de/gaspreise.

https://www.kfw.de/inlandsfoerderung/Unternehmen/Energie-Umwelt/F%C3%B6rderprodukte/Erneuerbare-Energien-%E2%80%93-Speicher-(275)/#1.

https://www.regelleistung.net/ext/?lang=en.

Safety Guidelines for Li-Ion Storage Systems. http://www.bves.de/wp2015/wp-content/uploads/2015/07/Safety_Guidelines_Li-Ion_Storage_Systems.pdf.

39

Integration of Renewable Energy by Distributed Energy Storages

Christian Doetsch and Anna Grevé

Fraunhofer Institute for Environmental Safety and Energy Technology UMSICHT, Osterfelder Strasse 3, 46047 Oberhausen, Germany

39.1 Introduction

To meet the GHG-emission reduction targets as well as the 1.5–2°C aim, a decarbonization of the global energy system is required (see Climate Summit COP21 in Paris, December 2015). This implies the substitution of fossil energy carriers by low carbon energy, which means less CO_2 from fossil fuel power plants and a higher share of renewable generation. Renewable energy from solar and wind shows a high additional potential for electricity generation, but they are variable and less predictable energy sources. By enabling the temporary balancing of supply and demand, energy storage has always been an important part of the energy system. Depending on the form of energy, which needs to be balanced, and the required storage period, different types of energy storages such as thermal, electrical, or virtual storages can be used.

39.2 Usage of Variable Renewable Energies and Induced Problems

Renewable energies are often called "variable renewable energies" due to the fact that technologies based on wind and solar only produce energy when the wind is blowing or the sun is shining. Therefore, energy storages seem to be the "silver bullet" for this obvious disadvantage of renewable energies. However, electrical storage technologies are still expensive and up to now only used in niches reasonable from the economic point of view. To understand the real challenges associated with integrating variable sources of electricity it is necessary to have a look at the whole electric system.

In developed countries, power grids were designed around the concept of large, controllable electric generators. Electric storage facilities (mostly pumped hydro) were only used for shifting energy from baseload power plants (e.g. nuclear) at nighttime to day peak-load and for bridging short fluctuations or unexpected power plant outages.

As the grid has practically no intrinsic storage capacity, the balance between electricity supply and demand must be maintained at all times to avoid blackouts. Therefore, the grid operator uses a three-phase planning process to ensure power plants produce the right amount of electricity at the right time and right location. The electricity supply is scheduled one day ahead, and then balanced with demand on an hour-by-hour and second-by-second basis.

Advances in Energy Storage: Latest Developments from R&D to the Market, First Edition.
Edited by Andreas Hauer.
© 2022 John Wiley & Sons, Ltd. Published 2022 by John Wiley & Sons, Ltd.

Nowadays, the share of renewable energies is increasing more (e.g. Germany) or less (e.g. France) rapidly. The main problem occurring with these variable energies is that the energy supply side becomes more and more unpredictable. Regional imbalances appear due to the fact that wind or sun is varying between hours but not too much in a region. That means there will be sometimes a huge surplus and sometimes a lack of energy at a regional level. This effect could be smoothened by enhanced grid systems, which allow for balancing demand and supply over hundreds of kilometers. On the other hand, in countries with a weak grid system, this effect becomes more challenging.

This potential imbalance problem is indicated by the share of variable renewable energies that doubles the problem: directly by more renewable energy power plants with variable power output and indirectly by lowering the amount of controllable power plants.

The development and forecast of renewable energies installed for different countries is displayed in Figure 39.1. There are countries like UK with an energy system based on fossil energies. There are other countries like France, which are focusing on nuclear energy. In addition, there are countries like Norway or Sweden, which have a long tradition in hydroelectric systems. Germany is facing another development, because it is in a very challenging transition, starting in the 1980s with a hard and soft coal based energy system (75% installed fossil power capacity) to a 60% renewable energy system. In 2050 the energy system in Germany will be mostly based on variable renewable energies (>80%). The current status (2020) is: 130 GWel RE vs. 80 GWel conventional power plant capacities. Additionally, 8.11 GWel of nuclear energy will fade out by 2022.

Due to the fact, that the most rapid change in the energy systems towards variable, renewable energies takes place in Germany (nearly 60% of installed power bases on

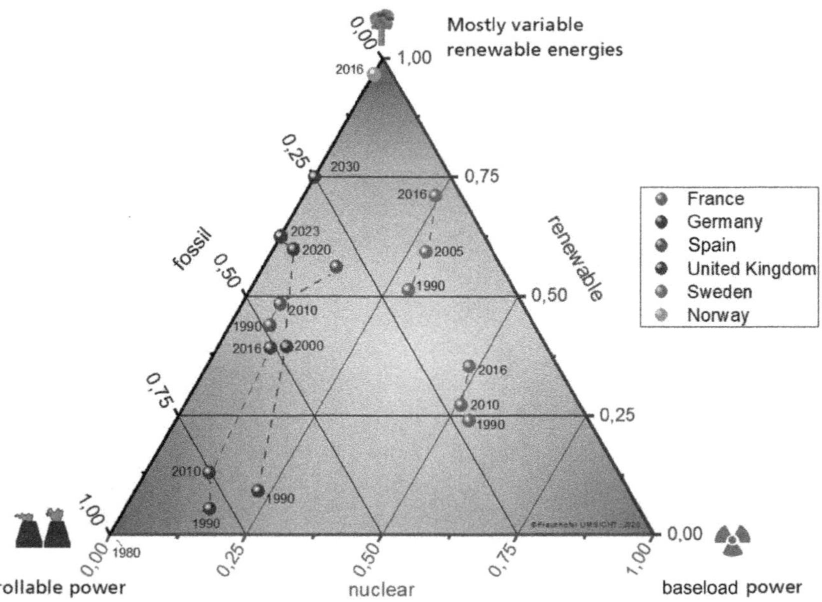

Figure 39.1 Ternary plot of installed electric power, clustered for renewable power (wind, solar, biomass, etc.) controllable power (fossil fueled) and nuclear power for baseload in different countries.

variable energies in 2020), it is a well suited example to show future challenges to balance the electric grid.

The increasing deployment of nuclear energy in the last century induced an increase of pumped hydro energy storages. Typical energy storage capacities are six full load hours, fitting to the day-night shift business case. Nowadays this storage pattern and business case is fading out due to high shares of solar energy with peak production in the daytime (Germany). Besides these hourly storages, there will be a need for long-term storages in the future, if the share of renewable energies is to exceed the 80% threshold.

Due to the fact that the most rapid change in the energy systems towards variable, renewable energies is taking place in Germany (more than 60% of installed power bases on variable energies in 2020), it is a well-suited example to show future challenges to balance the electric grid.

Figure 39.2 shows that there was on 23 August 2015 1.00 p.m. a share of more than 80% renewable energy and about 70% variable renewable energy in the German electric grid. Three days earlier (20 August 0:00 a.m.) the share of variable renewable energies was lower than 1%. This results in a gap of 42.7 GW (84% of total demand) or 12.4 GW (19% of total demand) which has to be closed by conventional power plants, energy storages, or other balancing means. Due to economic (power ramping could be sometimes more expensive than wasting energy) and technological reasons (must-run capacities to stabilize the grid) and some spatial mismatch the demand curve (grey) is lower than the production curve (grey area). This surplus leads to lower or sometimes negative energy prices. Figure 39.3 shows that the price decreases from more than 40 €/MWh to 5 €/MWh. In most cases, this surplus energy is exported to neighboring countries.

The technical need for energy storage will increase with a decreasing share of controllable power. This relationship will be influenced by grid structure and scale of balanced energy systems. In small scale systems or less connected systems the influence of individual demand peaks or productions peaks will be much more important than in bigger/well connected systems. This effect is shown in Figure 39.4 where the

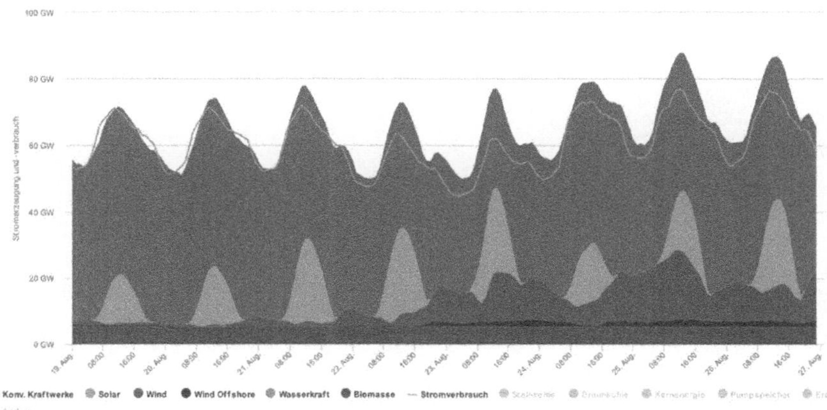

Figure 39.2 Electric energy demand and production in Germany: Week from 2015, 19 to 26 August 2015, which shows at hourly resolution a share of renewables between 15% and 80% (figures for 23 August 2015, 1.00 p.m.).

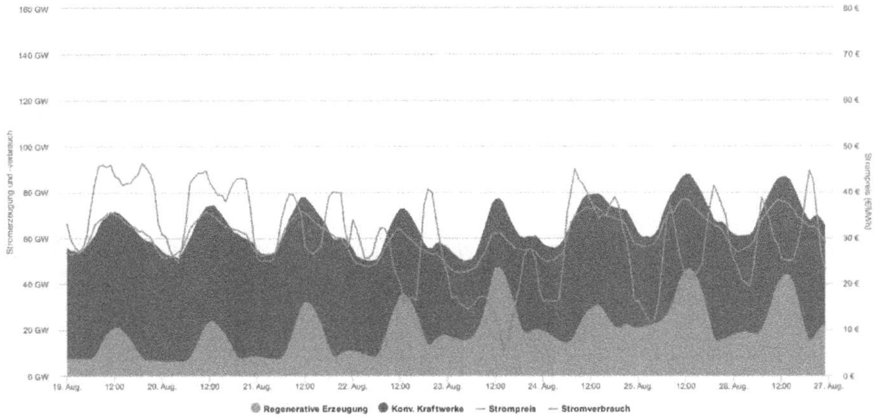

Figure 39.3 Electric energy prices compared to demand and production in Germany: Week from 2015, 19 to 26 August 2015, which shows price peaks between 5€/MWh up to 47€/MWh.

normalized power output (PV in Southern California) is shown over a day for one, two, six, and 25 PV plants. An advanced grid structure allows resolution of local imbalances in a broader region and to level some spatial fluctuations by this bigger system. This balancing effect has to be taken into account because it leads to lower energy storage demands in well-developed grids (likely in central Europe) or higher energy storage demands on small islands (e.g. Greek islands in the Mediterranean Sea).

This could be summarized to: An energy system with an advanced grid structure, more controllable power, and consolidation to a bigger system allows for lower balancing demand. But systems with a weak grid, less controllable power, or very local RE will need more balancing power.

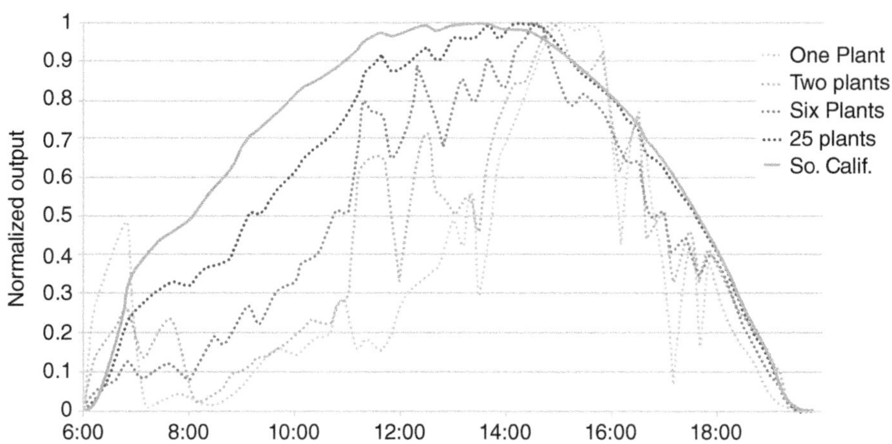

Figure 39.4 As the number of PV plants increases, their normalized, aggregated output becomes smoother (example Southern California for a partly cloudy day) [5].

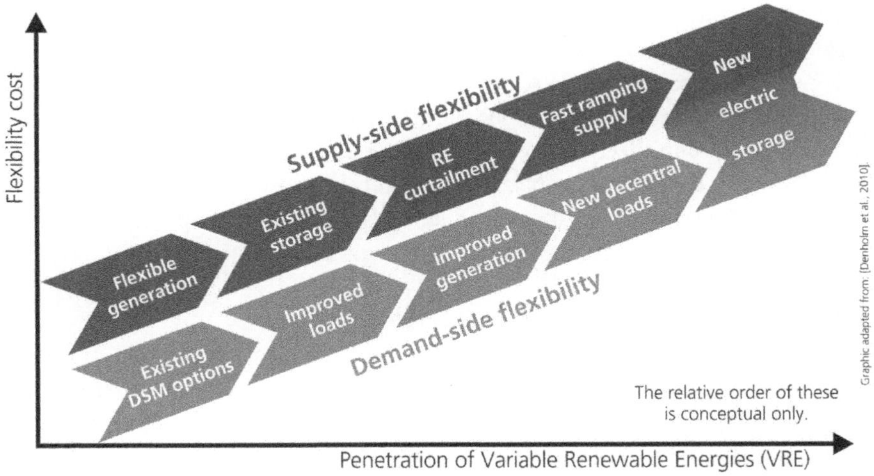

Figure 39.5 Conceptual figure that shows the principal relation between flexibility cost, penetration of variable renewable energies, and different balancing options on supply- and demand-side (Graphic adapted from [3]).

39.3 Energy Balancing Technologies and Options

There is more than one energy balancing technology. Although electrical energy storages are the most flexible and highly sophisticated solution, it is often not the most economic choice.

The different balancing options could be clustered in three main different groups. There is supply-side and the demand-side flexibility as a direct measurement. In addition, there is the indirect option of enhancing the grid to reach a higher level of consolidation and to lower the balancing demand. Figure 39.5 shows, in terms of quality (simplified), the different options

It is obvious that using both existing technologies/options (existing demand response, existing flexible generation, and existing storage) and curtailment of RE are solutions with the lowest cost. By raising penetration of variable renewable energies (VRE) the usable potential of these "cheap" flexibilities is fully exploited.

39.3.1 Demand-Side Flexibility/Demand Respond (DR) – Without New Electric Energy Storage

DR is often used for a wide range of demand-side based activities to level the load. The most common understanding is defined as: "Changes in electric usage by end-use customers from their normal consumption patterns in response to changes in the price of electricity over time, or to incentive payments designed to induce lower electricity use at times of high wholesale market prices or when system reliability is jeopardized" [2].

DR refers to a suite of demand-side options, including using more electricity when there is a surplus and using less when there is a scarcity [4]. In all cases, some ICT technology is needed to activate these options at the right time and to enable coordination with other options. In former times the ICT solution based on a simple central signal switching decentralized devices on and off. Today this could be organized economically by markets and price signals or technically by sensors reacting to voltage or frequency. The different options could be distinguished between improved loads and generation (by thermal storage) or new, additional loads.

39.3.1.1 Existing Demand-side Management (DSM)

DSM represents a huge potential of existing DR measurements, for example, pre-heating of buildings and water supplies, or pre-cooling of cold/frozen food storage, while delivering the same comfort to homeowners and service to industry. In most cases, load shifting is the main application. Advantages of DSM are low investments and high potentials. But connecting, organizing, and billing of all devices and customers could be difficult in relation to the small impact of each single device.

39.3.1.2 Improved Loads and Improved Generation

The core idea here is to improve existing heat pumps (HP) or decentralized combined heat and power units (μCHP), which are traditionally operated by heat demand with an additional thermal storage. This storage allows for more flexibility and operation modes based more on grid demand. Main advantages are higher power output, capacities, and cheap and efficient thermal storage devices. Depending of the size of the thermal storage load shifting could be realized anywhere between hours and days.

39.3.1.3 New Decentralized Loads

In contrast to DSM and improved loads/generation new decentralized loads means new flexibility options, for example, electric vehicles (EV) or power-to-heat allows for fuel switching – from gasoline or natural gas to (renewable) electricity – and due to intrinsic storage effects for new DR options. The intrinsic storage effect could be based on batteries (e.g. in EV) or thermal storage devices (power-to-heat). The main driver of this option is fuel switching to renewable (electric) energies. The gain of more flexibility is more a windfall effect. Sometimes there is a bivalent operations mode (i.e. two different energy sources) possible, which offers additional flexibility. All these means are more expensive than the first expected. Nevertheless, it will be an interesting option for the future, if the total energy system is moving more and more to electric sources.

39.3.2 Supply-Side Flexibility – Without New Electric Energy Storage

On the supply side, there are also traditional, improved, and new flexibility options.

39.3.2.1 Flexible Generation

Flexible generation means using existing power plants and operating them in the maximum range of their flexibility to meet the electric demand. There is less or no additional investments (CAPEX) necessary but flexible power outputs leads to higher OPEX (less efficiency) and less revenue in most cases. Additionally mechanical fatigue could rise which also increases OPEX through higher maintenance costs.

39.3.2.2 Existing Electrical Storage Devices

In many countries, there are lots of huge electric energy storage devices. But fast changing markets and/or changing energy systems due to high penetration of renewable energies destroys the former business model. While these storages are possibly not fitting perfectly to the new storage demand (capacity, power output, location, etc.) new business models have to be developed and realized. The main advantage of this flexibility option is the low additional costs. But alternatively it could be difficult to operate an unsuitable or wrongly located storage device economically.

39.3.2.3 Curtailment of Renewable Energies

The easiest way to balance a surplus of energy is switching off renewable energy production. But wasting electric energy is one disadvantage. Another disadvantage is that this is a unidirectional measurement because a gap of electric energy cannot be filled.

39.3.2.4 Fast Ramping Supply

Installing new (conventional) power plants (e.g. gas turbines) with special fast ramping options allows for more flexibility. The main disadvantage of this solution is very low full load hours. This leads to very low revenue to cover CAPEX.

39.4.3.3 New Electric Energy Storage

Building new electric energy storages allows for the highest flexibility but it is also a more expensive solution. There are different technologies currently available on the "storage market."

Pumped hydropower (PH) storage covers 99% of all storage capacities worldwide. This bulk storage is only usable in large-scale applications. Due to special environmental needs (hilly) and environmental impacts there have been less new storages built in recent years. Compressed Air Energy Storages (CAES) another bulk storage device is only realized in a few cases and with less efficiency. Additionally the usage is limited by special underground requirements (caverns). For all midsized and small applications, there are various battery types. These systems are modular that means applications could be realized from 1 kW to a few MW scale. Batteries are very flexible and have no special requirements in their environment. However, these technologies are mostly expensive (e.g. Li-batteries) or they have limited cycle life (lead-acid batteries) – compared to mechanical storages. Special features offer redox flow-batteries (RFB). These batteries have separate tanks for energy storage and a power conversion unit for charging and discharging. Therefore, there is a design flexibility for power output (MW) and storage capacity (MWh), which is not available with other batteries. This allows for special designs for different applications.

Supercaps and flywheels have minor relevance due to the fact that the energy capacity is too low to shift to variable renewable energies for hours.

39.4 Applications for Electric Energy Storages (Adapted from [4]).

There is a business case for a specific electric energy storage technology that not only depends on the technology itself but also strongly on the application and the regulatory framework. This section gives an overview of various services that can be

Table 39.1 Applications for energy storages.

Application
Energy shift by spot market trading (arbitrage)
Provision of balancing power
Provision of black-start services
Operational support for conventional power plants (ramping)
Guarantee of power quality and uninterrupted power supply
Electric supply capacity
T&D Deferral
Renewable energy time-shift
Increase of PV-self consumption
Time-of-use energy cost management
Peak-load reduction to minimize grid fees
Compensation of forecast errors (forecast hedging)
Active power balancing (micro-grids)

provided by different storage technologies depending on the respective requirements for each application. Most of the mentioned services are universal, but might vary slightly from country to country (see Table 39.1).

Energy shift by spot market trading (arbitrage). This takes advantage of the price spread between high and low spot market prices. Low electricity prices correlate with low residual load whereas high price levels are reached in times of high residual load. Therefore, charging electricity at low electricity prices, and discharging at high price levels leads to a smoothing of the residual load.

Provision of balancing power. Generation and consumption always have to be balanced to guarantee a constant grid frequency (e.g. 50 Hertz in the European interconnection). The demand for balancing power corresponds to the total sum of deviations between the reported schedules and actual customer consumption or actual power plant in feed. Balancing markets vary from country to country or even on a regional level. From the technological point of view, balancing power is the provision of a defined capacity, either positive or negative, available to the system operator that can be used to stabilize the grid frequency.

Provision of black-start services. Black-start services are required in case of power outages at power stations or parts of the electrical grid to restore operation. Black-start capability means that a power plant or device is able to start-up without an external energy supply.

Operational support for conventional power plants (ramping). As part of a so-called hybrid power plant, a storage system allows for faster load ramps and with that generates a more dynamic behavior compared with conventional power plants.

Guarantee of power quality and uninterrupted power supply. Energy storage systems with short response times can be used to guarantee voltage stability or uninterrupted power supply in case of a power failure.

Electric supply capacity (backup supply). Additional electric supply capacities can be required during peak load times.

T&D Deferral. To avoid or to defer grid extensions energy storage devices might be used for congestion relief by storing electricity when the maximum load capacity is exceeded.

Renewable energy time-shift. Renewable energy time shift describes the temporal shift of electricity generated from renewable energies from low-price to high-price phase (see also: arbitrage).

Increase of PV-self consumption. In times where electricity generation from PV-plants (household level) exceeds the respective demand, electricity can be stored (and vice versa).

Time-of-use energy cost management. This takes advantage of the price spread between high and low purchase prices for the end-user. Electricity is stored in low price phases and discharged in high price times to reduce the purchase costs for the end-user.

Peak-load reduction to minimize grid fees. Grid fees contain a load-depending weighting factor in the capacity charge. To reduce this factor electricity can be stored in low-load times and discharged at peak load times.

Compensation of forecast errors (forecast hedging). Energy storages can be used to compensate forecast errors in balancing groups to minimize external balancing demands. Within the same balancing group, electricity can be stored at times when the electricity supply exceeds forecast and can be discharged when forecast exceeds supply.

Active power balancing (micro-grids). Besides unintentional balancing due to forecast errors (see compensation of forecast errors), energy storages can be used for short-, medium-, or long-term active power balancing (e.g. island grids).

39.5 Business Cases for Electric Energy Storages

Business cases for energy storage might vary from country to country, depending on the respective market structures and regulatory frameworks (e.g. taxes, grid fees). Taking Germany as an example, in the past, only providing balancing power (secondary control) through large-scale storages (pumped hydro or compressed air) used to be economically feasible. Future forecasts show that there will be a business case for providing primary control (balancing power) using large-scale batteries as well as for small-scale batteries to increase PV self-consumption.

In addition, mobile batteries in electric vehicles could be economically feasible when providing different types of balancing power (primary and secondary control, minutes reserve).

Some of the services described in Figure 39.6 are only relevant for large-scale storages such as pumped hydro or compressed air. So far, the focus to integrated fluctuating forms of energy is mostly on large, central most cost-effective energy storage technologies like pumped hydro or the conversion of surplus electricity into hydrogen. The potential of small, distributed energy storage technologies is mostly unexplored.

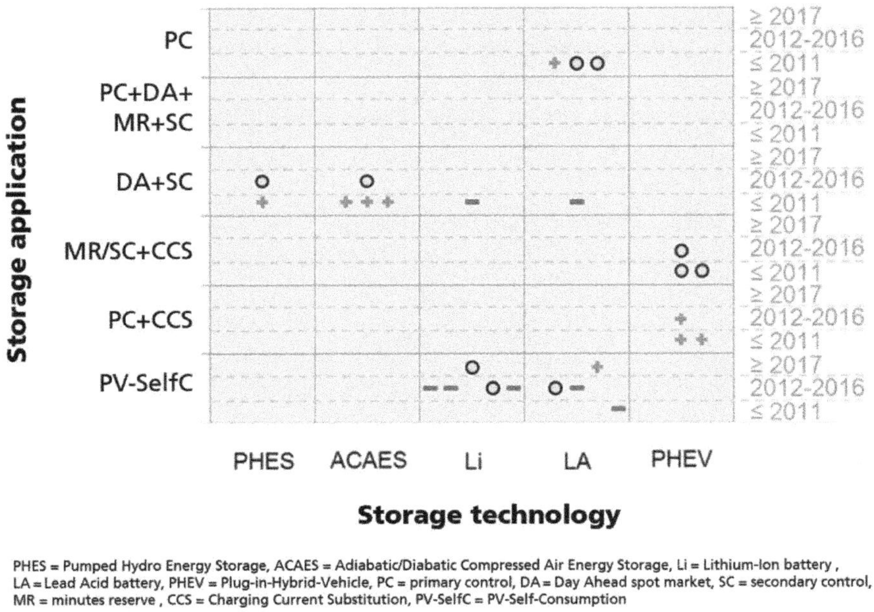

Figure 39.6 Overview of storage technologies and applications (+ business case, - no business case, o indifferent).

39.6 Distributed Storage Concepts

Distributed energy storages (DES) can be defined by their location within an energy system – either on the consumer side or directly at the local renewable energy source. DES can be electrical energy storages (EES) or thermal energy storages (TES) or a combination of both. Depending on their level of grid integration/interactions different cases can be considered (see Figure 39.7).

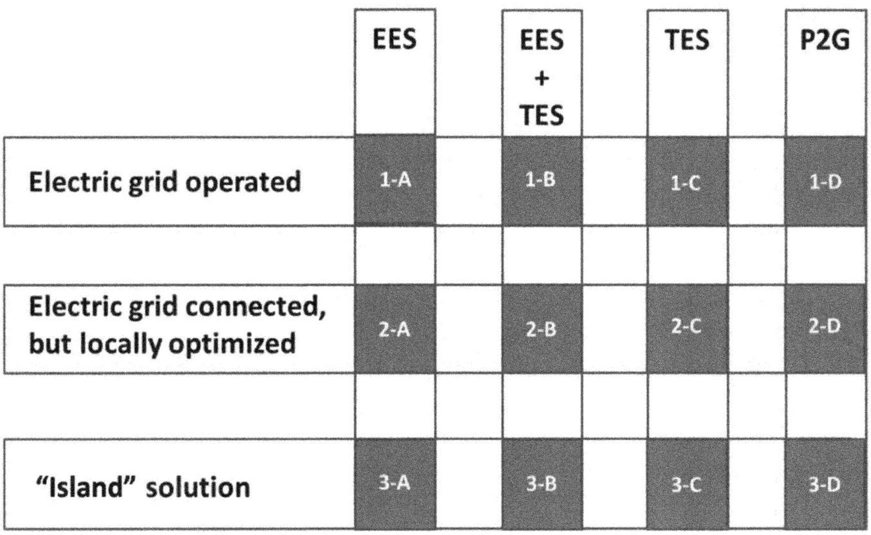

Figure 39.7 DES and level of grid integration (Example taken from IEA [7]).

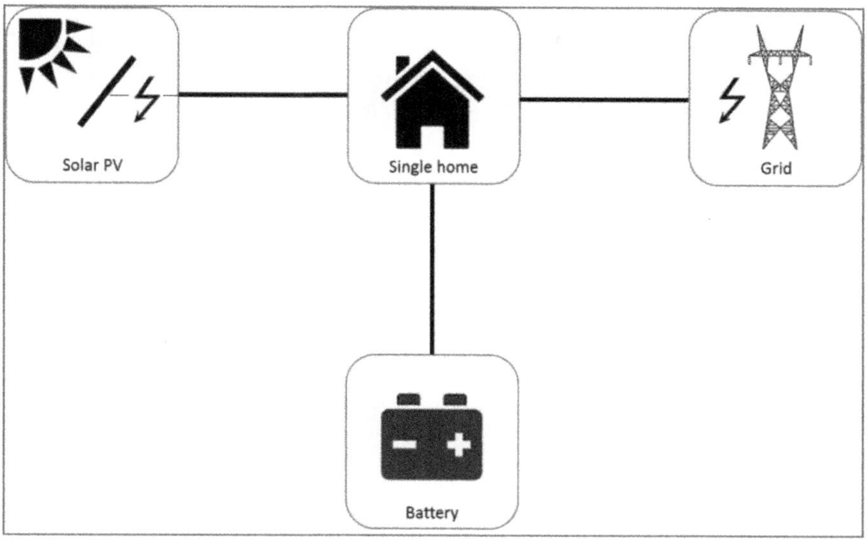

Figure 39.8 Increase of PV-self consumption (Example taken from IEA [7]).

DES systems with no connection to the distribution grid are referred to as "island solutions." In case the DES is grid-connected, it can be operated either grid-optimized or locally optimized. Examples for selected cases are given to show the contribution of DES in the integration of renewable energies.

Figure 39.8 shows a home battery combined with a PV-installation and connected to the grid. This scenario can be realized by any type of batteries, e.g. Li-Ion batteries or Lead-Acid batteries. In that case, electrical energy generated from PV can be stored in case generation exceeds demand and can be discharged when demand exceeds generation. By this, the degree of self-sufficiency can be increased and in the case of increasing electricity prices this can be economically feasible for the end-user.

Another business case for decentralized storages can be realized by thermal storage. The combination of a decentralized heat pump with a thermal storage allows for operating the heat pump in a grid-driven mode compared to a conventional heat-driven mode. This additional flexibility could be used to increase the self-consumption of energy from a roof mounted PV system. Another example is the combination of a micro-CHP (Combined Head and Power) system with a thermal storage in a household. In this case the micro-CHP unit could be operated more on the electric demand, so that the owner has the chance to buy less energy from the utility company. However, in both cases the thermal energy "shift" is limited to hours or a day, due to size limitations of a thermal storage (e.g. hot water tanks).

39.7 Summary

The decarbonization of the energy systems by (variable) renewable energies requires measures for grid balancing. The balancing demand is increasing with the share of VRE. Energy storages (electrical, thermal) are not the only means to meet this balancing demand; there are other options like curtailment and existing DS management. While energy storages offer the highest flexibility, other options might be economically feasible. After using this "low-hanging fruits," energy storages will be a broader solution in the near future. Some technologies are already competitive in defined applications (e.g. distributed household electric storage combined with roof-mounted PV). For bigger markets, most storage technologies are still too expensive. Additional R&D is still needed to reduce costs, but it must be focused. Due to different energy system structures, the optimal role for storage varies widely across regions. From the economic point of view, power markets are currently often ill-equipped to compensate storage for the suite of services they provide.

References

[1] Aggarwal, S. and Orvis, R., (2016). *Grid flexibility: Methods for modernizing the power grid*. Energy Innovation Policy & Technology LLLC. www.energyinnovation.org. March 2016.

[2] Balijepalli, M. and Pradhan, K. (2011). *Review of demand response under smart grid paradigm*. IEEE PES Innovative Smart Grid Technologies.

[3] Denholm, P., Ela, E., Kirby, B., and Milligan, M. (January 2010). *The Role of Energy Storage with Renewable Electricity Generation*. National Renewable Energy Laboratory.

[4] Kanngießer, A. (2013). Entwicklung eines generischen Modells zur Einsatzoptimierung von Energiespeichern für die techno-ökonomische Bewertung stationärer Speicheranwendungen. PhD thesis, Dortmund.

[5] Lew, D., Brinkman, G., Ibanez, E., et al. (2013). *Western Wind and Solar Integration Study Phase 2 (WWSIS-2)*. NREL/TP-5500-55588. Golden, CO: National Renewable Energy Laboratory.

40

Thermal Storages and Power to Heat

Per Alex Soerensen

PlanEnergi, Jyllandsgade 1, 9520 Skoerping Denmark

40.1 Introduction

Power has been used for the production of heat for decades, but in this chapter, the topic is the use of power to heat for the (P2H) regulation of intermittent electricity production from solar and wind. The technologies and systems described are technologies for the production of heat and hot water to be utilized in buildings and district heating systems. The technology includes heat pumps, electric boilers, and heat storages developed and introduced on the market for these purposes.

The chapter will also include concepts and examples of present and future integration of power to heat in the energy systems.

The power to heat concept was introduced from an analysis of energy systems with high fractions of intermittent electricity production from solar and wind. In periods with electricity production higher than consumption a cheap way of regulation is to produce heat for hot water and heating and store it for later use.

40.2 Why Power to Heat?

Replacement of fossil fuels with Renewable Energy Sources (RES) has been an ongoing process in many countries. In this process, electricity production from intermittent sources (solar and wind) will increase. In this situation, it is possible to benefit from trading electricity with neighboring regions and countries or to try to make the electricity production and consumption and thus the energy system more flexible. The feasibility of different ways of balancing the electricity system was calculated for the Danish energy system in 2006 [1]. The result suggested that it was feasible for the Danish society to include the Combined Heat and Power (CHP) plants to balance fluctuating wind power. Furthermore, there are major advantages in equipping small CHP plants as well as large CHP plants with heat pumps. By doing so, it would be "possible to increase the share of wind power from the present 20 to 40% without causing significant problems of imbalance between electricity consumption and production" [1].

The calculations showed that the investment in increased flexibility was profitable and the feasibility of wind power was improved.

Advances in Energy Storage: Latest Developments from R&D to the Market, First Edition.
Edited by Andreas Hauer.
© 2022 John Wiley & Sons, Ltd. Published 2022 by John Wiley & Sons, Ltd.

However, that was for a Danish energy system with 1400 MW of decentralized gas-fueled CHP plants producing district heating and with a heating system where district heating covered more than 50% of the heat market.

40.2.1 The German Energy System

In the German energy system, district heating covers only 15% of the heat consumption [2]. The conditions for power to heat will therefore differ from the Danish conditions because the possibilities of using CHP plants connected to district heating systems for energy balancing are reduced.

The Energy Balanced Demand (EBD) for Germany is described by [3]. The imbalances between supply and demand are calculated in a model where Germany is divided into 146 energy regions. For each region, regional load curves of supply and demand are set up based on actual demands and actual supply technologies in the regions and actual grid transmission capacities. To predict installed capacities of fluctuating electricity production EBD was calculated for 2020 (40% fluctuating electricity production), 2030 (64.7% fluctuating electricity production), and 2050 (86.2% fluctuating electricity production).

The results for 2020, where conventional power plants are still in use is a negative EBD (energy surplus) of 18 023 TWh (terrawatt hour). The EBD is primarily in the northern part of Germany. Figure 40.1 shows the sorted annual EBD and Figure 40.2 shows the geographical spreading of energy regions with negative EBD (dark) and regions in balance (light).

With increasingly fluctuating electricity production, the EBD will of course be extended if the energy system is not changed. But, the energy surplus in periods with fluctuating electricity production is much higher than the additional demand in periods with low fluctuating electricity production. Thus, electricity to a large extent has to be converted to other purposes (power to heat for instance) or the fluctuating production has to be turned off.

[2] A computer simulation model REMod-D (**R**enewable **E**nergy **Mod**el-**D**eutschland) that models a possible future German energy system for all sectors including all renewable energy converters, storage components, secondary energy converters, and loads for a whole year based on an hourly simulation. A generic optimizer is applied in order to identify system compositions with minimal overall annual cost [2]. The model is further developed in [4]. The total system is illustrated in Figure 40.3.

The result of the optimization is a future German energy system including a.o. 15 GW_{heat} electrically driven heat pumps connected to district heating and block heating if CO_2 emissions are to be reduced by 85% by 2050. Power to heat is thus an important part of an optimized future German energy system.

40.2.2 The Danish Energy System

As mentioned, more than 50% of Danish heat supply is covered by district heating and more than 50% of the heat production for district heating is produced by CHP plants. In the CEESA project [5] a group of 24 Danish researchers simulated scenarios with the entire Danish energy system covered 100% by renewable energy. The simulation showed that the integration of sectors is very important. Wind power in the proportion

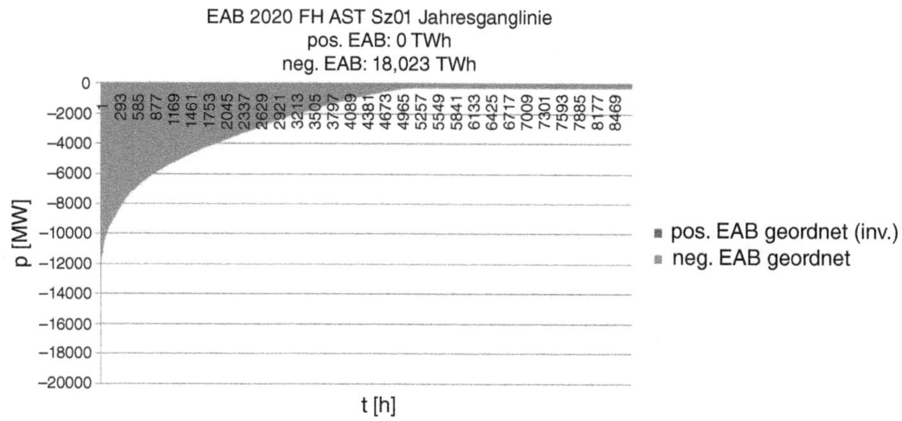

Figure 40.1 Sorted annual EBD for Germany in 2020 with 40% fluctuating electricity production [3].

20–25% can be integrated without significant changes to the Danish energy system through flexible operation of the CHP plants because these normally include a thermal storage covering three days of summer load. With more than 20–25% wind power, the most feasible next step in the integration is to install large heat pumps. With more than 40–45% wind power the transport sector needs to be integrated into the energy system by introducing electric vehicles (EV) and also flexible demands in households and the industry sector will be needed. The last step will be electrolyzers converting power to hydrogen for the production of methane/DME/methanol.

A Sankey diagram of CEESA 2050 100 RES can be seen in Figure 40.4. Of the fluctuating electricity 24%/14.2 TWh is utilized as power to heat.

The Danish reality is that 42% of electricity production came from wind power in 2015 and almost no large-scale heat pumps were implemented. Nearly 500 MW implemented capacity of electric boilers was enough to prevent imbalances in the electric grid, but electricity was exported to neighboring countries at very low prices because of the missing heat pumps.

40.3 Technologies for Power to Heat

Technologies for conversion of power to heat are thermal storages, electric boilers, and electric driven heat pumps.

Thermal storages can, if connected to district heating systems, be short-term accumulation tanks (normally steel tanks for diurnal use) or can be long-term or even seasonal storages depending on the flexibility demand in the electricity system. Thermal storages are described in [6]. In Section 40.4 two cases are described, one with long-term pit heat storage and one with diurnal accumulation tank.

Thermal storages in individual heating systems can be water tanks in the buildings and may be combined with storage capacity in the building, for instance the storage capacity in floors with floor heating.

The efficiency of electric driven heat pumps is several times higher than the efficiency of electric boilers so heat pumps are preferable for converting power to heat. If

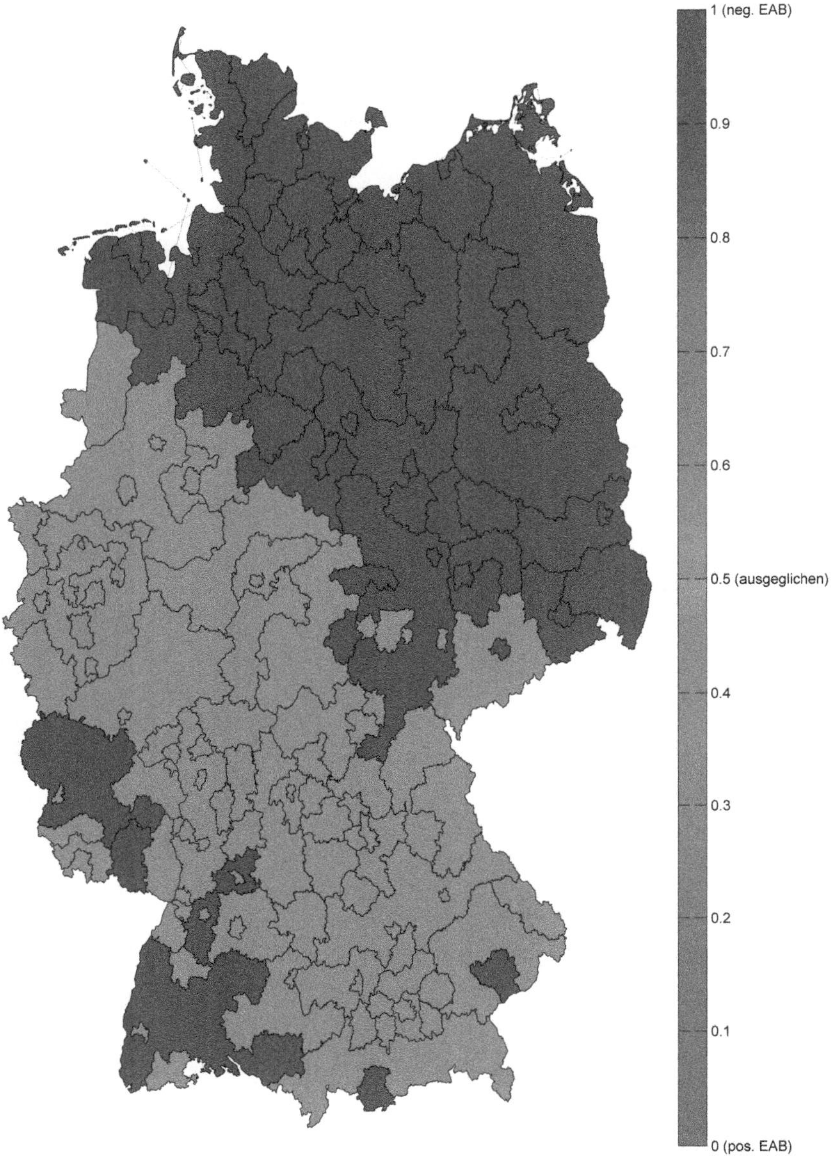

Figure 40.2 Geographical spread of negative EBD (dark) and regions in balance (light) for Germany in 2020 [3].

the heat pump heats up a building and only runs in periods with fluctuating electricity production there must be a backup system (for instance a gas boiler) or the heat pump must be oversized and connected to a thermal storage.

In district heating and cooling plants, a thermal storage is often already installed and the existing production capacity (CHP using natural gas, gas boilers, biomass boiler, etc.) works as backup. Furthermore, the price for a large heat pump per installed kW in district heating and cooling systems is remarkably lower than for single buildings. The efficiency is higher, the possible forward temperature is higher, and the integration in the electricity system is cheaper and more efficient.

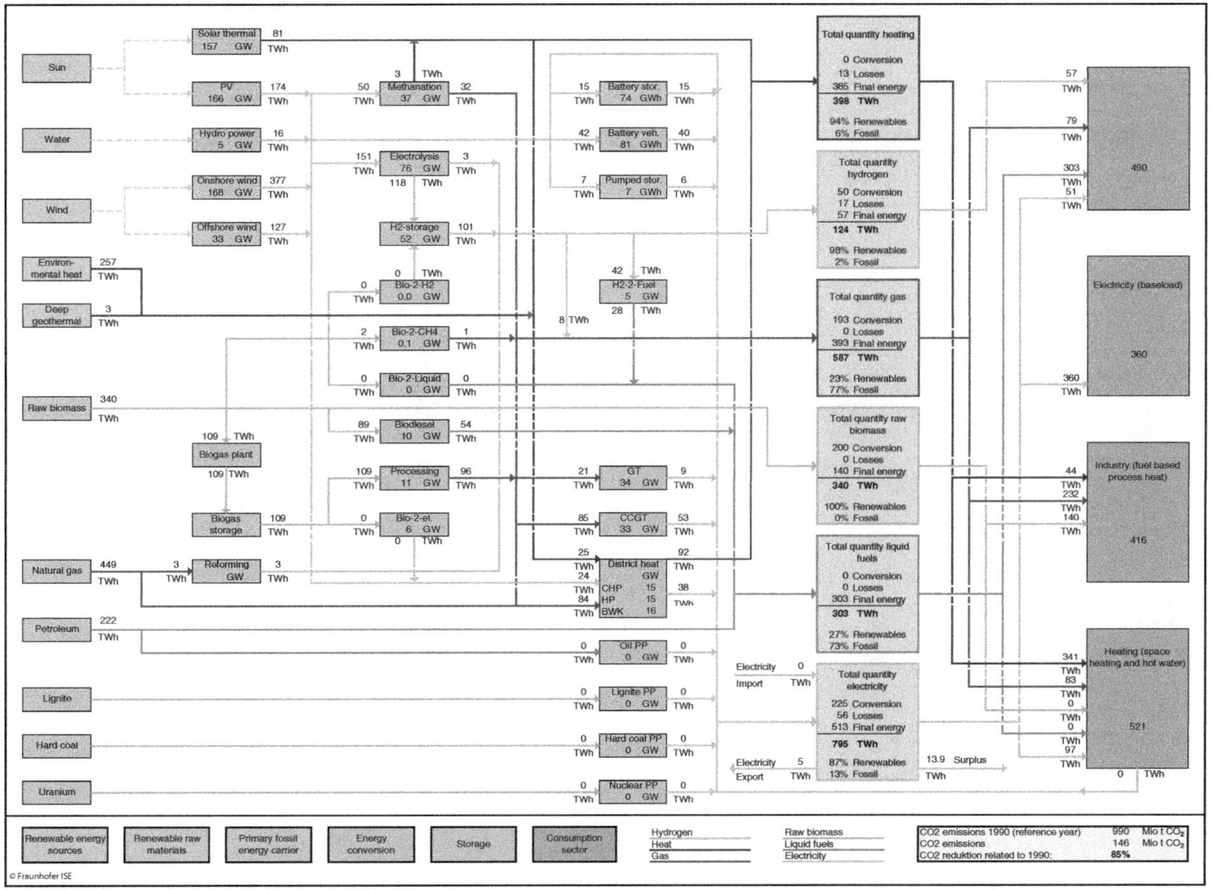

Figure 40.3 Scheme of a future German energy system as presented in the REMO-D model [4].

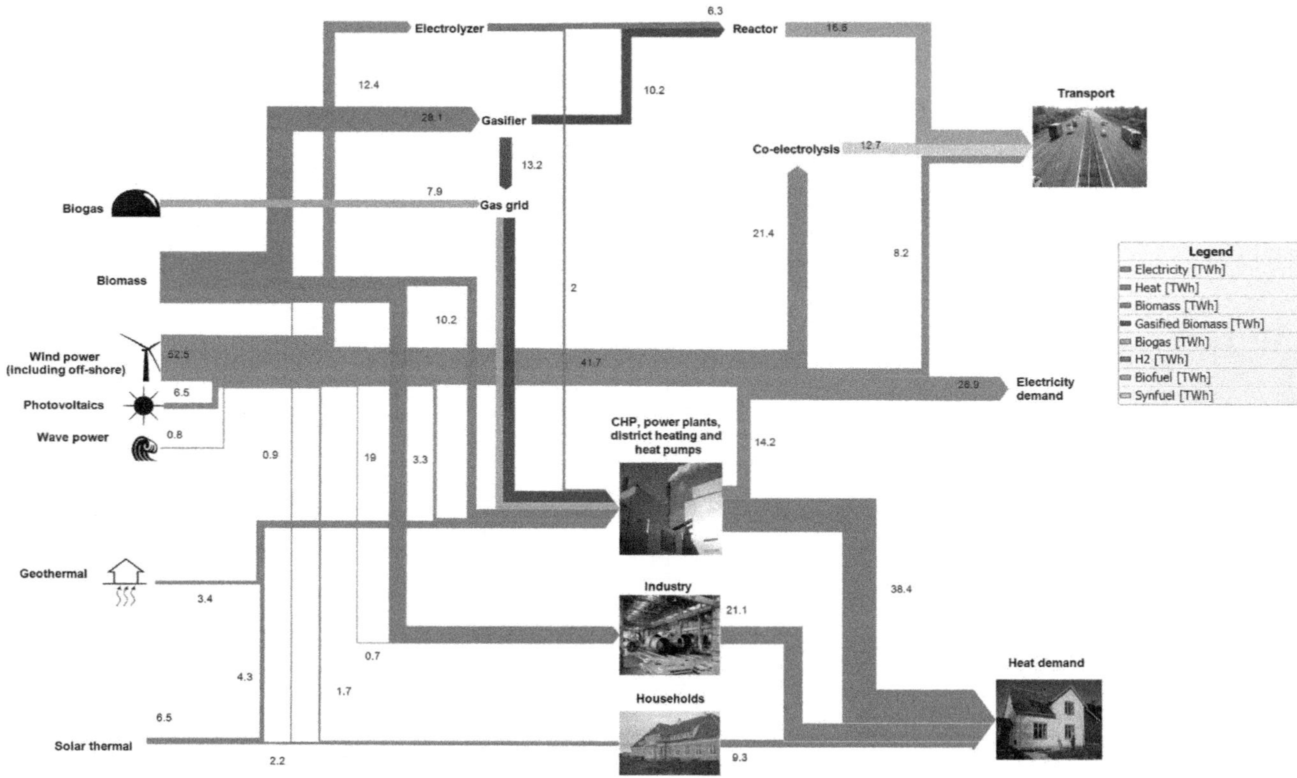

Figure 40.4 Sankey diagram of the CEESA 2050 100% renewable energy scenario [5].

These are the reasons why the best power to heat solutions are combined with district heating. Disadvantages are the heat loss in the district heating pipes and the investment in the pipes.

40.3.1 Large-scale Heat Pumps

For compression heat pumps, the practical heat output is usually three to five times (the coefficient of performance (COP)) the drive energy. This factor depends on the efficiency of the specific heat pump, the temperature of the heat source and the heat sink, and the temperature difference between the heat source and the heat sink. The energy flow is illustrated in the Sankey diagram in Figure 40.5.

Heat sources can be from the ambient air, surface water, ground water, sewage water, ground (soil), and excess heat from industries or fluegas.

Table 40.1 gives details COP, lifetime, regulation ability, and costs data for large-scale heat pumps.

40.3.2 Small-scale Heat Pumps

Heat sources for small-scale heat pumps can be ambient air, ventilation air, ground (soil), or ground water.

Table 40.2 gives COP, lifetime, regulation ability, and data for small-scale heat pumps.

40.3.3 Electric Boilers

Electric boilers are devices in the MW size using electricity for the production of hot water for industrial or district heating purposes. They are usually installed as peak load units in the same way as oil and gas boilers. The boilers are electrical resistance boilers up top 1–2 MW and electrode boilers (power is fed directly in to electrodes in the boiler) for larger systems.

Lifetime, regulation ability, and costs data can be found in Table 40.3.

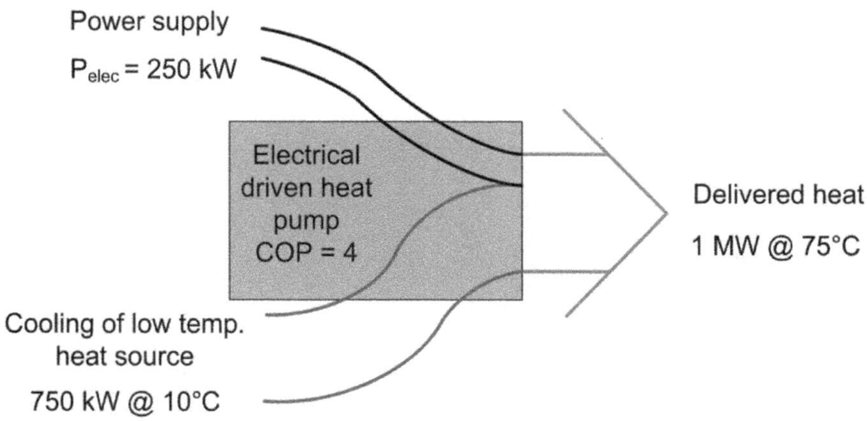

Figure 40.5 The electrical power consumption of 250 kW enables the heat pump to utilize 750 kW from a low temperature heat source at 10°C. Thus delivering 1 MW at 75°C (COP is 4). *Source*: [6].

Table 40.1 Data for large-scale heat pumps.

Technology	40 Electrical compression heat pumps – district heating									
	2015	2020	2030	2050	Uncertainty (2020)		Uncertainty (2050)		Note	Ref
Energy/technical data					Lower	Upper	Lower	Upper		
Heat generation capacity for one unit (MW_{heat})	4	4	4	4	3	6	3	10		[2]
Total efficiency, net (%), name plate	N/A	N/A	N/A	N/A	N/A	N/A	N/A	N/A		
Total eff., net (%), annual average, ambient heat source, no dev. in supply temp.	350	360	380	410	350	380	350	450	A, F, J, K	[3]
Total eff., net (%), annual average, ambient heat source, reduced supply temp.	350	400	480	600	350	450	350	700	A, B, F, J	[2, 3]
Total eff., net (%), annual average, waste heat 20°C, reduced supply temp.	440	500	600	740	440	600	440	850	A, B, F, J	[2, 3]
Total eff., net (%), annual average, waste heat 40°C, reduced supply temp.	700	900	1200	1800	700	1200	700	2000	A, B, F, J	[2, 3]
Electricity consumption for pumps, etc. (% of heat gen)	10	9	7	4	4	10	3	6	I, M	[2]
Forced outage (%)	0	0	0	0	0	1	0	1	G	[2]
Planned outage(weeks per year)	0.5	0.5	0.5	0.5	0	1	0	1	H	[2]
Technical lifetime (years)	25	25	25	25	15	30	15	30		[2]
Construction time (years)	0.5	0.5	0.5	0.5	0.3	0.7	0.3	0.7	C	
Space requirement (1000m2 per MW_{heat})	0.02	0.02	0.02	0.02	0.01	0.04	0.01	0.04		[1]
Regulation ability										
Primary regulation (% per 30 seconds)	10	10	10	10	10	25	10	30	D	[2]
Secondary regulation (% per minute)	20	20	20	20	20	40	20	40	D	[2]

Minimum load (% of full load)	10	10	10	10	10	10	10	10	D	[2]
Warm start-up time (hours)	0	0	0	0	0	1	0	1		[2]
Cold start-up time (hours)	6	6	6	6	1	12	1	12	E	[4, 5]
Environment										
SO_2 (g per GJ fuel)	0	0	0	0	0	0	0	0		
NO_X (g per GJ fuel)	0	0	0	0	0	0	0	0		
CH4 (g per GJ fuel)	0	0	0	0	0	0	0	0		
N2O (g per GJ fuel)	0.0	0.0	0.0	0.0	0.0	0.0	0.0	0.0		
Financial data										
Nominal investment (M€ per MW_{heat})	0.70	0.66	0.59	0.53	0.50	1.00	0.50	1.00	A, L	
-of which equipment (%)	50	50	50	50	30	70	30	70		[2]
-of which installation (%)	50	50	50	50	30	70	30	70		[2]
Fixed O&M (€/MW_{heat}/year)	2000	2000	2000	2000	1000	3000	1000	3000		[2]
Variable O&M (€/MWh_{heat})	8.4	8.2	8.8	8.2	4.3	9.1	6.5	11.9		
-of which is electricity costs (€/MWh_{heat})	6.4	6.4	7.1	6.6	2.8	7.1	5.0	9.9	M	
-of which is other O&M costs (€/MWh_{heat})	2.0	1.8	1.7	1.6	1.5	2.0	1.5	2.0	F	[2]

References

[1]Teknologisk Institut, 2015.
[3]Drejebog til store varmepumpeprojekter i fjernvarm esystemet, 2014, see http://www.planenergi.dk/varmepumper.
[4]Udredning vedrørende varmelagringsteknologier og store varmepumper til brug i fjernvarmesystemer, 2013.
[8]Innoterm, 2015.
[10] Johnson Controls, 2015.

Notes:

A Actual development within COP optimization and reduced investment cost depends on the development in fuel and electricity prices.
B The large potential for higher COP factors is primarily caused by lower supply temperatures in the future (40–80°C in 2015 and 30–60°C in 2050).
C The development within construction time will depend on future production figures and standardization of plants.
D The regulation ability of large heat pumps will depend on the future markets for regulation services.
E Cold start of time is starting a heat pump where stand-by heating has not been applied.
F Operation at part load will usually increase COP but increase variable O&M costs.
G May vary depending on availability of heat source.
H May vary depending on specific type, heat source, etc.
I The auxiliary electricity is not included in the total efficiency.
J The total efficiency net annual average is calculated using the practical COP.
K Average value for ambient heat sources. For air it will be lower and for sea and lake water it will be around the average value, whereas for groundwater the value will be higher. It is weighted so that the heat pump produces 60–70% of the demand of the district heating system. The supply temperature from the heat pump is fairly constant, since it is mixed with water from other production units in the months with the highest heat demand, when the supply temperature in the network is typically increased.
L Including heat uptake and buildings.
M The costs of auxiliary electricity consumption is calculated using the following electricity prices in €/MWh: 2015: 64, 2020: 71, 2030: 102, 2050: 165. These prices include production costs and transport tariffs, but not any taxes or subsidies for renewable energy.

Table 40.2 Data for air to water heat pump for a new multi-family house.

Technology	Heat pump, Air-to-water, new apartments									
					Uncertainty (2020)		Uncertainty (2050)			
Energy/technical data	**2015**	**2020**	**2030**	**2050**	**Lower**	**Upper**	**Lower**	**Upper**	**Note**	**Ref**
Heat production capacity for one unit (kW)	160	160	160	160	140	200	140	200		[2, 3]
Electricity generation capacity for one unit (kW)	0	0	0	0	0	0	0	0		
Expected share of space heating demand covered by unit (%)	100	100	100	100	70	100	70	100	C	
Expected share of hot tap water demand covered by unit (%)	100	100	100	100	70	100	70	100	C	
Heat efficiency, annual average, net (%), floor heating	440	450	460	480	440	470	450	520	B, D	[5, 6, 7]
Total efficiency, annual average, net (%), floor heating	390	400	405	410	390	410	400	490	B, E	[1, 5, 6, 7]
Heat efficiency, annual average, net (%), radiators	420	430	440	460	420	460	430	500	B, D	[5, 6, 7]
Total efficiency, annual average, net (%), radiators	370	380	390	405	370	400	380	480	B, E	[1, 5, 6, 7]
Auxiliary Electricity consumption (kWh/year)	10 000	10 000	10 000	10 000	8000	12 000	8000	12 000	F	[1, 4]
Technical lifetime (years)	15	15	15	15	12	20	12	20		[4, 6]
Regulation ability										
Change in capacity within 1 minute (%)	100	100	100	100	20	100	20	100		[4, 5]
Warm start-up time (hours)	0	0	0	0	0	0	0	0		[4, 5]
Cold start-up time (hours)	0	0	0	0	0	0	0	0		[4, 5]
Environment										
SO_2 (g per GJ fuel)	N/A	N/A	N/A	N/A	N/A	N/A	N/A	N/A		
NO_X (g per GJ fuel)	N/A	N/A	N/A	N/A	N/A	N/A	N/A	N/A		
CH_4 (g per GJ fuel)	N/A	N/A	N/A	N/A	N/A	N/A	N/A	N/A		

(Continued)

Table 40.2 (Continued)

N_2O (g per GJ fuel)	N/A	N/A	N/A	N/A	N/A	N/A	N/A	N/A		
Particles (g per GJ fuel)	N/A	N/A	N/A	N/A	N/A	N/A	N/A	N/A		
Financial data										
Specific investment (1000€/unit)	75	71	63	57	60	100	50	100		[6, 7]
–hereof equipment (%)	60	60	60	60	50	85	50	85	A	[6]
–hereof installation(%)	40	40	40	40	15	50	15	50		[6]
Possible additional specific investment (1000€/unit)										
Fixed O&M (€/unit/year)	1640	1650	1866	2411	1070	2850	1820	3980		
–of which is electricity costs (€/unit/year)	640	710	1020	1650	570	850	1320	1980	F	
–of which is other O&M costs (€/unit/year)	1000	940	846	761	500	2000	500	2000		[6]
Variable O&M (€/MWh)	0.50	0.47	0.42	0.38	0.2	1.0	0.2	1.0		[6]

References

[1]Sparenergi.dk, 2015.
[2]Energistyrelsen, 2015, personal communication.
[3]Energinet.dk, 2015, personal communication.
[4]Teknologisk Institut, 2015, personal communication.
[5]Cronborg, 2015, personal communication.
[6]Drejebog til store varmepumpeprojekter i fjernvarmesystemet, Energistyrelsen 2015.
[7]Udredning vedrØrende varmelagringsteknologier og store varmepumper til brug i fjernvarmesystemer, Energistyrelsen 2013.

Notes:

A Including domestic hot water storage tank and auxiliary equipment.
B Weighted by seasonal heat demand of average climate zones according to EN14825 and domestic hot water production according to EN16147.
C The share of heating demand and hot tap water demand covered depends on how the heat pump is operated in the total heat energy system in the house.
D Heat efficiency values refers to the COP of the heat pump.
E Total efficiency values are including auxilliary electricity consumption.
F The cost of auxiliary electricity consumption is calculated using the following electricity prices in EUR/MWh: 2015: 64, 2020: 71, 2030: 102, 2050: 165. These prices include production costs and transport tariffs, but not any taxes or subsidies for renewable energy.

Table 40.3 Data for large-scale electric boilers.

Technology	41 Heat only generation tech (boilers, heat pumps, geothermal) Electric boilers, 400 or 690 V, 0.06-5 MW; 10 or 15 kV, >10 MW									
	2015	2020	2030	2050	Uncertainty (2020)		Uncertainty (2050)		Note	Ref
Energy/technical data					Lower	Upper	Lower	Upper		
Heat generation capacity for one unit (MW)	5				1	25	1	25		
Total efficiency, net (%), name plate	98	99	99	99	98	99	99	99		[1]
Total efficiency, net (%), annual average	98	99	99	99	98	99	99	99		[1]
Electricity consumption for pumps, etc. (% if heat gen)	0.5	0.5	0.5	0.5	0.1	0.5	0.1	0.5		[1]
Forced outage (%)	1	1	1	1	0.5	1	0.5	1	E	[1]
Planned outage (weeks per years)	0.2	0.2	0.2	0.2	0.2	0.2	0.2	0.2	E	[1]
Technical lifetime (years)	20	20	20	20	20	20	20	20		[1]
Construction time (years)	0.5	0.5	0.5	0.5	0.5	1	0.5	1		[1]
Regulation ability										
Primary regulation (% per 30 seconds)	100	100	100	100	100	100	100	100		[1]
Secondary regulation (% per minute)	100	100	100	100	100	100	100	100		[1]
Minimum load (% full load)	5									[1]
Warm start-up time (hours)	0.008									[2]
Cold start-up time (hours)	0.08									[2]
Environment										
SO_2 (g per GJ fuel)	Cf. the utilized electricity									
NO_X (g per GJ fuel)										
CH_4 (g per GJ fuel)										
N_2O (g per GJ fuel)										
Particles (g per GJ fuel)										

(Continued)

Table 40.3 (Continued)

Financial data										
Nominal investment (M€ per MW), 400/690 V; 1–5 MW	0.15	0.15	0.14	0.13	0.10	0.25	0.10	0.25	A	[1]
–of which equipment	0.12	0.12	0.11	0.10	0.08	0.20	0.08	0.20	B	[1]
–of which installation	0.03	0.03	0.03	0.03	0.02	0.05	0.02	0.05	D	[1]
Nominal investment (M€ per MW); 10/15 kV;>10 MW	0.07	0.07	0.06	0.06	0.02	0.17	0.02	0.17	A	[1]
–of which equipment	0.06	0.06	0.05	0.05	0.02	0.14	0.02	0.14	C	[1]
–of which installation	0.01	0.01	0.01	0.01	0	0.03	0	0.03	D	[1]
Fixed O&M (€/MW/year)	1100	1070	1020	920	1000	1100	900	1000	A	[1]
Variable O&M (€/MWh)	0.8	0.9	1.0	1.0	0.5	0.9	0.5	1.0		[1]
–of which is electricity costs (€/MWh)	0.3	0.4	0.5	0.8	0.1	0.4	0.2	0.8	F	
–of which is other O&M costs (€/MWh)	0.5	0.5	0.5	0.4	0.4	0.5	0.3	0.5	A	[1]
Technology specific data										
Startup costs (€/MW/startup)	0	0	0	0	0	0	0	0		[1]

Notes:

[A] The investment and O&M costs are assessed in relation to an approx. operation time of 500 hours/year.

[B] The installation at low voltage necessitates a transformer substation and expansion of the distribution board. Costs for these are included in the stated equipment costs.

[C] Electrode boilers at medium-high voltage are directly connected to the distribution grid. Costs for the distribution board are included in the equipment costs.

[D] The installation costs include costs for electrical integration and grid connection fees.

[E] The forced outage of electric boilers is very limited and typically well below 1%. The planned outage is typically limited to 1 day/year.

[F] The cost of auxiliary electricity consumption is calculated using the following electricity prices in €/MWh: 2015: 64, 2020: 71, 2030: 102, 2050: 102. These prices include production costs and transport tariffs, but not any taxes or subsidies for renewable energy.

References

1 Tjaereborg Industry (2016).

2 Parat(n.d). Brochure *High Voltage Electrode Boiler*, available at http://parat.no/media/201154/Electrode-Boiler-web.pdf.

40.4 Examples of Power to Heat Concepts

Until now, power to heat installations have almost exclusively been electric boilers. The capacity installed in Danish district heating systems reached nearly 500 MW in 2015, and the German capacity of electric boilers is on the rise. However, implementation of electrically driven heat pumps is difficult because of the boundary conditions in the two countries (resulting in electricity prices at a level that makes it difficult to make a feasible project). Nevertheless, a few projects have been implemented. In Section 40.4.1 an example with long-term thermal storage is described and in Section 40.4.2 an example with short-term storage is outlined.

40.4.1 The SUNSTORE® Concept (Long-term Thermal Storage)

The SUNSTORE® concept [8] consists of a large heat storage (pit heat storage, borehole storage, or tank storage), solar collectors to heat up the storage, a heat pump to use the storage as heat source (and at the same time extend solar production, reduce heat loss from the storage, and extend the storage capacity) combined with a CHP plant.

The concept makes uses electricity in the heat pump when needed and produces electricity when needed. The storage makes this flexibility possible and the solar collectors helps to reduce fuel consumption for heating. This concept was implemented in 2011–2012 at two district heating plants in Denmark at Braedstrup District Heating (with borehole storage) and Marstal District Heating (with pit heat storage). The arrangement for Marstal can be seen in Figure 40.6.

The district heating plant in Marstal consists of the following main components:

- 33 365 m^2 solar collectors
- 75 000 m^3 pit heat storage
- 10 000 m^3 pit heat storage
- 2100 m^3 steel tank

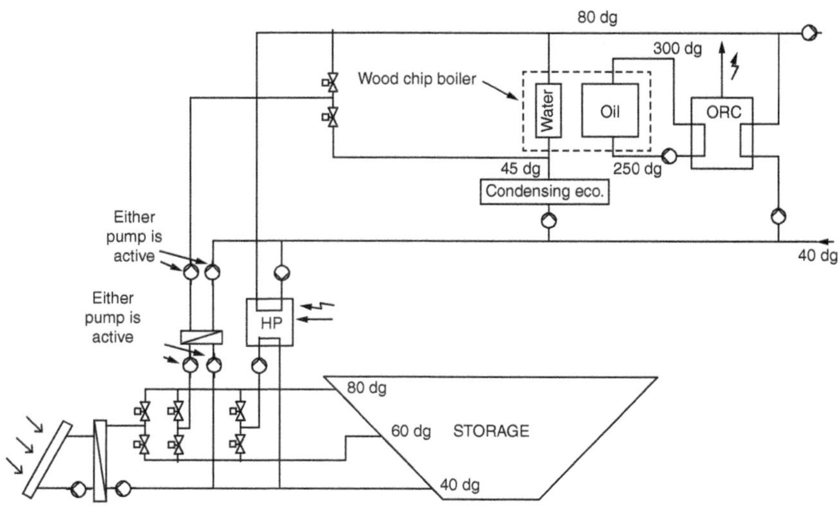

Figure 40.6 The SUNSTORE® concept in Marstal [9].

Figure 40.7 The energy plant in Marstal in December 2012.The floating lid of the storage can be seen left of center. *Source* Marstal District Heating.

- 4 MW biomass boiler with 750 kW ORC unit
- 1.5 MW_{th} electrical driven heat pump
- Bio oil boilers

The pit heat storage went into operation in June 2012 as a part of the project "SUNSTORE 4," which is supported by EC 7th Framework Program. The project enables the Marstal district heating plant to deliver heating to the consumers from 100% renewable energy sources where up to 50% is solar heat. The heat pump uses CO_2 as a refrigerant. Figure 40.7 shows the plant in Marstal with the city in the background.

40.4.2 Heat Pump Using Ground Water in Rye, DK (Short-term Thermal Storage)

The project "New heat pump and solar heating" was supported by the Danish program EUDP and started in 2011. The idea for the project is based on findings in the project "Naturgassens afløser" ("The replacement for natural gas") carried out for Region Central Denmark in 2010. In "Naturgassens afløser" it was found that the best solution for Rye District Heating to get weather changes in the natural gas price and thereby stabilize the heat production price was to implement a combination of a solar heat plant and a ground water based heat pump. The combination of solar heat (fuel free), heat pump (electricity consuming), and gas engines (electricity producing) in one plant is a combination that is expected to suit an overall energy system with fluctuating renewable electricity production very well and stabilize the heat production price. To be able to use a heat pump primarily when the electricity prices are low because of a high share of wind power the heat pump has to be larger than a heat pump running continuously. An investment in a larger heat pump was not economically feasible for the district heating company because the fluctuating electricity prices only represent a small part of the total electricity price including tax and transportation fees. This together with the fact that heat pumps have to be redesigned with a focus on district heating to be economical and technically optimized was the main reason for the need of support from the national Danish funding program EUDP. The design layout for Rye can be seen in Figure 40.8.

The district heating plant in Rye consists of the following main components:

- 2 combustion engines using natural gas
- 1 natural gas boiler

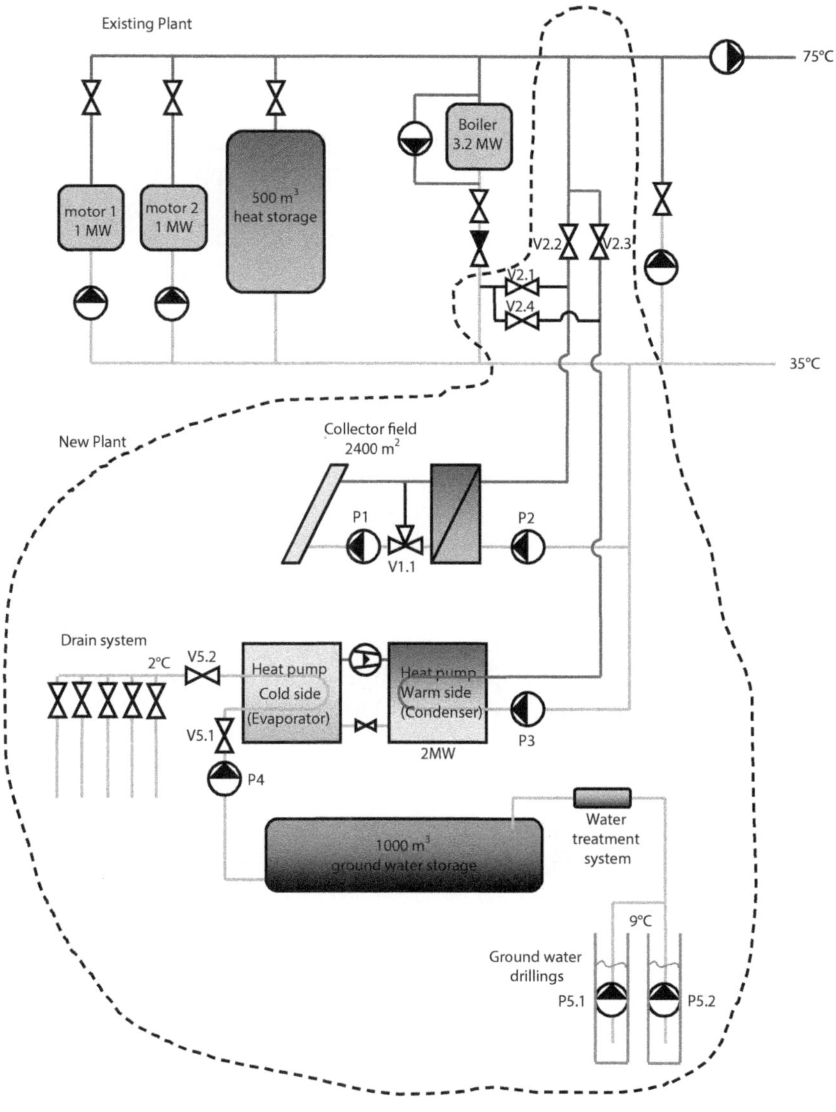

Figure 40.8 Principal diagram of the final plant for Rye [10].

- 1 accumulation tank
- 2,444 m^2 solar collectors
- 2 MW_{heat} electrical driven heat pump

The heat pump uses ammonium as the refrigerant. On average, the COP has been 3.74 in the period from 23rd February–31st December 2015. In the same period, solar covered 15.2% of the consumption and 59.4% of the heat production was covered by the heat pump.

40.5 The Future. Smart Energy Systems

In the future energy systems energy storages will be an essential part because production of fluctuating electricity from wind and solar is expected to increase to replace fossil fuels. This will increase the balance demand and the demand to store electricity. Since direct storage of electricity in for instance batteries for use in periods with low production of fluctuating electricity is expensive and since electricity to a large extent has to be converted to other purposes as explained in Section 40.2 the future energy systems will be hybrid systems, where electricity, heating, gas supply, and mobility are integrated.

The first feasible example of this is power to heat, where electricity is converted to, and stored as hot water for heating purposes. The next feasible steps will be the power to process heat in high temperature thermal storages and the integration of fluctuating electricity production in the mobility sector where car batteries in electrical vehicles are charged and discharged according to regulation demands.

To reach 80 to 100% RES in the energy system a total conversion of the transport sector is necessary meaning that production of transport fuels (hydrogen, methan, methanol, etc.) will be needed to cover transport, that cannot be covered by electricity directly. Therefore "power to gas" or "power to liquid fuels," where electrolyzers regulate electricity demand and electricity will be stored as gasses and liquid fuels will have to be added to the energy system in this phase.

Recently, a group of researchers from Aalborg University have introduced the Smart Energy System Concept, which is a cross-sectoral approach that makes use of synergies between the various energy sub-sectors. The structure is illustrated in Figure 40.9.

Conversion to Smart Energy Systems integrates the heating/cooling electricity and transport sectors and makes it possible to supply a society 100% with RES. The Smart Energy System uses technologies such as

Smart Electricity Grids to connect flexible demands such as heat pumps and EVs to intermittent renewable resources such as wind and solar power.

Smart Thermal Grids (District heating and cooling) to connect the electricity and heating sectors. This enables the utilization of thermal storage for creating additional flexibility and the recycling of heat losses in the energy system.

Smart Gas Grids to connect the electricity, heating, and transport sectors. This enables the utilization of gas storage for creating additional flexibility. If the gas is refined to a liquid fuel, liquid fuel storages can also be utilized.

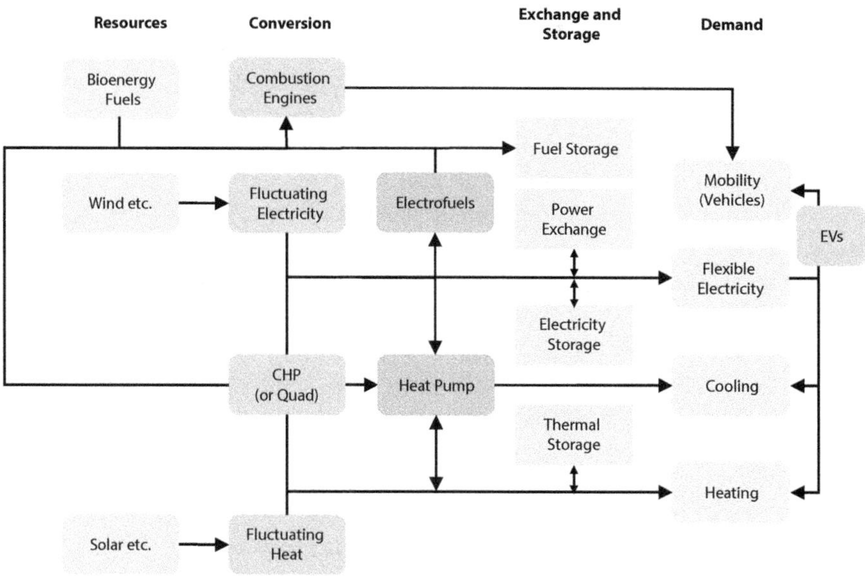

Figure 40.9 Smart Energy System structure [11].

Based on these fundamental infrastructures, a smart energy system is defined as an approach in which smart electricity, thermal, and gas grids are combined with storage technologies and coordinated to find synergies between them in order to achieve an optimal solution for each individual sector as well as for the overall energy system [11] [15].

In Smart Energy Systems not only power to heat, but also Power to Cold, Power to mobility and Power to Gas/liquid fuels are integrated

Acknowledgments

Thank you to Prof. Dr. Christian Doetsch, Fraunhofer UMSICHT, Dr. Andreas Palzer, Fraunhofer ISE, Prof. Dr Henrik Lund, Aalborg University, and Prof. Brian Vad Mathiesen, Aalborg University, for support.

References

[1] Lund, H. and Münster, E. (2006). Integrated energy systems and local energy markets. *Energy Policy* 34: 1152–1160.

[2] Henning, H.-M. and Palzer, A. (2014). A future German energy system with a dominating contribution from renewable energies. A holistic model based on hourly simulation. *Energy Technology* 2: 13–28.

[3] Doetsch, C., Droste-Franke, B., Mulder, G., and Scholz, Y. (May 2015). Electric energy storage – Future energy storage demand. IEA-ECES Annex 26. Final report. http://www.iea-eces.org/files/a5.1.1_2015_eces_26_final_report_public_rev16_web.pdf.

[4] Henning, H.-M. and Palzer, A. (2015). What will the energy transformation cost? Pathways for transforming the German energy system by 2050. Hg. v. Fraunhofer ISE. https://www.ise.fraunhofer.de/en/publications/veroeffentlichungen-pdf-dateien-en/studien-und-konzeptpapiere/study-what-will-the-energy-transformation-cost.pdf.

[5] Lund, H., Østergaard, P.A., Hvelplund, F., et al. (November 2011). Coherent energy and environmental system analysis. Department of Development and Planning, Aalborg University.

[6] Danish Energy Agency and Energinet.dk. Technology data for energy plants. Last update June 2017. https://ens.dk/sites/ens.dk/files/Analyser/technology_data_catalogue_for_energy_plants_-_aug_2016._update_june_2017.pdf.

[7] Danish Energy Agency and Energinet.dk. Technology data for individual heating plants and energy transport. Updated chapters, August 2016. https://ens.dk/sites/ens.dk/files/Analyser/technology_catalogue_individual_heating_plants_energy_transport_aug16.pdf.

[8] Soerensen, P.A., et al. (August 2014). SUNSTORE 4 final report. https://www.euroheat.org/wp-content/uploads/2016/04/SUNSTORE4_Report.pdf.

[9] Soerensen, P.A. and Jensen, M.V. (2013). The SUNSTORE® Concept for storing fluctuating electricity production from wind and solar. IRES. www.eurosolar.de/en/index.php/ires-conference-series/ires-2013.

[10] Jensen, M.V. (February 2016). New heat pump and solar heating. Final report. PlanEnergi. http://eudp-new.omega.oitudv.dk/sites/eudp-new.omega.oitudv.dk/files/slutrapporter/final_report_-_rye_kraftvarmevaerk_-_new_heat_pump_and_solar_heating_-_incl_app.pdf.

[11] Mathiesen, B.V., Lund, H., Hansen, K., Ridjan, I., et al. (2015). IDA's Energy Vision 2050. Aalborg University. Department of Development and Planning. https://vbn.aau.dk/en/publications/idas-energy-vision-2050-a-smart-energy-system-strategy-for-100-re.

Index

Advances in Energy Storage: Latest Developments from R&D to the Market, First Edition.
Edited by Andreas Hauer.
© 2022 John Wiley & Sons, Ltd. Published 2022 by John Wiley & Sons, Ltd.

b

C

g

h

n

o

q

r

S